	I B	II B	III A	IV A	V A	VI A	VII A	0
								2 **He** Helium 4.00260
			5 **B** Boron 10.81	6 **C** Carbon 12.011	7 **N** Nitrogen 14.0067	8 **O** Oxygen 15.9994	9 **F** Fluorine 18.998403	10 **Ne** Neon 20.179
			13 **Al** Aluminum 26.98154	14 **Si** Silicon 28.0855	15 **P** Phosphorus 30.97376	16 **S** Sulfur 32.06	17 **Cl** Chlorine 35.453	18 **Ar** Argon 39.948
28 **Ni** Nickel 58.69	29 **Cu** Copper 63.546	30 **Zn** Zinc 65.38	31 **Ga** Gallium 69.72	32 **Ge** Germanium 72.59	33 **As** Arsenic 74.9216	34 **Se** Selenium 78.96	35 **Br** Bromine 79.904	36 **Kr** Krypton 83.80
46 **Pd** Palladium 106.42	47 **Ag** Silver 107.868	48 **Cd** Cadmium 112.41	49 **In** Indium 114.82	50 **Sn** Tin 118.69	51 **Sb** Antimony 121.75	52 **Te** Tellurium 127.60	53 **I** Iodine 126.9045	54 **Xe** Xenon 131.29
78 **Pt** Platinum 195.08	79 **Au** Gold 196.9665	80 **Hg** Mercury 200.59	81 **Tl** Thallium 204.383	82 **Pb** Lead 207.2	83 **Bi** Bismuth 208.9804	84 **Po** Polonium (209)[a]	85 **At** Astatine (210)[a]	86 **Rn** Radon (222)[a]

metals ← → nonmetals

63 **Eu** Europium 151.96	64 **Gd** Gadolinium 157.25	65 **Tb** Terbium 158.9254	66 **Dy** Dysprosium 162.50	67 **Ho** Holmium 164.9304	68 **Er** Erbium 167.26	69 **Tm** Thulium 168.9342	70 **Yb** Ytterbium 173.04	71 **Lu** Lutetium 174.967
95 **Am** Americium (243)[a]	96 **Cm** Curium (247)[a]	97 **Bk** Berkelium (247)[a]	98 **Cf** Californium (251)[a]	99 **Es** Einsteinium (252)[a]	100 **Fm** Fermium (257)[a]	101 **Md** Mendelevium (258)[a]	102 **No** Nobelium (259)[a]	103 **Lr** Lawrencium (260)[a]

INSTRUMENTAL METHODS OF ANALYSIS

7 TH EDITION

INSTRUMENTAL METHODS OF ANALYSIS

Hobart H. Willard
LATE OF UNIVERSITY OF MICHIGAN

Lynne L. Merritt, Jr.
INDIANA UNIVERSITY

John A. Dean
UNIVERSITY OF TENNESSEE AT KNOXVILLE

Frank A. Settle, Jr.
VIRGINIA MILITARY INSTITUTE

WADSWORTH PUBLISHING COMPANY
Belmont, California
A Division of Wadsworth, Inc.

Chemistry editor: Jack Carey
Editorial assistant: Judy Walcom
Production editor: Vicki Friedberg
Managing designer: MaryEllen Podgorski
Print buyer: Karen Hunt
Designer: Gayle Jaeger
Copy editor: Carol Reitz
Art editor: Irene Imfeld
Technical illustrator: Mary Burkhardt
Compositor: Polyglot Pte Ltd
Cover: Harry Voight

Printed in the United States of America 19

7 8 9 10—99 98 97 96 95 94

Library of Congress Cataloging-in-Publication Data

Instrumental methods of analysis.

Includes bibliographies and index.
1. Instrumental analysis. I. Willard, Hobart Hurd, 1881–
QD79.I5I52 1988 543′.08 87-10601
ISBN 0-534-08142-8

PREFACE

During the preparation of this seventh edition, we were impressed by the explosive progress in the field of instrumental methods of analysis. Yet, on reviewing the first edition of this text, published in 1948, we found that most of the "modern" methods were already described there. The first edition included visual and photoelectric colorimeters, fluorescence, turbidimetry and nephelometry, spectrophotometry (including uv/visible and infrared), flame photometry, thermal conductivity, spectrography (including Raman), X-ray diffraction, radioactivity, thermal conductivity and other methods for gas analysis, refractometry and interferometry, the centrifuge, pH determinations (including the glass electrode), potentiometric titrations (including differential methods), conductometric titrations, electrolytic separations (including constant potential and constant current methods), polarography, and amperometric titrations.

Because visual colorimetry has faded from use, it is no longer treated in the text; also, the centrifuge is not considered as an analytical technique. Chromatography, however, is now a most valuable separation and quantitative technique; nuclear magnetic resonance is a very important and widely used method; and X-ray methods now make extensive use of fluorescence and absorption techniques.

It is apparent to us that, although the principles of most "modern" methods have been known for a long time, the explosion in the number of available instruments, the sensitivity and selectivity, the extremely low detection limits, and the ease of usage have been largely due to the advent of dedicated computers, the laser, and vastly improved electronics. Phenomena can now be observed and recorded in the nanosecond or even picosecond range. Large numbers of data points can be collected, stored, manipulated, and presented in seconds or fractions of seconds. Computers not only collect, store, and manipulate data but also control the whole instrument's operation.

The seventh edition has been restructured into four major sections: the transfer and processing of information (Chapters 2–4), spectroscopic methods (Chapters 5–15), separation methods (Chapters 17–20), and electrochemical methods (Chapters 21–24). The first chapter of each section provides a discussion of the basic principles of the methods discussed in the section. In addition to the four major sections, there are individual chapters on mass spectrometry (Chapter 16), thermal analysis (Chapter 25), and automated methods of analysis (Chapter 26). An introductory chapter, Chapter 1, has also been added to give a general overview of the role of instrumentation in analytical chemistry. All chapters have been thoroughly revised to include the most modern methods in general use today.

Chapter 21, "Introduction to Electroanalytical Methods of Analysis," has been thoroughly revised and updated. Section 21.4, on mass transfer by diffusion, includes Fick's laws and the Cottrell equation. Reversible, quasi-reversible, and irreversible reactions are carefully distinguished and discussed. Chapter 23, "Voltammetric Techniques," discusses the modern methods, which are distinguished from the classical polarographic, dc, and slow scan methods using a dropping mercury electrode.

Many examples have been added throughout the text, and summary sections are frequently included at the end of major sections. New material found in the appendixes includes a summary table of analytical methods, sources of laboratory experiments, and elemental detection limits for six different atomic spectroscopic methods.

The text contains much more material than can be covered in one semester and, perhaps, even in two semesters. Thus instructors must pick and choose the methods they consider most important for their particular classes. With the exception that the introductory chapters should be read before considering the later chapters in each section, the chapters are self-consistent and could be considered in any order.

To keep the text to a reasonable length, we have had to eliminate some methods, such as polarimetry and circular dichroism and electron spin resonance spectroscopy (except for a brief discussion in Chapter 5), as well as suggested experiments. The chapter on chemical analysis of surfaces in the sixth edition has been integrated, in part, into Chapters 14 and 16 on X-ray methods and mass spectrometry in the seventh edition.

We hope that this book will be useful not only as a text for courses in instrumental analysis in analytical chemistry but also as a short reference book. The Bibliography and Literature Cited sections at the end of each chapter contain references to many of the classical books and papers in the field as well as some of the most recent books, papers, and reviews.

We wish to express our appreciation to the following reviewers for their helpful comments: James Anderson, University of Georgia; Paul W. Bohn, University of Illinois, Urbana; Thomas Copeland, Northeastern University; Neil D. Danielson, Miami University; Arno Heyn, Boston University; John Lanning, University of Colorado; George Morrison, Cornell University; Gordon A. Parker, University of Toledo; John S. Phillips, University of Wisconsin; and Stanford Tackett, Indiana University of Pennsylvania.

Lynne L. Merritt, Jr.
John A. Dean
Frank A. Settle, Jr.

CONTENTS

4

COMPUTER-AIDED ANALYSIS 60

AN INTRODUCTION TO ABSORPTION AND EMISSION SPECTROSCOPY 97

ULTRAVIOLET AND VISIBLE SPECTROMETRY—INSTRUMENTATION 118

ULTRAVIOLET AND VISIBLE ABSORPTION METHODS 159

15

NUCLEAR MAGNETIC RESONANCE SPECTROSCOPY 422

MASS SPECTROMETRY 465

CHROMATOGRAPHY: GENERAL PRINCIPLES 513

18

GAS CHROMATOGRAPHY 540

HIGH-PERFORMANCE LIQUID CHROMATOGRAPHY: THEORY AND INSTRUMENTATION 580

HIGH-PERFORMANCE LIQUID CHROMATOGRAPHY: METHODS AND APPLICATIONS 614

INTRODUCTION TO ELECTROANALYTICAL METHODS OF ANALYSIS 656

POTENTIOMETRY 671

VOLTAMMETRIC TECHNIQUES 697

ELECTROSEPARATIONS, COULOMETRY, AND CONDUCTANCE METHODS 732

THERMAL ANALYSIS 761

PROCESS INSTRUMENTS AND AUTOMATED ANALYSIS 786

ABBREVIATIONS

absorption	Abs
alpha particle	α
alternating current	ac
American Society for Testing Materials	ASTM
American standard code for information interchange (computers)	ASCII
ampere	A
analog-to-digital converter	ADC, A/D
angstrom	Å
anodic	anod, a (subscript)
anodic stripping voltammetry	ASV
aqueous	aq
arithmetic logic unit	ALU
Association of Official Analytical Chemists	AOAC
asymmetry factor	AF
atmosphere	atm
atomic absorption spectrometry	AAS
atomic emission spectrometry	AES
atomic fluorescence spectrometry	AFS
atomic weight	at. wt.
attenuated total reflectance	ATR
Auger emission spectroscopy	AES
average	av (subscript)
backscatter	BS
barn	b
beta particle	β
binary coded decimal	BCD
boiling point	bp
bonded-phase chromatography	BPC
calorie	cal
capacitance	C
cathode ray tube	CRT
cathodic	cath, c (subscript)
centi- (prefix)(10^{-2})	c-
centimeter	cm
centipoise	cP
central processing unit	CPU
charge-coupled device	CCD
chemical ionization	CI
Christiansen effect detector	CED
circa	*ca.*
citrate	Cit
coherent anti-Stokes Raman spectroscopy	CARS
complementary metal oxide semiconductor	CMOS
Compton edge	CE
conductance	$1/R$
conductance-solids nebulizer	CSN
continuous flow analysis	CFA
continuous wave (laser)	CW
coulomb	C
counts per minute (second)	cpm (cps)
cubic centimeter	cm^3
curie	Ci
cycles per second (hertz)	Hz
cylindrical mirror analyzer	CMA
day	d
decibel	dB
degree Celsius	°C
degree Kelvin	K
deuterated triglycine sulfate	DTGS
deuteron	d
diameter	diam
differential scanning calorimeter	DSC
differential thermal analysis	DTA
digital-to-analog converter	DAC, D/A
digital voltmeter	DVM
diode transistor logic	DTL
direct current	dc
direct-current plasma	DCP
direct digital control	DDC
direct-injection enthalpimetry	DIE
direct memory access	DMA
disintegrations per minute (second)	dpm (dps)
disk-based operating system	DOS
dropping mercury electrode	dme, de (subscript)
dual-in-line package	DIP
dynamic mechanical analysis	DMA
dyne	dyn
effective aperture ratio	f/number
electromotive force	emf
electron	e, e^-
electron capture detector	ECD

electron impact	EI
electron probe	EP
electron spectroscopy for chemical analysis	ESCA
electron spin resonance	ESR
electron volt	eV
electrothermal atomic absorption spectrometry	EAAS
electrothermal vaporizer	ETV
energy-dispersive X-ray fluorescence analysis	EDXFS
equivalent weight	eq wt, equiv wt
erasable programmable read-only memory	EPROM
error function	erf
et alii (and others)	et al.
ethyl	Et
ethylenediamine-*N*,*N*,*N'*,*N'*-tetraacetate	EDTA, Y^{4-}
evolved gas analysis	EGA
evolved gas detection	EGD
exclusion chromatography	EC
exempli gratia (for example)	e.g.
exponential	exp
external	ext
farad	f
fast atom bombardment	FAB
fast Fourier transformation	FFT
field-effect transitor	FET
flame atomic absorption spectroscopy	FAAS
flame emission spectroscopy	FES
flame ionization detector	FID
flame photometric detector	FPD
flow-injection analysis	FIA
formal (concentration)	*F*
Fourier transformation	FT
Fourier transformation-ion cyclotron resonance	FT-ICR
Fourier transform-infrared	FTIR
free induction decay	FID
frequency	f
full width at half maximum	FWHM
gamma radiation	γ
gas (physical state)	*g*
gas chromatography	GC
gas chromatography–mass spectrometry	GC-MS
gas-liquid chromatography	GLC
gas-solid chromatography	GSC
gauss	G
Geiger-Müller	GM
geminal	*gem*
gram	g
hanging mercury drop electrode	HMDE
hertz	Hz
high performance liquid chromatography	HPLC
hour	hr
id est (that is)	i.e.
inch	in.
indicator	ind
inductance	L
inductively coupled (argon) plasma	ICP, ICAP
infrared	ir, IR
input/output	I/O
inside diameter	i.d.
integrated circuit	IC
integrated injection logic	IIL
internal	int
International Union of Pure and Applied Chemistry	IUPAC
ion exchange chromatography	IEC
ion microprobe mass analyzer	IMMA
ion scattering spectroscopy	ISS
ion-selective electrode	ISE
joule	J
kelvin (temperature)	K
kilo- (prefix)(10^3)	k-
kilocalorie	kcal
Kovats retention index	RI
laboratory information management systems	LIMS
large-scale integration	LSI
least significant bit	LSB
light emitting diode	LED
limiting	lim
linear variable differential transformer	LVDT
liquid (physical state)	liq, *l*
liquid chromatography	LC
liquid chromatography–mass spectrometry	LC-MS
liquid column chromatography	LCC
liquid-liquid (partition) chromatography	LLC
liquid-solid (adsorption) chromatography	LSC
liter	L (with prefixes), liter (alone)
local area network	LAN
logarithm (common, Briggsian, decadic)	log
logarithm (natural or Naperian)	ln
logical AND operation in Boolean algebra	· (center dot)
logical OR operation in Boolean algebra	+
low-angle laser light scattering	LALLS
lumen	lm
lysergic acid diethylamide	LSD
mass spectrometry	MS
maximum	max
medium-scale integration	MSI
mega- (prefix)(10^6)	M-
mercury cadmium telluride	MCT
meta-	*m-*
metal oxide semiconductor	MOS
metastable (state)	*m**, *m* (superscript)
meter	m
methyl	Me
micro- (prefix)(10^{-6})	μ-
micrometer (micron)	μm
microsecond	μsec
milli- (prefix)(10^{-3})	m-
milliampere	mA
milliequivalent	meq, mequiv
milliliter	mL
millimole	mM

million electron volts	MeV
minimum	min
minute	min
molar (concentration)	M
mole	mol
molecular weight	mol wt
monolayer	ML
most significant bit	MSB
multiple internal reflectance	MIR
nano- (prefix)(10^{-9})	n-
nanometer	nm
Naperian base	e
near infrared	NIR
near-infrared reflectance analysis	NIRA
negative	neg
nephelometric turbidity unit	NTU
neutron	n
normal (concentration)	N
normal hydrogen electrode (= SHE)	NHE
not AND (results of AND operation negated)	NAND
not OR (results of OR operation negated)	NOR
nuclear magnetic resonance	NMR, nmr
nuclear Overhauser effect	NOE
numerical aperture	NA
ohm	Ω
operating system	OS
operational amplifier	op amp
optical speed	f/number
optimum	opt
ortho-	*o-*
outside diameter	o.d.
oxidant	ox
oxide semiconductor field-effect transistor (MOSFET without metal gate)	OSFET
para-	*p-*
parent ion (mass spectrometry)	M
particle-induced X-ray emission	PIXE
parts per billion, volume	ppb, ng/mL
parts per billion, weight	ppb, ng/g
parts per million, volume	ppm, μg/mL
parts per million, weight	ppm, μg/g
pascal	Pa
percent	%
phenyl	ϕ, Ph
photoionization detector	PID
photomultiplier (tube)	PM, PMT
pico- (prefix)(10^{-12})	p-
poise	P
positron	β^+
potential	E
programmable peripheral interface	PPI
programmable read-only memory	PROM
propyl	Pr
proton	p
proton magnetic resonance	PMR
quantum (energy)	$h\nu$
quantum efficiency	QE

radian	rad
radio frequency	rf
random access memory	RAM
read-only memory	ROM
reciprocal ohm	mho, Ω^{-1}
reductant	red
reference	ref
refractive index	RI
reset-set	R-S
resistance	R
retention index (Kovats)	RI
reverse phase-ion pair partition	RP-IPP
revolutions per minute	rpm
sample and hold	S/H
saturated	satd
saturated calomel electrode	SCE
scanning Auger microprobe	SAM
scanning electron microscopy	SEM
second	sec
secondary ion mass spectrometry	SIMS
sigma	σ
small-scale integration	SSI
solid (physical state)	c, s
solvent (general)	S
specific gravity	sp gr
standard hydrogen electrode	SHE
standard temperature and pressure	STP
steradian	sr
support-coated open tubular (column)	SCOT
surface enhanced Raman spectroscopy	SERS
Système International	SI
temperature	t (Celsius), T (absolute)
tertiary	*tert-*
tesla	T
tetramethylsilane	TMS
thermal conductivity detector	TCD
thermionic emission detector	TED
thermogravimetric analysis	TGA
thermomechanical analysis	TMA
thermometric enthalpy titrimetry	TET
thousand electron volts	keV
torr (mm of mercury)	torr
transitor-resistor logic	TRL
triglycine sulfate	TGS
tritium	t, ^3H
ultraviolet	uv
universal asynchronous receiver transmitter	UART
universal synchronous/asynchronous receiver transmitter	USART
vacuum	vac
vacuum-tube voltmeter	VTVM
versus	vs.
very large-scale integration	VLSI
volt	V
volume	vol, V
volume per volume	v/v

volume per weight	v/w
wall-coated open tubular (column)	WCOT
watt	W
wavelength	λ
wavenumber	cm^{-1}
X-ray absorption edge	K edge, L edge, etc.
X-ray emission lines	$K\alpha$, $K\beta$, $L\alpha$, etc.
X-ray emission spectrometry	XES
X-ray energy level	K, L_{I}, L_{II}, L_{III}, etc.
X-ray fluorescence spectroscopy	XFS
year	yr
yttrium aluminum garnet	YAG

SYMBOLS

A absorbance; activity (radiochemistry); area; atomic weight

A_o amplifier gain; induced activity (at time zero)

A_p activity produced in a pulse

A_s activity at saturation

A_t activity of tagged material

a (subscript) anodic; specific absorptivity

a_i hyperfine coupling constant (ESR)

a_x activity of species x

B brightness of source

B magnetic field strength

b distance; grating spacing; optical path length; thickness

C concentration; capacitance

C_i integral capacitance

C_M concentration of solute in mobile phase

C_S concentration of solute in stationary phase

c (subscript) cathodic; velocity of light in a vacuum

D dielectric constant; diffusion coefficient

D^{-1} linear reciprocal dispersion

D_c concentration distribution ratio

D_M diffusion coefficient in mobile phase

D_S diffusion coefficient in stationary phase

d diameter; difference; distance; spacing

d_c cross section of column bore; diameter of collimating mirror

d_f effective thickness of stationary phase

d_{hkl} spacing of crystal planes

d_p particle diameter

E electrode potential; energy of a photon; potential of half-reaction; magnitude of electric vector

ΔE amplitude of voltage signal

E° standard electrode potential

$E_{1/2}$ half-wave potential

E_a anodic potential

E_b binding energy of an electron; core-electron binding energy

E_c cathodic potential

E_{CE} energy of Compton edge

E_i ionization energy

E_{ind} indicator electrode potential

E_j junction potential

E_k kinetic energy

$E_{\max}$ maximum energy

E_p peak potential

E_{ref} reference electrode potential

$\mathrm{E}_{\mathrm{rel}}$ relative error

e electronic charge

e^- Naperian base (natural logarithms, 2.718...)

e° solvent strength parameter

e_i input voltage

e_s voltage across input terminals of an amplifier

e_o output voltage

F faraday; fluorescence

F_c volume flowrate of mobile phase

F_y fractional error

$F(v)$ frequency-amplitude function

f focal length; fractional abundance; frequency; oscillator strength; secondary emission factor

$1/f$ low-frequency noise

Δf bandwidth of measurement frequencies

$f(\theta)$ geometric factor (fluorometers)

f/n aperture (n is a number)

$f(t)$ time-amplitude function

f_x activity coefficient of species x

G gain of photomultiplier tube

ΔG Gibbs free-energy change in a chemical process

ΔG° standard Gibbs free-energy change in a chemical process

g spectroscopic splitting factor

g_m, g_n statistical weight of atomic states

H magnetic field strength; plate height

ΔH enthalpy change in a chemical process; peak-to-peak separation

ΔH° standard enthalpy change in a chemical process

ΔH_s molal heat of solution

ΔH_v molal heat of vaporization

h height; Planck's constant

I current; radiant intensity; spin quantum number

I_o incident radiant energy; output intensity

I_p peak value of ac current amplitude

I_v emission line intensity

i angle of incidence; current

i- (prefix) iso-

i_a anodic component current

i_c charging current; cathodic component current

i_d current due to diffusive flux; diffusion-limited current
i_{lim} limiting current
i_p peak current
i_r residual current
J spin-spin coupling constant; joule
j inner quantum number; pressure gradient correction
K_a acid dissociation constant
K_{auto} autoprotolysis constant
K_d partition coefficient
K_f formation constant
K_i ionization constant (gaseous state)
K_{sp} solubility product
K_w ion product of water
k Boltzmann constant; force constant (infrared); general constant
k' partition ratio or capacity factor (chromatography)
$k_{\text{M/N}}$ selectivity ratio for solutes M and N
k_v absorption coefficient (optical)
L length or distance; quantum number for resultant angular orbital momentum; inductance
l orbital angular momentum of individual electron
M mass
M_I spin quantum number of nucleus
M_n number-average molecular weight
M_s angular momentum quantum number (electron)
M_w weight-average molecular weight
m demodulation factor; mass; mass of mercury (voltammetry); order number (optical); metastable state (superscript)
m^* metastable state
m^+ ionized mass fragment
m_{ex} degree of ac modulation
m/z, m/e mass-to-charge ratio
N noise; plate number (chromatography); total number of something
N_A Avogadro constant
N_b background count (counting rate)
N_{eff} effective plate number
N_j, N_m, N^* number of species in excited state
N_n, N_o number of species in ground energy state
N_{req} plates required
N_s sample counts (counting rate)
$N(E)$ energy distribution (Auger spectroscopy)
n electrons per molecule oxidized or reduced; faradays per mole of substance electrolyzed; principal quantum number; unshared p-electrons
n- semiconducting material containing an excess of negative charge carriers
n_{theor} theoretical plate number
P phosphorescence; polarization ratio; pressure; radiant power
ΔP pressure drop across a column
$P_F(t)$ fluorescence power at time t
P_i inlet gas pressure
P_M mass peak of parent compound
P_o incident radiant power; outlet gas pressure
p excited electron state; depolarization ratio (Raman); partial pressure (gaseous material); semiconducting material containing an excess of positive charge carriers
p° solute vapor pressure
Q flowrate; heat capacity; number of coulombs
R universal gas constant; resolution; resolving power (optical); Rydberg constant (X ray)
R_L load resistance
Rs resolution
r angle of refraction; count rate; radius; resolution (radiochemistry detectors)
r° programmed rate of temperature increase
r_D specific refraction
S electron spin; saturation factor (radiochemistry)
ΔS entropy change in a chemical process
ΔS° standard entropy change in a chemical process
S_1 first excited (singlet) electronic state
S_o ground electronic state
S/N signal-to-noise ratio
s standard deviation
s_b standard deviation of blank (background)
T absolute temperature; transmittance (optical)
T_1 first excited triplet state; spin-lattice (or longitudinal) relaxation time (NMR)
T_2 spin-spin (or transverse) relaxation time (NMR)
T_b boiling point
T_c column temperature (chromatography)
t time; prism base length
$t_{1/2}$ half-life
t_b counting time for background
t_g retention time in gradient elution
t_M transit time of nonretained solute (chromatography)
t_p time of solute passage through one plate
t_R retention time
t'_R adjusted retention time
t_s counting time for sample
u average linear velocity; reduced mass
V volume
V_g° specific retention volume (at 0 °C)
V_g volume of column occupied by gel matrix
V_i internal volume within porous particles
V_M volume of mobile phase
V_N net retention volume
V_R retention volume
V'_R adjusted retention volume
V_S cumulative internal volume within porous particles; volume of stationary phase
V_t total bed volume
v reduced velocity; scan rate; velocity; designation of vibrational levels
W physical slit width; weight; zone width
$W_{1/2}$ zone width at half peak height
W_b zone width at baseline
w effective aperture width; weight
w_L weight of stationary liquid phase

w_S weight of adsorbent phase
w_t weight of tagged material
X_C capacitive reactance
X_L inductive reactance
x distance; general designation of species
Z atomic number of an element; impedance
z charge on an ion in signed units of electronic charge; valence
z_+, z_- ionic charge
α polarizability; relative retention ratio
α_i degree of ionization
β blaze angle; buffer value; volumetric phase ratio
β_N Bohr magneton
γ activity coefficient; emulsion characteristic; surface tension; obstructive (or tortuosity) factor (chromatography)
Δ (prefix) symbol for finite change; spectral width
δ chemical shift (NMR); thickness of diffusion layer
ε molar absorptivity
ε_{tot} total porosity of column
η index of refraction; viscosity
η_D index of refraction (*D* line of sodium)
Θ cell constant (conductance)
θ angle; angle of diffraction
θ_c critical angle
θ_{vib} vibration level of diatomic molecule
2θ angular setting of diffraction angle (X ray)
$dx/d\lambda$ linear dispersion
$d\theta/d\lambda$ angular dispersion
κ specific conductance
Λ equivalent conductance
Λ_∞ equivalent conductance at infinite dilution column packing uniformity; decay constant; reduced column length
λ wavelength
λ_+, λ_- limiting equivalent ionic conductance
$\Delta\lambda$ bandpass
λ_{max} wavelength of an absorption maximum
μ ionic strength; linear absorption coefficient; magnetic moment of nucleus; known true value (statistics)
μ_B Bohr magneton
μ_e electron magnetic moment
μ_m mass absorption coefficient
μ_N nuclear magneton
μ/ρ mass absorption coefficient
v frequency; reduced velocity (chromatography)
$\bar{v}$ wavenumber
π pi; type of electron or bond
ρ density; resistivity
Σ summation symbol
σ reaction cross section; shielding constant (NMR, X ray); standard deviation of infinite population; gradient bandwidth
σ_{hkl} reciprocal lattice vectors
$4\sigma_v$ extra column broadening
τ mean emission lifetime; resolving time; time constant
Φ number of bombarding particles or flux
ϕ column flow resistance paremeter; photoluminescence efficiency; work function
ω angular frequency; chopping frequency; overpotential
ω_c angular velocity; cyclotron frequency
[ ] molar concentration of species within brackets

INSTRUMENTAL METHODS OF ANALYSIS

1 AN INTRODUCTION TO INSTRUMENTAL METHODS

The use of instrumentation is an exciting and fascinating part of chemical analysis that interacts with all the areas of chemistry and with many other fields of pure and applied science. Analyses of Martian soils, the body fluids of racehorses and Olympic athletes, the engine oil of commercial and military jet aircraft, and even the Shroud of Turin are examples of problems that require instrumental techniques. Often it is necessary to use several instrumental techniques to obtain the information required to solve an analytical problem.[1,2]

Analytical instrumentation plays an important role in the production and evaluation of new products and in the protection of consumers and the environment. This instrumentation provides the lower detection limits required to assure safe foods, drugs, water, and air. The manufacture of materials whose composition must be known precisely, such as the substances used in integrated circuit chips, is monitored by analytical instruments. The large sample throughputs made possible by automated instrumentation often relieve the analyst of the tedious tasks formerly associated with chemical analysis. Thus the analyst is freed to examine components of the analytical system, such as sampling methods, data treatment, and the evaluation of results.

1.1 TERMS ASSOCIATED WITH CHEMICAL ANALYSIS

It is necessary to distinguish between the terms *analytical technique* and *analytical method*.[3] A technique is a fundamental scientific phenomenon that has proved useful for providing information on the composition of substances. Infrared spectrophotometry is an example of an analytical technique. A method is a specific application of a technique to solve an analytical problem. The infrared analysis of styrene-acrylonitrile copolymers is an example of an instrumental method.

Two other terms associated with chemical analysis are *procedure* and *protocol*. A procedure is the written instructions for carrying out a method. The "standard methods" developed by the ASTM (American Society for Testing Materials) and the AOAC (Association of Official Analytical Chemists) are, in reality, standardized procedures. A procedure assumes that the user has some prior knowledge of analytical methodology, and thus it does not provide great detail, only a general outline of the steps to be followed. The procedure for infrared analysis of styrene-acrylonitrile copolymers involves the extraction of the residual styrene and acrylonitrile monomers from the polymer into carbon disulfide. The residual polymer is then dissolved and cast as a film directly on a sodium chloride plate. Both the carbon disulfide extract and the film are scanned to obtain absorbance measurements at wavelengths characteristic of styrene and acrylonitrile. The sample absorbances are compared with those of standards of known concentration. In contrast, the most specific description of a method is known as a protocol. The detailed directions must be followed, without exception, if the analytical results are to be accepted for a given purpose, such as an environmental analysis to meet Environmental Protection Agency (EPA) requirements or blood alcohol determinations for legal proceedings.

1.2 CLASSIFICATION OF INSTRUMENTAL TECHNIQUES

Most instrumental techniques fit into one of three principal areas: spectroscopy, electrochemistry, and chromatography (Table 1.1). Although several important techniques, including mass spectrometry and thermal analysis, do not fit conveniently into these classifications, these three areas do provide a basis for a systematic study of chemical instrumentation.

Advances in both chemistry and technology are making new techniques available and expanding the use of existing ones. Photoacoustic spectroscopy is an example of an emerging analytical technique. A number of existing techniques have been combined to expand the utility of the component methods. Gas chromatography–mass spectrometry (GC-MS) and inductively coupled plasma–mass spectrometry (ICP-MS) are examples of successful "hyphenated" methods (Table 1.2). The distribution of computer power to individual instruments has led to the widespread use of methods such as the Fourier transform to produce new techniques: Fourier transform infrared (FTIR) and pulsed nuclear magnetic resonance (carbon-13) spectroscopy.

The analyst should be aware of the functions performed by the computer(s) in a given analytical method.[4] These functions range from data acquisition and control to management of laboratory data systems. Although few analytical chemists actually develop programs and design hardware, they should understand the fundamental concepts of both computer hardware and software.

The chapters in this text that deal with specific techniques are organized to help the reader develop a general methodology for examining instrumental methods. This approach will allow the reader to investigate current techniques and methods as well as those that may appear in the future.

1.3 A REVIEW OF THE IMPORTANT CONSIDERATIONS IN ANALYTICAL METHODS

Although the instrument is often the most visible and exciting element of the analytical method, it is only one component of the total analysis. Before focusing on the role of instrumentation in an analytical method, the analyst should consider other steps important to the determination (Figure 1.1). The following discussion reviews the steps common to analytical methods and thus helps to place the role of instrumentation in proper perspective. A detailed discussion of the topics will not be attempted; they are covered in greater depth in the bibliographies and literature citations. Those topics that relate more closely to instrumentation are discussed in later chapters of this text.

The first task is to define the analytical problem.[1,2] When possible, this is done through direct interaction with the person(s) who desires the analysis. The analyst should determine the nature of the sample, the end use of the analytical results, the species to be analyzed, and the information required. Qualitative information may include elemental composition, oxidation states, functional groups, major components, minor components, and complete identification of all species present in the

TABLE **1.1** PRINCIPAL TYPES OF CHEMICAL INSTRUMENTATION

Spectroscopic techniques
Ultraviolet and visible spectrophotometry
Fluorescence and phosphorescence spectrophotometry
Atomic spectrometry (emission and absorption)
Infrared spectrophotometry
Raman spectroscopy
X-ray spectroscopy
Radiochemical techniques including activation analysis
Nuclear magnetic resonance spectroscopy
Electron spin resonance spectroscopy

Electrochemical techniques
Potentiometry (pH and ion selective electrodes)
Voltammetry
Voltammetric techniques
Stripping techniques
Amperometric techniques
Coulometry
Electrogravimetry
Conductance techniques

Chromatographic techniques
Gas chromatography
High-performance liquid chromatographic techniques

Miscellaneous techniques
Thermal analysis
Mass spectrometry
Kinetic techniques

Hyphenated techniques (see Table 1.2)
GC-MS (gas chromatography–mass spectrometry)
ICP-MS (inductively coupled plasma–mass spectrometry)
GC-IR (gas chromatography–infrared spectroscopy)
MS-MS (mass spectrometry–mass spectrometry)

sample. Quantitative data include the required accuracy and precision, range of expected analyte (substance being determined) concentrations, and detection limits for the analyte. Other considerations are the unique physical and chemical properties of the analyte, properties of the sample matrix, the presence of interferences that are likely to eliminate the use of certain analyte properties as measurement indicators, and finally an estimated cost of the analysis.[5] A major component of the cost is the time required for the analysis, both the instrument time and the labor required to perform the analysis. When appropriate, the costs of manual versus automated methods should be compared.

Once the problem has been defined, the next task is to select the appropriate method(s).[6,7] Factors to consider include the strengths and limitations of the

TABLE 1.2 THE STATE OF THE ART IN HYPHENATED TECHNIQUES

Initial technique / **Subsequent technique**	Gas chromatography	Liquid chromatography	Thin layer chromatography	Infrared	Mass spectrometry	Ultraviolet (visible)	Atomic absorption	Optical emission spectroscopy	Fluorescence	Scattering	Raman	Nuclear magnetic resonance	Microwaves	Electrophoresis
Gas chromatography	■	□		■	■	■		□	□			□	□	
Liquid chromatography	□	□	□	□	■	■	□	□	■	■	□	□		□
Thin layer chromatography	□	□	■	□	□	□		□	■		□			□
Infrared						□					□	□		
Mass spectrometry					■									
Ultraviolet (visible)				□					■	■	□			
Atomic absorption					□			□	□					
Optical emission spectroscopy					□		□		□					
Fluorescence					□	■	□	□	■	□	□		□	
Scattering						□		□	□		□			
Raman				□					□	□				
Nuclear magnetic resonance				□								■		
Microwaves					□									
Electrophoresis		□	□			■			■	■				■

NOTE: After personal communication from Tomas Hirschfeld.
■ Established.
□ Feasible in the state of the art.

technique when applied to the problem under consideration, restrictions placed upon the method by interferences present in the sample, and the quality of information obtained versus the cost of obtaining it. Appendix A gives a brief outline of the characteristics of commonly used analytical techniques.

Sampling is the next area to consider.[8] It is often the most important step in the entire analysis. What measures must be taken to obtain the samples required to provide the desired information? In some cases representative homogeneous samples are sought, whereas in other cases the heterogeneity of the sample is of primary interest. Do the field sampling and laboratory subsampling procedures assure the integrity of analytical results? Have proper procedures been used to store and preserve both samples and standards? Have all samples been properly labeled and recorded?

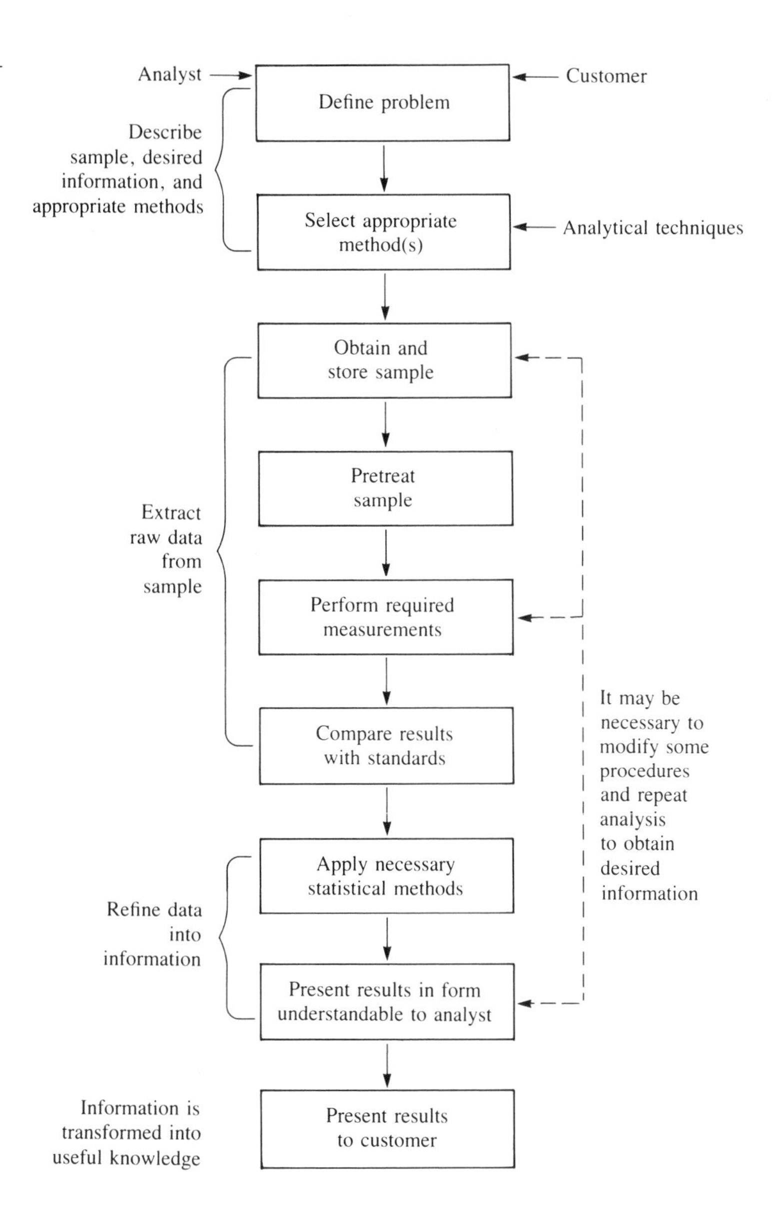

FIGURE **1.1**
Major steps in solving an analytical problem.

It is often necessary to perform some physical or chemical operations on the sample prior to the actual analysis.[9] These operations may reduce or remove interferences, bring the analyte concentration into the desired range for analysis, or produce species with quantitatively measurable properties from the analyte(s). Such operations include dissolution, fusion, separation, dilution, concentration, and chemical derivatization. Sophisticated instrumentation does not negate the need for fundamental laboratory skills; rather, it increases their importance. The proper cleaning, use, and knowledge of the tolerances of analytical balances, volumetric flasks, burets, pipets, and filtering apparatus are basic skills necessary in analyses.

Control of the chemical environment is often required to assure that the activities of the analytes remain constant during the measurement and to lessen the effects of interferences.[8] Methods such as controlling the atmosphere to which the sample is exposed, controlling the temperature of the sample, buffering the pH of sample solutions, and complexing the sample components are used for this purpose. Instrumental parameters, such as the amplitude and frequency of the input signal, detector sensitivity, and sampling rate, must be coordinated to measure the desired analyte under optimum conditions.

Once the measurement has been made, how is the desired precision for the analysis assured?[3,10] The analyst must select the method(s) of standardization best suited for the analysis at hand from the following: calibration ("working") curves, standard addition, internal standard, external standard, reference materials such as those provided by the National Bureau of Standards, "blind" samples, and control charts.

To assess the precision and to evaluate the results of analyses, the analyst must use statistical methods.[10,11] These methods include confidence limits, rejection of outlying points, regression analysis to establish calibration curves, tests for significance, and Gaussian and non-Gaussian distribution curves. In each analysis, the critical response parameters must be recognized and optimized if possible.[12] The simplex method and other procedures can be used to optimize the conditions for a given determination.

The clear, accurate presentation of results is an important requirement for any successful analytical method.[13] This involves maintaining a proper laboratory notebook, presenting data in graphical form if necessary, having a working knowledge of significant figures, being able to communicate the original problem and the procedures used to obtain the results, and summarizing the results. In today's laboratory, the analyst may use a computer-based word processor, spreadsheet, and laboratory information management system to prepare reports.

The object of every analysis is to extract the desired information from the sample and present it in a usable form. Instrumentation is only one component of the total method. The analyst should trace the flow of information through the entire process, not just the instrument component. This allows major sources of potential error to be identified.[14] Inadequate attention to any one of the areas outlined here can render the results meaningless, regardless of the potential analytical power of the instrumentation used.[15]

It is interesting to contrast instrumental with noninstrumental methods of analysis. Both have their strengths. Instrumental methods are usually faster and more sensitive after the necessary calibrations have been established. However, in many cases the classical gravimetric and volumetric methods are more accurate, even though much slower, than instrumental methods. Many instrumental methods offer improved detection limits over noninstrumental methods.

The following incident provides a comparison of the two types of analyses. A fresh, young Ph.D. was asked to do a determination of silica in a single sample. Having done his graduate research in X-ray fluorescence, he carefully prepared a series of standards and obtained a calibration curve. Later his supervisor and a coworker asked him for the results of his analysis. Ruefully, he had to admit that he had not yet run the analysis, yet his coworker had already obtained the results using the classical gravimetric method. Of course, the instrumental approach of the young Ph.D. would have eventually paid off, especially if there had been more samples involved.

1.4 BASIC FUNCTIONS OF INSTRUMENTATION

The purpose of chemical instrumentation is to obtain information from the substance being analyzed. In moving from the sample through the instrument to the output, the information (a chemical or physical quantity) is transformed. The number and complexity of the transformations are determined by the quality and quantity of data to be acquired from the sample under analysis.

Every analytical instrument may be divided into four basic components: a signal generator, an input transducer, an electronic signal modifier, and an output transducer. The signal results from the direct or indirect interaction of the analyte with some form of energy such as electromagnetic radiation, electricity, or thermal heating. Input transducers, also known as detectors, are devices that transform the physical or chemical property of the analyte into an electrical signal. Signal modifiers are electronic components that perform necessary and desirable operations, such as amplification and filtering, on the signal from the input transducer Finally, the output transducer converts the modified electrical signal into information that can be read, recorded, and interpreted by the analyst.

For example, in the spectrophotometric determination of copper in solution using dithizone as a reagent to form a red-violet complex, the radiation from a lamp is passed through a prism to obtain visible light at 525 nm to generate a signal that is absorbed by the copper complex. The amount of radiation absorbed at 525 nm is proportional to the concentration of the copper analyte. The radiant power is converted to an electrical current by a photodetector (input transducer). Electrical signal modifiers convert the current into a voltage and, after amplification and transformation to a logarithmic voltage, send it to the output transducer. This transducer may be a meter, a digital display, or a recorder.

Signal Generators

The signal used to transfer information from the analyte to the electrical modules of the instrument originates in the signal generator. Two general methods are used for signal generation: (1) application of an external signal to the sample and subsequent modification of this signal by the analyte as in absorption spectroscopy and (2) creation of a sample environment that allows the analyte to produce a signal as illustrated by potentiometric measurements. The signal generator is unique to each type of instrument. Its design requires an understanding of the physical properties of the instrument components, the chemical properties of the analyte, and the sample matrix.

Input Transducers[16]

Most input transducers are analog devices; that is, they measure physical and chemical properties continuously (Table 1.3). In most cases these devices produce analog electrical signals of voltage, current, or resistance. If the measured property is not continuous, the detector can be designed to give pulse outputs, such as in high-energy gamma radiation detectors. The quality and capabilities of the input transducer ultimately limit the overall performance of the instrument.

Signal Transformation Modules

The signal transformation module receives information from the detector, electrically converts it into a more meaningful form, and then presents it to the output transducer (Table 1.4). The detector used and the final form of the desired information determine the electronic composition of this module. Module components range from a single resistor for simple current-to-voltage conversion to a complex microprocessor that has a variety of signal-processing capabilities.

Output Transducers

The final instrument component, the output transducer, converts the modified electrical signal into information in a form useful to the analyst. This information

TABLE 1.3 INPUT TRANSDUCERS

Physical quantity measured	Input transducer	Electrical output
Concentration of electroactive species	Polarographic cell	Current
Ion activity in solution	Selective ion electrode	Voltage
Light intensity	Phototube	Current
	Photodiode	Current
Temperature	Thermistor	Resistance
	Thermocouple	Voltage

TABLE 1.4 ELECTRICAL SIGNAL TRANSFORMATIONS

Amplification	Digital-to-analog conversion
Analog-to-digital conversion	Filtering
Attenuation	Integration
Comparison	Rectification
Counting	Summation
Current-to-voltage conversion	Voltage-to-current conversion
Differentiation	Voltage-to-frequency conversion
Logarithmetic conversion	Antilogarithmetic conversion

TABLE 1.5 OUTPUT TRANSDUCERS

Alphanumeric printers	Oscilloscopes
Analog meters	Recorders, strip chart (y-t)
Digital meters	Recorders, x-y
Disks, hard	Tape cassettes
Diskettes, floppy	Video displays (cathode ray tubes)

may be displayed or recorded in either analog or digital form by a number of devices (Table 1.5).

Present instrumentation includes a relatively large number of signal generators and input transducers, fewer signal modifiers, and only a small number of different kinds of output transducers. An assortment of chemical and physical properties can be used to determine the analytes in a large number of sample environments. After the information from the analyte is converted to an electrical signal, the number of possible operations or modifications is limited, and there are even fewer types of output transducers. Specific input transducers are discussed in the appropriate chapters on instrumental methods. Output transducers are discussed in sections throughout the book when introduced as part of an instrument. After a brief introduction to the basic analog and digital solid-state circuit components in Chapter 3, some signal-modifying circuits are described in Chapter 4. Chapter 2 contrasts some of the electronic hardware and computer software techniques used to process electronic signals.

1.5 THE LITERATURE OF CHEMICAL INSTRUMENTATION[17]

The literature of analytical chemistry and the technical literature in general can be powerful aids to the analyst. A review of the literature associated with the analysis under consideration can save time-consuming operations in the laboratory. In many instances, the composition of industrial substances is partially or even fully revealed in patents or commercial literature, thus reducing or eliminating entirely the need for laboratory work.

Quantitative methods are abundant in the literature of analytical chemistry, and it is relatively simple to search this literature for the problem under consideration. Even if the literature does not provide a definite method for solving the problem, it may provide ideas for developing a new method.

Thus it is important for the analyst to become familiar with the literature and the methods of searching for pertinent information. Computer searching of the chemical literature is an important aspect of any contemporary analysis. It is now possible to conduct a computer search of *Chemical Abstracts* for abstracts written after 1968. It is important for the analyst to have a general understanding of the computer search procedures in order to select the combination of topics, authors, and so on that will provide either the desired information or an assurance that the specific information cannot be found. If the search is not well defined, the analyst will receive either irrelevant information or no information at all.

The literature of analytical chemistry is similar to that of other branches of chemistry. It may be classified by currentness and organization into three major groups: primary, secondary, and tertiary sources.

Primary sources contain the most recent original materials. The information is essentially unorganized, but in most cases it can be located by a computerized search provided it exists in a computerized information base. Manual searches may be necessary for older references. Primary sources include journals and other periodicals, government publications, dissertations, patents, and manufacturers' technical publications. Almost every article published in any chemical journal may contain information useful to the analyst. The methods used to identify and validate the purity of reactants and products are usually published as part of any paper that reports original research.

Periodicals of general interest to analytical chemists are *Analytical Chemistry, Analytica Chemica Acta, The Analyst* (*London*), *Talanta, Journal of the Association of Official Analytical Chemists,* and *Zeitschrift für Analytische Chemie.* Many articles of current interest are published in journals that cover particular fields within analytical chemistry. For example, developments in gas chromatography may be followed in the *Journal of Chromatography, Journal of High Resolution Chromatography, Chromatography Newsletter, Journal of Chromatographic Science,* and *Chromatographia.*

Secondary sources contain previously published materials in an accessible and useful format. Examples are reviews in periodicals and journals, bibliographies, tabular compilations, treatises, monographs, and textbooks. The function of secondary sources is to assemble widely scattered information into a concise, usable form. Materials in these sources are usually one to four years behind current developments because of the time required for collection and organization. One of the most useful comprehensive reviews is published annually in April by *Analytical Chemistry.* In even-numbered years recent developments in fundamental analytical techniques are covered, whereas in odd-numbered years the reviews are organized into applications such as air pollution, food, and coatings. Both types of reviews contain extensive bibliographies. A periodical devoted to reviewing analytical methods is *Critical Reviews of Analytical Chemistry.* The *Journal of Chemical Education* has a bimonthly feature dedicated to topics associated with chemical instrumentation.

In addition to the April reviews, *Analytical Chemistry* publishes several sections in the front, "A" (advertising) pages, that provide quick reviews of trends in instrumental methods and applications. Series with titles that begin with *Annual Reviews of* and *Advances in* provide ongoing summaries of progress in specific areas. Several trade periodicals such as *American Laboratory* and *Research and Development* provide similar articles. Treatises are available to provide information covering the entire spectrum of analytical techniques. The *Treatise on Analytical Chemistry,* edited by Kolthoff and Elving, is a concise, comprehensive, and systematic treatment of modern analytical chemistry, and *Comprehensive Analytical Chemistry* by Wilson and Wilson is another useful analytical treatise. The *Handbuch der Analytischen Chemie,* edited by Fresenius and Jander, is also an important reference.

A large number of monographs (an entire book devoted to a single topic) are available for all the methods of analytical chemistry. The materials are usually well organized but several years behind current developments. For any technique, it is most desirable to have the most current materials possible, but this is more critical in emerging techniques, such as liquid chromatography and inductively coupled plasma spectroscopy, than in the more mature methods, such as gas chromatography and atomic absorption spectroscopy. Selected monographs are cited in the Bibliography at the end of each chapter in this text.

Tertiary sources are publications designed to help the analyst use the primary and secondary sources and to keep up with developments in laboratory equipment, chemicals, and instrumentation. Abstracting journals are an important tertiary source. *Chemical Abstracts* and *Scientific Citation Index* are two important sources for analytical chemists. Lab guides or "Yellow Pages," published annually by *Analytical Chemistry, American Laboratory,* and *Research and Development,* are examples of this type of literature. Such sources include specific methods, procedures, and protocols for chemical analysis published by the American Society for Testing Materials (ASTM), the Association of Official Analytical Chemists (AOAC), and government regulatory agencies such as the Environmental Protection Agency (EPA) and the Food and Drug Administration (FDA).

A final word on the future of chemical literature: The electronic medium (computers and communications facilities) will provide rapid delivery of machine-searchable information. Laboratories will have the equipment to provide the analyst quick access to current information needed to solve problems. The analyst need only become familiar with the methods of computerized information retrieval.

1.6 IMPORTANT CONSIDERATIONS IN EVALUATING AN INSTRUMENTAL METHOD

The instrumental methods listed in Table 1.1 are grouped into three major divisions: spectroscopy, electrochemistry, and chromatography. The first chapter in each division of this text is devoted to general principles. The other chapters are dedicated to specific instrumental techniques. All chapters that discuss specific instrumental methods contain material from the following six major areas.

1. *How the method "works" (general theory).* The fundamental physical and chemical principles involved in the technique and the functions of the instrumental components are presented. The object of this section is to represent the transformation of information from one form to another as it passes from the sample through the instrument to the output device.
2. *Advantages and limitations of the method.* The capabilities and limitations of the method are featured in the second section, along with discussions of the presentation and interpretation of data. Major emphasis is on quantitative analysis and the limitations of the method, including types of samples handled, accuracy, precision, and limits of detection.
3. *Illustrative instrumentation.* This section describes several systems representative of current instruments. Diagrams and the accompanying discussion consider the practical aspects of the method, such as the relative cost, the training required to operate the instrument and interpret the results, and the time required for an analysis.
4. *Applications.* A general overview of the major areas for the application of the method is presented. Specific examples illustrate the utility of the method.

5. *Problems.* Problems in the quantitative and qualitative analysis of analytical data are given at the end of most chapters.
6. *Bibliography.* Bibliographical references have been selected to provide additional information at a suitable level. These references include general articles and books on each method as well as specific articles that give details on various applications, thus providing an important supplement to the material in each chapter.

BIBLIOGRAPHY

HIRSCHFELD, T., *Anal. Chem.*, **52**(2), 132A (1980).

LAITINEN, H., AND G. EWING, *A History of Analytical Chemistry,* the Division of Analytical Chemistry of the American Chemical Society, Washington, DC, 1977.

MOSSOTTI, V., "The Informational Structure of Analytical Chemistry," Chap. 1, *Treatise on Analytical Chemistry,* 2nd ed., I. Kolthoff, V. Mossotti, and P. Elving, eds., Part I, Vol. 4, John Wiley, New York, 1984.

PHILLIPS, J., *Anal. Chem.*, **53**(13), 1463A (1981).

RICHARDSON, J., AND R. PETERSON, eds., *Systematic Materials Analysis,* Vol. I, Chap. 1, Academic, New York, 1974.

WHAN, R., coordinator, *Metals Handbook,* 9th ed., Materials Characterization, Vol. 10, American Society for Metals, Metals Park, Ohio, 1986.

LITERATURE CITED

1. GRASSELLI, J., *The Analytical Approach,* American Chemical Society, Washington, DC, 1983.
2. SIGGA, S., *Survey of Analytical Chemistry,* McGraw-Hill, New York, 1968.
3. TAYLOR, J., *Anal. Chem.*, **55**(6), 600A (1983).
4. DESSY, R., *The Electronic Laboratory: Tutorials and Case Histories,* American Chemical Society, Washington, DC, 1985.
5. MASSART, D., *Trends in Anal. Chem.*, **1**(15), 348 (1982).
6. STROBEL, H., *J. Chem. Educ.*, **61,** A53 and A89 (1984).
7. RODGERS, J., *Am. Lab.*, **13**(3), 84 (1981).
8. TAYLOR, J., AND B. KRATOCHVIL, *Anal. Chem.*, **53,** 924A (1981).
9. LAITINEN, H., AND W. HARRIS, *Chemical Analysis,* 2nd ed., McGraw-Hill, New York, 1975.
10. TAYLOR, J., *Anal. Chem.*, **53,** 1588A (1981).
11. MCDOUGALL, D., et al., *Anal. Chem.*, **52,** 2242 (1980).
12. ECKSCHLAGER, K., AND V. STEPANEK, *Anal Chem.*, **54**(11), 1115A (1982).
13. HORWITZ, W., *Anal. Chem.*, **50**(6), 521A (1978).
14. ENKE, C., *Anal. Chem.*, **43**(1), 69A (1971).
15. PLEVA, M., AND F. SETTLE, *J. Chem. Educ.*, **62**(3), A85 (1985).
16. EWING, G., "Transducers," Chap. 4, *Treatise on Analytical Chemistry,* 2nd ed., I. Kolthoff and P. Elving, eds., Part I, Vol. 4, John Wiley, New York, 1984.
17. MELLON, M., *Chemical Publications, Their Nature and Use,* 5th ed., McGraw-Hill, New York, 1982.

2 MEASUREMENTS, SIGNALS, AND DATA

2.1 INTRODUCTION

A signal may be defined as the output of a transducer that is responding to the chemical system of interest. The signal may be divided into two parts, one caused by the analyte(s) and the other caused by other components of the sample matrix and the instrumentation used in the measurement. This latter part of the signal is known as noise.

Although the ability to separate significant data-containing signals from meaningless noise has always been a desirable property of any instrument, it has become imperative with the demand for increasingly sensitive measurements. The amount of noise present in an instrument system determines the smallest concentration of analyte that can be accurately measured and also fixes the precision of measurement at larger concentrations. Noise reduction (or signal enhancement) is a primary consideration in obtaining useful data from measurements that involve either weak signal sources, such as carbon-13 nuclear magnetic resonance spectroscopy, or trace amounts of analyte, as in voltammetry.

The two principal methods of enhancing the signal are (1) the use of electronic hardware devices, such as filters, or equivalent computer software algorithms to process signals from the measurement as they pass through the instrument and (2) postmeasurement mathematical treatment of data. Among the more useful postmeasurement methods are statistical techniques. In addition to signal enhancement, these techniques aid in identifying sources of error and determining precision, while providing a method for an objective comparison of results. This chapter will present some common noise-reduction techniques and briefly review important statistical methods typically used in the treatment of instrumental data.

2.2 SIGNAL-TO-NOISE RATIO

As concentrations decrease to trace levels or as signal sources become weak, the problem of distinguishing signals from noise becomes increasingly difficult, resulting in decreased accuracy and precision in measurements. The ability of an instrument system to discriminate between signals and noise is usually expressed as a signal-to-noise ratio (S/N), where

$$\frac{S}{N} = \frac{\text{average signal amplitude}}{\text{average noise amplitude}}$$

in the case of dc signals. An increase in the S/N ratio usually indicates a reduction in noise and thus a more desirable measurement. Once the physical or chemical quantity of interest is converted to an electrical signal, the S/N ratio cannot be increased by simple amplification alone, since each increase in the magnitude of the signal is accompanied by a corresponding increase in the value of the noise. Thus higher S/N ratios are usually obtained by electronic hardware devices (filters, lock-in amplifiers, etc.) or software algorithms (ensemble averaging, boxcar averaging, Fourier transformations, etc.) designed to reduce the contribution of the noise or to extract the signal from the noise. The hardware components will be discussed in more detail in Section 2.5 and in Chapter 3; the software algorithms are examined in Section 2.6.

2.3 SENSITIVITY AND DETECTION LIMIT[1]

A number of parameters, including the S/N ratio, affect the sensitivity of a particular instrumental method. Physical and chemical properties of the analyte, the response of the input transducer to the analyte, and the composition of the sample matrix are some of the more important factors that determine sensitivity. Sensitivity is defined as the ratio of the change in the instrument response (I_o, output signal) to a corresponding change in the stimulus (C, concentration of the analyte):

$$S = \frac{dI_o}{dC} \tag{2.1}$$

Slopes of calibration curves are used to determine the sensitivity values (Figures 2.1 and 2.2). It is usually desirable to maximize the sensitivity value unless one wishes to extend the instrument's range of response without diluting the sample.

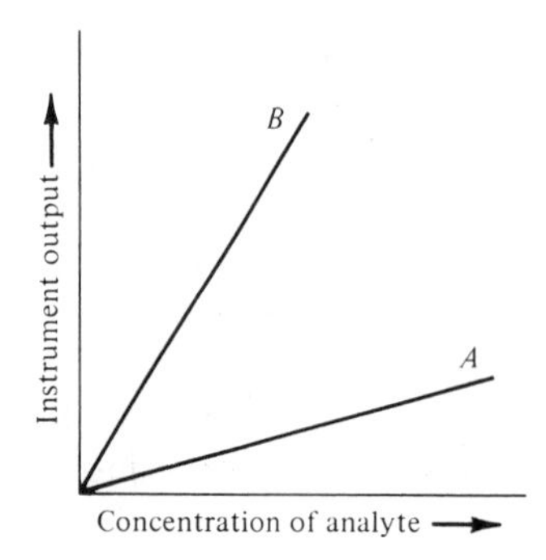

FIGURE 2.1
Linear response.

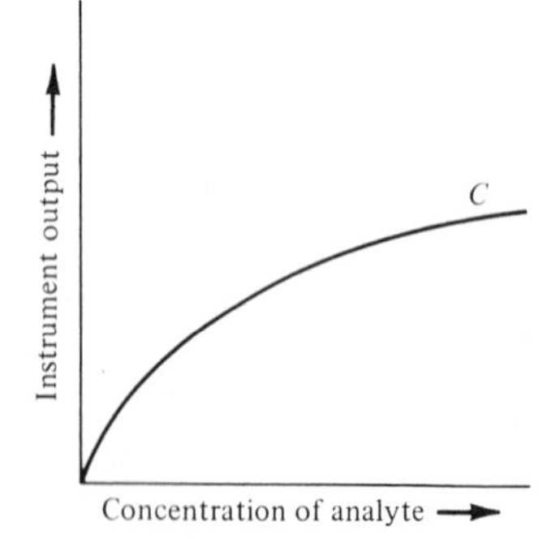

FIGURE 2.2
Nonlinear response.

Figure 2.1 shows a linear response (constant sensitivity) over the entire range of measured concentrations for both substances A and B. From the slopes of the curves, we see that the sensitivity of the method is much greater for substance B than for substance A. The nonlinear response in Figure 2.2 indicates a constantly changing value for sensitivity as a function of concentration. Measurements of substance C become less sensitive with increasing concentration. Sensitivity may also be expressed as the concentration of analyte required to cause a given instrument response. For example, in atomic absorption spectroscopy, sensitivity is expressed as concentration in micrograms per milliliter of analyte that produces an absorbance of 0.0043 absorbance unit (1.0% absorption). When comparing different techniques or instruments, one should be alert to the procedures used by the practitioners to arrive at sensitivity values.

As the concentration of the analyte approaches zero, the signal disappears into the noise and the detection limit is exceeded. The detection limit is most generally defined as the concentration of analyte that gives a signal, x, significantly different from the "blank" or "background" signal, x_B. This definition leaves the analyst with considerable freedom to define the phrase *significantly different*. When working with analytes in trace amounts, the analyst is confronted with two problems: reporting an analyte present when in fact it is absent and reporting an analyte absent when it is present. The literature of analytical chemistry has defined this difference to be an analyte concentration that produces a signal two times the standard deviation of the blank signal. Current guidelines define the detection limit as

$$x - x_B = 3s_B \tag{2.2}$$

where x is the signal with minimum detectable analyte concentration, x_B is the signal of the blank, and s_B is the standard deviation of the blank readings.

A comparison of sensitivity and detection limit is illustrated in Figure 2.3. It shows the results of atomic absorption analysis of solutions that contain equal concentrations of elements A and B. The sensitivities of the instrument system are identical for both metals. In the case of element B, little noise is present and thus the detection limit for this metal is considerably lower than that for element A. In the analysis of B, if reserve amplification is available, the signal amplitude can be increased (with a corresponding increase in noise), thus increasing the sensitivity for element B in this determination.

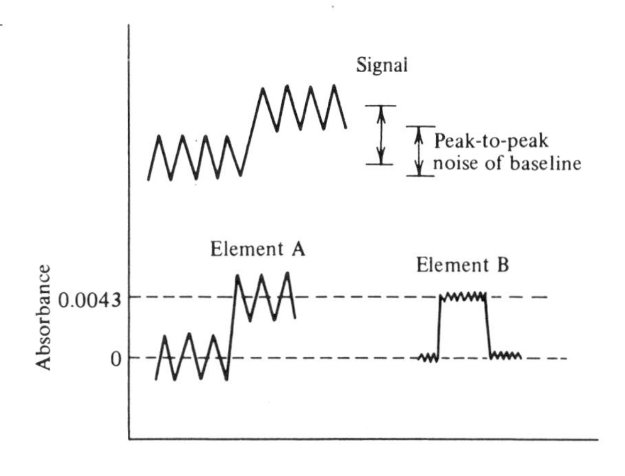

FIGURE 2.3
Detection limit for $S/N = 2$.

2.4 SOURCES OF NOISE[2]

It is important for the analyst who uses a particular instrumental method to be aware of the sources of noise and the instrument components used to minimize this noise because noise determines both the accuracy and detection limits of any measurement. Noise enters a measurement system from environmental sources external to the measurement system (Figure 2.4), or it appears as a result of fundamental, intrinsic properties of the system. It is usually possible to identify the sources of environmental noise and to either reduce or avoid their effects on the measurement. Such is not the case with fundamental noise because it arises from the discontinuous nature of matter and energy. Thus, fundamental noise ultimately limits accuracy, precision, and detection limits in every measurement.

The major kinds of noise associated with solid-state electronic devices are thermal, shot, and flicker.

Fundamental Noise

Thermal Noise. Noise that originates from the thermally induced motions in charge carriers is known as thermal noise. It exists even in the absence of current flow and is represented by the formula

$$V_{av} = \sqrt{4kTR\,\Delta f} \tag{2.3}$$

where V_{av} is the average voltage due to thermal noise, k is the Boltzmann constant, T is the absolute temperature, R is the resistance of the electronic device, and Δf is the bandwidth of measurement frequencies. Since thermal noise is independent of the absolute values of frequencies, it is also known as "white noise."

Methods for reducing thermal noise are suggested by Equation 2.3. Sensitive radiation detectors are often cooled to minimize this noise. Narrowing the frequency bandwidth of the detector is another way to reduce thermal noise, provided the frequencies important to the measurement of interest are not excluded. For example, if data-containing signals in the region between 10 and 20 kHz have a S/N ratio of 10 with a detector of $\Delta f = 1$ MHz, reducing the detector bandwidth by a

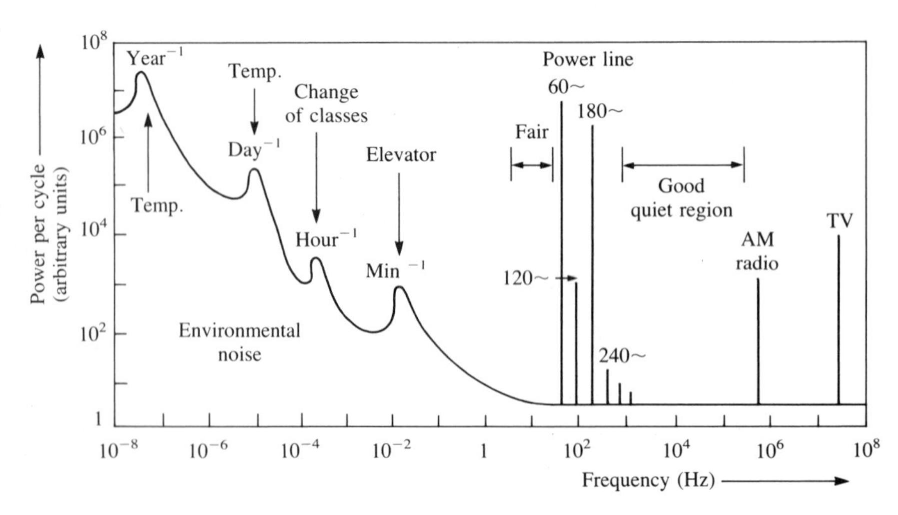

FIGURE **2.4** Pictorial representation of environmental noise in a typical location as a function of frequency. Note the $1/f$ character at low frequencies. [After T. Coor, *J. Chem. Educ.*, **45,** A540 (1968). Courtesy of the *Journal of Chemical Education.*]

factor of 100 to $\Delta f = 10$ kHz increases the S/N ratio by a factor of 10. It should be noted that this reduction in bandwidth is accompanied by a decrease in the intensity of the transmitted signal. Thermal noise is sometimes referred to as Nyquist noise after the physicist who derived Equation 2.3, or Johnson noise commemorating the engineer who first measured it.

Shot Noise. The magnitude of shot noise is much smaller than that of thermal noise and can therefore often be ignored. This kind of noise originates from the movement of charge carriers as they cross the $n - p$ junctions or arrive at electrode surfaces. Because these motions involve the movements of individual charge carriers, variations of current due to shot noise are random. Shot noise is signal dependent:

$$i_{av} = \sqrt{2Ie\,\Delta f} \tag{2.4}$$

where i_{av} is the shot noise, I is the intensity of the signal, e is the charge on the electron, and Δf is the measurement frequency bandwidth. This equation suggests that shot noise can be a problem at large signal values. Like thermal noise, shot noise is proportional to the square root of the measurement bandwidth, Δf, and can therefore be minimized by reducing the bandwidth.

Flicker Noise. The third kind of fundamental noise, flicker noise, is observed for low-frequency signals. Although the physical origins of this noise are not well understood, it can be represented by the following empirical equation:

$$V_{av} = \sqrt{\frac{KI^2}{f}} \tag{2.5}$$

where K is a constant depending on factors such as resistor materials and geometry, I is the dc current, and f is the frequency. Flicker noise predominates in measurements from 0 Hz (dc) up to about 300 Hz; it is due primarily to the contribution of the $1/f$ term. Although all solid-state devices are subject to flicker noise, field-effect transistors (FETs) seem to be affected less than bipolar devices. Flicker noise in amplifier systems is commonly referred to as drift. In sensitive measurements flicker noise may be eliminated by avoiding the use of low frequencies (including dc).

Environmental Noise

Environmental noise involves the transfer of energy from the surroundings to the measurement system and typically occurs at specific frequencies or a relatively narrow frequency of bandwidths. Two of the most common sources of environmental noise are the electric and magnetic fields produced by 60-Hz electrical transmission lines. This noise occurs not only at 60 Hz but also at frequencies corresponding to the harmonics (120, 180, 240,... Hz). Other sources of environmental noise are reflected radiant energy, mechanical vibration, and electrical interaction between different instruments. Reduction or elimination of this kind of noise involves shielding the circuits and wires used in signal transmission from external sources of energy. Proper grounding of all instruments and the transmission of signals at frequencies well removed from those of environmental noise are specific techniques for minimizing this noise.

2.5 HARDWARE TECHNIQUES FOR SIGNAL-TO-NOISE ENHANCEMENT

To avoid losing data, the signal from the input transducer (see Section 1.4) should be sampled at a rate twice that of the highest frequency component of the signal according to the Nyquist sampling theorem (see Section 2.6). Adherence to this theorem is important to obtain reliable results from either hardware or software S/N enhancement methods.

Filtering

Although amplitude and the phase relationship of input and output signals can be used to discriminate between meaningful signals and noise, frequency is the property most commonly used. As discussed in the previous section, white noise can be reduced by narrowing the range of measured frequencies; environmental noise can be eliminated by selecting the proper frequency. Three kinds of electronic filters are used to select the band of measured frequencies: low-pass filters that allow the passage of all signals below a predetermined cutoff frequency, high-pass filters that transmit all frequencies above a given cutoff point, and bandpass filters that combine the properties of the other two filters to pass only a narrow band of frequencies (Figure 2.5). The simplest filters are composed of passive circuit elements (resistors, R, capacitors, C, and inductance coils, L) with the transmitted frequencies determined by values of the individual circuit components (Figure 2.6). Bandpass filters can be designed using operational amplifiers.

Integration

Integration of dc signals for precisely limited time periods is a powerful way to reduce white noise. The coherent (nonrandom) signal adds directly with respect to

FIGURE **2.5** Filter types: (a) low-pass, (b) high-pass, and (c) active or bandpass.

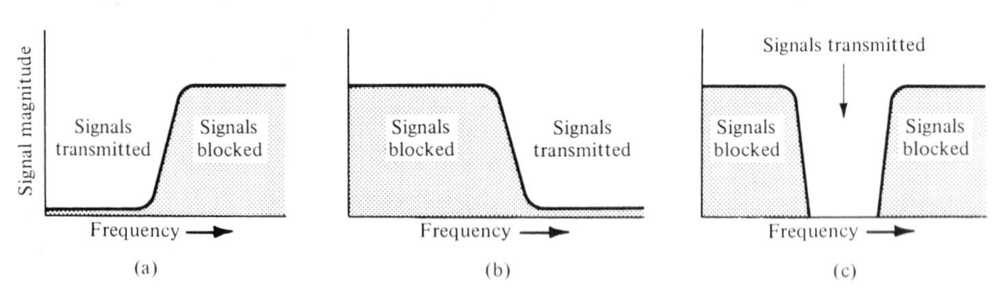

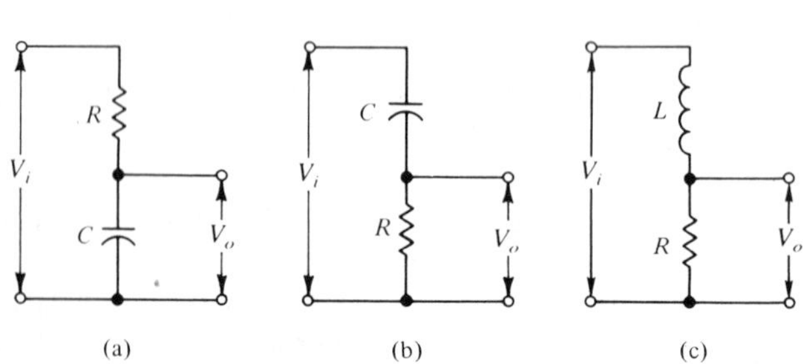

FIGURE **2.6** Passive filters: (a) low-pass RC filters, (b) high-pass LC filter, and (c) low-pass LR filter.

the integration time, whereas the random noise adds as the square root of the integration time; therefore, the S/N ratio increases with the square root of the integration time. Although a simple RC filter (Figure 2.6a) can be used to integrate signals, an operational amplifier with a capacitor in the feedback loop usually serves as a hardware integrator (see Section 3.2). Analog-to-digital converters such as voltage-to-frequency or dual slope devices have built-in S/N enhancement as a result of the integration techniques used in the signal conversion circuits.

Modulation/Demodulation

If the signal and noise cannot be separated by filtering, it is often advantageous to shift the signal of interest away from the noise frequency. To accomplish this, the signal is first transposed onto a carrier wave that has a desirable frequency, then it is transmitted to an amplifier tuned to the frequency of the carrier signal, and finally the original signal is recovered from the carrier wave. The first process is known as modulation; the final one as demodulation. Modulation/demodulation techniques can be used to process a signal in a region of minimum noise and also to discriminate between signal and noise on the basis of the signal's unique modulation configuration relative to the random pattern of the noise. This technique can be used, for example, to relocate signals away from dc where flicker noise is at its maximum. Any property of the carrier wave can be modulated by signals impressed upon it. Common examples are both amplitude and frequency modulation used in radio broadcasting and in optical spectrophotometers. The chopper (an electrical or mechanical device used to generate a signal alternating between the sample and reference measurements at a frequency selected to minimize noise) in a spectrophotometer should be located as close to the source of radiation as possible, since modulation removes only the noise arising after the chopper.

Active Filtering (Tuned Amplifiers)

Even when the signal is processed in a relatively noise-free environment, some noise will always be passed because of the bandwidth necessary to transmit the signal and the difficulty of obtaining and holding a match between signal frequencies and the filter bandpass. The lock-in or phase-sensitive amplifier offers a solution to these problems. Using a combination of signal frequency and phase relationships, it discriminates between both flicker and white noises. The functional components of a lock-in amplifier include a modulator (chopper), a multiplier, and a low-pass filter (Figure 2.7).

The data-containing signal at frequency f is superimposed onto the carrier wave frequency f_0 to produce a modulated signal, $f_0 + \Delta f$, that is then transmitted to an electronic device known as a multiplier. Simultaneously, a reference signal, modulated at the same frequency as the carrier signal and held in a constant phase relationship with the carrier wave, is sent to the multiplier. Under these conditions of identical frequencies and constant phase relationship, the multiplier can synchronously demodulate the combination of carrier and reference signals to yield a waveform at $2f_0 + \Delta f$, where Δf is the frequency containing the desired information. Since Δf is usually low-frequency data, it can be extracted from $2f_0$ by a low-pass filter. The bandwidth (of transmitted frequencies) can be adjusted by varying the RC time constant of the low-pass filter. These operations result in the transformation of the original spectrum of information frequencies Δf, centered about the carrier frequency f_0, to the same spectrum containing Δf centered at dc (0 Hz). As long as the carrier and reference waveforms occur at the same frequency and have a constant phase relationship (zero phase difference in the example of Figure 2.7), then the desired information minus the noise will appear in the transformed spectrum. This method of noise reduction is limited to data-containing signals that are periodic or can be modulated in such a way as to be made periodic. When this is not possible, as in the case of rapidly changing signals, other signal enhancement techniques must be used. Since the final low-pass filtering step is centered at dc, some flicker noise may still persist in lock-in amplifier systems.

Phase-sensitive detection is often used in spectrophotometers to achieve an increased S/N ratio. In atomic absorption spectroscopy, for example, the major sources of noise are the light source (hollow-cathode lamp) and the flame. Simultaneously chopping a reference light beam and the hollow-cathode light beam (Figure 2.7) produces two signals that have identical frequencies and a constant phase difference (180° in this example). Random noise in the flame, detector, and amplifier is minimized in the output of the lock-in amplifier. Noise that originates in the hollow-cathode source is not removed from the final signal because the lamp input is not modulated by the chopper. To remove the lamp noise, the lamp power supply must also be modulated to produce a periodic signal.

Boxcar Integrators

The boxcar integrator is a relatively simple method of signal enhancement for repetitive signals. It periodically samples the same portion of a signal for a fixed period of time and then averages the samples using a low-pass RC filter. This triggerable, gated integrator is a versatile measurement device. It provides S/N enhancement for the portion of the signal that is sampled. This technique has found wide application in instruments that require pulsed signal detection. It is best used for S/N ratio reduction in repetitive signals, although it can be used for more complex variable input waveforms.

When compared to the average value of a single pulse, boxcar integration gives S/N enhancement equal to the square root of the number of pulses integrated. Since

FIGURE **2.7** Application of a lock-in amplifier to an atomic absorption spectro-photometer.

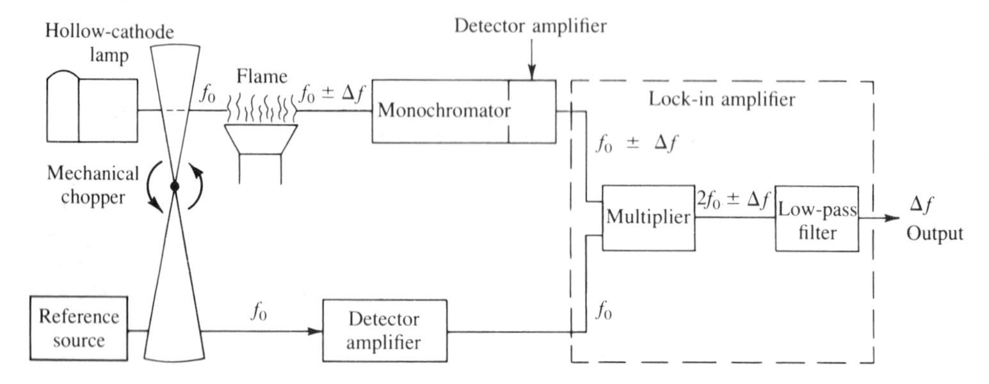

noise accumulates during the sampling time, further increases in the S/N ratio result from the shortened total sampling time of the boxcar method as compared to the time required to average a single pulse. As in the case of the phase-lock amplifier, the sampling frequency should be carefully selected to avoid interference from environment noise frequencies and their harmonics.

2.6 SOFTWARE TECHNIQUES FOR SIGNAL-TO-NOISE ENHANCEMENT[3–5]

The increased use of instruments that contain built-in microcomputers has increased the importance of software techniques for data acquisition and signal-to-noise enhancement. Operations such as filtering, linearization, and attenuation, formerly accomplished by hardware devices, are now achieved by software resident in the microcomputer component of the instrument. Software operations offer the advantages of flexibility and diversity. For example, a variety of software filters can be implemented by changing computer algorithms, whereas considerable effort may be required to change hardware filters. Nevertheless, in situations where the computer cannot execute the required function at a satisfactory rate, implementation with hardware components is necessary.

The minimum hardware required for software signal-processing functions is analog signal conditioning circuits and an analog-to-digital component as well as the microcomputer chips (see Section 4.3). The rates of sampling the analog data and of the analog-to-digital conversions must be fast enough to provide adequate resolution of the analog signal and thus ensure minimum loss of information. Although the resolution increases with the sampling rate, the upper limit of resolution is determined by the speed of the computer and the memory available for data storage. The minimum frequency required for accurate sampling, known as the "Nyquist frequency," should be twice that of the highest frequency component found in the data set. Each data point requires two coordinates, frequency and amplitude. If the sampling occurs at a rate less than the minimum, it is not clear which frequencies correspond to a given amplitude. If the sampling frequency significantly exceeds this minimum frequency, no additional information is transferred and the noise may increase because of the larger frequency bandwidth associated with faster sampling rates. Sampling rates corresponding to the fundamentals and harmonics of known environmental noise frequencies should be avoided.

Once the data are in digital form, a variety of software enhancement techniques may be used to increase the signal-to-noise ratio. Although these software techniques are readily available and widely used, caution should be exercised in their applications. The analyst should understand the advantages of each technique as well as potential problems such as undersampling, oversmoothing, and the time required to apply the technique to a set of data points.

Digital Filtering Technique[5–7]

Three of the most commonly used software signal enhancement techniques are boxcar averaging, ensemble averaging, and weighted digital filtering.

Boxcar Averaging. This software technique is the implementation of the hardware boxcar integration described in the previous section. In this method a group of closely spaced digital data points depicting a slowly changing analog signal is replaced by a single point representing the average of the group (Figure 2.8). Since this technique is well suited for applications in which the analog signal changes slowly with time, boxcar averaging can often be implemented in real time (averaging occurs simultaneously with the acquisition of the data). In this mode of operation, one group (boxcar) of points can be acquired and averaged before the next boxcar of data arrives. Enhancement of the S/N ratio can be calculated by the following equation:

$$S/N = \sqrt{n}\,(S/N)_0 \tag{2.6}$$

where $(S/N)_0$ is the signal-to-noise ratio of the untreated data and n is the number of points averaged in each boxcar. The effect of increasing n on the S/N ratio and signal resolution is shown in Figure 2.9.

Boxcar averaging can also be used for very rapidly changing signals when a short delay can be precisely controlled and the desired sampling interval is too fast for the available instrumentation—for example, fast laser spectroscopy in the nanosecond time frame.

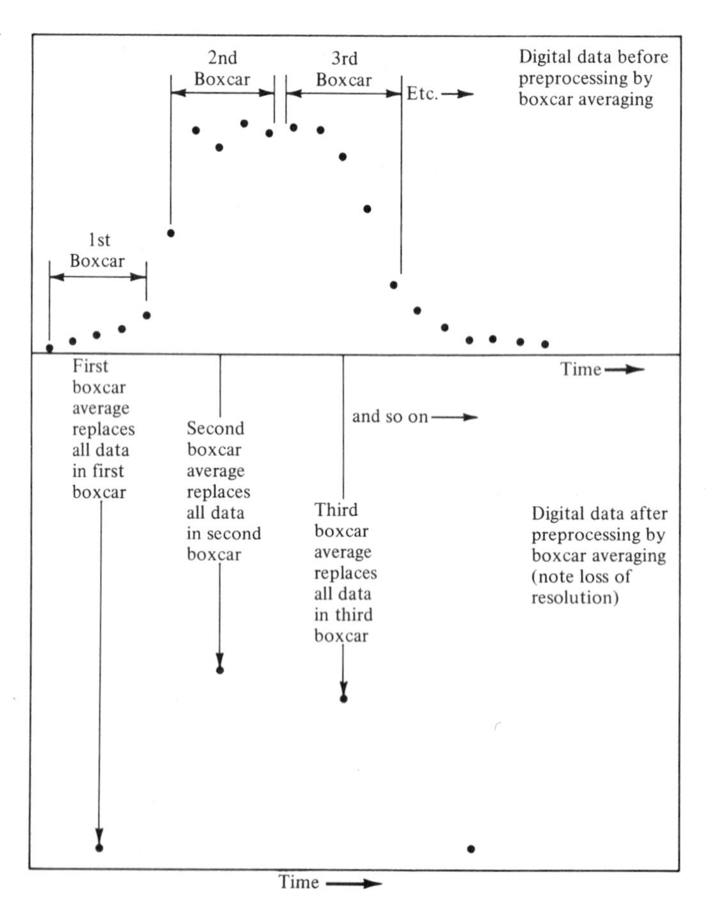

FIGURE **2.8**
Digital data before and after preprocessing by boxcar averaging. [After G. F. Dulaney, *Anal. Chem.*, **47,** 27A (1975). Courtesy of the American Chemical Society.]

Ensemble Averaging. This technique complements the boxcar method because it can be applied to signals that are changing rapidly. The results of n repeated sets of measurements of the same phenomenon are added and the sum is divided by n to obtain an average scan. If each set of measurements is recorded in the same way, the data contained in the measurements will sum coherently, whereas the random noise should average to a value smaller than the enhanced signal. To the extent that n represents a normal statistical distribution, the resulting S/N will be increased by a factor of n over that of a signal scan (Figure 2.10).

Computer-managed ensemble averaging has been used to extract small signals from background noise in instrumental techniques such as C-13 nuclear magnetic resonance spectroscopy. The principal liability of this technique is the time required to obtain a significant increase in the S/N ratio—100 scans to obtain an order-of-magnitude increase in the S/N ratio.

Smoothing (Weighted Digital Filtering). In digital filtering each of the data points to be averaged contributes equally to the calculation of the average (Figure 2.11). Assigning different weights to points as a function of their position relative to the central point can produce more realistic filtering. Adjustable filtering parameters include the mathematical smoothing function, the number of points and their positions relative to the central point in the moving average, and the number of times the data are processed by the smoothing function. Although this signal enhancement technique offers optimum flexibility in the choice of filter algorithms, the possibility of signal distortion is also great. The amount of time involved in weighted digital filtering usually requires that the method be applied after all the

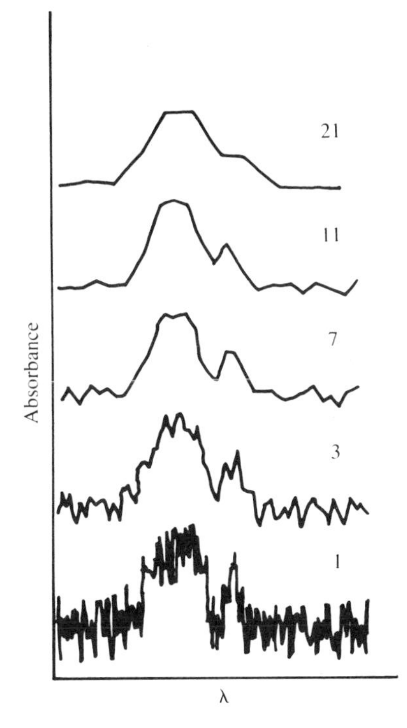

FIGURE **2.9**

Effect of boxcar averaging on a simulated noisy spectrum. The number of points included in the boxcar is given at the right of each plot. [After R. Thompson, *J. Chem. Educ.*, **62**(10), 866 (1985). Courtesy of the *Journal of Chemical Education.*]

FIGURE **2.10**

Effect of ensemble averaging on a simulated noisy spectrum. The number of times the signal was summed is given at the right of each plot. [After R. Thompson, *J. Chem. Educ.*, **62**(10), 866 (1985). Courtesy of the *Journal of Chemical Education.*]

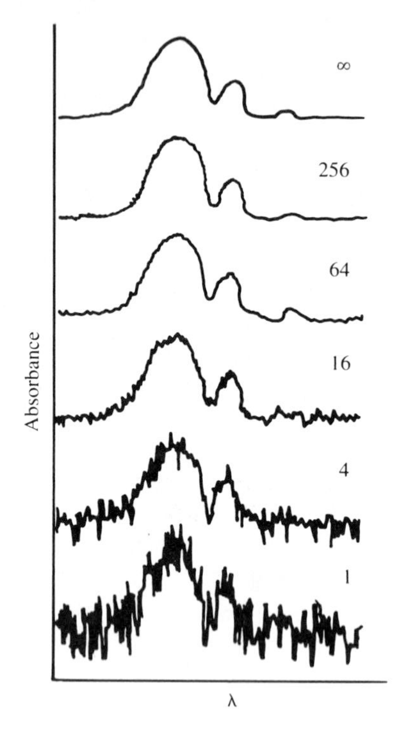

FIGURE **2.11**

Effect of a moving-average smooth on a simulated noisy spectrum. The number of points included in the smoothing function is given at the right of each plot. [After R. Thompson, *J. Chem. Educ.*, **62**(10), 866 (1985). Courtesy of the *Journal of Chemical Education.*]

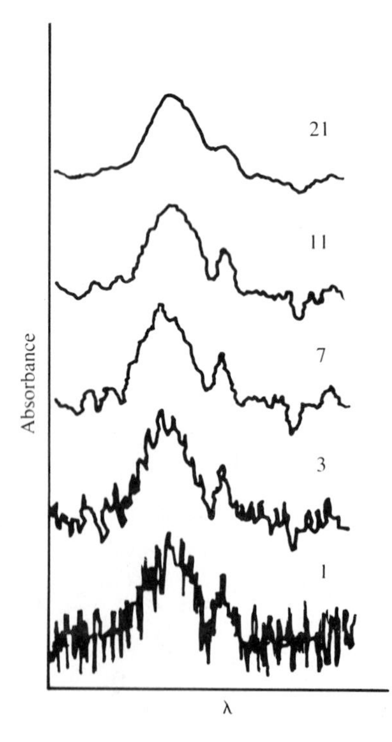

data have been acquired; in other words, extensive software digital filtering is not usually performed in real time.

Combining the three software signal enhancement techniques discussed thus far can produce an algorithm that is more useful than any individual technique. For example, an 8-point boxcar average coupled with an ensemble average of nine scans and a single postrun application of a 7-point filter would give a S/N increase of 22.4. More than 500 scans would have to be averaged to achieve the same result. Even if the ensemble and boxcar techniques were combined, 64 scans would be necessary.

A summary of signal enhancement techniques is given in Table 2.1. The central processing unit (CPU) time referred to in the table is the amount of computer time required to perform the calculations for a given process.

Fourier Transformations.[8–10] The mathematical operations known as Fourier transformations (FT) provide a powerful method of S/N enhancement. Applications of this technique in instrumental analysis usually fall into one of two general categories. The first involves the use of FT to produce spectroscopic methods that are much faster than conventional frequency domain methods. Data are rapidly collected in the time domain and then converted by FT to the conventional frequency domain. Since the time required for a single scan is greatly reduced, the time required for ensemble averaging is also decreased. Thus, the efficiency of ensemble averaging is improved by FT spectroscopy. Second, transformations of conventional signals may be multiplied by appropriate conditioning functions to achieve digital filtering and other useful signal modifications. A second mathematical operation, an inverse Fourier transformation, is required to restore the conditioned signal to its original form. Although a computer software algorithm, known as a fast Fourier transformation (FFT), has reduced the execution time by orders of magnitude over previous transform algorithms, software FFT signal conditioning techniques remain slower than hardware methods.

Two methods of data representation are the frequency-amplitude function, $F(\nu)$, and the less common time-amplitude function, $f(t)$. Both functions contain the same physical information but are expressed in different formats. The functions, known as a Fourier transform pair, are related by the following equations:

$$F(\nu) = \int_{-\infty}^{\infty} f(t) e^{-i(2\pi)\nu t} \, dt \tag{2.7}$$

$$f(t) = \int_{-\infty}^{\infty} F(\nu) e^{i(2\pi)\nu t} 2\pi \, d\nu \tag{2.8}$$

If, for example, $f(t) = A \cos 2\pi\nu_0 t$, then $F(\nu)$ is a single line at ν_0 with amplitude A (Figure 2.12a). If $f(t)$ is a square wave, then $F(\nu)$ broadens into the form shown in Figure 2.12b. Addition of three component waveforms, each occurring at a different frequency, produces the $f(t)$ pattern shown in Figure 2.13. The corresponding Fourier transformation, $F(\nu)$, consists of three lines located at the frequencies of the original waveforms. In this example, $f(t)$ represents the data produced by an optical interferometer or pulsed nuclear magnetic resonance (NMR) signal source, and $F(\nu)$ is the conventional frequency spectrum.

In Fourier transform spectroscopy the data are rapidly generated in the time domain [$f(t)$] form by either an interferometer (in the case of optical spectroscopy)

TABLE **2.1** DATA TREATMENT BY FILTERING, SMOOTHING, AND AVERAGING

Technique	Function	*S/N* improvement	Time required	Advantages	Disadvantages
Boxcar averaging (software)	Low pass filtering	Proportional to: (number of samples in box)$^{1/2}$	Number of PTS (number of samples $\times$ T_{conv}) CPU-time, small	Fast; useful in real time	Signal must slew slowly with respect to sampling rate; resolution lowered; some phase distortion
Ensemble averaging	*S/N* ratio enhancement	Proportional to: (number of scans)$^{1/2}$ 2	Number of PTS $\times$ T_{conv} $\times$ number of scans CPU-time, small	Useful even when $S/N < 1$ averages all random components regardless of f; negligible phase distortion	Signal must be stable; repetitive; noise must be random
Unweighted digital filter	Low pass filtering	Proportional to: (number of PTS-in-window)$^{1/2}$	Postrun CPU time; large	Capable of better resolution than boxcar; minimal phase distortion	
Weighted digital filter	Low pass, high pass, or bandpass filtering	Proportional to: (number of PTS-in-window)$^{1/2}$	Postrun CPU time; very large	Any filtering imaginable can be implemented	Slow; filter must have appropriate shape and width or distortion will occur
Analog filter hardware	Low pass, high pass, or bandpass filtering	Depends on components	Small	Fast	Possible phase and amplitude distortion

NOTE: After D. Binkley and R. Dessy, *J. Chem. Ed.*, **56,** 152 (1979). (Courtesy of the Journal of Chemical Education.)
PTS is the number of points; T_{conv} is the conversion time; and CPU is the central processing unit.

or a pulsed nuclear magnetic resonance signal. The resulting data are in the form of superimposed waves and include all the frequencies of the spectral range of the instrument. For the computer to have sufficient data to perform the necessary Fourier transformation, the Nyquist sampling function must be obeyed. Computer implementation of the transformation of the digitized data from time domain $f(t)$ to frequency domain $F(\nu)$ is carried out by means of a summation over a finite number of points N:

$$F(\nu_j) = \sum_{k=1}^{N} f(t_k)e^{-i(2\pi)\nu_j t_k}; \qquad j = 0, 1, 2, \ldots \tag{2.9}$$

where N is the total number of data points and ν_j is the jth component frequency out of a total of M frequencies. The results, $F(\nu)$, appear as a conventional frequency spectrum. Signal-to-noise enhancement is achieved by using the time saved by rapid

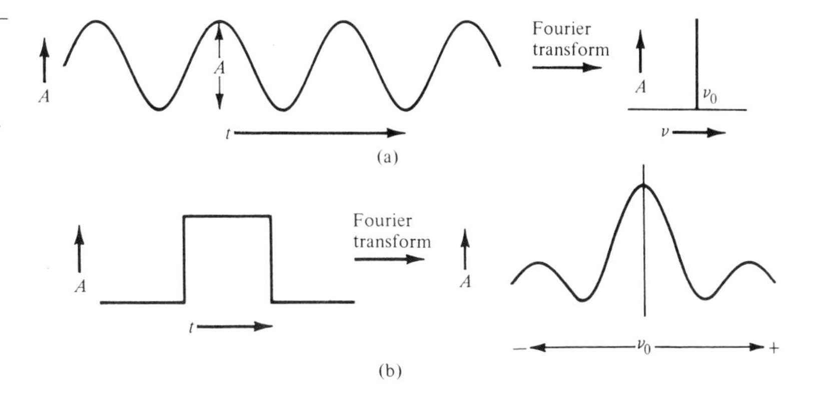

FIGURE 2.12
Simple Fourier transform pairs: (a) cosine function and (b) square-wave function.

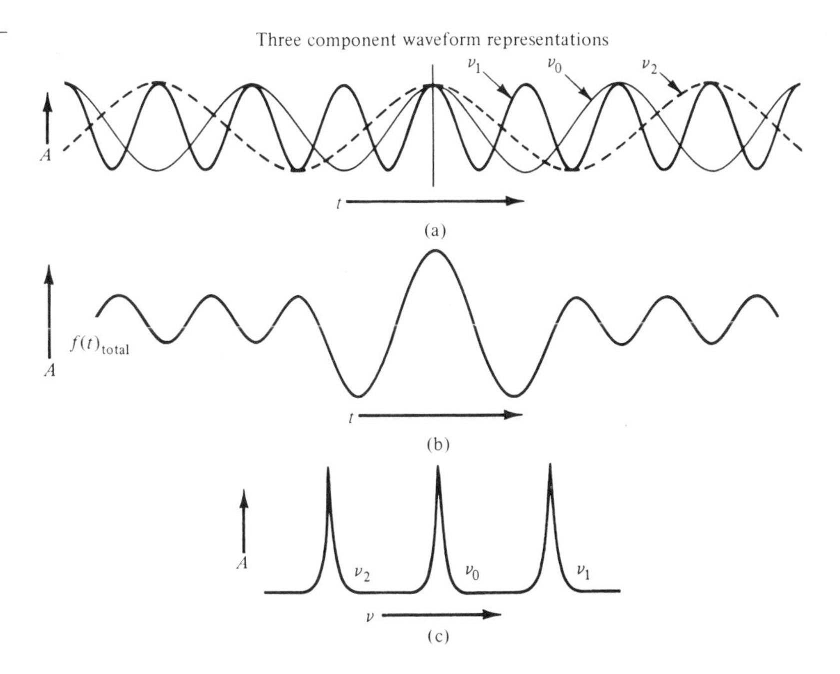

FIGURE 2.13
Fourier representation of data: (a) individual $f(t)$'s, (b) resultant $f(t)$, and (c) Fourier transform to $F(\nu)$.

scanning to make multiple scans that are then treated by signal averaging techniques. For example, in carbon-13 NMR spectroscopy, approximately 1000 scans are required to extract the weak carbon-13 signal from the noise. The FT technique requires 15 min to complete the scans, whereas the normal continuous-wave method would require 25 hr to obtain the same information.

The second application of FT is signal conditioning, illustrated by digital filtering. Data contained in an amplitude-frequency spectrum (Figure 2.14a) may be filtered by transforming them to an amplitude-time spectrum using a Fourier transformation (Figure 2.14b). The resulting waveform is then multiplied by an appropriate mathematical filter function to obtain the desired frequency response (Figure 2.14c). Finally an inverse Fourier transformation regenerates the filtered amplitude-frequency spectrum (Figure 2.14d). This method allows a variety of filter functions to be implemented with the appropriate software. Many of these filters would be impossible to implement using hardware devices. The major limitation of the software method is its slow speed relative to hardware techniques. In numerous situations where the speed of signal conditioning is not the limiting factor, the flexibility of software methods provides a great advantage over hardware techniques.

2.7 EVALUATION OF RESULTS

Total control of experimental variables is usually difficult and often impossible. Sampling methods, analysts' techniques, and instrument responses are potential sources of error. Statistical methods provide a means for objectively evaluating the source and amount of error in analytical methods. The common phrase *within experimental error* is meaningless if the magnitude of this error is not defined through the use of statistical techniques.

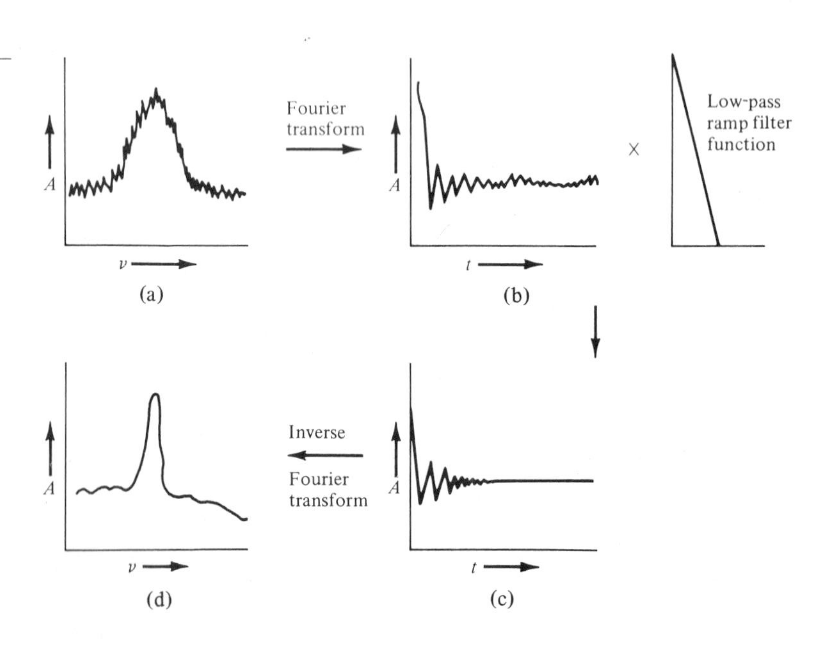

FIGURE **2.14**
Application of Fourier transforms in digital filtering.

Types of Errors[11]

To obtain reliable results from an analytical method, sources of error must be identified and either eliminated or minimized. Errors may be classified as one of two types, random (indeterminate) or systematic (determinate).

Since the intrinsically uncertain nature of the measurement technique is the source of random error, this kind of error occurs in every analysis. Thermal, shot, and flicker noise, discussed earlier in this chapter, are sources of random error. The magnitude of the random error is usually small and can therefore be minimized by filtering methods (either hardware or software).

The second kind of error, systematic or procedural error, causes results to deviate from the expected values in a constant manner. Sources include improper instrument calibration procedures, insufficient purity of reagents, and improper operation of the measurement instrument. This kind of error cannot be reduced by the application of statistical methods. Systematic errors may often be identified and minimized by modifying the analytical procedure.

Expression of Error

Error may be expressed in absolute terms as the difference between an analytical result, x, and the known true value, μ:

$$d = \mu - x \tag{2.10}$$

When this difference is expressed as an unsigned number, it is known as an absolute error. Because the absolute error represents a difference between the result and the true value, it must be expressed in the same units as these quantities. An absolute error has no significance when separated from the result or true value. For example, an absolute error of 5.1 μg/mL may be acceptable in an analysis of a sample containing 511 μg/mL lead but unacceptable in a sample containing 1.7 μg/mL lead.

The relative error, E_{rel}, is used to determine the accuracy of measurement and is typically expressed as the percentage of the known true value:

$$E_{rel} = \frac{d}{\mu}; \qquad \% E_{rel} = \frac{d}{\mu} \times 100 \tag{2.11}$$

Since the relative error is a dimensionless number, it can be used to determine the accuracy of results as well as to compare the accuracies of results expressed in different units.

The following example illustrates the use of absolute and relative errors for the comparison of results. Analyses of lead and zinc in a sample yield the following results: Pb = 653 μg/mL, d_{Pb} = 4.3 μg/mL; and Zn = 4.5 μg/mL, d_{Zn} = 0.15 μg/mL. Therefore,

$$\% E_{Pb} = \frac{4.3}{653} \times 100 = 0.65\%; \qquad \% E_{Zn} = \frac{0.15}{4.5} \times 100 = 3.3\%$$

It is obvious that the lead determination gives the more accurate result because it contains the smaller relative error, even though the absolute error is much greater than that of the zinc determination.

Precision and Accuracy

Accuracy may be defined as the agreement of a measurement with the known true value for the quantity being measured. Precision is concerned with the ability to reproduce the same values for a set of parallel observations. These terms are contrasted in Figure 2.15. The shots in Figure 2.15a are both accurate and precise, whereas the shots in Figure 2.15b are precise but not accurate. The situation in Figure 2.15b usually indicates the presence of a systematic error—in this case, perhaps poor alignment of the sighting mechanism on the rifle. While the accuracy of a measurement is determined by many factors, the precision is often limited by noise alone.

Precision and Significant Figures

Evaluation of an analytical method to discover the source and magnitude of errors requires careful acquisition and processing of data as well as appropriate statistical methods. Initial data must be reported with a precision that is indicated by the number of significant figures. Subsequent operations and calculations involving these data must preserve the correct number of significant figures so that the results give a true indication of both the accuracy and precision of the analysis. Moreover, unless the proper number of significant figures is maintained, the results of any statistical treatments are meaningless.

Significant figures usually have economic importance; the more significant figures reported in a measurement, the more costly the analysis. Factors that contribute to increased cost include more expensive instrumentation, higher-quality reagents, and increased expenditure of time and effort in performing the analysis. Once experimental observations have been recorded using the appropriate number of significant figures, care must be taken to ensure that the results of subsequent mathematical operations correctly represent the precision of the original measurements. It is impossible to increase the precision of experimental measurements by simple arithmetic operations. Likewise, care should be taken to assure that the number of significant figures is not improperly reduced in performing these simple operations. Readily available calculators and computers present the temptation to ignore the rules for handling significant figures and thus to produce results that nullify the experimental methods used to obtain the original data.

If the limit of error of a measurement is known, it should be represented as follows: 3946 ± 11. If the error is not stated, the last digit of the measured value is

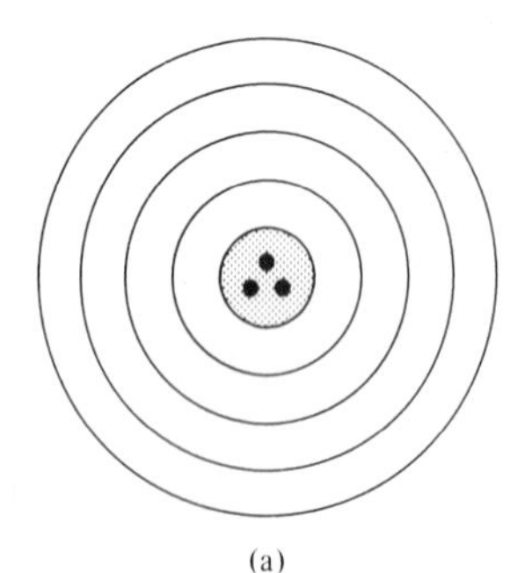

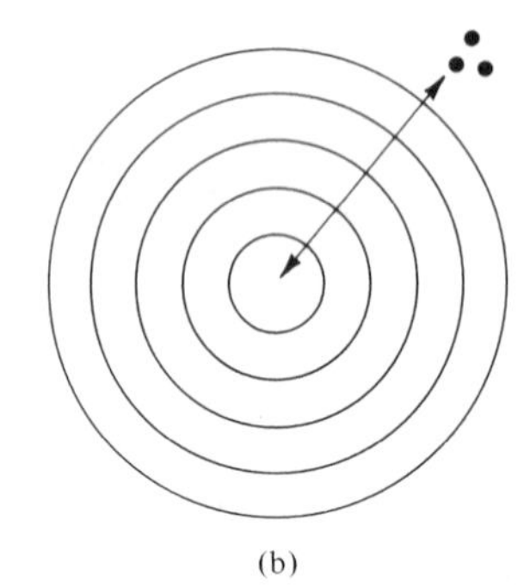

FIGURE 2.15
Accuracy versus precision: (a) shots accurate and precise and (b) shots precise only.

assumed to be uncertain and therefore determines the number of significant figures. Thus the value 3946 contains four significant figures and may represent a value between 3945.5 and 3946.5. The accepted rule for preserving the correct number of significant figures in multiplication and division is that the answer should contain only as many significant figures as the smallest number of figures in any factor (unless the factor is exactly known). For example, the product of 493.15 × 32 is 15,780.8, but the rule requires the answer to be rounded off to 16,000. This value should be unambiguously reported as 1.6×10^4. In addition and subtraction, the answer should be rounded off to the first column that has an uncertain digit. For example, 487. 3751 g, the sum of 394.5 g, 91.47 g, and 1.4051 g, becomes 487.4 g according to this rule.

Statistical Methods and Their Applications[12,13]

Data analysis can be said to be concerned with the study of populations and variation. If each measurement is thought of as an individual value, then repetition of the measurement produces a cluster or aggregate of values known as the population. Infinite repetition will generate the parent population or universe. Although the "true" value of a measured quantity may never be known, statistics provides a means of determining the features of the parent population from a few repetitive individual measurements. The three major functions of statistics are: (1) to determine the properties of the aggregate population, (2) to study the variations among individual measurements and the variations between the values of individual measurements and the average values of the aggregate, and (3) to reduce a large amount of data to a more easily comprehensible form.

A detailed discussion of elementary statistics is beyond the scope of this text. Excellent discussions of this topic are found in a number of references cited in the Bibliography at the end of this chapter. The precision of an analytical method is usually indicated by its confidence limit (Figure 2.16)—that is, the range of values about the mean that includes a specified percentage of the total observations. Precision increases with decreasing value of the standard deviation.

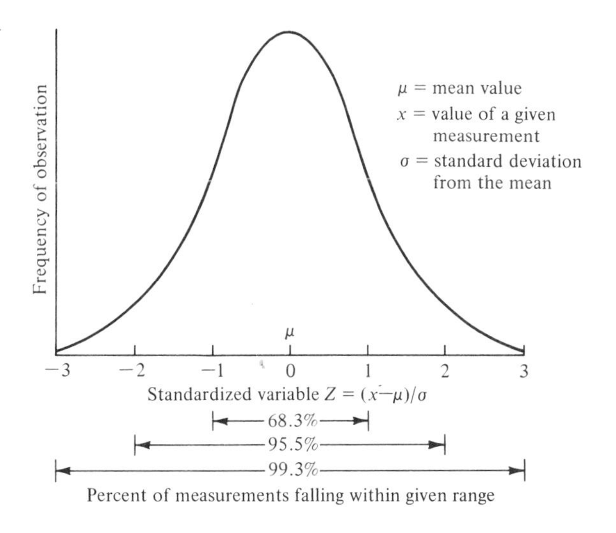

FIGURE **2.16**
Percent of measurements falling within a given range (assuming a Gaussian distribution).

Other statistical methods that are routinely used in analyses are the following:

1. The Q-test for outlying measurement (detects gross errors)
2. The t-test for significance in the difference between two means (useful in comparing results from different instruments, methods, or samples of supposedly the same composition)
3. The t-test for accuracy of results (for comparing the results of several methods against a known true value)

In addition to individual tests for significance of results, control charts are useful monitoring tools in laboratories where large amounts of data from a given method are generated over extended periods of time. The results of individual samples falling outside established confidence limits may be rejected. Trends in results may also be detected and corrective measures taken.

Although these and other statistical methods cannot solve all the problems involved in error analysis, they do provide the analyst with a systematic, objective approach to these problems. Statistical analysis of data should determine when differences among results have exceeded reasonable limits and have become large enough to imply the existence of real variations in sample compositions or analytical techniques.

2.8 ACCURACY AND INSTRUMENT CALIBRATION

Proper calibration (standardization) of instruments is essential in obtaining accurate analyses. The choice of a calibration technique is affected by the instrumental method, instrument response, interferences present in the sample matrix, and number of samples to be analyzed. Three of the most commonly used calibration techniques are the analytical or working curve, the method of standard additions, and the internal standard method.

Analytical Curve

In the analytical (working) curve technique, a series of standard solutions containing known concentrations of the analyte are prepared. These solutions should cover the concentration range of interest and have a matrix composition as similar to that of the sample solutions as possible. A blank solution containing only the solvent matrix is also analyzed, and the net readings—standard solution minus blank (background)—are plotted versus the concentrations of the standard solutions to obtain the working calibration curve (Figure 2.17). If a nonlinear plot results, as is often the case, electronic hardware or computer software can be used to compensate for the curvature and produce an output that is a linear function of concentration. The number of standard solutions analyzed in nonlinear regions should be increased to maintain accurate analysis of unknown samples.

Linearity may also be achieved in some analyses by varying instrumental parameters. In spectrophotometric analysis, changing the wavelength used to obtain absorption readings may produce a more linear working curve. It is of utmost importance to record all instrumental parameters used in obtaining data for the calibration curve because small variations in these parameters can affect the slope of the calibration curve. The calibration curve must be checked periodically using solutions of known concentration to detect any changes in instrument response.

Method of Standard Additions[14]

When it is impossible to suppress physical or chemical interferences in the sample matrix, the method of standard additions may be used. The instrument response must be a linear function of the analyte concentration over the concentration range and must also have a zero intercept (zero signal for zero concentration). A small amount of analyte solution of known concentration is added to a portion of a previously analyzed sample solution, and the analysis is repeated using identical reagents, instrument parameters, and procedures. If an instrument response R_x is obtained from a sample solution of unknown concentration x, and an instrument response R_0 is obtained from the sample solution to which a known concentration a of analyte has been added, then x can be calculated from the following equations:

$$R_x = kx \tag{2.12}$$

$$R_0 = k(a + x) \tag{2.13}$$

$$x = \frac{a \times R_x}{R_0 - R_x} \tag{2.14}$$

Readings must be corrected for any background signal. It is always advisable to check the result with at least one other standard addition. Additions of analyte equal to twice and to half the amount of analyte in the original sample are optimum statistically. All solutions should be diluted to the same final volume so that any interferent in the sample matrix will have an identical effect on each solution. Sufficient time must elapse between addition of the standard and the actual analysis to allow equilibration of the added standard with any matrix interferents.

FIGURE **2.17** Calibration curves for the determination of dimethylsulfoxide in aqueous solutions by (a) flame photometric detector and (b) flame ionization detector. [After M. O. Andrea, *Anal. Chem.*, **52,** 152 (1980). Courtesy of the American Chemical Society.]

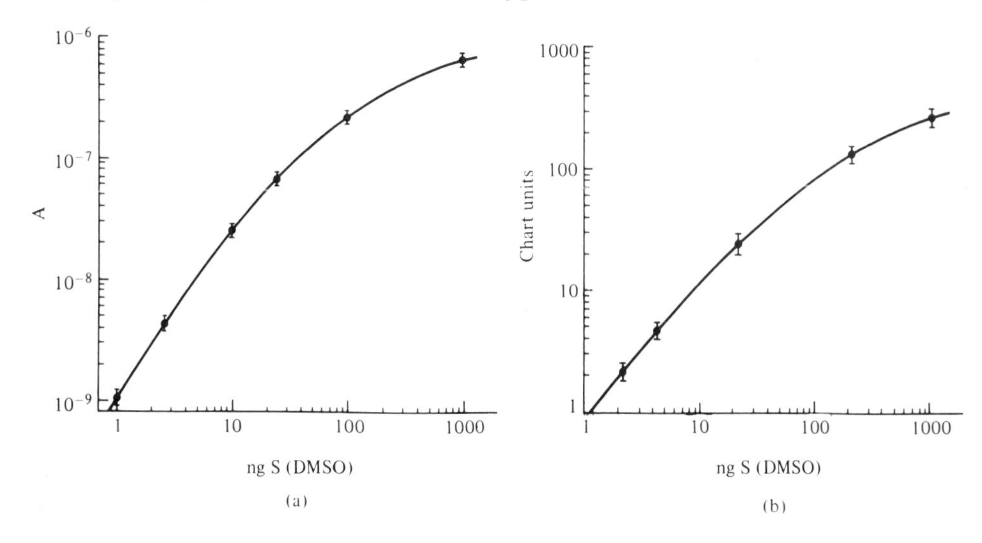

A graphic solution using the standard addition method is shown in Figure 2.18. The concentration scale (x-axis) is determined by the concentrations of analyte added to the sample solutions, and thus the unknown concentration is given by the point at which the extrapolated line intersects the concentration axis.

The method of standard addition is widely used in electroanalytical chemistry to obtain results that are more accurate than those obtained using calibration curves. Since the unknown and standard solutions are measured under identical conditions, matrix-sensitive voltammetric techniques such as anodic stripping voltammetry rely almost exclusively on standard additions for quantitative results. Atomic absorption and flame emission spectrophotometry use this method with complex sample matrices where viscosity, surface tension, flame effects, and other properties of the sample solution cannot be accurately reproduced in calibration solutions. Results from standard additions can also provide a systematic means of identifying sources of error in analysis, such as depletion of test reagents, a defective instrument, or inaccurate standard solutions.

Method of Internal Standard

An internal standard is used to minimize differences in the physical properties of a series of sample solutions that contain the same analyte. In this method, a fixed quantity of a pure substance is added to samples and standard solutions alike. The responses of the analyte and internal standard, each corrected for background, are determined, and the ratio of the two responses is calculated. If the parameters that affect the measured responses are controlled, the response of the internal standard line will be constant, since the concentration of the internal standard is fixed. If, however, one or more of the parameters that affect the measured responses vary, the responses of the analyte and the internal standard should be affected equally. Thus the response ratio (analyte to internal standard) depends only on the analyte concentration. A plot of the response ratio as a function of the analyte concentration yields a calibration curve. The standard must be added at the beginning of an analysis to allow for dissolution, mixing, and any other reactions to occur before a measurement is made. All equilibria must become established (some may be quite time dependent). The addition of standards to a dissolved sample can give misleading results if the possible reactions between the standard substance and the components are not considered.

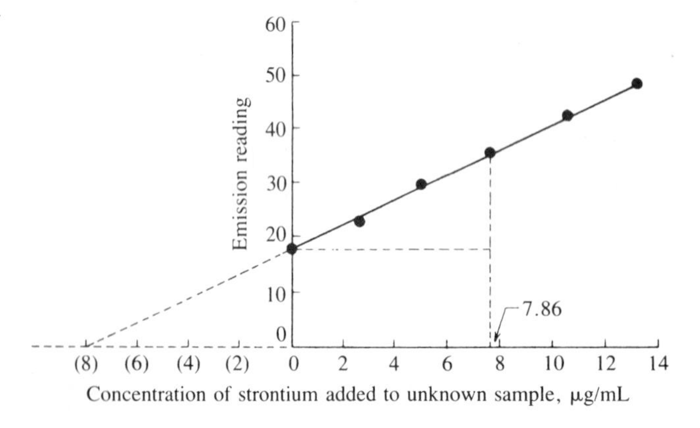

FIGURE **2.18**
Graphic representation of the standard addition method of evaluation.

The internal standard should be a substance, similar to the analyte, with an easily measurable signal that does not interfere with the response of the analyte. It should respond in a manner similar to the analyte to any variables that may affect the detector response. The concentration of the internal standard should be of the same order of magnitude as that of the analyte in order to minimize error in calculating the response ratios. This method is used extensively in gas chromatographic and atomic absorption analyses and to a lesser extent in infrared and emission spectroscopic determinations.

Isotopic Dilution

This is a special case of the method of internal standards that is used for quantitative determinations in radiochemical and mass spectral analysis. This technique measures the yield of a nonquantitative process, or it enables an analysis to be performed where no quantitative isolation procedure is known. To the unknown mixture that contains a compound with an inactive element P is added a known weight W_1 of the same compound tagged with the radioactive element P*. The specific activity A_1 of the tagged compound of weight W_1 is known. A small amount of pure compound is isolated from the mixture and the specific activity A is measured. The amount isolated need only be a very small fraction of the total amount present, merely a sufficient quantity for weighing or determining accurately. The extent of dilution of the radiotracer shows the amount W of the inactive element (or compound) present, as given by the expression

$$W = W_1\left(\frac{A_1}{A - 1}\right) \tag{2.15}$$

The method can be applied in mass spectrometry where A and A_1 represent the intensities of peaks that contain different isotopes (not necessarily radioactive) of a given element in a specific compound. The method has been used in the analysis of complex biochemical mixtures.

Comparison of Methods

Each of the methods has its advantages and limitations in quantitative analysis. If the analysis involves a large number of samples in a matrix with a known general composition, then the use of a calibration curve is favored. Standard addition is generally used when only a few samples are to be analyzed in a complex matrix. If the composition of the sample matrix is complex and the analysis includes a number of samples, then the method of standard additions may be the procedure of choice. Analysis that would otherwise require difficult quantitative separations may be performed using isotopic dilution.

2.9 CHEMOMETRICS[15,16]

Many of the techniques discussed in this chapter belong to an area known as chemometrics, the application of mathematical and statistical methods to chemical measurements in order to acquire chemical information on individual samples. These methods provide improved signal resolution and calibration by extracting increased information from the measurements. Major subdivisions of chemometrics are statistics, resolution, calibration, signal processing, modeling and parameter estimation, optimization, factor analysis, pattern recognition, image analysis, library searching of spectra, graph theory and structural handling, and artificial intelligence. Topics in this area are assuming increased importance as computer software becomes the critical interface between instruments and the resulting chemical information.

The practical application of the methods of chemometrics involves a variety of computer hardware and software. Until recently, scientific software was produced by instrument companies and computer manufacturers to enhance and support their products and by individual scientists for the implementation of specific tasks. Currently a number of scientific software applications packages are available from commercial software houses, publishing companies, and computer manufacturers.[16,17] The functions performed by these packages range from real-time data acquisition through statistics to expert systems.

With the emergence of smart instruments (see Chapter 4) controlled by built-in optimization algorithms, the costs of extracting more information from existing data must be weighed against the cost of producing additional data by performing more measurements, which often requires more sampling. For example, the experimental design of measurements affects the amount of information obtained from a given analytical method.[18] A well-designed measurement system should yield more information from a given amount of data than a poorly designed system. An application of this principle is the use of recursion to calculate the parameters for a calibration function. After each sample measurement, a measurement is made on a standard, the parameters of the calibration function are re-estimated, and these values are fed back to the experimental design function to update the calibration of the instrument.

Effective use of hyphenated methods (see Section 1.2) often involves one or more chemometric methods. GC-MS combines the ability of gas chromatography, an excellent quantitative technique, to separate the components of complex mixtures with mass spectrometry, a powerful qualitative technique, to identify the components. As long as the components are well separated on the column of the gas chromatograph, it is easy to obtain the desired analytical information. However, in many cases involving complex sample mixtures, good separation of components may be obtained only after time-consuming optimization of GC parameters, with some risk that the necessary separation will not be achieved. In this situation, the limits of traditional analytical methods may have been reached. Application of the mathematical tool known as multivariant analysis allows the number of components in each peak of the chromatogram of a poorly separated mixture to be estimated. In addition, the mass spectrum of each component of a given chromatographic peak can also be estimated provided there are fewer than four components in a single peak.

Chemometrics can combine principles from applied mathematics with computer technology and instrumentation to form powerful analytical methods. For example, a GC-MS instrument system capable of acquiring data at rates of megasamples per second requires efficient information-processing techniques to produce meaningful results from massive amounts of raw data. The mathematical tools in chemometrics provide the means to convert raw data into information,

information into knowledge, and ultimately knowledge into intelligence. In the GC-MS analysis of dioxins, the data from the instrument are used to determine the identities and amounts of the dioxins present in the sample (information). The information can be used to gain improved knowledge of the operation of the system from which the sample was taken, the relationships between the operating conditions of incinerators, and the amounts of dioxins in their emissions. Finally, this knowledge can be used in the intelligent operation of incinerators to minimize dioxin emissions, possibly using an expert system to monitor and control the incinerator.

PROBLEMS

1. Distinguish between the sensitivity and detection limit.

2. In which of the following measurements has the detection limit been reached? Show calculations to justify your answer.

Measurement	Analyte signal	Blank signal
1	1.52 + 0.05	1.38 + 0.07
2	0.94 + 0.03	0.81 + 0.02

3. What types of noise can be reduced by (a) reducing the bandwidth of measurement frequencies, (b) reducing the temperature of the measurement, and (c) reducing the frequency of the measurement?

4. Calculate the increase in the S/N ratio of a measurement by increasing the integration time from 1.0 sec to 5.0 sec. When is it not advisable to increase integration times as a means to improve the S/N ratio?

5. What types of noise can be reduced by (a) hardware filters, (b) integration of the signal, and (c) modulation/demodulation?

6. What advantages do the following hardware signal enhancement techniques offer over other hardware devices: (a) active filters and (b) boxcar integrators?

7. Explain how the software signal enhancement techniques of boxcar averaging and ensemble averaging complement each other.

8. (a) Calculate the increase in the S/N ratio by ensemble averaging 200 repetitive scans of a spectrum. (b) Discuss the limitations of ensemble averaging in signal enhancement.

9. How are fast Fourier transformations used to reduce noise?

10. Explain the advantages of fast Fourier transformations over ensemble averaging as a signal enhancement technique.

11. What types of experimental error are minimized by hardware and software filtering techniques? Explain.

12. Explain, using an example, how an accurate measurement could contain a large absolute error.

13. What is the real significance of significant figures?

14. A metal naphthenate sample, ashed and diluted to a fixed volume, gave an absorbance reading of 29. Solutions B and C, containing the same quantity of unknown solution plus 25 and 50 $\mu g/mL$ of barium, gave readings of 53 and 78, respectively. Calculate the quantity of barium in the original sample.

15. Given the accompanying data for determining the amount of ethyl alcohol in a wine sample using acetone, calculate the following: (a) the weight of alcohol in the sample and (b) the weight % of alcohol in the sample.

Calibration mixture

Component	Weight	Area
Acetone	1.18 g	45,000 units
Ethyl alcohol	0.42 g	18,500 units
Water	6.12 g	No peak detectable with flame ionization detector

Sample mixture

Acetone	1.50 g	60,100 units
Ethyl alcohol		24,200 units

16. A fermentation broth was known to contain some Aureomycin. To a 1000.0-g portion of the broth was added 1.00 mg of Aureomycin containing carbon-14 (specific activity = 150 counts/min/mg). From the mixture, 0.20 mg of crystalline Aureomycin was isolated, which had an activity of 400 counts in 100 min. Calculate the weight of Aureomycin per 1000 g of broth.

17. Select a journal article describing an analytical method and point out the chemometric methods used to obtain useful information about the analyte from the sample.

BIBLIOGRAPHY

CURRIE, L., "Sources of Error and the Approach to Accuracy in Analytical Chemistry," Chap. 4, *Treatise on Analytical Chemistry,* 2nd ed., I. Kolthoff and P. Elving, eds., Part I, Vol. 1, John Wiley, New York, 1978.

MANDEL, J., "Accuracy and Precision: Evaluation and Interpretation of Analytical Results," Chap. 5, *Treatise on Analytical Chemistry,* 2nd ed., I. Kolthoff and P. Elving, eds., Part I, Vol. 1, John Wiley, New York, 1978.

MEYERS, S., *Data Analysis for Scientists and Engineers,* John Wiley, New York, 1975.

MILLER, J., AND J. MILLER, *Statistics for Analytical Chemistry,* Halsted, New York, 1984.

MOSSOTTI, V., "Informational Structure of Analytical Chemistry," Chap. 1, *Treatise on Analytical Chemistry,* 2nd ed., I. Kolthoff, V. Mossotti, and P. Elving, eds, Part I, Vol. 4, John Wiley, New York, 1984.

PIERCE, T., AND B. HOHNE, eds., "Analytical Chemistry," Chaps. 22–28, *Artificial Intelligence: Applications in Chemistry,* American Chemical Society, Washington, DC, 1986.

TAYLOR, J. K., *Quality Assurance of Chemical Measurements,* Lewis, Chelsea, MI, 1987.

LITERATURE CITED

1. LONG, G., AND J. WINEFORDNER, *Anal. Chem.,* **55**(7), 713A (1983).
2. COOR, T., *J. Chem. Educ.,* **45,** A540 (1968).
3. THOMPSON, R., *J. Chem. Educ.,* **62**(10), 886 (1985).
4. SHUKLA, S., AND J. RUSLING, *Anal. Chem.,* **56**(12), 1347A (1984).
5. SAVISTY, A., AND M. GOLAY, *Anal. Chem.,* **36,** 1627 (1964).
6. DULANEY, G., *Anal. Chem.,* **47,** 25A (1975).
7. BINKLEY, D., AND R. DESSY, *J. Chem. Educ.,* **56,** 148 (1979).
8. BECKER, E., AND T. FARRAR, *Science,* **178,** 361 (1972).
9. MARSHAL, A., AND A. COMISAROW, *Anal. Chem.,* **47,** 491A (1975).
10. ISENHOUR, T., et al., *Anal. Chem.,* **53**(7), 889A (1981).
11. HARRIS, W., *Am. Lab.,* **31** (1978).
12. YOUDEN, W., *Statistical Methods for Chemists,* John Wiley, New York, 1951.
13. BAUER, E., *A Statistical Manual for Chemists,* 2nd ed., Academic, New York, 1971.
14. BADER, M., *J. Chem. Educ.,* **57,** 703 (1980).
15. KOWALSKI, B., et al., *Anal. Chem.,* **58**(5), 294R (1986).
16. SHARAF, M., D. ILLMAN, AND B. KOWALSKI, *Chemometrics,* John Wiley, New York, 1986.
17. BORMAN, S., *Anal. Chem.,* **57**(9), 983A (1985).
18. VANDEGINSTE, B., *Pure and Appl. Chem.,* **55**(12), 2007 (1983).

ELECTRONICS: FUNDAMENTALS AND APPLICATIONS OF SOLID-STATE DEVICES

This chapter provides an overview of the basic principles and circuit components of semiconductor technology used in modern instrumentation. No attempt has been made to review ac and dc circuits or older electronic devices such as vacuum tubes. References cited in the Bibliography at the end of this chapter contain reviews of fundamental electronics.

3.1 SEMICONDUCTOR COMPONENTS

The origins of contemporary integrated circuit (IC) chips can be traced to the transistor, a small low-power amplifier developed more than 30 years ago to replace the large power-hungry vacuum tube. To understand the general concepts of transistor operation, it is necessary to consider briefly the properties of semiconductor materials.

Semiconductor devices are made by introducing controlled numbers of "impurity atoms" into a crystal of semiconductor material (silicon or germanium) by a process known as doping. The impurity atoms usually belong to either group V or group III of the periodic chart and possess atomic radii that allow them to replace individual silicon or germanium atoms without disrupting the crystalline structure. If a group V element, such as phosphorus, is introduced into a pure semiconductor crystal, the resulting material will contain an excess of mobile carrier electrons and is known as an *n*-type semiconductor.

Conversely, doping with boron, a group III element, yields a *p*-type semiconductor that has a deficiency of carrier electrons. Each electron deficiency is known as a hole, which is not really a particle but merely the absence of an electron at a position where one would normally be found in a pure lattice of silicon atoms. A hole possesses a positive electric charge and can therefore carry electric current. Holes move through semiconductor material in much the same way that bubbles move through a liquid medium and almost as rapidly as a carrier electron.

The principal active components of electronic circuits are transistors and diodes. The diode is the simpler device that allows current to flow in one direction (Figure 3.1a) but not the other (Figure 3.1b). There are two major types of transistors, bipolar and field-effect transistors (FETs). These devices perform two electronic functions, switching and amplification. In nonlinear digital circuits, a transistor typically functions as a high-speed electronic switch, which is opened or closed depending on the applied voltages. In linear circuits, transistors operate as power amplifiers to increase the output voltage or current proportionally to the input.

Bipolar transistors, first constructed in 1948, consist of two *p-n* junctions formed by sandwiching a thin slice of one kind of semiconducting material between two thicker sections of the other semiconducting material (Figure 3.2). The FET was conceived in the early 1930s but was not fabricated in quantity until 30 years later. The metal oxide semiconductor field-effect (MOSFET) transistor is widely used in microelectronic circuits (Figure 3.3). Complementary MOS (CMOS) transistors represent a further advance in transistor technology. CMOS devices offer the advantages of reduced power consumption and excellent noise immunity compared with MOS components. The difference in fabrication methods allows roughly four times more MOS or CMOS transistors to be fitted on a given area than bipolar ones. However, bipolar transistors retain the major advantage of higher response speed.

Transistors may be used as discrete units or as components of a microelectronic circuit. The advent of microelectronics has not affected the functions of the basic components—namely, transistors, resistors, and so on. The major difference is that all these components are available as an electrical functional unit fabricated on a single small IC chip. Many problems of circuit design are solved within the IC, thus simplifying the design, operation, and maintenance of instrumentation.

FIGURE 3.1 Schematic representation of diode operation.

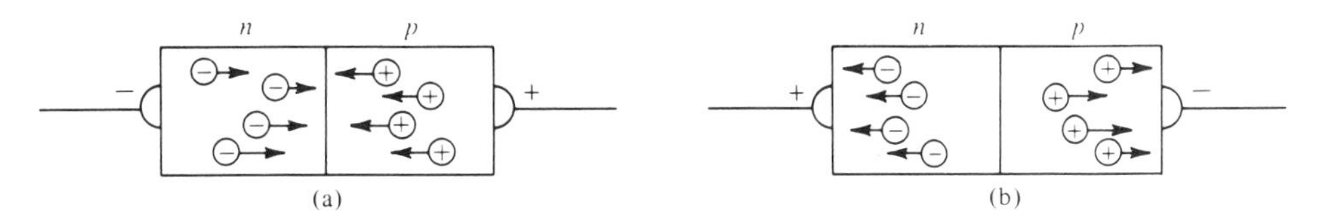

FIGURE 3.2 Transistor, *p-n-p* type.

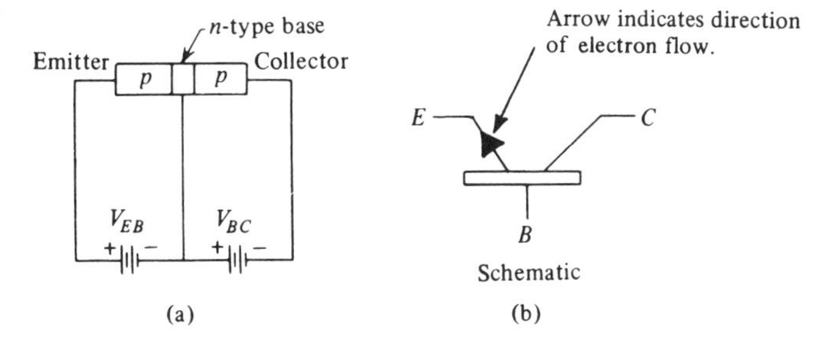

FIGURE 3.3 Enhancement-mode *n*-MOFSET. (a) *n*-Carriers conducting with positive gate voltage; (b) schematic.

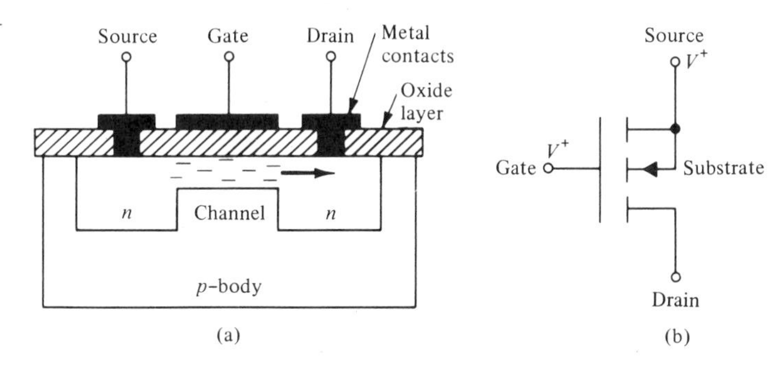

ICs may function in a linear or nonlinear manner. The output of a linear IC is directly proportional to the input. Linear IC applications include many types of amplification, modulation, and voltage regulation. The operational amplifier is the most important type of linear IC. Nonlinear ICs include all digital ICs and other circuits where there is not a linear relationship between the input and output signals. Digital ICs, the most important type of nonlinear ICs, usually use some form of bistable (on/off) operation. These ICs are common in computer circuits and in other digital applications such as counters, calculators, and digital data communication equipment.

ICs must be placed in a protective housing and have connections to the outside world. There are three methods of packaging ICs in containers (Figure 3.4): the TO-5 glass metal can, the ceramic flat pack, and the dual-in-line ceramic or plastic flat packs known as dual-in-line packages (DIPs). The popular, less expensive plastic DIP packages can have 14, 16, 18, 24, or 40 connecting pins. A minimum of two pins is required for connecting the IC to the power supply. The remaining connections are available for use as terminals for input and output signals.

3.2 OPERATIONAL AMPLIFIERS[1]

The operational amplifier (op amp) circuit is the basic component of most analog signal modifiers. It is a high-gain amplifier that, when connected to other electrical components, can perform a variety of mathematical operations on analog signals. The first op amp circuits were composed of vacuum tube amplifiers and other necessary elements such as diodes, resistors, and capacitors. Discrete transistors later replaced tubes, and finally entire operational amplifier circuits became available on IC chips. About one-third of all ICs are op amps; more than 2000 types are commercially available. These facts attest to the importance and diversity of op amp applications.

In addition to the obvious disparity in size, there are several other important differences between an IC op amp and earlier versions. An IC op amp consumes little power and operates at relatively low voltages (approximately 12 to 15 V) with power supplies that need not be highly regulated. The IC op amps have very low drift with

FIGURE 3.4

Types of IC packages. (Courtesy of Texas Instruments.)

both temperature and time and can withstand short circuits on the output without damage. The major disadvantages of the IC op amps are the limited operational voltage range (approximately 12 to 15 V), relatively low output current (less than 10 mA), and low input impedance (less than 1 MΩ). New generations of IC op amps have overcome the problems of low output current and low impedance. For high-voltage outputs, discrete transistors must still be used but can be controlled by op amps.

Operational Amplifier Characteristics

The desirable characteristics of op amps are derived from two properties of these devices: high-gain dc amplification and the provision for external feedback (returning a fraction of the output signal to the input). This combination of high gain and feedback allows the output to be independent of the internal parameters of the op amps. Thus the output signal can be made to depend primarily on the components in the external feedback circuit. In addition, the properties of high input impedance, low output impedance, and good response to high-frequency input signals are required for op amps to function properly.

Basic Operational Amplifier Circuits

Schematically the op amp is represented by a triangle with inputs and outputs (Figure 3.5a). Although the noninverted input is most often connected to ground (Figure 3.5b), it may be connected to an active signal input for certain applications. The gain ($G = V_o/V_i$) measured in this configuration is known as open-loop gain.

When the op amp is used for instrumental applications of signal transformation, it is normally used in the closed-loop configuration where the impedances Z_1 and Z_2 generally consist of resistors and/or capacitors external to the op amp (Figure 3.6). Since the input impedance of the op amp itself is large, i_a is practically zero and can be neglected. Thus $i_1 = i_2$.

Furthermore, normal operation of the operational amplifier requires that the two inputs be at essentially the same voltage, $V_d = (V_+ - V_-) \cong 0$. The allowable

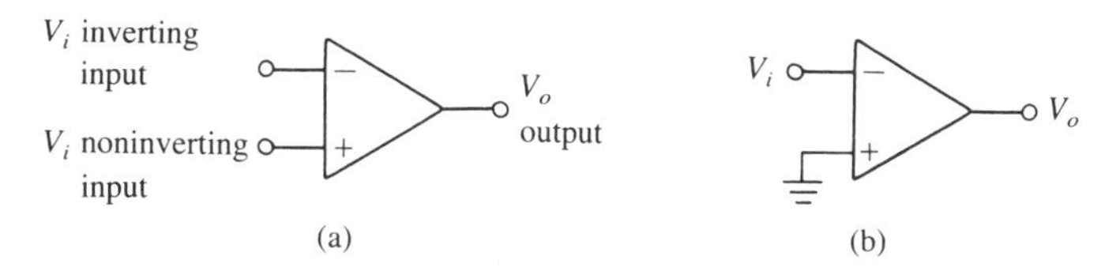

FIGURE 3.5
Operational amplifier representations.

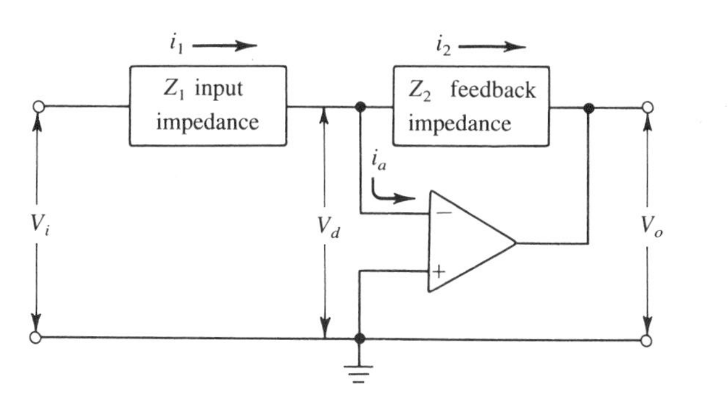

FIGURE 3.6
Operational amplifier with external components.

difference at the inputs then depends on the gain G and the limiting output voltage V_o. With a differential input, $V_o = G(V_+ - V_-)$, that has a maximum of approximately 10 V and a very large gain, the input difference, $(V_+ - V_-)$, must be close to zero.

Under these conditions (Figure 3.6),

$$i_1 = \frac{1}{Z_1}(V_i - V_d) = \frac{V_i}{Z_1} = i_2 = \frac{1}{Z_2}(V_d - V_o) = \frac{-V_o}{Z_2} \tag{3.1}$$

and therefore

$$V_o = -\frac{Z_2}{Z_1} V_i \tag{3.2}$$

The op amp is the basis for many signal-modifying circuits. Table 3.1 lists some applications that indicate how the composition of Z_1 and Z_2 determines the relationship between the input and output voltage signals. These circuits were initially developed as components for analog computers but have since found extensive use in instrumentation. The current polarographic instrumentation (see Chapter 23) is an excellent example of the use of op amp circuits to (1) generate required voltage ramps, (2) provide the accurate voltages required for precise measurements, and (3) amplify the resulting current.

Important Operational Amplifier Parameters

Regardless of the function of an op amp, its performance and reliability are determined by a number of parameters that appear as specifications for a particular device. The more important of these are listed here.

Input offset voltage. The voltage that must be applied across the input terminals to drive the output voltage to zero.

Input offset current. The current at the inputs necessary to make the output voltage zero. (This current is usually taken as the difference in the currents at the two inputs.)

Input bias current. The average of the two currents required to obtain a zero output voltage.

Slew rate. The rate at which the output voltage changes to the maximum saturated value (V/μsec).

Drift. The gradual change in offset voltage, bias current, and output voltage as a function of time and temperature.

Unity gain frequency bandwidth. The response of the op amp to signals is usually given as a Bode plot of the log of the open-loop gain versus the log of the frequency of the signal (Figure 3.7). The amplification is attenuated (reduced) at both high and low frequencies. The difference between f_1 and f_2 in Figure 3.7 is known as the unity gain frequency bandwidth, the region where maximum stable amplification is obtained. It should be noted that there is no low frequency rolloff for dc amplifiers.

TABLE 3.1

BASIC OPERATIONAL AMPLIFIER APPLICATIONS

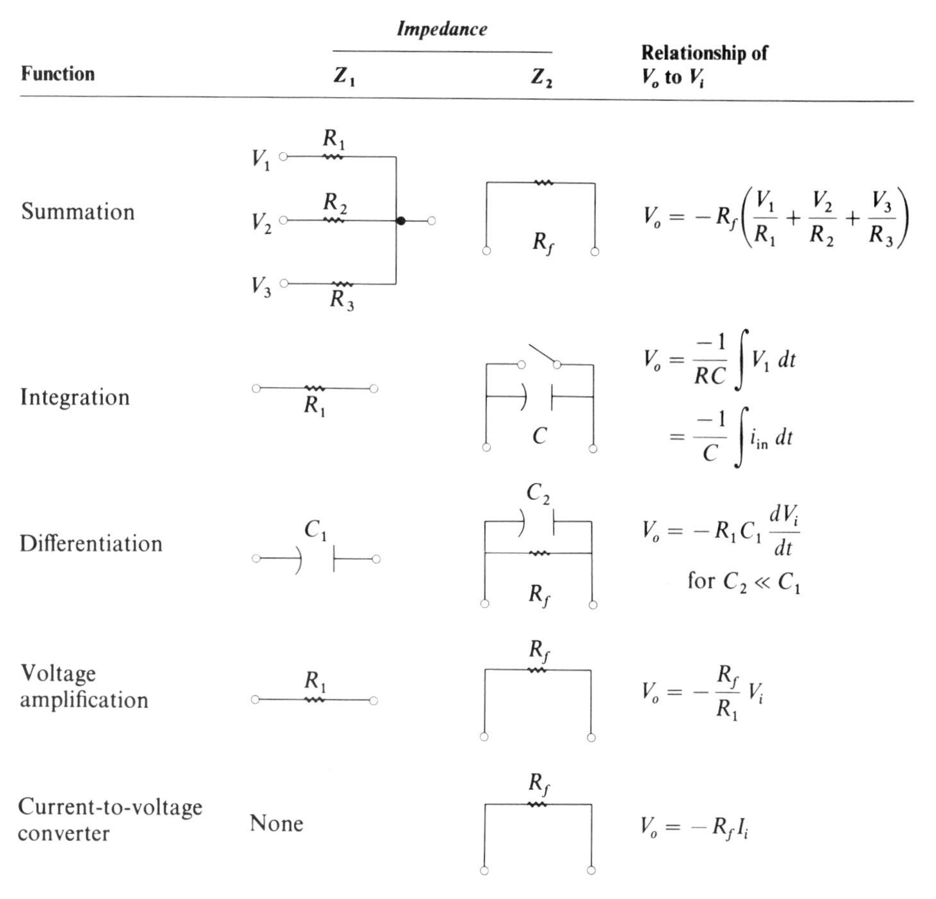

Function	Impedance Z_1	Impedance Z_2	Relationship of V_o to V_i
Summation	V_1, R_1; V_2, R_2; V_3, R_3	R_f	$V_o = -R_f\left(\frac{V_1}{R_1} + \frac{V_2}{R_2} + \frac{V_3}{R_3}\right)$
Integration	R_1	C	$V_o = \frac{-1}{RC}\int V_1\,dt = \frac{-1}{C}\int i_{in}\,dt$
Differentiation	C_1	C_2, R_f	$V_o = -R_1 C_1 \frac{dV_i}{dt}$ for $C_2 \ll C_1$
Voltage amplification	R_1	R_f	$V_o = -\frac{R_f}{R_1} V_i$
Current-to-voltage converter	None	R_f	$V_o = -R_f I_i$

NOTE: Z_1 and Z_2 refer to Figure 3.6. V_o is output voltage; V_i is input voltage; t is time.

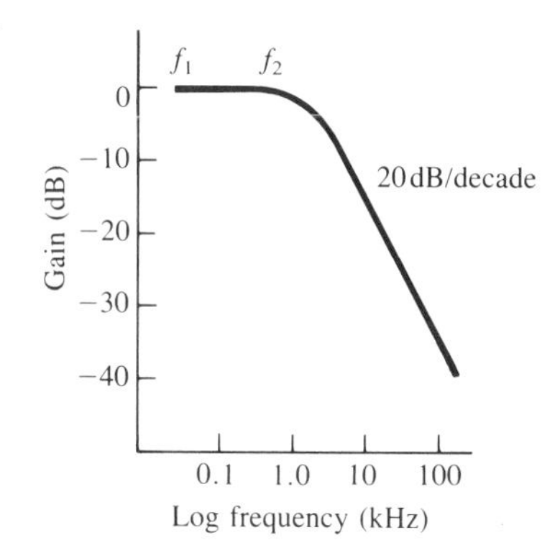

FIGURE 3.7
Bode plot.

Measurement of Current and Voltage

Obtaining precise measurements of currents and voltages from input transducers is one of the most important functions of op amps. As feedback-stabilized amplifiers, they can increase the magnitude of input signals so that these signals may be more precisely registered by signal modifiers and output transducers.

A combination of the basic op amp summing and scaling circuits is used to measure current (Figure 3.8). An adjustable resistance, R, controls the degree of amplification (the sensitivity). Provision is also made for adding a signed current from a controlled source to the input current at summing point P. This current, known as either the summing or bucking current, is used to adjust the level of the baseline, such as the dark current correction for photomultiplier output currents.

If the voltage drop across two points in a circuit is to be precisely determined, no current should flow between these points during the measurement. Voltage followers are used to isolate voltage measurement circuits in which no current flow is desirable from signal-modifying circuits where current flow is necessary (Figure 3.9a). These devices transfer voltage signals at the input to the voltage output without causing a current flow in the input circuit. The voltage follower op amp circuit has a gain of unity and provides a noninverted output voltage. The basic op amp properties of high input impedance and low output impedance are responsible

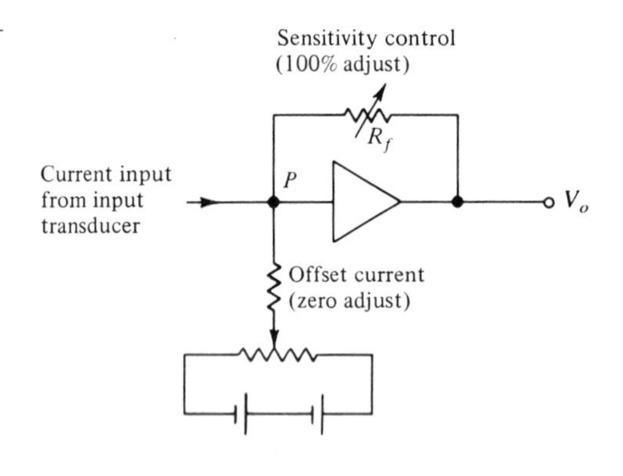

FIGURE 3.8
Current amplifier.

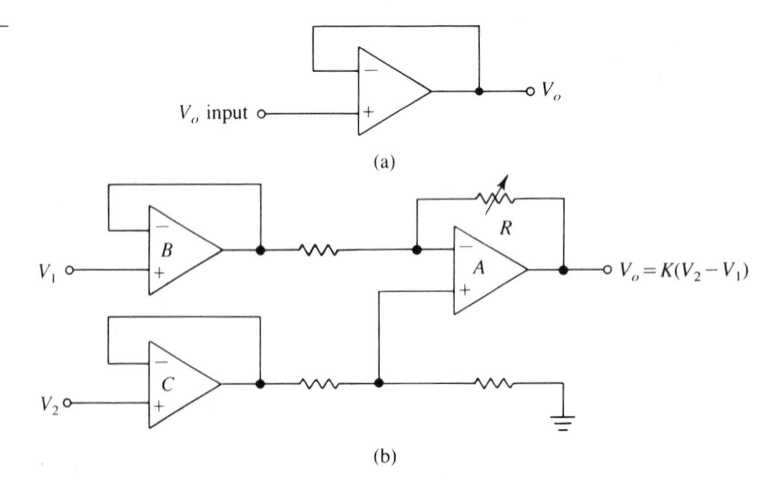

FIGURE 3.9
Voltage followers: (a) simple voltage follower and (b) instrument amplifier.

for the isolating (buffering) action of the voltage follower. This circuit is used to obtain accurate electrochemical cell potentials in the absence of electrolytic current flow.

Voltage followers *B* and *C* are attached to the inputs of a difference amplifier *A* to obtain a circuit known as an instrument amplifier (Figure 3.9b). This differential amplifier circuit is used to retrieve millivolts of analog data from volts of common-mode interference (signals of equal amplitude appearing simultaneously at the two inputs). It isolates the inputs from the outputs and thus protects the amplifier from high-voltage inputs and the device being measured from current leakage. Noise levels are also reduced. The gain, usually from 1 to 1000, may be adjusted by a single variable resistor *R* (Figure 3.9b). Properties of low drift, excellent linearity, and good noise rejection make the instrument amplifier a natural choice for extracting and amplifying low-level signals in the presence of high common-mode-noise voltages. They are widely used as transducer amplifiers for thermocouples, current shunts, and selective ion electrode meters.

Control of Current and Voltage

Circuits for the precise control of current and voltage may be constructed by using op amps. The potentiostat, a circuit that provides a constant, accurately known voltage, is illustrated in Figure 3.10a. The output voltage, which is equal to the reference voltage, V_{ref}, remains constant while allowing a considerable variation in the current output of the op amp. This circuit could be used to provide a constant potential across a varying load resistance—for example, the voltage control of a polarograph.

An amperostat or constant-current source assures that the current through the load, such as an electrolysis cell, is equal to that in the input circuit, $V_{ref/R}$ (Figure 3.10b). Neither the current in the feedback circuit nor the voltage across the load may exceed the maximum output values of the op amp. The current in the electrolysis cell remains constant as the voltage across the load changes. This circuit and the potentiostat circuit are used extensively in electrochemical instrumentation.

Active Filters (Tuned Amplifiers)

Sometimes it is desirable to amplify a signal that occurs at a given frequency or range of frequencies while suppressing signals of all other frequencies (see Chapter 2). Op amps may be used to make active bandpass filter circuits (Figure 3.11a). The output of this circuit (Figure 3.11b) shows a maximum gain A_o at a frequency f_o over

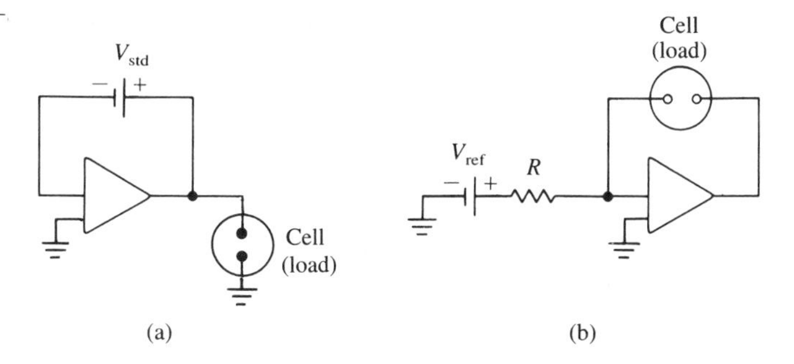

FIGURE 3.10
Voltage and current controllers: (a) constant-voltage source (potentiostat) and (b) constant-current source (amperostat).

a frequency range Δf (where Δf is defined by the maximum allowable gain, usually 3.0 dB below A_o). Values of circuit components and the operational parameters determine the magnitudes of A_o, f_o, and Δf. Amplifiers tuned to the frequency of a mechanical chopper can be used to improve signal-to-noise ratios in spectrometers.

At frequencies less than 10 kHz, active filters offer a distinct advantage over passive filters (circuits that contain only passive elements such as resistors, capacitors, and inductors). To operate at these low frequencies, passive filters require large inductors, which are expensive, bulky, and sluggish in their response to input signals.

Voltage Comparators

Basic Level Tester. A simple but important circuit is the voltage comparator (Figure 3.12). The relative magnitude of two analog inputs controls the digital output of a difference amplifier. Voltage comparators can be designed to give a voltage level equivalent to digital logic 1 at the output when the voltage on the noninverting (+) input exceeds the voltage on the inverting (−) input. When the inverting input voltage exceeds the noninverting voltage, the output voltage drops to a level corresponding to digital logic 0. If a constant reference voltage is applied to one input, the device can be used as a voltage level detector. Comparators are widely used as components of pulse counters and signal modifiers.

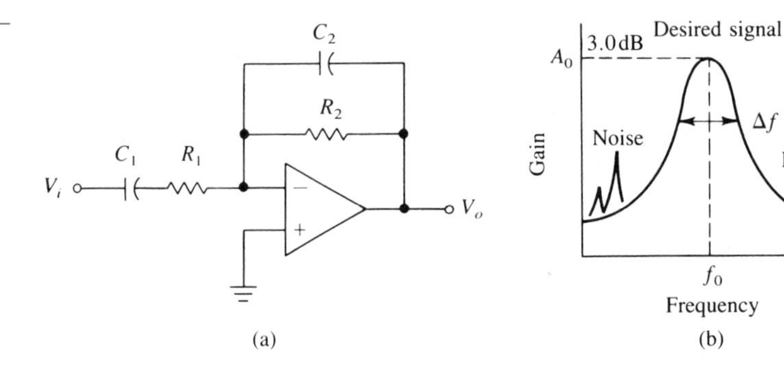

FIGURE 3.11 Active filter: (a) circuit and (b) response curve.

FIGURE 3.12 Voltage comparator: (a) schematic and (b) voltage curve shapes.

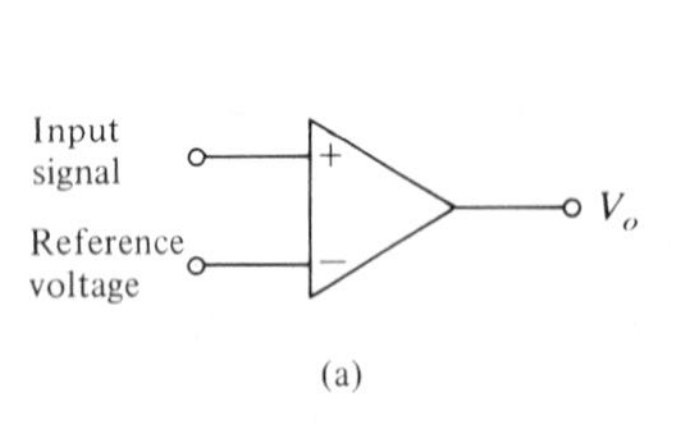

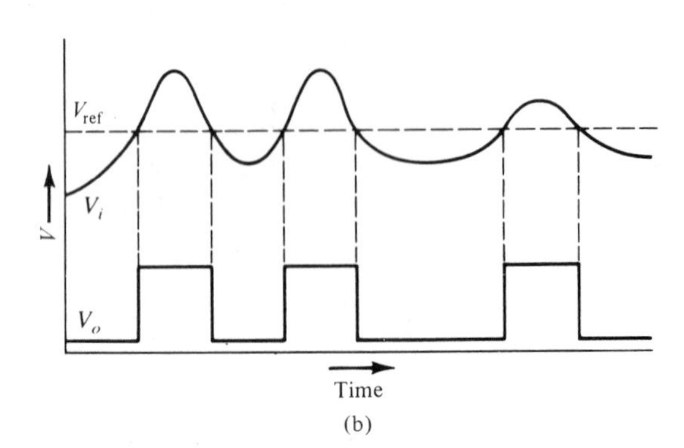

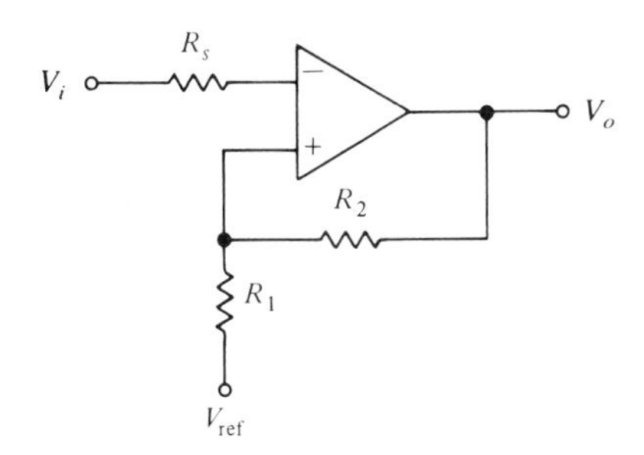

FIGURE 3.13
Schmitt trigger circuit.

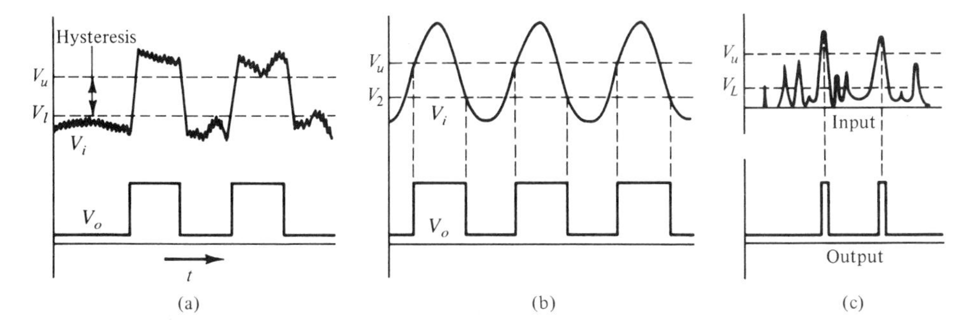

FIGURE 3.14
Applications of the Schmitt trigger: (a) noise reduction, (b) square-wave generator, and (c) pulse counting.

Schmitt Trigger. If the voltages on the two inputs of an op amp voltage comparator are equal, then the output will change state randomly in response to the noise present on both inputs. The Schmitt trigger circuit is used to eliminate problems associated with noisy input signals (Figure 3.13). Two voltage levels, V_u and V_l, are produced by V_{ref}, R_1, and R_2 of the circuit. The output of the trigger changes state from 1 to 0 when $V_i > V_u$ and from 1 to 0 when $V_i < V_l$ (Figure 3.14a). The voltage difference, $V_u - V_l$, known as the hysteresis voltage, should be larger than the magnitudes of expected noise levels.

The major advantage of the Schmitt trigger comparator is its noise immunity, which is equal to the voltage width of the hysteresis. Once this device changes state, small noise pulses do not cause reswitching. The advantage of the comparator is its ability to narrowly define a single-voltage comparator level (Figure 3.15). In addition to eliminating noise, the applications of a Schmitt trigger include square-wave generators (Figure 3.14b) and pulse counters (Figure 3.14c).

3.3 PRIMARY DIGITAL CIRCUITS[2]

Digital circuits were the first type of integrated circuit to be produced. They are currently the majority of nonlinear ICs and have found major applications in instrumentation for control and data processing. Logic functions generated by ICs are used to control instrument operation. For example, any one of three conditions may terminate a titration: depressing the off button, reaching the equivalence point on the titration curve, or emptying the buret. An IC that contains the appropriate circuits can be used to implement this logic in the laboratory. Initially data processing performed by internal instrumental ICs was limited to relatively simple functions, such as counting and preparing data for digital displays and computer interfaces. More complex data processing was done by computers external to the instrument.

Both the control and data processing functions have been expanded by designing instruments that contain microprocessor ICs. The increasing number of "smart" instruments indicates the operational advantages and economic practicality of replacing conventional circuits and mechanisms with microprocessors. Chapter 4 discusses microprocessor applications in more detail.

Digital ICs, like ordinary light switches, are binary devices that exist in only two states: off and on. Their inputs and outputs can therefore have one of two voltage levels. Logic low (off) is about at ground, whereas logic high (on) is a few volts positive. The term 1 usually refers to logic high and the term 0 to logic low.

Basic Logic Gates

There are three basic gates representing the fundamental logic functions: AND, OR, and NOT. Each of these gates is represented by a logic function. The operation of a gate is summarized by a truth table, which gives the output values as functions of various combinations of input values. The logic symbols and truth tables are given for the basic gates in Table 3.2.

In digital circuit diagrams, the symbols for logic gates are used whenever possible for simplicity. A single logic gate is a circuit containing several transistors and resistors. Integrated circuit chips may contain a few logic gates in the case of small-scale integrated (SSI) circuits (Figure 3.16) or several hundred logic gates in the case of large-scale and very large-scale integrated (LSI and VLSI) circuits.

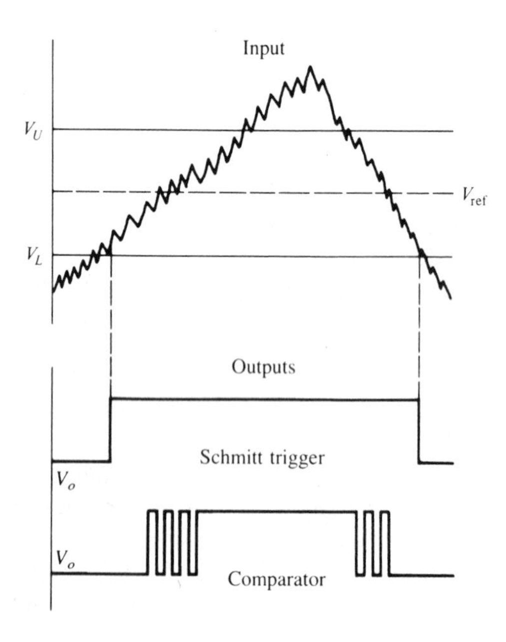

FIGURE **3.15**
Schmitt trigger output versus simple comparator output.

TABLE 3.2 BASIC LOGIC FUNCTIONS

Function	Symbol and Boolean expression	Truth tables
AND	$T = A \cdot B$	A B T 0 0 0 1 0 0 0 1 0 1 1 1
OR	$T = A + B$	A B T 0 0 0 1 0 1 0 1 1 1 1 1
NOT	$T = \bar{A}$	A T 1 0 0 1
NAND	$T = \overline{A \cdot B}$	A B T 0 0 1 1 0 1 0 1 1 1 1 0
NOR	$T = \overline{A + B}$	A B T 0 0 1 1 0 0 0 1 0 1 1 0

FIGURE 3.16 7400 Series TTL NOR gate ICs. (a) Quad 2-input NOR gates; (b) two 4-input NOR gates. (Courtesy of Signetics Corp.)

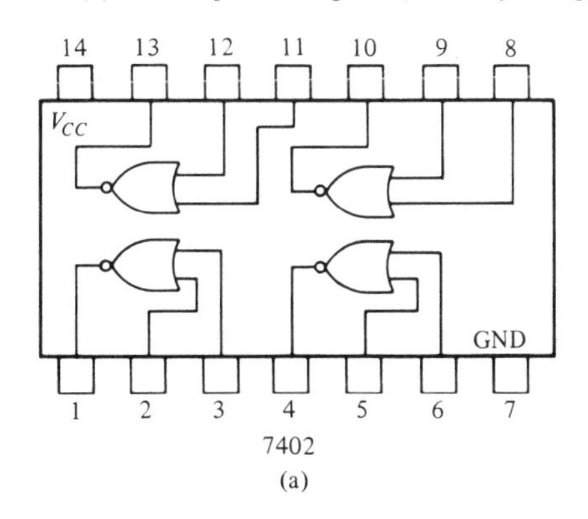

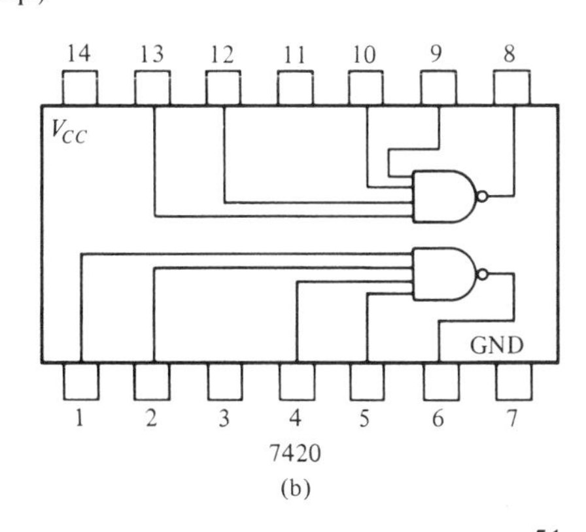

Sequential Logic Devices

The outputs of simple logic gates and the MSI circuits discussed in the preceding section are dependent only on the current state of the inputs. Another important class of digital circuits, known as sequential devices, has outputs that depend on a sequence of input states. These devices have the capacity to "remember" a succession of input states leading to the final output. The simplest sequential device is a flip-flop. MSI circuits that perform the functions of counting, storing, and manipulating digital data are composed primarily of flip-flops.

The simplest flip-flop circuit is the reset-set (RS) flip-flop shown in Figure 3.17. The truth table for this device indicates that a low or a logic 0 value at the S input "sets" the Q output to high or a logic 1. A logic 0 value at the R input "clears" the Q output to a logic 0. Simultaneous logic values of 1 at both the R and S inputs leave the logic value unchanged, whereas simultaneous values of logic 0 produce an ambiguous result. Thus the RS flip-flop serves as an elementary memory, remembering which input, R or S, was most recently set to logic 0.

The clocked RS flip-flop (Figure 3.18) can respond to changing logic values at the R and S inputs only when the T (clock) input is set to a value of logic 1. The truth table indicates that the logic of this device is the reverse of the simple RS flip-flop. A useful modification of the clocked RS flip-flop is the D flip-flop (Figure 3.19c). This device, also known as a data latch, transfers the logic value at input D to output Q when the clock (T) input is a logic 1. This transfer is not possible when the clock value is logic 0. The D flip-flop is useful in transferring digital data from one device to another under the control of the T input. A series of D flip-flops in parallel function as a single unit if their T inputs are connected to a single control device. This combination is known as a buffer or latch.

The JK flip-flop (Figure 3.19b) provides more flexibility in operation than the RS or D flip-flop. A logic 1 is required for the device to respond to inputs at J and K. If J and K are both set to logic 1, the complementary outputs Q and $\bar{Q}$ change state with each successive clock pulse. A logic 0 at the J input prevents a 0-to-1 logic transition at the Q output, whereas a logic 0 at the K input prevents a 1-to-0 logic transition at the Q output. The S (set) and R (clear) inputs override all the other input

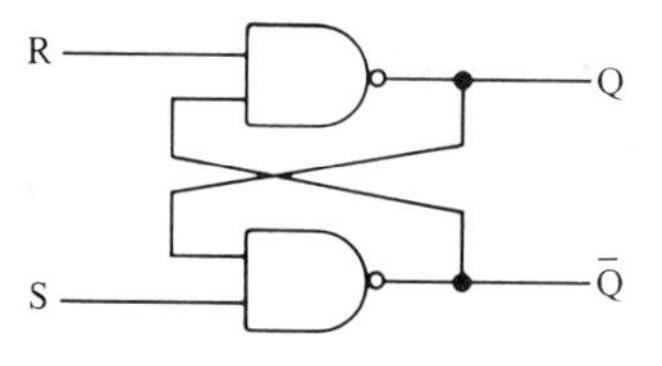

R	S	Q
0	0	Indeterminate
0	1	0
1	0	1
1	1	Original value of Q

FIGURE 3.17
RS flip-flop.

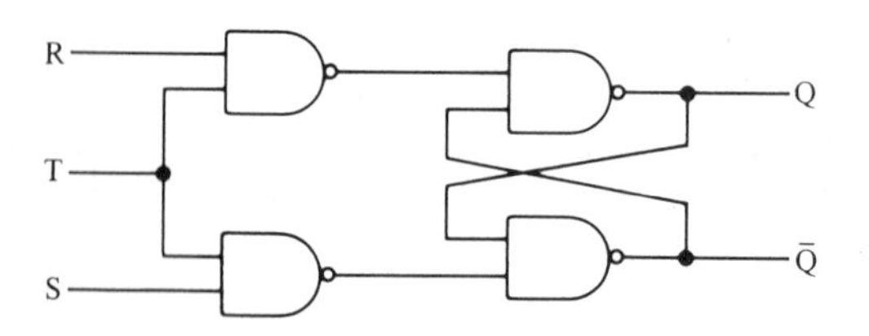

R	S	Q
0	0	Original value of Q
0	1	1
1	0	0
1	1	Indeterminate

FIGURE 3.18
Clocked RS flip-flop.

functions; a logic 0 at the clear input "clears" the value of Q to 0, and a logic value of 0 at the set input "sets" the value of Q to 1. JK flip-flops are integral components of digital counting and timing circuits (see Chapter 4).

Counters and Registers

Flip-flops are grouped together functionally as either registers or counters. A register can store data on the outputs of its component flip-flops. It is a simple device consisting of a group of interconnected flip-flops. Counters can perform more complex functions than registers for they contain both flip-flops and logic gates. Applications for counters include controlling sequences of operations and recording the number of events as a function of time. The output of a counter's flip-flops at any given time is called the state of the counter. The sequence of states can be simple, such as recording a series of consecutive pulses, or more complex, like counting up or down from a preset initial value.

Two common types of registers are the storage register (Figure 3.20) and the shift register (Figure 3.21). Registers may be used as temporary storage areas for digital data. Movement of data to and from a register requires a strobe pulse at the

FIGURE 3.19 Flip-flop representations: (a) gated RS flip-flop, (b) JK flip-flop, and (c) D flip-flop.

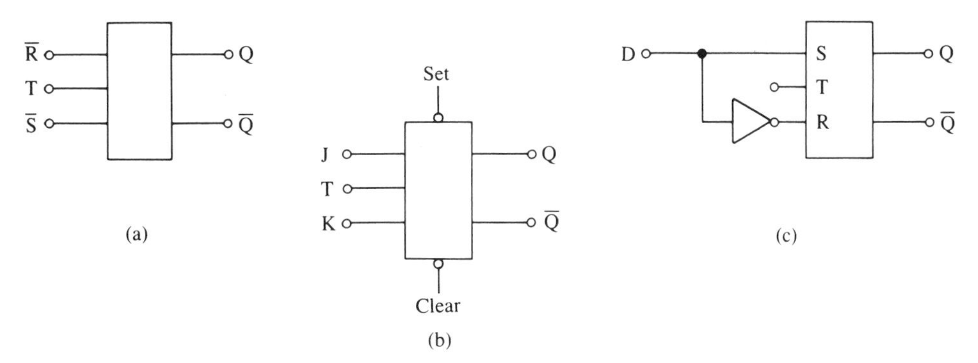

FIGURE 3.20 (a) 7475 Quad D latch; (b) schematic.

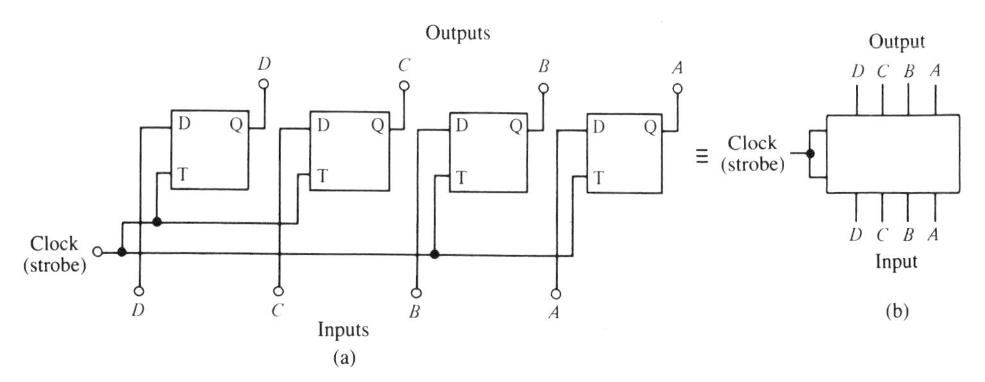

"clock" input of the register. This pulse causes the Q output of each flip-flop to assume the logic level of its respective input (Figure 3.20). When the strobe line is cleared to 0, the Q output values are held (latched) in the state that existed immediately prior to the strobe pulse. As long as there is no pulse on the clock inputs, the outputs may not change with the inputs. The operation of this register may be described as parallel data in/parallel data out, since all the flip-flops change state simultaneously—that is, synchronous action.

The 4-bit shift register (Figure 3.21) allows a sequence of digital pulses on a single input line (serial data) to be passed through the register. The data are moved sequentially right to left from one flip-flop to the next by a succession of clock pulses. After four clock pulses, the original input bit is on the D output. The next clock pulse pushes this data bit out of the register. The parallel outputs indicate the state of the flip-flops after each clock pulse. This register is used as a buffer to delay and store serial data and as a serial-to-parallel data converter. Buffers are registers for the temporary storage of small amounts of digital data. The concepts of serial and parallel transmission of data are discussed in Section 4.3. Eight-, 12-, and 16-shift registers are commonly used in instrumental circuits.

A simple binary counter can be constructed by connecting the input of one flip-flop to the input of the next (Figure 3.22). The first flip-flop will toggle (change its Q output logic level) with every logic level 1 input. The Q output of the second flip-flop will toggle on every other input pulse, the third on every fourth pulse, and so on. A binary count at the Q outputs (A through D) results from the "toggling" traveling through the series of flip-flops. This method of rippling pulses through a series of sequential flip-flops is known as asynchronous counting.

All the counter outputs can be reset to 0 by one of two methods. A logic level low (0) on the clear line simultaneously clears all the outputs of the JK flip-flops to 0 (Figure 3.22). The outputs are also cleared to 0 after every 16 consecutive logic 1 input pulses have been registered by the counter. Thus, this particular counter is said to have a modulus of 16 (counts from 0 to 15).

The counter rate of a ripple counter is limited by its propagation delay time.

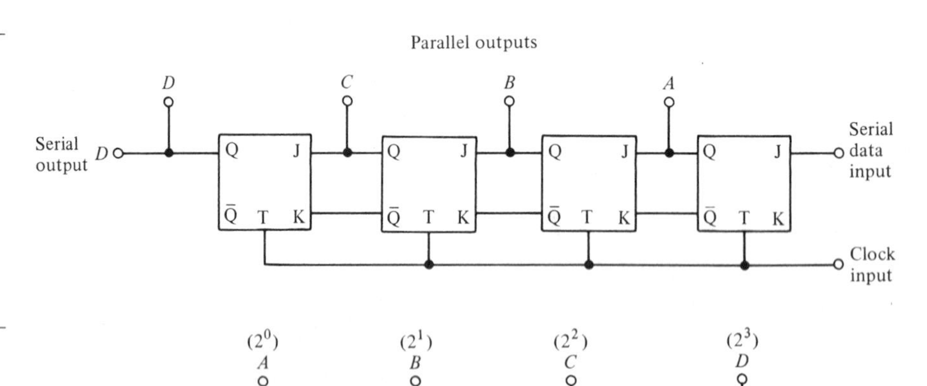

FIGURE 3.21
Four-bit shift register.

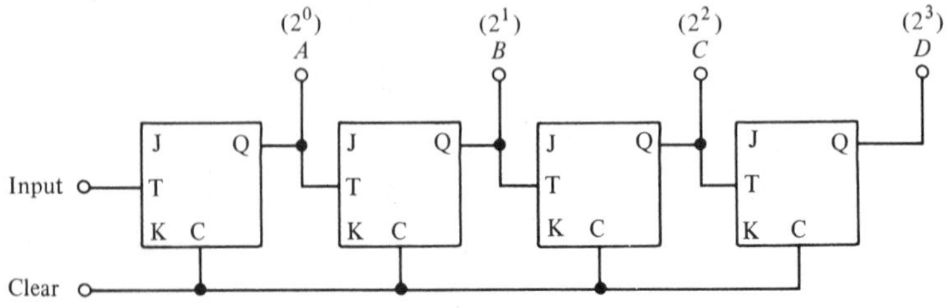

FIGURE 3.22
Simple modulo-16 counter.

This time is the sum of the delay times of the individual flip-flops that make up the counter. Delay times of individual flip-flops—that is, the time between a change in clock input and the resultant change to Q output—are typically 80 nsec per flip-flop. Thus the total propagation delay of a modulo-16 counter containing four flip-flops is 0.32 μsec, which limits the counting rates to 3.0 MHz.

The propagation delay is reduced by the synchronous binary counter, which allows all four flip-flops to change state simultaneously. The input pulse is applied simultaneously to the clock inputs of all the component flip-flops. Logic circuits constructed from flip-flops and auxiliary gates prevent transitions at Q outputs until the appropriate count is reached. The maximum propagation time for the synchronous counter is therefore the delay time of a single flip-flop. Thus the maximum counting rate of a synchronous modulo-16 counter is four times that of the equivalent asynchronous counter.

Binary coded decimal (BCD) counters are used in instruments where the convenience of readout in decimal data is desired. Four interconnected flip-flops form one stage of an asynchronous BCD counter (Figure 3.23). This flip-flop combination represents one digit of a decimal number. Table 3.3 lists the state of each flip-flop Q output as a function of the input count. Each digit requires one stage

FIGURE **3.23** Asynchronous BCD counter.

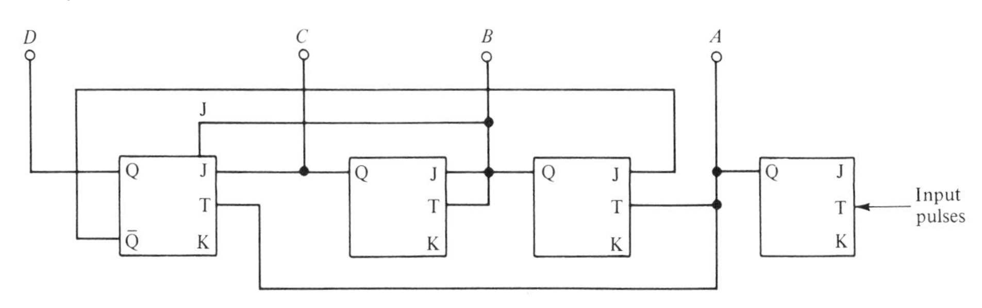

TABLE **3.3** BCD COUNTER OUTPUTS AS A FUNCTION OF COUNT NUMBER

Input count	*BCD counter outputs* D	C	B	A
0	0	0	0	0
1	0	0	0	1
2	0	0	1	0
3	0	0	1	1
4	0	1	0	0
5	0	1	0	1
6	0	1	1	0
7	0	1	1	1
8	1	0	0	0
9	1	0	0	1

of four flip-flops; thus four stages could represent any number from 0 to 9999. A commonly used IC containing a single stage is the 7490 decade counter.

Counters and latches may be combined to obtain the total count at a given time without interrupting the counting process—for example, in a digital wristwatch. Two ICs, a 7490 decade (BCD) counter, and a 7475 quad D latch, can be connected to obtain the desired counting function (Figure 3.24). A high (logic 1) on the strobe line causes each D flip-flop to assume the state of its respective counter flip-flop. Therefore the number in the counter at the time of the strobe pulse is stored (latched) on the outputs of the latch while the counter continues counting. The four flip-flops of the 7475 latch operating synchronously form a data register.

The data on the outputs of the 7475 latch may be the input to a decoder/driver IC producing a readout display. The 7447 BCD to 7-segment decoder/driver is connected into the total circuit. The decoder transforms the BCD input into output to light the proper light emitting diode (LED) segments of the display. BCD counters may be connected (cascaded) to produce any number of decimal digits.

The pulse counter uses a comparator or Schmitt trigger to detect pulses with voltage magnitudes greater than a present voltage level (Figure 3.25). The output of the comparator is then counted and displayed using digital ICs. Diodes at the inputs protect the comparator from high-voltage signals. This signal modifier could be part of a photon-counting instrument. If the output of the photomultiplier detector is connected to the noninverting input of the comparator, the photon count can be read on the digital display.

Connecting Digital IC Packages

A widely used technique of connecting ICs is that of tri-stating. Tri-state devices have a third input state in addition to the usual 1 and 0 logic levels; this third state is

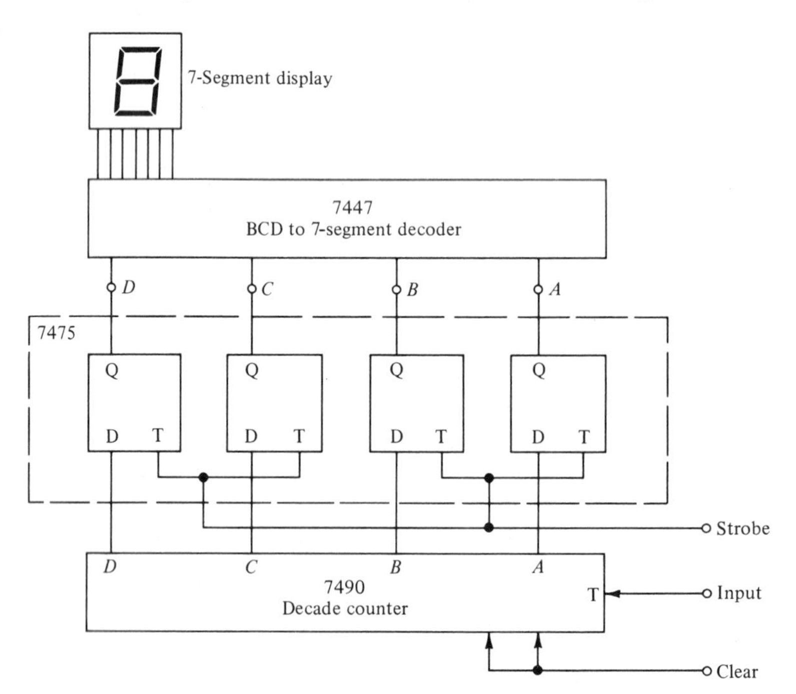

FIGURE 3.24
Latched BCD counter and 7-segment display.

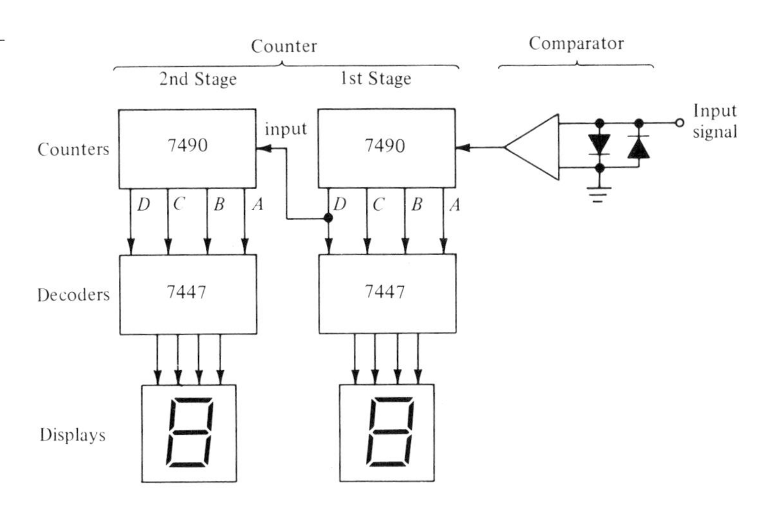

FIGURE 3.25
Pulse counter.

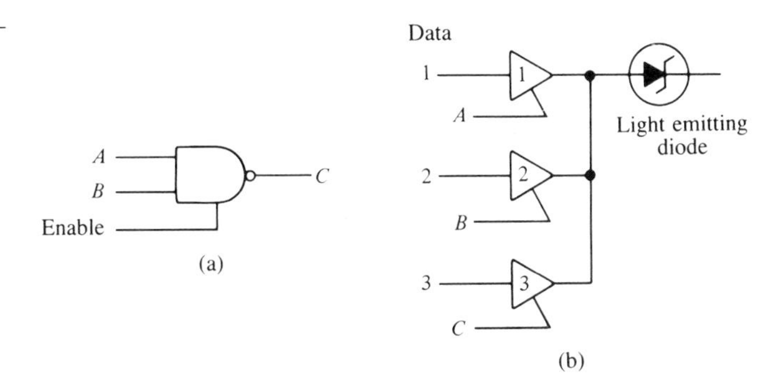

FIGURE 3.26
Tri-state devices: (a) tri-state 2-input NAND gate and (b) application of tri-state drivers.

equivalent to an open circuit or disconnect. It is controlled by the logic level applied to the output enable control (Figure 3.26a). A logic 1 on the enable input allows the results of the NAND operation to appear at the output of the gate, whereas a logic 0 at the enable input results in an open circuit at the gate output. Tri-stating does not affect the normal function of the gate or the signal levels that appear at the inputs and outputs.

The utility of tri-state devices is illustrated in Figure 3.26b. The status (logic state) of each driver input could be determined by sequentially connecting each driver output to the light emitting diode (LED) display using *A*, *B*, and *C* tri-state output enable controls. Tri-stating is widely used for transmitting digital data from different sources over shared lines to a centralized signal modifier.

3.4 DEVELOPMENT OF INTEGRATED CIRCUITS

Since the production of the first planar transistor in 1959, the number of elements in advanced integrated circuits has doubled each year. Circuits containing more than 2^{18} (262,144) elements on a single chip are now available. This increase in packing density, still far from the limits imposed by the laws of physics, is predicted to

TABLE 3.4 COMPONENT PACKING DENSITIES

Date	Components per IC	Circuit type	Application
1960	1	Discrete elements	Transistors
1962	10–64	Small-scale integration (SSI)	Gates, flip-flops, op amps
1964	64–1024	Medium-scale integration (MSI)	Counters, voltage-to-frequency converters
1969	1024–100,000	Large-scale integration (LSI)	Microprocessors, communications circuits
1974	$>100{,}000$	Very large-scale integration (VLSI)	Microcomputers
1988	10^7	Ultra large-scale integration (ULSI)	Third generation microcomputers

continue for IC technology. Table 3.4 presents the chronological development of IC technology, classifies ICs according to component packing density, and indicates typical applications for each kind of IC.

The cost of executing a given electronic operation has decreased as the space necessary to perform the operation has decreased. A NAND gate costing approximately $10 in 1961 is currently priced at less than 10¢. Op amp prices have likewise decreased during the same time period.

There are a number of reasons for this price decrease. Mass production of a complex circuit contained in a single high-density IC is less expensive than production of an equivalent circuit from several interconnected, lower-density IC packages. The smaller number of external interconnections reduces labor and materials costs. Interconnections of the IC within the package are more reliable than solder or connectors outside the package, thus reducing maintenance costs. Fewer external interconnections also mean that less intermediate testing is necessary during production.

Reducing the number of IC packages contained in an instrument results in less power consumption. Savings are therefore possible in power transformers, cooling fans, support racks, and cabinets. Instruments that incorporate high-density MSI and LSI packages in their design are smaller, have less rigorous power requirements, and require less control of operating environments than instruments that do not contain high-density ICs.

PROBLEMS

1. Distinguish between bipolar and MOS (metal oxide semiconductors) with respect to (a) construction, (b) advantages, and (c) limitations.

2. Using Table 3.1, draw an operational circuit for each of the following: (a) to integrate the areas under gas chromatographic peaks, (b) to determine the equivalence point of a potentiometric titration, (c) to convert the output of an amperometric titration to voltage, (d) to generate a voltage ramp for a polarograph, and (e) to sum currents necessary for a digital-to-analog conversion.

3. Calculate the minimum voltage that can be detected at the input of an operational amplifier with a gain of 1000 and a maximum output voltage of 12 V.

4. (a) Draw a symbol representing a 3-input NAND gate. (b) Construct a truth table for this gate.

5. Distinguish between the components in the following pairs: (a) a counter and a register, (b) asynchronous and synchronous counters, (c) a D latch and a JK flip-flop, (d) an incremental counter and a decremental counter, and (e) a decade counter and a modulo-16 counter.

6. (a) What are the three states of the output of a tri-state device? (b) Why is it necessary for an integrated circuit chip to be connected to the data bus of a computer with tri-state outputs?

BIBLIOGRAPHY

BROPHY, J., *Basic Electronics for Scientists,* 4th ed., McGraw-Hill, New York, 1983.

DIEFENDERFER, A., *Principles of Electronic Instrumentation,* 2nd ed., Saunders, Philadelphia, 1979.

HIGGINS, R., *Electronics with Digital and Analog Integrated Circuits,* Prentice-Hall, Englewood Cliffs, NJ, 1983.

HOROWITZ, W., AND A. HILL, *The Art of Electronics,* Cambridge University Press, New York, 1980.

MALMSTADT, H., C. ENKE, AND S. CROUCH, *Electronics and Instrumentation for Scientists,* Benjamin/Cummings, Reading, MA, 1981.

PIEPMEIR, E., "Analog Electronics," Chap. 2, *Treatise on Analytical Chemistry,* 2nd ed., I. Kolthoff, V. Mossotti, and P. Elving, eds., Part I, Vol. 4, John Wiley, New York, 1984.

VASSOS, B., AND G. EWING, *Analog and Digital Electronics for Scientists,* 3rd ed., John Wiley, New York, 1985.

For additional references to electronic hardware, consult the manufacturers' data books. In this rapidly developing field some of the most current applications are found in the literature from Analog Devices, Burr Brown, Intel, Motorola, National Semiconductor, RCA, Signetics, and Texas Instruments.

LITERATURE CITED

1. RIDGEWAY, T., S. BRIMMER, AND H. MARK, "Operational Amplifiers and Analog Circuit Analysis," Chap. 3, *Treatise on Analytical Chemistry,* 2nd ed., I. Kolthoff and P. Elving, eds., Sec. E, Part I, Vol. 4, John Wiley, New York, 1984.
2. ENKE, C., et al., *Anal. Chem.,* **54**(2), 367A (1982).

4 COMPUTER-AIDED ANALYSIS

It is evident from the previous chapters that the computer has become a basic tool in instrumental analysis. Microprocessors have become essential components of nearly every type of modern instrumentation from pH meters to mass spectrometers. In addition to larger mainframe and minicomputer systems, personal microcomputers with powerful, easy-to-use software are available for individual use in the laboratory. This chapter provides an introduction to computers and their applications. The concepts and jargon fundamental to the use of computers in instrumental analysis are presented.

4.1 INTRODUCTION

Digital computers have become integral components of modern methods of analysis. Applications of these devices to analytical instrumentation have increased with advances in computer technology. Initially computers were used to automate conventional calculations and existing instruments. Later new measurement methods were developed that were possible only through the use of computerized instrumentation and high-speed data processing techniques. Most recently microcomputers are influencing instrument design as well as analytical methods. To understand the role of a computer in a specific instrumental method, it is necessary to consider the interactions among instrument, computer, and analyst. Important combinations of interactions are off-line, on-line, in-line, and intra-line.[1]

Off-Line

Initial applications of digital computers to instrumentation used the off-line configuration (Figure 4.1). Computer programs for processing instrument output data are written in an analyst-oriented language, such as FORTRAN or BASIC.

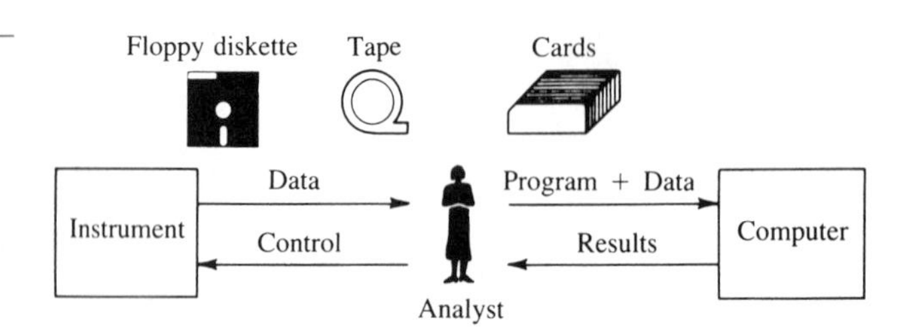

FIGURE 4.1 Off-line computer configuration.

Data are collected from the instrument, transferred to a suitable input medium (magnetic tape, floppy diskette, or punched cards), and then submitted with the appropriate program as a job to the computer. Jobs are processed sequentially by the computer operating in batch mode. The final results appear at an output device such as a line printer. The time required to obtain results depends on the length and priority of the job as well as the total number of jobs to be processed by the computer.

Although small computers can be operated off-line, this configuration is generally implemented with large computers in situations that require complex calculations, manipulation of sizable amounts of data, or a combination of the two. The efficient use of larger computers restricts their operation from being interrupted by an individual analyst or instrument for the purpose of data input or control. Because there is no direct communication between the computer and the instrument, data must be transferred indirectly under the control of the analyst. Thus a computer running in an off-line environment cannot respond to the instantaneous needs of a specific instrument or analyst, but rather processes jobs in the order submitted.

On-Line

By the late 1960s advances in electronics resulted in a compact, moderately priced minicomputer than could be operated on-line with instrumentation (Figure 4.2). In this configuration the computer is linked directly to instruments through an electronic interface. A single minicomputer dedicated to one or more instruments can perform specific tasks such as the acquisition and processing of data as well as instrument control functions. The analyst interacts with both the computer and the instrument to obtain and process data, control instrument operation, and retrieve results. Analyst-instrument interaction can be reduced by increasing the number of instrument control inputs and data outputs interfaced to the computer.

An on-line computer directly interfaced to an instrument can operate in a real-time mode. In this mode the computer can respond instantaneously to data acquired from the instrument. Computations, control functions, and output of information occur rapidly enough to improve the dynamic operation of the instrument.

A variety of analytical instrumentation has been interfaced to on-line computers. Gas chromatography uses this technique for rapidly reducing large amounts of data taken from one or more instruments to concise, accurate results. Dedicated minicomputers have decreased the time necessary to obtain results from X-ray diffraction studies of absolute compound configurations from months to hours. Another beneficiary is high-resolution mass spectrometry. Fourier transform nuclear magnetic resonance and infrared spectroscopy, methods that require the

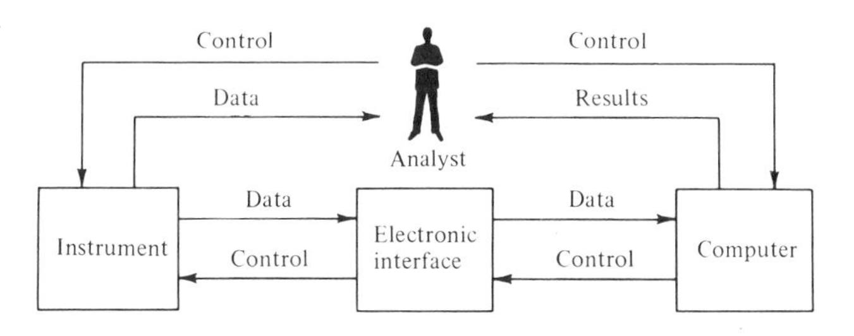

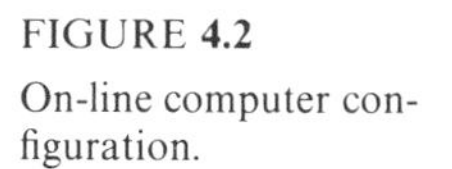
FIGURE 4.2
On-line computer configuration.

rapid execution of complex mathematical transformation functions, would be impossible without on-line computers.

In-Line

In the 1970s microcomputers became available as low-priced integrated circuit chips. The combination of reduced size and decreased price resulted in the replacement of minicomputers with microcomputers for some on-line applications. More important, however, instrument design was modified to include microcomputers as internal components. When the computer becomes an integral, dedicated part of the packaged instrument, the configuration is known as in-line (Figure 4.3).

Although microprocessors were originally designed to replace hardwired logic control circuits in relatively simple machines, such as cash registers, microwave ovens, and sewing machines, later generations of these devices are capable of data processing tasks in addition to instrument control. In-line microcomputers were introduced initially into chromatographs (gas and liquid) and spectrophotometers (infrared, visible, and ultraviolet) where precise control of instrument parameters and the ability to perform repetitive analyses are required.

In-line microcomputers can be controlled by programs stored in read-only memory (ROM). These programs are placed in the computer by the manufacturer and cannot be altered by the analyst. It is, however, possible to substitute or upgrade prestored programs (firmware) by changing ROM chips, thus avoiding extensive hardware modifications. The instrument designer may alter data handling functions, modify input and output formats, and even redefine instrument controls as new instrument applications emerge.

Intra-Line

The role of the computer in instrumentation continues to evolve. It is no longer a single device interfaced to an instrument for data acquisition and elementary control functions. Several microcomputers distributed within a single instrument can constitute subsystems that have the capability to change the nature of the chemical measurement system (Figure 4.4). The hardware components of microcomputer subsystems can include basic microprocessors, buses linking system components, programmable support chips (see Section 4.3), data domain conversion chips (see Section 4.3), and microprocessor-controlled robot arms (see Chapter 26). Important software elements are programs for data acquisition and processing, instrument control, database management, spreadsheet analysis, graphics displays and hardcopy, word processing, and, in some cases, artificial intelligence, including expert systems.

Dedicated microcomputers have replaced both hardwired circuits and more general minicomputers because their cost-performance characteristics are superior to either of the other approaches. The use of microcomputer control and logic

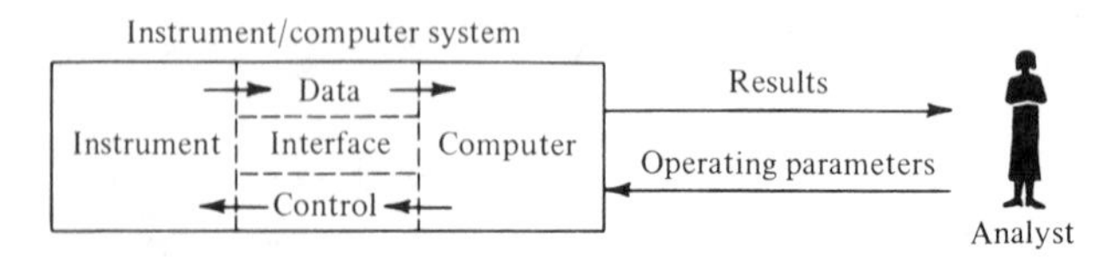

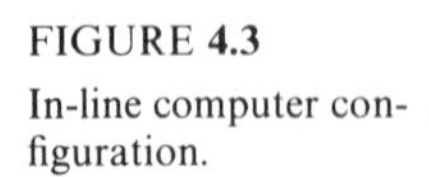
FIGURE 4.3
In-line computer configuration.

FIGURE 4.4 Intra-line computer configuration.

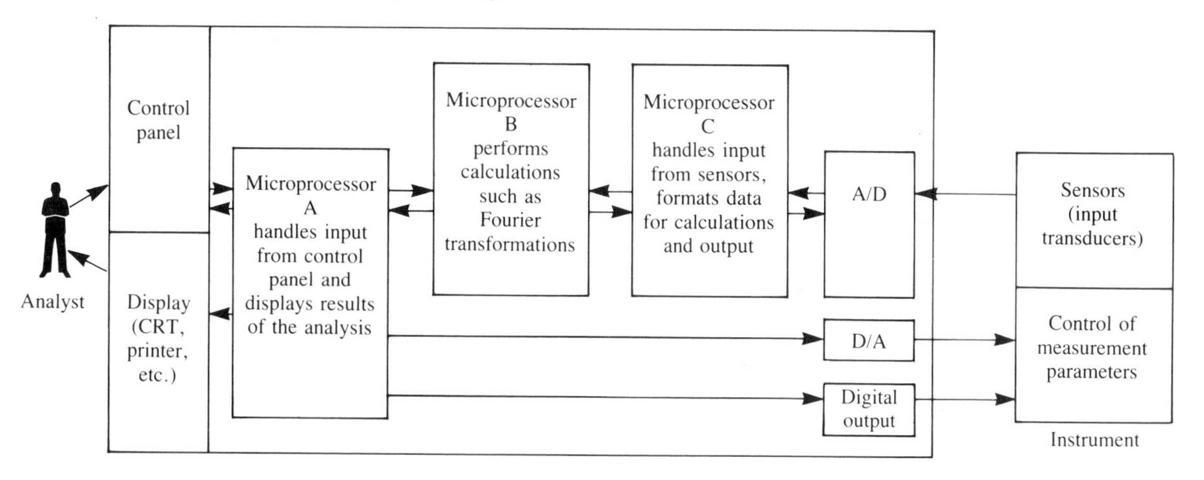

software ensures flexibility in instrument system modification and future expansion. ROM firmware provides the desirable fixed functions and dedicated characteristics of a hardwired electronic circuit.

Advantages resulting from the replacement of conventional instrumental electronic hardware with microcomputers include:

1. Increased reliability (decreased maintenance) due to substitution of electrical and mechanical components with sequences of programmed instructions
2. More complete and reliable analysis by prompting the analyst to enter and check all variables
3. Improved accuracy of results with automatic, periodic instrument calibration
4. Easier troubleshooting and maintenance by built-in diagnostic tests that check the functions of instrument components
5. Improved precision of results using digital signal processing
6. Ease of communication with other devices and computers external to the instrument

4.2 COMPUTER ORGANIZATION—HARDWARE

Although digital computers differ widely in data processing rates, memory sizes, and word lengths, every computer comprises five basic functional units connected by signal pathways known as buses. These units are the central processing unit, the arithmetic-logic unit, input, output, and memory. The heart of a computer is the central processing unit (CPU), which contains the control unit (CU) and the arithmetic-logic unit (ALU) (Figure 4.5).

Arithmetic-Logic Unit

The ALU performs the arithmetic and logic operations on data presented to it. Data are processed in the form of binary words, each word containing a specified number of binary bits. A bit consists of either a one or a zero. All operations are, therefore, performed using the principles of Boolean algebra. Arithmetic operations include addition and subtraction; logic operations involve "ANDing," ORing," and shifting all the bits of a word to the left or right. Although the number of basic operations performed by any computer is limited, the rapid execution of a series of these instructions, known as an algorithm, can produce many useful functions. Common algorithms include such functions as multiplication, integration, and manipulation of data arrays.

Control Unit

The CU is responsible for coordinating the operation of the entire computer system. Specifically, it generates and manages the control signals necessary to synchronize the flow of data on all buses with the operation of the functional units. The control unit also fetches, decodes, and executes successive instructions (a program) stored in the memory unit.

Central Processing Unit

The control unit is usually physically linked to the arithmetic-logic unit that it manages; the combination of CU and ALU is known as the central processing unit (CPU). Two critical parameters that are needed to evaluate CPU operation are the minimum time required to carry out specific types of instructions and the number of bits in the instructions, memory addresses (program counters), and data processed by the CPU. These parameters determine the rates at which data can be acquired and processed. The first generation of CPU chips, represented by the Motorola 6502, Intel 8080, and the Zilog Z80, processed data and instructions in 8-bit words.

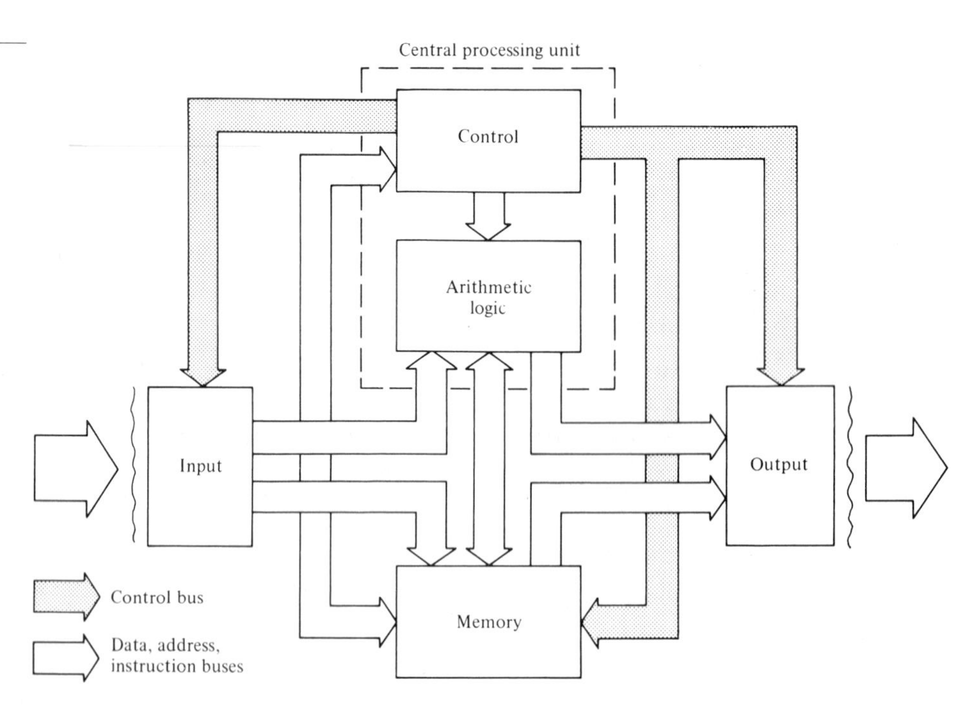

FIGURE **4.5**
Basic computer organization.

They used 16-bit program counters to directly address 64 kilobytes of memory. The next generation of CPU chips, DEC LSI-11, Intel 8088/8086, and the Motorola 68000/68008 series, had the ability to process 16-bit and, in some cases, 32-bit instructions and data as well as to address larger amounts of memory directly. The current or third generation of CPU chips includes the Motorola 68020, DEC MicroVAX, and Intel 80286/80386. Personal or lab computers designed around these chips have the memory access, speed, and control needed for applications requiring multiuser, multitasking operations.[2]

Memory Units

The memory unit is used by the CPU for rapid storage and recall of information. It is linked to the CPU by address lines, data lines, and control lines. Address lines carry binary information required to locate specific parts of the memory, data lines carry the information between memory locations and the CPU, and control lines direct the sequence of data transfers. Electronic memories currently available include ferrite core, bipolar, metal oxide semiconductor, magnetic bubble, and charge-coupled devices. Storage capacity, cost per bit of storage, reliability, and access time are important memory characteristics. Access time is the time required to read or write data at any storage location. Storage capacity is typically expressed as the number of storage locations, the number of bits in a given location that match the computer word length. Thus 1024 locations (known as 1K of memory) of 16-bit words contain twice as many bits as 1K of 8-bit words. The cost per bit is roughly related to the complexity of the total memory; the more auxiliary electronic components required, the higher the price. In general, faster memories are more expensive than slower ones. Memory access time is important because it often limits the overall operating speed of the computer.

Memories can be classified as volatile or nonvolatile; volatile memories lose their data when power is removed, whereas nonvolatile memories do not. There are memories that allow information to be transferred in both directions between the CPU and memory storage locations, read/write memories, and ROM (read-only memory) that permits transfer in only one direction—from memory to the CPU. When the access time of a read/write memory is independent of the position of the storage location in memory, the memory is known as random access memory (RAM). Although ferrite core memory can be classified as nonvolatile RAM, the terms RAM and ROM are usually applied to MOS- and CMOS-type memories. RAM is volatile and ROM is nonvolatile. Once a program is entered into ROM, it may never be altered by any means. An erasable, programmable ROM, EPROM, may be erased and reprogrammed. Both ROM and EPROM programming require special hardware.

The memory types discussed thus far are known as internal memories, since they are linked to the CPU by internal buses for rapid access to information. Information may also be stored external to the computer through a variety of peripheral devices and then transferred to the computer under the control of the CPU. External memory devices include magnetic disks (floppy and hard), magnetic tape, and, most recently, compact disks.

Two types of information are contained in memory: instructions and data. Sequences of instructions that direct the computer to perform specific tasks are known as programs. Under the supervision of the CPU, each successive instruction is fetched from a memory location and deposited in a special register of the CPU, where it is decoded and finally executed by the computer. Data required by the program are processed by the ALU. Although data may be transferred to and from the computer in a number of formats, they are ultimately processed by the CPU as binary information.

Input/Output Units

The remaining two modules, the input and output (I/O) units, provide the computer with its links to the outside world and thus are important considerations in interfacing instrumentation. Input units supply data to the ALU or, in some situations, directly to the internal memory. External sources of data include keyboards, instruments, and external memory devices. Output units transfer data from the ALU or internal memory to external devices, such as light emitting diodes (LEDs), printers, video terminals, plotters, external memory units, and control devices (stepper motors, relay switches, and so on). Thus a single computer control unit may supervise the operations of many individual peripheral I/O devices.

Buses

Buses are composed of the individual wires that link the component parts of a computer system and are classified according to the type of information they transmit (Figure 4.5). Buses that connect the parts of the CPU are known as internal buses, whereas buses that join the CPU to memory, to peripheral I/O devices, or to other computers are designated as external buses (Figure 4.6). The exact number of wires constituting a given bus is determined by the architecture of the computer and the specific function of the bus.

The information transmitted by buses is (1) data or instructions (*what*), (2) addresses of devices or memory (*where*), and (3) control signals (*when*). The number of lines in a data bus is usually determined by the number of bits in the word processed by the computer. For example, a computer that is capable of processing 16-bit words has a 16-line data bus. The popular IBM-PC is an exception, with a 16-bit internal data bus (Intel 8086/8088) but only an 8-bit external data bus. Both the origin and destination of the information transmitted by the data bus are designated by the signals on the address bus. Information can be sent to or received from CPU

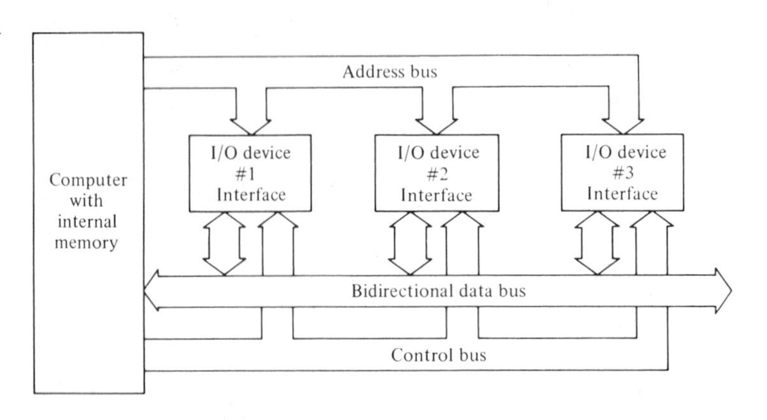

FIGURE **4.6**
External bus structure.

registers, memory locations, or peripheral devices. An address bus with 16 lines can directly address a total of 2^{16} (65,536) individual memory locations, registers, or devices. The signals sent from the CPU to the control bus manage the flow of information on the data bus.

Internal buses transmit instructions or data in a parallel manner; that is, signals on all lines of the bus are transmitted simultaneously. Parallel buses allow high-speed communications and require relatively simple interfacing hardware components. External buses can be either parallel or serial. Serial buses transmit the bits of data sequentially and thus need fewer lines than parallel buses. However, serial data transmission is slower and requires more complex interfacing hardware. Situations involving communication between the computer and a remote device usually involve serial buses. Standardization of bus architectures and protocols (the programmed response of the computer to messages from external devices) can simplify the task of computer interfacing. The RS-232 serial bus has become the most popular serial bus.[3] There are several parallel bus architectural standards, including the smart buses comprising the interfaces on the popular Apple II and IBM-PC microcomputers, the IEEE Standard 488-1975, the S-100, the Multibus, and the STD bus and, most recently, the NuBus.[4]

Information on the data bus is transmitted through the interface under the management of signals from the CPU appearing on the control bus. Designation of the sending and receiving devices is accomplished through the use of signals from the CPU sent to the address bus (Figure 4.6). Once a device has been assigned an address and hardwired to the interface, it is important to coordinate device addresses with the software required for the operation of the interface. This software must take into account the address of each device, methods of generating control signals, and responses to signals that come from the instrument components and other devices on the bus. The design of the hardware interface must therefore be coordinated with the development of the operating software.

Program instructions that enable the computer to communicate with devices attached to the interface bus are known as input/output (I/O) instructions. An I/O instruction contains the address code of the specific device to be serviced by the computer. Execution of an I/O instruction by the computer causes the address code to appear on the address bus simultaneously with an I/O signal on the appropriate control bus line. This combination of address and control signals is used to direct the flow of data to or from the device specified by the address code.

Microprocessors and Microcomputers

Although the basic components of microcomputers are generally the same as those of larger computers, there are differences in architecture and jargon. The distinction between a microcomputer and a microprocessor should be established. A microprocessor includes the central processing unit (CPU) with its arithmetic and control modules, a number of registers used for temporary storage, and connection to all the buses used for interfacing. In a microcomputer the memory and external circuitry necessary for timing the operation of the CPU are attached to the microprocessor.

Initially microcomputers were constructed from large-scale integrated (LSI) chips, including the microprocessor chip. Single very large-scale integrated (VLSI) circuits containing all the components of a microcomputer are now available. Both RAM and ROM memories are available, with the ROM chips containing instructions for the operation of the smart instrument and the RAM chips providing temporary storage of data and results. Microcomputers usually operate at speeds slower than larger computers, although separate chips, known as co-processors, can increase the speed of microcomputer systems. The addition of chips dedicated to specific tasks such as Fourier transformations of data sets can also increase the speed of a microcomputer system.

4.3 CIRCUITS FOR INTERFACING COMPUTERS TO INSTRUMENTS

The ideal computerized instrument system links the instrument directly to the computer and places the capabilities of the entire system at the disposal of the analyst. In a modern laboratory the tasks associated with chemical analyses are distributed among several types of computers. One or more microcomputers are dedicated to the operation of a single smart instrument, other independent micro- or minicomputers serve as communication controllers and intermediate data processors, and a large host computer has overall control of the results from all the instruments in one or more laboratories.

At the lowest level, the number of computer-related components of a smart instrument depends on the tasks required of the instrument. In addition to the hardware that makes up the instrument-computer interface, a variety of standard peripheral devices are available to facilitate analyst-computer interaction, storage of data and programs, display of results, and communication with external devices (Figure 4.7). In a smart instrument, each electronic component is linked to a

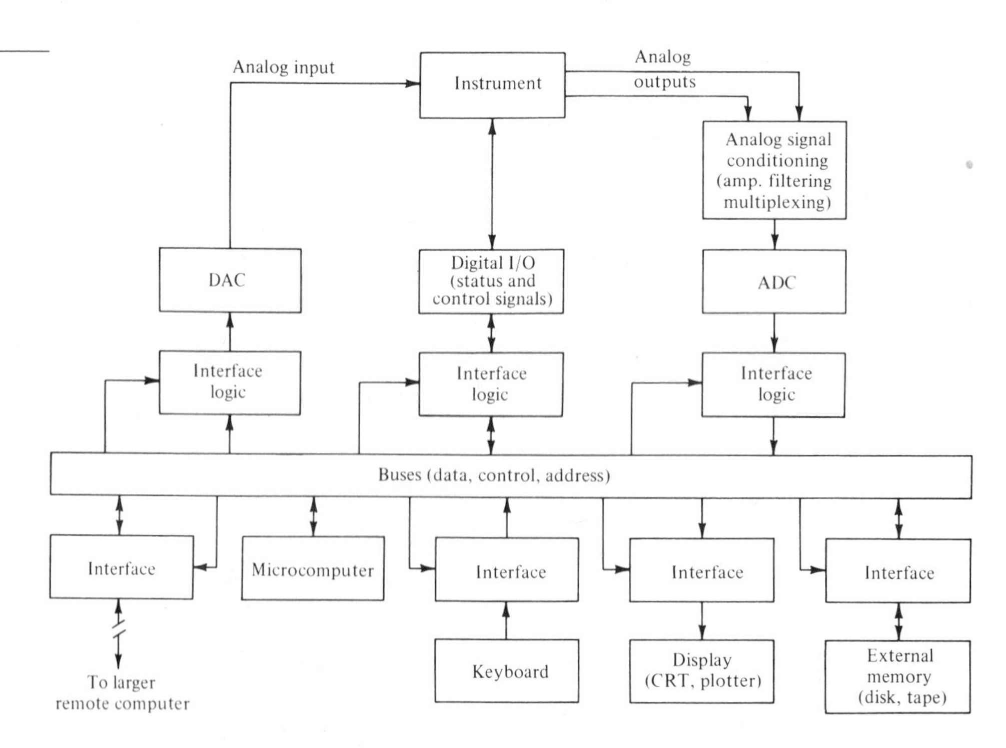

FIGURE 4.7
Microcomputer instrument system.

communication bus. Software routines, known as drivers, control the interaction among these components. Since standard peripheral devices such as terminals, printers, and disk drives represent a major expense, it is important to achieve efficient utilization in the overall operation of a laboratory.

Computer-Instrument Interface[5–7]

The focal point of a computerized instrument system is the interface between the computer's buses and the instrument's control and output signal lines (Figure 4.7). Two general methods have been used in computer-instrument interfacing. The older on-line method involves a separate microcomputer system with a general-purpose interface that can be connected to a variety of instruments that have only limited facilities for data processing.[8] This method is used for the experimental design of instrumentation and in situations where it is difficult to justify the expense of several smart instruments. Several commercial packages that include both hardware and software are available for interfacing instruments to popular microcomputer systems. By contrast, the in-line method places most of the interfacing functions within the instrument but also allows communication with external devices using standard serial and parallel buses.

The interface must provide the facilities for transmitting digital information to and from the computer. If the data from the instrument are in analog form, they must be converted. If the data are already in digital format, the interface needs only to control the flow of information along the data bus. While many modern instruments can be controlled directly by digital data from the computer, some still require analog control signals. In these cases, a digital-to-analog converter becomes an essential part of the interface. The individual components of interfaces are discussed in the following sections.

Multiplexers/Demultiplexers and Decoders

Multiplexers are often referred to as selectors because they function in a manner similar to mechanical selector switches. A 4-input multiplexer "selects" one of the four inputs to appear at the output (Figure 4.8a). Data on the multiplexer input lines

FIGURE **4.8** Four-input multiplexer. (a) Mechanical switch, (b) NAND gate, and (c) truth table for NAND multiplexer.

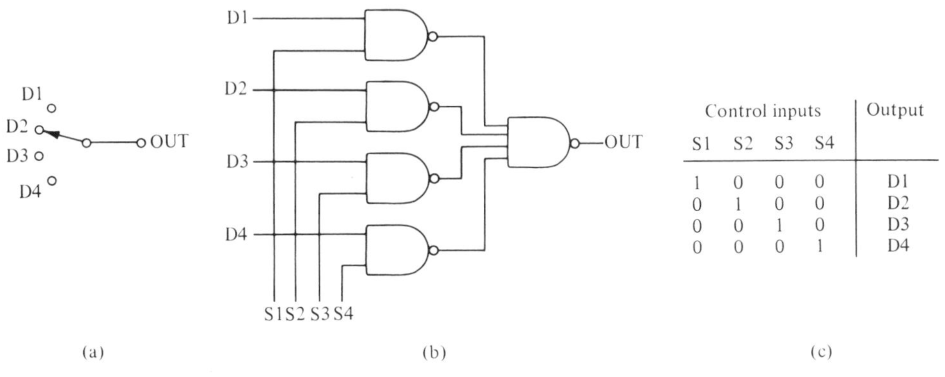

Control inputs				Output
S1	S2	S3	S4	
1	0	0	0	D1
0	1	0	0	D2
0	0	1	0	D3
0	0	0	1	D4

(D1 and D2 in Figure 4.8b) may be switched individually to the output using digital control switches S1 through S4. Signals on the input data lines may be either digital or analog, depending on the design of the multiplexer chip. Addition of gating logic within the chip allows the number of input lines to be increased and the number of control lines to be reduced. Figure 4.9a gives the block diagram of an 8-input multiplexer with three control lines. A 3-bit binary code at inputs S1, S2, and S3 determines which data input signal appears at the output (Figure 4.9b).

Two time-dependent factors must be considered when multiplexing. If the switching frequency of the multiplexer is too high, some input data will not be transferred to the output. According to the Nyquist sampling theorem (see Chapter 2), an input signal must be sampled at a rate at least twice that of the highest frequency component of the signal. For example, if the maximum frequency component of an input signal is 200 Hz, the switching frequency for this channel (input line) should be 400 Hz (channel sampled every 2.5 msec) in order to prevent any loss of data. At the other extreme, data will also be lost if the switching frequency is too low.

The other time-dependent multiplexing consideration is the settling time. Since switching is not an instantaneous process, a finite amount of time is required for the multiplexer's output to reach the value of the input once the channel has been selected. Semiconductor switching devices have shorter settling times and thus higher switching frequencies than electromechanical (reed relays) and mechanical switches. In general, the switching times of TTL family ICs are faster than those of either MOS or CMOS ICs (see Chapter 3). Another major limitation of semiconductor switching devices is the requirement of two power supplies, usually +15 and −15 V. Signal inputs and outputs may not exceed these limits without damaging the device.

The opposite of a multiplexer is a demultiplexer or decoder. A binary decoder produces a unique logic level output for each combination of binary inputs (Figure 4.10). It is a common practice in digital IC circuit design to make the unique output a logic level 0. A demultiplexer may be viewed as a multiplexer with its input and output functions reversed. Multiplexers/demultiplexers and decoders are used to implement logic functions in control devices and to route data signals through instruments and communication equipment.

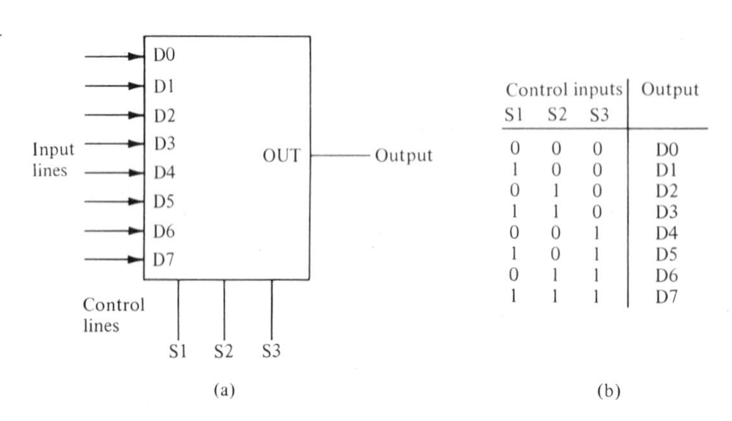

Control inputs S1	S2	S3	Output
0	0	0	D0
1	0	0	D1
0	1	0	D2
1	1	0	D3
0	0	1	D4
1	0	1	D5
0	1	1	D6
1	1	1	D7

FIGURE 4.9
Eight-input multiplexer: (a) block diagram and (b) truth table.

Digital-to-Analog Converters

Analog output can be produced from digital input using a summing op amp circuit (Table 4.1). For example, a 3-bit digital-to-analog converter (DAC) can be made by applying the total current from a resistor ladder to the input of an op amp (Figure 4.11). The drivers and diodes accurately set the voltage levels applied to the resistor ladder.

Three digital data bits are simultaneously applied to inputs A, B, and C, producing an output voltage V_o given by the following equation:

$$V_o = 8\,V_i\left(\frac{A}{4} + \frac{B}{2} + \frac{C}{1}\right) \tag{4.1}$$

where V_i is the voltage level of the logic 1 state from any of the binary input bits A, B, or C. Note that the relative values of the resistors determine that A will be the least

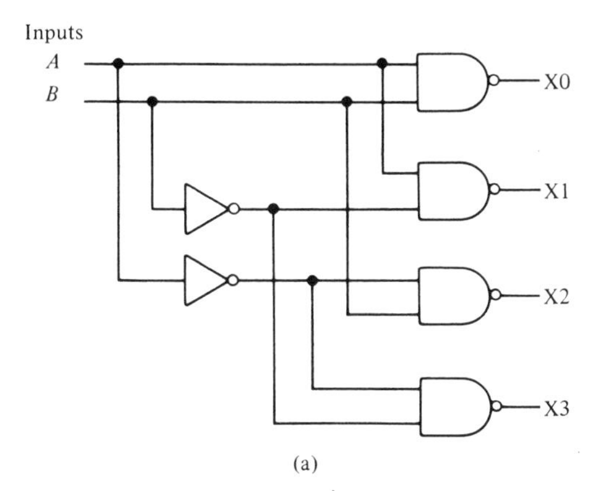

Inputs		Outputs			
A	B	X0	X1	X2	X3
0	0	1	1	1	0
0	1	1	1	0	1
1	0	1	0	1	1
1	1	0	1	1	1

(b)

FIGURE **4.10** Two-bit binary decoder: (a) logic circuit and (b) truth table.

TABLE **4.1** DAC OUTPUT AS A FUNCTION OF INPUT

Binary data word			
C	***B***	***A***	V_{out} **(Volts*)**
0	0	0	0.0
0	0	1	0.5
0	1	0	1.0
0	1	1	1.5
1	0	0	2.0
1	0	1	2.5
1	1	0	3.0
1	1	1	3.5

* The logic 1 state is assumed to produce a signal that is ~0.5 V due to diode limiting. The feedback resistor of the output op amp multiplies the output by 8.

significant bit and C the most significant bit. The operational amplifier sums the input currents and converts them to a scaled output voltage (Table 4.1). The range of the scale is determined by the value of the feedback resistor. The number of input bits may be increased by expanding the weighted resistance ladder. DACs with 4 to 16 bits are common.

A monolithic 4-bit DAC contained on a single chip latches a 4-bit data word onto the chip inputs (Figure 4.12). When the strobe line undergoes a 1-to-0 transition, the latch presents a new data word to the DAC inputs. The strobe pulse also triggers a monostable, which delays the signal before it changes the logic state of the end-of-conversion pulse. This delay allows the DAC to perform its conversion, after which the end-of-conversion output indicates that the analog output has reached a stable value (settled). The DAC is then ready to make another conversion.

Important DAC parameters are resolution, accuracy, settling time, and analog outputs associated with the least significant and most significant bits (LSB and MSB). The resolution is determined by the number of input bits the converter will handle; for example, a 10-bit DAC has 2^{10} or 1024 output levels and therefore a

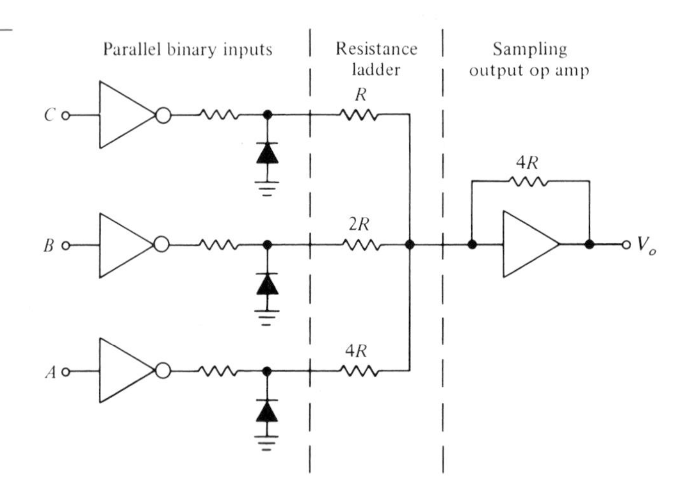

FIGURE **4.11**
Three-bit digital-to-analog converter.

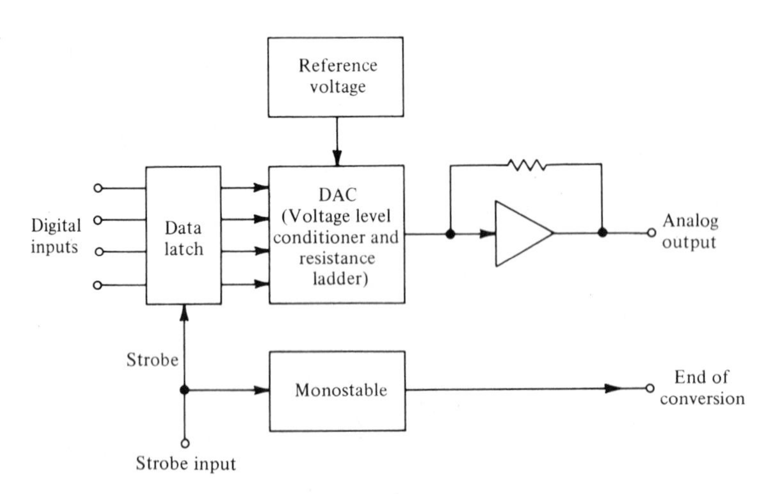

FIGURE **4.12**
Complete 4-bit DAC with control inputs.

resolution of 1 part in 1024. Accuracy is a measure of the deviation of the analog output voltage from its predicted value for a given input combination. The reference voltage is critical in determining accuracy. The settling time is the time required for a new, stable output value to be registered in response to a change in digital inputs.

DACs are used to operate any device that requires an analog input signal and that is interfaced to a digital source such as a computer. Applications include the display of digital data stored in computer memory on analog peripherals, such as plotters and graphics terminals, and the control of chemical instrumentation using computers to generate analog voltage signals, such as ramps and waves.

Analog-to-Digital Converters

An analog-to-digital converter (ADC) produces a digital representation of an analog input signal. Since most input transducers initially produce analog signals, ADCs find many applications in digital instrumentation. Resolution and sampling rate are two important parameters in interfaces that require analog-to-digital converters. The resolution of an ADC is defined as the smallest change in analog signal that can be observed at the digital output. Resolution is determined by the number of bits in the digital signal: 10 bits, 1 part in 1024; 16 bits, 1 part in 65,536. Thus for an A/D converter with a maximum range from 0 to 5 V dc and 12 bits of output, the smallest change that can be detected in the digital output is 5/4096 = 4.88 mV. Sample rates should be selected so as not to accumulate unnecessary data or bias the results. The Nyquist frequency rule (see Section 2.6) for sampling rates should be applied when necessary. The rule states that the rate of sampling must be greater than twice the highest frequency component of the A/D input.

The most important characteristics of ADCs are speed and accuracy of conversion. These parameters are determined by the particular method used for the analog-to-digital conversion, which, like digital-to-analog conversions, may be achieved by several methods. The voltage comparator is the heart of any ADC. The analog voltage to be converted is compared with a variable reference voltage. When the two voltages are equal, the comparator changes state and a digital output is registered. ADCs vary in the methods used to generate variable reference voltages and in the components used to register output data.

Counter Converters. The simplest ADC is the counter converter (Figure 4.13), which uses a binary up-counter latched to a DAC to produce a linear voltage ramp at the reference input of the converter. When the ramp voltage equals the analog input voltage, the comparator changes state, stopping the counter and allowing the digital outputs to be recorded by an output device. The size of the count is proportional to the magnitude of the analog input voltage. For example, if the 4-bit ADC (Figure 4.13) is set to cover analog inputs ranging from 0 to 1.0 V and a voltage of 0.40 V is placed on the input, the digital output will be 0110, corresponding to a value of 7/16 times the full-scale output. When the digital outputs have been recorded, the counter is reset to zero by a start-conversion pulse and the process is repeated. The conversion time is directly proportional to the magnitude of the analog input voltage, a limiting factor in most applications of computer interfacing.

Successive Approximation Counter. The popular successive approximation converter replaces the simple up-counter with a digital pattern generator and more complex control logic. This converter uses a series of logical guesses (approximations) to determine the digital equivalent of the analog input. Table 4.2 lists the sequence of approximations that a 4-bit ADC might use to digitize an analog input signal. A 4-bit binary number results from the series of approximations, with the most significant bit corresponding to the result of the first approximation, the next bit matching the result of the second approximation, and so on. Thus the magnitude of the analog signal in this example yields a binary 1101_2 corresponding to 13_{10}. This means that the analog input voltage is 13/16 of the full-scale voltage of this DAC.

Successive approximation counters require N approximations for an N-bit conversion. Ten approximations would therefore be required to determine the value of an analog input to an accuracy of 1 part in 1024 (2^{10}). The previously described counter ADC could require up to $2N$ trials to produce an N-bit digital number. Successive approximation converters are widely used because they have fast conversion times, which are independent of the magnitude of the analog input signal. The input signal for both the counter and successive approximation converters is usually in the high-level voltage range, 0–10 V maximum.

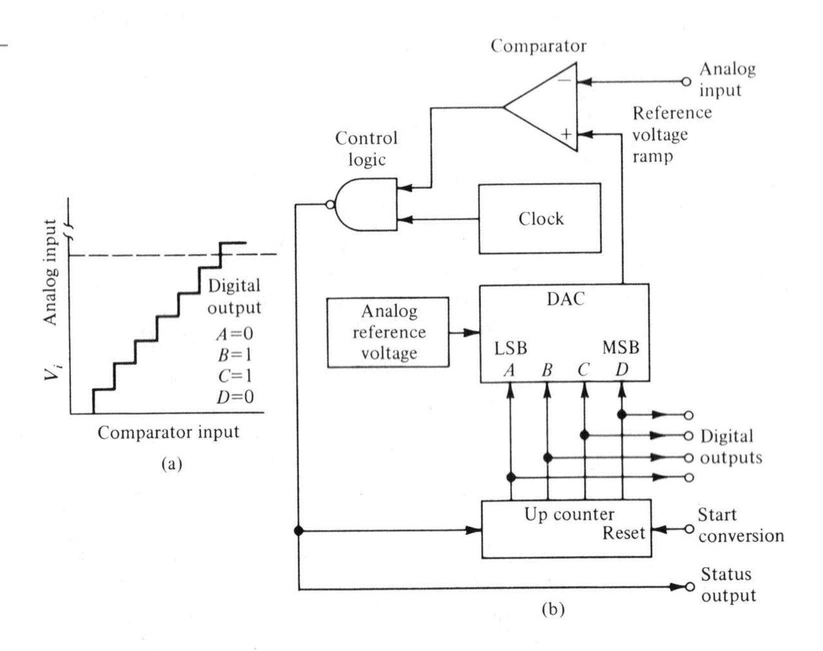

FIGURE **4.13**
Four-bit counter ADC: (a) example input and output and (b) composite circuit diagram.

TABLE **4.2**

EXAMPLE OF A SUCCESSIVE APPROXIMATION BY A 4-BIT ADC

Approximation	**Answer**	**Logic state**	**Weighted base-10 value**
$\# \geq 8$?	Yes	1×2^3 (MSB)	8
$\# \geq 12$?	Yes	1×2^2	4
$\# \geq 14$?	No	0×2^1	0
$\# = 13$?	Yes	1×2^0 (LSB)	1
			13

Voltage-to-Frequency Converters. In situations that require the conversion of small analog input signals (millivolt magnitudes), low-level ADCs are used. These converters are much slower (by factors of 100–1000) than the successive approximation converters, and they trade speed for the ability to represent accurately low-level input signals. These low-level ADCs integrate the analog signal and then compare it to a reference voltage to produce the digital output.

The voltage-to-frequency converter is a simple converter for transforming the analog input voltage into a series of digital pulses whose frequency is proportional to the magnitude of the input (Figure 4.14). The circuit, composed of resistor R, capacitor C, and op amp A, is used to produce an integral of the input voltage at the inverting input of the comparator (operational amplifier B). When the voltage of the integral output exceeds the value of the reference voltage, the comparator changes state, which turns on transistor Q1 and allows the capacitor to discharge. When the output of op amp A reaches ground, op amp B again changes state, turning off the transistor and allowing C to recharge again. Thus a series of pulses is generated at the output of op amp B. The rate at which the integrator op amp A charges is a function of the input analog voltage signal. The larger the input voltage, the faster the integrator ramp and, therefore, the higher the frequency of output pulses.

The output pulses go to the input of a binary counter, which counts pulses for a period of time determined by the monostable. The start-of-conversion input pulse resets the counter to zero and initiates the counting time interval by triggering the monostable. The monostable output, connected to one input of gate G, controls the counting time for pulses.

Dual Slope Converters. Dual slope ADCs (Figure 4.15) increase the accuracy of conversion over that of voltage-to-frequency converters by measuring a single integral rather than a series of integrals. An analog input voltage, V_1, is applied to the input of the integrator for a fixed period of time, t (Figure 4.16). At the end of this time a constant reference voltage, V_{ref}, of opposite polarity to V_i is applied to the

FIGURE **4.14** Four-bit voltage-to-frequency converter.

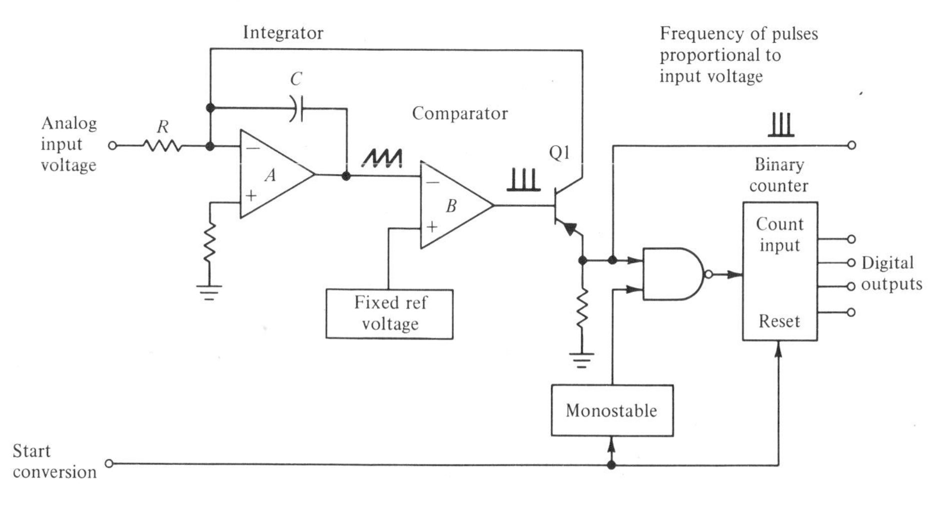

input of the integrator by switch S_1. The counter is also cleared to zero at the end of time period t. It then begins counting up, recording the time (t_1) required for the reference voltage ramp (slope = V_{ref}/RC) to reach zero. When the comparator senses this zero crossing, the counter is halted, its output is registered, and the analog

FIGURE 4.15 Dual-slope ADC.

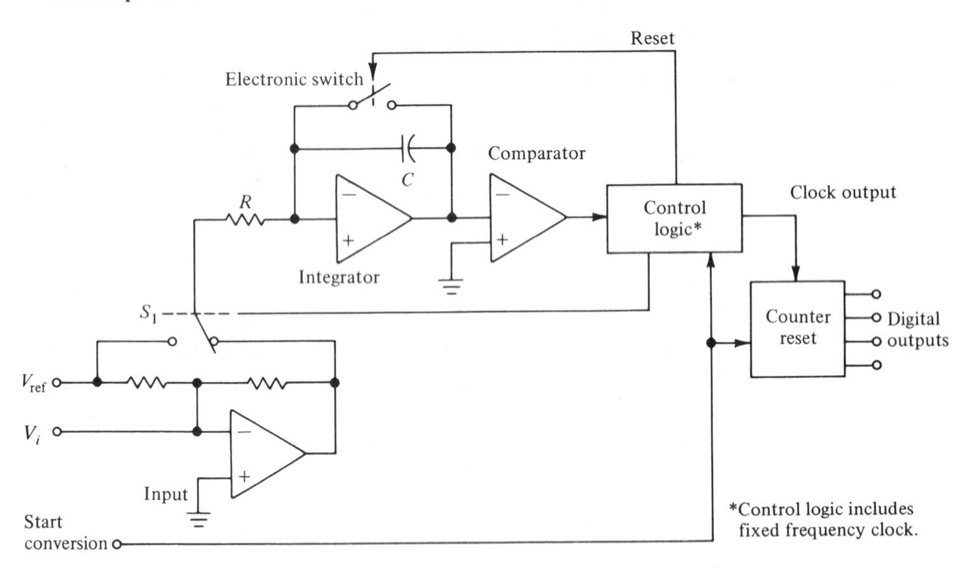

FIGURE 4.16 Operation of dual slope ADC.

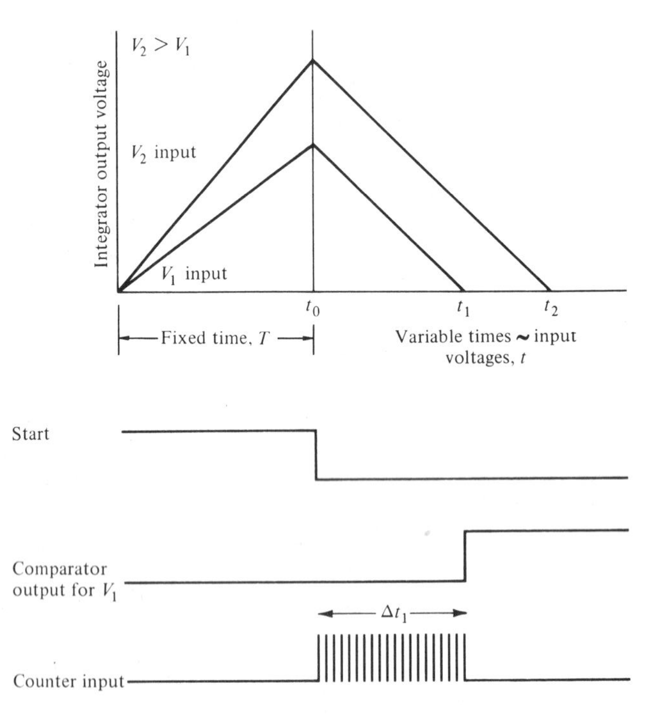

circuitry is reset by switches S_1 and S_2. A larger analog input voltage V_2 requires a proportionally longer time, t_2, to cross the zero axis (Figure 4.16).

Since the total charge gained by the capacitor C and V_{in} applied to the integrator is equal to the charge lost by the capacitor with V_{ref} applied,

$$\frac{V_i t}{R} = \frac{V_{ref} t_1}{R} \tag{4.2}$$

$$t_1 = \frac{t}{V_{ref}} V_i \tag{4.3}$$

If t and V_{ref} are constants for a given ADC, then the output of the counter (t_1) will be a binary representation of the analog input voltage (V_i). The conversion time must be greater than $2t$, and, therefore, the conversion frequency is limited to less than $(2t)^{-1}$ conversions per second.

Sample-and-Hold Amplifiers

These devices sample the analog signal and retain it at the input of the ADC for the time required to make the conversion. A sample-and-hold amplifier can be considered as a pair of voltage followers connected through an electronic switch (Figure 4.17a). The device has an analog input, a control input, and an analog output, which is in one of two operating modes: the sample mode in which the

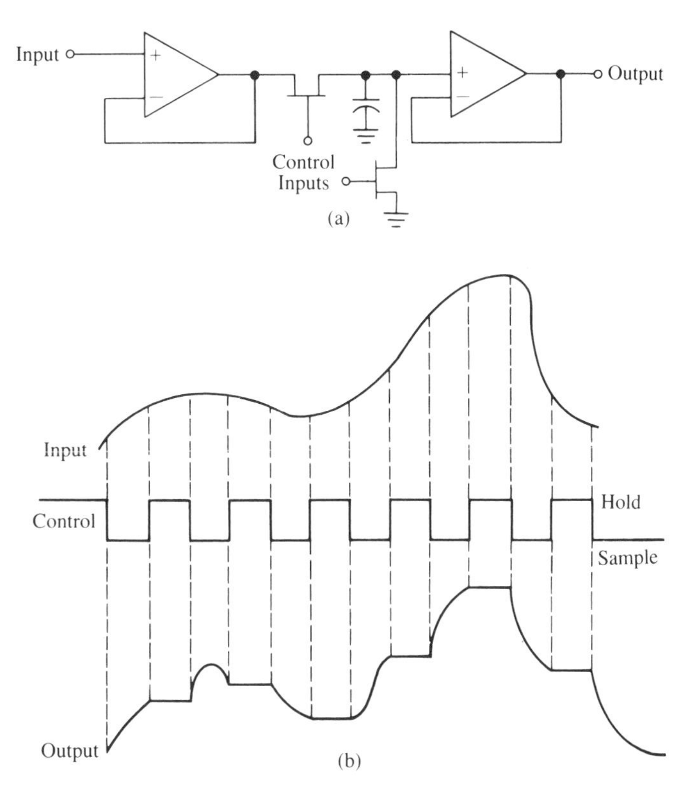

FIGURE **4.17**
(a) Sample-and-hold amplifier; (b) S/H output as a function of input.

output tracks the input or the hold mode in which the output retains the value of the signal at the time of the mode change (Figure 4.17b). This circuit behaves as an analog switch, which can sample the instantaneous input voltage and retain it at a constant dc level.

The track-and-hold amplifier is similar to the sample-and-hold amplifier, differing only in the relative amount of time it spends in the sample and hold modes. If the sampling time is long compared with the hold time, the device is known as a track-and-hold amplifier. Conversely, if the sampling time lasts only for the short period necessary to fully charge the capacitor (Figure 4.17b), the device is classified as a sample-and-hold amplifier.

In addition to their sampling function, these devices can amplify or attenuate the analog signal to make it compatible with the input requirements of the ADC, thus eliminating the need for separate sampling and scaling circuits. These amplifiers may also be wired to extend their applications. If the control switch is always closed, a sample-and-hold device may function as a conventional amplifier with excellent operating parameters. Several sample-and-hold amplifiers may be multiplexed using their control switches, since an open switch effectively disconnects the output of an amplifier.

Hybrid Data Acquisition Systems[4]

The individual circuits described in the preceding sections can be combined into a single system for the purpose of acquiring data from analog devices and converting them to a digital format to be processed by computers. A typical data acquisition system combines active filters for noise reduction, a multiplexer for sampling data from a number of devices, a sample-and-hold amplifier, an ADC, and finally a latch for temporary storage of the digitized data before transfer to the data bus of the computer (Figure 4.18). The system may be constructed from several medium- or large-scale integrated circuit chips or it may be contained on a single very large integrated circuit chip (Figure 4.19). This 80-pin chip includes all the component circuits necessary to multiplex and convert analog signals ranging from 0 to 5 V into equivalent digital outputs. Sampling rates can be varied from 18 kHz with 12-bit resolution to 40 kHz with 8-bit resolution. A low-drift instrumentation amplifier with a selection of gains from 2 to 1000 can be configured to accept either 8-channel differential or 16-channel single-ended inputs from analog devices.

Differential input devices use common rejection to eliminate grounding problems and to minimize environmental noise. Each input device is connected to

FIGURE **4.18** Configuration of a typical data acquisition system.

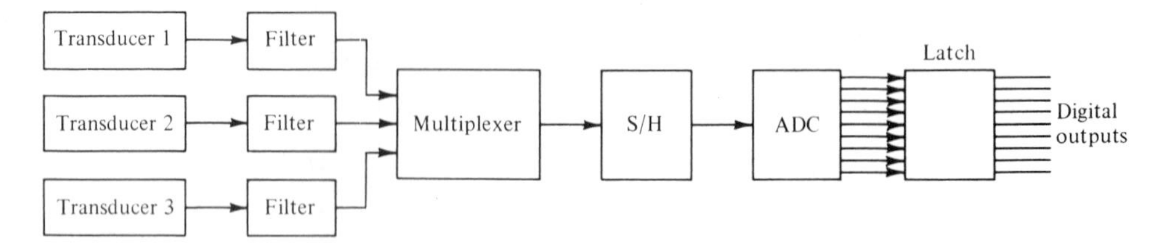

the data acquisition system with two wires that are subjected to the same environmental conditions. The system measures the difference in potential between these two wires as input voltage and rejects the ground potential between the signal source and the system chip. Furthermore, any voltage caused by sources of environmental noise common to both wires is rejected.

This system provides the complete signal conditioning required for interfacing an input transducer directly to a computer. The analog input signal channel is selected by the multiplexer; the signal is amplified by the instrumentation amplifier and transformed to a digital output using the S/H coupled to the dual slope ADC. Tri-state outputs allow the data to be conveniently transferred to the data input of a computer (Figure 4.20). Digital control signals from the computer can be used to select the analog input channel, to set the gain of the instrumental amplifier, and to transfer data from the ADC output to the computer.

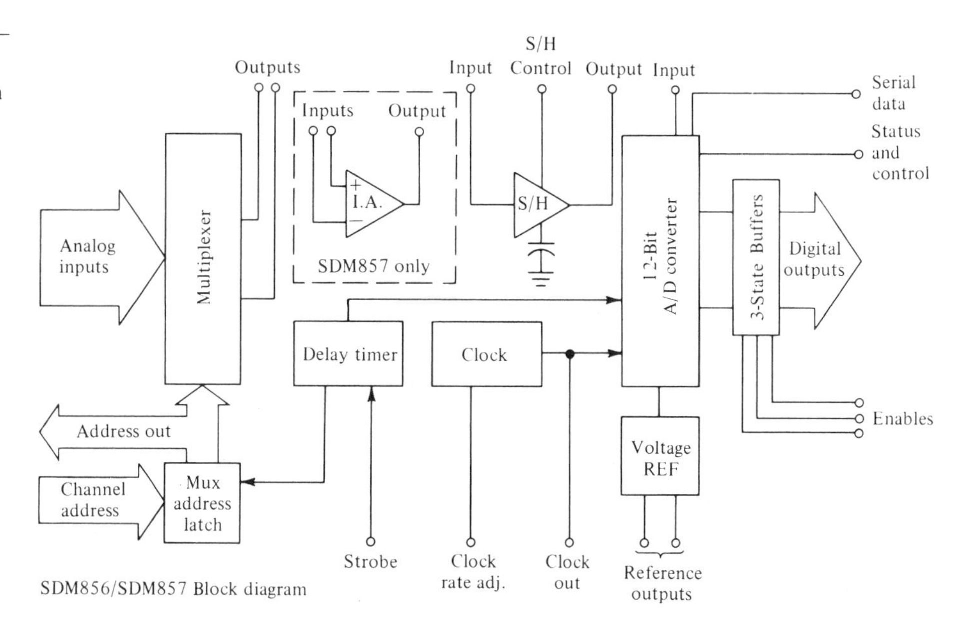

FIGURE **4.19**
Hybrid data acquisition system. (Courtesy of Burr-Brown.)

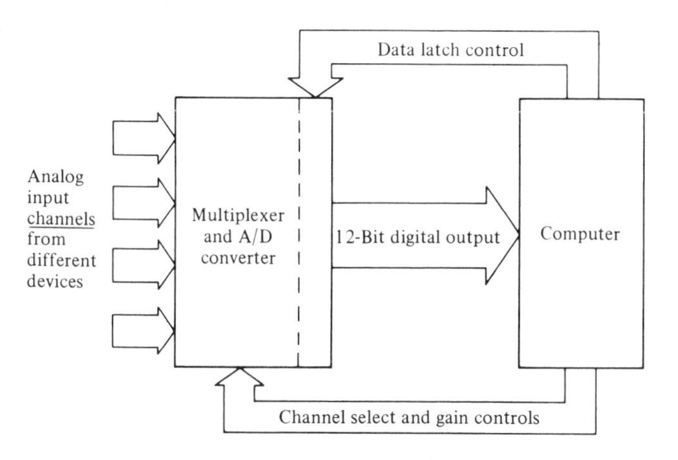

FIGURE **4.20**
Hybrid data acquisition system as a computer interface.

Universal Asynchronous Receiver/Transmitter

The universal asynchronous receiver/transmitter chip (UART) provides a programmable digital communications center on a single 40-pin chip (Figure 4.21). The transmitter register accepts bit parallel data and produces a bit serial output. The receiver register transforms bit serial input to bit parallel output. The two registers may operate independently or in series (full- or half-duplex, respectively). Other selectable operating parameters include the number of data bits in the transmitted code, the presence or absence of parity bits for error checking, and the transmission rate (bits per second). Transmitter buffer empty and receiver buffer full are two of the available status signals.

UARTs are used to link instruments with output devices, controllers, or computers at remote locations. The digital serial data are commonly transmitted over a pair of twisted wires. This inexpensive chip has greatly facilitated applications that require digital data transmission, such as remote monitoring and feedback control.

More recently, a serial port capable of both asynchronous and synchronous operation has been developed. It is called a USART (universal synchronous/asynchronous receiver transmitter) and operates in a manner similar to the UART.

Programmable Peripheral Interface

The programmable peripheral interface (PPI) LSI chip contains three digital I/O ports that can be controlled by software commands from the microcomputer (Figure 4.22). Each port can transfer 8 bits of information simultaneously to or from the computer. In addition to the three I/O ports, the PPI chip has a control port and chip control logic circuitry. The configuration of each port (whether it is receiving data from an instrument and sending them to the computer or transmitting data from the computer to an instrument) is controlled by commands programmed into the computer. These commands are transferred directly to the control port and decoded to set up specific configurations of the I/O ports. For example, the BASIC command OUT 6, 34 sends a control word (34) to the control port—in this example, device 6. Circuits on the chip read the data, 34, decode it, and configure the three ports accordingly (ports A and B output, port C input in this example). Each I/O port on the chip is addressed by the computer as a separate device and can transfer

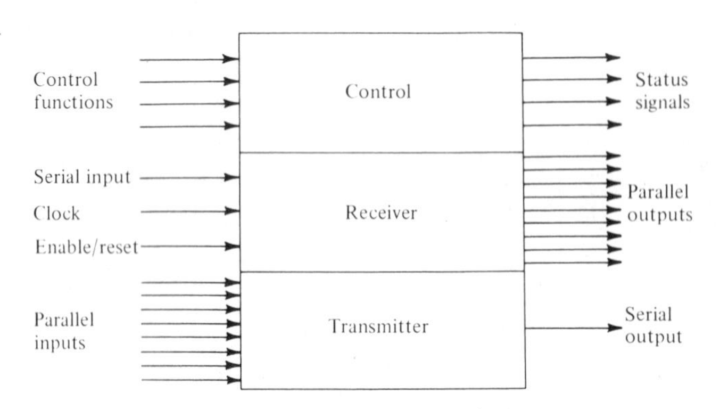

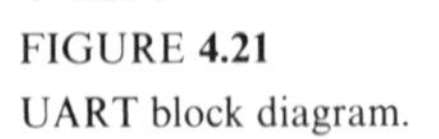
FIGURE 4.21
UART block diagram.

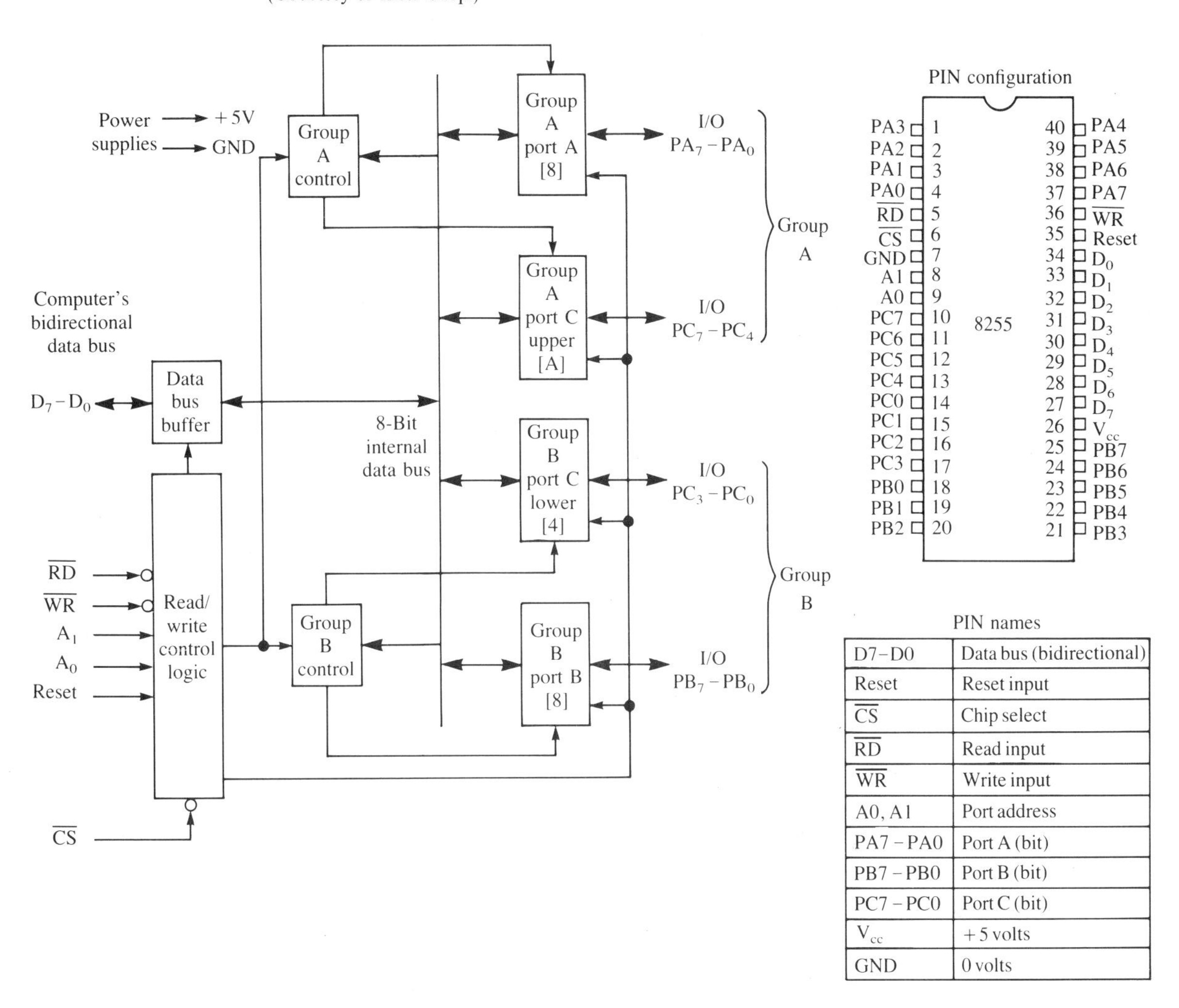

D7–D0	Data bus (bidirectional)
Reset	Reset input
$\overline{CS}$	Chip select
$\overline{RD}$	Read input
$\overline{WR}$	Write input
A0, A1	Port address
PA7 – PA0	Port A (bit)
PB7 – PB0	Port B (bit)
PC7 – PC0	Port C (bit)
V_{cc}	+5 volts
GND	0 volts

FIGURE **4.22** Block diagram and pin descriptions for the 8255 programmable peripheral interface chip. (Courtesy of Intel Corp.)

data independent of the other two I/O ports. Because I/O operations are fundamental to every microcomputer, the PPI chip is widely used as a universal parallel I/O device.

4.4 COMPUTER ORGANIZATION—SOFTWARE

While computer hardware costs have decreased approximately 30% each year, software costs have increased by 15%. Therefore, a discussion of software development is necessary to complement the preceding discussion of hardware.

Programming can be carried out at various levels, from the lowest, which is machine-oriented binary code, to the highest, which is machine independent. A comparison of programming a simple arithmetic addition at three different levels is given in Table 4.3.

Machine Language

The central processing unit of any computer responds to a set of binary coded instructions. The number of different instructions in the set and the binary codes for specific instructions depend on the internal architecture of the computer, which in turn varies with different manufacturers and models. Sequences of these binary coded instructions are known as machine language programs. All higher level programs must ultimately be translated into a series of machine language instructions, known as object programs. Writing low-level machine language (object) programs is a tedious, time-consuming task and should be avoided whenever possible.

Assembly Language

The next level of programming is assembly language. It is composed of a set of mnemonics (groups of letters and numbers) that represent machine language instructions, one assembly instruction for each machine instruction (Table 4.3). Programming at the assembly language level is easier and faster than machine language programming. A program known as an assembler is necessary to convert programs written in assembly language to machine language programs. Assembly language programs can be written, translated to object code, and executed on a single computer, provided an assembler program is resident in the computer's memory.

High-Level Languages[9]

Languages used at the highest level of programming have been developed for convenient, efficient use by humans. The detailed composition and arrangement of the machine code become the responsibility of a translator program. High-level languages are usually algebraic in nature, with each line of the program, known as source code, producing several lines of machine-readable object code (Table 4.3). Transportability (the machine independence) is a major factor in the convenience of high-level languages. The source code for a given program can be run on different

TABLE **4.3** COMPARISON OF PROGRAMMING AN ADD INSTRUCTION

Instruction language		
High level (FORTRAN or BASIC)	**Assembly (mnemonic)**	**Machine (binary code)**
D = B + C	MOV A, M	01111110
	INR M	00110100
	ADD M	10000110
	INR M	00110100
	MOV M, A	01110111

brands of computers provided that a suitable translator program is available for each computer. It is important to note that whereas high-level languages are usually machine independent, translator programs are not.

BASIC and FORTRAN are high-level languages with mature language support and extensive user bases. Pascal and Ada provide powerful structures within a relatively rigid framework. Forth[10] and C permit the user a great deal of flexibility in the design and implementation of special features. LISP and Prolog are useful in the manipulation of nonnumeric data and are therefore valuable in artificial intelligence and robotics.

Applications Packages

The analytical chemist no longer needs to spend large amounts of time developing programs. Commercial software, written in high-level languages and transparent to the user, is available for almost every laboratory task from data acquisition to the preparation of final reports. Labtech Notebook is an example of an integrated, general-purpose software package for data acquisition, process control, monitoring, and data analysis (Figure 4.23). The package can transmit data to other commercially available data analysis and management programs, such as Lotus 1-2-3 or Symphony, for additional data processing, formatting, and final report generation. A variety of interfacing hardware can be supported by the Notebook package. Two sophisticated routines are included for data processing: nonlinear curve fitting and fast Fourier transforms (FFTs). The analyst interacts with the Notebook package through a series of menus and thus no programming is required.

Other commercially available interfacing systems include both hardware and software components. Examples are ASYST from Data Translation, Inc., MEASURE from Lotus, LABSOFT from Cyborg Corp., and ADALAB from Interactive Microware, Inc. Finally, many current instrument systems that contain microcomputers include extensive applications packages for processing and storing data as well as displaying results in a variety of formats.

FIGURE 4.23 A functional diagram of Labtech Notebook. (Courtesy of Laboratory Technologies Corp.)

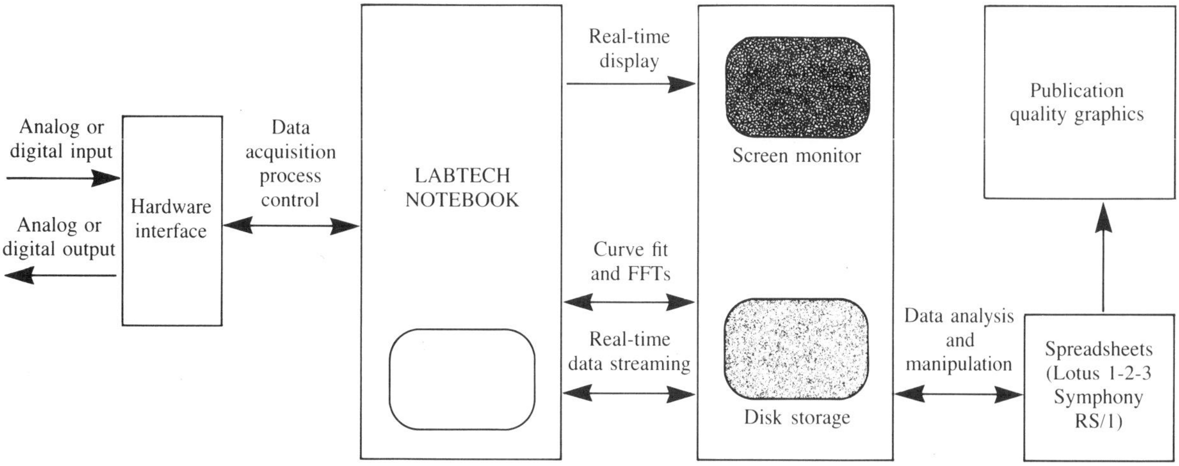

Translators

There are two kinds of high-level translator programs: compilers and interpreters. Both translators produce machine language object code but differ in their method of operation. A compiler translates the entire high-level program into object code. The resulting object program may not be executed until translation is complete. Programs are typically compiled in an off-line configuration; they may then be executed in off-line, on-line, or in-line configurations. Since compilers do not optimize the number of instructions required to perform a given task, object programs generated by a compiler are from two to five times longer than object programs produced from assembly language code written by skilled programmers.

Interpreters translate high-level programs line by line and execute each series of resulting machine language instructions before interpreting another line of source code. Programs must be interpreted and executed by the same computer. Because a program cannot be interpreted by one computer and executed by another, the interpreter software must be resident in the memory of the computer that executes the program. Interpreters (for example, BASIC) are typically used with on-line computers that respond to the programmer in real time.

Both interpreter and compiler programs suffer several disadvantages; both require large amounts of memory. Once an object program in machine language has been prepared by a compiler, it may be stored and quickly recalled for execution at any time. In contrast, the use of an interpreter requires that the high-level language source program be reinterpreted and executed, one line at a time, *each* time the program is run. No permanent object program is available for storage. Thus the use of an interpreter slows program execution and can become a serious limitation in laboratory applications that require high-speed data acquisition and manipulation. Interpreters, however, allow maximum flexibility in operator-computer interaction, including error and caution messages during program preparation and execution. In general, if memory costs are of little importance and relatively slow operation is acceptable, an interpreter language is the best choice for programming smaller computers. However, the method of programming in the future may be efficient compilers that generate compact object programs.

Operating Systems[11]

The overall operation of any computer system is controlled by software known as the operating system (OS). The OS or DOS (disk-based operating system) directs the CPU in coordinating the hardware and software functions of the entire system in a laboratory environment. Control of I/O devices, sequencing of programs to be run, and selection of languages for program development are some of the functions performed by the OS. The extensive interaction between the OS and any given high-level language requires that the language be compatible with the OS running on the computer. It is usually possible to run several OSs on a single type of computer. When purchasing software for programming in high-level languages and also applications packages, the user must match the package with both the computer hardware and the OS.

Popular operating systems for personal computers are MS/DOS and PC/DOS from Microsoft, CP/M and MP/M, single-user and multiuser systems from Digital Research, p-System from Softech, and RT-11, RSX-11/M, and RSTS from Digital Equipment Corp. These OSs are transportable from one brand of personal or laboratory computer to another within certain limits. UNIX is another popular OS developed at Bell Labs for use with a variety of computer systems.

Software Control of the Computer-Instrument Interface

A software programming strategy is required to control devices interfaced to a computer bus. Three strategies are polling (programmed I/O), interrupt, and direct memory access (DMA).

Polling or programmed I/O is the simplest method. After connecting the interface of each device to the three computer buses and assigning each device an address code, a program known as a polling loop is written to question sequentially each device to determine whether it requires service (Figure 4.24). The computer periodically asks a device: "Do you require service?" The device responds with a yes or no. A positive response causes the computer program to jump to a subroutine called a device handler that services the responding device. Whenever a negative response is received, the computer proceeds to poll the next device in sequence.

In actual practice, polling is implemented by the computer through a logic test of the status (1 or 0) of a specified data bit (flag) from the device. A typical service action is the transfer of a word or block of data to or from the device. The process of the computer questioning a device and receiving information in return is called handshaking. A service routine for a given device may elicit a number of handshaking exchanges to ensure the proper transfer of data. Two advantages of the polling method are minimal interface hardware and straightforward, simple software. The major disadvantage of this strategy is its demand on the computer's time. Each time a polling loop is entered, the status of each device in the loop must be checked even though it may not require service. If the time spent in the polling loop is objectionable for a given application, one of the other programming strategies may be implemented; if not, polling is the simplest technique to use.

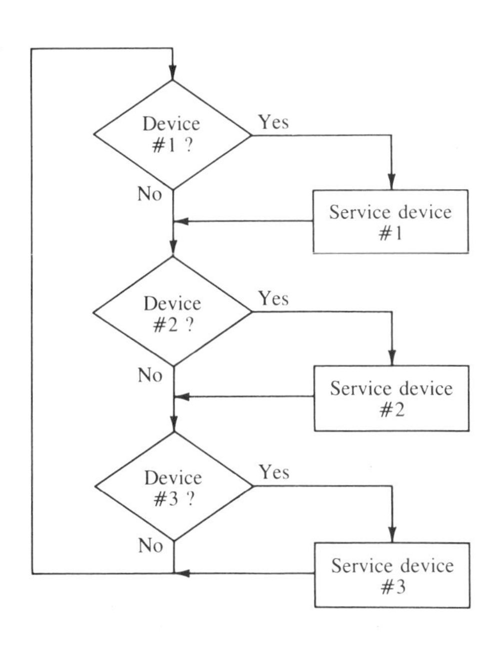

FIGURE 4.24
Polling method of input/output for three devices.

In situations where computer response time to interfaced devices is critical or where it is not desirable to have the computer spend a large fraction of its time in polling loops, the interrupt driven technique may be a suitable alternative. Each interfaced device is given the ability to initiate a request for service from the computer. The interrupt line on the control bus of the computer is used for this purpose. When an appropriate signal from a device is applied to the interrupt line, the computer breaks away from its current task, determines the device requesting service, and performs the specified service routine. Upon completion of this routine, it returns to the task it was executing when the interrupt signal was received. If several devices initiate interrupt signals simultaneously, the computer services them in order of preassigned priorities (priority interrupts).

Interrupts provide fast response to I/O devices and are therefore required in real-time systems that must provide the fastest possible response time to external conditions. This technique of I/O handling also promotes the efficient use of computer time by scheduling the computer to work on less important (background) tasks while waiting for interrupts. The major disadvantages of this method are extra interface hardware, more time required to service devices, and complex software.

The final technique, direct memory access (DMA), can be implemented on many computers by adding DMA hardware controllers. Upon receiving the correct interrupt signal from an I/O device, the DMA hardware controls the transfer of data directly between the device's interface and the computer's memory at a much faster rate than the previous two methods. Thus less computer time is required for servicing I/O devices. DMA is typically used with fast I/O devices such as multichannel analyzers, disks, and video terminals. This method is usually more expensive and adds significantly to the complexity of the computer system. However, once implemented, DMA requires little computer time for data transfer.

4.5 DATA REPRESENTATION

One of the major problems to be solved in any computer application is data transformation. Input data from peripheral devices and instruments must be converted to binary information required for the operation of all digital computers. Likewise, data transmitted from the computer to instrument or output devices must be understood by the device or analyst. Converters that perform the necessary data transformations are comprised of either hardware or software, or a combination. They are commercially available in standard devices or software packages.

Binary information can represent both fixed point (integer) and floating point numbers, alphanumeric characters, and computer instructions. The computer's interpretation of a group of binary bits depends on the architecture of the computer and the context in which the information appears.

In unsigned integer arithmetic, a computer with an *n*-bit word can represent $2n$ positive integers; for example, an 8-bit computer can represent numbers from 0 to 256_{10}. The subscript denotes the base of the number. For example, 256_{10} means 256 to the base 10, while 256_8 indicates 256 to the base 8. However, many computers use "two's complement arithmetic," which requires that both positive and negative numbers be represented. In fixed point notation the most significant bit is used to represent the sign of the number: 0 for positive integers and 1 for negative numbers. An 8-bit computer could represent signed numbers from -128_{10} through 0 to $+127_{10}$.

Floating point notation is used to expand the range of numbers that can be represented by a computer. For example, if a 16-bit computer can represent 65,536 integers directly, how can it handle the number $150{,}137_{10}$? Floating point notation solves this problem by representing the mantissa (number part) and the exponent (base part) with two or more 16-bit binary computer words. In this example, three 8-bit bytes are used to represent the mantissa, and one byte is reserved for the exponent. The first bit of the initial byte is used to designate the sign of the number and the first bit of the 4th byte the exponent; 0 indicates + and 1 indicates −. The number 4.519×10^{12} (4519×10^{9}) is

$$\overbrace{00000000}^{\text{1st byte}}\ \overbrace{00010001}^{\text{2nd byte}}\ \overbrace{10100111}^{\text{3rd byte}}\quad \text{(mantissa)}$$

$$\overbrace{00001001}^{\text{4th byte}}\qquad\qquad\qquad\qquad \text{(exponent)}$$

Floating point subroutines are necessary to perform arithmetic and logic operations on floating point data. These routines are available as either hardware or software components of the computer system, with hardware implementation having the advantage of speed. The range of numbers that can be processed by the popular 16-bit microcomputers is 10^{-38} to 10^{+38}. The faster execution times of the larger 32- and 64-bit computers is due in large part to the ability of these machines to process floating point data more efficiently than smaller 8- and 16-bit machines. Although binary numbers consisting of many digits are processed easily by computers, these numbers are cumbersome for human programmers to handle. The digits of binary numbers are therefore bunched together in groups of three or four and represented in octal or hexadecimal notation respectively. Table 4.4 gives examples of the

TABLE **4.4**

BINARY NUMBER EQUIVALENTS

Binary	Octal	Hexadecimal	Decimal
1	1	1	1
10	2	2	2
11	3	3	3
111	7	7	7
1000	10	8	8
1001	11	9	9
1010	12	A	10
1011	13	B	11
1100	14	C	12
1101	15	D	13
1110	16	E	14
1111	17	F	15
10000	20	10	16
10001	21	11	17
11011	33	1B	27

relationship among these three notations used for coding machine language programs.

4.6 THE AUTOMATED LABORATORY

Smart Instruments

The use of microcomputers as essential components of current instrumentation means that these instruments can adjust operating parameters during a series of measurements to maximize the information content of the data acquired. These instruments can perform calibrations during an analysis and make appropriate adjustments in either the data or the operating parameters. Many systems execute error-checking procedures and report any problems to the analyst.

Although examples of smart instruments are found in every succeeding chapter of this text, the microcomputer and hardware interface components are not discussed in any detail. The functions of these components in a smart atomic absorption instrument system are given in the following section.

A Smart Atomic Absorption Spectrophotometer[12]

Incorporation of in-line microcomputers into the design of atomic absorption (AA) spectrophotometers minimizes human error and effort, while maximizing performance and analytical throughput (Figure 4.7). Until the appearance of computerized instruments, the rate of multielement sample analyses was limited by the time required for the analyst to change lamps and readjust operating parameters. The accuracy and precision of results were limited by the narrow dynamic quantitative range of the instrument. Microcomputer data processing results in a linearized working range of up to four orders of magnitude. This means that chromium at a concentration of 90 $\mu g/mL$ can be determined with the same calibration and under the same conditions used to detect 0.01 $\mu g/mL$. The Perkin-Elmer Model 5000 atomic absorption spectrometer can handle up to 50 samples and determine six different elements automatically. All control knobs are replaced by keyboards consisting of numeric and functional keys that allow entry of operating parameters such as wavelength, slitwidth, and flowrates of fuel and oxidant (Figures 4.25 and 4.26). Data from the analysis of samples as well as the operational parameters are continually monitored and displayed on the control panel. Entering a set of parameters for a given analysis is a simple task. For example, inputting the desired wavelength from the keyboard and depressing the PEAK function key causes the grating in the monochromator to move to the position appropriate for the selected wavelength and to automatically center on the peak of the line from the lamp. The analyst selects the desired lamp on the multilamp turret by entering the position number and touching the LAMP # key. The lamp current in milliamperes is then entered and the LAMP MA key depressed. When the automatic burner control unit is activated, normal flows for an air/acetylene flame are immediately available. The FLAME ON/OFF button ignites the flame; gas flow adjustments are entered through the keyboard (Figure 4.26). Digital flow control offers more precision in resetting a previously optimized gas flow than do manually operated needle valves or pressure restrictor systems

A set of optimized operating parameters for the analysis of an element (known as a method) can be entered into the RAM memory from the keyboard. At any subsequent time these parameters can be recalled and used to prepare the instrument for the analysis of the specified element. Since the amount of RAM memory available for storage is both limited and volatile, permanent storage methods are available on magnetic cards or a tape cassette. Up to six method programs may be read from disks and stored in memory for sequential execution by the instrument system. Background correction is controlled by the microcomputer, which selects the continuum source appropriate for the desired wavelength. Mechanical and electrical components required for operation of the optical system have been reduced by the use of microcomputer software.

FIGURE **4.25** Keyboard/display panel of an atomic absorption spectrometer. (Courtesy of Perkin-Elmer Corp.)

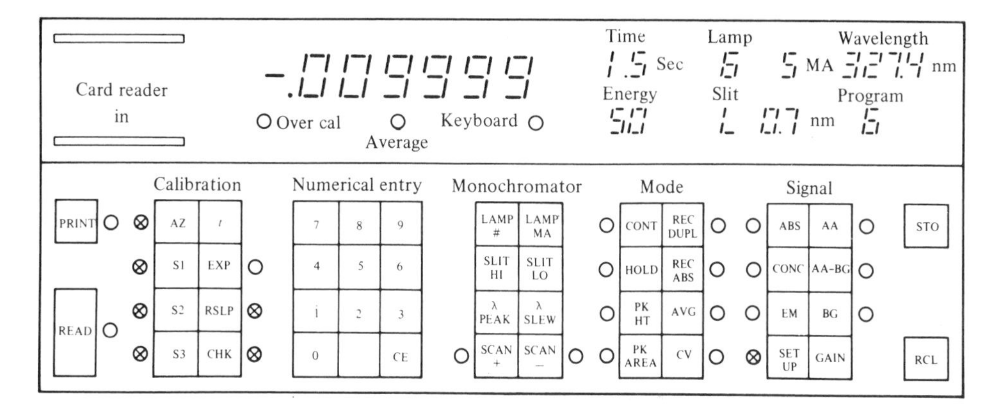

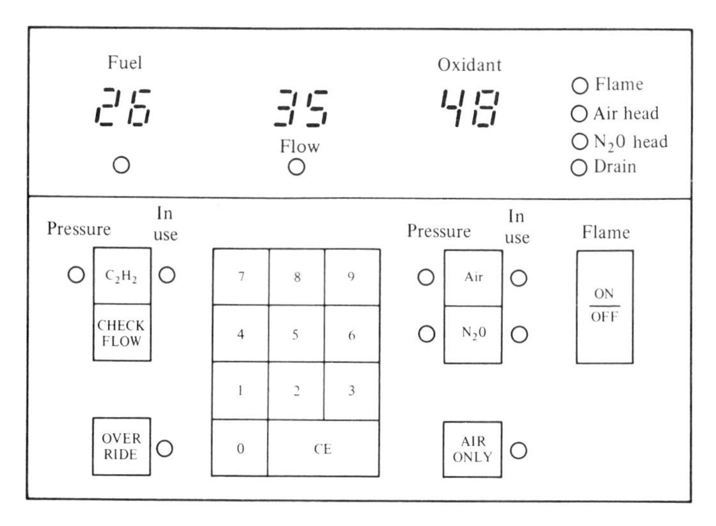

FIGURE **4.26** Control keyboard/display of flame parameters. (Courtesy of Perkin-Elmer Corp.)

The instrument provides integrated readings in units of absorbance, concentration, or emission intensity. These readings can be updated continuously or held on the instrument display. Integration times are selected on the keyboard, from 0.2 to 60 sec. Peak height or integrated peak area can be measured with electrothermal (flameless) furnaces. The instrument can be used for flame emission as well as atomic absorption spectrometry.

Calibration of the instrument to read directly in concentration units requires only two steps. First the operator enters the concentrations of the standards. The standards are then analyzed in the same sequence. Zero absorbance is set by aspirating a blank containing only solvent and depressing the ZERO button. If the calibration curve is known to be linear, only one standard is required. For nonlinear working curves, two or three standards are used, depending on the degree of nonlinearity. An algorithm stored in ROM memory then linearizes the instrument's response with respect to sample concentration over a specified range.

To recalibrate in the middle of a run, a standard is rerun and the RESLOPE key is depressed. This causes the instrument to be recalibrated with the same program used for the original calibration. The analyst can use a fixed expansion rather than diluted standards by entering the desired expansion value (from 0.01 to 100) on the keyboard and depressing the EXP button.

Analytical accuracy can be improved by the averaging feature of the instrument. By entering the number of readings to be averaged and pressing the AVG key, the average is displayed on the readout. Finally, the coefficient of variance and the standard deviation can be obtained by depressing appropriate keyboard buttons.

After the burner head is replaced with the graphite furnace head and the furnace control unit that contains a separate microcomputer, low-level samples can be analyzed. Furnace operation is controlled by the following programmed parameters: time intervals necessary to obtain and hold temperatures for dry, ash, and atomize cycles; the temperature for each cycle; times at which sample data are collected; time when AUTO ZERO control is activated; and the flowrate of furnace inert gas stream. A baseline feature makes it possible to store the difference between an electronic zero and a blank reading, thus permitting the true blank to be subtracted automatically. The use of this feature is demonstrated by the difficult analysis of barium in calcium chloride, using the graphite furnace (Figure 4.27). Without the background correction the calcium matrix would have caused an

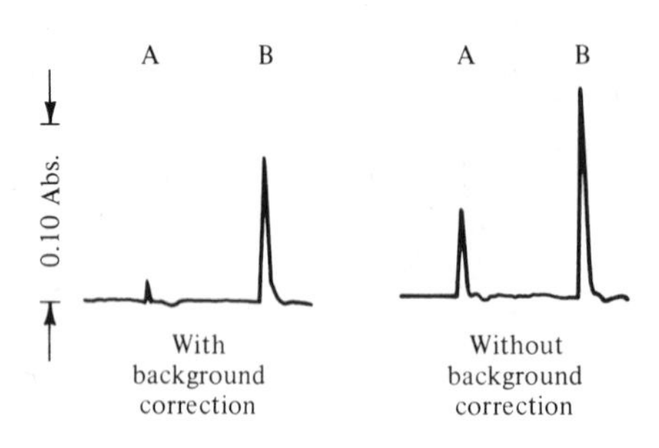

FIGURE 4.27
Determination of barium in calcium chloride with a graphite furnace using computerized background correction. [After R. D. Edigot, *Am. Lab.*, p. 75 (February 1978). With permission.]

erroneously high value of barium. As in the case of the flame spectrometer, method programs can be recorded on magnetic cards and reentered for subsequent analyses.

The computerized spectrophotometer is programmable not only from the keyboard and magnetic cards but also from an external terminal through a two-way RS232 communications interface. This standard interface permits external devices such as teletypes, CRT terminals, and computers to activate instrument functions. Using this capability, a tape cassette can store a large number of analytical methods for transfer later to the instrument. Data can be taken directly from the instrument's A/D converter at the chopping frequency and sent through the RS232 interface to an external computer for specialized data processing.

Local Area Laboratory Networks[13]

A local area network (LAN) may be loosely defined as a common link that provides for communication over relatively short distances among an assortment of computers, terminals, printers, disks, instruments, and other electronic devices. The primary function of LANs in the lab environment is the distribution of computing power to obtain efficient collection, processing, and storage of data. Coordination of LAN hardware with appropriate software packages is critical. A variety of transmission technologies are available for implementing LANs over distances of 500 m to approximately 50 km at rates of 10K bits/sec to 10M bits/sec. LAN transmissions may be either parallel or serial.

A laboratory computer system that uses several types of communications buses is shown in Figure 4.28. The 32-bit minicomputer controls the database laboratory management system, which receives the results from the lab data system. This minicomputer system contains a large amount of disk storage and devices for off-line archiving of results. Several terminals and a high-speed printer are available for manipulation and inspection of the database. One RS232 line is used to link the minicomputer to the 16-bit laboratory microcomputer with multitasking capabilities. In addition to being the hub of a star network, this lab microcomputer is also the terminus of an IEEE-488 parallel bus and a baseband LAN. A baseband LAN is capable of transmitting signals representing a broad range of frequencies simultaneously. These signals may carry both digital and analog information used by computers, telephones, and video devices.

The laboratory microcomputer system has a 10-megabyte hard disk, two 1-megabyte floppy disk drives, a printer, and a medium resolution graphics display. Five RS232 lines (ports) allow serial communication with the dedicated microcomputers in the nuclear magnetic resonance spectrometer (NMR), the Fourier transform infrared spectrometer (FTIR), a density meter, a gas chromatography data station, and the more remote minicomputer. The microcomputer can also communicate with several liquid chromatographs attached to the IEEE-488 parallel bus. Finally, the larger laboratory microcomputer can interact with several smaller microcomputers through a baseband LAN to acquire data and control several pH meters and balances, an atomic absorption spectrometer, an ultraviolet/visible spectrometer, and a temperature control system. Each small microcomputer is an 8-bit system with keyboard and graphics capabilities. A 20-megabyte hard disk is accessible to all the small microcomputers on the baseband system and the larger laboratory microcomputer.

The combination of several types of LANs, star, bus, and baseband units to form a single network provides redundancy and alternate storage of data. Two premises govern the design of this system:

1. Distributed processing results in faster response times and less downtime for any individual unit.
2. Because software cannot overcome hardware design limitations, highly flexible hardware allows the optimization of software for many applications within the system.

Laboratory Information Management Systems[14–16]

Current computerized instrumentation is capable of producing a great deal of scientific information. The analyst can be overwhelmed with results from these

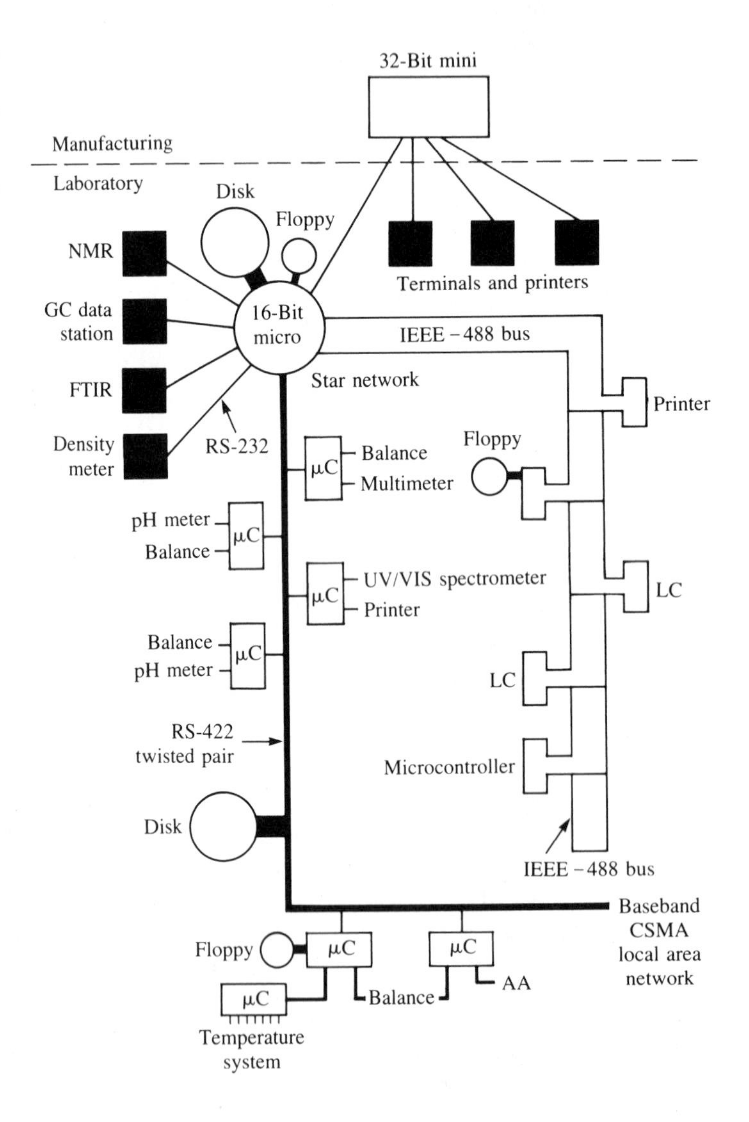

FIGURE 4.28
A local area network for linking the components of an analytical laboratory. μC = microcomputers. [After *Anal. Chem.*, **54**(12), 1296A (1982). Courtesy of American Chemical Society.]

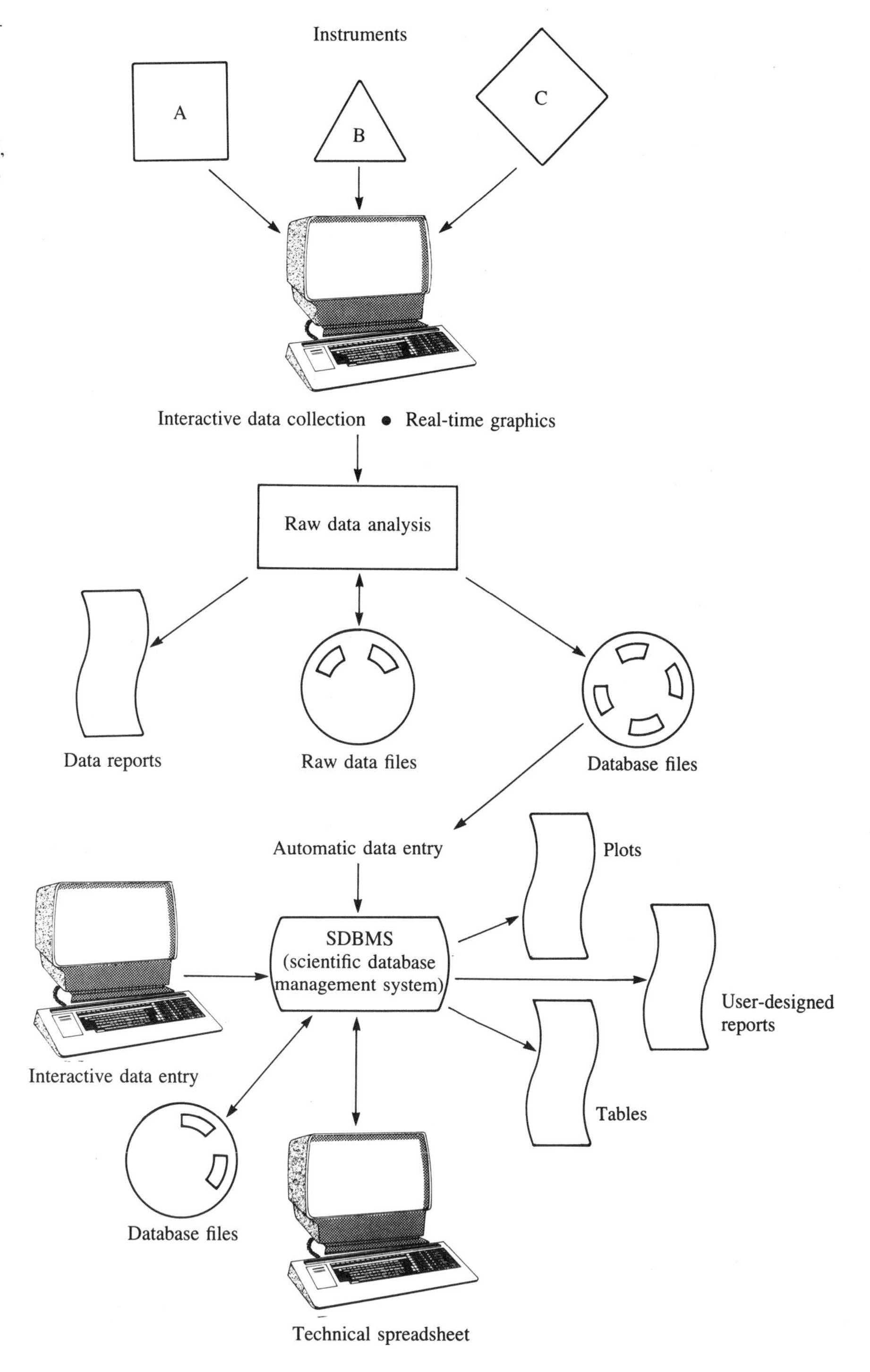

FIGURE **4.29**
Diagram of the LMS-1100 laboratory microcomputer system. [After M. Harder and P. Koski, *Am. Lab.*, **15**(9), 28 (1983). With permission.]

instruments, which are high-speed, automatic devices with the ability to perform many calculations that could not be obtained on earlier instruments. The computer that is responsible for this increased flow of information can also be used to assist the analyst in managing it. Applications packages known as database management programs were developed to meet the needs of the business community. These programs have been modified for use in analytical laboratories and are known as laboratory information management systems (LIMS).

Laboratory database management involves the creation, maintenance, and retrieval of a structured collection of analytical results for future study or comparison with another body of structured results. In laboratories that produce large volumes of data, such as quality control or clinical laboratories, the structuring, comparison, and reporting of data can consume large amounts of time. In addition the manual transfer of data from instrument to intermediate storage medium and into a final report can result in errors. Thus two immediate advantages of LIMS are a time savings and a reduction in data transmission errors.

A typical LIMS tracks each sample from log-in to completion of the required laboratory analyses. The system generates sample labels, receipts for the person submitting the sample, and work lists for laboratory personnel. The data from each analysis can be captured directly from instruments interfaced directly to the LIMS computer or entered manually from a video terminal (Figure 4.29). Once these data are stored on the computer's disk, they can be accessed by the LIMS for many different purposes, such as monitoring the progress of a sample through the laboratory or alerting personnel to abnormal results.

FIGURE **4.30** Diagram of the LMS scientific database management system (SDBMS). [After M. Harder and P. Koski, *Am. Lab.*, **15**(9), 28 (1983). With permission.]

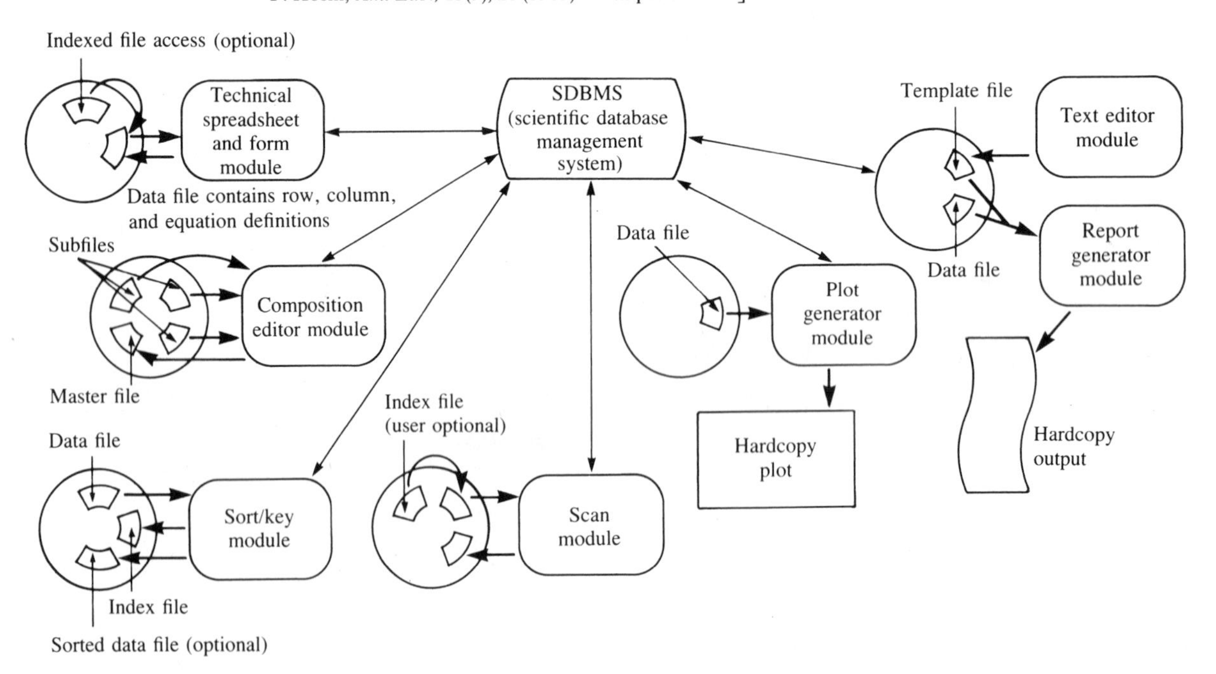

The most important function of a LIMS is the preparation of reports summarizing analytical results (Figure 4.30). A well-designed system allows the user to quickly prepare programs that generate properly formatted data and text. Desirable features of such a system are word processing (text preparation), list processing (sort, merge, and select), and a mathematics package that allows the user to calculate reported values from stored data. An archiving function is required to prepare and send data to devices that can store long-term, off-line information. It may be desirable to store condensed data, such as chromatographic retention times and areas, rather than entire sets of raw data, or entire chromatograms, in order to save storage space.

Some terms common to all database management systems can best be understood by drawing analogies with conventional methods of storing information. A file may be considered to be a manila folder that contains information on a particular set of samples. Each page in the folder is a record, and the information on each page is listed in fields. Thus, the ID number, origin, and results of each analysis would be the fields that constitute the record for a given sample.

The complex software of LIMS allows the user to access stored data in a variety of ways. One powerful access method involves relational databases that allow the user to construct new arrangements of data from several files or records in order to produce additional correlations or comparisons.

PROBLEMS

1. Briefly describe the role of each of the five basic units of the computer as they function with an instrument that is interfaced to the computer.

2. Describe the utility of the information carried by each of the three types of computer buses.

3. Distinguish between each of the following: (a) microcomputers and microprocessors, (b) serial and parallel data transmission, (c) multiplexing and demultiplexing, (d) ADCs and DACs, and (e) track-and-hold and sample-and-hold amplifiers.

4. If an 8-bit DAC has a 5.00-V full-scale output, what voltage results when 136_8 is sent to the input?

5. If a 10-bit DAC has a 10.0-V full-scale input, what output (binary and base 8) will result from an input signal of 6.23 V?

6. Calculate the resolution of the following ADCs: (a) a 4-bit converter and (b) a 12-bit converter.

7. Discuss the advantages and limitations of the following types of ADCs: (a) successive approximation counters, (b) voltage-to-frequency converters, and (c) dual slope converters.

8. Discuss the functions of the following in an instrument-computer interface: (a) UARTs, (b) PPIs, and (c) buffers.

9. Discuss the specific advantages of programs written at the following levels to instrument-computer interfacing: (a) machine language, (b) assembly language, (c) BASIC, FORTRAN, or Pascal, and (d) applications packages such as Lotus 1-2-3, LABSOFT, or ADALAB.

10. Distinguish between (a) a compiler and an interpreter, (b) an operating system and an applications package, and (c) polling and interrupts.

11. Perform each of the following conversions: (a) 001101111_2 to octal, hexadecimal, and decimal; (b) 362_8 to binary, hexadecimal, and decimal; (c) $4C9_{16}$ to binary, octal, and decimal; and (d) 1245_{10} to binary, octal, and hexadecimal.

12. Describe how smart instruments can interact through a LAN under the direction of a LIMS to provide an efficient flow of analytical information.

BIBLIOGRAPHY

BARKER, P., *Computers in Analytical Chemistry,* Pergamon, Elmsford, NY, 1983.
CARR, J., *Elements of Microcomputer Interfacing,* Reston, Reston, VA, 1984.
COFFORN, J., AND W. LONG, *Practical Interfacing for Microcomputer Systems,* Prentice-Hall, Englewood Cliffs, NJ, 1983.
CURRICK, A., *Computers and Instrumentation,* Heyden, Philadelphia, 1979.
DESSY, R., ed., *The Electronic Laboratory: Tutorials and Case Histories,* American Chemical Society, Washington, DC, 1985.
ELVING, P., V. MOSSOTTI, AND I. KOLTHOFF, eds., *Treatise on Analytical Chemistry,* 2nd ed., Part I, Vol. 4, Chaps. 7–9, John Wiley, New York, 1984.
RATZLAFF, K., *Computer-Assisted Experimentation,* John Wiley, New York, 1987.
SHEINGOLD, D., ed., *Analog-digital Conversion Handbook,* Prentice-Hall, Englewood Cliffs, NJ, 1986.
STONE, H., *Microcomputer Interfacing,* Addison-Wesley, Reading, MA, 1982.

For additional references to electronic hardware, consult the Bibliography in Chapter 3 and also the manufacturers' data books. In this rapidly developing field some of the most current applications are found in literature from Analog Devices, Burr-Brown, Intel, Motorola, National Semiconductor, RCA, Signetics, and Texas Instruments.

LITERATURE CITED

1. RATZLOFF, K., *Am. Lab.,* **10**(2), 17 (1978).
2. DESSY, R., *Anal. Chem.,* **58**(1), 78A (1986).
3. CAMPBELL, J., *The RS232 Solution,* SYBEX Inc., Berkeley, CA, 1984.
4. STOCKWELL, M., AND D. LARSEN, *Am. Lab.,* **16**(9), 41 (1984). See also Ushijima, D., *Macworld,* **4,** 129 (1987).
5. WOODWARD, W., F. WOODWARD, AND C. REILLY, *Anal. Chem.,* **53**(11), 1251A (1981).
6. LISCOUSKI, J., *Anal. Chem.,* **54**(7), 849A (1982).
7. DESSY, R., *Anal. Chem.,* **58**(6), 678A (1986).
8. BORMAN, S., *Anal. Chem.,* **57**(9), 983A (1985).
9. DESSY, R., *Anal. Chem.,* **55**(6), 650A (1983).
10. MACINTYRE, F., *Am. Lab.,* **17**(2), 18 (1985).
11. DESSY, R., *Anal. Chem.,* **55**(8), 883A (1983).
12. KRAMER, R., *Am. Lab.,* **15**(9), 46 (1983).
13. DESSY, R., *Anal. Chem.,* **54**(11), 1167A (1982).
14. DESSY, R., *Anal. Chem.,* **55**(1), 70A (1983).
15. LONG, E., *Am. Lab.,* **16**(9), 96 (1984).
16. MERRER, R., *J. Chem. Educ.,* **62**(5), A149 (1985), and **62**(6), A173 (1985).

AN INTRODUCTION TO ABSORPTION AND EMISSION SPECTROSCOPY

Spectroscopy is the measurement and interpretation of electromagnetic radiation absorbed, scattered, or emitted by atoms, molecules, or other chemical species. This absorption or emission is associated with changes in the energy states of the interacting chemical species and, since each species has characteristic energy states, spectroscopy can be used to identify the interacting species. As will be shown, quantitative information may also be obtained.

This chapter considers the fundamental principles of spectroscopy, whereas the next ten chapters describe specific techniques that use these principles for the identification and quantitative determination of chemical substances. The techniques and instrumentation required for different regions of the electromagnetic spectrum are so diverse that each region has given rise to seemingly quite different methods, although all are related by the fundamental principles discussed in this chapter.

5.1 THE NATURE OF ELECTROMAGNETIC RADIATION

Figure 5.1 shows a portion of an electromagnetic wave traveling through space in the x direction at the speed of light, approximately 3.00×10^8 m sec^{-1} in a vacuum. As the name indicates, there are both electrical and magnetic components of the wave at right angles to each other.

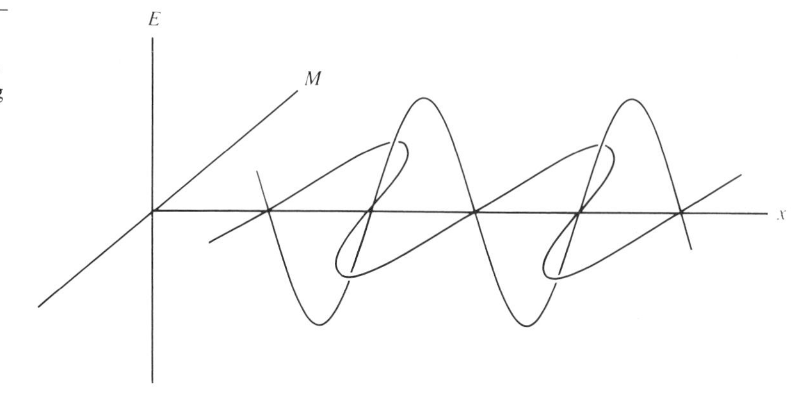

FIGURE 5.1
A portion of an electromagnetic wave traveling in the x direction, showing the electrical component, E, and the magnetic component, M.

For most discussions of the applications of spectroscopy to chemical analysis, electromagnetic radiation may be considered as an electromagnetic wave traveling at the speed of light. However, some properties of electromagnetic radiation are better described by considering the radiation as consisting of discrete particles (quanta) of energy, known as photons, that also travel at the speed of light. According to the Heisenberg uncertainty principle, it is impossible to measure simultaneously and exactly both the wave and particle properties of a photon. It is useful, however, to keep both properties in mind. It is often convenient to think of photons as particles of energy radiating from a source and characterized by an electromagnetic wave. The wavelength of the radiation, λ, can be visualized as the distance between maxima of either the electrical or magnetic component—that is, from crest to crest in Figure 5.2. In the figure the distance AB is one wavelength. Associated with wavelength is frequency, ν, the number of waves that pass a fixed point, such as P, in a unit of time. When a photon passes a particular region of space, the electric field in that region oscillates with the frequency ν. Wavelength and frequency are related to the energy of a photon, E, by Planck's constant h, 6.62×10^{-34} J sec, and c, the velocity of light in a vacuum (3.00×10^8 m sec^{-1}):

$$E = h\nu = \frac{hc}{\lambda} \tag{5.1}$$

Only frequency is truly characteristic of a particular radiation. Radiation is propagated through matter at velocities of less than c because of interactions between the electric vector and the bound electrons of the medium. The index of refraction of a medium, η, is the ratio of the speed of light in a vacuum to the speed of light in the medium. The index of refraction is also a function of wavelength; longer wavelengths have a smaller index of refraction in a transparent medium than do shorter wavelengths. When radiation of a particular wavelength enters matter, its velocity decreases but its frequency remains constant. In the ultraviolet, visible, and infrared regions of the spectrum, the velocity of radiation in air is within 0.1% of the velocity in a vacuum, making it satisfactory to use Equation 5.1 to interrelate wavelength and frequency.

Wavenumbers, $\bar{\nu}$, are sometimes used instead of frequency. They are calculated as follows:

$$\bar{\nu} = \frac{1}{\lambda} \tag{5.2}$$

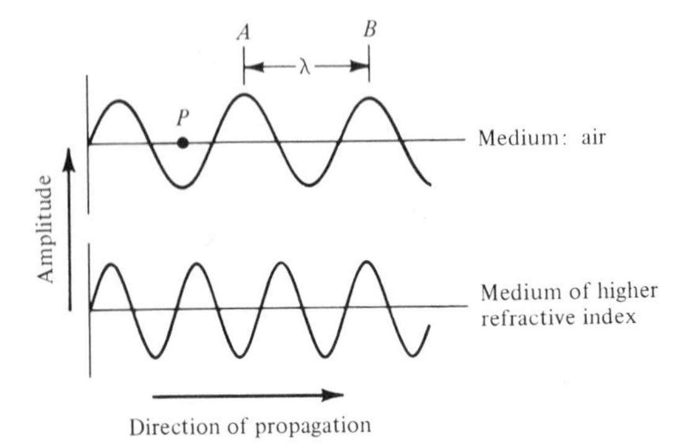

FIGURE 5.2
Some characteristics of electromagnetic radiation.

Wavenumbers, $\bar{\nu}$, in units of cm^{-1}, express the number of waves that occur per centimeter; this number is directly proportional to the frequency, but it should be emphasized that it is not a frequency:

$$\nu = c\bar{\nu} \quad \text{or} \quad \bar{\nu} = \frac{\nu}{c} \tag{5.3}$$

A beam that carries radiation with a very small wavelength spread approximating one discrete wavelength is said to be monochromatic. A polychromatic beam contains radiation with a wide distribution of wavelengths. Two waves can combine to interfere with each other, either constructively or destructively. In Figure 5.3 the two waves have the same wavelength, frequency, and amplitude but are out of phase by 180°. When the amplitudes are added at the same position in space, they destructively interfere to cancel exactly each other. If they were in phase, they would constructively interfere to produce a wave that has twice the amplitude of each component at all spatial positions.

Both the amplitude and frequency of the electromagnetic radiation are important properties in spectrochemical measurements. Since available detectors do not have fast enough frequency responses, however, the amplitude cannot be measured. Instead, the radiant power, P, which is proportional to the square of the wave amplitude, is determined. Radiant power is the amount of energy transmitted in the form of electromagnetic radiation per unit time. Radiant power is given by Equation 5.4, where E is the energy of a photon and ϕ is the photon flux, the number of photons per unit time:

$$P = E\phi = h\nu\phi \tag{5.4}$$

Although radiant power is widely referred to as intensity, intensity is strictly defined as the radiant power from a point source per unit solid angle, usually given as watts per steradian.

A plane polarized beam of radiation is one in which the electric vector is confined to a single plane; the magnetic vector is therefore restricted to an orthogonal plane—that is, a plane at right angles to the plane of the electric vector and the axis of propagation. Unpolarized radiation has waves in many planes. Polarization can be produced in several ways—for example, by selective absorption of radiation in specific planes as the radiation passes through certain substances.

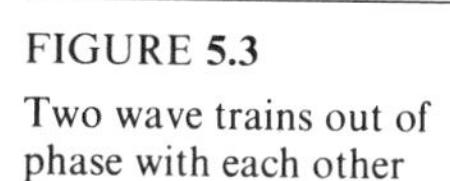

FIGURE 5.3
Two wave trains out of phase with each other

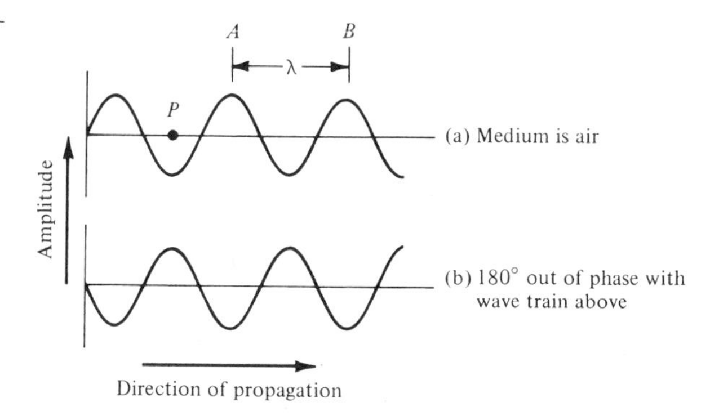

FIGURE 5.4 Schematic diagram of the electromagnetic spectrum. Note that the wavelength scale is nonlinear.

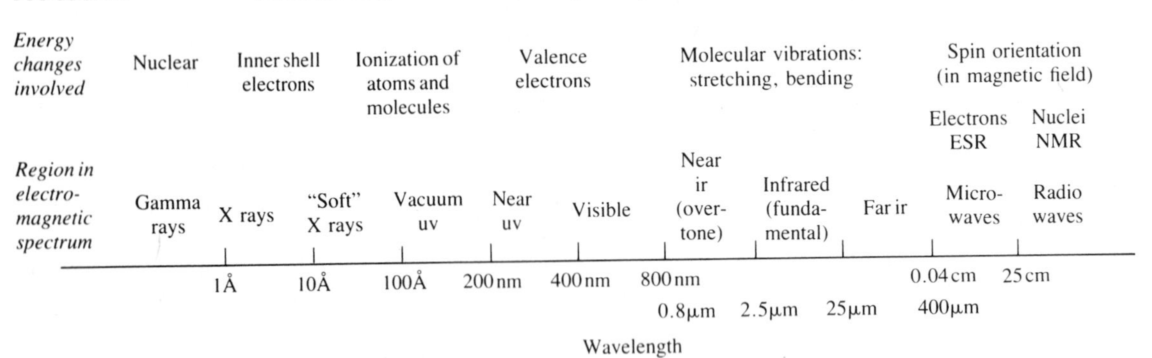

5.2 THE ELECTROMAGNETIC SPECTRUM

The interaction of matter and radiation takes place throughout the entire electromagnetic spectrum, the name given to the broad range of radiations that extends from cosmic rays with wavelengths as short as 10^{-9} nm all the way up to radio waves longer than 1000 km. Within these extremes and moving from short to long wavelengths are gamma rays, X rays, far, middle, and near ultraviolet rays, the visible light portion of the spectrum, infrared rays, and microwaves. The nature of all these radiations is the same, and all move with the speed of light. They differ only in frequency and wavelength and in the effects they can produce in matter.

The chemical and physical effects of various types of radiation are quite different, and these differences can be understood in terms of the various energies of the photons. In the radio-frequency range, the energy of one photon is very low, and the energy transitions are concerned with the reorientation of nuclear spin states of substances in a magnetic field. In the slightly higher-energy microwave region, there are changes in electron spin states for substances with unpaired electrons when in a magnetic field. In the infrared region, absorption causes changes in rotational and rotational-vibrational energy states, whereas absorption of radiation in the visible and ultraviolet regions causes changes in the energy of the valence electrons accompanied, in the case of molecules, by rotational-vibrational changes. X rays cause the ejection of inner electrons from matter, and, at the high-energy end, gamma rays can cause changes in the nucleus. The various regions in the electromagnetic spectrum are displayed in Figure 5.4 along with the nature of the changes brought about by the radiation.

5.3 ATOMIC ENERGY LEVELS

According to quantum theory, atoms can exist only in discrete potential energy levels. The potential energy of an atom depends on the electronic configuration. Transition of outer electrons between fixed energy levels leads to the emission or absorption of radiation at discrete energies. The frequency of radiation is proportional to the change in potential energy involved and is given by Equation

5.1. Each transition accounts for the presence of a specific frequency of radiation and hence the presence of a spectral line in either absorption or emission.

Grotrian developed a graphical way to present atomic energy levels and electronic transitions that is almost universally used. These diagrams permit the representation of spectral terms and transitions as shown in Figure 5.5 for several elements. The vertical axis is an energy axis, and the energy levels, or terms, are shown as horizontal lines. Absorption is represented by an upward line between adjacent energy levels. A spectral emission line results from a transition from a higher energy level to a lower one. The vertical distance representing a transition is a measure of the energy of the transition. The energy of the transition from the ground, or lowest, electronic state to the first excited state is great enough that only a few atoms with labile valence electrons have absorption spectra in the visible region. Absorption by atoms is limited to a comparatively few resonance lines. When an atom is excited by absorption of radiation or by collision with excited electrons, ions, or molecules, at atmospheric pressure, it normally remains in an excited state

FIGURE **5.5** Atomic term diagram for Li, Na, K, and Mg. Wavelengths are given in angstroms. (IP = ionization potential in eV.)

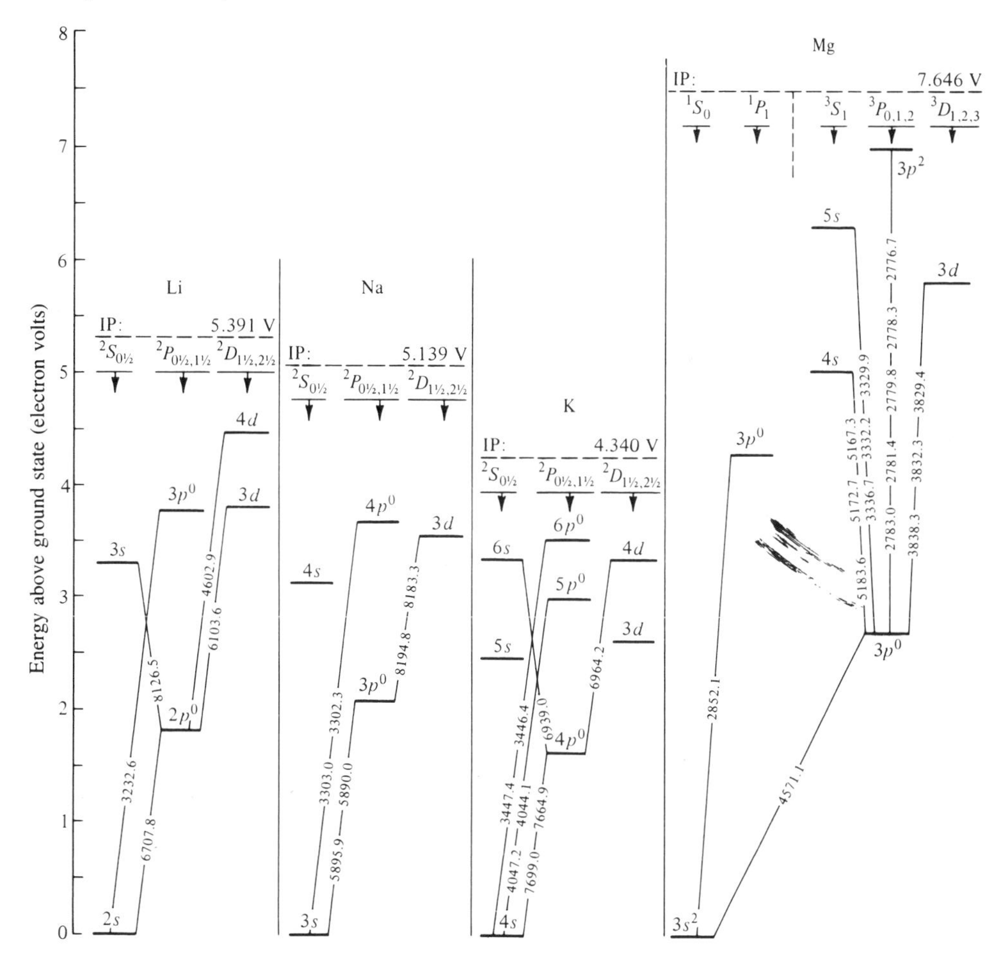

for only a very short time, approximately 10^{-9} sec, before it loses all or part of its excitation energy by collisions or by the emission of a photon.

The major contributions to the energy terms are associated with a principal quantum number $n = 1, 2, 3, \ldots$. For example, when the 3*s* valence electron of sodium is excited to 4*s*, 5*s*, 6*s*, and larger orbitals, the series of energy levels labeled ^2S results. The superscript refers to the multiplicity as explained below. The separation of levels decreases as the value of *n* increases. The continuum (ionization level) starts at $n = \infty$. The resultant orbital angular momentum or eccentricity of electronic orbitals that have the same principal quantum number accounts for somewhat smaller energy differences, which are classified as different series. The resultant orbital angular momentum is the vector sum of the orbital angular momenta of the individual electrons, $l = 0, 1, 2, 3, \ldots$ (symbol: *s*, *p*, *d*, *f*,...), and is represented by the quantum number $L = 0, 1, 2, 3, \ldots$ (series symbol: *S*, *P*, *D*, *F*,...). For example, a ^2P series results when the single valence electron of sodium is excited to *p* states of higher orbits.

The total angular momentum, the result of the different possible combinations of electron orbital momentum and electron spin angular momentum, accounts for still smaller differences, which distinguish the slightly different terms of multiplets. For example, each of the energies in the ^2P series is actually a pair of levels of slightly different energy depending on whether the total angular momentum is $\frac{1}{2}$ or $\frac{3}{2}$. The $^2S \leftrightarrow {}^2P$ transitions appear as doublets. The multiplicity is the number of different values of the total angular momentum for a particular resultant orbital angular momentum. The *s* terms, though singlets, always have the same multiplicity index as the *p*, *d*, and *f* terms that belong to them in a transition.

Selection rules are statements of transition probabilities that are based on experience or quantum-mechanical calculation. On the basis of selection rules, transitions are called allowed or forbidden. Selection rules are not absolute but define the transitions with the highest probability of occurrence. For example, the selection rule $\Delta L = \pm 1$ means that the spectral transitions that are most likely to occur are those between neighboring term series. In the case of multiplet terms, the inner quantum number *j* takes one of the two values $l + \frac{1}{2}$ or $l - \frac{1}{2}$. The selection rule for *j* is $\Delta j = 0, \pm 1$ ($j = 0 \rightarrow j = 0$ is not allowed).

The ionization energy or ionization limit of an atom is the energy to which the various series converge. The kinetic energy of the ejected electron is not quantized, and energy levels beyond this limit are continuous. The energy levels of the ion that remains correspond more closely to those of the preceding element in the periodic table than to those of the parent atom.

5.4 MOLECULAR ELECTRONIC ENERGY LEVELS

The electrons of a molecule can also be excited to higher energy states, and the radiation that is absorbed in the process, or the energy emitted in the return to the ground state, can be studied. The energies involved are generally large, 200–600 kJ mol^{-1}; consequently, the electronic spectra of molecules are usually found in the ultraviolet or visible region of the spectrum. In molecular spectra, transitions between electronic states are accompanied by transitions between rotational and vibrational energy levels that impart fine structure to the electronic absorption bands. As a result, the spectra of molecules are much more complicated than those of atoms. Molecular vibrations and rotations may reveal a great deal about molecular structure. Homonuclear diatomic molecules, which do not have vibration-rotation spectra in the infrared region, show vibrational and rotational structure in their electronic spectra.

Molecular electronic energies are represented by potential energy curves (or surfaces) in which the potential energy of each electronic state is plotted as a function of internuclear distance. Two-dimensional diagrams are inadequate, except in the case of diatomic molecules, but they are nevertheless useful in describing general phenomena that are observed. Examples of electronic energy levels are shown in Figure 5.6.

An electronic energy level is a physically stable state of a molecule when the potential energy curve has a minimum. Most molecules have excited electronic states that are not stable. Excitation to these states leads to dissociation, and the spectrum corresponding to these transitions is continuous. In some cases a stable excited state and dissociation occur as a radiationless transition between the two states.

Electronic states of simple molecules may be characterized by molecular quantum numbers that are derived from the quantum numbers of the component atoms. For example, a state is characterized by the resultant orbital angular momentum quantum number $\Lambda = 0, 1, 2, \ldots$ (symbols: Σ, Π, Δ, ...), and the multiplicity of a state is denoted by a superscript preceding this designation as in the case of atoms.

According to the Pauli exclusion principle, the spins of two electrons in the same orbit are opposite to each other—that is, paired. A molecule with an even number of electrons has all its electrons paired and is said to be in a singlet state. Whether the molecule is in the ground state or excited state, as long as the electrons are paired, the molecule is in a singlet state, $S_0, S_1, S_2, \ldots$.

To absorb radiation, a molecule must first interact with the radiation. This interaction must occur within approximately 10^{-15} sec, the period of oscillation of

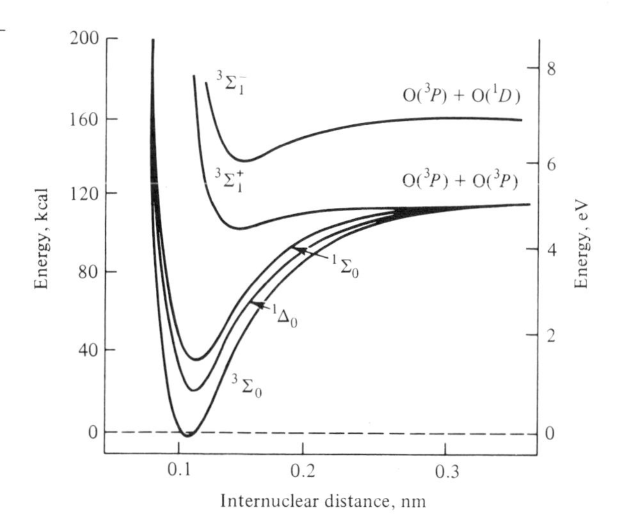

FIGURE **5.6**
Potential-energy curves for some of the electronic states of the O_2 molecule.

the electromagnetic wave. Consequently, exchange can occur only by interaction with the potential-energy component of the molecule's total energy via the movement of electrons. Born and Oppenheimer pointed out that because electrons move much more rapidly than nuclei, it is a good approximation to assume that, in an electronic transition, the nuclei do not change their positions. Therefore, an electronic transition is represented by a vertical line in an energy diagram. Furthermore, the electronic levels inside molecules are quantized so that the absorption bands can occur only at definite values corresponding to the energies required to promote electrons from one level to another.

Since a molecule vibrates, even when it is in the lowest vibrational energy level, a range of internuclear distances must be considered. In the lowest energy state the most probable internuclear distance is that corresponding to the equilibrium position. For the higher energy states the most probable configuration is at the ends of the vibration, where the atoms must stop and reverse their direction. For a solute molecule surrounded by solvent molecules, transitions are expected to have a greater probability of starting near the midpoint of the lowest vibrational level of the ground electronic state and proceeding to the $v = 2$ vibrational level of the excited electronic state. Transitions to other vibrational levels of the excited state occur with lower probabilities. Thus as Figure 5.7 shows, an electronic transition in absorption may show a series of closely spaced lines corresponding to different vibrational (and rotational) energies of the upper state.

Absorption terminates when the solute molecule arrives in any one of several possible vibrational levels in an excited electronic state that is still surrounded by the ground state of the solvent molecules. An excited molecule can return to its ground state by any of several paths shown in Figure 5.7. The favored route is the one that minimizes the lifetime of the excited state. Within 10^{-12} sec the electronically excited solute molecule drops to the lowest vibrational level of the lowest excited singlet

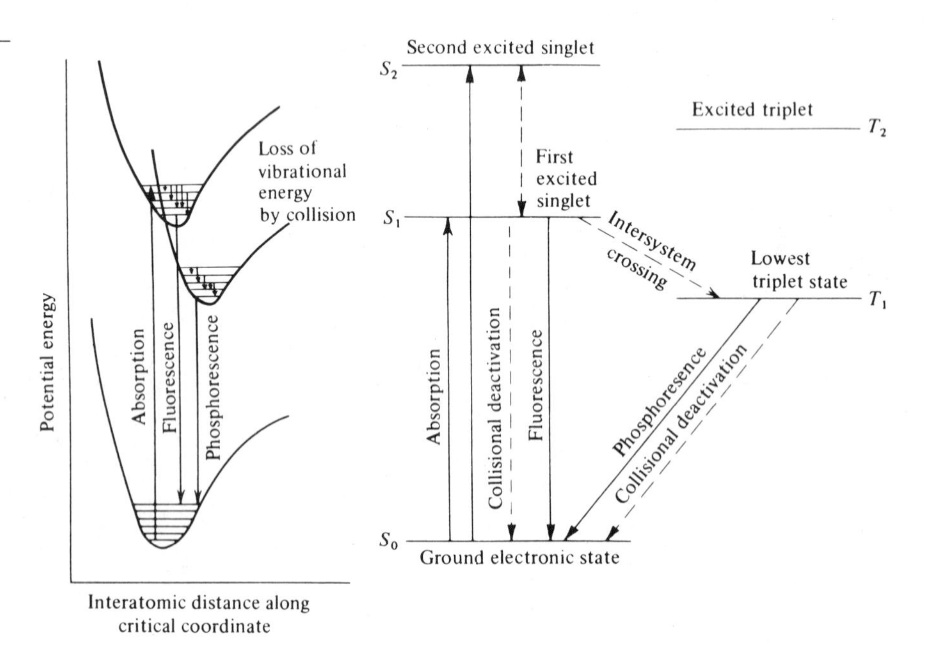

FIGURE 5.7
Schematic energy-level diagram for a diatomic molecule.

state by means of radiationless processes. The excess energy is transferred to other molecules through collisions as well as by partitioning the excess energy to other possible modes of vibration or rotation within the excited molecule. The solvent molecules reorient themselves to a state of equilibrium compatible with the new molecular polarity. The vibrational and solvent relaxation processes are accompanied by a loss of thermal energy.

There is another route by which molecules in higher excited states can reach the lowest vibrational state of a lower electronic level. Where the vibrational levels from different excited states overlap and have the same potential energy, internal conversion (a radiationless process) can occur. The excited molecule proceeds from the higher electronic state to the lowest vibrational level of the lower excited state via a series of vibrational relaxations, an internal conversion, and further relaxations. This process is not well understood; it occurs by direct vibrational coupling between electronic states and by quantum-mechanical tunneling.

When molecules reach the lowest vibrational level of the lowest excited singlet state, the radiation of fluorescence can occur when the electron returns to any of the vibrational levels of the ground electronic state. Each transition involves radiation of a specific wavelength. This radiative process ($S_1 \rightarrow S_0$) has a short natural lifetime (10^{-9} to 10^{-7} sec) so that in many molecules it can compete effectively with other processes capable of removing the excitational energy, such as internal conversion or intersystem crossing. The consequences of this mechanism are twofold. First, the fluorescence spectrum approximately mirrors the absorption (or excitation) spectrum. However, because the molecules relax to lower vibrational levels in the excited state and because of the solvent reorientation in the excited state and the ground state, the electromagnetic radiation corresponding to fluorescence is of lower energy than the exciting radiation and therefore appears at longer wavelengths. Second, although the intensity of the fluorescence spectrum depends on the excitation wavelength, its spectral pattern is independent of the excitation wavelength. Of course, if the absorption process leads to an electronic state in which the energy exceeds the bond strength of one of the solute's linkages, then excitation energy is lost by molecular dissociation before fluorescence can occur.

If the potential energy curve of the excited singlet state crosses that of the triplet state, some excited molecules may pass over to the lowest triplet state via an intersystem crossing that involves vibrational coupling between the excited singlet state, S_1, and the triplet state, T_1. A triplet state is one in which all the electrons in the molecule are paired except two. Although singlet-triplet transitions are forbidden, the internal conversion from the excited singlet to the triplet state may occur with some probability, since the energy of the lowest vibrational level of the triplet state is lower than that of the singlet state. The probability of intersystem crossing is greater when the potential energy curves cross at the lowest point on the excited singlet curve. After indirect occupation of the triplet state is achieved, the molecule undergoes a vibrational relaxation and solvent reorientation to arrive at the lowest vibrational level of the lowest excited triplet state. From this state, electromagnetic radiation can be emitted, or internal conversion can occur. If a radiative transition occurs, it is called phosphorescence. Phosphorescence takes place with low probability, since another spin reversal must occur. Consequently, the triplet state persists for a relatively long average lifetime. The decay time of phosphorescence is similar to the lifetime of the triplet state, approximately 10^{-4} to 10 sec. Spin-orbit coupling, which is a magnetic perturbation capable of flipping spins, is believed to be the main source of phosphorescence transitions back to the ground singlet state. The long lifetime of the triplet state greatly increases the probability of collisional transfer of energy with solvent molecules, which is very efficient in solution at room temperature and is often the main pathway for the loss of triplet state excitation energy. Because of this solvent interaction, phosphorescence in solutions is rarely observed at room temperature but can easily be observed by dissolving the solute in a solvent that freezes to form a rigid glass at the temperature of liquid nitrogen. Phosphorescence of some solutes can be observed at room temperature if the solute is adsorbed on a solid matrix. In summary, phosphorescence is influenced by vibrational relaxation and solvent reorientation in the excited singlet state, the triplet state, and the ground state, as well as intersystem crossing. Accordingly, the electromagnetic energy that corresponds to phosphorescence is still lower than fluorescence and appears at longer wavelengths.

In both fluorescence and phosphorescence the lower-energy photon is emitted in an arbitrary direction and at wavelengths longer than the excitation wavelength. Based on these phenomena, one has the twin techniques of spectrofluorometry and spectrophosphorimetry, which offer some unique advantages not possessed by absorption spectrophotometry for the study of chemical systems.

5.5 VIBRATIONAL ENERGY LEVELS

Molecular motion with the next lower energy after electronic transitions is vibration of the atoms of the molecule with respect to one another. Vibrational energies are usually an order of magnitude less than electronic energy. The simplest case involves the vibration of the atoms of a diatomic molecule. In a diatomic molecule the only vibration is the stretching of the bond between the two atoms represented by the two-dimensional potential energy diagram in Figure 5.7 (lowest electronic state). The allowed energies for the vibrational levels of a diatomic molecule, as given by quantum-mechanical theory, are

$$e_{\text{vib}} = \left(v + \frac{1}{2}\right)\frac{h}{2\pi}\sqrt{\frac{k}{u}} \qquad v = 0, 1, 2, \ldots \tag{5.5}$$

where k is the force constant that measures the force required to stretch a bond by a given distance (that is, the stiffness of the chemical bond) and u is the reduced mass for the two atoms, m_1 and m_2:

$$u = \frac{m_1 m_2}{m_1 + m_2} \tag{5.6}$$

Equation 5.5 indicates a pattern of energy levels with a constant spacing. At normal temperatures practically all molecules are in the ground electronic state. Except in cases where there are vibrational levels of very low energy, molecules are also usually in the vibrational ground state ($v = 0$).

Coupling with electromagnetic radiation occurs if the vibrating molecule produces an oscillating dipole moment that can interact with the electric field of the radiation. Homonuclear diatomic molecules like H_2, O_2, or N_2, which have a zero dipole moment for any bond length, fail to interact. However, the dipole moment of polar molecules like HCl can be expected to be some function, usually unknown, of

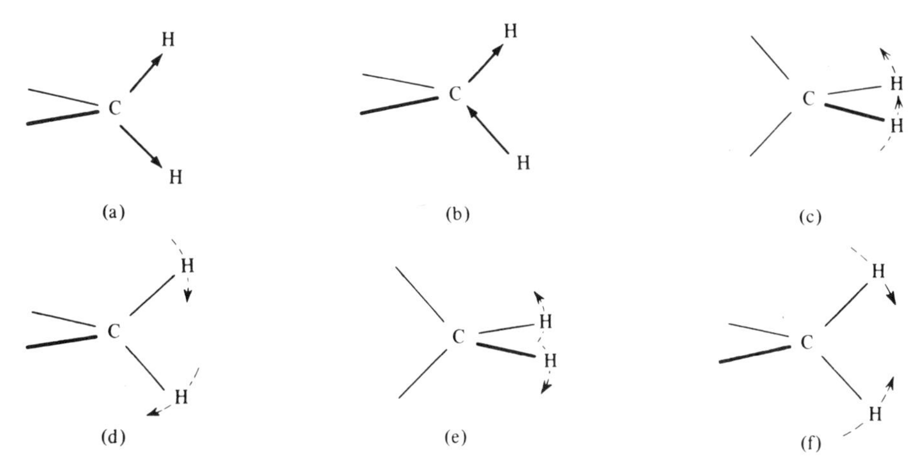

FIGURE 5.8
Vibrational modes of the H—C—H group: (a) symmetrical stretching, (b) asymmetrical stretching, (c) wagging or out-of-plane bending, (d) rocking or asymmetrical in-plane bending, (e) twisting or out-of-plane bending, and (f) scissoring or symmetrical in-plane bending.

the internuclear distance. The vibration of such molecules leads to an oscillating dipole moment, and a vibrational spectrum that lies in the infrared region is observed.

Even when there is interaction between a vibrating molecule and radiation, a further selection rule applies that restricts transitions resulting from the absorption or emission of a quantum of radiation by the relation $\Delta v = \pm 1$. Only $\Delta v = 1$ pertains to absorption spectroscopy. The vibrational frequency is greater for atoms with smaller reduced masses and for larger forces that restore the atoms to their equilibrium position (Equation 5.5). Thus motions that involve hydrogen atoms are found at much higher frequencies than are motions that involve heavier atoms. For multiple bond linkages, the force constants of double and triple bonds are roughly two and three times those of single bonds, and the absorption position becomes approximately two and three times higher in frequency. Interaction with neighboring atoms or groups may alter these values somewhat, as do resonating structures, hydrogen bonds, and ring strain.

The vibrational modes for a methylene group are illustrated in Figure 5.8. In a symmetrical group such as methylene there are identical vibrational frequencies. For example, the asymmetrical vibration in Figure 5.8(b) occurs in the plane of the paper and also in the plane at right angles to the paper. In space these two are indistinguishable and said to be one "doubly degenerate" vibration. In the symmetric stretching mode in Figure 5.8(a) there is no change in the dipole moment as the two hydrogen atoms move equal distances from the carbon atom. Such vibrations with no change in the dipole moment are infrared inactive. However, for the asymmetric vibrations, there is a change in the dipole moment because during these vibrations the centers of highest positive charge (hydrogen) and negative charge (carbon) move in such a way that the electrical center of the group is displaced from the carbon atom. These vibrations, involving changes in dipole moments, are observed in the infrared spectrum of the methylene group.

When a three-atom system is part of a larger molecule, it is possible to have bending or deformation vibrations. These vibrations imply the movement of atoms out from the bonding axis. Four types are distinguished.

1. *Deformation* (*or scissoring*). The two atoms connected to a central atom move toward and away from each other with deformation of the valence angle.

2. *Rocking or in-plane bending.* The structural unit swings back and forth in the symmetry plane of the molecule.
3. *Wagging or out-of-plane bending.* The structural unit swings back and forth in the plane perpendicular to the molecule's symmetry plane.
4. *Twisting.* The structural unit rotates back and forth around the bond that joins it to the rest of the molecule.

Splitting of bending vibrations due to in-plane and out-of-plane vibrations is found with larger groups joined by a central atom. An example is the doublet produced by the *gem*-dimethyl group. Bending motions produce absorption at lower frequencies than fundamental stretching modes.

Molecules composed of several atoms vibrate not only according to the frequencies of the stretching modes and bending motions, but also at overtones of these frequencies. When one bond vibrates, the remainder of the molecule is also involved. The harmonic (overtone) vibrations have a frequency that represents approximately integral multiples of the fundamental frequency. A combination band is the sum, or the difference, of the frequencies of two or more fundamental or harmonic vibrations. The unique quality of an infrared absorption spectrum arises largely from these bands, which are characteristic of the entire molecule. The intensities of overtone and combination bands are usually much smaller than those of fundamental bands.

The intensity of a fundamental vibrational absorption band is proportional to the square of the rate of change of the dipole moment with respect to the displacement of the atoms. In some cases, the magnitude of the change in dipole moment is quite small, producing only weak absorption bands as in the relatively nonpolar —C≡N group. By contrast, the large permanent dipole moment of the >C=O group causes strong absorption bands, often the most distinctive feature of an infrared spectrum.

5.6 RAMAN EFFECT

When a beam of electromagnetic radiation impinges on a particle that is small with respect to the wavelength of the radiation, the electrons of the particle are in an intense, alternating field caused by the electric and magnetic components of the radiation. The electrons of the particle oscillate with the frequency of the incident radiation and thereby produce electromagnetic radiation of the same frequency as the incident radiation but emanating from the particle in all directions. This appears to be scattered radiation and is known as Rayleigh scattering. If, however, the polarizability of the particle, usually a molecule, changes rather than remains constant, then the intensity of the scattered radiation varies accordingly. Polarizability is related to the ease of separation of charges in an external electrical field.

If one or more of the normal modes of vibration of a molecule involves changes in the polarizability, then the scattered radiation contains this vibrational frequency superimposed upon the frequency of the incident radiation. This is the Raman effect discovered by Sir C. V. Raman in 1928. Actually the scattered wave contains three frequencies, that of the incident radiation and the incident radiation frequency plus and minus that of the Raman-active vibration. For example, in the symmetric stretching mode of carbon dioxide, there is no change in the dipole moment as the two negative centers move equal distances in opposite directions from the positive center. However, the electron cloud around the molecule alternately elongates and contracts, changing the polarizability accordingly. In such a case, the changing molecular polarizability causes a modulation of the scattered light at the vibrational frequency.

In the Raman effect this mechanism is used by irradiating the sample with an intense monochromatic beam of radiation. The incident wavelength does not have to be one that is absorbed by the molecule, although it may be, as in the resonance Raman effect. Through the induced oscillating dipole(s) that it stimulates, the radiation leads to the transfer of energy with the rotational and vibrational modes of the sample molecules.

Most collisions of the incident photons with the sample molecules are elastic—that is, Rayleigh scattering where radiation is scattered in all directions by interaction with atoms in its path. However, about one in every million collisions is inelastic and involves a quantized exchange of energy between the scatterer and the incident photon to give weak scattered lines that are separated from the exciting line by frequencies equal to the vibrational frequencies of the scatterer as explained earlier. In the quantum-mechanical representation of the origin of Raman lines, the incident photon elevates the scattering molecule to a quasi-excited state whose height above the initial energy level equals the energy of the exciting radiation (Figure 5.9). This quasi-excited state then radiates energy in all directions except

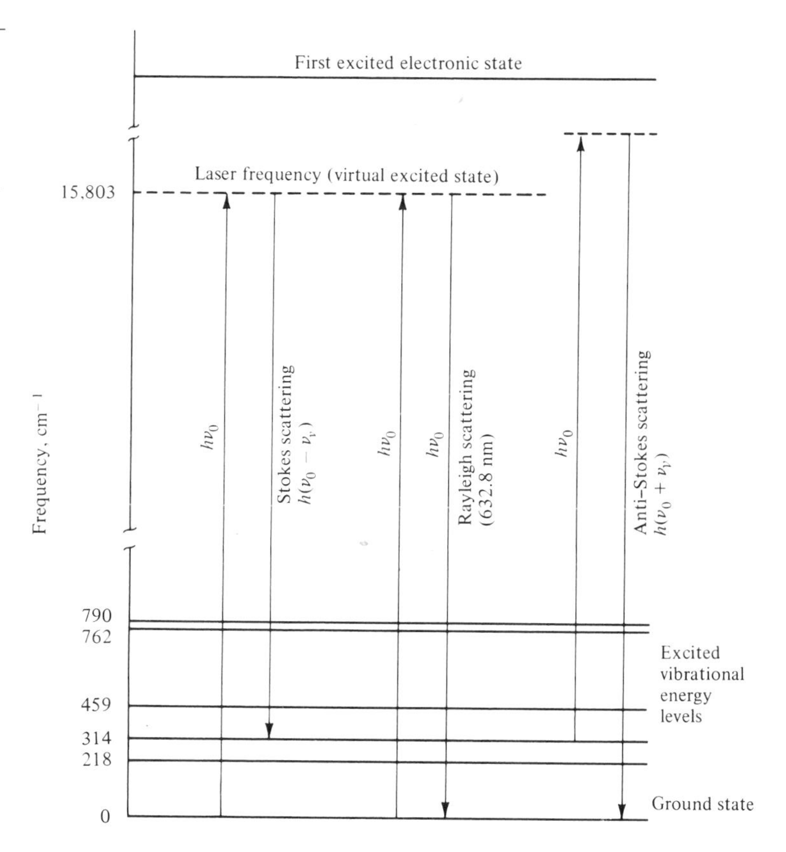

FIGURE **5.9**
Energy interchange involved in Rayleigh and Raman scattering; the molecule involved is CCl_4 and the source is a He-Ne laser.

along the line of action of the dipole—that is, the direction of the incident radiation. On the return to the ground electronic level, a vibrational quantum of energy may remain with the scatterer; if so, there is a decrease in the frequency of the reemitted radiation. If the scattering molecule is already in an excited vibrational level of the ground state, a vibrational quantum of energy may be abstracted from the scatterer, leaving it in a lower vibrational level and thus increasing the frequency of the scattered radiation. For either case, the shift in frequency of the scattered Raman radiation is proportional to the vibrational energy involved in the transition. Thus the Raman spectrum occurs as a series of discrete frequencies shifted symmetrically above and below the frequency of the exciting radiation and in a pattern characteristic of the molecule (Figure 5.10). The shift is independent of the frequency of the incident radiation; however, the intensity of the scattered radiation for a given vibration increases with the fourth power of the frequency of the incident radiation. The Raman lines usually studied are those on the low-frequency side of the incident radiation, the Stokes lines, which are more intense than the lines on the higher-frequency side, the anti-Stokes lines. By convention the positions of Raman lines are expressed as wavenumbers, but more correctly they are wavenumber differences.

For complex molecules, vibrational Raman spectra have the general appearance of the corresponding infrared absorption spectra, and often the same

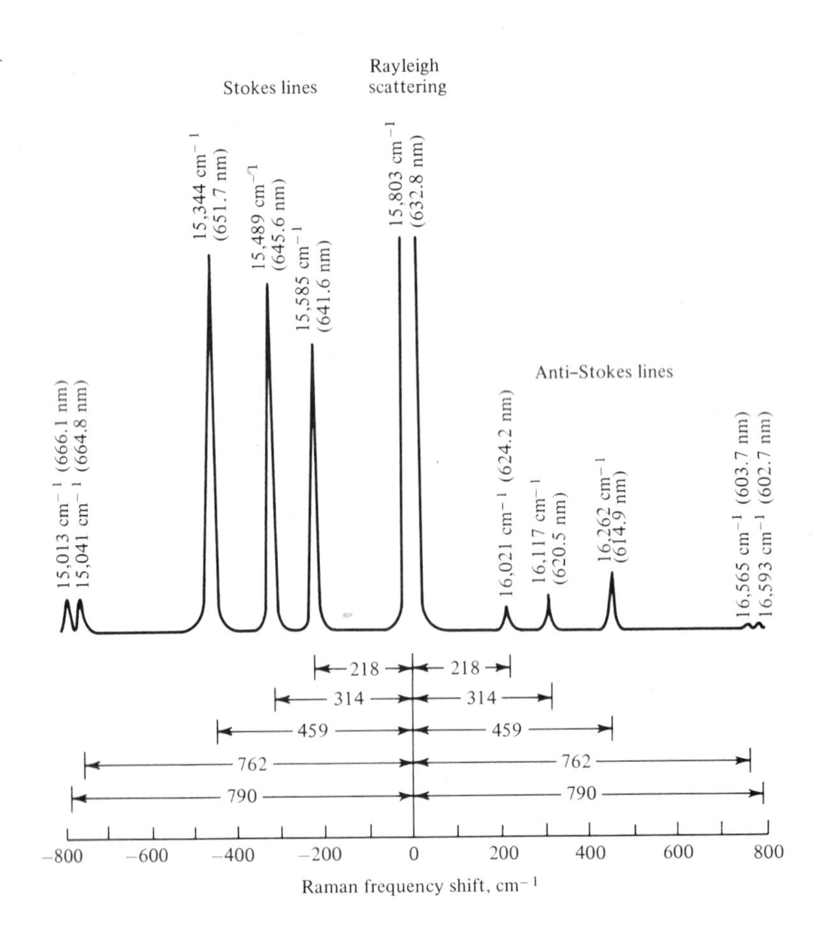

FIGURE **5.10**

Raman spectrum of CCl_4 obtained with a He-Ne laser.

vibrational energy-level separation shows up as a spectral line in both spectroscopic methods. When molecules with a center of inversion are considered, however, a dramatic difference between the two spectral techniques becomes evident. A homonuclear diatomic molecule exhibits a Raman vibrational spectrum because the molecule becomes more polarizable when it is lengthened than when it is shortened. For molecules with a center of symmetry, Raman spectra provide information on the symmetric vibrations of molecules and infrared absorption spectra on the antisymmetric vibrations. Raman spectral information thus complements the data obtained from infrared spectra for these systems.

5.7 LASERS

A laser provides an almost ideal monochromatic source of narrow linewidth. It emits radiant energy that is coherent, parallel, and polarized. A laser beam can be kept as a very slim cylinder only a few micrometers in cross section. Laser operation involves three principles: stimulated emission, population inversion, and optical resonance. Stimulated emission occurs when a photon strikes an excited atom or molecule and thereby causes that atom or molecule to emit its photon prematurely. This can occur only when the impinging photon has exactly the energy of the "stored" photon that would ultimately have been emitted spontaneously. The resulting emission falls precisely in phase with the electromagnetic wave that triggered its release and is identical in wavelength. Thus the incoming photon is now joined by a second photon from the excited atom or molecule, resulting in a gain or amplification of photons and giving a perfectly coherent (in-phase) beam of radiation.

The precedence of stimulated emission over spontaneous emission is the basis for achieving laser action. For this to occur, a population inversion must take place; that is, there must be more molecules (or atoms) in the excited state than in the ground state of the lasing material. This is possible only for a multilevel system, as illustrated in Figure 5.11 for the He-Ne laser. A helium atom, excited by an electrical discharge to a long-lived triplet level, can lose its energy only by collision with another atom that has a comparable energy level available. One excited state of neon lies only 313 cm^{-1} below the excited helium state. As a result, a radiationless excitation of neon can occur on collision of the excited helium and ground-state neon atoms. The excited level of neon has a relatively long lifetime before it spontaneously decays to the ground state with the emission of photons at 1153 and 632.8 nm. Thus it is possible to build up a larger population of neon atoms in the excited state than in the ground state (population inversion). The spontaneous emission of a single photon at either of these wavelengths can trigger a whole cascade of similar photons by the process of stimulated emission. The formation of the population inversion by optical irradiation of a system is known as optical pumping. Population inversion is relatively easy to accomplish with many organic molecules, since their energy levels constitute a multilevel system.

Optical resonance is the third principle essential to laser operation. Resonance is achieved by placing the lasing medium in a cavity situated between a pair of parallel, plane mirrors, as shown in Figure 5.12. Making the spacing between the mirrors an integral multiple of the desired wavelength means that there will be a buildup of energy at the desired wavelength. If the resulting radiation is collinear with the optic axis and the frequency falls within the bandwidth of one of the discrete optical frequencies, it is reflected back and forth through the cavity. Growth of the wave continues, and, if the gain on repeated passages through the lasing medium is sufficient to compensate for losses within the cavity, a steady wave is built up. Any wave that is inclined at an angle to the long axis of the cavity is lost after only a few reflections, or perhaps without ever striking one of the mirrors. If one of the mirrors is semitransparent, a portion of the wave can escape through it, constituting the output of the laser. The output is radiation of low divergence, all of the same frequency, and all in phase. A crystal of $LiIO_3$ or KH_2PO_4, internal to the laser cavity, may be used to double the frequency or to multiply the frequency by other integers.

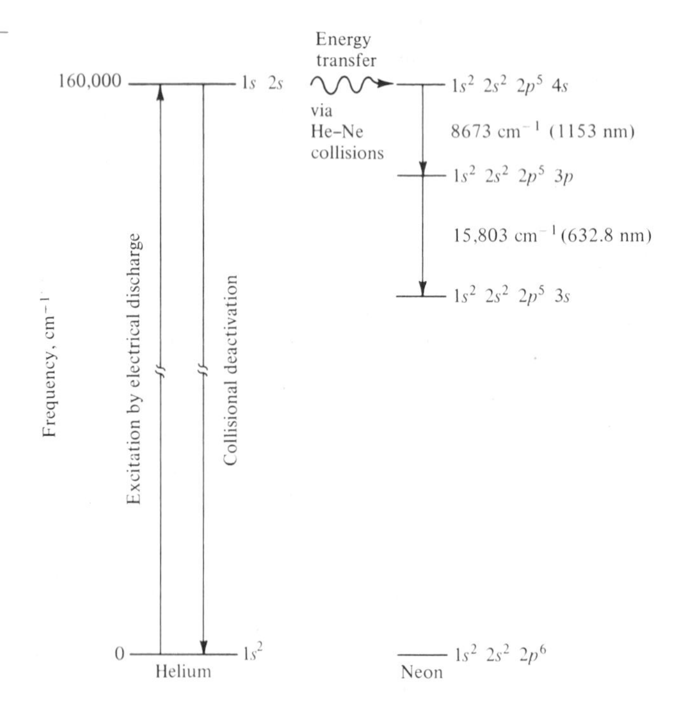

FIGURE 5.11
Energy levels involved in the He-Ne laser.

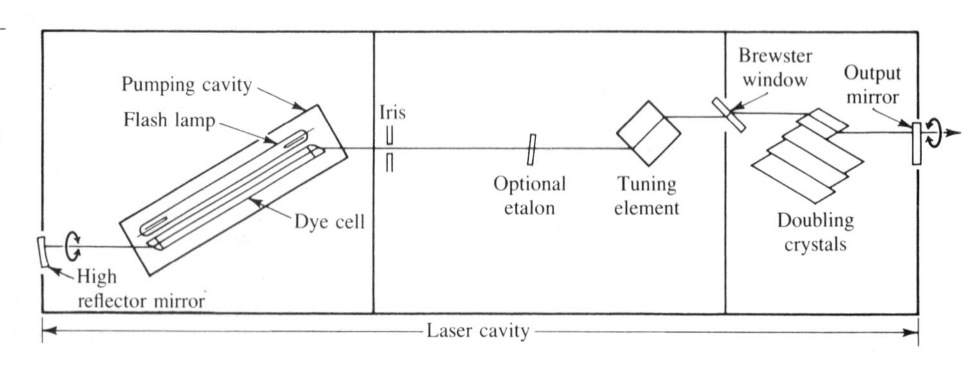

FIGURE 5.12
Optical schematic of a tunable dye laser. (Courtesy of Chromatix Inc.)

To pump the laser, flash lamps, inert gas arc lamps, or another laser has been used. Flash lamps are frequently used and have two configurations: coaxial and linear. In either configuration, high voltage from a low-inductance capacitor is rapidly pulsed through the lamp. Coaxial arrangement offers very high pulse power but low repetition rates. Linear flash lamps do not provide peak powers comparable to those of coaxial flash lamps, but they can operate at higher repetition rates to yield the same average power. All pulsed systems generate copious amounts of radio-frequency interference. Proper design and construction can reduce radio-frequency interference to acceptable levels, and equipment must be shielded to prevent upsetting nearby electronics.

The He-Ne laser line at 632.8 nm is favorably located in the spectrum where the least amount of fluorescent problems appear in routine analyses. An argon laser has intense lines at 488.0 and 514.5 nm; coupling with krypton adds two other major lines at 568.2 and 647.1 nm. The Ar-Kr laser is ideal for many experiments. The chances are that with at least one of the exciting lines, problems of photodecomposition, fluorescence, or absorption can be successfully circumvented. Tunable dye lasers, whose output frequency can be varied over a short range, are almost indispensable for studies that involve resonance Raman experiments. Several dyes are necessary to cover the entire wavelength range accessible to dye lasers. The appropriate dye solution is continuously pumped through the laser cavity. Prisms and gratings have been used to tune dye lasers.

5.8 NUCLEAR SPIN BEHAVIOR

In addition to charge and mass, about half of the known isotopes possess nuclear spin, or angular momentum. The spinning charge of the nucleus generates a magnetic field, and associated with the angular momentum is a magnetic moment. These nuclei resemble a tiny bar magnet, the axis of which is coincident with the axis of spin. When placed in a powerful, uniform magnetic field, such nuclei are acted upon by a torque and tend to assume an allowed orientation with respect to the external magnetic field. The field aligns the spinning nuclei against the disorienting tendencies of thermal processes. However, the nuclei do not align perfectly parallel (or antiparallel) to the field. Instead their spin axes are inclined to the field and, like the top of a gyroscope, precess about the direction of the external field. Each pole of the nuclear magnet sweeps out a circular path in the xy-plane, as shown in Figure 5.13. Increasing the strength of the external field only makes the nuclei precess faster. By applying a second, much weaker radio-frequency (rf) field at right angles to the uniform magnetic field, the nuclei can be made to undergo a transition to a higher energy level. When the frequency of the rotating component of this second rf field reaches the precession frequency, the spinning nuclei absorb energy and flip into a higher energy level. For protons and other nuclei with a spin of $\frac{1}{2}$, this means an orientation antiparallel to the uniform field. The resonance frequency, ν, that effects the transitions between energy levels is derived by equating the Planck quantum of energy with the energy of reorientation of a magnetic dipole:

$$\Delta E = h\nu = \frac{\mu H_0}{I} \tag{5.7}$$

where H_0 is the uniform magnetic field, μ is the magnetic moment of the nucleus, and I is the spin quantum number in units of $h/2\pi$. There are $2I + 1$ possible orientations and corresponding energy levels. Nuclei with $I = \frac{1}{2}$ give the best resolved spectra because their electric quadrupole moment is zero. They act as though they were spherical bodies possessing a uniform charge distribution that circulates over their surfaces. These nuclei include ^{1}H, ^{13}C, ^{19}F, and ^{31}P. Nuclei with spins of 1 or greater also possess nuclear electric quadrupole moments and so are readily disturbed by molecular electric field gradients. The result is a shortening of the spin lifetime in a given state and a "smearing out" of the spectroscopic signal. The electric quadrupole moment measures the nonspherical electric charge distribution of a nucleus.

As indicated in Equation 5.7, the frequency of the resonance absorption varies with the value of the applied field. For example, in a magnetic field of 14,092 gauss (G), protons precess at 60 MHz and this is the frequency required to flip their spins. In a field of 23,490 G, the precessional frequency rises to 100 MHz. Also, since the strength of the absorption signal is roughly proportional to the square of the magnetic field strength, higher values of field strength lead to a stronger signal.

The difference between the two energy levels for a proton is not very large compared with thermal energies, only about 0.04 J. Consequently, thermal agitation diminishes the slight excess of nuclei in the lower energy state. At normal temperatures and with a magnetic field of 14 kG, only about two protons of each million are in excess in the lower energy state. If the populations of the two energy states become equal, the system is said to be saturated and no absorption signal is

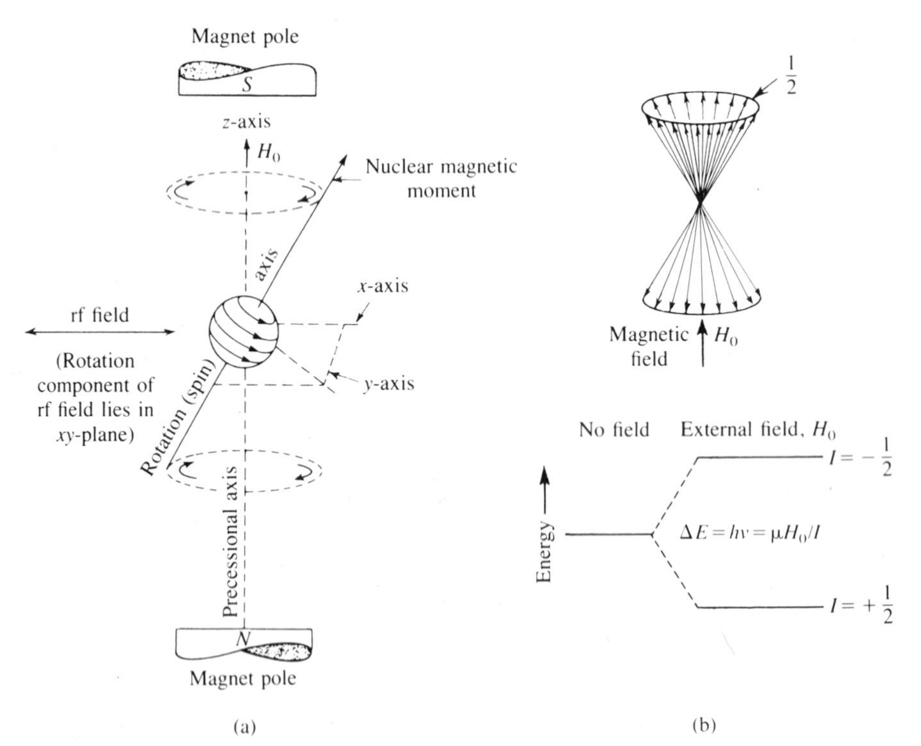

FIGURE 5.13
(a) Spinning nucleus in a magnetic field.
(b) Energy-level diagram for a nucleus (lower) and the nuclear orientation (upper).

observed. Thus mechanisms for replenishing the number of nuclei in the lower energy state are important in nuclear magnetic resonance (NMR).

Energy absorbed and stored in the higher energy level can be dissipated and the nuclei returned to the lower energy level by a process called spin-lattice relaxation. It is brought about by interaction of the spin with the fluctuating magnetic fields produced by the random motions of neighboring nuclei (called the "lattice," whether the material is crystalline, amorphous, or fluid). In solids and viscous liquids, the relaxation time is on the order of hours, but in typical organic liquids and dilute solutions, the time is in the range of 1–20 sec.

5.9 ELECTRON SPIN BEHAVIOR

The electron, like the proton and other spinning nuclei, is a charged particle with a spin and, hence, a magnetic field. It spins much faster than nuclei and thus has a much stronger magnetic field. A magnetic moment is associated with the spinning charge and, therefore, the electron behaves like a magnet with its poles along the axis of rotation. Since the electron has a magnetic moment due to its spin and also one associated with its circulation in its atomic orbit, the electron has a total magnetic moment equal to the vector sum of these two magnetic moments. The ratio of the total magnetic moment to the spin value is a constant for a given environment and is called the gyromagnetic ratio or spectroscopic splitting factor for that particular electron. The facts that these ratios differ for various atoms and environments and that local magnetic fields depend on the structure of the material result in characteristic spectra and lead to electron spin resonance spectroscopy.

In the absence of an external magnetic field, the free electron may exist in one of two states, $+\frac{1}{2}$ or $-\frac{1}{2}$, of equal energy and thus is degenerate. The imposition of an external static magnetic field, H_0, removes the degeneracy and causes the electron to precess. Two energy levels are established, as shown in Figure 5.14. The lower energy state has the spin magnetic moment aligned in the direction of the applied magnetic field and corresponds to the quantum number $M_s = -\frac{1}{2}$. The difference in energy between the two levels is given by

$$\Delta E = h\nu = g\beta_N H_0 \tag{5.8}$$

where g is the spectroscopic splitting factor and β_N is the Bohr magneton. Transitions from one state to the other can be induced by subjecting the electron to electromagnetic radiation in the microwave range, around 10,000 MHz. The interaction that causes the transitions is between the magnetic dipole of the electron and the oscillating magnetic field that accompanies the electromagnetic radiation. Thus when microwaves travel down a rectangular waveguide, they produce a

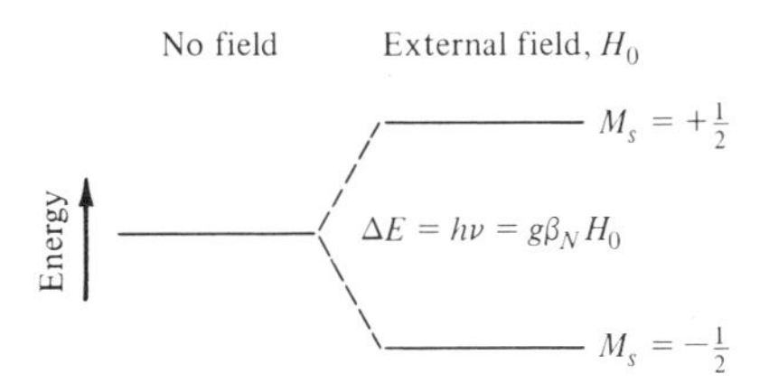

FIGURE 5.14
Energy-level diagram for an unpaired electron.

rotating magnetic field at any fixed point that can serve to flip over electron magnets in matter. When the magnetic field is expressed in kilogauss, the resonance frequency in megahertz for a free electron is given by

$$\nu = 2800H_0 \tag{5.9}$$

The typical energy involved is about 4 J mol^{-1}.

5.10 X-RAY ENERGY LEVELS

X-ray emission and absorption spectra are quite simple and all elements have a similar pattern. This relative simplicity of X-ray spectra is explained by the fact that the spectra result from transitions between energy levels of the innermost electrons in the atom. There are only a few electrons in these inner shells and the resulting energy levels are limited, thus giving rise to only a few permitted transitions. There is only one K shell ($n = 1$). The L electrons ($n = 2$) are grouped according to their binding energy into three sublevels: L_{I}, L_{II}, and L_{III}. The complete M shell ($n = 3$) consists of five sublevels.

The lines of heavier elements fall at higher energy positions in the spectrum. The relationship between the frequency of a given line and the atomic number of the element, Z, is

$$\nu = R(Z - \sigma)^2\left(\frac{1}{n_2^2} - \frac{1}{n_1^2}\right) \tag{5.10}$$

where R is the Rydberg constant, n_2 and n_1 are the electron quantum numbers, and σ is a shielding constant (approximately 1 for K electrons). The frequency of a given X-ray line therefore increases approximately as the square of the atomic number of the element involved.

A typical example of the X-ray energy levels that are involved in absorption and emission is shown in Figure 5.15. Absorption by electrons in an inner shell requires an energy that is at least equal to that required to eject an electron from the atom. As the energy of the incident radiation is increased, there is successive ionization first of electrons in the outermost shells, here the M shells, then of electrons in the L shells as the discrete L_{III}, L_{II}, and L_{I} absorption edge energies are progressively exceeded, and finally in the ionization of the K shell electron. X-ray spectra show no absorption lines but only an absorption edge because the electron is ejected completely from the atom.

X-ray emission lines result from electrons in outer shells falling into the vacant orbit formed when an inner electron has been ejected from the atom. Lines of significant intensity are those for which the selection rules $\Delta l = \pm 1$ and $\Delta j = 0, \pm 1$ are both satisfied. That is, an X-ray line is emitted if the rules give the difference in the initial and final states for the electron transition that fills the hole created by the absorption step. To illustrate from Figure 5.15, transitions from the L_{II} and L_{III} levels are permitted when the electron vacancy initially occurs in the K level, but the transition from the L_{I} level is not permitted because for it Δl is zero. The permitted transitions give rise to the $K\alpha_2$ and $K\alpha_1$ emission lines, respectively. Permitted transitions from the M_{II} and M_{III} sublevels give rise to the $K\beta_3$ and $K\beta_1$ lines in emission, respectively. Of course, the electron vacancies created when an L electron

FIGURE 5.15 Energy-level diagram of cadmium for X-ray transitions.

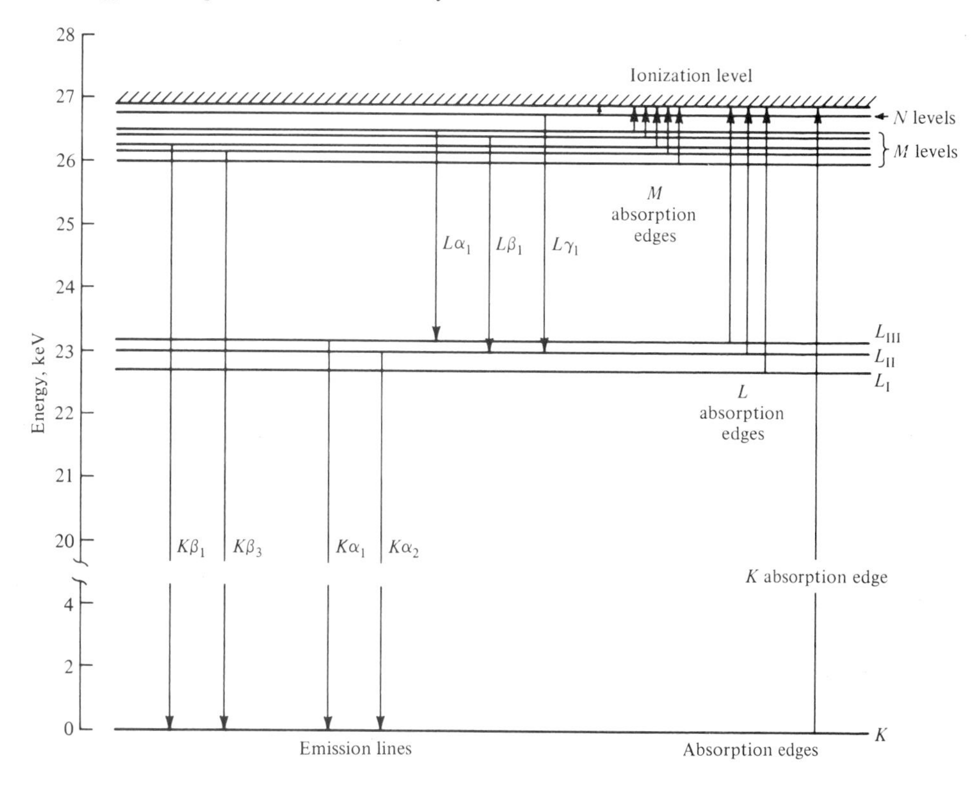

falls back to the K shell lead to a series of L emission lines, with the electron originating from the M or other outer shell, and so on for successively existing outer electron shells.

BIBLIOGRAPHY

BAUMAN, R. P., *Absorption Spectroscopy,* John Wiley, New York, 1965.

CROOKS, J. E., *The Spectrum in Chemistry,* Academic, New York, 1978.

DIXON, R. N., *Spectroscopy and Structure,* Methuen, London, 1965.

HERZBERG, G., *Molecular Spectra and Molecular Structure,* Van Nostrand Reinhold, New York, Vol. 1, 2nd ed., 1950; Vol. 2, 1945; Vol. 3, 1966; Vol. 4 (with K. P. Huber), 1979.

MEEHAN, E. J., "Optical Methods. Emission and Absorption of Radiant Energy," Chap. 1, *Treatise on Analytical Chemistry,* 2nd ed., P. J. Elving, E. J. Meehan, and I. M. Kolthoff, eds., Part I, Vol. 7, Wiley-Interscience, New York, 1981.

PETERS, D. G., J. M. HAYES, AND G. M. HIEFTJE, *Chemical Separations and Measurements,* Chap. 18, Saunders, Philadelphia, 1974.

SVELTO, O., AND D. C. HANNA, *Principles of Lasers,* 2nd ed., Plenum, New York, 1982.

THYAGARAJAN, K., AND A. K. GHATAK, *Lasers, Theory and Applications,* Plenum, New York, 1981.

WEBER, M. J., ed., *CRC Handbook of Laser Science and Technology,* Vol. 1, *Lasers and Masers,* Vol. 2, *Gas Lasers,* CRC Press, Boca Raton, FL, 1982.

WRIGHT, J. C., AND M. J. WIRTH, "Principles of Lasers," *Anal. Chem.,* **52,** 1087A (1980).

ULTRAVIOLET AND VISIBLE SPECTROMETRY—INSTRUMENTATION

In this chapter, design and operational features of characteristic instruments suitable for use in absorption spectrophotometry of the visible and ultraviolet regions of the spectrum are considered. The instrument modules are shown in schematic form in Figure 6.1. A source of radiation must be provided, with each spectral region having its own requirements. All spectrophotometers have some way to discriminate between different radiation frequencies through the use of filters, prisms, or gratings. Nondispersive methods, such as Fourier transform methods, are not widely used in the visible and ultraviolet regions, although some instruments are now available for this region. Since Fourier transform methods are so widely used in the infrared region, however, discussion of these methods is deferred to Chapter 11.

The sample absorbs a portion of the incident radiation; the remainder is transmitted to a detector, where it is changed into an electrical signal and displayed, usually after amplification, on a meter, chart recorder, or some other type of readout device.

6.1 RADIATION SOURCES[1,2]

Radiation sources in absorption spectrophotometry have two basic requirements. First, they must provide sufficient radiant energy over the wavelength region where absorption is to be measured. Second, they should maintain a constant intensity over the time interval during which measurements are made. If the intensity is low in the region where absorption is measured, the wavelength interval that passes through the sample must be relatively wide to obtain the necessary energy throughput. This wide range of wavelengths may cause errors in absorptivity measurements. Generally, in the ultraviolet and visible regions of the spectrum, intensity is not a problem. In design considerations, however, it must be remembered that the

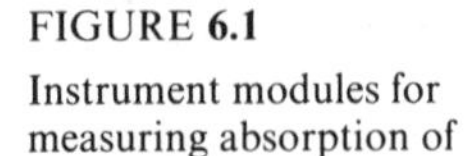

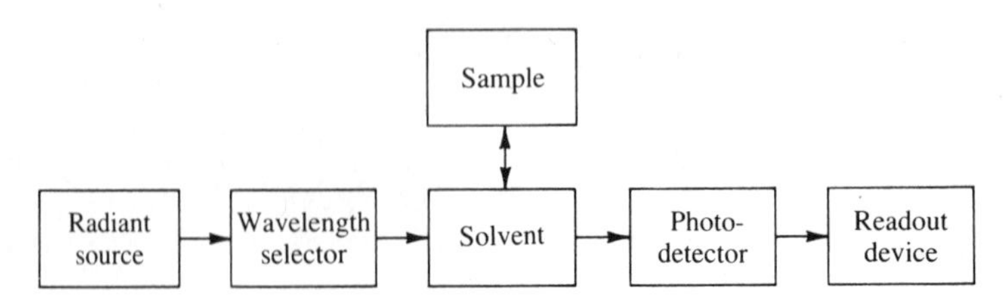

FIGURE 6.1
Instrument modules for measuring absorption of radiation.

flux density of the radiation (in joules/cm^2) varies as the square of the distance from the source for uncollimated and unfocused beams.

Lasers are not generally used in commercially available instruments in the ultraviolet and visible regions, since they are costly and since any one laser, even the tunable dye lasers, covers only a limited range of wavelengths (see the discussion of lasers in Chapter 5). However, for some special applications such as thermal lens spectrophotometry, to be discussed in the next chapter, lasers are necessary because of their very high spectral intensity. Spectral intensities, W m^{-2} sr^{-1} Hz^{-1}, of some radiation sources are as follows:[3] the sun and thermal sources, less than 10^{-16}; synchrotron light source at Brookhaven National Laboratory, about 10^{-6}; He-Ne (CW) laser, about 10^6; and pulsed lasers, 10^{14}.

Hydrogen or Deuterium Discharge Lamps

Work in ultraviolet regions of the spectrum is done mainly with hydrogen or deuterium discharge lamps operated under low pressure (approximately 0.2–0.5 torr) and low voltage (approximately 40 V dc) conditions. Heated cathodes provide the essential function of maintaining the discharge. The discharge has negative temperature versus resistance characteristics, so a current-regulated power supply is required. A vital feature of these lamps is a mechanical aperture between the cathode and the anode, which constricts the discharge to a narrow path. Normally the anode is placed close to the aperture, which creates an intense ball of radiation about 0.6–1.5 mm on the cathode side of the opening. The use of deuterium in place of hydrogen slightly increases the size of the radiating ball but enhances brightness—that is, luminance (candelas per unit area)—three to five times. Increased collection efficiency can be achieved by positioning the lamp arc at one of the foci of an elliptical reflector. This results in the collection of more than 2π steradians, which translates into more than 60% collection efficiency compared with typical housings, which offer a maximum of 10% collection efficiency. At less than 360 nm these discharge lamps provide a strong continuum that fulfills most needs in the ultraviolet region. With fused silica envelopes, work to about 160 nm is feasible. At wavelengths longer than about 380 nm, the discharge has lower intensity and emission lines superimposed on the continuum, thus generally negating its use in the visible region.

Incandescent Filament Lamps

Measurements above 350 nm and into the near infrared to 2.5 μm are usually made with incandescent filament lamps, which give a continuous spectrum over the range. In these lamps a wire filament, generally tungsten, is heated to incandescence by an electric current. The filament is enclosed in a hermetically sealed bulb of glass filled with an inert gas or a vacuum. Filaments are usually coiled to increase their emissivity, efficacy, and mean luminance. Incandescent lamps are rugged, low-cost units sufficiently bright for nearly all absorption work in the visible and near-ultraviolet regions.

Tungsten-halogen lamps are a special class of incandescent lamps with iodine added to normal filling gases. The envelope is fabricated of quartz to tolerate higher lamp operating temperatures of 3500 K. The iodine combines chemically at the bulb wall with sublimed tungsten. The resulting WI_2 migrates back to the hot filament where it decomposes and tungsten is redeposited. The cycle is repeated, continuously cleaning the bulb. These lamps maintain more than 90% of their initial light output throughout their life.

The spectral distribution of an incandescent filament is basically that of a blackbody radiator. Therefore, measurements very far from the peak wavelength are susceptible to stray light effects due to the much higher intensity radiation of other wavelengths. The tungsten lamp emits most of its energy in the near infrared, with a maximum at about 1000 nm, and drops off very rapidly in the ultraviolet region to $\frac{1}{100}$ of that value at about 300 nm. Only about 15% of the radiant energy falls within the visible region, with a lamp at an apparent color temperature (the radiation is equal to that of a perfect blackbody at that temperature) of about 2850 K (Figure 6.2). Often a heat-absorbing filter or cold dichroic mirror (a multilayer interference beam splitter that transmits long-wavelength radiation and reflects short-wavelength radiation) is inserted between the lamp and sample holder to remove the infrared radiation without seriously diminishing the radiant energy at shorter wavelengths. The glass envelope absorbs strongly below 350 nm. Incandescent lamps are important sources in spectrometric applications because of their excellent stability, rather than because of their spectral radiance.

Source Stability

High short-term stability of an incandescent or discharge source is required for single-beam spectrophotometers. The intensity of radiation from an incandescent or discharge source is proportional to the lamp voltage raised to some power that is larger than unity (three or four for incandescent lamps). To stabilize the photocurrent within 0.2%, which represents attainable spectrophotometric precision, the source voltage for incandescent lamps has to be regulated within a few thou-

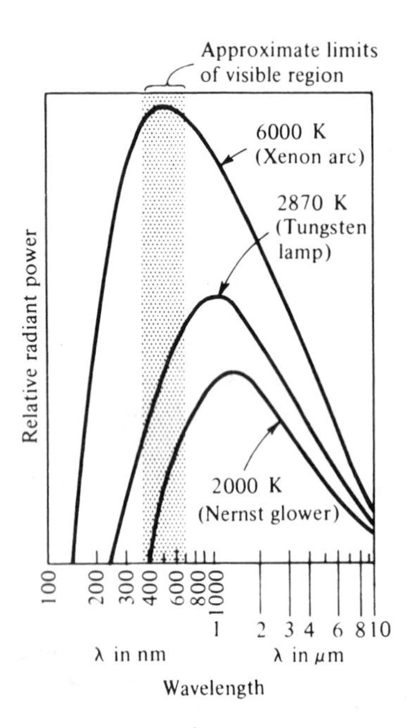

FIGURE **6.2**
Spectral distribution curves of radiant energy sources.

sandths of a volt. Source stability is achieved by using constant-voltage transformers and electronic voltage regulators.

By placing a second detector in the optical path and sampling a portion of the radiant energy, the monitored signal may be used to correct the lamp output. This is achieved by feeding the signal back to a programmable power supply and either increasing or decreasing the output current as required. In this manner the optical ripple can be reduced to 0.1% peak to peak over the short term. This order of stability is impossible to obtain with gas discharge or arc lamps.

Modulation or Pulsing Modes of Operation

A feedback loop within the source power supply allows the power supply to be modulated or pulsed (see the discussion of modulation in Chapter 2). Modulation with an external voltage source allows one to modulate or program the lamp system to follow a sine, square, or ramp function within the limits set by the various lamp operating parameters. By using optical feedback and external modulation, it is possible to obtain very little distortion of the optical signal. Since the lamp is now incorporated in the overall feedback path, any nonlinearities in the lamp characteristics are compensated. Such low levels of distortion are almost impossible to achieve using mechanical choppers because of the exact shape required for the chopper openings in relation to the beam geometry.

The pulsing mode of operation varies from a modulation mode in that the lamp is raised to a level well above the normal lamp operating conditions. The constant base current is set at a low value and this current is briefly increased during the pulse. This pulsing to a level above normal lamp operating conditions results in an optical output many times that normally attainable. The greatest increase is experienced in the ultraviolet region, the least in the infrared. The minimum pulse duration is 300 μsec; the maximum can be several seconds or longer, limited by the type of lamp used. Both pulsing and modulation result in decreased lamp life.

6.2 WAVELENGTH SELECTION

Spectrophotometric methods usually require the isolation of discrete bands of radiation. As we will see in the next chapter, Beer's law, which is the basis of quantitative work, is based on the assumption of monochromatic radiation. Additionally, in an emission mode, the most favorable signal ratio between background and the analytical emission lines must be selected. To isolate a narrow band of wavelengths, filters or monochromators, or both, are used. Now some nondispersive methods, particularly Fourier transform methods, are becoming very important. As mentioned earlier, these methods are, at present, more commonly used in other regions of the spectrum, especially in the infrared, and will be discussed in that context.

Filters

Filters provide high radiation throughput, approximately 50%–80% efficiency. Assembly of filter instruments is relatively easy for perhaps as many as five wavelengths. The bandpass, the total range of wavelengths transmitted, when interference filters are involved can equal that achieved with 0.25-m grating mounts.

Absorption Filters. Absorption filters derive their effects from bulk interactions of radiation within the material. Some types rely on selective scattering, and in others true ionic absorption predominates. Transmission is a smoothly decreasing function of thickness described by the exponential law of absorption (see Chapter 7).

Absorption filters are produced in a variety of host materials: gelatin, glass, liquid, and plastic. Glass filters are extensively used in automated chemical analysis equipment and colorimetry. The scattering type depends on scattering crystals formed within the glass mass through a reduction and thermal treatment. Shorter wavelengths are scattered and absorbed, while longer wavelengths are unaffected. The other absorption-type filter attains its selectivity through ions or molecules in true solution absorbing specific bands of radiation. Cut-on and cutoff (or sharp-cut) filters are widely used as blocking filters to suppress unwanted spectral orders from interference filters and diffraction gratings. One series consists of sharp cutoff filters that pass long wavelengths, the red and yellow series; the other series consists of long-wavelength cutoff filters, the blue and green series. Composite glass absorption filters are constructed from two sharp-cutoff filters, as shown in Figure 6.3. The spectral separation between 50% cut-on and cutoff points of the transmittance versus wavelength curve are from 20 to 70 nm. This range of wavelengths is known as the *bandwidth* or full width at half maximum (FWHM). The total range of

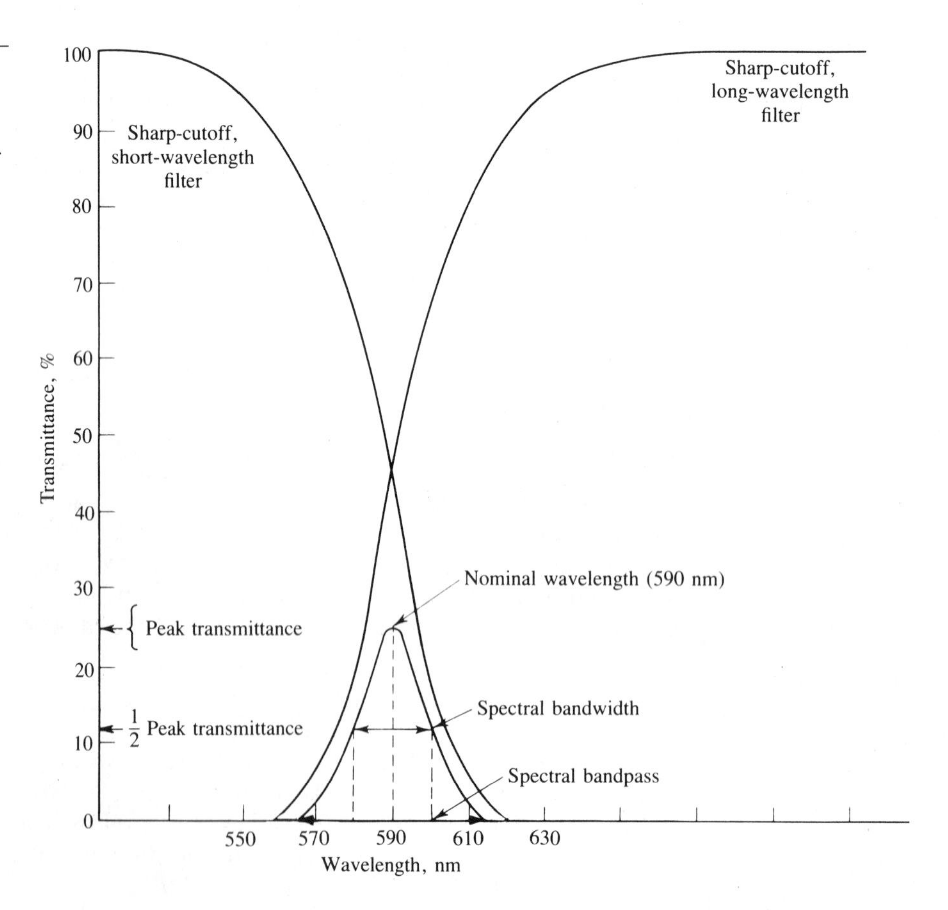

FIGURE **6.3**
Spectral transmittance characteristics of a composite glass absorption filter and its components.

wavelengths transmitted is the *bandpass*. Peak transmission is 5%–20%, decreasing with improved spectral isolation. Although such filters do not approach the extremely narrow bandpass and high peak transmission typical of the multilayer, all dielectric, interference filter, they have the advantage of relative insensitivity to the input angle.

There is a wide variety of plastic filters, both sharp-cut and intermediate bandwidth types. Plastic filters may be produced either by bulk colorants introduced into the basic batch or through subsequent dye treatments of clear base stock. Cut-on types, unlike their glass counterparts, exhibit no fluorescence in the visible region.

Interference Filters. Interference filters, as the name implies, are based on the phenomenon of optical interference (see Chapter 5). A simple two-interface (Fabry-Perot) filter consists of a dielectric spacer film (CaF_2, MgF_2, or SiO) sandwiched between two parallel, partially reflecting metal films, usually of silver (Figure 6.4). The thickness of the dielectric film is controlled to be only one, two, or three half-waves thick. These are referred to as first-, second-, or third-order filters, respectively.

A portion of the incident radiation normal to the filter (beam 1) passes through (beam 2), while another portion (beam 3) is reflected from surface B back to surface A. A portion of this reflected radiation is again reflected from surface A through the dielectric layer and exits as beam 4 parallel (actually coincident) to beam 2. Thus the path traveled by beam 4 is longer than that of beam 2 by twice the product of the dielectric spacer thickness and its refractive index. When the layer thickness b is half the wavelength of the radiation to be transmitted in the refractive index η of the dielectric, beams 2 and 4 are in phase and interfere constructively. The expression for central wavelengths at which full reinforcement occurs is

$$\lambda = \frac{2\eta b}{m} \tag{6.1}$$

where m is the order number. Since partial reinforcement occurs for other path differences, the filter actually transmits a band of radiant energy. Furthermore, the angle of incidence of the radiation must be 90°. Any phase shifts on reflection are ignored. The bandwidth is 10–15 nm, full width at half maximum (FWHM) transmission; the maximum transmission is usually 40% with this type of filter.

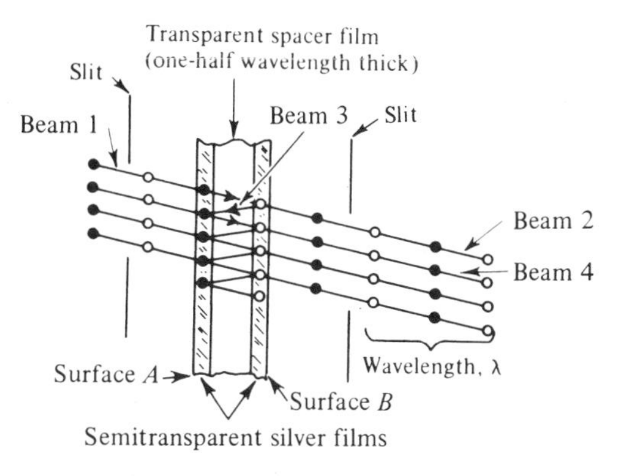

FIGURE 6.4
Schematic of an interference filter and path of light rays through the filter.

EXAMPLE 6.1

A dielectric layer of $\eta = 1.35$ that is 185 nm thick provides a first-order filter at a central wavelength of 500 nm. This filter also passes harmonic bands centered at 250 nm in the second order and 167 nm in the third order.

Unwanted transmission bands can be eliminated by using appropriate cut-on or cutoff absorption filters as one of the protecting glass covers. An excellent blocking filter is an interference filter whose first order matches the desired band when higher-order central wavelengths of the primary filter are used. Second- and third-order bands are narrower, and most interference filters are arranged to transmit one of these. However, the bandwidth and bandpass become shorter as the order is increased. Transmissions of Fabry-Perot filters are normally 10–100 times higher than those of a monochromator with an equivalent bandwidth. Band center wavelength, the nominal wavelength, varies with both the temperature of the filter and the angle of incidence of radiation that is to be filtered. The transmission band shifts to longer wavelengths with increasing temperature for most filters. In the visible spectrum the shift is approximately 0.01 nm/°C. With increasing angle of incidence, the band center wavelength shifts to shorter wavelengths.

A *wedge filter* consists of a wedge-shaped slab of dielectric deposited between the semireflecting metallic layers. A continuously variable transmission interference is obtained. At each point along the length of the filter a different wavelength is transmitted because the thickness changes. Different wavelengths are isolated either by moving the wedge past a slit assembly or by passing a slit assembly along the filter. A circular wedge filter can also be constructed and used to continuously sweep variable wavelengths past the slit assembly.

By replacing the metal films with a stack of all-dielectric films, as is done in *multilayer filters,* vastly improved performance is obtained. Since the absorption of the dielectric layers is very nearly zero, a considerable variety of bandwidths is produced, with high transmission being maintained at the same time as low background. The reflecting stack of films is comprised of alternating high index of refraction and low index of refraction layers, each one-quarter-wave optical thickness. When a train of waves strikes a multilayer optical coating, the beam divides at each film interface into a series of reflected and transmitted components. As shown in Figure 6.5, as wavelengths are scanned away from the central wavelength in either direction, rather quickly the layers are no longer a quarter-wavelength thick for the radiation and general transmission occurs. These transmission "wings" must be eliminated over the spectral region covered by the detector through the use of auxiliary blocking filters applied as a separate sandwich component or more directly as the multilayer substrate.

Extremely complex layer designs with 5 to 25 layers have evolved, and now filters of this type can be made with bandwidths less than 0.1 nm in particular wavelength regions (1–5 nm is usual) and with background transmission less than 10^{-6}. Fully blocked filters with a peak transmission of 80% are not uncommon and 55%–60% of the incident radiation is considered standard. Multilayer interference filters are available over the wavelength region from 180 nm in the ultraviolet to 35 μm in the infrared.

A serious problem can arise if the incident radiation is either highly convergent or divergent. First, the wavelength being transmitted at any point on the filter is obviously a function of the angle of incidence of the radiation, and one observes both broadening of the bandwidth and lowering of the peak transmission. Second, although only the effective index is usually considered, in actuality both the high and low index layers are shifting independently and a mismatch in effective layer thickness ensues with consequent detrimental effects.

A long-wave pass type of interference filter (also called a *dichroic filter*) is designed to transmit wavelengths longer than some desired cutoff wavelength. Customarily the cutoff is referred to as that wavelength at which the filter first reaches 5% absolute transmission. Properties of prime interest to the user are the steepness of rise from the region of low transmission to that of high transmission, the transmission in the passband, and the background transmission. Usually long-wave pass filters exhibit a slope of less than 5%, although steeper slopes can sometimes be achieved at the expense of smooth transmission in the passband. The passband transmission is generally not flat but has an oscillation because of secondary reflection bands. The long wavelength that the filter transmits is usually governed by the inherent absorption of the substrate material chosen, since in most instances the film materials are nonabsorbing over a greater wavelength range. Background transmission is usually 10^{-3} or less.

Monochromators

A monochromator consists, in general, of (1) an entrance slit that provides a narrow optical image of the radiation source, (2) a collimator that renders the radiation emanating from the entrance slit parallel, (3) a grating or prism for dispersing the incident radiation, (4) a collimator to reform images of the entrance slit on the exit slit, and (5) an exit slit to isolate the desired spectral band by blocking all of the dispersed radiation except that within the desired range. Typical optical arrangements are considered after each of these elements is discussed.

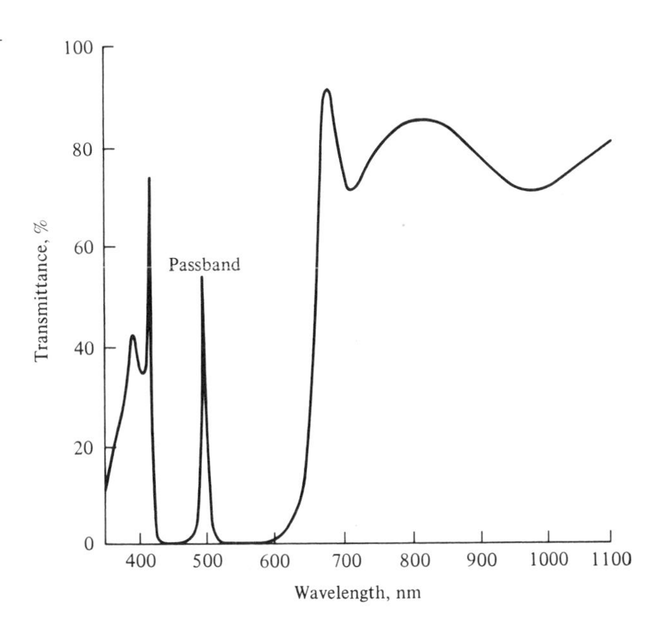

FIGURE **6.5**
Transmittance of a multilayer interference filter with passband at 500 nm.

The primary function of a monochromator is to provide a beam of radiant energy of a given nominal wavelength and spectral bandwidth. The spectral output of any monochromator used with a continuous radiation source, regardless of focal length and slit width, consists of a wavelength range with an average wavelength of the value indicated on the monochromator wavelength display. A secondary function of the monochromator is the adjustment of the energy throughput. The luminous flux that emerges from the exit slit can be varied by adjusting the slit width. However, since slit width also controls the spectral bandwidth, excessively wide slits, with consequent large spectral bandwidths, cause deviations from Beer's law (see Chapter 7). Excessively small slit width results in low-energy throughput and affects analytical sensitivity as a result of signal-to-noise degradation caused by the high photomultiplier dynode voltages that would be required for sensing the weaker signals.

The basic requirements imposed on a monochromator are design simplicity, resolution, spectral range, purity of exiting radiation, and dispersion. Each of these quantities is discussed in subsequent sections of this chapter. The choices of a radiation source and a detector are closely interrelated.

Mirrors or lenses are used to collect radiation from a source and direct it to the monochromator entrance slit. An effective collecting arrangement based on an off-axis ellipsoidal mirror is illustrated in Figure 6.6. When the source is placed at one focus of the mirror and the entrance slit at the other focus, radiation gathered from a large solid angle is focused on the slit.

Monochromator Performance. The performance of a monochromator involves three related factors: resolution, light-gathering power, and purity of the radiation output. The resolution depends on the dispersion and the perfection of the image formation, whereas the purity is determined mainly by the amount of stray or scattered radiation. Large dispersion and high resolving power in monochromators are necessary to measure accurately emission spectra with discrete lines or sharp absorption bands, whereas emission bands and the usual broad absorption bands show up with instruments of medium dispersion.

Dispersion. Dispersion is defined as the spread of wavelengths in space—that is, the separation of a mixture of wavelengths into component wavelengths. This is accomplished in a monochromator with a prism (refraction phenomenon) or a grating (diffraction phenomenon). Linear reciprocal dispersion, D^{-1}, is defined as the range of wavelengths over a unit distance in the focal plane of a mono-

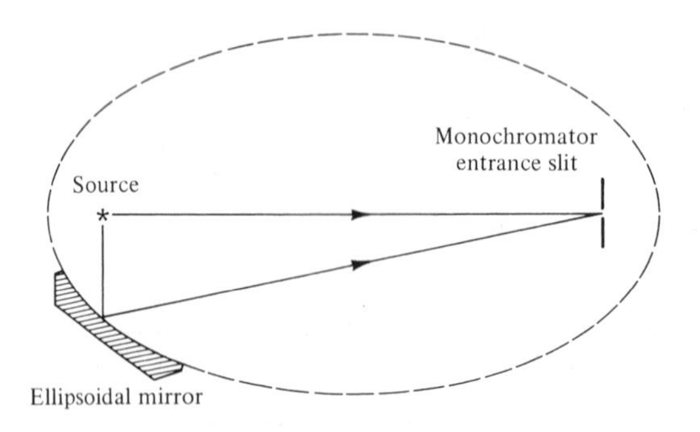

FIGURE **6.6**
Off-axis ellipsoidal mirror for gathering light from radiation sources.

chromator, or

$$D^{-1} = \frac{d\lambda}{dx} \tag{6.2}$$

The dimensions are in nanometers per millimeter. The relationship between angular dispersion, the angular range $d\theta$ over which a waveband $d\lambda$ is spread, and linear dispersion is

$$D = f\left(\frac{d\theta}{d\lambda}\right) \tag{6.3}$$

where f is the focal length of the monochromator.

Resolution. The resolution, or resolving power, of a monochromator is its ability to distinguish as separate entities adjacent spectral features, perhaps absorption bands or emission lines. Resolution is determined by the size and dispersing characteristics of the grating or prism, the optical design of which the dispersing device is a part, and the slit width of the monochromator. In recording spectrophotometers, resolution is also a function of the recording system and the scan speed.

The widely used definition for resolution, R, is

$$R = \frac{\bar{\lambda}}{d\lambda} = w\left(\frac{d\theta}{d\lambda}\right) \tag{6.4}$$

where $\bar{\lambda}$ is the average wavelength between two lines just resolved, $d\lambda$ is the wavelength difference between the lines, and w is the effective aperture width. Definitions aside, just what is the criterion for calling resolved two adjacent spectral lines? In practice any of several answers may be given. The most liberal statement is that peaks are resolved when the intensity falls at least 10% between them—a situation perhaps suitable for qualitative identification but hardly satisfactory for quantitation. Usually one would prefer a valley that extends to the background (baseline) between two discrete emission lines without any stipulation concerning line intensities; that is, the bases of the slit functions of the two lines may touch but not overlap. In this case, the resolution is the same as the base width of one slit function, as given in Equation 6.7. The resolution of an actual monochromator is generally less than the theoretical value because of optical aberrations, imperfections, diffraction effects, and other deleterious effects. The effect of spectral bandwidth on observed band shapes is shown in Figure 6.7. Too little resolution depresses the peak heights, and the observed bandwidth of the peaks increases. Separation of the two bands is less well defined, and uncertainties arise because of increased peak height dependence on minor changes in slit width. On the other hand, with too much resolution, unnecessary noise is superimposed upon the signal with no noticeable improvement in peak height or separation of the two bands. Obviously, there is an optimum spectral bandwidth. When the spectral bandwidth is one-tenth of the true or natural bandwidth of the peak, deviation from the true peak height is less than 0.5%.

The optical system of a monochromator produces an inherent curvature in the slit image that becomes evident when the slit height exceeds about 3 mm. In precision instruments the loss of resolution on lengthening slits is sometimes diminished by giving the entrance slit a radius of curvature that opposes the one produced optically.

Radiation-Gathering Power. With narrow spectral bandwidths, spectral features quite close together can be resolved. However, the signal-to-noise ratio is important. Sufficient radiation must reach the detector to enable the signal to be distinguished above the background. The radiation-gathering power of the instrument is critical in this case. The f/number, or speed of a spectrometer, is an indication of the ability of the collimator mirror to collect radiation that emerges from the entrance slit. It is expressed by

$$f/\text{number} = \frac{f_c}{d_c} \tag{6.5}$$

where f_c and d_c are the focal length and diameter of the collimating mirror, respectively. The smaller the f/number, the greater the radiation-gathering ability.

Slits. In practical spectrophotometry the monochromator module is not capable of isolating a single wavelength of radiation from the continuous spectrum emitted by the source. Rather, a definite band of radiation is passed by the monochromator.[4] This finite band arises from the slit distributions. The entrance or aperture of a monochromator is a long, narrow slit whose width is generally adjustable. Slits longer than 3 mm may have curved sides, but we shall speak, for simplicity, of the bright rectangle of radiation formed by the entrance slit when illuminated by the source. Inside the monochromator, the rays diverge from the entrance slit and illuminate the collimator mirror, which renders the rays parallel and focuses them on the dispersing element. Leaving the collimator, the parallel set of rays is a broadened version of the entrance slit. This rectangle of radiation must be large enough to illuminate the entire side of the prism or the length of the grating. In turn, the dispersing device separates the incident polychromatic radiation into an

FIGURE **6.7** Effect of spectral bandwidth on observed absorption band shapes of cytochrome *c*. Spectral bandwidths are (1) 20 nm, (2) 10 nm, (3) 5 nm, (4) 1 nm, and (5) 0.08 nm.

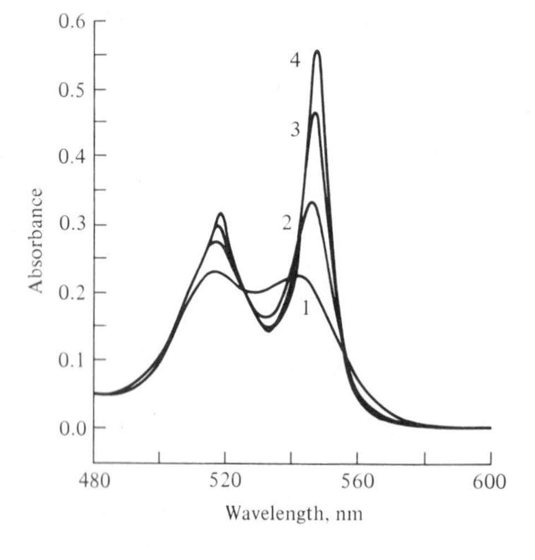

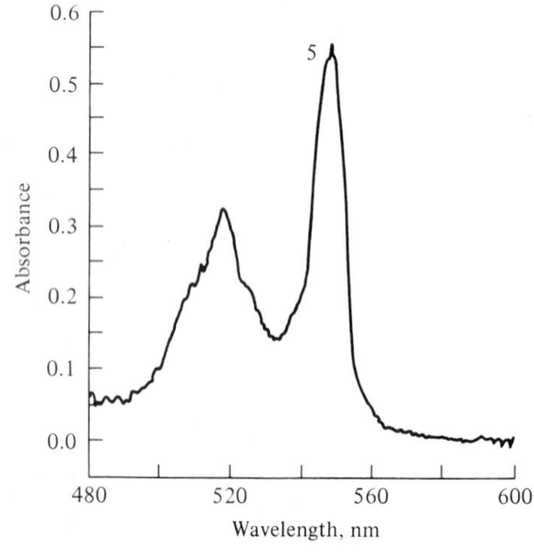

array of monochromatic rectangles, each of which leaves the dispersing device at a slightly different angle. The monochromatic rectangles overlap. The dispersed beam is intercepted by a second collimating mirror identical to the first (or a segment of the first collimator), which is used to focus and reduce each rectangle to an image of the entrance slit. These final images fall in a plane called the *focal plane*, in which a stationary exit slit is located. The distance between the second collimator and the exit slit is called the *focal length* of the monochromator.

Two small 45° mirrors enable the input and output to be in line. By removing one or both of these mirrors, parallel or right-angle configurations of input and output can easily be achieved. Assuming that the radiation is of equal intensity (irradiance) throughout the image, one can use the analysis of Buc and Stearns[5] and Hogness et al.[6] for dependence of intensity at the exit slit on wavelength setting when using symmetrical slit distributions. Referring to Figure 6.8, let ABCD be the image of a monochromator band at the exit plane and let EFGH be the dimensions of the exit slit. The numbers 1, 2, 3, ... correspond to positions of wavelength setting. Now plot the fraction of radiation transmitted by the system as a function of the position of the image of the entrance slit as it moves along the focal plane and passes over the exit slit when the dispersing device is pivoted as in wavelength scanning. Up to the point where the leading edge BD of the image reaches position 1, no radiation is transmitted by the exit slit. When BD reaches position 2, half the total radiation is passed. At the nominal wavelength, position 3, the exit slit aperture is filled and the transmitted intensity is 100% of that available. As the image reaches position 4, the transmittance of the system again falls to one-half, and at position 5 it falls to zero. This triangular plot of transmitted intensity versus the wavelength, which describes the position of the image, is the slit distribution versus the wavelength observed for monochromatic radiation on scanning. For continuous radiation a monochromator also gives a triangular intensity pattern, but the abscissa now represents the range of wavelengths passed at the setting of the central wavelength.

Generally slits are characterized only by their width. The spectral bandwidth may be rigorously defined as the wavelength interval of the radiation that leaves the exit slit of a monochromator between limits set at a radiant power level halfway between the continuous background and the peak of an absorption band of negligible intrinsic width. More simply, the *spectral bandwidth* is the difference in

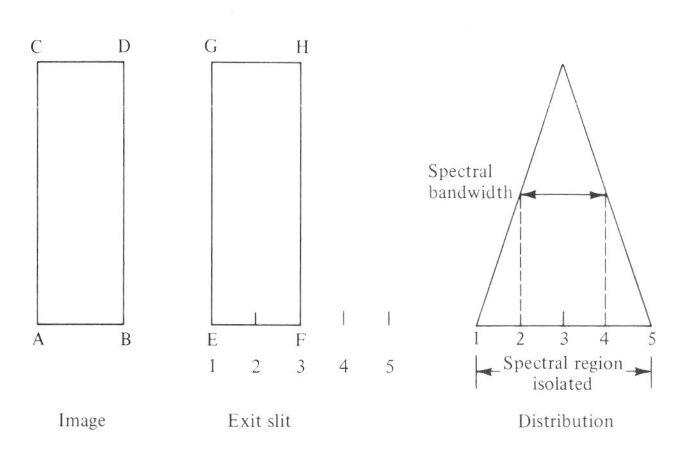

FIGURE **6.8**
Slit distribution function when the image size and the exit slit are identical.

wavelength between the points where the transmittance is half the maximum, or the bandwidth containing 75% of the radiant energy that leaves the monochromator (see Figure 6.8). The spectral region isolated (along the baseline, or abscissa), the *bandpass*, is the sum of the image width and the exit slit width in wavelength units. In Figure 6.8, the spectral region isolated—that is, the bandpass—corresponds to the distance from point 1 to 5.

Bandwidth can be expressed in terms of the physical width of the slits, W, and the reciprocal linear dispersion of the monochromator:

$$\text{bandwidth} = WD^{-1} \tag{6.6}$$

The bandpass, $\Delta\lambda$, of the slit function is given by

$$\Delta\lambda = 2WD^{-1} \tag{6.7}$$

for the case shown in Figure 6.8 where the image of the entrance slit and the exit slit widths are equal.

EXAMPLE 6.2

Using Equation 6.6 with slits 0.1 mm wide in a monochromator whose linear reciprocal dispersion is 1.6 nm/mm, the bandwidth would be 0.16 nm. Under the same conditions, two spectral lines separated in wavelength by 0.6 nm would be 0.38 mm apart in the focal plane of the monochromator at the exit slit (Equation 6.2).

The choice of slit width is basically a trade-off between intensity and resolution. For scanning molecular spectra, the slit width should be adjusted so that the spectral bandwidth is about one-tenth the natural bandwidth of the spectral feature to be recorded. For atomic line spectra, the lines recorded are actually slit functions (intensity versus frequency curves for the radiation emerging from the exit slit) with half-width equal to one spectral bandwidth. Thus the choice of slit width depends on the separation of the spectral lines or the isolation of the desired analytical line from adjacent spectral features.

Mirrors and Lenses. Within a monochromator the necessary collimating and focusing are performed by front-surface mirrors. By eliminating lenses, chromatic aberrations are eliminated. Chromatic aberration refers to the differing focal lengths for radiation of different wavelengths.

When radiation passes from one medium into a second with a different index of refraction, part of the radiation is reflected at the boundary surface. This loss for normal incidence of the beam and for pure dielectrics may be expressed as

$$\text{reflection loss} = \left(\frac{\eta_2 - \eta_1}{\eta_2 + \eta_1}\right)^2 \tag{6.8}$$

Since the first medium is normally air ($\eta_1 = 1$), the loss becomes a direct function of the substrate index alone. The loss for high-index materials, such as are customarily used in the infrared region, is quite severe and can lead to extreme attenuation in multicomponent systems. Even for low-index glasses the total loss can be serious if several optical components are involved in the optical path. Examination of a typical spectrophotometer reveals many glass-to-air interfaces in the mirrors and windows. A reflection loss of 4% occurs at each glass/air interface ($\eta_2 = 1.52$ for glass).

The reflection from a glass surface is modified by the application of either dielectric or metallic thin-film coatings.[7] In many cases a single layer of optical thickness equal to one-quarter of the wavelength of the radiation concerned is used. The normal reflectance at incidence of such a quarter-wave film is given by

$$\text{reflection loss} = \left(\frac{\eta_f^2 - \eta_a\eta_g}{\eta_f^2 + \eta_a\eta_g}\right)^2 \tag{6.9}$$

where η_f is the refractive index of the film, η_a that of the surrounding medium, and η_g that of the substrate. Application of antireflection coatings to each glass surface in the optical path reduces the reflectance to approximately 0.2% per surface.

Prisms as Dispersive Devices. The action of a prism depends on the refraction of radiation by the prism material. The dispersive power depends on the variation of the refractive index with wavelength. A ray of radiation that enters a prism at an angle of incidence i is bent toward the normal (vertical to the prism face), and at the prism-air interface, it is bent away from the vertical as depicted in Figure 6.9. To minimize astigmatism of a prism and achieve the best definition, the prism should be illuminated by parallel radiation with the slit parallel to the prism edge, and it should pass through the prism symmetrically so that the incident and emergent beams form equal angles to the faces; the prism is then at minimum deviation. The image of the entrance slit is projected onto the exit slit as a series of images arranged next to one another, caused by radiation of shorter wavelengths being more strongly bent than radiation of longer wavelengths. A nonlinear wavelength scale results. For a medium quartz prism monochromator of focal length 600 nm, typical values of reciprocal linear dispersion are 0.6 nm/mm at 230.0 nm, 1.04 nm/mm at 270.0 nm, 1.56 nm/mm at 310.0 nm, 2.9 nm/mm at 370.0 nm, 5.4 nm/mm at 450.0 nm, and 12.0 nm/mm at 600.0 nm. Flint glass provides about threefold better dispersion than quartz or fused silica and is the material of choice for the near-infrared/visible region of the spectrum. Fused silica or quartz is required for work in the ultraviolet region.

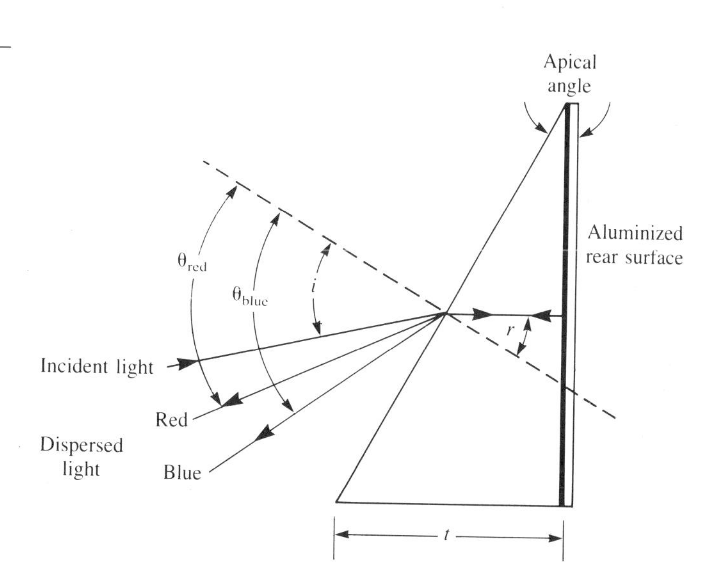

FIGURE 6.9
The prism as a dispersing medium. Littrow-type mounting: i is the angle of incidence, r is the angle of refraction, t is the base of the prism, θ (theta) is the angle of deviation, and the apical angle is 30°.

The resolving power of a prism is given by

$$R = t\left(\frac{d\eta}{d\lambda}\right) \tag{6.10}$$

where t is the base length of the prism. The resolving power is limited by the prism's base length and the dispersive power of the material, $d\eta/d\lambda$. The latter is not constant for a prism but increases from long wavelengths to shorter wavelengths. This requires a knowledge of the refractive index of the dispersing material and its rate of change as a function of wavelength, or the linear dispersion as a function of wavelength. A graph that supplies this information should be provided by the vendor for each instrument.

One of the most widely used prism monochromator designs is the Littrow mounting shown in Figure 6.10. The particular advantages of the Littrow mount are a high degree of dispersion in a compact arrangement, a single collimator mirror that serves to collimate the entrance beam and focus the dispersed beam, and the

FIGURE **6.10** Littrow mounting in three configurations. For wavelength selection: (a) 30° prism rotates, (b) Littrow mirror rotates, and (c) diffraction grating rotates.

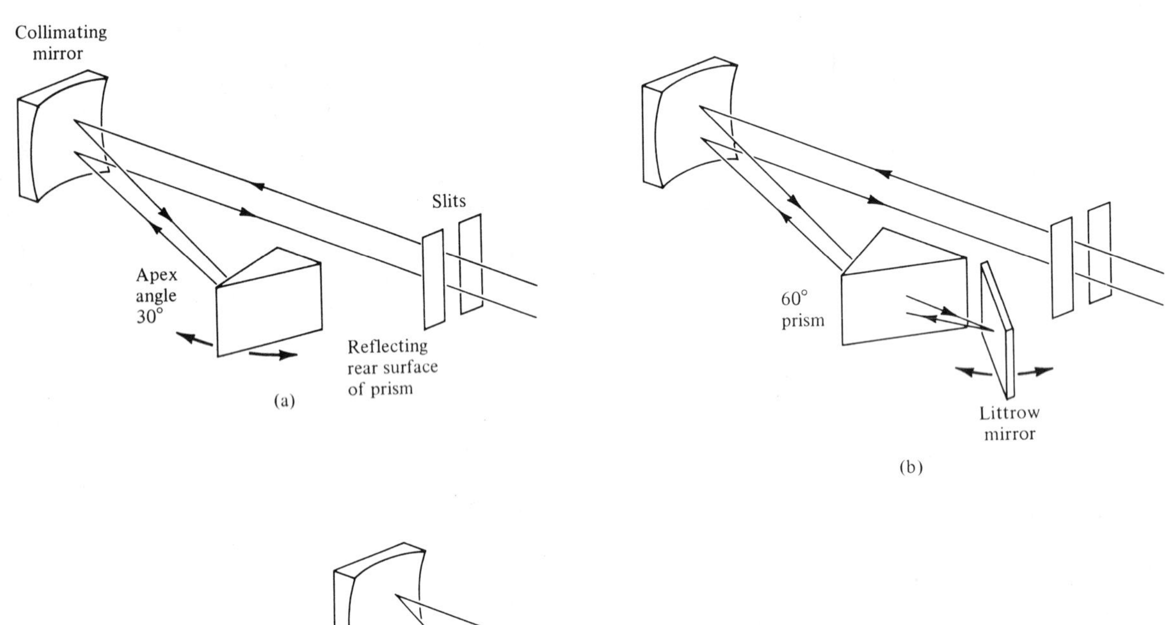

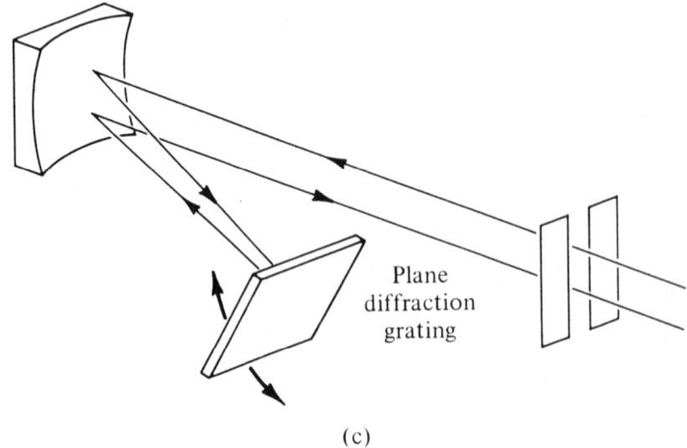

avoidance of double refraction if an anisotropic material like quartz is used. Prisms are mostly found today in double monochromators, particularly in a prism-grating double monochromator where the prism serves as an "order sorter" as well as the dispersing device in the first monochromator.

The Littrow mount accommodates several design variations: a 30° prism backed by a reflecting surface, a 60° plane grating and separate Littrow mirror, and a plane grating, discussed in the next section. The source illuminates a condensing mirror that brings the reflected beam to a focus on the plane of the entrance slit of the monochromator. The image of the entrance slit is collimated by the parabolic mirror and directed onto the dispersing device. The refracted or diffracted beam is sent back to the same collimating mirror, but at a different height, and the beam is then focused onto the exit slit, which selects a portion of the dispersed spectrum for transmission onto the sample and onto the detector. The upper and lower portions of the same slit assembly are used as entrance and exit slits, thus providing perfect correspondence of slit widths. The slit system can be fixed or continuously adjustable. In the 30° arrangement, the prism is rotated by means of a mount connected to the wavelength drive. In the 60° arrangement, the Littrow (plane) mirror behind the prism reflects the beam and returns it through the prism a second time, thus doubling the dispersion. The mirror is turned through a small angle to obtain the different wavelengths at the exit slit.

Gratings as Dispersive Devices. A grating is, in essence, an array of a very large number of equidistant slits that reflect or transmit radiation. Only at certain definite angles is radiation of any given wavelength in phase. At other angles the waves from the slits destructively interfere.

Virtually all gratings are of the reflection type. A conventional diffraction grating consists of parallel, equally spaced grooves ruled by a properly shaped diamond tool directly into a highly polished surface. The quality of a grating is closely connected to the degree of precision with which the straightness, the parallelism, and the equidistance of the grooves are controlled. The profile of the grooves, shown in Figure 6.11, is determined by the purpose for which the grating is intended. This must be maintained constant from the first to the last groove over the required ruling distances of 10 to 25 cm. Ruling a master grating is a slow, arduous

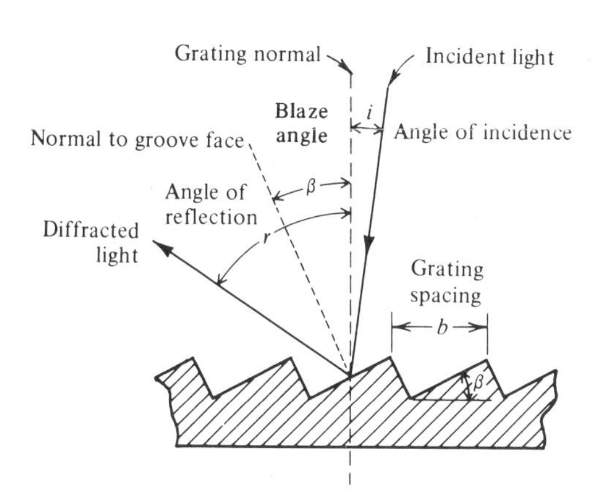

FIGURE **6.11**
Cross section of a diffraction grating showing the "angles" of a single groove, which are microscopic on an actual grating.

process that requires experience, skill, and unlimited patience. Only the ability to produce several replicas from the master has allowed ruled diffraction gratings to become widely available.

The process of replicating a master grating involves (1) applying a film of parting agent to the master, (2) vacuum depositing a layer of aluminum, and (3) attaching a glass or quartz base to the aluminum layer with epoxy cement. After an appropriate time interval the replica is separated.

Each groove of the grating has a broad face exposed to the incident radiation. The radiation incident on each groove is diffracted (that is, spread out) over a range of angles. At certain angles reinforcement, or constructive interference, occurs, as stated by the grating formula

$$b(\sin i \pm \sin r) = m\lambda \tag{6.11}$$

where b, the grating constant, is the distance between adjacent grooves, i is the angle of incidence, r is the angle of diffraction, and m is designated the order. A positive sign applies in the grating formula when incident and diffracted beams are on the same side of the grating normal. Rulings number from 20 grooves/mm in the far infrared to 3600 or more grooves/mm for the visible and ultraviolet regions.

The grating formula shows that the incident energy is diffracted into several orders, shown in Figure 6.12. The partition of energy into these different orders depends on the groove profile. Modern ruled gratings have an inclined groove profile (blazed) and, as a consequence, can concentrate most of the incident energy into a single order. The groove angle is controlled so that a maximum of energy is dispersed, or concentrated, into the wavelength (angular) region over which use is intended. As an approximation, efficiency peaks in the direction and at the wavelength where the angle of diffraction is equal to the angle of specular reflection from the face of the grating groove. This angle is called the *blaze angle* and the wavelength, the *blaze wavelength*. Most gratings are used in the first order as this

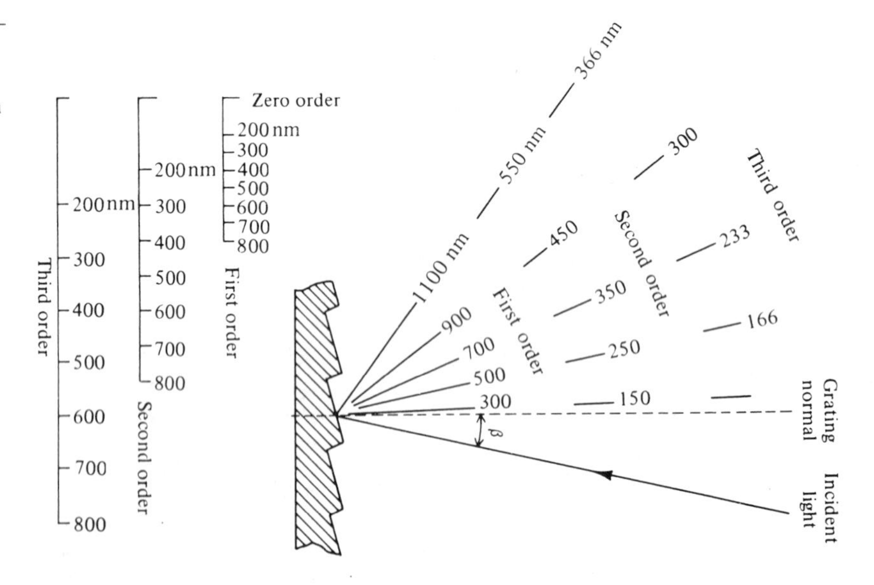

FIGURE **6.12**
Overlapping orders of spectra from a reflection grating.

gives high efficiency over a wide range. A portion of the incident radiation is simply reflected specularly by the grating and forms the zeroth order of the direct image.

EXAMPLE 6.3

To illustrate the use of the grating equation, the primary angle at which radiation of 300 nm is diffracted at normal incidence ($i = 0°$) by a grating ruled 1180 grooves/mm in the first order is given by

$$\sin r = \frac{m\lambda}{b} - \sin i$$

$$= \frac{1 \times 3.00 \times 10^{-5}\ \text{cm}}{(1/11{,}800)\ \text{cm}} - 0$$

$$= 0.354$$

The angle that has this sine is 20.8°. The second-order beam of 150 nm appears at the same angle. Some kind of filtering becomes necessary to prevent the overlap of orders. It may take the form of wavelength cutoff by the detector, a bandpass filter, or cross dispersion with another grating or prism.

In a Littrow mount the angle of incidence equals the angle of diffraction. Whenever the two angles are equal, as is usually the case, the grating equation becomes

$$m\lambda = 2b \sin i \tag{6.12}$$

The useful wavelength range of a grating monochromator is essentially determined by the fact that the efficiency, expressed as the ratio of reflected to incident radiation, drops on either side of the blaze wavelength. Maximum efficiency at the blaze wavelength is less than 70%. Intensity drops to half of the maximum intensity at the blaze wavelength when the diffraction angle is about half the blaze angle on the short-wavelength side or about three times the blaze angle on the long-wavelength side. This is equivalent to a range in the first order extending from two-thirds the blaze wavelength to twice the blaze wavelength.

The bandpass is limited by groove density, focal length, and slit width. As an example, a 0.25-m mount with a 1200-grooves/mm grating and a 1.0-mm slit offers about a 4-nm bandpass. Bandpass can be improved by limiting the entrance and exit slits but at a sacrifice in radiation throughput. In this respect, a 2400-groove/mm holographic plane grating (see below) can improve the bandpass to 2 nm with no change in throughput or in scattered radiation.

In grating monochromators the noise level (unwanted radiation) has two different names and origins: ghosts and stray radiation. Ghosts are related to periodic errors in the position of the grooves arising either from the ruling engine screw or from periodic vibrations that originate outside the machine during the ruling process. Stray (scattered) radiation originates in nonperiodic errors in the spacing of the grooves and in nonperfect planarity of the grooves' reflective surface. However, the scattered radiation level of a grating monochromator is generally lower than that of a filter photometer. Precision ruled gratings or plane holographic gratings may be required for more stringent requirements.

Angular dispersion, $d\theta/d\lambda$, of a grating, used in the autocollimating (Littrow) mode, is given by the expressions

$$\frac{d\theta}{d\lambda} = \frac{m}{b \cos r} \tag{6.13}$$

$$\frac{d\theta}{d\lambda} = \left(\frac{2}{\lambda}\right) \tan r \tag{6.14}$$

The equations are equivalent. In the plane of the exit slit, the linear dispersion is

$$\frac{dx}{d\lambda} = \frac{2f \tan r}{\lambda} = \frac{mf}{b \cos r} \tag{6.15}$$

Since cos r is virtually constant for reflection angles up to 20°,

$$\frac{d\lambda}{dx} \approx \frac{b}{mf} \tag{6.16}$$

Thus a grating monochromator has nearly constant dispersion throughout the spectrum and, consequently, a linear scale for wavelength. This feature is one of the most important advantages of gratings over prisms.

When the order m is regarded as fixed, large dispersion is obtained through the use of gratings with a large number of grooves per millimeter. From Equation 6.14 it is clear that for a given wavelength, dispersion is a function of only tan r. Changes in spacing and numbers of grooves have no effect on resolution and dispersion when a grating is used at a given angle.

Echelle (French for "step" or "ladder") *gratings* are composed of a grating with only a few grooves per millimeter (typically 80 grooves/mm) but used in a very high order—for example, the 75th order. A grating with so few lines and used in such a high order has a very severe overlapping of orders. Another grating or prism that disperses at right angles to the echelle is used to disperse the radiation before it falls on the echelle, thus giving a spectrum in two dimensions. The main echelle grating has a very high linear dispersion (Equation 6.13) and resolving power (Equation 6.17) due to the high order, m. An echelle spectrometer is shown in Figure 10.14.

In a grating monochromator the effective aperture is simply the width of an individual groove, b, multiplied by the total number of rulings, N, and by cos r, or bN cos r. Assuming that the angle between the incident and diffracted rays is small, the theoretical resolution of any grating is described by either of two formulas:

$$R = \frac{\lambda}{d\lambda} = mN = \frac{2Nb \sin r}{\lambda} \tag{6.17}$$

$$R = \frac{2W}{\lambda} \sin r \tag{6.18}$$

where r is the angle between the diffracted ray and the grating normal and W is the width of the ruled portion of the grating. Both equations are correct and equivalent; the second is written for the autocollimating (Littrow) mode of operation. Equation 6.17 implies that for a given order, m, the resolution increases with the number of lines ruled on the grating. Equation 6.18 makes clear that for a grating of width W and at a given wavelength, resolution is purely a function of the sine of the diffraction angle. The latter is typically a 10° to 15° angle for ordinary gratings, but many echelle gratings operate at 63°.

The ruled area of a grating should be large enough to intercept all of the incident radiation even when the grating is turned to its extreme angular position. Any smaller area decreases the useful radiation in the spectrum by wasting that which misses the grating and increasing the stray radiation.

As pointed out, the production of a master grating is a long, tedious process and is quite expensive. A method of producing gratings by photographic means using lasers has been developed in recent years. The gratings are known as *holographic gratings.*[8]

To fabricate regular holographic gratings, two collimated beams of monochromatic laser radiation are used to produce interference fringes in a photosensitive material that has been deposited on optically true glass. The glass is either plane or concave. The portion of the photoresist exposed to the laser beams where they constructively interfere is then washed away, creating a groove structure in relief. The grating is then coated with an appropriate reflective layer and can be used in the same manner as a ruled grating. Because holographic gratings are recordings of optical phenomena, they have absolutely no ghosts and a much lower stray light level than ordinary gratings. Holographic gratings can be produced with as many as 6000 grooves/mm in sizes up to 600 × 400 mm. By changing the wavelength of the laser beams, one obtains different groove spacings; the limiting factor is the availability of different wavelength lasers. When used with parallel beams of incident radiation, there are no aberrations. Even with nonparallel beam configurations, aberrations can sometimes be eliminated.

A concave holographic grating, when illuminated through an entrance slit, creates on rotation a perfect or nearly perfect image on a fixed exit slit. The concave diffraction grating is in itself a monochromator. No other optical components are required, thus eliminating all other elements and alignment of those components with respect to the grating. Compared with a plane ruled grating, the efficiency of a holographic grating generally is lower but is very uniform throughout the spectrum.

High-throughput systems that incorporate holographic gratings have been built from $f/5$ to $f/1$. Accompanying bandpasses of 0.4–4 nm are available in a physical configuration smaller than most 0.25-m grating systems. All dimensions and geometries are worked out by computer by the grating manufacturer and are provided to the customer. Although the grating can be used in a mounting that rotates, the simplest mount for applications involving a finite number of wavelengths is a spectrographic mount in which the grating remains fixed and a photodetector is positioned in the plane of the exit slit for each wavelength of interest. This provides an optical system with no moving parts that is able to monitor one or several wavelengths simultaneously. This type of mounting will be encountered in the discussion of spectrographic instruments (Chapter 10) and instruments for monitoring liquid chromatographic effluents (Chapter 19).

Grating Monochromator Systems. Popular grating systems usually involve some variation of the Ebert or Littrow monochromator configuration. Although continuously tunable, they are designed to pass only a single wavelength at a time. Therefore they are less useful with multichannel detection systems, since they generally vignette—that is, fade into the surrounding background, leaving no sharp edge to the image. This effect reduces the intensities of noncentral wavelengths because the optics are not physically large enough to prevent a loss of radiation at the edges. An f/number increase reduces vignetting. Radiation lost through vignetting is not completely out of the picture; instead it reappears as stray radiation, limiting the signal-to-noise performance of the spectrometer.

Ebert Mounting. The optical features of the Ebert mount are shown schematically in Figure 6.13. Entrance and exit slits are on either side of the grating. A single concave spherical mirror is used as a collimating and focusing mirror. Radiation entering the monochromator strikes the left side of the mirror and is collimated and reflected to the grating. The diffracted radiation goes to the right half of the same mirror and is focused on the exit slit. Normally the use of off-axis mirrors introduces astigmatism into an image. Astigmatism results when the vertical lines of an object are focused at a different point than the horizontal lines. The astigmatism of the first reflection from the mirror in the Ebert mounting is almost entirely compensated by the reflection from the opposite half of the mirror in the second reflection. Since the entrance and diffracted beams use different portions of the mirror, no scattering into the optical path results from the mirror. The Ebert design offers a high-aperture unit in a relatively small package. The wavelength is selected by simple pivoting of the grating about its vertical axis. The angle between the incident and diffracted rays remains constant. Focus is unaffected. A sine-bar drive produces a direct wavelength readout on a linear scale. A cosecant-bar drive provides a linear wavenumber scale.

Fastie suggested an "under-over" design in which the entrant beam passes below the grating and the diffracted emergent beam passes above. The Fastie-Ebert design is commonly used for small spectrophotometers. It offers high aperture, usually $f/3$ to $f/5$, has little or no astigmatism, and is compact.

Czerny-Turner Mounting. In the side-by-side Czerny-Turner mount, shown in Figure 6.14, two concave mirrors replace the single large mirror of the Ebert mounting. All advantages of the Ebert mounting pertain to this one also. The Czerny-Turner design has a low aperture, usually $f/6$ to $f/10$, and may be more difficult to align. It is generally used in more expensive systems for better stray radiation and resolution characteristics.

Littrow Grating Mounting. The Littrow configuration requires a single mirror for focusing and collimating. It is commonly used for small spectrophotometers. The Littrow mount is usually $f/3$ to $f/5$. In this autocollimating configuration (see

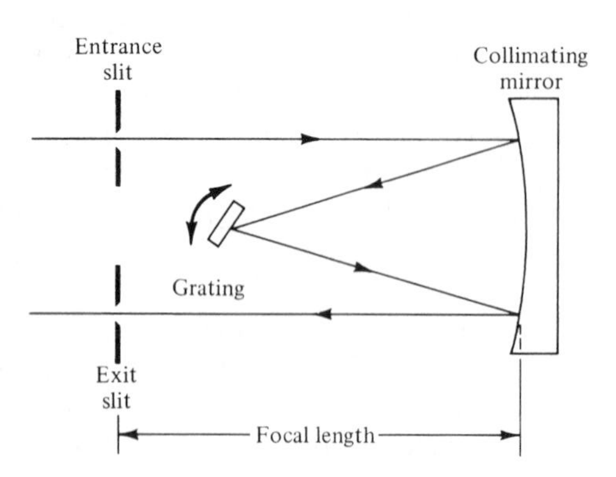

FIGURE **6.13**
The Ebert mounting.

Figure 6.10), the grating's axis is parallel to and in the plane of the grooves. In addition the axis of rotation intersects the radiation at right angles. As a result, angles of incidence and diffraction are on the same side of the grating normal and are nearly equal so that the grating equation becomes $m\lambda = 2b \sin i$. This mounting ensures high grating efficiency, good spectral purity, small aberrations, and compactness.

In a Littrow mounting the source illuminates a single mirror on its upper portion. The beam is collimated and directed to the grating where the beam is dispersed. The diffracted beam is sent back to the lower portion of the same mirror where it is focused on the exit slit. The upper and lower portions of the same slit assembly are used as entrance and exit slits, thus providing a perfect correspondence of slit widths. The slit width can be fixed or continuously variable.

Other Grating Mountings. Many other mountings for grating monochromators have been used. Among these are the Seya-Namioka, the Wadsworth, the Eagle, and the Rowland circle mounting. Some of these are discussed in Chapter 10.

Double Monochromation. A double monochromator consists of two dispersing systems used in series. The intervening slit is simply the exit slit of the first monochromator and the entrance slit of the second. Both wavelength drives must track together perfectly. In some double monochromators, different dispersing elements are used in each half; usually the first is a prism and the second a grating. The quartz prism acts as an order sorter to present a single order to the second grating monochromator. Sampling geometry is double beam.

With two identical monochromators used sequentially, the dispersion and resolution are approximately doubled, and stray radiation is greatly reduced. For example, if the intensity of the stray radiation in each monochromator is 0.1% of that of the primary beam, the usual situation in the ultraviolet and visible regions, then the double monochromator reduces this to 0.0001% (0.1% of 0.1%). In the near-infrared region, the stray radiation reduction is less, about 0.1% overall. Of course,

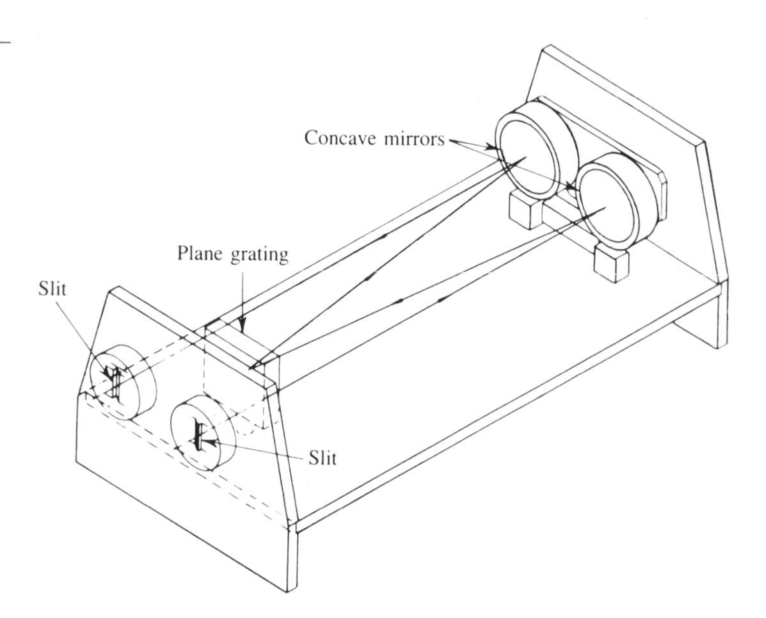

FIGURE **6.14**
The Czerny-Turner mounting.

the transmission factor of the double monochromator is about half that of either monochromator by itself. If desired, the increased resolution can be traded off for a gain in energy throughput. For example, by opening the slits to twice the width needed if only one monochromator was used, a resolution comparable to that of a single monochromator can be secured and, at the same time, a quadrupling of energy at the exit slit is secured in the double unit.

It is also possible to get some of the effect of two dispersing elements by passing the radiation through a single dispersing element twice. Such a double-pass monochromator is shown in Figure 6.21. This arrangement is considerably less expensive than a true double monochromator, but it is also less effective in reducing scattered radiation.

6.3 CELLS AND SAMPLING DEVICES

For the most accurate work the cells (cuvettes) that contain the sample and reference solutions should have perfectly parallel windows and be so positioned that the windows are perfectly perpendicular to the beam of radiation. Cylindrical cells are sometimes used in inexpensive instruments but, if used, they must be very carefully positioned so as to be reproducible each time. The cells used in ultraviolet-visible spectrophotometry are usually 1 cm in path length, but cells are available from 0.1 cm or even less to 10 cm and more. Transparent inserts are available to shorten the path length of the most common length cells from 1 cm to 0.1 cm. In the infrared region sample cells are generally much shorter, 0.1 mm to 1 mm, because most infrared solvents absorb appreciably (see Chapter 11).

The cells must be constructed of a material that does not absorb radiation in the region of interest. Fused silica or quartz is transparent from about 190 nm in the ultraviolet region to about 3–4 μm in the infrared region. Silicate glasses can be used from about 350 nm to 2–3 μm. Inexpensive plastic cells are available for use in the visible region. Special materials are necessary for the infared region as discussed in Chapter 11.

To obtain good absorbance data, the cells used should be matched. Matched pairs are obtainable from the manufacturers, but it is best to check the match by running identical samples in each cell and to reserve one cell for the sample solution and the other for the solvent. The cells should be scrupulously cleaned before and after each use. Nitric acid or aqua regia is recommended for cleaning. Dichromate solutions should not be used because of their tendency to be absorbed by glass. After cleaning, the cells should be thoroughly rinsed and dried, but never in an oven or over a flame, since the path length may be changed by such treatment. The faces of the cells should never be touched when handling them because fingerprints, grease, and lint may cause marked changes in the transmittance of the cells.

Several automated sample changers are available for use when large numbers of routine samples are to be investigated. Consult the manufacturer's literature for descriptions of sample changers and automatic sampling devices.

Fiber Optics

Fiber optic bundles are composed of numerous strands of glass or plastic fused at the ends. A single fiber transmits radiation; a bundle of fibers can transmit both radiation and images. Fiber optics transmit radiation by total internal reflection. Total reflection occurs whenever a ray traveling in a transparent medium strikes an interface with another medium of lower refractive index, provided that the angle of incidence is larger than a certain value known as the critical angle. This angle is defined by a ray that is refracted parallel to the interface. Thus any optical fiber must consist of a high refractive index core surrounded by a lower index sheath.

Radiation picked up within a limited cone or acceptance angle at one end of the fiber emerges from the other end at the same angle. A measure of this cone or numerical aperture (NA) is given by

$$\mathrm{NA} = \eta_3 \sin \theta_w = (\eta_1^2 - \eta_2^2)^{1/2} \tag{6.19}$$

where η_1 is the refractive index of the core material, η_2 the index of the sheath, η_3 the index of the surrounding medium, and θ_w the maximum acceptance half-angle.

In glass fibers the two glasses are chosen not only to give the required numerical aperture, but also to have compatible melting points and expansion coefficients. Typical fibers have a core index of 1.64 and a sheath index of 1.53, giving a numerical aperture of 0.54, which is equivalent to a full angle of 66°. Typical fiber diameters range from 10 to 150 μm. In plastic fibers a typical core is polymethylmethacrylate ($\eta = 1.49$) surrounded by a transparent polymer sheath with an index of 1.39. Fiber diameters range from 0.01 to 1.0 mm i.d.

Fiber optics are useful when the material to be investigated cannot readily be contained in a standard cell—for example, when the absorbance of a flowing solution is to be measured. The incident beam can be carried to the solution by the fiber optics, and the transmitted beam can then be brought back to the spectrophotometer by another fiber optic bundle. Fiber optics are also sometimes used to bring radiation, after dispersion, to impinge on one element of a linear diode array (see below).

6.4 DETECTORS[9,10]

A detector is a transducer that converts electromagnetic radiation into an electron flow and, subsequently, into a current flow or voltage in the readout circuit. Many times the photocurrent requires amplification, particularly when measuring low levels of radiant energy. There are single-element detectors such as solid-state photodiodes, photoemissive tubes, and photomultiplier tubes, and multiple-element detectors such as solid-state array detectors. Important characteristics of any type of detector are spectral sensitivity, wavelength response, gain, and response time.

Photoemissive Tubes

Vacuum photoemissive tubes are simply photocathode-anode combinations contained in an evacuated envelope. The typical single-stage vacuum phototube contains a radiation-sensitive cathode in the form of a half cylinder of metal coated on its receiving surface with a radiation-sensitive layer, usually an alkali metal, and an anode wire located along the axis of the cylinder or a rectangular wire that frames the cathode. The assembly is shown in Figure 6.15, along with a simple phototube circuit.

The photocathode operates on the principle that electrons are emitted from certain materials in direct proportion to the number of photons that strike their surface. For optimum efficiency a photocathode surface must have the highest possible absorption coefficient for the incident radiation. The surface material must also have a low work function in order to extend its spectral coverage to longer wavelengths. A photoemissive detector cannot respond to radiation with photons that have an energy below the work function, since the work function represents the energy necessary to just eject an electron from the surface of the material.

The transmission photocathode (shown in Figure 6.17b for a photomultiplier tube), which is superior from an electron-optical point of view, must be simultaneously thick enough to absorb most of the incident radiation and thin enough to allow the generated photoelectrons to traverse it while retaining enough energy to overcome the work function barrier at the vacuum interface. Similarly, in the reflection or opaque photocathode (shown in Figure 6.17a), those photons that are absorbed too deeply within the material generate photoelectrons that can no longer get to the surface and escape.

The spectral sensitivity for several types of photocathode materials is shown in Figure 6.16. Cathode quantum efficiency (QE), the parameter in the illustration, is the average number of photoelectrons emitted from the photocathode divided by the number of incident photons. There are essentially 11 different chemical compositions of photocathodes. These are offered in either semitransparent or opaque forms, which yield different responses. Either physical type can be combined with about ten different window materials to generate most of the commercial offerings. The most sensitive cathode compositions are the bialkali types (K-Cs-Sb), which yield tube responses in the millions of amperes per watt range. Red response is produced by the Ag-O-Cs type, which is usable to 1.1 μm. A series of Ga-In-As compositions have excellent sensitivities out to 1.1 μm. To reach the ultraviolet end of the spectrum, excellent response is obtained by coupling virtually any type of cathode with an appropriately transparent window. Cs-Te cathodes, the solar-blind type (i.e., insensitive to radiation in the visible and near-ultraviolet regions of the

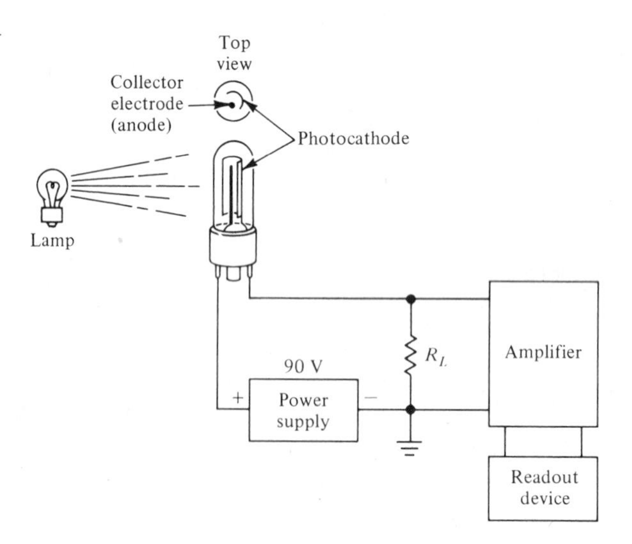

FIGURE **6.15**
Photoemissive tube and its accessory circuit.

spectrum), operate from 120 to 350 nm. Flat responses are available in the Ga-As series; this composition generates a tube usable from 200 to 940 nm, while varying in radiant response by only 10% from 440 to 880 nm. At short wavelengths the sensitivity of the detector is impaired by absorption by the envelope material. The short-wavelength limit is about 350 nm with a glass envelope or 200 nm with silica. Thus the shape of the spectral response is a function of the cathode composition and envelope material.

When radiation strikes the photocathode, photoelectrons are ejected and are drawn to the positive anode, creating a current. All the electrons are collected by maintaining the anode at +90 V relative to the cathode. The photoelectric current flows through the load resistance, R_L, developing the signal voltage, $e_s = iR_L$. The load resistance in the external circuit is normally the input resistor in an amplifier circuit. The resulting current that flows in the external circuit is directly proportional to the rate of photoelectron emission, which is proportional in turn to the incident radiation flux.

When a photoemissive tube is operated in darkness, a current still flows. This dark current is traceable to thermionic emission, field emission, ohmic leakage, and emission caused by the natural radioactivity of ^{40}K in the glass envelope. Photoemissive tubes are limited in sensitivity by this dark current and by the very low level of current produced by low radiation levels. Although currents as small as 10 pA may be amplified easily, there is great difficulty in amplifying lower currents, which may be smaller than the ohmic leakage across the tube envelope that shunts the load resistance. The time constant of the amplifier circuit (time required to reach $1/e$ of its final response) may become sufficiently large to degrade the effective response time of the detector.

The primary source of noise in photoemissive detectors at room temperature is generally shot noise (see Chapter 2). Shot noise is caused by the fundamentally discontinuous (quantized) nature of radiation and electrical current. Photons arrive

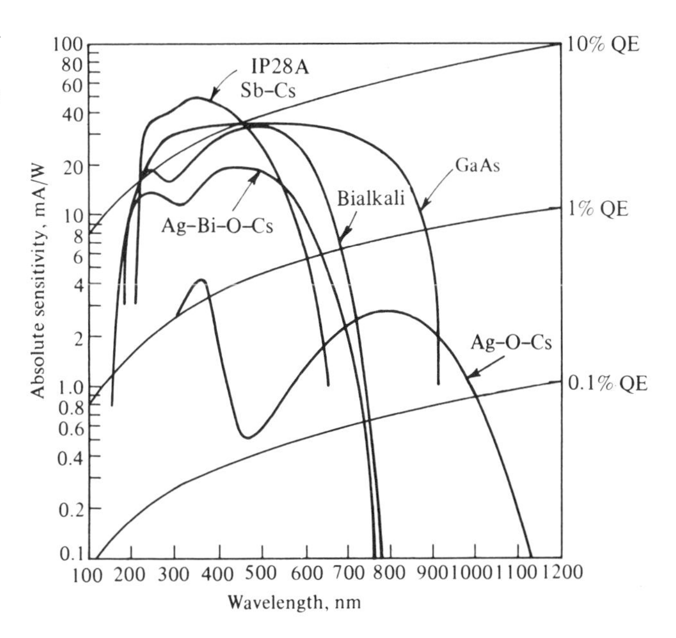

FIGURE **6.16**
Spectral response curves of selected photoemissive surfaces. (Courtesy of Radio Corporation of America.)

at the cathode randomly even though the overall intensity of the radiation is constant. Thus photoelectrons are emitted from the cathode and arrive at the anode also randomly, and yet the long-term rate of photoelectron pulses at the anode is constant and proportional to the radiant power. Thermionic electrons emitted from the cathode also occur randomly.

For an accuracy of 1% or better, a calibration curve is required to overcome the slight nonlinearity in dependence of photocurrent on illumination unless a suitable null method can be used. The gain is, of course, unity. The time constant is about 150 psec. Photoemissive tubes are useful mainly for following low-repetition-rate, high-intensity sources.

Photomultiplier Tubes[11]

The electron multiplier tube, or photomultiplier tube (PMT) as it is commonly called, is a combination of a photoemissive cathode and an internal electron-multiplying chain of dynodes. The two popular designs, the circular cage and in-line configurations, are shown in Figure 6.17. Incident radiation ejects photoelectrons from the cathode. The emitted photoelectrons are focused by an electrostatic field and accelerated toward a curved electrode, the first dynode, coated with a compound (BeO, GaP, or CsSb) that ejects several electrons when subjected to the impact of a high-energy electron. The rounded shape of the dynodes converges the electrons in one dimension, whereas the field-forming ridges at or near the dynode ends converge the electrons in the second dimension. Repeating this electron-multiplying process over successive dynodes maintained at higher voltages produces a current avalanche that finally impinges on the anode. Internal current amplification, or gain, is thereby achieved. The tube output may, of course, be further amplified. To prevent deterioration of dynode surfaces by local heating effects and to prevent tube fatigue, the anode current must be kept below 1 mA. This in turn requires that the voltage between the final dynode and the anode be restricted to 50 V or less. Response time is about 0.5 nsec with the GaP dynode surface; it is about 1–2 nsec with the other materials.

A resistor chain divides the operating voltage so that a potential difference of 75–100 V exists between adjacent dynodes. The most important interelectrode potential in a photomultiplier tube is the cathode-to-first-dynode voltage, which should always be maintained at the value recommended by the manufacturer of the tube. This can be done by using a constant-voltage diode (Zener diode) (see Chapter 3) of the proper voltage in place of the cathode-to-first-dynode divider resistor.

Ideally, the total gain, G, of a photomultiplier tube that has n stages and the secondary emission factor f per stage is $G = (f)^n$. The exact value of f depends on both the nature of the dynode secondary-emitting material and the imposed electrical potential. For older dynode materials the value of f has ranged from 3 to 10. With GaP coatings it can easily reach 50, so that the total number of dynode stages can be reduced while still achieving high gain. As a result of their large internal amplification, photomultiplier tubes can be used only at low power levels of about 10^{-14}–10^{-4} lumens. (A lumen is the luminous flux per steradian from a point source that has an intensity of 1 candela. A candela is the luminous intensity of $\frac{1}{60}$ of a square centimeter of projected area of a blackbody at the freezing point of platinum. A lumen is about 0.00147 W.) The ability to change the sensitivity over a wide range simply by changing the supply voltage is a unique advantage of photomultiplier tubes.

Since the dark current also produces an amplified current at the output, it sets a lower limit to the radiant power that can be directly detected. Thermal dark current is minimized by cooling the photomultiplier tube. Because the dark current represents a fairly steady component, it may be offset automatically by a potentiometer zeroing arrangement or be subtracted directly from the output signal.

Photomultiplier tubes with high gains are sensitive enough to detect pulses arising from individual photons that fall on the photocathode at low radiation levels. Photon counting generally results in a higher signal-to-noise ratio than measurement of the average output current at these low power levels. To count individual photons, a wide-band amplifier is used in series with a pulse-height discriminator (see Chapter 3) to eliminate low-amplitude spurious pulses that originate primarily from thermal emission from the dynodes, primarily the first

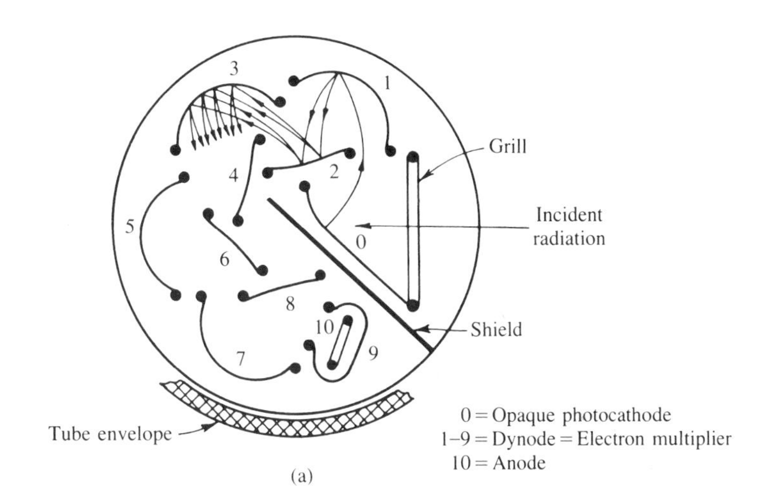

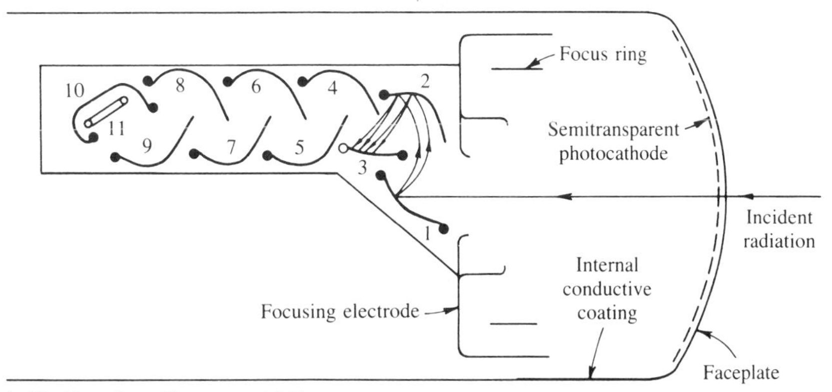

FIGURE **6.17**
Photomultiplier design: (a) the circular-cage multiplier structure in a "side-on" tube and (b) the linear-multiplier structure in a "head-on" tube. (Courtesy of Radio Corporation of America.)

dynode. These pulses are amplified to a lesser extent than pulses that originate at the photocathode. The transmitted pulses are counted with a high-speed digital counter and a timer. The signal-to-noise ratio is equal to the square root of the count rate. If there is an appreciable number of background counts due to dark currents then the signal-to-noise ratio is given by

$$S/N = \left(\frac{R^2\tau}{R + r}\right)^{1/2} \tag{6.20}$$

where R is the total count rate, r the dark current count rate, S/N the signal-to-noise ratio, and τ the counting time.

Photodiodes

Photodiodes operate on a completely different principle from the detectors discussed earlier. The construction of a planar-diffused silicon $p - n$ junction diode is shown in Figure 6.18. The process starts with a very high-resistivity intrinsic silicon material. Very shallow p and n diffusions are made in the top and bottom surfaces, respectively, and the top surface is covered with a protective SiO_2 layer. Metal contacts formed on the top and bottom surfaces provide electrical connections. The diffused p regions determine the junction and optically active area. A photon must reach the active (or intrinsic) area to produce current flow in the external circuit (see also Chapter 3).

A *p-n* semiconductor junction is reverse biased so that no current flows. When photons interact with the diode, electrons are promoted to the conduction band where they can act as charge carriers. Thus the generated current is proportional to the incident radiant power. Most of the devices detect only visible and near-infrared radiation. Diode responsivity is typically 250–500 mA W^{-1} across the visible spectrum. This is at least an order of magnitude more than vacuum photoemissive tubes but many orders of magnitude less than photomultiplier tubes, which can have very high gains as previously noted. The output photocurrent is linear up to 10 decades versus the illumination level. Typical spectral output is shown in Figure 6.19.

In general, the speed of the photodiode is limited by the time constant formed between the amplifier input impedance and the intrinsic shunt capacitance of 2–5 pF. To keep the capacitance as low as possible, extremely small devices are

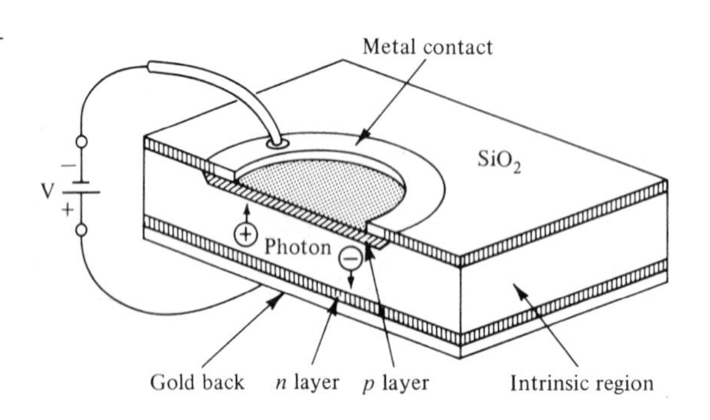

FIGURE **6.18**
Construction of a planar-diffused *p-n* junction photodiode.

used and the optical signal is coupled to the devices by a lens or by fiber optics. Rise times vary from less than 1 nsec to more than 10 μsec.

Many tiny photodiodes can be assembled in a linear or a two-dimensional array in which each diode gathers a signal simultaneously. A tiny capacitor is coupled to each diode and charged to a given level before illumination of the diode. On illumination of the diode, charge leaks off the capacitor. After the signals are obtained, each element of the array is scanned, the charge loss is recorded, and the capacitor is recharged. Thus one- and even two-dimensional data are obtained. For example, each diode may be placed so that it obtains the intensity of a narrow-wavelength range of a dispersed spectrum or the intensity of light falling on a two-dimensional picture. Solid-state arrays offer excellent stability. There is little spreading of charge from one diode to an adjacent diode, an important factor when sharp differences in intensity are encountered.

Linear photodiode arrays are commercially available with 128 to as many as 4096 elements per array.[12] The arrays may be as small as 13×2.5 μm spaced 15 μm apart. Each diode and its accompanying capacitor are connected to a MOS field-effect transistor (FET) switch, which is controlled by a clock. At the start pulse that initiates a scan, each diode element, in turn, is charged to its full reverse-biased potential. Radiation that falls on the diode then begins to discharge it. At the next cycle, each diode element is again recharged to its full potential and an analog-to-digital converter records the charge required, which is proportional to the integrated radiant power received by the diode in the interim. The digital values for each individual diode are stored in a memory unit. Several scans of each diode can be summed in memory. The diodes are charged sequentially. The integration time is the time between start pulses and is the same for all diodes; however, it takes a finite time to scan the total array, thus each diode records the integrated radiant power that falls on it at a slightly different time of the scan. Radiation that falls on the spaces between diodes is divided proportionally between the two adjacent diodes and, due

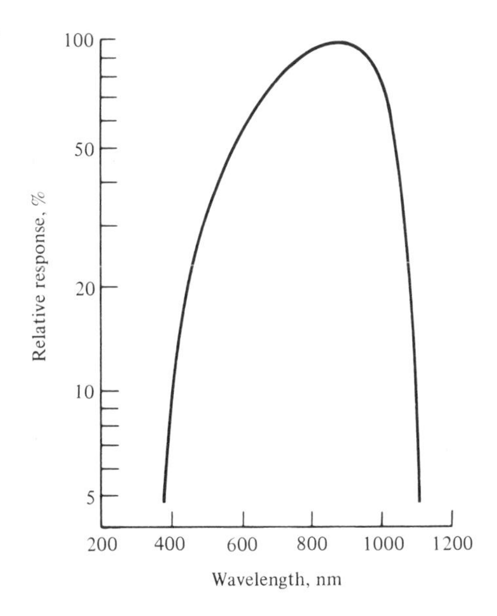

FIGURE **6.19**
Spectral response of a typical *p-n* photodiode.

to the discrete placement of the diodes, increases the bandpass and decreases somewhat the spectral resolution of the array compared with a scanning spectrophotometer with the same entrance slit width and an exit slit width equal to the spacing of the diodes in the array.

Another variation of photodiode array construction is the charge-coupled device (CCD). In this device, the readout of each diode in the array is triggered by a transfer pulse applied to the FET to simultaneously transfer each diode's charge into a corresponding stage of an analog shift register. Then the charges are clocked through the shift register to appear sequentially at an output. Thus the diodes all record the integrated radiant power that falls on them during the same time period.

6.5 READOUT MODULES

In the simpler instruments, direct-current (dc) signals are produced that are amplified by dc amplifiers and read on analog meters, recorders, digital voltmeters, or displays of computer systems. High-gain dc amplifiers are subject to significant drift and offset errors, however. The presence of low-frequency (l/f) noise in the signal seriously restricts the extent to which the signal-to-noise ratio can be improved by simple low-pass filtering. For these reasons, it is often desirable to modulate the signal and thereby transform it into some alternating-current (ac) frequency high enough to avoid the drift and l/f noise problems. After amplification by an ac amplifier, the signal is converted back into dc by a demodulator or rectifier because all the commonly used readout devices require dc signals. A current is usually converted to a voltage before display or recording (see Chapter 3).

Modulation is usually performed by interrupting (chopping) the beam of radiation that strikes the detector by means of a rotating sector disk (see Chapter 2) or by modulating the beam of radiation as previously described. If a chopper is used, it may be in various places in the optical system depending on the application. Typical placements are shown in Figures 6.21 and 6.23.

6.6 INSTRUMENTS FOR ABSORPTION PHOTOMETRY

Historically, scanning monochromators have been used with single-channel detectors on either single or mechanically complex double-beam systems for absorbance measurements over the 190–800 nm spectral range. These spectrometer systems often are used with analog calculation of percent transmittance and absorbance. They also may be interfaced to a microprocessor for simplified data reduction. Now a new generation of instruments incorporating recent advances in multichannel detector technology, concave holographic gratings, and microcomputers is appearing with heretofore unattainable capabilities.

The essentials of an analytical instrumental system are shown in Figure 6.1. It consists simply of a source, focusing optics, an unknown or standard sample holder, a wavelength isolation device, and a detector with amplifier and readout system. From an engineering standpoint, it is desirable that this type of system be detector limited; that is, the limiting factor should be the noise generated by the detector.

Anything that can be done to increase signal levels at the detector is therefore desirable. The measure of performance is usually defined as the precision or photometric accuracy.

In terms of construction one recognizes the difference between single-beam and double-beam radiation paths and whether the photometer module is direct reading or uses a balancing circuit. Special features include double monochromation and dual wavelength systems. In the final selection of an instrument, things to consider are initial cost, maintenance, flexibility of operation, resolution characteristics, wavelength range, accuracy, and perhaps auxiliary equipment to expand into other areas of application. Some instrument designs have concentrated on performing sequential measurements rapidly and automatically through the use of rapid sampling accessories. The accessory units fill and evacuate the spectrometer sample cells and make it possible to analyze many samples per hour on a routine basis. Spectrophotometers are available to perform reaction-rate analyses or to analyze solutions for several ingredients simultaneously. Many of the present advantages of spectrophotometers are direct results of the capabilities of the microprocessors used in automating the instruments. For example, one commercial instrument is a completely digital, interfaced spectrometer with a programmable statistical calculator that provides for unattended data acquisition, storage, and calculation of first- and second-derivative spectra, peak location, and peak area integration.

Instruments for absorption photometry may be classified as spectrophotometers (incorporating a monochromator) or filter photometers. Dispersive spectrometers are expensive but offer the advantage of continuous wavelength selectability. They are typically limited by radiation throughput, light scatter, and size. A monochromator is an inefficient device for transferring optical energy, and energy loss is serious. For example, if an exit slit width of 0.1 nm with a height of 10 mm is used, the resultant aperture to the detector is on the order of 10^{-6} mm^2. To sense the small amount of energy that falls on the detector, a photomultiplier tube is generally required. The noise of the photomultiplier tube further degrades the signal-to-noise ratio and places severe demands on the amplifier and recording system if high precision is to be obtained.

Filter photometers offer an economic advantage over dispersive instruments, as well as increased luminosity, particularly when the interference-type filter is used. Of course, filters must be available for all wavelengths desired. If an analysis requires a measurement at a wavelength for which no filter is available, less than maximal sensitivity must be accepted. Higher signal-to-noise ratios can usually be obtained than with a monochromator. Less sensitive detectors can also be used, since the radiation levels are higher. For some routine or repetitive analyses, the stability of filters offers advantages. Push button selection of the filters is possible, and there is not much danger of an incorrect wavelength setting. In spite of these advantages, modern dispersive instruments have so many more advantages and are so versatile that filter photometers are not often found in analytical laboratories at present.

Single-Beam Instruments[13]

The simplest type of absorption spectrometer is based on single-beam operation in which a sample is examined to determine the amount of radiation absorbed at a given wavelength. The results are compared with a reference (usually the solvent alone) obtained in a separate measurement. Changes in source intensity and detector sensitivity with wavelength generally limit single-beam instruments to measurements at one wavelength per comparison. The instruments are thus primarily used for the quantitative determination of a single component when a large number of similar samples are to be analyzed. In using a single-beam instrument, the absorption maximum of the analyte must be known in advance. The wavelength is then set to this value, the reference material (solvent blank) is positioned into the radiation path, and the instrument is adjusted to read 0% transmittance when a shutter is placed so as to block all radiation from the detector and to read 100% transmittance when the shutter is removed. After these adjustments have been made, the sample is placed in the path, the transmittance is read, and the concentration is determined by the use of either a calibration curve for the material or proper algebraic methods. Depending on the instrument used, the readout may be made directly by observation of the deflection of a meter, by a "null-balance" scale setting, by a recorder trace, or on a digital printout. An obvious requirement of single-beam instruments is a high degree of stability of both the source and the detector system because fluctuations cause errors in comparison of the two readings, sample and solvent. In routine work where errors of $\pm 1\%$ are insignificant, however, instrumental stability is usually not a limiting factor.

Filter Photometers. A relatively inexpensive, nonscanning absorption photometer can be designed around a set of filters. Such instruments are adequate for many methods, especially for absorbing systems with broad absorption bands. With a large energy throughput provided by the filter and a large aperture (often $f/1$), an amplifier is often unnecessary. A filter photometer is operated as described for single-beam instruments.

Spectrophotometers. A modern version of a single-beam spectrophotometer is shown in Figure 6.20. A small fraction of the energy from the source is split off by the beam splitter and used as a comparison beam to compensate for variations in lamp output, line-voltage fluctuations, and electronic instability. Two radiation sources are used, a quartz-halogen lamp for the visible range and a deuterium lamp for the ultraviolet. The dispersing medium is a Bausch and Lomb holographic grating. A photomultiplier tube is the detector throughout most of the ultraviolet-visible region and a silicon solid-state sensor is used in the red region of the spectrum. The Spectronic 1001 has a range of 190 to 950 nm and a spectral bandwidth of 2 nm. There are two built-in microprocessors, one for the main instrument functions and one to control the monochromator. Readout is on an alphanumeric display on the instrument, an optional printer, or, by means of an interface, an optional computer. Readout can be in transmittance, absorbance, or concentration (after calibration with standards). The microcomputer provides for many operations such as absorbance ratios or differences, statistical calculations, and multiple-wavelength (up to 12) measurements on a sample. Many accessories are available, including holders for several sample cells, holders for semi-micro flow-through cells, temperature-controlled or thermostatted cells, long-path cells, interfaces for strip chart recorders and computers, and automatic sampling equipment.

Double-Beam Instruments[14]

In double-beam instruments, a monochromatic beam of radiation is split or chopped into two components, usually of equal radiant power. One beam passes through the sample and the other through a reference solution or blank. However, the radiant power in the reference beam varies with the source energy, monochromator transmission, reference material transmission, and detector response, all of which vary with wavelength. If the output of the reference beam can be kept constant, then the transmittance of the sample can be recorded directly as the output of the sample beam. There are several ways to keep the reference output constant: (1) create a feedback loop to regulate photodetector sensitivity via dynode voltage, (2) control the monochromator slit width by means of servomotors and mechanical slit drives, and (3) position an optical wedge in the radiation path to

FIGURE **6.20** Schematic optical arrangement of the Spectronic 1001 split-beam spectrometer. (Courtesy of Milton Roy Company.)

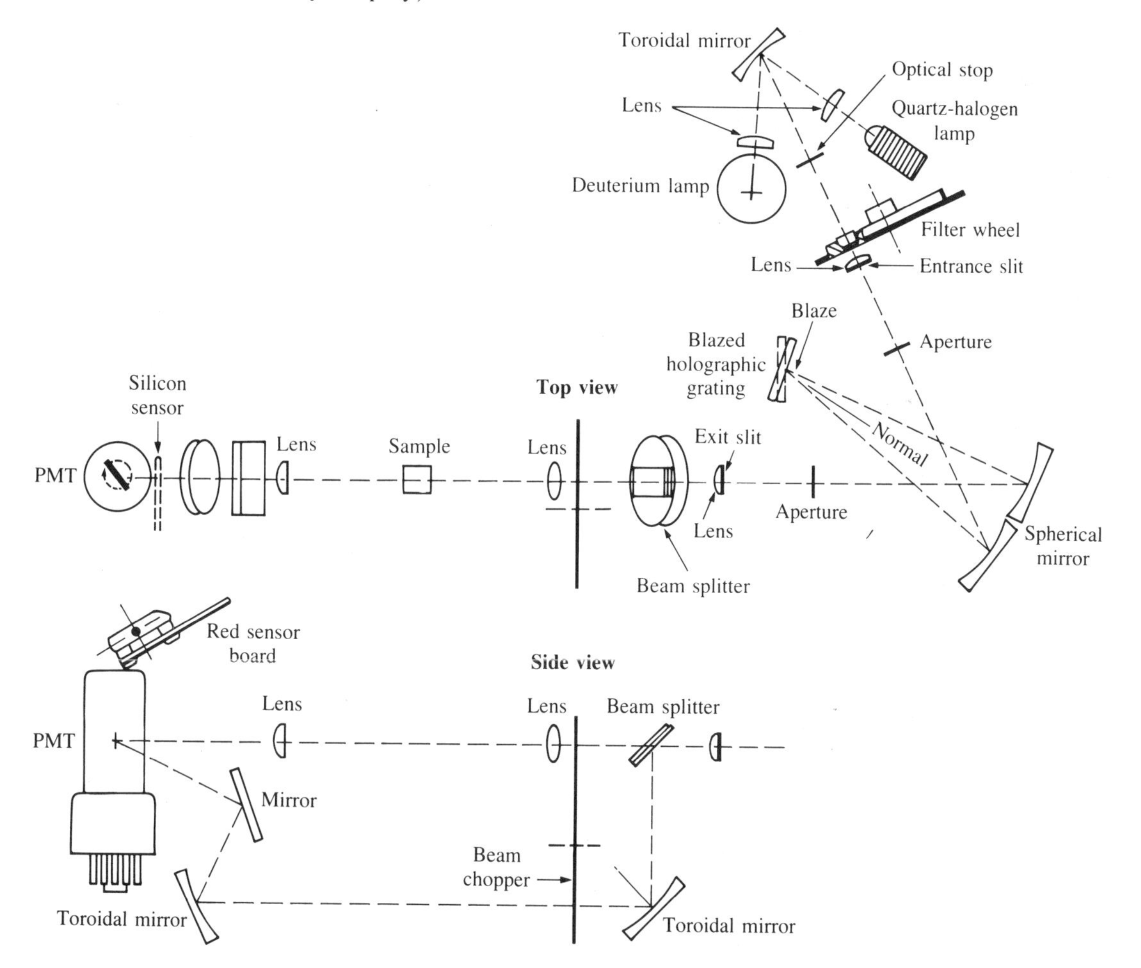

automatically increase or decrease the amount of radiation that reaches the detector.

Automatic gain control is the least expensive of the three modes, since it involves only electronic circuitry and no mechanical components. It provides constant slit width and thus constant resolving power during the scan when a grating monochromator is used. However, the noise level of the photodetector varies with gain and thus is not constant throughout the scan.

Automatic slit control is much more costly, since complicated mechanical slit drives must be incorporated into the instrument. However, photodetector gain and therefore noise remain constant. As the slit width varies, the resolution varies accordingly.

Optical wedge systems have intermediate utility. A less complicated mechanical drive system is needed than for slit drives. Photodetector gain as well as resolution remains constant throughout the scan.

Except for the optical wedge system, which is used in optical null procedures described later, it is not necessary to control the power of the beam through the reference solution in true double-beam operation. In double-beam operation, absorption spectra are automatically corrected for instrument response as a function of wavelength if the ratio P/P_0 is continuously measured. Source instability and amplifier drift affect both beams similarly and the effects should cancel. Solvent absorption is automatically subtracted by placing solvent in the reference beam. Another benefit is the ability to place two samples into the instrument and have the absorbance of one subtracted from the other, which reduces manual data manipulation. In such measurements only spectral differences are recorded.

Scanning Double-Beam Spectrophotometers. Spectrophotometers of this type feature a continuous change in wavelength. One beam permanently accommodates a reference or blank and the other the sample. An automatic comparison of the transmittance of sample and reference is made while scanning through the wavelength region. The ratio of sample to reference, after conversion to absorbance values, if desired, is plotted as a function of wavelength on a recorder. Automatic operation eliminates many time-consuming manual adjustments, especially for qualitative analysis where complex absorption curves are to be traced over a wide spectral range.

In an optical arrangement denoted *double-beam-in-time,* shown in Figure 6.21, radiation from either the visible or ultraviolet source (A) enters the grating monochromator in the Czerny-Turner configuration through the entrance slit (C). Broad-band filters contained in a filter wheel (B) are automatically indexed into position at the required wavelengths to reduce the amount of stray radiation and unwanted orders from the diffraction grating. The grating is actually two gratings ruled back-to-back so that by rotation the proper one for the region being investigated can be brought into position. The beam is passed through intermediate slits (D_1 and D_2) and returned to the grating for a second dispersion. The optical beam that emerges through the slit (E) is then directed alternately through the sample and reference cells (G_1 and G_2) by a rotating sector mirror (chopper) (F) and corner mirrors. The open part of the sector passes the beam directly to one path, while the mirrored part of the sector reflects the beam to the second path. This is the basis of the double-beam-in-time designation. Each beam, consisting of a pulse of radiation separated in time by a dark interval, is then directed onto a photomultiplier tube (H) for the ultraviolet-visible region or a lead sulfide detector (I) for the near-infrared region in a time-sharing procedure.

The Varian 2300 UV-Vis-NIR Ratio Recording/Scanning Double-Beam Spectrophotometer has a wavelength range of 185 to 3152 nm, ten scan speeds ranging from 0.01 to 10 nm/sec, and spectral bandwidths from 0.07 to 3.6 nm in the ultraviolet-visible range and from 0.16 to 14.4 nm in the near-infrared region. The detectors, sources, and grating are under automatic control. An 18-character digital printer is included. Built-in microprocessors permit the selection of output formats, automatic zeroing, data reduction, statistical treatment of the data, concentration routines with linear or nonlinear curves, kinetics, and many other methodologies. Many accessories are available, including thermostatted cells, automatic five-cell programmer, adapters for microcells, and a reflectance attachment.

In the *optical null* procedure, the two chopped beams fall on a single detector. If intensities are identical, the amplifier has a dc output. Any difference in intensities results in an ac signal at the chopping frequency. The unbalance signal is further

FIGURE **6.21** Schematic diagram of a double-beam spectrophotometer (Varian 2300) with dual-source, double-pass, double-sided grating, Czerny-Turner monochromator. (Courtesy of Varian Associates, Inc.)

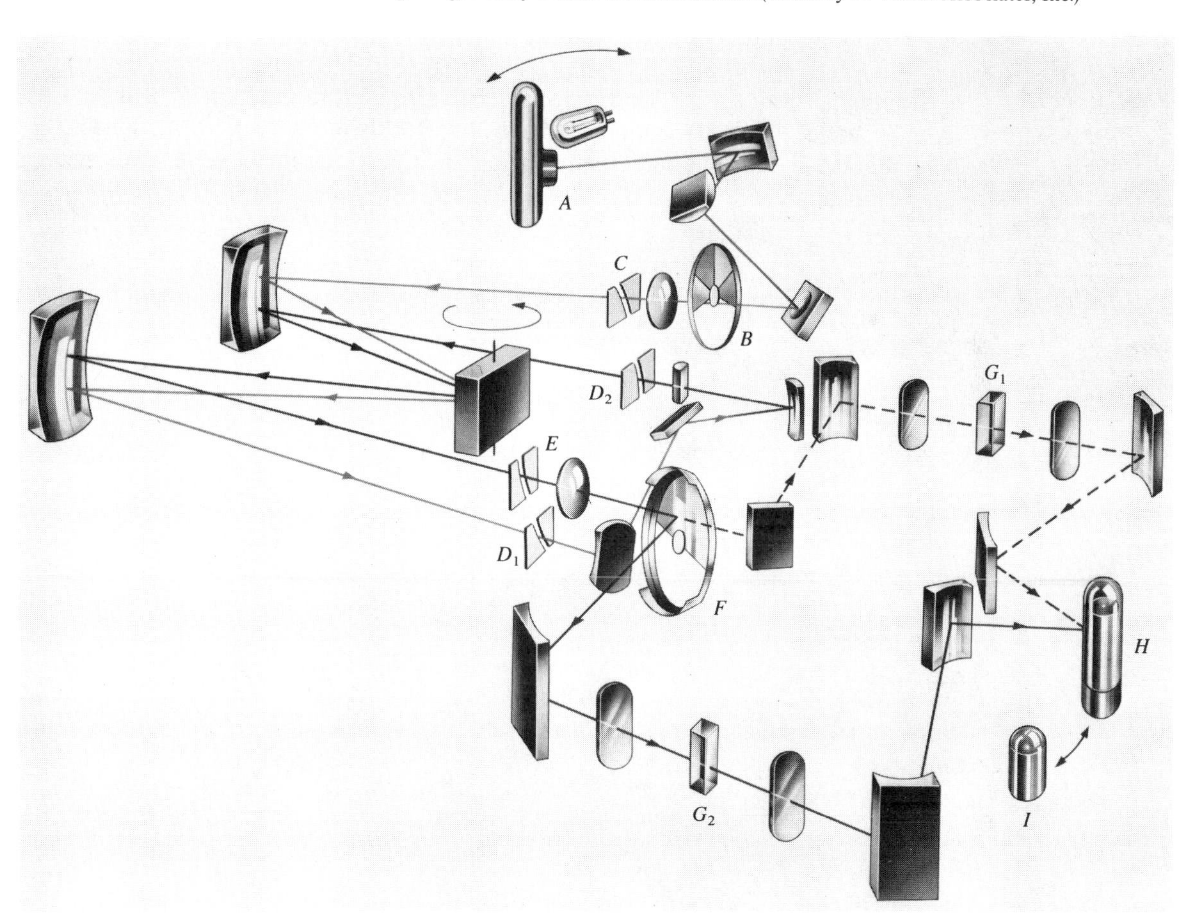

amplified and used to drive an optical attenuator into or out of the reference beam. The fraction of open space the attenuator gives to the reference beam compared with the space given when the reference sample is in both beams corresponds to the percent transmittance of the sample. A link between the servomotor that drives the attenuator and the recorder pen provides the transmittance trace. The optical null procedure has more use in infrared instruments where there are energy-limited situations and detectors have a slower response time. It will be discussed further in Chapter 11.

Double-Wavelength Spectrophotometer.[15,16] Double-wavelength spectrophotometry refers to the photometric investigation of a material by passing radiation of two different wavelengths through the same sample before reaching the detector, as shown in Figure 6.22. The sample is positioned close to the detector to compensate for any turbidity or scattering. Radiation from the source passes into a Czerny-Turner monochromator through slit S_1 to form an image of the mask on gratings G_1 and G_2. The dispersed radiation of both monochromators is focused by the collimating mirror M_3, split, and chopped. Bilateral optical attenuators O_1 and O_2, situated at the pupil image of each beam, are used to vary the radiant power of each beam continuously to compensate for power differences. Mirrors $M_{7\text{-}1}$ and $M_{7\text{-}2}$ converge the two beams through a single cuvette in the sample compartment. The time-separated sample, reference, and zero signals are compared.

Double-wavelength spectrophotometry provides information from two wavelengths per unit time. All other factors being equal, the resultant data should be more useful than data from a double-beam absorbance measurement. This is the fundamental principle underlying the application of double-wavelength spectro-

FIGURE **6.22** Optical schematic of a double-wavelength spectrophotometer. (Courtesy of Perkin-Elmer Corp.)

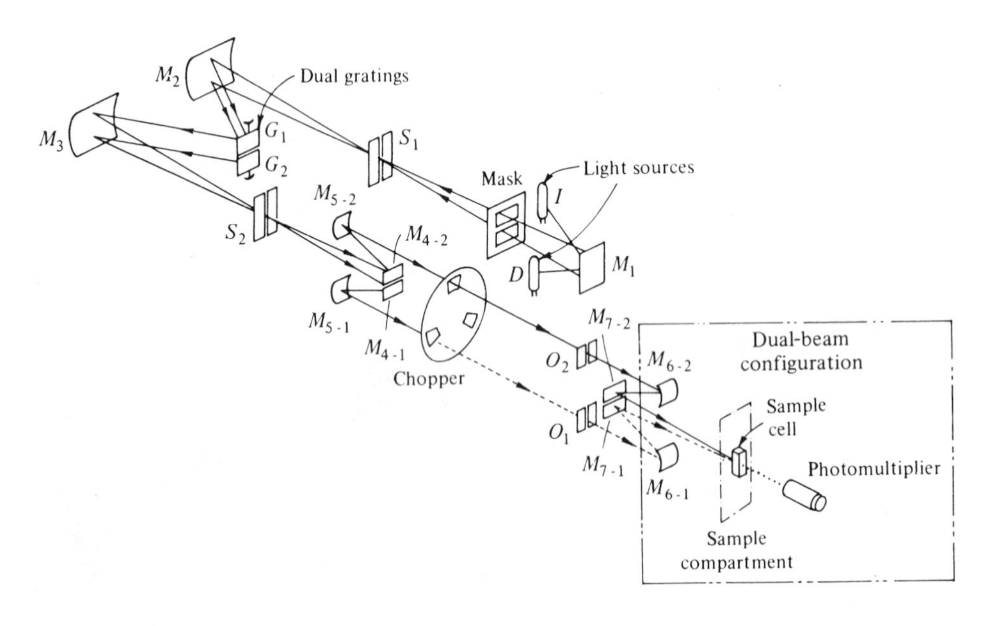

photometry. The measurement compensates for the presence of one parameter, be it an interfering impurity, a scattering sample, or an indistinct shoulder on the side of an absorption band.

The sample and reference beams may be set at different wavelengths (fixed) chosen such that the absorbance of an interfering component at both wavelengths is the same. This mode is used when the interfering substance strongly overlaps the analyte or when the effect of turbidity must be minimized.

The output of two fixed, but different, wavelengths may be measured independently. This mode is particularly suitable in reaction kinetic studies where the absorbance changes of two species can be monitored simultaneously. When compounds of protein or nucleic acid origin are studied, the relative absorbances at 254 and 280 nm are often used. Proteins absorb more strongly at 280 nm, and nucleic acids absorb more strongly at 254 nm.

In a third mode, the reference beam is fixed and the sample beam scans. This mode is used to determine the spectral characteristics of a highly turbid sample. Radiation is scattered in a random manner by turbidities. Both wavelengths suffer scattering to about the same extent and the difference between the two beams largely eliminates the effect of scattered radiation.

Derivative absorption measurements can be made by scanning with the two monochromators operating with a fixed, small wavelength difference between them. This mode is useful when the analyte in a two-component system absorbs on the side of the absorption band of the interfering component. Derivative spectrophotometry is discussed in Chapter 7.

Diode-Array, Reversed-Optics Spectrophotometer.[17] In the reversed-optics technique all wavelengths pass through the sample. In the HP 8450A instrument shown in Figure 6.23, the beam of radiation from the visible or ultraviolet lamp is directed through one of the five sample positions by a beam director mirror, returned by

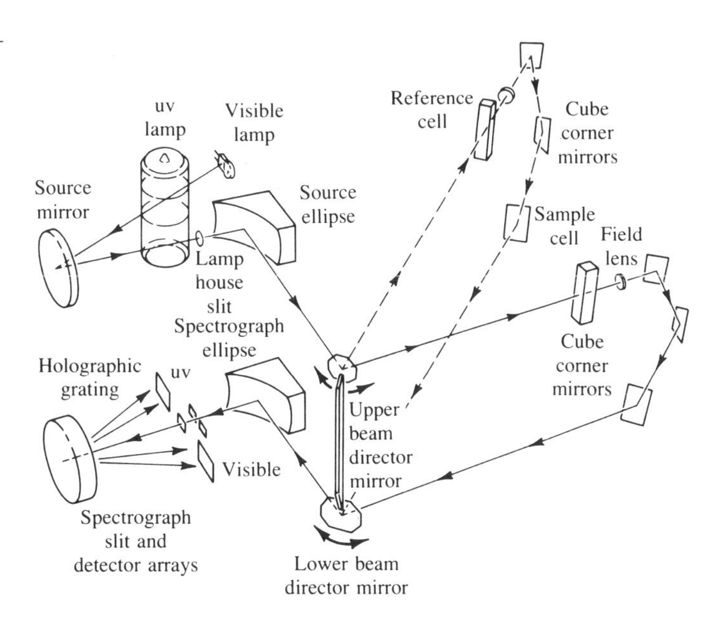

FIGURE **6.23**
Schematic diagram of a rapid scanning spectrophotometer (Hewlett-Packard HP 8450A) using the reversed-optics configuration. (Courtesy of Hewlett-Packard.)

means of a set of corner mirrors to another section of the beam director mirror, and then projected onto the entrance slit of the monochromator. A precision servomechanism controls the position of the beam director to better than 3 sec of arc and can switch from one sample position to the next in less than 146 msec. From two to all five of the cell positions can be used. The microcomputer remembers which sample positions are in use and directs the beam director accordingly. The beam that enters the slit of the monochromator is exactly centered on the 0.06-mm slit by the beam director by feedback from light reflected by the slit edges onto two photodiodes, one receiving light reflected from one slit edge and the other from the other slit edge.

The radiation that passes through the slit is dispersed by two holographic gratings on one substrate, giving two independent spectra that are focused on two photodiode arrays, one for the visible region and the other for the ultraviolet region. Each diode array consists of 211 diodes (but only 201 are used) covering the ranges of 200–400 nm in the ultraviolet region and 400–800 nm in the visible region; thus each diode in the uv range covers 1 nm and in the visible region, 2 nm. During the lamp turn-on procedure the gain of each individual diode is set at one of 16 values so that the analog-to-digital converter output does not exceed $\frac{15}{16}$ of its total range when converting the diode signal to digital values for storage and measurement.

The instrument passes radiation through the sample and reference cells twice in each 1-sec measurement period and also measures the dark current of each detector at least once each second. To compensate for any inequalities in sample cells, all cells should be filled with solvent and a balance measurement made. The instrument then measures and remembers the balance spectrum and uses this to correct subsequent measurements. The absorbance is calculated as follows:

$$A = -\log\left(\frac{S - D}{R - D} - B\right)$$

where S and R are the radiant powers that reach the detector after passing through the sample and reference solutions, respectively; D is the dark current, and B is the balance correction. The microcomputer allows storage of up to seven complete spectra or 12 partial spectra of standard substances and multicomponent analyses of an unknown that contains up to 12 components by adding together the individual standard spectra to produce the best match of the unknown spectrum using a regression analysis technique.

Results can be displayed on a video screen, printed, or recorded by an external computer or disk in many ways—for example, transmission, absorbance, first or second derivatives with respect to wavelength or time, concentration, smoothing, and statistical analyses. In addition attachments are available for reflectance measurements, autosampling, liquid chromatography flow-through cells, large path cells, and temperature-controlled cells.

In a reversed-optics configuration, fluorescence is detected at its own wavelength rather than at the wavelength of sample absorption. Thus measurements of absorbance at an isolated electronic transition are accurate. There is a deviation of the baseline at the wavelength of fluorescence. Scattering by small particles is compensated by the multicomponent technique using as standards a solution of the scattering material and a solution of the material to be determined. Although the sample is exposed to the total, undispersed radiation, decomposition is minimized by the very short time required to make each measurement.

PROBLEMS

1. Assuming that the limits of the visible spectrum are approximately 380 and 700 nm, find the angular range of the first-order visible spectrum produced by a plane grating that has 900 grooves/mm with the light incident normally on the grating.

2. For each individual plane reflection grating, supply the missing information in the following table.

Grating	Grooves/mm	Wavelength (nm)	Reflection angle
A		600	26.4°
B	1300		8.1°
C	1300	300	
D		500	9.0°
E	590		18.0°
F		1.2 μm	8.5°

3. Calculate the thickness of the dielectric spacer required for individual interference filters whose nominal wavelength is to peak in the first order at these values: (a) 410 nm, (b) 460 nm, (c) 580 nm, and (d) 750 nm. Assume that the dielectric material is magnesium fluoride ($\eta = 1.38$).

4. An interference filter is peaked at 625.0 nm in the second order. What are the third-order and the first-order passbands?

5. A third-order, 500-nm interference filter is desired. (a) From what adjacent transmission bands must it be isolated? (b) What type of absorption filter is suitable? (c) How thick a dielectric layer is required if magnesium fluoride ($\eta = 1.38$) is the dielectric material?

6. If magnesium fluoride ($\eta = 1.38$) is applied as a film to a glass mirror ($\eta = 1.52$ for the glass), to what value is the reflectance in air reduced at the wavelength of green light (525 nm)? More sophisticated multilayer coatings form a beamsplitter. A film of zinc sulfide ($\eta = 2.37$) applied to the glass raises the reflectance and transmittance to what values?

7. Assume that the image of the entrance slit is one-half the size of the exit slit in case A. What is the slit distribution function? In case B, the image size is twice the exit slit size. What slit distribution function is now obtained? Finally, what is the slit distribution function if either the image width approaches zero for a finite exit slit size, or the exit slit width approaches zero for a finite image size? For the last model, ignore the increase in diffraction that occurs as the slits are narrowed.

8. The yellow sodium line at 589.3 nm is actually a doublet of 0.59-nm peak separation. (a) What is the minimum number of lines that a grating must have to resolve this doublet in the first order and in the fourth order? (b) What must be the spectral bandpass to achieve baseline resolution? (c) What slit width is necessary if the monochromator has a reciprocal linear dispersion of 1.6 nm/mm in the first order?

9. Assuming ideal characteristics, would it be feasible to use an interference filter with a bandwidth of 8.0 nm to isolate the chromium emission line at 357.9 nm from the iron emission line at 372.0 nm?

10. The specifications of a commercial spectrophotometer are diffraction grating with 600 grooves/mm and focal length 330 mm. (a) What is the reciprocal linear dispersion for light of 500 nm? (b) Could the two emission lines of hydrogen, 656.28 and 656.10 nm, be resolved using a slit width of 0.010 mm?

11. Show mathematically why double-beam methods do not eliminate errors caused by improper dark current compensation. How would an extraneous factor have to affect the P/P_0 ratio if it can be eliminated by double-beam operation?

BIBLIOGRAPHY

ALTERNOSE, I. R, et al., "Evolution of Instrumentation for UV-Visible Spectrophotometry," F. A. Settle, ed., *J. Chem. Educ.*, **63,** A216, A262 (1986).

BAIR, E. J., *Introduction to Chemical Instrumentation,* McGraw-Hill, New York, 1962.

HARTLEY, M. C., *Diffraction Gratings,* Academic, London, 1982.

HOWELL, J. A., AND L. G. HARGIS, "Ultraviolet and Light Absorption Spectrometry," *Anal. Chem.,* **58,** 108R (1986).

MEEHAN, E. J., in *Treatise on Analytical Chemistry,* 2nd ed., P. J. Elving, E. J. Meehan, and I. M. Kolthoff, eds., Part I, Vol. 7, John Wiley, New York, 1981.

OLSEN, E. D., *Modern Optical Methods of Analysis,* McGraw-Hill, New York, 1975.

STROBEL, H. A., *Chemical Instrumentation,* 2nd ed., Addison-Wesley, Reading, MA, 1973.

LITERATURE CITED

1. HELL, A., *Anal. Chem.,* **43,** 79A (1971).
2. LEWIN, S. Z., "Luminous Gas Light Sources," *J. Chem. Educ.,* **42,** A165 (1965).
3. "Workshop on the Laser-Atomic Frontier," *Anal. Chem.,* **58,** 644A (1986).
4. ALMAN, D. H., AND F. W. BILLMEYER, JR., "A Review of Wavelength Calibration Methods for Visible-Range Photoelectric Spectrophotometers," *J. Chem. Educ.,* **52,** A281, A315 (1975).
5. BUC, G. L., AND E. I. STEARNS, *J. Opt. Soc. Am.,* **52,** 458 (1945).
6. HOGNESS, T, R., F. P. ZSCHEILE, JR., AND E. A. SIDWELL, JR., *J. Phys. Chem.,* **41,** 379 (1937).
7. BAUMEISTER, P., AND G. PINCUS, "Optical Interference Coatings," *Sci. Am.,* pp. 59–75 (December 1970).
8. FLAMAND, J., A. GRILLO, AND G. HAYAT, *Am. Lab.,* **7,** 47 (May 1975).
9. LYTLE, F. E., *Anal. Chem.,* **46,** 545A (1974).
10. TALMI, Y., "TV-Type Multichannel Detectors," *Anal. Chem.,* **47,** 658A, 697A (1975).
11. SACKS, R. D., "Emission Spectroscopy," Chap. 6, *Treatise on Analytical Chemistry,* 2nd ed., P. J. Elving, E. J. Meehan, and I. M. Kolthoff, eds., Part I, Vol. 7, John Wiley, New York, 1981.
12. JONES, D. C., *Anal. Chem.,* **57,** 1057A (1985).
13. LOTT, P. F., *J. Chem. Educ.,* **45,** A185, A273 (1968).
14. LOTT, P. F., *J. Chem. Educ.,* **45,** A89, A169 (1968).
15. PORRO, T. J., *Anal. Chem.,* **44,** 93A (1972).
16. SELLERS, R. L., G. W. LOWRY, AND R. W. KANE, *Am. Lab.,* **5,** 61 (March 1973).
17. WILLIS, B. G., D. A. FUSTIER, AND E. J. BONELLI, *Am. Lab.,* **13,** 62 (June 1981).

7 ULTRAVIOLET AND VISIBLE ABSORPTION METHODS

Analytical applications of the absorption of radiation by matter can be either qualitative or quantitative. The qualitative applications of absorption spectrometry depend on the fact that a given molecular species absorbs radiation only in specific regions of the spectrum where the radiation has the energy required to raise the molecules to some excited state. A display of absorption versus wavelength (or frequency) is called an absorption spectrum of that molecular species and serves as a "fingerprint" for identification. A brief discussion of the qualitative aspects of absorption spectrophotometry is found in the last section of this chapter and, in more detail, in Chapter 11 on infrared spectrophotometry. The quantitative aspects are considered first.

7.1 FUNDAMENTAL LAWS OF PHOTOMETRY

The radiant power of a beam of radiation is proportional to the number of photons per unit time. Absorption occurs when a photon collides with a molecule and raises that molecule to some excited state. Each molecule can be thought of as having a cross-sectional area for photon capture, and photons must pass within this area to interact with the molecule. The cross-sectional area varies with wavelength and represents, in effect, a probability that photons will be captured by any given molecule. The rate of absorption as a beam of photons passes through a medium depends on the number of photon collisions with absorbing atoms or molecules per unit time. If the number of absorbing molecules is doubled, by doubling either the length of the path of radiation through the medium or the concentration of absorbing species, the rate of absorption of photons doubles. Likewise, doubling the beam power doubles the number of photons that pass through the medium in unit time and doubles the number of collisions with absorbing molecules in unit time when the number of absorbing molecules remains constant.

If a parallel beam of monochromatic radiation of radiant power, P_0, traverses an infinitesimally small distance, dx, of an absorber, the decrease in power, $-dP$, is given by Equation 7.1, since the number of absorbing species is proportional to the thickness but is independent of P:

$$-dP = k'P\,dx \tag{7.1}$$

The proportionality constant, k', depends on the wavelength of the radiation. The concentration, C, is assumed constant. Separating the variables in Eq. 7.1 gives

$$\frac{-dP}{P} = -d(\ln P) = k'\,dx \tag{7.2}$$

which is a mathematical statement of the fact that the radiant power absorbed is proportional to the thickness traversed. Now if it is stipulated that P_0 is the radiant power at $x = 0$ and that P represents the radiant power of the transmitted (unabsorbed) radiation that emerges from the absorbing medium at $x = b$, Equation 7.2 can be integrated along the entire path:

$$-\int_{P_0}^{P} d\ln P = k' \int_0^b dx \tag{7.3}$$

obtaining

$$\ln P_0 - \ln P = \ln\left(\frac{P_0}{P}\right) = k'b \tag{7.4}$$

Equation 7.4 is known as Lambert's law (or Bouguer's law, since Bouguer really established this relationship several years before Lambert, but Bouguer's publication was not generally known) and states simply that, for parallel, monochromatic radiation that passes through an absorber of constant concentration, the radiant power decreases logarithmically as the path length increases arithmetically.

The dependence of radiant power on the concentration of absorbing species can be developed in a parallel manner if the wavelength and the distance traversed by the beam in the sample remain constant. In this case the number of absorbing molecules that collide with photons is proportional to the concentration, C. Thus

$$-dP = k''P\,dC \tag{7.5}$$

and separation of variables followed by integration from $C = 0$ to $C = C$ yields

$$\ln\frac{P_0}{P} = k''C \tag{7.6}$$

This relationship is known as Beer's law. If both concentration and thickness are variable, the combined Lambert-Beer law (often known simply as Beer's law) becomes

$$\ln\frac{P_0}{P} = kbC \tag{7.7}$$

Replacing natural logarithms by base-10 logarithms and calling the new constant a in accordance with accepted practice give

$$\log\frac{P_0}{P} = abC \tag{7.8}$$

Absorbance, A, is defined as

$$A = \log\frac{P_0}{P} = abC \tag{7.9}$$

Transmittance, T, is defined as

$$T = \frac{P}{P_0} \tag{7.10}$$

so that

$$A = \log \frac{1}{T} = -\log T \tag{7.11}$$

Percent transmittance is merely $100T$.

The proportionality constant a in Equations 7.8 and 7.9 is known as the absorptivity if C is given in grams of absorbing material per liter and b is in centimeters; therefore, a has the units liter g^{-1} cm^{-1}. If concentration is expressed in molar concentration and b is in centimeters, the proportionality constant is called the molar absorptivity and is designated as ε. Thus the Lambert-Beer law can also be written as

$$A = \log \frac{P_0}{P} = \varepsilon b C \tag{7.12}$$

where ε is in units of liter $mole^{-1}$ cm^{-1}.

As radiation passes through a cell (reference or sample), some radiation is reflected at each surface where there is a change in the refractive index. Instead of measuring P_0 as the radiant power that impinges on the cell, reflection losses can be almost entirely compensated by taking P_0 as the radiant power transmitted through a cell that contains only pure solvent. The sample to be measured should be placed in a cell as nearly identical as possible. Cells should be free of scratches, dirt, and fingerprints, all of which scatter radiation. Turbidities in the solvent or the sample also scatter radiation.

A plot of absorbance versus concentration should be a straight line passing through the origin, as shown in Figure 7.1. Readout scales and meter scales on spectrophotometers are usually calibrated to read absorbance as well as transmittance.

Absorption of radiation by molecules at specific wavelengths is frequently used for quantitative analyses owing to the direct relationship between absorbance and concentration described earlier. The sensitivity of spectrometric analysis is dictated by the magnitude of the absorptivity and the minimum absorbance that can be measured with the required degree of certainty.

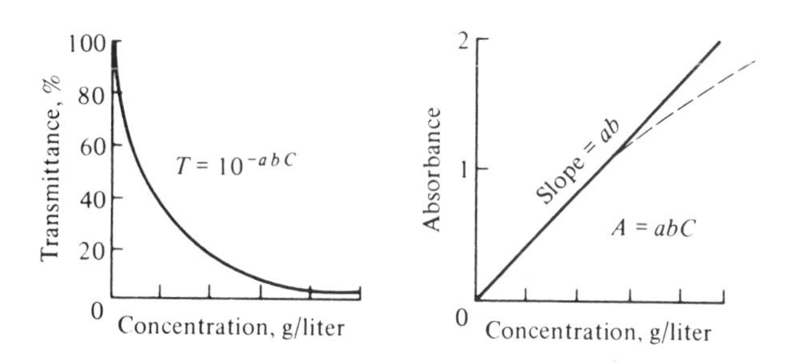

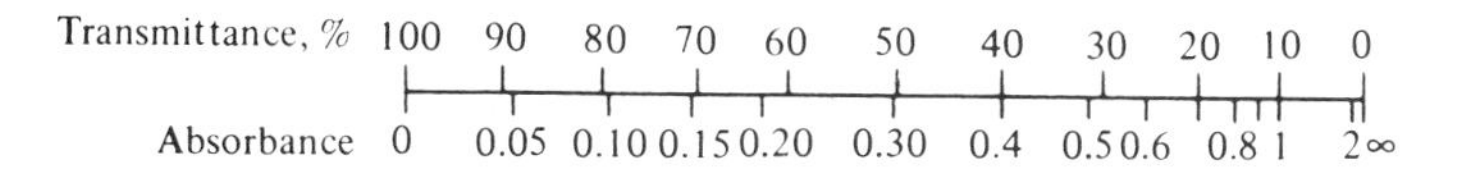

FIGURE 7.1
Representation of Beer's law and comparison between scales in absorbance and transmittance.

EXAMPLE 7.1

If the molar absorptivity for iron(II)-1,10-phenanthroline complex is 12,000 liter mol^{-1} cm^{-1} and the minimum detectable absorbance is 0.001, then, for a 1.00-cm path length, the minimum molar concentration that can be detected is

$$C = \frac{A}{\varepsilon b} = \frac{0.001}{(12{,}000 \text{ liter mol}^{-1}\text{ cm}^{-1})(1.00\text{ cm})} = 1.20 \times 10^{-7}M$$

Deviations from Beer's Law

Deviations from Beer's law fall into three categories: real, instrumental, and chemical. *Real deviations* arise from changes in the refractive index of the analytical system. Kortum and Seiler point out that Beer's law strictly applies only at low concentrations.[1] It is not absorptivity that is constant and independent of concentration but the expression

$$a = a_{\text{true}} \frac{\eta}{(\eta^2 + 2)^2} \tag{7.13}$$

where η is the refractive index of the solution. At concentrations of $10^{-3}M$ or less, the refractive index is essentially constant, but at high concentrations, the refractive index varies considerably and so does absorptivity. This does not rule out quantitative analyses at high concentrations, since bracketing the unknown by standard solutions or the use of a calibration curve can provide sufficient accuracy. The refractive index effect may be encountered in high-absorbance differential spectrophotometry, to be discussed later in this chapter.

Instrumental deviations arise primarily from the finite bandpass of filters or monochromators. The derivation of Beer's law assumed monochromatic radiation, but truly monochromatic radiation is approached in only specialized line emission sources. If absorptivity is essentially constant over the instrumental bandpass, then Beer's law is followed within close limits.

Actually a system appears to follow Beer's law if the *rate of change* of absorptivity versus wavelength is constant over the wavelength interval passed by the wavelength selector provided that the nominal wavelength setting is very reproducible. Thus, in Figure 7.2, the average absorptivity on the side of the absorption band is nearly independent of the bandpass at a constant nominal wavelength, whereas at the peak of the absorption band the average absorptivity decreases as the bandpass increases. On the other hand, since absorptivity versus wavelength changes rapidly with wavelength on the sides of an absorption band but much less rapidly at the peak, measurements at the peak are less sensitive to the nominal wavelength setting. The sensitivity to concentration is, of course, greatest at the peak of an absorption band because the absorptivity is a maximum at this point. Thus, for most quantitative determinations the peak wavelength is selected and the instrumental controls are not changed during the calibration and measurement or, alternatively, the spectra of the standards and unknowns are recorded in the region of the maximum with constant slit width and scanning speed and the absorbances at the maxima are read from the curve.

Departure from Beer's law is most serious for wide bandpasses and narrow absorption bands, and is less significant for broad bands and narrow bandpasses. It is the ratio between the spectral slit width and the bandwidth of the absorption band that is most important. Often the deviation from Beer's law becomes evident at higher concentrations—that is, at higher absorbances. On a plot of absorbance versus concentration, the curve bends toward the concentration axis. This lack of adherence to Beer's law in the negative direction, the broken line in Figure 7.1, is undesirable because of the rather large increase in the relative concentration error.

Chemical deviations from Beer's law are caused by shifts in the position of a chemical or physical equilibrium involving the absorbing species. Consider, for example, the following equilibria:

$$2CrO_4^{2-} + 2H^+ \rightleftarrows 2HCrO_4^- \rightleftarrows Cr_2O_7^{2-} + H_2O \tag{7.14}$$

The dichromate ion absorbs in the visible region at 450 nm. Upon diluting a dichromate solution, the equilibrium shifts to the left. The equilibrium can be controlled by converting all the chromium to CrO_4^{2-} by making the solution 0.05*M* in potassium hydroxide. Beer's law is then followed. Chromium should not be expected to follow Beer's law in a highly acidic solution because the dimerization step is dependent on concentration even at a constant pH. In general, if an absorbing species is involved in a simple acid-base equilibrium, Beer's law fails unless the pH and ionic strength are kept constant. In such situations the pH should be adjusted to at least three units more or less than the pK_a value of the monoprotic acid. Alternatively, the wavelength corresponding to an *isosbestic* (sometimes called isobestic) point can be used (Figure 7.3). An isosbestic point is a wavelength where the molar absorptivity is the same for two materials that are interconvertible. When the absorbing species is a complex ion, the concentration of free ligand must be constant and large in comparison to the amount of the analyte present, usually a 100-fold excess.

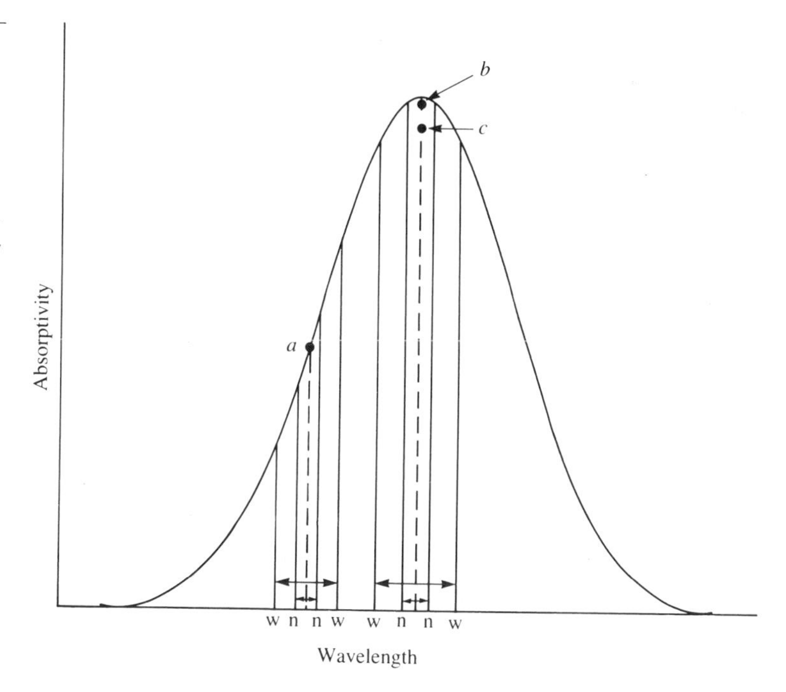

FIGURE 7.2
Changes in average absorptivity with narrow (*n-n*) and wide (*w-w*) wavelength invals. No apparent change is observed where the curve is linear (*a*). Average absorptivity decreases for wide band (*c*) as compared with narrow band (*b*) where the slope of the curve changes rapidly.

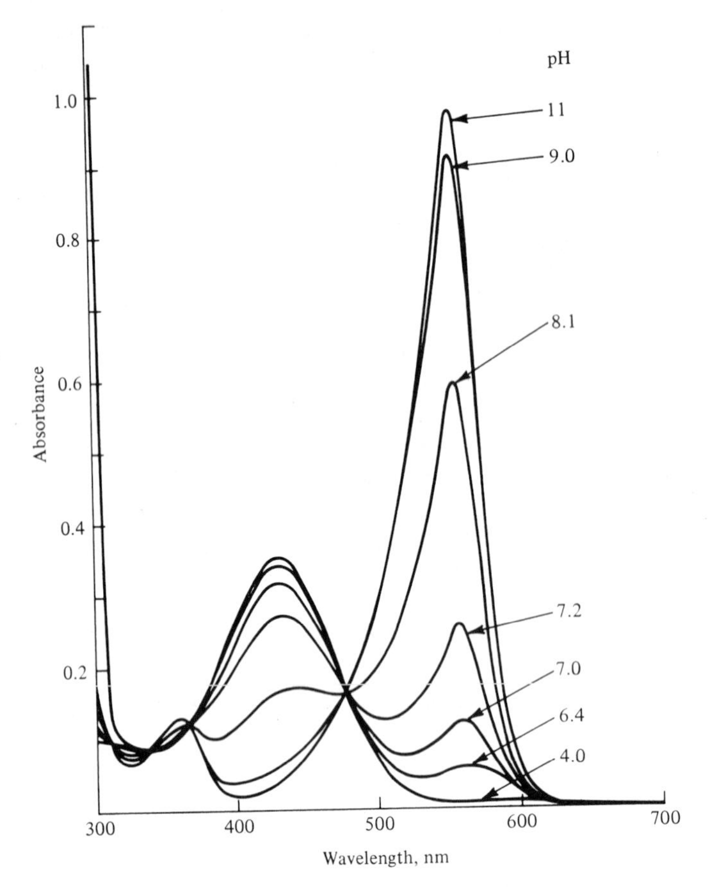

FIGURE 7.3
Chemical equilibrium between two solution components, the conversion of phenol red ($pK_a = 7.9$) from the yellow (acidic) to the red (basic) form. Absorption maxima are at 433 and 558 nm, respectively, for the acidic and basic forms. Isosbestic points are recorded at 338, 367, and 480 nm.

7.2 SPECTROPHOTOMETRIC ACCURACY

Accuracy means the nearness of a measurement to its true value. *Precision*, discussed in the following section, describes the reproducibility of results. A third term, *photometric linearity*, is defined as the ability of a photometric system to yield a linear relationship between the radiant power incident upon its detector and the readout.

Accuracy and linearity often are confused because an instrument with high photometric accuracy must also have good photometric linearity; however, accuracy and linearity are not synonymous. Linearity, at its simplest, is based on the additivity of empirical (not absolute) absorbance values. Absolute accuracy is based on the relationship between the measured transmittance and some absolute change in radiant power at the detector. Accuracy generally is determined by measuring the radiant energy flux through calibrated apertures mounted in the sample beam.

Numerous parameters affect photometric accuracy.[2] The spectral bandwidth selected and the observed bandwidth of the recorded absorption band are directly related. If the slits are too wide, the absorbance peak height is depressed (see above and Figure 7.2) and the observed bandwidth is greater than the natural bandwidth, giving erroneous absorbance values. When the spectral bandwidth is 10% or less of the natural bandwidth, the ratio of the observed peak height to the true peak height is at least 99.5%. Thus, if possible, the bandwidth should be decreased until there is no further change in the recorded spectra, as was shown in Figure 6.7. However, to maintain the best signal-to-noise ratio, the slits should not be narrowed more than necessary. If one is working with wide-band instruments, it is possible that a spectral bandwidth cannot be selected that fully resolves the absorption bands.

Too rapid a scan is another source of error when using scanning spectrophotometers. When the scan rate is too fast for a given spectral bandwidth and pen response period (the time required for the pen to traverse full scale), the recorded absorbance spectrum is shifted in the direction of the scan and the band is broadened and depressed. To obtain a "true" spectrum, the scan speed must be operated no faster than 0.4 of the natural bandwidth per pen response period.

A third source of error is wavelength inaccuracy. The degree of error varies with the magnitude of the wavelength inaccuracy and with the natural bandwidth. Wavelength calibration standards are available. Didymium glass is often used. It has sharp, multiple absorption bands in the 400–800-nm region. The emission lines of a mercury arc in quartz are also useful in the region 200–1100 nm.

Stray radiation can be a major problem.[3,4] In general, stray radiation becomes serious at the extreme ends of an instrument's spectral range where the photodetector sensitivity is low or the radiation beam that reaches the detector is weak. Principal sources of stray radiation are scattering by dust or smudges on the optical surfaces; optical surface imperfections of mirrors, lenses, and cuvette walls; scattering by diffraction at the slit edges (a natural phenomenon that occurs no matter how sharp the slit edges are made); reflections from interior surfaces of the monochromator that can be attenuated by the proper positioning of matte black baffle surfaces; and scattering by a grating that acts as a simple mirror instead of a perfect diffracting surface. As a result of any or all of these, a small fraction of undispersed radiant energy passes through the exit slit of the monochromator and reaches the detector. The absolute intensity of this stray radiation is constant at a given wavelength at any given time, but it can increase as the instrument ages and the optical materials in the monochromator deteriorate. The measurement of the amount of stray radiation in a spectrophotometer is discussed by Kaye[5,6] and Sharpe.[7] Kaye proposes two methods to measure the stray radiant power more accurately than the ASTM (American Society for Testing Materials) method.[8] The easiest method uses a blocking filter and a sample with a narrow absorption band and high absorbance. For details, see the references by Kaye.

Stray radiation affects photometric accuracy by causing apparent deviations from Beer's law. Positive deviations occur if the stray radiation is absorbed, and negative deviations occur if it is not absorbed. The latter is usually the case, and observed absorbances are reduced markedly as the stray radiation increases and the absorbance of the solution increases. The apparent absorbance, assuming that none of the stray radiation, P_S, is absorbed by the sample or reference material, is given by

$$A_{\text{apparent}} = \log\left(\frac{P_0 + P_S}{P + P_S}\right) \tag{7.15}$$

EXAMPLE 7.2

If an instrument has stray radiation of 1% at a particular wavelength, the instrument can never read more than two absorbance units regardless of the sample concentration, since from Equation 7.15:

$$A_{\text{apparent}} = \log\left(\frac{P_0 + 0.01P_0}{0 + 0.01P_0}\right) = \log\left(\frac{1.01}{0.01}\right) \approx 2$$

With a double monochromator it is possible to achieve stray radiation levels of less than 10^{-6} relative to the true signal. In the past, low-cost spectrophotometers were usually limited to a dynamic range of 0.001 to 2.000 absorbance units. However, recent improvements in stray radiation filters and circuit design extend this range to 3.000 absorbance units. The ability to measure high absorbance with good accuracy helps to increase sample throughput because it avoids the use of calibration curves and reduces the number of reruns necessitated by samples that exceed the upper limit of the instrument's range.

The attainment of low levels of stray radiation does not in itself guarantee good accuracy. In addition, it is necessary to adjust the spectrophotometer reading very accurately to 0% transmittance when the beam is blocked at the detector. The benefits of stray radiation levels as low as 0.01% cannot be realized unless the error of the 0% T adjustment is 0.0005% T or less.

The lower limit of the dynamic range of a spectrophotometer is determined by instrument stability and noise, which are primarily dependent on the electromechanical stability of the entire system, including the optics as well as the electronics. Very rigid mounts for optical components and the minimization of circuit susceptibility to ambient temperature and line voltage fluctuations permit measurements of low-absorbance solutions with scale expansion of the absorbance signal.

7.3 PHOTOMETRIC PRECISION

The ultimate precision of a photometric measurement is determined by instrumental noise—that is, the statistical fluctuations of the signal that reaches the detector, which in turn is a function of the type of detector used. A quantitative analysis should be conducted within the range of transmittance for which a given uncertainty in transmittance, ΔT, or the equivalent quantity, ΔP (the noise associated with the radiant energy reaching the detector), causes the least uncertainty in concentration, ΔC. Instruments with detectors subject only to Johnson noise, noise independent of radiant power, display optimum precision in one range, and spectrophotometers with photoemissive detectors limited by shot noise perform best in a somewhat different concentration range.

Uncertainty of the transmittance setting for most instruments is on the order of 0.01–0.002 of the total scale; the latter value is considered a practical limit in ordinary work. The actual value of ΔT is ascertained by measuring the transmittance of perhaps 30 portions of a solution and calculating the standard deviation of the measurements. Each measurement should include the operations of emptying, refilling, and repositioning the cuvette in its holder. Usually ΔT is taken as twice the average standard deviation of the replicate readings in order to include the uncertainty involved in setting the scale to 0% T with the beam of radiation occluded and to 100% T with the pure solvent. With modern spectrophotometric instruments capable of scale expansion or with digital readout devices, the influence of reading error on the precision of the readout signal is considerably lessened or made completely negligible with respect to noise.

Relative Concentration Error

The two extremes to be considered in determining the relative concentration error of photometric measurements are (1) the uncertainty in determining P—namely, δP—is independent of P, and (2) the uncertainty that δP varies with P because of shot noise, the statistical variation in the number of photons that reach the detector. The latter is the more usual situation in modern uv-visible photometers that use photomultiplier tubes with low radiant powers incident on the photomultiplier tube. The former case arises in older instruments with higher radiant powers, where the photocurrent is proportional to P, the output is measured on a scale linear in P, and the uncertainty in measurement arises primarily from reading the scale.

The derivation of the relative concentration error for the first case above is as follows. Assuming that Beer's law is obeyed, rearranging the expression for the law gives

$$C = \frac{A}{ab} = \frac{1}{ab} \log \frac{P_0}{P} = \frac{-\log T}{ab} \tag{7.16}$$

Differentiation of Equation 7.16 gives

$$dC = \frac{-0.434}{ab}\left(\frac{dP}{P}\right) \quad \text{or} \quad \frac{dC}{dT} = \frac{-0.434}{Tab} \tag{7.17}$$

Replacing the quantity ab by its equivalent from Beer's law and rearranging give

$$\frac{dC}{C} = \frac{-0.434}{A}\left(\frac{dP}{P}\right) \quad \text{or} \quad \frac{dC}{C} = \frac{0.434}{\log T}\left(\frac{dT}{T}\right) \tag{7.18}$$

Thus, the relative concentration error, dC/C, depends inversely on the product of the absorbance and transmitted radiant power.

The transmittance at which the propagation of error is smallest is found by differentiating Equation 7.18 and setting the derivative equal to zero:

$$\frac{d}{dP}\left(\frac{dC}{C}\right) = \frac{-dT}{2.30}\frac{d}{dP}\left(P \log \frac{P_0}{P}\right)^{-1} = \frac{dT}{2.30}\left(P \log \frac{P_0}{P}\right)^{-2}\left(\log \frac{P_0}{P} - 0.434\right) = 0 \tag{7.19}$$

This yields the nontrivial solution

$$\log \frac{P_0}{P} = 0.434 = A \quad \text{or} \quad T = 0.368$$

The minimum error then becomes

$$\frac{dC}{C} = \frac{-0.434}{(0.368)(0.434)} dT = -2.72\, dT \tag{7.20}$$

Equation 7.20 is strictly true only when differentials are involved. Thus, it is reasonable to say that a 0.1% error in transmittance produces a 0.27% error in sample concentration. A plot of the relative concentration error (and also the error due to path length) as a function of absorbance (Figure 7.4) is rather flat between an absorbance of 0.3 and 0.6 so that careful adjustment of the solution concentration to read 36.8% $T(A = 0.434)$ is of little value.

The minimum relative concentration error for most modern uv-visible photometers where shot noise predominates, the second case above, is determined in a similar manner. The noise (dP or dT) is proportional to the square root of the radiant power—that is, $dT = k\sqrt{P}$. Replacement of dT by $k\sqrt{P}$ in Equation 7.19 yields

$$\frac{dC}{C} = -0.434k\left(\sqrt{P}\log\frac{P_0}{P}\right)^{-1} \tag{7.21}$$

Differentiation of Equation 7.21 gives

$$\frac{d}{dP}\left(\frac{dC}{C}\right) = 0.434k\left(\sqrt{P}\log\frac{P_0}{P}\right)^{-2}\left[\frac{1}{2}\left(\frac{1}{\sqrt{P}}\log\frac{P_0}{P}\right)\frac{-\sqrt{P}}{P}0.434\right] \tag{7.22}$$

The minimum for this function is when

$$\log\frac{P_0}{P} = 2(0.434) = 0.868 = A \quad \text{or} \quad T = 0.135$$

For detectors where shot noise predominates, the region of minimum error extends over a very wide range of absorbance values from 0.3 to 2.0 ($T = 0.5$ to 0.01), as shown in Figure 7.4. Thus more concentrated solutions may be used, which, in turn, reduces errors in solution preparation and cuvette matching. Clean windows and scratch-free cuvette faces are less significant.

The foregoing derivations apply to single-beam spectrophotometers. The noise level for double-beam instruments is the quadratic sum of the noise levels of both the sample beam and the reference beam. The calculations for a double-beam instrument therefore are more complex. The final results are identical, however.

Sample Handling

Since all cells (cuvettes) have slight imperfections, reflection and scattering losses change as different parts of the cuvette face are exposed to the beam of radiation. Therefore, it is important to reposition a cuvette as precisely as possible when duplicating an analysis.

To clean cuvettes (and other optical surfaces), lens paper soaked in spectrograde methanol, which is held by a hemostat or a similar device, should be used. When cuvette faces are cleaned in this manner a film of methanol is left, which quickly evaporates to leave the faces free of contaminants. If maximum precision is required

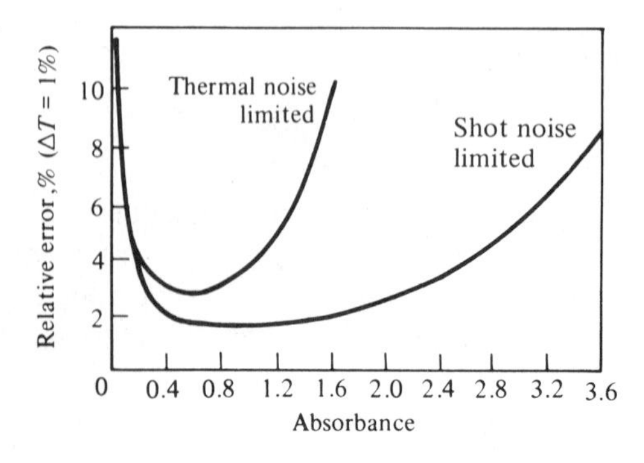

FIGURE 7.4
Relative concentration error, $\Delta C/C$, in percent, for a constant transmittance error.

when duplicating photometric measurements, the best method is to leave the cuvette in place and change the solution by means of a syringe. Reproducibility is almost twice as good as that obtained when removing the cuvette. With instrumentation capable of precisely reading a 0.0001 absorbance unit change, very small differences are measurable.

7.4 QUANTITATIVE METHODOLOGY

In developing a quantitative method for determining an unknown concentration of a given species by absorption spectrophotometry, the first step is to choose the absorption band at which absorbance measurements are to be made. An ultraviolet–visible absorption spectrum of the species to be determined is obtained either from the literature or experimentally, preferably by means of a scanning double-beam spectrophotometer. From inspection of the absorption spectrum, a suitable absorption band is selected. Absorptivity at any given wavelength is constant and an inherent characteristic of the absorbing species. The path length is made constant by using carefully matched cuvettes.

The numerical value of the absorptivity determines the slope of the analytical curve and influences the concentration range over which determinations can be made. When there are several absorption bands of suitable absorptivity, the band selected should favor wavelength regions that correspond to a relatively high output of the radiation source and high spectral sensitivity of the photometer. The absorption band should not overlap absorption bands of the solvent or possible contaminants, including excess reagents that might be in the sample.

Although many organic compounds absorb quite strongly, only a limited number of inorganic ions do, and it is the normal procedure of inorganic absorption spectrophotometry to add a reagent species to the solution of the inorganic ion that reacts with it and, in the process, bring about a marked change in the spectral absorption characteristics of the reagent. The formation of metal-inorganic complexes is well known. For example, since the iron(II) ion is very weakly colored, a complexing agent, 1,10-phenanthroline, is added to form a highly colored complex that is suitable for the determination of very small amounts of iron. A few moments' reflection also brings to mind possibilities among organic compounds. For example, although alcohols have no absorption spectra between 200 and 1000 nm, the treatment of an alcohol with phenyl isocyanate yields the corresponding phenyl alkyl carbamate, which absorbs at about 280 nm. Semicarbazones display maxima that are shifted to longer wavelengths by 30–40 nm, with an average increase in molar absorptivity of 10,000 compared with the original carbonyl compound. Conversely, the strong absorption of anthracene can be eliminated by a Diels-Alder reaction with 1,2-dicyanoethylene.

There are numerous compilations of spectrophotometric methods of analysis. Tables of recently proposed methods for metals, nonmetals, and organic compounds are found in the comprehensive reviews by Hargis and Howell.[9,10] Analysis of drugs is discussed by Sunshine[11] and organic compounds and drugs by Pesez and Bartos.[12]

Although very few reactions are specific for a particular substance, many reactions are quite selective, or can be rendered selective through the introduction of masking agents, control of pH, solvent extraction, adjustment of the oxidation state, or prior removal of interferents. Both the color-developing reagent and the absorbing product must be stable for a reasonable period of time. It is often necessary to specify that the color comparisons be made within a definite period of time, and it is always advisable to prepare standards and unknowns on a time schedule.

Solvents

Solvents used in spectrophotometry must meet certain requirements to assure successful and accurate results. The solvent chosen must dissolve the sample, yet be compatible with cuvette materials. Solubility data for common substances are available in reference works such as the *Merck Index* and *Lange's Handbook of Chemistry*. The solvent must also be relatively transparent in the spectral region of interest. To avoid poor resolution and difficulties in spectrum interpretation, a solvent should not be used for measurements near or below its ultraviolet cutoff—that is, the wavelength at which absorbance for the solvent alone approaches one absorbance unit. Ultraviolet cutoffs for solvents commonly used are given in Table 7.1.

Once a solvent is selected based on physical and spectral characteristics, its purity must be considered. The absorbance curve of a solvent, as supplied, should be "smooth"—that is, have no extraneous impurity peaks in the spectral region of interest. Solvents especially purified and certified for spectrophotometric use are available from suppliers.

Selection of Analytical Wavelength

When filter photometers are used, the proper filter can be selected while the calibration curve is prepared. A series of standard solutions is prepared, including a blank. Using one filter at a time, a series of calibration curves is plotted. The filter that permits the closest adherence to linearity over the widest absorbance interval and yields the largest slope, but with a small or zero intercept, is the best choice. Naturally, if a spectrophotometer is available, the wavelength of maximum absorbance, λ_{max}, is quickly ascertained from a wavelength scan, and this wavelength usually is the most suitable.

Simultaneous Spectrophotometric Determinations

When no region can be found free from overlapping spectra of two chromophores (groups that produce color in a compound), it is still possible to devise a method based on measurements at two or more wavelengths. Two dissimilar chromophores have different powers of radiation absorption at some or several points in their absorption spectra. If, therefore, measurements are made on an unknown solution at two wavelengths where the absorptivities of the two components are different, it is possible to set up two independent equations and solve them simultaneously for the two unknown concentrations. First, one should obtain spectra of each pure component and then select two wavelengths where the difference in molar absorptivities is maximal. For the system illustrated in Figure 7.5, $(\varepsilon_1/\varepsilon_2)_{\lambda 1}$ and $(\varepsilon_2/\varepsilon_1)_{\lambda 2}$ are maximal. Neither of these wavelengths need necessarily coincide with

TABLE 7.1

ULTRAVIOLET CUTOFFS OF SPECTRO-GRADE SOLVENTS (ABSORBANCE OF 1.00 IN A 10.0-mm CELL VS. DISTILLED WATER)

Solvent	Wavelength (nm)	Solvent	Wavelength (nm)
Acetic acid	260	Hexadecane	200
Acetone	330	Hexane	210
Acetonitrile	190	Isobutyl alcohol	230
Benzene	280	Methanol	210
1-Butanol	210	2-Methoxyethanol	210
2-Butanol	260	Methylcyclohexane	210
Butyl acetate	254	Methylene chloride	235
Carbon disulfide	380	Methyl ethyl ketone	330
Carbon tetrachloride	265	Methyl isobutyl ketone	335
1-Chlorobutane	220	2-Methyl-1-propanol	230
Chloroform (stabilized with ethanol)	245	*N*-Methylpyrrolidone	285
Cyclohexane	210	Nitromethane	380
1,2-Dichloroethane	226	Pentane	210
1,2-Dimethoxyethane	240	Pentyl acetate	212
N,*N*-Dimethylacetamide	268	1-Propanol	210
N,*N*-Dimethylformamide	270	2-Propanol	210
Dimethylsulfoxide	265	Pyridine	330
1,4-Dioxane	215	Tetrachloroethylene (stabilized with thymol)	290
Diethyl ether	218	Tetrahydrofuran	220
Ethanol	210	Toluene	286
2-Ethoxyethanol	210	1,1,2-Trichloro-1,2,2-trifluoroethane	231
Ethyl acetate	255	2,2,4-Trimethylpentane	215
Ethylene chloride	228	*o*-Xylene	290
Glycerol	207	Water	191
Heptane	197		

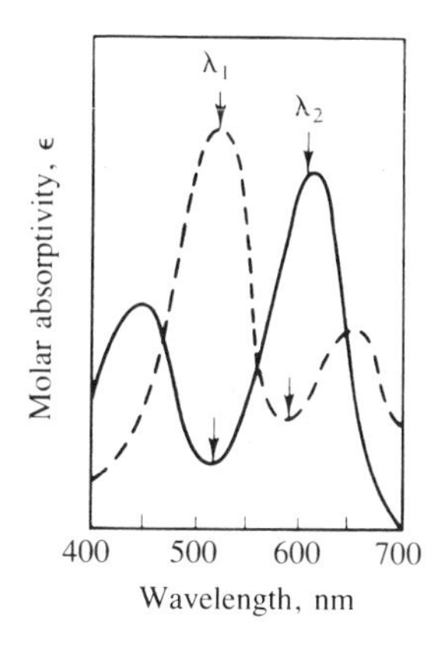

FIGURE 7.5

Simultaneous spectrophotometric analysis of a two-component system. Selection of analytical wavelengths is indicated by arrows.

an absorption maximum for either component. Now two simultaneous equations are written:

$$C_1(\varepsilon_1)_{\lambda 1} + C_2(\varepsilon_2)_{\lambda 1} = A_{\lambda 1} \tag{7.23}$$

$$C_1(\varepsilon_1)_{\lambda 2} + C_2(\varepsilon_2)_{\lambda 2} = A_{\lambda 2} \tag{7.24}$$

The equations are solved for the concentration of each component.

EXAMPLE 7.3

The absorbance of a 0.000100M solution of dye A in a 1.00-cm cell is 0.982 at 420 nm and 0.216 at 505 nm. The absorbance of a 0.000200M solution of dye B is 0.362 at 420 nm and 1.262 at 505 nm. The absorbance of a mixture of the two dyes is 0.820 at 420 nm and 0.908 at 505 nm. Using Equation 7.12, the value of ε_A at 420 nm is

$$(\varepsilon_A)_{420} = \frac{0.982}{1.000 \times 10^{-4}M \times 1 \text{ cm}} = 9820$$

Likewise, $(\varepsilon_A)_{505} = 2160$, $(\varepsilon_B)_{420} = 1310$, and $(\varepsilon_B)_{505} = 6310$. For the mixture, using Equations 7.23 and 7.24:

$$0.820 = 9820C_A + 1310C_B$$

$$0.908 = 2160C_A + 6310C_B$$

From the first equation above,

$$C_A = \frac{0.820 - 1310C_B}{9820}$$

and substituting this into the second equation gives

$$0.908 = 2160\left(\frac{0.820 - 1310C_B}{9820}\right) + 6310C_B$$

and $C_B = 0.0001208M$. Substituting this result into the equation for C_A gives $C_A = 0.0000674M$.

It is also possible, and in fact desirable if conditions permit, to make measurements at more wavelengths than there are unknowns. Such a set of equations is said to overdetermine the situation. The best values for the unknown concentrations are then determined by the method of least squares. In favorable situations it is possible to determine a dozen or more unknown concentrations provided only that wavelengths can be found that are sufficiently distinct for each component and that Beer's law holds for each component at each wavelength. Simultaneous determinations rest on the assumption that the total absorbance at each wavelength is the sum of the absorbances of each component.

Computer software programs can handle multicomponent mixtures and deviations from Beer's law. Options involving concentration calculations include linear least squares with forced zero intercept, a method that uses several stored standards and the origin to construct the calibration curve. Another option is ordinary least squares in which the computed calibration curve does not necessarily pass through the origin. Also available are second-order least-squares-best curves with or without a zero intercept (a method that handles deviations from Beer's law because of nonlinearity at high absorption) and multicomponent analysis of 12 or more components using stored standard curves to construct and fit a synthetic spectrum that best matches that of the unknown mixture. Caruana and colleagues have developed a fast algorithm for the resolution of overlapping spectral bands.[13]

7.5 DIFFERENTIAL OR EXPANDED-SCALE SPECTROSCOPY

In the *ordinary* spectrophotometric method, two adjustments are required before the actual measurement of standards and unknowns is made. First the zero point of the transmittance scale must be adjusted to read zero with no radiation reaching the detector. This is done by placing an occluder in the sample beam; the occluder, which may simply be a shutter, represents a completely opaque species. The second manipulation brings the adjustment of the 100% transmittance onto the scale by placing pure solvent in the sample beam and balancing the instrument to read 100% (or full scale). After these operations, the instrument is capable of measuring any radiant power that falls between total darkness and the power passing through the pure solvent. To complete the analysis, the transmittance of at least one solution of known concentration, but preferably more, is measured to establish the proportionality between absorbance and concentration from Beer's law; then the transmittance of the unknown solution is measured.

In the derivation of the relative concentration error, the incident radiant power, P_0, was considered a constant and was disregarded in finding the optimum conditions. Setting $(d/dP)(dC/C) = 0$ means that P_0 should be infinite. However, under ordinary spectrometric methods, the maximum value of P_0 is limited by the length of the potentiometer slidewire, since one end corresponds to zero radiant power and the other end to the full power at zero concentration, or P_0. This is an artificial requirement. The transmittance scale can also be calibrated—for example, by using two reference solutions that contain the absorbing species in different concentrations. Two neutral density filters would be equally suitable. The only condition is that one reference absorber must transmit more radiant energy than the sample to be measured and the other reference absorber must transmit less radiant energy than the sample. From these possible alterations of the ordinary method, three differential or scale-expansion techniques arise that can be used to increase precision: the high-absorbance method, the trace-analysis method, and the maximum-precision method.

In the *high-absorbance* method the dark current is still measured using a shutter to occlude the radiation beam while adjusting the scale to read 0% T. To make the 100% T adjustment, however, a finite reference solution replaces the pure solvent. The reference solution is more dilute than the unknown. For example, if the sample shows a 20% transmittance by the ordinary method and a standard reference solution reads 36%, the latter solution is used to set the instrument to read 100% T. A threefold scale expansion is accomplished, as illustrated in Figure 7.6(II). The unknown now reads 55.6% T relative to the reference solution, since $20/x = 36/100$ and $x = 55.6$.

To compensate for the lower amount of transmitted radiation that reaches the detector in the scale-expansion method, the instrument must have one or more provisions for reserve sensitivity. One approach is to increase the amplification of the detector output without, however, increasing the noise component of the signal. A second method involves increasing the slit width if such an increase is compatible with required spectral purity and the natural bandwidth of the absorption band being measured. Stray radiation in the instrument limits the usefulness of the high-absorbance method.

The artificial P_0 does increase the accuracy of the measurement and thus justifies the expanded scale that automatically results. The relative concentration error becomes finite at the high end of the transmittance scale. In fact, the error becomes dependent in a pronounced manner on the absorbance of the reference standard used to make the 100% T setting, as illustrated in Figure 7.7. The error function is given by

$$\frac{dC}{C} = \frac{0.434\,\Delta T}{\left(\frac{P}{P_0}\right)\left(\log\frac{P_0}{P}\right)_{\text{sple}} + \left(\log\frac{P_0}{P}\right)_{\text{ref}}} = \frac{0.434\,\Delta T}{(TA)_{\text{sple}} + A_{\text{ref}}} \tag{7.25}$$

where the subscripts *sple* and *ref* refer to the sample and reference solutions, respectively. The position of minimum error gradually shifts to 100% T on the transmittance scale as the reference concentration increases; this amounts to comparing the reference and unknown at the same scale setting.

The optical path of the cuvettes must be known with a precision equal to the best precision expected for the differential method, or else one cuvette must be used for all measurements. To minimize volumetric errors, aliquots should be taken by weight. The precision attainable by the high-absorbance method may approach 0.01%. A calibration curve is often desirable, since increasing the sensitivity may cause some deviations from Beer's law. For interesting discussions of the differential method and a rigorous treatment of the error function, the reader is referred to papers by Bastian,[14–16] Hiskey,[17,18] and Reilley and Crawford.[19,20]

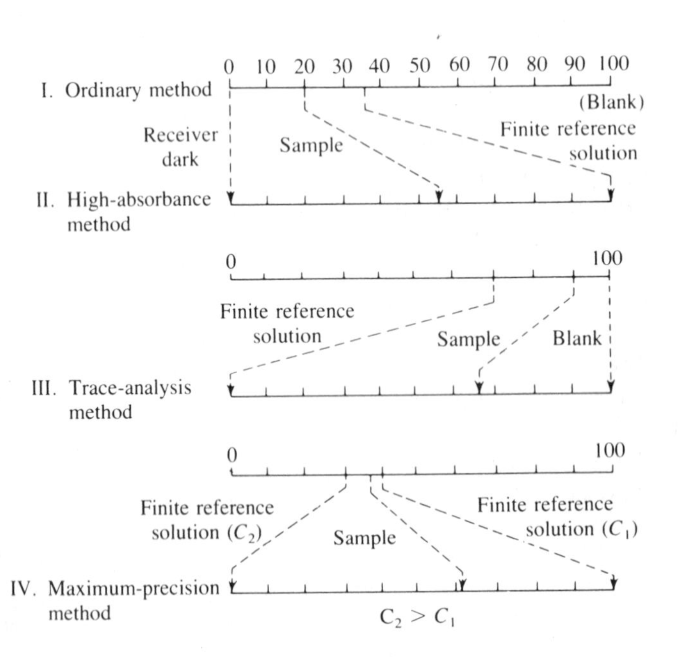

FIGURE 7.6
Differential (or expanded-scale) spectrophotometry.

In the *trace-analysis* method a large increase in sensitivity is achieved by "arranging" a positive deviation from Beer's law. The opaque reference standard is abandoned. The transmittance scale is set to 100% T with the solvent blank as in the ordinary method, but the zero energy is "faked" by placing in the radiation beam a reference solution, a screen, or a neutral density filter. Whichever is used should transmit a finite amount of radiant energy, but an amount less than that of the most concentrated sample solution. For example, following Figure 7.6(III), suppose that in the ordinary method a standard reference shows 70% transmittance and the sample reads 90%. Using the 70% T reference in the beam to make the 0% T setting results in a severalfold scale expansion, and the sample now reads 67% of the full scale.

To use this method, the instrument must have a zero-suppression (dark current, or "bucking") control, with which the response obtained through a finite reference solution can be made to read 0% T. A calibration curve must be constructed because the sensitivity increase is achieved at the expense of a positive deviation from Beer's law. Interaction between the controls used for the 0% T and the 100% T scale settings necessitates making several trials before the transmittance scale is adjusted. The error at the low-transmittance end of the transmittance scale now becomes finite. The expression for the relative concentration error becomes

$$\frac{dC}{C} = \frac{0.434(1 - T_{\text{ref}})}{TA} \Delta T \tag{7.26}$$

The *maximum-precision* method is simply a combination of the preceding two methods. Both ends of the transmittance scale are calibrated with standard reference solutions spaced around transmittance values that lie within that relatively flat portion of the curve of relative concentration error. In this method a standard

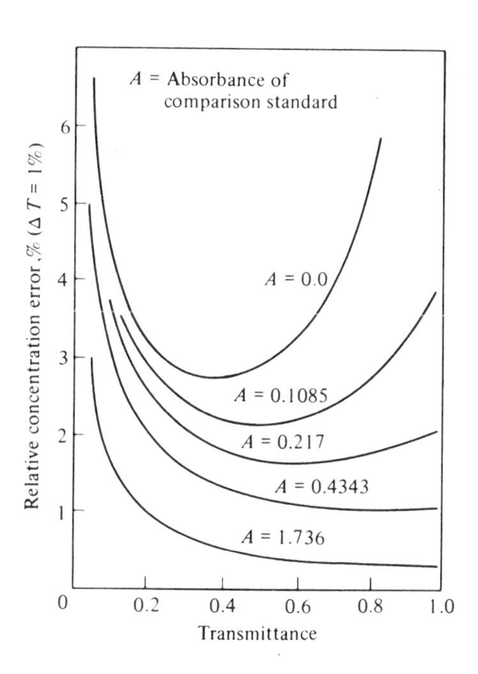

FIGURE **7.7**
Plot of the relative concentration error. [After C. F. Hiskey, *Anal. Chem.*, **21,** 1440 (1949). Courtesy of *Analytical Chemistry*.]

solution with a concentration somewhat higher than the unknown is used to set the scale at 0% *T*. A second standard solution that is more dilute than the unknown is placed in the beam of radiation for the 100% *T* adjustment. Figure 7.6(IV) illustrates the procedure for reference solutions that read 30% transmittance (used to set the 0% *T*) and 40% (used to make the 100% *T* setting). The maximum-precision method improves an already favorable transmittance reading (at 36.3% *T*) by making the reading ten times more precise at 63.0% *T*. A calibration curve is necessary because there are positive deviations from Beer's law. The instrument requirements for this method are the combined requirements of the previous two methods, and the limitations likewise tend to be a combination. The relative concentration error is given by

$$\frac{dC}{C} = \frac{0.434(T_{100\%\,\mathrm{ref}} - T_{0\%\,\mathrm{ref}})}{TA} \Delta T \tag{7.27}$$

In theory, the precision of this method can be increased by making the difference between the two reference solutions small. This assumes, however, the availability of an instrument with very high sensitivity and stability.

To sum up the four methods, in the ordinary method both ends of the transmittance scale are set precisely at 0% and 100% *T*. In the high-absorbance method, the 100% *T* setting is offset by using a reference instead of a blank; in the trace-analysis method, the 0% *T* setting is offset by using a reference solution instead of an occluder; and in the high-precision method, both ends of the scale are offset by using reference solutions.

7.6 DIFFERENCE SPECTROSCOPY

In difference spectroscopy two samples are used, one in the reference beam and the other in the sample beam of a double-beam spectrophotometer. The recording, usually made with pen range expansion, is the difference in transmittance (or absorbance) of the two samples. Common features in the two spectra cancel. Usually the concentration of absorbing material in the two samples is identical but some solution parameter such as pH is different. The method is used in toxicology laboratories for the analysis of drugs. Barbiturates, for example, have an absorption maximum at about 260 nm in 0.45*M* NaOH and a maximum at 250 nm or below at pH 10.3.[11] Using the pH 10 solution as reference, the spectrum from 230 to 280 nm is recorded. This spectrum is very characteristic of the particular barbiturate and thus provides a means of identification. Biochemists frequently use difference spectroscopy to study the conformation of globular proteins in solution by varying the solvent (adding alcohol) or changing the temperature of one sample. An enzyme-catalyzed reaction can be measured in the presence of nonenzyme-catalyzed reactions by placing the substrate in both beams. The nonenzyme reactions are canceled out, and the enzyme-catalyzed reaction produces a difference signal when enzyme is added to one sample.

Recording accurate difference spectra requires a spectrophotometer with low stray radiation levels, good resolution, good photometric linearity, repeatability, and an overall system designed to minimize noise with samples that significantly attenuate the signal in both reference and sample channels. The solutions must be clean—that is, free from particulate matter and dissolved gases that may form bubbles during measurement. The cuvettes must be identical, capable of exact positioning, and, of course, clean. It may be necessary to carry out reactions in the cuvettes without removing and replacing them. Schlieren effects due to localized areas of different densities must be avoided. Gentle stirring of the solutions during measurement may be necessary to assure homogeneity.

7.7 DERIVATIVE SPECTROSCOPY[21]

In derivative spectroscopy the first or higher derivative of absorbance or transmittance with respect to wavelength is recorded versus the wavelength. In a derivative spectrum the ability to detect and to measure minor spectral features is considerably enhanced. This enhancement of characteristic spectral detail can distinguish between very similar spectra and follow subtle changes in a spectrum. Moreover, it can be of use in quantitative analysis to measure the concentration of an analyte whose peak is obscured by a larger overlapping peak due to something else in the sample (and thus avoid prior separations), as shown in Figure 7.8. In this particular example, if one draws the best guess of the tangent to the spectrum with an interfering band as shown by the broken lines, the reading of 0.4 for the absorbance of the analyte is far too low, whereas a guess at the maximum of the analyte peak gives an absorbance of 1.9, far too high. Referring to the derivative spectrum in the lower right of Figure 7.8, one takes as the measure of the analyte

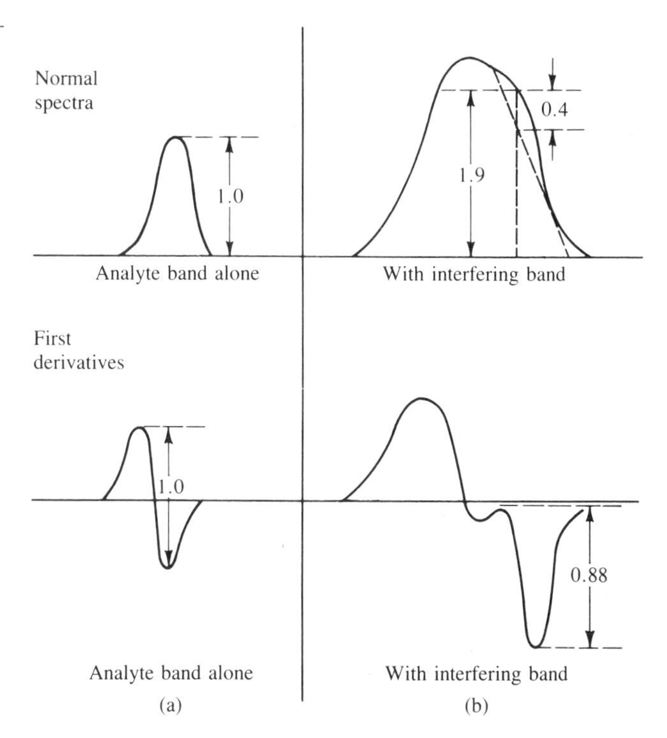

FIGURE **7.8** First-derivative spectrometry for the quantitative measurement of the intensity of a small band (a) alone and (b) obscured by a broader overlapping band. [Reprinted with permission from T. C. O'Haver, *Anal. Chem.*, **51**, 91A (1979). Copyright 1979 American Chemical Society.]

absorbance the vertical distance between the adjacent maximum and minimum of the first derivative. Now the estimate of the analyte absorbance is low by only 12%. Whenever the interfering band is broader than the analyte band by at least a factor of two, it is usually advantageous to base the measurement on the derivative spectra.

If two substances, X and Y, absorb in the same spectral region, the absorbances are additive—that is, $A = A_X + A_Y$, and each substance follows Beer's law, then $P = P_0 10^{-A}$. The first derivative of P with respect to λ then becomes

$$\frac{dP}{d\lambda} = 10^{-A}\frac{dP_0}{d\lambda} - 2.303P_0\,10^{-A}\left[bC_X\left(\frac{d\varepsilon}{d\lambda}\right)_X + bC_Y\left(\frac{d\varepsilon}{d\lambda}\right)_Y\right] \tag{7.28}$$

If P_0 does not vary significantly with wavelength in the spectral region of interest, as is usually the case, then $dP_0/d\lambda$ can be neglected. If, further, the molar absorptivity of one component, say ε_Y, varies only slightly with λ in the spectral region under investigation, then $bC_Y(d\varepsilon/d\lambda)_Y$ can be neglected, and Equation 7.28 reduces to

$$\frac{dP}{d\lambda} = -2.303PbC_X\left(\frac{d\varepsilon}{d\lambda}\right)_X \tag{7.29}$$

Under the same conditions, the differentiation of $A = A_X + A_Y$ with respect to λ yields

$$\frac{dA}{d\lambda} = bC_X\left(\frac{d\varepsilon}{d\lambda}\right)_X \tag{7.30}$$

Thus the first derivative is directly proportional to the concentration of X provided that P_0 and ε_Y do not vary appreciably with wavelength in the spectral range measured.

If the interfering substance, Y, has an absorption band close to X, the first derivative curves are distorted (see Figure 7.8b) unless the two absorption bands are nearly identical in λ_{max} and bandwidth, in which case the first-derivative curve is of no help.

Second-derivative spectra are also sometimes useful. At λ_{max}, the second derivative is directly proportional to concentration. For adequate sensitivity, however, $d^2\varepsilon/d\lambda^2$ must be large. This requires a narrow absorption band, 4 nm or less; thus the second-derivative method is most useful for atomic and gas molecular spectra.

A variety of different experimental techniques have been used to obtain derivative spectra. If the spectrum has been recorded digitally or is otherwise available in computer-readable form, then the differentiation can be done numerically. Alternatively, the derivative spectra may be recorded directly in real time, either by wavelength modulation or by obtaining the time derivative of the spectrum when the spectrum is scanned at a constant rate. In the latter case, a quite simple electronic differentiator is used. Most modern uv/vis spectrophotometers are capable of recording derivative spectra. Dual-wavelength spectrophotometers can obtain first-derivative spectra by scanning the spectrum with a small, constant difference between the two wavelengths.

7.8 PHOTOMETRIC TITRATIONS[22,23]

The change in absorbance of a solution may be used to follow the change in concentration of a radiation-absorbing constituent during a titration. The absorbance is directly proportional to the concentration of the absorbing constituent. A plot of absorbance versus titrant consists, if the reaction is complete, of two straight lines that intersect at the end point. For reactions that are appreciably incomplete, extrapolation of the two linear segments of the titration curve establishes the intersection and end-point volume. Possible shapes of photometric titration curves are shown in Figure 7.9. Curve (a), for example, is typical of the titration where the titrant alone absorbs, as in the titration of arsenic(III) with bromate-bromide, where the absorbance readings are taken at the wavelength where the bromine absorbs. Curve (b) is characteristic of systems where the product of the reaction absorbs, as in the titration of copper(II) with EDTA. When the analyte is converted to a non-absorbing product—for example, titration of *p*-toluidine in butanol with perchloric acid at 290 nm—curve (c) results. When a colored analyte is converted to a colorless product by a colored titrant—for example, bromination of a red dyestuff—curves similar to (e) are obtained. Curves (d) and (f) might represent the successive addition of ligands to form two successive complexes of different absorptivity.

Photometric titrations have several distinct advantages over a direct photometric determination. The presence of other absorbing species at the analytical wavelength, as in curve (a) of Figure 7.9, does not necessarily cause interference, since only the change in absorbance is significant. However, the absorbance of nontitratable components (color or turbidity) must not be intense because, if so, the

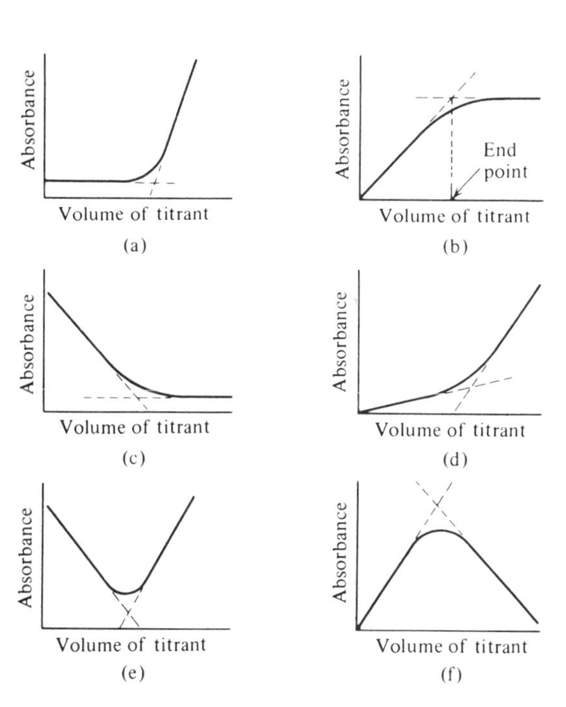

FIGURE 7.9
Possible shapes of photometric titration curves.

absorbance readings are limited to the undesirable upper end of the absorbance scale unless the slit width or amplifier gain can be increased. Only a single absorber needs to be present among the reactant, titrant, or products. This extends photometric methods to a large number of nonabsorbing constituents. Precision of 0.5% or better is attainable because a number of pieces of information are pooled in constructing the segments of the titration curve.

The analytical wavelength is selected on the basis of two considerations: (1) avoidance of interference by other absorbing substances and (2) need for a molar absorptivity that causes the change in absorbance during the titration to fall within a convenient range. Often the chosen wavelength lies well apart from an absorption maximum.

Volume change is seldom negligible, and straight lines are obtained only if there is correction for dilution. This is done simply by multiplying the measured absorbance by the factor

$$A_{\text{corrected}} = \frac{V + v}{V} A_{\text{observed}} \tag{7.31}$$

where V is the initial volume and v is the volume of titrant added up to any point. If the correction is not made, the lines are curved downward toward the volume axis and erroneous intersections are obtained. Use of a microsyringe and a relatively concentrated titrant is desirable. Stray-radiation error also affects the linearity of the titration curve. The upper limit of concentration permissible is found by delivering measured portions of a colored substance known to obey Beer's law from the buret into a beaker of transparent liquid. After correcting for dilution, the plot of absorbance versus concentration is a straight line up to the absorbance value where the stray-radiation error becomes detectable.

Areas of particular applicability are for solutions so dilute that the indicator blank is excessive by other methods, or when the color change is not sharp due perhaps to titration reactions that are incomplete in the vicinity of the equivalence point, or when extraneous colored materials are present in the sample. Ordinarily there is no difficulty in working with solutions of either high or low ionic strength or in nonaqueous solvents. One of the attractive features of photometric titrations is the ease with which the sensitivity of measurements can be changed simply by changing the wavelength or the length of the cell path. When self-indicating systems are lacking, an indicator can be deliberately added, but in a relatively large amount to provide a sufficient linear segment on the titration curve beyond the equivalence point.

Variations on the method of photometric titrations are used to determine other physical or chemical properties of substances. One example is the determination of the dissociation constants of weak acids or bases such as pH indicators. A constant amount of indicator is added to a series of buffer solutions that range in pH values from ± 2 or more units about the pK_a value of the indicator, and the absorbance of one form or the other of the indicator is measured. The absorbance is then plotted against the pH of the solutions. The limiting values of the absorbance at high and low pH represent the absorbance of the "basic" and "acidic" forms of the indicator. Using intermediate pH values where the indicator is changing color, the pK_a is given by $pK_a = \text{pH} + \log[\text{acid}]/[\text{base}]$. The relative concentrations of acidic and basic forms of the indicator at any given pH are $[\text{acid}] = |A_{\text{pH}} - A_{\text{base}}|$ and $[\text{base}] =$

$|A_{pH} - A_{acid}|$, where A_{pH} is the absorbance at the pH value and A_{acid} and A_{base} are the absorbancies of the completely acidic and basic forms of the indicator, respectively.

Another example is the determination of the composition of complex ions. In the mole-ratio method the concentration of the metal ion is held fixed and the concentration of the reagent is increased stepwise. On a graph of absorbance versus moles of reagent added, the intersection of the extrapolated linear segments determines the ratio: moles of reagent per moles of metal. The graph obtained resembles Figure 7.9b. In the method of continuous variations, the sum of the molar concentrations of the two reactants is kept constant as their ratio is varied. The abscissa of the extrapolated peak corresponds to the ratio present in the complex. The graph obtained resembles Figure 7.9f. In the slope-ratio method, two series of solutions are prepared. In the first series various amounts of metal ion are added to a large excess of the reagent, whereas in the second series different quantities of reagent are added to a large excess of metal ion. The absorbance of the solutions in each series is measured and plotted against the concentration of the variable component. The combining ratio of the components in the complex is equal to the ratio of the slopes of the two straight lines.

All one needs to carry out photometric titrations in the visible region is a light source, a series of narrow-bandpass filters, a titration vessel (which can be an ordinary beaker), a receptor, and a buret or other titrant delivery unit. The entire unit is then housed in a light-tight compartment. Photometers or spectrophotometers with a provision for the inclusion of a suitable titration vessel from 5- to 100-mL capacity are suitable. It is imperative that the titration vessel remain stationary throughout the titration. By the use of Vycor beakers and an appropriate spectrophotometer, titrations may be conducted in the ultraviolet region. Provision for magnetic stirring from underneath or some type of overhead stirrer is desirable; otherwise, manual agitation after the addition of each increment of titrant is necessary. For the transmission of photometric end points, the ends of two fiberoptic light pipes can be located facing each other across an area below the buret tip in the titrating vessel. One pipe conducts dispersed light to the sample and the other conducts the transmitted light to the photodetector.

7.9 OTHER TECHNIQUES

Photoacoustic Spectroscopy[24]

Photoacoustic spectroscopy, sometimes also called optoacoustic spectroscopy, is, as the names imply, a combination of optical methods with acoustical detection of the signal. When radiation, modulated at an acoustical frequency, is absorbed by a substance, the radiant energy is converted to heat. The heat causes a gas or liquid surrounding the sample to expand and contract at the modulation frequency. These acoustical waves are detected by a microphone.

A typical experimental arrangement is shown in Figure 7.10. The radiation source can be the output from a monochromator that furnishes radiation in the ultraviolet, visible, or infrared region; a pulsed, tunable laser; or a Fourier transform infrared spectrometer. The radiation must be pulsed at an acoustical frequency, usually 50–1200 Hz. A Fourier transform infrared spectrometer, if used as a radiation source, must have a low mirror velocity to produce modulation of the radiation at an acoustical frequency.

The photoacoustical cell is filled with an optically transparent gas, often air or helium. The gas must be free of CO_2 and water vapor when operating in the infrared region, since these impurities are strong infrared absorbers. The cell volume is kept small, usually less than 1 cm^3, to preserve the strength of the acoustical signal. For solid samples, the radiation is incident on the sample from above so that the material may be held in place by gravity, thus avoiding another interface. Liquids and gases may be investigated by the photoacoustical method; in these cases the sample fills the entire cell.

The sensitive microphone (or hydrophone in the case of liquid samples) must be responsive to the modulation frequency of the radiation. The entire cell must be isolated from vibrations. The signal from the microphone is amplified by a low-noise preamplifier and then transmitted to the signal processing unit, which further amplifies the signal and presents a recording of photoacoustical signal intensity versus frequency of radiation.

Radiation reflected by the sample produces no signal, since it is not absorbed by the gas and merely exits through the transparent windows of the cell. The photoacoustical signal is linearly dependent on the concentration of absorbing analyte over a rather considerable range of concentration.

The photoacoustical method is useful for the investigation of weakly absorbing fluids. It is easier to detect and measure a weak signal than to measure a small difference between two strong signals, as is the case in absorption measurements.

Photoacoustical measurements require almost no sample preparation and, furthermore, sensitive samples can be easily transferred and sealed in the cell under an inert atmosphere. A major disadvantage of the photoacoustical method is that the 100% absorption background spectrum is the energy output function of the spectrometer. This 100% absorption spectrum is determined by measuring the spectrum of a completely absorbing substance such as carbon black. The difference

FIGURE **7.10** Schematic diagram of a photoacoustic spectrometer.

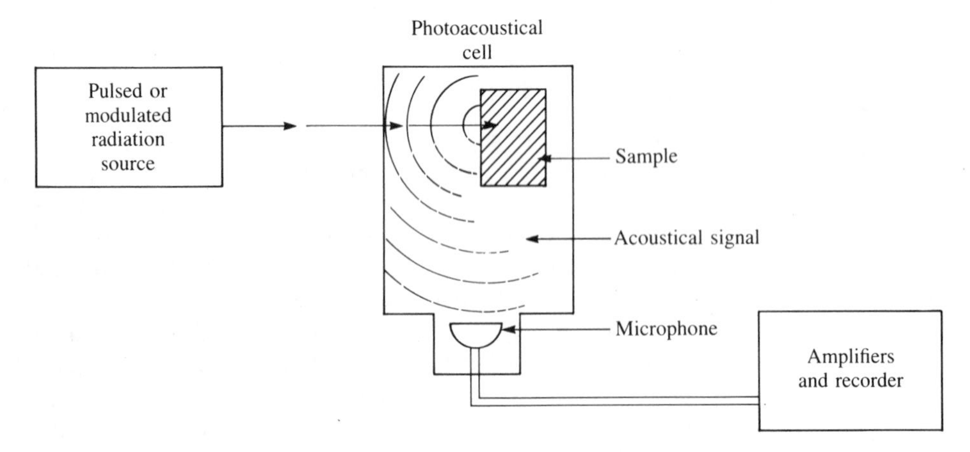

between the intensity from the sample and this background spectrum must be considered when the absorbance of the sample is desired.

Photoacoustical spectroscopy has been applied to the investigation of such substances as coal, semiconductors, plastics, foods, and pharmaceuticals. It is especially useful for solid samples that cannot be ground to fine powders (see the KBr pellet technique discussed in Chapter 11) or that are changed by grinding. It has also been applied to the qualitative investigation and quantitative determination of thin-layer chromatograms.

Thermal Lens Spectroscopy[25–27]

When a solution is irradiated with a laser beam of a frequency that is absorbed by a solute, the uneven cross-sectional intensity distribution of the laser beam gives rise to a greater amount of radiation absorbed in the center of the beam path in the solution than at the sides. The subsequent heating produced by absorption of the radiation causes the refractive index of the solution to decrease more in the center than at the sides of the volume of solution irradiated by the beam. The result is a negative (concave) lens effect, which causes the laser beam to diverge. If a small limiting aperture is placed in the beam center after the sample cell, the decrease in intensity can be detected by a phototube or photodiode. The decrease in intensity depends sensitively on the sample absorbance. Thermal lens spectroscopy can measure absorbances as low as 10^{-7}. A reduction in the precision of the method can result from several causes including excess reagent, solvent absorption, matrix effects, and impurities.

Flow Injection Analysis[28,29]

An interesting method of sampling and photometric measurement has recently been developed and is becoming widely used for routine work. This method, known as flow injection analysis (FIA), involves injection of a measured amount of liquid sample into a moving, nonsegmented stream of reagent. As the injected sample plus reagent carrier proceeds down a narrow tube, propelled by a constant flow pump, the sample develops a colored species, which is detected by a flow-through photodetector or a spectrophotometer equipped with flow-through cells. The signal generated by the colored species is a sharp peak with a trailing edge. The height of the peak is proportional to the concentration of analyte. The sample peak may be detected in about 15 sec after injection in a well-designed system and then, after about another 15 sec to clear the reactant out of the detector, another sample can be injected. The whole system can be made very compact. Samples as small as 30 μL and reagent volumes of less than 1 mL per analysis are possible. (See also Chapter 26.)

Spectra of Solids

Absorption measurements on nonclear solutions and solids are precluded with standard absorption spectrophotometers; yet these materials are found in the real world. With the proper instruments or accessories it is possible to obtain useful data from turbid liquids, powders, and opaque and translucent solids.

Reflection occurs whenever a light ray encounters a boundary between two media. The light reflected from the first surface of contact is called the specular (gloss, sheen) component. These encounters are repeated over and over in granular or fibrous structures where a light beam encounters a new interface every few millionths of a centimeter. These repeated encounters result in thorough diffusion such that the surface tends to appear uniformly bright in all directions. This is the diffuse component that is responsible for color where color exists. Particle size plays an important role. Whiteness and reflectance increase as the diameter of the particles is reduced to about half the wavelength of the incident light. At very small particle diameters, less than one-fourth the light wavelength, scattering takes over and diffuse reflectance falls off.

The determination of specular reflection (gloss), diffuse reflectance, and color measurements is very important in many industries such as paint, textile, and paper, but it is beyond the scope of this text. Color measurements are discussed by Hunter.[30]

7.10 CORRELATION OF ELECTRONIC ABSORPTION SPECTRA WITH MOLECULAR STRUCTURE

When molecules interact with radiant energy in the visible and ultraviolet regions, the absorption of energy consists in displacing an outer electron in the molecule. Rotational and vibrational modes are found combined with electronic transitions. Broadly, the spectrum is a function of the whole structure of a substance rather than of specific bonds. No unique electronic spectrum is found; this is a poor region for product identification by the "fingerprint" method. Information obtained from this region should be used in conjunction with other evidence to confirm the identity of a compound—for example, the previous history of a compound, its synthesis, auxiliary chemical tests, and other spectroscopic methods. On the other hand, electronic absorption often has a very large magnitude. Molar absorptivity values frequently exceed 10,000, whereas in the infrared region they rarely exceed 1,000. Thus dilute solutions are more easily measured in visible-ultraviolet spectrophotometry.

We will consider only those molecules capable of absorption within the wavelength region from 185 to 800 nm. Compounds with only single bonds involving σ-valency electrons exhibit absorption spectra only below 150 nm and are discussed only in interaction with other kinds. In covalent saturated compounds that contain heteroatoms, like nitrogen, oxygen, sulfur, and halogen, unshared *p*-electrons are present in addition to σ-electrons. Excitation promotes a *p*-orbital electron into an antibonding σ^*-orbit—that is, a $p \rightarrow \sigma^*$ transition, such as occurs in ethers, amines, sulfides, and alkyl halides. In unsaturated compounds absorption results in the displacement of π-electrons. Molecules that contain single absorbing groups, called *chromophores*, undergo transitions at approximately the wavelengths indicated in Table 7.2.

Molecules with two or more isolated chromophores absorb radiation of nearly the same wavelength as a molecule that contains only a single chromophore of a particular type, but the intensity of the absorption is proportional to the number of that type of chromophore in the molecule. Chromophores do not interact appreciably unless they are linked to each other directly; interposition of a single

TABLE 7.2 ELECTRONIC ABSORPTION BANDS FOR REPRESENTATIVE CHROMOPHORES

Chromophore	System	λ_{max}	ε_{max}	λ_{max}	ε_{max}	λ_{max}	ε_{max}
Ether	—O—	185	1,000				
Thioether	—S—	194	4,600	215	1,600		
Amine	—NH$_2$	195	2,800				
Thiol	—SH	195	1,400				
Disulfide	—S—S—	194	5,500	255	400		
Bromide	—Br	208	300				
Iodide	—I	260	400				
Nitrile	—C≡N	160					
Acetylide	—C≡C—	175–180	6,000				
Sulfone	—SO$_2$—	180					
Oxime	—NOH	190	5,000				
Azido	>C=N—	190	5,000				
Ethylene	—C=C—	190	8,000				
Ketone	>C=O	195	1,000	270–285	18–30		
Thioketone	>C=S	205	Strong				
Esters	—COOR	205	50				
Aldehyde	—CHO	210	Strong	280–300	11–18		
Carboxyl	—COOH	200–210	50–70				
Sulfoxide	>S → O	210	1,500				
Nitro	—NO$_2$	210	Strong				
Nitrite	—ONO	220–230	1,000–2,000	300–4,000	10		
Azo	—N=N—	285–400	3–25				
Nitroso	—N=O	302	100				
Nitrate	—ONO$_2$	270 (shoulder)	12				
	—(C=C)$_2$— (acyclic)	210–230	21,000				
	—(C=C)$_3$—	260	35,000				
	—(C=C)$_4$—	300	52,000				
	—(C=C)$_5$—	330	118,000				
	—(C=C)$_2$— (alicyclic)	230–260	3,000–8,000				
	C=C—C≡C	219	6,500				
	C=C—C=N	220	23,000				
	C=C—C=O	210–250	10,000–20,000			300–350	Weak
	C=C—NO$_2$	229	9,500				
Benzene		184	46,700	202	6,900	255	170
Diphenyl				246	20,000		
Naphthalene		220	112,000	275	5,600	312	175
Anthracene		252	199,000	375	7,900		
Pyridine		174	80,000	195	6,000	251	1,700
Quinoline		227	37,000	270	3,600	314	2,750
Isoquinoline		218	80,000	266	4,000	317	3,500

methylene group, or *meta*-orientation about an aromatic ring, is sufficient to insulate chromophores almost completely from one another. However, certain combinations of functional groups afford chromophoric systems that give rise to characteristic absorption bands.

A great deal of "negative" information may be deduced regarding molecular structures. If a compound is highly transparent throughout the region from 220 to 800 nm, it contains no conjugated unsaturated or benzenoid system, no aldehyde or keto group, no nitro group, and no bromine or iodine. If the screening indicated the presence of chromophores, the wavelength(s) of maximum absorbance is ascertained and tables are searched for known chromophores. Further information may be deduced from the shape, intensity, and detailed location of the bands. Finally the absorption spectrum is compared with the spectra in standard compilations of ultraviolet-visible spectra.[31–34] Often structural details are inferred from the close resemblance of a compound's spectrum with that of a compound of known and related structure; for example, in petroleum ether, the spectra of toluene and chlorobenzene are similar.

7.11 TURBIDIMETRY AND NEPHELOMETRY

Turbidity is an expression of the optical property of a sample that causes radiation to be scattered and absorbed rather than transmitted in straight lines through the sample. Scattering is elastic so that both incident and scattered radiation have the same wavelength. Turbidity is caused by the presence of suspended matter in a liquid. A scattering center is actually an optical inhomogeneity in an otherwise homogeneous medium. An atom, molecule, thermal density fluctuation, colloidal particle, or suspended solid can produce an optical inhomogeneity that results in scattering of radiation. The intensity of the perpendicularly polarized component of scattered radiation and the parallel component is a function of the relative refractive index, the size parameter, and the angle of observation (relative to the incident radiation), as well as the concentration of scatterer. When there is no molecular absorption in the sample, the refractive index is the conventional value. The size parameter ($\alpha = 2\pi r/\lambda$) involves the radius of the scattering center and the wavelength of the incident radiation. This ratio determines the phase distribution of the scattered radiation around the scattering center. The phase distribution shapes the scattering envelope and determines the resulting angular distribution of the scattered radiation. When the size parameter is smaller than one-tenth the wavelength of the incident radiation and the refractive index of the particle is not greatly different from that of the surrounding medium, the scattering envelope is symmetrical and is called Rayleigh scattering. As the size parameter becomes approximately one-fourth the wavelength of the incident radiation, scattering is concentrated in the forward direction. Ultimately, for particles larger than the wavelength of incident radiation, the radiation intensity scattered by the particle depends in a very complicated manner on the angle of scattering, but a large amount is scattered in the forward direction. This is known as Mie scattering. For very large particles, there is no wavelength dependence. Light-scattering theory is complicated by other sample parameters such as particle shape, molecular absorption, sample concentration, and size distribution of scatterers. Consequently, the relationship between any measurable indication of scattered radiation intensity and concentration of scatterer is not simple. Analytical determinations must be empirical. In fact, differences in the physical design of an instrument cause differences in the measured values for turbidity, even though the same calibration material was used for each instrument.

There are two methods for measuring the turbidity of a sample, turbidimetry and nephelometry. A turbidimeter measures the amount of radiation that passes through a sample in the forward direction, analogous to absorption spectrophotometry. A nephelometer measures the amount of radiation scattered by a turbid sample. These measurements are made at an angle to the direction of the beam of radiation through the sample, usually 45° or 90°, analogous to fluorometry.

Standards

The suspension that results from accurately weighing and dissolving 5 g of hydrazinium (2+) sulfate ($N_2H_4 \cdot H_2SO_4$) and 50 g of hexamethylenetetramine in 1 L of distilled water is defined as 4000 nephelometric turbidity units (NTU). After standing 48 hr the insoluble polymer (formazin) formed by the condensation reaction develops a white turbidity. This turbidity can be prepared repeatedly with an accuracy of $\pm 1\%$. The mixture can be diluted to prepare standards of any desired value.

Instrumentation[35,36]

For the measurement of very small amounts of turbidity (sample transmittance greater than 90%), as in water and waste-water analysis, the nephelometric method is the choice. A typical flow-through instrument with a 90° detection angle is shown in Figure 7.11. Although this angle is not the most sensitive to concentration, it is probably the least sensitive to variations in particle size. It also affords a simple optical system that is relatively free from stray radiation. If the sample is entirely free of scatterers, no scattered radiation reaches the photodetector and the indicating meter reads zero. Increasing turbidity gives an increase in the meter reading. A linear response is obtained at around zero and extends to a certain turbidity, after which the response begins to level off. Further increases in turbidity cause a decrease in response, and finally the instrument goes blind at a high turbidity value. Both sensitivity and linearity are functions of the path traversed by the scattered radiation. Whereas sensitivity increases as the path length increases, linearity is sacrificed at high concentrations because the sample becomes increasingly opaque and the radiation cannot penetrate. The shorter the radiation path in the nephelometer, the higher the turbidity that can be measured; but sensitivity is lost at low concentrations. This tradeoff is eliminated with an adjustable path length. Stray radiation becomes an added complication with the use of a short path length. Any scratches, imperfections in the cell windows, dirt, films, or condensation on the walls scatters radiation, some of which usually reaches the detector and gives a positive error to the turbidity measurement.

To overcome the problem of stray radiation caused by cell windows in a nephelometer when attempting to measure very low turbidities, the surface scatter instrument was designed (Figure 7.12). When a very narrow beam of radiation strikes the surface of the sample at a very small angle, part of the beam is reflected by

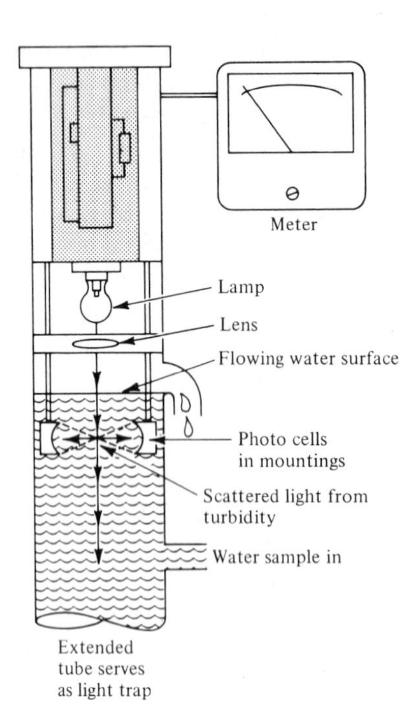

FIGURE 7.11
Low-range turbidimeter. (Courtesy of Hach Chemical Co.)

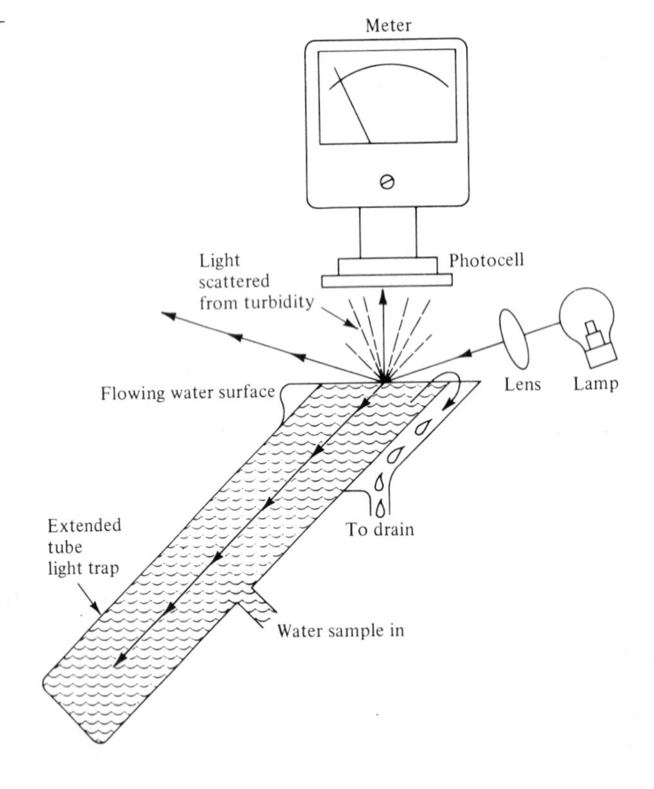

FIGURE 7.12
Surface scatter turbidimeter. (Courtesy of Hach Chemical Co.)

the water surface and escapes to a light trap. The remaining portion enters the sample at approximately a 45° angle. If particles of turbidity are present, radiation scattering occurs and some of the scattered radiation reaches the detector located slightly above the sample. These instruments are capable of accurately measuring trace turbidities in hundredths of NTU. The surface scatter approach can also be used for high turbidities. These instruments provide a practical way to continuously monitor turbidity in water and monitor many industrial operations.

Applications of nephelometry and turbidimetry are widely varied. Some determinations involve systems that are turbid prior to entering the analytical laboratory, such as in the determination of suspended material in waters. Measurements of the clarity of beverages and pharmaceuticals are typical of this simple kind of nephelometric determination. This is essentially an appearance measurement designed to evaluate the amount of haze, or cloudiness, in a sample. Clarity and sparkle are important characteristics of product quality; the presence of suspended materials, even in amounts so small as to be invisible at the bottling point, results after bottling and storage in an unsightly and unpalatable sediment. Determining the suitability of industrial process waters and the clarity of boiler feed waters and condensates is another typical everyday example of turbidity measurement.

PROBLEMS

1. In the transmission curve of an NaCl plate, opaque below the cutoff at 450 cm^{-1}, stray radiation is noted at 379 cm^{-1}. A similar tracing obtained using a CaF_2 plate that is opaque below 900 cm^{-1} shows no stray radiation. From the difference in the sample cutoff wavelengths, what was the origin of the stray radiation?

2. Calculate the effect of stray radiation on absorbance, assuming no stray radiation is absorbed by the sample, when the true absorbance is 0.5, 1.0, 1.5, and 2.0. Assume three levels of stray radiation: 0.1%, 1.0%, and 5.0%. Calculate the observed absorbance at each level of stray radiation (or the percent change in absorbance).

3. With a certain filter photometer and using a 510-nm filter and 2.00-cm cuvettes, the reading on a linear scale for P_0 was 85.4. With a $1.00 \times 10^{-4} M$ solution of a chromophore in the cuvette, the value of P was 20.3. Calculate the molar absorptivity.

4. The simultaneous determination of titanium and vanadium, each as their peroxide complex, can be done in steel. When 1.000-g samples of steel were dissolved, colors developed, and diluted to 50 ml exactly, the presence of 1.00 mg of Ti gave an absorbance of 0.269 at 400 nm and 0.134 at 460 nm. Under similar conditions 1.00 mg of V gave an absorbance of 0.057 at 400 nm and 0.091 at 460 nm. For each of the following samples, 1.000 g in weight and ultimately diluted to 50 mL, calculate the percent titanium and vanadium from the following absorbance readings:

Sample	A_{400}	A_{460}
1	0.172	0.116
2	0.366	0.430
3	0.370	0.298

Sample	A_{400}	A_{460}
4	0.640	0.436
5	0.902	0.570

5. The absorption spectra for tyrosine and tryptophan in $0.1M$ NaOH are shown in the illustration below. Select appropriate wavelengths for the simultaneous determination of each component.

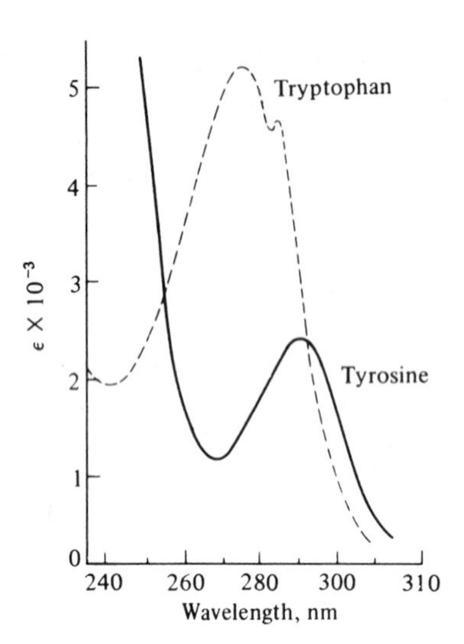

6. A mixture of sodium acetate and *o*-chloroaniline solution, 10 mL of each, was titrated in glacial acetic acid at 312 nm with a $0.1102M$ $HClO_4$ solution. Sodium acetate does not absorb in the ultraviolet portion of the spectrum, but it is a stronger base than *o*-chloroaniline. The results in the table were obtained (corrected for dilution). Plot the results and calculate the concentrations of sodium acetate and *o*-chloroaniline in the original aliquots.

Volume of titrant (mL)	Absorbance	Volume of titrant (mL)	Absorbance
0.00	0.68	8.25	0.37
1.00	0.68	8.50	0.32
2.00	0.68	8.75	0.26
3.00	0.68	9.00	0.20
4.00	0.67	9.25	0.14
5.00	0.66	9.50	0.09
6.00	0.63	10.50	0.02
7.00	0.56	11.00	0.02
8.00	0.42	11.50	0.02

7. In a photometric titration of magnesium with $0.00145M$ EDTA at 222 nm, the following procedure was used. All reagents except the magnesium-containing solution were placed in the titration cell, and the slit width was adjusted to give zero absorbance. The following readings were observed after additions of the standard EDTA:

Absorbance	EDTA added (mL)	Absorbance	EDTA added (mL)
0.000	0.00	0.429	0.60
0.014	0.10	0.657	0.80
0.200	0.40	0.906	1.00

At this point, the magnesium solution was added and the absorbance fell to zero. The titration was continued with the following results:

Absorbance	EDTA added (mL)	Absorbance	EDTA added (mL)
0.000	1.00	0.360	3.00
0.020	1.50	0.580	3.20
0.065	2.00	0.803	3.40
0.160	2.50	1.000	3.60
0.240	2.80	1.220	3.80

Plot the results, explain the curves obtained, and calculate the number of micrograms of magnesium in the sample.

8. Graph the data and determine the acid dissociation constants for these materials.

	p-Nitrophenol $\varepsilon \times 10^{-3}$			*Papaverine (cation form)* $\varepsilon \times 10^{-4}$	
pH	317 nm	407 nm	pH	239 nm	251 nm
3.0		0.33	2.0	3.36	5.90
4.0	9.72	0.33	3.0	3.36	5.90
5.0	9.72	0.50	4.0	3.39	5.83
6.0	9.03	1.66	5.0	3.48	5.63
6.2	8.61	2.28	5.6	3.86	5.19
6.4	8.19	3.99	5.8	3.93	4.91
6.6	7.36	5.14	6.0	4.30	4.61
6.8	6.39	7.22	6.2	4.61	4.15
7.0	5.55	9.16	6.4	4.86	3.71
7.2	4.45	11.65	6.6	5.22	3.30
7.4	3.61	13.40	6.8	5.46	2.77
7.6	2.92	15.00	7.0	5.66	2.51
7.8	2.08	16.90	7.4	6.03	2.00
8.0	1.81	17.50	8.0	6.27	1.63
9.0	1.39	18.33	11.0	6.43	1.56
10.0	1.39	18.33	12.0	6.44	1.56

9. Determine the acid dissociation constant for each indicator from the absorbance at the wavelength of maximum absorbance, measured as a function of pH. The ionic strength is 0.05.

Bromophenol blue $\lambda_{max} = 592$ *nm*		*Methyl red* $\lambda_{max} = 530$ *nm*		*Bromocresol purple* $\lambda_{max} = 591$ *nm*	
Absorbance	**pH**	**Absorbance**	**pH**	**Absorbance**	**pH**
0.00	2.00	2.00	3.20	0.00	4.00
0.18	3.00	1.78	4.00	0.24	5.40
0.58	3.60	1.40	4.60	0.66	6.00
0.98	4.00	0.92	5.00	0.87	6.20
1.43	4.40	0.48	5.40	1.13	6.40
1.75	5.00	0.16	6.00	1.37	6.60
2.10	7.00	0.00	7.00	1.72	7.00
				2.00	8.00

10. A series of chromium(III) nitrate solutions were measured according to the ordinary method with 0% T set with the phototube darkened and 100% T set with the pure solvent. From the results obtained at 550 nm, (a) calculate the relative concentration error for each measurement, assuming $\Delta T = 0.004$, and (b) plot the results as relative concentration error versus absorbance. Calculations should be done for the two limiting cases: shot-noise limited and Johnson-noise limited. Assume $k = 0.004$ for the spectrometer in the shot-noise case.

Concentration (M)	Absorbance	Concentration (M)	Absorbance
Blank	0	0.0300	0.357
0.0050	0.060	0.0400	0.476
0.0100	0.119	0.0500	0.595
0.0150	0.179	0.0600	0.714
0.0200	0.238	0.0800	0.952
0.0250	0.298	0.1000	1.190
		0.1100	1.309

11. From the series of solutions used in Problem 10, the 0.0500M solution was used to set the 100% T reading. The results in the table were obtained. Assuming $\Delta T = 0.004$, plot the relative concentration error versus transmittance.

Concentration (M)	Transmittance	Concentration (M)	Transmittance
0.0500	1.000	0.110	0.230
0.0600	0.775	0.120	0.162
0.0700	0.584	0.130	0.120
0.0800	0.453	0.140	0.097
0.0900	0.380	0.150	0.074
0.1000	0.295		

12. In using the low-absorbance method a 0.100M solution of chromium(III) nitrate was used as the standard with which the scale was set at 0% T by means of the zero-suppressor (dark-current) control. Pure solvent was used to set the 100% T point. The results in the table were obtained. Assuming $\Delta T = 0.004$, plot the relative concentration error versus transmittance.

Concentration (M)	Transmittance	Concentration (M)	Transmittance
Blank	1.000	0.0500	0.197
0.0100	0.745	0.0600	0.130
0.0200	0.546	0.0700	0.077
0.0300	0.420	0.0800	0.037
0.0400	0.286	0.0900	0.012

13. With the maximum precision method, a 0.0500M chromium(III) nitrate solution was used to set the 100% T point, and a 0.100M solution was used to set the 0% T point. The results in the table were obtained. Assuming $\Delta T = 0.004$, plot the relative concentration error versus transmittance.

Concentration (M)	Transmittance	Concentration (M)	Transmittance
0.0500	1.000	0.0800	0.247
0.0600	0.695	0.0900	0.128
0.0700	0.437	0.100	0.000

14. The absorbances of a series of phosphate solutions, using the phosphatovanadomolybdate complex at 420 nm, are shown in the following table when using a 5.0-mg phosphate solution as reference standard (actual $A = 1.075$). (a) Determine the increase in precision for the measurement of the 5.2-mg phosphate solution as compared with the normal method. (b) Estimate the relative concentration error for the 5.6-mg phosphate solution when the 6.0-mg phosphate solution is used to set the zero-transmittance reading in addition to the 5.0-mg phosphate solution being used for the 100% transmittance reading.

mg P_2O_5/100 mL	Absorbance
5.0	0.000
5.2	0.046
5.4	0.092
5.6	0.138
5.8	0.184
6.0	0.230
6.2	0.276

15. A series of solutions is prepared in which the amount of iron(II) is held constant at 2.00 mL of $7.12 \times 10^{-4}M$, while the volume of $7.12 \times 10^{-4}M$ 1,10-phenanthroline is varied. After dilution to 25 mL, absorbance data for these solutions in

1.00-cm cuvettes at 510 nm are as shown in the following table. (a) Evaluate the composition of the complex. (b) Estimate the value of the formation constant of the complex.

1,10-Phenanthroline (mL)	Absorbance
2.00	0.240
3.00	0.360
4.00	0.480
5.00	0.593
6.00	0.700
8.00	0.720
10.00	0.720
12.00	0.720

16. The method of continuous variation was used to investigate the species responsible for the absorption at 510 nm when the indicated volumes of $6.72 \times 10^{-4} M$ iron(II) solution were mixed with sufficient $6.72 \times 10^{-4} M$ 1,10-phenanthroline to equal a total volume of 10.00 mL, after which the entire system was diluted to 25 mL. Cuvettes were 1.00 cm. (a) Determine the composition of the complex. (b) Calculate the molar absorptivity of the complex.

Iron(II) (mL)	Absorbance	Iron(II) (mL)	Absorbance
0.00	0.000	5.00	0.565
1.00	0.340	6.00	0.450
1.50	0.510	7.00	0.335
2.00	0.680	8.00	0.223
3.00	0.794	9.00	0.108
4.00	0.680	10.00	0.000

17. Evaluate the composition of the iron(II)-1,10-phenanthroline complex with an absorption peak at 510 nm on the basis of the following absorbance data obtained, after dilution to 25 mL, in 1.00-cm cuvettes.

Iron(II) constant at 5.00 mL of $7.00 \times 10^{-4} M$		*Ligand constant at 10.00 mL of $2.10 \times 10^{-3} M$*	
$7.00 \times 10^{-4} M$ ligand (mL)	***A***	**$7.00 \times 10^{-4} M$ iron(II) (mL)**	***A***
1.00	0.177	0.50	0.177
2.00	0.235	1.00	0.352
3.00	0.352	1.50	0.530
4.00	0.470	2.00	0.706
5.00	0.585	2.50	0.883

18. The dissociation of the complex between thorium and quercetin can be expressed as $ThQ_2 \rightleftarrows Th + 2Q$ (omitting formal charges). For a solution that was $2.30 \times 10^{-5} M$ in thorium and contained a large excess of quercetin, sufficient to ensure that all the thorium is present as the complex, the absorbance was 0.780. When the same amount of thorium is mixed with a stoichiometric amount of quercetin, the absorbance was 0.520. Calculate (a) the degree of dissociation and (b) the value of the formation constant of the complex.

BIBLIOGRAPHY

BAUMAN, R. P., *Absorption Spectroscopy*, John Wiley, New York, 1962.

BURGESS, C., AND A. KNOWLES, eds., *Techniques in Visible and Ultraviolet Spectrometry*, Vol. 1, Chapman and Hall, New York, 1981.

FORBES, W. R., Chap. 1, *Interpretive Spectroscopy*, S. K. Freeman, ed., Reinhold, New York, 1965.

JAFFE, H. H., AND M. ORCHIN, *Theory and Applications of Ultraviolet Spectroscopy*, John Wiley, New York, 1962.

MCDONALD, R. S., *Anal. Chem.*, **56,** 361R (1984).

MEEHAN, E. J., Chap. 2, *Treatise on Analytical Chemistry*, 2nd ed., P. J. Elving, E. J. Meehan, and I. M. Kolthoff, eds., Part I, Vol. 7, John Wiley, New York, 1981.

OLSEN, E. D., *Modern Optical Methods of Analysis*, McGraw-Hill, New York, 1975.

SILVERSTEIN, R. M., G. C. BASSLER, AND T. C. MORRILL, *Spectrometric Identification of Organic Compounds*, 4th ed., John Wiley, New York, 1981.

WEST, W., ed., *Chemical Applications of Spectroscopy*, Vol. IX, Part 1, 2nd ed., John Wiley, New York, 1968.

LITERATURE CITED

1. KORTUM, G., AND M. T. SEILER, *Angew. Chem.*, **52,** 687 (1939).
2. ERICKSON, J. O., AND T. SURLES, *Am. Lab.*, **8,** 41 (June 1976).
3. COOK, R. B., AND R. JANKOW, *J. Chem. Educ.*, **49,** 405 (1972).
4. SLAVIN, W., *Anal. Chem.*, **35,** 561 (1963).
5. KAYE, W., *Anal. Chem.*, **53,** 2201 (1981).
6. KAYE, W., *Am. Lab.*, **15**(11), 18 (1983).
7. SHARPE, M. R., *Anal. Chem.*, **56,** 339A (1984).
8. ASTM, *Standard Methods of Estimating SRE*, E-387-84 (1984).
9. HARGIS, L. G., AND J. A. HOWELL, *Anal. Chem.*, **56,** 225R (1984); **52,** 302R (1980).
10. HOWELL, J. A., AND L. G. HARGIS, *Anal. Chem.*, **54,** 171R (1982); **50,** 234R (1978).
11. SUNSHINE, I., *Handbook of Spectrophotometric Data of Drugs*, Chemical Rubber Co., Cleveland, 1981.
12. PESEZ, M., AND J. BARTOS, *Colorimetric and Fluorometric Analysis of Organic Compounds and Drugs*, Dekker, New York, 1974.
13. CARUANA, R. A., R. B. SEARLE, T. HELLER, AND S. I. SHUPACK, *Anal. Chem.*, **58,** 1162 (1986).
14. BASTIAN, R., *Anal. Chem.*, **21,** 972 (1949).
15. BASTIAN, R., R. WEBERLING, AND F. PALILLA, *Anal. Chem.*, **22,** 160 (1950).
16. BASTIAN, R., *Anal. Chem.*, **23,** 580 (1951).
17. HISKEY, C. F., *Anal. Chem.*, **21,** 1440 (1949).
18. HISKEY, C. F., J. RABINOWITZ, AND J. G. YOUNG, *Anal. Chem.*, **22,** 1464 (1950).
19. CRAWFORD, M., *Anal. Chem.*, **31,** 343 (1959).
20. REILLEY, C. N., AND C. M. CRAWFORD, *Anal. Chem.*, **27,** 716 (1955).
21. O'HAVER, T. C., *Anal. Chem.*, **51,** 91A (1979).
22. GODDU, R. F., AND D. N. HUME, *Anal. Chem.*, **26,** 1679, 1740 (1954).
23. HEADRIDGE, J. B., *Photometric Titrations*, Pergamon, New York, 1961.
24. MCCLELLAND, J. F., *Anal. Chem.*, **55,** 89A (1983).

25. HARRIS, J. M., AND M. J. DOVICHI, *Anal. Chem.*, **52,** 695A (1980).
26. SKOGERBOE, K. J., AND E. S. YEUNG, *Anal. Chem.*, **58,** 1014 (1986).
27. PHILLIPS, C. M., S. R. CROUCH, AND G. E. LEROI, *Anal. Chem.*, **58,** 1710 (1986).
28. BETTERIDGE, D., *Anal. Chem.*, **50,** 832A (1978).
29. RUCICKA, J., *Anal. Chem.*, **55,** 1040A (1983).
30. HUNTER, R. S., *Off. Dig. Feder. Soc. Paint Technol.*, **35,** 350 (1963).
31. LANG, L., *Absorption Spectra in the Ultraviolet and Visible Regions,* Academic, New York, 1961.
32. Sadtler Research Laboratories, *Ultraviolet Reference Spectra,* Philadelphia, updated continuously.
33. PHILLIPS, J. P., D. BATES, H. FEUER, AND B. S. THYAGARAJAN, *Organic Electronic Spectral Data,* John Wiley, New York, 1986.
34. SILVERSTEIN, R. M., G. C. BASSLER, AND T. C. MORRILL, *Spectrometric Identification of Organic Compounds,* 4th ed., John Wiley, New York, 1981.
35. SURLES, T., J. O. ERICKSON, AND D. PRIESNER, *Am. Lab.*, **7**(3), 55 (March 1975).
36. WENDLANDT, W. W., *J. Chem. Educ.*, **45,** A861, A947 (1968).

FLUORESCENCE AND PHOSPHORESCENCE SPECTROPHOTOMETRY

Luminescence is the term applied to the reemission of previously absorbed radiation. This chapter is concerned with molecular photoluminescence, in which photons of electromagnetic radiation are absorbed by molecules, raising them to some excited state and then, on returning to the ground state, the molecules emit radiation—that is, luminesce. Photoluminescence includes fluorescence and phosphorescence. In fluorescence the energy transitions do not involve a change in electron spin. Phosphorescence involves a change in electron spin and is therefore much slower than fluorescence. The theory was discussed in Chapter 5.

In nearly all cases, absorption involves excitation from the lowest vibrational level of the ground state of a molecule to a higher vibrational level of the first excited singlet state. Upper vibrational levels quickly relax to the lowest vibrational level of the particular electronic level, since vibrational relaxation occurs in about 10^{-12} sec, much faster than other energy dissipation processes. On reversion to the ground electronic level luminescence may result, with the molecule usually ending up in an excited vibrational level. Because the vibrational levels of both ground and excited states are similar, the fluorescence spectrum often is a sort of mirror image of the exciting absorption spectrum (Figure 8.1). The lifetime of an excited singlet state is usually 10^{-9}–10^{-6} sec and fluorescence lifetimes fall in this range. The lifetime is

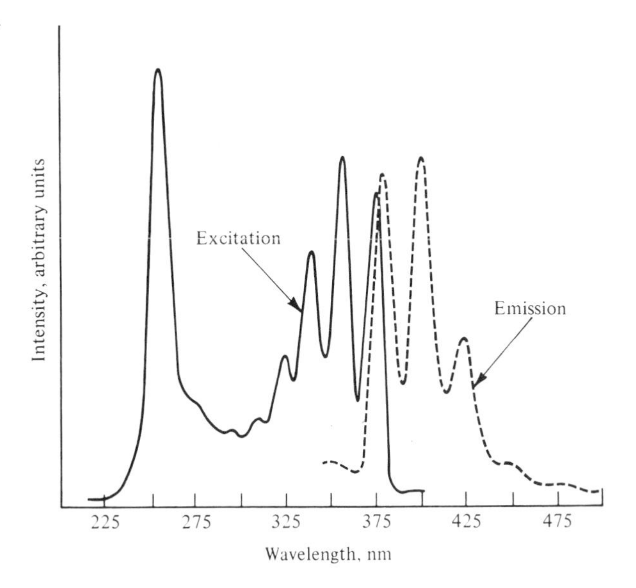

FIGURE **8.1**
Excitation and fluorescence emission spectra of 0.3 μg/mL anthracene in methanol.

defined as the time required for the population of the excited state to decrease to $1/e$ of its original value after the excitation source is turned off.

For a molecule to phosphoresce, the excited singlet state produced by the absorption of radiation must change to the triplet state by intersystem crossing, which involves a change in electron spin, a "forbidden" transition. This merely means that the probability of this transition is low and also the rate is slow. In order to emit a photon and return to the ground state, the electron spin of an electron in the molecule must again change. Phosphorescence lifetimes are therefore much longer than fluorescence lifetimes, being 10^{-4}–10 sec. Because other deactivating processes are generally much faster, phosphorescence is usually rare unless special precautions are taken to slow down the competing processes. Phosphorescence is seen more frequently in rigid media at low temperatures.

For any molecule to photoluminesce it must first be raised to an excited state by the absorption of radiation. The luminescence spectrum always has a longer wavelength (lower energy) than the exciting wavelength because the excitation process involves an amount of energy equal to the electronic change plus a vibrational energy increase, whereas the luminescence involves the electronic energy change minus a vibrational energy change. Phosphorescence wavelengths are longer than fluorescence wavelengths if both exist for a molecule because the first triplet state must lie at or below the energy level of the first singlet state if intersystem crossing is to occur.

This chapter emphasizes fluorescence spectroscopy, which is the more widely used of the two photoluminescence methods, although new analytical methods are being developed using phosphorescence.[1]

Fluorescence spectroscopy has assumed a major role in analysis, particularly in the determination of trace contaminants in our environment, industries, and bodies, because for applicable compounds fluorescence gives high sensitivity (in the low parts per trillion) and high specificity. High sensitivity results from the difference in wavelengths between the exciting and fluorescence radiation. This results in a signal contrasted with essentially zero background; it is always easier to measure a small signal directly than a small difference between two large signals as is done in absorption spectrophotometry. High specificity results from dependence on two spectra, the excitation and emission spectra, and the possibility of measuring the lifetimes of the fluorescent state. Two compounds that are excited at the same wavelength but emit at different wavelengths are readily differentiated without the use of chemical separation techniques. Likewise two compounds may fluoresce at the same wavelength but require different excitation wavelengths. Also, a fluorescent compound in the presence of one or more nonfluorescent compounds is readily analyzed fluorometrically even when the compounds have overlapping absorption spectra. Even nonfluorescent or weakly fluorescent compounds can often be reacted with strong fluorophores, enabling them to be determined quantitatively (see the discussion of derivatization techniques in Chapter 19). The phenomenon of fluorescence itself is subject to more rigorous constraints on molecular structure than is absorption. Many drugs possess rather high quantum efficiencies for fluorescence—for example, quinine and lysergic acid diethylamide (LSD). As little as 1 ng/mL of the latter can be detected in a 5-mL sample of blood plasma or urine. Carcinogens, such as benzopyrene, are easily determined fluorometrically in air pollution analyses.

To ascertain the excitation and emission spectra (Figure 8.1) the following procedure is commonly followed. The excitation monochromator is varied until fluorescence occurs; often this can simply be viewed visually. The excitation monochromator is then set at this wavelength (or at any point within the excitation wavelength band) and the emission monochromator is allowed to scan, recording the emission spectrum. The emission monochromator is then set at the wavelength at which maximum fluorescence occurred, the excitation monochromator is allowed to scan, and the excitation spectrum is recorded. In turn, the final emission spectrum is obtained by setting the excitation wavelength and again scanning with the emission monochromator. For analytical applications the emission spectrum is used. Often an excitation spectrum is first made to confirm the identity of the substance and to select the optimum excitation wavelength.

8.1 STRUCTURAL FACTORS[2–6]

Fluorescence is expected in molecules that are aromatic or contain multiple-conjugated double bonds with a high degree of resonance stability. Both classes of substances have delocalized π-electrons that can be placed in low-lying excited singlet states. In polycyclic aromatic systems where the number of π-electrons available is greater than in benzene, these compounds and their derivatives are usually much more fluorescent than benzene and its derivatives. Substituents strongly affect fluorescence. Substituents that delocalize the π-electrons, such as $—NH_2$, $—OH$, $—F$, $—OCH_3$, $—NHCH_3$, and $—N(CH_3)_2$ groups, often enhance fluorescence because they tend to increase the transition probability between the lowest excited singlet state and the ground state. Electron-withdrawing groups that contain $—Cl$, $—Br$, $—I$, $—NHCOCH_3$, $—NO_2$, or $—COOH$ decrease or quench the fluorescence completely. Thus aniline fluoresces but nitrobenzene does not.

Molecules with a nonbonding pair of valence electrons—for example, an amine with a lone pair on its nitrogen atom—often fluoresce. Such electrons can be promoted without disruption of bonding. In general, a delocalized π-system must also be part of this type of molecule to ensure fluorescence.

Molecular rigidity reduces the interaction of a molecule with its medium and thus reduces the rate of collisional deactivation (internal conversion, see Figure 5.7). This reduced rate of deactivation by nonradiative processes leads to a greater probability of luminescence. For example, fluorescein and eosin are strongly fluorescent, but a similar compound, phenolphthalein, which is nonrigid and in which the conjugate system is disrupted, is not fluorescent. Given a series of aromatic compounds, those that are the most planar, rigid, and sterically uncrowded are the most fluorescent. In this same sense, substances fluoresce more brightly in a glassy state or in viscous solution. The probability of intermolecular energy transfer between the fluorescer and other molecules tends to be reduced at low temperature and in a medium of high viscosity in which the rotational relaxation time of the fluorescer is much longer than the lifetime of the excited state.

Fluorescence intensity and wavelength often vary with the solvent. Solvents capable of exhibiting strong van der Waal's binding forces with the excited-state species prolong the lifetime of a collisional encounter and favor deactivation.

Solvents that have molecular substituents such as Br, I, NO_2, or —N=N— groups are undesirable because the strong magnetic fields that surround their bulky atomic cores promote spin decoupling of electrons and triplet state formation, giving rise to marked fluorescence quenching, although these same solvents may promote phosphorescence. Indole illustrates the wavelength shifts that may occur in different solvents. Although the excitation wavelength remains at 285.0 nm in each solvent, the wavelength of maximum fluorescence is 297.0 nm in cyclohexane, 305.0 nm in benzene, 310.0 nm in 1,4-dioxane, 330.0 nm in ethanol, and 350.0 nm in water.

Changes in the system pH, if it affects the charge status of the chromophore, may influence fluorescence. Both phenol and anisole fluoresce at pH 7, but at pH 12 phenol is converted to the nonfluorescent anion, whereas anisole remains unchanged. Similarly, aniline fluoresces in the visible region at pH 7 and 12, but the protonated cation is nonfluorescent at pH 2. These observations can be explained by comparing the resonance forms of anions and cations. For example, aniline in acid solution has the positive charge fixed at the nitrogen atom, and the anilinium ion has only the same resonance forms as benzene, which fluoresces only in the ultraviolet region. In neutral or basic solution, however, aniline has three additional resonance structures, resulting in a more stable excited singlet state and a longer wavelength of fluorescent radiation. In fact, some substances are so sensitive to pH that they are used as indicators in acid-base titrations. The merit of such indicators is that they can be used in turbid or intensely colored systems.

Frequently, weakly fluorescent or nonfluorescent aromatic compounds are strongly phosphorescent. Usually this indicates the involvement of $n - \pi^*$ absorption transitions. Such $n - \pi^*$ excited states have smaller energy differences between the excited singlet and triplet levels and have longer excited-state lifetimes. These conditions favor the population of the triplet state and lead to phosphorescence. Carbonyl-substituted aromatic compounds frequently exhibit this behavior.

The formation of chelates with metal ions, in general, also promotes fluorescence by promoting rigidity and minimizing internal vibrations. The introduction of paramagnetic metal ions, such as copper(II) and nickel(II), gives rise to phosphorescence but not fluorescence in metal complexes. By contrast, magnesium and zinc compounds show only strong fluorescence. Generally only those cations that are diamagnetic when coordinated and that are nonreducible form fluorescent complexes. The transition metals with unfilled outer *d*-orbitals quench fluorescence completely. On the other hand, whereas paramagnetic species quench fluorescence, they strongly promote intersystem crossing so that those cations are observed to promote phosphorescence. Dissolved oxygen is a strong quencher of fluorescence, since oxygen is paramagnetic. Oxygen must be removed from solutions by degassing with nitrogen or other inert gas before fluorescence measurements are made.

Phosphorescence lifetimes are also affected by molecular structure. Unsubstituted cyclic and polycyclic hydrocarbons and their derivatives that contain CH_3, NH_2, OH, COOH, and OCH_3 substituents have lifetimes of 5–10 sec for most benzene derivatives and 1–4 sec for many naphthalene derivatives. The nitro group diminishes the intensity of phosphorescence and the lifetime of the triplet state to about 0.2 sec. Aldehydic and ketonic groups diminish the lifetime to about 0.001 sec. The introduction of bulky substituents that force a planar configuration to become nonplanar markedly shortens lifetimes.

8.2 PHOTOLUMINESCENCE POWER AS RELATED TO CONCENTRATION

The quantitative relationship between the fluorescent power (or phosphorescent power) and concentration is derived from the following considerations. The fluorescent power is proportional to the number of molecules in excited states, which, in turn, is proportional to the radiant power absorbed by the sample. Thus

$$P_F = \Phi_F(P_0 - P) \tag{8.1}$$

where P_F is the radiant power of fluorescence, Φ_F is the fluorescence efficiency or quantum yield of fluorescence, P_0 is the radiant power incident on the sample, and P is the radiant power emerging from the sample. $P_0 - P$ is therefore the radiant power absorbed by the sample. The fluorescence efficiency, Φ_F, is simply the ratio of the number of photons emitted as fluorescence to the number of photons absorbed:

$$\Phi_F = \frac{\text{photons emitted as fluorescence}}{\text{photons absorbed}} \tag{8.2}$$

Applying Beer's law to Equation 8.1, one obtains

$$P_F = \Phi_F P_0(1 - e^{-\varepsilon bc}) \tag{8.3}$$

and expanding Equation 8.3 in a power series yields

$$P_F = \Phi_F P_0 \varepsilon bc\left[1 - \frac{\varepsilon bc}{2!} + \frac{(\varepsilon bc)^2}{3!} - \cdots + \frac{(\varepsilon bc)^n}{(n+1)!}\right] \tag{8.4}$$

If εbc is small, only the first term in the series is significant and Equation 8.4 can be written as

$$P_F = \Phi_F P_0 \varepsilon bc \tag{8.5}$$

which is in error by only 2.5% if εbc is as large as 0.05.

Equation 8.5 indicates a linear relationship between fluorescent power and concentration provided that εbc is small—that is, when the concentrations are very dilute. At higher concentrations the relationship becomes nonlinear, and at high concentrations where all the radiation is absorbed by the sample and none is transmitted ($P = 0$), the limit is

$$P_F = P_0 \Phi_F \tag{8.6}$$

and there is no dependence on concentration.

In practice only a portion of the fluorescence is observed by the detector, since fluorescence radiation is emitted in all directions. Thus the detector response is decreased by two factors: $f_{(\Phi)}$, a geometrical factor representing the solid angle of fluorescing radiation subtended by the detector, and $g_{(\lambda)}$, the efficiency of the detector for the fluorescent wavelength. The detector response, F, is then

$$F = P_0 f_{(\Phi)} g_{(\lambda)} \Phi_F \varepsilon bc \tag{8.7}$$

A similar expression can be written for phosphorescence.

Of particular interest in Equation 8.5 is the linear dependence of fluorescence on the excitation power. This means sensitivity can be increased by working at high excitation powers to give large signal-to-noise ratios. Since the source intensity can change from time to time, fluorescence signals are not measured as absolute parameters; rather, they are expressed in terms of relative fluorescence. All measurements are made relative to reference standards of known concentration. All readings must be corrected for background fluorescence. A major advantage of photoluminescence instrumentation is the ability to adjust the sensitivity of instruments, so that after a standard of known concentration is measured, the instrument can be readily adjusted to read directly in concentration. In photoluminescence photometry, the b term in Equation 8.5 is not the path length of the cell, but the solid volume of the beam defined by the excitation and emission slit widths together with the beam geometry. Therefore, slit widths are the critical factor and not the cell dimension (Figure 8.2)

A plot of fluorescence (or phosphorescence) versus concentration, shown in Figure 8.3, is often found to be linear over two or more decades, but there are limiting factors—namely, the blank fluorescence, quenching, and absorbance of exciting radiation by the solvent or by too high concentrations of solute. The minimum detectable quantity of an analyte is generally limited by the magnitude of

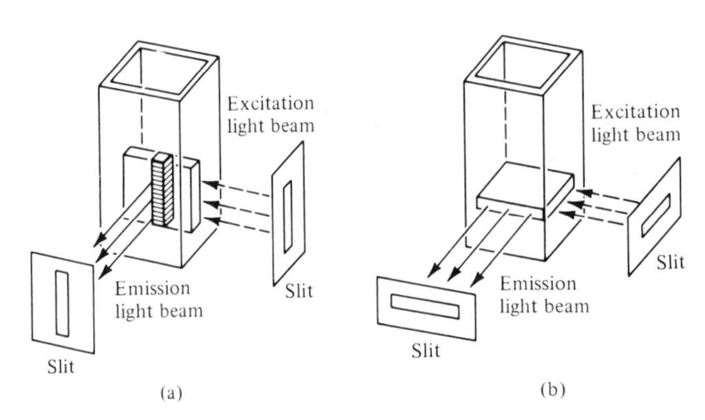

FIGURE **8.2**
Illumination of a sample solution for a given spectral slit width: (a) conventional configuration and (b) horizontal beam focused on the sample cell after image rotation, an arrangement requiring only 0.6 mL of sample. (Courtesy of Perkin-Elmer Corp.)

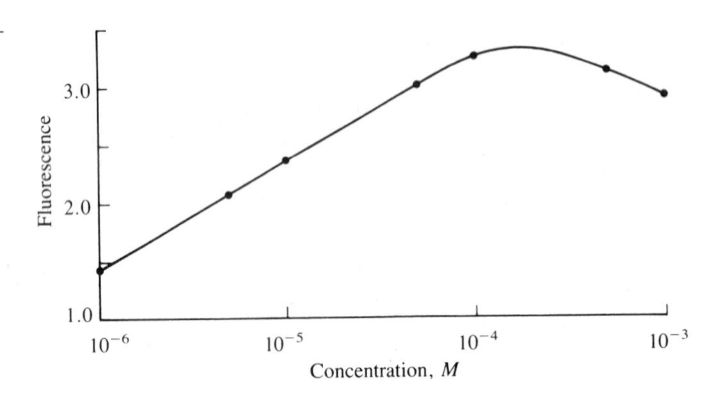

FIGURE **8.3**
Fluorescence/concentration graph for the coenzyme NADH in distilled water solution. The linear portion of the curve extends from about 10^{-4} to $10^{-8}M$.

the blank. Solvent fluorescence and light scattering produce signals that at some point obscure the fluorescence of the analyte.

According to the general expression for fluorescence, a sharp negative deviation is exhibited at high absorbance values and concentrations (Equation 8.3). Although a sufficiently dilute concentration has been stated as a concentration that has an absorbance less than 0.05, a stricter limitation puts the value at 0.02 or less. There is a slight curvature in the calibration curve between 0.02 and 0.05 absorbance units. At higher concentrations self-quenching and self-absorption may also be responsible for negative deviations. Self-quenching results when fluorescing molecules collide and lose their excitation energy by radiationless transfer. Collisional impurity quenching leads to loss of fluorescence because an excited complex forms between the excited analytical species and a ground-state impurity molecule and there is subsequent nonradiative energy loss. Dissolved oxygen, being paramagnetic, is a particularly serious offender. Heavy atoms or paramagnetic species strongly affect the rate of intersystem crossing, which, in turn, alters the quantum efficiency for fluorescence or phosphorescence. To avoid quenching in most fluorometric procedures, paramagnetic species must be rigorously excluded from the sample solution. Energy transfer quenching occurs when an impurity is present whose first excited singlet state is at an energy less than that of the excited singlet state of the analytical species. A nonradiative transfer occurs followed by a further radiationless loss by the impurity. Aromatic substances are prime offenders in this category. "Spec-pure" solvents are essential for careful work.

If the solution under investigation contains, besides the fluorescing molecules, a solute that absorbs either the exciting or the fluorescent radiation, the measured fluorescent power is reduced. To minimize this effect, the sample can be diluted to reduce the interfering absorbance provided that the concentration of the analyte is not reduced to an intolerably low level. Another method of minimizing this interference is to view the fluorescence near the front surface of the cell. Likewise, the method of standard additions may be used, since the absorbance of the interfering component remains constant.

If the analyte is too concentrated, the fluorescence versus concentration curve (Figure 8.3) may have a maximum and then actually show a decrease in fluorescent power with increasing concentration. This behavior is due to attenuation of the exciting radiation as it passes through the cell and results in a faster decrease in the exciting power than an increase in fluorescent power in the region of the cell observed by the detector (see Figure 8.2a). Again, observation of the fluorescence from the front part of the cell rather than at 90° can help alleviate this problem. It is imperative in quantitative determinations to be aware of this problem, since a given fluorescent power can correspond to two values of concentration. One should dilute the sample and note whether the fluorescence increases or decreases on dilution to determine on which side of the maximum the solution lies. Another phenomenon that occurs with some solutes in high concentrations is the formation of a complex between the excited-state molecule and another molecule in the ground state, an *excimer*. The excimer may then dissociate into two ground-state molecules with emission of fluorescent radiation but at longer wavelengths than the normal fluorescence. This phenomenon is also sometimes called self-quenching. Dilution helps reduce this effect, since excimer concentration varies as the square of the solute concentration.

8.3 INSTRUMENTATION

A generalized luminescence instrument is illustrated in Figure 8.4. It consists of (1) a source of radiation, (2) a primary filter or excitation monochromator, (3) a sample cell, (4) a secondary filter or emission monochromator, (5) a photodetector, and (6) a data readout device. In contrast to ultraviolet-visible instrumentation, two optical systems are necessary. The primary filter or excitation monochromator selects specific bands or wavelengths of radiation from the source and directs them through the sample in the sample cell. The resultant luminescence is isolated by the secondary filter or emission monochromator and directed to the photodetector, which measures the power of the emitted radiation. For the observation of phosphorescence, a repetitive shutter mechanism or electronic delay system is required.

In the following discussion, fluorescence instruments are categorized as filter fluorometers (also called fluorimeters, but this is not the official IUPAC recommendation), spectrofluorometers, and compensating spectrofluorometers. The prefix "spectro" implies that at least one dispersive monochromator is used in the instrument, usually as the excitation monochromator. A spectrofluorometer has the advantage over a filter instrument of being able to measure the spectral distribution of fluorescence emission (emission spectrum) and the variation in emission spectral radiance with excitation wavelength (excitation spectrum).

Sample Cell Geometry

There are four arrangements for illuminating and viewing the sample: the right-angle (90°) method, the frontal (37°) method, the rotating-cell method, and the straight-through (transmission) method. Figure 8.5 illustrates the first three methods. The right-angle geometry is used almost exclusively in commercial instruments. The 90° geometry is efficient because none of the sample cuvette surfaces directly illuminated by the excitation beam are viewed by the emission monochromator; hence, no cuvette fluorescence (from the trace uranium content of some glasses and from quartz) or radiation reflected from these surfaces enters the emission monochromator. Scattered radiation originates only from the bulk of the

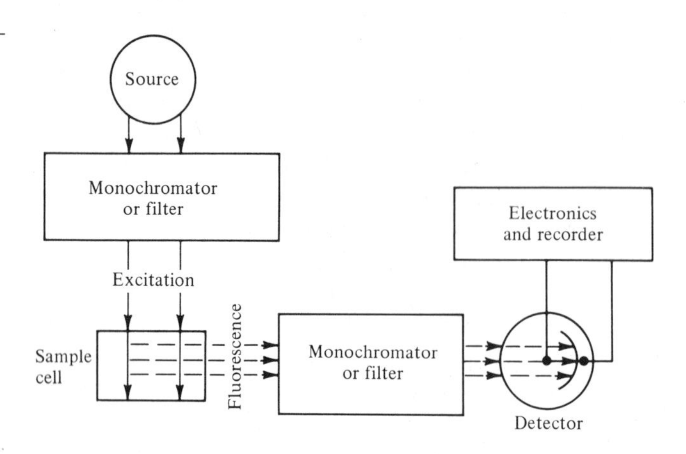

FIGURE 8.4
Basic components of fluorescence instrumentation.

solution itself. In the 90° viewing mode, the excitation radiation does pass through a fairly long solution path so that there is an upper concentration limit observed before attenuation of the exciting radiation disrupts the linear relationship between luminescent power and solute concentration. The arrangement of entrance and exit slits, as shown in Figure 8.2, influences the effective path length (or critical volume viewed).

The frontal method of cell illumination is used primarily for semiopaque materials or solids, or for solutions that are highly absorbing. The disadvantage of this configuration is that the emission monochromator views directly illuminated cell surfaces with their own residual fluorescence and unavoidable reflected radiation. Reflected radiation is minimized by the choice of 37° as the take-off angle.

The straight-through method is seldom used, although it formerly enjoyed some popularity with fluorescing pellets after the fusion fluxes were cooled in the determination of uranium with LiF-Na_2CO_3.

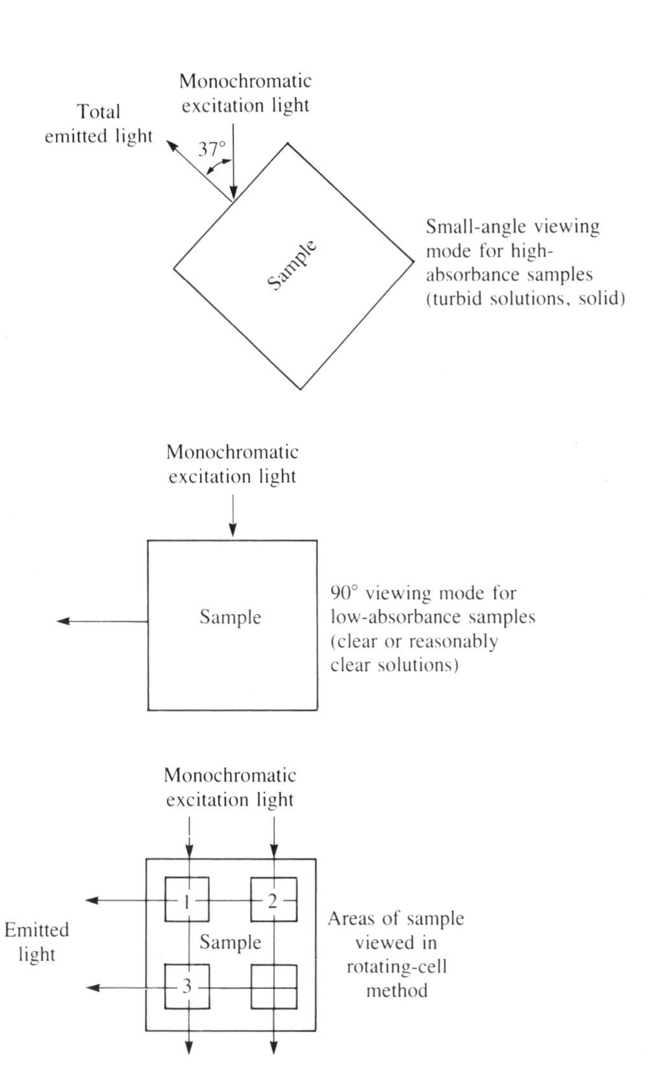

FIGURE **8.5**
Viewing modes in fluorescence.

The rotating-cell method has recently been developed and presents a novel way of evaluating and correcting the fluorescent power for primary (exciting) beam absorption and secondary (emission) beam absorption.[7] The cell is rotated so that three measurements are taken in the positions illustrated in Figure 8.5. The difference in the power between positions 1 and 2 gives a measure of the absorption of fluorescence radiation by the sample. The difference between positions 1 and 3 gives a measure of the absorption of exciting radiation by the sample. Using the corrections obtained by the cell rotation method, fluorescence/concentration graphs linear to absorbances as high as 2.7 have been obtained.

Sources

The primary factors to consider when selecting a radiation source for luminescence instrumentation are lamp intensity, the wavelength distribution of emitted radiation, and stability. Of particular interest from Equation 8.5 is the linear dependence of the luminescent signal on the power of the exciting radiation. Consequently, it is advantageous to use a source as powerful as possible. Scanning spectrofluorometers require radiation sources that emit continuously over a wide spectral range. Filter fluorometers or spectrofluorometers, used specifically for analytical measurements at fixed wavelengths, may use atomic spectral line sources.

High-pressure dc xenon arc lamps are used in nearly all commercial spectrofluorometers. The xenon lamp emits an intense and relatively stable continuum of radiation that extends from 300 to 1300 nm. Several strong emission lines lie between 800 and 1100 nm. The spectral output approximates that of a blackbody radiator with the equivalent color temperature of about 6000 K (see Figure 6.2). During lamp operation the arc discharge is compressed within the narrow gap between the electrodes. Arc flicker determines the short-term stability, about 0.3%. Long-term stability is 1% drift per hour and is limited by electrode wear and arc wander. In contrast with mercury arcs, the spectral distribution is not as strongly dependent on operating gas pressure and voltage.

A xenon flash lamp is a compact, low-cost source. The sample is excited by a high-energy flash produced by the discharge of a charged capacitor through a lamp filled with xenon. By making the flash repetitive, ac methods of amplification can be used and full advantage taken of the high peak flash intensity. A 0.8-mm-diameter capillary flash lamp has obvious advantages for microcell and continuous-flow arrangements, since the image produced at the sample position is about 2 mm wide and 18 mm high.

Low-pressure mercury vapor lamps are most frequently used in filter fluorometers. The fill gas pressure is quite low (approximately 10 torr). The resulting arc discharge is spatially very diffuse and much less intense than that of a high-pressure arc. The stability is generally better. The lamps may be coated with a phosphor to emit a more nearly continuous spectrum, or they may have a clear bulb of ultraviolet-transmitting material to permit use of the individual mercury lines that appear at (intensity) 253.7 (very strong), 313 (m), 365 (m), 404.7 (m) 407.8 (w), 435.8 (s), 546.1 (s), 577.0 (m), and 579.1 (m) nm. Interference filters are used to select individual mercury lines for excitation, or bandpass filters are used to select for several lines. The discrete line emission output of the mercury lamp shows considerable sensitivity for a material whose excitation spectrum coincides with a mercury emission line but less sensitivity for a more strongly fluorescent compound, which is excited more efficiently with another excitation wavelength but not one appearing in the mercury lamp output.

Clearly, a laser would be an acceptable source provided monochromatic radiation or very high intensities were needed for excitation. Ultratrace inorganic ion determination is one area where laser-excited fluorescence is applicable.[8]

Filter Fluorometers

Filter fluorometers offer the advantage of lower cost and convenience for repetitive, quantitative determinations. In this class of instruments, excitation and emission wavelengths are chosen by absorption or interference filters. For repetitive routine analyses, the lack of versatility is not a major drawback, and the high sensitivity, if the excitation wavelength is at an emission line of the source, may be a distinct advantage. The small size of filters, as compared with monochromators, results in a compact instrument.

A filter fluorometer usually consists of a mercury lamp as an excitation source, a primary filter to transmit the desired excitation wavelength, and a sample cuvette. A photomultiplier tube measures the fluorescence emission. The secondary filter between the sample and the photodetector is selected to transmit the fluorescence and to absorb scattered excitation radiation. If flow cells are incorporated into the sample compartment, they can be part of continuous-flow instrumentation such as high-performance liquid chromatographic (HPLC) detection systems (see Chapter 19).

The use of optical filters results in very high excitation radiation levels and efficient detection, but sacrifices the selectivity obtained by the more precise selection of excitation and emission wavelengths with monochromators. Excitation filters are generally bandpass types, which transmit a rather broad band of wavelengths, although interference filters are used to isolate single mercury lines. Off-band transmission, and hence stray radiation level, is usually quite high for interference filters compared with absorption filters. This is attributable to unavoidable pin holes and defects in the film coatings. Emission filters are usually of the sharp-cutoff type, which pass long wavelengths and attenuate shorter wavelengths. A point to remember is that bandpass filters have more than one transmission band or window. Photodetectors often have a low-level response over a large wavelength range. This may result in a high stray-radiation signal. Sharp-cutoff filters, particularly glass filters, frequently fluoresce themselves.

Filter fluorometers almost always use a single-beam arrangement with source intensity control to lessen the effects of fluctuations and drift in source intensity and detector response. A variety of ingenious ways for obtaining the monitoring channel have been devised. A ratio system is shown in Figure 8.6. The optical path of the reference beam is shown on the right side of the diagram. A portion of the lamp radiation passes through the primary optical system and is attenuated by means of the reference aperture disk before reaching the reference photodetector. The electrical signal from the reference photodetector and the sample signal from the sample photodetector are fed into a solid-state electronic divider, which computes the ratio of sample-to-reference signals.

For a filter fluorometer (or any spectrofluorometer operated at constant excitation and emission wavelengths), Equation 8.5 is reduced to $P = kc$. The constant k may be established by calibration with standards. The dial or meter reading is proportional to concentration. Measurements may be extended to extremely low concentrations by increasing the sensitivity of the photodetector or the intensity of the radiation source.

Fluorescence measurements are usually made by reference to some arbitrarily chosen standard. The standard is placed in the instrument and the circuit is balanced with the reading scale at some chosen setting. Without readjusting any circuit components, the standard is replaced by known solutions of the analyte and the fluorescence of each is recorded. Finally, the fluorescence of the solvent and cuvette alone is measured to establish the true zero concentration reading. Some fluorometers are equipped with a zero-adjust circuit. A plot of fluorescence readings against the concentration of the reference solutions furnishes the calibration curve. Some commonly used fluorescence standards are rhodamine B in ethylene glycol, quinine hydrogen sulfate in $0.1N$ H_2SO_4, tryptophan in water, and anthracene in cyclohexane or ethanol. Glass reference filters are also suitable.

Spectrofluorometers

The greatest analytical scope of fluorescent analysis is achieved by replacing filters with grating monochromators to give scannable wavelength selection throughout the region 200–800 nm, the most useful region for the fluorescence technique. Another advantage of spectrofluorometry is that the analyst sees scattered radiation by comparing the sample spectrum with a blank spectrum and examining for distortion of the fluorescence peak or the presence of additional peaks. A filter photometer cannot optically resolve the various Rayleigh (scattered radiation from small particles—the frequency is the same as that of the incident radiation, see Section 5.6) and Raman (scattered radiation by molecules, which has frequencies somewhat higher and lower than the incident radiation, see Section 5.6 and Chapter 12) scatter peaks, and the analyst does not observe their presence.[9]

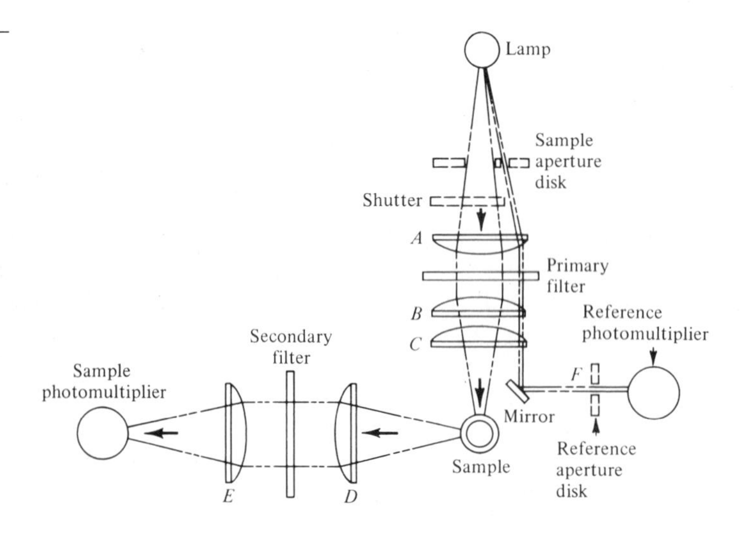

FIGURE **8.6**

Optical diagram for Farrand Ratio Fluorometer. (Courtesy of Farrand Optical Co., Inc.)

Spectrofluorometers usually incorporate grating monochromators of the Czerny-Turner type with a 0.25-m focal length and $f/4$ or $f/5$ aperture. The monochromators use gratings with 600 grooves/mm, blazed (in the first order) for 300 nm in the excitation unit and for 500 nm in the emission unit. Filters are used to block out higher-order diffracted radiation. The basic concepts of a spectrofluorometer are illustrated in Figure 8.7. The excitation monochromator is located between the radiation source and the sample, and the emission monochromator is between the sample and the photomultiplier tube. For quantitative analysis, one selects the desired excitation and emission wavelengths and compares the relative fluorescent powers of standard and unknown samples. For good spectral selectivity and ability to resolve spectral fine structure, the emission monochromator should be able to resolve two lines 1.0 nm apart.

Instruments that measure the characteristics of the sample without a continuous comparison to a reference standard have the same major drawbacks that appear in attempting to record absorption spectra with a single-beam instrument (Figure 8.8a). If the source is unstable and varies in intensity, it can create false peaks in the spectrum. Equally serious is the variation of the sensitivity of the photodetector with regard to wavelength. The term *uncorrected* is applied to this type of spectrofluorometer because the excitation and emission spectra presented are a combination of the true spectra of a compound and various instrumental artifacts. When excitation spectra are plotted, no attempt is made to hold P_0 constant. Excitation spectra are hence a composite of the true excitation spectra, the spectral distribution of lamp output with wavelength, and excitation monochromator efficiency with wavelength. Likewise, emission spectra are a composite of the true emission spectra, the spectral distribution of detector response, and emission monochromator efficiency with wavelength. In certain regions of the spectrum these instrumental factors become dominant. Despite these drawbacks,

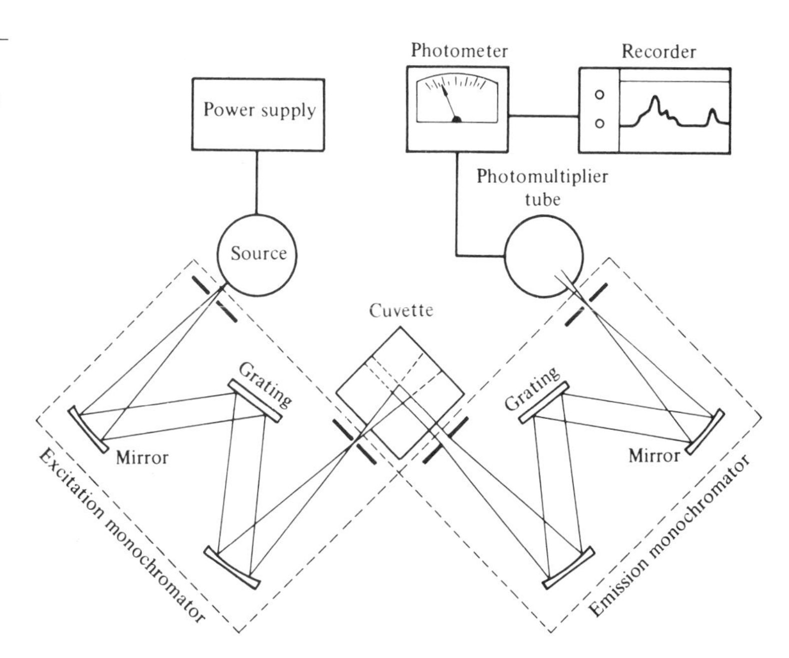

FIGURE **8.7**
Schematic diagram of a fluorescence spectrophotometer.

the uncorrected spectrofluorometer is used for quantitative analyses with the convenience that wavelengths are selected and used for rough comparisons of spectra from other laboratories and direct comparison within one laboratory.

Instrumental limitations caused by the instability of the xenon source are overcome by the *ratio mode* operation (Figure 8.8b). A small fraction of the exciting radiation is directed to a reference photodetector, which is chosen primarily for wide wavelength response. The emitted radiation is isolated by the emission photodetector especially selected for low dark current and high sensitivity. The output signal of the reference phototube is amplified and fed into the photomultiplier dynode voltage control circuit, where it is used as a monitor to control the dynode voltage of both photomultiplier tubes in an inverse relationship. As the excitation radiation increases or decreases in power due to fluctuations in the xenon lamp, there is a corresponding increase or decrease in relative fluorescence. The reference detector is the component in the system with the necessary range and speed of response to monitor the excitation radiant power. Since the signal is monitored after the excitation monochromator, excitation spectra are compensated for wavelength-dependent energy fluctuations. It does not provide a true corrected excitation spectrum, however.

As the chopper rotates, all radiation is blocked from one or the other photomultiplier tube. Any signal from a detector during the period when the chopper blocks radiation from it is dark current inherent in the detector. The two signals from the photomultipliers, now separated in time, are fed into a differential

FIGURE **8.8** Different operational modes in fluorescence instrumentation shown schematically.

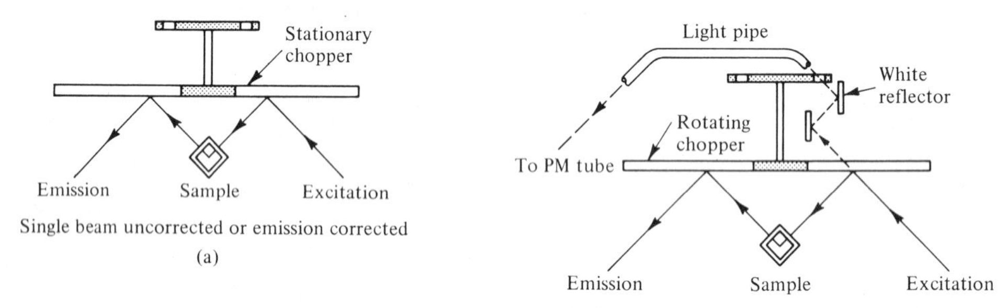

Single beam uncorrected or emission corrected
(a)

Ratio mode. Uncorrected or emission corrected.
(b)

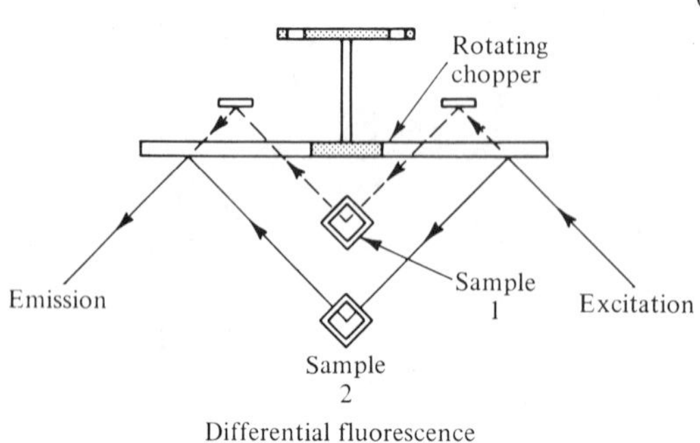

Differential fluorescence
(c)

amplifier, which effectively subtracts dark current and presents only the difference signal to the recorder.

Double Monochromation. The optical diagram of an instrument that uses double monochromation is shown in Figure 8.9. The optical section of the instrument comprises four grating monochromators. Two are used in tandem for dispersing excitation energy and the other two are used for examining the fluorescence spectra. An optical filter inserted into the radiation path between the xenon source and the excitation monochromator rejects second-order excitation when excitation wavelengths longer than 400 nm are used. The double monochromator drastically reduces scattered radiation and also permits the use of larger slits to achieve the same resolution as in conventional monochromators that have smaller slit widths. Larger slits allow more radiation to pass through the excitation monochromator to excite the sample. This is very advantageous for samples at low concentrations because a maximum amount of energy is made available without any sacrifice in resolution.

Double-Beam Fluorescence Spectrophotometer.[10] *Double beam* refers to an optical-electronic system that permits the use of a second optical beam for essentially simultaneous compensation of the sample fluorescence by comparison with a reference sample (Figure 8.8c). Double-beam spectrophotometry has increased convenience over single-beam measurements because reference sample compensation can be performed by the instrument automatically and accurately. Figure 8.10 shows the optical schematic of one instrument. The design concept utilizes a classic time-sharing electro-optical system with equivalent optical paths for sample and reference beams and an optical chopper to create an ac signal as well as to split the beam. The arc of a 150-W xenon lamp is focused on the entrance slit of the

FIGURE **8.9** Optical diagram of a fluorescence spectrophotometer that uses double monochromation. G, gratings; M, mirrors; S, slits. (Courtesy of Baird-Atomic.)

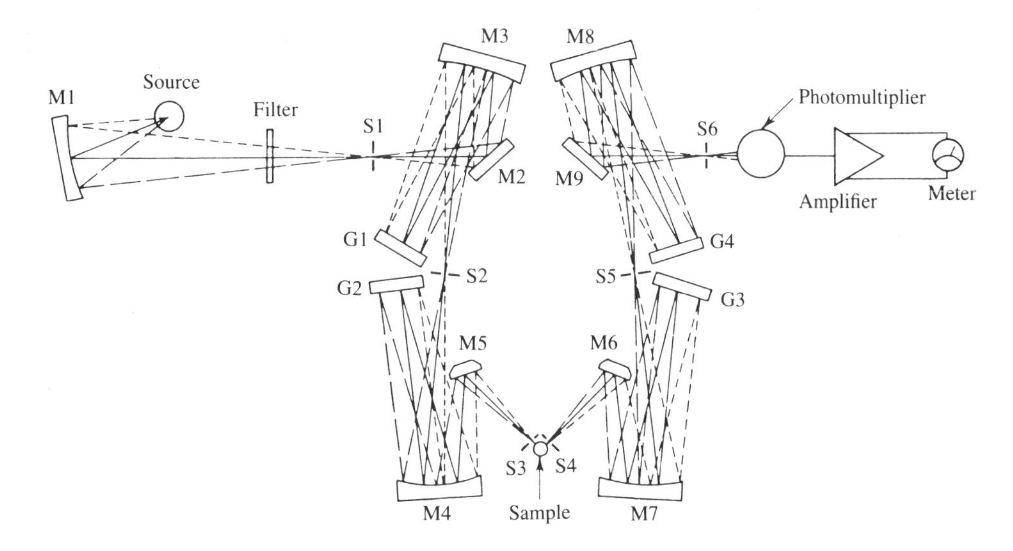

entrance monochromator after the radiation is optically chopped with a bow-tie-shaped disk. A sector mirror located beyond the exit slits of the excitation monochromator is driven by a shaft common to the chopper.

In the fluorescence double-beam mode the emitted radiation is viewed at right angles to the exciting radiation for both the sample and reference beams. The radiation from each beam is focused alternately on the entrance slit of the emission monochromator using common or equivalent front-surface mirrors. The lattice mirror is a half-aluminized, half-transmitting flat mirror that reflects the sample beam and transmits the reference beam to the emission monochromator. Both beams that emerge from the exit slits of the emission monochromator are imaged on the photomultiplier detector 180° out of phase.

When the chopper disk is open the first time, the sector mirror reflects the beam to the reference position. When the chopper disk is open the second time, the sector mirror is also open and therefore allows radiation to pass to the sample position. The remaining time of each revolution of the chopper measures the optical zero during the opaque portion of the disk.

The double-beam technique provides greater analytical precision, convenience, and speed over comparable single-beam techniques by canceling out the interfering scatter and fluorescence of reagents. It also enables small differences to be measured between two very similar fluorescent samples.

Complete correction for the excitation spectrum is made by recording the ratio of the output signal from the sample detector to that of a reference detector that monitors the emission of a quantum counter. The quantum counter is a special cell that contains a high concentration of a fluorescent material, often rhodamine B (3 g/liter in ethylene glycol). Over a considerable range of exciting wavelengths, all incident energy is absorbed and a fixed quantum fraction is reemitted at essentially fixed wavelength (640 nm). A portion of this emitted beam is then allowed to fall alternately on the same photomultiplier that is used to measure the fluorescence emission from the sample. Signals caused by the light from the quantum counter and the sample are electrically separated and used to operate a ratio

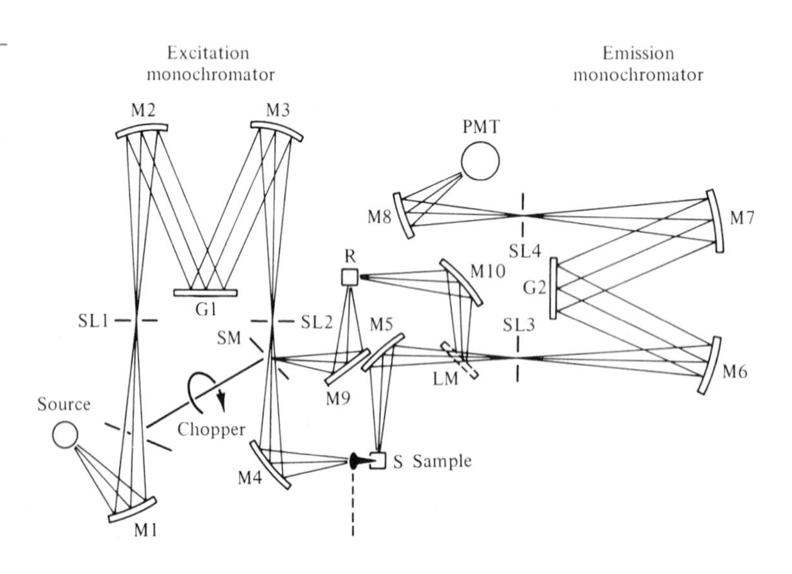

FIGURE **8.10**
Optical diagram of a double-beam fluorescence spectrophotometer. SM, sector mirror; G, grating; LM, lattice mirror; M, mirror; PMT, photomultiplier tube; R, reference cuvette; SL, entrance and exit slits. (Courtesy of Perkin-Elmer Corp.)

recorder. Since the light emitted from the quantum counter is proportional to the quanta falling on the sample, corrected excitation spectra are obtained automatically.

The *corrected spectra computer* is a microcomputer that automatically stores correction factors when rhodamine B or known reference samples are used.[11] During the calibration mode the operator must only scan the spectrophotometer in a prescribed manner. For subsequent wavelength scans, in either corrected excitation or corrected emission modes, the information stored in memory during the calibration mode scan is recalled and used to correct automatically the spectral data. In other words, an error curve is generated and stored for both the excitation and emission spectra.

Corrected fluorescence measurements yield useful information concerning energy transfer studies (inter- and intramolecular), energy of the Stoke's shifts, and studies of macromolecules. Because a corrected fluorescence excitation spectrum is identical to an absorption spectrum, but at 1000 times the sensitivity, it is used advantageously for trace analysis applications that normally use ultraviolet–visible absorption spectrophotometry.

Total Luminescence Spectroscopy[12]

Total luminescence spectroscopy is a method for measuring and displaying spectral intensity as a function of *all* useful excitation and emission wavelengths. A computer-controlled spectrofluorometer is used to scan the emission spectrum of a sample repeatedly as the excitation wavelength is increased in small increments. Intensity values are recorded and plotted on a digital plotter. The display is a three-dimensional representation comprising a series of closed, continuous, equal-intensity contours (see Figure 8.11). It provides spectral data much as a topographical map displays height above sea level on a plane surface. The result is a graphic pattern in which spectral peaks, anomalies, and symmetries are readily observed. In many cases the known signature patterns of specific compounds are immediately discerned or deduced. For complex mixtures in which overlapping patterns prevent the identification of individual components, considerable information is still available. Classes of constituent compounds are often identified, differences between closely similar compounds are noted, and changes in a particular mixture, sampled at different times or under different conditions, become readily apparent.

Once the data are stored in the computer, many different types of mathematical manipulations are possible. One involves the subtraction of the solvent background. When contour maps of individual components are available and there is a set of calibrated matrices, it is straightforward to solve for the concentration of the individual components using a set of simultaneous equations. Several different pairs of excitation/emission wavelengths are required to solve for the concentration of the individual components in the mixture. The maximum emission power for a component with the least interference from the others is selected. The total luminescence contour of the mixture aids in the determination of excitation/emission wavelength pairs that maximize the response to the desired materials while minimizing the response to other materials. When a bank of standard spectra is available, for quantitation, once a complete spectrum has been measured, the spectral contribution of one component at a time is subtracted from the mixture,

FIGURE **8.11** Front (a) and back (b) views of the three-dimensional excitation fluorescence spectrum of Quaker State Super Blend Motor Oil. [Reprinted with permission from J. S. Siegel, *Anal. Chem.*, **57,** 934A (1985). Copyright 1985 American Chemical Society.]

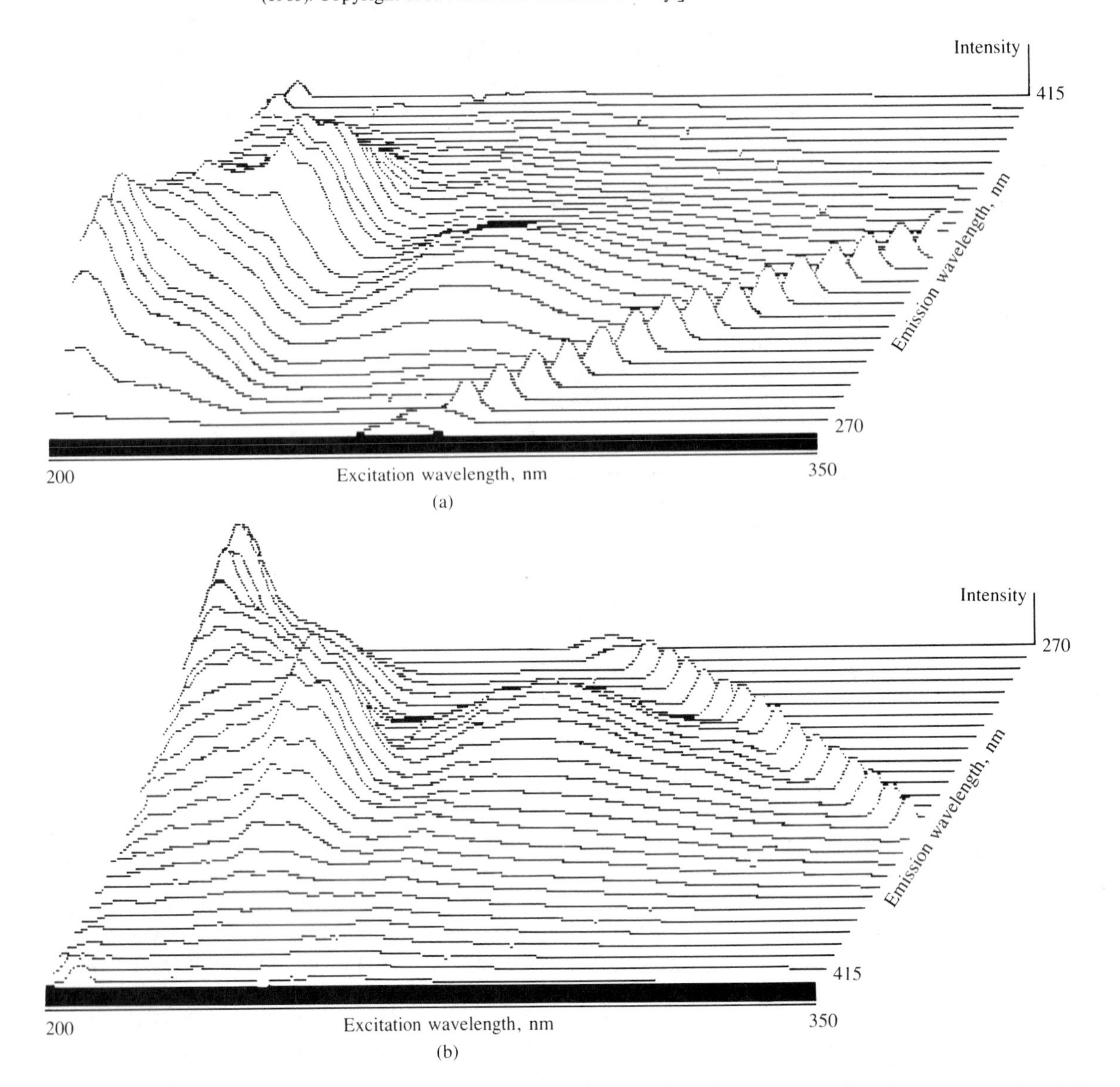

eventually resulting in a flat contour, which confirms that all components of the mixture are accounted for.

8.4 FLUORESCENCE LIFETIME MEASUREMENTS[13,14]

As mentioned in the introduction to this chapter, fluorescence lifetimes are on the order of 10^{-9} to 10^{-6} sec. Modern instrumentation permits the measurement of such short lifetimes and therefore adds another selectivity parameter to luminescence measurements. Two different techniques are used for such measurements, time-resolved and phase-resolved.

In the time-resolved method, the sample is excited by a short-duration pulse of radiation and the resulting luminescent power is measured as a function of time after the pulse ceases. This method is discussed in the following section, which deals with phosphorescence where relatively simple instrumentation can measure the long lifetimes involved. With laser pulses and modern electronic circuits, the very short lifetimes of fluorescence are measurable. Most luminescence decay curves are apparently first-order kinetically and thus,

$$P_F(t) = P_F(0)e^{-kt} \tag{8.8}$$

where $P_F(t)$ is the fluorescence power at time t, $P_F(0)$ is the power at time zero, and k is the first-order rate constant. Since the radiation source gives a finite pulse, however short, and the electronics have a finite response time, a mathematical process is needed to deconvolute the measured decay curve from the measurements obtained.[15]

A second time-resolved method is the time-correlated, single-photon counting method. In this case the source is pulsed repetitively. Two photomultiplier tubes are placed so that one observes the radiation source and the other observes the fluorescing sample. The first photon from the source to reach its photomultiplier tube starts a time-to-amplitude converter, and the first photon from the sample stops the converter. The output of the converter is a voltage pulse proportional to the time interval. By measuring a large number of pulses and fluorescence responses, sorting and counting the voltage outputs in a multichannel analyzer, one obtains a plot of power versus time for the fluorescing material, since the time between the pulse and photon release is a function of the fluorescence lifetime. Since only one photon per pulse is counted, the pulse intensity can be very low. The more photons counted, the better the results.

Another approach to time-resolved fluorometry is cross correlation.[16] In this technique a randomly modulated source is used. Such a source should approach "white light"; that is, it should contain all frequencies, thus appearing noisy. The radiation from a fluorescing material irradiated with this source appears noisy too, but, in reality, each frequency component is represented and modified by the fluorescing molecule. The population of the excited states of a molecule follows low-frequency fluctuations, cannot follow high-frequency fluctuations, and partially follows mid-frequency fluctuations depending on the lifetime of the excited state. A continuous-wave (CW) laser is a good source because the laser output contains apparent amplitude modulations at discrete frequencies related to the laser's mode spacing—that is, at $c/2L$, $2c/2L$, $3c/2L, \ldots, nc/2L$, where c is the velocity of light and L is the distance between the laser's mirrors. For many lasers, L is about 1 m so the modes are multiples of 150 MHz up to 4.5 GHz or higher for some types of lasers.

With a microwave spectrum analyzer, the ratio of fluorescent radiant power to source power at each frequency is measured and plotted. The resulting plot of fluorescent radiant power versus frequency is the frequency-domain equivalent of the time-domain exponential decay. Thus it is a Lorentzian curve, the Fourier transform of an exponential. Either a Fourier transform is made on the data or, by plotting the curve with appropriate abscissa units, the half-life is read as the reciprocal of the Lorentzian half-width.

Another approach to cross correlation is to split the laser beam into two parts, with one part going to the fluorescing sample and the other going to a photodetector on a movable track. When this photodetector is moved, a variable delay is introduced in detecting the laser pulses. If the waveforms observed by the two detectors are multiplied, the product is zero except during the time when the laser pulse strikes its detector. At that time, the product of the outputs is a sample of the fluorescence power at the time delay introduced in the direct laser pulse beam. Thus the decay curve is traced directly by smoothly moving the detector in the laser beam section.

In the phase-resolved method of lifetime determination, the sample is excited by a continuous but sinusoidally modulated radiation source. The radiant power of the source can be represented by

$$P_F(t) = A(1 + m_{ex} \sin 2\pi f t) \tag{8.9}$$

where A is the dc power (average) of the exciting beam, m_{ex} is the degree of ac modulation—that is, the ratio of the amplitude of the ac component to the dc component, and f is the frequency of modulation. The emitted fluorescence as a function of time, $P_F(t)$, is phase shifted and also partially demodulated to an extent dependent on the lifetime of the fluorescing species (Figure 8.12). The radiant power of the emitted fluorescence as a function of time is given by

$$P_F(t) = A'(1 + m_{ex} m \sin(2\pi f t - \Phi)) \tag{8.10}$$

where A' is the dc component of the fluorescent emission, Φ is the phase-shift angle, and m is the demodulation factor:

$$m = \cos \Phi \tag{8.11}$$

FIGURE **8.12**

Phase shift of fluorescent samples when excited by sinusoidal radiation: (a) short lifetime and (b) longer lifetime. Broken line is excitation radiation.

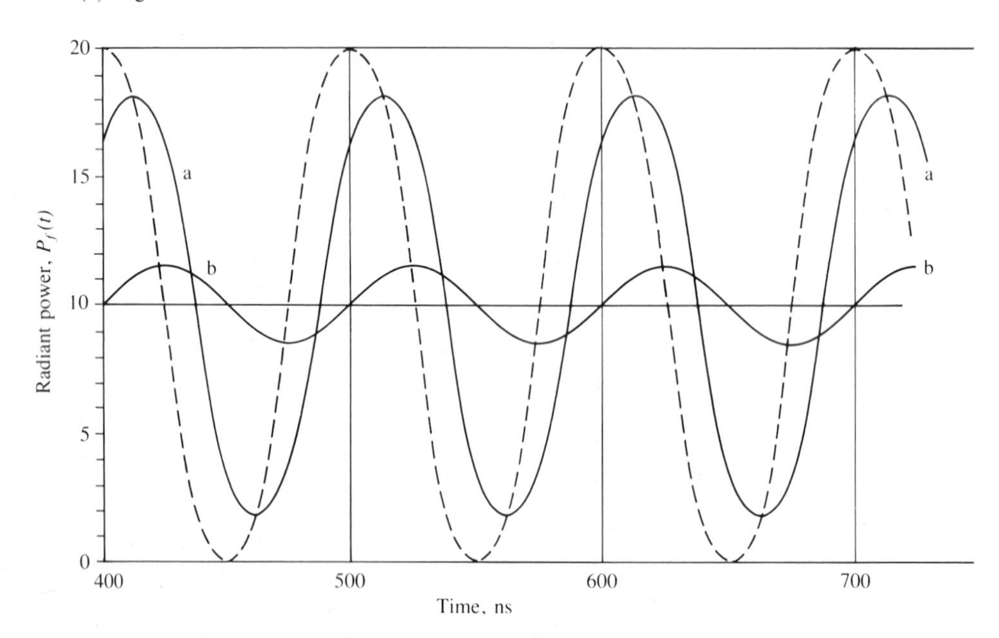

The demodulation factor and phase shift are, in practice, measured relative either to a scattering solution that has a lifetime, τ, equal to zero or to a reference solution with known lifetime. The fluorescence lifetime is calculated from the phase shift or the demodulation according to Equation 8.12 or Equation 8.13:

$$\tau = \frac{1}{2\pi f} \tan \Phi \tag{8.12}$$

$$\tau = \frac{1}{2\pi f} \left(\frac{1}{m^2} - 1 \right)^{1/2} \tag{8.13}$$

Several methods of modulating radiation are available, but the most versatile is to pass a laser beam through an electrooptically modulated Pockels cell. A Pockels cell is a noncentrosymmetric crystal that changes its optic axis when a voltage is applied across it. If plane-polarized radiation passes through the crystal parallel to the optic axis, it is transmitted without diminution and, if not parallel, the intensity is reduced by a factor proportional to $\cos^2 \theta$, where θ is the angle between the plane of polarization of the incident radiation and the optic axis of the crystal. Thus polarized light from a laser beam can be varied in intensity from zero when the axis is at 90° to 1 when the axis is parallel to the polarization. The detected emission signal is compared with a reference signal of the same frequency. In a method known as phase-resolved fluorescent spectroscopy (PRFS), the detector is controlled to measure the signal during only half of the modulation cycle; however, the half measured can be shifted at will by offsetting the detector by a variable phase angle, Φ_D. When $\Phi_D = \Phi$, where Φ is the phase angle of the fluorescent radiation, a maximum signal results, and at $\Phi_D = \Phi \pm 90°$, a null signal is obtained. This furnishes a means of distinguishing several components with different fluorescence lifetimes and, therefore, different phase angles.

With either method of lifetime measurement, it is possible to analyze multicomponent mixtures. First, the measured emission curves are broken down into the individual component curves. For the time-resolved method, a curve-stripping procedure might be used in which the longest lifetime component is determined from the measurements after shorter lifetime components have decayed. This curve is extrapolated back to time zero and subtracted from the total curve, the next longest lifetime curve is determined, subtracted, and so on. In the phase-resolved method, several alternatives are used to separate component spectra. One of the best is to set up a series of simultaneous equations by measuring standards and the multicomponent solution at a series of detector phase angles, Φ_D. Since power is proportional to concentration, the proportionality constant is determined at each phase angle for each known component from the standard solutions. The number of measurements (phase detector angles) must equal or exceed the number of components.

8.5 INSTRUMENTATION FOR PHOSPHORESCENCE MEASUREMENTS

Instrumentation for phosphorescence investigations is identical to that described for fluorescence measurements with the addition of a radiation interrupter and provision for immersion of the sample in a Dewar flask for liquid nitrogen temperatures. Excitation radiation from a xenon source, after dispersion by the excitation monochromator, is admitted to the sample via a fixed slit system and a rotating can chopper (Figure 8.13) (or a set of slotted disks with equally spaced ports) that permits excitation of the sample and periodic out-of-phase measurement of phosphorescence. This allows measurement of the phosphorescent signal without interference from scattered radiation and short-lived fluorescence. The sample cuvette is a small Dewar flask made of fused silica and silvered, except in the region where the optical path traverses the Dewar. The solvent frequently used is a mixture of diethyl ether, isopentane, and ethanol in a volume ratio of 5:5:2. When cooled to liquid nitrogen temperature, it gives a clear transparent glass.

The resolution time of the instrument is the length of time between the cutoff of each pulse of excitation radiation admitted to the sample and the clearing of the optical path by the second opening to allow the phosphorescence emission to enter the emission monochromator. This time is a function of the motor speed, the size and spacing of the openings, and the relative radial positions of the ports to one another. Decay curves can be recorded if the detector circuit is equipped with an oscillograph. Time resolution in phosphorimetry is considerably improved by the use of a microsecond-duration pulsed source in place of a rotating can. In this case, the photomultiplier detector is gated so as to be off during the flash in order to avoid overloading the detector from large scattered radiation levels. A pulsed radiation source provides fast cutoff of the excitation radiation and allows phosphorescence decay times down to milliseconds to be observed and recorded.

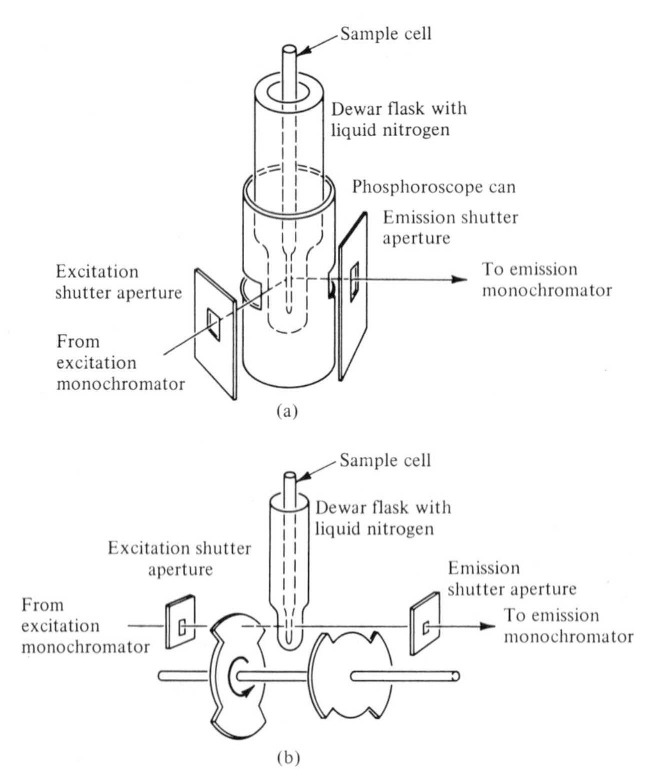

FIGURE **8.13**
Schematic diagram of the interruption of excitation radiation and phosphorescence emission: (a) rotating can device and (b) rotating shutter. [Reprinted with permission from T. C. O'Haver and J. D. Winefordner, *Anal. Chem.*, **38**, 602 (1966). Copyright 1966 American Chemical Society.]

Despite the obvious advantages of time resolution, phosphorimetry is not as widely utilized as fluorometry. This is due to the added complexity of phosphorimetric instrumentation, the smaller number of species that phosphoresce, and the inconvenience of having to cool the sample to obtain adequate phosphorescence quantum efficiencies.

8.6 ROOM TEMPERATURE PHOSPHORESCENCE

Recently several substances have been observed that have measurable phosphorescent intensities at room temperatures when adsorbed on a solid.[17,18] The processes that lead to phosphorescence at room temperature when certain molecules are absorbed on solid surfaces are not clearly understood. Different solid supports, such as silica gel, filter paper, and thin-layer chromatography papers, influence the luminescence. Some molecules phosphoresce better on one support than another, or they may fluoresce on one type of support and phosphoresce on another. Heavy ions, such as Pb^{2+}, Hg^{2+}, or I^-, tend to promote phosphorescence and quench fluorescence of some compounds. Likewise, pH has a major effect on the luminescence of some compounds.

Asafu-Adajaye and Yue have developed techniques for the determination of many pesticides and toxic substances without separation using different substrates, different heavy ions, pH control, and different emission and excitation wavelengths.[19] Both fluorescence and phosphorescence were used in the determinations.

Ramasamy and colleagues point out that quantum efficiencies for the luminescence of substances adsorbed on solid surfaces at room temperature are much lower than for the same substances in solution at low temperatures.[20] They also point out that oxygen still retains its strong quenching effect, and, therefore, one should remove oxygen by flowing nitrogen over the solids. Nevertheless, the convenience of working at room temperatures rather than very low temperatures makes room-temperature phosphorescence an attractive option.

8.7 COMPARISON OF LUMINESCENCE AND ULTRAVIOLET-VISIBLE ABSORPTION METHODS

Luminescence is usually the method of choice for quantitative analytical purposes if applicable. Fewer substances luminesce than absorb radiation in the ultraviolet or visible region. In fact, a substance has to absorb before it can emit luminescent radiation. Thus the selectivity of luminescent methods is greater than absorption methods. Furthermore, luminescence is more selective because both the emission and absorption spectra can be obtained. If commercial instrumentation for measuring luminescent lifetimes becomes available, a third factor would be available for discrimination among compounds.

Luminescence measurements are usually more sensitive than absorption methods. In absorption spectrophotometry, concentration is proportional to absorbance, which is the logarithm of the ratio between large quantities (incident and transmitted radiant power, P_0 and P). In fluorescence and phosphorescence methods, concentration is directly related to the luminescent radiant power, which can be measured at right angles to the incident radiation. This means that luminescent power is measured against a very small background. It is always easier to measure a small signal against a very small or zero background than to measure a small difference between large signals.

Luminescence, nevertheless, is not as widely used in analytical laboratories probably because many fewer compounds luminesce and because luminescence instruments are more complicated and expensive than absorption instruments. If both emission and absorption spectra are measured, two monochromators are required.

PROBLEMS

1. A 1.00-g sample of a cereal product was extracted with acid and treated to isolate the riboflavin plus a small amount of extraneous material. The riboflavin was oxidized by the addition of a small amount of $KMnO_4$, the excess of which was removed by H_2O_2. The solution was transferred to a 50-mL volumetric flask and diluted to the mark. A 25-mL portion was transferred to the sample cuvette and the fluorescence was measured. Initially the fluorometer had been adjusted to read 100 scale divisions with a solution of quinine bisulfate. The solution read 6.0 scale divisions. A small amount of solid sodium dithionite was added to the cuvette to convert the oxidized riboflavin back to riboflavin. The solution now read 65 scale divisions. The sample was discarded and replaced in the same cuvette by 24 mL of the oxidized sample plus 1 mL of a standard solution of riboflavin that contained 0.500 μg/mL of riboflavin. A small amount of solid sodium dithionite was added. The solution read 94 scale divisions. Calculate the micrograms of riboflavin per gram of cereal.

2. Solutions of varying amounts of aluminum were prepared, 8-quinolinol was added, and the complex was extracted with chloroform. The chloroform extracts were all diluted to 100 mL and compared in a fluorometer. The readings in the table were obtained. Plot the fluorometer reading versus the aluminum concentration. Over what concentration range is Equation 8.6 valid?

Aluminum (μg/50 mL)	Fluorometer reading	Aluminum (μg/50 mL)	Fluorometer reading
2	10	12	53
4	19	14	60
6	28	16	66
8	37	18	71
10	45		

3. From the spectrum for phenanthrene shown in the figure, select the optimum wavelength of excitation and for fluorescence emission. Do this (a) assuming only a filter fluorometer is available and (b) when a grating monochromator spectrofluorometer is available.

PROBLEM 3

Excitation and emission spectra of phenanthrene. *A*, excitation; *F*, fluorescence; *P*, phosphorescence.

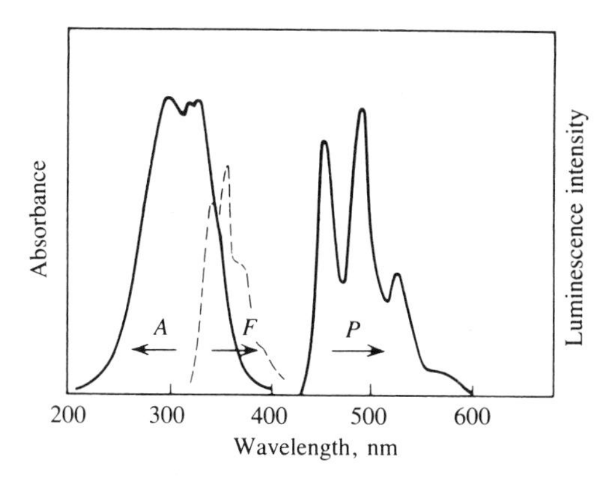

4. From the spectra for phenanthrene and naphthalene (see the figure), devise a method of analysis for each component assuming (a) only a spectrofluorometer is available and (b) both a spectrofluorometer and a phosphorescence spectrometer are available.

PROBLEM 4

Excitation and emission spectra of naphthalene. *A*, excitation; *F*, fluorescence; *P*, phosphorescence.

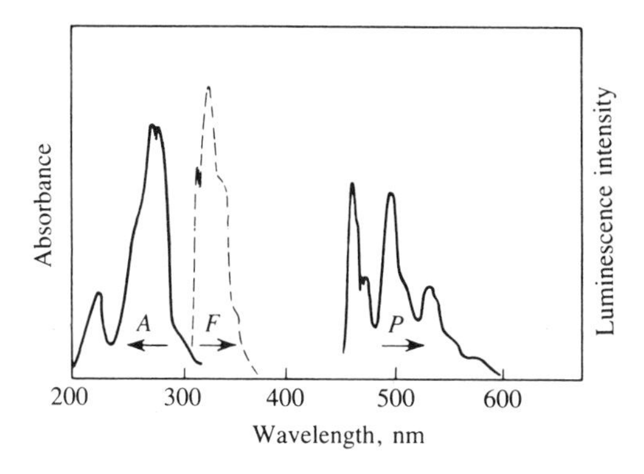

5. (a) Although early optical designs involved front-face fluorescence from solution cells, why is this arrangement impractical? (b) What two phenomena does the right-angle arrangement for solutions introduce?

6. (a) How can Raman and Rayleigh scatter be identified when included in a fluorescence emission spectrum? (b) What is the limiting effect of Raman scatter when added to the sample measurement? (c) When Raman scatter does interfere, what four alternatives does the analyst have for eliminating it? (d) Raman shifts of various solvents are as follows: cyclohexane (2880 cm^{-1}), water (3380 cm^{-1}), ethanol (2920 cm^{-1}), and chloroform (3020 cm^{-1}). At what wavelength would the Raman effect appear for each solvent when the excitation wavelength is one of these mercury emission lines: 254, 313, 365, or 436 nm?

7. 3,4-Benzopyrene is a potent carcinogen often found in polluted air. A sensitive method for its determination is to measure its fluorescence in sulfuric acid solution. The excitation wavelength is 520 nm and the emission wavelength is 548 nm. Ten liters of air were drawn through 10.00 mL of dilute sulfuric solution. One milliliter of this sulfuric acid solution gave a reading of 33.3 in a fluorometer. Two different standard solutions of 3,4-benzopyrene containing 0.750 μg and 1.25 μg per 10.00 mL of sulfuric acid gave readings of 24.5 and 38.6 when 1.00 mL was placed in the same cell used for the unknown. A blank sample with no 3,4-benzopyrene gave a reading of 3.5. Calculate the weight of 3,4-benzopyrene in 1 liter of air.

8. From the phosphorescence plus fluorescence spectrum and the phosphorescence spectrum alone, the following peaks were obtained for trivalent metal chelates of dibenzoylmethane. Construct approximate energy-level diagrams for the fluorescence and phosphorescence transitions to the vibrational levels in the ground electronic state.

	Metal ion		
	Al	Sc	Y
λ_{ex} for F (nm)	417.5	425.0	429.0
λ_{ex} for P (nm)	478.5	485.0	491.0
Phosphorescence (nm)	478.3	484.0	491.0
	495.0	500.0	508.0
	512.0	515.0	525.0
	530.8	537.0	548.0
	548.0	556.0	580.0
	575.5	573.5	
Phosphorescence plus fluorescence (nm)	417.1	434.0	428.3
	440.0	450.0	452.0
	467.6	484.0	489.0
	481.0	515.0	508.0
	495.0	520.0	525.0
	512.0	537.0	548.0
	530.3	~556.0	~569.0
	~550.0		

9. From the response of the fluorescent samples to sinusoidal excitation shown in Figure 8.12, calculate the lifetimes of the fluorescence. What are the demodulation factors for each material?

BIBLIOGRAPHY

BOWMAN, R. L., "History and Development of the Spectrophotofluorometer," *Fluorescence News*, **3** (1), 1 (1974).
DEMAS, J. N., *Excited State Lifetime Measurements*, Academic, New York, 1983.
GUILBAULT, G. G., *Practical Fluorescence: Theory, Methods and Techniques*, Dekker, New York, 1973.
HERCULES, D. M., "Some Aspects of Fluorescence and Phosphorescence," *Anal. Chem.*, **38,** 29A (1966).
LAKOWICZ, J. R., *Principles of Fluorescence Spectroscopy*, Plenum, New York, 1983.
LOTT, P. F., AND R. J. HURTUBISE, "Instrumentation for Fluorescence and Phosphorescence," *J. Chem. Educ.*, **51,** A315, A358 (1974).

PASSWATER, R. A., *Guide to Fluorescence Literature,* Vols. I–III, Plenum, New York, 1967.
SCHULMAN, S. G., ed., *Molecular Luminescence Spectroscopy,* Vol. I, John Wiley, New York, 1985.
SEITZ, W. R., "Luminescence Spectrometry (Fluorimetry and Phosphorimetry)," Chap. 4, *Treatise on Analytical Chemistry,* 2nd ed., P. J. Elving, E. J. Meehan, and I. M. Kolthoff, eds., Part 1, Vol. 7, John Wiley, New York, 1981.
WEHRY, E. L., ed., *Modern Fluorescence Spectroscopy,* Plenum, New York, 1981.

LITERATURE CITED

1. HURTUBISE, R. J., *Anal. Chem.,* **55,** 669A (1983).
2. SCHULMAN, S. G., *Fluorescence News,* **7**(4), 25 (1973); **7**(5), 33 (1973).
3. WEHRY, E. L., *Fluorescence News,* **6**(1), 1 (1971).
4. WILLIAMS, R. T., AND J. W. BRIDGES, *J. Clin. Pathol.,* **17,** 371 (1964)
5. BECKER, R. S., *Theory and Interpretation of Fluorescence and Phosphorescence,* Wiley-Interscience, New York, 1969.
6. WEHRY, E. L., "Effect of Molecular Structure and Molecular Environment on Fluorescence," in *Practical Fluorescence: Theory, Methods and Techniques,* G. G. Guilbault, ed., Dekker, New York, 1973.
7. ADAMSONS, K., A. TIMNICK, J. F. HOLLAND, AND J. E. SELL, *Anal. Chem.,* **54,** 2186 (1982); *Am. Lab.,* **16,** 16 (1984).
8. WRIGHT, J. C., AND F. J. GUSTAFSON, *Anal. Chem.,* **50,** 1147A (1978).
9. PASSWATER, R. A., *Fluorescence News,* **7**(3), 17 (1973).
10. PORRO, T. J., AND D. A. TERHAAR, *Anal. Chem.,* **48,** 1103A (1978).
11. DICESARE, J. L., AND T. J. PORRO, *Trends in Fluorescence,* **1,** 16 (1978).
12. GIERING, L. P., *Industrial Res. Dev.,* p. 134 (September 1978).
13. CLINE LOVE, L. J., AND L. A. SHAVER, *Anal. Chem.,* **48,** 364A (1976).
14. MCGOWN, L. B., AND F. Y. BRIGHT, *Anal. Chem.,* **56,** 1400A (1984).
15. JANNSON, P. A., ed., *Deconvolutions with Applications in Spectroscopy,* Academic, Orlando, FL, 1984.
16. HIEFTJE, G. M., AND G. R. HAUGEN, *Anal. Chem.,* **53,** 755A (1981).
17. VO-DINH, T., *Room Temperature Phosphorimetry for Chemical Analysis,* John Wiley, New York, 1984.
18. HURTUBISE, R. J., *Solid Surface Luminescence Analysis,* Dekker, New York, 1981.
19. ASAFU-ADAJAYE, E. B., AND S. Y. YUE, *Anal. Chem.,* **58,** 539 (1986).
20. RAMASAMY, S. M., Y. P. SENTHILNATHAN, AND R. J. HURTUBISE, *Anal. Chem.,* **58,** 612 (1986).

FLAME EMISSION AND ATOMIC ABSORPTION SPECTROSCOPY

9.1 INTRODUCTION

The absorption and emission of radiant energy by atoms provide powerful analytical tools for both quantitative and qualitative analysis (see Chapter 5). Flame emission spectroscopy (FES) has been used since the early 1900s. In the 1960s, atomic absorption spectroscopy (AAS) was developed as an analytical method. Most recently, new sources for plasma emission spectroscopy offer capabilities that complement FES and AAS for many analyses.

Table 9.1 summarizes atomic spectroscopic methods. This chapter will discuss methods that use combustion flames and a single nonflame method, electrothermal AAS. Chapter 10 will present emission spectroscopic methods that use nonflame excitation sources. Two major limitations apply to all atomic spectroscopic methods: (1) their limited ability to distinguish among oxidation states and chemical environments of the analyte elements and (2) their insensitivity to nonmetallic

TABLE 9.1 PRIMARY METHODS FOR ATOMIC SPECTROSCOPY

Method (abbreviation)	Energy source	Measured quantity
Emission	***Source of excitation***	
Flame emission spectroscopy (FES)	Flame (1700–3200 °C)	Intensity of radiation
Atomic fluorescence spectroscopy (AFS)	Flame (1700–3200 °C)	Intensity of scattered radiation
Electric arc	Plasma from dc arc (4000–6500 °C)	Intensity of radiation
Electric spark	Plasma from ac spark (4500 °C)	Intensity of radiation
Inductively coupled plasma (ICP)	Argon plasma produced by induction from high-frequency magnetic field (6000–8500 °C)	Intensity of radiation
ICP-AFS	Same as ICP	Intensity of scattered radiation
Direct-current argon plasma	Argon plasma produced by a dc arc (6000–10,500 °C)	Intensity of radiation
Absorption	***Source of atomization***	
Flame atomic absorption spectroscopy (FAAS)	Flame (1700–3200 °C)	Absorption of radiation
Electrothermal absorption spectroscopy	Electric furnace (1200–3000 °C)	Absorption of radiation

elements. The latter limitation has been overcome for selected nonmetals by modifying the optical components to extend the range of usable wavelengths farther into the ultraviolet region.

The relationship between analyte concentration and the measured signal is different for the methods discussed in this chapter. Flame emission spectroscopy and atomic fluorescence spectroscopy (AFS) are emission methods and therefore the intensity of the emitted radiation is directly proportional to the concentration. In AAS determinations, the absorbance is measured and the concentration of the analyte is related to the signal by the Lambert-Beer law, usually called Beer's law.

The Role of Combustion Flames in FES and AAS

Combustion flames provide a means of converting analytes in solution to atoms in the vapor phase freed of their chemical surroundings. These free atoms are then transformed into excited electronic states by one of two methods: absorption of additional thermal energy from the flame or absorption of radiant energy from an external source of radiation.

In the first method, known as flame emission spectroscopy (FES), the energy from the flame also supplies the energy necessary to move the electrons of the free atoms from the ground state to excited states. The intensity of radiation emitted by these excited atoms returning to the ground state provides the basis for analytical determinations in FES.

In the second method, atomic absorption spectroscopy (AAS), the flame that contains the free atoms becomes a sample cell. The free atoms absorb radiation focused on the cell from a source external to the flame. As in all absorption spectroscopic methods, the incident radiation absorbed by the free atoms in moving from the ground state to an excited state provides the analytical data. A major modification of AAS was the replacement of the flame by an electrothermal furnace (electrothermal atomic absorption spectroscopy, EAAS), resulting in improved detection limits for many elements.

In both FES and flame AAS, the sample solution is introduced as an aerosol into the flame, where the analyte ions are converted into free atoms. Once formed, the free atoms are detected and determined quantitatively by FES or AAS. If the source of exciting radiation is placed at right angles to both the flame and the optical axis of the spectrometer, a method known as atomic fluorescence spectroscopy (AFS) results. This technique has dramatically improved detection limits for certain elements. At this time only one atomic fluorescence instrument is available commercially, an inductively coupled plasma (ICP-AFS) unit from Baird Corporation.

Although inductively coupled plasma and dc plasma are finding increased use (see Chapter 10), FES and AAS remain the workhorses for many analyses (Table 9.2). FES still offers the lowest detection limits for the alkali metals and good limits for many other elements. Flame AAS, the method of choice for many elements in the 1960s and 1970s, now provides the lowest detection limits for only a few elements. It is, however, widely used for many routine determinations. Electrothermal AAS provides the best known detection limits for a large number of elements (see Appendix G).

TABLE **9.2** "BEST" DETECTION LIMITS FOR ATOMIC SPECTROSCOPIES (NANOGRAMS PER MILLILITER)

Li 0.001 A,G	Be 0.003 D,E											B 0.1 E	C 44 E	N * E	O * E	F * E
Na 0.004 A,D	Mg 0.0002 D											Al 0.01 D	Si 0.005 D	P 0.3 D	S 10 D,E	Cl * E
K 0.004 A,D	Ca 0.0001 E	Sc 0.4 E	Ti 0.03 E	V 0.06 D,E	Cr 0.004 D	Mn 0.0005 D	Fe 0.01 D	Co 0.008 D	Ni 0.05 D,E	Cu 0.005 D	Zn 0.0003 C,D	Ga 0.01 D	Ge 0.1 D	As 0.02 B,D	Se 0.02 B,C,D	Br * E
Rb 0.02 A	Sr 0.002 E	Y 0.04 E	Zr 0.06 E	Nb 0.2 E	Mo 0.02 D		Ru 30 B,E,F	Rh 0.1 D	Pd 0.05 D	Ag 0.001 D	Cd 0.0002 D	In 0.006 D,G	Sn 0.03 D	Sb 0.08 B,C,D	Te 0.002 B	I 3 D,E
Cs 0.02 A,D	Ba 0.01 D,E		Hf 10 E	Ta 5 E	W 0.8 E	Rs 6 D,E,F	Os 0.4 E	Ir 0.5 D	Pt 0.2 D	Au 0.01 D	Hg 0.001 B,C	Ti 0.01 D	Pb 0.007 D	Bi 0.01 B,D		
		La 0.1 E	Ce 0.4 E	Pr 10 E	Nd 0.3 E		Sm 1 E	Eu 0.06 A,E	Gd 0.4 E	Tb 0.1 E	Dy 4 E,F	Ho 0.7 D	Er 0.3 D,E	Tm 0.2 E	Yb 0.01 D,E	Lu 0.1 E
			Th 3 E		U 1.5 E											

A. Flame emission spectrometry
B. Flame atomic absorption
C. Atomic fluorescence spectrometry
D. Electrothermal atomic absorption
E. ICP
F. DCP
G. Laser-assisted ionization

* Spectra from these elements have been observed in the ICP, but detection limits are not available in a comparable form.

SOURCE: After M. Parsons and S. Major, *Appl. Spectrosc.*, **37** (5), 411 (1983). With permission.
NOTE: See Appendix G for a more complete listing.

Flame Spectrometric Methods

Even though the characteristics of free-atom formation in the flame and problems associated with interferences are common to the three flame spectrometric methods (FES, AAS, and AFS), each method has its own instrumentation, unique advantages, and definite limitations.

9.2 INSTRUMENTATION FOR FLAME SPECTROMETRIC METHODS[1,2]

The basic components of flame spectrometric instruments are discussed in this section. These components provide the following functions required in each method: (1) deliver the analyte to the flame, (2) induce the spectral transitions (absorption or emission) necessary for the determination of the analyte, (3) isolate the spectral lines required for the analysis, (4) detect the increase or decrease in intensity of radiation at the isolated lines(s), and (5) record these intensity data.

Pretreatment of Sample[3]

Flame AAS and FES require that the analyte be dissolved in a solution in order to undergo nebulization (see the next section). The wet chemistry necessary to dissolve the sample in a matrix suitable for either flame method is often an important component of the analytical process. The analyst must be aware of substances that interfere with the absorption or emission measurement (Section 9.4). When these substances are in the sample, they must be removed or masked (complexed). Reagents used to dissolve samples must not contain substances that lead to interference problems.

Sample Delivery

The device that introduces the sample into the flame or plasma plays a major role in determining the accuracy of the analysis. The most popular sampling method is nebulization of a liquid sample to provide a steady flow of aerosol into a flame. An introduction system for liquid samples consists of three components: (1) a nebulizer that breaks up the liquid into small droplets, (2) an aerosol modifier that removes large droplets from the stream, allowing only droplets smaller than a certain size to pass, and (3) the flame or atomizer that converts the analyte into free atoms.

Nebulization

Pneumatic nebulization is the technique used in most atomic spectroscopy determinations. The sample solution is introduced through an orifice into a high-velocity gas jet, usually the oxidant. The sample stream may intersect the gas stream in either a parallel or perpendicular manner (Figure 9.1a). Liquid is drawn through the sample capillary by the pressure differential generated by the high-velocity gas stream passing over the sample orifice. The liquid stream begins to oscillate, producing filaments. Finally, these filaments collapse to form a cloud of droplets in the aerosol modifier or spray chamber. In the spray chamber the larger droplets are removed from the sample stream by mixer paddles or broken up into smaller droplets by impact beads (Figure 9.1b) or wall surfaces. The final aerosol, now a fine mist, is combined with the oxidizer/fuel mixture and carried into the burner (Figure 9.2).

A typical distribution range of droplet diameters is shown in Figure 9.3. Droplets larger than about 20 μm are trapped in the spray chamber and flow to waste. The distribution of drop sizes is a function of the solvent as well as the

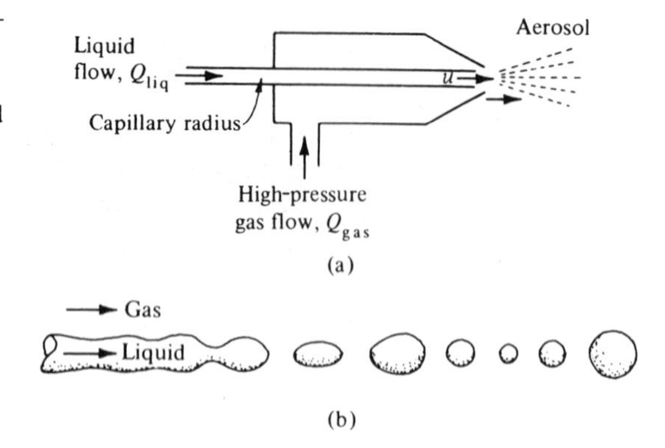

FIGURE **9.1**
(a) Construction of pneumatic nebulizer and (b) breakdown of liquid filament into droplets.

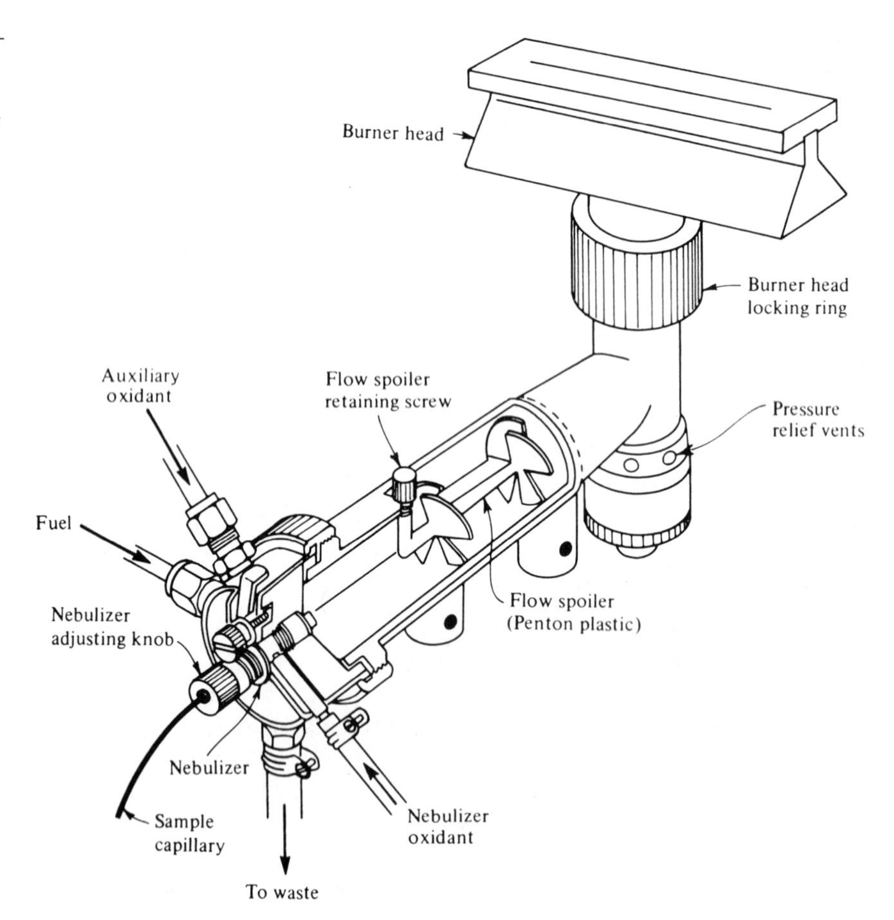

FIGURE **9.2**
Slot burner and expansion chamber. (Courtesy of Perkin-Elmer Corp.)

components of the sampling system. In AAS only a small percentage (usually 2% or 3%) of the nebulized analyte solution reaches the burner.

The diameter of the aerosol droplets produced by the nebulizer is determined by the physical properties of the sample solution as described in the following equation:[4]

$$d_S = \frac{585}{v}\left(\frac{\gamma}{\rho}\right)^{0.5} + 597\left[\frac{\eta}{(\gamma\rho)^{0.5}}\right]^{0.45} \cdot 1000\left[\frac{Q_l}{Q_g}\right]^{1.5} \quad (9.1)$$

where d_s is the Sauter mean diameter of the aerosol droplets (the diameter of the drop with a volume-to-surface-area ratio the mean of the distribution), v is the velocity difference between gas and liquid flows (m/sec), γ is the surface tension of the liquid (dyne/cm), ρ is the liquid density (g/mL), η is the liquid viscosity (poise), and Q_l and Q_g are the volume flow rates of the liquid and gas, respectively (mL/sec). The equation has proved valuable in predicting trends for aerosol generation. Unanticipated changes in viscosity and surface tension are avoided by keeping the sample and standard matrices identical and by avoiding total acid or salt concentrations greater than about 0.5%.

Atomization

The atomization step must convert the analyte within the aerosol into free analyte atoms in the ground state for AAS, FES, and AFS analysis. Very small sample volumes (5–100 μL) or solid samples can be handled by flameless electrothermal methods.

Flame Atomizers. The sequence of events involved in converting a metallic element, M, from a dissolved salt, MX, in the sample solution to free M atoms in the flame is depicted in Figure 9.4. After the aerosol droplets containing MX enter the flame, the solvent is evaporated, leaving small particles of dry, solid MX. Next, solid MX is converted to MX vapor. Finally, a portion of the MX molecules are dissociated to give neutral free atoms. These M atoms are the species that absorb

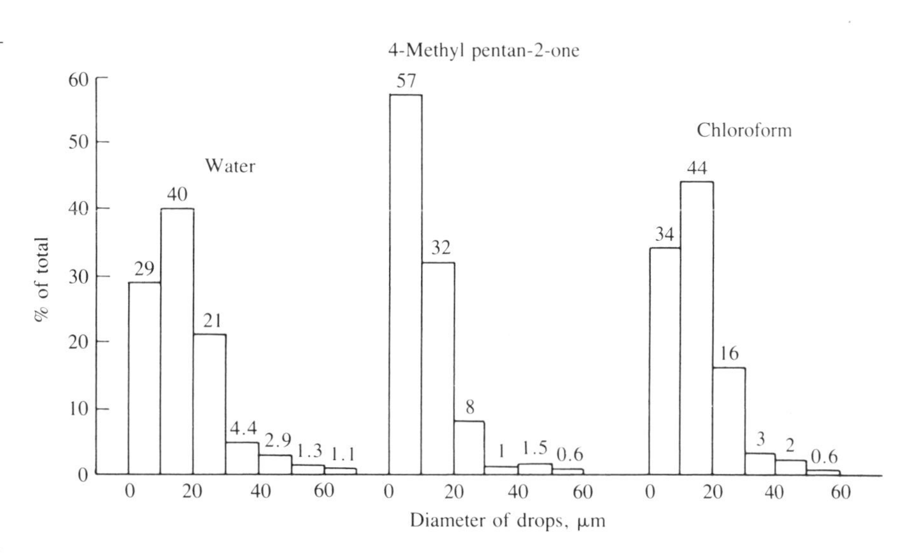

FIGURE **9.3**

Representation of a drop-size distribution from a pneumatic nebulizer. [After J. A. Dean and W. J. Carnes, *Anal. Chem.*, **34**, 192 (1962). Courtesy of American Chemical Society.]

radiation in AAS and AFS, and they are the potential emitting species in FES. The efficiency with which the flame produces neutral analyte atoms is of equal importance in all the flame techniques.

If the events proceed vertically from the top down in Figure 9.4, the efficiency of free-atom production is high. Processes that branch horizontally interfere with the production of free analyte atoms. These processes include: (1) excitation and emission of radiation by MX(*g*) molecules, (2) reaction of M(*g*) atoms with flame components at high temperatures to produce molecules and ions that also absorb and emit radiation, and (3) formation of M^{+x} ions, which, in addition to reducing the efficiency of free-atom production, complicate the analysis by adding lines to the spectrum.

The flame remains the most generally useful atomizer for atomic spectroscopy despite the developments in electrothermal atomization. A satisfactory flame source must provide the temperature and fuel/oxidant ratio required for a given analysis. The maximum operating temperature of the flame is determined by the identities of the fuel and oxidant (Table 9.3), whereas the exact flame temperature is fixed by the fuel/oxidant ratio. In addition, the spectrum of the flame itself should not interfere

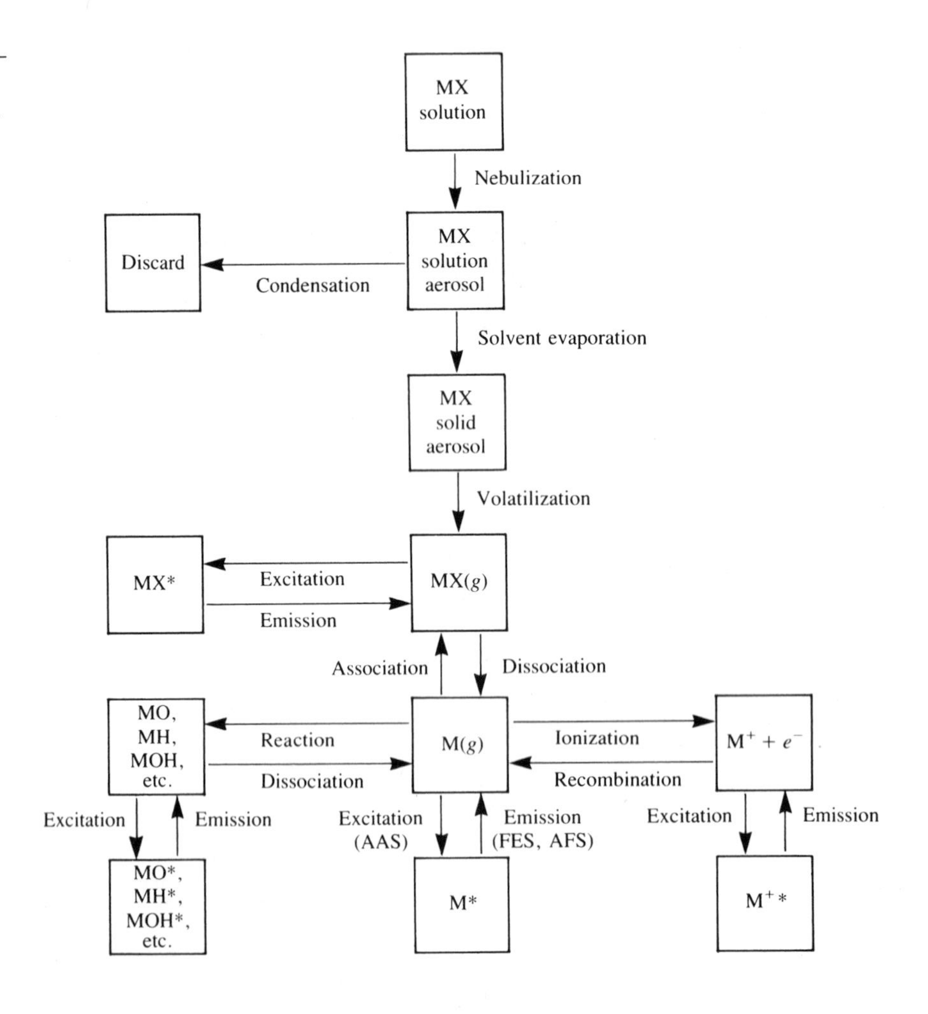

FIGURE 9.4
Flame atomization processes for the salt MX. Asterisk (*) indicates excited state.

with the emission or absorption lines of the analytes. Components of the flame gases limit the usable range to wavelengths longer than 210 nm.

Flames are not uniform in composition, length, or cross section. The structure of a premixed flame, supported on a laminar flow burner, is shown in Figure 9.5. (Laminar flow is defined as a mode of gas flow in which the lines of flow are approximately parallel and change smoothly, if at all, in time and space.) Emerging from region *A*, the unburned hydrocarbon gas mixture passes into a region of free heating about 1 mm thickness (region *B*). In this region, the mixture is heated by energy (conduction and radiation) from region *C*. Diffusion of radicals into region *B* initiates combustion. Flame gases travel upward from the reaction zone with velocities of 1–10 m/sec. Gases that emerge from region *C* consist mainly of CO_2,

TABLE **9.3**

CHARACTERISTICS OF COMMON PREMIXED FLAMES

Fuel	**Oxidant**	**Temperature*** **(°C)**	**Burning velocity†** **(cm sec^{-1})**
Acetylene	Air	2400	160–266 (160)
Acetylene	Nitrous oxide	2800	260
Acetylene	Oxygen	3140	800–2480 (1100)
Hydrogen	Air	2045	324–440
Hydrogen	Nitrous oxide	2690	390
Hydrogen	Oxygen	2660	900–3680 (2000)
Propane	Air	1925	43

* Stoichiometric mixture.
† Values in parentheses are probably the ones most applicable to laboratory burners.

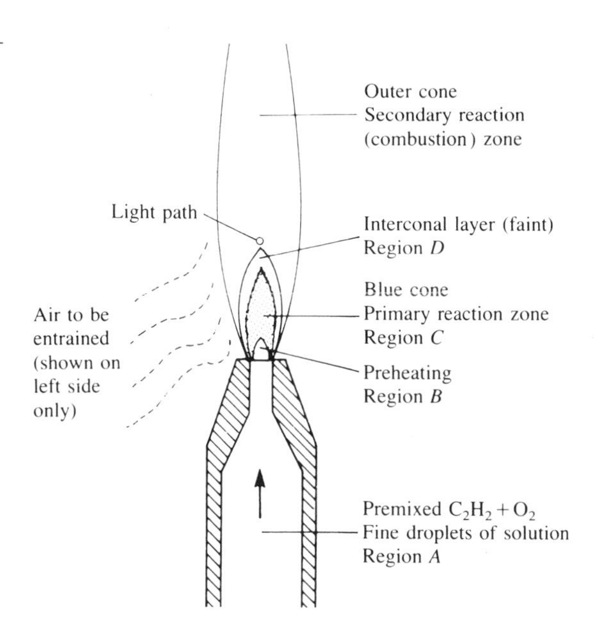

FIGURE **9.5**
Schematic structure of a laminar flow flame.

CO, H_2O, and N_2 if air is used as an oxidant. Lesser amounts of H_2, H, O, OH, and NO are also found at this point. The specific composition of the gases is dependent on the composition of the initial mixture. In region *C*, the concentration of radicals (C_2, CH, H_3O^+, HCO^+) is too high for the gases to achieve thermal equilibrium. The conditions in this region are reducing and the intense emission of radiation from flame components can create noise problems at the detector. As a result, this region is rarely used for AAS unless refractory oxides must be reduced (e.g., BO_2) to obtain the atomic emission.

As the gases reach region *D* (Figure 9.5), the interconal zone, they approach thermal equilibrium. This region is cooler and more oxidizing than the primary reaction zone (region *C*). Oxidation is completed in the outer cone with the assistance of surrounding air. The conditions in region *D* are optimum for most AAS measurements.

The temperature of the flame (Table 9.3) determines its utility in both AAS and FES. The exact temperature depends on the fuel/oxidant ratio and is generally highest for a stoichiometric mixture. Fuel-rich flames are usually cooler. Optimum temperatures vary for AAS and FES but in both cases depend on the excitation and ionization potentials of the analyte (Table 9.4). Temperatures high enough to cause ionization of the analyte atoms are usually undesirable in both methods unless an ionization buffer (an easily ionizable element added to the matrix to suppress ionization of the analyte) is added to the sample.

The concentration of unexcited and excited atoms in a flame is determined by the fuel/oxidant ratio. It varies in different parts of the flame. Studies of atomic or molecular distributions within the flame envelope have been made by measuring absorption, emission, or fluorescence as the flame is moved vertically (or horizontally) relative to the light path of the optical system. Figure 9.6 shows the distribution of free atoms obtained by absorption measurements in a 10-cm long acetylene/air

TABLE **9.4**

PERCENT IONIZATION OF SELECTED ELEMENTS IN FLAMES*

Element	**Ionization potential (eV)**	**Acetylene/ air, 2400 °C**	**Acetylene/ oxygen, 3140 °C**	**Acetylene/ nitrous oxide, 2800 °C**
Lithium	5.391	0.01	16.1	
Sodium	5.139	1.1	26.4	
Potassium	4.340	9.3	82.1	
Rubidium	4.177	13.8	88.8	
Cesium	3.894	28.6	96.4	
Magnesium	7.646		0.01	6
Calcium	6.113	0.01	7.3	43
Strontium	5.694	0.01	17.2	84
Barium	5.211	1.9	42.3	88
Manganese	7.43			5

* Partial pressure of metal atoms in the flame is assumed to be 1×10^{-6} atm for acetylene/air and acetylene/oxygen flames, and approximately 10^{-8} atm for the acetylene/nitrous oxide flame.

flame. Contours are drawn at intervals of 0.1 absorption unit with maximum absorbance at the center. Different elements exhibit different free-atom flame profiles. Neither the area of observation nor the fuel/oxidant ratio is critical for copper, whereas for molybdenum, the region of maximum free-atom concentration is sharply localized. The height of the maximum free-atom concentration is the point at which the increased atomization with flame height is just balanced by the rate of decrease in the concentration of free atoms through dilution by the flame gases and the formation of oxides and hydroxides.

The distribution pattern obtained by measuring atomic emission often differs dramatically from that obtained from absorption measurements for a given element. For example, the emission lines of boron (249.7 nm) and antimony (259.8 nm) are either absent or very weak in the outer mantle of a stoichiometric flame, but they appear in high concentrations in the reaction zone of a fuel-rich flame. A fuel-rich acetylene flame (ratio of fuel to oxidant exceeds that needed for stoichiometric combustion) provides the reducing atmosphere necessary for the production of a large free-atom population of those elements that have a tendency to form refractory oxides. Incandescent carbon particles present in fuel-rich flames cause luminosity that produces a high background. Thus the position of observation and the fuel/oxidant ratio must be optimized for each element in both FES and flame AAS methods.

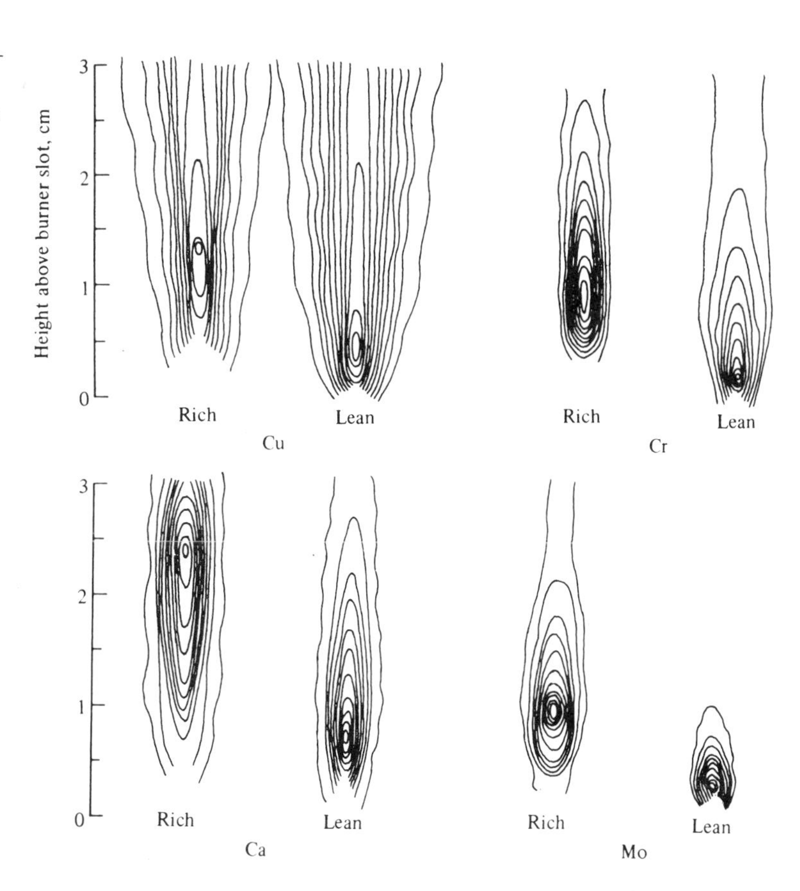

FIGURE **9.6**
Distribution of atoms in a 10-cm air/acetylene flame. Fuel-rich and fuel-lean results are shown. Contours are drawn at intervals of 0.1 absorbance unit with maximum absorbance in center. [After C. S. Rann and A. N. Hambly, *Anal. Chem.*, **37,** 879 (1965). Courtesy of American Chemical Society.]

In FES, the acetylene/air flame is used for practically all determinations involving alkali metal elements. The higher flame temperature of the acetylene/nitrous oxide flame is required for analysis of the alkaline earth metals as well as Ga, In, Tl, Cu, Co, Cr, Ni, and Mn. The hotter flame allows the sensitive analysis of additional elements whose refractory oxides are not reduced to the atomic state in acetylene/air flames. The acetylene/nitrous oxide flame is unique in combining high temperature with a propagation rate and analyte residence time not much greater than those of the cooler acetylene/air flame. Special burner heads (5 cm slot length, 0.5 mm width) and control units for igniting and extinguishing the flame are required to eliminate the risk of flashback into the spray chamber and consequent explosion. Proper safety precautions should be taken in operating all flame instruments, both FES and AAS units.

Shielding a flame with a sheath of inert gas, blown around the outside of the flame, causes an elongation of the interconal zone. This elongation reduces noise and provides a wider range of excitation conditions over a small flame volume. Shielding is particularly useful for multielement FES determinations if there is a means for adjusting the burner height (relative to the observation light path). Shielded flames are also used in AFS but usually not required in AAS.

In flame AAS, the fuel/oxidant ratio and observation height are chosen to provide the maximum number of free atoms while minimizing interferences from emission, ionization, or compound formation. For elements where stable (refractory) oxide, carbide, or nitride formation reduces the concentration of free atoms in the flame, two approaches can increase the number of free atoms: (1) use a cooler, reducing acetylene/air flame to minimize the formation of the thermal stable species or (2) use a hotter acetylene/nitrous oxide flame to dissociate these stable species into free atoms (in spite of the risk of decreasing free atoms through ionization).

Electrothermal Atomization.[5,6] As we have seen, the nebulizing system used with flames wastes the sample, and the residence time of free analyte atoms in the portion of the flame where they absorb radiation from the external source is short. Electrically heated devices, such as graphite furnaces and carbon rod analyzers, are common flameless atomizers that complement flame AAS. The most attractive features of electrothermal AAS are (1) high sensitivity (analyte amounts of 10^{-8} to 10^{-11} g absolute), (2) the ability to handle small sample volumes of liquids (5–100 μL), (3) the ability to analyze solid samples directly without pretreatment (in most cases), and (4) low noise from the furnace. Matrix effects from components in the sample other than the analyte are usually much more severe than those encountered in flame AAS and thus the precision, typically 5%–10%, compares unfavorably with that of flame AAS. Electrothermal atomizers are much more difficult to use than flame atomizers, although computer control of the electrical heating cycles has reduced many operational difficulties.

Electrothermal AAS has increased sensitivity because the production of free analyte atoms by an electrically heated carbon atomizer is more efficient than with a flame atomizer. If samples of equal analyte concentration are atomized by an electrothermal and by a flame device, a much higher concentration of analyte per unit time appears in the optical path of the electrothermal atomizer. However, electrothermal atomizers can maintain the relatively high concentration of free atoms for only a brief time, whereas flame atomizers can produce a signal as long as the sample solution is being atomized. Any variation in the rate of atomization during the comparatively short time available for measurement of radiation absorption in the furnace can lead to serious analytical errors. It is therefore much easier to reproduce results using a flame atomizer because of the longer measurement times for minimizing signal variations by averaging.

An apparatus for electrothermal atomization consists of three components: the workhead, the power unit, and the controls for the inert gas supply. The workhead replaces the burner-nebulizer assembly in the AAS spectrometer. The power unit supplies the operating current at the proper voltage to the workhead. Computer control at the workhead provides the reproducible heating conditions necessary for analysis. The gas control unit provides for metering and control of the flow of inert gas around the exterior and through the workhead during the analysis. This inert environment prevents destruction of the graphite at high temperatures by air oxidation. Sometimes hydrogen gas is introduced into the tube if the unit uses a hydrogen diffusion flame during the reductive ashing cycle.

The heated graphite analyzer, shown in Figure 9.7, consists of a hollow graphite cylinder 28 mm long and 8 mm in diameter, positioned in such manner that the radiation from the external source (hollow-cathode tube) passes through the center of the cylinder. The interior of the cylinder is coated with pyrolytic graphite. Electrodes at the end of the cylinder are connected to a low-voltage, high-current power supply capable of delivering up to 3.6 kW to the cylinder walls. Liquid samples are introduced with a microsyringe through the small opening in the top of the cylinder. Solid samples can be introduced through the end of the tube with a special sampling spoon or on a microdish made of tungsten.

A metal housing surrounding the furnace is water-cooled to allow the entire atomizer unit to be rapidly restored to ambient temperature after each sample has been atomized. Inert gas, usually argon, enters the graphite cylinder at both ends and exits through the sample introduction port at the center of the tube. This gas flow ensures that matrix components vaporized during the ashing step are quickly removed and that nothing is deposited on the inner wall of the tube where subsequent vaporization during the atomizing step could produce a large background absorption signal. Removable quartz windows at each end of the tube prevent ambient air from entering. The inert gas that flows around the exterior of the tube is controlled by a separate valve.

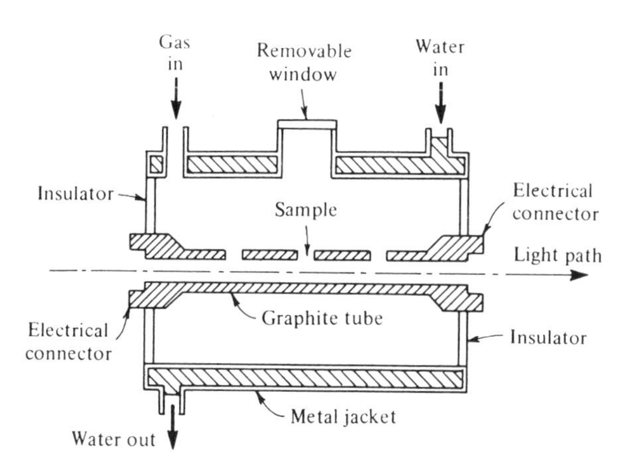

FIGURE 9.7

Cross section of a heated graphite atomizer. (Courtesy of Perkin-Elmer Corp.)

A miniature version of the furnace is the carbon rod atomizer (Figure 9.8). It consists of a three-piece tube or cup unit. The workhead that contains the small furnace is supported between two graphite electrodes inserted in water-cooled terminal blocks. The furnace itself is 9 mm long and 3 mm in diameter with a maximum sample capacity of 10 μL for smooth tubes and 25 μL for tubes with grooves (or threads). The central unit can be replaced with a vertical cup held between the two electrodes. This version is useful for solid samples or samples that require preliminary chemical treatment performed directly in the cup. All units are coated with pyrolytic graphite. In normal use the carbon rod atomizer is protected from oxidation by a sheath of inert gas directed onto the rod from a chimney beneath. When hydrogen gas is added to generate a reducing environment, the gas ignites spontaneously when the cup or tube reaches incandescence.

A thin graphite plate, known as the L'vov platform, added to the bottom of the graphite tube allows for better control of atomization (Figure 9.9). The sample is placed on the plate and is heated by radiation from the walls so that the temperature increase of the plate is delayed relative to that of the tube walls and the gaseous vapor inside the tube. With this device, the walls and vapor can reach a steady-state

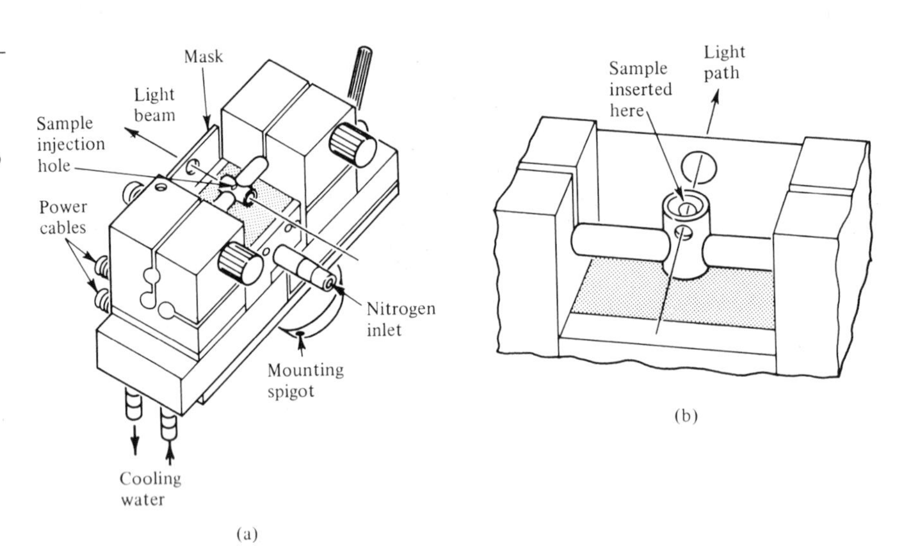

FIGURE **9.8**
Carbon rod atomizer: (a) horizontal rod version and (b) vertical cup version. (Courtesy of Varian Associates, Inc.)

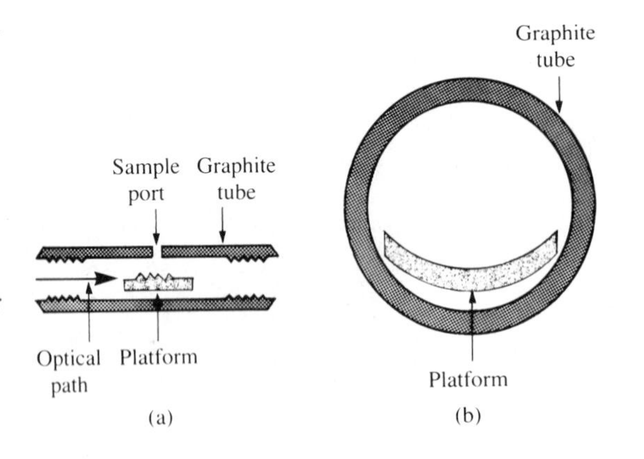

FIGURE **9.9**
(a) The modified L'vov platform. Side view of the platform position within the graphite tube. Tube dimensions, 28 × 6 (i.d.) mm. Platform dimensions, 7 × 5 mm. (b) The modified L'vov platform, end view. [After S. R. Koirtyohann and M. A. Kaiser, *Anal. Chem.*, **54**(14), 1518A (1982), by permission.]

temperature before the sample is atomized. This technique requires fast recording of the absorbance signal, since the peaks appear and disappear rapidly.

After insertion or injection of the sample into the electrothermal atomizer, a heating sequence is initiated to take the sample through three steps: dry, ash or char, and atomize (Figure 9.10). The control of heating is independently programmable, with a temperature ramp and isothermal hold time available for each step. In the drying cycle, the sample is heated for 20–30 sec at 110–125 °C to evaporate any solvent or extremely volatile matrix components. The sample that remains after this step appears as a slight stain or crust on the interior of the graphite tube or rod. The ash or char cycle is performed at an intermediate temperature selected to carry out the necessary processes, volatilization of higher-boiling matrix components and pyrolysis of matrix materials such as fats and oils that will crack and carbonize. This step often converts the analyte to a different chemical state. The obvious problem at this stage is loss of analyte if the ashing temperature is too high or is maintained for too long.

In the final stage, the optimum maximum power is applied to raise the furnace unit to the selected atomization temperature. In this step the analyte residue is dissociated and volatilized into free atoms responsible for the observed absorption. The transient signal produced by the brief period of absorption provides the output to the recorder or microcomputer data processing system. Both peak height and peak area have been used to determine the concentration of analyte.

The proper temperature and timing parameters must be carefully selected for each step in the electrothermal process. The identity of the analyte and the composition of the sample matrix are the most important factors in choosing these parameters. The evaporization of solvent in the drying cycle must be smooth and gentle to avoid mechanical losses by foaming or splattering. The progress of the drying step can be observed by monitoring the absorption signal without background correction. The escaping vapors should produce a smooth curve without the appearance of humps or spikes, which indicate that the heating rate is too rapid. The ashing cycle is followed in the same way; usually no analyte is lost

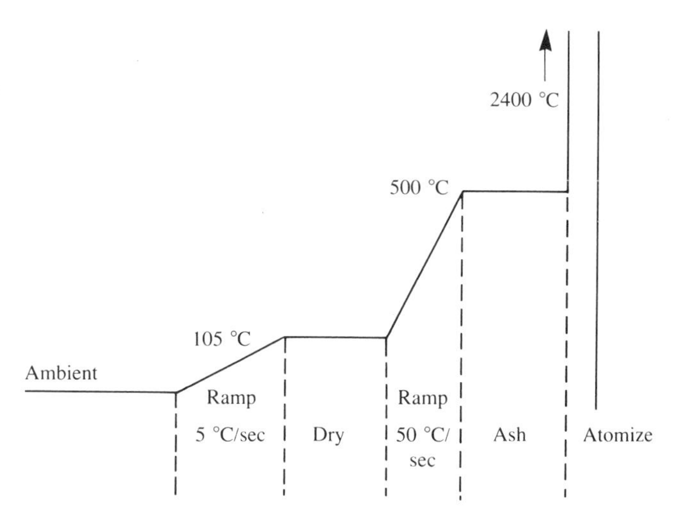

FIGURE **9.10**

Temperature profile with ramp heating. (Courtesy of Pye Unicam.)

until a specific temperature is reached, and then the analyte signal appears in the form of a sharp peak (Figure 9.11). Most organic materials pyrolyze at around 350 °C, leaving a residue of amorphous carbon. At this temperature, a stream of air or oxygen may be introduced into the furnace to convert this carbon residue into carbon dioxide.

A wide variety of samples can be handled directly with little or no pretreatment required. These include organic solvents, viscous liquids, liquids that contain a high concentration of dissolved solids, and pulverized solids. The ashing step destroys organic components, thus eliminating the need for pretreatment of these samples. In contrast to flame AAS, the lifetime of the free atoms in the optical path is short, approximately $\frac{1}{100}$ sec or less. The absorption signal must be measured rapidly, requiring fast response from both the detector and recording system. Care should be exercised in replacing a flame atomizer with an electrothermal atomizer. In many cases the response time of an instrument adequate to measure the relatively stable signal from analyte absorption in the flame is not fast enough to measure the transient signal from an electrothermal atomizer.

The deterioration of the atomizer surface with use creates two major problems that directly affect analytical results. First, the peak height for a given concentration of analyte decreases with use, and second, the electrical resistance of the atomizer changes. These effects are minimized by preheating the graphite to form a hard pyrolytic layer on the surface. This coating prevents the sample from soaking into the graphite and thus gives more reproducible atomization for carbide-forming elements. The operational lifetime of the tube or furnace surface is also prolonged.

Accurate temperature sensors are required for control of the power source. Recently, silicon photodiodes have been used for sensing and controlling temperatures of 600–3000 °C. These diodes have the rapid response times needed for the precise reproduction of heating conditions. The response times of thermocouples are usually too slow for tight control of heating.

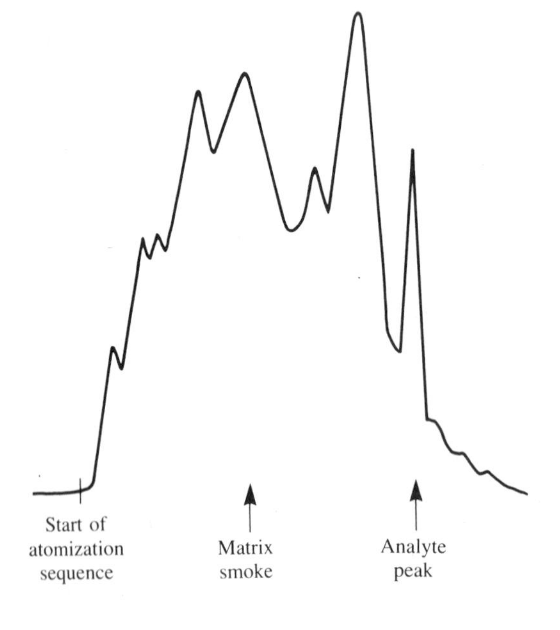

FIGURE **9.11**
Appearance of analyte peak after matrix smoke. (Courtesy of Pye Unicam.)

Electrothermal atomizers lead to background absorption due to incomplete combustion or decomposition of molecules during the final atomization step. In addition, residual ash and solvent materials may be present and contribute to the background through partial decomposition. This background absorption can be large and is often variable, depending on the analyte matrix. Thus it is necessary to provide a means for background correction. Recently, very rapid heating systems have been shown to reduce matrix effects.

Substitution of the graphite furnace by a tungsten ribbon creates a less expensive atomizer with a reduction in electrical power consumption and elimination of water cooling (Figure 9.12).[7] The tungsten atomizer has the following advantages over conventional graphite furnaces: (1) reduction in matrix interferences due to the faster heating rates of tungsten (6000 °C/sec), (2) elimination of memory effects and interference because tungsten is less porous than any form of graphite, and (3) increased sensitivity and lower detection limits. In most applications the sample is injected directly into the cell that contains the heated tungsten filament. Arsenic, selenium, and mercury are analyzed by using nickel, silver, and selenium, respectively, as matrix modifiers, and then injecting the analyte solutions directly. This technique eliminates the need for hydride-generation and cold-vapor attachments for the analyses of As, Se, and Hg.

Chemical Vaporization.[8] For elements (As, Bi, Ge, Sb, Se, Sn, and Te) that form volatile, covalent hydrides, the technique of hydride formation and subsequent analysis improves the atomization efficiency by quantitatively transferring the analyte as a hydride into the vapor phase. Electrodeless discharge lamps are often used as sources for this technique to provide the higher intensity required for elements such as As and Se. Chemical vaporization offers improved detection limits for the elements listed above as well as for the direct determination of elemental mercury.

Most AAS instrument manufacturers offer apparatus for the chemical pretreatment of samples to generate a volatile product that is then subjected to AAS measurement. The apparatus consists of a vapor-generation unit in which a metallic hydride or mercury vapor is produced. The vapors are then injected into the atomizer in a stream of insert gas. Commonly used atomizers for hydrides are either

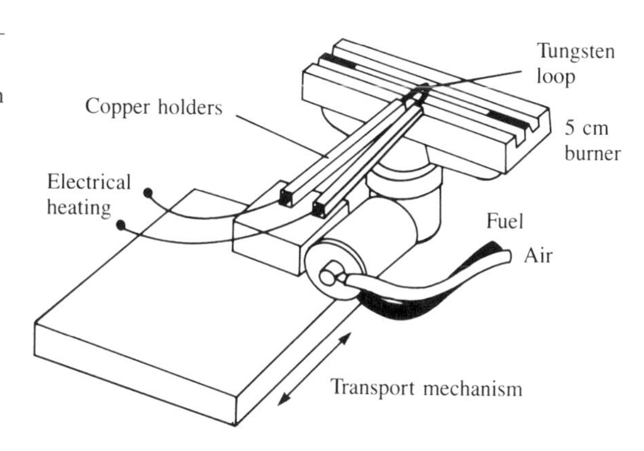

FIGURE **9.12**
Use of a tungsten ribbon instead of a carbon furnace avoids the need for high electrical current. The standard filament reaches 2400 °C at 6300 °C/sec.

an air/hydrogen flame or a relatively low-temperature, tube-type quartz furnace. Gaseous hydrides may be generated with sodium borohydride, dispensed in pellet form, as the reducing agent added to an acidic solution. The atoms in mercury vapor generated by heating or chemical reaction may be analyzed directly without further atomization.

Spectrometers

The spectrometers used to isolate the wavelengths of interest for a specific determination are described in Sections 9.4 and 9.5. These components are similar to those used in visible and ultraviolet instruments discussed in Chapter 6.

Detector-Readout Unit

Since the wavelengths of the analytical resonance lines fall in the ultraviolet-visible portions of the spectrum, the most common detectors for AAS, AFS, and FES are photomultiplier (PM) tubes (see Chapter 6). Care must be taken never to exceed the saturation limit of the PM tube by flooding it with light, too much radiation from the flame, or a high concentration of sample components, even though these signals are eliminated by modulation in the final signal readout. When the electronic noise in the detector-amplifier system can be reduced to a negligible level, scale expansion can be used profitably. With this technique, the zero point is displaced off-scale by applying a potential opposite to the signal arriving from the detector. Thus, the full scale of the readout device can be used as the upper end of a greatly expanded scale. In AAS, the reading for small decreases in transmitted radiation may be increased manyfold.

In a single-beam instrument, noise originating in the flame can be minimized by a chopper that modulates radiation from the sample, P, and reference P_0, so that the alternating beams strike the detector at a frequency determined by the speed of the beam chopper (see Section 2.5). Current from the PM tube is converted to a voltage using a high resistance in the feedback loop of an operational amplifier. After P and P_0 have been demodulated, the absorbance is calculated by taking the difference, $\log_{10} P_0 - \log_{10} P = \log_{10}(P_0/P)$. Sample absorbance may be measured and averaged for different periods of time. These integration times typically range from 0.1 to 90 sec. Long integration times are desirable for optimizing the S/N ratio. Integrating dc signals for precisely limited time periods is a powerful method of reducing noise (see Sections 2.5 and 2.6).

Although many instruments have the computational capability to correct for nonlinear calibration curves in regions of high analyte concentrations, this function is most easily performed by microcomputer-controlled instruments. In the linear region, data from one standard and a blank are sufficient to define the relationship between concentration and absorbance. In nonlinear regions, three standards and a blank are usually sufficient to permit a software curve-corrector algorithm to function accurately (Figure 9.13). It is important that these algorithms have a threshold control to allow the analyst to adjust the calibration curve in only the region of nonlinearity. Even though the software is capable of fitting curves in the nonlinear regions, it is unwise to run samples with absorbances that fall in the upper portion of the curved segment of the calibration curve. In this region the absorbance

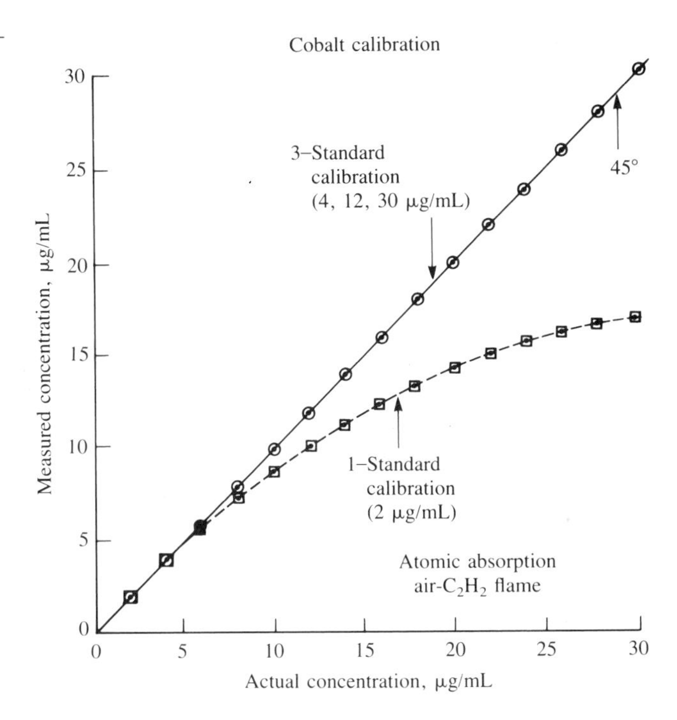

FIGURE 9.13
Calibration curves. (Courtesy of Perkin-Elmer Corp.)

change per unit concentration (sensitivity) becomes small and therefore the uncertainty of the measurement becomes large.

9.3 FLAME EMISSION SPECTROMETRY[9]

In flame emission spectrometry, the sample solution is nebulized (converted into a fine aerosol) and introduced into the flame where it is desolvated, vaporized, and atomized, all in rapid succession. Subsequently, atoms and molecules are raised to excited states via thermal collisions with the constituents of the partially burned flame gases. Upon their return to a lower or ground electronic state, the excited atoms and molecules emit radiation characteristic of the sample components. The emitted radiation passes through a monochromator that isolates the specific wavelength for the desired analysis. A photodetector measures the radiant power of the selected radiation, which is then amplified and sent to a readout device, meter, recorder, or microcomputer system. A typical FES instrument is shown in Figure 9.14.

The radiant power of the spectral emission line that appears at frequency v, I_v, is determined by the number of atoms that simultaneously undergo the spectral transition associated with the emission line. It is given by the expression

$$P_v = \frac{V \cdot A_t \cdot h \cdot v \cdot N_0 \cdot g_u \cdot e^{-E/kT}}{B(T)} \tag{9.2}$$

where V is the flame volume (aperture ratio) viewed by the detector, A_t is the number of transitions each excited atom undergoes per second, N_0 is the number of free analyte atoms present in the electronic ground state per unit volume (which is proportional to the concentration of analyte in the sample solution nebulized), g_u is the statistical weight of the excited atomic state, k is the Boltzmann constant, T is the absolute temperature, $B(T)$ is the partition function of the atom over all states, and E is the energy of the excited state. Equation 9.2 indicates that the higher the flame temperature, the greater the number of atoms in the excited state. The ratio of excited atoms to ground-state atoms under conditions of thermal equilibrium is given in Table 9.5 for selected emission lines of some commonly determined elements.

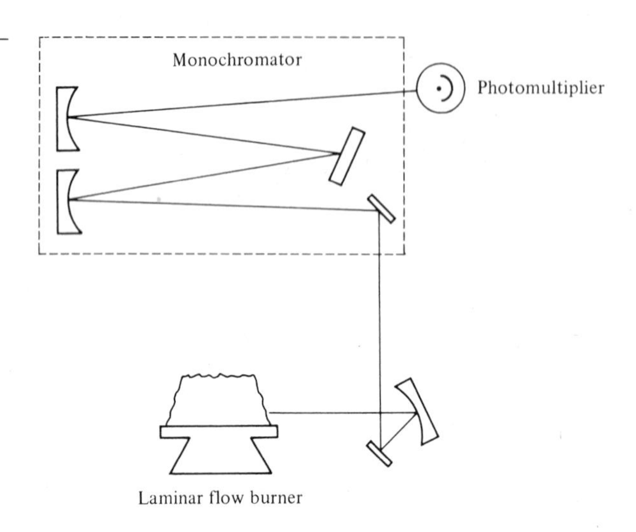

FIGURE **9.14**
Schematic arrangement of a flame emission spectrophotometer.

TABLE **9.5**

VALUES OF N^*/N_0 FOR VARIOUS RESONANCE LINES

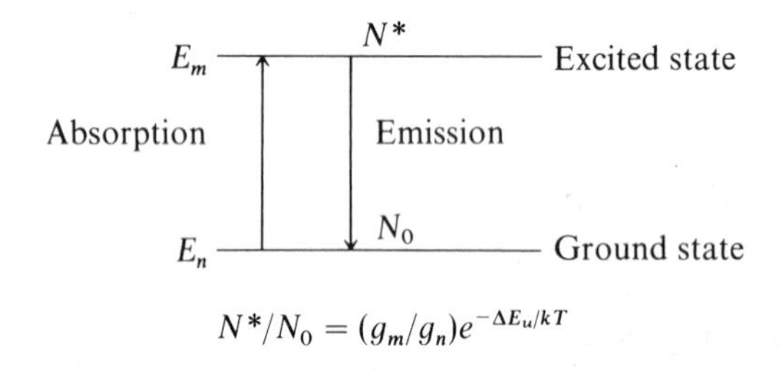

$$N^*/N_0 = (g_m/g_n)e^{-\Delta E_u/kT}$$

Resonance line		g_m/g_n	ΔE **(in eV)**	N^*/N_0 **2000 K**	N^*/N_0 **3000 K**
Cs	8521	2	1.45	4.44×10^{-4}	7.24×10^{-3}
Na	5890	2	2.10	9.86×10^{-6}	5.88×10^{-4}
Ca	4227	3	2.93	1.21×10^{-7}	3.69×10^{-5}
Fe	3720		3.33	2.29×10^{-9}	1.31×10^{-6}
Cu	3248	2	3.82	4.82×10^{-10}	6.65×10^{-7}
Mg	2852	3	4.35	3.35×10^{-11}	1.50×10^{-7}
Zn	2139	3	5.80	7.45×10^{-15}	5.50×10^{-10}

A grating spectrometer, equipped with a laminar flow burner and a good detection-readout system, serves equally well for FES and flame AAS because both require the measurement of the intensity of selected wavelengths of radiation that emerge from the flame. The wavelengths usually fall into the visible or ultraviolet region and the resulting photons are detected by photomultiplier tubes (see Chapter 6). FES requires a monochromator capable of providing a bandpass of 0.05 nm or less in the first order. Slits should be adjustable to allow for greater radiant power for situations where high resolution is not required. Spectrometers of 0.33–0.5 m focal length with adjustable slits meet these requirements. The instrument should have sufficient resolution to minimize the flame background emission and to separate atomic emission lines from nearby lines and molecular fine structure. In contrast, emission band spectra from molecular species show up more clearly with instruments of low dispersion. The ability to scan a portion of a spectrum is often a desirable instrumental feature. Proper positioning of the flame to ensure sampling of the optimum flame zone is important. The best entrance optics design just fills the monochromator optics with a solid angle of radiation. At a high aperture ratio, the limit of detection is restricted, not by the shot noise of the photodetector, but by the instability of the flame and the flicker noise of emission from the matrix.

Background correction in FES can be accomplished with a dual-channel instrument (see Section 9.6). One channel is tuned to the emission line of the analyte and the other is set to a nearby wavelength where analyte emission is not observed, but where background emission from the flame or continuum is measured. The analytical signal is the difference in the intensities from the two wavelengths. The Zeeman method of correction (Section 9.6) can also be applied to FES measurements.

9.4 ATOMIC ABSORPTION SPECTROMETRY[10,11]

The absorption of radiation by atoms in the sun's atmosphere was first observed in 1814. However, it was only in 1953 that an Australian physicist, Alan Walsh, demonstrated that atomic absorption could be used as a quantitative analytical tool in the chemical laboratory. Today AAS is one of the most widely used methods in analytical chemistry.

The AAS phenomenon can be divided into two major processes: (1) the production of free atoms from the sample and (2) the absorption of radiation from an external source by these atoms. The conversion of analytes in solution to free atoms in the flame was discussed earlier in this chapter.

The absorption of radiation by free atoms (those analyte atoms removed from their chemical environment but not ionized) in the flame involves a transition of these atoms from the highly populated ground state to an excited electronic state (see Section 5.3). Although other electronic transitions are possible, the atomic absorption spectrum of an element consists of a series of resonance lines, all originating with the ground electronic state and terminating in various excited states. Usually the transition between the ground state and the first excited state, known as the first resonance line, is the line with the strongest absorptivity. The absorptivity for a given element decreases as the energy difference between the ground state and the excited states increases. All other factors being equal, if an analysis requires high sensitivity, the first resonance line of the analyte is used.

The wavelength of the first resonance line for all metals and many metalloids is longer than 200 nm, the short wavelength limit for operation in the conventional ultraviolet region. The first resonance line for most nonmetals falls into the vacuum ultraviolet region below 185 nm and, therefore, cannot be measured with conventional spectrometers. Thus AAS instrumentation finds wide application for the analysis of metals and metalloids. The optical systems of AAS instruments can be modified to detect resonance lines of nonmetals (<200 nm), but these modifications add a significant expense to the instrument and are not commonly used.

For AAS to function as a quantitative method, the width of the line emitted by the narrow-line source (see Section 9.3) must be smaller than the width of the absorption line of the analyte in the flame (Figure 9.15). The shape of the spectral line emitted by the source is a critical parameter in AAS. The flame gases are considered as a sample cell that contains free, unexcited analyte atoms capable of absorbing radiation at the wavelength of the resonance line emitted by the external source. Unabsorbed radiation passes through a monochromator that isolates the resonance line and then into a photodetector that measures the power of the transmitted radiation. Absorption is determined by the difference in radiant power of

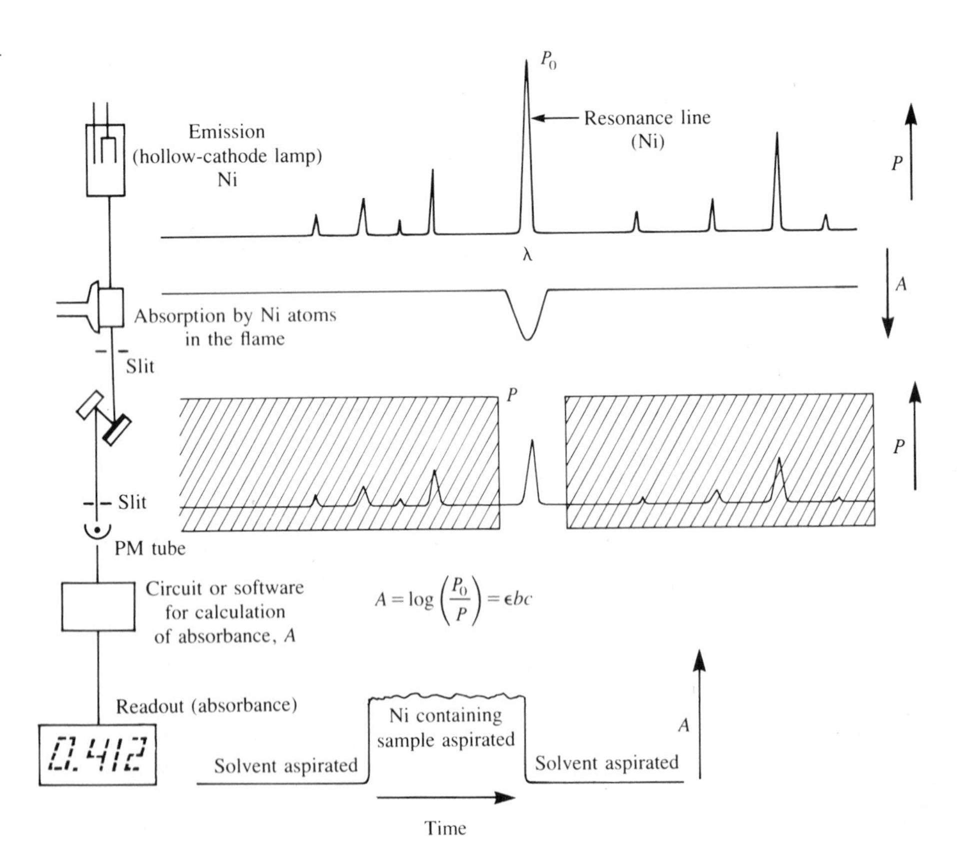

FIGURE **9.15**
Atomic absorption measurements and results.

the resonance line in the presence and absence of analyte atoms in the flame. Instrumentation for AAS is shown in Figure 9.16.

Transitions from the ground state to the first excited state occur when the frequency of incident radiation from the source is exactly equal to the frequency of the first resonance line of the free analyte atoms. Part of the energy of the incident radiation, P_0, is absorbed. The transmitted power, P, may be written

$$P = P_0 e^{-(k_\nu \cdot b)} \quad (9.3)$$

where k_ν is the absorption coefficient of the analyte element and b is the average thickness of the absorbing medium—that is, the horizontal path length of the radiation through the flame. Around the center of the resonance line there is a finite band of wavelengths caused by absorption line broadening within the flame and also broadening associated with the emission source. The two principal causes of line broadening are Doppler (see p. 248) and Lorentz, or pressure broadening. Lorentz broadening is caused by collisions of the absorbing atoms with other molecules or atoms present in all flames. These collisions cause the analyte atoms to have a small range of energies centered on the resonance frequency, resulting in a broadening of the resonance line.

For Equation 9.3 to be valid, the bandwidth of the incident radiation from the source absorbed by the analyte atoms must be narrower than the absorption line of the analyte. This means that the line width of the primary radiation source must be less than 0.001 nm, the usual width of resonance lines found in the absorption spectra of free atoms (Figure 9.15). This requirement on the width of resonance lines emitted from the source arises because all but the most expensive monochromators have bandpasses greater than 0.01 nm. Walsh demonstrated that a hollow cathode made of the same element as the analyte emits lines that are narrower than the corresponding atomic absorption line width of the analyte atoms in the flame. This is the basis for current commercial AAS instrumentation.

Specifications for a typical atomic absorption spectrophotometer might include a 0.33–0.5-m focal length Czerny-Turner monochromator with a 64 × 64-mm grating ruled with 2880 grooves/mm and blazed at 210 nm to cover a range of

FIGURE **9.16** Optical diagrams of (a) a single-beam and (b) a double-beam atomic absorption spectrometer. (Courtesy of Instrumentation Laboratories, Inc.)

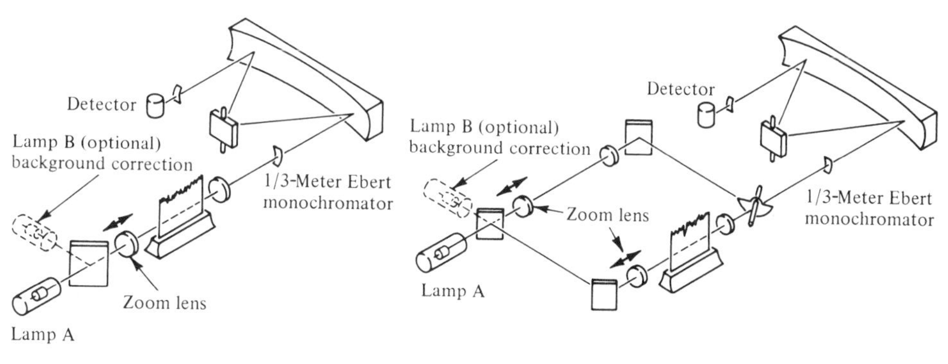

wavelengths of 190–440 nm. A second grating, ruled with 1440 grooves/mm and blazed at 580 nm, covers the range of 400–900 nm. The two gratings are often mounted back-to-back on a turntable. Spectral bandwidths should be 0.03–7 nm in the ultraviolet and 0.06–14 nm in the visible region. They should be keyboard- or switch-selectable. The geometry of the beam of radiation must be designed to provide optimum performance with both flame and electrothermal atomizers. Optical elements are needed to focus an image of the source lamp on the flame (or furnace), collect a small solid angle of radiation from the flame (or furnace), and provide sufficient radiant flux at the photodetector. Such a system minimizes the effect of radiation produced by thermal emission of analyte atoms in the flame. One design has a zoom lens system to produce optimum transmission of radiation for any resonance line.

The more sophisticated AAS system shown in Figure 9.17 is a two-channel, double-beam, microcomputer-controlled spectrophotometer capable of background correction in either one or two channels (see Section 9.6). The instrument can determine two elements simultaneously, thereby doubling the speed of single-element instruments. Alternatively, analytical accuracy can be improved by using the element in one channel as an internal standard (see Section 2.8). In another operating mode, the instrument can extend the analytical range of a given element; the same element is determined in both channels but with resonance lines of different sensitivity. It is also possible to determine the ratios between the concentrations of two elements, an internal standard and the unknown. This operating mode minimizes errors that result from fluctuations in flame conditions, aspiration rate, sample viscosity, and temperature of the sample solution.

The microcomputer contained in an AAS system (see Section 4.6) enables the instrument to calculate the best analytical curves in one or both channels using up to five standards, compute ratios, apply appropriate statistical techniques, and present

FIGURE **9.17** Optical schematic of a two-channel, double-beam atomic absorption spectrometer. (Courtesy of Instrumentation Laboratories, Inc.)

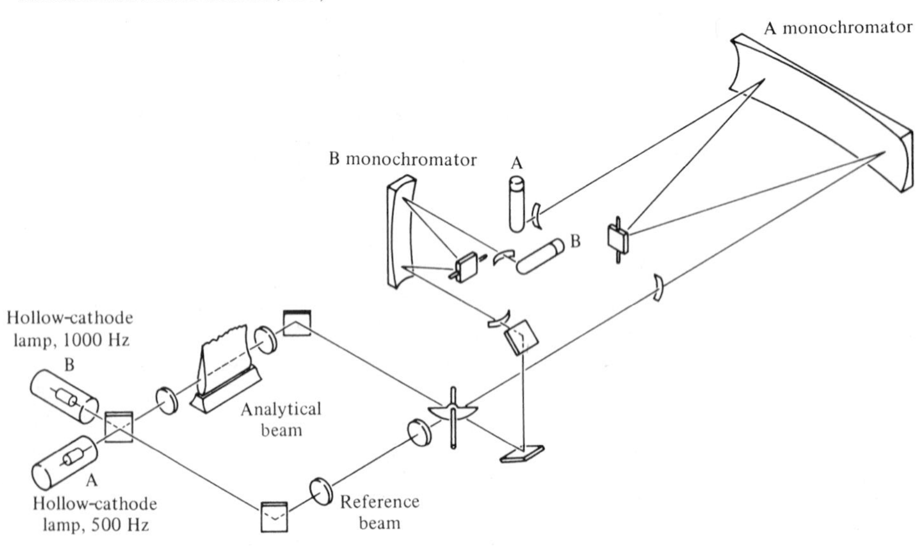

results in graphical or tabular form. In addition the microcomputer can be programmed to handle routine tasks associated with any determination: selection of the resonance wavelength, slit width, fuel and oxidant flows, burner height, lamp current, and electrothermal furnace operating parameters.[12] A deuterium lamp, used for background correction, can be applied independently to either or both channels (see Section 9.6).

The hollow-cathode lamp has become the source most often used in AAS. The electrodeless discharge lamp provides a more intense and stable source for certain elements in AAS; it is also the source of choice for atomic fluorescence measurements.

A hollow-cathode lamp has a Pyrex body with an end window of quartz. Within, an anode wire is positioned along the outside of the cylindrical cathode, as shown in Figure 9.18. The cathode is made of (or lined on the interior with) the element to be determined. The lamp is evacuated and filled with an inert gas, usually argon or neon, at a pressure of 4–10 torr. Lamps operate at currents below 30 mA and voltages up to 400 V. Discharge occurs between the two electrodes, ionizing some of the inert fill gas atoms. The cathode (40 mm i.d.) is bombarded by these energetic, positively charged, inert gas ions. The ions are accelerated toward the cathode surface by the electrical potential that exists between the two electrodes. When the positively charged ions collide with the negatively charged cathode surface, the metal atoms of the cathode are ejected (sputtered) into the gaseous atmosphere inside the lamp. Here, the metal atoms absorb energy by colliding with fast-moving filler gas ions, are elevated to excited electronic states, and finally return to the ground state by emitting radiation characteristic of the metallic cathode.

With lamps that contain elements with high ionization potentials, neon is used as the fill gas. Neon improves the intensity of the emitted resonance lines, since its ionization potential is greater than that of argon. Argon is usually used only when there is a neon line in close proximity to a resonance line of the cathode element. When the cathode is formed into a cylinder or cup, the discharge tends to concentrate in the cup. In this configuration more efficient sputtering and excitation occur. The addition of a protective shield (nonconductive) of mica around the outside of the cathode just behind the lip of the cup causes the lamp to radiate more intensely by preventing spurious discharges around the outside of the cathode.

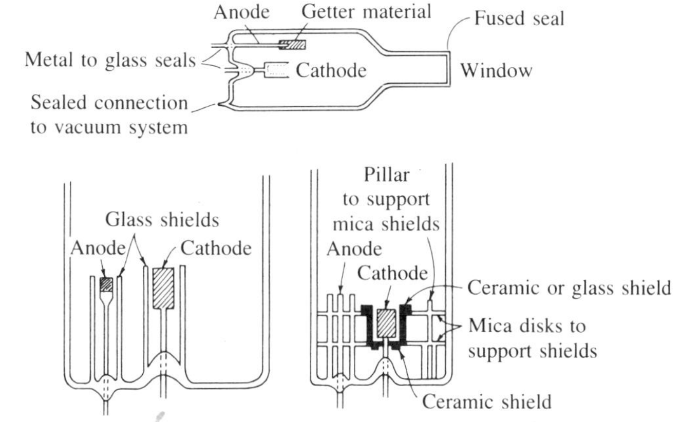

FIGURE **9.18**
Schematic diagram of shielded-type hollow-cathode lamp.

Cathode construction differs for various metals. If the metal is easily worked, the entire cathode may be made from it. If an expensive metal is involved, a thin liner of the metal is inserted into a copper cathode. (Consequently, copper resonance lines appear in the spectrum of many lamps.) When the metal's melting point is low, a cuplike carrier electrode is used. For hard or brittle metals, an alloy or sinter of pressed metal powder is used. Multielement lamps are also constructed in this fashion using metal powders.

As mentioned, the shape of the spectral line emitted by the source is an important parameter in AAS determinations. Although increasing the lamp current beyond a certain optimum value increases its output intensity, the sensitivity for many elements is reduced through line broadening or self-reversal. Doppler broadening is caused by the motions of the radiating atoms as a result of thermal activity at the hot cathode. For a given emission line, the line broadening is proportional to the square root of the temperature. Thus, to produce the required narrow emission lines, the temperature of the radiating plasma near the cathode must be kept as low as possible. This is done by keeping the lamp current small. Self-reversal or self-absorption is caused by the absorption of radiation by unexcited atoms in the source lamp. These unexcited atoms of the cathode material exist in the cooler portions of the lamp. Reduction of line intensity due to self-reversal is proportional to the length of the path traveled by emitted radiation in the lamp and the lamp current. Good lamp design can minimize this type of absorption by reducing the path length and reducing lamp current.

A liability of AAS is the need (usually) for a different hollow-cathode lamp for each element to be analyzed. On the positive side, these light sources with their narrow emission line widths provide virtual specificity for each element. Multielement lamps are available for a few combinations of elements. The cathode is made from sections or rings of the different elements or from an alloy or pressed powder that contains the elements blended so as to obtain emission lines of equal intensity from each element. During the lifetime of the lamp, the more volatile metal sputters at a faster rate and slowly covers the other elements. This leads to an increase in intensity of the volatile element and a steady decrease in the intensities of the other elements. In general, multielement lamps sacrifice some sensitivity when compared with single-element lamps.

Another approach in overcoming the single-element, single-lamp problem is the lamp turret accessory. Up to six individual lamps are located in the turret and are kept at the correct operating currents for the elements involved, thus eliminating delays caused by warming up the lamps. Computer-controlled rotation of the turret brings the lamp for the element selected into the source position and no further adjustment is required. This accessory is invaluable when several elements are to be determined in the same solution. All types of hollow-cathode lamps have finite operating lifetimes; they are often warranted for 5000 mA-hr.

For volatile elements, where reduced intensity and short lamp life are problems with hollow-cathode lamps, an electrodeless discharge lamp offers an alternative source. This lamp has also found application in AFS where its high-intensity output improves the determinations. Electrodeless discharge lamps are fabricated by sealing small amounts of the metal, iodine (or the more volatile metal iodide salt), and argon at a low pressure in a small quartz tube. This tube is placed inside a ceramic cylinder on which an antenna from a microwave generator (2450 MHz,

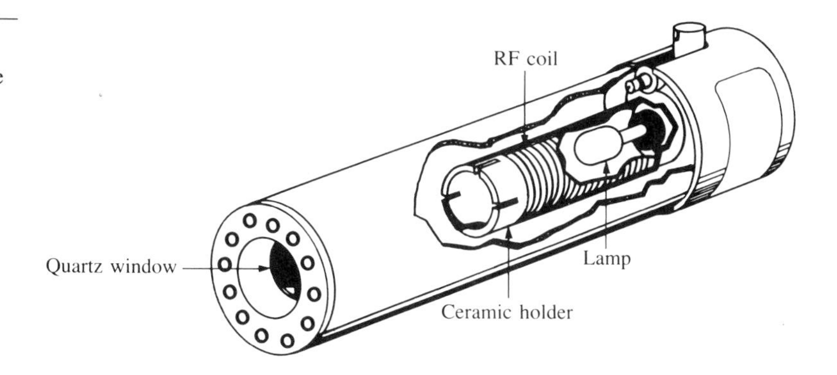

FIGURE 9.19
Electrodeless discharge lamp. (Courtesy of Perkin-Elmer Corp.)

200 W) is coiled (Figure 9.19). When an alternating field of sufficient power is applied, the coupled energy vaporizes and excites the atoms inside the tube, thus producing their characteristic emission spectrum. Lamp temperature is an important operating parameter; an increase in temperature of 130 °C can effect a 1000-fold increase in line intensity. Optimum temperature varies with different elements.

Electrodeless discharge lamps cannot be powered with hollow-cathode tube power supplies; a separate power supply must be provided. Warm-up periods of 30 min are required to stabilize the lamp output.

9.5 ATOMIC FLUORESCENCE SPECTROMETRY[13,14]

The basic principle of atomic fluorescence spectroscopy (AFS) is the same as that of molecular fluorescence described in Chapters 5 and 8. Free analyte atoms formed in a flame absorb radiation from an external source, rise to excited electronic states, and then return to the ground state by fluorescence. AFS offers the same advantages over AAS for trace analysis that molecular fluorescence has over ordinary absorption spectroscopy; the radiation from fluorescence is measured (in principle) against a zero background, whereas ordinary absorption measurements involve the ratio of two signals.

In AFS the exciting source is placed at right angles to the flame and the optical axis of the spectrometer (Figure 9.20). Some of the incident radiation from the

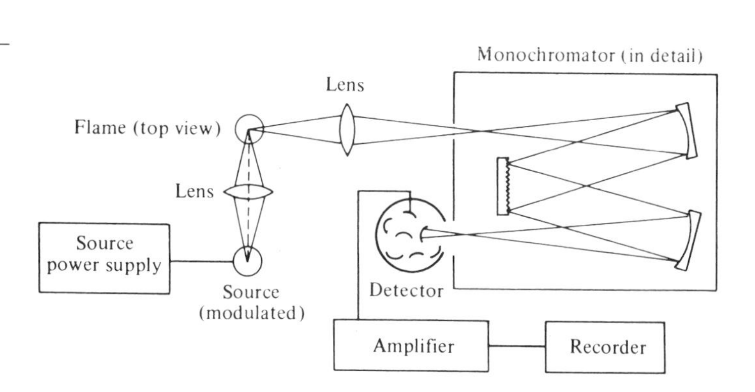

FIGURE 9.20
Schematic diagram of equipment for atomic fluorescence spectrometry.

source is absorbed by the free atoms of the test element. Immediately after this absorption, energy is released as atomic fluorescence at a characteristic wavelength upon the return of the excited atoms to the ground state.

The best burner system for AFS is probably a combination of acetylene/nitrous oxide and hydrogen/oxygen/argon using a rectangular flame with a premixed laminar flow burner (see Section 9.2). The flame should have a low background and a low quenching cross section, in addition to being efficient in producing a large free-atom population. Although the influence of flame background on detection limits in AFS is most severe with an unmodulated source and dc detection, the presence of intense flame background is a problem even in systems that use modulation. Even though the unmodulated flame background is not amplified directly with ac modulation, its presence results in noise at the output of the amplifier. A commercial instrument for AFS, the Baird atomic inductively coupled plasma system, is discussed in Section 10.4.

The intensity of the fluorescence is linearly proportional to the exciting radiation flux. When there is no analyte, only background radiation from the flame is detected. This difference between AFS and AAS is significant near the analyte detection limit and makes AFS the method of choice for trace analysis for selected metals. AFS exhibits its greatest sensitivity for elements that have high excitation energies.

9.6 INTERFERENCES ASSOCIATED WITH FLAMES AND FURNACES

Essentially the same interferences occur in FES and flame AAS but to different extents. The literature contains many contradictory statements and hasty generalizations based on inadequate measurements and misunderstandings. Interferences can be separated into four general classes: (1) background absorption, (2) spectral line interference, (3) vaporization interference, and (4) ionization effects.

Background Absorption

The background associated with flames is caused by the large number of species present in the flame that are capable of broad band absorption of radiation. The species include metal oxides, hydrogen molecules, OH radicals, and portions of fragmented solvent molecules. Radiation emitted from the hollow-cathode source is absorbed by these species as well as by the analyte atoms. This background absorption results in a direct interference and a corresponding error in analytical results.

In some respects, background absorption in AAS is a more serious problem than in FES because it is likely to go unnoticed. The monochromatic nature of the source negates the need for scanning. Even with automatic background correction, care must be exercised in obtaining measurements. In areas of high background absorption (50%–80%), the signal may be displaced into the high-absorption region (>90%) where measurement error is at a maximum.

A number of techniques correct for background absorption. A solution that contains none of the analyte can be prepared and its absorbance at the resonance line of the analyte determined. The absorbance of this "blank" solution is then used to correct the measurements obtained from sample solutions. This procedure requires that the matrix of the blank and sample solutions be similar and that the absorption of the blank is indeed a measure of the background absorption. In practice, it is difficult to prepare the solutions required to make accurate background corrections with this procedure.

A more precise technique involves measuring the absorbance at a wavelength close to the resonance line used in the determination. This absorbance measurement is used to correct the absorbance signal at the resonance line. This technique assumes that the variation due to the background absorption is insignificant over the small wavelength difference involved. It allows the correction to be made using only the sample solution. The problem with this technique is locating an appropriate resonance line close to the line used for the determination.

The use of a broad-band, continuous source of radiation in conjunction with the hollow-cathode line source is a popular technique for background correction. The resonance line of the hollow cathode is absorbed by both free analyte atoms and molecular species in the flame. The radiation from the continuous source, usually a deuterium lamp, is absorbed over the entire bandwidth of wavelengths passed by the monochromator, usually 0.1–0.2 nm, while the free analyte atoms absorb in a narrow range (0.005–0.001 nm). Thus, the decrease in the intensity of radiation from the continuous source is due almost entirely to absorption by the components of the background, whereas absorption by the free analyte atoms is negligible. In this situation, absorption of radiation from the deuterium lamp is a measure of the background and the background correction is made at the same nominal wavelength as the resonance line used for the AAS determination.

The use of a broad-band continuous source for background correction can be incorporated into either single-beam or double-beam optical configurations. The modulation frequency used to chop the beams from the radiation sources must provide for fast correction, up to three corrections per 10 msec. This rate of signal correction is required because both the sample and background may change rapidly, especially in the case of electrothermal atomization. At wavelengths in the visible region, additional problems arise because the deuterium source intensity is weaker than the output of many hollow-cathode sources that operate at optimum current levels. Substitution of the deuterium lamp with a 150-W xenon-mercury lamp provides adequate intensity from 200 to 600 nm.

The Zeeman effect on radiation emitted from hollow-cathode tubes has been used to correct for background.[15] If an intense, alternating magnetic field is applied to a hollow cathode, the lamp emits a single line at the resonance frequency and a doublet centered about the resonance frequency but shifted slightly in wavelength from the resonance line by Zeeman splitting (Figure 9.21). The signal from the resonance line is affected by both atomic and background absorption, whereas the signal from the nonresonance doublet is affected only by background absorption. The difference between these signals provides a measurement of the corrected atomic absorption. This technique offers the advantage of using a single radiation source and only one optical path through the atomizer. When separate sources are used, it is sometimes difficult to match exactly the optical paths. A difficulty associated with the technique is that the magnetic field can cause instability in the lamp, resulting in an irregular emission signal.

The pulsed hollow-cathode lamp background correction also uses a single radiation source.[16] Two absorption measurements are made, one with the lamp run at a normal low current value and a second with the lamp pulsed to a large current value. The first measurement indicates the absorbance due to both analyte atoms and background, whereas the second measurement indicates primarily background. The large current eliminates the resonance line due to self-reversal and thus provides a measurement of the background absorption. Substraction of these two absorption measurements yields a corrected value for atomic absorption. The lamp must not be pulsed to too high a current or else the tube-life will be shortened. This technique appears to be easier to implement and operate than the Zeeman method.

Spectral Line Interference

Spectral line interferences occur when a line of interest cannot be readily resolved from a line of another element or from a molecular band. Interference of this type is closely associated with the resolving power of the monochromator. Atomic line interferences are more serious in FES, even in those instruments that contain the best monochromators. Their bandpass is still an order of magnitude broader than the line profiles of cathode lamps (typically 0.005 nm) and absorbing atoms in AAS. Molecular spectral interference is also more severe in FES. Chemiluminescence (see Section 5.4) of molecular fragments formed in the flame gas is a major source of molecular interference. In flame AAS and AFS this interference can be minimized by amplitude modulation of the radiation source. No such possibility exists in FES.

Instances of serious spectral interference in FES involve the manganese triplet (403.1, 403.3, 403.5 nm), the gallium line (403.3 nm), the potassium doublet (404.4,

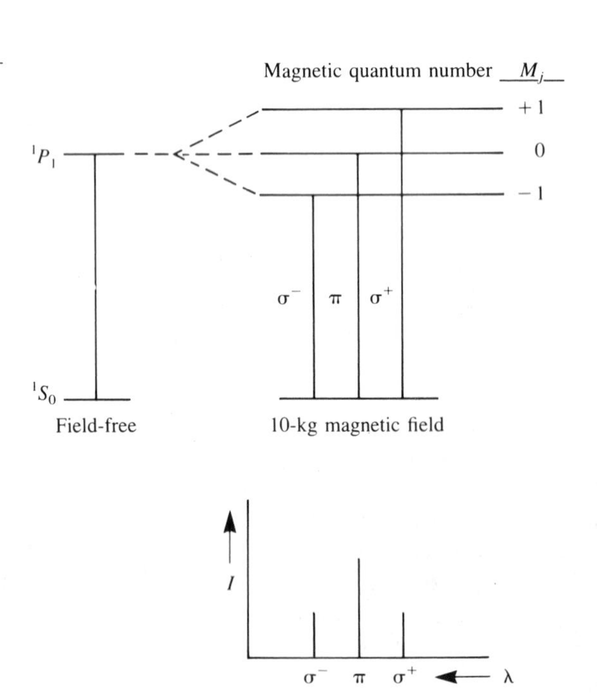

FIGURE 9.21
Zeeman splitting.

404.7 nm), and the lead line (405.8 nm). The band systems of CaOH extending from 543 to 622 nm interfere with the sodium doublet (589.0, 589.6 nm) and the barium line (553.6 nm).

Vaporization Interference

Interferences of this type arise when some component of the sample alters the rate of vaporization of salt particles that contain the analyte. They arise from a chemical reaction that changes the vaporization behavior of the solid, or as the result of a physical process in which the vaporization of the matrix controls the release of the analyte atoms.

Hotter flames tend to minimize vaporization interferences. Use of the acetylene/nitrous oxide flame is often justified because it is better for decomposing refractory phosphates, sulfates, silicates, and aluminates than the cooler acetylene/air flame. The shifting of equilibrium of reactions involving these thermally stable molecules, due to small uncontrolled changes in flame temperatures, influences AAS and FES to similar degrees. Thus, in situations involving thermally stable analyte molecules, AAS results are apparently as dependent on flame temperature as those obtained from FES.

Metal compounds in the flame are usually simple di- or triatomic molecules such as CaO or CaOH. Elements such as Na, Cu, Tl, Ag, and Zn are almost completely atomized in the flame; they do not form compounds with flame components in noticeable proportions. Alkaline-earth elements form monoxides unless very fuel-rich flames are used. Some metals such as Al and Ti form refractory oxides, which are extremely stable. As a result, the free-atom concentrations of these metals are negligible in flames of stoichiometric composition and moderate temperature. However, these metals may be analyzed in the reducing environment of a fuel-rich acetylene/nitrous oxide flame.

Releasing agents provide a chemical means for overcoming some vaporization interferences by pretreatment of sample solutions. In calcium AAS determinations, a few hundred parts per million of lanthanum or strontium are often added to solutions to minimize interference due to phosphates. At these concentrations, lanthanum or strontium preferentially binds phosphate, thus releasing the calcium for atomic absorption. Calcium is also released from phosphate by complexing the calcium with EDTA. Once in the flame the EDTA is destroyed. The calcium does not reform with the phosphorus entities in the flame gases. Thus, a releasing agent may either combine with the interfering substance or deny the analyte to the interfering substance by mass action. In either situation the analyte is left free to vaporize in the flame.

When the physical characteristics (viscosity and surface tension) of the sample and calibration solutions are different, an interference occurs that affects the rate of nebulization of analyte atoms into the flame (see Equation 9.1) and therefore the rate of vaporization of analyte. Errors caused by this type of interference are minimized by the method of standard additions (see Section 2.8).

Ionization Interference

At elevated flame and furnace temperatures, atoms with low ionization potentials become ionized (Table 9.5). Any ionization reduces the population of both the ground state and the excited state of neutral free atoms, thus lowering the sensitivity of the determination. This problem is readily overcome by adding an excess (ca. 100-fold) of a more easily ionized element such as K, Cs, or Sr to suppress ionization in both sample and calibration solutions. The more easily ionized atoms produce a large concentration of electrons in the vapor. These electrons, by mass action, suppress the ionization of analyte atoms. Thus, suppressant should be added to samples that contain variable amounts of alkali metals analyzed by acetylene/air flames to stabilize free-electron concentrations. The addition of suppressants is even more important in analyses that require the hotter acetylene/nitrous oxide flames.

9.7 APPLICATIONS

Flame Emission Spectroscopy

Most applications of both FES and flame AAS have been the determination of trace metals, especially in liquid samples. It should be remembered that FES offers a simple, inexpensive, and sensitive method for detecting common metals, including the alkali and alkaline earths, as well as several transition metals such as Fe, Mn, Cu, and Mn. FES has been extended to include a number of nonmetals: H, B, C, N, P, As, O, S, Se, Te, halogens, and noble gases. FES detectors for P and S are commercially available for use in gas chromatography (see Chapter 18).

FES has found wide application in agricultural and environmental analysis, industrial analyses of ferrous metals and alloys as well as glasses and ceramic materials, and clinical analyses of body fluids. FES can be easily automated to handle a large number of samples. Array detectors interfaced to a microcomputer system permit simultaneous analyses of several elements in a single sample.

Atomic Absorption Spectroscopy

AAS has been used for trace metal analyses of geological, biological, metallurgical, glass, cement, engine oil, marine sediment, pharmaceutical, and atmospheric samples. As in FES, liquid samples usually present few problems in pretreatment. Thus, most solid samples are first dissolved and converted to solutions to facilitate analysis. Gas samples are generally pretreated by scrubbing out the analyte and analyzing the scrubbing solution or adsorbing the analytes on a solid surface and then leaching them into solution with appropriate reagents. Direct sampling of solids may be accomplished using an electrothermal furnace.

The chemistry involved in the pretreatment of samples is a vital component of both FES and AAS determinations. In trace analysis, the analyst must be alert to possible sources of sample contamination such as storage containers, impurities in pretreatment reagents and solvents, and incomplete removal of prior samples from the nebulizer system. Careful attention must be given to minimizing contamination from room dust and contact with an analyst's skin or clothing and laboratory glassware.

The detection limits and sensitivities provide a means of comparing the quantitative characteristics of atomic spectroscopic methods for a given element (see Section 2.3).[17] Table 9.2 summarizes the detection limits for FES, flame AAS, electrothermal AAS, and AFS. ICP and DCP methods are discussed in Chapter 10.

9.8 COMPARISON OF FES AND AAS[18,19]

A comparison of the analytical performances of AAS and FES reveals that the two methods complement each other in many respects. FES is better for determinations that involve alkali, alkaline-earth, and rare earth elements as well as Ga, In, and Tl. Flame AAS permits Ag, Al, Au, Cd, Cu, Hg, Pb, Te, Sb, Se, and Sn to be detected with high sensitivity. The performance of both methods for other elements is similar. The choice of the method depends on the matrix to be analyzed and the sensitivity and the selectivity desired. In routine analyses it is reasonable to combine both methods. FES, however, has one important advantage in that it permits simultaneous quantitative, multielement analyses. AFS exhibits its greatest sensitivity for elements with analytical lines at shorter wavelengths and thus competes more strongly with AAS.

Alkemade has compared FES and flame AAS on a theoretical basis.[20] He showed that AAS can be more sensitive for a given element only if the brightness of the lamp exceeds that of a blackbody at the temperature of the flame, with both measured at the wavelength of the analytical line. For nonthermal sources (electrical discharges), such as hollow-cathode lamps, the spectral radiance of the lamp may be much greater than that of the flame, permitting better performance of AAS at shorter wavelengths. The reader should consult Alkemade's paper for details of commonly accepted fallacies concerning AAS. The higher temperatures of inductively coupled and dc plasmas provide the necessary excitation energies at all wavelengths.

For most elements, the flame is a primary source of noise in FES or flame AAS signals. Since background intensity is considerably greater in FES than in AAS, the flame noise may influence detection limits more strongly in FES. In AAS, the lamp used as a primary radiation source is an additional source of noise. There is no consistent difference in the reproducibility of results between the two methods. Relative standard deviations of 0.5%–1.0% are obtained in favorable cases. Except for work near the detection limit, the instability of the nebulizer and flame is the major contribution to the scattering of results in both methods.

For many elements electrothermal AAS provides lower detection limits than either FES or flame AAS. The analytical trade-offs for these lower limits of detection are a decrease in precision and an increase in the time required for analysis.

PROBLEMS

1. For the analysis of cement samples, a series of standards was prepared and the emission intensity for sodium and potassium was measured at 590 and 768 nm, respectively. Each standard solution contained 6300 μg/mL of calcium as CaO to compensate for the influence of calcium on the alkali readings. The results are shown in the table. For each cement sample 1.000 g was dissolved in acid and diluted to exactly 100 mL. Calculate the percent of Na_2O and K_2O.

Concentration ($\mu g/mL$)	*Emission reading* Na_2O	K_2O
100	100	100
75	87	80
50	69	58
25	46	33
10	22	15
0	3	0
Cement A	28	69
Cement B	58	51
Cement C	42	63

2. In Problem 1, what contributed to the emission reading of the blank at the analytical wavelength for sodium, but did not for potassium? A small quartz spectrometer was used to obtain the results. Would the blank reading be larger, smaller, or the same if a filter photometer equipped with glass absorption filters had been used? [See *Anal. Chem.*, **21,** 1296 (1949).]

3. Boron gives a series of fluctuation bands due to the radical BO_2 that lie in the green portion of the spectrum. Although the overlapping band systems present a problem in the measurement of the flame background, the minimum between adjacent band heads can be used. These results were obtained:

Boron present ($\mu g/mL$)	*Emission reading* 518-nm peak	505-nm minimum
0	36	33
50	44	36
100	52	39
150	60.5	42.5
200	68.5	45.5

What are the concentrations of boron in these unknowns?

A	45	36.5
B	85	65
C	66	50

[See *Anal. Chem.*, **27,** 42 (1955).]

4. A calibration curve for strontium, taken at 460.7 nm, was obtained in the presence of 1000 $\mu g/mL$ of calcium as CaO and also in the absence of added calcium. These results are shown in the table. (a) Graph the calibration curve on rectilinear graph paper and also on log-log paper. (b) What might be the cause of the upward curvature in the region of low concentrations on the rectilinear graph when calcium is absent? (c) Why does the addition of calcium straighten the calibration curve and increase the net emission reading for strontium?

Strontium present (μg/mL)	*Emission reading* No calcium	Calcium added
0	0	13
0.25	2	18.5
0.5	6	24
1.0	16	36
2.5	44	70
5.0	94	125
7.5	150	181
10.0	200	238

5. Calculate the iron content in a diethyldithiocarbamate extract using the following data:

Absorbance units Blank	Sample	Iron added (μg/200 mL)
0.0020	0.0090	None
0.0214	0.0284	2.00
0.0414	0.0484	4.00
0.0607	0.0677	6.00

6. To illustrate the effect of aqueous-organic solvents on droplet size, calculate the mean droplet diameter for (a) water, (b) 50% methanol–water, and (c) 40% glycerol–water. Pertinent data follow.

System	Surface tension (dyne/cm)	Viscosity (dyne/cm^2)	Density (g/cm^3)	Velocity of aspirating gas (m/sec)	Q_{air}/Q_{liq}
Ethanol, 50%	28	0.029	0.934		
Glycerol, 40%	68.6	0.039	1.102	279	2540
Methanol, 50%	30.6	0.027	0.946	198	9540
Methyl isobutyl ketone	24.6	0.0051	0.801		
Water	73	0.010	1.00	198	6400

7. For (a) water, (b) 50% (v/v) ethanol–water, and (c) methyl isobutyl ketone as solvents, plot the droplet diameters for solution flowrates ranging from 0.1 to 5 mL/min. As values for a typical nebulizer, assume the velocity of the aspirating gas to be 333 m/sec and Q_{gas} to be 8.5 liter/min. Other data are given in Problem 6.

8. Calculate the fraction of cesium atoms ionized in a flame at 2000 K when the total cesium concentration in the flame gases is (a) 10^{-4} atm, (b) 10^{-6} atm, and (c) 10^{-7} atm.

9. Calculate the fraction of lithium atoms ionized in a premixed laminar flame at 3000 K when the concentration sprayed into the flame is (a) $10^{-2}M$, (b) $10^{-3}M$, and (c) $10^{-4}M$.

10. What individual amounts of (a) cesium, (b) rubidium, (c) potassium, or (d) lithium should be added to a flame at 2500 K and at 2800 K to suppress the ionization of a solution containing 0.23 μg/mL of sodium?

	K_i (atm)	
Element	**2500 K**	**2800 K**
Li	1.48×10^{-9}	2.63×10^{-8}
Na	4.8×10^{-9}	7.40×10^{-8}
K	1.8×10^{-7}	2.08×10^{-6}
Rb	3.9×10^{-7}	3.98×10^{-6}
Cs	1.45×10^{-6}	1.32×10^{-5}

11. From the absorbance traces shown in the figure for arsenic by atomic absorption, estimate (a) the signal-to-noise ratio, (b) the sensitivity of the method, and (c) the detection limit.

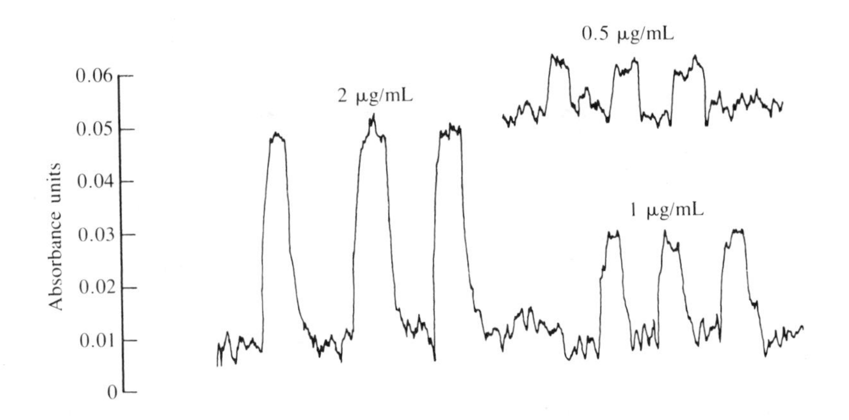

12. State two advantages and a major limitation of each of the following methods: (a) flame AAS, (b) electrothermal AAS, (c) FES, and (d) AFS.

13. List three techniques that are used to correct for background absorption in flames and furnaces. Discuss the advantages and limitations of each technique.

14. Define each of the following terms: (a) nebulization, (b) laminar flow burner, (c) refractory compound, (d) ionization buffer, (e) L'vov platform, (f) hydride generator, (g) resonance line, and (h) Lorentz broadening.

15. List three types of interferences associated with flames and furnaces and illustrate each. Give one method used to correct for or eliminate each type of interference.

16. What are the major steps in the atomization of an analyte using an electrothermal furnace? Explain the physical or chemical processes that occur in each step.

BIBLIOGRAPHY

ALKEMADE, C., et al., *Metal Vapors in Flames,* Pergamon, New York, 1982.

DEAN, J., *Flame Photometry,* McGraw-Hill, New York, 1960.

DEAN, J., AND T. RAINS, eds., *Flame Emission and Atomic Spectrometry: Theory,* Vol. 1, 1969; *Components and Techniques,* Vol. 2, 1971; *Elements and Matrices,* Vol. 3, 1975, Dekker, New York.

EBDON, L., *An Introduction to Atomic Absorption Spectroscopy—A Self Teaching Approach,* Heydon, Philadelphia, 1982.

HOLCOMBE, J., AND T. RETTBERG, *Anal. Chem.,* **58**(5), 124R (1986).

L'VOV, B., *Analyst,* **112,** 355 (1987).

MAYER, B., *Guidelines to Planning Atomic Spectrometric Analysis,* Elsevier, New York, 1982.

OTTAWAY, J., AND A. URE, *Practical Atomic Absorption Spectroscopy,* Pergamon, New York, 1983.

SACKS, R., "Emission Spectroscopy," Chap. 3, *Treatise on Analytical Chemistry,* 2nd ed., P. J. Elving, E. J. Meehan, and I. M. Kolthoff, eds., Part I, Vol. 7, John Wiley, New York, 1981.

WELTZ, B., *Atomic Absorption Spectroscopy,* Verlag Chemie, New York, 1976.

LITERATURE CITED

1. SYTY, A., *Crit. Rev. Anal. Chem.,* **4,** 155 (1974).
2. BROWNER, R., AND A. BOORN, *Anal. Chem.,* **56**(7), 786A, 875A (1984).
3. MCCELLAN, B., "Water," chapter in Vol. 3, *Flame Emission and Atomic Absorption,* J. Dean and T. Rains, eds., Dekker, New York, 1975.
4. NUKIYAMA, S., AND Y. TANASAWA, *Experiments on the Atomization of Liquids in an Air Stream,* E. Hope, translator, Defense Research Board, Department of National Defense, Ottawa, Ontario, Canada, 1950.
5. KOIRTYOHANN, S., AND M. KAISER, *Anal. Chem.,* **54**(14), 1515A (1982).
6. STURGEON, R., *Anal. Chem.,* **49,** 1255A (1977).
7. BERNDT, H., AND MESSERSCHMIDT, J., *Anal. Chem. Acta,* **136,** 407 (1982).
8. ROBBINS, W., AND J. CARUSO, *Anal. Chem.,* **51,** 889A (1979).
9. PICKETT, E., AND S. KOIRTYOHANN, *Anal. Chem.,* **41,** 28A (1969).
10. WALSH, A., *Spectrochim. Acta.,* **7,** 108 (1955).
11. SLAVIN, W., *Anal. Chem.,* **54,** 685A (1982).
12. LIDDEL, P., *Am. Lab.,* **17**(5), 100 (1985).
13. SYCHRA, V., V. SVOBODA, AND I. RUBESKA, *Atomic Fluorescence Spectroscopy,* Van Nostrand Reinhold, London, 1975.
14. VAN LOON, J., *Anal. Chem.,* **53** (2), 333A (1981).
15. BROWN, S., *Anal. Chem.,* **49,** 2187 (1977).
16. SMITH, S., AND G. HIEFTJE, *Appl. Spectrosc.,* **37,** 419 (1983).
17. PARSONS, M., S. MAJOR, AND A. FORSTER, *Appl. Spectrosc.,* **37**(5), 411 (1983).
18. KAHN, H., *Ind. Res./Dev.,* **24**(2), 156 (1982).
19. SLAVIN, W., *Anal. Chem.,* **58**(4), 589A (1986).
20. ALKEMADE, C., *Appl. Opt.,* **7,** 1261 (1968).

ATOMIC EMISSION SPECTROSCOPY WITH PLASMA AND ELECTRICAL DISCHARGE SOURCES

10.1 INTRODUCTION

Atomic emission spectroscopy (AES) has long been a standard method for metal analysis. Recent developments in noncombustion plasma sources have expanded the number of applications and generated new interest in AES. There are few restrictions on the samples; they may range from alloys to ores or from ashes of organic materials to atmospheric dust. As with atomic absorption analysis, a major limitation is the inability of AES to distinguish among the oxidation states of a given element.

In atomic emission spectroscopy, a minute part of the sample is vaporized and thermally excited to the point of atomic emission. The energy required for these processes is supplied by an electric arc or spark, or more recently by a laser or plasma composed of inert gas. It is convenient to distinguish these sources from the chemical flame sources used to obtain emission spectra as described in Chapter 9. This chapter discusses the advantages and limitations of both types of sources for specific analyses. A comparison of atomic spectroscopic methods is given in Table 10.1.

The atomic spectrum emitted by a sample is used to determine its elemental composition. The wavelength at which the intensity measurement is made identifies the element, whereas the intensity of the emitted radiation quantifies its concentration. The fundamentals of atomic spectroscopy are discussed in Chapter 5.

10.2 INSTRUMENTATION

Instrumentation for AES may be divided into three major components: (1) the sampling device and source, (2) the spectrometer, and (3) the detector and readout device. The sampling device and source depend on the type of sample and the analytical data desired. As the emission spectrum emerges from the source it is focused onto the spectrometer's entrance slit, where it is dispersed into its component wavelengths. The optical elements permit site selection within the source's discharge and thus eliminate the need to move the entire source assembly. The use of mirrors, properly placed, also eliminates the chromatic and spherical aberrations encountered with lens optics. At the exit slit(s) of the spectrometer, the radiation is sensed by a photodetector. In older instruments photographic film was used to record the spectra, but most modern systems use photomultiplier tubes or diode arrays linked directly to computer-driven data processing systems.

Sampling Devices and Sources

No single source is best for all applications. The analyst should have available a wide variety of sources that can be selected in accordance with the requirements of the desired analysis. Factors that influence the selection of an excitation source are the concentration of the elements being determined, the vapor pressures or volatilities of these elements, the excitation potentials of the atomic lines used in the analysis, and the physical condition of the sample. In general, for solid samples arc excitation is more sensitive, while spark sources are more stable. Plasma sources are the choice for solutions and for gaseous samples; their sensitivity enables trace analyses to be carried out at the parts-per-billion level.

TABLE **10.1** ADVANTAGES AND DISADVANTAGES OF ATOMIC SPECTRAL TECHNIQUES

	Method							
	FAES	**FAAS**	**EAAS**	**DCPS**	**ICPS**	**AFS**	**Arc**	**Spark**
Advantages								
Inexpensive instrument	X	X	X				X	X
Inexpensive maintenance	X	X	X				X	X
Wide dynamic range	X			X	X	X		
Low matrix interferences				X	X	X		
Low spectral interferences	X	X	X			X		
Multielement				X	X	X	X	X
Small sample capacity			X					
Disadvantages								
Expensive instrument				Moderately	X	Moderately		Moderately
Expensive maintenance				Moderately	X	X		
Limited dynamic range		X	X					
High matrix inteferences	X	X	X					
High spectral interferences				Moderately	X		Moderately	X
Single element	X	X	X					
Poor precision			X				X	

Electrical Discharges. Electrical discharges have long been used as excitation sources. When the discharge strikes the sample surface, it produces high current densities that volatilize a certain, small amount of sample. Atoms in the resulting vapor are then excited by collisions in the discharge plasma.

Direct-Current Arcs. The simplest electrical discharge is the dc arc between two solid electrodes. Typically one electrode supports the sample, while the other is the counter electrode. In the United States the anode is generally the sample-containing electrode, whereas in Europe the cathode is used as the sample holder.

The dc arc consists of a high-current (5–30 A), low-voltage (10–25 V) discharge usually operating in air. Arc temperatures range from 4000 to 6000 K. Excitation of the sample atoms by a dc arc is thus both thermal and electrical in origin. The resulting plasma of high-velocity ions, electrons, and atoms produces the atomic emission spectra. The energy available for excitation varies along the length of the arc discharge. Near the electrodes the energy of the plasma is greatest and the sample is quickly vaporized into the high-temperature region. In this region, the excitation lines are caused almost entirely by thermal energy.

Electrode configurations for handling a variety of sample types are shown in Figure 10.1. A popular configuration consists of a counter electrode and a lower cup electrode that contains the sample (Figure 10.1c). High-purity graphite is a popular electrode material because of its desirable physical and chemical properties. It is easily available in the required purity and is highly refractory, allowing the volatilization of high-boiling sample components. Chemically, it resists attack by strong acid or redox reagents. Its emission spectrum contains few lines, thus minimizing spectral interference. Solid samples, usually in powder form, are placed in the cup electrode either alone or mixed with graphite to enhance conductivity. Conductive metals are typically cast or machined into appropriately shaped electrodes that are used in place of the cup electrode (Figure 10.1a and b). Currently

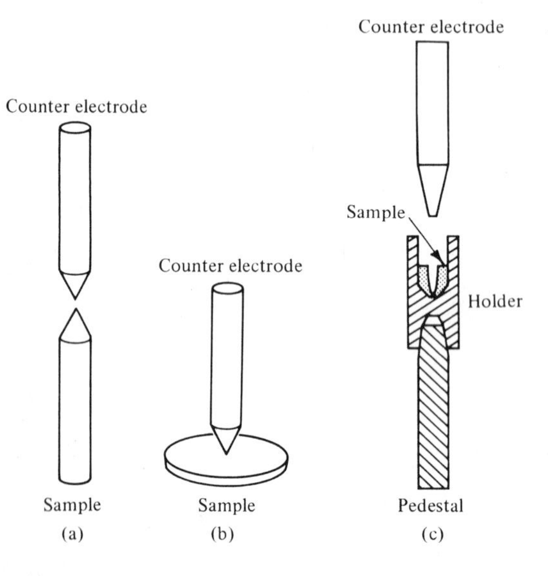

FIGURE **10.1**
Electrode configurations: (a) point-to-point, (b) point-to-plane, and (c) carrier distillation.

plasma sources (ICP and DCP) provide a much easier method of handling liquid samples and are replacing both arc and spark sources in these applications.

The arc produces spectral lines of high intensity, but the spectra produced contain only the more easily excitable lines, preferentially from neutral atoms. Arc discharges are therefore most useful for trace determinations. The arc does not produce results as reproducible as some of the other excitation methods because the discharge spot tends to wander about on the surface of the sample with resultant variations in excitation parameters of position and time.

Radiation emitted by a dc arc discharge is rich in both emission lines and molecular spectra. Because the electrodes are heated to high temperatures, they emit intense and spectrally continuous background radiation resembling emission from a blackbody source. Atomic line spectra are more complex than the spectra obtained from lower-temperature emission flame sources because the electrical and thermal energy of the arc is sufficient to populate many more of the higher electronic energy levels than the lower energy of the flame source. Arc spectra are further complicated by the many lines originating from ions of elements that have relatively low ionization energies. The molecular spectra of the cyanogen molecules, formed when an arc is operated between carbon electrodes in air, consist of a molecular band emission in the region 360–420 nm.

The dc arc source is better suited for qualitative or semiquantitative analyses, rather than for rigorous quantitative work. Because of its high sensitivity, the dc arc is the excitation source commonly selected for qualitative survey analyses of both trace and major elements, metals, alloys, plastics, rocks, minerals, soils, and biological materials such as tissue, bone, and fluids. Approximately 70 elements, mostly metals, can be satisfactorily determined using the dc arc technique.

An internal standard can be used to minimize the effects of arc instability and the sample matrix (see Section 2.8). The element selected as the internal standard must have vaporization and excitation characteristics that closely match the element being determined. Proper use of this technique can improve precision up to an order of magnitude.

Alternating-Current Arcs. Alternating currents can be used to produce an arc. Ac arcs are operated in either high-voltage (2200–4400 V) or low-voltage (100–400 V) regions. In both situations, the arc ceases at the end of each half-cycle as the applied voltage drops below the value needed to maintain it. In high-voltage ac arcs, the plasma spontaneously ignites during the next half-cycle when the applied voltage exceeds the voltage required for the dielectric breakdown of the gas between the electrodes. In low-voltage ac arcs, a low-voltage, high-current spark is used to initiate the arc in the next half-cycle. Electrode temperatures are lower for ac arcs than for dc arcs due to the intermittent presence of the arc.

The ac arc offers the major advantage of increased reproducibility (precision) over the dc arc because the cathode and anode spots form at different positions on the respective electrodes at a rate of 120 times/sec (60-Hz source). Thus electrode sampling with an ac arc is more statistically significant than sampling with the slow, erratic wandering of the dc arc. However, the sensitivity of the ac arc is usually less than that of the dc arc. In summary, the ac arc represents a compromise between stability and sensitivity for many elemental analyses.

High-Voltage, Alternating-Current Sparks. The high-voltage ac spark discharge is not as sensitive as the dc arc source but it provides the greatest precision and stability of all the electrical discharge sources. It is the method of choice for AES analysis of ferrous metals in industrial operations. Sparks contain high peak currents and power density, which result in the population of high-energy electronic levels of atoms and also more extensive ionization than found in arcs. Emission from both singly and doubly charged ions is common. The distribution of energy over a large number of excitation possibilities results in spectra that contain more lines than are found in spectra produced with arc sources. This renders qualitative analysis of ac sparks more tedious, however.

In metal analysis, the ac spark source is usually limited to analytes whose concentrations are greater than 0.01%, depending on the sensitivity of the element being determined. With solutions or the copper electrode method, spark excitation can be used successfully for determinations in the parts-per-million range.

The ac spark provides the same sampling advantages as the ac arc, mainly a sampling spot that moves rapidly about on the electrode surface. Conducting samples are usually ground flat and used as one electrode with a pointed graphite counter electrode. Powdered samples are mixed with graphite and pressed into a pellet, which is used as the plane electrode.

Microprobes[1]

The laser microprobe is well suited to the analysis of very small samples or tiny localized areas on larger samples. An optical pulsed ruby laser is focused via a conventional microscope onto a minute area of the sample after visual focusing has been used to select the sample area. The microprobe is shown in Figure 10.2. The intense heat of the laser vaporizes a small amount of the sample, leaving a hemispherical crater about 50 μm in diameter on the surface of the sample and produces a plasma. Once formed, the plasma absorbs radiation from the laser beam, thus stopping further vaporization at the sample surface. This absorption by the plasma means that the amount of sample vaporized in air at atmospheric pressures is independent of the total energy of the laser. If the pressure above the sample

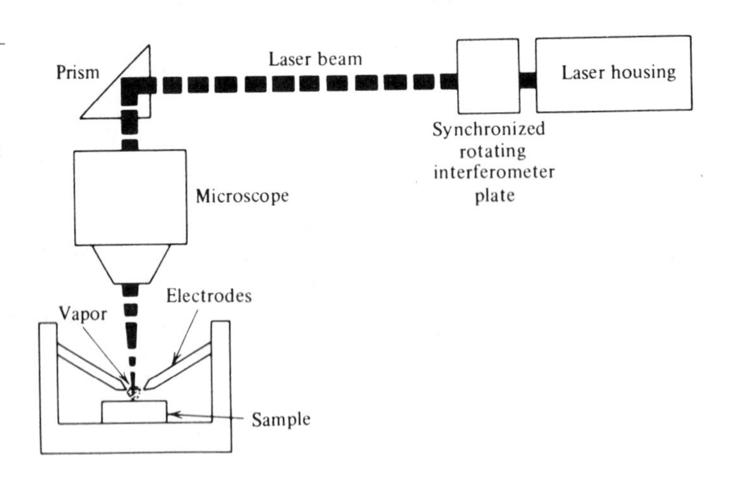

FIGURE **10.2**
Schematic diagram of a laser microprobe. (Courtesy of Jarrell-Ash Co.)

is reduced, the amount of sample vaporized and the crater size increase with increasing total laser energy.

Although emission spectra suitable for analyses can be produced directly with laser sources, cross excitation with a spark discharge results in improved resolution of spectral lines and an increased signal-to-noise (line/background) ratio. The vapor plume generated by the laser is passed between two closely spaced electrodes that are charged to a high voltage. The electrode gap breaks down, producing a spectrum similar to that produced by a spark source.

The laser microprobe furnishes a tool for examining the interior of individual cells and can even be used with living organisms, since it is essentially a nondestructive method because of the small amounts of sample consumed. It is also useful in analyzing the inclusion areas of alloys or corrosion spots on metal surfaces by those interested in these conditions. Resolution of the laser beam is currently limited to 50 μm by the lenses used to focus the laser beam. The intense energy of the laser beam focused on too small an area would shatter the lens or melt the cement that positions the lenses in their holders.

Noncombustion Plasma Sources[2]

The chemical and physical properties of noncombustion, flamelike plasmas offer performance and operational advantages over both traditional electrical discharge sources and combustion flame excitation sources. Noncombustion sources improve the accuracy, sensitivity, and precision for a large number of elements over that found with electrical discharge sources. Handling liquid and gaseous samples is much easier with these plasma flame sources. The higher temperatures and cleaner chemical environments of plasmas overcome two problems associated with combustion flames: lower temperatures and reactive chemical environments. Plasmas retain the convenience of liquid sample handling and the precision of excitation conditions characteristic of flame sources.

The linear dynamic concentration range of plasma emission sources is four or more orders of magnitude, in general much larger than the range found with FES. This increase in range is due to the narrow emission profile of the plasma, which reduces self-absorption. (Self-absorption is the absorption of photons by the same atoms or ions that emitted them.) Self-absorption leads to a reduction in the intensity of observed radiation, thus limiting the range of useful quantitative measurements. Spectra rich in atom and ion lines are produced by noncombustion plasma flame sources. Thus the analyst is not limited to analytical lines involving ground-state transitions but can select from first or even second ionization state lines. Spectrometers with superior resolving power are, however, required to isolate an analytical line from the many lines present in the spectrum.

Inductively Coupled Plasma Sources.[3] The inductively coupled argon plasma (ICAP or ICP) torch is a special type of plasma that derives its sustaining power by induction from a high-frequency magnetic field. The problem of igniting and burning a torch fueled by an inert gas (argon) is interesting. Initally argon gas passes through a 25-mm quartz tube and, upon emerging at the tip, is surrounded by an induction coil. An ac current flows through this coil at a frequency of around 30 MHz and power levels of about 2 kW (Figure 10.3).

The argon gas stream (the support gas) that enters the coil is initially seeded with free electrons from a Tesla discharge coil. These seed electrons quickly interact with the magnetic field of the coil and gain sufficient energy to ionize argon atoms by collisional excitation. Cations and electrons generated by the initial Tesla spark are accelerated by the magnetic field in a circular flow perpendicular to the stream that emerges from the tip of the torch (Figure 10.4). Reversal of the direction of the

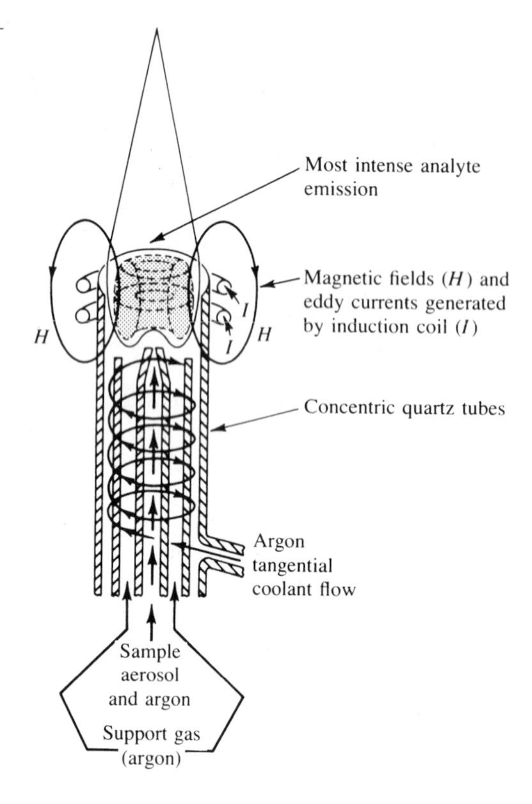

FIGURE **10.3**
Schematic configuration of an induction-coupled argon plasma (ICAP) torch.

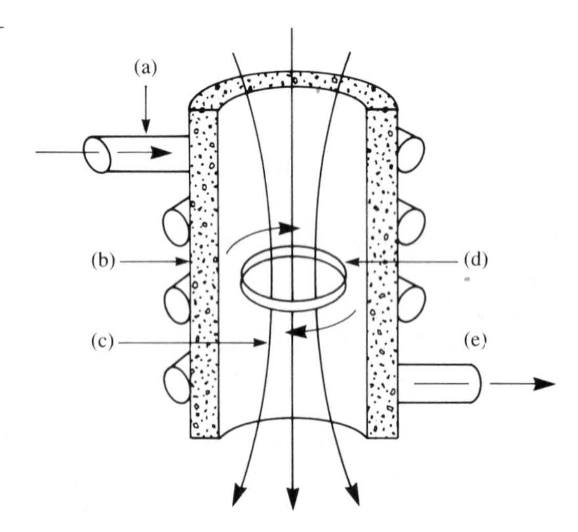

FIGURE **10.4**
Operation of an inductively coupled plasma source. (a) Passing current through a coil (b) wrapped around a quartz tube (c) sets up a magnetic field, (d) which causes an eddy current of ions and electrons (e) whose motion generates intense heat in a continuously ionized flow of gas.

current in the induction coil reverses the direction of the magnetic field applied to the mixture of atoms, ions, and electrons. The fast-moving cations and electrons, known as an eddy current, collide with more argon atoms to produce further ionization and intense thermal energy. A flame-shaped plasma forms near the top of the torch. Temperatures in the plasma range from 6000 to 10,000 K.

Power transfer between the induction coil and the plasma (mixture of fast-moving electrons and cations) is similar to a power transfer in a transformer. The induction coil is equivalent to a two-turn primary winding and the plasma is equivalent to a one-turn secondary winding.

The high temperatures produced by the plasma generally require a second stream of gas, usually argon, to provide a vortex flow of argon to cool the inside quartz walls of the torch. This flow also serves to center and stabilize the plasma. The unique characteristics of a high-frequency ICP facilitate the introduction of the sample in aerosol form through the innermost concentric tube of the torch. At frequencies of 27 MHz, a phenomenon known as the skin effect occurs, which gives the ICP plasma a toroidal shape (Figure 10.5). This plasma shaping lengthens the resident time (approximately 2 msec) of the sample in the interior, high-temperature zone of the plasma and increases the detection limits for many elements. Typical argon flow rates are 1 L/min in the sample carrier gas, 0–1 L/min for the support gas, and 15 L/min for the coolant stream.

A long, well-defined tail emerges from the high-temperature plasma on the tip of the torch. This tail is the spectroscopic source. It contains all the analyte atoms and ions that have been excited by the heat of the plasma. The optical window used for analysis falls just above the apex of the primary plasma cone (Figure 10.5) and just under the base of the flamelike afterglow. In this window, the high-background or current-carrying region of the plasma is excluded from the spectrometer slit.

There is no electrode contact in the ICP source. Therefore excitation and emission zones are spatially separated. As a result, there is a relatively simple background spectrum that consists of argon lines and some weak band emission from OH, NO, NH, and CN molecules. This low background, combined with a high signal-to-noise ratio of analyte emission, results in low detection limits, typically in the parts-per-billion range. The well-defined boundary between the tail that contains the excited analyte and the current-carrying portion of the inert plasma is one of the main reasons for the minimal interelement and matrix effects associated with ICP measurements. The high temperatures in the optical window (radiation

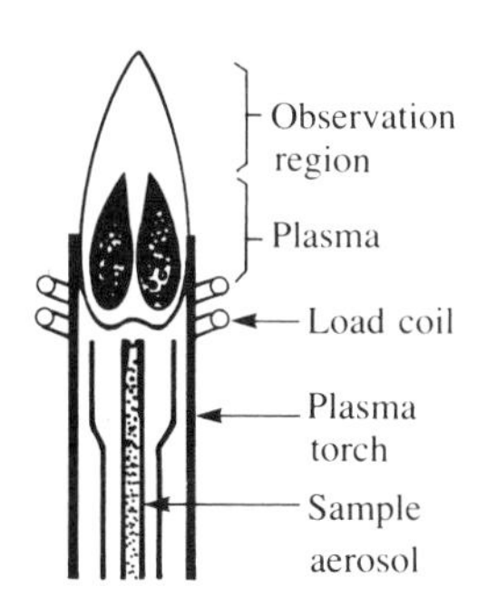

FIGURE **10.5**
The inductively coupled plasma, showing the injection of the sample aerosol into the toroidal (or annular) plasma. (After M. Walsh and M. Thompson, *A Handbook of Inductively Coupled Plasma Spectroscopy*, p. 2, Methuen, New York, 1983.)

zone) ensure the complete breakdown of chemical compounds and impede the formation of other interfering compounds.

For many elements, ion line emission from the ICP source is considerably more intense than neutral atom line emission. For calcium, the neutral atom line at 422.7 nm cannot even be observed relative to the intensity of the ion emission lines at 394.4 and 396.2 nm. This phenomenon is observed for many other elements, such as Ba, Be, Fe, Mg, Mn, Sr, Ti, and V, where ion lines provide the best detection limits.

Direct-Current Plasma Sources.[4,5] In direct-current plasma (DCP) sources, a high-velocity inert gas produces a high-temperature plasma and separates the excitation region from the analytical observation zone (Figure 10.6). In a typical dc arc discharge, the current density and energy available for atomic excitation are independent of the arc current. If the current is increased, the cross-sectional area of the arc current path increases proportionally with no increase in the arc temperature. To increase the current density and thus the plasma temperature, it is necessary to squeeze the plasma to decrease the current cross section. In DCP this is accomplished by cooling the edges of the plasma with a high-velocity inert gas vortex. The cooler outer areas of the plasma cannot support significant ionization, and therefore the current density in the channel region is increased due to the reduction in size of the arc channel. The high-velocity gas used to reduce the size of the plasma is also useful in separating the current channel from the observation zone. A DCP requires about 1 kW of power and consumes about 8 L/min of welder's grade argon.

Of particular significance is the DCP plasma's stability in the presence of varying solvent types, such as those that contain large amounts of dissolved solids (as high as 25%), organics, and high concentrations of acids or bases. The degree to which aqueous calibration standard matrices must be matched to sample matrices is governed only by the effect of viscosity and surface parameters on the rate of sample introduction (similar to ICP).

In the three-electrode DCP source, the plasma jet is formed between two spectroscopic carbon anodes and a tungsten cathode in an inverted **Y** configuration

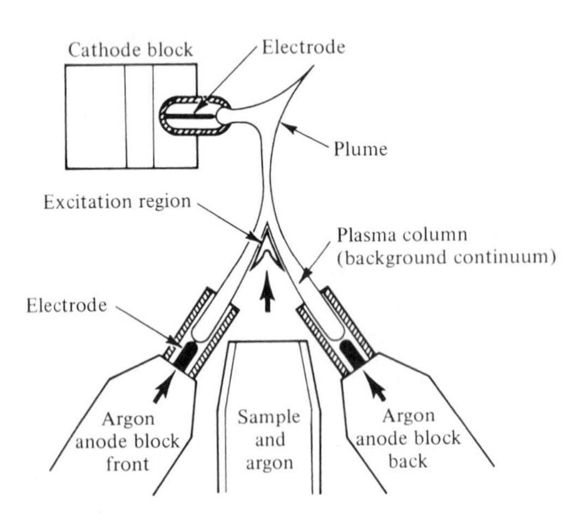

FIGURE **10.6**
Schematic of the dc argon plasma source in the **Y** configuration. (Courtesy of Beckman Instruments.)

(Figure 10.6). The sample excitation region and observation zone are centered in the crook of the **Y**, where the spectral contribution from the molecular bands of the plasma is minimal. The arc is initiated by moving the three electrodes into contact using argon-driven pistons and then withdrawing them. No spark is required in this process. The **Y** configuration stabilizes the position of the plasma and the sample excitation area and provides an effective means of separating the high-current plasma from the observation zone.

Once ignited, the plasma is sustained by a low voltage (400 V at 7.5 A) and produces temperatures as high as 9000 to 10,000 K. The excitation region at the juncture of the plasma is approximately 6000 K. Samples are nebulized and introduced into the excitation area in aerosol form. The entire cycle of desolvation, molecular dissociation, and excitation takes place in the extremely high-temperature plasma during the residence time of the sample. Operations are conducted in the inert atmosphere of high-velocity argon, nitrogen, or helium. Because the intense lines of excited atoms and ions of the analyte are observed in a region separated from the main plasma, an optimum signal-to-background noise is obtained, thus producing detection limits lower than those of ICP for many elements.

A major limitation of DCP sources is their inability to be incorporated into totally automated systems. The plasma-supporting electrodes are consumed by the arc and require reshaping after 2 hr of continuous operation. The relatively high consumption rates of expensive gases required for shaping the plasma are another limitation of DCP sources.

Introduction of Liquid Samples into Plasma Sources[6]

Liquid sample introduction is equally important in plasma methods. In all cases, the analyses can be only as good as the sample introduction. However, optimum conditions for sample introduction into plasma sources differ markedly from those for flame atomic absorption sources (see Section 9.2). The transport efficiency (percentage of analyte mass reaching the atomizer compared with that aspirated) for ICP is usually lower than that for AAS. The time required for washout of the analyte from ICP spray chambers is longer that that required for AAS due to the substantially lower gas and liquid flow rates (approximately 1 L/min and 1 mL/min, respectively) compared with those of AAS systems (typically 18 L/min and 6–8 mL/min). The problems of slow washout are aggravated by the broad linear range of ICP, up to five orders of magnitude. Thus it is possible that a concentrated sample of 2000 μg/mL could be followed by one containing 0.1 μg/mL of analyte. In this case, any analyte that remains from the more concentrated sample introduces a large error in the determination of the less concentrated sample.

There are three practical considerations when introducing liquid samples into plasma sources: (1) there is a maximum drop size for introducing the sample into the atomizer, (2) the rate of solvent flow must fall within a specific range of values, and (3) the analytical precision of an ICP system is strongly dependent on careful control of these parameters over the long term. Any significant change in either the drop size or the solvent loading that reaches the atomizer reduces both the accuracy and precision of the ICP system. Flames are generally far less susceptible to variations in solvent loading than plasmas, although the introduction of organic solvents can reduce the temperature of air/acetylene flames.

Pneumatic nebulation is the method used in most ICP and DCP determinations. These nebulizers are described in detail in Section 9.2. Systems for DCP sample introduction generally tolerate both suspended particles and high concentrations of dissolved solids much better than ICP systems. However, the excitation characteristics of DCP sources are greatly affected by high concentrations of easily ionized elements such as Na and Ca.

Two methods of special interest for use with plasma sources are high solids nebulizers and electrothermal vaporizers. The problem of nebulizer blockage associated with pneumatic nebulizers can be overcome with either cross-flow nebulizers (Figure 10.7) or Babington-type nebulizers. In the cross-flow nebulizer, clogging is avoided because the liquid sample stream can interact at right angles with the argon gas to produce the aerosol. The original Babington concept was developed for spraying paints and used a spherical surface with small holes in a circular pattern. In the maximum dissolved solids nebulizer (Figure 10.8), the liquid that flows over the outer surface of the sphere is nebulized by gas flowing from within the sphere through the small holes.

Both these nebulizers are inherently blockage free because of their design and operation. They are used with samples that contain suspended particles or when it is not convenient to totally dissolve the sample with acids or other reagents. They are also useful when the viscosity of samples varies widely. When using these devices it should be remembered that transporting the analyte to the plasma or flame does not ensure a proportional supply of atoms or ions. Analyses should be verified with standard reference materials whenever possible.

In ICP the electrothermal vaporizer (ETV) is used for direct sample introduction into the Ar gas stream in contrast to its use as an atomizer in AAS. The use

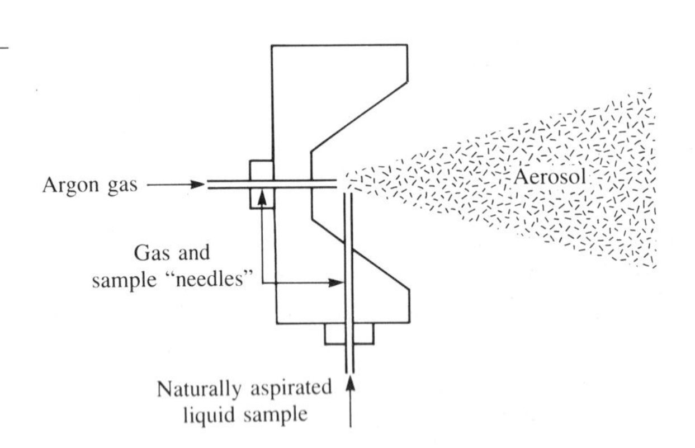

FIGURE **10.7**
Cross-flow nebulizer. (Courtesy of Applied Research Laboratories, Bausch and Lomb.)

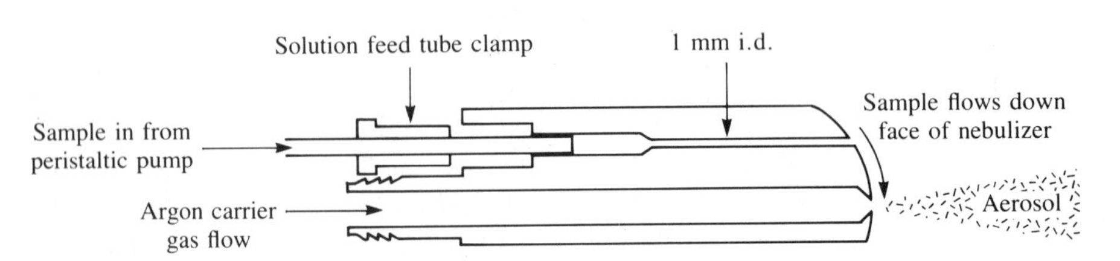

FIGURE **10.8**
Maximum dissolved solids nebulizer. (Courtesy of Applied Research Laboratories, Bausch and Lomb.)

of ETV with ICP allows microsampling while producing detection limits comparable to those obtained with furnace AAS. The other advantages of ICP, such as wide linear working range, freedom from interferences, and multielement analyses, are retained.

The open graphite rod device (Figure 10.9) is one of the most functional ETVs. When the sample contacts the electrically heated graphite rod, it is vaporized and swept into the plasma. The signals produced by the excitation of the analyte by the plasma are transient, and thus the direct-reading spectrometer detection circuitry must be modified. Otherwise, the plasma background rather than the analyte signal is observed during the extended integration period, reducing analyte detection limits.

Flow injection techniques (see Section 26.10) have several advantages over conventional methods of sample introduction for ICP. First, a relatively small volume of sample is necessary to produce a signal comparable to that produced with continuous nebulization. Second, because of the transient character of the signal produced with flow injection, the washout process starts much sooner than in a nebulizer. This results in an increased sample throughput.

A conductive-solids nebulizer (CSN) accessory for ICP sources combines the advantages of ICP analysis and spark-emission spectroscopy.[7] In this technique, the flat metal surface of the analyte material is placed on a spark stand, it is purged with argon to remove air, and then the spark source is switched to its preintegration condition. During this time the spark discharge remelts and homogenizes the sample

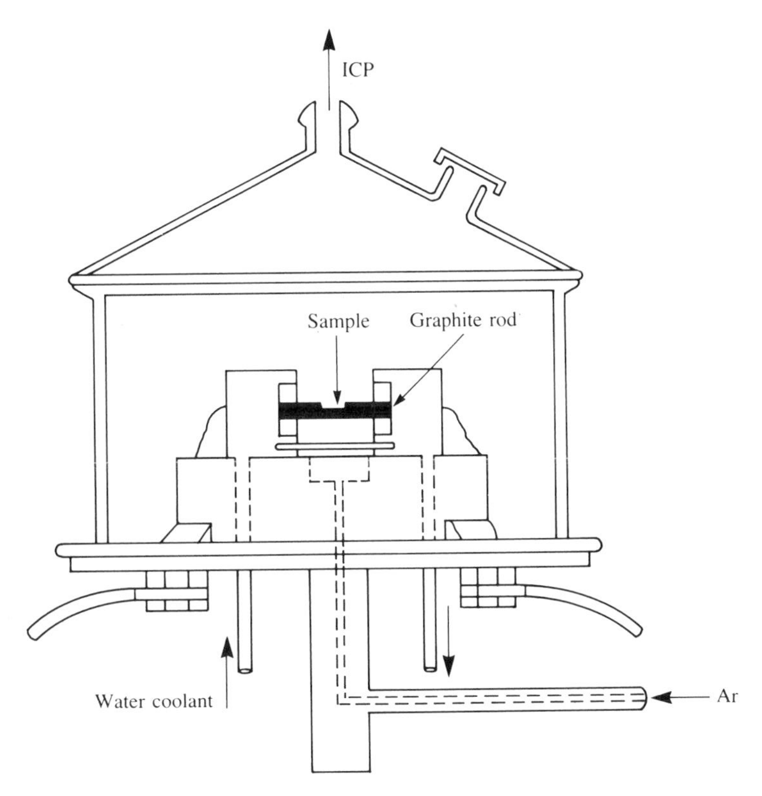

FIGURE **10.9**
Electrothermal vaporizer for ICP. [After R. F. Browner and A. W. Boorn, *Anal. Chem.*, **56**(7), 875A (1984).]

surface. Next, material from the sample surface is sputtered into a stream of argon. The resulting aerosol is swept into the plasma where it is vaporized and excited to obtain the emission spectrum. The signal integration period is usually less than 10 sec. This method offers the ease of solid sampling characteristics of spark-emission spectroscopy plus the improved linearity of response and reduced inter-element effects associated with ICP spectroscopy.

Hollow-Cathode Discharge Lamps as Emission Sources

Although hollow-cathode discharge lamps are widely used as radiation sources for both atomic absorption and atomic fluorescence spectrometry (see Section 9.5), their use in atomic emission spectroscopy has been limited despite some attractive characteristics. The major limitation of this emission source is the requirement of low-pressure operation. Since time-consuming pumpdown and other gas-handling operations are required for each analysis, these sources are not used in laboratories where rapid, high-volume analyses are required.

The hollow-cathode lamp consists of two coaxial cylinders (Figure 10.10a). The inner graphite cylinder is the cathode and contains the sample material. The discharge material is helium. The radiation is emitted from the negative glow, which is confined to the cathode cavity. At low temperatures, sample erosion is caused by sputtering due to ion bombardment. At higher temperatures, thermal volatilization becomes the predominant process. The relatively low temperatures of the discharge and the low operating pressures produce spectra that contain sharp, narrow lines compared with the lines obtained from a discharge source that operates at atmospheric pressure. These lamps are often used to determine elements with low boiling points in high-melting-point matrices. The lamp current can be ramped to obtain selective volatilization of lower-boiling elements. A number of nonmetals that have high excitation energies and analytical lines at wavelengths shorter than 220 nm have been determined using hollow-cathode sources.

The glow discharge lamp is similar to the hollow-cathode discharge lamp, differing only in the position of the sample (Figure 10.10b). The sample is not placed in the middle of the lamp but rather becomes one of the outer surfaces that seal the lamp. Atoms are released from the sample surface by cathode sputtering. Adjustment of operating conditions is critical, and thus the method requires more skill than other arc or spark techniques.

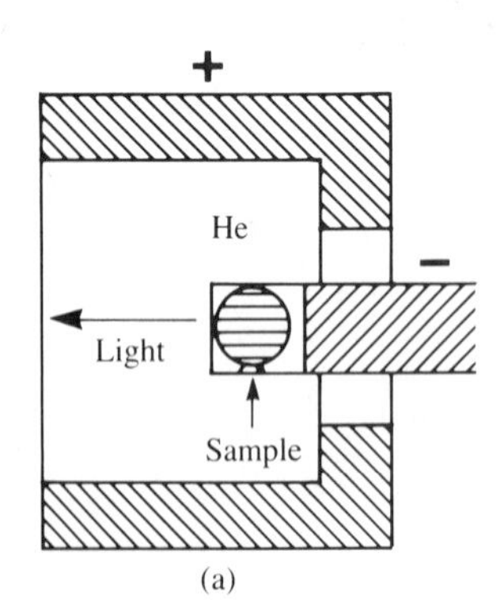

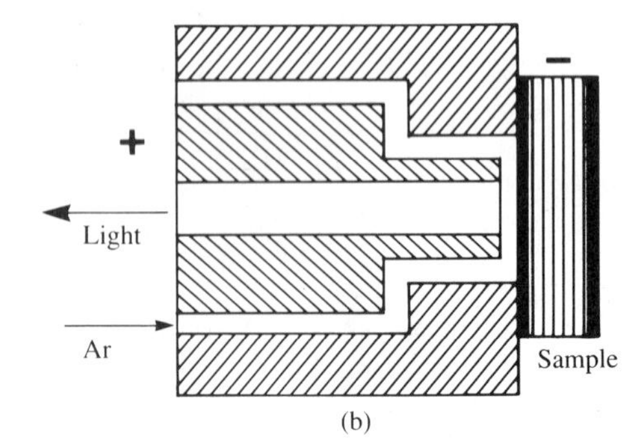

FIGURE **10.10**
(a) Hollow-cathode lamp (the sample is placed in the cathode cavity).
(b) Glow discharge lamp. (Courtesy of Applied Research Laboratories, Bausch and Lomb.)

Atomic Emission Spectrometers

In emission spectrometers the sensitivity is limited by noise originating from two major components, the flicker of the source and the dark current of the photomultiplier detector. Since the dark current noise remains constant, it is important to decrease the source noise of the spectral background by decreasing the signal bandpass until either the detector noise predominates or the spectral linewidth is reached. High-resolution, high-luminosity monochromators are necessary to isolate a spectral line from its background without loss of radiant power.

The optical components of emission spectrometers have undergone considerable change over the years. In the 1940s and 1950s, large dispersive monochromators were developed for ferrous metal analyses. These instruments have gradually given way to more compact, less expensive plane-grating, scanning monochromators.

Both concave and plane diffraction gratings are used as dispersive elements. A concave grating has focusing as well as dispersing capabilities (see Section 6.2). It requires no additional optical components, whereas the plane grating requires one or more lenses or mirrors to focus the image slit onto the focal plane. Another optical device is the echelle grating system, which is capable of resolution and dispersion an order of magnitude greater than a conventional grating of equal focal length.

Concave Grating Instruments. The spectrometer shown in Figure 10.11 is used for nonscanning, multielement analyses. It has a concave holographic grating mounted in a Rowland circle configuration. In this configuration, known more specifically as the Paschen-Runge mounting, the entrance slit, grating, and focal plane lie on the circumference of the Rowland circle. The Rowland circle has a radius of curvature half that of the grating. If the entrance slit and the grating are on the Rowland circle, the spectrum is focused on the circle. The grating is the only optical component between the entrance and exit slits. A focal length of 1 m assures good spectral dispersion and resolution.

Positioning of the photomultiplier tubes external to the spectrometer housing reduces secondary-array clutter and avoids secondary-array scattered radiation. The secondary optics consist of mirrors positioned behind the exit slits, which project and focus the radiation onto the cathodes of the photomultiplier tubes. For most determinations only one array of slits is used. However, for extended analyses of up to 60 elements, a second array of slits lying on the circle directly below the first is available. Individual exit slits are positioned to pass only spectral wavelengths for the elements being determined.

The Paschen-Runge mounting, positioned on the Rowland circle, is popular for large concave gratings used in routine analyses (Figure 10.11). Its major advantage is the ability to cover a large range of wavelengths. It suffers from the limitations of nonlinear dispersion and the dependence of optical speed on the position of the exit slit relative to the grating. (Optical speed refers to the amount of radiation collected by the detector during a fixed time period.)

The more compact Eagle mounting (Figure 10.12) has become popular in spite of the rather complicated adjustments needed to change the wavelength region. All the components lie on one side of a Rowland circle, giving a configuration that resembles a Littrow mount. Astigmatism is slight. A typical instrument with an

Eagle mounting (1.5 m focal length and fixed slit widths of 10, 20, and 50 μm) provides a dispersion of 1.6 nm/mm in the first order. It covers a spectral range of 225–625 nm.

The Wadsworth mounting (Figure 10.13) is the only common grating configuration that does not use the Rowland circle. It requires a collimating mirror to illuminate the concave grating with parallel radiation. The optical speed of the arrangement is high, since the grating is used at about half the image distance. The linear dispersion is about half that of Rowland-circle mountings. Most other mountings are astigmatic and thus produce optical aberrations in certain spectral regions. The smaller aberrations produced by the stigmatic character of Wadsworth mountings result in sharp, intense spectral lines. Instrument size is a major disadvantage of Wadsworth mountings. A 1.5-m spectrometer, ruled 600 lines/mm, can cover a range of 225–725 nm with an average linear dispersion of 1.09 nm/mm in the first order.

FIGURE **10.11** Schematic diagram of nonscanning (direct-reader) spectrometer using a holographic concave grating in the Rowland circle configuration. (Courtesy of Applied Research Laboratories, Bausch and Lomb.)

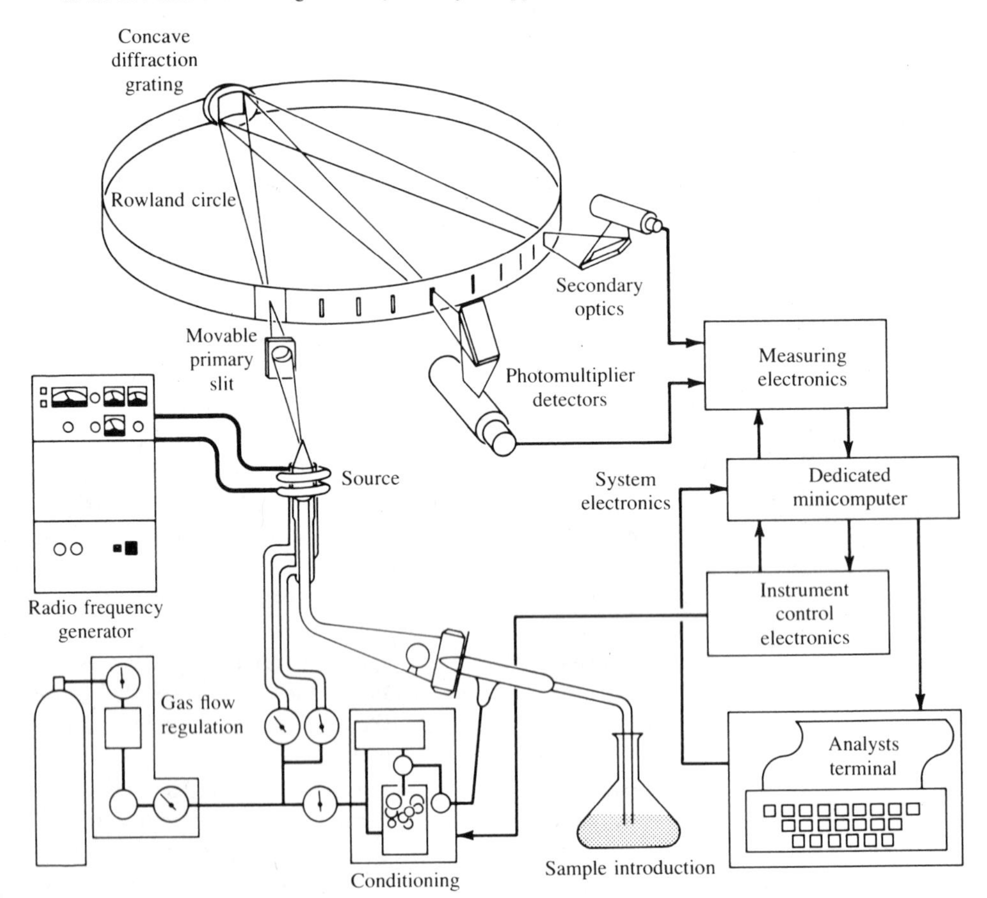

Plane-Grating Instruments. Plane gratings are used in most of the grating AES instruments produced today. In these instruments, the grating serves only as the dispersing element, and therefore a pair of concave mirrors is usually required to image the entrance slit onto the focal plane. Different spectral regions are focused on the camera detector's film or different wavelengths on the exit slit by rotating the grating. A device known as a sine bar drive is used to obtain rapidly an accurate linear wavelength readout of the position of spectral lines on the focal plane.

The Ebert mounting (Figure 6.13) is used in almost all large (3-m focal length) spectrometers that contain plane gratings. This mounting is nearly stigmatic and achromatic, so that light of all wavelengths is brought to focus on the detector without changing the detector-to-mirror distance. This makes it easy to change wavelengths by simply rotating the grating. Higher orders are easily accessible. Standard gratings have 600 or 1200 lines/mm, resulting in resolution ranging from

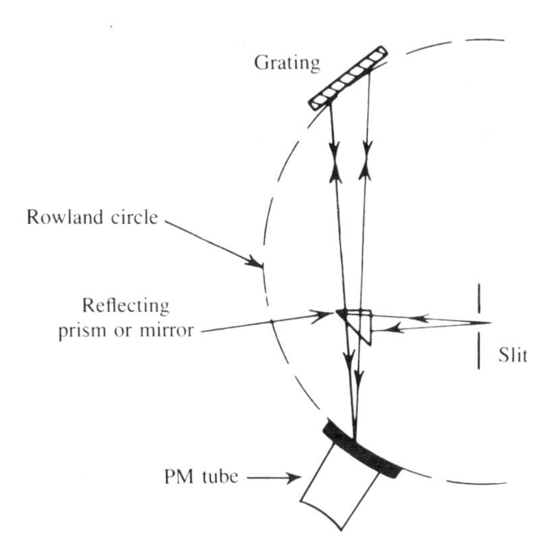

FIGURE **10.12**
Eagle mounting for a concave grating.

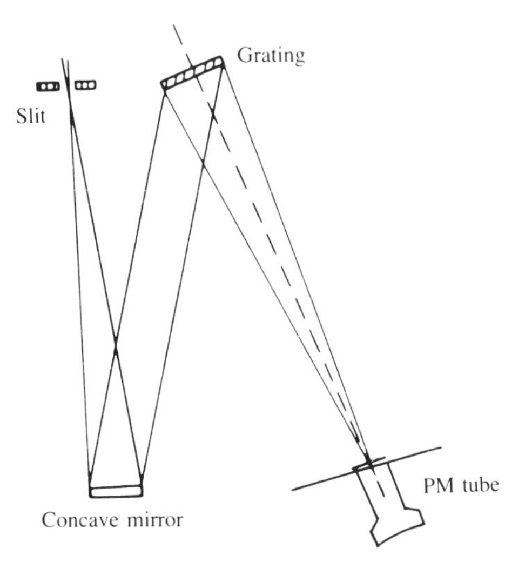

FIGURE **10.13**
Wadsworth mounting for a concave grating.

0.51 nm/mm in the first order to 0.07 nm/mm in the third order. The spectral region covered is 180–3000 nm. A device known as an order sorter is an accessory available for some instruments. This device, a fore-prism arrangement placed between the source and the entrance slit of the spectrometer, positions various grating orders one above another on the photographic plate of the detector.

Current AES instruments have plane-grating, scanning monochromators. An example of this type is the ARL scanning spectrometer, which uses a Czerny-Turner plane-grating monochromator with a 1 m focal length (Figure 6.14). With slits of 20 μm and a 1500-lines/mm grating, the resolution is 0.022 nm/mm over the wavelength range 190–670 nm. The rotation of the grating by a computer-controlled stepper motor permits the programmed selection of desired wavelengths and the sequential examination of spectral lines. It is possible to record an entire spectrum or to collect intensity data for specific lines. The computer not only controls the collection of data but also checks the analytical parameters for each line analyzed, makes the necessary background correction, recalibrates the wavelength scale of the instrument, performs required calculations, and prints the results in report form.[8]

Echelle Spectrometers.[9–11] Echelle gratings provide excellent dispersion and resolution over a wide range of wavelengths in a relatively compact instrument. The optical configuration for a modified Czerny-Turner mount is shown in Figure 10.14. The different orders appear in horizontal lines, with the longer wavelengths of the lowest order appearing at the bottom and the highest order at the top. Here the prism cross-dispersing element separates wavelengths in the vertical direction, while the echelle grating separates them horizontally. Each line is equivalent to a segment (order) of a conventional high-resolution spectrum and successive lines are adjacent segments (successive orders). The free spectral wavelength covered by one order varies from 1.8 nm at 200 nm to 11.1 nm at 500 nm for an echelle with 79 grooves/nm, blaze angle 63°26′, and 128 mm wide.

FIGURE **10.14** (a) Typical configuration for an echelle grating/prism spectrometer. (b) Echelle spectral pattern; each horizontal spectral pattern is a grating order. (Courtesy of Beckman Instruments.)

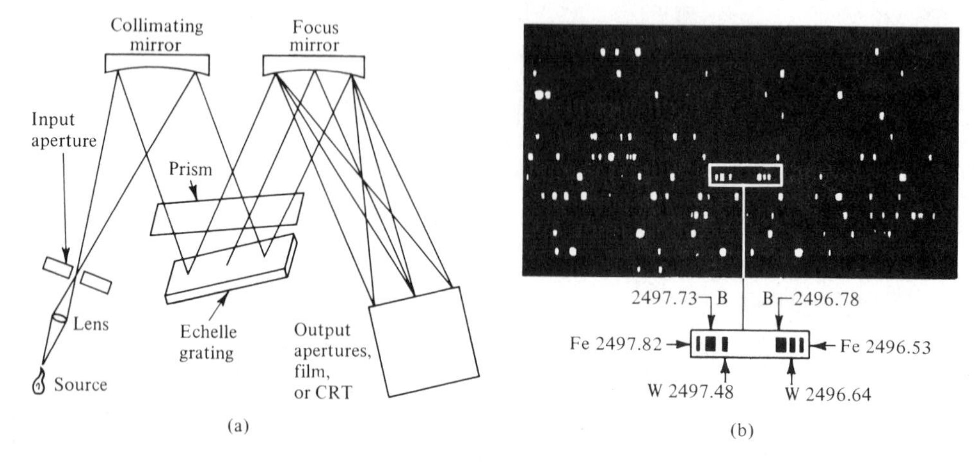

The two-dimensional display pattern permits high dispersion of all wavelengths from 190 to 800 nm in a compact array, typically 10 × 13 mm. A one-dimensional linear output would require a distance of 2 m for the same coverage. Since the spectrometer operates in multiple orders, the angular change in the radiation dispersed in any one order is relatively small. Therefore, all wavelengths are measured at their optimum blaze angle. By operating at the blaze angle, maximum energy throughput is obtained. This increases the efficiency and sensitivity of the total system. Another feature that leads to increased energy throughput is the short focal length required, typically 0.75 m. As a result, low levels of light intensity, normally lost in instruments with longer focal lengths, are detectable with echelle gratings.

One factor that can limit the energy throughput of echelles is the slit height, generally less than 1 mm. This is necessary to prevent overlap of the vertically displaced orders on the detector array. However, this reduction in throughput is somewhat offset by the favorable grating efficiency and the high order of the echelle. For a specified spectral resolution, an echelle spectrometer can have much wider slits than a medium-resolution conventional spectrometer.

Significant improvements in detection limits have been achieved with echelle instruments, particularly with high-background "atom reservoirs" such as those found in some plasma sources or the nitrous oxide/acetylene flame of AAS. Spectral interferences, such as CaOH band emission on the Ba 553 nm emission line, or Fe 249.782 nm and W 249.748 nm lines on the B 249.773 nm atomic emission line, can be eliminated with the use of a high-resolution echelle monochromator. This practical property is of great importance in emission spectroscopy when high-temperature electrical discharges or noncombustion plasma flame sources are used.

In qualitative analysis, the echelle instrument can operate in the photographic mode as a high-resolution spectrograph. A special camera attaches directly to the instrument and the results are displayed as a two-dimensional array on photographic film. In contrast to conventional spectrometers, two wavelength settings are required when echelle instruments are used with photomultiplier detectors. Both the order number and the coordinate or wavelength in the order must be selected. Sometimes a spectral line appears in more than one order. In this case it becomes necessary to select the order that provides the highest relative signal intensity. For example, the Ca 228.8 nm line exhibits the following intensities (in parentheses) for the specified order number: 110 (11), 111 (28), 112 (100), 113 (39), and 114 (5).

Echelle instruments have widespread applications in multielement determinations that use a two-dimensional matrix of photomultiplier tubes (Figure 10.15) or scanning diode arrays. With the incorporation of a microcomputer into the system, data from each detector in the array are collected, calculations required for quantitative analysis are performed, and the final report is generated automatically. One position in the detector array can be reserved for the background and another for the value of the internal standard.

Detectors and Readout Devices

Detectors used in emission instrument systems fall into two categories, photographic emulsions and photoelectric transducers. Most recently mass spectrometers are being used as detectors for ICP sources. Emulsions have the advantages of integrating impinging radiation over the entire time of the exposure and recording all spectral features simultaneously over a wide range of wavelengths. These characteristics allow weak spectral lines to be detected by using long exposure times, and they also provide a relatively inexpensive means of obtaining permanent records of emission spectra. Limitations of emulsions are long exposure times, the need to process and analyze spectra, nonlinear response to radiation intensity, low precision and sensitivity in quantitative determinations, and a limited dynamic range.

Photoelectric transducers have replaced emulsions in recent years. These are usually photomultiplier tubes and solid-state photodiodes. The devices exhibit linear response to radiant energy over a dynamic analyte concentration range of five to seven orders of magnitude. Energy data are obtained quickly with excellent precision and sensitivity. Limitations of these detectors are the initial higher cost and the ability to measure only a single spectral line at any given instant. This latter disadvantage has been addressed with multichannel PM arrangements and photodiode arrays. Interfacing these devices to computer systems for data collection and processing as well as measurement control is a relatively simple task. Microcomputers have become vital components of current emission instruments.

Photographic Detection. Photographic materials consist of a light-sensitive emulsion coated on a glass plate or plastic film. The emulsion contains light-sensitive crystals of silver halides suspended in gelatin. On exposure, the silver halide crystals that receive radiation form a latent image. Subsequent chemical treatment converts the exposed silver halide crystals into a black deposit of silver at the site of the latent image. After development, the emulsion is fixed in a solution that dissolves (by complexation) the unexposed silver halides. Finally, the photographic material is washed thoroughly to remove the chemicals used in developing and fixing. The entire series of operations follows rigidly controlled conditions with respect to time, temperature, and chemicals. These operations are best carried out with automated processing equipment.

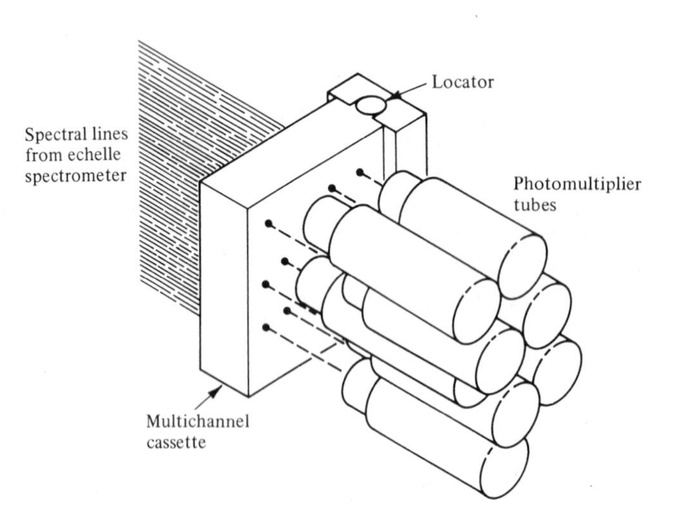

FIGURE **10.15**
Components in direct-reader approach to multielement echelle spectrometry. (Courtesy of Beckman Instruments.)

Devices known as microdensitometers are used to obtain the optical density or transmission data required for quantitative analysis of the line images on the photographic emulsion. These devices vary somewhat in their design, but they are all based on a simple optical system that projects a visible light beam through a small region of the emulsion. As the emulsion is moved in front of the light beam or the beam is moved across the emulsion, a photodetector measures the relative intensity of the light transmitted through the emulsion. The resulting power values are plotted or printed directly, or digitized for further processing by a computer system. These power data are expressed in units of either percent transmission or absorbance.

In quantitative analysis, either the ratio of powers of the analyte line to the internal standard line is plotted against the concentration, or the logarithm of the ratio is plotted against the logarithm of the concentration to establish a calibration curve. Once this curve has been established, similar unknown samples are rapidly analyzed. A new working curve is established for each type of sample and whenever a new component is introduced into the sample matrix.

The spectrum from a low-current dc iron arc is often used for the calibration of photographic instruments. This spectrum contains a large number of easily identifiable lines dispersed over the entire visible and ultraviolet regions.

Photoelectric Detectors. Since the radiation involved in atomic emission spectroscopy lies in the visible and ultraviolet regions of the spectrum, the detectors discussed in Chapter 6 can be used. The photomultiplier (PM) tube and the junction photodiode are the two transducers used almost exclusively in current AES instrument systems. Photodiodes are used in applications where the detection of low radiant power is not required. Their low cost, small size, and low power requirements relative to PM tubes make them attractive transducers.

The exit slit of the monochromator must isolate radiation from individual lines by precise positioning of the slit and accurate control of the slit width. Alternatively, a more complicated mirror system within the spectrometer may be used to sequentially focus radiation of varying wavelengths onto the exit slit. The radiant power of the spectral line that impinges on the slit is converted by the PM tube into current that is then used to charge a capacitor-resistor circuit.

Two modes of integration are used to average the PM signal, both with integration times of 25–40 sec. In the older method, the capacitor voltage is the quantity read out. The voltage across the capacitor at the end of the sampling time is a function of the accumulated charge (detector current $\times$ integration time). Thus the capacitor voltage is proportional to the time integral of the radiant power. The radiant powers are now compared with operational amplifier circuits that output the voltage ratio from two lines.

In integration with constant time, a voltage-to-frequency circuit is used to convert the capacitor voltage into electrical pulses of equal height (Figure 4.14). By counting the pulses generated during the integration period, a number proportional to the radiant power is obtained. This measuring technique has the advantages of a large dynamic range, extending over six decades, and built-in noise reduction. The output is usually in digital form, ready for computer processing.

Direct-reading, computerized spectrometers are usually calibrated with a high- and low-concentration standard. In real analyses, the relationship between the analytical signal, R, and the solution concentration, C, is described by

$$R = C \times \tan(A) + R_0 \tag{10.1}$$

where R_0 is the background intensity and A is the slope of the calibration curve. Rearrangement of this equation, in which C_0 is the concentration equivalent of the background (the concentration of analyte required to give a line-to-background ratio of unity), gives

$$C = R \times \cot(A) - C_0 \tag{10.2}$$

After the calibration is completed, the instrument is ready for analytical samples. Typically, sample analysis, data output, and any associated data management functions are completed within 2 min after sample introduction. Since the value of C is dependent on the spectral line intensity and the analyte element, calibration can diagnose problems associated with spectrometer performance and preparation of standard solutions. The precision of results is generally better than 0.5% relative. The reproducibility over an 8-hr period is good.

Mass Spectrometers. The use of mass spectrometers as detectors for ICP sources is discussed in detail in Section 16.13. The mass spectrometer systems used for this purpose can both process data and generate reports. The detection limits for many elements are significantly lower (ca. 10%) than those observed for ICP-AES systems with photographic or photoelectric detectors. ICP-MS systems also permit the determination of the isotopic composition of samples.

Instrument Configurations for Multielement Analysis

Detection systems used for multielement analysis are either sequential or simultaneous. In the former system, dispersed radiation is transmitted sequentially to a single detector. This is generally accomplished by scanning monochromators or rotating filters. Scanning monochromators have a grating or mirror usually controlled by a computer-driven stepper motor. Linear scanning, rapid slewing to preselected wavelengths, and manual operation are possible with these devices.

The most common simultaneous or parallel systems are multichannel with one or more detectors for each analyte element. The multiple PM tubes positioned on the circumference of a Rowland circle (Figure 10.11), the PM multichannel module used with an echelle spectrometer (Figure 10.15), and linear photodiode arrays are examples of simultaneous detection systems. Multielement simultaneous analysis is used effectively in situations that require routine determinations. Instruments used in such analyses, sometimes called quantometers, have widespread applications, especially in the ferrous metals industries.

For less routine analyses involving greater control of optical parameters, instruments with greater flexibility are required.[12] If sample sizes are adequate, sequential multielement instruments provide the flexibility needed. The sequential system shown in Figure 10.16 employs a double monochromator to minimize interference from stray radiation. Computer-controlled stepper motors provide fast movement of the grating to the desired wavelength. The plasma plume observation height is optimized automatically for each element by moving mirror M1 and lens

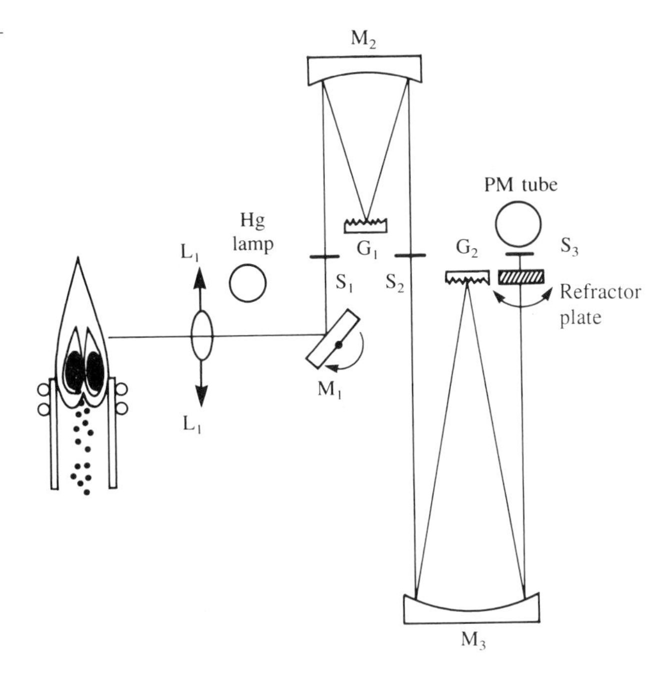

FIGURE 10.16
Optical diagram of a double monochromator for sequential multi-element analysis. (Courtesy of Instrumentation Laboratory.)

L1 in tandem. Two additional features add accuracy and precision: (1) a built-in mercury lamp and (2) a refractor plate. The 254 nm line of a mercury lamp is used to automatically calibrate the grating scan mechanism daily and to provide a reference wavelength at the beginning of each scan. The plate is used to fine tune the analytical wavelength during peak search and measurement as well as during background measurements.

A monochromator enclosed in a vacuum is used to determine sulfur, phosphorus, and boron in steel in the vacuum ultraviolet (190 nm) region of the spectrum. Although the conventional ultraviolet region of the spectrum usually provides the best analytical lines for phosphorus and boron, common components of steel interfere spectrally with both of these elements in this region.

10.3 TYPICAL APPLICATIONS[13,14]

AES has been used in the (1) analysis of ferrous and nonferrous alloys; (2) determination of trace metal impurities in alloys, metals, reagents, and solvents; (3) analysis of metals in geological, environmental, and biological materials;[15] and (4) water analysis. ICP instruments have been coupled with mass spectrometers to provide a powerful analytical technique (see Chapter 16).

Simultaneous ICP optical emission spectrometers are used for routine quantitative analysis of wear metals in lubricating oils. When the metal components of any mechanical system are in moving contact, minute metallic particles are generated by wear. If oil is used to lubricate the system, these particles are suspended in the oil and are usually too small to be trapped by filters. When the rate of wear is low (normal conditions in a well-lubricated system), both the number and size of the metallic particles are small. With increasing wear, the rate of particle generation and the particle size increase.

Analysis of lubricating oils identifies the surface from which the particles were abraded and the rate of wear. It is possible to predict from abnormal wear rates when mechanical failure will occur. Various makes and models of equipment have different metal contact surfaces; examples with various engine parts are lead/tin and lead/copper for Mercedes Benz bearings, aluminum for Caterpillar bearings, chromium for Volvo piston rings, and jet aircraft engines.

10.4 ICP ATOMIC FLUORESCENCE SPECTROSCOPY[16]

In atomic fluorescence spectroscopy (AFS), an excitation source external to the atomization cell is used to produce the fluorescence emission signal (see Section 9.5). This technique is a hybrid of AAS (external radiation source) and AES (emission signal proportional to the concentration of the analyte) and incorporates the advantages of both. AFS combines simplicity of operation and freedom from spectral interferences from AAS with simultaneous multielement analysis and large, linear dynamic ranges from AES. Other desirable features of AFS are optical stability, small baseline drift, and no requirement for background correction.

The plasma/AFS spectrometer consists of up to a dozen hollow-cathode lamps (HCL) and the same number of PM tubes (Figure 10.17). Each HCL/PM tube module is dedicated to the determination of a single element. Radiation from a given HCL is directed into the plasma. Part of the atomic fluorescence produced by the interaction of the analyte atoms in the plasma with the radiation from the HCL is directed to the PM tube corresponding to the HCL element. An optical interference filter and lens are used to exclude background radiation and focus the selected analyte line on the appropriate PM tube. Each element module is independent of the other modules and is adjusted to interact with the plasma at a height selected to obtain the optimum analytical signal. During a multielement analysis, the HCLs of the modules are pulsed at a frequency of about 500 Hz and the detection electronics are synchronously gated. Thus at any given instant, only one fluorescence signal is produced and detected. AFS detection limits are comparable to those obtained with flame AAS.

10.5 COMPARISON OF METHODS: ICP VERSUS AAS[17–19]

Although all the methods discussed in Chapters 9 and 10 have applications (see Table 10.1), AAS (including flame, thermal, and fluorescence techniques) and ICP have emerged as the two major analytical methods of atomic spectroscopy. The two primary advantages of ICP over AAS are the larger dynamic concentration range, up to 10^5 for many elements, and the relatively simple application to analyses that permits the simultaneous determination of several elements in a sample. Although the sensitivities of AAS techniques are equal or better than those for ICP at low concentrations, the AAS techniques become less sensitive at higher

FIGURE **10.17** Block diagram for ICP atomic fluorescence spectrometer. (Courtesy of Baird Corp.)

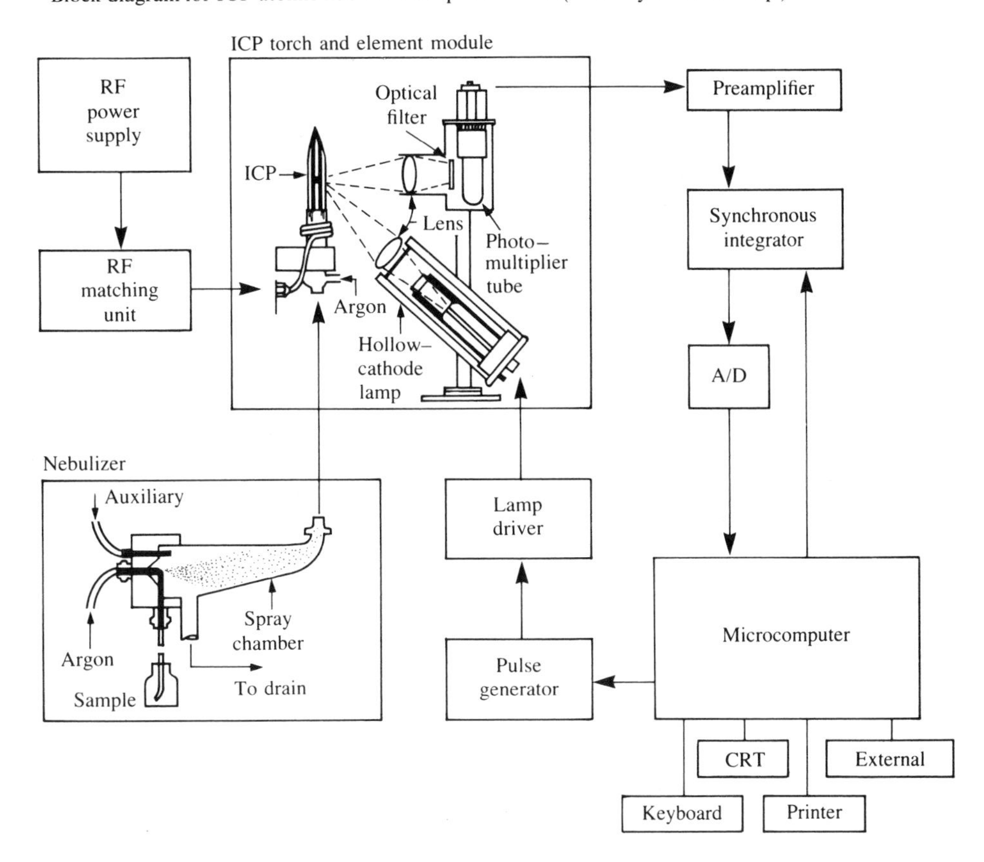

concentrations. Multielement ICP instruments have the great advantage that a sample needs to be aspirated only once. Only one or two elements can be determined simultaneously using AAS. Thus a multielement analysis with AAS requires repeated aspirations, adjustment of instrument parameters, and proper tracking of the samples.

Two limitations of ICP relative to AAS techniques are greater sensitivity to matrix effects and poorer powers of detection for some elements, particularly nonmetals that are difficult to excite. Recent advances in sample introduction and excitation sources have reduced these problems in many determinations. It is often desirable to have both AAS and ICP instruments available to provide complementary analytical methods.

For methods development and analyses of a few samples of one type, flame AAS is much simpler and less expensive than ICP. For most elements, ICP cannot match the detection limits attainable with furnace or fluorescence AAS. In general, AAS instruments are less expensive than ICPs by a price factor of $\frac{1}{2}$ to $\frac{1}{5}$. The cost of argon is a major operating expense. At a typical flow rate of 20 L/min, a low-pressure tank of liquid argon will last 3 weeks at 8 hr/day. At a price of $250 per tank, the yearly cost of argon is approximately $4000. Both AAS and ICP instruments generate heat, which adds to the cost of laboratory air conditioning. In situations where the method has been developed and there is a heavy load of similar samples requiring multielement determinations, ICP offers a definite advantage.

A final comparison of the best detection limits, those equal to or within a factor of four of the best detection limits reported in the literature, is shown in Table 9.2. It is interesting to note that flame emission atomic spectroscopy, the "granddaddy" of all emission methods, still affords the best detection limits for the alkali metals and is competitive for many other elements. Flame AAS, *the* method of the 1960s and 1970s, provides the best detection limits for only a few elements. However, it is still the method of choice for many elemental analyses. Electrothermal AAS is usually the most sensitive technique for elements where it has been used (55 out of 56 elements). It should be noted that whereas EAAS is the most sensitive technique for many elements, it is limited by poor precision and its single-element capability.

ICP and DCP offer overall improvement in detection limits for many elements over the single-element flame techniques. In practice ICP and DCP appear to have essentially equal detection limits for many elements. Their multielement capabilities mean that detection limits have often been measured at compromised conditions.

PROBLEMS

1. A sample of an unknown light-metal alloy was placed on a spark stand and a spectrum was recorded. The spectrogram had lines at the following wavelengths: 643.8, 518.4, 517.3, 481.0, 472.2, 468.0, 383.8, 383.2, 382.9, 361.1, 346.6, and 340.3 nm, plus many lines of aluminum. What elements, besides aluminum, are present? (*Hint:* The *CRC Handbook of Chemistry and Physics* contains lists of wavelengths.)

2. In the spectrographic determination of lead in an alloy, using a magnesium line as an internal standard, the results shown in the table were obtained. (a) Prepare a calibration curve on a log-log paper. (b) Evaluate the concentrations for solutions A, B, and C.

	Densitometer reading		
Solution	**Mg**	**Pb**	**Concentration of lead (mg/mL)**
1	7.3	17.5	0.151
2	8.7	18.5	0.201
3	7.3	11.0	0.301
4	10.3	12.0	0.402
5	11.6	10.4	0.502
A	8.8	15.5	
B	9.2	12.5	
C	10.7	12.2	

3. The following emission signals were obtained from ICP analysis of 25-μL aliquots of whole blood spiked with Mn standard solution. The blood samples were first diluted 10-fold with 0.1*M* HCl. Using the method of standard additions, calculate the concentration (μg/mL) of Mn in the blood sample.

Solution	Intensity reading
Sample	25.3
Blank	5.2
Sample + 0.005 μg/mL Mn	55.2
Sample + 0.010 μg/mL Mn	85.4
Sample + 0.020 μg/mL Mn	165.3

4. A water sample containing trace amounts of zinc is analyzed using an ICP with a photomultiplier tube detector. A calibration sample containing 1.4 ppm of zinc gives a signal of 124.5 units. If the background signal is 8.2 units and the concentration equivalent of the background is 0.02 ppm, calculate the concentration of zinc in a sample that gives a signal response of 94.5 units.

5. For each of the following electrical discharge sources list the major strengths and limitations: (a) dc arcs, (b) ac arcs, and (c) ac sparks.

6. What advantages does an ICP source offer over an electrical discharge source? Do electrical discharge sources have any advantages over ICP sources? Explain.

7. Discuss the differences in nebulization required for flame AAS and ICP atomizers.

8. Discuss the advantages and limitations of each of the following optical configurations used in AES: (a) Paschen-Runge mounting, (b) Wadsworth mounting, and (c) echelle grating.

9. What are the unique advantages of the following detectors used in AES: (a) photographic emulsions, (b) photomultiplier tubes, and (c) photodiode arrays?

BIBLIOGRAPHY

BARNES, R., ed., *Emission Spectroscopy,* Halsted, New York, 1976.

FASSEL, V., *Anal. Chem.,* **51,** 1290A (1979).

GROVE, E., ed., *Applied Atomic Spectroscopy,* Vols. 1 and 2, Plenum, New York, 1978.

KELIHER, P., et al., *Anal. Chem.,* **58**(5), 334R (1986).

MONTASER, A., AND D. GOLIGHTLY, *Inductively Coupled Plasmas in Analytical Atomic Spectrometry,* VCH Publishers, Deerfield Beach, FL, 1986.

SACKS, R., "Emission Spectroscopy," Chap. 6, *Treatise on Analytical Chemistry,* 2nd ed., P. J. Elving, E. J. Meehan, and I. M. Kolthoff, eds., Part I, Vol. 7, John Wiley, New York, 1981.

SCHRENK, W., *Analytical Atomic Spectroscopy,* Plenum, New York, 1975.

TOLG, G., *Analyst,* **112,** 365 (1987).

WALSH, M., AND M. THOMPSON, *A Handbook of Inductively Coupled Plasma Spectroscopy,* Methuen, New York, 1983.

LITERATURE CITED

1. BRECH, F., *Analysis in Instrumentation,* Vol. 6, p. 215, Plenum, New York, 1969.
2. SKOGERBOE, R., AND G. COLEMAN, *Anal. Chem.,* **48,** 611A (1976).
3. FASSEL, V., AND R. KNISELY, *Anal. Chem.,* **46,** 1110A, 1155A (1974).

4. REEDNICK, J., *Am. Lab.*, **11**(3), 53 (1979).
5. ZANDER, A., *Anal. Chem.*, **58,** 1139A (1986).
6. BROWNER, R., AND A. BOORN, *Anal. Chem.*, **56**(7), 786A, 875A (1984).
7. DALAGER, P., J. GOULTER, AND D. TASKER, *Res. Dev.*, **4,** 114 (1985).
8. KAHN, H., AND D. CHASE, *Am. Lab.*, **15**(8), 46 (1983).
9. LOWEN, E., "The Echelle Story," *Leeman Notes (Bausch and Lomb)*, Leeman Labs, Inc., Lowell, MA, 1980.
10. MACDONALD, J., *Am. Lab.*, **15**(9), 90 (1983).
11. KELIHER, P., AND C. WOHLERS, *Anal. Chem.*, **48,** 333A (1976).
12. NYGAARD, D., D. CHASE, AND D. LEIGHTY, *Res. Dev.*, 172 (February 1984).
13. EISENTRAUT, K., et al., *Anal. Chem.*, **56**(2), 1087A (1984).
14. BROEKAERT, J., *Trends Anal. Chem.*, **1**(11), 249 (1982).
15. BURNS, D., et al., *Anal. Chem.*, **57**(9), 1048A (1985).
16. VAN LOON, J., *Anal. Chem.*, **53**(2), 333A (1981).
17. KAHN, H., *Res. Dev.*, **24**(2), 156 (1982).
18. PARSONS, M., S. MAJOR, AND A. FORSTER, *Appl. Spectrosc.*, **37**(5), 411 (1983).
19. VAN LOON, J., *Anal. Chem.*, **52**(8), 955A (1980).

11 INFRARED SPECTROMETRY

The infrared region of the electromagnetic spectrum extends from the red end of the visible spectrum to the microwave region. The region includes radiation at wavelengths between 0.7 and 500 μm or, in wavenumbers, between 14,000 and 20 cm^{-1}. The spectral range used most is the mid-infrared region, which covers frequencies from 4000 to 200 cm^{-1} (2.5 to 50 μm). Infrared spectrometry involves examination of the twisting, bending, rotating, and vibrational motions of atoms in a molecule (Figure 5.8). The fundamental theory was discussed in Chapter 5. Upon interaction with infrared radiation, portions of the incident radiation are absorbed at specific wavelengths. The multiplicity of vibrations occurring simultaneously produces a highly complex absorption spectrum that is uniquely characteristic of the functional groups that make up the molecule and of the overall configuration of the molecule as well.

Atoms or atomic groups in molecules are in continuous motion with respect to one another. The possible vibrational modes in a polyatomic molecule can be visualized from a mechanical model of the system, shown schematically in Figure 11.1. Atomic masses are represented by balls, their weights being proportional to the corresponding atomic weights and arranged in accordance with the actual space geometry of the molecule. Mechanical springs, with forces that are proportional to the bonding forces of the chemical links, connect and keep the balls in balance. If the model is suspended in space and struck, the balls appear to undergo random

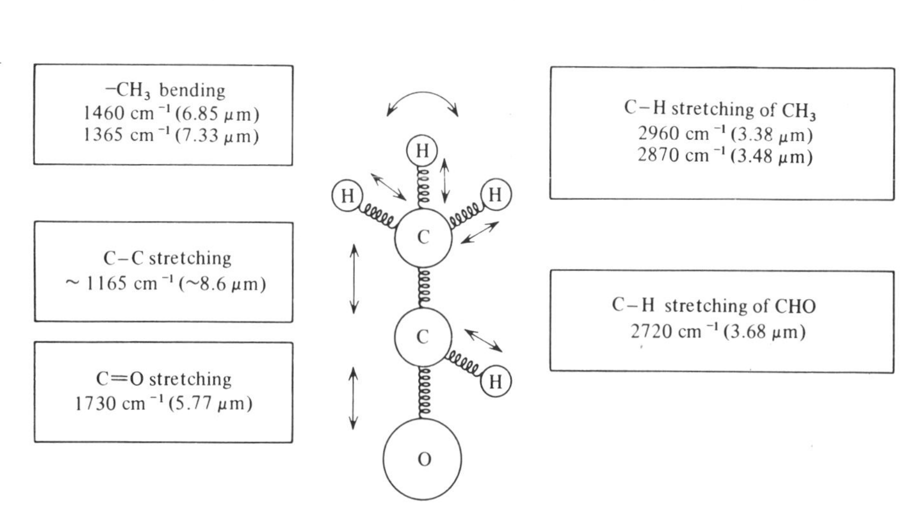

FIGURE 11.1
Vibrations and characteristic frequencies of acetaldehyde.

chaotic motions. However, if the vibrating model is observed with a stroboscopic light of variable frequency, certain light frequencies are found at which the balls appear to remain stationary. These represent the specific vibrational frequencies for the motions.

11.1 CORRELATION OF INFRARED SPECTRA WITH MOLECULAR STRUCTURE

The infrared spectrum of a compound is essentially the superposition of absorption bands of specific functional groups, yet subtle interactions with the surrounding atoms of the molecule impose the stamp of individuality on the spectrum of each compound. For qualitative analysis, one of the best features of an infrared spectrum is that the absorption or the *lack of absorption* in specific frequency regions can be correlated with specific stretching and bending motions and, in some cases, with the relationship of these groups to the rest of the molecule. Thus, when interpreting the spectrum, it is possible to state that certain functional groups are present in the material and certain others are absent. With this one datum, the possibilities for the unknown sometimes can be narrowed so sharply that comparison with a library of spectra of pure compounds permits identification. Since most organic courses now include a discussion of the relationship of infrared spectra to molecular structure, the discussion in this text will be limited to a very general overview.

Near-Infrared Region

In the near-infrared (NIR) region, which meets the visible region at about 12,500 cm^{-1} (0.80 μm) and extends to about 4000 cm^{-1} (2.50 μm), there are many absorption bands that result from harmonic overtones of fundamental and combination bands often associated with hydrogen atoms. Among these are the first overtones of the O—H and N—H stretching vibrations near 7140 cm^{-1} (1.40 μm) and 6667 cm^{-1} (1.50 μm), respectively, and combination bands that result from C—H stretching and deformation vibrations of alkyl groups at 4548 cm^{-1} (2.20 μm) and 3850 cm^{-1} (2.60 μm). The absorptivity of NIR bands is from 10 to 1000 times less than that of mid-infrared bands. Thicker sample layers (0.5–10 mm) compensate for these smaller molar absorptivities. Because the absorptivities are so low, the NIR beam penetrates deeper into a sample in reflectance techniques, giving a more representative analysis. Furthermore, minor impurities are less troublesome in both reflectance and transmission methods than in the mid-infrared region.

The NIR region is accessible with quartz optics, and this is coupled with greater sensitivity of near-infrared detectors and more intense radiation sources. The near-infrared region is often used for quantitative work in addition to qualitative identification. Water has, for example, been determined in glycerol, hydrazine, Freon, organic films, acetone, and fuming nitric acid. Absorption bands at 2.76, 1.90, and 1.40 μm are used depending on the concentration of the test substance. Where interferences from other absorption bands are severe or where very low concentrations of water are being studied, the water can be extracted with glycerol or ethylene glycol prior to measurement.

Near-infrared spectrometry is a valuable tool for analyzing mixtures of aromatic amines. Primary aromatic amines are characterized by two relatively intense absorption bands near 1.97 and 1.49 μm. The band at 1.97 μm is a combination of N—H bending and stretching modes, and the one at 1.49 μm is the first overtone of the symmetric N—H stretching vibration. Secondary amines exhibit an overtone band but do not absorb appreciably in the combination region. These differences in absorption provide the basis for rapid quantitative analytical methods. The analyses are normally carried out on 1% solutions in CCl_4 using 10-cm cells. Background corrections are obtained at 1.575 and 1.915 μm. Tertiary amines do not exhibit appreciable absorption at either wavelength. The overtone and combination bands of aliphatic amines are shifted to about 1.525 and 2.000 μm, respectively. Interference from the first overtone of the O—H stretching vibration at 1.40 μm is easily avoided with the high resolution available with near-infrared instruments.

Near-infrared reflectance spectra find wide acceptance in the food and grain industry for the determination of protein, fat, moisture, sugar, oils, iodine numbers (unsaturation), and so on.[1] NIR reflectance spectra have also been used for the determination of substances in wood, components of polymers, and even geological exploration from aircraft. The intensities are weak but measurable and reproducible.

The composition of the material to be determined must be restricted to a finite number of known samples. The computer of the instrument is "trained" by measuring and recording the spectra of these known samples at a large number of wavelengths. As many as 100 knowns may be used. The computer then develops a set of equations, which are later used to analyze the unknown samples. The number of samples and wavelengths required and calculation methods have been discussed by Honigs and colleagues.[2,3] The method is fast and useful in routine analyses. For many substances sample preparation is extremely simple. The sample is merely poured into a cell and inserted into the sample chamber of the instrument. The NIR reflectance method really depends on pattern recognition by the computer and, if there are unexpected substances in the unknown, false determinations can be made.

Mid-Infrared Region[4]

Many useful correlations have been found in the mid-infrared region (Figure 11.2). This region is divided into the "group frequency" region, 4000–1300 cm^{-1} (2.50–7.69 μm), and the fingerprint region, 1300–650 cm^{-1} (7.69–15.38 μm). In the group frequency region the principal absorption bands are assigned to vibration units consisting of only two atoms of a molecule—that is, units that are more or less dependent on only the functional group that gives the absorption and not on the complete molecular structure. Structural influences do reveal themselves, however, as significant shifts from one compound to another. In the derivation of information from an infrared spectrum, prominent bands in this region are noted and assigned first. In the interval from 4000 to 2500 cm^{-1} (2.50–4.00 μm), the absorption is characteristic of hydrogen stretching vibrations with elements of mass 19 or less. The C—H stretching frequencies are especially helpful in establishing the type of compound present; for example, C≡C—H occurs around 3300 cm^{-1} (3.03 μm), aromatic and unsaturated compounds around 3000–3100 cm^{-1} (3.33–3.23 μm),

FIGURE 11.2 Some characteristic infrared absorption bands. Band positions are given for dilute solution in nonpolar solvents. Intensities are expressed as strong (s), medium (m), weak (w), and variable (v). (Courtesy of American Cyanamid Company.)

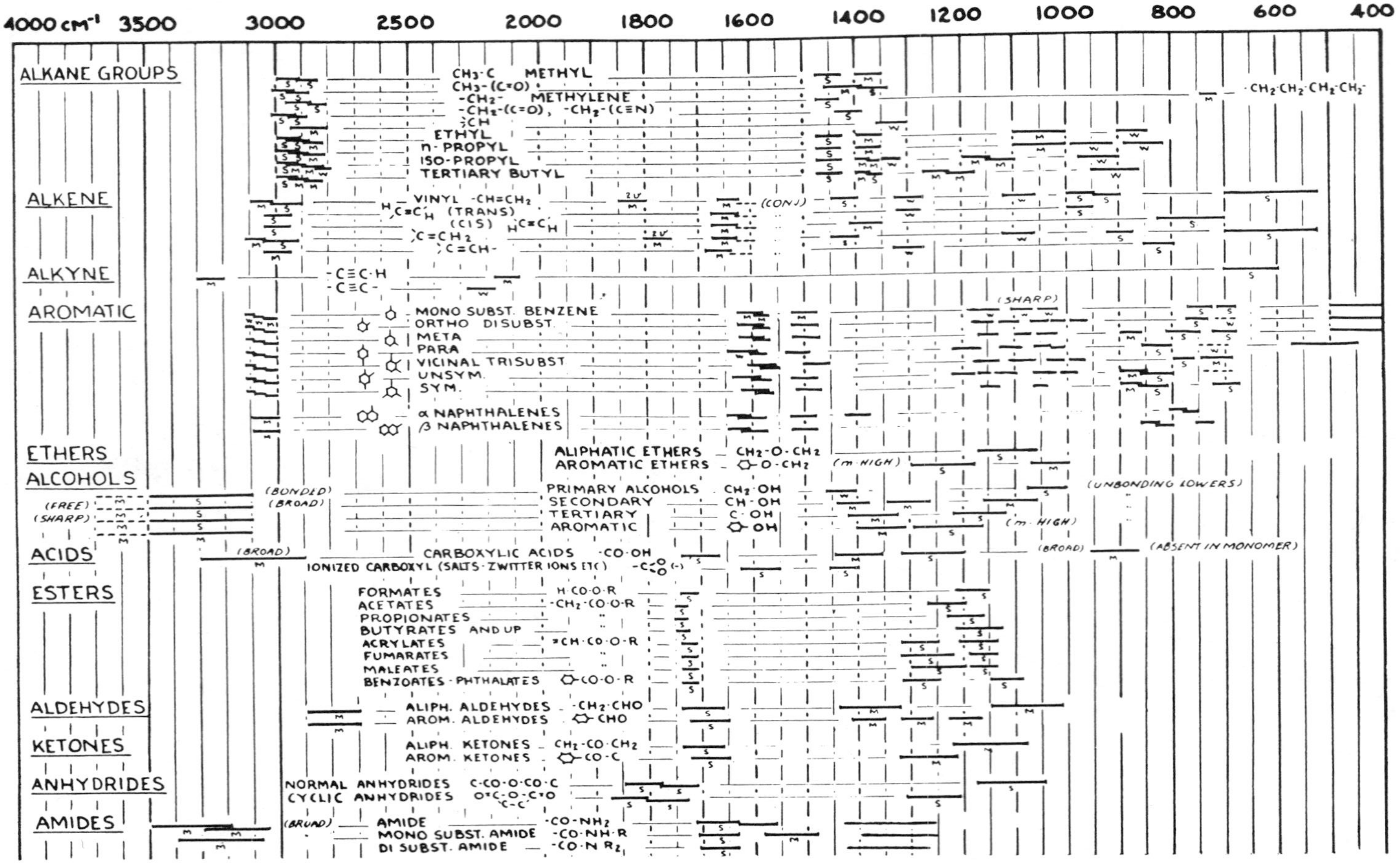

4000 cm⁻¹ 3500 3000 2500 2000 1800 1600 1400 1200 1000 800 600 400

ALKANE GROUPS
CH_3·C METHYL
CH_3-(C=O)
-CH_2- METHYLENE
-CH_2-(C=O), -CH_2-(C≡N)
CH
ETHYL
n-PROPYL
ISO-PROPYL
TERTIARY BUTYL
·CH_2·CH_2·CH_2·CH_2·

ALKENE
VINYL -CH=CH_2
HC=CH (TRANS)
(CIS) HC=CH
C=CH_2
C=CH-
(CONJ)

ALKYNE
-C≡C·H
-C≡C-

AROMATIC
MONO SUBST. BENZENE
ORTHO DISUBST.
META
PARA
VICINAL TRISUBST
UNSYM.
SYM.
α NAPHTHALENES
β NAPHTHALENES
(SHARP)

ETHERS
ALIPHATIC ETHERS CH_2-O-CH_2
AROMATIC ETHERS ⌬-O-CH_2
(m HIGH)

ALCOHOLS
(FREE) (SHARP)
(BONDED) (BROAD)
PRIMARY ALCOHOLS CH_2·OH
SECONDARY CH·OH
TERTIARY C·OH
AROMATIC ⌬-OH
(UNBONDING LOWERS)
(m HIGH)

ACIDS
(BROAD)
CARBOXYLIC ACIDS ·CO·OH
IONIZED CARBOXYL (SALTS·ZWITTER IONS ETC) -$CO_2^{(-)}$
(BROAD)
(ABSENT IN MONOMER)

ESTERS
FORMATES H·CO·O·R
ACETATES -CH_2·CO·O·R
PROPIONATES "
BUTYRATES AND UP "
ACRYLATES =CH·CO·O·R
FUMARATES "
MALEATES "
BENZOATES·PHTHALATES ⌬-CO·O·R

ALDEHYDES
ALIPH. ALDEHYDES -CH_2·CHO
AROM. ALDEHYDES ⌬-CHO

KETONES
ALIPH. KETONES CH_2·CO·CH_2
AROM. KETONES ⌬-CO·C

ANHYDRIDES
NORMAL ANHYDRIDES C·CO·O·CO·C
CYCLIC ANHYDRIDES O=C-O-C=O

AMIDES
(BROAD)
AMIDE -CO·NH_2
MONO SUBST. AMIDE -CO·NH·R
DI SUBST. AMIDE -CO·NR_2

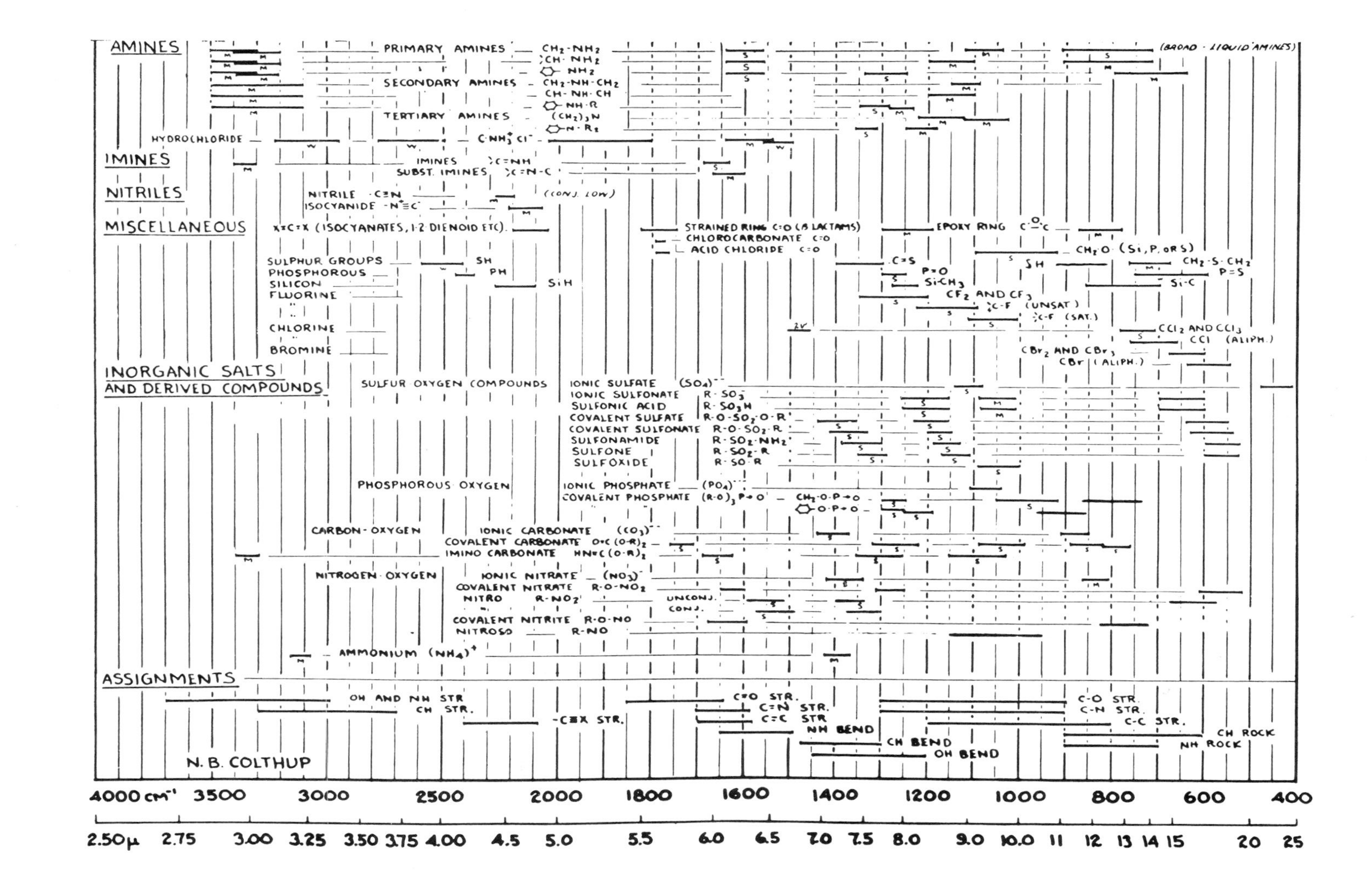

AMINES
PRIMARY AMINES
CH2-NH2
CH-NH2
NH2
SECONDARY AMINES
CH2-NH-CH2
CH-NH-CH
NH-R
TERTIARY AMINES
(CH2)3N
N-R2
HYDROCHLORIDE
C-NH3+ Cl-
(BROAD - LIQUID AMINES)
IMINES
IMINES C=NH
SUBST. IMINES C=N-C
NITRILES
NITRILE -C≡N
ISOCYANIDE -N+≡C-
(CONJ. LOW)
MISCELLANEOUS
X=C=X (ISOCYANATES, 1-2 DIENOID ETC.)
STRAINED RING C=O (β LACTAMS)
CHLOROCARBONATE C=O
ACID CHLORIDE C=O
EPOXY RING
SULPHUR GROUPS
PHOSPHOROUS
SILICON
FLUORINE
CHLORINE
BROMINE
SH
PH
SiH
C=S
P=O
Si-CH3
CF2 AND CF3
C-F (UNSAT.)
C-F (SAT.)
CH2-O (Si, P, OR S)
SH
CH2-S-CH2
P=S
Si-C
CCl2 AND CCl3
CCl (ALIPH.)
CBr2 AND CBr3
CBr (ALIPH.)
INORGANIC SALTS AND DERIVED COMPOUNDS
SULFUR OXYGEN COMPOUNDS
IONIC SULFATE (SO4)--
IONIC SULFONATE R-SO3-
SULFONIC ACID R-SO3H
COVALENT SULFATE R-O-SO2-O-R
COVALENT SULFONATE R-O-SO2-R
SULFONAMIDE R-SO2-NH2
SULFONE R-SO2-R
SULFOXIDE R-SO-R
PHOSPHOROUS OXYGEN
IONIC PHOSPHATE (PO4)---
COVALENT PHOSPHATE (R-O)3 P→O
CH2-O-P→O
O-P→O
CARBON-OXYGEN
IONIC CARBONATE (CO3)--
COVALENT CARBONATE O=C(O-R)2
IMINO CARBONATE HN=C(O-R)2
NITROGEN-OXYGEN
IONIC NITRATE (NO3)-
COVALENT NITRATE R-O-NO2
NITRO R-NO2
UNCONJ.
CONJ.
COVALENT NITRITE R-O-NO
NITROSO R-NO
AMMONIUM (NH4)+
ASSIGNMENTS
OH AND NH STR
CH STR.
-C≡X STR.
C=O STR.
C=N STR.
C=C STR
NH BEND
CH BEND
OH BEND
C-O STR.
C-N STR.
C-C STR.
CH ROCK
NH ROCK
N. B. COLTHUP
4000 cm-1 3500 3000 2500 2000 1800 1600 1400 1200 1000 800 600 400
2.50μ 2.75 3.00 3.25 3.50 3.75 4.00 4.5 5.0 5.5 6.0 6.5 7.0 7.5 8.0 9.0 10.0 11 12 13 14 15 20 25

and aliphatic compounds at 3000–2800 cm^{-1} (3.33–3.57 μm). When coupled with heavier masses, the hydrogen stretching frequencies overlap the triple-bond region.

The intermediate frequency range, 2500–1540 cm^{-1} (4.00–6.49 μm), is often called the *unsaturated* region. Triple bonds, and very little else, appear from 2500 to 2000 cm^{-1} (4.00–5.00 μm). Double-bond frequencies fall in the region from 2000 to 1540 cm^{-1} (5.00–6.49 μm). By judicious application of accumulated empirical data, it is possible to distinguish among C=O, C=C, C=N, N=O, and S=O bands. The major factors in the spectrum between 1300 and 650 cm^{-1} (7.69 and 15.38 μm) are single-bond stretching frequencies and bending vibrations (skeletal frequencies) of polyatomic systems that involve motions of bonds linking a substituent group to the remainder of the molecule. This is the fingerprint region. Multiplicity is too great for assured individual identification of the bands, but collectively the absorption bands aid in identifying the material.

Far-Infrared Region

The region 667–10 cm^{-1} (15.0–1000 μm) contains the bending vibrations of carbon, nitrogen, oxygen, and fluorine with atoms heavier than mass 19, and additional bending motions in cyclic or unsaturated systems. The low-frequency molecular vibrations in the far-infrared are particularly sensitive to changes in the overall structure of the molecule; thus the far-infrared bands often differ in a predictable manner for different isomeric forms of the same basic compound. The far-infrared frequencies of organometallic compounds are often sensitive to the metal ion or atom, and this too can be used advantageously in the study of coordination bonds. Moreover, this region is particularly well suited to the study of organometallic or inorganic compounds whose atoms are heavy and whose bonds are inclined to be weak.[5]

Compound Identification

Much information about a compound or mixture can be obtained before running an infrared spectrum. Such information should include the physical state and appearance, solubility, melting point, flame tests, and especially the history of the sample. It is helpful to ascertain, if possible, whether the sample is pure or a mixture of components. After the infrared spectrum is run, if the compound is organic the hydrogen stretching region is usually investigated first to determine whether the sample is aromatic, aliphatic, or both. Further examination of the group frequency region is then undertaken to try to establish the functional groups present *or absent*.

In many cases the interpretation of the infrared spectrum on the basis of characteristic frequencies is not sufficient to identify positively a total unknown, but perhaps the type or class of compound can be deduced. One must resist the tendency to overinterpret a spectrum—that is, to attempt to interpret and assign all of the observed absorption bands, particularly those of moderate and weak intensity in the fingerprint region. Once the category is established, the spectrum of the unknown is compared with spectra of appropriate compounds for an exact match. If the exact compound is not in the file, particular structure variations within the category may assist in suggesting possible answers and eliminating others. Several collections of spectra[6,7] and characteristic group frequencies[8,9] are available.

11.2 INSTRUMENTATION

Infrared instrumentation is divided into two classes, dispersive and nondispersive. The dispersive instruments use a prism or grating and are similar to ultraviolet-visible dispersive spectrometers except that, in the infrared region, different sources and detectors must be used. Nondispersive spectrometers may use interference filters, tunable laser sources, or an interferometer in the very popular Fourier transform infrared (FTIR) spectrometer. In the infrared region, front-surfaced mirrors are used, since glass and quartz, used in lenses, are opaque to infrared radiation.

It is convenient to divide the infrared region into three segments, with the dividing points based on instrumental capabilities (Table 11.1). Different radiation sources, optical systems, and detectors are needed for the different regions. The standard dispersive infrared spectrometer is a filter-grating or prism-grating instrument covering the range from 4000 to 650 cm^{-1} (2.50–15.38 μm). Grating instruments offer high resolution, which permits the separation of closely spaced absorption bands, accurate measurements of band positions and intensities, and high scanning speeds for a given resolution and noise level. Radiation from a source emitting in the infrared region is interrupted (chopped, pulsed, or modulated) at a low frequency, often 10–26 Hz, and is passed alternately through the sample and the reference cells placed before the monochromator. This minimizes the effect of stray radiation emanating from the sample and cell before it reaches the detector, a serious problem in most of the infrared region. The temperature and relative humidity in the room that houses the instrument should be controlled, since NaCl and KBr used in cells are hygroscopic. Modern spectrometers generally have attachments that permit speed suppression, scale expansion, repetitive scanning, and automatic

TABLE 11.1 COMPONENTS OF DISPERSIVE INFRARED SPECTROMETERS

	Region of electromagnetic spectrum		
	Near-infrared	**Mid-infrared**	**Far-infrared**
Wavenumber (cm^{-1})	12,500 – 4000	4000 – 200	200 – 10
Wavelength (μm)	0.8 – 2.5	2.5 – 50	50 – 1000
Source of radiation	Tungsten filament lamp	Nernst glower, Globar, or coil of Nichrome wire	High-pressure mercury-arc lamp
Optical system	One or two quartz prisms or prism-grating double monochromator	Two to four plane diffraction gratings with either a foreprism monochromator or infrared filters	Double-beam grating instruments for use to 700 μm; interferometric spectrometers for use to 1000 μm
Detector	Photoconductive cells	Thermopile, thermistor, pyroelectric, or semiconductor	Golay, pyroelectric

control of slit, period, and gain. Often these are under the control of a microprocessor. Accessories such as beam condensers, reflectance units, polarizers, and microcells can usually be added to extend versatility or accuracy. Radiation sources and detectors will be discussed in detail before dispersive and nondispersive spectrometers.

Radiation Sources

In the region beyond 5000 cm^{-1}, blackbody sources without envelopes commonly are used. The same spectral characteristics cited for the tungsten incandescent lamp apply to these as well (Figure 6.2). Unfortunately, the emission maximum lies in the near-infrared. A fraction of the shorter-wavelength radiation is present as stray radiation, and this is particularly serious for long-wavelength measurements.

A closely wound *Nichrome coil* raised to incandescence by resistive heating is a suitable source. A black oxide film forms on the coil, giving acceptable emissivity. Temperatures up to 1100 °C are reached. The Nichrome coil requires no water cooling and little or no maintenance and gives long service. This source is recommended where reliability is essential, such as in nondispersive process analyzers and inexpensive spectrometers or filter photometers. Although simple and rugged, this source is less intense than other infrared sources.

A hotter and therefore brighter source is the *Nernst glower*, which has an operating temperature as high as 1500 °C. Nernst glowers are constructed from a fused mixture of oxides of zirconium, yttrium, and thorium molded in the form of hollow rods 1–3 mm in diameter and 2–5 cm long. The ends of the rods are cemented to short ceramic tubes to facilitate mounting; short platinum leads provide power connections. Nernst glowers are fragile. They have a negative temperature coefficient of resistance and must be preheated to be conductive. Therefore, auxiliary heaters are provided as well as a ballast system to prevent overheating once the glower becomes conductive. A glower must be protected from drafts, but at the same time adequate ventilation is needed to remove surplus heat and evaporated oxides and binder. The energy output is predominantly concentrated between 1 and 10 μm, with relatively low energy beyond 10 μm. Radiation intensity is approximately twice that of Nichrome and Globar sources except in the near-infrared.

The *Globar*, a rod of silicon carbide 6–8 mm in diameter and 50 mm long, has characteristics intermediate between heated wire coils and the Nernst glower. It is self-starting and has an operating temperature near 1300 °C. The temperature coefficient of resistance is positive and may be conveniently controlled with a variable transformer. Its resistance increases with the length of time used, so provision must be made for increasing the voltage across the unit. It is often encased in a water-cooled brass tube, with a slot provided for the emission of radiation. The spectral output of the Globar is about 80% that of a blackbody radiator. In comparison with the Nernst glower, the Globar is a less intense source below 10 μm, the two sources are comparable out to about 15 μm, and the Globar is superior beyond about 15 μm. It finds some use out to about 50 μm.

In the very far-infrared, beyond 50 μm (200 cm^{-1}), blackbody-type sources lose effectiveness, since their radiation decreases with the fourth power of the wavelength. High-pressure mercury arcs, with an extra quartz jacket to reduce thermal loss, give intense radiation in this region. Output is similar to that from blackbody sources, but additional radiation is emitted from a plasma, which enhances the long-wavelength output.

Tunable diode lasers have some special uses in infrared spectrometry, especially in nondispersive instruments for process monitoring where only one substance is being determined. A diode laser emits a band of very narrow wavelength. Most tunable diode lasers must be operated at cryogenic temperatures and the nominal wavelength depends on the temperature. Tuning over a narrow wavelength range is accomplished by varying the diode current.

Detectors[10]

At the short-wavelength end, below about 1.2 μm, the preferred detection methods are the same as those used for visible and ultraviolet radiation. The detectors used at longer wavelengths can be classified into two groups: (1) thermal detectors, in which the infrared radiation produces a heating effect that alters some physical property of the detector, and (2) photon detectors, which use the quantum effects of the infrared radiation to change the electrical properties of a semiconductor.

Thermal Detectors. The active element in any thermal detector is as small as possible to maximize its temperature change for any level of infrared radiant energy. For the same reason, the element is blackened and thermally insulated from its substrate. When radiation ceases, the element returns to the temperature of the substrate, with a decay time determined by the finite thermal conductance of the insulation. Material properties affected in thermal detectors include an expansion of a solid, gas, or fluid (Golay cell), electrical resistance (thermistor), voltage induced at the junction of two dissimilar materials (thermocouple and thermopile), and electric polarization (pyroelectric).

Thermal detectors can be used over a wide range of wavelengths (Figure 11.3), which includes both visible and infrared radiation, and they operate at room temperature. Their main disadvantages are slow response time (milliseconds) and lower sensitivity relative to other types of detectors. Their response times set an upper limit to the frequency at which the radiation can be usefully modulated, chopped, or pulsed. The total mass represented by the receiver, absorbing material, and temperature-sensing element must heat during each half-cycle when radiation strikes the detector and cool when the detector is occluded.

A *thermocouple,* fabricated from two dissimilar metals like bismuth and antimony, produces a small voltage proportional to the temperature of the junction. The surface that receives the incident radiation is coated with a metal oxide, such as gold or bismuth black, which has little thermal mass. A *thermopile* consists of several (often six) thermocouples connected in series so their outputs add. Half of the junctions are called "hot" and make up the active element. Alternate junctions, the "cold" ones, are thermally bonded to the substrate and remain at a relatively stable temperature. Thin-film techniques have miniaturized thermopiles. The entire assembly is mounted in an evacuated enclosure with an infrared-transmitting window so that conductive heat losses are minimized (Figure 11.4). Thermopiles offer the simplest and most direct means for converting radiant energy into an electrical signal. Frequency response is flat below 35 Hz. Response time is about

30 msec. To prevent the faint signals from being lost in the stray (noise) signals picked up by the lead wires, a preamplifier is located as close as possible to the detector.

A *thermistor* functions by changing resistance when heated. To minimize noise and drift, an infrared thermistor–bolometer detector contains two closely spaced thermistor flakes; the *therm*ally sensitive res*istors* are sintered oxides of manganese, cobalt, and nickel, which have a high-temperature coefficient of resistance (approximately 4%/°C). One of these 10-μm-thick flakes is an active detector, while the other acts as a compensating (or reference) detector. The active flake is coated

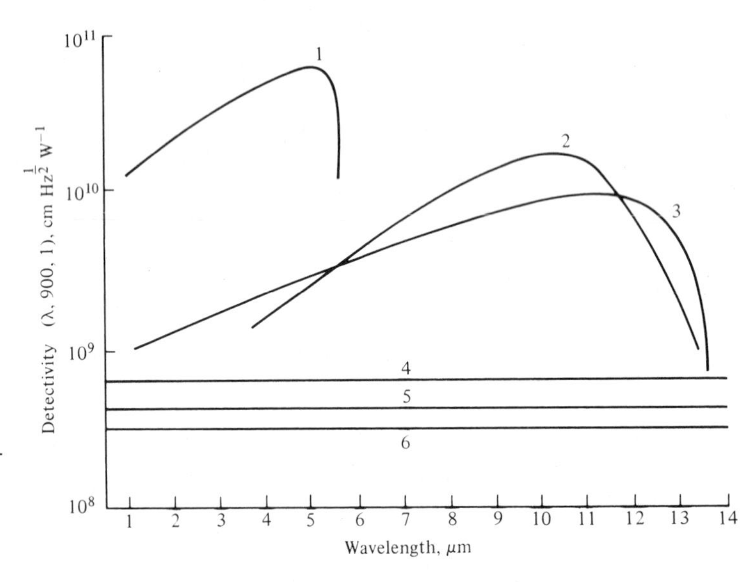

FIGURE 11.3
Wavelength response of some infrared detectors: (1) InSb at 77 K, (2) PbSnTe at 77 K, (3) PbSnTe at 4.2 K, (4) pyroelectric at 300 K, (5) thermistor at 300 K, and (6) thermopile at 300 K. Detectivity is obtained by irradiating the detector with monochromatic power at the same wavelength as that where the detector produces its peak output, chopping frequency is 900 Hz, and noise is measured in a 1-Hz bandwidth. (Courtesy of Barnes Engineering Co.)

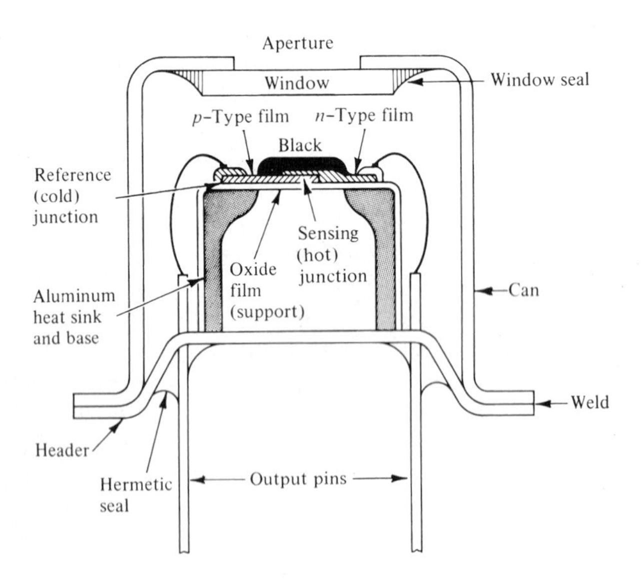

FIGURE 11.4
Schematic construction of a thermocouple (or thermopile) in cross section.

with black material to increase its infrared absorption, whereas the compensating flake is optically shielded to prevent its exposure to the incident infrared radiation. The flakes are separately mounted on an insulating substrate that is placed on a heat sink. When connected in a bridge circuit, the shielded wafer compensates for ambient temperature changes. A steady bias voltage is applied across the detector elements. The drift due to ambient temperature change tends to cancel because both the active and compensating flakes are connected in series. By judicious choices of substrate material and the thickness of the flakes, detectors are constructed with time constants on the order of a few milliseconds. In general, however, the response time and responsivity must be traded. Greater thermal contact assures faster response time, but since good thermal contact also prevents the flake from reaching higher temperature, the detector responsivity decreases.

A *pyroelectric* detector contains a noncentrosymmetrical crystal, which, below its Curie temperature, exhibits an internal electric field (or polarization) along the polar axis. The electric field results from the alignment of electric dipole moments. Heat resulting from radiation absorption produces thermal alteration of the crystal lattice spacing, which, in turn, changes the value of the electric polarization. The surfaces of the crystal normal to the polarization axis then develop a polarization charge. If electrodes are applied to these surfaces (one of the electrodes must be infrared-transparent) and connected through an external circuit, free charge will be brought to the electrodes to balance the polarization charge, thus generating a current in the external circuit. The detector is a thin plate of pyroelectric material between two electrodes, forming a capacitor. The high impedance is reduced by mounting the subassembly in an enclosure that contains a field-effect transistor connected as a source follower and a matched load resistor. Pyroelectric materials include triglycine sulfate (TGS), deuterated triglycine sulfate (DTGS), $LiTaO_3$, $LiNbO_3$, and some polymers. Generally TGS and DTGS show superior features but have limited use because of their hygroscopic nature and low Curie points (approximately 49 °C). For ease in handling and higher Curie points, $LiTaO_3$ or $LiNbO_3$ is frequently used. With a 100-MΩ load resistor, the response time is 1 msec and responsivity (ratio of detector output to incident radiation) is about 100. With a 1-MΩ load resistor the response time is 10 μsec and responsivity is 1. The absence of a bias voltage means that there is minimal low-frequency noise to interfere with low-frequency scanning signals.

Unlike other thermal detectors, the pyroelectric effect depends on the rate of change of the detector temperature rather than on the temperature itself. This allows the pyroelectric detector to operate with a much faster response time and makes these detectors the choice for Fourier transform spectrometers where rapid response is essential. It also means that this type of detector responds only to changing radiation that is chopped, pulsed, or otherwise modulated; it ignores steady background radiation.

The *Golay* pneumatic detector, shown in Figure 11.5, uses the expansion of a gas as the measuring device. The unit consists of a small metal cylinder closed by a rigid blackened metal plate (2-mm square) at one end and by a flexible silvered diaphragm at the other end. The chamber is filled with xenon. Radiation passes through a small infrared-transmitting window and is absorbed by the blackened plate. Heat, conducted to the gas, causes it to expand and deform the flexible diaphragm (mirror). To amplify distortions of the mirror surface, light from a lamp inside the detector housing is focused on the mirror, which reflects the light beam onto a phototube. With the flexible mirror in its rest position, an image of half the Moiré grid falls on the other half so that no light passes through. Flexing of the mirror moves the image of the grid laterally so that varying amounts of light can reach the phototube. In an alternate arrangement, the rigid diaphragm is used as one plate of a dynamic condenser. A perforated diaphragm a slight distance away serves as the second plate. The distortion of the solid diaphragm relative to the fixed plate alters the separation and hence the capacity. Response time is approximately 20 msec. The Golay detector has a sensitivity similar to that of a thermocouple. It is significantly superior as a detector for the far-infrared region. Since the angular aperture is 60°, the detector must be used with a system of condensing mirrors to concentrate the incident radiation.

Photon Detectors. The more sensitive detectors rely on a quantum interaction between the incident photons and a semiconductor. The result produces electrons and holes. This is the internal photoelectric effect. A sufficiently energetic photon that strikes an electron in the detector can raise that electron from a nonconducting state to a conducting state. The excitation of electrons requires a definite minimum-energy photon. The detectors thus exhibit a sharp cutoff toward the far-infrared region. As conductors, electrons contribute to the current flow in one of several ways depending on the configuration of the semiconductor. These are referred to as photovoltaic or photoconductive devices.

In a *photoconductive detector,* consisting of a homogeneous semiconductor chip, the presence of electrons in the conduction band lowers the chip's resistance. Intrinsic hole-electron pairs are created by raising an electron from the valence band to the conduction band of the semiconductor. Extrinsic excitation refers to electrons raised from or to impurity doping levels within the forbidden band of the semiconductor. In either case a bias current or voltage registers this change at the output.

Photovoltaic detectors generate a small voltage when exposed to radiation. InSb detectors use a diffused *p-n* junction (Chapter 2) in single-crystal indium antimonide. The *p*-type InSb is in a thin layer over the *n*-type material. Radiation is incident upon the *p*-type surface. Photons that have sufficient energy generate hole-electron pairs that are then separated by the internal field existing at the *p-n* junction. The result is a voltage that can be electronically processed as required by the application. The valence-to-conduction band gap of InSb is 0.23 eV at liquid nitrogen temperature.

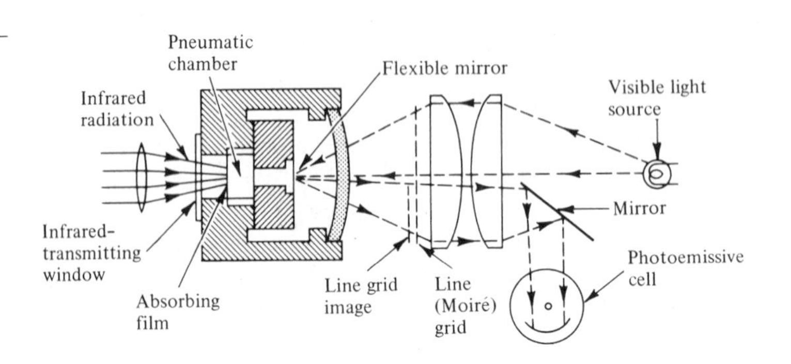

FIGURE 11.5
Golay pneumatic infrared detector.

This accounts for the detector's sensitivity cutoff at a wavelength of 5.5 μm. The detector forms an integral part of its Dewar-type vacuum bottle cooling unit (Figure 11.6). The cooling well has sufficient volume to permit the use of either liquid nitrogen (4 hr/charge) or Joule-Thomson coolers with compressed nitrogen gas. Single-element detectors, as well as mosaic arrays, are available. The signal output is linear with irradiance over nine orders of magnitude. The time constant is less than 1 μsec.

Lead tin telluride detectors extend spectral sensitivity to considerably longer wavelengths than InSb. Two types are available. The first is cooled by liquid nitrogen and has optimum sensitivity throughout the 5–13-μm region. The second type is liquid-helium-cooled and has optimum performance in the 6.6–18-μm region. Mercury cadmium telluride (MCT) detectors are even more efficient than lead tin telluride detectors and about five times more sensitive than pyroelectric detectors. When these detectors are used with a current mode amplifier, the speed of response is not compromised; detector sensitivity and response times as fast as 20 nsec can be obtained. Bias currents are not required, resulting in very low low-frequency noise.

Dispersive Spectrometers

Most dispersive spectrometers are double-beam instruments in which two equivalent beams of radiant energy are taken from the source. By means of a combined rotating mirror and radiation interrupter, the source is flicked alternately between the reference and sample paths. In the optical-null system, the detector responds only when the intensity of the two beams is unequal. Any imbalance is corrected by an attenuator (an optical wedge or comb shutter) moving in or out of the reference beam to restore balance. The recording pen is coupled to the attenuator. Although very popular, the optical-null system has serious faults. Near zero transmittance of the sample, the reference-beam attenuator moves in to stop practically all radiation in the reference beam. Both beams are then blocked, no energy is passed, and the spectrometer has no way of determining how close it is to the correct transmittance value. The instrument will go dead. On the other hand, in the mid-infrared region the

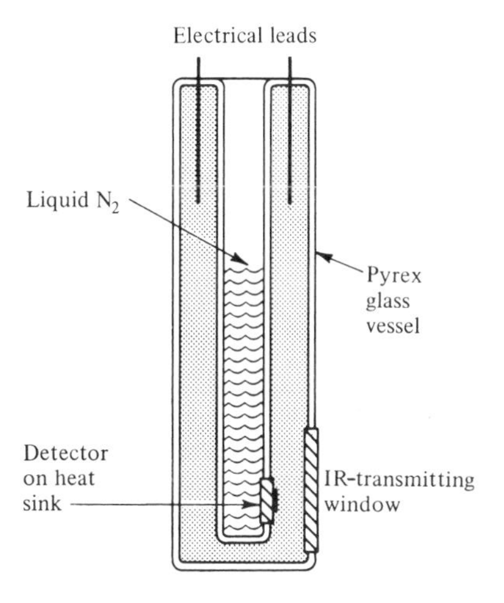

FIGURE **11.6**
Simplified construction diagram of indium antimonide detector cooled by liquid nitrogen.

electrical beam-ratioing method is not an easy way to avoid the deficiencies of the optical-null system. To a large extent it is trading optical and mechanical problems for electronic problems.

Monochromators that use prisms for dispersion utilize a Littrow 60° prism-plane mirror mount (Figure 6.15). Mid-infrared instruments have a sodium chloride prism for the region 4000–650 cm^{-1} (2.50–15.38 μm), with a potassium bromide or cesium iodide prism and optics for the extension of the useful spectrum to 400 cm^{-1} (25 μm) or 270 cm^{-1} (37 μm), respectively. Quartz monochromators, designed for the ultraviolet–visible regions, extend their coverage into the near-infrared (to 2500 cm^{-1}, or 4.0 μm).

Plane reflectance grating monochromators dominate today's dispersive instruments. To cover the wide wavelength range, several gratings with different ruling densities and associated higher-order filters are necessary. This requires some complex sensing and switching mechanisms for automating the scan with acceptable accuracy. Because of the nature of the blackbody emission curve, a slit programming mechanism must be used to give nearly constant energy and resolution as a function of wavelength. The principal limitation is energy. The resolution and signal-to-noise ratio are limited primarily by the emission of the blackbody source and the noise-equivalent power of the detector. Two gratings are often mounted back-to-back so that each need be used only in the first order. The gratings are changed at 2000 cm^{-1} (5.00 μm) in mid-infrared spectrometers. Grating instruments incorporate a sine-bar mechanism to drive the grating mount when a wavelength readout is desired, and a cosecant-bar drive when wavenumbers are desired. Undesired overlapping orders are eliminated with a fore-prism or by suitable filters.

The optical arrangement for a filter-grating spectrometer is shown in Figure 11.7. The filters are inserted near a slit or slit image when the required size of the filter is not too large. The circular variable filter is simple in construction. It is frequently necessary to use gratings as reflectance filters when working in the far-infrared region in order to remove unwanted second and higher orders from the radiation incident on the far-infrared grating. For this purpose small plane gratings are used, which are blazed for the wavelength of the unwanted radiation. The grating acts as a mirror, reflecting the desired radiation into the instrument and diffracting the shorter wavelengths out of the beam. A grating acts like a good mirror to wavelengths longer than the groove spacing.

Probably the most elegant filter is a prism because it provides a narrow band of wavelengths with high efficiency over a relatively broad spectral range. The prism and grating must track together over consecutive grating orders. Radiation from the parabolic mirror enters the fore-prism, where it is dispersed so that only a relatively narrow band of wavelengths is allowed to fall on the grating. The resolution of the prism can be quite low because it need only exclude the adjacent orders, but for higher orders from the grating the interval between orders gets successively narrower. Thus it is preferable to use more than one grating and confine their application to lower orders.

The use of microprocessors has alleviated many of the tedious requirements necessary to obtain usable data. Integrated scan controls allow the operator to select a single recording parameter, such as scan time, slit setting, or pen response. The microprocessor automatically optimizes these and other conditions. Even under high-resolution conditions where the noise might obscure spectral detail, the operator can improve the S/N ratio by selecting the appropriate multiplier. When this occurs, the instrument automatically changes the scanning parameters to optimize conditions. The discussion by Smith[11] on "trading rules" for spectrometers should be read by everyone before running a spectrum.

Single-beam photometers have the capability for accurate measurement in quantitative analysis. Until recently they were not extensively used, primarily because of the need for extensive reduction of the data produced. However, a built-in microcomputer results in the combined system's programmability as well as the virtually instantaneous reduction of data and availability of results.

Nondispersive Spectrometers

One type of nondispersive spectrometer uses filters to isolate the wavelengths desired. Such instruments are useful in cases where the same determination is done repetitively, as in industrial process control. The optical schematic of a very simple infrared analyzer is shown in Figure 11.8. Different wavelengths are selected by three circular, variable interference filters that cover the spectral range 2.5–14.5 μm. These filters are mounted on the shaft of a high-resolution potentiometer, which senses its position. This position information is compared with a fixed value from the microcomputer in the feedback loop of the dc servomotor that drives the filter wheel to the desired position. The instrument has a Nichrome wire source and a $LiTaO_3$ pyroelectric detector. A $f/1.5$ optical system with the low-resolution filter yields a high S/N ratio, allowing the detection of absorption as small as 0.0001 absorbance unit.

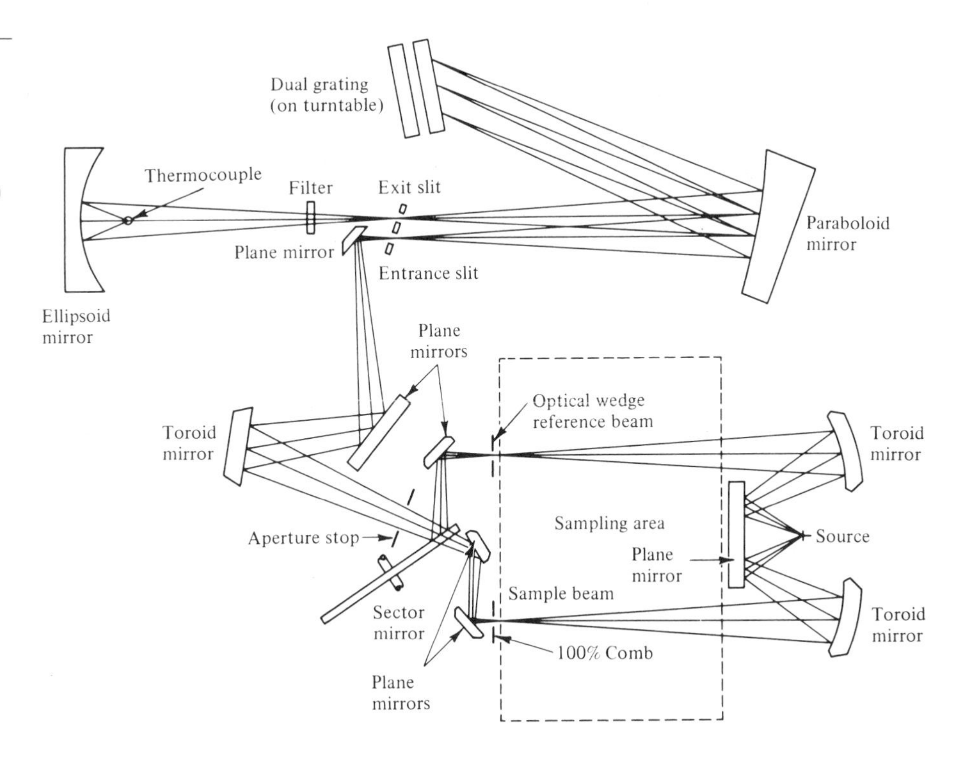

FIGURE 11.7
Optical schematic of a filter-grating, double-beam infrared spectrophotometer. (Courtesy of Perkin-Elmer Corp.)

Simple filter infrared analyzers have been designed around the same optical schematic shown in Figure 11.8. Wavelength is selected through the use of interchangeable, slide-mounted, fixed-wavelength interference filters. Several filters may even be mounted on a filter wheel for convenient selection. Equipped with a micro flow-through cell designed for continuous monitoring, these infrared analyzers can be coupled to liquid chromatographs for monitoring effluents, or they can be used for the quantitative determination of fiber finishes and lubricating oils, or dissolved hydrocarbons in water after extraction into CCl_4 or Freon solvent. Strongly absorbing liquids can be handled with a fixed flow-through internal reflection cell. Such instrumentation is used for the selective monitoring of compounds in on-line process control applications (see Chapter 26).

Fourier Transform Infrared (FTIR) Spectrometer[3,12]

Infrared radiation can be analyzed by means of a scanning Michelson interferometer (Figure 11.9). This consists of a moving mirror (4), a fixed mirror (3), and a beamsplitter (*C*). Radiation from the infrared source (*B*) is collimated by a mirror (2), and the resultant beam is divided at the beamsplitter; half the beam passes to a fixed mirror (3) and half is reflected to the moving mirror. After reflection, the two beams recombine at the beamsplitter and, for any particular wavelength, constructively or destructively interfere, depending on the difference in optical paths between the two arms of the interferometer. With a constant mirror velocity, the intensity of the emerging radiation at any particular wavelength modulates in a regular sinusoidal manner. In the case of a broadband source the emerging beam is a complex mixture of modulation frequencies that, after passing through the sample compartment, is focused onto the detector (*G*). This detector signal is sampled at precise intervals during the mirror scan. Both the sampling rate and the mirror velocity are controlled by a reference signal incident upon a detector (*E*), which is produced by modulation of the beam from the helium-neon laser (*A*).

The resulting signal from detector *G* is known as an interferogram (stored in memory 1) and contains all the information required to reconstruct the spectrum via the mathematical process known as Fourier transformation (see Section 2.6). For

FIGURE 11.8 Infrared analyzer, single-beam: (a) optical schematic and (b) circular interference filter drive system. (Courtesy of Foxboro/Wilks, Inc.)

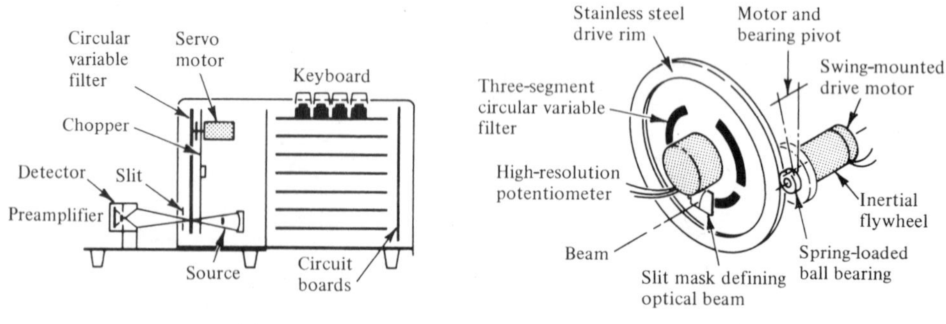

FIGURE 11.9 Infrared Fourier transform interferometric spectrometer: (a) optical path diagram and (b) block diagram of the instrument's functions. (Courtesy of Nicolet Instrument Corp.)

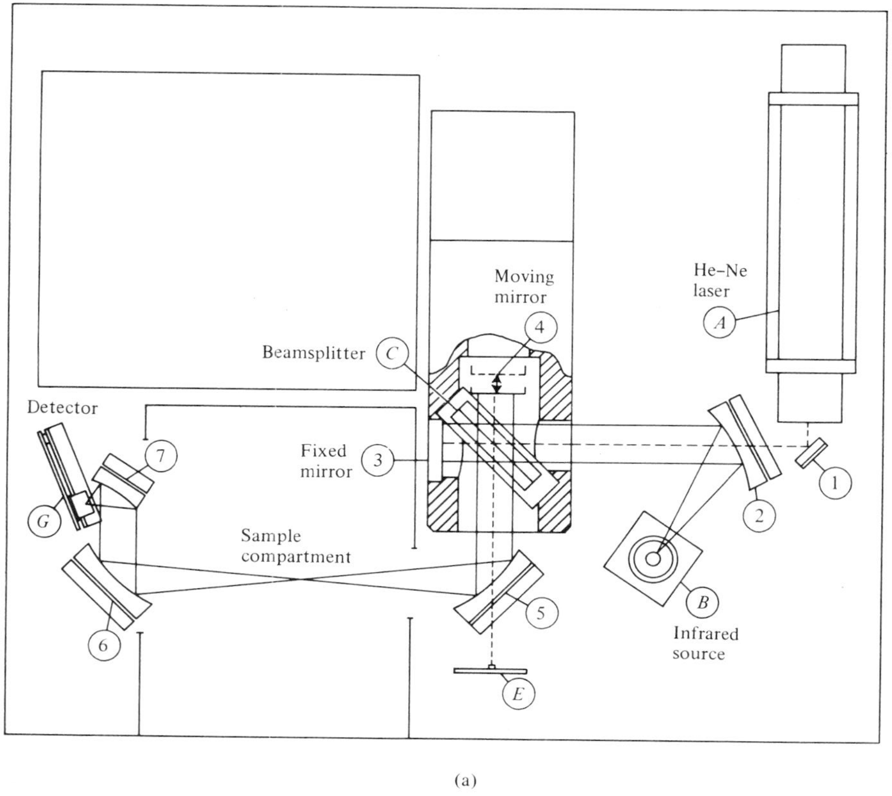

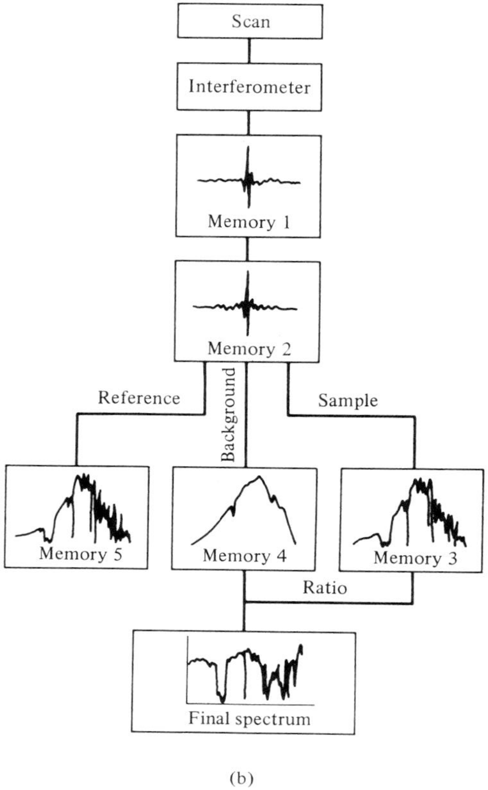

spectroscopy, Equation 2.7 may be written as

$$P_{(x)} = \int_{-\infty}^{+\infty} P_{0(\bar{v})} \cos(2\pi x \bar{v})\, d\bar{v} \tag{11.1}$$

where $P_{(x)}$ is the intensity of the total beam at the detector, $P_{0(\bar{v})}$ is the intensity of the source at frequency $\bar{v}$ (in cm^{-1}), x is the mirror displacement (in cm), and the summation is over all frequencies or as many as are in the radiation that reaches the detector. Only cosine waves are needed, since x is measured from the position where both arms of the interferometer are equal, at which point all waves constructively interfere; that is, all wave intensities are at a maximum at $x = 0$. The Fourier transform or spectrum as a function of frequency is then given by the equivalent of Equation 2.6, or

$$P_{(\bar{v})} = \int_{-\infty}^{+\infty} P_{(x)} \cos(2\pi x \bar{v})\, dx \tag{11.2}$$

The transform is carried out by a computer, which is an essential part of the spectrometer, by an appropriate algorithm such as the Cooley-Tukey transform algorithm.[13]

This technique has several distinct advantages over the conventional dispersive techniques. There is only one moving part involved, mirror 4, mounted on a frictionless air bearing. Dispersion or filtering is not required so that energy-wasting slits are not needed. This is a major advantage, particularly with energy at a premium in the far-infrared. The use of a helium-neon laser as a reference results in near absolute frequency accuracy, better than 0.01 cm^{-1} over the range 4800–400 cm^{-1}. Because all wavelengths are simultaneously detected throughout the scan, the scanning interferometer achieves the same spectral signal-to-noise ratio as a dispersive spectrometer in a fraction of the time (Felgett's advantage).

The automatic process between the initiation of the scan and the final plot is shown in Figure 11.9b. The interferogram recorded with each scan is stored in memory 1. This interferogram is then automatically aligned with, and added to, the averaged interferograms in memory 2. At the same time annotation of the plot is begun in preparation for the final spectrum. After a number of scans, often 32 (approximately 60 sec), this averaged interferogram is Fourier transformed to produce a single-beam spectrum, which, in the standard sample mode, is stored in memory 3. This single-beam spectrum is then ratioed against the stored background (run once a day) in memory 4 and the resulting "double-beam" spectrum is plotted on the high-speed digital plotter. Memory 5 is additional space available for storage of a reference spectrum that would be used in the spectral subtraction technique. This memory is also used to store a newly measured spectrum while maintaining the sample and background spectra for further manipulation. The time from insertion of the sample to the completed plot is about 2 min.

In Equations 11.1 and 11.2, the summation should extend from $-\infty$ to $+\infty$. Of course, this is impossible. If a Fourier series is suddenly terminated (i.e., a "boxcar" truncation), however, the resulting transform will have some wiggles or "ringing" especially on the wings of an intense peak. To minimize these effects the series may be gradually terminated by multiplying the data by some apodizing function such as a triangular function, trapezoidal function, or the Norton-Beer function. Resolution is, however, decreased somewhat by the use of an apodizing function.

Resolution of a Fourier transform spectrometer is given by

$$R = \frac{1}{\Delta_{max}} \text{ cm}^{-1} \tag{11.3}$$

where R is the maximum attainable resolution in wavenumbers and Δ_{max} is the maximum retardation—that is, twice the distance of the mirror movement in centimeters. Besides the decrease in resolution due to the application of an apodizing function, resolution is adversely affected by divergence of the beam and instability of the mirror motion, in either speed or alignment, resulting in poorer resolution than that predicted by Equation 11.3.

Since the computer uses digital data, the photodetector signal must be converted to digital form by an analog-to-digital converter. The data must be sampled at precise points and the resultant data stored. The data must be sampled at least twice every wavelength, so the sampling interval must be half the minimum wavelength to be observed; that is, $\bar{\nu}_{max} = 1/2\Delta x = 1/\lambda_{min}$. If the sampling intervals are longer than this value, the high-frequency waves will be "aliased" or folded back into the spectrum.

Felgett's advantage, mentioned earlier, results because an interferometer measures all wavelengths simultaneously. Thus an interferometer-spectrometer, when compared with a sequential dispersive spectrometer, carries out N measurements, where N is the number of resolution elements of the dispersive spectrometer, in the same time as one complete scan on the sequential dispersive spectrometer. The signal is N times as strong and the noise is $\sqrt{N}$ as great, so the signal-to-noise advantage (Felgett's advantage) is $\sqrt{N}$. Another advantage of interferometric instruments is known as Jacquinot's advantage. This is due to the increased energy throughput. The interferometer has a large circular entrance aperture rather than smaller entrance slits like a dispersive instrument, but the interferometer does lose half the radiation at the beamsplitter.

Fourier transform infrared spectrometers are still more expensive than sequential dispersive instruments due to the precision needed for mirror movement and the computer that is also required. However, dispersive instruments are now often equipped with computers that can record and remember spectra, plot absorbance, transmittance, or some function of absorbance, subtract one spectrum from another, and do other functions, and this adds to their price. The power of FTIR instruments has led to much competition in the instrumentation industry, resulting in an increased availability of commercial instruments and a decrease in price. Low-cost FTIR instruments are beginning to appear. Fourier transform spectrometers are faster than dispersive instruments and therefore are especially useful in situations that require fast, repetitive scanning. They are used in recording the output of gas or liquid chromatographs or obtaining kinetic data.

11.3 SAMPLE HANDLING

Infrared instrumentation has reached a remarkable degree of standardization as far as the sample compartment of various spectrometers is concerned. Sample handling itself, however, presents a number of problems in the infrared region. There is no rugged window material for cuvettes that is transparent and also inert over this region. The alkali halides are widely used, particularly sodium chloride, which is transparent at wavelengths as long as 625 cm^{-1} (16.00 μm). Cell windows are easily fogged by exposure to moisture and require frequent repolishing. Silver chloride is often used for moist samples or aqueous solutions, but it is soft, is easily deformed, and darkens on exposure to visible light. Teflon has only C—C and C—F absorption bands. For frequencies less than 600 cm^{-1}, a polyethylene cell is useful. Infrared transmission materials are compiled in Table 11.2. Materials of high refractive index produce strong, persistent interference fringes.

Gases

In the analysis of gases, the usual path length is 10 cm. When this is too short to measure the spectra of minor components or substances encountered in trace analysis, a variable-path cell provides path lengths in steps of 1.5 m for 20-, 40-, and 120-m cells. The path is folded using internal gold-surfaced mirrors and gold-plated or stainless steel metal components. Further gains in sensitivity can be realized by increasing the pressure of the gas sample in the cell to 10 atm; however, pressure broadening of absorption bands is troublesome in quantitative work. Long-path

TABLE **11.2**

INFRARED-TRANSMITTING MATERIALS

Material	Wavelength range (μm)	Wavenumber range (cm^{-1})	Refractive index at 2 μm
NaCl, rock salt	0.25–17	40,000–590	1.52
KBr, potassium bromide	0.25–25	40,000–400	1.53
KCl, potassium chloride	0.30–20	33,000–500	1.5
AgCl, silver chloride*	0.40–23	25,000–435	2.0
AgBr, silver bromide*	0.50–35	20,000–286	2.2
CaF_2, calcium fluoride (Irtran-3)	0.15–9	66,700–1,110	1.40
BaF_2, barium fluoride	0.20–11.5	50,000–870	1.46
MgO, magnesium oxide (Irtran-5)	0.39–9.4	25,600–1,060	1.71
CsBr, cesium bromide	1–37	10,000–270	1.67
CsI, cesium iodide	1–50	10,000–200	1.74
TlBr-TlI, thallium bromide-iodide (KRS-5)*	0.50–35	20,000–286	2.37
ZnS, zinc sulfide (Irtran-2)	0.57–14.7	17,500–680	2.26
ZnSe, zinc selenide* (vacuum deposited) (Irtran-4)	1–18	10,000–556	2.45
CdTe, cadmium telluride (Irtran-6)	2–28	5,000–360	2.67
Al_2O_3, sapphire*	0.20–6.5	50,000–1,538	1.76
SiO_2, fused quartz	0.16–3.7	62,500–2,700	
Ge, germanium*	0.50–16.7	20,000–600	4.0
Si, silicon*	0.20–6.2	50,000–1,613	3.5
Polyethylene	16–300	625–33	1.54

* Useful for internal reflection work.

gas cells are intended for measurements in the range of a few parts per million and lower, concentration ranges encountered with problems in air pollution, air monitoring, process instrumentation, and purity determinations. When working in a spectral region where water vapor or carbon dioxide absorption occurs, a dual-cell system for compensation is desirable.

Liquids and Solutions

Samples that are liquid at room temperature are usually scanned in their neat form or in solution. The sample concentration and path length are chosen so that the transmittance lies between 15% and 70%. For neat liquids, this represents a very thin layer, about 0.001–0.05 mm thick. For solutions, concentrations of 10% and cell lengths of 0.1 mm are most practical. Unfortunately, not all substances are soluble to a reasonable concentration in a solvent that is nonabsorbing in regions of interest. When possible, the spectrum is obtained in a 10% solution in CCl_4 in a 0.1-mm cell in the region 4000–1333 cm^{-1} (2.50–7.50 μm) and in a 10% solution of CS_2 in the region 1333–650 cm^{-1} (7.50–15.38 μm). Transparent regions of selected solvents are shown in Figure 11.10. To obtain the spectra of polar materials that are insoluble in CCl_4 or CS_2, chloroform, methylene chloride, acetonitrile, and acetone are useful solvents. Sensitivity is gained by going to longer path lengths if a suitably transparent solvent can be found. In a double-beam spectrometer a reference cell of the same path length as the sample cell is filled with pure solvent and placed in the reference beam. Moderate solvent absorption, now common to both beams, is not observed in the recorded spectrum. However, solvent transmittance should never fall to less than 10%.

The possible influence of a solvent on the spectrum of a solute must not be overlooked. Particular care should be exercised in selecting a solvent for compounds that are susceptible to hydrogen-bonding effects. Hydrogen bonding through an —OH or —NH group alters the characteristic vibrational frequency of that group; the stronger the hydrogen bonding, the more the fundamental frequency is lowered. To differentiate between inter- and intramolecular hydrogen bonding, a series of spectra at different dilutions, yet with the same number of absorbing molecules in the beam, must be obtained. If, as the dilution increases, the hydrogen-bonded absorption band decreases while the unbonded absorption band increases, the bonding is intermolecular. Intramolecular bonding shows no comparable dilution effect.

Infrared solution cells are constructed with windows sealed and separated by thin gaskets of Teflon or copper or lead that have been wetted with mercury. The whole assembly is securely clamped together and permanently mounted in a stainless steel holder. If copper or lead is used, as the mercury penetrates the metal, the gasket expands to produce a tight seal. Cells are provided with tapered fittings to accept the needles of hypodermic syringes for filling. Each cell is labeled with its precise path length as measured by interference fringes. An unassembled (demountable) cell consists of two window pieces and a Teflon fitting (which also accommodates a syringe for filling). The Teflon forms a leak-proof seal when the cell is slipped into a mount and knurled nuts are turned down until finger tight.

Flow-through cells are useful for the continuous analysis of liquids. In repetitive sampling applications, quantitative accuracy is increased and sample handling facilitated. Coupling the infrared instrument to a chromatograph with a micro flow-through cell makes it possible to monitor a column effluent on a functional group basis.

The variable-path-length cell consists of a cylindrical, Teflon-lined, stainless steel chamber with parallel windows. The path length is continuously adjustable from 0.005 to 5 mm and reproducible to within 0.001 mm. A vernier and scale provided on the cylinder permit the cell thickness to be read to within ± 0.0005 mm. The accuracy of the thickness settings is ± 0.001 mm or 1%, whichever is larger.

The variable-path-length liquid cell serves two very useful purposes in the infrared laboratory. The first is solvent compensating and differential analysis. When filled with solvent and mounted in the reference beam of the spectrometer, its path length may be adjusted to compensate for unwanted solvent absorption in the sample beam. In differential analysis, two liquids are examined that differ only slightly in the concentration of the minor component. Careful adjustment of the

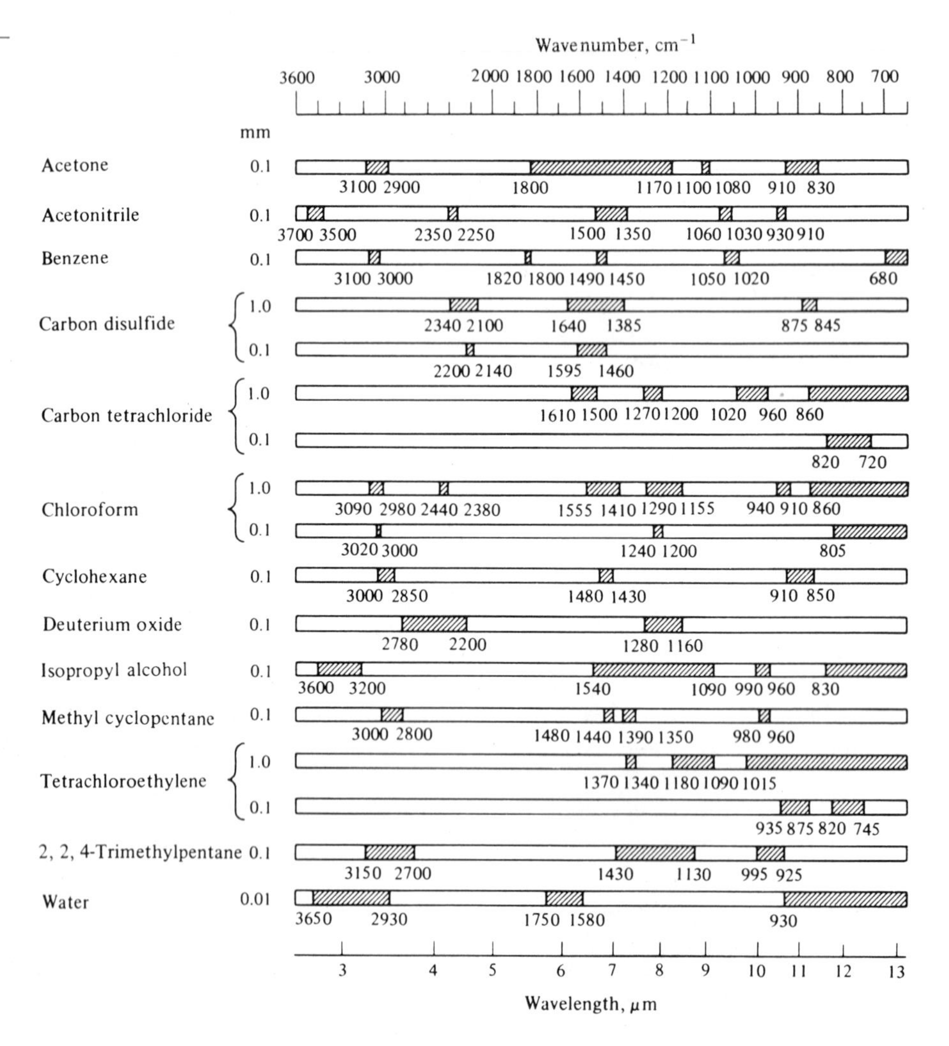

FIGURE 11.10
Transmission characteristics of selected solvents. The material is considered transparent if the transmittance is 75% or greater. Solvent thickness is given in millimeters.

variable cell enhances the spectrum of the minor component in the fixed path cell in the sample beam.

The second application uses the variable-path-length cell in the sample beam. In order for weaker absorption bands of many liquids to appear with the proper intensity, the stronger bands must be totally absorbing and their exact location thus obscured. In such situations, the variable-path-length cell is precisely adjusted to provide just the right intensity for the exact location of both weak and strong bands. The variable-path-length cell eliminates the requirement for a large collection of different-thickness fixed-path-length cells.

The minicell is an economical approach for obtaining qualitative infrared spectra of liquids. It consists of a threaded, two-piece plastic body and two silver chloride cell window disks of special design. The windows fit into one portion of the cell. The second portion of the cell is then screwed in to form the seal. The AgCl cell windows each contain a 0.025-mm circular depression (also available in 0.100-mm depression). The rim of the window is flat and the circumference is beveled to ensure proper sealing. Because AgCl flows slightly under pressure, a tight seal is formed. The circular depression in each window enables the cell path to be varied: the windows can be (1) placed back-to-back for a conventional smear, (2) arranged with one back to the circular depression for 0.025-mm path length, or (3) positioned with facing circular depressions for 0.050-mm path length.

Films

The spectra of liquids not soluble in a suitable solvent are best obtained from capillary films. A large drop of the neat liquid is placed between two infrared-transmitting windows, which are then squeezed together and mounted in the spectrometer in a suitable holder. Plates need not have a high polish, but they must be flat to avoid distorting the spectrum.

For polymers, resins, and amorphous solids, the sample is dissolved in any reasonable volatile solvent, the solution poured onto a rock-salt plate, and the solvent evaporated by gentle heating. If the solid is noncrystalline, a thin homogeneous film is deposited on the plate, which then can be mounted and scanned directly. Sometimes polymers can be "hot pressed" onto plates.

Mulls

Powders, or solids reduced to small particles, can be examined as a thin paste or mull. A small amount of the sample is thoroughly ground in a clean mortar until the powder is very fine. After grinding, the mulling agent is introduced in a small quantity just sufficient to take up the powder. The resulting mixture approximates the consistency of toothpaste. The mixture is then transferred to the mull plates, and the plates are squeezed together to adjust the thickness of the sample. Sample thickness is adjusted so that the strongest bands display 60% to 80% absorption.

Multiple reflections and refractions off the particles are reduced by grinding the particles to a size an order of magnitude less than the analytical wavelength and by surrounding the particles with a medium whose index of refraction is a close match. Liquid media include mineral oil or Nujol, hexachlorobutadiene, perfluorokerosene, and chlorofluorocarbon greases (fluorolubes). The latter are used when the absorption by the mineral oil masks the presence of C—H bands. For qualitative analysis the mull technique is rapid and convenient, but quantitative data are difficult to obtain even when an internal standard is incorporated into the mull. Polymorphic changes, degradation, and other changes may occur during grinding.

Pellet Technique

The pellet technique involves mixing the finely ground sample (1–100 μg) with potassium bromide powder and pressing the mixture in an evacuable die at sufficient pressure (60,000–100,000 psi) to produce a transparent disk. Potassium bromide becomes quite plastic at high pressures and flows to form a clear disk. Grinding and mixing are conveniently done in a vibrating ball mill. Other alkali halides are also used, particularly CsI or CsBr for measurements at longer wavelengths. Good dispersion of the sample in the matrix is critical. There must be no moisture. Freeze-drying the sample is often a necessary preliminary step.

KBr wafers can be formed, without evacuation, in a Mini-Press. Two highly polished bolts, turned against each other in a rugged stainless steel cylinder, produce a clear wafer from the KBr placed between the bolts. Pressure is applied with wrenches for about 1 min to 75–100 mg of powder. Some substances may decompose or change form under the heat and pressure of forming a pellet.

By the pellet technique and using an internal standard, quantitative analyses are readily performed, since an accurate measurement can be made of the ratio of weight of sample to internal standard in each disk or wafer.

Cell Thickness

One of two methods is used to measure the path length of infrared absorption cells: the *interference fringe* method or the *standard absorber* method. The interference fringe method is ideally suited to cells whose windows have a high polish. With the empty cell in the spectrometer on the sample side and no cell in the reference beam, the spectrometer is operated as near as possible to the 100% line. Enough spectrum is run to produce 20 to 50 fringes. The cell thickness, b, in centimeters, is calculated from the equation

$$b = \frac{1}{2\eta}\left(\frac{n}{\bar{\nu}_1 - \bar{\nu}_2}\right) \tag{11.4}$$

where n is the number of fringes (peaks or troughs) between two wavenumbers, $\bar{\nu}_1$ and $\bar{\nu}_2$, and η is the refractive index of the sample material. If measurements are made in wavelength (micrometers), the equation is

$$b = \frac{1}{2\eta}\left(\frac{n\lambda_1\lambda_2}{\lambda_2 - \lambda_1}\right) \tag{11.5}$$

where λ_1 is the starting wavelength and λ_2 is the final wavelength between which the fringes are counted. The fringe method also works well for measuring film thickness.

The standard absorber method may be used with a cell in any condition and with cavity cells (minicells) whose inner faces do not have a finished polish. The 1960-cm^{-1} (5.10-μm) band of benzene may be used for calibrating cells that are shorter than 0.1 mm in path length, and the 845-cm^{-1} (11.83-μm) band for cells

0.1 mm or longer in path length. At the former frequency, benzene has an absorbance of 0.10 for every 0.01 mm of thickness; at 845 cm^{-1}, benzene has an absorbance of 0.24 for every 0.1 mm of thickness.

Internal Reflectance

The scope and versatility of infrared spectrometry as a qualitative analytical tool have been increased substantially by the technique of internal reflectance (also known as attenuated total reflectance, ATR). When a beam of radiation enters a plate (or prism) surrounded by or immersed in a sample, it is reflected internally if the angle of incidence at the interface between sample and plate is greater than the critical angle (which is a function of refractive index). On internal reflection, all the energy is reflected. However, the beam appears to penetrate slightly beyond the reflecting surface and then return. The depth of penetration is approximately one wavelength and is given by

$$d_p = \frac{\lambda_1}{2\pi(\sin^2\theta - \eta_{sp}^2)^{1/2}} \qquad (11.6)$$

where d_p is the depth of penetration in wavelength units, λ_1 is the wavelength of the radiation in the prism (i.e., $\lambda_1 = \lambda/\eta_p$, where η_p is the refractive index of the prism material), $\eta_{sp} = \eta_s/\eta_p$, where η_s is the index of refraction of the sample, and θ is the angle of incidence of the beam. The angle, θ, must be greater then the critical angle, θ_c, for total reflection to occur and should actually be about 0.2 degrees larger than the calculated critical angle because the index of refraction of a substance undergoes a marked change in the vicinity of an absorption band. The critical angle, θ_c, is given by

$$\theta_c = \sin^{-1}\left(\frac{\eta_2}{\eta_1}\right) \qquad (11.7)$$

where η_2 and η_1 are the indices of refraction of the sample and prism materials, respectively.

When a material is placed in contact with the reflecting surface, the beam loses energy at those wavelengths where the material absorbs due to an interaction with the penetrating beam. This attenuated radiation, when measured and plotted as a function of wavelength, is an absorption spectrum characteristic of the material and is similar to an infrared spectrum obtained in the normal transmission mode. Internal reflection both simplifies infrared sampling and opens new areas of practical investigation. Using the technique, qualitative infrared absorption spectra are easily obtained from most solid materials without the need for grinding or dissolving or making a mull.

Most internal reflectance work is done by means of an accessory that is readily inserted in, and removed from, the sampling space of a conventional infrared spectrometer (Figure 11.11a). The accessory consists of a mirror system that sends the source radiation through the attachment and a second mirror system that directs the radiation into the monochromator. The mirrors are front-surfaced aluminum and accommodate the full width of the instrument beam. Sample holders are mounted in any of three positions to change the external angle. The three standard positions are 30°, 45°, and 60°. The length-to-thickness ratio of the plate determines the number of reflections after the angle of incidence is selected. Plate dimensions vary from 0.25 to 5 mm in thickness and 1 to 10 cm in length. Twenty-five internal reflections are standard for a 2-mm plate. Parallelism and flatness of sampling surfaces and surface polish are critical. Where the best performance is required, it is desirable to have matching optical units in both beams of the monochromator. In this way a much flatter P_0 curve is obtained, since absorptions such as those in the reflector plates and from Teflon "0" rings in liquid cells cancel and there is minimal interference from atmospheric absorption.

In the single-pass plate, radiation is introduced through an entrance aperture that consists of a simple bevel at one end of the plate and, after propagation via multiple internal reflections down the length of the plate, leaves through an exit aperture either parallel or perpendicular to the entrance aperture. The angle of the bevel determines the interior angle of incidence. This type of plate is useful for bulk materials, thin films, and surface studies. In the double-pass plate, radiation enters as before, propagates down the length of the plate, is totally reflected at the opposite end by a surface perpendicular to the plate length, and returns to leave the plate via the exit aperture at the same end as the entrance aperture. The free end of the plate can be dipped into liquids or powders or placed in closed systems.

The apparent depth to which the radiation penetrates the sample is only a few micrometers and is independent of sample thickness. Consequently, ATR spectra can be obtained for many samples that cannot be studied by normal transmission methods. These include samples that show very strong absorptions, resist preparation as thin films, are characteristic only as thick layers, or are available only on a nontransparent support. Aqueous solutions are handled without compensating for very strong solvent absorptions. Samples that contain suspended matter, such as dispersed solids or emulsions that produce high backgrounds in transmission

FIGURE 11.11 Internal reflection attachment: (a) three-position pin plate for 30°, 45°, and 60° plates and (b) variable-angle model. (Courtesy of Foxboro/Wilks, Inc.)

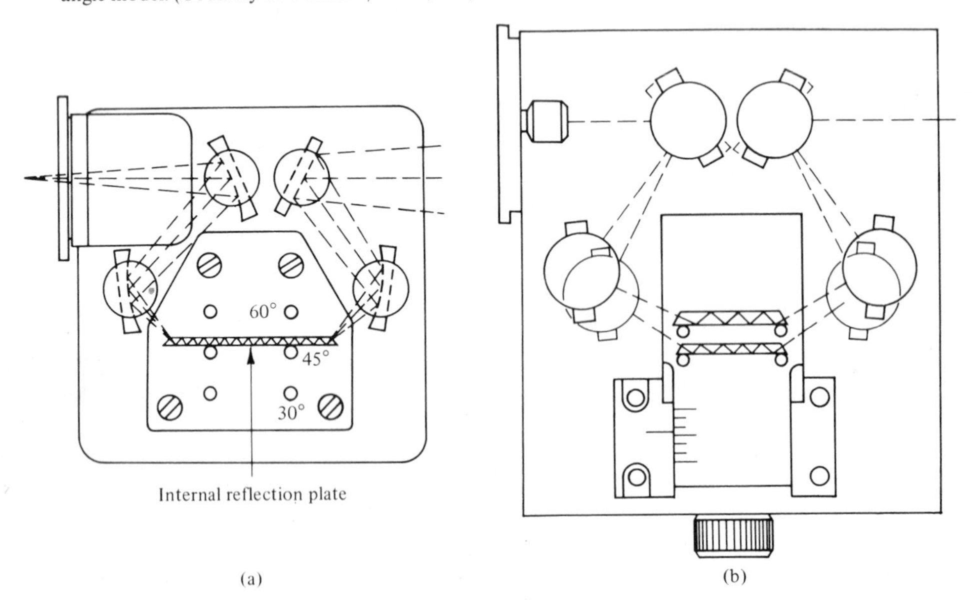

spectra, give better results by multiple internal reflectance. Various sample holders are available. Some samples self-adhere to the reflector plate. In other cases, some pressure is required to bring the sample in contact with the plate. Liquids are analyzed using a germanium reflector plate that has both the required insolubility and the short effective path length. The heated sample holder is a modified solid sample holder equipped with two wafer heaters and operates to 250 °C in order to study thermal effects *in situ* or as a means of heat-softening various plastic samples to improve optical contact.

A CIRCLE cell, a long cylinder with tapered conical ends, is also available. Such a cell is well suited to FTIR instruments because of the match between the circular beam of the FTIR and the shape of the cell. The CIRCLE cell is especially useful for low-volume, liquid flow-through cells.

The appearance and intensity of a reflectance spectrum depend on the difference of the indices of refraction between the reflector plate (Table 11.2) and the rarer medium containing the absorber, and the internal angle of incidence. Thus a reflector plate with a relatively high index of refraction should be used. The material that performs most satisfactorily for most liquid and solid samples is KRS-5, thallium(I) bromide iodide. Its refractive index is high enough to permit well-defined spectra of nearly all organic materials, although it is soluble in basic solutions. AgCl is recommended for aqueous samples because of its insolubility and lower refractive index; germanium could also be used but it is brittle. Zinc selenide (Irtran IV) has been used but it too is brittle and it releases H_2Se in acid solutions. By varying the angle of incident radiation from 30° to 60°, the depth of penetration of energy into the sample may be changed. A commercial unit is shown in Figure 11.11b; it has a scissor-jack assembly linking the four mirrors and sample platforms in a pantograph system. At steep angles (near 30°) the depth of penetration into the sample is considerably greater (perhaps an order of magnitude greater) than at grazing angles near 60°. Being able to vary the angle of incidence and thus the depth of penetration of the radiation is of considerable significance in the study of surfaces. Much can be learned about the surfaces of a film or plastic in which chemical additives are suspected of migrating to the surface, or where a surface has been exposed to synthetic weathering.

Photoacoustic Spectroscopy

Photoacoustic spectroscopy was discussed in Chapter 7 for ultraviolet-visible measurements. Photoacoustic methods are also applicable in the infrared region of the spectrum. The sampling depth for photoacoustical spectrometry depends on two factors: the transparency of the material to the radiation and the thermal diffusivity (thermal conductivity divided by the product of sample density times specific heat). A sample can be optically transparent or opaque and thermally thick or thin. Thus it is difficult to compare the sampling depth of photoacoustic spectrometry with attenuated total reflectance. However, for many polymers ATR samples to a depth of approximately one wavelength and photoacoustic spectroscopy in the infrared region (at the same wavelength) samples to a depth of 10–30 μm.[14]

Microscopes

Special microscopes are available with parabolic front-surfaced mirrors as the beam condenser to focus the infrared radiation on a very small area and to collect the transmitted or reflected radiation and with auxiliary visual optical elements to view the area irradiated. These infrared microscopes are very useful in investigating very small samples such as fibers and contaminants in heterogeneous mixtures. Microscope attachments small enough to fit into the normal sample compartment of FTIR instruments are available. The mercury cadmium telluride (MCT) detector cooled to liquid nitrogen temperature is recommended as the detector in the FTIR instrument, since its size can be reduced to match the image of the illuminated sample. Reducing the size of the MCT detector reduces the noise correspondingly so that the signal-to-noise ratio of the spectrometer is maintained constant. The microscope attachment can also be used with macro samples by focusing on a thin edge of a sample or on a portion of a wedge-shaped section. A high-pressure diamond cell can be used to prepare thin samples of some materials. Picogram quantities of substances can be identified by the microscopic technique and femtogram quantities can be detected.[15]

11.4 QUANTITATIVE ANALYSIS

The application of infrared spectroscopy as a quantitative tool varies widely from one laboratory to another. However, the use of high-resolution grating instruments materially increases the scope and reliability of quantitative infrared work. With the advent of Fourier transform infrared spectrometers with their increased signal-to-noise ratio, quantitative work is even more precise. Quantitative infrared analysis is based on Beer's law. Apparent deviations arise from either chemical or instrumental effects. In many cases scattered radiation makes the direct application of Beer's law inaccurate, especially at high values of absorbance. Since the energy available in the useful portion of the infrared region is usually quite small, it is necessary to use rather wide slits in the monochromator of dispersive instruments to increase the signal-to-noise ratio. This causes a considerable change in the apparent molar absorptivity. Therefore, molar absorptivities should be determined empirically. Because the refractive index changes rapidly near an absorption band, dilute solutions, less than 2%, should be used. It is always advisable, and usually essential, to create a calibration curve from known standard solutions when determining concentrations of unknowns or to use an internal standard. Sometimes the spectrometer is set at the wavelength of maximum absorbance of the material to be determined and not changed throughout the measurement of knowns and unknowns. Sometimes the spectrometer is scanned in a narrow region near the wavelength of maximum absorption and the transmittance readings are taken from the maximum of the tracing.

The baseline method involves the selection of an absorption band of the substance under analysis that does not fall too close to the bands of other matrix components (Figure 11.12). The value of the incident energy, P_0, is obtained by drawing a straight line tangent to the spectral absorption curve at the position of the sample's absorption band and then measuring the distance from this tangent to the 0% T line. The transmitted power, P, is measured at the point of maximum absorption of the analyte. The value of the absorbance, $\log(P_0/P)$, is then plotted against concentration for a series of standard solutions, and the unknown concentration is determined from this calibration curve.

Many possible errors are eliminated by the baseline method. The same cell is used for all determinations. All measurements are made at points on the spectrum that are defined by the spectrum itself; thus, there is no dependence on wavelength settings. The use of such ratios eliminates changes in instrument sensitivity, source intensity, and adjustment of the optical system.

Pellets from the disk technique can be used in quantitative measurements. Uniform pellets of similar weight are essential, however, for quantitative analysis. Known weights of KBr are taken plus known quantities of the test substance, and from the measurements of absorbance on these knowns a calibration curve is constructed. The disks are weighed and their thickness measured at several points on the surface with a dial micrometer. The disadvantage of measuring pellet thickness is overcome by using the internal standard method. Potassium thiocyanate makes an excellent internal standard. It should be preground, dried, and then reground with dry KBr to make a concentration of about 0.2% by weight of thiocyanate. The final mix is stored over phosphorus pentoxide. A standard calibration curve is made by mixing known weights of the test substance with a known weight of the KBr-KCNS mixture and then grinding. The test substance is usually around 10% of the total weight. The ratio of the thiocyanate absorption at 2125 cm^{-1} (4.70 μm) to a chosen absorption band of the test substance is plotted against the concentration of the test substance.

The difference method discussed in Section 7.6 is useful in infrared quantitative methods when greater accuracy than obtainable by ordinary quantitative methods is desired. A series of standards is prepared with concentrations extending from just above to just below the concentration of the unknown. The unknown is placed in the sample beam, the standards, one after another, are placed in the reference beam (using the same cell), and the differences are plotted against concentration. The concentration of the unknown is that concentration on the curve where the difference is zero.

Multicomponent mixtures are quantitatively determined by infrared spectroscopy just as in ultraviolet-visible spectroscopy (see Section 7.4). In case Beer's law appears not to be followed exactly, as is often the case in infrared methods where peaks may overlap significantly and baselines cannot be accurately determined, a series of standard mixtures is run containing known amounts of the substances to be

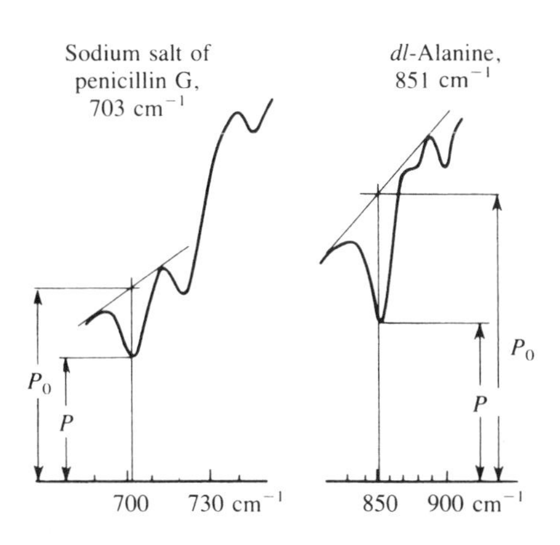

FIGURE **11.12**
Baseline method for calculation of the transmittance ratio in quantitative analysis.

determined.[16,17] A series of proportionality constants between concentration and absorbance is obtained. These proportionality constants are not necessarily the same as εb in Beer's law. The resulting equations are solved by matrix algebra methods using a computer. There must be at least as many known mixtures as there are proportionality constants, normally n, where n is the number of components to be determined. None of the mixtures should be algebraically equivalent. There must also be measurements at as many wavelengths as there are unknowns to be determined, and these wavelengths should be chosen to give as little overlapping of peaks as possible. It is also possible to overdetermine a system by measuring absorbances at more wavelengths than there are unknowns and to obtain the best solution by combining least-squares and matrix methods.

For quantitative measurements, the single-beam system has some fundamental characteristics that can result in greater sensitivity and better accuracy than double-beam systems. All other things being equal, a single-beam instrument has a greater signal-to-noise ratio. There is a factor-of-two advantage in looking at one beam all the time rather than two beams, each half the time. The S/N ratio increases only as $\sqrt{2}$, however. Thus, in any analytical situation where background noise is appreciable, the single-beam spectrometer or, of course, a Fourier transform infrared spectrometer is superior.

PROBLEMS

1. What would be the frequency of the fundamental absorption if its first overtone was observed at 1820 cm^{-1}?

2. The molecular heterotope $^{35}Cl^{37}Cl$ has a fundamental band at 554 cm^{-1} in the gaseous state. Where would one expect the first and second overtones? What window material would be suitable?

3. Assuming a simple diatomic molecule, obtain the frequencies of the absorption band from the force constants given here. Refer to Chapter 5 for the theory. Compare your answers with the tabulated positions in Figure 11.2.

a. $k = 5.1 \times 10^5$ dyne cm^{-1} for C—H bond in ethane
b. $k = 5.9 \times 10^5$ dyne cm^{-1} for C—H bond in acetylene
c. $k = 4.5 \times 10^5$ dyne cm^{-1} for C—C bond in ethane
d. $k = 7.6 \times 10^5$ dyne cm^{-1} for C—C bond in benzene
e. $k = 17.5 \times 10^5$ dyne cm^{-1} for C≡N bond in CH_3CN
f. $k = 12.3 \times 10^5$ dyne cm^{-1} for C═O bond in formaldehyde

4. The apparent specific absorptivities are given for various infrared absorbers. Calculate the minimum liquid concentrations determinable (mg/mL) in 0.025-mm cells (for an absorbance reading of 0.0043).

a. $\alpha = 900$ for $CHCl_3$ at 1215 cm^{-1}
b. $\alpha = 1320$ for CH_2Cl_2 at 1259 cm^{-1}
c. $\alpha = 4900$ for C_6H_6 at 1348 cm^{-1}
d. $\alpha = 6080$ for $COCl_2$ at 1810 cm^{-1}
e. $\alpha = 4400$ for $CH_2ClCOCl$ at 1821 cm^{-1}
f. $\alpha = 1010$ for water at 1640 cm^{-1}

5. The presence of ethylene in samples of ethane is determined by using the absorption band of ethylene at 1443 cm^{-1} (6.93 μm). A series of standards gave the following data:

Percent ethylene	0.50	1.00	2.00	3.00
Absorbance	0.120	0.240	0.480	0.719

Calculate the percentage of ethylene in an unknown sample that had an absorbance of 0.412 when the same cell and instrument were used.

6. Estimate the minimum concentration detectable ($A = 0.0043$) in 0.050-mm cells for each of the following compounds, given their molar absorptivities: (a) phenol at 3600 cm^{-1}, $\varepsilon = 5000$; (b) aniline at 3480 cm^{-1}, $\varepsilon = 2000$; (c) acrylonitrile at 2250 cm^{-1}, $\varepsilon = 590$; (d) acetone at 1720 cm^{-1}, $\varepsilon = 8100$; and (e) isocyanate (in polyurethane foam) monomer at 2100 cm^{-1}, $\varepsilon = 17{,}000$.

7. (a) Calculate the thickness of the four cells from their interference fringe patterns shown in the figure.

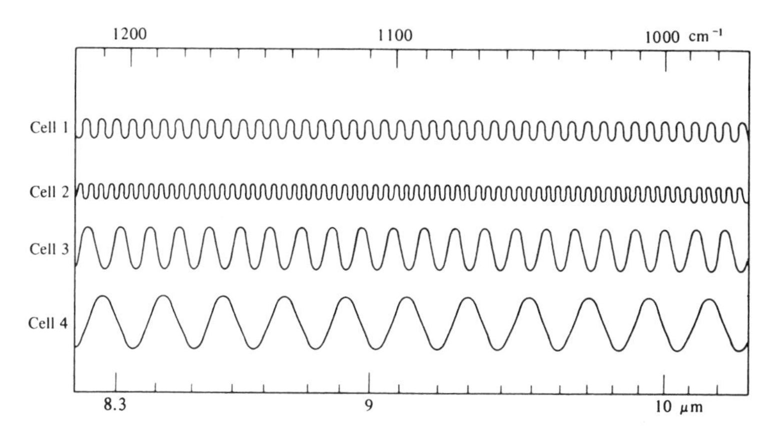

(b) Calculate the path length of the infrared absorption cell from the transmittance curve for benzene shown.

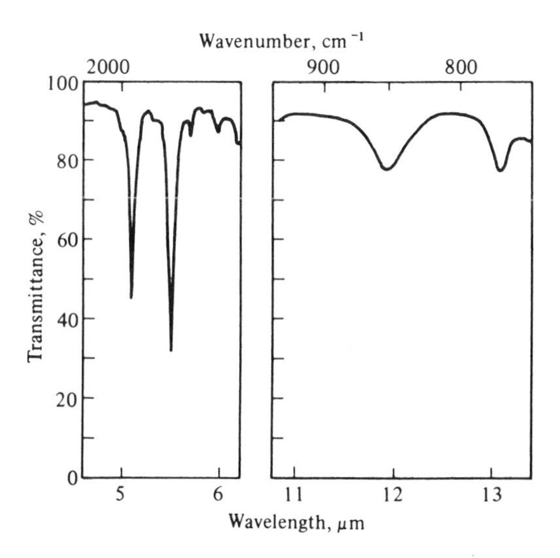

8. Identify the particular xylene from the following infrared data:

Compound A: absorption bands at 767 and 692 cm^{-1} (13.04 and 14.45 μm)
Compound B: absorption band at 792 cm^{-1} (12.63 μm)
Compound C: absorption band at 742 cm^{-1} (13.48 μm)

9. A bromotoluene, C_7H_7Br, has a single band at 801 cm^{-1} (12.48 μm). What is the correct structure?

10. Deduce the structure of the compound with the molecular formula $C_6H_{12}O$ whose infrared spectrum is shown in the figure.

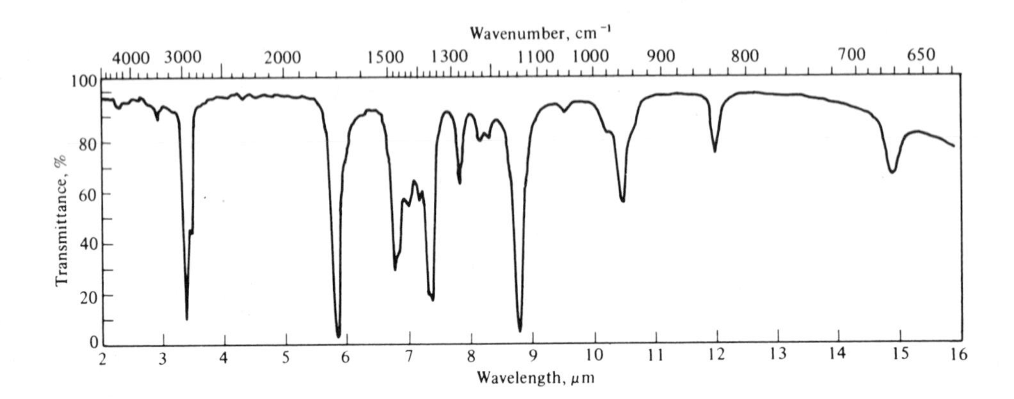

11. The upper curve in the figure is part of the spectrum of a 2.000% solution of compound X in CCl_4. The lower curve is a solution of unknown concentration of the same compound in CCl_4. Determine the concentration of compound X in the unknown solution.

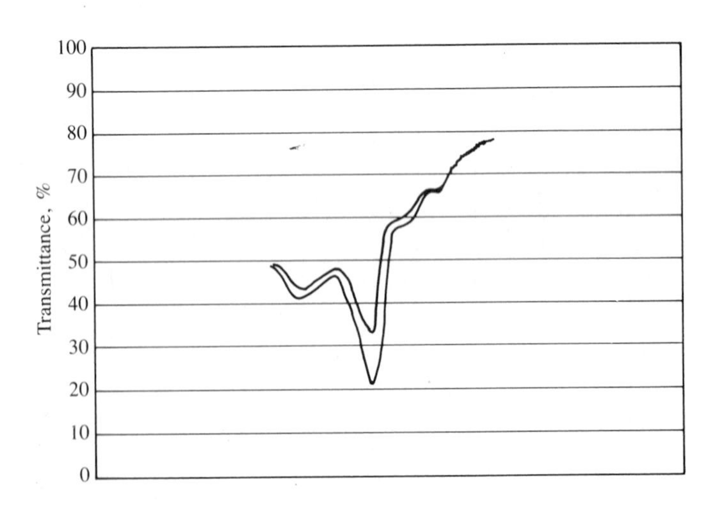

12. The concentration of mixtures of three isomeric compounds, 1, 2, and 3, is to be determined. Amounts of the pure isomers are dissolved in a solvent that does not absorb in the infrared region where the compounds absorb. The pertinent data for calibration are shown in the table. A 2.00% solution of the unknown mixture gave an absorbance of 0.270 at λ_1, 0.120 at λ_2, and 0.732 at λ_3. Calculate the concentration of each isomer in the pure mixture.

	Percent in standard	$A_{\lambda 1}$	$A_{\lambda 2}$	$A_{\lambda 3}$
Compound 1	1.50	0.426	0.051	0.008
Compound 2	2.50	0.040	0.280	0.060
Compound 3	2.00	0.002	0.008	0.304

13. The data in the table were obtained for mixtures of two compounds, X and Y, dissolved in a nonabsorbing solvent. A solution of an unknown mixture of compounds X and Y had absorbances of 0.184 at 1770 cm^{-1} and 0.256 at 1733 cm^{-1}. Calculate the percentage of each compound in the solution.

Frequency (cm^{-1})	**Percent X**	**Percent Y**	A
1770	1.000	1.000	0.212
1770	2.000	1.000	0.341
1733	2.000	3.000	0.598
1733	1.000	3.000	0.541

BIBLIOGRAPHY

BAUMAN, R. P., *Absorption Spectroscopy,* John Wiley, New York, 1962.

BELL, R. J., *Introductory Fourier Transform Spectroscopy,* Academic, New York, 1972.

BRAME, E. G., AND J. G. GRASSELLI, eds., *Infrared and Raman Spectroscopy,* Vol. 1, Parts A, B, and C, *Practical Spectroscopy Series,* Dekker, New York, 1977.

COATES, J. P., J. M. D'AGOSTINO, AND C. R. FRIEDMAN, "Quality Control by Infrared Spectroscopy," *Am. Lab.,* **18**(11), 82 (1986), **18**(12), 40 (1986).

COLTHUP, N. B., L. H. DALY, AND S. E. WIBERLEY, *Introduction to Infrared and Raman Spectroscopy,* 2nd ed., Academic, New York, 1975.

FERRARO, J. R., AND L. J. BASILE, eds., *Fourier Transform Infrared Spectroscopy,* Academic, New York, 1982.

GRIFFITHS, P, R., AND J. A. DE HASETH, *Chemical Infrared Fourier Transform Spectroscopy,* John Wiley, New York, 1986.

HARRICK, N. J., *Internal Reflection Spectroscopy,* John Wiley, New York, 1967.

NAKAMOTO, K., *Infrared and Raman Spectra of Inorganic and Coordination Compounds,* 4th ed., John Wiley, New York, 1986.

REIN, A. J., AND S. K. MORRIS, "Design of an IBM AT-Based FTIR Spectrometer," *Am. Lab.,* **18**(9), 86 (1986).

SILVERSTEIN, R. M., G. C. BASSLER, AND T. C. MORRILL, *Spectrometric Identification of Organic Compounds,* 4th ed., John Wiley, New York, 1981.

SMITH, A. L., "Infrared Spectroscopy," Chap. 5, *Treatise on Analytical Chemistry,* 2nd ed., P. J. Elving, E. J. Meehan, and I. M. Kolthoff, eds., Part I, Vol. 7, John Wiley, New York, 1981.

WILLIAMS, D. H., AND I. FLEMING, *Spectroscopic Methods in Organic Chemistry,* 2nd ed., McGraw-Hill, New York, 1973.

LITERATURE CITED

1. BORMAN, S. A., *Anal. Chem.,* **56,** 933A (1984).
2. HONIGS, D. E., G. M. HIEFTJE, H. L. MARK, AND T. HIRSCHFELD, *Anal. Chem.,* **57,** 2299 (1985).
3. HONIGS, D. E., G. M. HIEFTJE, AND T. HIRSCHFELD, *Appl. Spectrosc.,* **38,** 844 (1984).
4. STEWART, J. E., in *Interpretive Spectroscopy,* S. K. Freeman, ed., pp. 131–169, Van Nostrand Reinhold, New York, 1965.
5. LOW, M. J. D., *Anal. Chem.,* **41,** 97A (1969); *J. Chem. Educ.,* **47,** A163, A255, A348, A415, (1970).
6. POUCHERT, C. J., ed., *The Aldrich Library of Infrared Spectra,* 2nd ed., Aldrich Chemical Co., Milwaukee, 1975.
7. Sadtler Research Laboratories, *Catalog of Infrared Spectrograms,* Philadelphia (a continuously updated subscription service).
8. DEAN, J. A. ed., *Lange's Handbook of Chemistry,* 13th ed., McGraw-Hill, New York, 1985.
9. DEAN, J. A., ed., *Handbook of Organic Chemistry,* McGraw-Hill, New York, 1986.
10. EWING, G. W., *J. Chem. Educ.,* **48,** A521 (1971).
11. SMITH, A. L., *Infrared Spectroscopy,* pp. 285–292, *Treatise on Analytical Chemistry,* 2nd ed., P. J. Elving, E. J. Meehan, and I. M. Kolthoff, eds., Part I, Vol. 7, John Wiley, New York, 1981.
12. PERKINS W. D., *J. Chem. Ed.,* **63,** A5 (1986).
13. COOLEY, J. W., AND J. W. TUKEY, *Math. Comput.,* **19,** 297 (1962).
14. SAUCEY, D. A., S. J. SIMKO, AND R. W. LINTON, *Anal. Chem.,* **57,** 871 (1985).
15. KATON, J. E., G. E. PACEY, AND J. F. O'KEEFE, *Anal. Chem.,* **58,** 465A (1986).
16. KACOYANNAKIS, J. F., J. K. ALLEN, AND M. L. PARSONS, *Industrial Chemical News,* p. 32 (February 1983).
17. BROWN, C. W., *Anal. Chem.,* **54,** 1472 (1982).

12 RAMAN SPECTROSCOPY

When monochromatic radiation is scattered by molecules, a small fraction of the scattered radiation is observed to have a different frequency from that of the incident radiation; this is known as the *Raman effect*. Since its discovery in 1928, the Raman effect has been an important method for the elucidation of molecular structure, for locating various functional groups or chemical bonds in molecules, and for the quantitative analysis of complex mixtures, particularly major components. Although vibrational Raman spectra are related to infrared absorption spectra, a Raman spectrum arises in a different manner and thus often provides complementary information. Vibrations that are active in Raman scattering may be inactive in the infrared, and vice versa. A unique feature of Raman scattering is that each line has a characteristic polarization, and polarization data provide additional information about molecular structure.

12.1 THEORY

The Raman effect arises when a beam of intense monochromatic radiation passes through a sample that contains molecules that undergo a change in molecular polarizability as they vibrate. Recall that in order for a vibrational mode to be active in the infrared region, the vibration must cause a change in the permanent dipole moment of the molecule. The dipole moment is the product of the charge of the dipole and the charge separation distance. On the other hand, for a vibration to be active in the Raman effect, the *polarizability* of the molecule must change during the vibration. Polarizability is the value of the induced dipole moment divided by the strength of the field that causes the induced dipole moment. In other words, the electron cloud of the molecule must be more readily deformed in one extreme of the vibration than in the other extreme. It is strictly a quantum effect.

Classical electromagnetic theory predicts the Raman effect, although a quantum mechanical treatment is needed for a detailed explanation. According to the classical theory, the polarization, P, expressed as dipole moment per unit volume, is given by

$$P = \alpha E \tag{12.1}$$

where E is the magnitude of the electric vector of the electromagnetic field that acts on the molecule and α, the polarizability, is the proportionality constant. Since the magnitude of the electric vector of the electromagnetic field varies with time, t, in a sinusoidal manner,

$$E = E_0 \cos 2\pi\nu t \tag{12.2}$$

the polarization becomes

$$P = \alpha E_0 \cos 2\pi\nu t \tag{12.3}$$

The polarizability, α, consists of two parts: α_0, the polarizability when the atoms of a molecule are in their equilibrium positions, and a second term that is the sum of the polarizabilities of the molecule due to the various rotational and vibrational motions. Each term of this second part varies with the frequency associated with the particular rotation or vibration. Thus,

$$\alpha = \alpha_0 + \Sigma\left(\frac{\delta\alpha_n}{\delta r}\right) r_n \cos 2\pi\nu_n t \tag{12.4}$$

where α_n is the polarizability associated with the nth rotational or vibrational mode and r_n is the maximum displacement of the involved atoms.

Combining Equations 12.3 and 12.4 gives

$$P = E_0\alpha_0 \cos 2\pi\nu t + E_0\Sigma\left(\frac{\delta\alpha_n}{\delta r}\right) r_n \cos 2\pi\nu_n t \cos 2\pi\nu t \tag{12.5}$$

and

$$P = E_0\alpha_0 \cos 2\pi\nu t + \frac{1}{2} E_0\Sigma\left(\frac{\delta\alpha_n}{\delta r}\right) r_n \{\cos 2\pi(\nu - \nu_n)t + \cos 2\pi(\nu + \nu_n)t\} \tag{12.6}$$

The first term has the frequency of the incident radiation and is the Rayleigh scattering. The second term represents the Stokes $(\nu - \nu_n)$ and anti-Stokes $(\nu + \nu_n)$ Raman bands. Note that $\delta\alpha_n/\delta r$ must not be zero.

Most collisions of the incident photons with the sample molecules are elastic (Rayleigh scattering). According to the first term in Equation 12.6, the electric field produced by the polarized molecule oscillates at the same frequency as the passing electromagnetic wave so that the molecule acts as a source sending out radiation of that frequency in all directions. As shown in Figure 5.9, the incident radiation does not raise the molecule to any particular quantized level; rather, the molecule is considered to be in a virtual excited state. As the electromagnetic wave passes, the polarized molecule ceases to oscillate and returns to its original ground level in a very short time (approximately 10^{-12} sec).

A small proportion of the excited molecules (10^{-6} or less) may undergo changes in polarizability during one or more of the normal vibrational modes. According to the second term in Equation 12.6, this is the basis for the Raman effect. Usually incident radiation, ν_0, is absorbed by a molecule in the lowest vibrational state. If the molecule reemits by returning not to the original vibrational state, but to an excited vibrational level, ν_v, of the ground electronic state, the emitted radiation is of lower energy—that is, lower frequency $(\nu_0 - \nu_v)$—than the incident radiation. The difference in frequency is equal to a natural vibration frequency of the molecule's ground electronic state. Several such shifted lines (the *Stokes* lines) normally are observed in the Raman spectrum, corresponding to different vibrations in the molecule. This provides a richly detailed vibrational spectrum of a molecule (Figure 5.10).

A few of the molecules initially absorb radiation while they are in an excited vibrational state and decay to a lower energy level, so that the Raman frequency is higher than the incident radiation. These are the *anti-Stokes* lines. Thus the spectrum of the scattered radiation consists of a relatively strong component with frequency unshifted (Rayleigh scattering) corresponding to photons scattered without energy exchange, and the two components of the Raman spectrum: the Stokes lines and the anti-Stokes lines. Normally only the Stokes lines are considered in chemical analysis. These are more intense because, under usual circumstances, most molecules are initially in the lowest vibrational level.

In the usual Raman method the excitation frequency of Raman sources is selected to lie below most S–S^* electronic transitions and above most fundamental vibrational frequencies, but this need not be the case, as the next section will show.

Resonance Raman Spectroscopy

Thus far we have explored what might be called the ordinary Raman effect—that is, Raman scattering induced by excitation far removed from any electronic transitions. If, however, the excitation frequency falls near or becomes coincident with an electronic absorption, ν_{abs}, a very significant enhancement may occur such that

$$\text{intensity} \propto (\nu_{exc} - \nu_v)^4 \left(\frac{\nu_{abs}^2 + \nu_{exc}^2}{\nu_{abs}^2 - \nu_{exc}^2} \right)^2 \tag{12.7}$$

The advantage of the *resonance* Raman effect lies in its greater sensitivity and selectivity as a probe of chromophore structure. The enhanced Raman lines have intensities 10^2–10^6 times greater than normal Raman intensities. Consequently, resonance Raman spectra have low detection limits (10^{-6}–$10^{-8}\,M$) and are much simpler than normal Raman spectra, since only vibrational modes associated with the "chromophore" are enhanced.

The resonance Raman effect results from the promotion of an electron into an excited electronic-vibrational state, accompanied by immediate relaxation into a vibrational level of the ground state. The process is not preceded by prior relaxation to the lowest vibrational level of the excited state as in ordinary fluorescence. The distinction is shown in Figure 12.1. Consequently, the resonance Raman emission process is essentially instantaneous and the resulting spectra consist of narrow bands. Resolution is good, 10–20 cm^{-1} (0.3–0.6 nm at 500 nm). For molecules in solution, electronic states are broadened by many closely spaced vibrational states. Excitation with radiation anywhere within this continuum gives rise to the same resonance Raman spectrum with an intensity proportional to the absorption intensity.

Spectra may also be obtained using an excitation frequency just below the absorption band. Such spectra display smaller resonance enhancement, typically less than tenfold. These spectra are called preresonance Raman spectra.

Not all the normal Raman bands are equally enhanced. Only those vibrations that exhibit a large change in equilibrium geometry upon electronic excitation produce strongly resonance-enhanced Raman bands. In practical terms this means that two classes of vibrational modes produce intense resonance-enhanced spectra: totally symmetrical vibrations and those nontotally symmetrical vibrations that vibronically couple two electronic states. Since an electronic transition is often more or less localized in one part of a complex molecule, the resonance Raman effect provides highly detailed information about the vibrational modes of chromophores that have an absorption band near the wavelength of the incident radiation. This selectivity, for example, is quite apparent in the heme proteins where resonance Raman bands are due solely to vibrational modes of the tetrapyrrole chromophore.

Coherent Anti-Stokes Raman Spectroscopy (CARS)

Coherent anti-Stokes Raman spectroscopy (CARS) depends on the fact that when a molecule is polarized by an exciting field, the polarization contains terms that involve the square, cube, and so on of the field strength (nonlinear terms) as well as the first power (linear term). Because of the nonlinear terms, which become important only when the exciting field or fields are very strong as in laser excitation, the molecule can combine several photons to yield another photon of different but related frequency. In CARS the sample is irradiated with photons from a laser such as the Nd:YAG laser or a flash pumped dye laser with frequency ν_1, and simultaneously with photons from another tunable dye laser with frequency ν_2. The frequency of the tunable dye laser is varied. When the difference in the frequencies, $\nu_1 - \nu_2$, equals a Raman frequency, a new frequency, $\nu_3 = 2\nu_1 - \nu_2$, is generated. The new frequency, ν_3, is coherent with the generating frequencies—that is, laserlike—and is quite intense compared with ordinary Raman lines. It is detected in the forward direction and its intensity is plotted against the variable frequency, ν_2. The new frequency appears in the anti-Stokes region with respect to the main frequency, ν_1. Using this technique, the Raman frequencies of highly fluorescent dyes and biological materials have been measured. In the ordinary Raman method, the intense fluorescence would completely obscure the Raman lines from such samples.

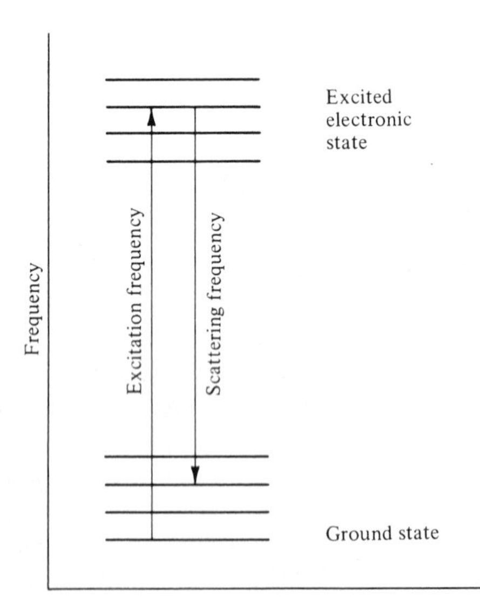

FIGURE **12.1**
Resonance Raman effect shown schematically.

Surface Enhanced Raman Spectroscopy (SERS)

Raman scattering is enhanced when the analyte is adsorbed on colloidal metallic surfaces. Silver, gold, and copper are the metals found to be most effective. For colloidal silver particles, the enhancement factors are typically 10^3–10^6 times the normal Raman intensities. Surface enhancement and resonance enhancement effects are roughly multiplicative, thus leading to large Raman signals and low detection limits, often in the range of 10^{-9}–$10^{-12}\,M$. Shony and colleagues have shown that colloidal silver is best prepared by reduction of Ag^+ with citrate and subsequent separation of the colloidal silver particles by sedimentation to obtain a fraction with particle sizes of 25–500 nm without bulk crystallites.[1]

The colloidal silver can be used in solution to adsorb the analyte or it can be used as a film on a glass microscope slide. The analyte is spotted on the prepared glass slide and the Raman effect is observed in the usual manner. The latter method is especially useful for analytes in water-immiscible solutions. Silver can also be coated on other substrates such as paper or quartz and on rough metallic surfaces with submicrometer protrusions. The basis of the enhancement effect is not clearly understood but is believed to be a combination of several electromagnetic and chemical effects.

Concentration is proportional to the power of the signal in surface enhanced Raman spectroscopy (SERS). For crystal violet the linear proportionality between concentration and signal power extends over a range of 10^{-7} to $10^{-10}\,M$. For 1-nitropyrene adsorbed on silver coated on a solid substrate, the proportionality holds over the range 10^{-4} to $10^{-6}\,M$.

12.2 INSTRUMENTATION

The primary function of a Raman spectrometer is clean rejection of the intense Rayleigh scattering and detection of the weak Raman-shifted components. A computer adds significantly to the power and versatility of the spectrometer. An intense monochromatic radiation source, sensitive detection, and high light-gathering power, coupled with freedom from extraneous stray radiation, must be built into a Raman spectrometer. A double holographic grating monochromator minimizes stray radiation from the unshifted laser excitation wavelength. Visual-type optics are used throughout, and the entire spectrum is covered by a single grating.

The laser Raman spectrophotometer, shown in Figure 12.2, consists of two basic units: the laser excitation unit and the $f/6.7$ spectrometer unit. The laser beam enters from the rear of the spectrometer into the depolarization autorecording unit and, after passing through this unit, it illuminates the sample. The Raman scattering, collected at 90° to the exciting laser beam, is focused on the entrance slit of a 0.5-m focal length, Czerny-Turner grating double monochromator. Immediately ahead of the spectrometer is a polarization scrambler, which overcomes grating bias caused by polarized radiation. A polarization analyzer is placed between the condenser lens and the polarization scrambler when the polarization of the Raman spectrum is measured. The spectrometer is arranged in a back-to-back configuration. The

Raman scattered radiation is dispersed using gratings with 1200 grooves/mm and finally passes through the monochromator exit slit and onto the photocathode of the photomultiplier tube. The photomultiplier tube is placed in a thermoelectric cooler (−30 °C), markedly lowering the dark current and reducing noise, thus providing a high sensitivity and favorable signal-to-noise ratio. In this particular instrument, the signals from the detector are amplified and counted with two photon-counting systems, one of which is for the detection of the ordinary Raman spectrum, while the other is for the calculation of the depolarization ratio (see Section 12.5). These signals are displayed graphically as a function of wavelength or frequency.

Photon counting has long been recognized as the most effective means of recovering low-level Raman signals. However, it suffers from limited range and high cost. On the other hand, inexpensive dc amplifiers are excellent for strong signals; they provide the required range but suffer from low signal-to-noise ratios for weak signals. One can combine the best of both these techniques by operating the amplifier as a photon counter with a built-in discriminator for low-level signals and, when the signal is strong enough, automatically switching in the dc system for the remaining levels of amplification.

The intermediate slit has a vertical mask that allows the height to be shortened when one is running a solid sample or using the transverse illumination technique for microsamples. Masking of the intermediate slit reduces the spectral background and enhances the signal-to-noise ratio.

FIGURE **12.2** Optical schematic of a laser Raman spectrophotometer. (Courtesy of Jeol, Ltd.)

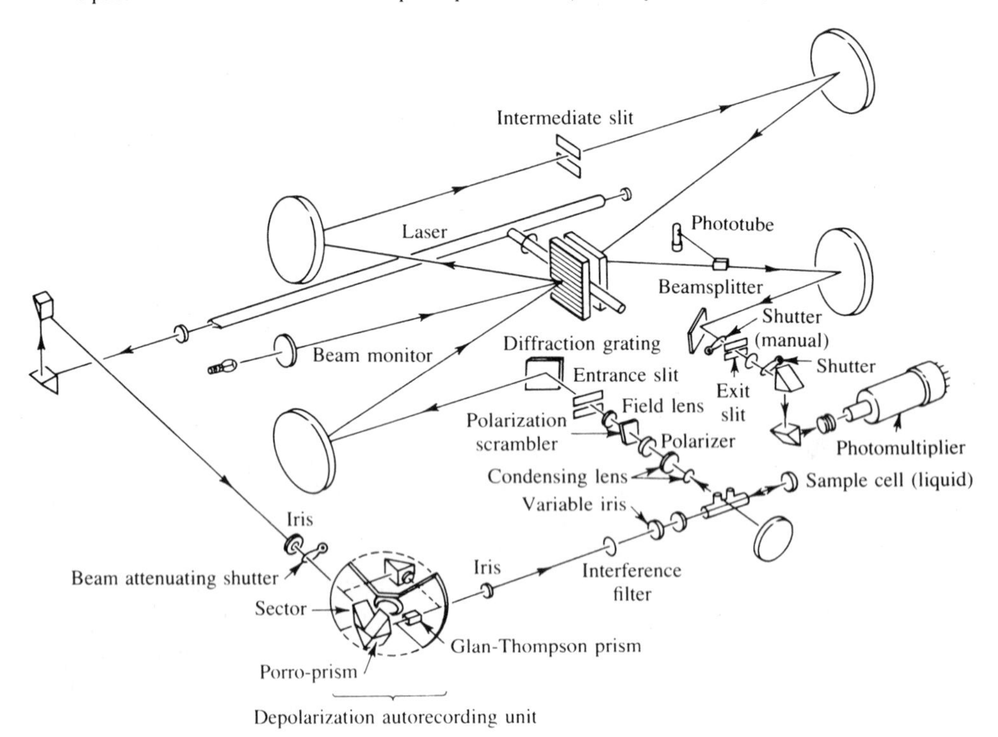

The operation of the Raman spectrometer, except for the selection of slit widths and heights, is controlled automatically. The slit width selector wheel is linked directly to the electronics, so after the scanning speed and slit width are determined, the time constant is automatically selected to exactly match the criterion: time constant = (spectral bandwidth)/(scanning speed × 4). A precise electronic master clock provides pulses to the stepping motor that operates the scanning system and to the stepping motor that operates the X-Y recorder. Through its solid-state divider circuits, the master clock maintains the correct ratio between these pulses so that scanning speeds can be varied independently from the range of the recorder.

The Jobin-Yvon instrument (Figure 12.3) combines the throughput of a double monochromator system with the rejection levels of a triple monochromator system. The *f*/8 monochromator is designed to reduce stray radiation by using two concave, aberration-corrected, holographic gratings with 2000 grooves/mm. Both gratings are on the same shaft; thus, the tracking error is zero. The scattered radiation from the sample is gathered by a high-aperture objective lens, which focuses the radiation onto the straight horizontal entrance slit, slit 1, of the first monochromator. The concave grating, G1, diffracts the beam and focuses the selected wavelength onto the exit slit, slit 2. The nearly monochromatic radiation that exits from the first monochromator through slit 2 is now focused onto the entrance slit, slit 3, of the second monochromator by mirrors, m_1 through m_5. These mirrors are not in the monochromator cavities and work in monochromatic radiation; therefore, they do not contribute to scattered stray radiation. Radiation that enters slit 3 is then diffracted and focused onto the exit slit, slit 4, of the second monochromator by the concave grating, G2. All four slits are individually controlled by the stepping motor. This permits complete versatility with the computer in order to obtain spectra at either constant slit width or constant bandpass.

Detectors

Photographic detection has given way almost entirely to photoelectric detection. The choice of phototube response depends on which laser line is used. The trialkali photocathode has about 7% quantum efficiency at 632.8 nm and falls in efficiency fourfold for every 1000 cm^{-1}. Raman shifts of 3700 cm^{-1} require response

FIGURE **12.3** Laser Raman spectrophotometer. (Courtesy of Instruments SA, Inc.)

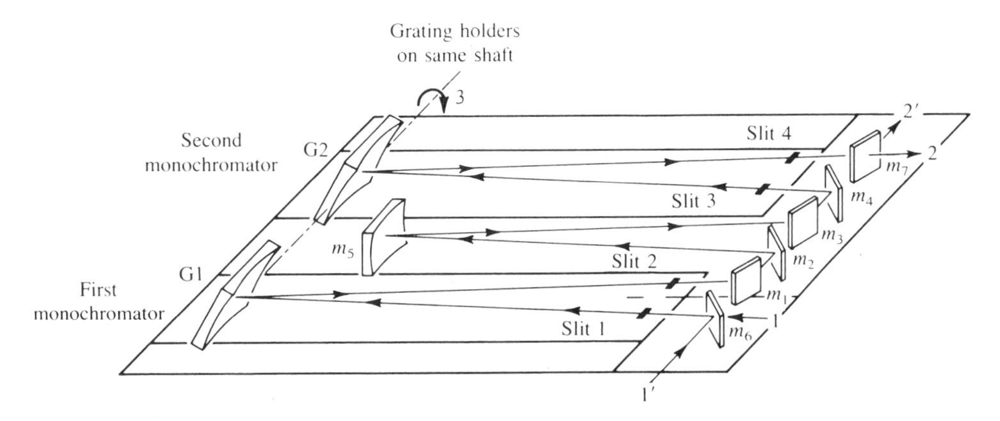

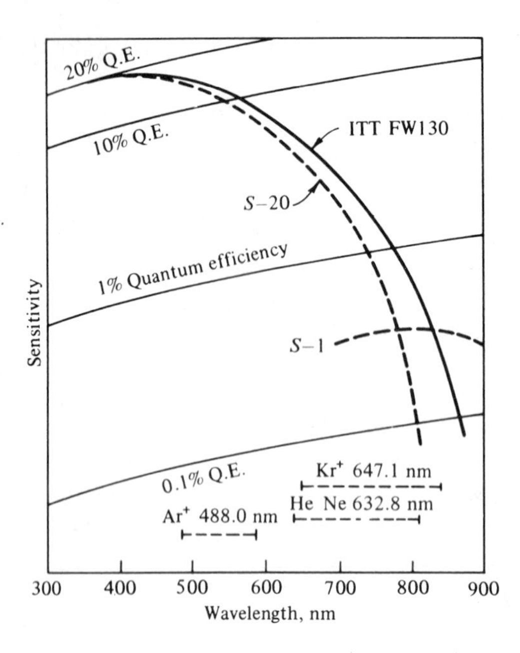

FIGURE 12.4
Sensitivity of several photomultiplier tubes. The dashed horizontal lines represent the range of 3500 cm^{-1}, which normally includes the Raman shift from the designated laser exciting lines.

to approximately 826.6 nm for the He-Ne laser. The extended red-sensitive multialkali cathode and gallium arsenide photocathode designs have much higher quantum efficiency in the red portion of the spectrum out to 900 nm. By contrast, photomultiplier tubes are near their peak sensitivity at 488.0 nm, one of the emission lines of the Ar-Kr laser. Obviously, laser selection and detector choice are interwoven, as shown in Figure 12.4.

12.3 SAMPLE HANDLING AND ILLUMINATION

The use of laser excitation allows Raman spectroscopy to be performed on specimens in almost any state: liquid, solution, transparent solid, translucent solid, powder, pellet, or gas. Liquids are usually examined with a single pass of the laser beam either axially or transverse to the neat liquid sample contained in a glass capillary tube. If the liquid is clear, the beam focused to a diffraction-limited point in a small sample passes through the liquid and is reflected back again for another pass. Considerable gain in Raman intensity is achieved beyond a single pass, permitting work with extremely small samples in the microliter or even nanoliter range. The volume of the focused He-Ne laser beam is about 8 nL at the diffraction-limited point. When greater volumes of sample are available, the exciting radiation may be passed several times through larger cells. Photo- or heat-labile materials are studied in spinning cells, which reduce exposure time in the laser beam. Water is a weak scatterer and therefore an excellent solvent for Raman work. This has important consequences in studies of biochemical interest and in the pharmaceutical industry. Other widely used solvents are carbon disulfide, carbon tetrachloride, chloroform, and acetonitrile. Their obscuration ranges are shown in Figure 12.5.

Gas samples are handled with powerful laser sources and efficient multiple passes or intralaser-cavity techniques, but they are difficult to study because of their

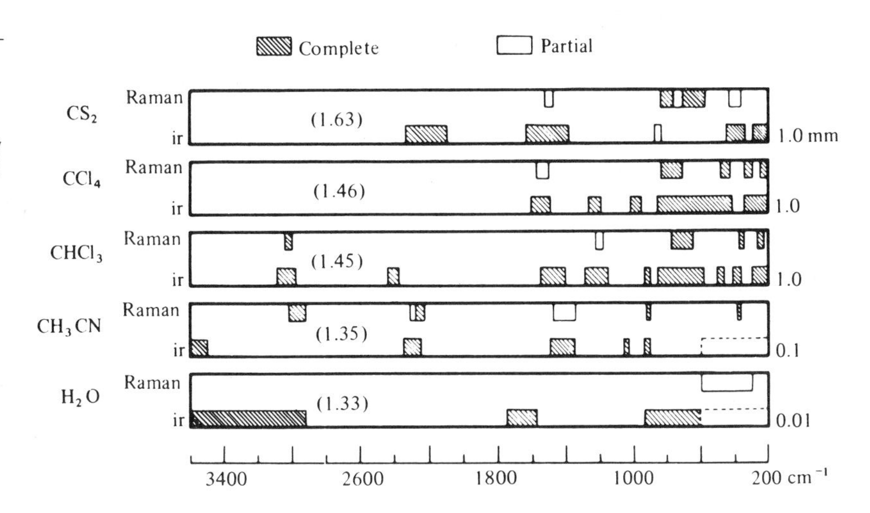

FIGURE 12.5
Obscuration ranges of the most useful solvents for Raman spectrometry in solution. The infrared obscuration at the indicated path lengths is given for comparison; the refractive index of the solvent is given in parentheses.

low scattering. Nevertheless, pollutant studies of emissions are being made from considerable distances from the point of emanation.

Powders are tamped into an open-ended cavity for front-surface illumination or into a transparent glass capillary tube for transverse excitation. Forward (180°) sample illumination provides higher collection efficiency (better S/N ratio); however, Raman lines in the low-frequency region are more easily observed with right-angle viewing because the ratio of Raman to Tyndall and Rayleigh scattering is improved. For a translucent solid, the laser beam is focused into a cavity on the face of the sample, cut into either a cast piece or a pellet formed by compression of powder. The cavity functions as a light trap, producing multiple scattering of the incident photons via the intrinsic reflectivity and transmittance of the specimen to the laser frequency. These arrangements are shown in Figure 12.6. A transparent solid sample preserves the directionality of the laser beam as it passes through the specimen, producing a scattering image that is collinear with the monochromator slit aperture for scattering observed at 90° to the direction of incidence. Bulk polymer samples and also single, very fine fibers are run intact. Polymerization studies on such individual fibers often yield information about their orientation and crystalline properties.

If fluorescence arises from impurities in the sample, one can clean the sample, often without a great deal of effort, and generally observe a rather marked improvement in the quality of the spectrum. Techniques such as gas chromatography fractionation, recrystallization, distillation, and filtration are suitable for this purpose. The so-called drench quenching technique also works quite well but is time-consuming in some cases. This technique simply involves soaking the sample in the laser beam until the luminescence background decays to some reasonable level. If the fluorescence arises from the sample itself, one has little choice but to select a different excitation line.

12.4 STRUCTURAL ANALYSIS

In Raman spectroscopy those vibrations that originate in relatively nonpolar bonds with symmetrical charge distributions and that are symmetrical in nature produce the greatest polarizability changes and are the most intense. Vibrations from

—C=C—, —C≡C—, —C≡N, —C=S, —C—S—, —S—S—, —N=N—, and —S—H bonds are readily observed. Raman lines are more characteristic than infrared absorption bands of the skeletal vibrations of finite chains and rings of saturated and unsaturated hydrocarbons.

The position of the symmetric ring stretching vibration of cyclic compounds is characteristic of the type and size of ring present in the compound. Aromatic compounds have particularly strong spectra. All have a strong ring deformation mode at 1600 ± 30 cm^{-1}. Monosubstituted compounds have an intense symmetric ring stretching vibration at about 1000 cm^{-1}, a strong in-plane hydrogen bending

FIGURE **12.6** Experimental arrangements for laser excitation of specimens in various physical forms.

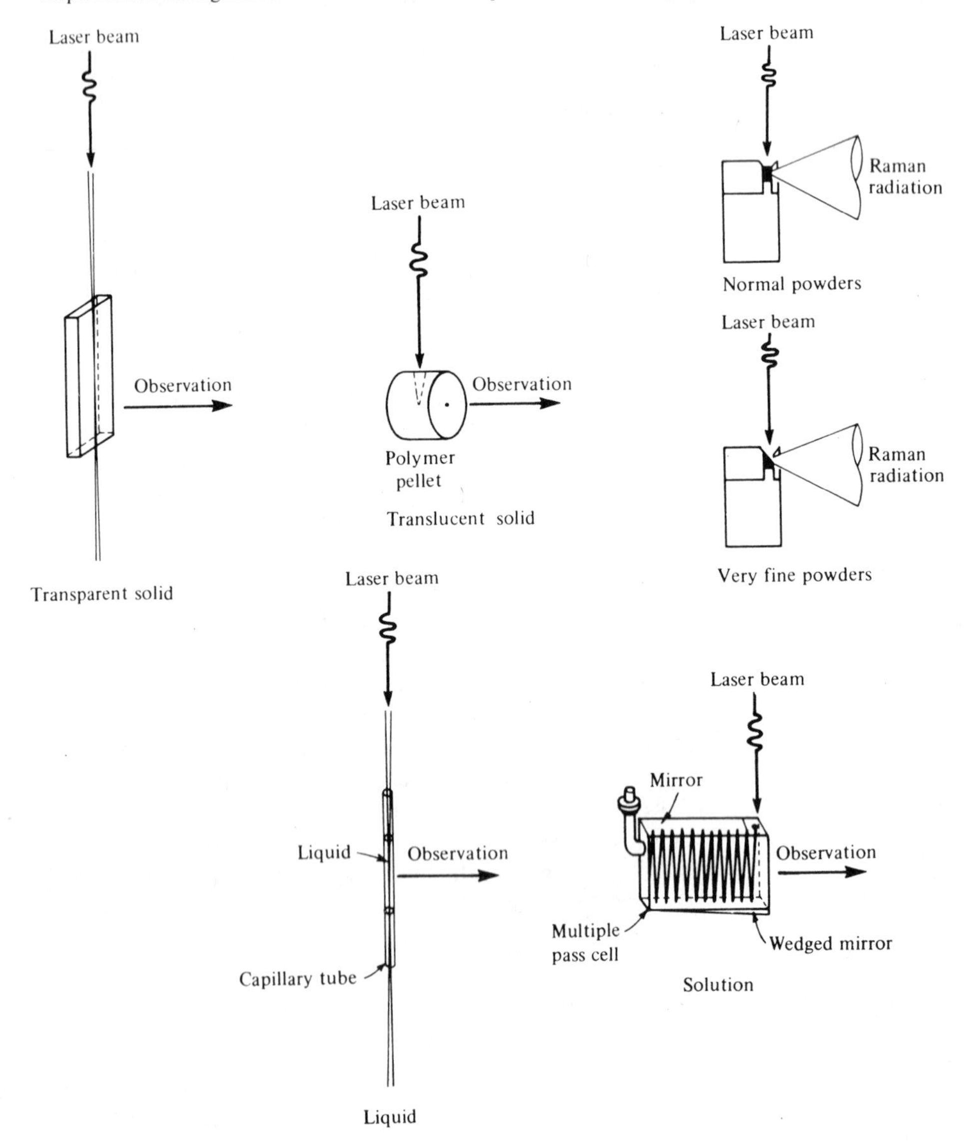

vibration at about 1025 cm^{-1}, and a weak depolarized in-plane bending vibration at about 615 cm^{-1} (Figure 12.7). *Meta*- and 1,3,5-trisubstituted compounds have only the line at 1000 cm^{-1}. *Ortho*-substituted compounds have a line at 1037 cm^{-1}, and *para*-substituted compounds have a weak line at 640 cm^{-1}.

The intense Raman band near 500 cm^{-1} is characteristic of the —S—S— linkage, and the band near 650 cm^{-1} derives from —C—S— stretching (Figure 12.8). Also, the —S—H stretching band near 2500 cm^{-1}, normally weak in the infrared, shows high intensity in the Raman spectrum.

Although the —C≡N stretch appears in both the infrared and Raman spectra, it is markedly diminished in intensity in the infrared region when an electronegative group such as chloride is α-substituted. The intensity is retained in the Raman spectrum.

FIGURE **12.7** Raman spectrum of styrene monomer.

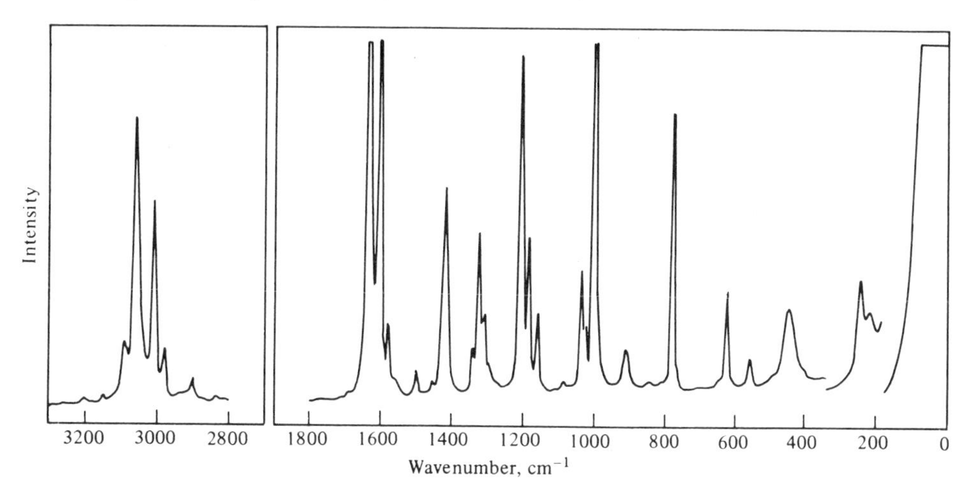

FIGURE **12.8** Raman spectrum of diethyldisulfide, $C_2H_5SSC_2H_5$.

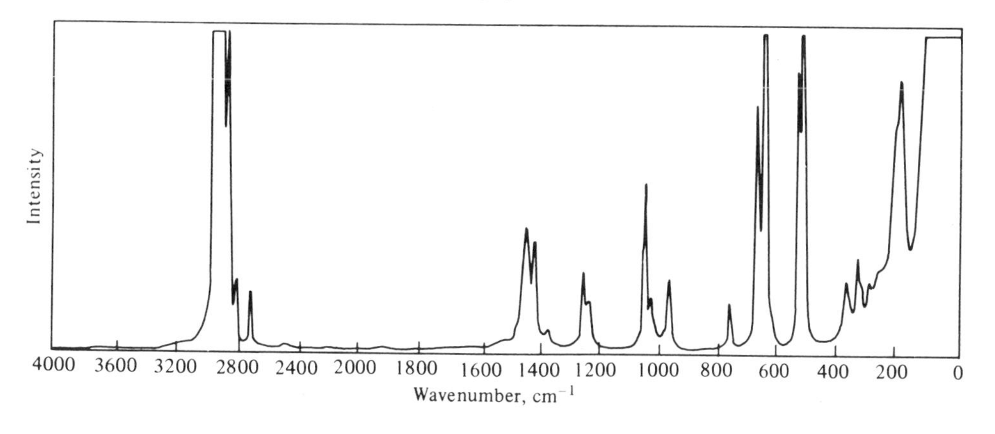

Whenever the 3300-cm^{-1} region in the infrared is badly obscured by intense OH absorption, Raman spectroscopy is helpful because the OH band is weak, whereas the NH and CH stretching frequencies still exhibit moderate intensity.

In the Raman spectrum the symmetrical methyl deformation frequency near 1380 cm^{-1} is sensitive to the environment of the methyl group. It is quite weak in alkyl compounds, but the band intensity is considerably enhanced when the methyl is attached to an aromatic ring and some types of double bonds.

Skeletal motions are very characteristic and highly useful for cyclic and aromatic rings, steroids, and long chains of methylenes. The spectrum in the region 800–1500 cm^{-1} is quite characteristic. In solid samples, sharpening and intensifying of certain bands appear to be a function of crystallinity.

For all molecules that have a center of symmetry, a band allowed in the infrared is forbidden in the Raman, and vice versa. In molecules with symmetry elements other than a center of symmetry, certain bands may be active in the Raman, infrared, both, or neither. For a complex molecule that has no symmetry, all of the normal vibrational modes are allowed in both the infrared and Raman spectra. The symmetry selectivity in Raman spectra results in a generally simpler spectrum than the corresponding infrared spectrum, a characteristic that is frequently useful. One other obvious difference is the tendency for peaks in a Raman spectrum to have a greater range of intensities, from very weak to very strong. These factors, plus the contrasting relative intensities for a given group, are the main basis for making more confident assignment of chemical structure through the combination of infrared and Raman data.

Raman spectroscopy has a distinct advantage in the detection of low-frequency vibrations; the lower limit is dictated by the nature of the sample. With gases, information can be taken to within 2 cm^{-1} of the exciting line. In less favorable cases, within 20–50 cm^{-1} is more typical. This corresponds to the far-infrared region where measurements are difficult. Most of the important vibrations in metal bonding of inorganic and organometallic compounds fall in the low-frequency region as a result of the large masses of the metal atoms. Raman spectroscopy has been applied to the analysis of strong acids and other aqueous solutions and to determination of the degree of dissociation of strong electrolytes and their corresponding activity coefficients.

The Raman technique has proved to be particularly valuable in the study of single crystals where the infrared technique has greater limitations on sample size and geometry. In addition, polarization data obtained from Raman spectra allow the unambiguous classification of fundamentals and lattice modes into the various symmetry classes. Although Raman spectroscopy will never challenge X-ray diffraction as a tool for quantitative structural analysis, it is the preferred technique when qualitative information is sufficient because it is faster and less expensive.

12.5 POLARIZATION MEASUREMENTS

When a polarized beam of radiation is incident upon a molecule, the induced oscillations of the electrons in the molecule are in the same plane as the electric vector of the electromagnetic wave. Thus the resultant emitted radiation tends to be polarized in the same plane.

If one assumes a Cartesian coordinate system around a sample placed at the origin with the incident radiation along the x-axis and observation of the Raman effect at 90°, say along the y-axis, two situations are possible for observing the depolarization of the incident radiation caused by the molecule. In one situation, the polarization of the incident radiation is changed and the variation in intensity of the emitted radiation is observed. In the other and most used situation, the polarization of the incident radiation is fixed and the polarization of the Raman radiation is observed in two different planes.

In the first situation mentioned, the incident radiation is polarized in the xy-plane (parallel illumination) and then in the xz-plane (perpendicular illumination). Unless the molecule depolarizes the incident radiation, the intensity viewed along the y-axis will be zero for the parallel illumination and a maximum for the perpendicular illumination. This is because, to observe an electromagnetic wave, the electric vector must have a component perpendicular to the line of motion of the wave.

In the second situation, the incident radiation is polarized with its electric vector in the xz-plane (perpendicular illumination). The resultant Raman radiation, observed along the y-axis, will have a greater intensity when viewed through a polarizer that passes radiation polarized in the yz plane than through the polarizer when oriented at 90° to pass the radiation polarized in the xy-plane. If the molecule does not depolarize the radiation, the intensity of the parallel radiation (xy-plane) will be zero.

The *depolarization ratio* is defined as

$$p = \frac{P_{\parallel}}{P_{\perp}} \tag{12.8}$$

In the second situation described above, $P_{\perp}$ is the radiant power observed when the electric vector of the Raman radiation is in the yz-plane, and $P_{\parallel}$ is the power of the Raman radiation polarized in the xy-plane. Since $P_{\perp}$ is always greater, the depolarization ratio may vary from near zero for highly symmetrical types of vibrations to a theoretical maximum of 0.75 for totally nonsymmetrical vibrations. In the first situation described—that is, when the polarization of the incident radiation is changed from parallel to perpendicular and the total emitted Raman radiation is measured for each orientation of the incident beam—$P_{\parallel}$ is the radiant power observed along y when the incident radiation is polarized in the xy-plane and $P_{\perp}$ is the observed power when the incident radiation is polarized in the xz-plane. In this case, the depolarization ratio, p, may vary from zero to a maximum of $\frac{6}{7}$, or 0.86.

When measuring depolarization ratios, one should check the instrument with known Raman bands. The 218-cm^{-1} band of carbon tetrachloride should have the maximum value of 0.75, and the 459-cm^{-1} band should have a value of 0.01 ± 0.001.

A depolarization unit consists of an analyzer prism and a depolarization compensator in the path of the Raman scattered radiation. With this accessory unit the analyzer prism is set at 0°, then at 90°, and the ratio of the band powers are measured. Polarization characteristics of the monochromator are eliminated by the polarization compensator, a quartz wedge, which completely scrambles the polarized Raman radiation that enters the spectrometer.

The instrument shown in Figure 12.2 automatically records both the depolarization ratios and the Raman spectrum at the same time. This is very useful for the detection of a weak Raman band overlapped by a strong band, as in the study of intermolecular interaction in solution, a quick analysis of substances, and the study of the polarized Raman spectrum of a single crystal. The laser beam is chopped at 21.4 Hz by the rotating-sector mirror, then alternately passed through two different illumination systems. One beam has the same polarization plane as the laser, and the other beam has a polarization plane perpendicular to the laser beam created by passage through a half-wave plate and a Glan-Thompson prism. The polarization plane is perpendicular to the axis of the beam. Raman scattering caused by the two beams is converted into electrical signals and recorded on a chart.

Any vibrational mode that is not totally symmetric has the maximum depolarization ratio. Totally symmetrical vibrations have depolarization ratios between zero and the maximum. Thus the measurement of depolarization ratios for Raman lines is useful in assigning the frequencies to specific vibrational modes of a molecule.

Depolarization is illustrated by the partial Raman spectrum of $CHCl_3$, shown in Figure 12.9. Nonspherical symmetry exists for the molecule. The antisymmetric C—Cl_3 bending vibration at 261 cm^{-1} and the C—Cl stretching mode at 760 cm^{-1} are depolarized. At 366 cm^{-1} the polarized C—Cl_3 symmetric bending mode appears, as does the polarized symmetric C—Cl stretching mode at 667 cm^{-1}. The C—H stretching mode is not shown.

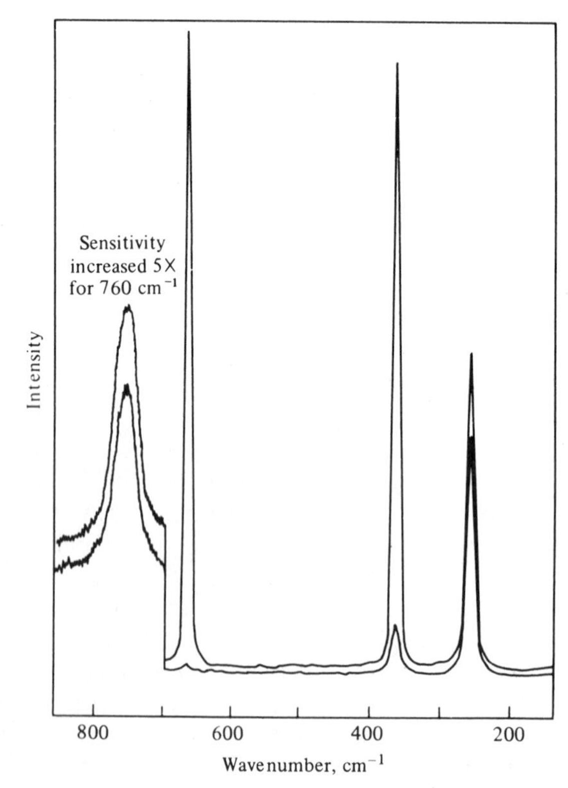

FIGURE **12.9**

Partial Raman spectrum of $CHCl_3$ illustrating depolarization. The lower trace is when the direction of polarization of the incident beam is perpendicular to the direction of observation; the upper trace is when the incident beam is polarized parallel to the direction of observation.

12.6 QUANTITATIVE ANALYSIS

The determination of the absolute radiant powers of Raman bands is even more difficult than the determination of the absolute radiant powers of infrared absorption bands. For this reason, the power of a Raman line is usually measured in terms of an arbitrarily chosen reference line, usually the line of CCl_4 at 459 cm^{-1}, which is scanned before and after the spectral trace of the sample. Scattering powers, or peak heights on the spectrum, are then converted to *scattering coefficients* by dividing the recorded height of the sample peak by the average of the heights of the dual traces of the CCl_4 peak. Both standard and sample must be recorded in cells of the same dimensions.

For quantitative analysis the power of Raman lines is directly proportional to the number of scatterer molecules and thus to the scattering coefficient. For mixtures in which the components all have the same molecular type, there is a direct proportionality between the scattering coefficient and the volume fraction of the compound present. For mixtures of dissimilar type, Raman shifts vary among the various compounds and a broad band is recorded at the position characteristic of these bond types. The area under the recorded peak is used as a measurement of scattering power.

Some examples of the applicability of the Raman method to the solution of otherwise difficult analytical problems are the determination of styrene monomer in styrene/butadiene latexes[2] and the determination of isobutane, nitrogen, and methane in mixtures of the three gases.[3]

12.7 COMPARISON OF RAMAN WITH INFRARED SPECTROSCOPY

Raman spectroscopy offers distinct advantages over the more direct infrared absorption measurement. Raman spectroscopy can be used to detect and analyze molecules with infrared inactive spectra, such as homonuclear diatomic molecules. For complicated molecules whose low symmetry does not forbid both Raman and infrared activity, certain vibrational modes are inherently stronger in the Raman effect and weaker in, or apparently absent from, the infrared spectrum. Raman activity tends to be a function of the covalent character of bonds. Hence a Raman spectrum reveals information about the backbone structure of the molecule, whereas the strong infrared features are indicative of polar segments.

Raman spectra can be used to study materials in aqueous solutions, a medium that transmits infrared radiation very poorly. Another advantage is the ability to examine the entire vibrational spectrum with one instrument, unlike infrared spectroscopy in which the far-infrared is usually scanned separately from the mid-infrared. For quantitative determinations, the Raman scattering power is directly proportional to the concentration, whereas in infrared spectroscopy, it is the logarithm of the ratio of incident to transmitted power—that is, the absorbance that is proportional to concentration. Finally, sample preparation for Raman spectroscopy is generally simpler than for the infrared.

The sensitivity of resonance Raman spectroscopy to only chromophore vibrational modes may be considered either a strength or a weakness. Spectra are generally greatly simplified and a series of molecules that contain slightly different chromophores give spectra that are easily distinguished. On the other hand, if a series of molecules contains the same chromophore with, for example, different aliphatic side chains, the resonance Raman spectra are nearly identical.

There are some shortcomings of the Raman technique. Both liquid and solid samples must be free from dust particles or the Raman spectrum may be masked by Tyndall scattering. The primary disadvantage of Raman spectroscopy is the fluorescent background that accompanies intense laser irradiation of many biological materials. Relative to the Raman signal, the background can be enormous, completely obliterating the spectrum. Even if one could observe the Raman spectrum superimposed on the fluorescence background, the noise contribution of the fluorescence emission degrades the signal-to-noise ratio of the Raman spectrum. Although the problem can be attacked by careful sample preparation, time-resolved spectroscopy, or coherent anti-Stokes Raman spectroscopy (CARS), there will always be experiments that remain difficult to perform.

PROBLEMS

1. What instrumental factors have led to a false but widely held belief that C—H stretching vibrations are weak in Raman spectroscopy?

2. Suppose one wishes to scan a Raman spectrum as fast as possible. The minimum time constant of the electronics is 0.1 sec, and the desired resolution (bandpass) is 8 cm^{-1}. (a) What should be the scan speed? (b) If the dispersion of a double monochromator with 1200-grooves/mm grating, working in the first order, is 0.55 nm/mm at the exit slit, what can the maximum slit opening be?

3. By what factor is the scattered intensity of a given band reduced in changing the excitation frequency from an argon laser (488.0 nm) to a neodymium-doped laser (1065 nm)? Ignore changes in detector response as well as grating and reflector efficiencies.

4. What are the relative intensities of a Raman line excited by a He-Ne laser at 632.8 nm and one excited by an argon laser at 488.0 nm?

5. When excited by the mercury line at 435.8 nm, the spectral trace of benzene has lines at 606, 850, 991, 1176, 1584, 1605, 3047, and 3063 cm^{-1}. At what wavelengths will these Raman lines appear if benzene is irradiated with (a) a He-Ne laser (632.8 nm), (b) an argon laser (488.0 and 514.5 nm), and (c) a krypton laser (568.2 and 647.1 nm)?

6. If unfiltered laser excitation from either the krypton or argon laser were used to excite the Raman spectrum of benzene (see Problem 5), to what extent would each suite of Raman lines overlap?

7. For carbon disulfide, all vibrations that are Raman active are infrared inactive and vice versa, whereas for nitrous oxide (N_2O), the vibrations are simultaneously Raman and infrared active. What can one conclude about the structures of N_2O and CS_2?

8. For CCl_4, four principal Raman lines appear at 218, 314, 459, and 791 cm^{-1}. None of the corresponding infrared frequencies is absorbed. The depolarization ratios are 0.75, 0.75, 0.01, and 0.72, respectively. What can be concluded about the symmetry of the molecule? Is its spatial configuration planar or tetrahedral?

9. For each unknown mixture of the trimethylbenzenes, compute the volume percent of each. The scattering coefficient of the pure compound at each analytical wavenumber is shown in the table.

	1,2,3-		*1,2,4-*		*1,3,5-*	
Mixture	**625 cm^{-1}**	**0.627**	**716 cm^{-1}**	**0.208**	**570 cm^{-1}**	**0.555**
A		0.209		0.069		0.185
B		0.251		0.054		0.189
C		0.157		0.077		0.211

10. Deduce the structure of the compound whose infrared and Raman spectra are shown in the figure. The molecular weight is 140.

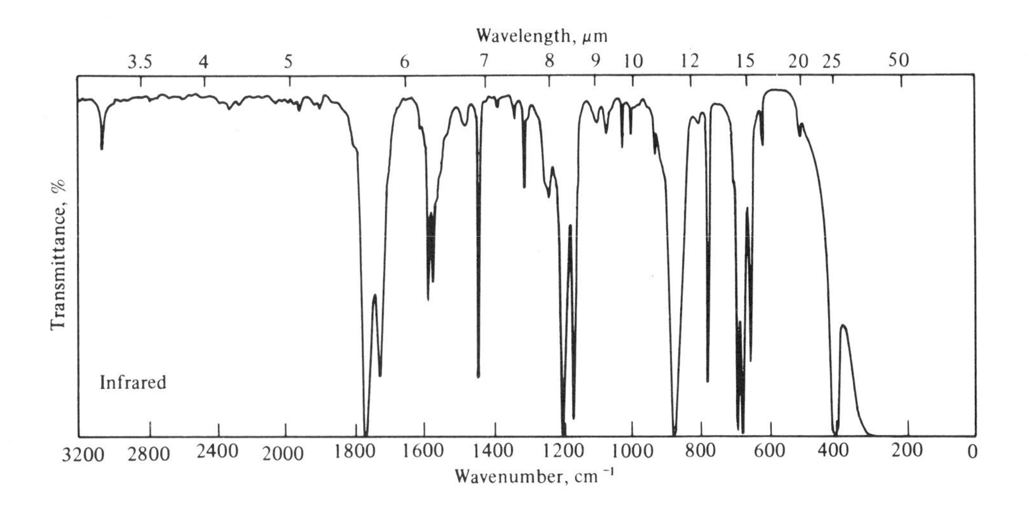

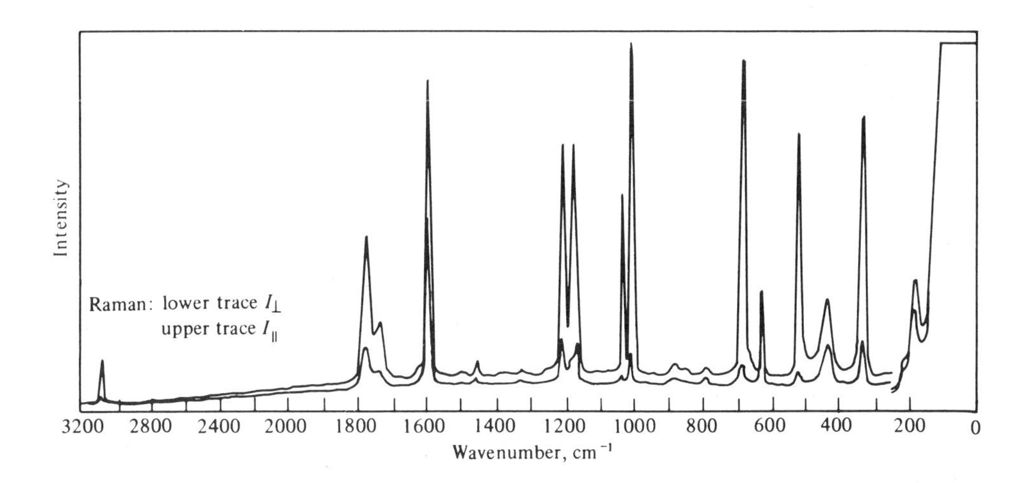

11. The molecular weight of the compound is 54. Using the spectra in the accompanying figure, determine the structure of the compound.

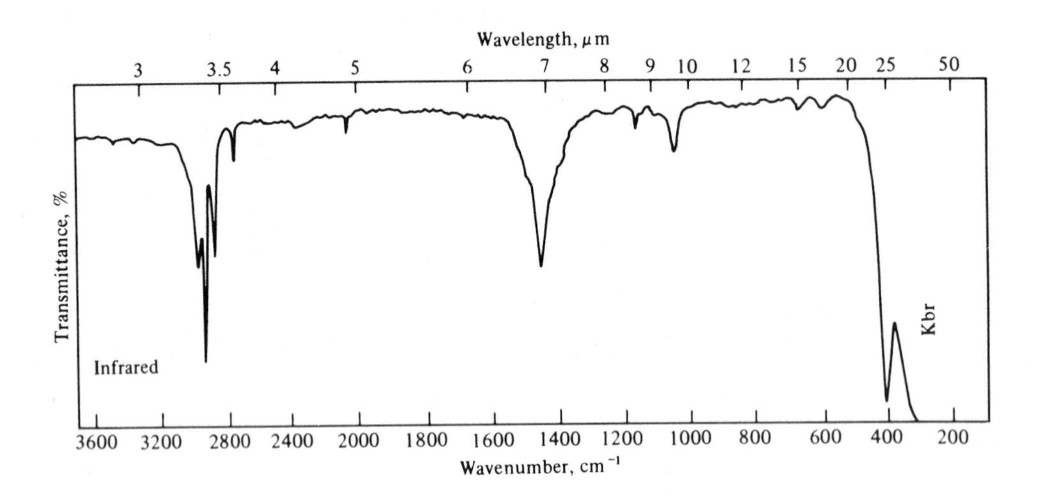

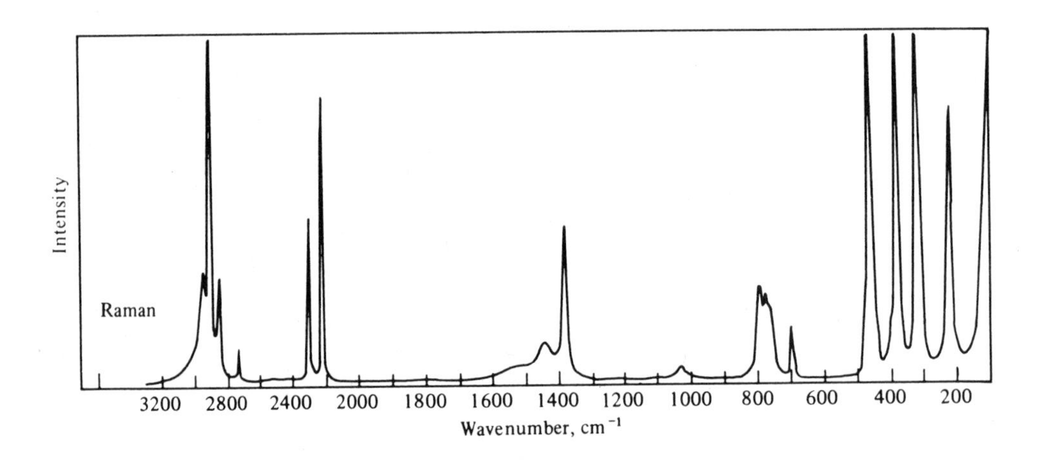

BIBLIOGRAPHY

ALLKINS, J. R., "Tunable Lasers in Analytical Chemistry," *Anal. Chem.*, **47,** 752A (1975).

CHONG, R. K., AND T. E. FURTALS, eds., *Surface Enhanced Raman Scattering,* Plenum, New York, 1982.

CLARK, R. J. H., AND R. E. HESTER, eds., *Advances in Infrared and Raman Spectroscopy,* Vol. 11, 1984; Vol. 12, 1986; John Wiley, New York.

COLTHUP, N. B., "Infrared and Raman Spectroscopy," in *Guide to Modern Methods of Instrumental Analysis,* T. H. Gouw, ed., Wiley-Interscience, New York, 1972.

GRASSELLI, J. G., M. K. SNAVELY, AND B. J. BULKIN, *Chemical Applications of Raman Spectroscopy,* John Wiley, New York, 1981.

GREEN, R. B., "Dye Laser Instrumentation," *J. Chem. Educ.*, **54,** A365 (1977).

HABER, H. S., "Fluorescence Problems in Raman Spectroscopy," *Am. Lab.*, p. 67 (November 1973).

HARVEY, A. B., "Coherent Anti-Stokes Raman Spectroscopy (CARS)," *Anal. Chem.*, **50,** 905A (1978).

MORRIS, M. D., AND D. J. WALLAN, "Resonance Raman Spectroscopy," *Anal. Chem.*, **51,** 182A (1979).

PARKER, F. S., *Applications of Infrared, Raman, and Resonance Raman Spectroscopy in Biochemistry,* Plenum, New York, 1983.

SPIRO, T. G., *Biological Applications of Raman Spectroscopy,* Vols. 1 and 2, John Wiley, New York, 1987.

STROMMEN, D. P., AND K. NAKAMOTO, *Laboratory Raman Spectroscopy,* John Wiley, New York, 1984.

TOBIN, M. C., *Laser Raman Spectroscopy,* Wiley-Interscience, New York, 1971.

WASHBURN, W. H., "Synergistic Use of Infrared and Raman Spectroscopy," *Am. Lab.*, p. 47 (November 1978).

LITERATURE CITED

1. SHONY, R-S., L. ZHU, AND M. D. MORRIS, *Anal. Chem.*, **58,** 1116 (1986).
2. WANAHECK, P. L., AND L. E. WOLFRAM, *Appl. Spectrosc.*, **30,** 542 (1976).
3. DILLER, D. E., AND R. F. CHANG, *Appl. Spectrosc.*, **34,** 411 (1980).

X-RAY METHODS

When an atom is excited by the removal of an electron from an inner energy level, it may return to its normal state by transferring an electron from some outer level to the vacant inner level. The energy of this transition either appears as X rays or is used to eject a second electron from the outer shell (Figure 13.1). This latter method is called the Auger emission process; observation of the energy of such an ejected electron leads to a method of analysis known as Auger emission spectroscopy (AES). This is discussed later in this chapter. If X rays are emitted, their wavelengths are characteristic of that element and their intensities are proportional to the number of excited atoms. Thus, X-ray emission methods can be used for both qualitative and quantitative measurements.

Atoms are excited in several ways: by direct bombardment of the material with electrons (direct emission analysis, electron probe microanalysis, and Auger emission spectroscopy), by bombardment with protons or other particles (particle-induced X-ray emission, PIXE), or by irradiation of the material with X rays of shorter wavelength. In the latter case, the exciting X radiation causes an inner electron to be ejected with subsequent emission of the characteristic X rays of the sample (fluorescence analysis). It is also possible to measure the energy of the electrons ejected from the sample, and this leads to a method of analysis known as electron spectroscopy for chemical analysis (ESCA.) This too is described later in this chapter.

Another method of X-ray analysis uses the differing absorption of X rays by different materials (absorption analysis). Major discontinuities in the absorption of X rays by an element occur when the energy of the X rays becomes sufficient to knock an electron out of an inner level of an atom.

Still another method of using X rays in analytical work is the diffraction of X rays from the planes of a crystal (diffraction analysis). This method depends on the wave character of the X rays and the regular spacing of planes in a crystal. Although diffraction methods are used for quantitative analysis, they are most widely used for qualitative identification of crystalline phases.

Since electrons quickly lose their energy by colliding with molecules, methods that involve electron bombardment or detection of emitted electrons must be carried out in a vacuum. Furthermore, since electrons cannot penetrate deeply into solid materials or escape from any significant depth when generated in a solid, such methods characterize the surface layers of solids and are useful for analyzing surfaces. Auger electron spectroscopy (AES) and electron spectroscopy for chemical analysis (ESCA), described later in this chapter, are carried out in a vacuum and usually characterize not more than the outer 2 nm of the surface under investigation.

In 1913 Moseley first showed the extremely simple relationship between atomic number, Z, and the reciprocal of the wavelength, $1/\lambda$, for each spectral line that belongs to a particular series of emission lines for each element in the periodic table. This relationship is expressed as

$$\frac{c}{\lambda} = a(Z - \sigma)^2 \tag{13.1}$$

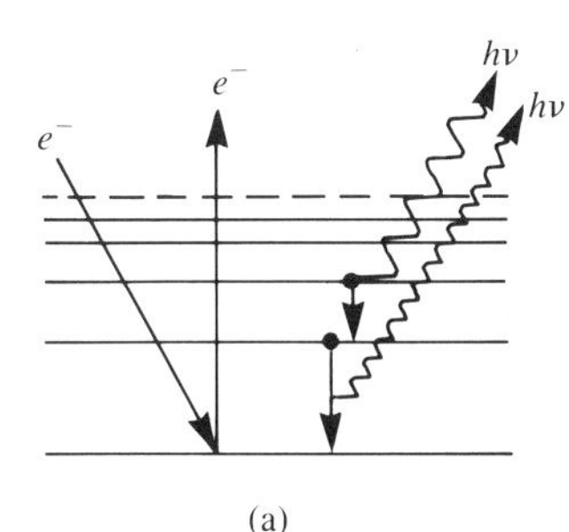

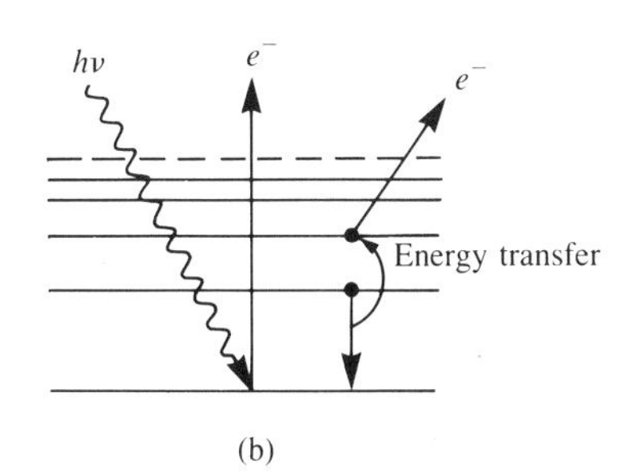

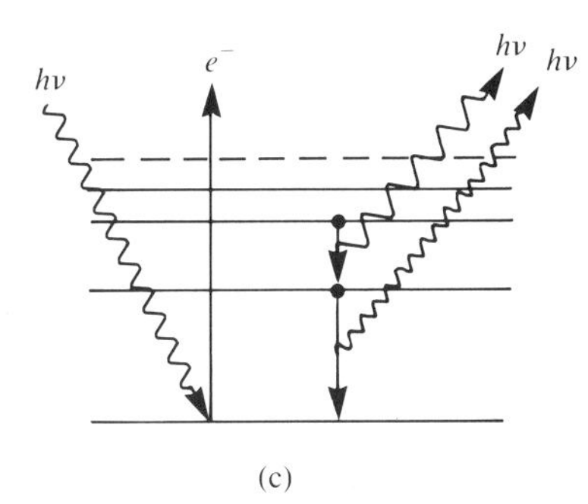

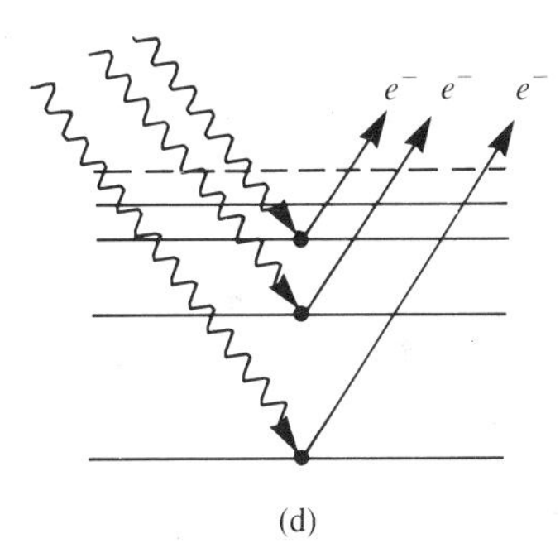

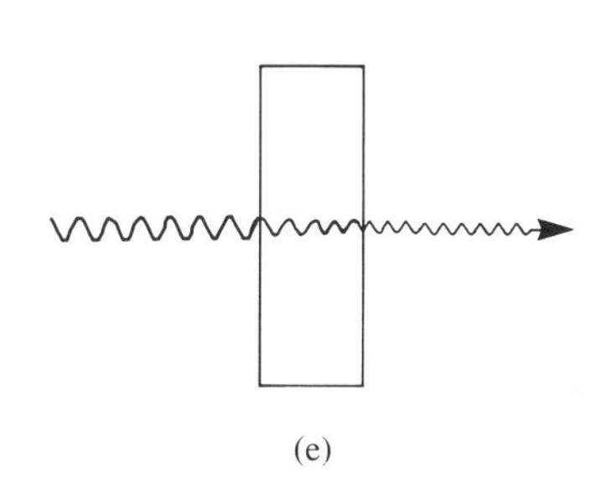

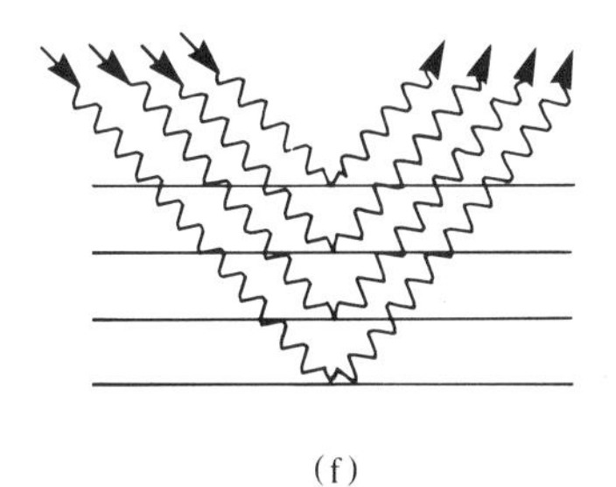

FIGURE **13.1**
Methods of X-ray analysis. Primary processes are shown on the left. (a) X-ray emission spectroscopy (XES): the primary electron beam ejects electrons from inner energy levels; secondary X rays are emitted as outer-level electrons fall into vacant inner levels. (b) Auger emission spectroscopy (AES): primary X rays eject electrons from inner energy levels; when electrons fall into vacant inner levels by nonradiative processes, excess energy ejects electrons from outer levels. (c) X-ray fluorescence spectroscopy (XFS): primary X rays eject electrons from inner energy levels; X rays are emitted as electrons fall from outer levels to vacant inner levels. (d) Electron spectroscopy for chemical analysis (ESCA): primary X rays eject electrons from inner energy levels; energy of the emitted electrons is measured. (e) X-ray absorption: the intensity of X rays is diminished as they pass through material; discontinuities in absorptivity occur when X rays have sufficient energy to eject electrons as in parts a–d. (f) X-ray diffraction: X rays are diffracted by the planes of a crystal.

where a is a proportionality constant and σ is a constant whose value depends on the particular series.

X-ray emission and absorption spectra are quite simple because they consist of very few lines compared with the emission (Chapter 10) or absorption spectra observed in the visible and ultraviolet regions (Chapters 6 and 7). This relative simplicity is because the X-ray spectra result from transitions between energy levels of the innermost electrons in the atom. There are only a few electrons in these inner levels, giving rise to only a few permitted transitions. There is only one principal energy level, $n = 1$ (the K level), which contains only two electrons. The second principal energy level electrons, $n = 2$ (the L level), are grouped according to their binding energy into three sublevels: L_{I}, L_{II}, and L_{III}. The complete third principal energy level, $n = 3$ (the M level), consists of five sublevels. X-ray emission or absorption spectra are dependent only on atomic number and not on the physical state of the sample or its chemical composition, except for the lightest atoms, because the innermost electrons are not involved in chemical binding and are not significantly affected by the behavior of the valence electrons.

13.1 PRODUCTION OF X RAYS AND X-RAY SPECTRA

An X-ray tube is basically a large vacuum tube containing a heated cathode (electron emitter) and an anode, or target (Figure 13.2). Electrons emitted by the cathode are accelerated through a high-voltage field between the target and cathode. On impact with the target, the stream of electrons is quickly brought to rest. The electrons transfer their kinetic energy to the atoms of the material that makes up the target. Part of the kinetic energy is emitted in a continuous spectrum of X rays covering a broad wavelength range, with a broad maximum in intensity and falling off to a definite short-wavelength limit (Figure 13.3). The broad range of X-ray photon energies is due to deceleration of the impinging electrons by successive collisions with the atoms of the target material. Consequently, the emitted X rays have longer wavelengths than the short-wavelength cutoff, λ_0, which is independent of the target element and depends only on the voltage across the X-ray tube. At the

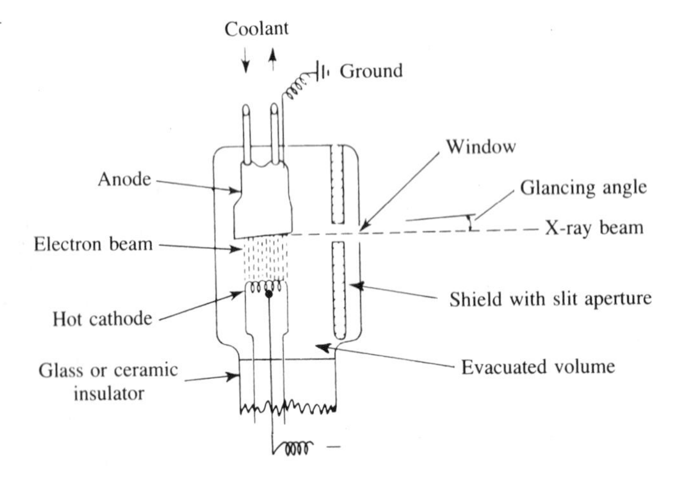

FIGURE 13.2
Schematic of an X-ray tube.

cutoff wavelength, all the energy of the electron is converted, at one impact, to a photon. The relationship between the voltage and λ_0, in angstrom units (Å), is given by the Duane-Hunt equation:[1]

$$\lambda_0 = \frac{hc}{eV} = \frac{12{,}393}{V} \tag{13.2}$$

where V is the X-ray tube voltage in volts, e is the charge on the electron, h is Planck's constant, and c is the velocity of light. An increase in the tube voltage results in an increase in the total energy emitted and a movement of the spectral distribution toward shorter wavelengths. The wavelength of maximum intensity is about 1.5 times the short-wavelength limit. The intensity of the spectrum increases with the atomic number of the target element.

If sufficiently energetic electrons are available, the transfer of energy from the impinging electron beam may eject an electron from one of the inner levels of the target atoms. Within each atom, the place of the ejected electron is promptly filled by an electron from an outer level whose place, in turn, is taken by an electron coming from still farther out. Another possibility is that the energy released from the first transition results in another electron being ejected from one of the outer levels of the atom. Thus the ionized atom returns to its normal state in a series of steps, in each of which an X-ray photon of definite energy is emitted or the excess energy is released by ejection of a second electron with a characteristic energy. These transitions give rise to the characteristic line spectrum of the material in the anode or of a specimen pasted on the target. When originating in an X-ray tube, these lines are superimposed on the continuum. The K series of lines is observed when an electron in the innermost K level ($n = 1$) is dislodged and electrons drop down from the L ($n = 2$) or M ($n = 3$) levels into the vacancy in the K level. Corresponding

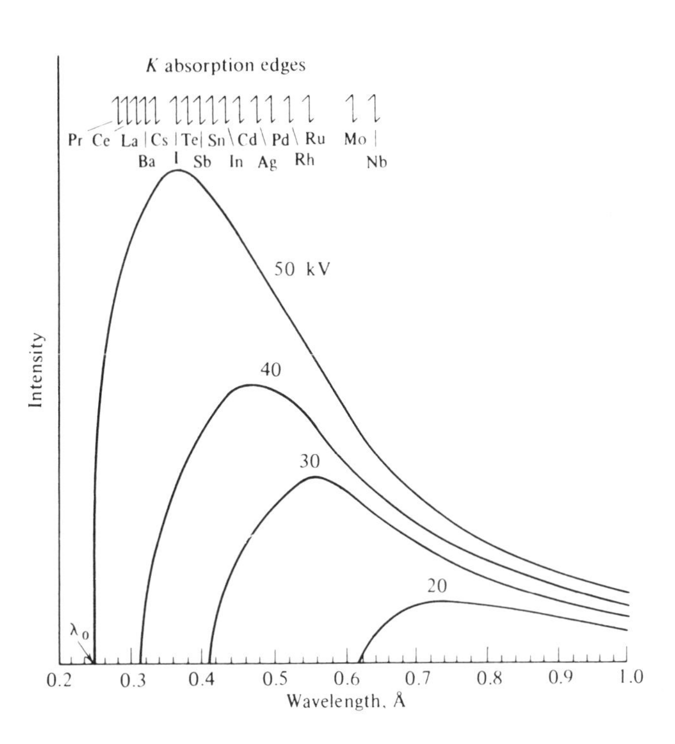

FIGURE **13.3**
X-ray continuum from a target operated at voltages specified. Along the top edge are indicated the wavelengths of K absorption edges for elements Z 41–59.

vacancies in the L levels are filled by electron transitions from outer levels and give rise to the L series. The K series consists of two prominent lines known as the $K\alpha$ and $K\beta$ lines, each of which is a closely spaced doublet. The $K\alpha$ doublet arises from transitions of electrons from L levels to the unfilled K level. The higher-energy $K\beta$ lines arise from transitions from the M levels to the K level. The L series consists of several closely spaced lines that result from transitions from M levels to unfilled L levels.

The characteristic X-ray spectrum of an element is also excited by the irradiation of a sample with a beam of X rays, provided that the primary X radiation is sufficiently energetic to remove an electron from an inner level of the element. Since the inner electron must be completely removed from the element, the energy required is greater than that of any emission lines in the element's spectrum; emission lines result when an electron falls into a vacant inner level from higher energy levels within the atom. When the energy of the exciting radiation just equals the energy required to remove an electron from the element, the exciting radiation is strongly absorbed; that is, there is a sharp rise in the absorption of the exciting radiation (Figure 13.4). This is known as an *absorption edge*. If an X-ray tube is used to produce the exciting radiation, λ_0, the short-wavelength cutoff must be equal to or shorter than the wavelength of the absorption edge, and thus there is a critical potential that must be applied to the tube.

EXAMPLE 13.1

To calculate the short-wavelength limit for an X-ray tube operated at 50 kV, use the equation

$$\lambda_0 = \frac{12{,}393}{50{,}000} = 0.2479 \text{ Å}$$

Upon irradiation with this energy, europium ($Z = 63$), whose K absorption edge lies at 0.255 Å, would emit its characteristic K series of lines, though with low intensity. However, the energy is insufficient for excitation of the K lines of gadolinium ($Z = 64$), whose K absorption edge lies at 0.247 Å.

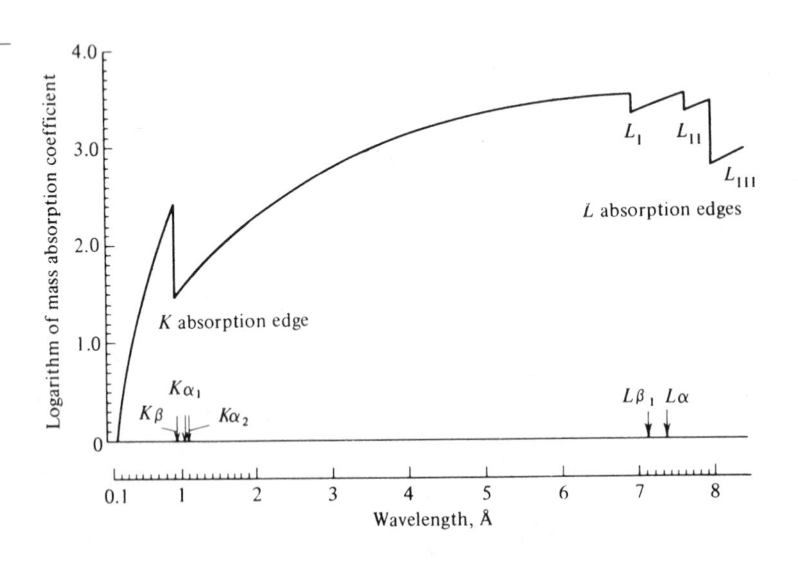

FIGURE 13.4
X-ray absorption spectrum of bromine. The characteristic emission lines of the K and L series are shown with arrows.

TABLE 13.1

CHARACTERISTIC WAVELENGTHS OF ABSORPTION EDGES AND EMISSION LINES FOR SELECTED ELEMENTS

Element	Minimum potential for excitation of K lines (kV)	K absorption edge (Å)	$K\beta$ (Å)	$K\alpha_1$ (Å)	L_{III} absorption edge (Å)	$L\alpha_1$ (Å)
Magnesium	1.30	9.54	9.558	9.889	247.9	251.0
Titanium	4.966	2.50	2.514	2.748	27.37	27.39
Chromium	5.988	2.070	2.085	2.290	20.7	21.67
Manganese	6.542	1.895	1.910	2.102	19.40	19.45
Cobalt	7.713	1.607	1.621	1.789	15.93	15.97
Nickel	8.337	1.487	1.500	1.658	14.58	14.57
Copper	8.982	1.380	1.392	1.541	13.29	13.33
Zinc	9.662	1.283	1.295	1.435	12.13	12.26
Molybdenum	20.003	0.620	0.632	0.709	4.912	5.406
Silver	25.535	0.484	0.497	0.559	3.698	4.154
Tungsten	69.51	0.178	0.184	0.209	1.215	1.476
Platinum	78.35	0.158	0.164	0.186	1.072	1.313

SOURCE: J. A. Dean, ed., *Lange's Handbook of Chemistry*, 13th ed., McGraw-Hill, New York, 1985.

As the wavelength of the incident radiation is decreased or the potential across an X-ray tube is increased, there is successive ionization: first of electrons in the M levels of the sample or target, then of electrons in the L levels as the L_{III}, L_{II}, and L_{I} absorption edges are progressively exceeded, and finally in the K level absorption edge. The K spectra are generally used for the detection and analysis of elements up to about neodymium ($Z = 60$). The L spectra are used from lanthanum to the *trans*-uranium elements when an X-ray tube that has a maximum rating of 50 kV produces the exciting radiation. The wavelengths of selected spectral lines and the absorption edges of some elements are shown in Table 13.1.

EXAMPLE 13.2

Consider a vacant orbital in a bromine atom produced by the ejection of an electron from the innermost K level of electrons. The energy required just to lift a K electron out of the environment of the atom must exceed the energy of the K absorption edge at 0.9181 Å, or

$$V = \frac{12{,}393}{0.9181} = 13{,}498 \text{ V} \quad (\text{or } 13.498 \text{ kV})$$

The wavelength of the K absorption edge is always shorter than that of the K emission lines. The $K\beta_1$ line at 0.934 Å arises when an electron drops from the M level; the $K\alpha_1$ and $K\alpha_2$ lines, a closely spaced doublet at 1.048 and 1.053 Å, arise from sublevels of slightly different energies within the L shell. In energy units, the $K\alpha_1$ line represents the difference: K edge minus L_{III} edge. Thus, for bromine whose L_{III} edge lies at 7.399 Å (or 1.675 kV),

$$K\alpha_1 = 13.498 - 1.675 = 11.823 \text{ kV} \quad \text{or } 1.048 \text{ Å}$$

The absorption and emission spectra for bromine are shown in Figure 13.4, and the energy level diagram is shown in Figure 13.5.

The bond character in molecules and solids affects the X-ray spectra of the light elements whose emission lines originate from the valence electron shell and even the lines and absorption edges from the next innermost shell. In general, relative to the lines of the free element, the lines of the atom in a compound are shifted toward shorter wavelengths if the atom has a positive charge, and toward longer wavelengths if it has a negative charge in the compound. Similar fine structure is observed at an absorption edge if a high-resolution spectrometer is used. Mean wavelengths and shifts in the emission lines for the different oxidation states of sulfur are given in Table 13.2.

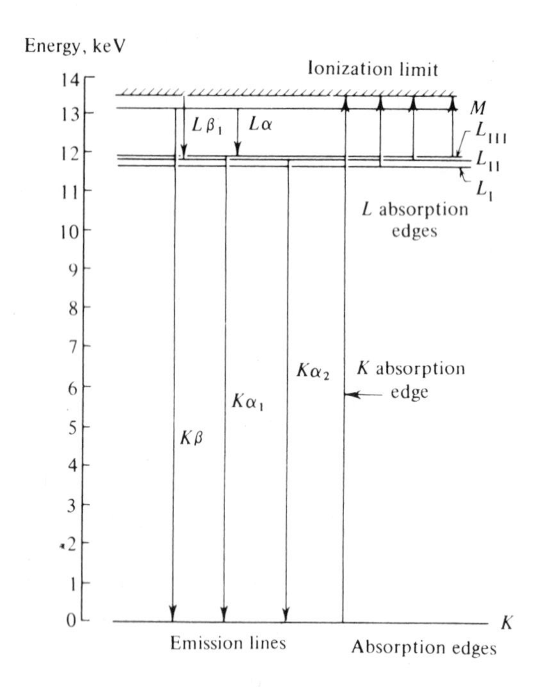

FIGURE **13.5**
Energy level diagram of bromine ($Z = 35$) showing the transitions that give rise to the absorption discontinuities and the emission lines.

TABLE **13.2**

MEAN WAVELENGTHS AND SHIFTS OF *K* LINES OF SULFUR FOR THE DIFFERENT OXIDATION STATES OF SULFUR

	λ (*XU**)		*Mean shift*	
Oxidation state	$K\alpha_1$	$K\alpha_2$	$\Delta\lambda$ (XU)	ΔE (eV)
S^{6+}	5358.08	5360.89	−2.76	+1.19
S^{4+}	5358.63	5361.47	−2.20	+0.95
S^{2+}	5360.13	5362.93	−0.72	+0.31
S^0	5360.83	5363.66	0	0
S^{2-}	5361.15	5363.99	+0.33	−0.14

* One angstrom ≡ 1002.02 XU, where 1 XU = 1/3029.45 the spacing of the cleavage planes of a calcite crystal, a former standard wavelength unit.
SOURCE: A. Faessler, "X-Ray Emission Spectra and the Chemical Bond," pp. 307–319 in *Proceedings of the Xth Colloquium Spectroscopicum Internationale,* Spartan Books, Washington, 1963.

Besides bombardment of the target with electrons, as in an X-ray tube, and irradiation of a target with energetic photons, as in the production of fluorescent radiation from a sample, X rays are also produced during the decay of certain radioactive isotopes. Many isotopes emit gamma rays, which are the same as short-wavelength X rays. Other isotopes decay by K capture. In this process the nucleus captures a K electron, thus becoming an element with one lower atomic number. The vacant K level is filled by electrons that fall in from outer levels, thus emitting characteristic X rays. An example of a radioactive isotope that decays by K capture and may be useful as a source of essentially monoenergetic X rays is ^{55}Fe. The K lines of manganese are emitted. (See Chapter 14 for a more complete description of radioactive isotopes.)

13.2 INSTRUMENTATION

Instrumentation associated with X-ray methods is outlined schematically in Figure 13.6. The components of this generalized instrument are discussed more fully in subsequent sections.

X-ray Generating Equipment

The modern X-ray tube is a high-vacuum, sealed-off unit, shown schematically in Figure 13.2, usually with a copper or molybdenum target, although targets of chromium, iron, nickel, silver, and tungsten are used for special purposes. The target is viewed from a very small angle above the surface. If the focal spot is a narrow ribbon, the source appears to be very small when viewed from the end, which leads to the sharper definition demanded in diffraction studies. For fluorescence work, the focus is much larger, about 5×10 mm, and is viewed at a larger angle (about 20°). Because the target becomes very hot due to collisions of high-energy electrons, it is cooled by water and is sometimes rotated when a very intense X-ray beam is generated. The X-ray beam passes out of the tube through a thin window of beryllium or a special glass. For wavelengths of 6–70 Å, ultrathin films (1 μm aluminum or cast Parlodion films) separate the X-ray tube from the remainder of the equipment, which must be evacuated or flushed with helium.

Associated equipment includes high-voltage generators and stabilizers. Voltage is regulated by regulating the main ac supply. Current regulation is achieved by

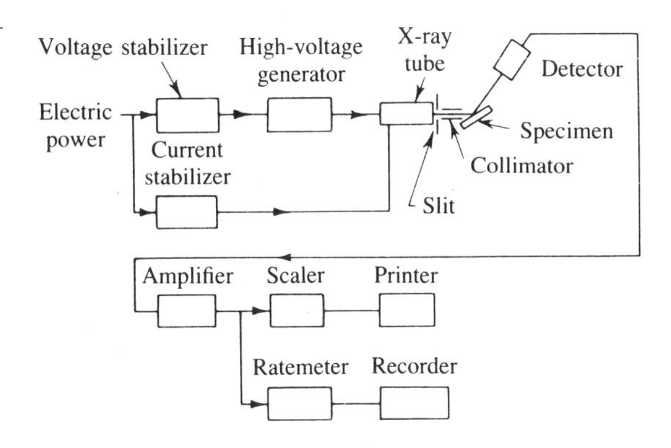

FIGURE 13.6
Instrumentation for X-ray spectroscopy.

monitoring the X-ray tube dc current and controlling the filament voltage. Either full-wave rectification or a constant high potential is used to operate the X-ray tube. In full-wave rectification the voltage reaches its peak value 120 times a second but persists at that value for only a small fraction of the time. A constant high potential, obtained through electronic filtering, increases the output of characteristic X rays from a specimen, particularly with elements that emit at short wavelengths. With a tube operated at 50 kV the gain is twofold for elements up to about atomic number 35 (Br), increasing to fourfold for atomic number 56 (Ba). Commonly, X-ray tubes are operated at 50 or 60 kV. Tubes with a 100-kV rating are available and extend the range of elements whose K series can be excited. The sensitivity of analyses is also greater because a higher voltage increases the intensity of lines emitted by the X-ray tube target.

Collimators

Radiation from an X-ray tube is collimated either by a series of closely spaced, parallel metal plates or by a bundle of tubes, 0.5 mm or smaller in diameter. In a fluorescence spectrometer (see Figure 13.22), one collimator is placed between the specimen and the analyzer crystal to limit the divergence of the rays that reach the crystal. The second collimator, usually coarser, is placed between the analyzer crystal and the detector. It is particularly useful at very low reflection angles for preventing radiation that has not been reflected by the crystal from reaching the detector. Increased resolution is obtained by decreasing the separation between the metal plates of the collimator or by increasing the length of the unit (usually a few centimeters), but this is achieved at the expense of diminished intensity.

Filters

When the wavelengths of two spectral lines are nearly the same and there is an element with an absorption edge at a wavelength between the lines, that element may be used as a filter to reduce the intensity of the line with the shorter wavelength. In X-ray diffractometry it is common practice to insert a thin foil of the proper metal in the X-ray beam to remove the $K\beta$ lines from the spectrum while transmitting the $K\alpha$ lines with a relatively small loss of intensity. Filters for the common targets of X-ray tubes are listed in Table 13.3. Background radiation (the continuum) is reduced

TABLE 13.3 FILTERS FOR COMMON TARGETS OF X-RAY TUBES

Target element	$K\alpha_1$ (Å)	$K\beta$ (Å)	Filter	K absorption edge filter (Å)	Thickness* (mm)	Percent loss of $K\alpha_1$
Mo	0.709	0.632	Zr	0.689	0.081	57
Cu	1.541	1.392	Ni	1.487	0.013	45
Cr	2.290	2.085	V	2.269	0.0153	51
	$L\alpha_1$ (Å)	$L\beta_1$ (Å)				$L\alpha_1$
Pt	1.313	1.120	Zn	1.283	—	—
W	1.476	1.282	Cu	1.380	0.035	77

* To reduce the intensity of the $K\beta$ line to 0.01 that of the $K\alpha_1$ line.

by the same method. Usually it makes no difference whether the filter is placed before or after the specimen unless the specimen fluoresces. If this is the case, the filter is placed at the entrance slit of the goniometer.

Analyzing Crystals

Virtually monochromatic X radiation is obtained by reflecting X rays from crystal planes. The relationship between the wavelength of the X-ray beam, the angle of diffraction, θ, and the distance between each set of atomic planes of the crystal lattice, d, is given by the Bragg condition:[2]

$$m\lambda = 2d \sin \theta \tag{13.3}$$

where m represents the order of the diffraction. The geometrical relationships are shown in Figure 13.7. For the ray diffracted by the second plane of the crystal, the distance $\overline{CBD}$ represents the additional distance of travel in comparison with a ray reflected from the surface plane. Angles CAB and BAD are both equal to θ. Therefore,

$$\overline{CB} = \overline{BD} = \overline{AB} \sin \theta \tag{13.4}$$

and

$$\overline{CBD} = 2\overline{AB} \sin \theta \tag{13.5}$$

where $\overline{AB}$ is the interplanar spacing, d. To observe a beam in the direction of the diffracted rays, $\overline{CBD}$ must be some multiple of the wavelength of the X rays so that the diffracted waves are in phase. Note that the angle between the direction of the incident beam and that of the diffracted beam is 2θ. In order to scan the emission spectrum of a specimen, the analyzing crystal is mounted on a goniometer, an instrument for measuring angles, and rotated through the desired angular region, as shown in the schematic diagram of a fluorescence spectrometer (Figure 13.22).

The range of wavelengths usable with various analyzing crystals is governed by the d-spacings of the crystal planes and by the geometric limits to which the goniometer can be rotated. The d value should be small enough to make the angle 2θ greater than approximately 8° even at the shortest wavelength used; otherwise excessively long analyzing crystals are needed to prevent the incident beam from entering the detector. A small d-spacing is also favorable for producing a larger

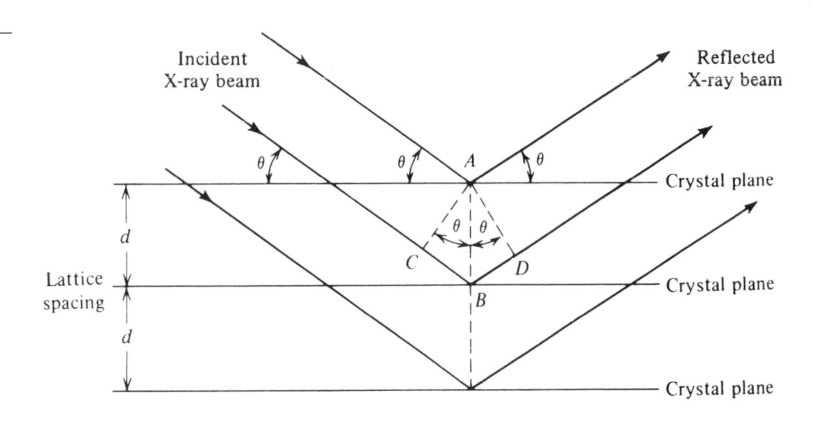

FIGURE 13.7
Diffraction of X rays from a set of crystal planes.

dispersion, $\delta\theta/\delta\lambda$, of the spectrum, as seen by differentiating the Bragg equation:

$$\frac{\partial\theta}{\partial\lambda} = \frac{m}{2d\cos\theta} \tag{13.6}$$

On the other hand, a small d value imposes an upper limit to the range of wavelengths that can be analyzed because at $\lambda = 2d$ the angle 2θ becomes 180°. Actually, the upper limit to which goniometers can be rotated is mechanically limited to a 2θ value of around 150°. For longer wavelengths a crystal with a larger d-spacing must be selected. Crystals commonly used are listed in Table 13.4. These crystals are all composed of light atoms; only sodium chloride, quartz, and the heavy metal fatty acids contain elements heavier than $Z = 9$ so that their own fluorescent X rays will not interfere with measurements. Higher-order reflections, m greater than 1, from the analyzing crystal result in the overlap of lines that originate from different elements.

Analyzing crystals have presented problems in the extension of X-ray analysis beyond a few angstroms. Potassium hydrogen phthalate crystals, due to their large interplanar spacings, d, have made the determination of magnesium and sodium more practical. To extend the analytical capabilities beyond 26 Å, multiple monolayer soap film "crystals" are used. So far, a lead stearate decanoate "crystal" has proved to be superior for elements $Z = 9$ to $Z = 5$. These heavy metal fatty acid "crystals" are prepared by repeatedly dipping an optical flat into a film of the metal fatty acid on water—that is, the Langmuir-Blodgett technique. Recently, multilayer

TABLE **13.4** TYPICAL ANALYZER CRYSTALS

Crystal	Reflecting plane	Lattice spacing d (Å)	*Useful range* (Å)	
			Maximum*	Minimum†
Topaz	303	1.356	2.62	0.189
Lithium fluoride	200	2.014	3.89	0.281
Aluminum	111	2.338	4.52	0.326
Sodium chloride	200	2.821	5.45	0.393
Calcium fluoride	111	3.16	6.11	0.440
Quartz	1011	3.343	6.46	0.466
Ethylenediamine d-tartrate (EDDT)	020	4.404	8.51	0.614
Ammonium dihydrogen orthophosphate (ADP)	200	5.325	10.29	0.742
Pyrolytic graphite	002	6.71	12.96	0.936
Gypsum	020	7.60	14.70	1.06
Mica	002	9.963	19.25	1.39
Lead palmitate		45.6	78.3	6.39
Strontium behenate		61.3	121.7	8.59

* Maximum $2\theta = 150°$, $m\lambda = 2d \sin 75°$.
† Minimum $2\theta = 8°$, $m\lambda = 2d \sin 4°$.

interference mirrors have been used for Bragg reflectors for long-wavelength X rays. These mirrors are made by vapor deposition or sputtering several layers of controlled thickness (several nanometers) alternately of some light-element atoms and heavy-element atoms on a silicon substrate. The substrate can be curved. Thus the mirror focuses the beam and collimating slits are not needed.

Detectors

Photographic Emulsions. Photographic film can be used to measure the intensity of radiation and is often used in diffraction studies. It is also used to measure the distribution of radioactive material in a thin section of a substance (autoradiography) or the distribution of different X-ray absorbers in a thin section of material (microradiography). Film badges are used to measure the total exposure of workers to ionizing radiation. The nature of the photographic process is discussed in Chapter 10. Suffice it to say here that with high-energy radiation such as X rays and radiation from radioactive nuclides, each particle of silver halide that absorbs radiation can be developed. Therefore, there is a direct linear relationship between blackening of the developed film and the intensity of radiation.

The Ionization Chamber. In the ionization chamber an electric field is applied between two electrodes across a volume of gas (air for alpha particles, krypton or xenon under pressure for X or gamma radiation) (Figure 13.8). The potential across the electrodes is adjusted to minimize recombination of the ion pairs without causing amplification by the production of additional ion pairs by energetic ions that collide with neutral gas molecules (Figure 13.9). An ionization chamber is an accurate, quick-acting detector even for weak radiation. The sample is placed outside the window or inserted in a well extending into the chamber volume. For each ionizing event, a *pulse ion chamber* produces an electronic pulse proportional to the number of electrons released by the ionizing radiation as these electrons are collected at the anode. *Current ion chambers* integrate the events and provide a dc current.

The Geiger Counter. The Geiger counter, also called a Geiger-Müller or GM tube, is shown schematically in Figure 13.10. A potential of 800–2500 V is applied to a central wire anode surrounded by a cylindrical cathode (a glass tube that has been silvered, or a brass cylinder). The two electrodes are enclosed in a gas-tight envelope typically filled to a pressure of 80 torr of argon gas plus 20 torr of methane or

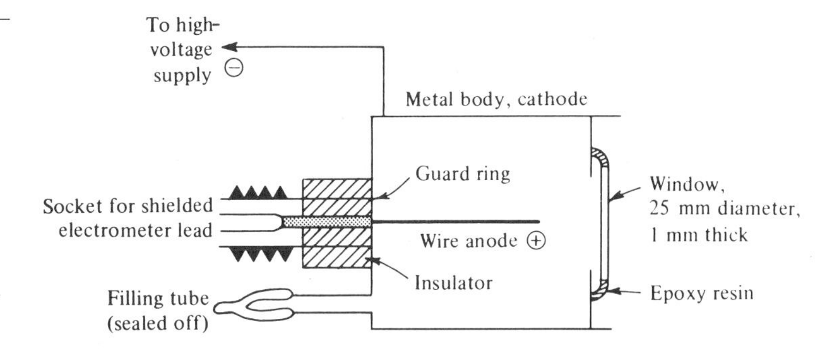

FIGURE **13.8**
Schematic diagram of an ionization chamber.

ethanol or 0.08 torr of chlorine. A thin end-window of mica, about 2.5 cm in diameter and 2–3 mg/cm^2 thick, or a glass wall in dipping counters is the point of entry of the radiation.

When an ionizing particle enters the active volume of the Geiger counter, collision with the filling gas produces an ion pair. This is followed by migration of

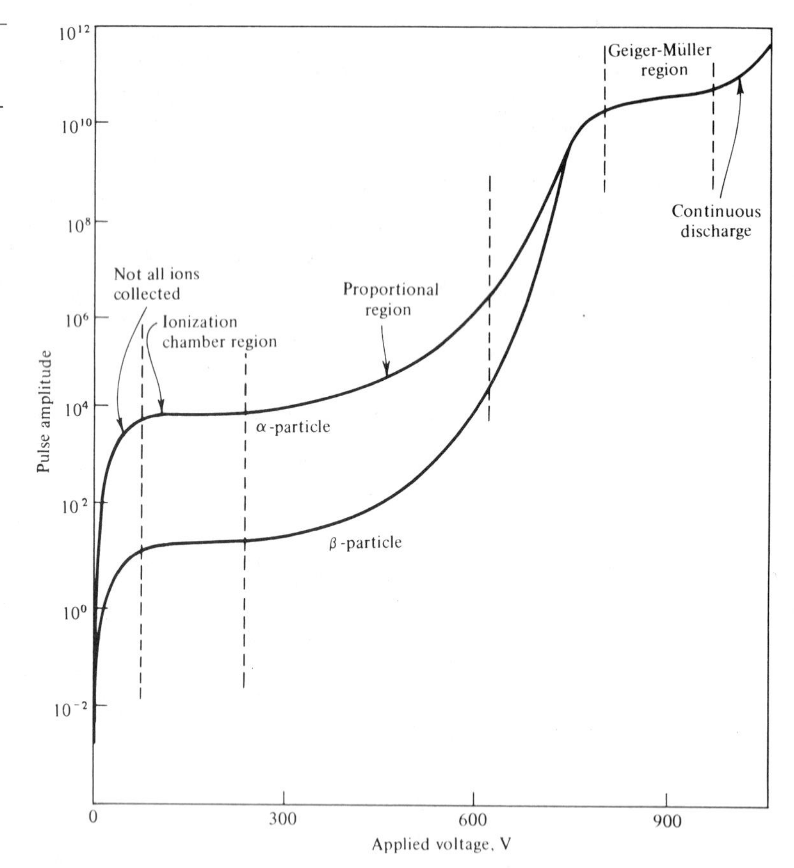

FIGURE 13.9
Pulse amplitude as a function of applied voltage for the ionization type of detectors.

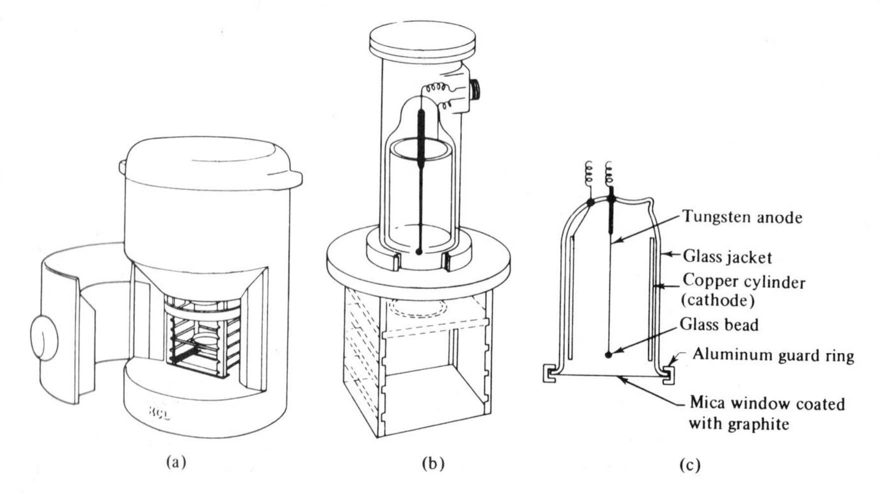

FIGURE 13.10
(a) End-window type of Geiger counter, (b) counter and sample holder with shielding removed, and (c) schematic of counter.

these charged particles toward the appropriate electrode under the voltage gradient. The mobility of the electron is quite high, and under the influence of the potential gradient it soon acquires sufficient velocity to produce a new pair of ions upon collision with another atom of argon. Under these conditions, which are repeated many times, each original ionizing particle that enters the active volume of the counter gives rise to an avalanche of electrons traveling toward the central anode. Photons, emitted when the electrons strike the anode, spread the ionization throughout the tube. These processes produce a continuous discharge, which fills the whole active volume of the counter in less than a microsecond. Each discharge builds up to a constant pulse of maximum amplitude (10 V) and 50–100 μsec duration. These pulses are counted precisely with the aid of scaling circuits or a ratemeter with no intermediate amplification. This is the principal advantage of the Geiger counter.

While the electron avalanche is collected on the anode, the positive ions, being much heavier, progress only a short distance on their way to the cathode. Their travel time is about 200 μsec. During most of this time their presence as a virtual sheath around the anode effectively lowers the potential gradient to a point where the counter is insensitive to the entry of more ionizing particles—the *dead time* of the counter. Because the halogen or organic gas molecules in the counter have a lower ionization potential than argon, after a few collisions the ions moving toward the cathode consist of only these lower energy particles. In contrast to argon ions, these positive ions, when discharged at the cathode, do not produce photons. Consequently, photons that could initiate a fresh discharge are prevented from forming and the counter is self-quenching. When the organic filling gas ions are discharged at the cathode, they decompose into various molecular fragments and eventually the quenching gas is exhausted. Counter life is limited to about 10^{10} counts. Because chlorine atoms merely recombine, this quencher gas is available for further use. A halogen-quenched counter has a life in excess of 10^{13} counts.

Counting rates are limited to about 15,000 counts/min because of the long dead time of 200–270 μsec and the large reduction in the true count rate as the maximum count rate is approached. This is due to two or more ionizing events that occur so close together in time that the counter cannot resolve them. Sensitivity for beta particles is excellent, but for X and gamma radiation the sensitivity is less than that of the scintillation counter described below. There is no possibility of using pulse height discrimination with a Geiger counter because the pulses all have the same amplitude, nor is there any practical correction for coincidence losses when a spectrum is scanned.

The argon-filled Geiger counter, using halogen as a quencher gas, has a sensitive volume large enough to detect nearly the entire large-area beam used in some X-ray optics. The tube is relatively insensitive to scattered short-wavelength radiation and thus its background intensity is low. Its quantum efficiency (the ratio of pulses produced per incident photon) is about 60%–65% in the range 1.5–2.1 Å, and decreases to 40% below 1.4 Å and above 2.9 Å (Figure 13.11).

Proportional Counters. When the electric field strength at the center electrode of an ionization chamber is increased above the saturation level, but under that of the Geiger region (Figure 13.9), the size of the output pulse from the chamber starts to increase but remains proportional to the initial ionization. A device operated in this fashion is called a proportional counter.

Pulse formation is identical with that described for Geiger counters, but gas amplification is approximately 1000 times less. Consequently, a preamplifier ($\times 10$) is needed and is mounted close to the detector to avoid reducing the pulse size through capacitance in connecting cables. In the proportional region few, if any, photons are released. Consequently, the total number of secondary electrons is proportional to the number of primary pairs produced by the original ionizing particle. Furthermore, the discharge is limited to the immediate environment of the entering ionizing particle and the path traversed by the ion pair plus their secondary electrons and positive ions. The dead time is thus very short, about 0.25 μsec. Multiplication factors from 10 to 10^5 are possible. They are dependent on applied voltage, gas pressures, and counter dimensions.

Proportional counters are useful for counting at extremely high rates, 50,000–200,000 counts/sec; the upper limit is imposed by the associated electronic circuitry. The signal produced is extremely small and requires both a preamplifier and a second stage of amplification before the signal is fed to a scaler. Excellent plateaus about 100 V long are obtained with slopes as low as 0.1% counting-rate change per 100 V (whereas values of less than 1% variation are uncommon with Geiger counters).

The proportional counter has about the same spectral sensitivity characteristics as the Geiger counter (Figure 13.11). The window material and thickness have a great influence on the spectral characteristics. Detector windows present challenging problems in work at very long wavelengths (to 70 Å) because these windows must be transparent to very low-energy photons and, in addition, must be capable of supporting atmospheric pressure. Typical windows are 1-μm aluminum foil dipped in Formvar (usable for sodium and magnesium X rays), 1-μm hydrocarbon (cast Formvar, Parlodion, or collodion) films, and 0.1-μm hydrocarbon films. The 0.1-μm films must be supported on a 70% optical transmission grid or on the 0.5-mm spacing blade on a flow detector collimator. Their use is required for X radiation

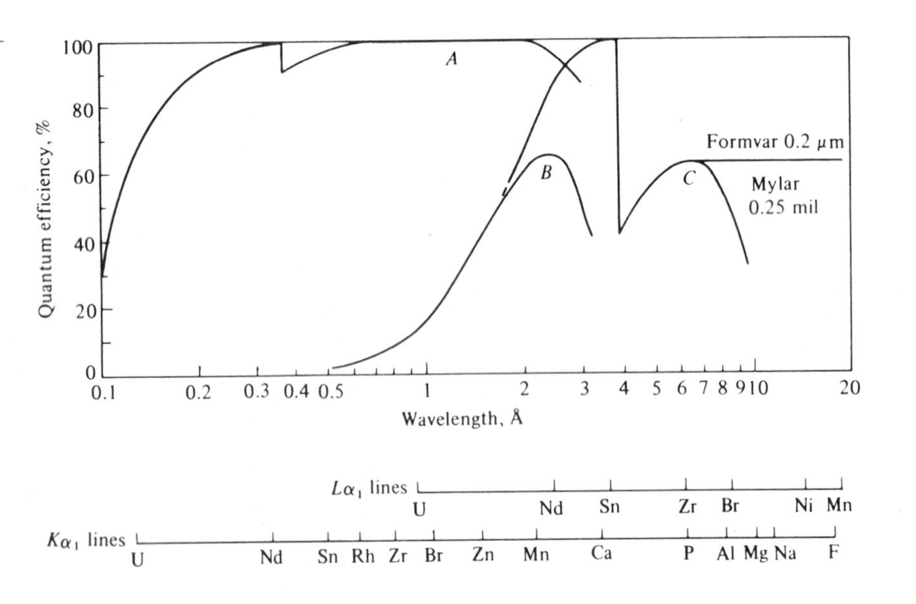

FIGURE 13.11
Quantum efficiencies of detectors commonly used in X-ray spectrometry. *A*, scintillation counter with Tl-activated, NaI scintillator. *B*, argon-filled Geiger counter with Be window. *C*, gas-flow proportional counter; 90% Ar, 10% CH_4; 6-μm Mylar window. Lower scales indicate the wavelengths of representative emission lines.

from oxygen, nitrogen, and boron. The lifetimes of unsupported, 1-μm films never exceed 8 hr.

A sample of radioactive material can actually be placed inside the active volume of a flow proportional counter (Figure 13.12), thus avoiding losses due to window absorption. With this type of counter, the chamber is purged with a rapid flow of counter gas, often 10% methane in argon, and a steady flow of gas is maintained during counting. The counter life is virtually unlimited, since the filling gas is constantly replenished. Such a counter is particularly suited for distinguishing and counting low-energy alpha and beta particles. For X-ray detection the flow proportional counter is equipped with an extremely thin window, usually 6-μm Mylar film, which naturally decreases the losses due to window absorption of very soft X rays. Its range extends to 12 Å, and it is the counter of choice for long-wavelength X radiation. Windows of 0.1-μm Formvar or "thin" nitrocellulose film (often supported by screens) extend the transmission to approximately 120 Å and 160 Å, respectively. Since the signal is proportional to the initial energy, pulse height discrimination may be used.

Scintillation Counters. *Scintillators* are chemicals used to convert radiation energy into light. When an ionizing particle is absorbed in any one of several transparent scintillators, some of the energy acquired by the scintillator is emitted as a pulse of visible light or near-ultraviolet radiation. The light is observed by a photomultiplier tube, either directly or through an internally reflecting optic fiber. The combination of a scintillator and photomultiplier tube is called a scintillation counter (Figure 13.13). A good match should exist between the emission spectrum of the scintillator and the response curve of the photocathode. The decay time for scintillators is very short: 250 nsec for a sodium iodide crystal, 20 nsec for anthracene, and 10 nsec for liquid organic systems. The signal from a scintillation counter is proportional to the energy dissipated by the radiation in the scintillator, so that this counter may be used with pulse height discrimination.

For counting alpha particles the best scintillator is a thin layer of silver-activated zinc sulfide, which may be coated on the envelope of the photomultiplier tube.

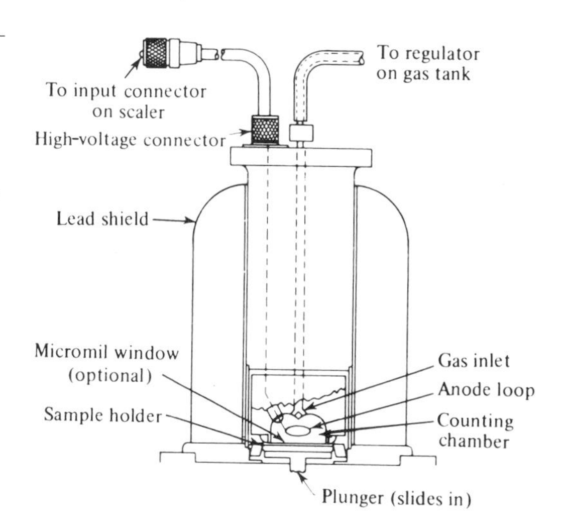

FIGURE **13.12**
Schematic diagram of a flow proportional counter mounted in a shield; the very thin window is optional. The sample is inserted into the active volume by means of the lateral slide holder. (Courtesy of Nuclear-Chicago Corp.)

Scintillation crystals of anthracene or stilbene (wavelength of emission: 445.0 and 410.0 nm, respectively) affixed by a good optical liquid to an end-window photomultiplier tube are suitable for beta particles of moderate and high energy. However, organic liquid scintillators are often preferred because of their shorter decay times. Low-energy beta emitters, such as ^{3}H, ^{14}C, and ^{35}S, are commonly counted by dissolving the compound that contains the radionuclide in the liquid scintillator solution (see Chapter 14).

To measure X rays, gamma radiation, or bremsstrahlung (continuous X radiation produced when electrons are decelerated by the electric fields of nuclei) from high-energy beta emitters, an inorganic scintillator such as a sodium iodide crystal doped with 1% thallium(I) iodide is best. This scintillator has a large photoelectric cross section, a high density, which provides a high probability of absorption, and a high transparency to its own radiation (the optical emission lines of thallium), which enables large thicknesses to be used for the absorption of X and gamma radiation. When such radiation interacts with a NaI(Tl) crystal, the transmitted energy excites the iodine atoms and raises them to higher energy states. When an iodine atom returns to its ground electronic state, this energy is reemitted in the form of an ultraviolet radiation pulse, which is promptly absorbed by the thallium atom and reemitted as fluorescent light at 410.0 nm. The crystal is sealed from atmospheric moisture and protected from extraneous light by an enclosure of aluminum foil, which serves also as an internal reflector. Such a scintillation counter has a nearly uniform and high quantum efficiency throughout the important X-ray region, 0.3–2.5 Å, and is usable to possibly 4 Å (Figure 13.11). Longer wavelengths are absorbed in the coating of the crystal.

The well-type scintillation crystal increases the counting efficiency by surrounding the sample with the detector crystal. The sample is placed in a well drilled into a crystal 5–10 cm in diameter; the size is chosen so that it contains the entire path of the ionizing particle or radiation and measures the total energy. The resolution of a NaI(Tl) counter spectrometer is relatively poor (peak width of 6%

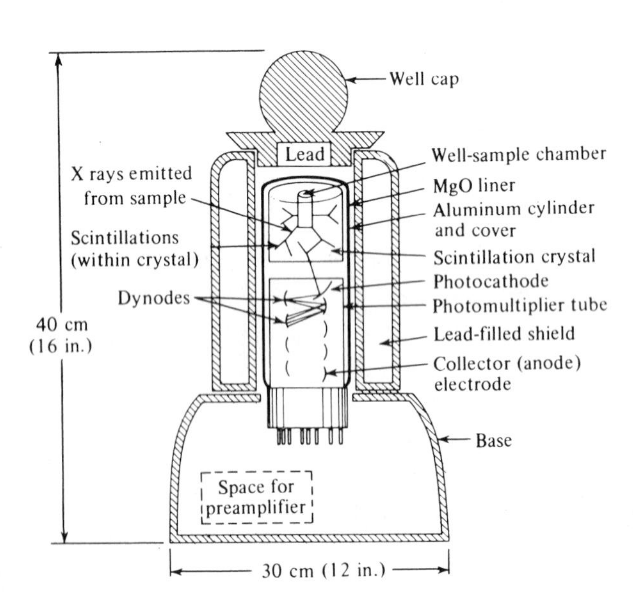

FIGURE **13.13** • Well-type crystal scintillation counter and shield.

at 1 MeV and 18% at 100 keV), but the efficiency approaches 100%, and the instrument is a multichannel device because the entire spectrum is recorded at one time.

Semiconductor Detectors. Semiconductor detectors have revolutionized X- and gamma-ray spectroscopy by providing energy resolution unattainable with previous detectors. In these detectors the charge carriers produced by ionizing radiation are electron-hole pairs rather than ion pairs. The ionizing radiation lifts electrons into the conduction band and these electrons travel toward the positive electrode with high mobilities. The positive charge travels in the opposite direction by successive exchanges of electrons between neighboring lattice sites. Two general types of semiconductor detectors will be discussed: the surface barrier silicon detector and the lithium-drifted silicon and germanium detectors.

A surface barrier detector consists of a *p-n* junction formed at the surface of a slice of silicon. At the junction there is a planar region where there are no charge carriers and no electric field. This region is called the *depletion region.* There is a thin depletion depth when no bias voltage is applied across the *p-n* junction. If a reverse bias is applied, the depletion depth is increased and is given by $d \cong 0.5\sqrt{\rho V}$ (in micrometers), where ρ is the silicon resistivity in ohm-centimeters and V is the bias in volts. Thus, with higher bias and higher resistivity, deeper depletion depths are formed. If charges are injected into the depletion region, they are swept out of it by the electric field, and a voltage pulse appears across the *p-n* junction. If energy measurements are to be made, the selection of a detector for a particular purpose requires selecting a depletion depth greater than or equal to the range of the particle of interest.

The lithium-drifted germanium detector consists of a virtually windowless Ge(Li) crystal (germanium doped with lithium), a vacuum cryostat maintained by cryosorption pumping, a liquid nitrogen Dewar, and a preamplifier. The Ge(Li) crystal is fabricated by drifting lithium ions (a donor) into and through *p*-type germanium. This is performed under the influence of a high electric field at 400 °C. This process results in the compensation of all acceptors within the bulk material, yielding a very high-resistivity (or intrinsic) region that acts like ultrapure germanium within the bulk material. The drifting process is discontinued while a layer of *p*-type germanium still remains (Figure 13.14). This intrinsic or compensated volume becomes the radiation-sensitive region. When ionizing radiation

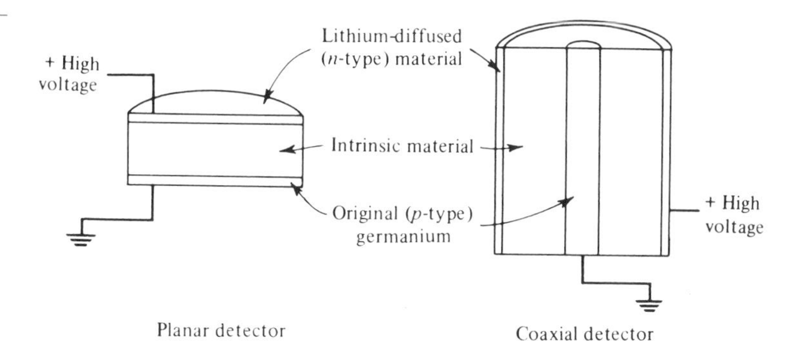

FIGURE **13.14**
Schematic diagrams of two common types of lithium-drifted germanium detectors.

enters the intrinsic layer, electron-hole pairs are created, and the charge produced is rapidly collected under the influence of the bias voltage. The completed detector must be maintained at 77 K at all times to prevent precipitation of the lithium, since the lithium drift process is not stable at room temperature. At this low temperature thermal noise is greatly reduced and the resolution capabilities are vastly increased. In recent years, extremely pure germanium crystals have been grown. These crystals do not require lithium drifting to produce "intrinsic" germanium. Consequently, the crystals need cooling only during use in order to reduce the noise. Since the energy difference between the valence and conduction bands is only 0.66 eV in germanium and 1.1 eV in silicon, some electrons jump into the conduction band at room temperature and cause noise in the detector. Reducing the temperature reduces the thermal energy of the electrons in the crystals and thus the number of electrons that have sufficient energy to jump into the conduction band without excitation by incident radiation.

Lithium-drifted silicon detectors are prepared in a similar manner by drifting lithium ions into *p*-type silicon. They have become quite popular in recent years. These detectors come in a multitude of sizes. Silicon detectors are preferred for X rays longer than 0.3 Å. For shorter wavelength, Ge detectors are necessary, since Si, due to its low atomic number, cannot absorb the X rays effectively in the depth of crystals available.

The energy resolution of semiconductor detectors as compared with gas-filled ionization chambers is intrinsically good because of the large number of electron-hole pairs formed per photoelectron absorbed. The average energy for electron-hole pair production is 2.95 eV for germanium and 3.65 eV for silicon. This is less than the 30 eV required for the production of an ion pair in rare gas-filled detectors and far less than the 500 eV required to produce a photoelectron in a NaI(Tl) scintillation detector. Thus, for a given amount of energy absorbed, about ten times as many electron-hole pairs are formed as ion pairs in gases and about 170 times as many electron-hole pairs as photoelectrons in scintillators. Since the relative resolution is proportional to the square root of the signal, the resolution of the Ge(Li) detector is about a factor of 13 better than the NaI(Tl) detector and three times better than an ionization or proportional detector (see the illustration that accompanies Problem 18 in Chapter 14). A typical figure for energy resolution of a Si(Li) semiconductor detector is about 155 eV for 2.10-Å X rays and ranges from 120 to 200 eV over the range 1–10 Å. This is the full width at half maximum (FWHM) of a peak in the energy spectrum. The rise time is about 10 nsec.

The lithium-drifted silicon detectors and "intrinsic" germanium detectors have increased the popularity of what has become known as energy-dispersive analysis. In energy-dispersive methods of X-ray analysis, the sample is irradiated with X rays, gamma radiation from a radionuclide source, or ions to produce secondary X radiation characteristic of the elements present in the sample. These characteristic X rays pass through a hollow shield onto a semiconductor detector. The purpose of the shield is only to prevent any X rays from hitting the edges of the detector where they may not be completely absorbed. The output of the detector is amplified by a preamplifier and then passed into a pulse height analyzer (Figure 13.15) and associated electronic circuits that eventually present counts versus X-ray energy and thus give qualitative evidence and even rough quantitative measures of the various elements in the sample.

This method, sometimes known as X-ray energy spectrometry (XES) or energy-dispersive X-ray fluorescence analysis (EDXFS), discussed later in this chapter, has the advantage of being able to give a complete qualitative analysis in one operation. Depending on the primary radiation source used, the method can be applied to bulk analyses or just to analyses of surfaces and thin films to a thickness of about 2 nm. Elements of atomic weight down to about carbon are detected. The apparatus is also simple, although the detector must be cooled to liquid nitrogen temperatures for best resolution and the whole must be contained in a vacuum when elements of low atomic number ($Z < 12$) are being sought.

Auxiliary Instrumentation

Detectors require auxiliary electronic equipment, including a high-voltage supply, an amplifier (often plus a preamplifier), a scaler, and a count-registering unit. The required stability of the high-voltage supply and the required sensitivity and linearity of the amplifier are dictated by both the detector and the application. The signal produced when an X-ray quantum is absorbed by proportional, scintillation, or semiconductor detectors is extremely small and requires both a preamplifier and a second stage of amplification before the signal can be fed to a scaler, recorder, or computer. To diminish noise pickup, the preamplifier is located immediately after the detector in the latter's housing.

A significant amount of radiation from natural radioactive elements and cosmic rays is always present in the vicinity of a detector. Insertion of the counter into a shield of lead 2–3 in. thick reduces the background counting rate appreciably (Figures 13.12 and 13.13).

Pulse Height Discrimination. Whenever the amplitude of the pulse is proportional to the energy dissipation in the detector, the measurement of pulse height is a useful tool for energy discrimination. Current pulses are first fed into a linear amplifier of sufficient gain to produce voltage output pulses in the amplitude range of 0–100 V. These amplified pulses are then sorted into groups (analyzed) according to their pulse heights.

One method of analyzing the pulse is with a single-channel analyzer. The baseline discriminator passes only those pulses above a certain amplitude and eliminates pulses below this amplitude. It is useful for excluding scattered radiation and amplifier noise. Pulses associated with a particular energy must be amplified sufficiently so that their amplitudes exceed the discriminator setting. In practice this is accomplished by adjusting a combination of the gain of the amplifier and the dc voltage applied to the detector.

A pulse height analyzer also contains a second discriminator called the window width, the channel width, or the acceptance slit. Now all pulses above the sum of the baseline and window setting are also rejected. Only pulses with an amplitude within the confines of these settings pass on to the counting stages. These operations are outlined schematically in Figure 13.15. With circuits for pulse height discrimination, it is possible to discriminate electronically against unwanted wavelengths of different elements. Discrimination between elements eight to ten atomic numbers apart is possible with a scintillation detector. A proportional detector, because of its narrower pulse amplitude distribution, can discriminate between elements four to

FIGURE 13.15

Block diagram of a single-channel pulse height analyzer.

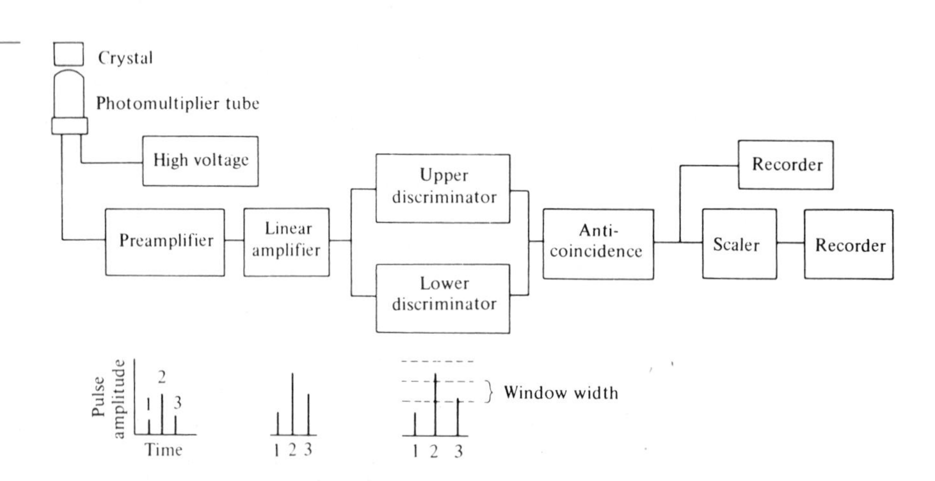

six atomic numbers apart. With semiconductor detectors, even better resolution is possible, one or two atomic numbers apart. For precise quantitative measurements, simple energy dispersion measurements, even with semiconductor detectors, are not sufficient. One must use wavelength dispersion from reflecting (Bragg) crystals, an energy dispersion detector, and pulse height discrimination. Unlike a filter, a pulse height analyzer is used to pass either line of superposed spectral lines, serving, in effect, as a secondary monochromator. It is particularly useful for rejecting higher-order scattered radiation from elements of higher atomic number when determining the elements of lower atomic number. Modern pulse height analyzers with ever-improving resolution make finer and finer nondispersive analyzers.

EXAMPLE 13.3

The use of a pulse height analyzer for Si $K\alpha_1$ radiation illustrates the step-by-step operations of a pulse height analyzer. A relatively pure sample of silicon is inserted into the sample holder (see Figure 13.22). Scanning with the goniometer from 106° to 110° provides the graph shown in Figure 13.16. From an ethylenediamine *d*-tartrate crystal, Si $K\alpha_1$ radiation is reflected at $2\theta = 108°$. Next, with the goniometer set manually at the peak of the silicon radiation, the distribution of pulses due to the silicon X-ray quanta is obtained by scanning the pulse amplitude baseline and using a 1-V window. The integral curve of intensity versus pulse amplitude is shown in Figure 13.17. From this information, the base of the pulse height discriminator would be set at 8.5 V and the window width at 13.0 V, since it is noticed that no pulses are detected until the upper-line setting approaches 21 V. In this example, the silicon radiation was peaked at 15 V. Naturally, if the silicon pulses were peaked at a lower or higher voltage, the window and baseline settings would be different. The peak distribution (in volts) is a function of the dc voltage on the counter and of the amplifier gain.

13.3 DIRECT X-RAY METHODS

The process of exciting characteristic spectra by electron bombardment was applied many years ago in the investigation of characteristic spectra of the elements by Siegbahn and others. In this manner the element hafnium was discovered by Von Hevesy and Coster in 1923. The specimen must be plated or smeared on the target of the X-ray tube. Disadvantages are that the X-ray tube must be evacuated each time the specimen is changed, a demountable target is required, and the heating effect of

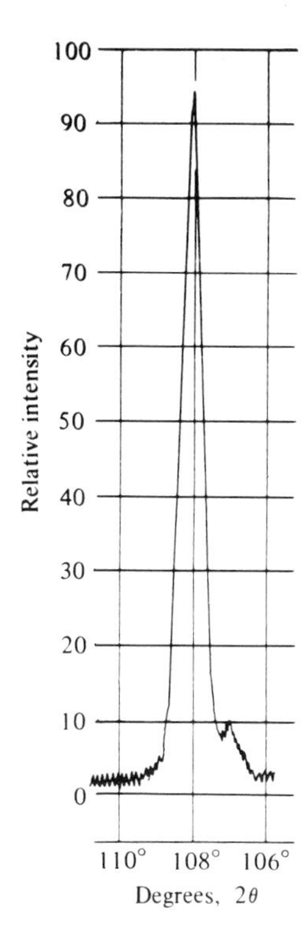

FIGURE 13.16
Relative intensity of the Si $K\alpha_1$ line as function of goniometer setting (2θ). Analyzing crystal: ethylenediamine *d*-tartrate.

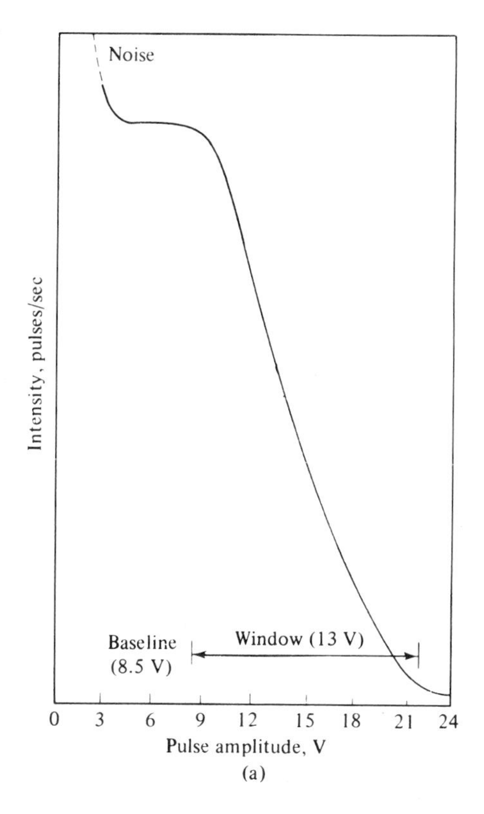

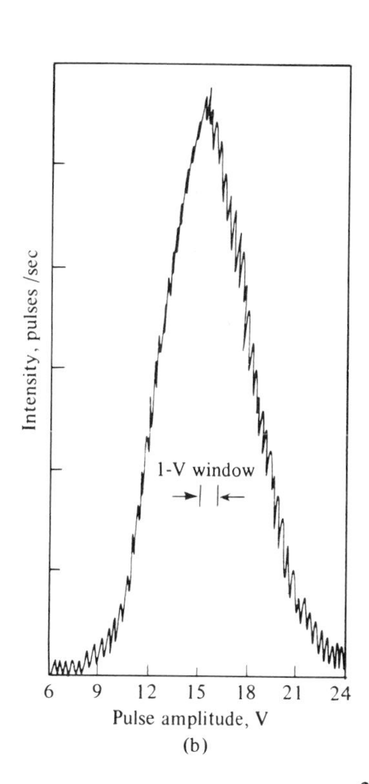

FIGURE 13.17
Pulse amplitude distribution of Si $K\alpha_1$ radiation: (a) integral curve and (b) differential curve. Goniometer set at 108°

the electron beam may cause a chemical reaction, selective volatilization, or melting of the sample coating. These difficulties virtually prohibit the large-scale application of the direct method to routine analyses except for electron probe microanalysis and Auger emission spectroscopy (AES), discussed later in this chapter.

Electron Beam Probe

Electron probe microanalysis, developed by Castaing in 1951, is a method for the nondestructive elemental analysis from an area only 1 μm in diameter at the surface of a solid specimen.[3] A beam of electrons is collimated into a fine pencil of 1-μm cross section and directed at the specimen surface exactly on the spot to be analyzed. This electron bombardment excites characteristic X rays essentially from a point source and at intensities considerably higher than with fluorescent excitation. The limit of detectability (in a 1-μm region) is about 10^{-14} g. The relative accuracy is 1%–2% if the concentration is greater than a few percent and if adequate standards are available.

Three types of optics are used in the microprobe spectrometer: electron optics, light optics, and X-ray optics (Figure 13.18). Of these, the most complex is the electron optical system, a modified electron microscope, which consists of an electron gun followed by two electromagnetic focusing lenses to form the electron beam probe. The specimen is mounted as the target inside the vacuum column of the instrument and under the beam. A focusing, curved-crystal X-ray spectrometer or a semiconductor detector with a pulse height discriminator is attached to the evacuated system with the focal spot of the electron beam serving as the source of X radiation. A viewing microscope and mirror system allow continuous visual observation of the exact area of the specimen where the electron beam strikes. Point-by-point microanalysis is accomplished by translating the specimen across the beam.

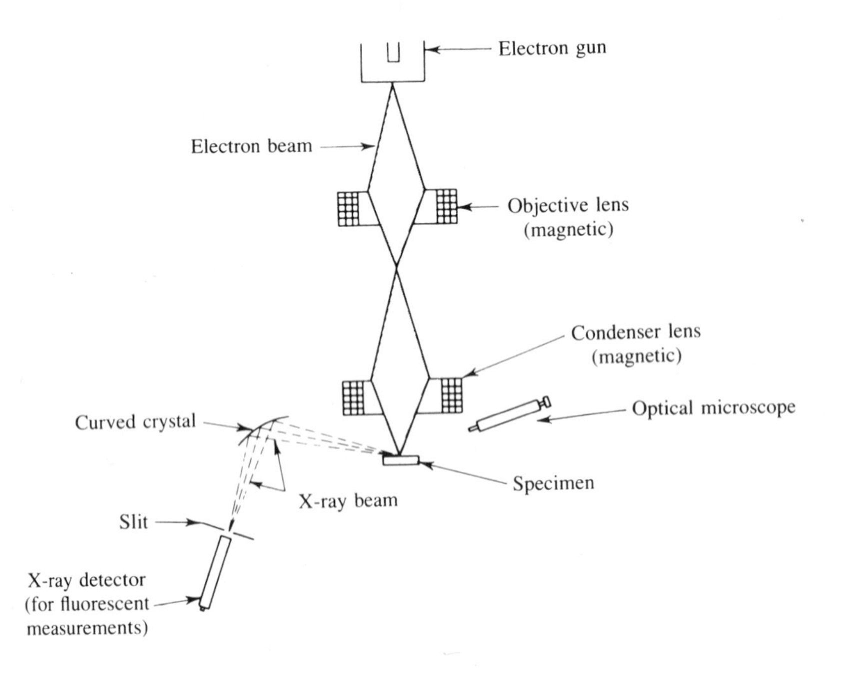

FIGURE 13.18
Schematic of an electron probe microanalyzer. The X-ray beam can be passed directly into the detector or reflected from the analyzer crystal.

The method is used to study the variations in concentration that occur near grain boundaries, the analysis of small inclusions in alloys or precipitates in a multitude of products, and corrosion studies where excitation is restricted to thin surface layers because the beam penetrates to a depth of only 1 or 2 μm into the specimen.

13.4 X-RAY ABSORPTION METHODS

Because each element has its own characteristic set of K, L, M, and other absorption edges, the wavelength at which a sudden change in absorption occurs is used to identify an element present in a sample, and the magnitude of the change determines the amount of the particular element present. The fundamental equation for the transmittance of a monochromatic, collimated X-ray beam is

$$P = P_0 e^{-(\mu/\rho)\rho x} \tag{13.7}$$

where P is the radiant power of P_0 after passage through x cm of homogeneous matter of density ρ and whose linear absorption coefficient is μ. Note the similarity to Beer's law discussed in Chapters 5 and 7. The parenthetical term μ/ρ is the mass absorption coefficient, often expressed simply as μ_m. It depends on the wavelength of the X rays and the absorbing atom; that is,

$$\mu_m = CZ^4\lambda^3 \frac{N_A}{W} \tag{13.8}$$

where N_A is Avogadro's number, W is the atomic weight, and C is a constant valid over a range between characteristic absorption edges. It is significant that the mass absorption coefficient is independent of the physical or chemical state of the specimen. In a compound or mixture it is an additive function of the mass absorption coefficients of the constituent elements; namely,

$$\mu_{mT} = \mu_{m1} W_1 + \mu_{m2} W_1 + \cdots \tag{13.9}$$

where μ_{m1} is the mass absorption coefficient of element 1 and W_1 is its weight fraction, and so on for all the elements present. Because only one element has a change in mass absorption coefficient at the edge, the following relationship is obtained for the ith element:

$$2.3 \log \frac{P}{P_0} = (\mu''_{m1} - \mu'_{m1}) W_i \rho x \tag{13.10}$$

where the term in parentheses represents the difference in mass absorption coefficient at the edge discontinuity. Thus, the logarithm of the ratio of beam intensities on the two sides of an absorption edge depends only on the change in mass absorption coefficients of the particular element in the beam; the product, ρx, is the mass thickness of the sample in grams per square centimeter. There is no matrix effect, which gives the absorption method an advantage over X-ray fluorescence analysis in some cases.

By analogy with absorption measurements in other portions of the electromagnetic spectrum, one would expect to obtain a representative set of transmittance measurements on each side of an absorption edge with an X-ray spectrometer and extrapolate to the edge. However, X-ray absorption spectrometers that provide a continuously variable wavelength of X radiation are not commercially available. Instead, only a single attenuation measurement is made on each side of the edge. A multichannel instrument is required.

The general procedure is illustrated by the determination of tetraethyl lead and 1,2-dibromoethane in gasoline. Four channels are required. One channel is used as a reference standard; the other three channels provide for the analyses for lead and bromine and a correction for variations of the C/H ratio and the presence of any sulfur and chlorine. Primary excitation is provided by an X-ray tube operated at 21 kV. The secondary targets for each channel are as follows, with the fluorescent X-ray lines used:

Channel 1: RbCl, Rb $K\alpha_1$

Channel 2: RbCl, Rb $K\alpha_1$

Channel 3: $SrCO_3$, Sr $K\alpha_1$

Channel 4: NaBr, Br $K\alpha_1$

The relationship between the pertinent absorption edges and the target fluorescent emission lines is shown in Figure 13.19. In operation, a reference sample is sealed in the sample cell in Channel 1; the sample to be analyzed is placed in the remaining channels. The exposure is started and automatically terminated when the integrated intensity in Channel 1 reaches a predetermined value (perhaps 100,000 counts in a

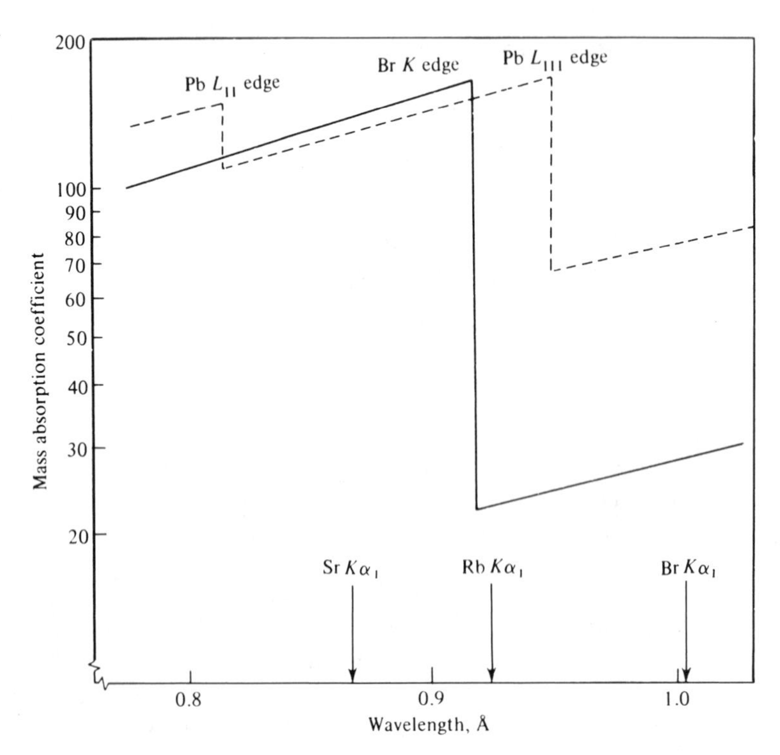

FIGURE **13.19**
Absorption edges and emission lines pertinent to the X-ray absorption analysis of lead tetraethyl and dibromomethane in gasoline.

time interval of 100 sec). The integrated intensities accumulated in the other channels are then recorded. Initially the four channels are adjusted to reach 100% transmittance with nominally pure gasoline. Results for bromine are computed from the difference in counts between Channels 3 and 2; for lead from Channels 2 and 4.

Radiography

Another application that uses the different absorbing powers of different elements toward an X-ray beam permits the gross structure of various types of small specimens to be examined under high magnification. Positions where there are elements that strongly absorb the X rays appear light, and positions where there are elements that do not absorb the X rays appear dark on a film placed behind the sample.

Clark and Gross developed a method that uses ordinary X-ray diffraction equipment.[4] Any of the common targets operated at 30–50 kV can be used. No vacuum camera is necessary. The microradiographic camera, shown in Figure 13.20, is designed to fit as an inset in the collimating system of commerical X-ray equipment. Special photographic film that has an extremely fine grain makes magnifications up to 200 times possible without loss of detail from graininess. Sample thicknesses vary from 0.075 mm for steels to 0.25 mm for magnesium alloys. Only a few seconds of exposure are necessary.

Various techniques are possible depending on the specimen. Biological specimens may be impregnated with a material of high molecular weight to characterize particular structures. Occasionally the necessary density variations are initially present in the sample. More often, various selective monochromatic wavelengths from different target elements must be used.

Another technique uses a tube with a very small focal spot as an X-ray source. If the focal spot approaches a point source, magnification in the microradiograph is obtained by simple geometry with considerable sharpness. Actually, focal spots as small as 5 μm in diameter can be obtained and magnifications up to 50 times or so are possible. The magnification is the ratio of the distance of film from the target to the distance of object from the target.

To avoid the use of photographic film, which requires time for development and later inspection, modern radiographic instruments often use a television camera covered with a film of material that fluoresces on exposure to X rays. The image from the camera is then monitored on a television screen. This method of X-ray fluoroscopic examination allows direct on-line examination of objects. The objects can also be manipulated so as to observe any characteristics in the most favorable position. If the object being inspected is immobile, the X-ray and television

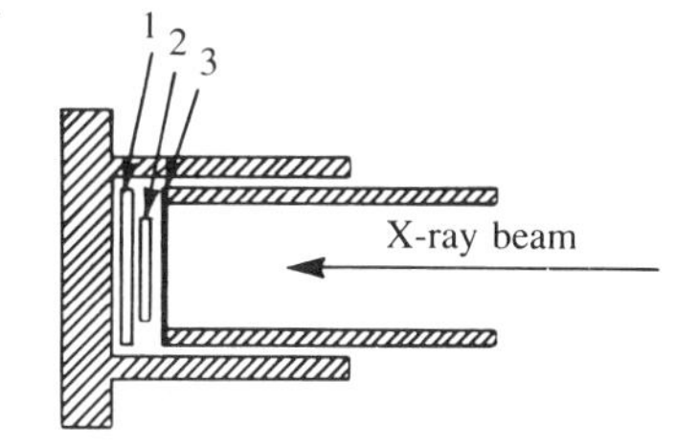

FIGURE **13.20**
Schematic of microradiographic camera: (1) film, (2) sample, and (3) black paper.

apparatus can be made mobile. Thus welds in heavy castings, boilers, pipelines, and so on can be inspected in situ. The security systems used in airports for inspecting luggage are another example of the X-ray fluoroscopic inspection method.

One of the earliest uses of X-ray absorption methods was in medicine for inspection of the human body. Bones contain a high proportion of calcium and therefore absorb X rays more strongly than other tissues. Both film methods and fluoroscopy are now used to observe the body. If blood or soft tissues are of interest, then contrast can be enhanced by ingestion or injection of more strongly absorbing substances such as barium compounds or iodine-containing chemicals. A recent development in medicine is *tomography*. In conventional radiography, the X-ray source, the object being investigated, and the detection device are held as still as possible in order to obtain a sharp image. In tomography, the source and the detection device are moved so as to keep the plane being inspected always in the same relative position between the source and detector. The shadows of other planes are displaced relative to the source and detector as the movement takes place, thus blurring or obliterating their images. The tomographic method gives three-dimensional detail.

Nondispersive X-ray Absorptiometer

The general arrangement of a nondispersive X-ray absorptiometer is shown in Figure 13.21. A tungsten target X-ray tube is operated at 15–45 kV. In the X-ray beam is a synchronous motor-driven chopper that alternately interrupts half of the X-ray beam. A variable-thickness aluminum attenuator (in the shape of a wedge) is placed between the chopper and the reference sample compartment. Duplicate reference and sample cells up to 65 cm long can be accommodated and those for liquids and gases arranged for continuous-flow process stream analyses if desired. Both halves of the X-ray beam fall on a common phosphor-coated photomultiplier tube, which is protected from visible light by a thin metallic filter.

In operation, a reference sample is placed in the appropriate cell and the specimen to be analyzed in the sample tube. The attenuator is adjusted until the absorption in the two X-ray beams is brought into balance. The change in thickness of aluminum required for different samples is a function of the difference in

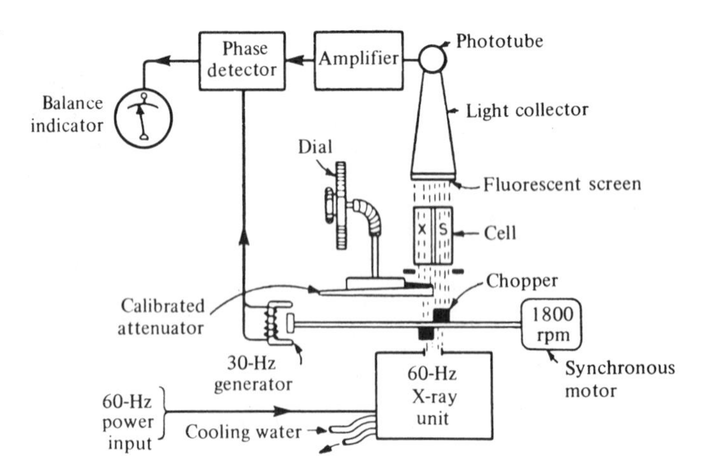

FIGURE **13.21**
Nondispersive X-ray absorptiometer. (Courtesy of General Electric Co.)

composition. Prior calibration enables a determination in terms of the solute in an unknown. Liquids are simplest to handle. With solids the thickness of solid specimens and the density of samples of powders must be uniform to a precision greater than that expected in the results.

Polychromatic absorptiometry can be used to determine chlorine in hydrogen. Sulfur in crude oil can be distinguished from the carbon-hydrogen residuum. Other examples are barium fluoride in carbon brushes, barium or lead in special glasses, and chloride in plastics and hydrocarbons. In fact, the method is applicable to any sample that contains one element that is markedly heavier than the others and when the matrix is essentially invariant in concentration.

13.5 X-RAY FLUORESCENCE METHOD

Characteristic X-ray spectra are excited when a specimen is irradiated with a beam of sufficiently short-wavelength X radiation. Intensities of the resulting fluorescent X rays are smaller by a factor of roughly 1000 than an X-ray beam obtained by direct excitation with a beam of electrons. Only the availability of high-intensity X-ray tubes, very sensitive detectors, and suitable X-ray optics renders the fluorescence method feasible. Beam intensity is important because it influences the time necessary to measure a spectrum. A certain number of quanta must be accumulated at the detector to reduce sufficiently the statistical error of the measurement. The detection limit of the analysis—that is, the lowest detectable concentration of a particular element in a specimen—depends on the peak-to-background ratio of the spectral lines. Relatively few cases of spectral interference occur because of the relative simplicity of X-ray spectra.

X-ray Fluorescence Spectrometers

Wavelength Dispersive Devices. The general arrangement for exciting, dispersing, and detecting fluorescent radiation with a plane-crystal spectrometer is shown in Figure 13.22. The specimen in the sample holder (often rotated to improve uniformity of exposure) is irradiated with an unfiltered beam of primary X rays, which causes the elements present to emit their characteristic fluorescence lines. A portion of the scattered fluorescence is collimated by the entrance slit of the goniometer and directed onto the plane surface of the analyzing crystal. The line radiations, reflected according to the Bragg condition, pass through an auxiliary collimator (exit slit) to the detector where the energy of the X-ray quanta is converted into electrical impulses, or counts.

The primary slit, the analyzer crystal, and secondary slit are placed on the focal circle so that Bragg's law is always satisfied as the goniometer head is rotated. The detector is rotated at twice the angular rate of the crystal. The analyzer crystal is a flat single-crystal plate, 2.5 cm wide and 7.5 cm long. The specimen holder is often an aluminum cylinder, although plastic material is used to examine acid or alkaline solutions. A thin film of Mylar supports the specimen, and an aluminum mask restricts the area irradiated (often a rectangle 18 mm × 27 mm). Intensity losses caused by the absorption of long-wavelength X rays by air and window materials are reduced by evacuating the goniometer chamber. Another method for reducing losses is to enclose the radiation path in a special boot that extends from the sample surface to the detector window and then displace the air by helium, which has a low absorption coefficient. Vacuum spectrometers are used where helium is scarce and for the elements boron ($Z = 5$) to sodium ($Z = 11$).

Focusing spectrometers that involve reflection from or transmission through a 10-cm or 28-cm curved crystal have been described.[5] Collimators are not required and the increase in intensity obtained by focusing the fluorescence lines makes the technique suitable for the analysis of small specimens. In the curved-crystal arrangement the analyzing crystal is bent to a radius of curvature twice that of the focal circle (Figure 13.23). A slit on the focusing circle acts as a divergent source of polychromatic radiation from the specimen. All the radiation of one wavelength that diverges from the slit is diffracted at a particular setting of the crystal, and the diffracted radiation converges to a line image at a symmetric point on the focusing circle. The angular velocity of the detector is twice that of the crystal and, as the two of them move along the periphery of the circle, the X-ray spectral lines are dispersed and detected just as in the flat-crystal arrangement.

To excite fluorescence, the primary radiation must obviously have a wavelength shorter than the absorption edge of the spectral lines desired. Continuous as well as characteristic radiation of the primary target serves the purpose. To get a continuous spectrum of short enough wavelength and sufficient intensity, one may calculate the required voltage of the X-ray tube from Equation 13.2, remembering that the wavelength of maximum intensity is approximately $1.5\lambda_0$. In qualitative analyses it is usually desirable to operate the X-ray tube at the highest permissible voltage to ensure that the largest possible number of elements in the specimen are excited to fluoresce. It also ensures the greatest possible intensity of fluorescence for each element in quantitative analyses. In two cases, however, the X-ray tube voltage should be made lower than the available maximum: (1) when it is desirable not to excite fluorescence of all elements in the specimen but rather to use selective excitation conditions, and (2) when very long-wavelength spectral lines are excited in

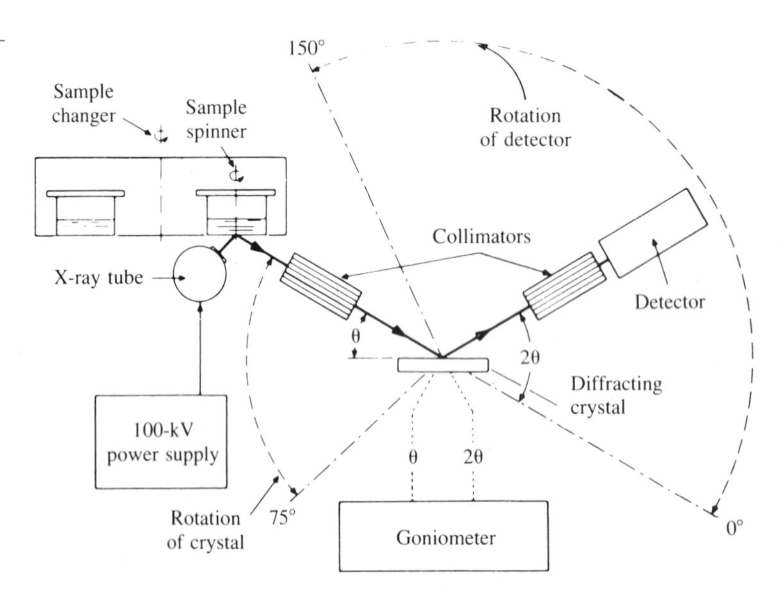

FIGURE **13.22**
Geometry of a plane-crystal X-ray fluorescence spectrometer. (Courtesy of Philips Electronic Instruments.)

order to minimize scattering of primary radiation through the system by holding down the intensity of the short-wavelength continuum. Besides X rays, electron bombardment as used in the scanning electron microscope (SEM), the electron probe (EP), and the transmission electron microscope also excites the characteristic fluorescent X rays of the elements present in a sample. For this reason, many scanning electron microscopes, electron probes, and transmission electron microscopes are now available with optional crystal analyzers and detectors or Si(Li) or germanium detectors and pulse height analyzers to permit better X-ray identification of elements. The most modern instruments are computer-controlled. The computer controls the scanning rate to scan rapidly where no lines exist and slowly in the regions of fluorescence peaks. It also records the output of the detector. The best instruments in terms of resolution also use energy-dispersive detectors, gas-filled ionization or proportional counters, or semiconductors, with pulse height discriminators for the detector output to improve resolution.

Certain radioisotopes are X-ray emitters and thus can be used as excitation sources for fluorescence. Since no high-voltage supply or high-vacuum equipment is necessary, radioisotope sources are often used in portable equipment for such applications as the monitoring of mine waters, stream pollution, and other field testing, especially for pollutants. X-ray-emitting isotopes also give monochromatic X rays without the continuous background involved in ordinary X-ray tubes. On the other hand, the radiation cannot be turned off and constant, bulky shielding is required. The intensity of these sources is usually low. Some sources may have a rather short half-life and need frequent replacement. Also care must always be exercised in disposing of the old sources.

Bombardment of materials by ions such as protons can also lead to the emission of fluorescent X-rays. Protons of several million electron volts (MeV) of energy from ion accelerators have a flux density about two orders of magnitude greater

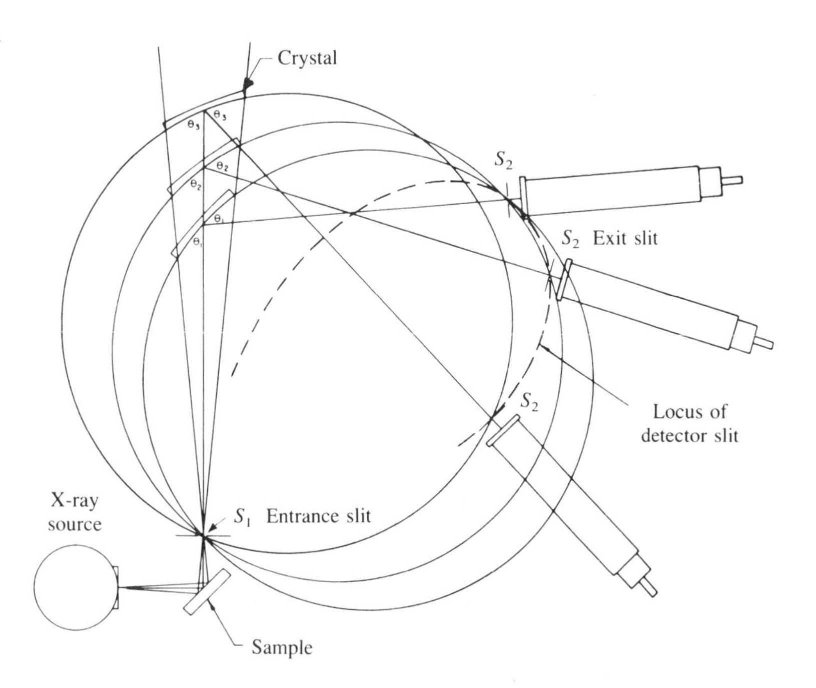

FIGURE 13.23
Focusing X-ray optics, using a curved analyzing crystal. (Courtesy of Applied Research Laboratories, Inc.)

than that from a standard X-ray tube. Protons do not penetrate deeply into matter and thus the use of protons or other ionized particles is particularly useful for very thin samples such as particulate samples collected on a thin substrate. This method of analysis is sometimes known by the acronym PIXE (particle-induced X-ray emission).

Energy Dispersion Spectrometers. For some samples where only a very few elements are present and their X-ray lines are widely separated in wavelength, the crystal analyzer may be eliminated and energy-dispersive detectors with pulse height discrimination used in its place. For concentrations greater than about 1% and elements separated by a few atomic numbers, energy dispersion analysis is very useful because the intensities are increased about 1000-fold. The intensity at the detector is drastically increased because no collimating slits are required, thus increasing the energy throughput, and the detector can be placed very close to the sample, again increasing the energy intercepted by the detector. With such an increase in radiation intensity, weaker primary sources can be used—for example, radioisotopes. The resolution of an energy dispersion instrument, however, is as much as 50 times less than the wavelength dispersion spectrometer using a crystal; thus, lines from nearby elements may overlap. As pointed out, for precise quantitative measurements, wavelength dispersion followed by energy-dispersive detectors and pulse height discriminators must be used to get sufficient resolution.

Analytical Applications

For qualitative analysis using a wavelength-dispersive spectrometer, the angle θ between the surface of the crystal and the incident fluorescence beam is gradually increased. At certain well-defined angles the appropriate fluorescence lines are reflected. In automatic operation the intensity is recorded on a moving chart as a series of peaks corresponding to fluorescence lines above a background that arises principally from general scattering. The angular position of the detector, in degrees of 2θ, is also recorded on the chart. Additional evidence for identification may be obtained from relative peak heights and the critical excitation potential. If one uses an energy-dispersive spectrometer, the output of the pulse height discriminator is recorded.

For quantitative analysis, the intensity of a characteristic line of the element to be determined is measured. The goniometer is set at the 2θ angle of the peak and counts are collected for a fixed period of time, or the time is measured for the period required to collect a specified number of counts. For major elements, 200,000 counts can be accumulated in 1 or 2 min. The goniometer is then set at a nearby portion of the spectrum where a scan has shown that only the background contributes. Background counts require much longer times; a very low background may require 10 min to acquire 10,000 counts. The net line intensity—that is, peak minus background—in counts per second is then related to the concentration of the element via a calibration curve.

Particle size and shape are important and determine the degree to which the incident beam is absorbed or scattered. Standards and samples should be ground to the same mesh size, preferably finer than 200 mesh. Errors from differences in packing density are handled by the addition of an internal standard to the sample.

Powders are pressed into a wafer in a metallurgical specimen press or converted into a solid solution by fusion with borax. Samples are best handled as liquids. If they can be conveniently dissolved, their analysis is greatly simplified and precision is greatly improved. Liquid samples should exceed a depth that will appear infinitely thick to the primary X-ray beam, about 5 mm for aqueous samples. The solvent should not contain heavy atoms. In this respect HNO_3 and water are superior to H_2SO_4 or HCl.

Before relating the intensity of fluorescent emission to the concentration of the emitting element, it is usually necessary to correct for matrix effects. Matrix dilution avoids serious absorption effects. The samples are heavily diluted with a material that has a low absorption, such as powdered starch, lithium carbonate, lampblack, gum arabic, or borax (used in fusions). The concentration, and therefore the effect, of the disturbing matrix elements is reduced, along with the measured fluorescence. However, the most practical way to apply a systematic correction is with an internal standard. Even so, the internal standard technique is valid only if the matrix elements affect the reference line and analytical line in exactly the same way. The choice of a reference element for use as an internal standard depends on the relative positions of the characteristic lines and the absorption edges of the element to be determined, the reference element, and the disturbing elements responsible for the matrix effects. If either the reference line or the analytical line is selectively absorbed or enhanced by a matrix element, the internal standard line to analytical line intensity ratio is not a true measure of the concentration of the element being determined. Preferential absorption of a line would occur if a disturbing element had an absorption edge between the comparison lines. The intensity of a line can be enhanced if a matrix element absorbs primary radiation and then, by fluorescence, emits radiation that, in turn, is absorbed by a sample element and causes the sample to fluoresce more strongly. Thus, if the matrix fluorescence lies between the absorption edges of the analytical and internal standard elements, selective enhancement might result. Many algorithms have been proposed for correcting for matrix effects and converting intensities to concentration.[6–9]

Matrix effects such as absorption and enhancement are often negligible when thin samples are used for analyses. A thin sample is defined as one in which the total mass per unit area, m, is given by

$$m \leq \frac{0.1}{\bar{\mu}} \tag{13.11}$$

where $\bar{\mu}$ is the effective average mass absorption coefficient. For excitation by X rays, the value of $\bar{\mu}$ is given by

$$\bar{\mu} = \mu_1 \operatorname{cosec} \theta_1 + \mu_2 \operatorname{cosec} \theta_2 \tag{13.12}$$

where μ_1 and μ_2 are the mass absorption coefficients for the exciting and fluorescent radiation and θ_1 and θ_2 are the angles of incidence and emergence, respectively. For these thin samples the count rate of the characteristic radiation, I_i, from an element is directly related to the element mass per unit area, m_i, in grams per square centimeter by

$$I_i = S_i m_i \tag{13.13}$$

where S_i is the calibration factor for element i. The National Bureau of Standards has developed two thin-glass reference samples for calibration of X-ray fluorescence spectrometers.[10] One reference sample contains Al, Si, Ca, V, Mn, Co, and Cu, and the other contains Si, K, Ti, Fe, Zn, and Pb. Thin samples are particularly useful for the elemental analysis of particulate matter collected on a filter, mesh, or membrane. Particulate matter from airborne or waste-water samples is an example of such a use.

The X-ray fluorescence method, inherently very precise, rivals the accuracy of wet chemical techniques in the analysis of major constituents. On the other hand, it is difficult to detect an element present in less than one part in 10,000. The method is attractive for elements that lack reliable wet chemical methods—for example, elements such as niobium, tantalum, sodium, and the rare earths. It often serves as a complementary procedure to optical emission spectrography, particularly for major constituents, and also for the analysis of nonmetallic specimens because the sample need not be an electrical conductor. To overcome air absorption for elements of atomic number less than 21, operating pressure must be 0.1 torr or less. Even so, below magnesium the transmission becomes seriously attenuated, although the method has been extended to boron. The ultimate limit of X-ray fluorescence (XRF) in absolute terms is about 10^{-8} g, whereas that of the PIXE method is around 10^{-12} g.

Simultaneous analysis of several elements is possible with automatic equipment. Instruments of this type have semifixed monochromators with optics mounted around a centrally located X-ray tube and sample position. Each crystal is adjusted to reflect one fluorescence line to its associated detector. A compatible recording unit permits both optical and X-ray units to be recorded with the same console.

13.6 X-RAY DIFFRACTION

Every atom in a crystal scatters an X-ray beam incident upon it in all directions. Because even the smallest crystal contains a very large number of atoms, the chance that these scattered waves would constructively interfere would be almost zero except for the fact that the atoms in crystals are arranged in a regular, repetitive manner. The condition for diffraction of a beam of X-rays from a crystal is given by the Bragg equation (Equation 13.3). Atoms located exactly on the crystal planes contribute maximally to the intensity of the diffracted beam. Atoms exactly halfway between the planes exert maximum destructive interference, and those at some intermediate location interfere constructively or destructively depending on their exact location but with less than their maximum effect. Furthermore, the scattering power of an atom for X rays depends on the number of electrons it possesses. Thus the position of the diffraction beams from a crystal depends only on the size and shape of the repetitive unit of a crystal and the wavelength of the incident X-ray beam, whereas the intensities of the diffracted beams depend on the type of atoms in the crystal and the location of the atoms in the fundamental repetitive unit, the unit cell. No two substances, therefore, have absolutely identical diffraction patterns when one considers both the direction and intensity of all diffracted beams. However, some similar, complex organic compounds may have almost identical patterns. The diffraction pattern is thus a fingerprint of a crystalline compound, and the crystalline components of a mixture can be identified individually.

Reciprocal Lattice Concept

Diffraction phenomena are interpreted most conveniently with the aid of the reciprocal lattice concept. A plane can be represented by a line drawn normal to the plane. The spatial orientation of this line describes the orientation of the plane. Furthermore, the length of the line can be fixed in an inverse proportion to the interplanar spacing of the plane that it represents.

When a normal is drawn to each plane in a crystal with a length inversely proportional to the interplanar spacing and the normals are drawn from a common origin, the terminal points of these normals constitute a lattice array. This is called the *reciprocal lattice* because the distance of each point from the lattice is reciprocal to the interplanar spacing of the planes that it represents. Figure 13.24 shows, near the origin, the traces of several planes in a unit cell of a crystal—namely, the (100), (001), (101), and (102) planes. [Planes in crystals are represented by the Miller indices (h, k, l), where h, k, and l are the reciprocals of the intercepts of the plane on the three crystal axes: x, y, and z. Thus the plane (100) intercepts the x-axis at $x = 1$ and is parallel to the y- and z-axes. A plane (112) intercepts the x-axis at $x = 1$, the y-axis at $y = 1$, and the z-axis at $z = \frac{1}{2}$.] The normals to these planes, also indicated, are called the reciprocal lattice vectors, σ_{hkl}, and are defined by

$$\sigma_{hkl} = \frac{\lambda}{d_{hkl}} \tag{13.14}$$

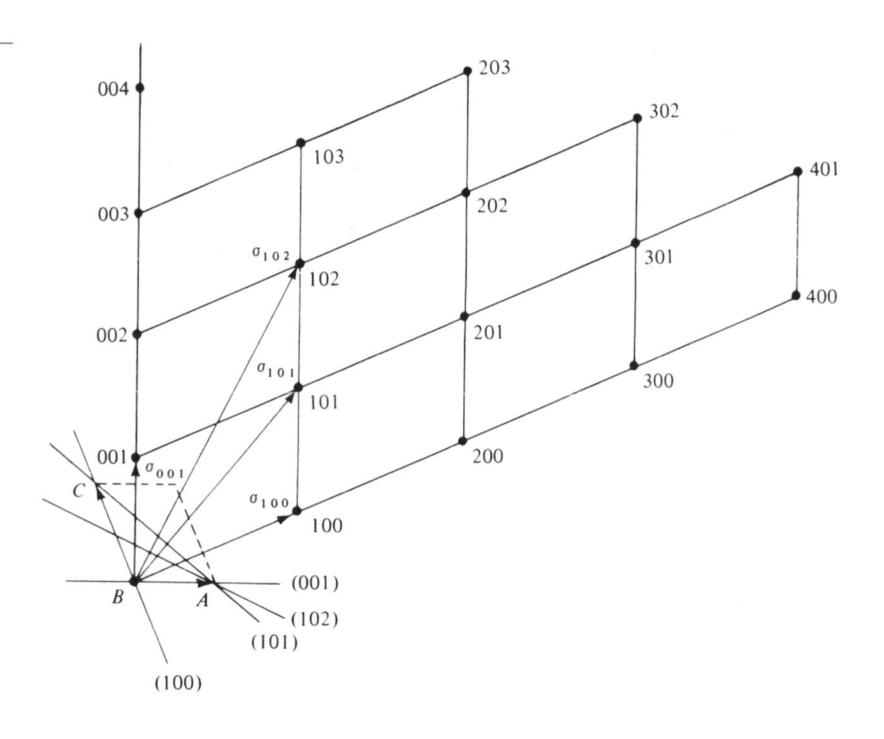

FIGURE **13.24**
Side view of several planes in the unit cell of a crystal, with the normals to the planes indicated.

In three dimensions, the lattice array is described by three reciprocal lattice vectors whose magnitudes are given by

$$a^* = \sigma_{100} = \frac{\lambda}{d_{100}} \tag{13.15a}$$

$$b^* = \sigma_{010} = \frac{\lambda}{d_{010}} \tag{13.15b}$$

$$c^* = \sigma_{001} = \frac{\lambda}{d_{001}} \tag{13.15c}$$

and whose directions are defined by three interaxial angles, α^*, β^*, and γ^*.

Writing the Bragg equation in a form that relates the glancing angle θ most clearly to the other parameters, one has

$$\sin \theta_{hkl} = \frac{\lambda/d_{hkl}}{2} \tag{13.16}$$

The numerator is taken as one side of a right triangle with θ as another angle and the denominator as its hypotenuse (Figure 13.25a). Because of the physical meaning of the quantities in Equation 13.14, the construction can be interpreted as shown in Figure 13.25b. The diameter of the circle ($\overline{ASO}$) represents the direction of the incident X-ray beam. A line through the origin of the circle, parallel to $\overline{AP}$ and forming the angle θ with the incident beam, represents a crystallographic plane that satisfies the Bragg diffraction condition. The line $\overline{SP}$, also forming the angle θ with the crystal plane and 2θ with the incident beam, represents the diffracted beam's direction. Then the line $\overline{OP}$ is the reciprocal lattice vector to the reciprocal lattice point P_{hkl} that lies on the circumference of the circle. The vector σ_{hkl} originates at the point on the circle where the direct beam leaves the circle. The Bragg equation is satisfied when and only when a reciprocal lattice point lies on the "sphere of reflection," a sphere formed by rotating the circle upon its diameter $\overline{ASO}$.

Thus, the crystal in a diffraction experiment can be pictured at the center of a sphere of unit radius, and the reciprocal lattice of this crystal is centered at the point where the direct beam leaves the sphere, as shown in Figure 13.26. Because the orientation of the reciprocal lattice bears a fixed relation to that of the crystal, if the crystal is rotated, the reciprocal lattice can be pictured as rotating also. Whenever a

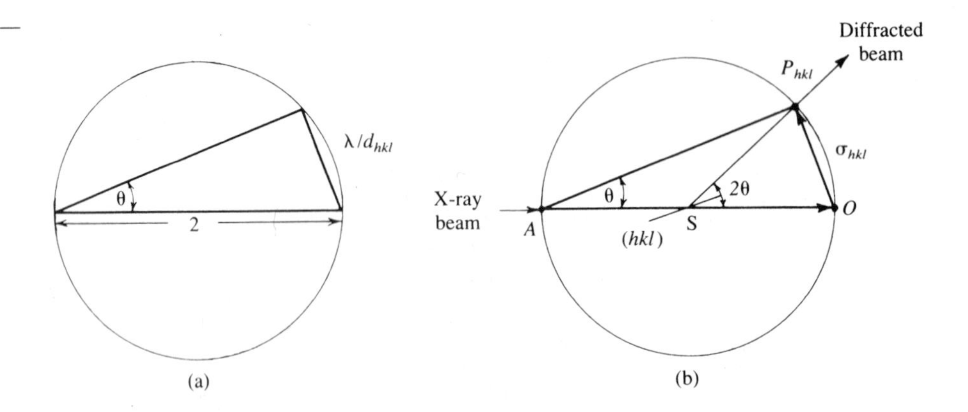

FIGURE **13.25**
Representation of the diffraction condition.

reciprocal lattice point intersects the sphere, a reflection emanates from the crystal at the sphere's center and passes through the intersecting reciprocal lattice point.

Diffraction Patterns

If the X-ray beam is monochromatic and a single crystal is the sample, there will be only a limited number of angles at which diffraction of the beam can occur. The actual angles are determined by the wavelength of the X rays and the spacing between the various planes of the crystal. In the *rotating-crystal method,* monochromatic X radiation is incident on a single crystal that is rotated about one of its axes. The reflected beams lie as spots on the surface of cones that are coaxial with the rotation axis. If, for example, a single cubic crystal is rotated about the (001) axis, which is the equivalent to rotation about the c^* axis, the sphere of reflection and the reciprocal lattice are as shown in Figure 13.27. The diffracted beam directions are determined by intersection of the reciprocal lattice points with the sphere of reflection. All the reciprocal lattice points that lie in any one layer of the reciprocal lattice perpendicular to the axis of rotation intersect the sphere of reflection in a circle. The height of the circle above the equatorial plane is proportional to the vertical reciprocal lattice spacing c^*. By remounting the crystal successively about different axes, one can determine the complete distribution of reciprocal lattice points. Of course, one mounting is sufficient if the crystal is cubic, but two or more may be needed if the crystal has lower symmetry.

In the *powder method*, the crystal is replaced by a large collection of very small crystals, randomly oriented, and a continuous cone of diffracted rays is produced. There are some important differences, however, with respect to the rotating-crystal method. The cones obtained with a single crystal are not continuous because the diffracted beams occur only at certain points along the cone, whereas the cones with

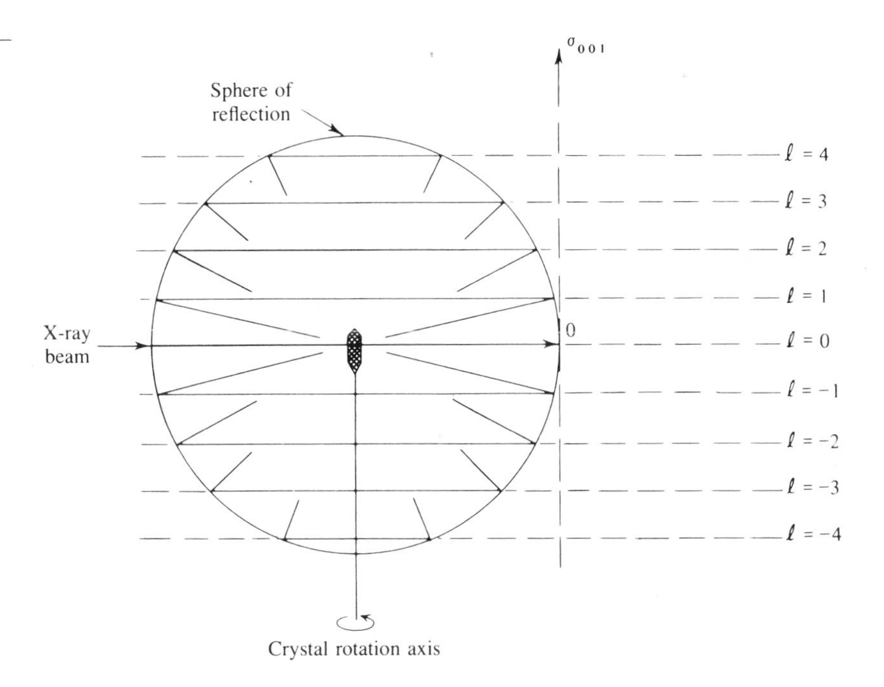

FIGURE **13.26**
Reciprocal lattice construction for a rotating crystal.

the powder method are continuous. Furthermore, although the cones obtained with rotating single crystals are uniformly spaced about the zero level, the cones produced in the powder method are determined by the spacings of prominent planes and are not uniformly spaced. The origin of a powder diagram is shown in Figure 13.27. Because of the random orientation of the crystallites, the reciprocal lattice points generate a sphere of radius σ_{hkl} about the origin of the reciprocal lattice. A number of these spheres intersect the sphere of reflection.

Automatic Diffractometers

Results are achieved rapidly and with much better precision when automatic diffractometers are used to record diffraction data. A diffractometer to record data from powdered samples is built much like the instrument shown in Figure 13.22. The X-ray tube furnishes the radiation directly. (Filters are generally used to get more nearly monochromatic radiation.) The diffracting crystal is replaced by the powdered or metallic sample. To increase the randomness of orientation of the crystallites, the sample may be rotated in its own plane—that is, the plane perpendicular to the bisector of the angle between the source and detector beams. Note also that as the sample is rotated in the other plane to sweep through various θ angles, the detector must be rotated twice as rapidly to maintain the angle 2θ with the irradiating beam.

Proportional, scintillation, or semiconductor detectors, with their associated circuitry, are far superior to photographic film in regard to the number of reflections per day that can be recorded. With them a precision of 1% or better is achieved. Even with the best darkroom and photometric procedures, the relative degree of blackness of each spot or cone on a film cannot be estimated with an accuracy of much more than 10%, and often the error in estimation is greater than this. The principal advantage of photographic film over counters is that it provides a means of recording many reflections at one time.

Automatic single-crystal diffractometers are quite complex. A cradle assembly provides a wide angular range for orienting and aligning the crystal under study. A precision diffractometer assembly allows the detector to traverse a spherical surface from longitude $-5°$ to $150°$ and from latitude $-6°$ to $60°$. The complete unit provides four rotational degrees of freedom for the crystal and two for the detector.

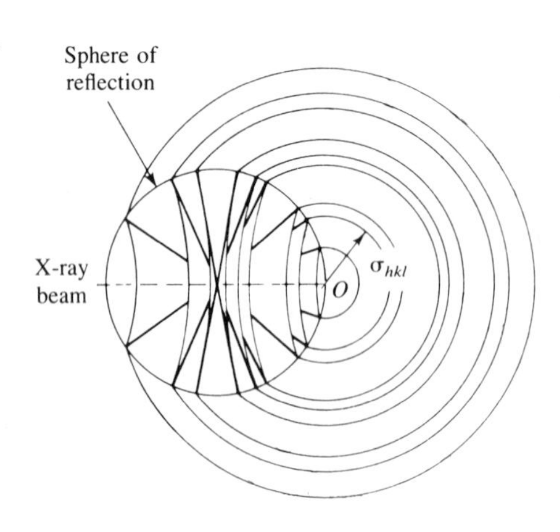

FIGURE 13.27
Origin of powder diffraction diagrams in terms of the series of concentric spheres generated from the reciprocal lattice points about the origin, O, of the reciprocal lattice and their intersection with the sphere of reflection.

Various crystal and counter angles are set on the basis of programmed information and the resulting diffraction intensity is measured as a function of angle. Single-crystal diffractometers controlled by computers that set the angles for the crystal and detector and record the data are widely used for gathering the information for crystal structure determinations.

Choice of X Radiation

Two factors control the choice of X radiation, as is seen by arranging the terms of the Bragg equation:

$$\theta = \sin^{-1}\left(\frac{\lambda}{2d}\right) \tag{13.17}$$

Because the ratio in parentheses cannot exceed unity, the use of long-wavelength radiation limits the number of reflections that are observed. Conversely, when the unit cell is very large, short-wavelength radiation tends to crowd individual reflections very close together.

The choice of radiation is also affected by the absorption characteristics of a sample. Radiation that has a wavelength just shorter than the absorption edge of an element contained in the sample should be avoided because then the element absorbs the radiation strongly. The absorbed energy is emitted as fluorescent radiation in all directions and increases the background. It is obvious, then, why one commercial source provides radiation sources from a multiwindow tube with anodes of silver, molybdenum, tungsten, and copper.

Specimen Preparation

Single crystals are used for structure determinations whenever possible because of the relatively large number of reflections obtained from single crystals and the greater ease of their interpretation. A crystal should be of such size, usually less than 1 mm, that it is completely bathed by the incident beam. Generally, a crystal is affixed to a thin glass capillary that, in turn, is fastened to a brass pin, as shown in Figure 13.28a.

When single crystals of sufficient size are not available or when the problem is merely the identification of a material, a polycrystalline aggregate is formed into a cylinder whose diameter is smaller than the diameter of the incident X-ray beam. Metal samples (metals usually consist of randomly oriented crystallites) are

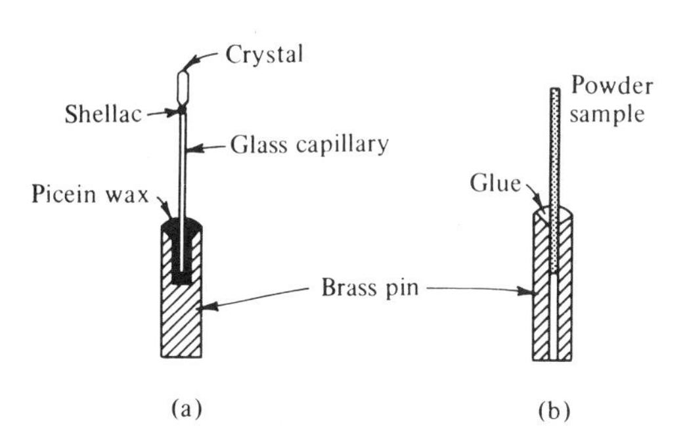

FIGURE **13.28**
Specimen mounts for X-ray diffraction: (a) single crystal and (b) powdered sample.

machined to a desirable shape, plastic materials can often be extruded through suitable dies, and all other samples are best ground to a fine powder (200–300 mesh) and shaped into thin rods after mixing with a binder (usually collodion) or tamped into a fine, uniform glass capillary. The mount is shown in Figure 13.28b.

Although liquids cannot be identified directly, it is frequently possible to convert them to crystalline derivatives that have characteristic patterns. Many of the classical derivatives are used, for example, in the identification of aldehydes and ketones as 2,4-dinitrophenylhydrazones, fatty acids as *p*-bromoanilides, and amines as picrate derivatives.

X-ray Powder Data File

If only the identification of a powder sample is desired, its diffraction pattern is compared with diagrams of known substances until a match is obtained. This method requires that a library of standard films be available. Alternatively, *d* values calculated from the diffraction diagram of the unknown substance are compared with *d* values of known substances stored on cards, microfiche, computer disks or tapes, or in book format with images of the data cards in the X-ray powder data file.[11] An index is available for the substances so listed. The cataloging scheme[12] used to classify different substances lists the three most intense reflections as the primary classification on cards or on computer-readable disks or tapes in the manner shown in Figure 13.29. The three most intense reflections are shown on the upper left corner of the card. The cards or files are then arranged in order of decreasing *d* values of the most intense reflections, based on 100 for the intensity of the most intense reflection observed. The file is continuously reexamined to eliminate errors and to remove deficiencies. Replacement cards for substances have a star in the upper right corner.

To use the file to identify a sample that contains one component, the *d* value for the most intense line of the unknown is looked up first in the file. Since there is probably more than one listing that contains the first *d* value, the *d* values of the next two most intense lines are matched against the values listed. Finally, the various

5-0628

d	2.82	1.99	1.63	3.258	NaCl	★
I/I_1	100	55	15	13	SODIUM CHLORIDE	HALITE

Rad. Cu λ 1.5405 Filter Dia.
Cut off I/I_1
Ref. Swanson and Fuyat, NBS Circular 539, Vol. II, 41 (1953)

Sys. Cubic S.G. O_H^5 – Fm3m
a_0 5.6402 b_0 c_0 A C
α β γ Z 4 Dx 2.164
Ref. Ibid.

εα nωβ 1.542 εγ Sign
2V D mp Color
Ref. Ibid.

An ACS reagent grade sample recrystallized twice from hydrochloric acid.
X-ray pattern at 26°C.

Replaces 1-0993, 1-0994, 2-0818

dÅ	I/I_1	hkl	dÅ	I/I_1	hkl
3.258	13	111			
2.821	100	200			
1.994	55	220			
1.701	2	311			
1.628	15	222			
1.410	6	400			
1.294	1	331			
1.261	11	420			
1.1515	7	422			
1.0855	1	511			
0.9969	2	440			
.9533	1	531			
.9401	3	600			
.8917	4	620			
.8601	1	533			
.8503	3	622			
.8141	2	444			

FIGURE **13.29**
X-ray data card for sodium chloride. (Courtesy of American Society for Testing Materials.)

cards or files involved are compared. A correct match requires that all the observed lines of the unknown and the file agree. It is also good practice to derive the unit cell from the observed interplanar spacings and to compare it with that listed in the file.

If the unknown is a mixture, each component must be identified individually. This is done by treating the list of d values as if they belonged to a single component. After a suitable match for one component is obtained, all the lines of the identified component are omitted from further consideration. The intensities of the remaining lines are rescaled by setting the strongest intensity equal to 100 and repeating the entire procedure.

The same results are obtained much easier if a computer is used to search the files and to match the d spacings found with the data on tapes or disks.

X-ray diffraction furnishes a rapid, accurate method for identifying the crystalline phases in a material. Sometimes it is the only method available for determining which of the possible polymorphic forms of a substance are present—for example, carbon as graphite or diamond. Differentiation among various oxides such as FeO, Fe_2O_3, and Fe_3O_4, or between materials present in such mixtures as KBr + NaCl, KCl + NaBr, or all four substances, is easily accomplished with X-ray diffraction, whereas chemical analysis would show only the ions present and not the actual state of chemical combination. Identifying the presence of various hydrates is another possibility.

Quantitative Analysis

X-ray diffraction is adaptable to quantitative applications because the intensities of the diffraction peaks of a given compound in a mixture are proportional to the fraction of the material in the mixture. However, direct comparison of the intensity of a diffraction peak in the pattern obtained from a mixture is fraught with difficulties. Corrections are frequently necessary for the differences in absorption coefficients between the compound being determined and the matrix. Preferred orientations must be avoided. Internal standards help but do not overcome the difficulties entirely.

Structural Applications

A discussion of the complete structural determination for a crystalline substance is beyond the scope of this text. It will suffice to point out that, with careful work, atoms can be located to a precision of hundredths of an angstrom or better.

In polymer chemistry a great deal of information can be obtained from an X-ray diffraction diagram. Fibers and partially oriented samples show spotty diffraction patterns rather than uniform cones; the more oriented the specimen, the spottier the pattern. Figure 13.30 shows the fiber diagram of polyethylene. The center row of spots in the pattern is called the equator, and the horizontal rows parallel to the equator are called the layer lines. The equatorial spots arise by diffraction from lattice planes that are parallel to the fiber axis. The layer line spots arise by diffraction from planes that intersect the fiber axis. The repeat distance along the polymer chain is calculated from the distances of the layer lines from the equator and their separation from one another. In the simplest cases the repeat distance corresponds to that of a fully extended chain of the known chemical composition.

Crystal Topography

There are several experimental diffraction techniques by which the microscopical defects in a crystal can be observed. Most crystals are far from perfect and exhibit regions (grains) with somewhat differing orientations, or they may contain individual defects such as dislocations or faults distributed throughout the crystal. Studies of these defects are important in understanding the nature of stress in metals, the nature and behavior of "doped" crystals used in semiconductors, the production of "perfect" crystals, and other phenomena.

Microradiographic methods are based on absorption, and the contrast in the images is due to differences in absorption coefficients from point to point. X-ray diffraction topography depends for image contrast on point-to-point changes in the direction or the intensity of beams diffracted by planes in the crystal.

One much-used method of X-ray diffraction topography is known as the Berg-Barrett method. The experimental arrangement is shown in Figure 13.31. The crystal is set so as to reflect the X rays at the Bragg angle for some plane. Geometric resolution of about 1 μm is achieved and single dislocations can be resolved. The contrast on the film is due to variations of the reflecting power from imperfections in the crystal.

Another method for X-ray diffraction topography is known as the Lang method. The experimental setup is shown in Figure 13.32. A ribbon X-ray beam is collimated to such a small angular divergence that only one characteristic wavelength is diffracted by the crystal. Simultaneous movement of the crystal and film allows a large area of the crystal to be investigated.

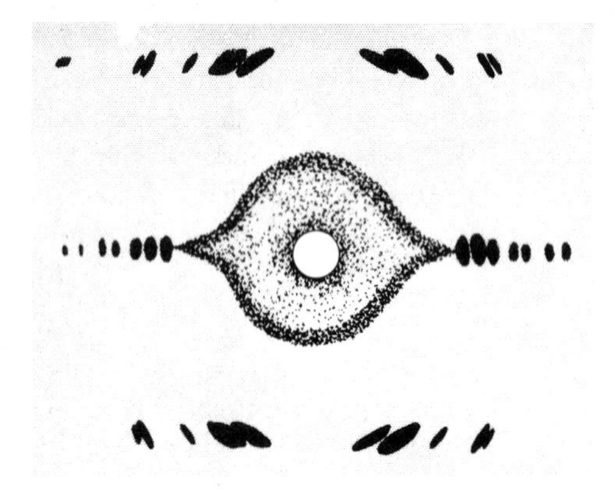

FIGURE **13.30**
Fiber diagram of polyethylene. [From A. Ryland, *J. Chem. Educ.*, **35,** 76 (1958). Reproduced by permission.]

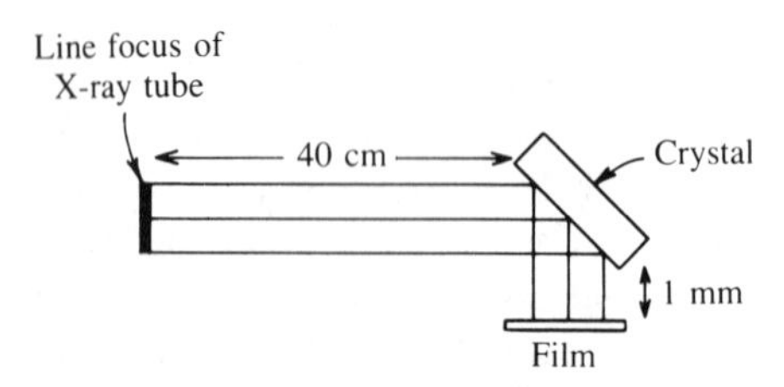

FIGURE **13.31**
Experimental arrangement for the Berg-Barrett method of X-ray diffraction topography.

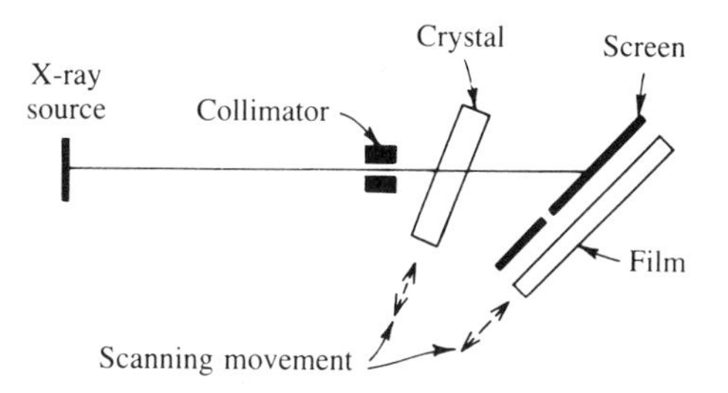

FIGURE 13.32
Experimental arrangement for the Lang method of X-ray diffraction topography.

13.7 AUGER EMISSION SPECTROSCOPY (AES)

AES measures electrons emitted from a surface, induced by electron bombardment (Figure 13.1b). The first step is ionization of an inner atomic level by a primary electron. Once the atom is ionized, it must relax by emitting either a photon (an X ray) or an electron (the nonradiative Auger process). In most instances nature chooses the Auger process. For example, a KLL Auger transition means that the K-level electron undergoes the initial ionization. An L-level electron moves in to fill the K-level vacancy and, at the same time, gives up the energy of that transition (L to K) to another L-level electron, which then becomes the ejected Auger electron. Other Auger electrons originate from LMM and MNN transitions. At this point the atom is doubly ionized. The energy of the ejected electron is a function only of the atomic energy levels involved in the Auger transition and is thus characteristic of the atom from which it came. A threshold energy, related to the transition energy, exists and a primary energy five to six times greater than the Auger energy maximizes the sensitivity to that particular transition. All elements except hydrogen and helium produce Auger peaks. Most elements have more than one intense Auger peak so that a recording of the spectrum of energies of Auger electrons released from any surface, compared with the known spectra of pure elements, enables a chemical analysis to be made.

Since an excited atom can lose energy either by the X-ray emission process or by Auger electron emission, the relative sensitivities of these two techniques are complementary when considering the relative abundance of X rays and Auger electrons after ionization of a particular level. In the light elements ($Z < 30$), the Auger process dominates, which makes AES relatively more sensitive. For heavier elements, the electron microprobe becomes more sensitive for transitions following ionization of inner shells. The sensitivity of AES is maintained, however, by utilizing Auger transitions between outer shells—for example, MNN—where the Auger process dominates.

Although the penetration of the primary electron beam may be as deep as 1 μm for high-energy electrons, the Auger electrons emitted are, on the average, of much lower energies. Electrons of such low energy must originate very close to the surface if they are to escape without being lost by inelastic scattering before reaching the surface. Typically, Auger electrons come from the first few atomic layers; the sampling depth is about 2 nm.

The sensitivity of the Auger technique is determined by the probability of the Auger transitions involved, the incident beam current and energy, and the collection efficiency of the energy analyzer. With a high-sensitivity cylindrical mirror analyzer

(CMA), discussed later, the detection limit for the elements varies between 0.1 and 1 atomic percent (or 10^{-3} of a monolayer). Because the electron beam can be focused to a small diameter (50 nm), it is possible to do spatial resolution or surface mapping of a sample. When operated in this manner, AES is usually referred to as the Auger microprobe. AES is traditionally run with high-intensity electron guns and low-resolution analyzers, resulting in fast analysis. Unfortunately, this limits one to primarily elemental information with little information on chemical bonding, such as one clearly obtains in the chemical shifts of electron scattering for chemical analysis (ESCA), to be discussed later in this chapter, or in the molecular fragments found in secondary ion mass spectrometry (SIMS), discussed in Chapter 16. For example, hydrocarbon compounds appear in AES only as a C peak, since H is not observed.

Figure 13.33 shows the Auger spectra of Ag, Cd, In, and Sb, in these cases the *MNN* transitions. The spectra are very similar; the only major difference is the shifts in energy from one element to the next. These shifts are of the order of 25 eV and, since the peak positions can be measured to an accuracy of ± 1 eV, there is no

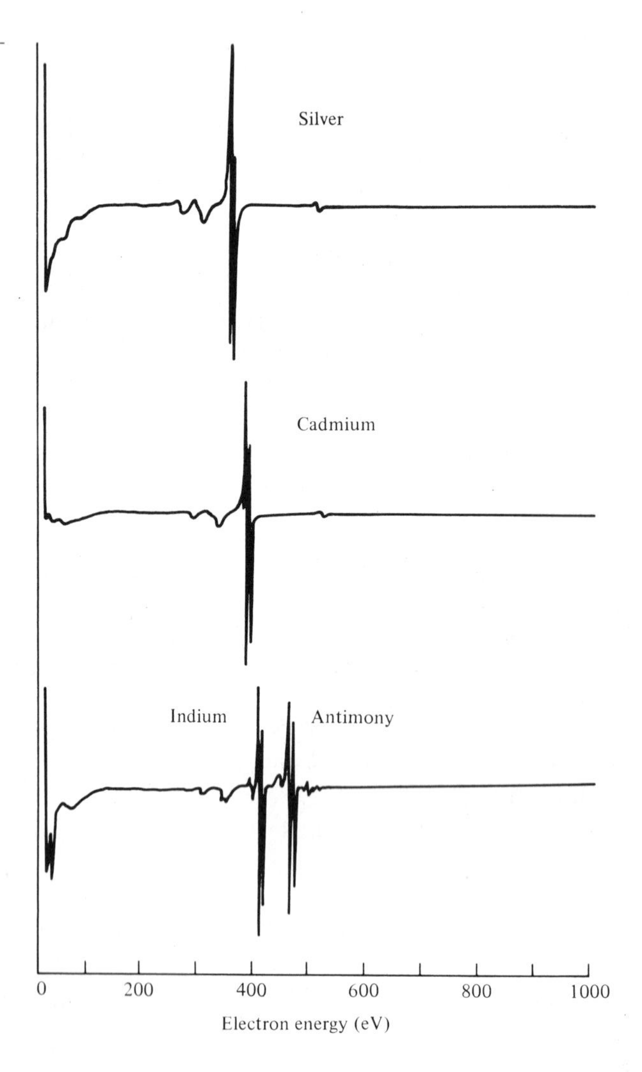

FIGURE 13.33
Auger spectra (differential form) from silver, cadmium, indium, and antimony; all are *MNN* transitions.

ambiguity in the identification of adjacent elements in the periodic chart. Auger spectra of all the elements lie between 50 and 1000 eV. The *KLL* transitions correspond quite nicely with tabulated X-ray energies. There are, however, some minor differences due to the energy Auger electrons must lose in escaping from the sample. The Auger lines are relatively broad due to the double uncertainty of the exact origins and destinations within a sublevel of both the electrons that fall into inner levels and those being expelled from outer levels. Lines due to *LMM* transitions tend to correlate with X-ray energies, but some lines begin to appear that the selection rules for photon emission forbid. As the atomic number increases, the Auger spectra become more complex and may overlap.

The strength of AES is its ability to give both a qualitative and quantitative nondestructive analysis of the elements in the immediate surface atomic layers from a very small area of a solid. When combined with a controlled removal of surface layers by ion sputtering, AES provides the means to solve some very important problems, such as corrosion on metallic surfaces. To provide ion sputtering for surface cleaning and/or depth profiling, the sample chamber is backfilled with argon and an electron impact source is used.

AES Instrumentation

The Auger spectrometer consists of an ultra-high vacuum chamber console, a sample carrousel and manipulator unit, and a combination electron gun/energy analyzer unit (Figure 13.34). Auxiliary equipment often includes a grazing incidence

FIGURE **13.34** Schematic diagram of an Auger spectrometer with computer control.

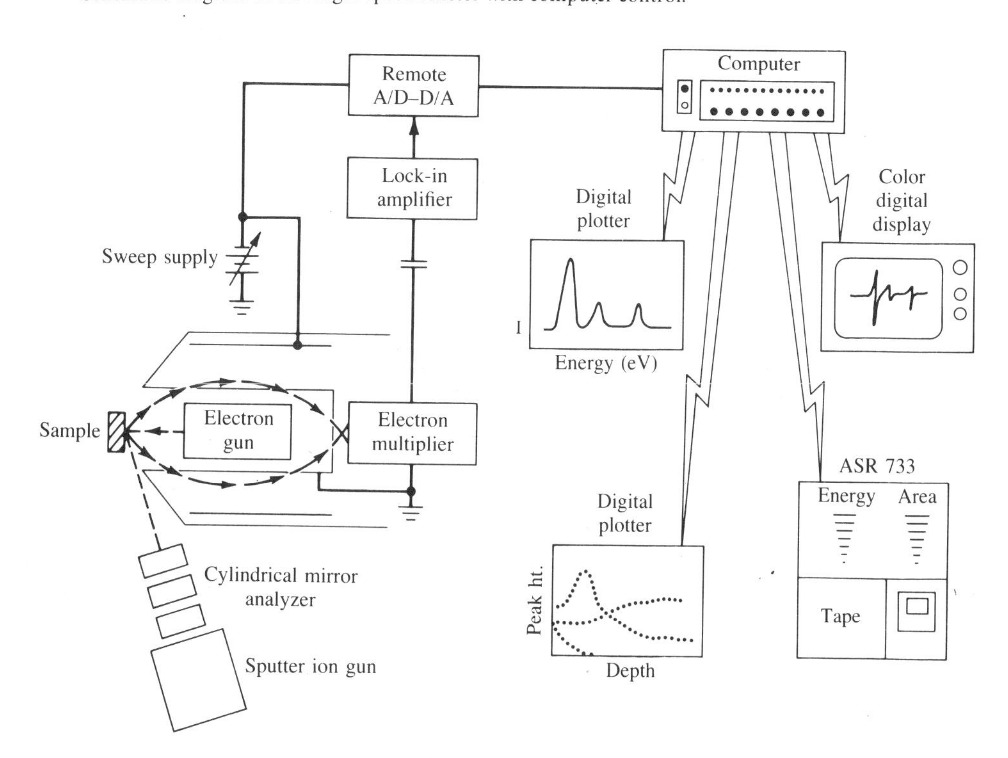

electron gun and a sputter ion gun for cleaning surfaces and profiling studies. Auger spectrometers are available as large-beam ($\sim$25-μm) depth-profiling instruments or as Auger microprobes with beam diameters of 5μm. If an instrument is to be used as a high-sensitivity depth profiler and also as a high lateral resolution microprobe, two electron guns may be required to realize the optimum use of each operating mode.

The energy distribution of electrons emitted from the target, N_E, is evaluated by scanning the negative voltage applied to the outer cylinder of the cylindrical mirror analyzer (CMA) (Figure 13.35). Thus as the voltage applied to the outer cylinder is scanned, the secondary electron distribution is measured by the current output of the analyzer. Because the Auger peaks are superimposed on a rather large continuous background of secondary electrons, it has become popular to differentiate electronically the N_E function. This is done by applying a small ac voltage on the dc energy voltage and synchronously detecting the in-phase component of the output current of the electron multiplier with a lock-in amplifier. Unfortunately, the use of the lock-in amplifier in data acquisition places significant limits on sensitivity and quantification of the data. The limitations are substantially reduced through the use of digital techniques. The digital storage of signal-averaged data with a high dynamic range makes it possible to use digital filters to remove high-frequency noise components from the signal. Signal averaging using multiple passes over a given energy range is used to ensure that the Auger peaks stand out above the noise. In this way, sensitivity is substantially improved without serious loss of energy resolution. Consequently, one now has a means for preserving the chemical shift information while improving sensitivity for elements present in low concentrations. In effect, this allows the choice between energy resolution (that is, oxidation state information as in ESCA) and sensitivity to elements in low concentrations to be made after the data are acquired.

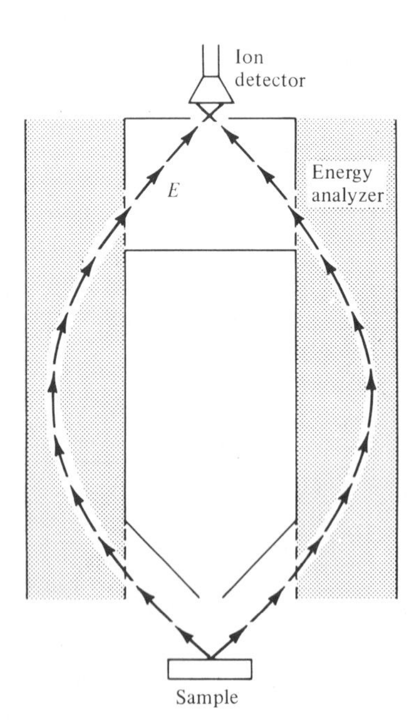

FIGURE 13.35
Cylindrical mirror analyzer.

Quantitative Analysis with AES

Quantitative analysis is possible in principle. The relative intensities of peaks depend on the following variables: inner-level ionization cross sections, Auger transition rates, and inelastic scattering of the emitted Auger electrons. With careful calibration, quantitative analysis of a homogeneous sample should be achieved with an accuracy of about 10%. Thus, although quantitative Auger spectroscopy is not very exact, it is still of significant value in surface analysis.

The difficulties in the quantification of Auger data are illustrated in depth-profiling analysis. In this operation the change in peak shape of the differentiated spectra distorts the conventional peak-height estimates of concentration. Changes in peak shape occur because of changes in the oxidation state of the emitting atom. This is particularly severe in cases where bonding orbitals are involved in the Auger transitions. Peak-shape changes also may be induced by matrix effects, since the energy loss mechanisms and their effect on peak shape may be different in different matrices.

To provide accurate quantitative information within the normal quantitative limits of AES, it is necessary to use the total Auger current. This is expressed by the area under the curve for the peak of interest, with background subtracted. By using undifferentiated data and integrating over the total number of Auger electrons in the peak, one still has a choice of the limits of integration. Now chemical shifts may be observed and a depth profile can be realized with no artifacts.

Scanning Auger Microprobe (SAM)

The scanning Auger microprobe has a finely focused, scanning electron beam as the probe for AES analysis of a surface. For a surface area, the SAM provides an electron micrograph, an Auger image of selected elements on a cathode ray tube display for observation or photographing, and a depth-composition profile. First, the cathode ray tube image is used to determine specific points or an area of interest on the sample surface. From the micrograph obtained, points of interest or an area up to 200 μm can be chosen for compositional analysis.

The image mode is useful because the positions of various elements in the sample are delineated in a few minutes. An *xy* recording of a line scan across a selected region of the image for a given element gives a relative quantitative measure of its concentration.

The thin-film analyzer mode of the SAM simultaneously sputter-etches a relatively large surface area (several millimeters in diameter) and multiplexes the Auger signals from a smaller area about 15 μm in diameter. Up to six selected Auger peaks can be recorded as the surface is etched. Depth resolution is about 10% of the total etched thickness. Elements are detected when present in quantities down to 0.1% of a monolayer.

At best, the spatial resolving power of SAM is 500 nm with tungsten thermionic emitters as the electron source. The use of LaB_6 emitters extends the resolving power to 100–200 nm because of their high electron optical brightness. Field emitters, which provide even higher brightness, are expected to allow one to analyze smaller areas in less time when they come into use.

Many surface composition problems involve compositional inhomogeneity, both in depth and across a material surface. Scanning Auger microscopy offers high resolution in the form of scanning electron microscope-type images, combined with elemental mapping capabilities. For example, graphite present in spherical nodules indicates that a specimen of cast iron is ductile cast iron. Examination of the carbon Auger images shows carbon to be present not only in the graphite nodules but also, at lower concentration, in regions between nodules. Carbon is not evident in the circular regions around nodules. The iron Auger image shows iron to be present everywhere in the specimen surface except in the graphite regions.

13.8 ELECTRON SPECTROSCOPY FOR CHEMICAL ANALYSIS (ESCA)

ESCA is concerned with the measurement of core-electron binding energies. A molecule or atom is bombarded with high-energy X rays, which cause the emission from sample atoms of inner-level electrons. All electrons whose binding energies are less than the energy of the exciting X rays are ejected (Figure 13.1d). The kinetic energies, E_k, of these photoelectrons are then measured by an energy analyzer. The core-electron binding energies E_b, are computed via the relationship

$$E_b = h\nu - E_k - \phi \tag{13.18}$$

where $h\nu$ is the energy of the exciting radiation and ϕ is the spectrometer work function, a constant for a given analyzer. Binding energies unambiguously define a specific atom. AES can be compared to ESCA because both originate from similar fundamental processes. In ESCA the ionizing source is an X-ray photon that ejects an inner-core electron, whereas in AES the electron ejection is caused by an impinging electron. The photoelectron escape depths are the same as those for Auger electrons of the same kinetic energy. The energy of the ejected electron (Auger electron or E_k) is thus characteristic of the atom involved and its chemical environment.

Although the incident X-ray photon may penetrate and excite photoelectrons to a depth of several hundred nanometers, only the photoelectrons from the outermost layers have any chance to escape from the material environment and eventually be measured. Most ESCA measurements of solids generate useful information from only the outer 2 nm of the surface layer.

The required sample size is a microgram or less. The sampling area is approximately 1 cm^2. Applicability to the second-row elements, including carbon, nitrogen, and oxygen, makes ESCA an important structural tool for organic materials. The detection limit depends on the particular element being measured, but ranges from 1% of a monolayer for light elements to 0.1% of a monolayer for heavy elements.

To determine the absolute binding energy, the work function of the sample and of the spectrometer must be known. Unfortunately, the work function is not independent of the physical state of the material; neither is it easy to measure or calculate. If the material being examined is a conductor and the material is in good electrical contact with the spectrometer, which presumably is itself a good conductor, the Fermi levels for the two are the same. Binding energies for different materials are then referred to this Fermi level. Although the electron ejected from the conducting material must overcome the work function of that material, it adjusts to the work function of the spectrometer on entering the spectrometer chamber. The correction, ϕ, in Equation 13.18 is, then, the work function of the material out of which the spectrometer is constructed and remains constant as long as the surface of the spectrometer chamber does not change.

Frequently only a relative binding energy (chemical shift) is desired in chemical studies. Here it is sufficient that ϕ be constant. For example, the chemical shift of a metal oxide relative to the metal is calculated from the measured kinetic energies:

$$\Delta E_{\text{oxide}} = E_{k(\text{metal})} - E_{k(\text{oxide})} \tag{13.19}$$

It is ΔE_{oxide}, the chemical shift, that gives chemical structural information.

Chemical Shift

The utility of ESCA for the chemist is the result of chemical shifts that are observed in electron binding energies. The binding energies of core electrons are affected by the valence electrons and therefore by the chemical environment of the atom. When the atomic arrangement surrounding the atom that ejects a photoelectron is changed, it alters the local (quantum) charge environment at that atomic site. This change, in turn, is reflected as a variation in the binding energy of *all* the electrons of that atom. Thus, not only the valence electrons but also the binding energies of the core electrons experience a characteristic shift. Such a shift is inherent to the chemical species that produces the results and thus provides the capability of chemical analysis. In a simple sense, the shifts of the photoelectron lines in an ESCA spectrum reflect the increase in binding energy as the oxidation state of the atom becomes more positive. In general, any parameter, such as oxidation state, ligand electronegativity, or coordination, that affects the electron density about the atom is expected to result in a chemical shift in electron binding energy.

A major portion of the strength of ESCA as an analytical tool lies in the fact that chemical shifts are observed for every element in the periodic chart except hydrogen and helium. Magnitudes of chemical shifts vary from element to element, and the sensitivity for a particular element varies with the photoelectron cross section. In general, the ESCA chemical shifts lie in the range 0–1500 eV. For instance, the position of the nitrogen 1*s* peak at a binding energy of 398 eV correlates very well with nitrogen being in a formal oxidation state of −1. To make this assignment one must refer to a catalog of reference nitrogen spectra or to correlation charts, such as Figure 13.36. One should be cautious about making structural assignments on the

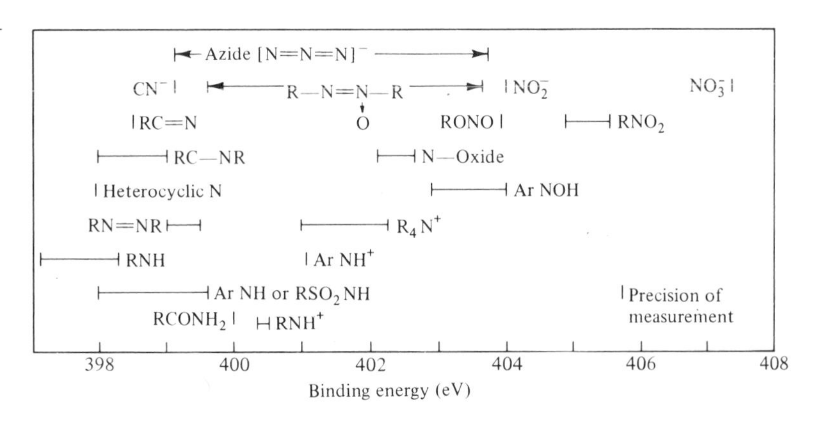

FIGURE **13.36** Correlation chart for nitrogen (1*s*) electron binding energies and organic functional groups.

basis of small ESCA binding energy shifts, since factors such as crystal potential energy differences and sample charging cause apparent shifts in magnitude of the observed chemical shifts.

In general, photoelectron peaks are narrower than the corresponding X-ray emission lines and in most cases vary from 1 to 3 eV (FWHM). Since the chemical shifts for a given element are on the order of 10 eV, ESCA is not a high-resolution method. The FWHM resolution of about 1.0 eV makes it possible to resolve electron energies that differ by 0.5–1.0 eV, and thus to measure chemical shifts that are of interest to most chemists. For example, Figure 13.37 shows the partial ESCA spectra of Cu_2O, CuO, and metallic copper. One can clearly distinguish between metallic copper and CuO, but it is difficult to decide between metallic copper and Cu_2O. However, by also observing the oxygen 1*s* spectrum (530.8 eV for Cu_2O and 530.1 eV for CuO), one can easily distinguish between metallic copper and its two oxidation states. The classic example of chemical shifts is the carbon 1*s* ESCA spectrum of ethyltrifluoroacetate shown in Figure 13.38. Each carbon atom is located in a different chemical environment and the ESCA spectrum contains four distinct photoelectron lines. The trifluorocarbon yields the photoelectron line at the highest binding energy, since the fluorine atoms withdraw electron density from the carbon atom most efficiently. In this example, the relative positions of the photoelectron lines reflect the relative electronegativities of the various substituents. The respective carbon photoelectron lines actually appear above their peaks in the spectrum.

ESCA Instrumentation

Instrumentation for ESCA, shown in Figure 13.39, involves a radiation source of sufficient energy to eject an electron from the sample. There must also be a device that collects the emitted electrons, counts them, and carefully measures their kinetic energy. A storage and display unit is usually included. Since it is necessary to ensure that the mean free path of the photoelectrons is large enough to allow them to traverse the distance from the sample to the detector without suffering energy loss, ESCA is a vacuum technique with a maximum operating pressure of about 5×10^{-6} torr.

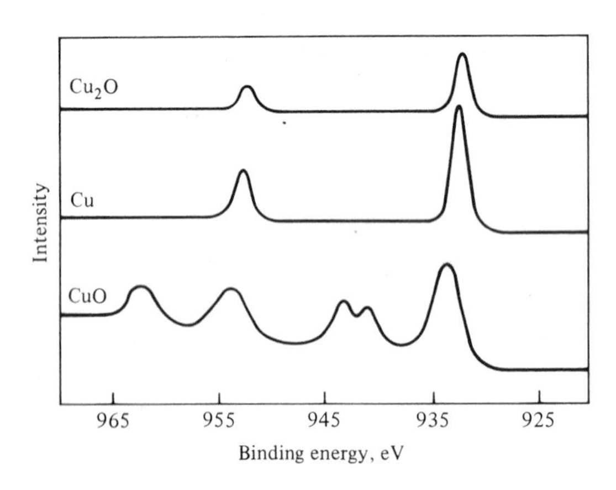

FIGURE **13.37**
ESCA spectra of Cu_2O, CuO, and metallic copper.

ESCA spectra can be obtained on solids, liquids, and gases. The physical form of the sample is not important. However, this is a vacuum technique; therefore, low-vapor-pressure solids are most easily run. A solid sample need only be placed on a probe that is appropriately positioned relative to the X-ray beam and the spectrometer slit. Liquids cannot be run as such but must be condensed onto a cryogenic probe and run in the condensed phase. Alternatively, liquids are vaporized and run in the gaseous state. To obtain spectra on the gaseous sample, the spectrometer must be equipped with a differential pumping system to prevent the pressure in the analyzer from rising above 10^{-4} torr.

Source. Soft X rays, such as Mg $K\alpha_{1,2}$ and Al $K\alpha_{1,2}$ with a FWHM of 0.75 and 0.95 eV, respectively, are usually used for photoelectron excitation. It is important to have at least two alternate sources to distinguish photoelectron peaks from Auger peaks. When a different X-ray source is used, the photoelectron peaks shift in kinetic energy but the kinetic energies of Auger peaks remain constant and appear at the same energy position in the spectrum. A source of small linewidth is advantageous if an element exists in several oxidation states in a sample and it is desired to extract the binding energies of each of these oxidation states.

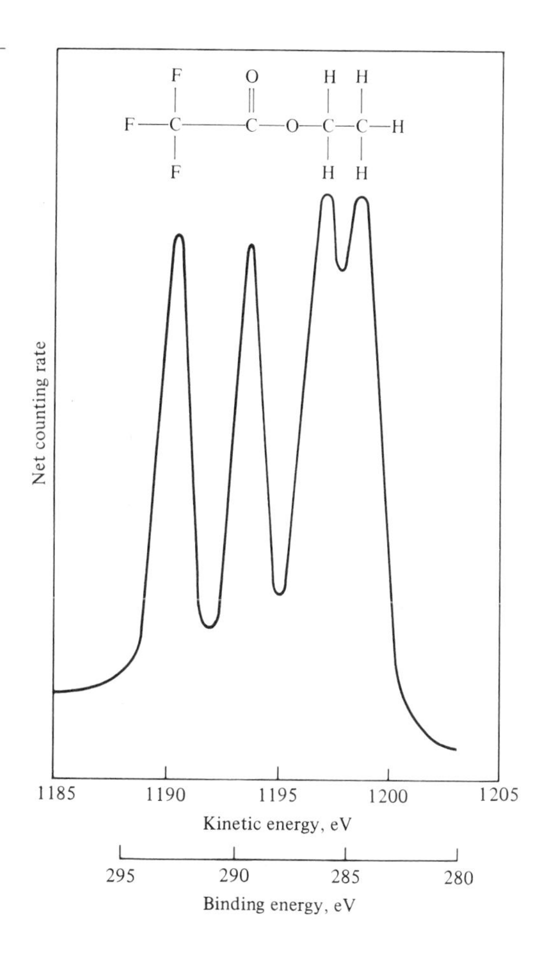

FIGURE **13.38**
ESCA spectrum of ethyltrifluoroacetate.

Two types of X-ray systems are used in commercial instruments. In one the sample is illuminated directly by the output of the source. This polychromatic type of source is quite simple and permits the use of different target materials and hence different energies of photoelectron excitation. The other system incorporates an X-ray monochromator to disperse the X radiation and thus provide monochromatic illumination of the sample surface. Spectral interferences and background are thereby reduced, but X-ray intensities that impinge on the sample surface are much lower.

The monochromatic source shown in Figure 13.39 places the X-ray anode, a spherically bent crystal disperser, and the sample on a Rowland circle. Only the X radiation from the anode is reflected to the sample by the crystal. The source radiation will be free from the X-ray satellite structure (arising from $K\alpha_{3,4}$ and $K\beta$ lines) that plagues the photoelectron spectra generated with polychromatic beams. Elimination of the broad background bremsstrahlung radiation improves the signal-to-background ratio and sharply reduces X-radiation damage in the sample. For example, radiation-induced changes, such as the reduction of Cu(II) to Cu(I) that occurs in 20 sec with bremsstrahlung radiation, occurs only after 10 hr with the crystal-reflection method.

FIGURE 13.39 Schematic of an ESCA spectrometer using a digital processor and crystal dispersion to achieve X-ray monochromatization. Commercial version is the Hewlett-Packard model 5950B.

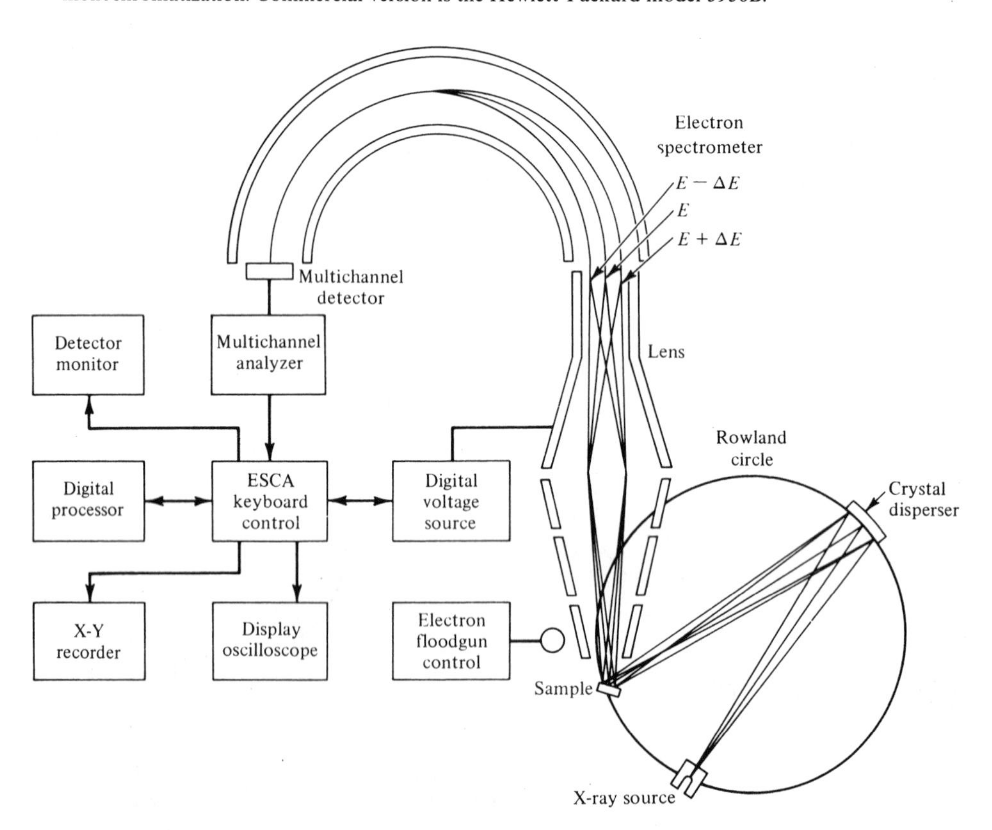

ESCA Electron Analyzers. The electron analyzers in ESCA instrumentation use the double-focusing principle. An electrostatic field sorts the electrons. As the field is varied, electrons of appropriate kinetic energy are focused at the detector. Initially, in the instrument illustrated, the photoelectrons are channeled through a series of four interconnected electrostatic lenses. This method for photoelectron collection is referred to as dispersion compensation, with the four lenses acting as selective apertures. In addition, retarding voltages are used to bring the photoelectrons into focus so that the inherent linewidths of the X-ray photons are removed. In addition, the lenses provide the mechanism for selecting and scanning the particular photoelectron energy of interest.

Detector. Both continuous channel and discrete dynode electron multipliers are used to count the electrons. The continuous channel detector counts electrons with high efficiency to very low energies and, compared with discrete dynode multipliers, is more stable to atmospheric and other gases.

Scan and readout systems are of either a continuous or incremental type. In the continuous mode the focusing field is increased continuously as a function of time as the signal from the detector is simultaneously monitored by a ratemeter. The focusing field and output from the ratemeter are synchronized to allow the accurate recording of spectra. In the continuous mode the energy region of interest is scanned only once. This is somewhat of a handicap, since there can be no signal averaging. Signal averaging becomes necessary when the signal is weak or the quantity of the sample is small.

The incremental scanning mode increases the field in a series of small steps, counting the signal during each increment. When the counting rate at each increment is plotted as a function of the focusing field, a spectrum is produced. Instrumentation that uses the incremental scanning mode has either a multichannel analyzer or a small dedicated computer to accumulate the data. In such systems, signal averaging is achieved by performing repetitive scans over the energy region of interest. The counting rate in each increment is added to the preceding one. When the system contains a dedicated computer, both energy scan and data acquisition are under its control. Usually several energy regions are scanned sequentially, with the computer storing the data until they are retrieved by the operator.

Scanning ESCA. ESCA in its earlier years was always thought of as a low-spatial-resolution technique because the specimen is excited by flooding the surface with X rays. These X rays could not be readily focused. Thus, elemental images could not be generated in a manner analogous to that of a scanning Auger microprobe where the source of excitation, a scanning electron beam, is focused to a small spot. However, if a focused electron beam is used to bombard a thin foil of aluminum that has a thin specimen mounted on the side opposite the beam, a localized source of Al $K\alpha$ X rays is produced in the aluminum foil (Figure 13.40). This causes a spatially localized source of photoelectrons to be created in the specimen (but there is also bremsstrahlung radiation). The result is an ESCA spectrum from an area less than 20 μm in diameter. If a scanning electron beam is used to create the X rays in the aluminum foil, two-dimensional photoelectron images are obtained.

A practical example of the use of surface analysis involved the study of a metallographically polished cast iron specimen. An ESCA survey spectrum indicated

FIGURE 13.40
Schematic arrangement for spatially resolved ESCA.

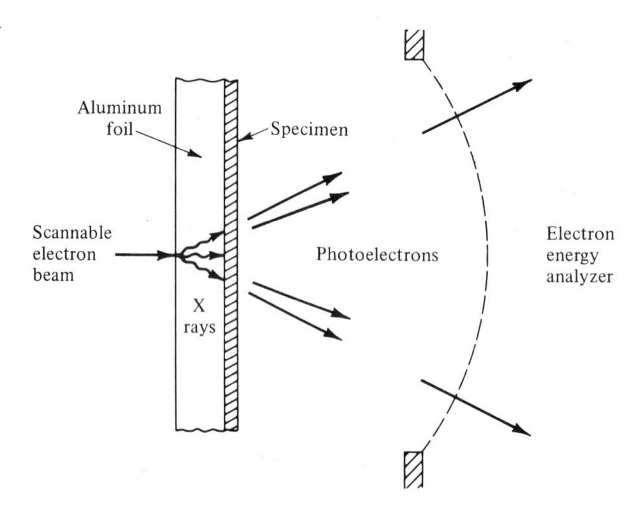

the presence of Fe, O, and C as major components of the material. High-resolution spectra of Fe and C showed the presence of both oxidized and reduced iron species and of at least two carbon species. Detailed analysis of the carbon binding energies revealed the presence of a carbide and a species with a binding energy characteristic of graphite. These conclusions are consistent with the fact that cast iron contains precipitated graphite as well as phases rich in iron carbide. The oxidized iron species probably was caused by surface oxidation of the specimen.

Quantitative Analysis

The intensity of a photoelectron line is proportional not only to the photoelectric cross section of a particular element, but also to the number of atoms of that particular element that are present in the sample. Analyses of mixtures are often accurate to $\pm 2\%$. An example involves measuring the intensities of photoelectron lines in spectra obtained from mixtures of MoO_2 and MoO_3 at binding energies that correspond to each oxide. No instrumental technique had existed that was capable of performing this analysis. Even though the surface of MoO_2 was significantly contaminated with MoO_3, a linear calibration curve resulted. Mixtures of PbO—PbO_2, Cr_2O_3—CrO_3, and As_2O_3—As_2O_5 can also be measured quantitatively Estimates of the total protein content of various grains can be made by measuring the intensities of the nitrogen and sulfur peaks.

PROBLEMS

1. What is the short-wavelength limit for a 60-kV X-ray tube? What is the atomic number of the element for which just insufficient energy is available for excitation?

2. Write the transition relations for each of these lines: (a) $K\alpha_2$, (b) $K\beta_1$, and (c) $L\alpha_1$.

3. Calculate the critical excitation potentials for the K and L series of the elements in the table.

Element	K absorption edge (Å)	L_{III} edge (Å)
Mg	9.54	247.9
Mn	1.895	19.40
Cu	1.380	13.29
Mo	0.620	4.912
W	0.178	1.215
Pt	0.158	1.072

4. Calculate the wavelengths of the $K\alpha_1$ lines for the elements in Problem 3.

5. For what elements will Mo $K\alpha_1$ prove sufficiently energetic to excite their L_{III} spectra? What is the limit for K spectra?

6. The measurements in the table were obtained with a proportional flow counter. Plot the results on graph paper and select an operating voltage.

Applied voltage (V)	Observed count rate (counts/min)	Applied voltage (V)	Observed count rate (counts/min)
1200	225,000	1600	231,000
1300	231,700	1700	231,500
1400	232,400	1800	233,000
1500	231,400	1900	271,000

7. Using a gamma emitter, the measurements in the table were obtained with an ionization chamber. Plot the results on graph paper and select an operating voltage.

Applied voltage (V)	Observed count rate (counts/min)	Applied voltage (V)	Observed count rate (counts/min)
1200	19,400	1470	41,000
1245	29,100	1500	40,900
1290	35,700	1545	41,100
1335	38,900	1590	41,500
1380	40,500	1635	42,200
1425	40,600	1680	45,100
		1725	48,700

8. What causes the discontinuity in the efficiency curve of the NaI scintillation counter and of the argon-filled flow proportional counter (Figure 13.11)?

9. Compute the goniometer setting (2θ) for the $K\alpha_1$ lines of Ti, Co, Ag, and Pt, when the analyzing crystal is (a) NaCl, (b) EDDT, or (c) gypsum.

10. Discuss four ways that might be used to separate interfering spectral lines.

11. For the determination of uranium in aluminum by measurement of U $L\alpha_1$ fluorescence, the counting relationships in the table were obtained. Determine the slope of the calibration curve (cps/% U) over each interval of uranium concentration and the counting time (in minutes) required to achieve a 1% precision at the 95% confidence level.

U(wt %)	Count rate (cps)
2	436
5	835
10	1262
15	1533
20	1720

12. Sulfur (0.4%–6.0%) has been determined in carbon materials by X-ray fluorescence using the S $K\alpha$ line (5.36 Å), which under a particular set of operating conditions gave 1 cps equivalent to 0.014% S. Background radiation is equivalent to 0.05% S. Select a proper analyzing crystal, goniometer setting, excitation conditions (X-ray tube voltage), and counting times to achieve results with a deviation that does not exceed 5% at the 95% confidence level.

13. The pulse amplitude distributions for the elements Mg, Al, Si, P, S, and Ca show a peak at 11.0, 13.0, 15.0, 17.4, 19.0, and 31.8 V, respectively. For each the width at one-half peak height is 2.5 V and the base width is 9.5 V. What baseline and window settings (in volts) would be used in the following situations: (a) the determination of Mg in the presence of P, (b) the determination of S in the presence of Mg, (c) the determination of P in the presence of Al, (d) the separation of calcium from all the others, and (e) the total Al plus Si in a sample, all other elements being absent?

14. Suggest methods for handling each pair of overlapping X-ray spectral lines whose wavelength in angstroms is enclosed in parentheses: (a) Mn $K\alpha_1$(2.103)–Cr $K\beta$(2.085), (b) Zn $K\alpha_1$(1.435)–Re $L\alpha_1$(1.433), and (c) Nb $K\alpha_1$(0.746)–W $L\alpha_1$(1.476). (Hint: second-order spectra must be considered.)

15. Graphically represent the following disturbing effects in the use of an internal standard element S for the determination of element E, with the disturbing element being D. Plot all absorption edges and emission lines: (a) selective absorption of S, (b) selective absorption of E, (c) enhancement of S, and (d) enhancement of E.

16. Strontium has been determined in sediments of oil-bearing formations using yttrium as an internal standard. The spectral characteristics of these elements are:

Sr $K\alpha_1$, 0.877; $K\beta$, 0.783; K edge, 0.770

Y $K\alpha_1$, 0.831; $K\beta$, 0.740; K edge, 0.727

(a) Compute the critical voltage for each element. (b) Using a LiF crystal, compute the Bragg angle (2θ) for the emission lines. (c) Calibration data are obtained as in the table. Plot intensity ratio versus concentration of strontium. Unknown samples gave these intensity ratios—Sr:Y—*A*, 0.8860; *B*, 0.7802; *C*, 0.6011.

Sr (wt %)	Measurement of Y $K\alpha_1$; time (sec) for 6400 counts	Measurement of Sr $K\alpha_1$; time (sec) for 6400 counts
0.0000	41.1	80.1
0.1000	40.1	60.4
0.2000	40.2	49.5
0.3000	40.0	41.6
0.4000	42.4	38.3

17. What is the difference in wavelength (and Bragg angle) between Hf $L\alpha_1$ and the second order of Zr $K\alpha_1$ (1.566 Å and 0.784 Å, in first order, respectively), with a LiF analyzing crystal? Suggest a method for eliminating the second-order Zr line.

18. Oil paintings have been tested for authenticity by examining the pigment composition with the electron probe and a hypodermic needle core. With an ammonium dihydrogen orthophosphate crystal (101 plane), $d = 10.62$, the characteristic X-ray spectra (and relative intensity) shown in the table were obtained. Was the painting produced before or after the time ($\approx$A.D. 1900) when titanium-white pigments became available?

Top white layer (2θ)	Bottom white layer (2θ)
27.6° (8)	34.0° (12)
30.0° (60)	36.0° (100)
58.0° (4)	71.0° (2)
63.0° (18)	

19. From the data given in Table 13.3, estimate the mass absorption coefficients of (a) Zr for the Mo $K\alpha_1$ radiation, (b) Ni for the Cu $K\alpha_1$ radiation, and (c) Cu for the W $L\alpha_1$ radiation.

20. Calculate the reduction in intensity of an X-ray beam from the Mo $K\alpha_1$ line (0.709 Å) that results from 1 mL of 1% TEL liquid in *n*-octane. The aviation mixture consists of 61.5% $Pb(C_2H_5)_4$ and 38.5% $C_2H_4Br_2$ per milliliter of TEL. The mass absorption coefficients are C, 0.64 cm^2/g; H, 0.38 cm^2/g; Pb, 140 cm^2/g; and Br, 0.79 cm^2/g. Density is 0.72 g cm^{-3}. Do the calculations for (a) an *n*-octane blank and (b) versus air in the reference path. The path length is 1.0 cm.

21. Repeat Problem 20 using the Cu $K\alpha_1$ line (1.542 Å). The mass absorption coefficients are C, 4.6 cm^2/g; H, 0.43 cm^2/g; Pb, 241 cm^2/g; and Br, 88 cm^2/g. Note the increase in absorption by lead and also by the carbon atoms.

22. Identify the emission lines and, from these, the base wire and plate metal in each sample. The spectrometer had a LiF crystal and the tungsten target operated at 50 kV.

Sample	Plate metal (2θ)	Base wire (2θ)
1	48.64°	41.30° (2nd)
2	99.87° (2nd)	63.88° (3rd)
3	31.19°	110.92° (2nd)
4	16.75°	110.92° (2nd)

23. A sample ground and pelleted with lithium carbonate and starch gave the following emission lines (2θ) with a LiF crystal. Identify each line.

111.0°, 100.2°, 57.6°, 48.4°, 45.1°, 44.0°, 40.4°

24. Suggest an X-ray method for each of these determinations: (a) the thickness of electroplated metal films, such as successive layers of Cu, Ni, and Cr on steel (chrome plate), (b) the thickness of SrO and BaO on evaporated electrode coatings, and (c) the concentration of fillers and impregnants, such as BaF_2 in carbon brushes.

25. Iron-55 decays by K-electron capture to stable ^{55}Mn with the attendant emission of K-line X-rays of Mn. The half-life of ^{55}Fe is 2.60 yr. For which elements would this isotope be a convenient source of X radiation for absorption analysis?

26. An unknown material was placed in the sample holder of an X-ray fluorescence unit that used a tungsten target tube operated at 60 kV to furnish the exciting radiation. A mica crystal was used in the analyzer. The lattice spacing of mica is 9.984 Å. Reflections were observed at angles (2θ) of 9° 34′, 12° 8′, 19° 12′, 24° 24′, and 38° 58′. Calculate the wavelength of the fluorescent lines and identify the elements present.

27. In the case of polyethylene, the repeat distance obtained from the X-ray diffraction pattern of a fiber diagram was 2.54 Å. What type of chemical structure is implied?

28. The full width at half maximum (FWHM) for a semiconductor detector is 155 eV at the Mn $K\alpha_1$ line at 2.102 Å. If the resolution of lines for quantitative purposes requires a separation of the peaks by at least three half-widths, could a semiconductor be used in the quantitative determination of Mn in the presence of Cr using the $K\alpha_1$ lines?

BIBLIOGRAPHY

BARR, T. L., "Applications of ESCA in Industrial Research," *Am. Lab.*, **10,** 65 (November 1978), p. 40 (December 1978).

BERTIN, E. P., *Principles and Practice of X-Ray Spectrometric Analysis,* 2nd ed., Plenum, New York, 1975.

BIRKS, L. S., *X-Ray Spectrochemical Analysis,* 2nd ed., Wiley-Interscience, New York, 1969.

BIRKS, L. S., *Electron Probe Microanalysis,* 2nd ed., Wiley-Interscience, New York, 1971.

BRIGGS, D., AND M. P. SEAH, eds., *Practical Surface Analysis by Auger and X-Ray Photoelectron Spectroscopy,* John Wiley, New York, 1983.

CARLSON, T. A., *Photoelectron Auger Spectroscopy,* Plenum, New York, 1975.

CZANDERNA, A. W., ED., *Methods of Surface Analysis,* Elsevier, New York, 1975.

DAVIS, L. E., N. C. MACDONALD, P. W. PALMBERG, G. E. RIACH, AND R. E. WEBER, *Handbook of Auger Electron Spectroscopy,* 2nd ed., Physical Electronics Industries, Eden Prairie, MN, 1976.

HARRIS, L. A., "Auger Electron Emission Analysis," *Anal. Chem.*, **40,** 24A (1968).

HERCULES, D. M., AND S. H. HERCULES, "Analytical Chemistry of Surfaces," *J. Chem. Ed.*, **61,** 402, 483, and 592 (1984).

HOFMANN, S., AND R. FRECH, "Auger Electron Spectroscopy," *Anal. Chem.*, **57,** 716 (1985).

JENKINS, R., *X-Ray Technology,* Norelco Reporter, **32,** Philips Electronic Instruments, Inc., Rahway, NJ, p. 22 (No. 1xr, August 1985).

JENKINS, R., R. W. GOULD, AND D. GEDCKE, *Quantitative X-Ray Spectrometry,* Dekker, New York, 1981.

LIEBHAFSKY, H. A., H. G. PFEIFFER, E. H. WINSLOW, AND P. J. ZEMANY, *X-Ray Absorption and Emission in Analytical Chemistry,* John Wiley, New York, 1960.

LIEBHAFSKY, H. A., H. G. PFEIFFER, E. H. WINSLOW, AND P. J. ZEMANY, *X-Rays, Electrons, and Analytical Chemistry,* Wiley-Interscience, New York, 1972.

LIEBHAFSKY, H. A., E. A. SCHWEIKERT, AND E. A. MYERS, "The Nature of X-Rays, Spectrochemical Analysis by Conventional X-Ray Methods, and Neutron Diffraction and Absorption," Chap. 13, *Treatise on Analytical Chemistry,* 2nd ed., P. J. Elving, E. J. Meehan, and I. M. Kolthoff, eds., Part I, Vol. 8, Wiley-Interscience, New York, 1986.

MARKOWICZ, A. A., AND R. E. VAN GRIEKEN, "X-Ray Spectrometry," *Anal. Chem.,* **58,** 279R (Fundamental Reviews), 1986 and earlier reviews.

MCMURDIE, H. F., C. S. BARRETT, J. B. NEWKIRK, AND C. O. RUUD, *Advances in X-Ray Analysis,* Vol. 21, Plenum, New York, 1978, and other volumes in this series.

MULLER, R. O., *Spectrochemical Analysis by X-Ray Fluorescence,* Plenum, New York, 1972.

NEWBURY, D. E., D. C. JOY, P. ECHLIN, C. E. FIORI, AND J. I. GOLDSTEIN, *Advanced Scanning Electron Microscopy and X-Ray Microanalysis,* Plenum, New York, 1986.

PFLUGER, C. E., "X-Ray Diffraction," *Anal. Chem.,* **50,** 161R (Fundamental Reviews), 1978 and earlier reviews.

SPROULL, W. T., *X-Rays in Practice,* McGraw-Hill, New York, 1946.

SWARTZ, W. E., JR., "X-Ray Photoelectron Spectroscopy," *Anal. Chem.,* **45,** 788A(1973).

TERTIAN, R., AND F. CLAISSE, *Principles of Quantitative X-Ray Fluorescence Analysis,* Heyden, London, 1982.

THOMPSON, M., M. D. BAKER, A. CHRISTIE, AND J. F. TYSON, *Auger Emission Spectroscopy,* Chemical Analysis, Vol. 74, John Wiley, New York, 1985.

WITTRY, D. B., "X-Ray Microanalysis by Means of Electron Probes," Chap. 61, *Treatise on Analytical Chemistry,* I. M. Kolthoff and P. J. Elving, eds., Part I, Vol. 5, Wiley-Interscience, New York, 1964.

LITERATURE CITED

1. DUANE, W., AND F. L. HUNT, *Phys. Rev.,* **66,** 166 (1915).
2. BRAGG, W. L., *The Crystalline State,* Macmillan, New York, 1933.
3. CASTAING, R., thesis, University of Paris, 1951.
4. CLARK, G. L., AND S. T. GROSS, *Ind. Eng. Chem., Anal. Ed.,* **14,** 676 (1942).
5. BIRKS, L. S., E. J. BROOKS, AND H. FRIEDMAN, *Anal. Chem.,* **25,** 692 (1953).
6. ROUSSEAU, R. M., *X-Ray Spectrom.,* **13**(3), 115 (1984).
7. ROUSSEAU, R. M., *X-Ray Spectrom.,* **13**(3), 121 (1984).
8. VREBOS, B., AND J. A. HELSEN, *X-Ray Spectrom.,* **14**(1), 27 (1985).
9. VREBOS, B., AND J. A. HELSEN, *Adv. X-Ray Anal.,* **28,** 31 (1985).
10. PELLA, P. A., D. E. NEWBURY, E. B. STEELE, AND D. H. BLACKBURN, *Anal. Chem.,* **58,** 1133 (1986).
11. "Powder Diffraction File," Sets 1–36, JCPDS (Joint Committee on Powder Diffraction Standards), International Centre for Diffraction Data, Swarthmore, PA, 1986. A continuing project with data organized in many ways (i.e., minerals, metals and alloys, inorganic phases, organic and organometallic compounds, etc.).
12. HANAWALT, J. D., H. W. RINN, AND L. K. FREVEL, *Ind. Eng. Chem., Anal. Ed.,* **10,** 457 (1938).

14 RADIOCHEMICAL METHODS

The phenomenon of radioactivity presents to the analyst a number of specific properties that are characteristic of a particular radionuclide. These properties include the type and energy of the radiation, the half-life, and the decay scheme. The fact that a given nuclide may be radioactive does not, in any way, affect its chemical properties before radioactive emission takes place, and after detection its fate is of no consequence. Radionuclides, most artificially produced, with suitable half-lives exist for all the elements except a very few (for example, oxygen, nitrogen, helium, lithium, and boron), and even of these, oxygen, lithium, and boron can be determined by activation techniques. In general, the chemical operations performed in radiochemistry are the standard techniques of analytical chemistry: precipitation, volatilization, solvent extraction, controlled potential electrolysis, and ion-exchange chromatography.

14.1 NUCLEAR REACTIONS AND RADIATIONS

A radionuclide is characterized by three factors: (1) its half-life, (2) the type of transition involved when it decays, and (3) the type and energy of the radiation emitted upon decay. Such information is essential for the recognition and understanding of the problems associated with the measurement of radionuclides.

Particles Emitted in Radioactive Decay

The heavy, naturally occurring radioactive elements, such as thorium and uranium, emit, among other products, doubly ionized helium particles known as *alpha particles*. Alpha particles have only a slight penetrating power, being stopped by thin sheets of solid materials and penetrating only 5–7 cm of air. Their energies are generally very high, however, and may exceed 10 MeV (million electron volts). As a result, the ionizing power of an alpha particle is high, and consequently they can generally be distinguished from beta or gamma radiation on the basis of pulse amplitude (see Chapter 13). The ionization chamber is the preferred detector.

A *beta particle* is a very energetic electron or positron. The feature of beta radiation that sets it apart from both alpha-particle and gamma radiation is the continuous distribution in energy from nearly zero energy to a maximum energy (E_{max}) that is characteristic of the specific beta transition. The maximum energy is often expressed as penetrating power or range. For example, a 0.5-MeV beta particle has a range of 1 m in air and produces about 60 ion pairs per centimeter of its path. Above 0.4 MeV beta particles have sufficient energy to penetrate the windows of most counting devices, and measurement is not difficult. Below this energy value, however, special techniques are required. Very thin window counters may be used. The sample may be introduced directly into the active volume of the counter, or it may be dissolved in a liquid scintillator.

Gamma radiation, actually high-energy photons, is monoenergetic. Gamma-ray spectra therefore consist of discrete lines. The penetrating power of gamma radiation is much greater than that of either alpha or beta particles, but the ionizing power is less. The sensitivity of the detector must be increased by using longer chambers, by gas fillings under pressure that have a high atomic number, or by thicker scintillator material. A filter of sufficient thickness to absorb all beta particles when inserted between the sample and detector permits the measurement of gamma radiation exclusively from a mixture of activities.

A long-lived positron-emitting nucleus may decay by capturing one of its own orbital K-electrons; this is called *K-capture* or *internal conversion*. The excess energy is emitted as a gamma ray. The resulting ion with a vacant K-orbital then emits X radiation characteristic of the new element. The daughter element is one atomic number less than its parent.

Interaction of Nuclear Radiation with Matter

Since the radiations from radionuclides are detected by means of their interactions with matter, a brief summary of these modes of interaction is presented here. Gamma rays lose energy on passage through matter by three essentially different mechanisms: *photoionization* (or the photoelectric effect), the *Compton effect*, and *pair production*.

Photoionization is important for heavy absorbing elements and for low gamma-ray energies. In this process gamma rays eject an inner electron from the atom, leaving a positive ion. The excess energy of the photon—that is, the difference between the binding energy of the electron and the energy of the photon—is imparted to the electron as kinetic energy (compare with X-ray absorption; see Chapter 13).

Gamma rays of intermediate energy can be scattered by electrons with binding energies negligible in comparison with the photon energy. This is the Compton effect. Upon interacting with an electron, part of the photon's energy is transferred to the electron. The electron is then ejected from the atom, and a new photon of lower energy proceeds from the collision in an altered direction. The Compton effect is important with light target elements and with gamma rays that have energies less than 3 MeV.

Pair production can occur at photon energies greater than 1.02 eV, the energy corresponding to twice the rest mass of the electron. A photon is annihilated (disappears) and an electron-positron pair is created. This process is important with heavy elements. Conversely, when a positron and electron meet, the two are annihilated, and two gamma rays with energies of 0.51 MeV each are produced.

The attenuation of gamma radiation is given by

$$P = P_0 e^{-\mu x} \tag{14.1}$$

where P is the radiant power of P_0, which is transmitted through an absorber of thickness x, and μ is the linear absorption coefficient. Often the energy of a gamma ray is expressed as the thickness of an absorber required to diminish the radiant power by half. This half-thickness is given by

$$x_{1/2} = \frac{0.693}{\mu} \tag{14.2}$$

Absorber thicknesses are frequently given in units of surface density (g/cm^2)—that is, ρx, where ρ is the density of the absorber. Equation 14.1 becomes

$$P = P_0 e^{-(\mu/\rho)\rho x} \tag{14.3}$$

where μ/ρ is known as the mass absorption coefficient.

Beta particles interact primarily with the electrons in material traversed by the particle. In the material, molecules may be dissociated, excited, or ionized. However, it is the ionization that is of primary interest in the detection of beta particles. As a beta particle is slowed down while moving through matter, the *specific ionization*—that is, the number of ion pairs produced per unit track length—increases and reaches a maximum near the end of the track. On the average, each ion pair produced represents a loss of 35 eV. Actually, the beta particle may lose a large part of its energy in a single interaction, but if it does, the ions produced have so much excess energy that they in turn produce additional ion pairs. The absorption of beta particles in matter follows approximately the exponential relation given by Equation 14.1 for gamma radiation, up to a certain thickness where the absorption finally exceeds that predicted by the exponential law and soon becomes infinite. This maximum thickness is known as the *range of the beta particle* (Figure 14.1). For energies of 0.5–3 MeV, the maximum energy of the beta particle is given by the following range-energy relationship:

$$\text{range (in mg/cm}^2\text{) of absorber} = 0.520E_{\max} - 0.090 \tag{14.4}$$

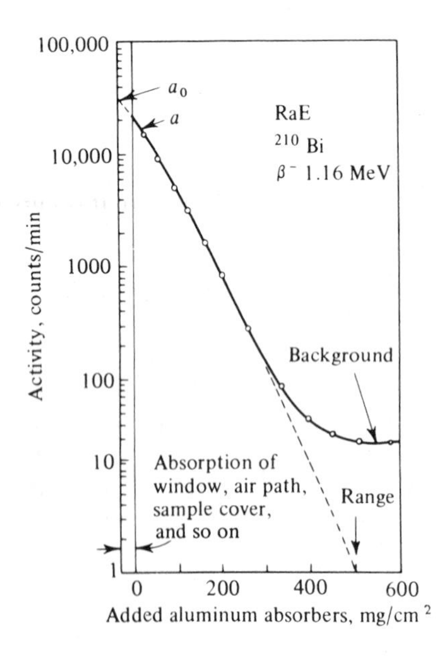

FIGURE **14.1**
Absorption of beta particles.

where the energy is expressed in millions of electron volts. In lower energy regions it is best to use a range-energy curve. Thus, unwanted beta radiation can be removed by an appropriate thickness of absorber, as calculated from Equation 14.4.

Radioactive Decay

The decay of a radionuclide follows the first-order rate law, which may be written in differential form as

$$\frac{dN}{dt} = -\lambda N \tag{14.5}$$

where N is the number of radionuclides that remain at time t, and λ is the characteristic decay constant (in time^{-1}). The activity A is related to N by the equation

$$A = \lambda N \tag{14.6}$$

and is usually the quantity observed or computed. The rate equation may be integrated to yield

$$A = A_0 e^{-\lambda t} \tag{14.7}$$

or

$$\ln A = \ln A_0 - \lambda t \tag{14.8}$$

where A_0 is the activity at some initial time and A is the activity after elapsed time t.

The time required for half of the radioactive material to disintegrate and release its energy (that is, to decay), the *half-life* of the radionuclide, is generally used in describing radioactive emitters; namely,

$$t_{1/2} = \frac{1}{\lambda} \ln \frac{A}{A/2} = \frac{0.693}{\lambda} \tag{14.9}$$

For example, strontium-90 has a half-life of approximately 30 yr. In 30 yr half of the nuclei will disintegrate and release their energy; in another 30 yr half of the remainder of the nuclei will decay. This process continues such that every 30 yr half of the remainder of the element decays and the other half remains for future decay. After ten half-lives only 0.1% of the radioactive nuclei remains. An accurate knowledge of the characteristic decay constant is essential when working with short-lived radionuclides to correct for the decay that occurs while the experiment is in progress. Decay schemes for selected radionuclides are shown in Figure 14.2.

A selection of radionuclides and their characteristics is given in Table 14.1. Nuclides with very short half-lives decay too rapidly to be generally useful; on the other hand, a long-lived nuclide is difficult to measure because disintegrations are too infrequent.

Units of Radioactivity

Activity is expressed in terms of the *Curie*, where 1 Ci is 3.700×10^{10} disintegrations (radioactive decay events) per second (dps). *Specific activity*, the activity per unit quantity of radioactive sample, is expressed by dps per unit mass or volume, or in units such as microcurie or millicurie per milliliter, per gram, or per millimole. The last is preferable for labeled compounds.

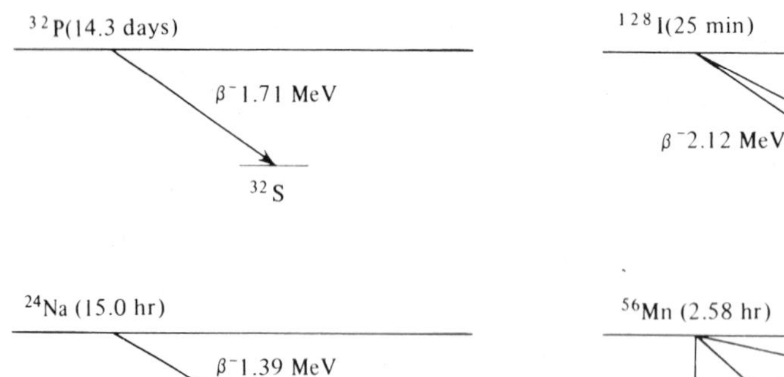

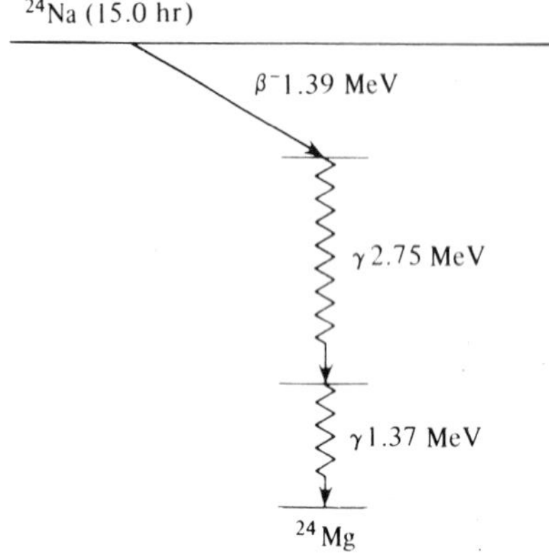

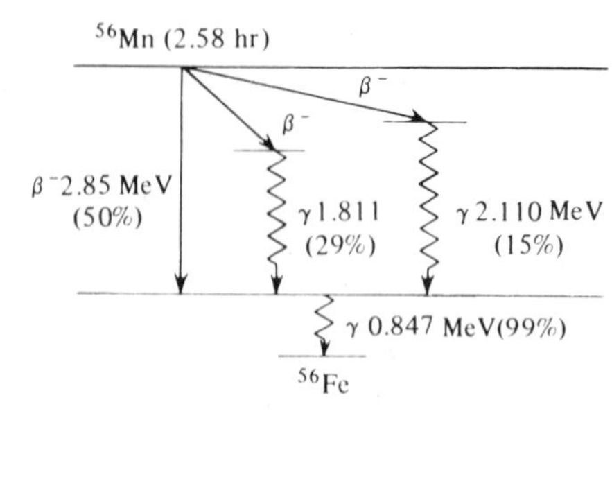

FIGURE **14.2**
Some decay schemes of radioisotopes.

14.2 MEASUREMENT OF RADIOACTIVITY

The best method for detecting and measuring radiation from nuclides in any particular situation depends on the nature of the radiation and the energy of the radiation or particles involved, as discussed in the preceding section. Specific detectors were discussed in Section 13.2.

The random nature of nuclear events requires that a large number of individual events be observed to obtain a precise value of the counting rate or the total sample count of a single sample. Several possible problems may arise in any measurement. (1) The ionizing particle may never reach the active volume of the detector; instead, it may be absorbed in the walls or air path. (2) Due to its *dead time* the detector may not have recovered from a previous event. (3) The ionizing particle may not produce an ion in the sensitive volume of the detector. (4) The detector may not be perfectly efficient, and the detector efficiency varies in different regions of the detector.

When resolution is the highest priority, the energies and intensities of the radiations are measured with a lithium-drifted germanium, Ge(Li), detector coupled to a multichannel, pulse height analyzer system. If sensitivity is paramount, a proportional counter equipped with a NaI(T1) scintillation crystal is the choice. These detectors were described in Chapter 13.

The Statistics of Radioactivity Measurements

Of primary interest in the measurement of radioactivity is the counting time (*preset time*) or the number of counts required (*preset count*) for a certain statistical error of the count. In practical work the true average count of any sample is not known, and it is necessary to substitute for it the observed counting rate, N_s/t_s,

TABLE **14.1** NUCLEAR PROPERTIES OF SELECTED RADIOISOTOPES

Radioisotope	Half-life	Target isotope	Natural abundance (%)	Thermal neutron cross section (barns)	Major radiations, energies in MeV (γ intensities, %)
^{3}H	12.26 yr				β^- 0.0186; no γ
^{14}C	5730 yr	^{13}C	1.108	0.0009	β^- 0.156; no γ
^{15}O	123 sec				β^+ 1.74; γ 0.511
^{22}Na	2.62 yr				β^+ 1.820, 0.545; γ 0.511, 1.275(100)
^{24}Na	14.96 hr	^{23}Na	100	0.53	β^- 1.389; γ 1.369(100), 2.754(100)
^{28}Al	2.31 min	^{27}Al	100	0.235	β^- 2.85; γ 1.780(100)
^{32}P	14.28 day	^{31}P	100	0.19	β^- 1.710; no γ
^{35}S	87.9 day	^{34}S	4.22	0.27	β^- 0.167; no γ
^{36}Cl	3.08×10^5 yr	^{35}Cl	75.53	44	β^- 0.714; γ 0.511
^{38}Cl	37.29 min	^{37}Cl	24.47	0.4	β^- 4.91; γ 1.60(38), 2.17(47)
^{40}K	1.26×10^9 yr		0.118	70	β^- 1.314; β^+ 0.483; γ 1.460(11)
^{42}K	12.36 hr	^{41}K	6.77	1.2	β^- 3.52; γ 0.31, 1.524(18)
^{45}Ca	165 day	^{44}Ca	2.06	0.7	β^- 0.252
^{51}Cr	27.8 day	^{50}Cr	4.31	17	γ 0.320(9); e^- 0.315
^{56}Mn	2.576 hr	^{55}Mn	100	13.3	β^- 2.85; γ 0.847(99), 1.811(29), 2.110(15)
^{55}Fe	2.60 yr	^{54}Fe	5.84	2.9	Mn X rays
^{59}Fe	45.6 day	^{58}Fe	0.31	1.1	β^- 1.57, 0.475; γ 0.143(1), 0.192(3), 1.095(56), 1.292(44)
^{60}Co	5.263 yr	^{59}Co	100	19	β^- 1.48, 0.314; γ 1.173(100), 1.332(100)
^{63}Ni	92 yr	^{62}Ni	3.66	15	β^- 0.067; no γ
^{65}Ni	2.564 hr	^{64}Ni	1.16	1.5	β^- 2.13; γ 0.368(5), 1.115(16), 1.481(25)
^{64}Cu	12.80 hr	^{63}Cu	69.1	4.5	β^- 0.573; β^+ 0.656; e^- 1.33, γ 0.511, 1.34(1)
^{65}Zn	245 day	^{64}Zn	48.89	0.46	β^+ 0.327; e^- 1.106; γ 0.511, 1.115(49)
^{69m}Zn	13.8 hr	^{68}Zn	18.6	0.10	γ 0.439(95); e^- 0.429
^{76}As	26.4 hr	^{75}As	100	4.5	β^- 2.97; γ 0.559(43), 0.657(6), 1.22(5), 1.44(1), 1.789, 2.10(1)
^{80}Br	17.6 min	^{79}Br	50.52	8.5	β^- 2.00; β^+ 0.87; γ 0.511, 0.61(7), 0.666(1)
^{80m}Br	4.38 hr				γ 0.037(36); e^- 0.024, 0.036, 0.047

TABLE **14.1** (continued)

Radioisotope	Half-life	Target isotope	Natural abundance (%)	Thermal neutron cross section (barns)	Major radiations, energies in MeV (γ intensities, %)
^{82}Br	35.34 hr	^{81}Br	49.48	3	β^- 0.444; γ 0.554(66), 0.619(41), 0.698(27), 0.777(83), 0.828(25), 1.044(29), 1.317(26), 1.475(17)
^{90}Sr }	27.7 yr				β^- 0.546; no γ
^{90}Y }	64.0 hr				β^- 2.27; no γ
^{110m}Ag	255 day	^{109}Ag	48.65	89	β^- 1.5; γ 0.658(96), 0.68(16), 0.706(19), 0.764(23), 0.818(8), 0.885(71), 0.937(32), 1.384(21), 1.505(11)
^{122}Sb	2.80 day	^{121}Sb	57.25	6	β^- 1.97; β^+ 0.56; γ 0.564(66), 1.14(1), 1.26(1)
^{124}Sb	60.4 day	^{123}Sb	42.75	3.3	β^- 2.31; γ 0.603(97), 0.644(7), 0.72(14), 0.967(2), 1.048(2), 1.31(3), 1.37(5), 1.45(2), 1.692(50), 2.088(7)
^{128}I	24.99 min	^{127}I	100	6.4	β^- 2.12; γ 0.441(14), 0.528(1), 0.743, 0.969
^{137}Cs }	30.0 yr				β^- 0.511, 1.176; γ 0.662(85)
^{137m}Ba }	2.554 min				γ 0.662(89); e^- 0.624, 0.656
^{198}Au	2.697 day	^{197}Au	100	98.8	β^- 0.962; γ 0.412(95), 0.676(1), 1.088
^{204}Tl	3.81 yr	^{203}Tl	29.5	11	β^- 0.766

SOURCE: J. A. Dean, ed., *Lange's Handbook of Chemistry*, 13th ed., McGraw-Hill, New York, 1985.

or the total counts, N_s, where t_s is the counting time. The error introduced by this substitution is small when dealing with the large number of counts used in the measurement of radioactivity. The deviation, q, from the true average count of a single sample, N_s, or background, N_b, is given by

$$q_s = \pm K\sqrt{N_s} \quad \text{or} \quad q_b = \pm K\sqrt{N_b} \tag{14.10}$$

where K is the number of standard deviations (Table 14.2) for a particular error or confidence limit.

Expressed as counting rate (counts per unit time), the deviation, Q, is

$$Q_s = \pm K\sqrt{\frac{N_s}{t_s}} \quad \text{or} \quad Q_b = \pm K\sqrt{\frac{N_b}{t_b}} \tag{14.11}$$

A radioactivity detector always shows some *background activity* due to natural sources of radiation, and any background is recorded simultaneously with the sample activity. The uncertainty in the net count of a single sample after subtraction of the background is

$$q_y = \sqrt{q_s^2 + q_b^2} \qquad (14.12)$$

The corresponding error in the net counting rate of a single sample is

$$Q_y = \sqrt{Q_s^2 + Q_b^2} \qquad (14.13)$$

$$Q_y = K\sqrt{\left(\frac{N_s}{t_s}\right)^2 + \left(\frac{N_b}{t_b}\right)^2} \qquad (14.14)$$

Because of the variation in the counting rate of the background, the uncertainty of the counting rate of the sample corrected for the background is always larger than the uncertainty of the recorded rate of the sample including the background.

When the background count is small in comparison with the count of the sample, the fraction of total time given to the counting of the background should be small. However, when the total count is close to the background, the time devoted to the background should approach that devoted to the sample. The optimum distribution of counting time between background and sample is given by

$$\frac{t_b}{t_s} = \sqrt{m\left(\frac{N_b}{N_s}\right)} \qquad (14.15)$$

where m is the number of experimental samples measured, not including the background.

To express the fractional error, F_y, of the sample corrected for background—that is, $N_s - N_b$, we use

$$F_y = \frac{K}{N_s - N_b}\sqrt{\left(\frac{N_s}{t_s}\right) + \left(\frac{N_b}{t_b}\right)} \qquad (14.16)$$

Preset Time or Preset Count

The counting time required for a sample to achieve a predetermined precision is obtained by combining Equations 14.11 and 14.15, squaring, and rearranging. The

TABLE **14.2**

TABLE OF CONSTANTS OF RELATIVE ERROR

Error (confidence limit)	Probability of occurrence (%)	K
Probable	50.0	0.675
Standard deviation (one sigma)	31.7	1.000
Nine tenths	10.0	1.645
Ninety-five hundredths	5.00	1.960
Two sigma	4.55	2.000
Ninety-nine hundredths	1.00	2.576
Three sigma	0.27	3.000

result is

$$t_s = \frac{K^2}{F_y^2}\left[\frac{N_s + N_b/c}{(N_s - N_b)^2}\right] \tag{14.17}$$

where $c = \sqrt{mN_b/N_s}$. The time required to count the background is obtained from Equation 14.15.

For a preset count, the counting time (Equation 14.17) is multiplied by the counting rate.

The number of counts required for a certain statistical precision is dependent on the relation of the counting rates of sample and background to each other (and not to their absolute values). As many counts must be taken of a sample that has a counting rate of 200 counts per minute (cpm) and a background of 20 cpm as of a sample that has 1000 cpm and a background of 100 cpm. Of course, the sample of 200 cpm must be counted five times as long as the sample counting 1000 cpm.

EXAMPLE 14.1

A sample gave a counting rate of 200 cpm in a 10-min counting period. The background gave a counting rate of 40 cpm in a 5-min counting period. What is the fractional error of the sample corrected for background when it is desired to achieve a 0.95 error? How much time should be devoted to counting the sample and the background individually?

From Equation 14.16, the fractional error of the sample corrected for background is

$$F_y = \frac{1.96}{200 - 40}\sqrt{\left(\frac{200}{10}\right) + \left(\frac{40}{5}\right)} = 0.066$$

There are five chances out of 100 that the error of the sample corrected for background will be greater than 6.6%. The counting time required for a fractional error of 0.066 is obtained by inserting the proper values into Equations 14.15 and 14.17:

$$\frac{t_b}{t_s} = \sqrt{\frac{40}{200}} = 0.45$$

$$t_s = \frac{(1.96)^2}{(0.066)^2}\left[\frac{200 + (40/0.45)}{(200 - 40)^2}\right] = 9.97 \text{ min}$$

$$t_b = 0.45 \times 9.97 = 4.46 \text{ min}$$

Coincidence Correction

In order to correct for counting losses at high counting rates caused by the finite resolution time of counters, two options are available. One method is to construct a calibration curve from a series of dilutions of a strong sample or from measurements of a series of standards of known strengths. A second procedure is to measure two samples separately and then measure the two samples together. The resolving time of the counter, τ, is given by

$$\tau = \frac{N_1 + N_2 - N_{1,2} - N_b}{2N_1 N_2} \tag{14.18}$$

where N_1 represents the counting rate of source 1 plus background and N_2 that of source 2 plus background, and $N_{1,2}$ is the counting rate of source 1 plus source 2, and the background N_b.

If, as is often true, the resolving time of the counter is longer than that of any other part of the measuring circuit, then the true counting rate, N_0, is related to the observed counting rate, N_s, and the resolving time, τ; namely,

$$N_0 = \frac{N_s}{1 - N_s\tau} \tag{14.19}$$

14.3 NEUTRON SOURCES

The chain-reacting pile is without peer as a tool for the general quantitative production of radioactivity by neutron activation because of the magnitude and spatial extent of the bombarding neutrons (thermal neutron flux) it is able to sustain. *Thermal* (or slow) *neutrons* are neutrons in a kinetic-energy equilibrium with the surrounding temperature. The pile is better than the cyclotron with respect to both intensity and the magnitude of the effective flux produced. Fluxes available range from 10^{11} to 10^{14} neutrons cm^{-2} sec^{-1}.

Preformed, portable generating sources, such as radionuclide cobalt-60 or californium-252 sources, can be useful in meeting many analytical requirements. Californium-252, available commercially as encapsulated source material, has a half-life of 2.6 yr and decays primarily by spontaneous fission with the emission of 10^{12} neutrons cm^{-2} sec^{-1}. In general, these are *fast neutrons*, with an average energy of about 2 MeV. Conversion into thermal neutrons is accomplished through collision with and elastic scattering from hydrogen atoms.

Pneumatic transfer systems permit the rapid, safe transfers of samples from an input station to the neutron source for irradiation. The samples then proceed to a decay station where unwanted, short-lived activities are allowed to decay. Next the samples proceed to a shielded counting station containing the detector, and finally to a dump for storage or disposal. In modern systems sample transfer and irradiation times are under computer control.

14.4 ACTIVATION ANALYSIS

Activation analysis is a technique of elemental analysis based on selectively inducing radioactivity in some atoms of the elements that make up the sample and then selectively measuring the radiations emitted by the radioactive atoms. Qualitative identification is achieved from the energy of the emitted spectrum. Quantitative determination is based on the intensity of radiation(s) characteristic of the particular element.

Neutron Activation Analysis

Neutron activation is the general term for irradiating material with neutrons to create radionuclides. Three steps are involved: (1) neutron bombardment of the sample, (2) recording the energy spectrum of the gamma or beta radiation produced, and (3) analysis of the significance of the spectrum features. The energies of the spectral peaks identify the elements present; the areas of the peaks define the quantities of each element (Figures 14.3 and 14.4).

The sample is placed in a neutron source long enough to produce enough radionuclide product that can be measured with the desired statistical precision.

Quite frankly, this requires some educated guesswork. The probability of a nuclear reaction is dependent on the nature of the target nuclide and the energy of the bombarding neutron. This probability or *cross section* is usually highest for the (n, γ) reactions with thermal neutrons. Most particle-emitting reactions have positive threshold energies and therefore have appreciable cross sections with only the more energetic neutrons. The production of radioactive atoms, N^*, is equal to the difference between the rate of formation and the rate of decay; that is,

$$\frac{dN^*}{dt} = \Phi\sigma N - \lambda N^* \tag{14.20}$$

where Φ is the number of bombarding particles, or *flux* (in $cm^{-2}\ sec^{-1}$), σ is the reaction cross section expressed in units of cm^2/target atom ($10^{-24}\ cm^2$/nucleus, also denoted *barns*), N is the number of target nuclei available, and λ is the characteristic decay constant. The number of target nuclei is given by

$$N = \frac{wN_A f}{M} \tag{14.21}$$

where w is the weight of the parent nuclide, N_A is the Avogadro constant, M is the atomic weight of the nuclide, and f is the fractional abundance of the target nuclide. Integration of Equation 14.20 over the time of irradiation yields the number of radioactive nuclei at the end of the irradiation period:

$$N^* = \frac{\Phi\sigma N}{\lambda}(1 - e^{-\lambda t}) \tag{14.22}$$

The *induced activity*, A_0, after irradiation for time t is the product λN^*. Substituting $0.693/t_{1/2}$ for the decay constant in the exponential term in Equation 14.22 yields

$$A_0 = \Phi\sigma N\left[1 - \exp\left(-\frac{0.693t}{t_{1/2}}\right)\right] \tag{14.23}$$

where t is the duration of the irradiation period. During the irradiation period some of the radionuclide produced decays. The term within the brackets of Equation 14.23 is the *saturation factor*, S. It represents the ratio of the amount of activity produced during the irradiation period to that produced in infinite time. At $t/t_{1/2}$ values of 1, 2, 3, 4, 5, 6,..., ∞, S has corresponding values of 0.50, 0.75, 0.87, 0.94, 0.97, 0.98,..., 1.00. The induced activity reaches 98% of the saturation value for irradiation periods equal to six half-lives.

EXAMPLE 14.2

Calculate the activity for a 10.0-mg sample of an aluminum alloy containing 0.041% manganese after a 0.50-hr irradiation in a flux of 5×10^{13} neutrons $cm^{-2}\ sec^{-1}$. Other necessary information is obtained from Table 14.1 for insertion into Equation 14.23:

$$A_0 = \frac{(0.00041)(0.0100\ g)(1.00)(6.02 \times 10^{23}\ \text{nuclei mol}^{-1})}{54.94\ g\ mol^{-1}}$$

$$\times (5 \times 10^{13}\ \text{neutrons cm}^{-2}\ sec^{-1})(13.3 \times 10^{-24}\ cm^2\ \text{nuclei}^{-1})$$

$$\times [1 - e^{-(0.693)(0.50\ hr)/2.58\ hr}]$$

$$A_0 = 3.68 \times 10^6\ \text{disintegrations/sec}$$

As the foregoing example illustrates, the duration of irradiation is determined by the neutron flux, the half-life of the radioactive product, and the sample matrix. In practice, a sample is seldom irradiated for longer than one or a few half-lives of the induced activity of interest because of the rapid asymptotic approach of the saturation factor to unity.

Short-lived activities are enhanced, relative to longer-lived activities, by the use of a short irradiation period, followed quickly by counting. Longer-lived activities are enhanced by the use of a longer irradiation period, followed by an appreciable delay for the decay of interfering short-lived activities, before counting.

EXAMPLE 14.3

In the determination of iron in aluminum alloys, the 1.29-MeV gamma of iron-59 was measured. A decay period of 1 wk before chemical processing allowed for the decay of the 15.0-hr sodium-24 formed from the reaction $^{27}Al(n, \alpha)^{24}Na$:

$$A = A_0 e^{-(0.693)(168)(15.0)} = 0.00235A_0$$

After an interval of 7 days (168 hr), the sodium-24 activity will have decayed to 0.00235 of its original value. The gamma radiation from sodium-24 at 1.369 MeV would no longer constitute an interference to iron.

Since there is usually a significant interval between the cessation of the irradiation and the measurement of the induced activity, the activity at any time after irradiation must be corrected for the intervening decay of the radionuclide being counted.

EXAMPLE 14.4

At 120 min after discharge from the reactor, all of the aluminum-28 (2.3 min half-life) and most of the other short-lived isotopes will have decayed to negligible activity. The activity of manganese-56, from Example 14.3, has dropped to 2.153×10^6 counts/sec. The activity at the moment of removal from the reactor is

$$2.153 \times 10^6 \text{ counts/sec} = A_0 e^{-(0.693)(2.0\text{ hr})/2.58\text{ hr}}$$

$$A_0 = 3.68 \times 10^6 \text{ counts/sec}$$

Of the two forms of the activation analysis method, this example represents the purely instrumental or nondestructive form. It involves only activation of the sample, followed by gamma-ray spectrometry of the activated sample.

If interferences from other induced activities prevent the purely instrumental detection of the activity of interest, the analyst may profitably resort to a postirradiation radiochemical separation procedure. After irradiation, the sample is dissolved and chemically equilibrated with a relatively large amount (but accurately known and typically about 10.0 mg) of the element or elements of interest. These are known as *carriers*. If high levels of nuclides with high specific activity are present, they are diluted with similar amounts of hold-back carriers of the particular nuclides. Then the element of interest is separated and purified by any suitable separation procedure. Finally it is counted. The amount of carrier element recovered is measured quantitatively so that the results can be normalized to 100% recovery. Thus it is possible to use fairly rapid, nonequilibrium separations. Recoveries of 50% or better are desirable, since the higher the recovery, the better the counting statistics.

Quantitative Analysis

In quantitative determinations the comparative method is used. Both samples and standards, each sealed in polyethylene or quartz containers, are irradiated in the same physical location and therefore under the same flux conditions. The relative activities are directly proportional to the respective concentrations of parent nuclide:

$$\frac{A_{\text{unknown}}}{A_{\text{standard}}} = \frac{N_{\text{unknown}}}{N_{\text{standard}}} \tag{14.24}$$

A gamma-ray spectrometer is used to count a series of activated samples and standards under exactly the same conditions, but at different decay times and perhaps for different lengths of time.

An illustrative gamma-ray spectrum (or pulse height spectrum) is shown in Figure 14.3. Manganese-56 follows the disintegration scheme outlined in Figure 14.2. The major feature in its spectrum is the sharp symmetrical full-energy or photopeak at 0.847 MeV, the result of total absorption of the gamma-ray energy by the detector. Two less intense photopeaks appear at 1.811 and 2.11 MeV. Any identification is confirmed by sequential measurements of the gamma spectrum and observation of the decay rates, which correspond to the half-life of the nuclide present.

The continuous curve below the full-energy peaks is the *Compton-continuum* region. The maximum energy (in MeV) that a gamma-ray photon can lose by Compton scattering, which is the elastic scattering of photons by electrons, is given by

$$E_{\text{CE}} = \frac{E}{1 + 0.511/2E} \tag{14.25}$$

where E_{CE} is the energy of the *Compton edge*. For the 0.847-MeV photopeak, E_{CE} is 0.627 MeV. Analogous Compton edges lie at 1.59 and 1.88 MeV for the less in-

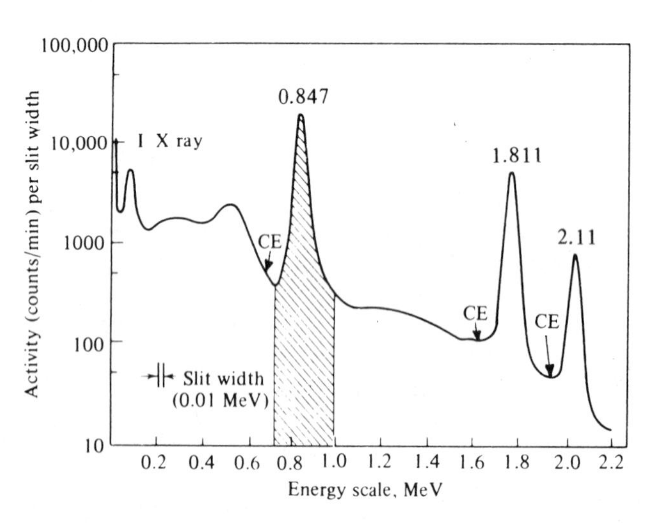

FIGURE **14.3**
Gamma spectrum of manganese-56. The area indicated under the photopeak at 0.847 MeV would be used in quantitative work. The letters *CE* indicate the Compton edges for each photopeak.

tense photopeaks. Less prominent features are a small peak at 0.511 MeV from an annihilation photon resulting from pair production in the shielding material by the gamma-ray peaks that exceed 1.02 MeV, peaks at full energy minus 0.511 MeV and full energy minus 1.02 MeV (twice 0.511 MeV), and several back-scatter peaks of various sizes and at various energies (usually less than 0.3 MeV).

The net area under each photopeak is directly proportional to the absolute gamma emission rate of the corresponding isotope. Usually the net photopeak area above the Compton continuum is computed using a "linear-base" approximation, as shown in Figure 14.4 for the chromium-51 and ruthenium-103 photopeaks. All counting rates must be corrected to the chosen reference time.

For the analysis of mixed gamma-emitting isotopes, the characteristic lower-energy portion of the gamma-ray spectrum of the most energetic full-energy peak is subtracted from the total spectrum. This step is repeated for each successively less energetic peak, as shown in Figure 14.4. These operations are called *spectrum stripping* and are performed automatically on the multichannel analyzer from standard curves stored in the memory with the aid of appropriate software programs. Finally, the multichannel analyzer adds together the counts in any groups of channels, such as those included within a photopeak.

When only the positron annihilation peak at 0.511 MeV is present in the spectrum, the sample must be counted twice, using different decay times following irradiation. The calculation of the contribution of the two elements is facilitated by knowledge of the different decay properties of the radionuclides.

Fast-Neutron Activation Analysis

For a few elements, activation by fast neutrons is more sensitive than activation with thermal neutrons. In some instances the product formed is more readily

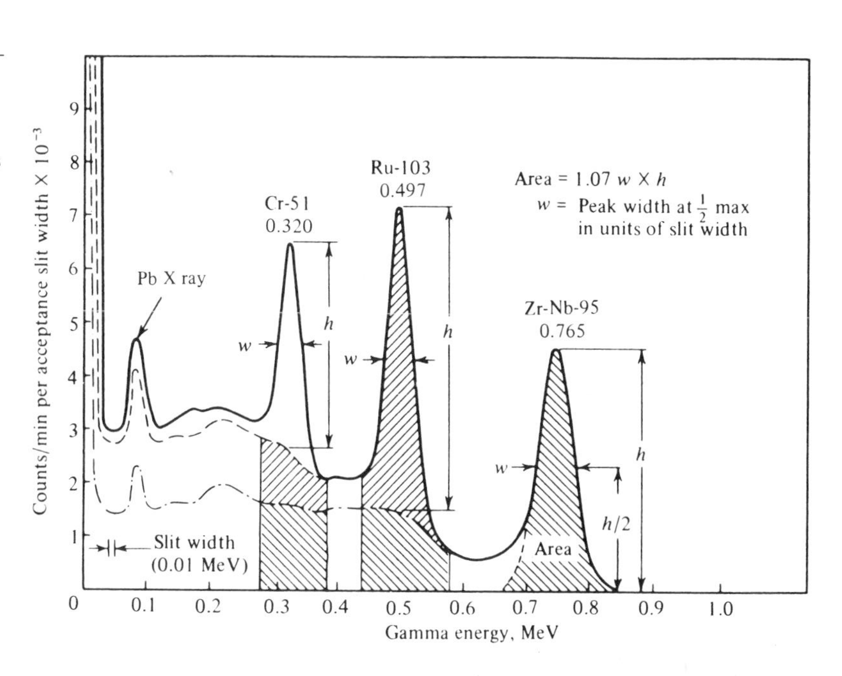

FIGURE **14.4**

Typical three-component gamma spectrum illustrating two methods of quantitative analysis: linear-base approximation and spectrum stripping.

detected because it is a gamma emitter instead of a pure beta emitter. In other cases the product has a more convenient half-life relative to interfering activities or is more easily separated chemically. The list of applicable elements includes N, O, F, Al, Si, P, Cr, Mn, Fe, Cu, Y, Mo, Nb, and Pb.

A fast-neutron (fission-spectrum) flux is present in the core of a nuclear reactor. Samples are wrapped with cadmium foil to prevent thermal neutrons from reaching the sample. Californium-252 is another source (see Section 14.3).

The diversity of reactions produced by 14-MeV neutrons often dictates the use of an accelerator-type neutron generator. The accelerator uses a target of metallic tritium and a positively charged deuterium beam to form neutrons with energies of about 14 MeV by the $^{3}H(d, n)^{4}He$ reaction. Pulsed generators provide very high, fast-neutron fluxes, well over 10^{16} neutrons cm^{-2} sec^{-1}, of about 30 msec duration. Pulsed operation favors the formation of short-lived (less than 50 sec) radioisotopes. The amount of a particular activity, A_p, produced in a pulse, relative to the amount of activity produced in the usual steady-state operation of the same reactor to saturation activity, A_s, is approximately $A_p/A_s = 70/t_{1/2}$. The nondestructive determination of oxygen is now frequently done by activation analysis with 14-MeV neutrons to form 7.35-sec ^{16}N via the $^{16}O(n, p)^{16}N$ reaction. The gamma radiation emitted by ^{16}N is mostly 6.13 MeV, which is exceptionally high, so that there are essentially no interferences except for fluorine, which also undergoes the $^{19}F(n, \alpha)^{16}N$ reaction. Total analysis time for oxygen is less than 1 min; the limit of detection is 1 $\mu g/g$ in a 10-g sample. For six elements, O, Si, P, Fe, Y, and Pb, the sensitivity is greater using fast rather than thermal neutrons.

Prompt Gamma-Ray Analysis

Prompt gamma-ray analysis involves the analysis of gamma rays emitted simultaneously with neutron absorption. Emission occurs in times shorter than molecular processes and at energies as high as 10 MeV. Thus the physical state of a sample has no effect on the emitted radiation. This high-energy radiation is much more penetrating than X rays and can be detected through process vessel walls. Since these gamma rays originate from a nuclear event, their energy and intensity are independent of the chemical state of the element(s) being monitored.

Prompt gamma-ray analysis is effective for H, B, Hg, Cl, Ti, Co, Mn, Ni, Cu, and Cd. One application is the determination of TiO_2 present as a slurry in ethylene glycol; titanium-49 emits a gamma ray with an energy of 1.381 MeV. Another example is the measurement of chlorine either as CCl_4 in hexane or as HCl in water. The matrix does little to impede the transmission and absorption of neutrons or the emission and transmission of gamma rays.

Summary of Activation Analysis

The technique is useful in analyzing for most elements in the periodic table. The chemical and physical states of the sample have little effect on the results. Activation analysis offers advantages over more traditional methods, especially in sensitivity and the elimination of complex sampling systems. Other advantages include elemental analysis of bulk contents, off-line or on-line use through container walls, limited effects of impurities, and quick response (1–10 min per analysis). It is the most common technique for establishing the elemental composition of geological and biological standard reference materials.

14.5 ISOTOPE DILUTION ANALYSES

The isotope dilution technique is the only feasible approach in the determination of a substance that cannot be quantitatively isolated in pure form for weighing or for determination by other methods. The technique is very valuable, particularly in biological studies, when exceedingly small amounts of substances are to be determined in the presence of hard-to-separate substances that interfere with other analytical methods. This technique also measures the yield of a nonquantitative process.

Several assumptions are implicit in the isotope dilution methods. The tagged substance should be chemically identical with the substance being determined. The radionuclide used for tagging must not undergo radiative self-decomposition, nor should it undergo exchange reactions or isotopic fractionation processes that would alter the isotopic composition of the tagged molecule. Uniform mixing must not be taken for granted. Occasionally it is necessary to break up any combination of analyte with another compound so that the analyte will be available for mixing with the tagged material.

The general principles of isotope dilution analyses are outlined below. In the *direct method*, the sample being analyzed contains substances of natural isotopic abundance. A tracer of unnatural isotopic composition (radioactive or stable) is added. (In isotope dilution with stable isotopes, the samples measured need to be free of contamination only with the element that enters into the isotope-ratio determination; other impurities introduce no error.)

Experimentally a known weight, w_t, of the tracer, whose specific activity, A_t, is known from a separate measurement, is added to the unknown sample. A small amount of the pure compound is isolated from the mixture by a suitable chemical separation method. Purity, not quantitative recovery, is the goal. The amount isolated needs to be only a sufficient quantity for determining the radioactivity accurately. Finally, the specific activity, A, of the material isolated after isotopic dilution (inactive element plus tracer) is measured under conditions comparable to those for the radiotracer alone.

The extent of dilution of the radiotracer shows the amount, w, of inactive element (or compound) present, as given by the equation

$$w = w_t\left(\frac{A_t}{A} - 1\right) \tag{14.26}$$

In *reverse isotope dilution*, a known weight of untagged material is added to a sample containing an unknown weight of tagged material of known specific activity. Thereafter the experimental method and calculations are the same as for direct isotope dilution.

The method has proved valuable in the analysis of complex biochemical mixtures, studies of human metabolism, and radiocarbon dating of archaeological and anthropological specimens.

14.6 LIQUID SCINTILLATION SYSTEMS

Liquid scintillation counting has as its primary application the counting of weak beta emitters, such as tritium, carbon-14, and phosphorus-32. However, other radionuclides can also be measured. These include higher energy beta emitters as well as alpha and gamma emitters. Isotopes with intermediate half-lives have low specific activity and require larger amounts of the isotope to give an adequate counting rate. Isotopes with short half-lives require prompt shipment and delivery from the source; those with half-lives longer than 12 hr can usually be obtained and used without difficulty.

Liquid Scintillators

Samples are usually counted by dissolving the compound containing a small amount of the radionuclide (called the *tracer*) in a liquid scintillator. The tracer labels the sample and allows the molecules to be traced or followed by the liquid scintillation spectrometer. The basic objective of the technique is to arrange for emitted beta particles to collide with the solvent molecules. The energy resulting from the collisions excites the solvent molecules and is transferred to other molecules until it is finally transferred to the scintillator molecules. The scintillator molecules absorb the energy, a portion of which is then emitted in the form of photons of visible light, called fluorescence. The photons are detected by a photodetector and converted into electrical energy for counting by the spectrometer.

The bulk solvent must efficiently transfer energy to a scintillator molecule and be capable of dissolving the scintillators and the sample material. Aromatic solvents, such as toluene or xylene, are favored because of their efficiency in energy transfer. 1,4-dioxane is used when large amounts of water are involved. Naphthalene is often added to improve the energy-transfer process and reduce quenching. Sometimes aqueous sample solutions are incorporated into a toluene-based system by adding a nonionic surfactant such as Triton X-100. Glycol ethers and alcohols are also used as secondary solvents to improve water miscibility and to allow counting at low temperatures.

The scintillator must be capable of absorbing light at one wavelength and reemitting it at a longer wavelength. Most scintillation spectrometers are sensitive to the fluorescent emission of the primary scintillator. However, if an older model is being used, it may be necessary to add a secondary scintillator. The latter absorbs the light emitted by the primary scintillator and emits it at an even longer wavelength. When used, it is added to the extent of one-tenth or less of the primary scintillator. The most popular primary scintillator is 2,5-diphenyloxazole (PPO). The most widely used secondary scintillators are 2,2′-*p*-phenylenebis(5-phenyloxazole), or POPOP, and 2,2′-*p*-phenylenebis(4-methyl-5-phenyloxazole). For the primary scintillator the fluorescence emission maximum lies in the range 360–365 nm, whereas that for POPOP lies around 410–420 nm. As shown in Figure 14.5, the POPOP absorption spectrum overlaps well with the PPO emission spectrum. Since POPOP has a very large molar absorptivity in the overlap region, only a very small amount of POPOP is necessary for nearly complete absorption of the PPO fluorescence. The POPOP fluorescence closely matches the response of the blue-sensitive photomultiplier tubes.

Liquid scintillation procedures can be extended to various samples. If radioactive carbon dioxide is being measured, the scintillator solution should contain a trapping agent, perhaps 1-amino-2-phenylethane. Quaternary ammonium hydroxides in methanol find use as tissue solubilizers. For materials that cannot be solubilized, a suitable gelling agent is used to prevent settling; this technique is called suspension counting. Finely divided amorphous silica is widely used as a gelling agent.

Color and Chemical Quenching and Luminescence

Two types of quenching affect the accuracy of results in liquid scintillation systems. In chemical quenching, some of the energy released by the emission of the beta particle during radioactive decay is absorbed by the liquid components of the sample. Therefore, the total amount of released energy is unavailable for producing photons.

In color quenching, colored material in the sample absorbs a portion of the photons emitted by the scintillator. This type of quenching is common in samples that have been prepared using colored biological specimens such as blood or plant

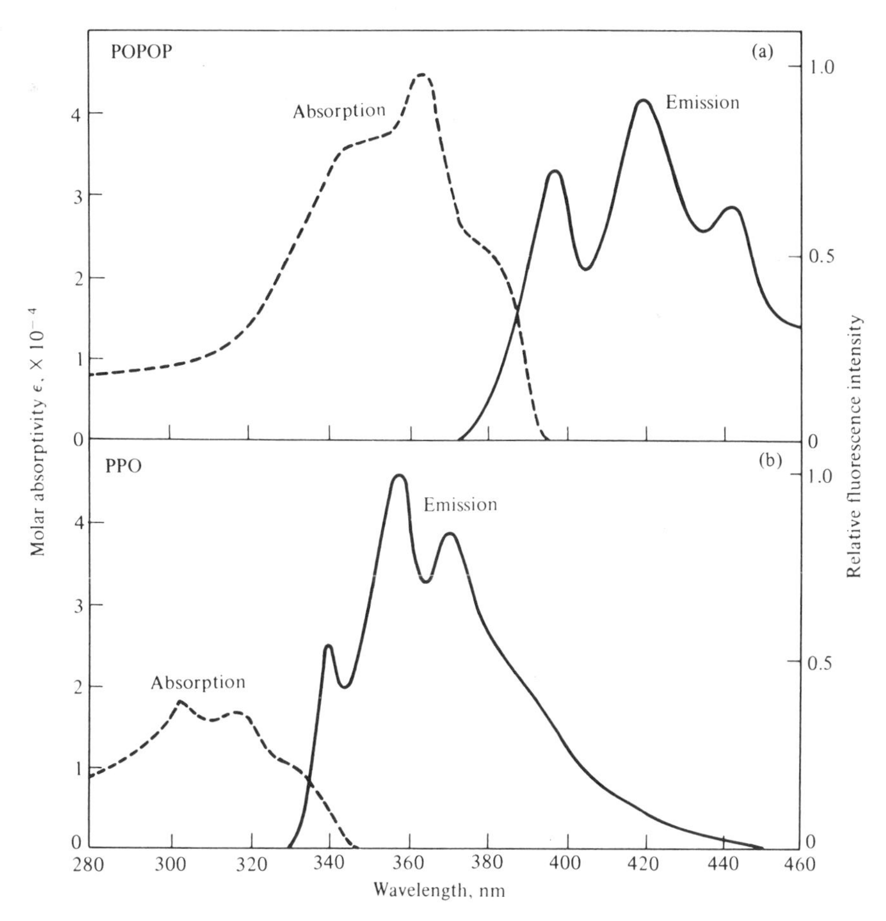

FIGURE 14.5
(a) Absorption and fluorescence emission spectra of POPOP.
(b) Absorption and fluorescence emission spectra of PPO.

tissue extracts. Color quenching can be reduced or eliminated by digestion with hydrogen peroxide and perchloric acid. For ^{3}H and ^{14}C, complete combustion of the sample to water and a soluble carbonate produces a simple counting environment.

Chemiluminescence and phosphorescence (see Chapter 8) give yet a third change in the spectrum. Luminescence correction is based on spectral pattern recognition by a computer and delayed-coincidence counting. The delayed-coincidence counts are normalized to yield chance-coincidence counts. The correct sample activity is obtained by spectral stripping.

Use of Cerenkov Radiation

The Cerenkov technique is applicable for high-energy beta emitters, such as phosphorus-32, without the use of a scintillator. Cerenkov counting is not subject to chemical quenching. Color quenching may be a problem but can often be overcome by bleaching; the by-products of the bleaching reaction do not cause chemical quenching.

When a charged particle, a beta particle in this case, travels with a velocity greater than that of light in a given medium, Cerenkov radiation results. The emission takes place in the violet-ultraviolet region. The solution should not contain materials that absorb in this wavelength region.

The threshold for the emission of Cerenkov radiation can be calculated as

$$\text{threshold (in MeV)} = 0.511\left[\frac{1}{(1 - 1/\eta^2)^{1/2}}\right] - 1 \qquad (14.27)$$

where η is the refractive index of the solvent. Water, with a refractive index of 1.3325, gives a threshold of 0.262 MeV. Even though the energy of the beta particles emitted in the decay of phosphorus-32 is a continuum (from 0 to 1.710 MeV), the average energy of 0.690 MeV is fairly high. Low-energy beta emitters that have their energy maximum between the threshold and about 0.600 MeV cannot be expected to be counted by this technique, since the efficiency will be extremely low.

Instrumentation

A diagram of a typical beta scintillation counter is shown in Figure 14.6. High-density polyethylene vials are economical; they should have background counts of less than 10 cpm (for tritium counting). Borosilicate glass vials are impervious to aromatic hydrocarbons and best for sample storage; the vials should have a low background count from potassium, less than 20 cpm when determining tritium.

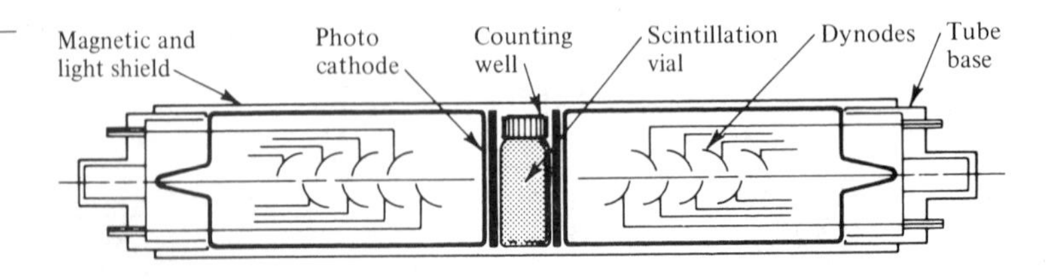

FIGURE **14.6**
Diagram of a typical beta scintillation counter showing only the counting well and photomultiplier tube detectors.

FIGURE 14.7
Energy spectrum of tritium and carbon-14.

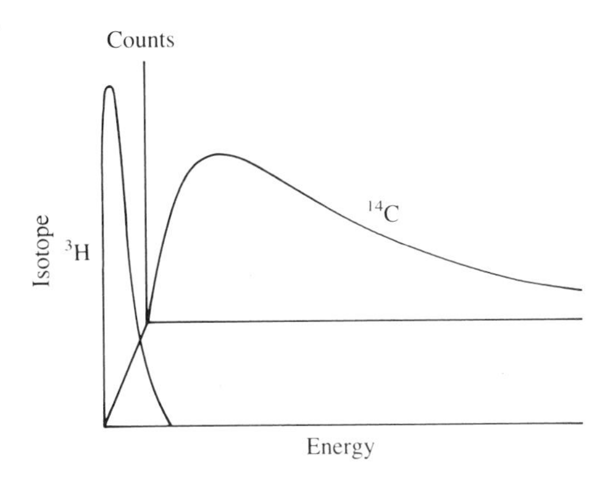

Instrumentation is available that uses a spectrum unfolding technique to separate completely the individual radionuclides of a dual labeled sample. This is a patented method for dual label analysis using one continuous energy region. The method allows the counting of each radionuclide at maximum efficiency and eliminates the need for spillover corrections; these are corrections for overlapping energy regions of the individual isotopes, such as tritium and carbon-14, whose energy curves are shown unfolded in Figure 14.7. Concern about the ratio of two radionuclides in a sample is thereby eliminated.

14.7 OTHER APPLICATIONS OF RADIONUCLIDES

The introduction of a radioactive-labeled material into a sample system, labeled reagents, or the use of radioactive tracers for procedure development finds application in analytical chemistry.

Tagging Compounds

Since the radioactive isotopes are chemically identical with their stable-isotope counterpart, they may be used to "tag" a compound. The tagged compound may then be followed through any analytical scheme, industrial system, or biological process. It is essential that a compound be tagged with an atom that is not readily exchangeable with similar atoms in other compounds under normal conditions. For example, tritium could not be used to trace an acid if it were inserted on the carboxyl group, where it is readily exchanged by ionization with the solvent.

In biological investigations the purity of the isolated material is important. One must be certain that the radioactivity of the compound isolated is due to the compound itself and not to some minor contaminant that may have a high specific activity. The usual chemical methods of purification of a tagged substance should be carried out in a preparation. On fractions thus purified, specific activities are determined. Lack of constancy of specific activity is evidence of radioactive or radiochemical impurities, and further purification is then required.

Radionuclides have been used to study errors that result from adsorption and occlusion in gravimetric methods and to devise methods of preventing coprecipitation, adsorption, and occlusion.

Analyses with Labeled Reagents

Radiometric methods that employ reagent solutions or solids tagged with a radionuclide have been used to determine the solubility of numerous organic and inorganic precipitates, or a radioreagent for titrations involving the formation of a precipitate. In this type of application it is necessary to establish the ratio between radioactivity and weight of the radionuclide plus carrier present. This may be established by evaporating an aliquot to dryness, weighing the residue, and measuring the radioactivity.

In solubility studies, the compound of interest is synthesized, using the radionuclide, and a saturated solution of the compound is prepared. A measured volume of the saturated solution is evaporated to dryness, and the radioactivity of the residue is determined. From the previously established relationship between weight and radioactivity, the amount of the compound present can be calculated.

The efficiency of an analytical procedure can be determined by adding a known amount of a radioisotope before analysis is begun. After the final determination of the element in question, the activity of the precipitate is determined and compared with the activity at the start.

Chemical yields need not be quantitative in an analytical procedure when the results are corrected by the recovery of the radionuclide. To the mixture of ions a known amount of radionuclide is added, or to a mixture of activities a known amount of carrier element is added, and then separated in the necessary state of chemical and radiochemical purity but without attention to yield. The isolated sample is determined by any suitable method, and the activity is measured.

PROBLEMS

1. A sample of ^{35}S contains 10 mCi. After 174 days, how many disintegrations per minute occur in the sample?

2. How much activity of the ^{32}P, the ^{131}I, and the ^{198}Au remains (a) after 14 days, (b) after 30 days, and (c) after 60 days?

3. Calculate the probable error and the 1σ and 2σ variations (in percent) for each of these total numbers of counts: (a) 3200, (b) 6400, (c) 8000, (d) 25,600, and (e) 102,400.

4. Compute the dead time of the Geiger counter and the corresponding counting losses from this information: sample A gave a rate of 9728 counts/min, sample B gave a rate of 11,008 counts/min, and together samples A plus B gave a rate of 20,032 counts/min.

5. Assuming that the dead time of a Geiger counter is 200 μsec and that there are no other counting losses, what is the efficiency of the counter for (a) 2500 ionizing particles per second, (b) 1000, (c) 200, and (d) 5?

6. What is the useful range of counting rates if the dead time of the detector is (a) 0.25 μsec, (b) 1.0 μsec, (c) 5 μsec, and (d) 270 μsec?

7. The decay of a particular halogen, subjected to several hours of irradiation, provided the data in the table. Plot the decay curve on semilog paper and analyze it into its components. What are the half-lives and the initial activities of the component activities? Can you identify the particular halogen?

Time (min)	Activity (counts/min)	Time (min)	Activity (counts/min)
10	1800	50	650
18	1400	60	550
24	1215	80	430
32	970	120	330
36	880	180	270
40	800	240	230

8. In a certain measuring arrangement, the beta particles of ^{136}Cs are absorbed as shown in the table. (Correction is made for gamma radiation.) Find the maximum energy of the beta radiation. What is the aluminum half-thickness?

Thickness of aluminum (mg/cm^2)	Activity (counts/min)	Thickness of aluminum (mg/cm^2)	Activity (counts/min)
0	10,000	53	270
12	4700	72	45
27	1700	85	10
41	730	100	10

9. To a crude mixture of organic compounds that contains some benzoic acid and benzoate was added 40.0 mg of benzoic acid-7-^{14}C (activity = 2000 counts/min). After equilibration, the mixture was acidified and extracted with an immiscible solvent. The extracted solid, following removal of the solvent, was purified by recrystallization of the benzoic acid to a constant melting point. The purified material weighed 60.0 mg and gave a rate of 500 counts/min. Compute the weight of benzoic acid in the crude mixture.

10. Argon ionization detectors, used in gas-liquid chromatography, utilize as radioisotope source ^{90}Sr and its daughter ^{90}Y. Estimate the range of the beta particles emitted in air and in iron (the material of construction of cell walls).

11. If a 10.0-mg sample of aluminum foil were irradiated for 30 min in a neutron flux of 5×10^{11} neutrons cm^{-2} sec^{-1}, how long should the sample be allowed to "cool" before chemical processing or counting so that the strong aluminum activity will have decayed to less than 1 count/min?

12. For the irradiation time and flux stated in Problem 11, what is the limit of detection (40 counts/sec) for traces of sodium as sodium-24 in "pure" aluminum foil after the aluminum activity has decayed to less than 1 count/min. Counting geometry is 100%. Assume no other activities are present and ignore corrections for the absorption of sodium beta particles by the aluminum foil.

13. What weight of sample should be taken for the activation analysis of an aluminum alloy that contains 0.019% zinc if the irradiation time is 62 hr with a flux of 5×10^{11} neutrons cm^{-2} sec^{-1}, followed by a cooling period of 24 hr? A rate of 1000 counts/min is desirable.

14. In a particular aluminum alloy, these elements are present in the following percentages: Cu, 0.30; Mn, 0.30; Ni, 0.59; Co, 0.0053. If all samples weighed 10.0 mg, how long should the irradiations be continued for the determination of each element? Assume a rate of 10,000 counts/min in a 5-min counting period is desirable after a cooling period of 0.7 day. Flux is 5×10^{11} neutrons/cm^2/sec.

15. An assay was done using thymidine-5′-triphosphate[^{32}P] with a specific activity of 22 Ci/mmol (supplied by vendor). The ^{32}P was counted at 97% efficiency. This particular assay used 15 μg of DNA; the detector output recorded 410,558 counts/min. All experiments with this lot of isotope were corrected back to a reference date 22 days earlier. Calculate the moles of thymidine per unit weight of DNA.

16. In the analysis of mixtures of sodium and potassium carbonates, the half-lives of ^{24}Na and ^{42}K are too nearly the same to permit a resolution of the composite gross decay curve if the total beta radiations were counted. Suggest a method for the analysis of this binary mixture using the beta radiations.

17. It took a weekend (48 hr) to count a set of samples for ^{32}P. If no half-life correction were made, what would be the error in counts per minute for the last sample as compared with the first one counted in the set?

18. From the gamma-ray spectrum of neutron-activated seawater taken with a Ge(Li) detector, identify the elements present. Photopeak energies are expressed in keV.

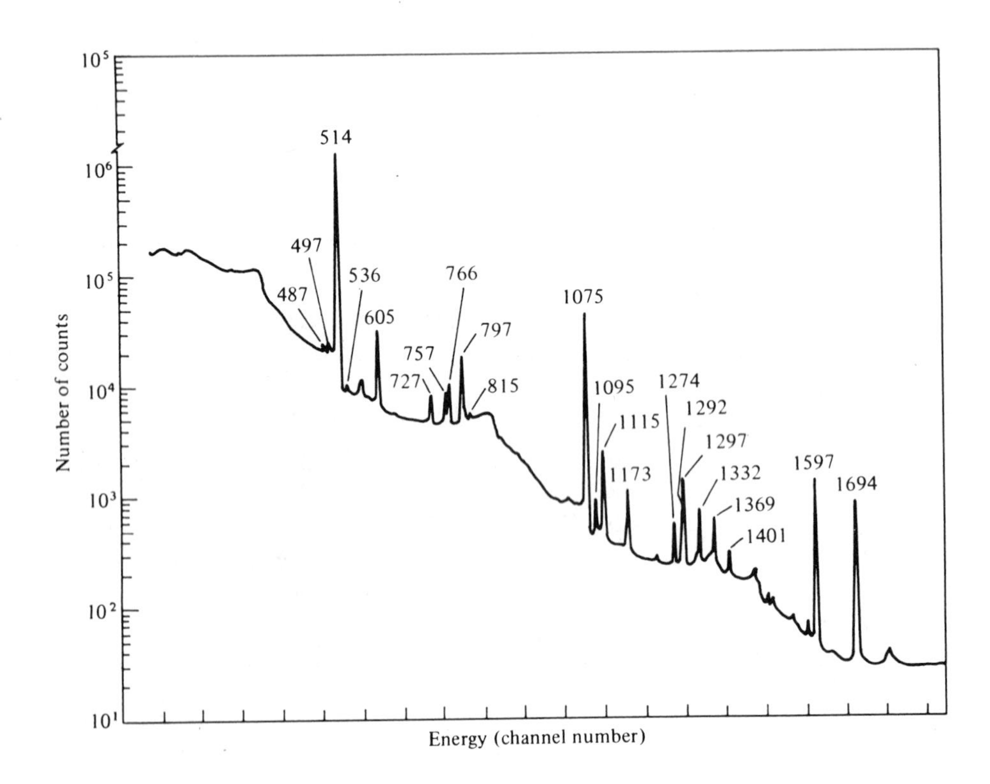

BIBLIOGRAPHY

BARKOUSKIE, M. A., "Liquid Scintillation Counting," *Am. Lab.,* **8,** 101 (May 1976).

BERNSTEIN, K., "Neutron Activation Analysis," *Am. Lab.,* **12,** 151 (September 1980).

CROUTHAMEL, C. E., AND R. R. HEINRICH, "Radiochemical Separations," Chap. 96, *Treatise on Analytical Chemistry,* I. M. Kolthoff and P. J. Elving, eds., Part I, Vol. 9, Wiley-Interscience, New York, 1971.

FINSTON, H. L., "Radioactive and Isotopic Methods of Analysis," Chap. 94, *Treatise on Analytical Chemistry,* I. M. Kolthoff and P. J. Elving, eds., Part I, Vol. 9, Wiley-Interscience, New York, 1971.

KATZ, S. A., "Neutron Activation Analysis," *Am. Lab.,* p. 16 (June 1985).

NUCLEAR MAGNETIC RESONANCE SPECTROSCOPY

In nuclear magnetic resonance (NMR) spectroscopy, the characteristic absorption of energy by certain spinning nuclei in a strong magnetic field, when irradiated by a second and weaker field perpendicular to it, permits the identification of atomic configurations in molecules. Absorption occurs when these nuclei undergo transitions from one alignment in the applied field to another one. The amount of energy required to cause a particular nucleus to realign depends on such factors as field strength, electronic configuration around the particular nucleus, anisotropy, type of molecule, and intermolecular interactions. The spectra obtained answer many questions such as (referring to specific nuclei): Who are you? Where are you located in the molecule? How many of you are there? Who and where are your neighbors? How are you related to your neighbors? The result is often the delineation of complete sequences of groups or arrangements of atoms in the molecule. Consequently, organic chemists have enthusiastically embraced NMR spectroscopy to identify and characterize molecules.

Analytical chemists, on the other hand, were more reluctant to accept NMR instrumentation. The sensitivity of NMR, compared with optical techniques, gas and liquid chromatography, and mass spectrometry, was lower by several orders of magnitude and usually precluded its use as a method for trace analyses. Also, the cost and complexity of maintaining standard operating conditions frequently turned the tide in favor of other methods. Now, however, new, high-powered superconducting magnets and a new family of NMR instruments, called Fourier transform NMR, built around a small, high-speed digital computer, have revolutionized the practice of NMR in organic chemistry and firmly entrenched NMR techniques in the analytical chemists' arsenal of weapons.

15.1 BASIC PRINCIPLES

The nuclei of certain isotopes have an intrinsic spinning motion around their axes. The spinning of these charged particles, or their circulation, generates a magnetic moment along the axis of spin (Figure 15.1). If the nuclei are placed in an external magnetic field, their magnetic moment can align with or against the field. Each individual nucleus spins around its axis, and the axis of the nuclear magnetic moment so generated precesses about the force line of the applied magnetic field as shown in Figure 15.2. The field aligns the spinning nuclei against the disordering tendencies of thermal processes. Increasing the strength of the field only makes the nuclei precess faster. The frequency of precession, ν_0, is known as the Larmor frequency of the observed nucleus.

Nuclear Magnetic Energy Levels

For a nucleus to be magnetic, it must possess spin angular momentum with an actual magnitude of $(h/2\pi)\sqrt{I(I+1)}$. The maximum observable value of the spin angular momentum is $hI/2\pi$, where I is the spin quantum number of the particular nucleus and h is Planck's constant. A spinning nucleus, like any spinning charge, generates a magnetic moment, μ, parallel to the axis of spin. The magnetic moment is quantized, as is the spin moment. The magnetic quantum number, m, has the values $-I, -I+1, \cdots, I-1, I$. The ratio of the magnetic moment to the angular moment, γ, is known as the magnetogyric ratio. Different nuclei have different values of γ. The maximum observable component of the magnetic moment, μ, has the value $m\mu/I$ or $\gamma Ih/2\pi$.

Nuclei with even mass number, A, and even charge, Z, behave as though they were nonspinning spherical bodies and have no magnetic moment. Nuclei such as ^{12}C and ^{16}O are of this type. These nuclei are not detectable by NMR. Nuclei with odd mass number, A, have half-integral spins, and nuclei with even A but Z odd have integral spins and, thus, these classes of nuclei give NMR signals.

In the class of nuclei with half-integral spins, those nuclei with $I = \frac{1}{2}$ behave like charged, spinning spherical bodies and give the best-resolved spectra. Among

FIGURE **15.1**

(a) A spinning nucleus generates a magnetic moment, μ, which precesses at the Larmor frequency, ν_0, around an external field, H_0, along the z-axis. (b) Precession of μ about H_0 when weak-rf magnetic field, H_1, rotating at frequency ν is in the xy-plane. (c) When $\nu = \nu_0$, the precessing nucleus absorbs energy and flips to antiparallel orientation with respect to H_0.

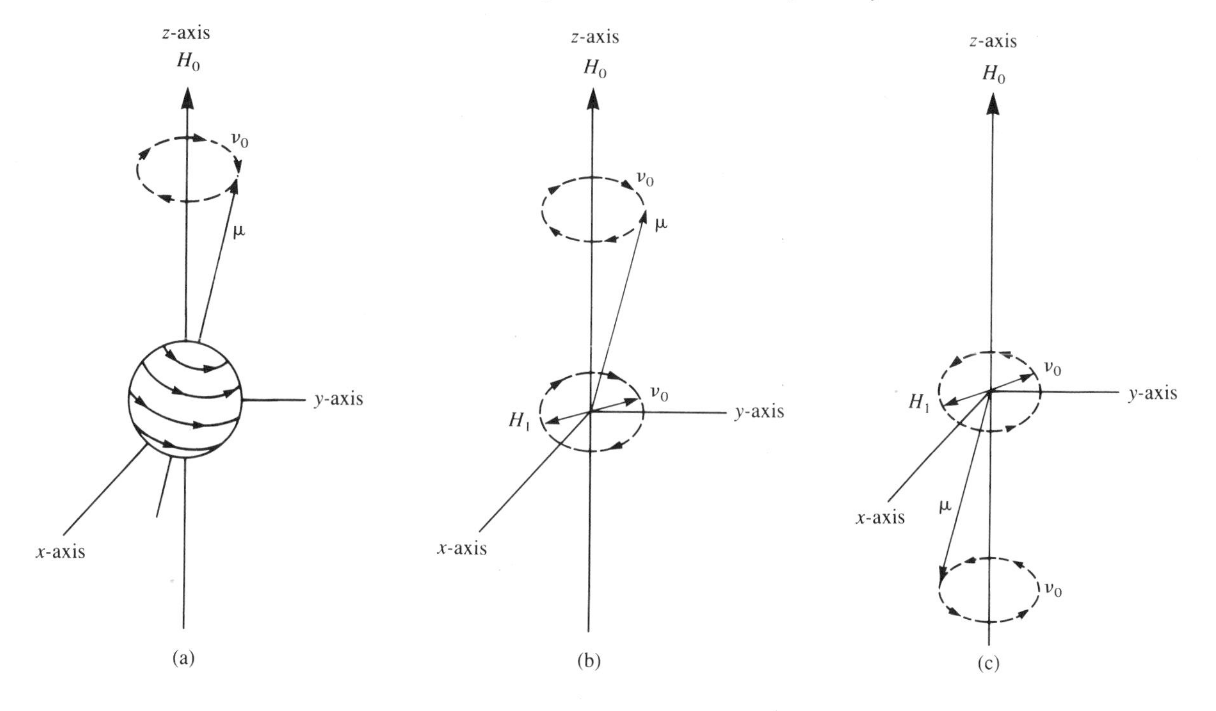

important nuclei with $I = \frac{1}{2}$ are ^{1}H, ^{13}C, ^{19}F, and ^{31}P. Nuclei of interest with $I = \frac{3}{2}$ are ^{7}Li, ^{11}B, and ^{35}Cl. Nuclei with integral spins of $I = 1$ include ^{2}H and ^{14}N.

Nuclei with spins greater than $\frac{1}{2}$ behave like nonspherical, charged rotating bodies. Some of these nuclei behave like oblate spheroids and others like prolate spheroids. Such nuclei often show line broadening in NMR experiments.

The fact that ^{12}C and ^{16}O do not have nuclear magnetic moments is perhaps advantageous, since organic compounds that contain only C, H, and O show only proton spectra without complications from interactions of the protons with carbon and oxygen. The natural abundance of ^{13}C, about 1.1%, is so low that its effects are not seen in ordinary proton NMR spectra; however, as shown later in this chapter, ^{13}C spectra are observed when necessary with modern sensitive instruments.

The magnetic axis of the nucleus assumes $2I + 1$ orientations with respect to the external magnetic field, and each orientation corresponds to a discrete energy level (with respect to that in a zero field) given by

$$E = \frac{-m\mu}{I} H_0 \tag{15.1}$$

where I is the spin number, m is the magnetic quantum number, E is the energy, μ is the magnetic moment of the nucleus, and H_0 is the external field strength in gauss (G). The spectrum of allowed values of m, therefore, is $I, I - I, \ldots, -I + I, -I$. Each value corresponds to a discrete orientation (and energy level). Hence, a

FIGURE 15.2 Nuclear orientation and energy levels of nuclei in a magnetic field for different spin numbers.

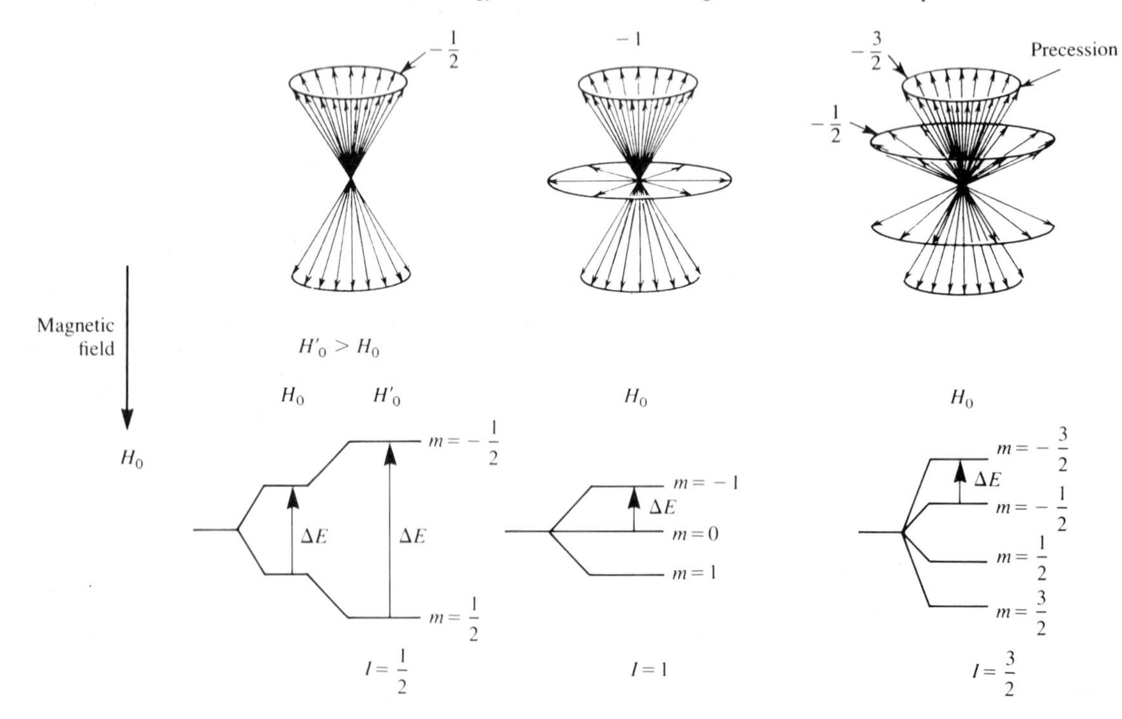

nucleus with spin $\frac{1}{2}$ has two orientations; with spin 1, three orientations; and so on (Figure 15.2).

At equilibrium the population of the various nuclear energy levels is predictable by use of a Boltzmann distribution:

$$\frac{n_{\text{upper}}}{n_{\text{lower}}} = e^{-\mu H_0/IkT} \qquad (15.2)$$

where k is the Boltzmann constant and T is the absolute temperature. For a magnetic nucleus of spin $\frac{1}{2}$ in a field of 14.09 kG, the distribution predicts a population ratio of 0.9999904 at room temperature. The lower energy level (orientation parallel to the applied magnetic field) is favored to the extent of approximately 9.6 excess nuclei out of every million. These nearly equal populations with such small energy differences between the levels make it easy to saturate the system—that is, equalize the populations—thus eliminating the signal. Therefore, relaxation processes that allow nuclei in the higher energy states to return to the lowest energy state are important. These processes are described later in this chapter.

Magnetic Resonance

The resonance frequency, ν_0, that effects the transition between energy levels is derived by equating the Planck quantum of energy with the energy of reorientation of a magnetic dipole (Equation 15.1). Since m can change by only ± 1,

$$\Delta E = h\nu_0 = \frac{\mu H_0}{I} \qquad (15.3)$$

When a constant magnetic field, H_0, is applied to a sample, the nuclear magnetic moments precess about the axis of this external field but without phase coherence; that is, the nuclear magnetic moments are randomly distributed around the cone of precession. A radio frequency (rf) current passing through a coil in the yz-plane (Figure 15.1) creates an alternating magnetic field, H_1, perpendicular to the plane of the coil—that is, along the x-axis. This alternating linear magnetic field may be thought of as the sum of two equal fields rotating in opposite directions in the xy-plane. Only the component rotating in the same direction as the precessing nuclear magnetic moments can interact with the nuclei.

When the frequency of the rf field, H_1, becomes equal to the Larmor frequency, ν_0, of the precessing magnetic moments of the nuclei, the rf field, H_1, and the spinning nuclei can exchange energy. If energy is transferred from H_1 to some of the spinning nuclei that are in the lower energy state, these nuclei flip their spin direction so that their magnetic moments precess against the applied external field.

Since there is a linear relationship between resonance frequency and the external applied magnetic field, H_0, NMR spectra are expressed as intensity of absorption versus resonance frequency, H_1, at fixed H_0, or against H_0 at fixed resonance frequency if the rf frequency, H_1, is fixed and the applied magnetic field, H_0, is varied.

For a proton, $\mu = 1.41 \times 10^{-30}$ J G^{-1} or 2.7927 nuclear magnetons. (Although the unit of magnetic field used in this chapter and commonly in NMR discussions is the gauss, G, the official IUPAC nomenclature is the tesla, T, which equals

10,000 G.) From Equation 15.3,

$$\nu_0 = \frac{\Delta E}{h} = \frac{(1.41 \times 10^{-30}\ \text{J G}^{-1})(14{,}092\ \text{G})}{(6.626 \times 10^{-34}\ \text{J sec})(\frac{1}{2})} = 60 \times 10^6\ \text{sec}^{-1}$$

or

$$\nu_0 = \frac{2.7927(5.05 \times 10^{-31}\ \text{J G}^{-1})(14{,}092\ \text{G})}{(6.626 \times 10^{-34}\ \text{J sec})(\frac{1}{2})} = 60 \times 10^6\ \text{sec}^{-1}$$

Thus, in a magnetic field of 14,092 G, the protons precess 60 million times per second, or 60 MHz. Consequently, 60 MHz is the resonance frequency required to flip some of the excess population in the lower energy state to the higher energy state.

Properties of nuclei frequently encountered in NMR spectroscopy are listed in Table 15.1. Because the strength of the absorption signal is roughly proportional to the square of the magnetic field, larger values of field strength lead to a stronger signal. One major area of progress in NMR spectroscopy has been the development of stronger magnetic field strengths through the use of super-conducting magnets.

Relaxation Processes

Upon irradiation of a particular nucleus, the rate of absorption of energy is initially greater than the rate of emission because of the slight excess of nuclei in the lower energy state. However, the absorption signal rapidly attains some finite value. Only if relaxation back to the lower energy state can occur at least as rapidly as absorption does the intensity of nuclear absorption at a given frequency remain constant. Otherwise, in a short time, the rf field equalizes the populations of the energy states, the spin system becomes saturated, and the absorption signal disappears. Since the difference in populations among the various energy levels is small, saturation is easy to bring about experimentally.

Two types of relaxation processes are involved. One, *spin-lattice* (or longitudinal) *relaxation*, is brought about by interaction of the spin with fluctuating magnetic fields produced by random motions of neighboring nuclei. The term spin-lattice relaxation originated from considering an experimental set of nuclei as a spin

TABLE **15.1** MAGNETIC RESONANCE PROPERTIES OF SELECTED NUCLEI

Isotope	**Magnetic moment, μ/μ_N***	**Relative sensitivity at constant H_0†**	***NMR frequency (MHz)*** At 14.09 kG	At 21.14 kG	At 23.49 kG
^{1}H	2.7927	100	60.000	90.000	100.000
^{2}H	0.8574	0.96	9.210	13.815	15.352
^{13}C	0.7024	1.59	15.086	22.629	25.147
^{19}F	2.6288	83.4	56.444	84.666	94.087
^{31}P	1.1317	6.64	24.288	36.432	40.485

* In multiples of the nuclear magneton, $eh/4\pi Mc$.

† Sensitivity relative to the proton, assuming equal numbers of nuclei and the same relaxation time ratio, T_2/T_1.

system embedded in a lattice of other nuclei and electrons. The energy transferred from the nucleus in an upper energy state is given to the lattice as extra translational or rotational energy. Relaxation occurs, in part, from the thermal motions of other nuclei. Their Brownian motion gives rise to magnetic fields that occasionally have a fluctuation whose frequency is equal to the precession frequency of the nucleus to be relaxed. These oscillating components therefore induce transitions and provide a mechanism by which nuclei lose their excess magnetic energy as thermal energy to the lattice. Basically, the relaxation process is first order and decreases exponentially with time. It is defined by the expression

$$(n - n_{eq})_t = (n - n_{eq})_0 e^{-t/T_1} \tag{15.4}$$

where n is the initial excess population in the lower energy level; n_{eq} is its equilibrium value in the presence of the rf field, H_1; and T_1 is a rate constant called the spin-lattice relaxation time. Operationally T_1 represents the time, t, required for the Boltzmann distribution to reach $1/e$ of its initial value in the presence of H_0. In solids and viscous liquids, the relaxation time is on the order of hours, but in typical organic liquids and dilute solutions the time is in the range 0.01–100 sec.

A second time constant, T_2, is assigned to the *spin-spin* (or transverse) *relaxation* process. In this process, nuclei exchange spins with neighboring nuclei by interaction of their magnetic moments. Although no spin energy is lost by this mechanism, the precessing nuclei lose phase coherence—that is, some precess faster and some slower—and the net magnetization in the xy-plane falls toward zero. Recording the dispersion mode signal (see Figure 15.5) as a function of time provides a curve from which T_2 can be calculated. Actually, in mobile liquids and gases where molecular diffusion and rotation are rapid processes, the interactions between adjacent magnetic moments average out and T_2 is not a significant factor in line broadening. In general, $T_2 \leq T_1$ and the linewidth is given approximately as

$$\Delta\nu_{1/2} \approx \frac{1}{T_1} \tag{15.5}$$

where $\Delta\nu_{1/2}$ is the linewidth at half-height in hertz. If T_1 is 1 sec, the linewidth at half-height would be 1 Hz.

Pulsed (Fourier Transform) NMR

The conventional NMR spectrometer scans the spectrum at a slow rate to avoid passing over a spectral line too rapidly, since the lines are usually narrow. The spectrometer spends most of its time recording background; only occasionally does it record the desired information. Efficiency and consequently the sensitivity of such a system are far from optimum. The time required to observe a NMR spectrum by the continuous-wave method is Δ/r (in seconds), where Δ is the range of frequencies (spectral width) that must be scanned to cover the chemical shift range and r is the resolution desired. For ^{13}C at 25 MHz, where Δ is typically about 5 kHz and the linewidths are about 1 Hz, one must scan the 5-kHz region at a rate of 1 Hz sec^{-1}, or slower. This requires a minimum time of 5000 sec (or 83 min).

If a spectrum is thought of as a large number of small increments in frequency, each increment being just large enough to contain a typical spectral line, and if these increments could be examined simultaneously, this would remove the constraint on the scanning rate. If the spectrum contained N increments, where N is just the spectral width (in hertz), the signal-to-noise ratio attainable would be improved by approximately the factor $(N)^{1/2}$, since the signals add linearly whereas the noise adds as the square root of the number of observations. For the proton NMR spectrum, this factor is typically around 30, whereas for ^{13}C, it can be as large as 100. This improvement in the signal-to-noise ratio, due to the fact that all frequencies are simultaneously observed, is known as Felgett's advantage (see the discussion of Fourier transform infrared spectrometers in Chapter 11).

Naturally the instrumentation must be modified appropriately to accomplish the simultaneous excitation of all the spectral lines and to sort the resulting information into the conventional representation of spectral lines. This is accomplished by applying a strong pulse of rf energy (H_1) to the sample for a very short time (1–100 μsec). This pulse is applied by a coil with its axis parallel to the x-axis of the spectrometer. A pulse of short duration contains a wide band of frequencies centered around the nominal frequency. Under the influence of the pulse of rf energy, the magnetic moments of all magnetic nuclei spiral away from the z-axis in the direction of the y-axis. If the pulse is of the proper strength and duration, the magnetic moments are tipped by 90° and come to rest parallel to the y-axis in the xy-plane. The angle through which the magnetic vector is tipped away from the z-axis, α, is given by

$$\alpha = \gamma H_1 \tau \tag{15.6}$$

where γ is the magnetogyric ratio, τ is the time of duration of the pulse, and H_1 is the field strength of the pulse. After the pulse has terminated, the restoring torque of the static field, H_0, causes a precession around the z-axis at the resonant frequencies. The free precession of the magnetic moments of the nuclei under the influence of only the static field induces decaying sinusoidal voltages in a coil of wire surrounding the sample (axis parallel to the x- or y-axis). The voltages decay partly because the nuclear magnetic moment vectors slowly spiral back up to their original positions with an exponential time constant, T_1, the spin-lattice relaxation time, and thus the projection of the vectors in the xy-plane (which generates the signal being measured) diminishes. After several time constants ($3T_1$ to $5T_1$), the nuclei will have regained equilibrium and a second pulse is applied to repeat the process. A second reason for the decay of the free induction signal is magnetic inhomogeneities near the precessing nuclei. Thus, some otherwise identically precessing nuclei begin to precess at slightly different rates and slowly lose phase coherence, thus diminishing the resultant sum of all the precessing vectors. The spin-spin relaxation time, T_2, is an important factor in this case.

The behavior of the magnetic moment vectors during a pulse sequence is shown in Figure 15.3. Figure 15.3a shows the magnetic moment vectors precessing around an external field, H_0, in the z direction. The vectors are randomly distributed around H_0; that is, there is no phase coherence. At the conclusion of a 90° pulse, H_1, applied by means of a coil with the axis along x, all vectors have been aligned parallel to y (Figure 15.3b). After the pulse is terminated, the vectors spiral back toward their original situation (Figure 15.3a) and gradually lose phase coherence. An intermediate stage is shown in Figure 15.3c. The signal observed by a coil with axis along the x direction (this can be the same coil used to generate the pulse) is shown in Figure 15.3d. Each group of equivalent nuclei leads to a signal similar to that shown but with its characteristic Larmor frequency. The signal received by the coil is the sum of all the signals. After Fourier transformation, the frequency domain signal appears as in Figure 15.3e.

The free induction decay (FID) time domain response and the conventional frequency domain representation of the NMR spectrum form a Fourier transform pair (see Chapter 2). The FID signal contains the sine waves of all frequencies corresponding to the Larmor frequencies of all nuclei "tipped" by the impulse. The frequency domain response is calculated from the time domain response, and vice versa. The response of the entire spin system is picked up in the normal manner (see Section 15.3), amplified, and detected in the spectrometer. The free induction decay

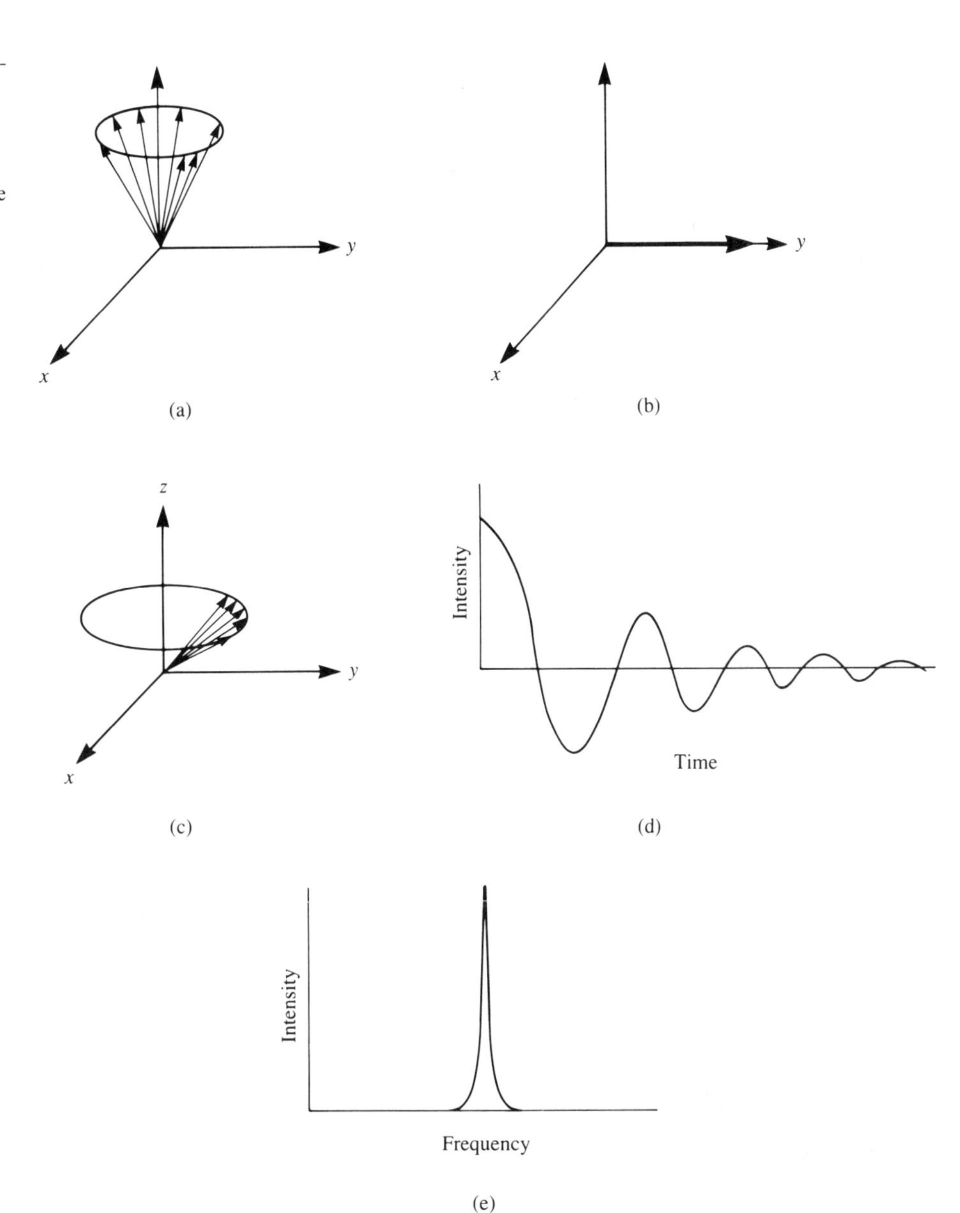

FIGURE **15.3**
Behavior of magnetic moment vectors in response to a 90° pulse. H_0 in $-z$ direction. (See text for explanation.)

signal following each repetitive pulse is digitized by a fast analog-to-digital (ADC) converter, and the successive digitized transient signals are coherently added in the computer until an adequate signal-to-noise ratio is obtained. Using the Cooley-Tukey algorithm, the computer then performs a fast Fourier transformation to the frequency domain and plots a normal spectral presentation of the NMR absorption versus frequency in 10–20 sec.

The sampling time during which data points must be collected to obtain the true NMR spectrum after Fourier transformation depends on the spectral width Δ. The time per data point (or dwell time) must be $(2\Delta)^{-1}$ sec/point. Thus, for a spectral width of 5 kHz, a dwell time of 100 μsec is required. Multiplying the dwell time by the number of data points, N, to be collected during the free induction decay yields the time required for recording the interferogram digitally—namely, $N/2\Delta$. For an 8-K interferogram (8192 points), the time is 0.82 sec. This is also the minimum repetition time between two pulses when several pulse interferograms must be accumulated in order to improve the signal-to-noise ratio. The resolution for this 8-K interferogram (in Hz) is $2\Delta/N$, or 1.23 Hz. Hence, for a full data table, $8192 = (\text{ADC rate}) \times (\text{acquisition time}) = 2\Delta \times (\text{acquisition time})$.

The pulsed Fourier transform technique makes possible the study of less sensitive nuclei, such as ^{13}C, ^{14}N, ^{15}N, ^{17}O, ^{31}P, and even unstable, radioactive nuclei with spins. Due to chemical shielding, each nucleus resonates within a range of Larmor frequencies, depending on the chemical environment. To rotate all nuclear spins within that range by the same angle, the strength of the rf pulse must meet the requirement: $\mu H_1/I \gg 2\pi\Delta$. Furthermore, the pulse width, t_p, must be much shorter than the relaxation times—that is, $t_p \ll T_1$ and T_2—so that relaxation is negligible during the pulse and all of the nuclei start to precess in phase at the termination of the pulse.

Proper selection of pulse sequences under computer control allows measurement of the various T_1 values. In one method, a 90° pulse is used to orient the spins. During a delay time, τ, a portion of these spins relax along the z-axis and some portion of the net nuclear magnetization along the z-axis (M_z) is reestablished. A second pulse flips the remaining vectors out of the xy-plane in a $-z$ direction, whereas those that have reconstituted M_z are flipped back along y where they produce a second signal. The amplitudes of the first, A_0, and the second, A_τ, signals are related to the delay time by

$$\log(A_0 - A_\tau) = \log A_0 - \left(0.434 \frac{\tau}{T_1}\right) \tag{15.7}$$

A plot of $\log(A_0 - A_\tau)$ versus τ enables T_1 to be calculated from the slope. These measurements provide additional information of value to chemists. T_1 depends on the average distance of the nucleus from magnetic neighboring nuclei and on the types of molecular motion that the functional group is undergoing. Internal rotations and segmental motions are detected in this way.

Wide-Line NMR

Wide-line spectra are those in which the observed width of the resonance line is as large as or larger than the major resonance shifts caused by differences in the chemical environment of the observed nucleus. Thus, wide-line NMR supplies information regarding the concentration and physical environment of an observed isotope, but not its chemical environment. It is applicable to solids as well as liquids. Sample sizes range from 0.1 to 50 mL.

An early and continuing application of wide-line NMR is for the quantitative analysis of materials for the particular isotope content from the integrated area under the NMR absorption band. It is a rapid, nondestructive method of analyzing for the proton content of fats and oils and of moisture in many types of materials. The determination of the fluorine content in plastics and chemical compounds is another area of application. Calibration with a standard is required, and accuracy and precision are limited by environmental factors.

The width and shape of the resonance lines are indicative of the physical environment of the isotope. In particular, the width reveals the degree of motional freedom of the isotope in its physical environment, valuable information with respect to high polymer chemistry and solid-state physics. If a magnetic nucleus exists in an isolated magnetic field, its spectrum appears as a single line. However, if two hydrogen nuclei are contiguous to each other in a rigid state, the respective hydrogen nucleus is affected by the local magnetic field caused by the dipole-dipole interaction, in addition to the external field. Stretched fluorocarbons (Teflon), for example, exhibit a doublet spectrum. If the temperature dependence of the linewidth is measured, the transition temperature for conformational change of the polymer can be obtained.

15.2 CONTINUOUS-WAVE NMR SPECTROMETERS

Continuous-wave NMR instrumentation involves six basic units: (1) a magnet to separate the nuclear spin energy states; (2) at least two rf channels, one for field/frequency stabilization and one to furnish rf irradiating energy; a third may be used for each nucleus to be decoupled; (3) a sample probe containing coils for coupling the sample with the rf field(s); (4) a detector to process the NMR signals; (5) a sweep generator for sweeping either the magnetic or rf field through the resonance frequencies of the sample; and (6) a recorder to display the spectrum. These are schematically shown in Figure 15.4. The spectrum is scanned by the field-sweep method or the frequency-sweep method. If the magnetic field, H_0, is held constant, which keeps the nuclear spin energy levels constant, then the rf

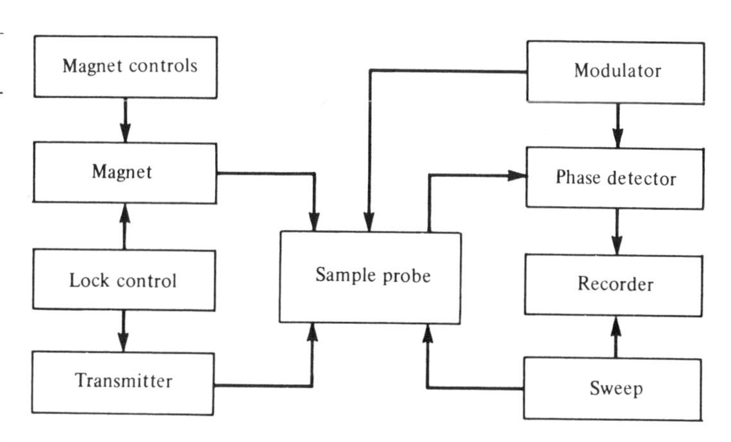

FIGURE 15.4
Block diagram of a high-resolution, continuous wave NMR spectrometer.

signal is swept (varied continuously over a spectral range) to determine the frequencies at which energy is absorbed; this is the *frequency-sweep* method. If the rf signal, H_1, is held constant, then the magnetic field is swept, which varies the energy levels, to determine the magnetic field strengths that produce resonance at the fixed resonance frequency; this is the *field-sweep* method.

The Magnet

The strength of the magnetic field, H_0, determines the Larmor frequency of any nucleus. Because chemical shifts and spectrometer sensitivity are field dependent, it is often desirable to operate at the highest field strength commensurate with homogeneity and stability. The stronger the magnetic field, the better the line separation of chemically shifted nuclei on the frequency scale (see Section 15.4). Since coupling constants remain unaffected by the magnetic field strength, multiplet overlapping decreases with increasing field strength, and homonuclear couplings become small compared with chemical shift differences. The homonuclear multiplets approach first-order systems assignable by following the multiplet rule. Moreover, the relative population of the lower-energy spin level increases with increasing field, leading to a corresponding increase in the sensitivity of the NMR experiment.

For high-resolution work the magnetic field over the entire sample volume must be maintained uniform in space and time. Effective homogeneity of the field is promoted by (1) the use of large pole pieces composed of a very homogeneous alloy, (2) the polishing of pole faces to optical tolerances, and (3) the use of a narrow pole gap—that is, a smaller cross section and consequently a compromise with decreased sensitivity. Shim coils are also used to iron out field inhomogeneities. These shim coils are manufactured of special material without soldering and encased in epoxy resin. They are mounted permanently in the pole cap covers of the magnet. Since inhomogeneities of a magnetic field take the form of gradients, corrections are made by creating small corrective fields that oppose these gradients. Corrections are available along the x-, y-, and z-axes, in the xy- and yz-planes, and for curvature in the z direction.

Permanent magnets are simple and inexpensive to operate but require extensive shielding and must be thermostatted to ± 0.001 °C. Commercial units that use electromagnets or permanent magnets operate at 14.09, 21.14, or 23.49 kG. An electromagnet requires elaborate power supplies and cooling systems, but these disadvantages are offset by the opportunity to use different field strengths to disentangle chemical shifts from multiplet structures and to study different nuclei.

Cryogenic superconducting solenoids produce homogeneous fields at 51.7 and 70.5 kG and, as a result, make possible high-resolution experiments with 220- and 300-MHz rf fields, respectively. Some instruments are now available with 93.9- and 117.4-kG fields and operate at 400 MHz and 500 MHz frequencies for proton resonances. In the cryogenic solenoids many turns of copper-clad, niobium-tantalum superconducting wire are immersed in a Dewar flask that holds liquid helium. The Dewar is surrounded by another that holds liquid nitrogen. Aided by shim coils positioned in the probe, the 220-MHz spectrometer achieves a resolution of 1.1 Hz (full linewidth at half-maximum amplitude) and a 55:1 signal-to-noise ratio for the ethylbenzene quartet in a 1% solution. The 300-MHz instrument achieves a resolution of 1.5 Hz and a 65:1 ratio under the same conditions.

The Probe Unit

The probe unit is the sensing element of the spectrometer system. It is inserted between the pole faces of the magnet in the *xy*-plane of the magnet-air gap by an adjustable probe holder. The probe unit houses the sample, the rf transmitter(s), output attenuator, receiver, and phase-sensitive detector. The sample is contained in a cylindrical, thin-walled, precision-bore glass tube that has an outer diameter of 5 mm. To average small magnetic field inhomogeneities in the *xz*-plane, an air-bearing turbine rotates the sample tube at a rate of 20–40 or so revolutions per sec. This spinning produces sidebands in the spectrum because the NMR peaks are modulated at the spinning frequency. Since most NMR samples, except for ^{13}C solids and for the wide-line technique, are liquid solutions, the sample tube is filled until the length/diameter ratio is about five, which approximates that of an infinite cylinder. For minute samples, a tube with a capillary bore that widens to a spherical cavity at the position of the rf coil is used. If the ^{1}H spectrum is being studied, an amount between a few micrograms and a few milligrams of sample is dissolved in a solvent such as CCl_4 or CS_2 or a solvent that has had all the protons replaced by deuterium atoms. Chloroform-*d*, acetone-d_6, and benzene-d_6 are commonly used. The deuterium serves two purposes. It replaces hydrogen nuclei that would otherwise generate a background solvent signal and would overwhelm the signals from the sample. In addition, the magnetic resonance response of the deuterium nuclei is used to lock the ratio of the magnetic field and frequency of the instrument over long periods of time. If the ^{13}C spectrum is desired, an amount of sample between a few milligrams and a few hundred milligrams is dissolved in one of the same deuterated solvents used for proton NMR. The deuterium in this case serves only the purpose of locking the spectrometer.

Two probe designs are used. A *single-coil* probe has one coil that not only supplies the rf radiation to the sample but also serves as a part of the detector circuit for the NMR absorption signal. To detect the resonance absorption and to separate the NMR signal from the imposed rf field, a rf bridge is used. The exciting signal is balanced against an equal-amplitude reference, with the modulation appearing as bridge unbalance and extractable in that form. Alternatively, the signal can be amplified and then subjected to diode detection to extract the resonance spectrum. *Crossed-coil* (nuclear induction) probes have two coils, one for irradiating the sample and a second coil mounted orthogonally for signal detection. The irradiating coil is split into halves with the sample inserted between. This coil is oriented with its axis perpendicular to the magnetic field (that is, along the *x*-axis). The detector coil is wound around the sample tube with its axis (the *y*-axis) perpendicular to both the field H_0 (*z*-axis) and the rf field (H_1) axis. Since magnetic resonance produces a net magnetization in the *xy*-plane, a current is generated at resonance in the receiver coils from an indirect coupling between the rf field and receiver coils, with the coupling produced by the sample itself. This design permits selective pickup of the resonance signal while virtually excluding the applied rf field.

Instrument Stabilization

The NMR spectrum is recorded directly on precalibrated chart paper. This demands that the ratio of rf frequency to field strength be very stable. To obtain an NMR peak with a linewidth of 0.1 Hz at 60 MHz requires an overall stability of

$0.1/(60 \times 10^6)$, or about two parts in 10^9. There is no problem with the rf frequency, but the magnet stability is only about 10^{-7}/hr. Independent stabilization of the two units is difficult. The field-frequency (H/ν) relation of Equation 15.3 is more reliably maintained by means of servo loops that lock them together. Since an NMR signal is an ac signal with two components 90° out of phase, the detectable signals are an absorption component and a dispersion component, as shown in Figure 15.5. By using a phase-sensitive detector that is finely tuned to sense only one component, either the absorption or the dispersion mode can be observed. Since NMR signals are usually observed in the absorption mode, the dispersion mode is available for field-frequency control and for measuring T_1, the spin-lattice relaxation time constant. The field-frequency control loop is produced in two ways. An external lock uses a signal from a separate, adjacent sample, whereas the internal lock uses the sample under analysis for the locking signal.

In the external lock system, an external reference nucleus is continuously irradiated at its resonance frequency and the resultant NMR dispersion signal is continuously monitored while the spectrum is being swept. If the frequency of irradiation exactly equals the frequency at the center of the dispersion signal, the NMR error signal is zero. If the magnetic field changes, the resonance condition is no longer fulfilled and an output signal is produced. The output error signal is amplified and fed back into a coil that decreases or augments the applied H_0.

In an internal lock system, a suitable reference such as tetramethylsilane (TMS) is added to the sample so that the analytical sample nuclei and reference nuclei experience exactly the same field. An automatic shim control continuously compensates for the several magnetic field gradients.

Since all variations in the magnetic field and rf frequency cannot be eliminated entirely and since it is the ratio of H_0 and ν that is important, all measurements of H_1 are made with respect to the internal standard, TMS, which gives a single intense peak at a higher field strength than most other protons. Resonances are normally stated in parts per million (δ) (see Equation 15.9) from the TMS peak and most occur at a lower H_0—that is, "downfield."

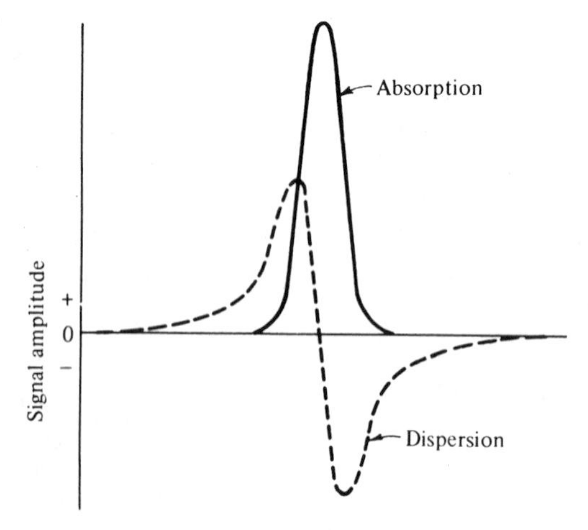

FIGURE **15.5**
Line shapes of the two observable NMR signals.

Sweeping Modes

To flip the rotating nuclear axes with respect to the magnetic field in the field-sweep method, a linearly polarized rf field is imposed at right angles (x-axis) to the magnetic field (z-axis). As pointed out, a linearly polarized oscillating magnetic field along the x-axis is equivalent to oppositely rotating circular magnetic fields in the xy-plane. Auxiliary coils, wound around the pole pieces, allow a sweep of the applied magnetic field to be made.

In the frequency-sweep mode, the magnetic field and the locking frequency are held constant while the portion of the spectrum of interest is scanned by sweeping the observing rf. It is then possible to introduce a second rf for the purposes of spin decoupling or spin tickling experiments (see Section 15.5). This second rf field is held constant at a fixed separation from the reference line in the spectrum, and is thus held at exact resonance for a chosen signal while the rest of the spectrum is examined by the rf field. The resultant decoupled spectrum is then observed by sweeping ν through any desired part of the spectrum.

In the field-sweep method, the frequency of the irradiating field, H_1, is kept constant while the magnetic field, H_0, is varied. If the instrument uses a permanent magnet, H_0 is varied by means of coils that carry a carefully regulated but variable dc current wound around the pole pieces of the magnet. For decoupling experiments, the second rf frequency, H_2, is varied as H_0 is varied so as to keep a constant difference between the two fields, thus assuring that the nucleus being decoupled is always irradiated at its resonance frequency.

The sweep range for protons is approximately from 2000 Hz downfield to 500 Hz upfield; for other nuclei, such as ^{13}C and ^{19}F, the sweep range must be increased to more than 10 kHz. At the usual sweep rate of about 1 Hz/sec, a ringing envelope-like pattern appears over the trailing edge of an absorption band. This arises from rapid sweeping through the resonance condition. The frequency of the rotating component of magnetization varies with the changing sweep field and, as a result, the induced signals are alternately in and out of phase with the applied rf field. The observation of ringing is a good indication of a homogeneous field.

Minimal-Type NMR Spectrometers

Among the families of continuous-wave NMR spectrometers, the minimal type has stressed reliability, ease of operation, and a cost/performance trade-off. This basic instrument often utilizes a permanent magnet of 14, 21, or 23 kG field strength and rf fields of 60, 90, or 100 MHz, respectively. Emphasis is on the measurement of proton NMR spectra. This family might be designated as a high-resolution, minimal-type NMR spectrometer lacking most of the auxiliary accessories found on the more sophisticated instruments.

A schematic diagram of a typical instrument is shown in Figure 15.6. The 60-MHz frequency for protons is synthesized by taking the fourth harmonic of a 15-MHz crystal oscillator. This signal is modulated by a low-frequency oscillator to create an NMR signal that can be easily amplified. The field sweep is generated by passing a variable but carefully regulated dc current through coils wound around the magnet pole pieces. The signal is amplified and presented on the y-axis of the recorder while the sweep is coupled to the x-axis. A separate lock/decoupler unit is available so that decoupling experiments can be performed or the frequency

FIGURE 15.6 Schematic circuit diagram of Varian EM-360 NMR spectrometer. (Courtesy of Varian Associates.)

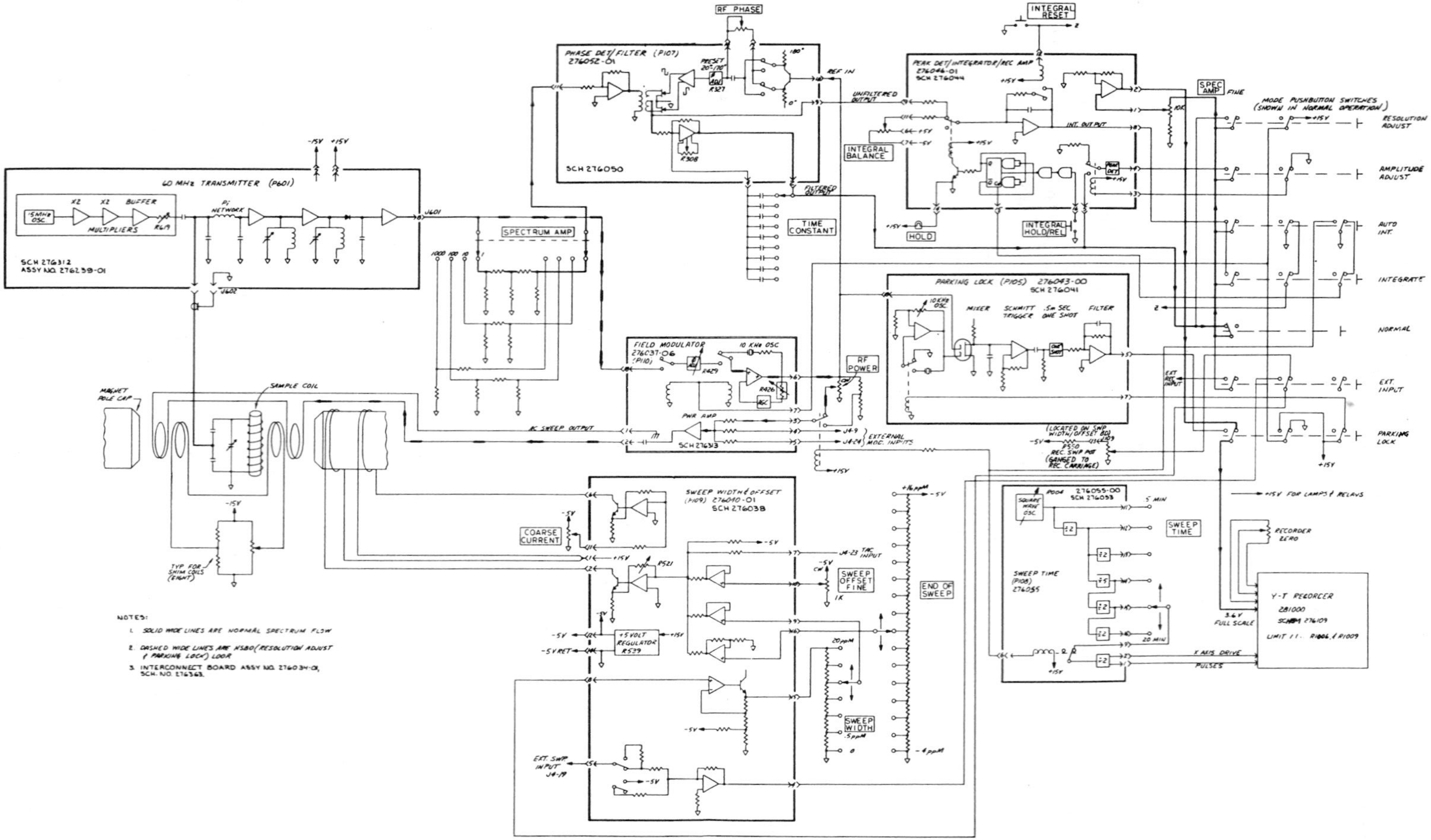

NOTES:
1. SOLID WIDE LINES ARE NORMAL SPECTRUM FLOW
2. DASHED WIDE LINES ARE K580 (RESOLUTION ADJUST & PARKING LOCK) LOOP
3. INTERCONNECT BOARD ASSY NO. 276034-01, SCH. NO. 276363.

of a reference signal can be used to stabilize (lock) the instrument over long periods of time.

Multipurpose NMR Spectrometers

The second family of NMR spectrometers is more diverse. These instruments are designed primarily for research and emphasis is on high performance and versatility, with cost being a secondary consideration. The high precision comes through the use of homonuclear and heteronuclear lock systems and frequency synthesizers. They are also characterized by high intrinsic sensitivity and the ability to study a variety of nuclei.

The strength of the magnetic field is quite important, since sensitivity, resolution, and the separation of chemically shifted peaks increase as the field strength increases. In addition, the complexities of spin-spin couplings are reduced at higher fields. These instruments use cryogenic solenoids and rf fields of 220, 300, or even 500 MHz.

Wide-Line NMR Spectrometers

The wide-line NMR spectrometer uses a frequency synthesizer to generate the rf field and a permanent magnet or a compact, light-weight electromagnet. Slowly varying scan voltages are directly injected in the regulator for the magnet power supply for the electromagnet. These signals cause a corresponding change in magnet energizing current, thereby creating a scan of the magnetic field. Sample probe temperatures may be varied over the range -170 to 200 °C. Sample tubes are 15 or 18 mm in outer diameter. The standard magnetic field is 9.4 kG for protons and 10.0 kG for ^{19}F; the rf field is 40 MHz.

Instruments are also available in which the rf applied field is continuously adjustable over a basic frequency range of 300 Hz to 31 MHz, usually in steps of 10 Hz. The continuously adjustable H_1 level allows a quantitative determination of T_1 relaxation values.

For signal detection a sweep unit generates sinusoidal audio-modulation voltages that have selectable frequencies of 20, 40, 80, 200, and 400 Hz. The output is amplified for simultaneous application to the probe modulation coils and to the x-axis of an oscilloscope. The sweep unit also provides a reference voltage with the same frequency as the selected modulation frequency. This is applied to a phase-sensitive detector to guarantee clear differentiation between absorption and dispersion modes. The detected signal contains a superimposed audio frequency. Following either selective or broad-band amplification, the signal can be recorded in the first derivative mode or, according to the sideband procedures, in the undifferentiated form. An integrating circuit measures the area under the absorption band.

15.3 PULSED FOURIER TRANSFORM NMR SPECTROMETER

An NMR spectrometer capable of pulsed Fourier transform measurements is a combination of a continuous-wave circuit, as found in conventional NMR spectrometers, a computer-controllable pulse generator, and a digital computer. A simplified block diagram of a pulsed Fourier transform unit is shown in Figure 15.7. The computer-controllable pulse programmer generates dc pulses. For ^{13}C NMR resonance, the output of a 25-MHz rf oscillator in the Fourier transform unit is fed to a rf gate and sent to the power amplifier units only when the gate is open. The digital pulse programmer controls gate on/off (that is, the rf input signal of the H_1 transmitter) and determines the pulse width, interval, and repetition rate. The timing of each pulse train is controlled by the clock pulse from the built-in, highly stable, crystal-controlled oscillator to assure high resettability. The widths of the resulting rf pulses are adjustable in 1-μsec steps for 90°, 180°, or other tip angles. In addition, various pulse sequences can be programmed. After the rf pulse is amplified, the intense rf pulse excites the sample to be investigated. The free induction decay signal is then filtered, amplified, detected by the phase-sensitive detector, and digitized. This signal is accumulated while all protons are simultaneously decoupled. For ^{13}C experiments, the ^2H signal of the deuterated solvent is used for the internal lock of the magnetic field. The lock signal is observed on the oscilloscope to adjust the magnetic homogeneity during signal accumulation. The oscilloscope is also used to monitor the free induction decay signal and for the quick display of accumulated signals and the Fourier transformed spectra to determine whether the signal-to-noise ratio is satisfactory.

The remainder of the operation is under computer control. The operator sets the values of the spectral width, acquisition time, number of transients, and pulse width. For the first look at a sample, these and other parameters are optimized. For repetitive or routine analyses, however, optimum parameters can be stored on a disk and the experiment set up simply by loading the parameter set into the computer. Acquisition of data is automatic and terminates when the number of transients requested is reached. A printout of the parameter set and the number of completed transients can be obtained for a permanent record. The fact that data are stored in digital form in the computer makes possible the numerical integration of the areas under each of the lines. This results in particularly accurate, drift-free integral values, and greatly decreases the requirements of operator skill. Spectra and integrals are plotted simply by entering the commands.

FIGURE **15.7** Block diagram of pulsed Fourier transform NMR spectrometer.

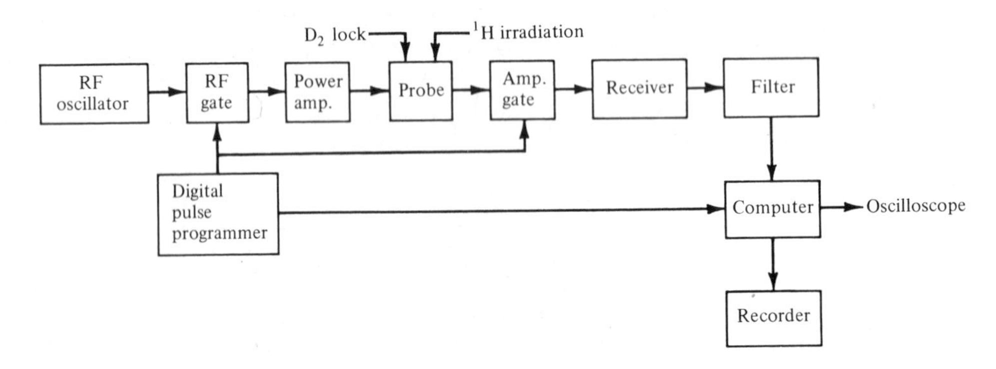

Magic Angle Spinning and Cross Polarization

No longer are chemists restricted to examining NMR spectra from liquid solutions. For nuclei, such as ^{13}C, line broadening for a polycrystalline material arises from the fact that the molecules are oriented in all possible directions. The shielding any particular nucleus receives from its electronic environment is thus a function of the orientation of the molecules that contain it. This type of line broadening due to chemical shift anisotropy can largely be removed by rotating the sample very rapidly (faster than 2 kHz) about an axis oriented at an angle of approximately 54.7° (the magic angle) with respect to the external magnetic field. The averaging that occurs is similar to tumbling in liquids. The quality of the NMR spectrum can approach that of the liquid-state spectrum. The angle is a property of the local fields that electrons exert on nuclei and of tensors that describe such behavior.

The accessory to accomplish NMR experiments with crystalline materials on 220-MHz superconducting NMR spectrometers includes a probe and a high-power, 100-W, 220-MHz amplifier capable of also performing dipole decoupling and cross polarization. Solids are machined into rotors or packed as powders into hollow rotors. Samples are introduced into the probe by dropping the rotor into a tube at the top of the magnet. The rotor automatically positions itself at the magic angle.

Because ^{13}C has a low natural abundance, NMR measurements on this nucleus generally require signal averaging. Cross-polarization techniques overcome the problems of long T_1 times and consequent limited sensitivity. The technique relies on the presence of a system of abundant nuclear spins (^1H) in order to observe the NMR signal from the dilute nuclear spin (^{13}C). The procedure consists of four basic timed sequences of rf pulses. The four-part procedure is (1) polarizing the ^1H spin system by applying a 90° rf pulse at the ^1H resonance frequency, (2) spin locking in the rotating frame by applying a 90° phase shift to the foregoing field, (3) establishing $^{13}C{-}^1H$ contact by applying a rf field at the ^{13}C resonance frequency, and (4) observing the ^{13}C free induction decay while the ^1H field is maintained for decoupling. The entire sequence is repeated many times until a suitable signal-to-noise ratio for ^{13}C is achieved. Details are given by Miknis and colleagues.[1]

15.4 SPECTRA AND MOLECULAR STRUCTURE

For most purposes, high-resolution NMR spectra are described in terms of *chemical shifts* and *coupling constants*. Two other parameters sometimes involved are the *spin-lattice* (T_1) and the *spin-spin* (T_2) *relaxation times* of the nuclei. Internal rotation, chemical exchange, and other rate processes affect relaxation times to produce pronounced temperature-dependent effects on the spectra. In solids, direct magnetic dipole-dipole interactions dominate. The nuclei are exposed to slightly different magnetic environments due to the different, but fixed, orientations of the molecules to the external and internal magnetic fields. Consequently, resonances occur over a range of frequencies and the lines are broad. In liquids and gases, the direct dipole-dipole interactions are averaged to zero by rapid intra- and intermolecular motions and narrow-line NMR spectra are observed.

Chemical Shifts

An important feature of high-resolution NMR spectra is *chemical shift*. In different chemical environments the same type of nucleus is shielded slightly from the applied field in a manner that depends on the distribution of the surrounding electrons. For a fixed external field, H_0, different screening factors cause slightly different resonant frequencies.

The magnitude of the effective field felt by each group of nuclei is expressed as follows:

$$H_{\text{eff}} = H_0(1 - \sigma) \tag{15.8}$$

where σ is a nondimensional shielding constant and may be either a positive or negative number. Thus, the protons at various sites in a molecule are spread out into a spectrum according to the values of their shielding parameters. A field or frequency sweep brings protons at each particular site into resonance one after another. The more the field induced by the circulating electrons shields the nucleus and opposes the applied field, the higher must be the applied field to achieve resonance if the field is varied, or the lower must be the resonance frequency if the frequency is varied. The specific locations of the shifted resonance frequencies are used to characterize the neighbors of a given nucleus. The value of the shielding constant depends on several factors, among which are the hybridization and electronegativity of the groups attached to the atom that contains the nucleus being studied. Shielding effects seldom extend beyond one bond length except with very strong electronegative groups.

Because NMR spectrometers with different field strengths are in use, it is desirable to express the position of resonance in field-independent units and with respect to the resonance of a reference compound. For proton spectra in non-aqueous media, the reference material is tetramethylsilane, $(CH_3)_4Si$, abbreviated TMS, whose position is assigned as exactly 0.0 on the δ scale. TMS contains 12 protons, but these are all chemically equivalent and therefore give rise to a single sharp signal at an H_0 (applied field) higher than most other proton resonances. The magnitude of the chemical shift is expressed in parts per million:

$$\delta = \frac{H_r - H_s}{H_r} \times 10^6 = \frac{\nu_s - \nu_r}{\nu_r} \times 10^6 \tag{15.9}$$

where H_r and H_s are the positions of the absorption lines for the reference and sample, respectively, expressed in magnetic units (gauss), and ν_r and ν_s are the corresponding frequencies expressed in hertz. A positive δ value represents a greater degree of shielding in the sample than in the reference.

Recommended reference materials for other nuclei include CS_2 or TMS for ^{13}C, trichlorofluoromethane (CCl_3F) for ^{19}F, NH_3 liquid for ^{15}N (or ^{14}N), TMS for ^{29}Si, and 85% phosphoric acid for ^{31}P. The numbers on the dimensionless (shift) scale downfield from the reference are designated positive.

Proton resonances from C—H bonds are located in the range $\delta = 0.9$–1.5 when only aliphatic groups are substituents. Protons in CH_3 groups usually appear at $\delta = 0.9$–1.0 when the bonds on the adjacent carbon atom are to H, CH, or CH_2 groups; CH_2 and CH protons are slightly farther downfield in that order. An adjacent, unsaturated bond shifts the resonance position of CH_3 to $\delta = 1.6$–2.7.

An adjacent oxygen atom markedly shifts proton signals downfield to $\delta = 3.2-3.4$ for aliphatic entities and to $\delta = 3.6-3.9$ for aryl—O—CH situations.

Many common groups produce special shielding effects because they allow the circulation of electrons in only certain preferred directions within the molecule. Figure 15.8 shows shielding (+) and deshielding (−) zones in the neighborhood of triple, double, and single bonds to carbon. In C=C and C=O double bonds, the deshielding zone extends along the bond direction; even C—C bonds show some deshielding in this direction. This anisotropy of the magnetic susceptibility of chemical bonds means that the shielding or deshielding of a neighboring proton in the molecule is dependent on its distance from the bond and its orientation with respect to that bond. Aromatic rings exhibit a strong anisotropic effect. When such compounds are placed in a magnetic field, the six π-electrons circulate in two parallel doughnut-shaped orbits on each side of the ring. The resulting magnetic field opposes H_0 in a cone-shaped zone of excess shielding that extends along the hexad axis, but it reinforces H_0 in a zone of deshielding that extends from the edge of the ring. In aromatic compounds the deshielding zone is more commonly occupied; thus, protons on aromatic rings appear at much lower fields ($\delta = 6-7$) than olefinic protons ($\delta = 5-6$). In acetylenes, the electron current circulates in such a way that the shielding zone extends along the bond direction and acetylenic protons appear at high fields ($\delta = 1.6-3.0$). Table 15.2 gives some proton chemical shifts; Table 15.3 gives selected ^{13}C chemical shifts. Other more complete tables for proton, ^{13}C, ^{15}N (or ^{14}N), ^{19}F, ^{29}Si, and ^{31}P chemical shifts and coupling constants can be found in handbooks edited by Dean.[2,3]

Processes that result in chemical exchange or conformational change and that are complete in 1–0.001 sec may give rise to spectra that are time averages in comparison with those expected in terms of instantaneous molecular conformations. If the exchange rate is high in comparison with the frequency of the chemical shifts and spin-spin couplings (see below), the local fields seen from the

FIGURE **15.8**

Shielding (+) and deshielding (−) zones in the neighborhood of triple, double, and single bonds to carbon and aromatic rings.

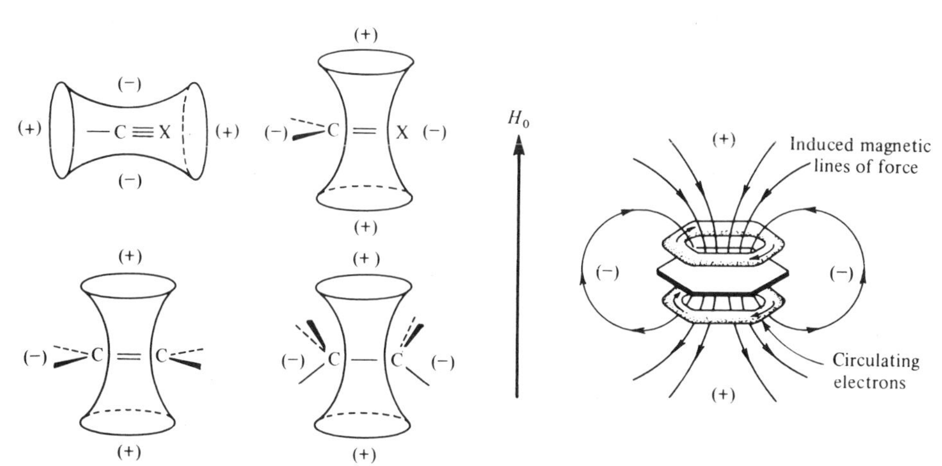

TABLE **15.2** PROTON CHEMICAL SHIFTS

Substituent group	Methyl protons (δ)	Methylene protons (δ)	Methine proton (δ)
HC—C—CH_2	0.95	1.20	1.55
HC—C—NR_2	1.05	1.45	1.70
HC—C—C=C	1.00	1.35	1.70
HC—C—C=O	1.05	1.55	1.95
HC—C—NRAr	1.10	1.50	1.80
HC—C—NH(C=O)R	1.10	1.50	1.90
HC—C—(C=O)NR_2	1.10	1.50	1.80
HC—C—(C=O)Ar	1.15	1.55	1.90
HC—C—(C=O)OR	1.15	1.70	1.90
HC—C—Ar	1.15	1.55	1.80
HC—C—OH (and OR)	1.20	1.50	1.75
HC—C—C≡CR	1.20	1.50	1.80
HC—C—C≡N	1.25	1.65	2.00
HC—C—SR	1.25	1.60	1.90
HC—C—OAr	1.30	1.55	2.00
HC—C—O(C=O)R	1.30	1.60	1.80
HC—C—SH	1.30	1.60	1.65
HC—C—(S=O)R and —SO_2R	1.35	1.70	
HC—C—NR_3^+	1.40	1.75	2.05
HC—C—O(C=O)CF_3	1.40	1.65	
HC—C—Cl	1.55	1.80	1.95
HC—C—O(C=O)Ar	1.65	1.75	1.85
HC—C—Br	1.80	1.85	1.90
HC—CH_2	0.90	1.30	1.50
HC—C=C	1.60	2.05	
HC—C≡C	1.70	2.20	2.80
HC—(C=O)OR (and NR_3)	2.00	2.25	2.50
HC—SR	2.05	2.55	3.00
HC—O—O	2.10	2.30	2.55
HC—(C=O)R	2.10	2.35	2.65
HC—C≡N	2.15	2.45	2.90
HC—CHO	2.20	2.40	
HC—Ar (and NR_2)	2.25	2.45	2.85
HC—SSR	2.35	2.70	
HC—(C=O)Ar	2.40	2.70	3.40
HC—SAr	2.40		
HC—NRAr	2.60	3.10	3.60
HC—SO_2R and —(SO)R	2.60	3.05	
HC—Br	2.70	3.40	4.10
HC—NR_3^+	2.95	3.10	3.60
HC—NH(C=O)R	2.95	3.35	3.85

TABLE 15.2 (continued)

Substituent group	Methyl protons (δ)	Methylene protons (δ)	Methine proton (δ)
HC—Cl	3.05	3.45	4.05
HC—OH and —OR	3.20	3.40	3.60
HC—NH_2	3.50	3.75	4.05
HC—O(C=O)R	3.65	4.10	4.95
HC—OAr	3.80	4.00	4.60
HC—O(C=O)Ar	3.80	4.20	5.05
HC—F	4.25	4.50	4.80
HC—NO_2	4.30	4.35	4.60
Cyclopropane		0.20	0.40
Cyclobutane		2.45	
Cyclopentane		1.65	
Cyclohexane		1.50	1.80
Cycloheptane		1.25	

Substituent group	Proton shift (δ)	Substituent group	Proton shift (δ)
HC≡CH	2.35	HO—C=O	10–12
HC≡CAr	2.90	HO—SO_2	11–12
HC≡C—C=C	2.75	HO—Ar	4.5–6.5
HAr	7.20	HO—R	0.5–4.5
HCO—O	8.1	HS—Ar	2.8–3.6
HCO—R	9.4–10.0	HS—R	1–2
HCO—Ar	9.7–10.5	HN—Ar	3–6
HO—N=C(oxime)	9–12	HN—R	0.5–5

SOURCE: J. A. Dean, ed., *Lange's Handbook of Chemistry,* 13th ed., McGraw-Hill, New York, 1985, in which additional proton chemical shifts may be found.
NOTE: Values are given on the officially approved δ scale. R = alkyl group; Ar = aryl group.

nucleus of the exchanging atom are averaged out to result in a single line, somewhat broader than normal.

Spin-Spin Coupling

Nuclei can interact with each other to cause mutual splitting of the otherwise sharp resonance lines into multiplets, called *spin-spin coupling*, sometimes also called *J coupling*. These multiplets arise because magnetic moments of nuclei interact with each other through the strongly magnetic electrons in the intervening bonds. The strength of the coupling, denoted by J, is given by the spacing of the multiplets and is expressed in hertz. Proton-proton couplings in aliphatic organic compounds (Table 15.4) are usually transmitted through only two or three bonds, although weak couplings are often transmitted further. In certain rigid structures with favorable geometry, coupling through four bonds may be reasonably large. When unsaturated systems occur between the protons, long-range coupling is enhanced.

TABLE 15.3

^{13}C CHEMICAL SHIFTS

Substituent group	Primary carbon	Secondary carbon	Tertiary carbon	Quaternary carbon
Alkanes				
C—C	−20 to 30	25 to 45	30 to 60	35 to 70
C—O	40 to 60	40 to 70	60 to 75	70 to 85
C—N	20 to 45	40 to 60	50 to 70	65 to 75
C—S	10 to 30	25 to 45	40 to 55	55 to 70
C—Halide	−37 to 35	−10 to 45	30 to 65	35 to 75
	(I) (Cl)	(I) (Cl)	(I) (Cl)	(I) (Cl)

Substituent group	δ	Substituent group	δ
Alkynes	70 to 100	Isocyanides	130 to 150
Alkenes	110 to 150	Carbonates	150 to 160
Aromatics	110 to 135	Oximes	155 to 165
C-substituted	125 to 145	Ureas	150 to 170
Heteroaromatics	115 to 140	Thioureas	165 to 185
C-α	135 to 155	Esters, anhydrides	150 to 175
Cyanates	105 to 120	Amides	160 to 180
Isocyanates	115 to 135	Acids, acyl chlorides	160 to 185
Thiocyanates	110 to 120	Aldehydes	175 to 205
Isothiocyanates	120 to 140	Ketones	175 to 225
Cyanides	110 to 130		

NOTE: Values are given on the δ scale, relative to TMS.

TABLE 15.4

PROTON SPIN COUPLING CONSTANTS

Structure	*J* (Hz)	Structure	*J* (Hz)
>C(H)(H)	12–15	CH_2—C≡C—CH<	0–3
>CH—CH< (free rotation)	6–8	>CH—C≡CH	0–3
>CH—OH (no exchange) (—NH)	5	H—C=C—H (ring) (3-member)	0–2
>CH—C(H)=O	1–3	(4-member)	2–4
H_t, H_g C=C H_c, H (*gem*)	0–3	(5-member)	5–7
(*cis*)	6–14	(6-member)	6–9
(*trans*)	11–18	(7-member)	10–13
		furan (2–3)	1.8
		(3–4)	3.5
		(2–4)	0–1
		(2–5)	1–2

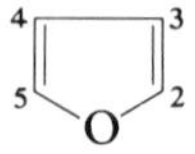

TABLE **15.4** (continued)

Structure		J (Hz)
H_c, H_t, H_g on C=C bearing >CH	(*cis*)	0.5–3
	(*trans*)	0.5–3
	(*gem*)	4–10
>C=CH—CH=C<		10–13
=CH—C=O (aldehydic H)		6
Cyclohexane (H_a, H_e)	(*a-a*)	8–10
	(*a-e*)	2–3
	(*e-e*)	2–3
Cyclopentane	(*cis*)	4–6
	(*trans*)	4–6
Cyclobutane	(*cis*)	8
	(*trans*)	8
Cyclopropane	(*cis*)	9–11
	(*trans*)	6–8
	(*hetero*)	4–6
Benzene (H, H)	(*o*)	6–10
	(*m*)	1–3
	(*p*)	0–1
Pyridine	(2–3)	5–6
	(3–4)	7–9
	(2–4)	1–2
	(3–5)	1–2
	(2–5)	0–1
	(2–6)	0–1
Pyrrole (N—H)	(1–2)	2–3
	(1–3)	2–3
	(2–3)	2–3
	(3–4)	3–4
	(2–4)	1–2

Structure		J (Hz)
Thiophene	(2–3)	5–6
	(3–4)	3.5–5.0
	(2–4)	1.5
	(2–5)	3.4
Benzene (F, H)	(*o*)	6–10
	(*m*)	5–6
	(*p*)	0–2
Benzene (CH_3, F)	(*o*)	2.5
	(*m*)	1.5
	(*p*)	0
	(2–5)	1–3
>C(H)F		45–52
>CH—CF<	(*gauche*)	0–12
	(*trans*)	10–45
H_t, H_g, H_c on C=C bearing F	(*gem*)	72–90
	(*cis*)	1–8
	(*trans*)	12–40
F—C=C—CH_3		2–4
H—C=C—CF<		0–6

In allylic systems, four-bond couplings reach a maximum of about 3 Hz when the angle between the plane that contains the olefinic protons and the C—H bond of the allylic carbon atom is about 90°. In H—C—C=C—C—H systems, five-bond couplings of about 3 Hz are observed. In acetylenes, allenes, and cumulenes, observable couplings are transmitted over many bonds, up to nine in polyacetylenes. In aromatic rings, couplings of protons in *ortho* positions (through three bonds) are 7–9 Hz, *meta* (four bonds) 2–3 Hz, and *para* (five bonds) 0.5–1.0 Hz.

Couplings depend also on geometry. The dihedral angle between planes determines the coupling of protons on adjacent carbon atoms (Figure 15.9). Adjacent axial-axial protons, displaying a dihedral angle of 180°, are strongly coupled, whereas axial-equatorial and equatorial-equatorial protons are coupled only moderately. *Trans* and *cis* protons on olefinic double bonds show $J_{trans}/J_{cis} \cong 2$, which is useful in assigning structures of geometrical isomers.

The number of lines in a multiplet is given by $2nI + 1$, where n is the number of nuclei producing the splitting. For protons, this becomes $n + 1$ lines. The relative intensity of each of the multiplets, integrated over the whole multiplet, is proportional to the number of nuclei in the group. Intensities of the peaks within a multiplet are given by simple statistical considerations and are proportional, therefore, to the coefficients of the binomial expansion. Thus, one neighboring proton splits the observed resonance into a doublet (1:1), two produce a triplet (1:2:1), three a quartet (1:3:3:1), four a quintet (1:4:6:4:1), and so on.

The magnitude of J is independent of the field strength, unlike the chemical shift. Thus, as H_0 increases, the multiplets move farther apart but the spacing of the peaks within each multiplet remains the same. The rato J/δ, where δ is the chemical shift difference between the two coupled nuclei, is the critical parameter that determines the appearance of the spectrum. When J/δ is 0.05 or less, the spectrum consists of well-separated multiplets with the theoretical intensities of the peaks, and the spectrum is said to be first-order. When $0.05 < J/\delta < 0.15$, the spectrum may appear first-order but peak intensities are no longer theoretical. When $J/\delta > 0.15$, the spectrum begins to deviate noticeably from the simple first-order appearance. New peaks may appear and intensities are no longer binomial because some spin states that were degenerate when J/δ was small split because the magnetic field mixes states. Ultimately, when the chemical shift difference, δ, vanishes, the multiplet will collapse to a singlet.

A strongly coupled system of three or more spins is difficult to unravel by inspection alone, although certain patterns become recognizable with experience. Use of 220-, 300-, or even 500-MHz NMR spectrometers is valuable. A higher field strength improves the ratio of chemical shift to coupling constant by causing the chemical shift to increase and spreading the NMR spectrum over a wider range (Figure 15.10). It reduces complicating second-order spectrum effects and produces a sensitivity gain in addition.

Other nuclei with spins of $\frac{1}{2}$ interact with protons (and each other) and cause

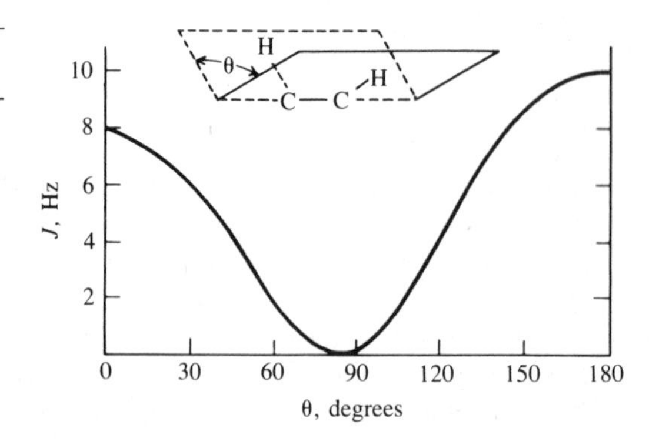

FIGURE **15.9**
Dependence of the coupling constant J on the dihedral angle θ in the saturated system H—C—C—H. [By permission from M. Karplus and D. H. Anderson, *J. Chem. Phys.*, **30,** 6 (1959); M. Karplus, *J. Chem. Phys.*, **30,** 11 (1959).]

FIGURE **15.10**

NMR spectra of acrylonitrile at 60, 100, and 220 MHz illustrate the primary advantage of operating at the highest attainable rf and magnetic field. (Courtesy of Varian Associates.)

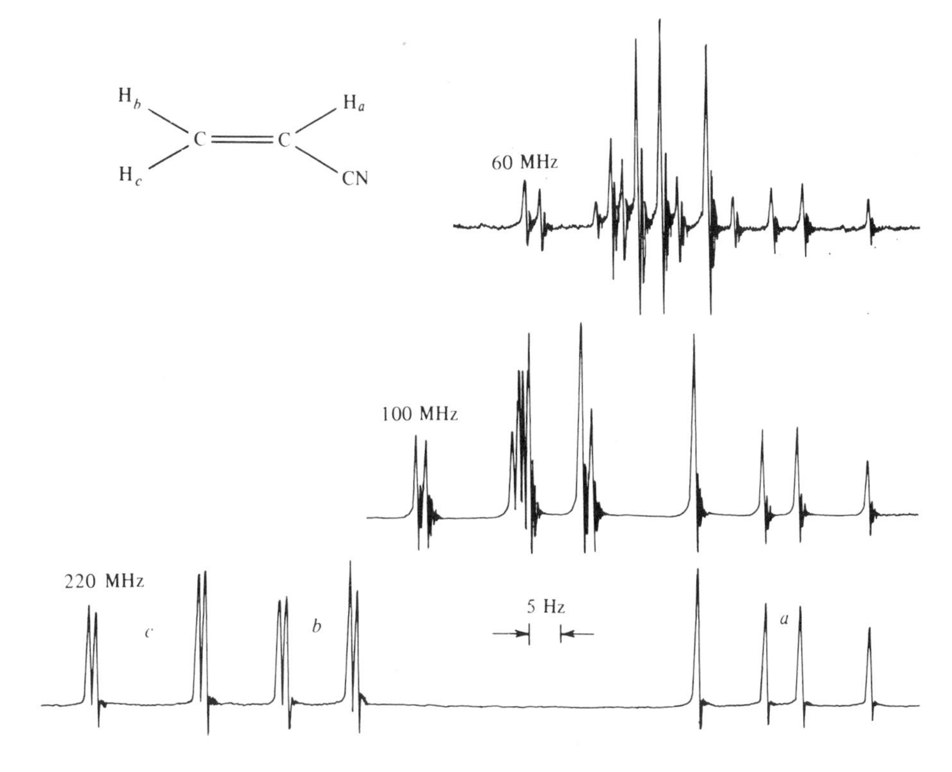

observable spin-spin coupling. Without deliberate isotopic substitution, significant numbers of only fluorine and phosphorus occur naturally. In fact, the presence of one of these elements may be deduced from an otherwise unexplained coupling effect. Usually J is larger than for most proton-proton couplings. The direct coupling of $^{13}C—H$ is often noticeable as sidebands on a proton NMR spectrum. These sidebands do not vary as the sample spinning rate is changed, as do the spinning sidebands. To a first approximation, the magnitude of the $^{13}C—H$ coupling is proportional to the percent of *s*-character in the C—H bond.

When a proton is coupled to a nucleus that has a nonzero quadrupole moment, the latter provides efficient spin-lattice relaxation (T_1 is decreased) and this is usually sufficient to decouple, completely or partially, the spin-spin interaction with the proton. Coupling of protons with chlorine, bromine, or iodine nuclei is not observed. In the case of ^{14}N, the decoupling is usually partially effective so that the 1:1:1 triplet that results from $^{14}N-^{1}H$ coupling is normally broad and featureless, except for ammonium ion in strongly acidic media.

15.5 ELUCIDATION OF NMR SPECTRA

Application of NMR to structure analysis is based primarily on the empirical correlation of structure with observed chemical shifts and coupling constants. Extensive surveys have been published.[4–8] These tabulations are used to predict the position of resonance lines for a postulated compound. These predictions are compared with the sample spectrum. Conversely, one searches compilations and published spectra to ascertain groups that might occur at the positions observed in the sample spectrum. One of the unique advantages of NMR is that spectra can often be interpreted without reference to data from structurally related compounds.

To distinguish labile protons from nonlabile protons, brief vigorous shaking with a few drops of D_2O generally results in a complete exchange of the labile protons by deuterium and collapse of their absorption signal. Deuterium does not absorb in the proton spectral region. Because of the smaller value of μ/I for deuterium, its coupling to hydrogen is much smaller (about one-sixth) than the corresponding H-H coupling and, therefore, the spin-spin couplings do not appear in the proton spectrum.

Double Resonance (or Spin Decoupling)

Nuclear magnetic double resonance, or spin decoupling, is achieved by irradiating an ensemble of nuclei not only with a rf H_1(i.e., ν_1) at resonance with the nuclei to be observed but additionally with a second, relatively strong rf field H_2 (i.e., ν_2) perpendicular to H_0 and at resonance with the nuclei to be decoupled. Decoupling is achieved when $\gamma H_2/2\pi \gg |J_{AX}|$, but H_2 is still sufficiently low in power that it does not cause significant changes in the other nuclei. The spin and magnetic moments of the nucleus irradiated with ν_2 are effectively averaged out over the possible spin states so that other nuclei do not interact with it; thus, the coupling is removed and any multiplets involved collapse into a single peak. By reversing the roles of irradiated and observed nuclei, the spin-coupled nucleus is unequivocally identified. Decoupling experiments are carried out to convert homonuclear ($^1H-^1H$) or heteronuclear ($^{19}F-^1H$, $^{13}C-^1H$) multiplets into singlets or less complex multiplets. A shorthand designation for these experiments uses brackets to indicate the nucleus being irradiated—for example, H–{H} or H–{X}.

Experimentally, in the frequency-sweep technique, the second rf field, ν_2, is varied until a characteristic change occurs in the lines of a particular multiplet. For example, if ν_2 is held at a fixed separation from the reference line (at H_5-H_{TMS} in Figure 15.11) and the spectrum is swept by ν_1 while the magnetic field and locking frequency are held constant, signals from protons H_3 change from a pair of quartets to a pair of triplets (upper spectrum of Figure 15.11), indicating that H_3 and H_5 have a small long-range coupling constant. Small splittings also disappear in the patterns of H_4 and H_6. In a single experiment the effect on all other resonances of irradiating a chosen resonance is observed.

In the field-swept technique of decoupling, the spectrum is swept while a fixed frequency difference $\Delta\nu$ is maintained between the observing ν_1 and irradiating ν_2 frequencies. Only those coupled resonances separated by the chosen frequency difference are observed. For each critical frequency difference a separate decoupling experiment is required.

In routine ^{13}C work usually all $^{13}C-^1H$ multiplets are decoupled for sensitivity and simplicity reasons. This is achieved when the decoupling field, H_2, covers the range of all proton Larmor frequencies. This is at least 1 kHz at about 90 MHz in a magnetic field, H_0, of 23 kG. Decoupling fields with large frequency ranges are realized either by a very large rf power, H_2, so that $\mu H_2/2\pi I = \Delta = 1$ kHz (for ^1H), or by application of broad-band decoupling in which the coherent proton radio frequency is modulated with "white" noise. If the frequency spread of this noise is greater than the range of proton frequencies and high enough power is used, the various C—H multiplets are collapsed, giving single lines for all carbons formerly coupled to protons. In the latter case, no detailed consideration is necessary for the selection of the proton radio frequency.

Decoupling increases the sensitivity of NMR measurements because the intensities of all multiplet lines in a coupled spectrum are accumulated in one singlet in the decoupled spectrum. This technique is especially useful when working with the low-abundance natural ^{13}C in samples.

The Nuclear Overhauser Effect

The nuclear Overhauser effect (NOE) involves saturation of one signal in the spectrum and observation of the changes in the intensities of other signals. It does not depend on the two nuclei being spin coupled. In ^{13}C–^{1}H experiments it arises from an intramolecular dipole–dipole relaxation mechanism. Although decoupling increases the sensitivity of NMR experiments because the intensities of all multiplet lines in a coupled spectrum are accumulated in one singlet signal in the decoupled spectrum, the intensity of the ^{13}C signal often increases much more than expected. This additional increase is the nuclear Overhauser effect enhancement. In a ^{13}C–^{1}H decoupling experiment, the transitions of ^{1}H are irradiated while the resonances of ^{13}C nuclei are observed. The irradiation of the proton frequencies results in equalization of the proton spin-state populations and this, in turn, causes changes in the normal population of the ^{13}C states with an increased fraction in the lower energy state, thus increasing the intensity of the ^{13}C lines. The theoretical maximum value of the nuclear Overhauser enhancement is 2.98 for ^{13}C–^{1}H experiments, but it will be less if there are other pathways for relaxation.

FIGURE **15.11** Frequency-swept spin decoupling at 100 MHz. (Courtesy of Varian Associates.)

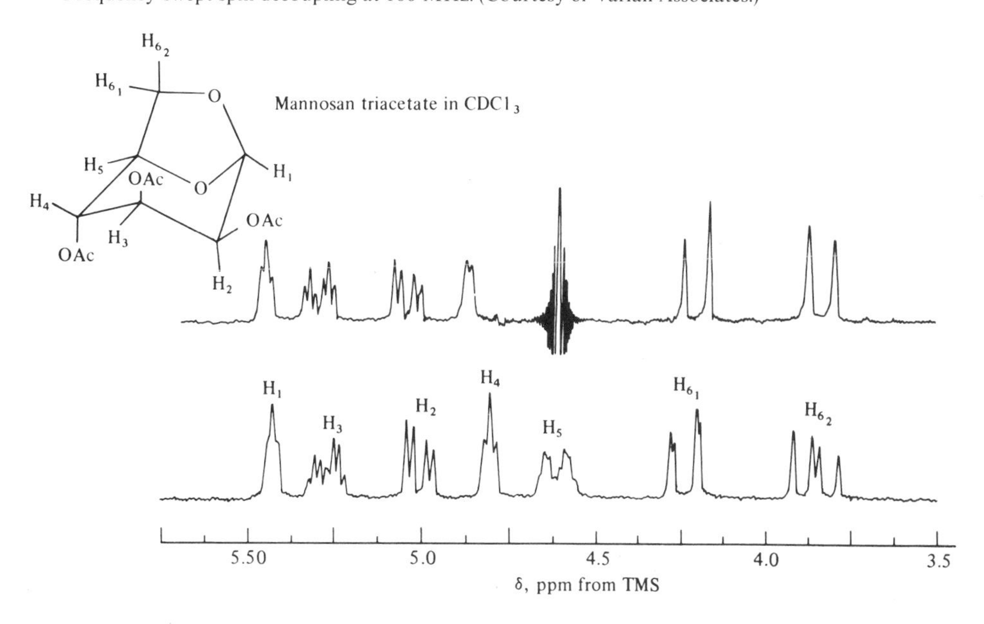

The nuclear Overhauser effect is a short-range effect and is proportional to r^{-6}, where r is the internuclear distance. The nuclear Overhauser effect is helpful in determining which groups are nearby in space. For $^1H-^1H$ experiments the NOE effect is usually not observed for nuclei separated by more than about 3.5 Å.

The extra intensity enhancement due to the nuclear Overhauser effect is welcome when obtaining qualitative NMR spectra because it means that fewer repetitions are needed to obtain the desired signal-to-noise ratio. However, the peak areas are no longer proportional to the concentration of specific nuclei, and therefore the NOE cannot be used in normal quantitative work. Furthermore, the NOE is variable among nonequivalent nuclei in a sample. This means that the relaxation times, T_1, are variable and in pulsed FT-NMR experiments the pulses must not be repeated until all the nuclei have completely relaxed if the intensities are to be proportional to the concentrations of the various nuclei. Either the nuclear Overhauser effect enhancement factors are made all equal to zero by gating the decoupler on only during the acquisition period and off during the delay period, or all NOE enhancement factors are assumed to have equal value if the carbon atoms being studied are all protonated and restricted in their motions by being part of the framework of a fairly large molecule. However, if nonprotonated carbon atom signals must be integrated, the required delay times may be very long and the analysis time may become prohibitive. In such cases it often proves useful to add a relaxation agent such as chromium(III) acetylacetonate, which quenches the nuclear Overhauser effect for all carbons and shortens all T_1 values enough so that pulses can be repeated at about 2-sec intervals.

Spin Tickling

In *spin tickling* a weak irradiating field, ν_2, is used, a field whose bandwidth is only slightly larger than the width of individual lines. The effect of irradiating one line of a multiplet in the spectrum is to split other lines in the same spectrum coupled to it. Hidden lines in a complex region of the spectrum are located by irradiating lines one at a time while looking at the effect on other regions of the spectrum. Frequency-swept spin tickling is illustrated in Figure 15.12. In this spectrum, one of the olefinic protons produces the set of four lines centered around 6.3δ. The proton coupling of 14 Hz is midway between the normal *cis* and *trans* couplings of 10 and 17 Hz, respectively. While "sitting on" the peak around 646 Hz from TMS, an external audio oscillator, ν_2, is slowly varied so as to sweep ν_2 through the low-field pattern of signals from 7 to 9δ. A dip in the signal at 646 Hz occurs when ν_2 is 777.6 or 764.2 Hz larger than the frequency of the oscillator used to lock to the TMS signal. Likewise, the recorder pen is positioned on the peak at about 627 Hz and the splitting effects are noted when ν_2 is 737.5 and 723.7 Hz larger than the TMS lock frequency. Thus, the four hidden lines are located and their positions in the spectrum are indicated by the arrows. From the large spacings the average value of phosphorus coupling to this proton is 40.3 Hz, as compared with the *cis* coupling constant of 13.5 Hz, and agrees with *trans* phosphorus–hydrogen coupling.

Solvent Influence and Shift Reagents

The solvent may affect the screening of particular protons by several possible mechanisms such as van der Waals interaction, anisotropy of the susceptibilities of the surrounding molecules, and the reaction field of the medium and specific solute-solvent interactions. Polyhalogenated hydrocarbon solvents tend to cause the largest shifts, up to 0.5 ppm downfield. This is thought to be due to van der Waals interactions leading to a decrease in the diamagnetic shielding of the nucleus. In Figure 15.13 the complex pattern has been resolved (see the upper trace) by the addition of 50% benzene to the CCl_4 solution in which the NMR spectrum was initially taken. Benzene solvent molecules associate with electron-deficient sites in solute molecules. Because benzene is highly anisotropic, different protons in the solute experience shielding or deshielding depending on their orientations to the benzene ring. In this particular example, the ring protons are shielded (located above the benzene ring) and appear at higher field.

An isolated carbonyl group induces a solvent effect that changes sign near a plane drawn through the carbonyl carbon atom and perpendicular to the carbonyl

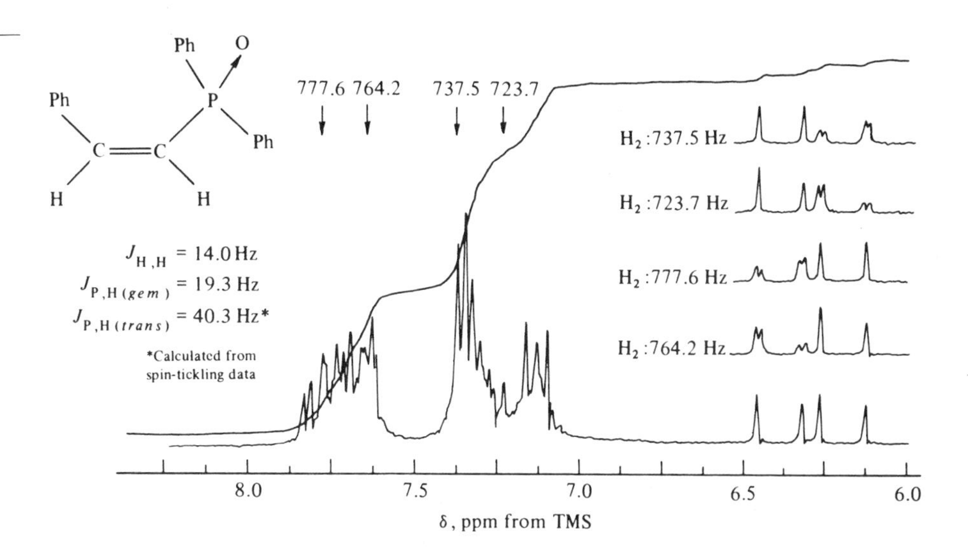

FIGURE **15.12**
Frequency-swept spin tickling at 100 MHz. (Courtesy of Varian Associates.)

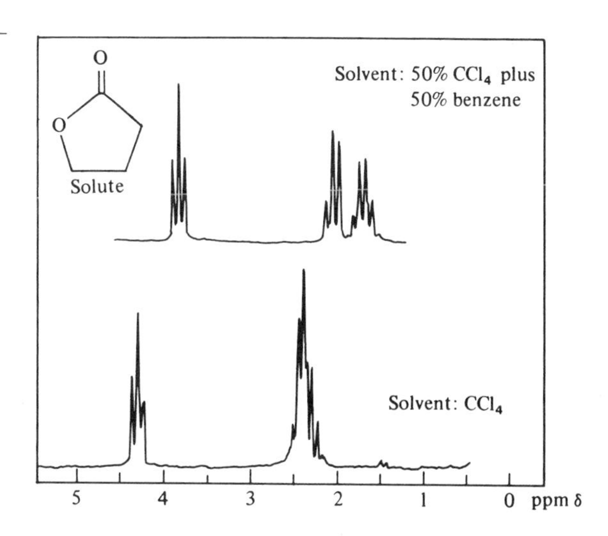

FIGURE **15.13**
Solvent effect on an NMR spectrum.

group. The corresponding shift relative to pyridine changes sign near a plane drawn through the α-carbon atoms. Thus, in the case of an equatorial proton adjacent to the carbonyl group in a cyclohexanone ring, the shift is small or zero in benzene and negative in pyridine, whereas for axial protons or methyl groups, the shift is positive in benzene and small or zero in pyridine.[9]

Proton resonance peaks are spread across a broader range of magnetic field strength by the addition of a paramagnetic compound to the solution being studied. The most commonly used shift reagents are lanthanide fluorinated β-diketones. They function by acting as Lewis acids, forming a complex with the substance under investigation that acts as a nucleophile. Induced shifts are attributed to a pseudocontact, or dipolar interaction, between the shift reagent and the nucleophile. The most commonly used metal chelates are those of Eu(III) and Yb(III), which normally induce downfield shifts, and Pr(III), which induces upfield shifts. Shift reagents give a resolution of peaks comparable to the resolution achievable with 100- or even 220-MHz spectrometers (Figure 15.14). The fastest and easiest technique for obtaining induced shifts is to add a few milligrams of shift reagent directly to the nucleophile dissolved in solvent. Increments of shift reagent are added until sufficient resolution is attained. There is essentially a linear dependence of the chemical shifts upon added shift reagent. The most frequently used solvents are chloroform and carbon tetrachloride. Once shift reagents have been used to re-

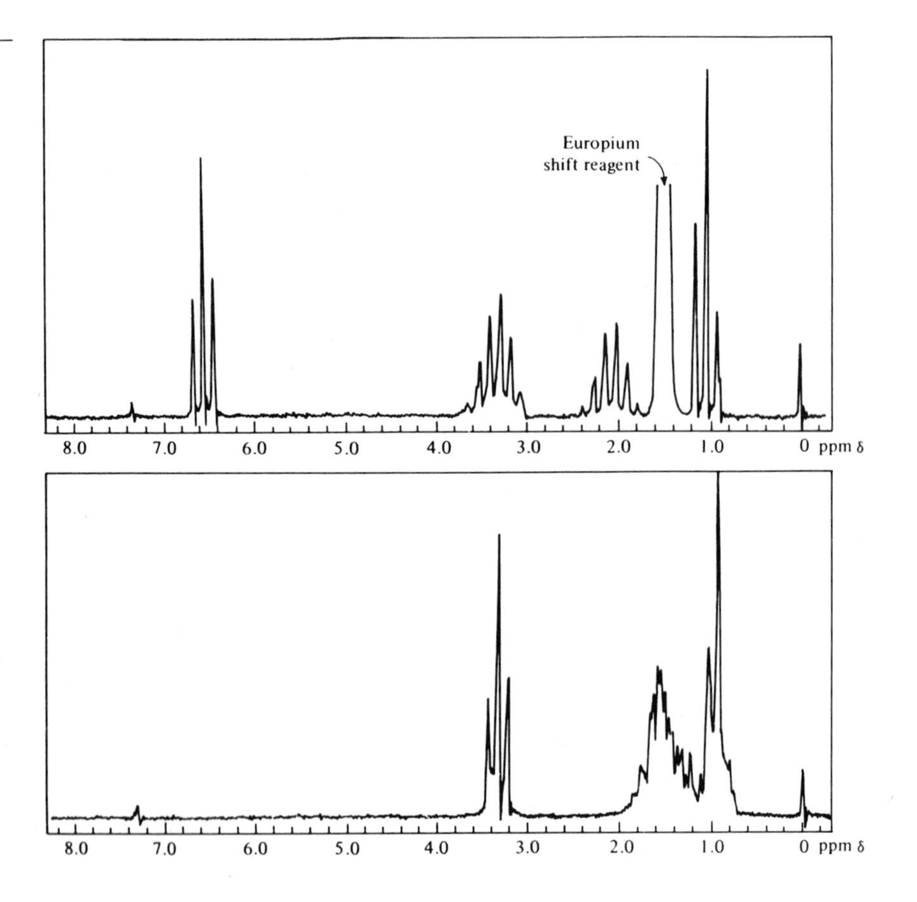

FIGURE **15.14**
NMR spectra of di-*n*-butyl ether (1.0×10^{-4} moles) in 0.5-mL CCl_4 alone (*lower trace*) and with 5.0×10^{-5} moles of tris-(1,1,1,2,2,3,3-heptafluoro-7, 7-dimethyl-4,6-octanedione) europium(III) added (*upper trace*).

solve the NMR peaks of a compound, spin-decoupling experiments become possible.[10–12]

Two-Dimensional Fourier Transform NMR

In recent years, a number of pulse sequences have been developed that permit the presentation of spectra as a function of two independent frequency parameters such as ^{13}C chemical shifts versus the J or spin-spin couplings, correlations between nuclei based on the nuclear Overhauser effect, or chemical shifts correlated with coupling constants. The experiments involve two pulses (which may result in different tip angles) separated by a variable time period and observation of the free induction decay signal. Thus the sequence is pulse, evolution, pulse, detection. Sometimes the data obtained in a series of acquisitions are reordered in a different sequence to form a new set of free induction decay signals and then these new FIDs are Fourier transformed. The presentation of NMR data in two-dimensional form often greatly simplifies the interpretation of complex spectra. Many different pulse sequences have been proposed and used, although the theoretical basis for all is not yet completely understood.

Analysis of NMR Spectra

If two or more nuclei in a compound have identical chemical shifts, they are said to be *chemical shift equivalent*. Nuclei that have identical coupling constants to all other magnetic nuclei in the molecule are said to be *magnetically equivalent*. Two or more nuclei related by some element of symmetry are chemical shift equivalent. Also nuclei that exchange by some process in 10^{-3} sec or less are chemical shift equivalent. To illustrate the difference between magnetic and chemical shift equivalency, consider the two compounds: (1) *o*-dibromobenzene and (2) *m*-dibromobenzene. In compound 1, hydrogens H_1 and H_4 are related by a plane of

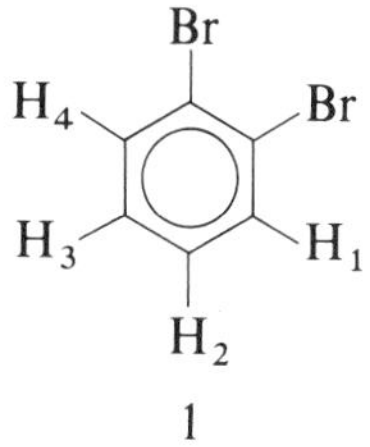

Br, Br, H_4, H_3, H_1, H_2
1

Br, Br, H_4, H_3, H_1, H_2
2

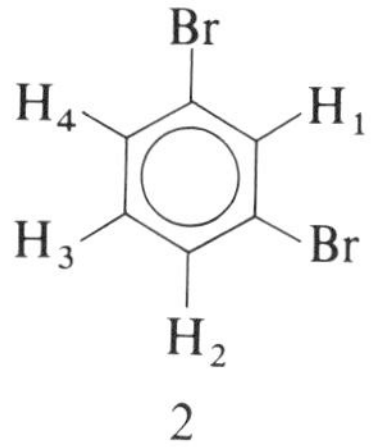

symmetry passing between the two bromine atoms and H_2 and H_3. Hydrogens H_2 and H_3 are also related to each other. Thus there are two pairs of chemical shift equivalent nuclei. These nuclei, however, are not magnetically equivalent, since the coupling constant between H_1 and H_2, for example, is an *ortho* coupling constant whereas that between H_4 and H_2 is a *meta* coupling constant.

On the other hand, compound 2 with a plane of symmetry passing through hydrogens H_1 and H_3 contains a pair of magnetically equivalent nuclei, H_2 and H_4, because these nuclei are related by the symmetry plane and the coupling constants are identical with respect to each of the other magnetic nuclei. Hydrogen H_2 is *meta* to H_3, as is H_4. Likewise, H_2 and H_4 are both *meta* to H_1 (with a bromine in between).

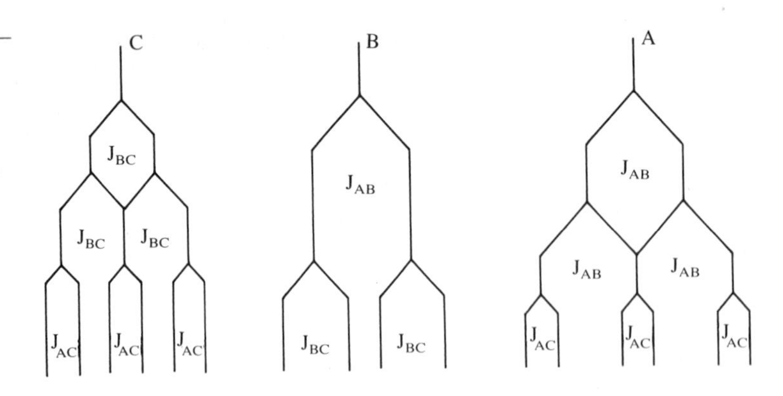

FIGURE 15.15
Diagram of an AB_2C spectrum with chemical shifts ν_A, ν_B, and ν_C and coupling constants J_{AB}, J_{AC}, and J_{BC}. (Not drawn to scale.)

A method of notation has been developed to classify these spin systems. Chemical shift equivalent nuclei are given the same letter of the alphabet. Nuclei with chemical shift differences comparable to the coupling constants are given letters at one end of the alphabet, whereas those with large differences are given letters at the other end. Chemical shift equivalent nuclei that are not magnetically equivalent are given the same letter but are distinguished by primes. Magnetically equivalent nuclei are given the same letter and the number of such nuclei is indicated by a subscript. Thus in the two examples given, *o*-dibromobenzene is designated as AA′BB′, and *m*-dibromobenzene is AB_2C.

Nuclei that have identical chemical shifts and coupling constants to all other nuclei by virtue of fast rotation about a single bond are magnetically equivalent. Thus bromoethane is an A_3B_2 system and difluoromethane is an A_2X_2 system.

The type of NMR spectrum observed is strongly dependent on the spin system. Coupling constants between magnetically equivalent nuclei are not observed, and such spectra usually contain fewer lines than spectra involving only chemical shift equivalent nuclei. As pointed out, first-order spectra are obtained only when $J/\delta < 0.05$. These spectra can usually be analyzed easily. Other spectra usually require computer-aided computations to sort out the observed lines and to determine the coupling constants.

As an example of a simple first-order spectrum, consider an AB_2C spectrum such as the *m*-dibromobenzene case above. The hydrogens are coupled to each other. Suppose that the chemical shifts are large compared with the coupling constants. The spectrum can then be diagramed as in Figure 15.15. Note that two multiplets are really triplets with each line further split into a doublet, whereas the lines caused by the B_2 nuclei consist of two doublets with each of the relevant coupling spacings repeated twice.

Overlapping spectral features are a source of difficulty in interpretation. The distribution of protons shown by the integration curve is helpful. (See the following section.)

15.6 QUANTITATIVE ANALYSIS AND INTEGRATION

The area under an absorption band is proportional to the number of nuclei responsible for the absorption. A device for electronically integrating the absorption signal is a standard item on most commercial spectrometers. The integral is represented as a step function; the height of each step is proportional to the number of nuclei in that particular region of the spectrum. Accuracy is typically within $\pm 2\%$. For quantitative analysis, a known amount of a reference compound can be included with the sample. The NMR signal of the reference compound preferably should contain a strong singlet lying in a region of the NMR spectrum unoccupied by sample peaks. Exact phasing out of the dispersion signal is crucial in integration. From the two peak areas, A_{unk} and A_{std}, and the weight of the internal standard taken, W_{std}, the amount of the unknown present is calculated by

$$W_{unk} = W_{std} \times \frac{N_{std}}{N_{unk}} \times \frac{M_{unk}}{M_{std}} \times \frac{A_{unk}}{A_{std}} \tag{15.10}$$

where N's are the numbers of protons in the groups giving rise to the absorption peaks, and M's are the molecular weights of the compounds.

Whenever the empirical formula is known, the total height (in any arbitrary units) divided by the number of protons yields the increment of height per proton. Lacking this information, but deducing the assignment of a particular absorption band, one calculates the increment per proton from the height for the assigned group divided by the number of protons in the particular group. Unfortunately, there is no way of handling overlapping bands. Provided that at least one resonance band from each component of a mixture is free from extensive overlap by other absorptions, NMR quantitative analysis should be possible.

EXAMPLE 15.1

An NMR spectrum shows three single peaks located at +440, +300, and +120 Hz in a field of 60 MHz with TMS as reference. The integral heights are 4.2, 1.7, and 2.5 units, respectively. Without knowledge of the empirical formula, the integral heights bear the ratio 5:2:3 and, since no splitting is observed, this must mean a group with five protons (probably an aromatic ring from the chemical shift), a methylene group, and a methyl group, each not coupled with one another. If one knew that the empirical formula was $C_9H_{10}O_2$, dividing the total integral height of 8.4 units by the number of protons gives 0.84 unit as the increment per proton.

EXAMPLE 15.2

Returning to the information given in Example 15.1, these structures are present:

C_6H_5— —CH_2— —CH_3

From the position of the phenyl group resonance and the lack of multiplet structure, the benzyl group is suggested. Placing an oxygen next to the phenyl ring would shift upfield the *ortho* and *para* proton resonances; likewise, a carbonyl group would shift the *ortho* proton resonances downfield. The position of the isolated methyl group resonance suggests an adjacent double bond, either a carbonyl or phenyl ring. The latter is impossible because a methylene group must be accommodated. Two structural candidates remain as possibilities: benzyl methyl ketone and benzyl acetate (if one can eliminate the presence of sulfur). In the former compound the methylene proton resonances would be expected at δ 3.6; in the latter at δ 5.1. The observed position is δ 5.0, and the compound is benzyl acetate.

EXAMPLE 15.3

The NMR spectrum shows a single peak at δ 6.83, a quadruplet at δ 4.27, and a triplet at δ 1.32. For the multiplets the coupling constant is about 7 Hz. The integrator readings are in the ratio 1:2:3, respectively. The empirical formula is $C_8H_{12}O_4$. The upfield methyl group is split into a triplet by an adjacent methylene group, which, in turn, is split into a quadruplet—the typical ethyl pattern. This is confirmed by the integrator readings. Assignment of the low-field group is not so simple. It is not quite in the location expected for a benzene ring, nor does it contain sufficient numbers of protons. However, an olefinic proton absorption is a possibility, although further shielding is indicated. Returning to the methylene group, its downfield position could be due to an adjacent —O—(C=O)— structure. If the olefinic proton were alongside the carbonyl group, its absorption position would be reasonable. Summarizing the available information, one obtains

$$CH_3—CH_2—O—\overset{\overset{\displaystyle O}{\|}}{C}—CH=$$

which is exactly half the empirical formula. The complete structure is either diethyl fumarate or diethyl maleate. Having gotten this far, one would take an NMR spectrum of each compound and find that in diethyl maleate the low-field single peak occurs at δ 6.28. Coupling between the olefinic protons collapsed because $\delta_2 + \delta_1 = 0$.

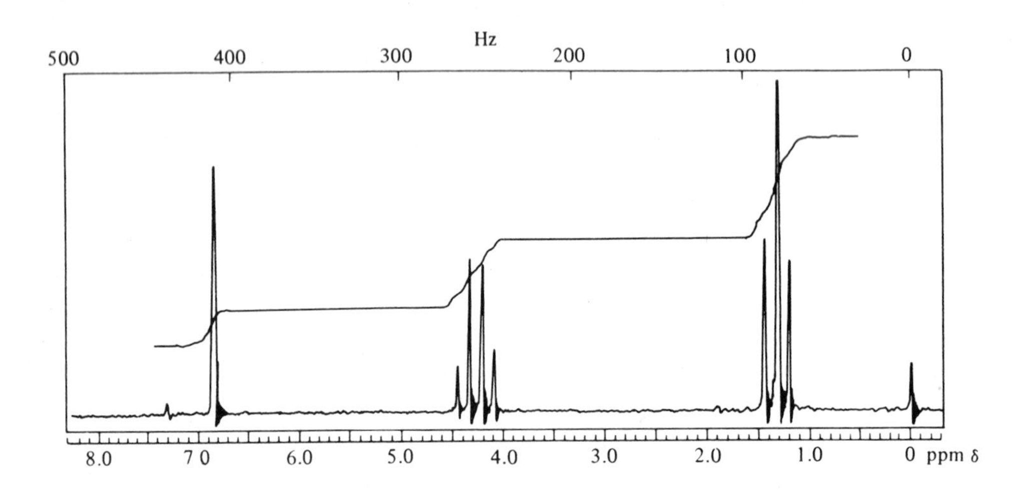

EXAMPLE 15.4

The NMR spectrum for the compound with empirical formula $C_7H_{16}O$ shows a symmetrical heptet centered around δ 3.78 and an unsymmetrical doublet with individual peaks at δ 1.18 and δ 1.12. The integrated areas are 2 units for the heptet and 24 + 6 units for the doublet. The absence of peaks at low field indicates the absence of aldehyde, unsaturation, and probably hydroxyl groups. The heptet indicates a probable isopropyl group whose methine proton is split by six methyl protons. Its field position indicates an adjacent oxygen. In turn, the methine proton splits the *gem*-methyl protons into a doublet. A logical presumption is an isopropyl ether group, with the oxygen atom serving to isolate the methine proton from the remainder of the molecule. Using the heptet proton as the divisor, the integrator readings indicate nine unassigned protons in a single peak superimposed on the low-field wing of the methyl doublet in the isopropyl group. This could mean only three isolated methyl groups—that is, a *tert*-butyl group. The compound is *tert*-butyl isopropyl ether.

The determination of the ethylene chain length in commercial nonionic detergents provides an example of quantitative analysis. Approximately 10% solutions of each surfactant in CCl_4 are prepared, and the spectrum and several repeat integrals of each are run (Figure 15.16). The ratio of the ethylene oxide proton signal to the cetyl chain signal is calculated and, since the number of protons in the cetyl chain is known, the calculation of the number of O—CH_2—CH_2—O units is a matter of simple proportion. For each detergent the total analysis time is about 10 min; the alternative is a lengthy titration procedure.

Impure samples are assayed using another compound as the internal standard. The NMR spectrum of

$$\text{(isopropylphenyl)}\!-\!O\!-\!\overset{\overset{O}{\|}}{C}\!-\!\overset{\overset{CH_3}{|}}{N}\!-\!\overset{\overset{O}{\|}}{C}\!-\!CH_2\!-\!S\!-\!P(OCH_3)_2$$

has a singlet peak due to the N—CH_3 protons at $\delta = 3.40$ and a doublet due to the —(C=O)—CH_2—S—P— group at $\delta = 4.3$. Benzyl benzoate, which gives a well-resolved peak at $\delta = 5.4$, is an excellent internal standard.

Products from the air oxidation of *p*-cymene yield a complex mixture. The NMR spectrum, with peak assignments indicated, is shown in Figure 15.17, which also shows the position of the internal standard line from benzyl benzoate.

The determination of water in liquid N_2O_4 in the less than 0.1 weight % range illustrates the usefulness of NMR in trace analysis and in a system for which calibration standards are not available. At −10 °C the exchange of protons between dissolved H_2O, HNO_2, and HNO_3 molecules is rapid so that a single NMR peak is observed for the exchanging protons. As little as 0.03% H_2O can be detected with a relative standard deviation of 7.5% based on a single spectral scan. Accuracy is improved by using the accumulated average of 50 scans. With this procedure, 0.01% H_2O (100 μg/mL) can be determined. Benzene is used as an internal standard.

Residual H_2O in samples of high-purity D_2O is determined by the method of standard additions, using standard microvolumetric techniques. The resulting NMR integrals are plotted versus the weight percent of added H_2O and extrapolated to zero integral.

A well-known assay of a pharmaceutical formulation by NMR is that of aspirin, phenacetin, and caffeine mixtures.[13] The procedure takes about 20 min and does not require the separation of individual constituents. A separate analytical

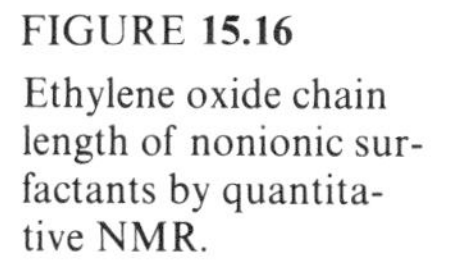

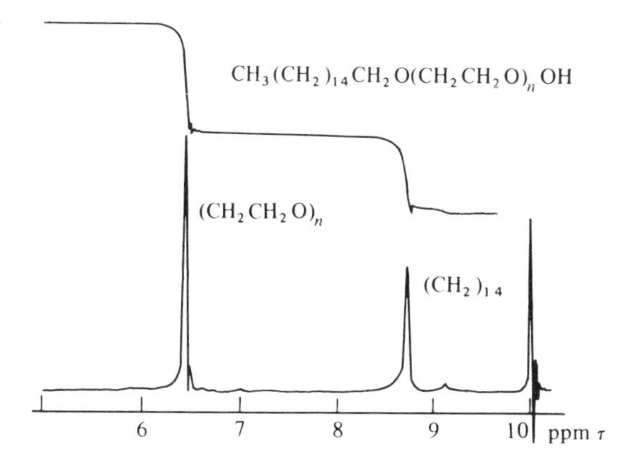

FIGURE **15.16**
Ethylene oxide chain length of nonionic surfactants by quantitative NMR.

FIGURE 15.17

NMR spectrum of mixtures of products obtained from the oxidation of *p*-cymene, with benzoyl benzoate added as internal standard. Integrals are not shown. (Courtesy of International Scientific Communications, Inc.)

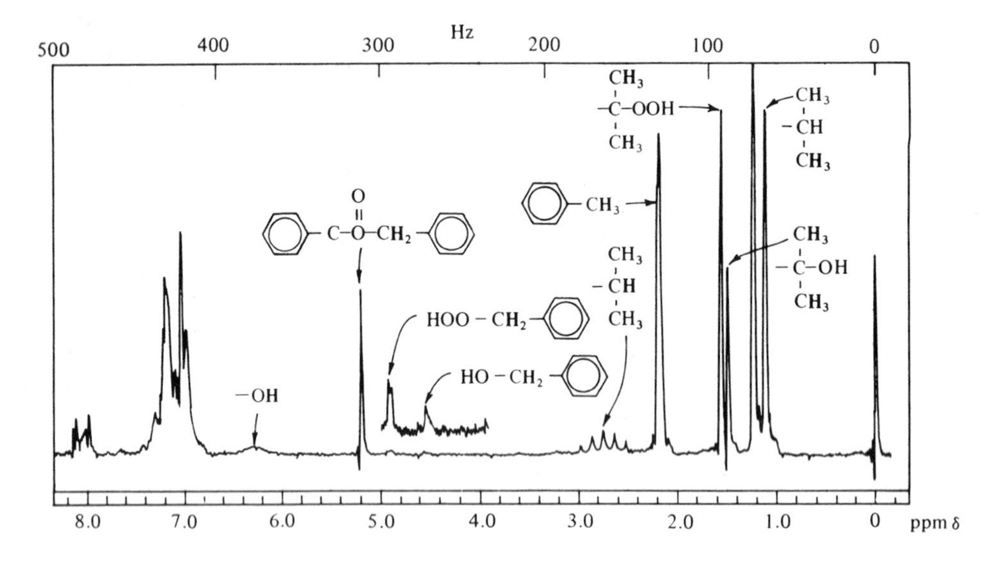

peak of known origin is present for each component; no calibration curves are required, since the integrals for the protons that give rise to the various analytical peaks are constant and unaffected by solvent or solute interactions. For aspirin the sharp peak at about $\delta = 2.3$, which represents the ester methyl group, is used. For phenacetin the amide methyl group at $\delta = 2.1$ is preferable; the quartet at $\delta = 4.0$ is suitable, although one of the methyl peaks of caffeine overlaps the quartet at $\delta = 3.9$ and a correction is necessary. For caffeine, the two methyl resonances at $\delta = 3.4$ and 3.6 are used.

15.7 NMR IMAGING IN MEDICINE[14–17]

In Chapter 13, it was pointed out that X rays could be used to generate a two- or three-dimensional image of an object such as the human body. Such an imaging procedure is known as X-ray tomography. Unfortunately the soft tissues of the body show little contrast to X rays and exposure to X rays is detrimental to health. On the other hand, nuclear magnetic resonances are very sensitive to changes in the physical and chemical environments of the nuclei, especially of protons, which are the most abundant and widely distributed in the body, and the energy involved in NMR techniques is so low that health risks are eliminated. Therefore NMR imaging methods for the human body are now being rapidly developed and improved and are widely used.

In normal NMR methods considered to this point, the uniformity of the externally applied magnetic field has been emphasized. If one adds to the externally applied, homogeneous magnetic field another field with a gradient in the same direction, then protons in the sample will resonate at differing frequencies depending on the strength of the field at their location. Protons in regions where the total field is lower precess at lower Larmor frequencies than those protons in regions of higher applied fields. Thus one has added a spatial marking mechanism. By observing the NMR signal from a sample or body in the gradient field from many different angles and, at the same time, measuring the signal as a function of frequency, one can reconstruct the two- or three-dimensional image of the body. The most modern NMR imaging techniques now use pulse sequences and Fourier transform techniques to speed up the process and improve contrast in the resultant images—for example, to measure areas or volumes where T_1 or T_2 is different. For additional information on NMR imaging, see the Bibliography at the end of this chapter.

PROBLEMS

1. At 43 °C the NMR spectrum of liquid acetylacetone shows a peak at δ 5.62 (37 units on the integrator) and a peak at δ 3.66 (19.5 units), plus additional peaks that do not concern us. Calculate the percent enol composition.

2. A hydrocarbon sample shows NMR bands over the interval δ 1.0–5.5. Benzophenone used as an internal standard shows NMR bands in the δ 6–7 region. The relative integrals were 228 and 184 units for 0.8023 g of benzophenone and 0.3055 g of sample, respectively. Calculate the percent hydrogen in the sample.

3. Phenol formaldehyde resins include a class prepared with excess phenol, called novolacs, that consist of phenolic nuclei linked by methylene bridges at positions *ortho* or *para* to the hydroxyl group. The integration of the spectrum gives the ratio of aromatic to methylene protons as 30 to 18. Calculate the average chain length and the average molecular weight.

4. Chlorination of *o*-cyanotoluene gave *o*-cyanobenzyl chloride in about 50% yield. Attempted fractionation of the washings and mother liquor gave a liquid whose proton magnetic resonance spectrum gave three singlets upfield from the aromatic ring signals as follows (integral units in parentheses): δ 2.52 (13), δ 4.72 (20), and δ 7.01 (10). (a) Assign the NMR signals. (b) From the signal intensities, deduce the relative molar proportions and the proportions by weight of the three constituents in the liquid mixture.

5. The NMR proton spectrum of the liquid diketene, $C_4H_4O_2$, shows two signals of equal intensity. What structure is consistent with this information?

6. A sample is believed from its mass spectrum to be either of the following dicyanobutenes.

$$CH_3{-}CH{=}C\begin{matrix} \diagup CH_2CN \\ \diagdown CN \end{matrix} \qquad\qquad NC{-}CH{=}C\begin{matrix} \diagup CH_2CN \\ \diagdown CH_3 \end{matrix}$$

I   II

What characteristic in the NMR spectrum would identify each isomer?

7. The phosphorus resonance of phosphonic acid and phosphinic acid is reported to be a doublet in the former and a triplet in the latter compound. Write the structures of the two acids.

8. Addition of methyldichlorosilane to vinyl acetate gives an adduct whose likely structure is

$$CH_3\text{—}Si(Cl)_2\text{—}CH_2\text{—}CH_2\text{—}O\text{—}CO\text{—}CH_3$$

or

$$CH_3\text{—}Si(Cl)_2\text{—}CH(CH_3)\text{—}O\text{—}CO\text{—}CH_3$$

The NMR spectrum shows two bands with clearly resolved triplet splitting. Which structure is supported by the NMR evidence?

9. On the basis of the two peaks of equal strength found in the NMR spectrum of the sodium salt of Feist's acid in D_2O, which structure is correct?

HOOC—C——CH—COOH (ring C bearing CH_3; C=C double bond in ring)   HOOC—CH——CH—COOH (ring C bearing $=CH_2$)

10. Deduce the structure of the compound with the spectrum shown whose empirical formula is $C_{10}H_{11}NO_4$.

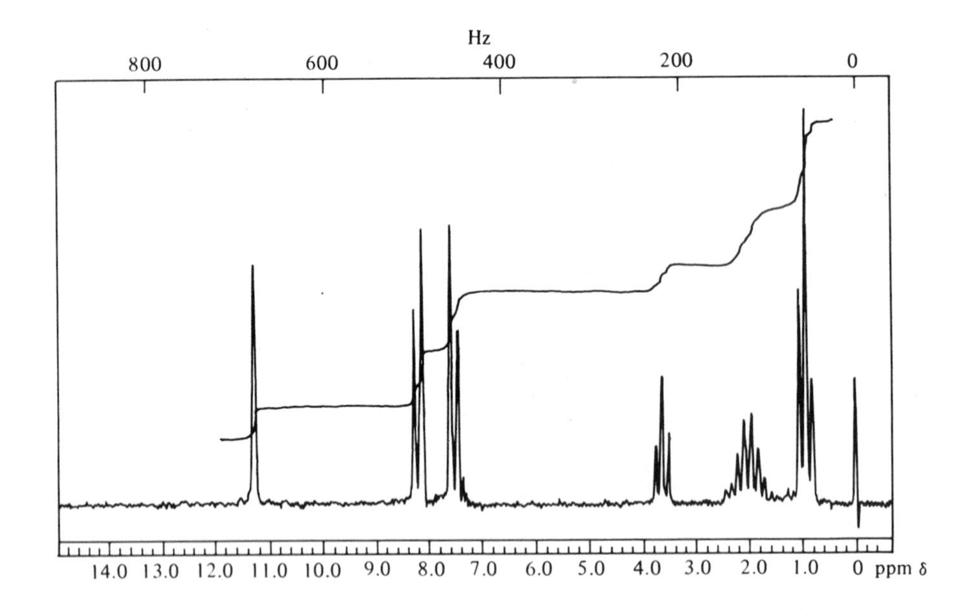

11. The NMR spectrum contains a single peak at δ 3.58 and another single peak at δ 7.29. Integrated intensities are 8 and 20 units, respectively. From mass spectral information, the compound (mol wt 246) is known to contain two sulfur atoms. Deduce its structure.

12. The NMR spectrum contains single peaks at δ 7.27, δ 3.07, and δ 1.57. The empirical formula is $C_{10}H_{13}Cl$. Deduce the structure of the compound.

13. Overlapping multiplets always present a challenge in unraveling a NMR spectrum, such as the one for the compound $C_6H_{11}BrO_2$ shown. Deduce the structure. Ignore the small benzene peak at δ 7.32.

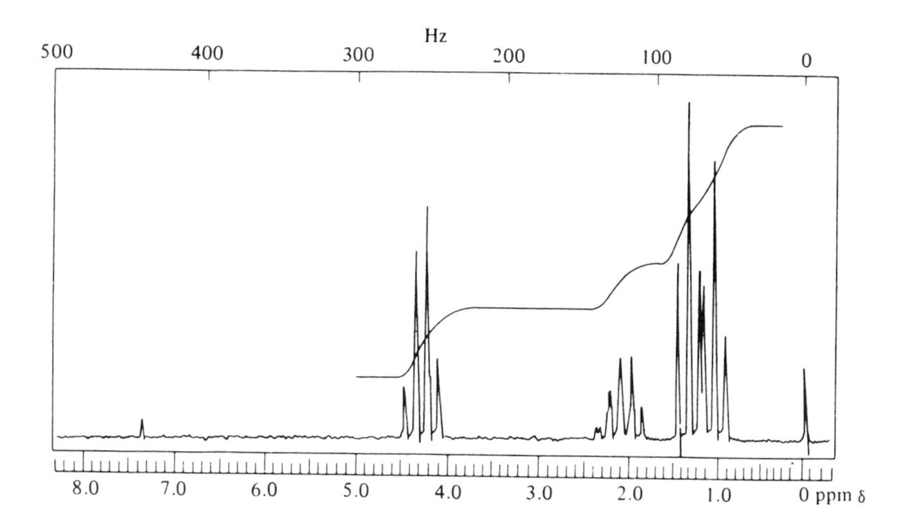

14. Deduce the structure of the compound $C_8H_{14}O_4$ from the NMR spectrum shown. Be cognizant of the requirement for spin-spin coupling.

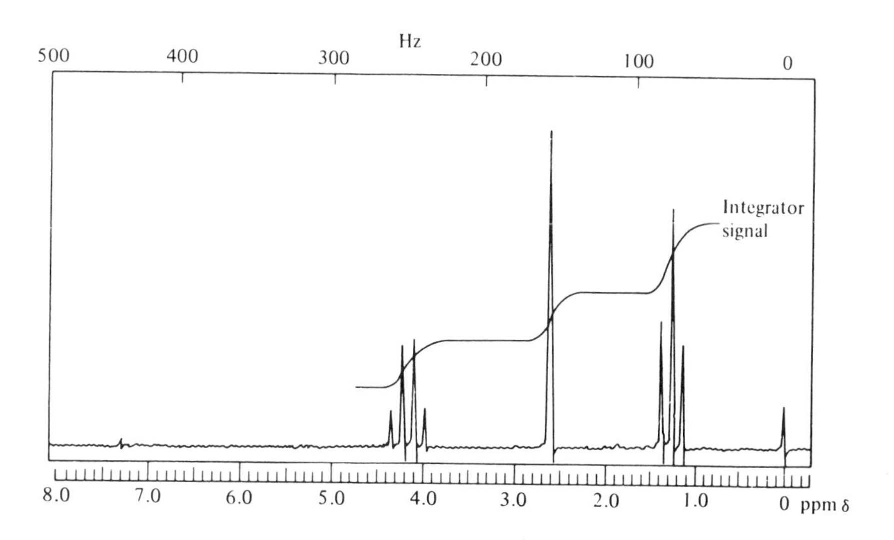

15. Chart the spin-spin couplings involved in the spectrum of methyl methacrylate shown. The lower trace is an expanded scale. The resonance of external benzene was 418.4 Hz from TMS, the internal standard, in a rf field of 60 MHz.

PROBLEM 15

The proton NMR spectrum of methyl methacrylate (liquid) taken on a Varian model A-60 spectrometer. Internal standard was TMS; external standard was benzene.

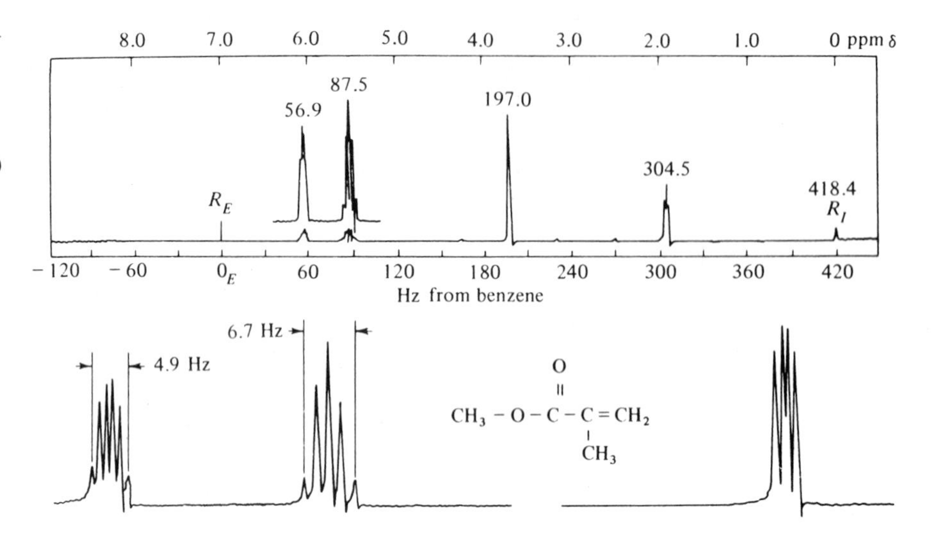

16. Suggest a method for obtaining the resonance frequency of ^{11}B (19.3 MHz) on an NMR spectrometer equipped with a 5-MHz crystal oscillator when $H_0 = 14.09$ kG.

17. In the double resonance procedures, why cannot a simple sweep of the static magnetic field be used to generate the spectrum?

18. Estimate the nuclear Overhauser effect sensitivity enhancement for (a) decoupling protons from ^{13}C NMR spectra, and (b) decoupling ^{13}C from ^{1}H spectra.

19. The NMR spectrum of a polyester dissolved in trifluoroacetic acid gave multiplets centered around δ 2.0 (45), δ 4.75 (114), and δ 8.0 (133). The integrator readings for the peaks are given in parentheses. The peaks at δ 4.75 include all protons immediately adjacent to oxygen atoms. (a) Determine the mole percent of the three following components in the polymer. (b) Determine the weight percent of each component.

1. —O—C(=O)—C$_6$H$_4$—C(=O)—O—

2. —O—CH_2—CH_2—O—

3. —O—CH(—CH_2—CH_2—)(—CH_2—CH_2—)CH—O—

BIBLIOGRAPHY

ALLERHAND, A., AND S. R. MAPLE, "Ultra-High Resolution NMR," *Anal. Chem.,* **59,** 441A (1987).

BECKER, C. D., *High Resolution NMR,* Academic, New York, 1980.

BUDINGER, T. F., AND P. C. LAUTERBUR, "Nuclear Magnetic Resonance Technology for Medical Studies," *Science,* **226,** 288 (1984).

COX, R. H., AND D. E. LEYDEN, in *Treatise on Analytical Chemistry,* 2nd ed., P. J. Elving, M. M. Bursey, and I. M. Kolthoff, eds., Part I, Vol. 10, Chap. 1, John Wiley, New York, 1983.

FARRAR, T. C., "Selective Sensitivity Enhancement in FT-NMR," *Anal. Chem.,* **59,** 679A, 749A (1987).

FARRAR, T. C., AND E. D. BECKER, *Pulse and Fourier Transform NMR,* Academic, New York, 1971.

HARRIS, R. K., *Nuclear Magnetic Resonance Spectroscopy,* John Wiley, New York, 1986.

JACKMAN, L. M., AND F. A. COTTON, eds., *Dynamic Nuclear Magnetic Resonance Spectroscopy,* Academic, New York, 1975.

JACKMAN, L. M., AND S. STERNHELL, *Applications of Nuclear Magnetic Resonance Spectroscopy in Organic Chemistry,* Pergamon, New York, 1969.

JELINSKI, L. W., "Modern NMR Spectroscopy," *Chem. Eng. News,* p. 26 (November 5, 1984).

KASLER, F., *Quantitative Analysis by NMR Spectroscopy,* Academic, New York, 1973.

LEVY, G. C., AND D. J. CRAIK, "Recent Developments in Nuclear Magnetic Resonance Spectroscopy," *Science,* **214,** 291 (1981).

LEVY, G. C., R. LICHTER, AND G. NELSON, *Carbon-13 Nuclear Magnetic Resonance Spectroscopy,* 2nd ed., John Wiley, New York, 1980.

LEYDEN, D. E., AND R. H. COX, *Analytical Applications of NMR,* John Wiley, New York, 1977.

LOMBARDO, A., AND G. C. LEVY, in *Treatise on Analytical Chemistry,* 2nd ed., P. J. Elving, M. M. Bursey, and I. M. Kolthoff, eds., Part I, Vol. 10, Chap. 2, John Wiley, New York, 1983.

MACOMBER, R. S., "A Primer on Fourier Transform NMR," *J. Chem. Educ.,* **62,** 213 (1985).

MASON, J., ed., *Multinuclear NMR,* Plenum, New York, 1986.

MULLEN, K., AND P. S. PREGOSIN, *Fourier Transform NMR Techniques: A Practical Approach,* Academic, New York, 1976.

POPLE, J. A., W. G. SCHNEIDER, AND H. J. BERNSTEIN, *High-Resolution Nuclear Magnetic Resonance,* McGraw-Hill, New York, 1959.

WILLIAMS, D. A. R., AND D. J. MOWTHORPE, *Nuclear Magnetic Resonance Spectroscopy,* John Wiley, New York, 1986.

LITERATURE CITED

1. MIKNIS, F. P., V. J. BARTUSKA, AND G. E. MACIEL, "Cross-Polarization ^{13}C NMR with Magic-Angle Spinning," *Am. Lab.,* **11,** 19 (November 1979).
2. DEAN, J. A., ed., *Lange's Handbook of Chemistry,* 13th ed., McGraw-Hill, New York, 1985.
3. DEAN, J. A., ed., *Handbook of Organic Chemistry,* McGraw-Hill, New York, 1986.
4. *The Aldrich Library of NMR Spectra,* Edition 7, Vols. 1 and 2, Aldrich Chemical Co., Milwaukee, 1983.
5. SADTLER RESEARCH LABORATORIES, *Nuclear Magnetic Resonance Spectra,* Philadelphia, a continuously updated subscription service.
6. SILVERSTEIN, R. M., G. C. BASSLER, AND T. C. MORRILL, *Spectrometric Identification of Organic Compounds,* 4th ed., John Wiley, New York, 1981.

7. VARIAN ASSOCIATES, *High Resolution NMR Spectra Catalog,* Vol. 1, 1962; Vol. 2, 1963. Palo Alto, CA (out of print).
8. ASAHI RESEARCH CENTER CO., LTD., Japan, ed., *Handbook of Proton-NMR Spectra and Data,* Academic, Orlando, FL, 1985.
9. RONAYNE, J., AND D. H. WILLIAMS, *J. Chem. Soc. Sect. B,* 535 (1967).
10. KIME, K. A., AND R. E. SIEVERS, "A Practical Guide to Uses of Lanthanide NMR Shift Reagents," *Aldrichimica Acta,* **10,** 54(1977).
11. SIEVERS, R. E., ed., *Nuclear Magnetic Resonance Shift Reagents,* Academic, New York, 1973.
12. WENZEL, T. J., AND J. LAIA, *Anal. Chem.,* **59,** 562 (1987).
13. HOLLIS, D. P., *Anal. Chem.,* **35,** 1682 (1963).
14. SMITH, S. L., *Anal. Chem.,* **57,** 595A (1985).
15. BAX, A., AND L. LERNER, *Science,* **232,** 960 (1986).
16. DUMOULIN, C. L., *Spectroscopy,* **2,** 32, (1987).
17. SOCHUREK, H., *National Geographic,* **171,** 2 (1987).

MASS SPECTROMETRY

Mass spectrometry, one of the most generally applicable of all the analytical tools, provides qualitative and quantitative information about the atomic and molecular composition of inorganic and organic materials. The first mass spectrometer dates back to the work in England of J. J. Thompson in 1912 and of F. W. Aston in 1919, but the instrument that served as a model for more recent ones was constructed in 1932. The mass spectrometer produces charged particles that consist of the parent ion and ionic fragments of the original molecule, and it sorts these ions according to their mass/charge ratio. The mass spectrum is a record of the relative numbers of different kinds of ions and is characteristic of every compound, including isomers. High-resolution mass spectrometry can provide the elemental composition of the molecular and fragment ions. In many cases, together with data from infrared, ultraviolet-visible, and nuclear magnetic resonance spectra, an experimenter can arrive at a definite identification or structure assignment of compounds.

The main advantages of mass spectrometry as an analytical technique are its increased sensitivity over most other analytical techniques and its specificity in identifying unknowns or confirming the presence of suspected compounds. The enhanced sensitivity results primarily from the action of the analyzer as a mass/charge filter to reduce background interference and from the sensitive electron multipliers used for detection. Sample size requirements for solids and liquids range from a few milligrams to subnanogram quantities as long as the material can exist in the gaseous state at the temperature and pressure in the ion source. The excellent specificity results from characteristic fragmentation patterns, which can give information about molecular weight and molecular structure. In addition, a mass spectrometer is an essential adjunct to the use of stable isotopes in investigating reaction mechanisms and in tracer work. Also, mass spectrometry has contributed greatly to a more detailed understanding of kinetics and mechanisms of unimolecular decomposition of molecules.

Interfacing other equipment with a mass spectrometer is often desirable. Coupled techniques have increased in importance in the past few years. Discussions involving gas chromatography and liquid column chromatography coupled with mass spectrometry are deferred until these topics are considered in Chapters 18 and 19, respectively. Considered in this chapter are the coupling of two or three mass analyzers to form tandem mass spectrometers such as the double-focusing sector spectrometers and MS-MS quadrupole systems. Another coupled technique discussed is mass spectrometry of inductively coupled plasmas (ICP-MS), which was introduced in Chapter 10.

16.1 SAMPLE FLOW IN A MASS SPECTROMETER

First let us examine in a general way a typical mass spectrometer and the principles upon which it depends. Functionally, all mass spectrometers perform three basic tasks: (1) creating gaseous ion fragments from the sample, (2) sorting these ions according to mass (strictly mass/charge ratio), and (3) measuring the relative abundance of ion fragments of each mass.

There is no universal mass spectrometer. Certain designs and configurations lend themselves to the solution of specific problems better than others. A block diagram of the components of a mass spectrometer is shown in Figure 16.1. The essential parts are: (1) sample inlet system(s), (2) ion source, (3) ion acceleration and mass (ion) analyzer, (4) ion-collection system, (5) data handling system, and (6) vacuum system. These components will be described in more detail in subsequent sections.

The operation of the spectrometer requires a collision-free path for the ions. To achieve this, the pressure in the spectrometer should be less than 10^{-6} torr (1 torr = 133.3 pascals or newtons per square meter). Volatile and gaseous samples are introduced through a leak into the ionization chamber from the inlet system. Once formed, ions are sorted into discrete mass/energy arrays based on three properties that can be determined with relative ease: energy, momentum, and velocity. A measurement of any two of these allows the mass/charge ratio to be determined. For many years the conventional way of doing this was to use energy and momentum: accelerating the ions in an electric field to a given energy and dispersing them in a magnetic field according to their momentum. Other combinations are also used to sort the ion fragments. For example, energy and velocity are used in time-of-flight mass spectrometers; momentum and velocity are used in the Fourier transform mass spectrometer. An electronic detector with accompanying signal processor and readout device completes the general components of a mass spectrometer.

Whatever the specific type of mass spectrometer, and these will be discussed in detail in later sections along with the various ion sources and detectors, mass spectrometry not only is useful for qualitative and quantitative identification but is also an extremely powerful tool for the evaluation of molecular structure, as described in Section 16.9.

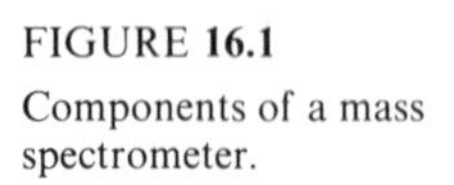
FIGURE **16.1**
Components of a mass spectrometer.

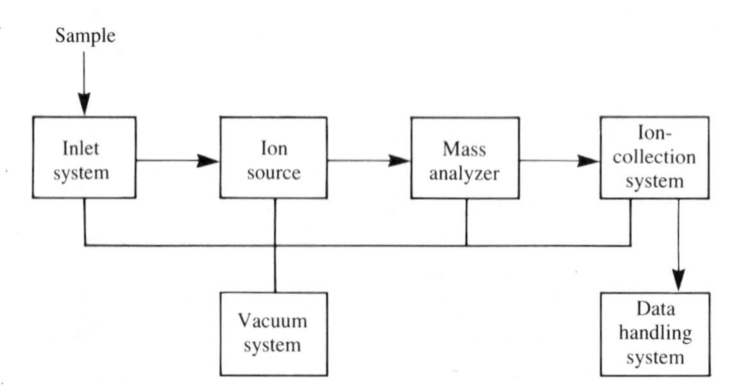

16.2 INLET SAMPLE SYSTEMS

To handle various types of samples and also standards as required, the inlet sample system is usually fitted with several sample inlets (Figure 16.2).

Handling Gas Samples

The introduction of gases merely involves transfer of the sample from small containers of known volume (about 3 mL) coupled to a mercury manometer. The sample is then expanded into a reservoir (perhaps 3–5 L) immediately ahead of the sample "leak."

Introduction of Liquids

Liquids are introduced in various ways: (1) by break-off devices, (2) by touching a micropipet to a sintered glass disk under a layer of molten gallium, or (3) by hypodermic needle injection through a silicone rubber septum. The low pressure in the reservoir draws in the liquid and vaporizes it instantly.

Solids

Solids with a very low vapor pressure can be introduced directly into an entrance to the ion chamber on a silica or platinum probe and then volatilized by gently heating the probe until sufficient vapor pressure is indicated by the total ion current indicator or by the appearance of a spectrum.

Often a simple chemical reaction suffices to convert a nonvolatile compound into a derivative that still retains all the important structural features but now has

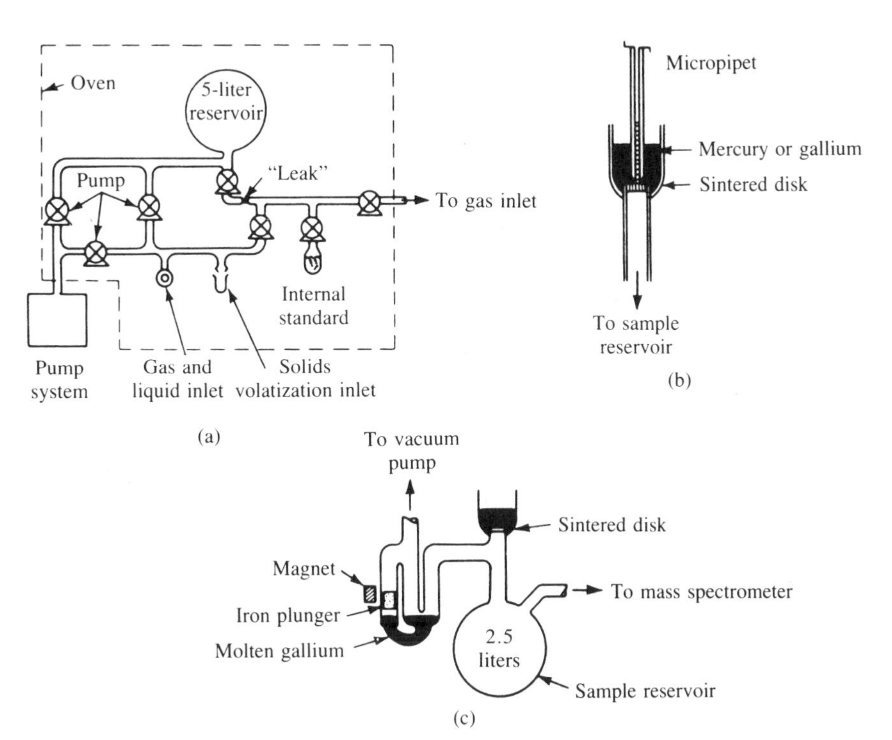

FIGURE 16.2
(a) Inlet sample system for a mass spectrometer. (b) Introduction of liquids through a sintered disk. (c) Magnetically actuated, gallium cutoff valve.

sufficient vapor pressure. Suggestions for accomplishing this may be found in Section 18.2

Heated Inlet Systems

Heated inlet systems extend the usefulness of mass spectrometry to polar materials, which tend to be adsorbed on the walls of the chambers at room temperature, and to less volatile compounds insofar as they have a vapor pressure on the order of 0.02 torr at the temperature of the sample reservoir, usually 200 °C. The temperature is limited by the materials of construction and by thermal degradation of the sample. Above 200 °C most compounds that contain oxygen or nitrogen are thermally decomposed.

The Molecular "Leak"

From the inlet system the gases diffuse through a molecular leak into the ion source. The leak is a pinhole restriction about 0.013–0.050 mm in diameter in a gold foil. The preferred type of flow into the ion source depends on the purpose for which the instrument is intended. For analytical work, conditions for molecular flow are usually used in which collisions between molecules and the walls are much more frequent than collisions between molecules themselves. In isotope studies viscous flow is preferred in which a gas molecule is more likely to collide with other gas molecules than with the surfaces of the container; thus, there is no tendency for various components to flow differently from the others and adversely affect the relative concentrations of an isotope cluster. Viscous flow occurs when the mean free path is smaller than the tube diameter, and molecular flow when it is greater than the tube diameter.

In on-stream analysis the sample must be admitted to the instrument at or near atmospheric pressure. Consequently, it is necessary to drop the pressure by a viscous flow system to a range in which molecular flow can be achieved by the use of leak perforations of reasonable size. In one system, gas is drawn by an auxiliary mechanical pump through a pair of viscous leaks. The leaks are so proportioned that the pressure intermediate between them is about 1 torr. The sample is admitted to the mass spectrometer through a perforated foil from the region of intermediate pressure between the two leaks.

16.3 IONIZATION METHODS IN MASS SPECTROMETRY[1]

Ionization methods in mass spectrometry are divided into two categories. First, in gas-phase ionization techniques, the analyst is dealing with those substances that are volatile or volatilizable via specific, quantitative derivatization procedures. The sample is vaporized outside the ion source. This is followed by ionization in the gas phase using unimolecular (electron impact or field ionization) or bimolecular (chemical ionization) methods. Combinations of ion sources offer dual capabilities within one source housing (Figure 16.5). Second, in desorption techniques (field desorption, ^{252}Cf desorption, fast-atom or ion bombardment, and laser desorption), ions are formed from samples in the condensed phase inside the ion source. The thermospray method is discussed in Section 19.9.

Ion sources have the dual function of producing ions without mass discrimination from the sample and accelerating them into the mass analyzer with a small spread of kinetic energies prior to acceleration. Double-focusing mass analyzers must be used with sources that produce a large spread of energies in order to obtain sufficient resolution.

In all source designs there must be an ion withdrawal and focusing system, as shown in Figure 16.3, in which the ions are removed electrostatically from the chamber and accelerated toward the mass analyzer. Several pairs of focusing elements and slits then control the direction, shape, and width of the ion beam.

Electron-Impact Ionization

The electron-impact ion source is the most commonly used and highly developed ionization method (Figure 16.3). Once past the molecular leak, the neutral molecules are in a chamber that is maintained at a pressure of 0.005 torr and a temperature of 200 ± 0.25 °C.

An electron gun is located perpendicular to the incoming gas stream. Electrons emitted from a glowing filament are drawn off by a pair of positively charged slits through which the electrons pass into the body of the chamber. An electric field maintained between these slits accelerates the electrons. The ionizing electrons from the cathode are formed into a tight helical beam by a small magnetic field, on the order of 100 G, which is confined in the ionization region. The number of electrons is controlled by the filament temperature, whereas the energy of the electrons is controlled by the filament potential. Ions are formed by the exchange of energy during the collision of the electron beam and sample molecules. This results in a Franck-Condon transition producing a molecular ion, which is usually in a high state of electronic and vibrational excitation.

The filament potential is variable. A range of 6–14 V is used in molecular weight determinations when it is desirable to avoid fragmentation. A source operating at 70 V, the conventional operating potential, provides sufficient energy to ionize and cause the characteristic fragmentation of sample molecules. Fragmentation can provide positive identification of an unknown. At 70 V the appearance of the spectrum is nearly independent of the electric field, and reproducibility for quantitative work is thereby obtained.

The positive ions formed in the ionization chamber are drawn out by a small electrostatic field between the large repeller plate (charged positive) behind them and

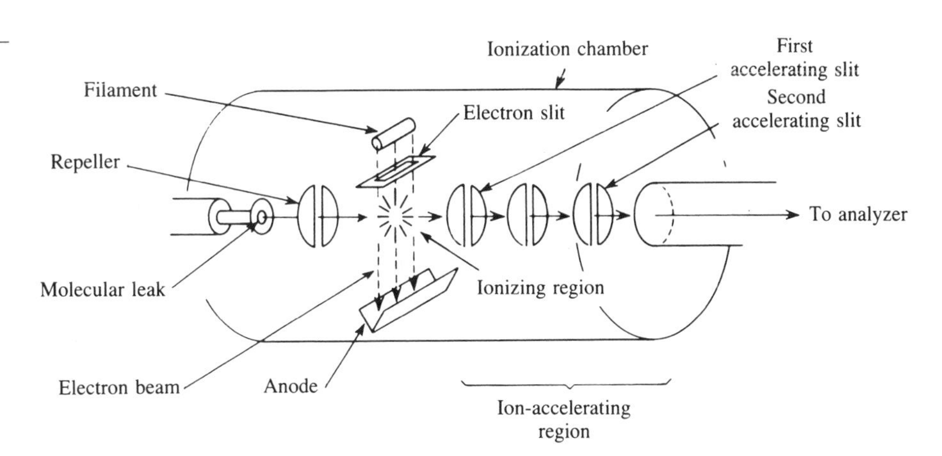

FIGURE 16.3
Electron-impact ion source and ion-accelerating system.

the first accelerating slit (charged negative) ahead of them (Figure 16.3). The positive charge on the repeller plate does not affect the un-ionized molecules as they enter the ionization chamber. A strong electrostatic field between the first and second accelerating slits of 400–4000 V accelerates the ions of masses m_1, m_2, m_3, and so on, to their final velocities. The ions emerge from the final accelerating slit as a collimated ribbon of ions with velocities and kinetic energies given by

$$zV = \tfrac{1}{2}m_1v_1^2 = \tfrac{1}{2}m_2v_2^2 = \tfrac{1}{2}m_3v_3^2 = \cdots \quad (16.1)$$

where z is the charge of the ion.

Chemical Ionization[2]

Chemical ionization results from ion-molecule chemical interactions involving a small amount of sample with an extremely large amount of a reagent gas.

Positive Chemical Ionization. A two-part process occurs. In the first step a reagent gas (methane in our example) is ionized by electron-impact ionization in the source:

$$\text{[electron impact]}\ CH_4 + e^- = CH_4^+ + 2e^-(CH_3^+, CH_2^+, \ldots) \quad (16.2)$$

The electron energy must be 200–500 V to ensure penetration of the ionization electrons into the active volume. Primary reagent ionization is followed by second-order processes in which the primary ion reacts with additional reagent gas molecules to produce a stabilized reagent ion plasma:

$$\text{[secondary ions]}\quad CH_4^+ + CH_4 = CH_5^+ + CH_3 \quad (16.3)$$

$$CH_3^+ + CH_4 = C_2H_5^+ + H_2 \quad (16.4)$$

The second part of the chemical ionization process occurs when a reagent ion (CH_5^+ or $C_2H_5^+$) encounters a sample molecule (MH). The reagent ion and sample molecule react via any of several modes:

$$\text{[proton transfer]}\quad CH_5^+ + MH = CH_4 + MH_2^+ \quad (16.5)$$

$$\text{[hydride abstraction]}\quad CH_3^+ + MH = CH_4 + M^+ \quad (16.6)$$

$$\text{[charge transfer]}\quad CH_4^+ + MH = CH_4 + MH^+ \quad (16.7)$$

Negative Chemical Ionization.[3] Negative chemical ionization is the counterpart of positive chemical ionization. Typical negative ion-forming reactions are:

$$\text{[resonance capture of electron]}\quad M + e^- = M^- \quad (16.8)$$

$$\text{[dissociative capture of electron]}\quad RCl + e^- = R^{\bullet} + Cl^- \quad (16.9)$$

$$H_2O + e^- = H^{\bullet} + OH^- \quad (16.10)$$

Also ion-molecule reactions occur between negative ions formed in the ion source and neutrals. These include charge transfer, hydride transfer, and anion-molecule adduct formation. The reactions of the hydroxide ion are mimicked by methoxide and amide reagents. Fluorocarbons produce the fluoride ion.

Reagent Gases. Different reagent gases are specific for certain functional groups and provide excellent control of sample ion fragmentation. When the interest is primarily to determine the molecular weight of a compound or to confirm it as one of a small set of compounds, a low-energy reactant, such as *tert*-$C_4H_9^+$ (from iso-C_4H_{10}), is frequently used. An even weaker protonating agent such as NH_4^+ generates $(M + H)^+$ and $(M + NH_4)^+$ ions, useful in characterizing polyhydroxy compounds like sugars. Deuterium oxide is used to determine the presence of active hydrogen. Oxygen and hydrogen are used as reagent gases in negative ion chemical ionization–mass spectrometry. Competition between localized chemical-ionization-induced reactions at various sites in a molecule produces structural information that is often absent from the electron-impact spectrum. For example, with NO as the reagent gas, all alcohols give ions at $(M - 17)$, but only primary and secondary alcohols give additional ions at $(M - 1)$ and $(M + 30 - 2)$.

Argon, helium, and nitrogen, as reagent gases, produce fragmentation patterns essentially identical to electron-impact ionization but with increased sensitivity. Helium, because of its use as a carrier gas in gas-liquid chromatography, has been used as a reagent gas for charge exchange reactions (illustrated by Equation 16.7).

Chemical Ionization Unit. For an effective *chemical ionization unit* the basic physical requirements are a tightly enclosed source housing, a high-speed pumping system, and a differential pumping barrier between the source and analyzer regions (Figure 16.4). This allows pressures inside the source to reach 0.5–4.0 torr, while the

FIGURE **16.4** Combination chemical ionization (CI) and electron-impact (EI) ionization source. (Courtesy of Varian Associates.)

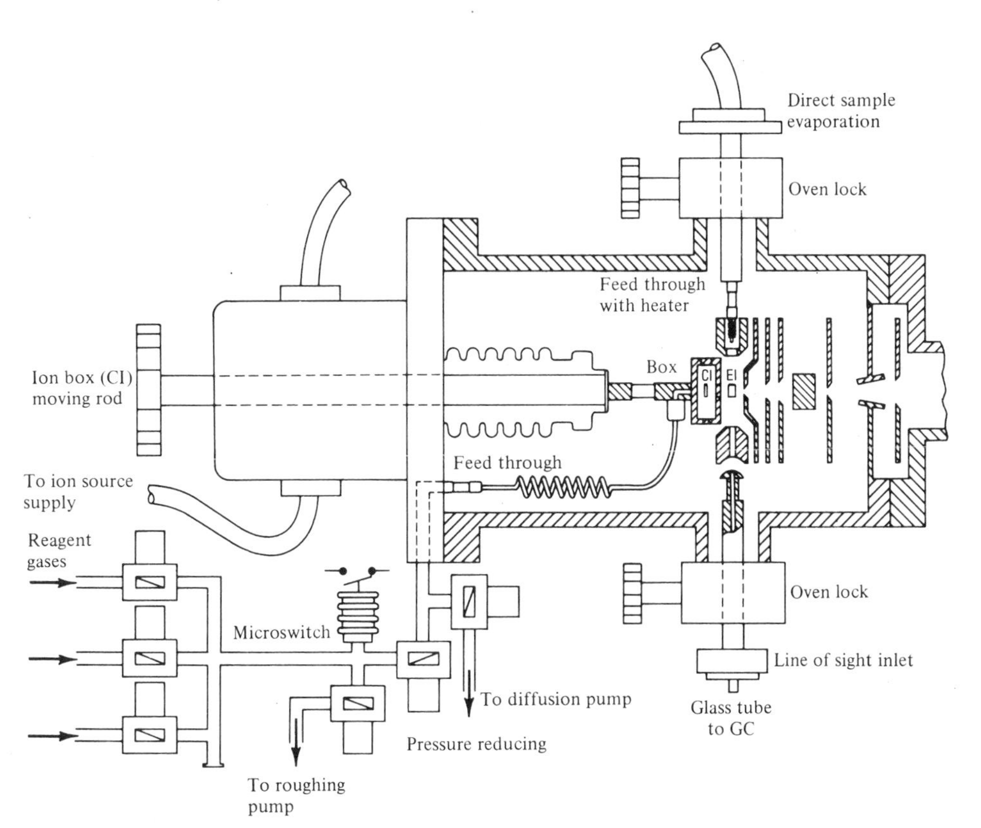

pressures outside the source are about four orders of magnitude less. With the unit illustrated in Figure 16.4, electron-impact spectra are obtained by simply turning off the reagent gas and admitting the sample at normal electron-impact pressures.

Field Ionization[4]

The application of a very strong electrical field induces the emission of electrons. When a molecule is brought between two closely spaced electrodes in the presence of a high electrical field (10^7–10^8 V/cm), it experiences an electrostatic force similar to that on the plates of a charged condenser. If the metal surface (anode) has the proper geometry (a sharp tip, cluster of tips, or a thin wire) and is under high vacuum (10^{-6} torr), this force can be sufficient to remove an electron from the molecule without imparting much excess energy. In the absence of electron collisions there is little or no fragmentation of the ion. Sensitivities are an order of magnitude less than those of electron-impact ionization.

In a field ionization source shown in Figure 16.5, a thin needle of a few micrometers diameter constitutes the anode that is located 1 mm away and immediately behind the exit slit of the ion chamber. The exit slit serves as the cathode and opposite field-forming member. The remainder of the source consists of the focusing slits common to all ion sources. The field ionization unit can be combined with the electron-impact source in a single assembly.

Field Desorption[5,6]

In field desorption a wire (similar to that used in field ionization) is dipped into or has deposited on it a solution of the sample under study. The emitter wire is then heated with an electric current while it is at the high-voltage conditions used in field ionization. The material is evaporated into a source that may be a chemical ionization plasma or an electron-impact unit. Due to the nonhomogeneity of the wire size and directional orientation, the surface ions generated during the field desorption have considerable energy distribution and thus spread over a small angle.

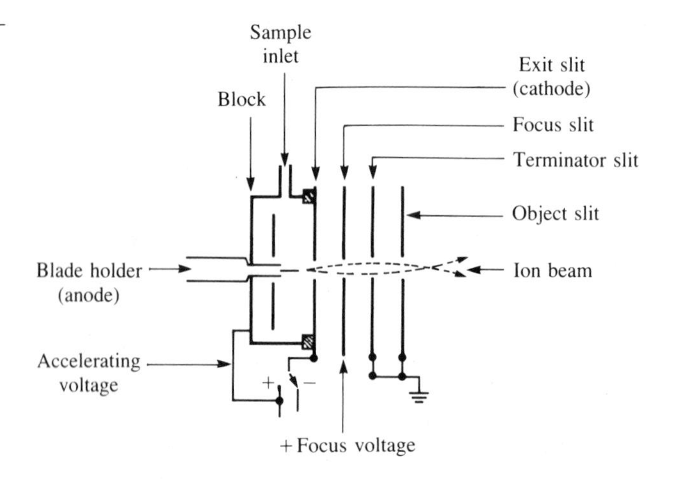

FIGURE **16.5**
Field ionization source.

Fast Atom Bombardment[7–9]

The fast atom bombardment equipment is shown schematically in Figure 16.6. Usually a relatively high pressure of argon gas is placed between the ionizer and the sample. Argon gas is ionized by a hot filament-type ion source, then accelerated in an electrostatic field, and focused into a beam that bombards the sample. The Ar^+ provides high kinetic energy as well as undergoing charge exchange with the solvent in which the sample is dissolved. Without the intentional collisions, the sample would be primarily bombarded by ions as in secondary ion mass spectrometry (see Section 16.14).

Solvents, such as glycerol, monothioglycerol, carbowax, or 2,4-dipentylphenol, easily dissolve organic compounds yet do not evaporate in a vacuum. Since a monolayer of material is completely sputtered in a matter of seconds, it is essential that the sample surface be continuously renewed. The bombarding argon beam ionizes the solvent—for example, glycerol (G). Glycerol ions react with the surrounding glycerol molecules to produce $(G_n + H)^+$ as reactant ions analogous to the chemical ionization method discussed in an earlier section. The sample undergoes proton transfer, hydride transfer, or ion-pair reaction with the reactant ion $(G_n + H)^+$ to produce pseudomolecular ions such as $(M + H)^+$, $(M - H)^-$, and $(M + G + H)^+$. These ions are then extracted by a slit lens system designed to collect ions emitted approximately normal to the sample surface and direct them into the mass analyzer.

Ion Bombardment

The incident beam of monoenergetic noble gas ions from an ion gun (often using argon) is directed vertically onto a properly positioned sample. When the ion beam strikes the surface, secondary ions are released. If monoatomic, these secondary ions are analyzed and detected as such. Molecular ions can dissociate into positive and negative ions. The secondary ion mass spectrum (SIMS) therefore consists of both

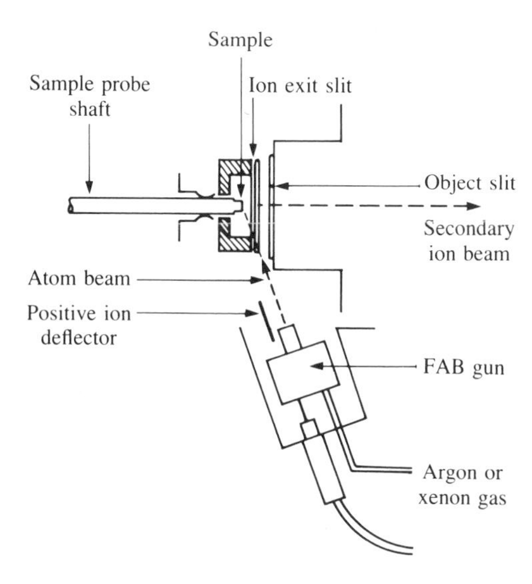

FIGURE **16.6**
Schematic diagram of the fast atom bombardment (FAB) ion source.

those secondary ions that are stable to such dissociation and the ionic fragments of those that are not.

The ion source consists of a cylindrical grid with an external filament. Ions formed by electron bombardment of a noble gas inside the grid are extracted axially from one end and focused on the target by an electrostatic aperture lens system. The duoplasmatron, shown in Figure 16.24 but minus the magnetic field, is one type of ion source. The ion gun produces a nominal 1-mm-diameter ion beam throughout an operating range of 300–3000 eV. Beam diameter can be decreased in steps to a minimum of 0.1 mm. Such convenient beam selection permits the rapid transition from localized surface analysis with the small beam to a large-area, higher sensitivity analysis where spatial resolution is not critical.

Plasma Desorption[10]

Plasma desorption produces molecular ions from samples coated on a thin foil when a highly energetic fission fragment from californium-252 "blasts through" from the opposite side of the foil (Figure 16.13). The fission of the californium-252 nucleus is highly exothermic, and the energy released is predominantly carried away by a wide range of fission fragments, which are heavy atomic ion pairs. The ion pair fission fragments depart in opposite directions. A single fission fragment passing through a thin film can desorb many positive and negative ions, electrons, and up to 2000 neutral molecules. It is the interaction of these fission fragments as they pass through matter that forms the basis of the plasma desorption process. According to the theory of the interaction of high-energy, charged heavy ions with matter, atoms and molecules of the sample matrix perceive the swiftly moving ion passing through the matrix as equivalent to a short burst of photons. The electromagnetic radiation induces electronic excitation, which quickly dissipates into a complex spectrum of atomic, molecular, and matrix excitation.

Laser Desorption[11–14]

Laser desorption methods involve the interaction of a pulsed laser beam with the sample to produce both vaporization and ionization. The *microprobe technique* uses a laser beam that is focused to a very small spot on the back side of a thin metal foil that holds a thin film of sample. Ions emerge on the front side from a small cratered hole in the foil. The *bulk analysis technique* uses less focused beams and larger samples. The laser beam produces a microplasma that consists of neutral fragments together with elementary molecular and fragment ions, largely protonated/deprotonated species that have predominantly unit charge. Since the laser pulse lasts only a few microseconds, suitable mass analyzers are limited to time-of-flight and Fourier transform spectrometers.

Thermal (Surface) Ionization

The thermal or surface ionization source is useful for inorganic solid materials. Samples are coated on a tungsten ribbon filament and then heated until they evaporate. When an atom or molecule is evaporated from a surface (at approximately 2000 °C) it has a certain probability of being evaporated as a positive ion. This probability is predictable and is a function of the ionization potential, E_i, of the sample and the work function, ϕ, of the filament material. The relationship for the ratio of ions, n^+, to neutral species, n^0, is given by the Langmuir-Saha equation:

$$\frac{n^+}{n^0} = \exp\left[\frac{z(\phi - E_i)}{kT}\right] \tag{16.11}$$

where z is the electronic charge.

Summary of Ionization Methods

Little or no fragmentation is desirable when an analysis of a mixture of compounds is needed and the list of possible components is limited. On the other hand, fragmentation can provide positive identification of an unknown. Some compounds do not give a molecular ion in an electron-impact source because of the excess ionization energy imparted to the molecule during the ionization step. This is a disadvantage of the electron-impact source.

The general absence of carbon-carbon cleavage reactions from the chemical ionization spectra means that they provide little skeletal information. The small amount of fragmentation provides a sensitivity increase up to 100 times for the chemical ionization process because most of the ionization is concentrated in the molecular ion. Sensitivity is enhanced even more by high cross sections for the chemical ionization process and long ion residence times. To provide unambiguous confirmation of the sample identity, required in library searches, it would be desirable to use chemical ionization/electron-impact ionization alternately (Figure 16.7). In general, chemical ionization spectra are intermediate in their extent

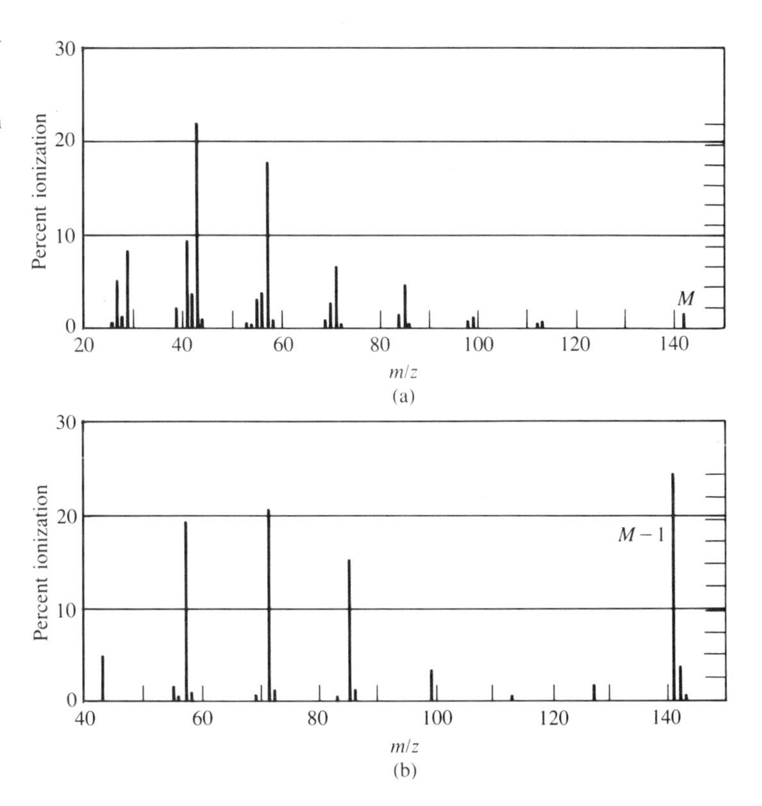

FIGURE 16.7
(a) Electron-impact spectrum compared with (b) chemical ionization spectrum.

of fragmentation between the very simple field ionization spectra and the more complex electron-impact spectra. Both positive and negative chemical ionization are well suited for the analysis of polar compounds in the lower mass range, which can be adjusted up to at least 1000 daltons (amu). Spectra contain abundant ions related to molecular weight as well as fragment ions eminently suitable for structure determination. Background is generally lower than that produced by secondary ion mass spectrometry (SIMS; see Section 16.14) or fast ion bombardment (FAB) ionization. Chemical ionization produces very sensitive quantitative analysis in the femtomole to picomole sample range.

Field desorption is a valuable technique for studying surface phenomena, such as adsorbed species and trapped samples, and the results of chemical reactions on surfaces. It is also an appropriate method for large lipophilic polar molecules.

When working with polar substances (usually of higher molecular weight) and salts, problems arise. These materials may be unstable when heated to the temperature at which they vaporize and may undergo molecular change when derivatization procedures are attempted. Field desorption, californium-252 desorption, fast atom or ion bombardment, and laser desorption techniques deal rather effectively with these classes of substances. Samples may be bulk solids, liquid solutions, thin films, or monolayers. The available energies range from approximately 0.1 eV (field desorption) through thousands of electron volts (fast atom bombardment and secondary ion mass spectrometry) to millions of electron volts (^{252}Cf plasma and laser pulses).

In the fast atom bombardment method the even-electron species $(M + \mathrm{H})^+$ and $(M - \mathrm{H})^-$ predominate, whereby the sample forms a stable ion by the addition or loss of a proton. Molecules that are alkali metal salts form similar stable species by the addition or loss of the alkali metal cation. There is also a significant tendency to produce metastable ions. The use of a liquid matrix makes it possible to ionize large polar molecules simply and without any prior chemical derivatization.

Thermal (surface) ionization is appropriate for inorganic compounds that generally have low ionization potentials (3–6 eV). Surface ionization is especially useful in determining isotope ratios in inorganic compounds for geochemical applications or in studies of elements involved in nuclear chemistry. There is no ionization of the background gases in the mass spectrometer. On the other hand, surface ionization is inefficient for organic compounds whose ionization potentials usually lie in the range 7–16 eV.

16.4 MASS ANALYZERS

The function of the mass analyzer is to separate the ions produced in the ion source according to their different mass/charge ratios. After leaving the ion source, the ions are accelerated by a system of electrostatic slits (Figure 16.3) and then they enter the particular analyzer. For some purposes the analyzer should be able to distinguish between minute mass differences, whereas discrimination between integral mass numbers suffices for other applications. Resolving power is discussed in the last portion of this section.

Single-focusing mass analyzers focus ions of different mass/charge ratios in different directions—that is, spread out in the plane of the exit slit. Since the accelerating potential experienced by an ion during its transit of the accelerating slits depends on where in the source it is formed, there is a lack of uniformity in the kinetic energy of particles of a given species as they leave the ion source. In a single-focusing mass analyzer, these variations cause a broadening of the ion beam that reaches the collector and a loss of resolving power. To achieve high resolution, tandem mass spectrometers are required.

Magnetic-Deflection or Sector Mass Analyzer

The magnetic-deflection or sector-type system involves a magnetic field that causes ions to be deflected along curved paths. Initially, a set of electrostatic slits accelerates the ions from the source, forms them into a narrow beam, and directs them toward the magnetic field. The velocity of the ions is controlled by the potential, V, applied to the slits. Subsequently the ions are diverted into circular paths by a magnetic field parallel to the slits and perpendicular to the ion beam. A stable, controllable magnetic field, H, separates the components of the total ion beams according to momentum. By this means, the individual ion beams are separated spatially and each has a unique radius of curvature (ion trajectory), r, according to its mass/charge (m/z) ratio. Only ions of a single m/z value will have the proper trajectory leading to the exit slit ahead of the detector. By changing the magnetic field strength, ions with differing m/z values are brought to focus at the detector slit.

The ion velocity, v, in the magnetic field is given by the equation

$$\tfrac{1}{2}mv^2 = zV \tag{16.12}$$

or, solving for the velocity term,

$$v = \sqrt{\frac{2zV}{m}} \tag{16.13}$$

As the ions enter the magnetic field, they experience a force at a right angle to both the magnetic lines of force and their line of flight. Equating the centripetal and the centrifugal forces,

$$\frac{mv^2}{r} = Hzv \tag{16.14}$$

Or, rearranging the equation, the radius of curvature of the flight path is proportional to its momentum and inversely proportional to the strength of the magnetic field:

$$r = \frac{mv}{zH} \tag{16.15}$$

Eliminating the velocity term between Equations 16.13 and 16.15, we get

$$r = \frac{1}{H}\sqrt{2V\left(\frac{m}{z}\right)} \tag{16.16}$$

Ions accelerated through a uniform electrostatic field and then deflected through a uniform magnetic field therefore have different radii of curvature of their orbits.

Only those ions that follow the path that coincides with the arc of the analyzer tube in the magnetic field are brought to a focus on the exit slit where the detector is located. Ions with other mass/charge (m/z) ratios strike the analyzer tube (which is grounded) at some point, are neutralized, and are pumped out of the system along with all other nonionized molecules and uncharged fragments. Thus, the magnetic field classifies and segregates the ions into beams, each with a different m/z ratio, where

$$\frac{m}{z} = \frac{H^2 r^2}{2V} \tag{16.17}$$

To obtain the mass spectrum, either the accelerating voltage or the magnetic field is varied. Each m/z ion from light to heavy is successively swept past the detector slit at a known rate. Development of the fully laminated magnet coupled with an increased sector radius and correction for curvature aberrations and image rotation has enabled longer slits to be used to obtain high sensitivity while maintaining a specified resolution. Both fast scan rates and high mass range can be achieved.

In sector instruments the ion source, the collector slit, and the apex of the sector-shaped magnetic field are collinear, as shown in Figure 16.8 for the 60°-sector instrument. Notice that the ion source and collector are completely removed from the magnet region. This isolation permits the use of unusual and diverse ion source constructions, and the use of conventional electron multipliers for ion detection. Since the ion transit time from the accelerator slits to the magnetic field is significant, peaks that result from metastable transition products are often observed.

Double-Focusing Sector Spectrometers

In a single-focusing magnetic sector instrument there is a lack of uniformity of ion energies, since the accelerating potential experienced by an ion depends on where in the source it is formed. The resulting spread in ionic energies produces a spread in their radii of curvature in the magnetic field. The result is peak broadening and low to moderate resolution. Magnetic/electrostatic sector instruments use magnetic and electric fields to disperse ions according to their momentum and translational energy. An electrostatic deflection field is incorporated between the ion source and the mass analyzer. Ions are accelerated out of the source and are collimated into a

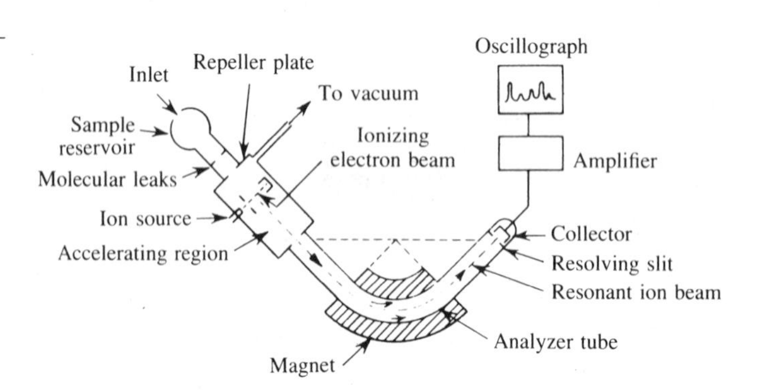

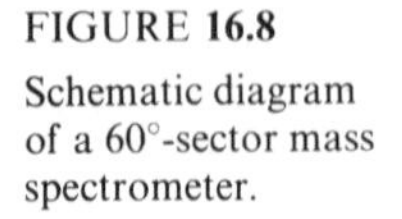

FIGURE **16.8**
Schematic diagram of a 60°-sector mass spectrometer.

narrow beam by a set of slits. As the ions pass through the electrostatic sector, they are dispersed according to their translational energy. Only those ions that have the correct translational energy pass through the slits at the end of the electrostatic sector. Finally, the magnetic sector disperses the ions according to their momentum. A mass spectrum is obtained by scanning the magnetic field strength to bring ions with different m/z ratios sequentially to focus at the detector.

The Mattauch-Herzog geometry involves an electrostatic sector with 31° 50′ angular deflection followed by a beta slit that determines the energy bandpass of the energy filter (Figure 16.9).[15] This geometry uses a cylindrical electrostatic analyzer as an energy filter that allows a band of energies to pass into the magnetic sector. Ions are focused thereby for both velocity and direction. Focusing is accomplished by acceleration or deceleration of the ions as they enter the electrostatic field. Positive ions that travel closer to the positive plate are slowed down, whereas those that travel closer to the negative plate are accelerated. Following the intermediate slit, a monitor assembly is allowed to intercept a fixed fraction of the total ion beam. The ions then enter a homogeneous magnetic field, are deflected through 90°, and come to a focus along a plane that is nearly parallel to the exit face of the magnet. The ions that form this focal image have been separated according to their m/z ratio, as dictated by Equation 16.15 and

$$v = \sqrt{\frac{2zV}{m}} \pm \Delta v \tag{16.18}$$

The mass spectral resolution depends on the effect of the spread in the velocity, Δv, being small compared with the effects of the accelerating voltage and magnetic field strength. For this reason the accelerating voltage is kept high (20–25 kV), as is the magnetic field strength (15 kG). At these values, ions of mass 6–240 daltons are displayed simultaneously along a focal plane 25 cm long. The Mattauch-Herzog geometry lends itself to the use of image-forming detectors, such as channel electron multiplier arrays. Resolution is at least 20,000.

Another version of a double-focusing mass spectrometer is shown in Figure 16.10. The ideal shapes and arrangements of the electrostatic and magnetic fields were determined by computer calculation of the ion beam trajectory, taking

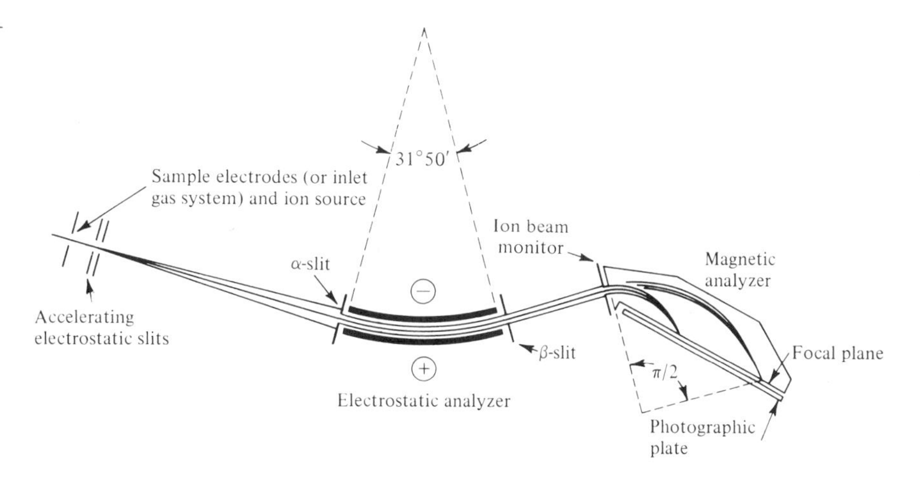

FIGURE **16.9**
Schematic diagram of a double-focusing mass spectrometer showing basic Mattauch-Herzog geometry.

into account the fringing field. Thus aberrations are eliminated from the ion optical geometry. Three-dimensional focusing is achieved with a quadrupole lens placed between the electrostatic and magnetic fields. A quadrupole lens is also placed in front of the electrostatic analyzer. The detector is a secondary electron multiplier. In this configuration, ions of only one m/z ratio are sharply in focus at any given combination of field strengths.

Quadrupole Mass Analyzer

A quadrupole field is formed by four electrically conducting, parallel rods (Figure 16.11). Opposite pairs of electrodes are electrically connected. One diagonally opposite pair of rods is held at $+U_{dc}$ volts and the other pair at $-U_{dc}$ volts. A rf oscillator supplies a signal to the first pair of rods that is $+V \cos \omega t$ and a rf signal retarded by 180° ($-V \cos \omega t$) to the second pair. The equipotential surfaces in the region between the four rods appear as oscillating hyperbolic potentials.

Ions from the ion source are injected into the quadrupole array through a circular aperture. A range of energies can be tolerated. As the ions proceed down the longitudinal z-axis, they undergo transverse motion in the x- and y-planes perpendicular to the longitudinal axis. The dc electric fields tend to focus positive ions in the positive plane and defocus them in the negative plane. As the superimposed rf field becomes negative during part of the negative half-cycle of the alternating field, positive ions are accelerated toward the electrodes and achieve a substantial velocity. The following positive half-cycle has an even greater influence on the motion of the ion, causing it to reverse its direction (away from the electrode) and accelerate even more. The ions exhibit oscillations with increasing amplitudes until they finally collide with the electrodes and become neutral particles. The lighter

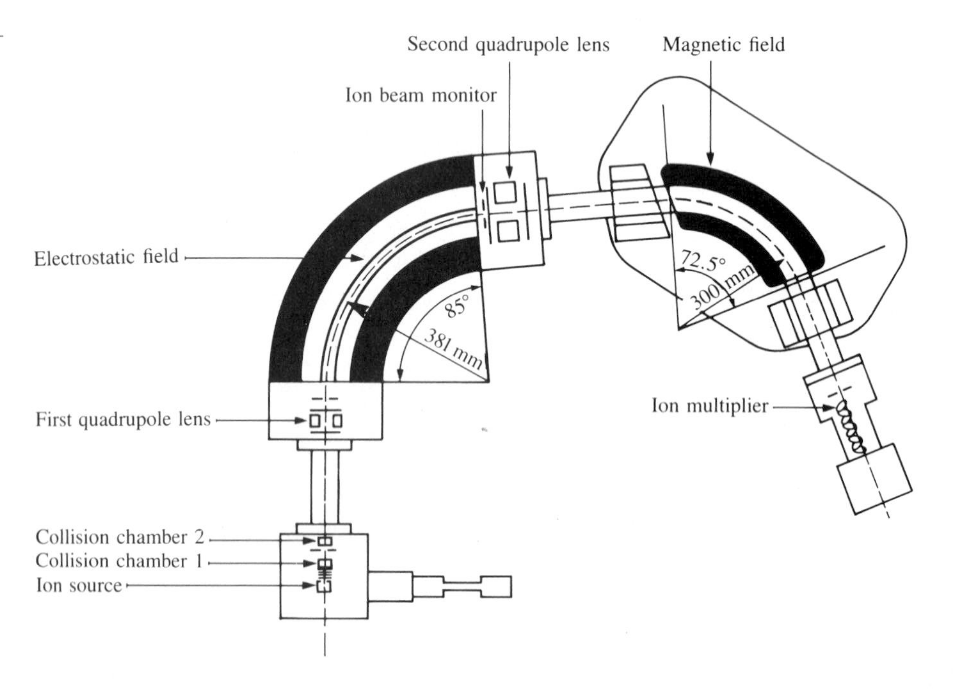

FIGURE **16.10**
Ultrahigh-resolution double-focusing mass spectrometer. (Courtesy of Jeol, Ltd.)

the ion in mass, the smaller the number of cycles before it is collected by the electrode.

By controlling the ratio V_{dc}/V_{rf}, the field can be established to pass ions of only one m/z ratio down the entire length of the quadrupole array. For example, an ion of mass 800 daltons with an energy of 10 eV requires 129 μsec to pass through a 20-cm quadrupole filter. By simultaneously ramping the dc and rf amplitudes, ions of various m/z ratios are allowed to pass through the mass filter to the detector and an entire spectrum can be produced. Using an intuitive approach, if one scans at a rate such that the transmission character of the mass filter changes by no more than 0.1 dalton during the transit time of a particular ion, a maximum scan rate of 780 daltons/sec can be reached before resolution, peak shape, and intensity are significantly sacrificed. Under these conditions 90% of the ion current at m/z 800 is transmitted to the detector.

The quadrupole analyzer is not restricted to the detection of monoenergetic sources. Ions are accepted within a 60° cone around the axis. The quadrupole analyzer therefore does not require focusing slits, which results in higher sensitivity. The resolution is a function of the number of cycles an ion spends in the field. Increasing the rod length (usually 5–20 cm) increases the resolution and the capability to handle ions of higher energies. If the rf frequency is increased, the length of the analyzer can be reduced. Rod diameters are also a factor: increasing the rod diameter increases the sensitivity by a large factor, whereas decreasing the diameter increases the mass range.

The quadrupole mass analyzer is well suited for the registration of negative ions, since the analyzer does not discriminate between the polarity of the ions. Simultaneous pulsed positive and negative ions produced in a conventional chemical ionization source are alternately pulsed from the source, with the appropriate potentials, through a quadrupole analyzer to two electron multipliers, one for positive and one for negative ions. For collection of negative ions, the first

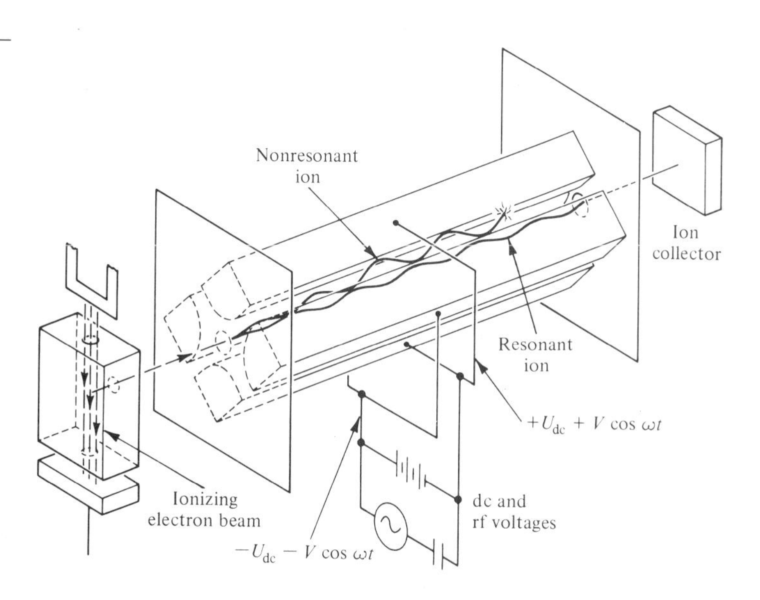

FIGURE **16.11**
Quadrupole mass analyzer.

dynode of the multiplier must be supplied with a positive potential. The advantage here is that some compounds have negative quasimolecular ion spectra that are one to three orders of magnitude more intense than their positive ion spectra.

Time-of-Flight Spectrometer

The time-of-flight (TOF) mass spectrometer operates in a pulsed rather than a continuous mode. The ions are formed in the source as a discrete ion packet, initially consisting of ions with a variety of m/z ratios, that is pulsed (via acceleration slits at the exit of the ion source) into the field-free region of the flight tube, 30–100 cm long (Figure 16.12).

The essential principle of time-of-flight mass spectrometry is that if ions with different masses are all accelerated to the same kinetic energy, each ion acquires a characteristic velocity that depends on its m/z ratio. The ion beam reaches drift energy, typically 2700 eV, in less than 2 cm. Upon leaving the accelerating slits, the ions enter and move down a field-free region with whatever velocity they have acquired. Each ion has a kinetic energy, zV, as expressed by Equation 16.12. Because all ions have essentially the same energy at this point, their velocities are inversely proportional to the square roots of their masses. As a result, ions with different m/z ratios spatially separate as they travel down the flight tube. Ions of high velocity (low m/z ratios) speed on ahead and arrive at the detector before the heavier ions of lower velocity (high m/z ratios). Hence the original beam becomes separated into "wafers" of ions according to their mass/charge ratio. The wafers of ions impact sequentially on the flat cathode of the ion detector.

For TOF mass spectrometers, the m/z ratio of an ion and its transit time t (in microseconds) through a flight distance L (in centimeters) under an acceleration voltage V are given by

$$t = L\sqrt{\left(\frac{m}{z}\right)\left(\frac{1}{2V}\right)} \quad \text{or} \quad \frac{m}{z} = \frac{2Vt^2}{L^2} \tag{16.19}$$

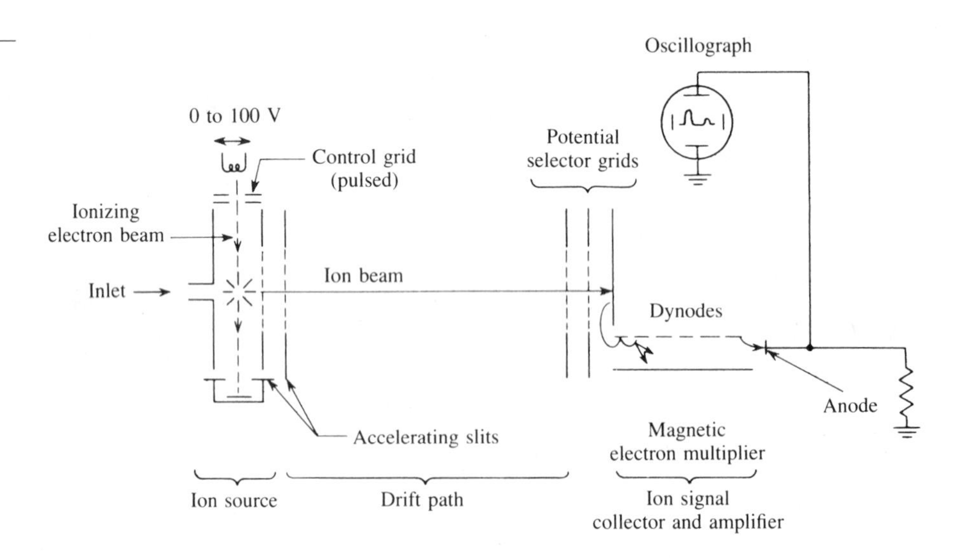

FIGURE **16.12**
Schematic diagram of a time-of-flight mass spectrometer.

Instruments are capable of single-shot detection. With a 100-MHz transient recorder, data can be measured every 10 nsec.

By using two or more ions of known mass in the TOF spectrum as mass calibration points, the exact values of L and V need not be known. The mass calibration is based on the exact conditions of the instrument during the entire period of the measurement. This is not an extrapolation procedure because the measurement is digital rather than analog.

Isotope peaks will be unresolved in plasma desorption measurement (without energy selection prior to analysis by the time interval digitizer). However, this has no effect on the evaluation of the centroid of the distribution, which yields the isotopically averaged m/z value.

Californium-252 Plasma Desorption TOF Mass Spectrometer[10]

The general features of a TOF measurement in conjunction with californium-252 plasma desorption are portrayed in Figure 16.13. The californium-252 source is mounted behind the sample. When a fission event occurs, the two fission fragments are hurled in opposite directions, one toward the sample and the other toward a fission fragment detector. The fission fragment detector produces an electronic pulse that gives a precise time marker for the occurrence of a fission event. An ion detector located 50 cm away and in a line of sight with the surface of the fission track is sensitive to the desorbed ions and photons that are emitted. This sensitivity is enhanced by accelerating the ions toward the detector using an electric field (10–20 kV). Each ion that the desorbed ion detector senses initiates the formation of an electronic pulse that is precisely synchronized with the arrival time of the ion at the detector. This flurry of activity each time a fission track is formed in the sample is followed by a quiescent period of about 50 μsec before another fission track is formed in the sample.

Since the time intervals between the initiation of the fission track and the arrival of the ions at the ion detector can be measured precisely, the time of flight of each ion and hence its velocity and mass are measured; the kinetic energy of each ion is known from the electric field potential between the sample foil and acceleration grid.

The electronic pulse from the desorbed ion detector is used to start a fast electronic clock that "ticks" at the rate of 50 million times per second. The clock circuity determines whether this was the first, second, third, or nth ion to be detected from the current fission track formed in the sample and also how long it took for that ion to travel from the site of the fission track to the secondary ion detector located 50 cm away. The electronic marvel that performs these measurements is the time interval digitizer. Its function parallels that of the magnet in a magnetic sector mass spectrometer. At 10 kV, a m/z ion takes about 30 μsec to make the 50-cm trip. If the ions were monoenergetic and there were no instrumental aberrations, the peak width would be 78 psec (the limit of precision of the time interval digitizer). This corresponds to a mass resolution of 200,000. Since the upper time limit of the digitizer is 320 msec, the theoretical upper mass limit would be a 10^{12}-m/z ion if it could be desorbed and detected. Thus the plasma desorption TOF instrument is excellent for studies of high-molecular-weight biomolecules.

Resolving Power

The most important parameter of a mass analyzer is its resolving power. Mass peaks of ions have no natural linewidth so that the breadth of a peak is representative of the mass analyzer performance. Recorded ion peaks are usually gaussian in shape. Two peaks of equal height, h, separated by a mass difference, Δm, are said to be resolved when the height, Δh, of the valley between them is 10% or less of the peak height. For this condition, $\Delta m/m$ equals the peak width at a height that is 5% of the individual peak height. The resolving power is then given by the value $m/\Delta m$, where m is the mass of the lighter peak. This definition of resolving power is called "10% valley resolution." Occasionally a definition of resolving power based on $\Delta h/h = 0.5$ is used. These two definitions are illustrated in Figure 16.14.

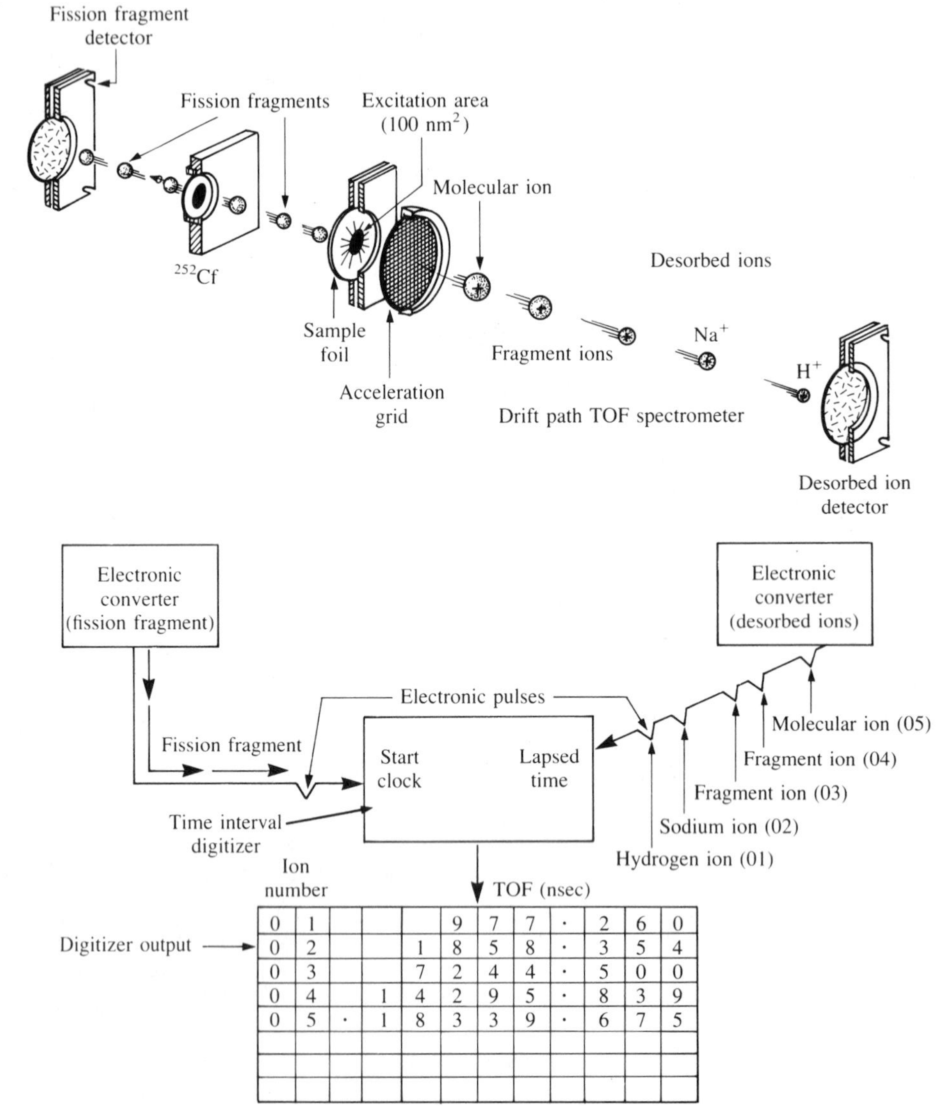

FIGURE **16.13**
Schematic diagram of the ^{252}Cf plasma desorption TOF mass spectrometer and how the data are converted to digital information. [Reprinted with permission from R. D. Macfarlane, *Anal. Chem.*, **55,** 1250A (1983). Copyright 1983 American Chemical Society.]

Mass spectrometric applications govern the required resolving power. A resolution of 1 part in 200 adequately distinguishes between m/z 200 and m/z 201; adjacent peaks in a mass spectrum are generally distinguishable if the intensity ratio is not greater than 10 to 1. The need for high resolution arises when it is necessary to determine accurate masses of molecular or fragment ions of moderate to great complexity. From an accurate mass of the molecular ion, the molecular formula of the compound can often be uniquely determined. For example, the nominal masses of dihydronaphthalene ($C_{10}H_{10}$) and phthalazine ($C_8H_6N_2$) are 130, whereas their exact masses are 130.0783 and 130.0531, respectively. By the 10% valley definition, a resolution of 5159 would be adequate for this mass difference of 0.0252 dalton. To distinguish $CH_2Cl_2^+$ of mass 83.9534 from $CDCl_2^+$ of mass 83.9518, a resolution of 52,500 is necessary.

Comparison of Mass Analyzers

A magnetic sector instrument with medium resolution has a mass range of 2500 at 4-kV ion energy. Mass resolution is continuously variable up to 25,000 (10% valley definition). The 90°-sector instrument is able to record metastable peaks that can aid in structural elucidation. Single-focusing mass spectrometers are directional focusing but not velocity focusing for ions of a given mass, so their resolving power is limited because of the spread in kinetic energy of the ions that leave the acceleration chamber. A considerable improvement in resolving power is achieved with double-focusing sector instruments by passing the ions through an electric field prior to their deflection by the magnetic field. Resolving power lies in the range of 100,000.

The quadrupole mass analyzer is ideal for coupling a gas chromatograph with a mass spectrometer. Rod contamination is minimized by the use of a low-resolution prefilter placed in front of the main quadrupole mass analyzer. This combination of instruments is discussed in Chapter 18. Quadrupole mass analyzers show a linear variation of mass in focus with the quadrupole field voltage so that it is particularly easy to fit such a spectrometer with an automatic mass calibration unit.

The TOF mass analyzer is excellent for kinetic studies of fast reactions and for the direct analysis of effluent peaks from a gas chromatograph. The instrument is unique in its ability to register as molecular ions even those molecular ions that

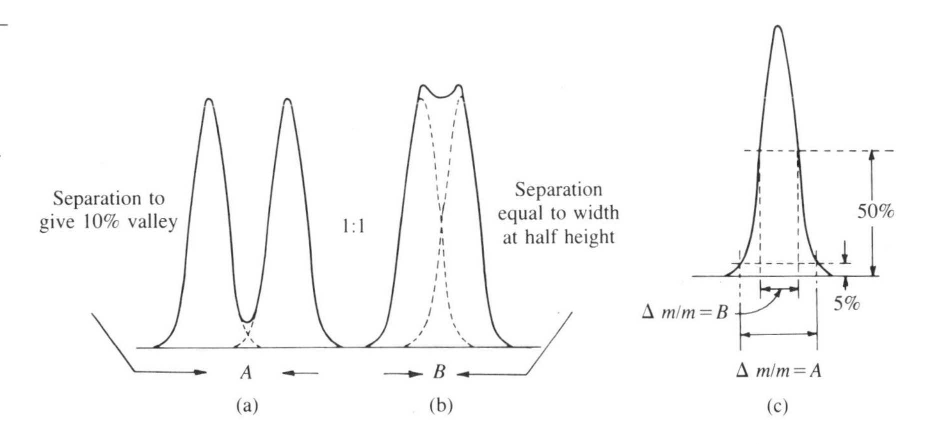

FIGURE **16.14**
(a) Resolution power equal to 10,000 by 10% *valley definition*; that is, m/z signals with separation $\Delta m/m = A$ equal signal width at 5% height points, (c).
(b) Resolution based on *width at half height definition*; that is, m/z values drawn with separation $\Delta m/m = B$ equal signal width at 50% height points, (c).

decompose in the flight tube; thus, sensitivity is not degraded by metastable decompositions of heavy ions. Resolution is limited below 1000 by kinetic energy spreads of the ions. Resolution can be improved by lengthening the flight tube; however , this degrades transmission efficiency and overall sensitivity.

16.5 ION-COLLECTION SYSTEMS

Resolved ion beams, after passage through a mass analyzer, sequentially strike a detector. Several types of detectors are available. The electron multiplier is most commonly used.

Faraday Cup Collector

The Faraday cup collector is a simple and effective means of monitoring ion current in the focal plane of the mass spectrometer. It consists of a cup with suitable suppressor electrodes (to suppress secondary ion emission) and guard electrodes, as shown in Figure 16.15. Currents as low as 10^{-15} A may be detected.

Electron Multiplier

For ion currents less than 10^{-15} A, an electron multiplier is necessary (Figure 16.16). The ion beam strikes the conversion dynode. It is a metal plate that converts impinging ions to electrons. Either positive or negative ions are accelerated by the constant high voltage of the conversion dynode, and they are converted into electrons and/or positively charged ions when they strike the plate. The currents created are then further multiplied by the electron multiplier in either of two configurations. A discrete dynode multiplier has 15 to 18 individual dynodes coated with a metal oxide that has high secondary electron emission properties. The dynodes are arranged in either venetian blind or box-and-grid fashion. Secondary electrons, emitted by the dynodes, are constrained by a magnetic field to follow circular paths, causing them to strike successive dynodes. The magnetic field is produced by a number of small permanent magnets. Continuous dynode multipliers consist of a leaded glass that contains a mixture of metal oxides that is drawn into a hollow tube to form the multiplier channel. The tube is curved to prevent ion feedback. Electrons cascade down the tube attracted by the voltage characteristics established by the inherent resistivity of the glass. For either type of electron

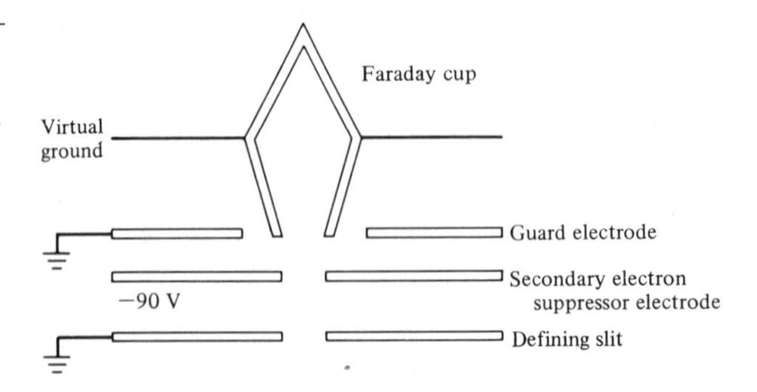

FIGURE 16.15 Schematic diagram of a Faraday cup collector.

multiplier the gain ranges from 10^5 to 10^7. The limiting factor is either the system noise level or the system background.

Channel Electron Multiplier Array

A channel plate is composed of a regular (usually hexagonal) close-packed array of channels in a flat plate of semiconducting material. Typical pore diameters lie in the range 10–25 μm. The length-to-diameter ratio determines the gain characteristics of the device, with a ratio of 40 giving a gain of 10^3 electrons per initial ion. The plate is about 1 mm thick. The inside of each pore, or channel, is coated with a secondary electron emissive material; thus each channel constitutes an independent electron multiplier. To achieve higher gain, two plates can be operated in tandem, with the output of one plate forming the input of the next. A schematic of a channel electron multiplier array is shown in Figure 16.17. Energetic particles enter the first channel plate, where they collide with the wall and produce secondary electrons.

Photographic Plate

A spectrometer of the Mattauch-Herzog or similar design allows the simultaneous focusing of ions in a plane and so the mass spectrum can be recorded on a photographic plate. A photographic plate can give greater resolution than an

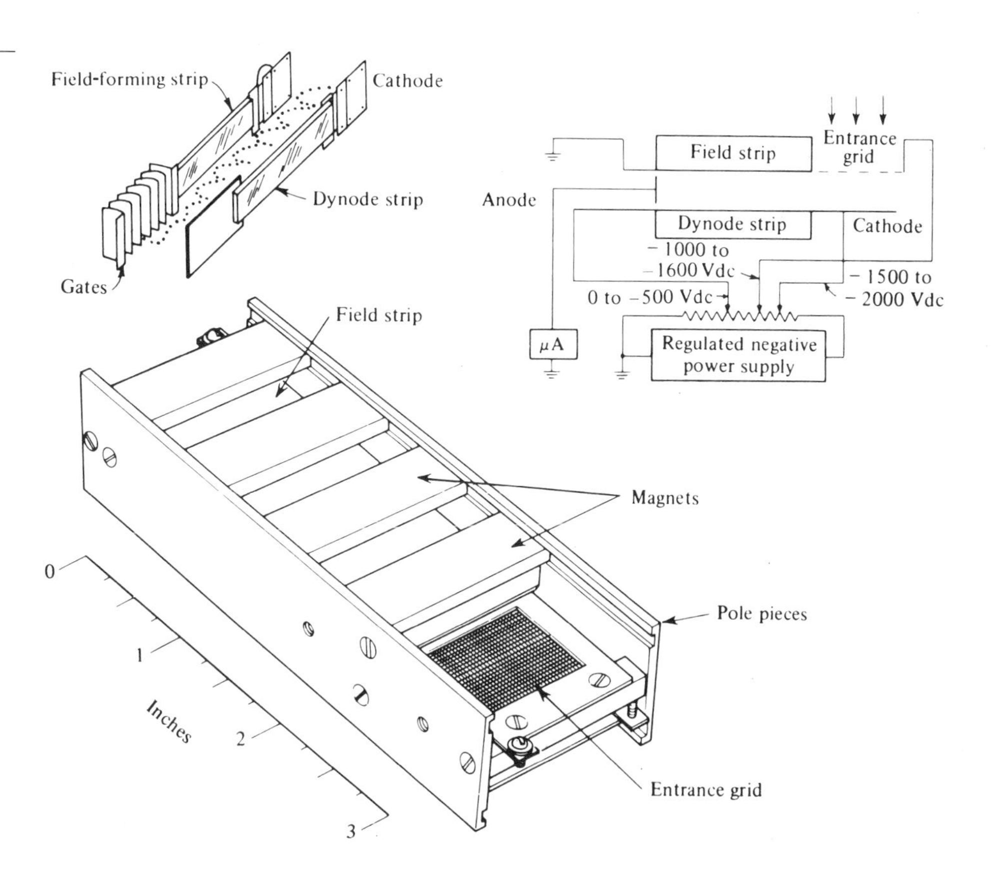

FIGURE **16.16**
Electron multiplier phototube and typical electrical circuit for operating the tube.

FIGURE 16.17 Multichannel electron multiplier arrays. (a) Schematic construction of a multichannel array. (b) Operation of electron amplification. (c) Schematic arrangement of the tandem type. Actually a slice angle (5°, 8°, or 13°) is selected to prevent a primary ion from passing through the channel and to prevent ionic feedback between two plates.

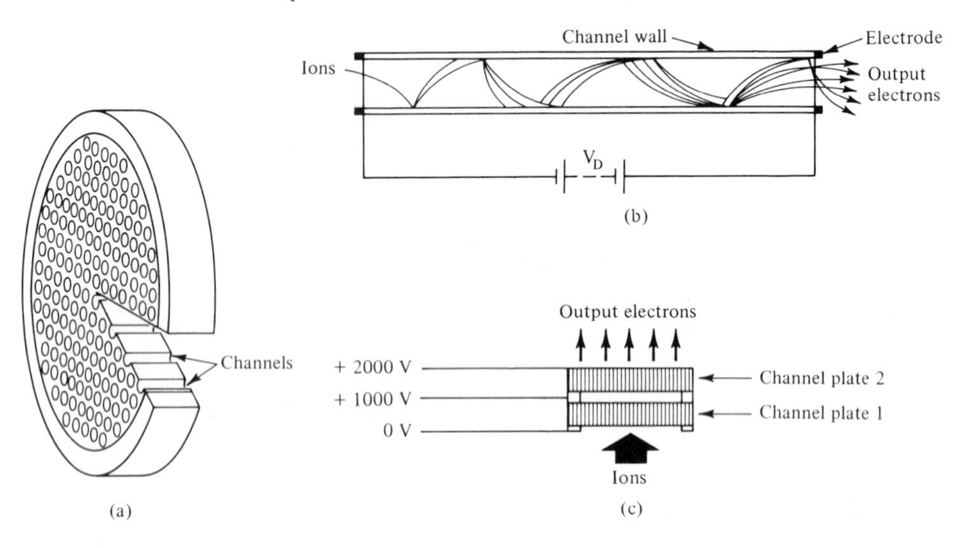

electrical detector. It is more cumbersome, though. Photographic emulsions were discussed in Chapter 10. Because the photographic plate is a time-integrating device, it can provide the highest sensitivity of any detector. Spectra from extremely small samples, samples with low vapor pressure, and ions with a short life can be detected. All of these might be missed with an electrical detector.

16.6 VACUUM SYSTEM

For the operation of a mass spectrometer, the ion source, the mass analyzer, and the detector must be kept under high vacuum conditions of 10^{-6}–10^{-7} torr (1.3×10^{-4} to 1.3×10^{-5} Pa). Both the speed at which the instrument is operational after cleaning or opening to atmospheric pressure and the efficiency of maintaining a high vacuum are related to the capacity and speed of the vacuum system. Most systems use a combination of oil diffusion pumps to maintain a high vacuum and backing rotary pumps to reduce the initial pressure to approximately 0.001 torr. However, oil diffusion pumps are being replaced more and more by turbomolecular pumps. Turbomolecular pumps contain no working fluid; the pumping effect is purely mechanical. Therefore background spectra are practically nonexistent and thus eliminate problems due to accidental venting.

For gas chromatographic–mass spectrometer systems, differential pumping is recommended. One pumping system is connected to the ion source part. Only a small hole is provided for the ion transmission between the ion source and the mass analyzer. A second pumping system keeps a very low pressure in the analyzer part of the mass spectrometer.

16.7 DATA HANDLING

On modern mass spectrometers the data are digitized and collected on magnetic tape or stored in the memory of a computer for subsequent processing. Such a system permits rapid accumulation of the wealth of data generated. On request, the dedicated microcomputer reconstructs the mass spectrum.

Electrical detection of the mass spectrum allows operation of the ion-collection system in one of several modes. Scanning the mass spectrum across the detector is an extremely inefficient way to collect information and is seldom done. Peak switching allows the detector to view selected regions of the spectrum, bypassing those that contain no information or unwanted information. In this mode the magnetic field is varied to scan the spectral regions. Resolution generally suffers in peak switching because the slits must be widened to allow for slight inaccuracies in field settings while still viewing a portion of the mass spectrum that includes the mass number of interest. In the selective-ion mode the instrument simply monitors a single mass number to achieve low limits of detection by long-term integration of the signal.

16.8 ISOTOPE-RATIO SPECTROMETRY

The ability of the mass spectrometer to precisely measure isotope ratios has allowed medical researchers to study body functions using isotope-labeled tracers. Stable isotopes are used to "tag" compounds and thus serve as tracers to determine the ultimate fate of the compound in chemical and biological reactions.[16] The mass spectrum displays amounts of the added isotope in the fragment ions as well as in the parent ion. Thus, the position of the tracer isotope in the molecule can often be determined without laborious chemical degradation techniques.

A further extension of isotope-ratio mass spectrometry is in geochemistry. Precise age dating is based on the rate of decay of radioactive nuclides (see Chapter 14). Knowing the decay rate of uranium-238 to lead-206, potassium-40 to argon-40, and rubidium-87 to strontium-87, and measuring the ratio of the isotopes of one of these pairs, one can determine the age of minerals and rocks.

For the precise determination of isotope ratios of gases H/D, $^{13}C/^{12}C$, $^{15}N/^{14}N$, $^{18}O/^{16}O$, and $^{34}S/^{32}S$, isotope-ratio mass spectrometers are used. One design is shown in Figure 16.18. After leaving the ion source and passing through special electrostatic focusing lenses, the emerging ion beam is projected at an angle of 26.5° with an energy of 10 kV into the field of the 90° separating magnet from which it emerges at the same angle. This arrangement doubles the dispersion and at the same time focuses the ion beam stigmatically. The field is supplied by a permanent magnet of 0.325 tesla (3250 G) induction. Mass setting is effected by varying the ion accelerating voltage. A mass scale generated by the computer facilitates setting to the desired mass numbers. For special applications that require a higher mass range, an electromagnet with 1 tesla can be substituted.

The double Faraday collector for masses 2 and 3, located in the middle section of the analyzer system, handles the H_2/HD analysis. The ion-collector system has a working resolution of 25 (10% valley definition). The ion-collector system for carbon, nitrogen, oxygen, and sulfur has a working resolution of 200 (10% valley definition).

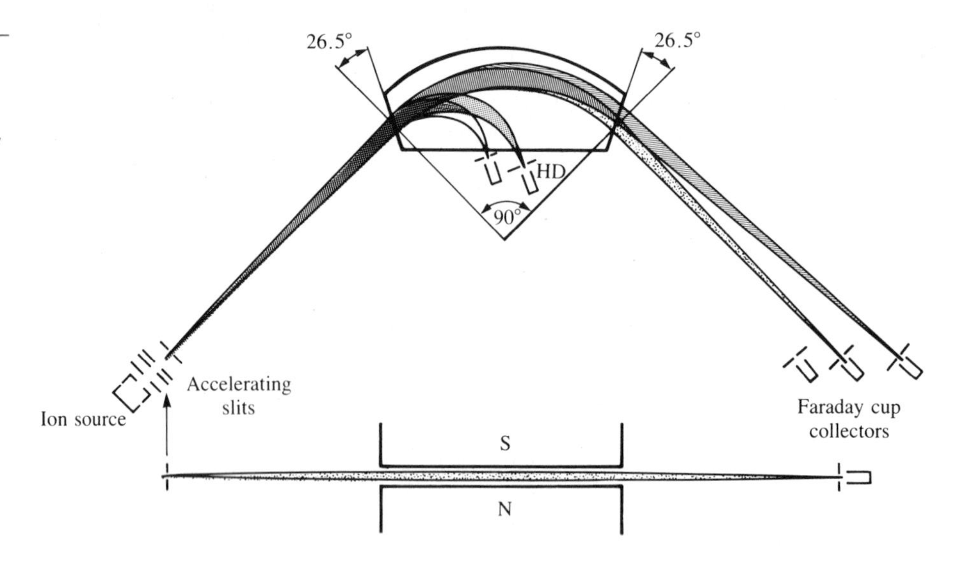

FIGURE **16.18**
Schematic diagram of an isotope-ratio mass spectrometer. (Courtesy of Finnigan MAT.)

The inlet system is designed for measuring the sample and standard alternately. For balancing the ion currents of sample and standard, the size of the reservoirs can be varied by means of stepping motors. An automatic system carries out balancing and at the same time adjusts the reservoir volumes (between 1 and 20 mL) such that the ion current of the more abundant isotope becomes optimum. For measurements on SO_2 the inlet system can be heated up to 80 °C together with the ion source housing.

16.9 CORRELATION OF MASS SPECTRA WITH MOLECULAR STRUCTURE

The ionization efficiency of a mass spectrometer source (see Section 16.3) must be high so that a large portion of the neutral sample particles present are converted to ions. High efficiency is particularly important for the analysis of nanogram quantities of sample material and trace impurities in solids.

When bombarded by electrons in the electron-impact method of ionization, every substance ionizes and fragments uniquely. A molecule may simply lose an electron or it may fragment into two smaller units, an ionized fragment and a neutral particle, the sum of whose masses equals their precursor. A molecular (or parent) ion is generally observed in considerable intensity when the gaseous molecules are bombarded with electrons of energy just sufficient to cause ionization, but not bond breakage, about 8–14 eV for most organic molecules. As the electron energy is increased further, the molecular ion is formed with excess energy in its electronic and vibrational degrees of freedom. Because energy between the bonds redistributes rapidly, all of the bonds are affected simultaneously. As soon as the excess energy over the ground-state energy possessed by the molecular ion becomes equal to the dissociation energy of some particular bond, the appropriate fragment ions are formed. Although the peak intensities are extremely sensitive to ionizing voltage at low values of the ionizing voltage, the relative peak intensities become fairly constant once the ionization voltage exceeds 50 eV. At higher ionizing energies the total production of ions is higher, but the net effect of higher overall intensity and the resultant severe fragmentation is an increase in relative intensity of the fragment peaks at the expense of the parent peak.

Sometimes both electron-impact and chemical ionization spectra (or field ionization spectra) are needed for a given compound. The chemical ionization or field ionization spectrum might show a molecular ion, and the fragmentation of an electron-impact spectrum might identify the class of compound. Calibration of a mass spectrometer must be done in the electron-impact mode; perfluoroalkanes so often used as markers would deactivate emitter wires used in field ionization or field desorption, and chemical ionization would provide few, if any, fragment ions.

Molecular Identification

In the identification of a compound, the most important information is the molecular weight. The mass spectrometer is unique among analytical methods in being able to provide this information very accurately, often to four decimal places when a high-resolution mass spectrometer is used. At ionizing voltages ranging from 9 to 14 V in the electron-impact mode of ionization, it is assumed that no ions heavier than the molecular ion form. Therefore, the mass of the heaviest ion, exclusive of isotopic contributions, gives the nominal molecular weight with low-resolution mass spectrometers and the exact molecular weight with high-resolution instruments.

The number of possible molecular formulas is restricted by a study of the relative abundances of natural isotopes for different elements at masses one or more units larger than the parent ion (Table 16.1). Observed values are compared with those calculated for all possible combinations of the naturally occurring heavy isotopes of the elements. For a compound $C_wH_xO_zN_y$, a simple formula allows one to calculate the percent of the heavy isotope contributions from a monoisotopic peak, P_M, to the P_{M+1} peak:

$$100\frac{P_{M+1}}{P_M} = 0.015x + 1.11w + 0.37y + 0.037z$$

Tables of abundance factors have been calculated by Beynon and Williams for all combinations of C, H, N, and O up to mass 500.[17] Table 16.2 illustrates the spectral peak contributions at nominal mass 135 for a few of the compounds that have a parent peak (or possibly a fragment peak) of mass 134. Once the empirical formula is established with reasonable assurance, hypothetical molecular structures are written. For this purpose one can use the entries in the formula indices of Beilstein and *Chemical Abstracts*. If the molecular weight of the parent peak can be measured to several decimal places, the possible molecular structures are further restricted.

Compounds that contain chlorine, bromine, sulfur, or silicon are usually apparent from prominent peaks at masses 2, 4, 6, and so on, units larger than the nominal mass of the parent or fragment ion. The abundance of heavy isotopes is treated in terms of the binomial expansion $(a + b)^m$, where a is the relative abundance of light isotopes, b is the relative abundance of heavy isotopes, and m is the number of atoms of the particular element present in the molecule. When two elements with heavy isotopes are present, the binomial expansion $(a + b)^m(c + d)^n$ is used.

If the mass of the parent ion is measured with a high-resolution mass spectrometer, the number of possible empirical formulas can be still further restricted. Because the masses of the elements are not exactly integral multiples of a unit mass, a sufficiently accurate mass measurement alone enables the elemental composition of the ion to be determined. For combinations of C, H, N, and O, the relationship is

$$\frac{\text{exact mass difference from nearest integral mass} + 0.0051z - 0.0031y}{0.0078} = \text{number of H's}$$

TABLE **16.1**

(A) ABUNDANCES OF SOME POLYISOTOPIC ELEMENTS (%)

Element	% Abundance	Element	% Abundance	Element	% Abundance
^{1}H	99.985	^{16}O	99.76	^{33}S	0.76
^{2}H	0.015	^{17}O	0.037	^{34}S	4.22
^{12}C	98.892	^{18}O	0.204	^{35}Cl	75.53
^{13}C	1.108	^{28}Si	92.18	^{37}Cl	24.47
^{14}N	99.63	^{29}Si	4.71	^{79}Br	50.52
^{15}N	0.37	^{30}Si	3.12	^{81}Br	49.48

(B) SELECTED ISOTOPE MASSES

Element	Mass	Element	Mass
^{1}H	1.0078	^{31}P	30.9738
^{12}C	12.0000	^{32}S	31.9721
^{14}N	14.0031	^{35}Cl	34.9689
^{16}O	15.9949	^{56}Fe	55.9349
^{19}F	18.9984	^{79}Br	78.9184
^{28}Si	27.9769	^{127}I	126.9047

SOURCE: Lange's Handbook of Chemistry, 13th ed., J. A. Dean, ed., McGraw-Hill, New York, 1985.

TABLE **16.2**

HEAVY-ISOTOPE CONTRIBUTIONS TO PARENT PEAK OF MASS 134 (%) AND EXACT PARENT MASSES

Empirical formula	P_{M+1} peak	Exact mass
$C_5H_{10}O_4$	5.72	134.1316
$C_5H_{14}N_2O_2$	6.47	134.1778
$C_6H_{14}O_3$	6.83	134.1748
$C_8H_6O_2$	8.82	134.1342
$C_9H_{10}O$	9.93	134.1774
$C_9H_{12}N$	10.30	134.2005
$C_{10}H_{14}$	11.03	134.2206

For example, a crystalline solid that contains only C, H, and O gave the mass 134.0368 for the molecular ion. Thus,

$$\frac{0.0368 + 0.0051z}{0.0078} = 6\text{H's} \quad \text{when } z = 2 \text{ oxygen atoms}$$

and the empirical formula is $C_8H_6O_2$. One substitutes integral numbers for z (oxygen) and y (nitrogen) until the divisor becomes an integral multiple of the numerator within 0.0002 mass unit.

Two general rules aid in writing formulas. (1) If the molecular weight of a C, H, O, and N compound is even, so is the number of hydrogen atoms it contains; if the molecular weight is divisible by four, the number of hydrogen atoms is also divisible by four. (2) When nitrogen is known to be present in any compound of C, H, O, As, P, S, Si, and the halogens that have an odd molecular weight, the number of nitrogen atoms must be odd.

Once the exact molecular formula has been decided, the total number of rings and double bonds can be determined by the formula

$$\tfrac{1}{2}(2w - x + y + 2)$$

when covalent bonds make up the molecular structure. A benzene ring has four (one ring and three double bonds); a triple bond has two.

Metastable Peaks

A one-step decomposition process may be indicated by an appropriate *metastable* peak in the mass spectrum. Metastable peaks arise from ions that decompose in the field-free region after they are accelerated out of the ion source but before they enter the analyzer. Their lifetime is about 10^{-6} sec. A metastable ion transition takes the general form:

$$\text{original ion} \rightarrow \text{daughter ion} + \text{neutral fragment}$$

The metastable peak m^* appears as a weak, diffuse (often humped-shape) peak, usually at a nonintegral mass, given by

$$m^* = \frac{(\text{mass of daughter ion})^2}{\text{mass of original ion}}$$

In a spectrum that is linear with respect to mass values, the distance of m^* below the daughter ion is of similar magnitude to the distance of the daughter ion below the original ion. The foregoing relationship holds only for ions that decompose in a small portion of the accelerating region and is more frequently observed with 60°- and 90°-sector instruments where a field-free region exists after the accelerating slits and before the magnetic analyzer. Of course, the absence of a metastable peak from the spectrum does not preclude a particular decomposition.

Mass Spectra and Structure

The mass spectrum of a compound contains the masses of the ion fragments and the relative abundance of these ions often plus the parent ion. The uniqueness of the molecular fragmentation aids in structural identification. Because sufficient molecules are present and dissociated for the probability law to hold, the dissociation fragments always occur in the same relative abundance for a particular compound. The mass spectrum becomes a "fingerprint" for each compound because no two molecules are fragmented and ionized in exactly the same manner on electron bombardment. There are sufficient differences in these molecular fingerprints to permit the identification of different molecules in complex mixtures. Usually the complexity of mixtures is lessened by carrying out preliminary separations.

To a considerable extent the breakdown patterns are predictable. Conversely, the size and structure of a molecule can often be reconstructed from the fragment ions in the spectrum of a pure compound. For example, Table 16.3 indicates the relative abundance of the significant fragments produced from three isomeric octanes. In the structural formulas the asterisk indicates the bond that is broken in the most probable process of fragmentation, and the plus sign indicates the next most probable process. Favored sites for bond rupture in the molecule parallel chemical bond lability. The mass 114 corresponds to the parent ion formed by the loss of a single electron from the parent compound; mass 99 corresponds to the loss of a methyl group plus an electron; mass 71, to the loss of a propyl group plus an electron; mass 57, to the loss of a butyl group; and mass 43, to the loss of a pentyl group.

It is usual practice in reporting mass spectra to normalize the data by assigning the most intense peak (the so-called base peak) a value of 100. Other peaks are reported as percentages of the base peak.

When working from a spectrum, it is advisable to tabulate the prominate ion peaks, starting with the highest mass, and also to record the probable group(s) lost to give these ion peaks. Common fragment ions are listed in the literature,[18] also possible structures for the positively charged ion fragments. Finally, the mass spectral features are predicted from available correlation data. These features are then checked against the actual spectrum.

Usually only one bond is cleaved. In succeeding fragmentations a new bond is formed for each additional bond that is broken. When fragmentation is accom-

TABLE **16.3** MASS SPECTRAL PATTERN OF TRIMETHYLPENTANES

	Relative abundances (%)		
Mass/charge ratio	**2,3,3-Trimethylpentane** C C + | C—C * C + C—C | C	**2,2,4-Trimethylpentane** C C | | C—C * C + C—C | C	**2,3,4-Trimethylpentane** C C | | C—C * C * C—C | C
114	0.1	0.02	0.3
99	3	5	0.1
71	1	1	40
57	70	80	9
43	15	20	50

SOURCE: H. W. Washburn, H. F. Wiley, S. M. Rock, and C. E. Berry, *Ind. Eng. Chem. Anal. Ed.*, **17**, 75 (1945).

panied by the formation of a new bond as well as by the breaking of an existing bond, a *rearrangement* process is said to have occurred. The migrating atom is almost exclusively hydrogen. Six-membered cyclic transition states are most important, although alternative ring sizes also operate.

Some general features of the mass spectra of compounds can be predicted from the following general rules for fragmentation patterns.

1. Cleavage is favored at branched carbon atoms: tertiary > secondary > primary, with the positive charge tending to stay with the branched carbon (carbonium ion).
2. Double bonds favor cleavage beta to the bond (but see rule 8).
3. A substance that has a strong parent peak often contains a ring; the more stable the ring, the stronger the parent peak.
4. Ring compounds usually contain peaks at mass numbers characteristic of the ring.
5. Saturated rings lose side chains at the alpha carbon. The peak corresponding to the loss of two ring atoms is much larger than for the loss of one ring atom.
6. In alkyl-substituted ring compounds, cleavage is most probable at the bond beta to the ring if the ring has a double bond next to the side chain.
7. A hetero-atom induces cleavage at the bond beta to it.
8. Compounds that contain a carbonyl group tend to break at this group, with the positive charge remaining with the carbonyl portion.

The presence of Cl, Br, S, and Si is easy to deduce from the unusual isotopic abundance patterns of these elements. These and other elements, such as P, F, and I, are also detectable from the unusual mass differences they produce between some fragment ions in the spectrum.

For linear alkanes the peak at mass 43 is always very large for compounds larger than butane. The presence of a large peak at mass 43 suggests the presence of higher alkyl groups. In the same way, other characteristic low-mass fragment ions can be used to identify certain terminal groups:

1. Mass 30 is always large for primary amines.
2. Masses 31, 45, 59, and so on indicate the presence of oxygen as an alcohol or ether.
3. Masses 19 and 33 indicate an alcohol.
4. Mass 66 indicates a monobasic carboxylic acid.
5. Masses 77 and 91 indicate the presence of a benzene ring.

Look for small-mass neutral fragments lost from the molecular ion:

1. Loss of mass 18 (H_2O) is indicative of alcohols, aldehydes, and ketones.
2. Loss of masses 19 (F) and 20 (HF) is indicative of fluorides.
3. Loss of mass 27 (HCN) indicates aromatic nitriles or nitrogen heterocycles.
4. Loss of mass 29 indicates either CHO or C_2H_5; an infrared spectrum can assist in the selection.
5. Loss of mass 30 indicates either CH_2O or NO and an infrared spectrum should be consulted.
6. Loss of masses 33 (HS) and 34 (H_2S) characterizes thiols.

7. Loss of mass 42 could indicate either CH_2CO via rearrangement from methyl ketone or an aromatic acetate or an aryl —$NHCOCH_3$ group.
8. Loss of mass 43 is indicative of either C_3H_7 or CH_3CO.
9. Loss of mass 45 indicates either —COOH or OC_2H_5

EXAMPLE 16.1

A solid, melting at 33 °C, has this mass spectrum; in the table peak intensities are given as percent of the base peak. Metastable peaks appear at 46.5, 53.5, 67.9, 106.3, 121.7, and 147.9 mass units. Insofar as possible, deduce the structure of the compound.

m/z	Percent of base peak	m/z	Percent of base peak
65	30	155	61
91	100	172	18
92	32	200	78
107	11	201	10.6
108	10	202	5.8

The isotopic cluster at 202, 201, 200 amu strongly suggests the presence of one sulfur atom (natural abundance of sulfur-34 is 4.22%). The peak at 200 amu is very strong, suggesting the presence of a ring and probably indicating that the parent has a molecular weight of 200 amu. The metastable peaks provide this information:

147.9: $200^+ = 172^+ + 28$; probable loss of $CH_2{=}CH_2$, maybe CO

121.7: $200^+ = 156^+ + 44$; probable loss of $CH_2{=}CHOH$

106.3: $108^+ = 107^+ + 1$; loss of H

67.9: $172^+ = 108^+ + 64$; probable loss of SO_2

53.5: $155^+ = 91^+ + 64$; probable loss of SO_2

46.5: $91^+ = 65^+ + 26$; loss of $HC{\equiv}CH$, probably from tolyl ring structure via a rearrangement

The sulfur atom is apparently present as a SO_2 group. The isotopic cluster at 201 indicates not more than nine carbon atoms: $(10.6 - 0.76)/1.10 \approx 9$. The 0.76 is the contribution of sulfur-33 to the peak at 201 amu.

The base peak at 91 amu strongly suggests a tolyl structure, as does the metastable peak at 46.5 amu. The loss of 45 mass units in the transition from 200 to 155 amu suggests the presence of an ethoxy group, which is confirmed by the metastable peaks at 147.9 and 121.7. (Both are the result of rearrangement processes; note that the daughter ion has an even mass from an even-mass parent.) The abundance of the peak at 155 amu suggests that it could originate only from the molecular ion. The ethoxy group is linked to the SO_2 group; note the loss of 64 mass units from both peak 172 (after loss of an ethylene group via rearrangement) and peak 155 (after loss of an ethoxy group by cleavage). Now the pieces are a phenyl group with an attached methyl and an ethyl sulfonate group. The ring substitution cannot be ascertained from the mass spectrum. However, ethyl 4-toluenesulfonate has a melting point of 33 °C.

The identification of relatively simple molecules by mass spectrometry is not difficult if the interpreter is armed with a few pertinent bits of information derived from other spectral sources. It is understandable that the diagnostic task becomes somewhat more demanding as the molecular intricacy increases. The successful application of mass spectrometry to the structural elucidation of highly complex materials involves obtaining and interpreting the spectra of molecules with related structures. Similarities and differences between the spectrum of an unknown compound and that of a known can then be used to arrive at a structural assignment.

16.10 QUANTITATIVE ANALYSIS OF MIXTURES

The system used in quantitative analysis by mass spectrometry is basically the same as that used in infrared or ultraviolet absorption spectrometry. Spectra are recorded for each component. Consequently, samples of each compound must be available in a pure state. From inspection of the individual mass spectra, analysis peaks are selected on the basis of both intensity and freedom from interference. If possible, monocomponent peaks (perhaps molecular-ion peaks) are selected. The sensitivity is given in terms of the height of the analysis peak per unit pressure. This is obtained by dividing the peak height for the analysis peak by the pressure of the pure compound in the sample reservoir of the mass spectrometer.

Calculation of the sample composition is simplified if the components of the mixture give at least one peak whose intensity is entirely due to the presence of one component. The height of the monocomponent peak is measured and divided by the appropriate sensitivity factor to give its partial pressure. Then division by the total pressure in the sample reservoir at the time of analysis yields the mole percent of the particular component.

If the mixture has no monocomponent peaks, simultaneous linear equations are then set up from the coefficients (percent of base peak) at each analysis peak. There will be one equation for each compound in the mixture with *n* terms (unless one or more terms are zero) when *n* components are in the mixture. Take, for example, the analysis of a mixture of butyl alcohols whose individual mass spectra are tabulated in Table 16.4. Four equations are written using the selected four mass peaks and the analytical intensity of each peak found with the sample:

$$90.58x_1 + 1.47x_2 + 1.02x_3 + 2.46x_4 = M_{56} = 126.7$$

$$0.26x_1 + 100.00x_2 + 17.78x_3 + 4.98x_4 = M_{59} = 301.5$$

$$6.59x_1 + 0.59x_2 + 100.00x_3 + 5.03x_4 = M_{45} = 322.6$$

$$0.79x_1 + 0x_2 + 0.29x_3 + 9.06x_4 = M_{74} = 14.8$$

To achieve greater speed in computation, the matrix of coefficients is inverted, yielding a set of equations in terms of each unknown and the analytical masses:

$$\text{(butyl)}\quad x_1 = 110.70M_{56} - 1.625M_{59} - 0.744M_{45} - 28.77M_{74}$$

$$(\textit{tert}\text{-butyl})\quad x_2 = 1.39M_{56} + 100.08M_{59} - 17.67M_{45} - 45.53M_{74}$$

$$(\textit{sec}\text{-butyl})\quad x_3 = -6.83M_{56} - 0.489M_{59} + 100.31M_{45} - 53.56M_{74}$$

$$\text{(isobutyl)}\quad x_4 = -9.39M_{56} + 0.157M_{59} - 3.17M_{45} + 1108.0M_{74}$$

Peak intensities from the mixture spectrum are substituted into the inverse matrix equations, yielding the number of divisions of base peak due to each component. Division by the appropriate sensitivity factor (Table 16.5) yields the partial pressure of each component. Each partial pressure is divided by the total computed pressure, yielding mole percent. The sum of the partial pressures determined in this way should equal the total sample pressure. A discrepancy would indicate an unsuspected component or a change in the operating sensitivity.

TABLE **16.4**

MASS SPECTRAL DATA (RELATIVE INTENSITIES) FOR THE BUTYL ALCOHOLS

	Percent of base peak (italic)			
m/z	**Butyl**	***sec*-Butyl**	***tert*-Butyl**	**Isobutyl**
15	8.39	6.80	13.30	7.47
18	2.18	0.23	0.49	2.05
27	50.89	15.87	9.87	42.20
28	16.19	2.98	1.67	5.94
29	29.90	13.94	12.65	21.17
31	*100.00*	20.31	35.53	63.10
33	8.50			53.40
39	15.63	3.36	7.70	19.03
41	61.57	10.13	20.82	55.68
42	32.36	1.64	3.32	60.46
43	61.36	9.83	14.45	*100.00*
45	6.59	*100.00*	0.59	5.03
55	12.29	2.06	1.55	4.35
56	90.58	1.02	1.47	2.46
57	6.68	2.74	9.02	3.89
59	0.26	17.78	*100.00*	4.98
60		0.64	3.26	0.57
74	0.79	0.29		9.06

SOURCE: A. P. Gifford, S. M. Rock, and D. J. Comaford, *Anal Chem.*, **21**, 1026 (1949).

TABLE **16.5**

ANALYSIS OF A MIXTURE OF BUTYL ALCOHOLS*

Component	**Value of x**	**Sensitivity divisions/ 10^{-3} torr**	**Partial pressures, 10^{-3} torr**	**Mol%**
Butyl	$x_1 = 12{,}871$	1151	11.18	24.4
tert-Butyl	$x_2 = 23{,}976$	2093	11.46	25.0
sec-Butyl	$x_3 = 30{,}555$	2698	11.33	24.8
Isobutyl	$x_4 = 14{,}234$	1205	11.81	25.8

SOURCE: A. P. Gifford, S. M. Rock, and D. J. Comaford, *Anal Chem.*, **21**, 1026 (1949).
* Mass peaks used: 45, 56, 59, and 74.

An outstanding feature of quantitative mass spectrometric analysis is the large number of components that can be handled with no need for fractionation or concentration. Mixtures that contain as many as 30 components can be accommodated, and quantities of material as low as 0.001 mol % can be detected in hydrocarbon mixtures. Calculations are generally performed by the computer component of the mass spectrometer system. More complex mixtures, covering a wide boiling range, may require a rough or simple distillation before analysis. Precision normally falls within the range ± 0.05–1.0 mol %.

16.11 FOURIER TRANSFORM MASS SPECTROMETRY[19–21]

Fourier transform mass spectrometry (FT-MS) is a technique in which mass analysis is performed by detecting the cyclotron frequencies of ions—in other words, the cyclic motion of ions in a uniform magnetic field. It evolved from ion cyclotron resonance mass spectrometry, which suffered from several limitations—namely, restricted mass range, low mass resolution, and slow scanning speeds. Comisarow and Marshall,[22] who are responsible for the original demonstration of FT-MS, made use of new technology and operating principles that overcame many of these limitations. Their method involves the temporal separation of ion formation, excitation, and detection. Direct observation of induced image currents in the walls of the analyzer cell allows short measurement times of tens to hundreds of milliseconds. Experimentally this is achieved in FT-MS by the simultaneous time domain observation of signals from all excited ions, followed by Fourier transformation of the data thus obtained to yield a frequency domain (mass) spectrum. A mass spectrum produced by this technique is the result of resolving different cyclotron frequencies of ions with different masses rather than resolving ion trajectories or drift velocities. The fact that ion formation and detection are separated in time, not space, makes the system mechanically simple and easy to maintain and operate. There are no slits or ion optic lenses to adjust. Negative and positive ions behave exactly alike, except that their orbits are in the opposite direction.

Ions produced by a selected ionization mode enter a cubical cavity, roughly 2 cm on an edge, through a hole in the trap plate. The ions are trapped inside the cubical cavity (analyzer cell) that is situated between the pole caps of an electromagnet, or a superconducting solenoid magnet (Figure 16.19). The ions are trapped in the direction parallel to the magnetic field by a shallow electrostatic field. For positive ions a potential well of this type is produced by a positive voltage on the two side plates and a negative voltage on the upper plate, lower plate, and the two end plates. The ions are constrained by the magnetic field strength, B (in tesla), to move in circular orbits that have a characteristic cyclotron frequency, ω_c (in hertz), which depends only on the ion's m/z value:

$$\omega_c = zB/2\pi m = 1.536 \times 10^7 \,(B/m) \qquad (16.20)$$

The cyclotron frequency is independent of the velocity of the ion. One consequence of this for mass spectrometry is that an ensemble of ions that have the same m/z ratio may have greatly different position coordinates and velocities and yet still have the same cyclotron frequency. Only the radius of the cyclotron orbits is affected by the velocity of the ion.

To detect the ions of interest, they must be excited into coherent orbital motion. The radii of the cyclotron orbits are expanded by subjecting the ions to an alternating electric field. When the frequency of the sine wave signal generator is the same as the cyclotron frequency, the ion is steadily accelerated to a larger and larger radius of gyration r given by

$$r = \frac{V_p t}{2dB} \tag{16.21}$$

where V_p is the voltage of the rf excitation signal, t is the duration of the rf pulse, and d is the separation of the two electrodes. When the frequency of the signal generator is not equal to the cyclotron frequency, the ion motion is not perturbed significantly.

To accelerate the ions, an alternating electric field is established perpendicular to the magnetic field by connecting a sine wave signal generator to the upper (transmitter) plate of the analyzer cell (Figure 16.19). The lower (transmitter) plate (which is analogous to electrode 2 in Figure 16.20a) is connected to the input of an amplifier. This is done by applying a very fast-frequency sweep voltage immediately following the ion-formation event in the electron-impact source or after a delay time in the chemical ionization mode. For example, the excitation energy can be set to cover the cyclotron frequencies of all the ions of interest. The coherent motion of the excited ions induces image currents in the form of a transient decay in the receiver circuit. This transient ion excitation decay is a time-domain signal containing all the frequencies of all the excited ions in the cell.

Orbiting ions generate characteristic alternating currents, which are detected by receiver plates. In one configuration, a packet of positive ions might be undergoing coherent cyclotron motion between two electrodes that are connected through a resistor to ground. As the ions move away from the first electrode and closer to the second, the electric field of the positive ions causes electrons in the external circuit to flow through the resistor and accumulate on the second electrode. On the other half of the cyclotron orbit, the electrodes leave the second electrode and accumulate on the first electrode as positive ions approach.

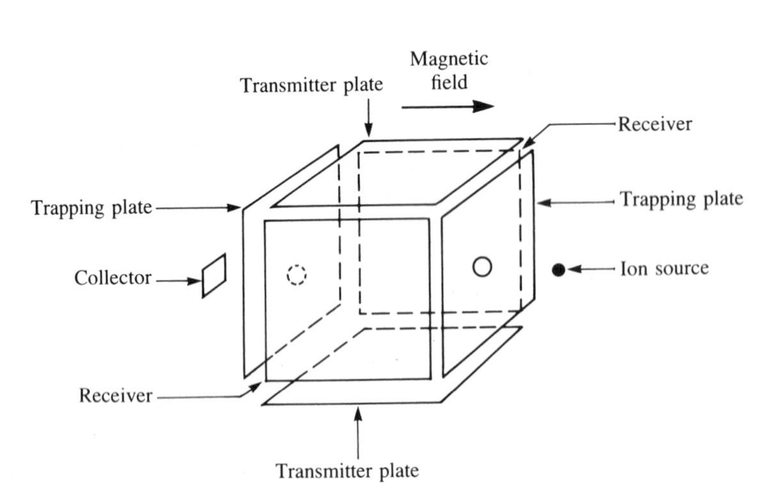

FIGURE **16.19**
The analyzer cell of a Fourier transform mass spectrometer (semi-exploded view). Magnet pole faces are located above and below the transmitter plates.

Negative ions behave exactly the same as positive ions except that their orbits are in the opposite direction. Negative ion spectra are obtained by using a negative trap voltage. Alternating the polarity of the computer-controlled trap voltage on successive scans while signal averaging results in simultaneous negative ion and positive ion spectra.

Electrons move in the external circuit (from electrode 1 to electrode 2 in Figure 16.20a) as positive ions move to the right. Electrode 1 is at ground potential and electrode 2 is connected through a resistor to ground. As the packet of positive ions moves closer to electrode 2, electrons are attracted to it by the electric field of the ions. A small negative voltage develops on electrode 2 as the electron current, called the image current, flows through the resistor, and a small ac voltage develops across the resistor. Cyclotron motion produced by the magnetic field gives rise to alternating ion image currents at the same frequency as the cyclotron frequency of the ion with a specific m/z value. The amplitude of the current is proportional to the number of the specific m/z ions in the analyzer cell. The ions are made to oscillate between the two electrodes. This makes it possible to detect the ions without ever having them collide with the electrodes. After the excitation pulse is turned off, the signal of the image current induced by the accelerated ions is amplified, digitized, and stored in a computer. Fourier transform mass spectrometry is unique in that increased measurement time increases both sensitivity and resolution. For a mixture of different ions, the signal at the output of the amplifier is a composite transient signal with a frequency spectrum that is related to the mass spectrum. The Fourier transform of the transient signal is automatically computed to yield mass spectra. The detection sensitivity is the same for low-mass ions as it is for high-mass ions.

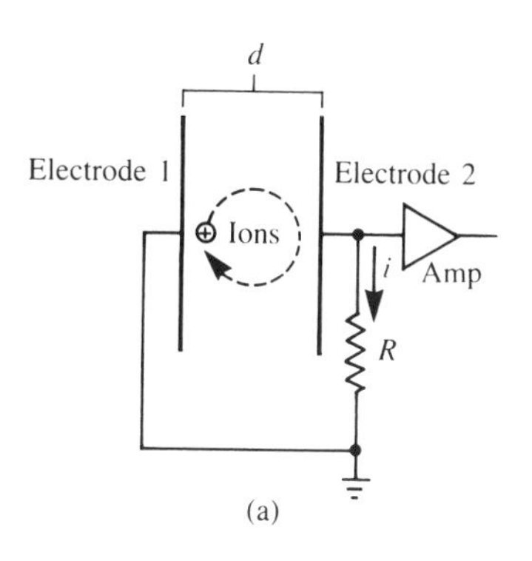

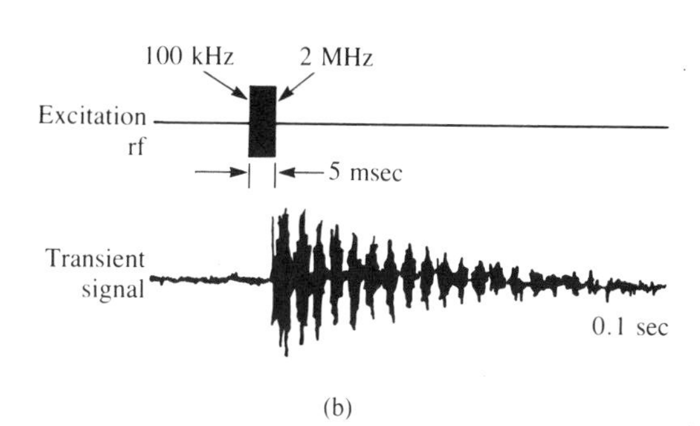

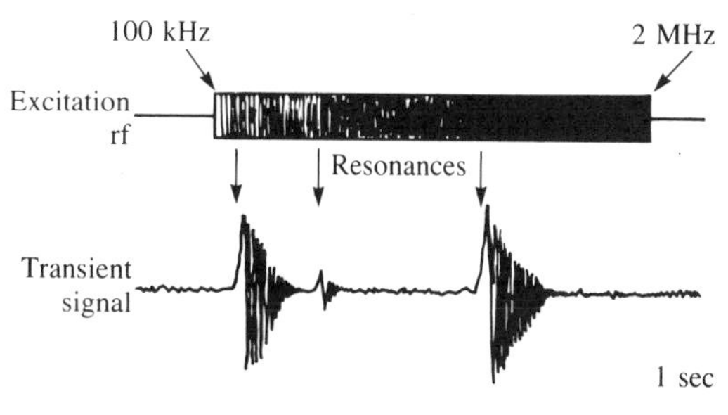

FIGURE **16.20**
(a) Ion image current. (b and c) Methods of signal detection in Fourier transform-mass spectrometry. (b) Fourier transform–ion cyclotron resonance (FT-ICR). (c) Rapid-scan ion cyclotron resonance. [After W. D. Bowers, R. L. Hunter, and R. T. McIver, Jr., *Ind. Res/Dev.*, p. 124 (November 1983). Copyright 1983 *Research & Development* magazine.]

Fourier Transform–Ion Cyclotron Resonance (FT-ICR)

To detect a signal the ions must be accelerated to move coherently in phase. There are two methods for accomplishing this: FT-ICR and rapid-scan ICR. In FT-ICR, cyclotron resonance is induced by exposing the ions to a rf electric field pulse with a frequency that is varied linearly during the irradiation period of a few milliseconds. The frequency sweep must cover the entire range of cyclotron frequencies of interest. The brief rf pulse accelerates all ions in the analyzer cell and produces a complex transient signal that is a composite of all the cyclotron frequencies in the cell (Figure 16.20b).

When the frequency of the applied sine wave signal is the same as the cyclotron frequency of an ion, a resonance condition is established. The ion is steadily accelerated to a larger radius (orbit) of gyration. This effect provides the basis for mass spectrometry because ions that have a different cyclotron frequency are not accelerated.

Rapid-Scan Ion Cyclotron Resonance

In this technique ions stored in the analyzer cell are subjected to a frequency-swept excitation rf (or "chirp") signal that scans more slowly across the spectrum, typically in 1 sec. When the frequency of the excitation signal matches the cyclotron frequency of an ion, that ion absorbs energy from the circuit and is accelerated. A transient signal is induced in the amplifier (the chirps). Later in the scan, when resonance is established with an ion of different cyclotron frequency, that ion, too, is accelerated and detected (Figure 16.20c). Excitation and detection occur simultaneously.

One feature of the rapid-scan method is a temporal separation of the transient signals. With this separation, the signals are digitized at a much slower rate than Fourier transform–ion cyclotron resonance. Rapid-scan ion cyclotron resonance is well suited for acquiring a wide-range mass spectrum, since the digitized data are stored directly on magnetic disks. Modern microelectronics enable the transformation to be made on single transient decay waveforms in the millisecond time domain. This time scale accommodates scanning speeds compatible with chromatography.

The capabilities of FT-MS are best established with respect to obtaining high-resolution measurements. Separation of C_6D_6 and C_6H_{12} molecular ions at m/z 84 is achieved with $m/\Delta m = 220{,}000$ at a field strength of 1.2 tesla. Mass resolution is directly proportional to magnetic field strength and inversely proportional to mass at a constant pressure. The theoretical maximum resolution obtainable at about 10^{-8} torr is about 800,000, even for measurements greater than 1000 daltons.

16.12 TANDEM MASS SPECTROMETERS[23–28]

Tandem mass spectrometry (MS-MS) uses a wide range of mass analyzers in a series arrangement. Each instrument configuration has certain characteristics that in a particular application have specific advantages over another configuration. Double-focusing sector spectrometers, discussed in Section 16.4, are one example. Two other examples follow.

Triple Quadrupole Mass Spectrometry[29,30]

A triple quadrupole system produces a mass spectrum from ions formed initially in a normal mass spectrum. Ions formed from the sample in the ion source are separated by mass dispersion in the first quadrupole unit that functions as a mass filter (Figure 16.21). A specific fragment ion of any mass that appears in a compound's normal mass spectrum can be selected with the first mass analyzer. This particular parent ion is further fragmented by collision with a target gas in the reaction chamber, a rf-only quadrupole, which is pressurized with a collision gas. The rf-only quadrupole collision chamber also focuses scattered ions. The mass spectrum of the resulting daughter ions is obtained by scanning with the mass analyzer that is the third quadrupole unit. In this manner a complete three-dimensional fragmentation map may be obtained by recording the mass spectrum of each fragment ion of a parent compound. Figure 16.22 is an example of such a map for cyclohexane. The normal electron-impact mass spectrum is displayed along the diagonal. The subsequent fragmentation of each parent ion is shown toward the rear and left of the plot.

Quadrupole FT-MS System

In the quadrupole FT-MS instrument, the positive and negative ions produced in the source are focused into a beam and injected into a quadrupole mass analyzer. Only the ion(s) of interest is transmitted. Next the transmitted ions are accelerated into a superconducting solenoid magnet, trapped in an ion cyclotron resonance cell, and detected by Fourier transform mass spectrometry.

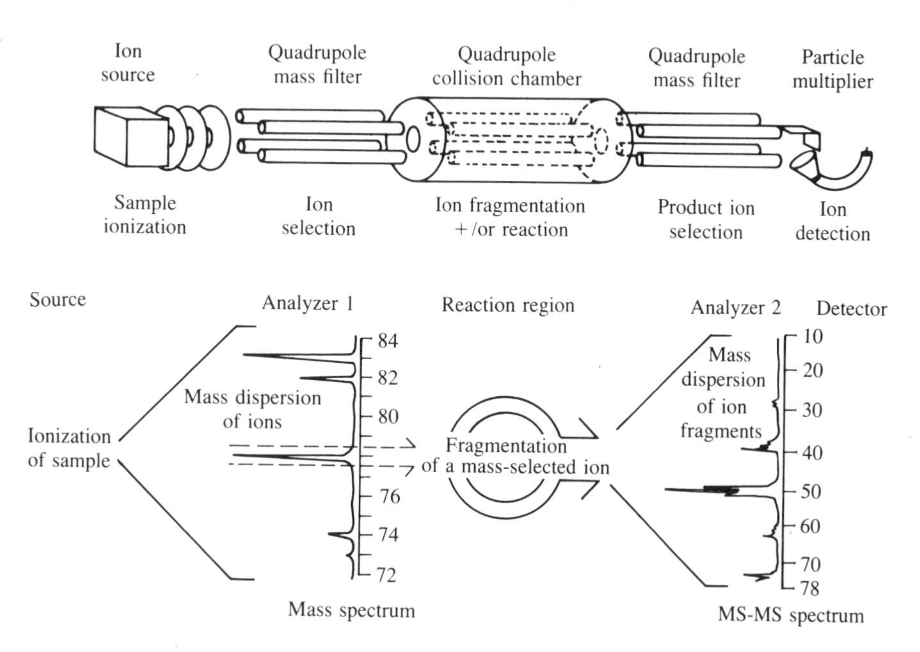

FIGURE **16.21**
Conceptual diagram of the triple quadrupole mass spectrometer showing each component and its function. [Reprinted with permission from R. A. Yost and C. G. Enke, *Anal. Chem.*, **51,** 1251A (1979). Copyright 1979 American Chemical Society.]

FIGURE **16.22**

Three-dimensional fragmentation map for cyclohexane. [Reprinted with permission from R. A. Yost and C. G. Enke, *Anal. Chem.*, **51,** 1252A (1979). Copyright 1979 American Chemical Society.]

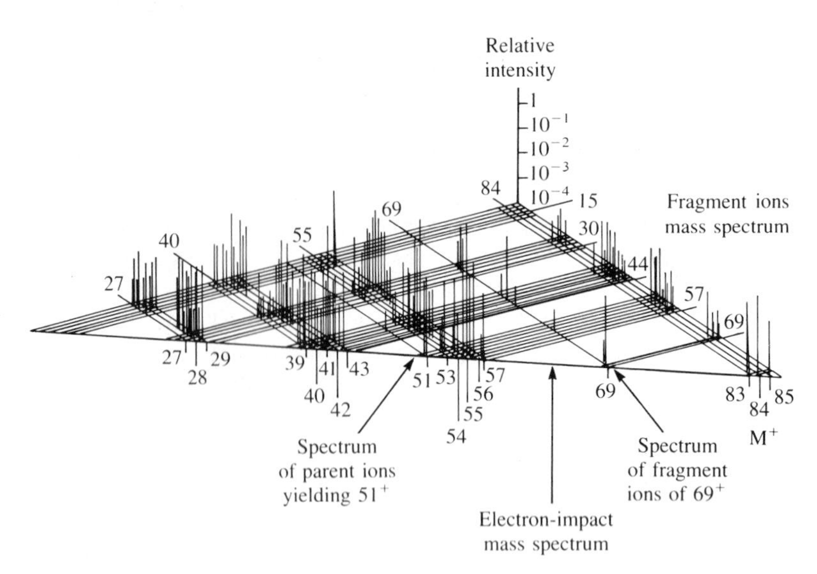

16.13 INDUCTIVELY COUPLED PLASMA–MASS SPECTROMETRY (ICP-MS)[31–33]

In ICP-MS, the sample is introduced into the ICP unit (see Chapter 10) by conventional ICP sample introduction methods, which include nebulization, hydride generation, and electrothermal vaporization. A portion of the ionized gas from the tail flame of the ICP is introduced into the vacuum system of the mass spectrometer, as shown in Figure 16.23. The plasma torch is positioned horizontally to achieve this introduction more easily. The plasma tail flame flows around the tip of a water-cooled nickel-alloy cone called the sampler. Gas from the ICP is extracted through a small aperture (0.4–0.8 mm) drilled in the cone and into the first (expansion) stage that operates at a pressure of about 1 torr. The central orifice of the conical skimmer is located behind the sampler at an appropriate position to transmit as much of the sampled beam as possible into a second vacuum chamber. The pressure here is low enough (about 5×10^{-4} torr) and the mean free path long enough for ion lenses to collect, focus, and transmit the ions to the mass analyzer. Transmitted ions are detected with a channel electron multiplier that is generally operated in the pulse counting mode—that is, detecting individual ions in order to discriminate against rf background from the quadrupole mass analyzer and ICP. To be observed by the mass analyzer, an ion must be present as such in the ICP and survive the extraction process, or be formed by chemical reactions during the extraction process.

The instruments have a quadrupole mass analyzer and yield essentially unit mass resolution over a mass range up to $m/z = 2000$. The multichannel scaling data system is normally set with a data acquisition memory group of 1024 channels, a dwell time per channel of 1 msec, and 60 separate sweeps. The quadrupole control is set for the first mass and the mass range required, and its scan is synchronized with each sweep of the scaling data system. A complete mass spectrum can be integrated and recorded in slightly longer than 1 min. Detection limits in ICP-MS are generally

FIGURE **16.23** Schematic diagram of an ICP-MS instrument showing the sampling interface. [Reprinted with permission from R. S. Houk, *Anal. Chem.*, **58,** 97A (1986). Copyright 1986 American Chemical Society.]

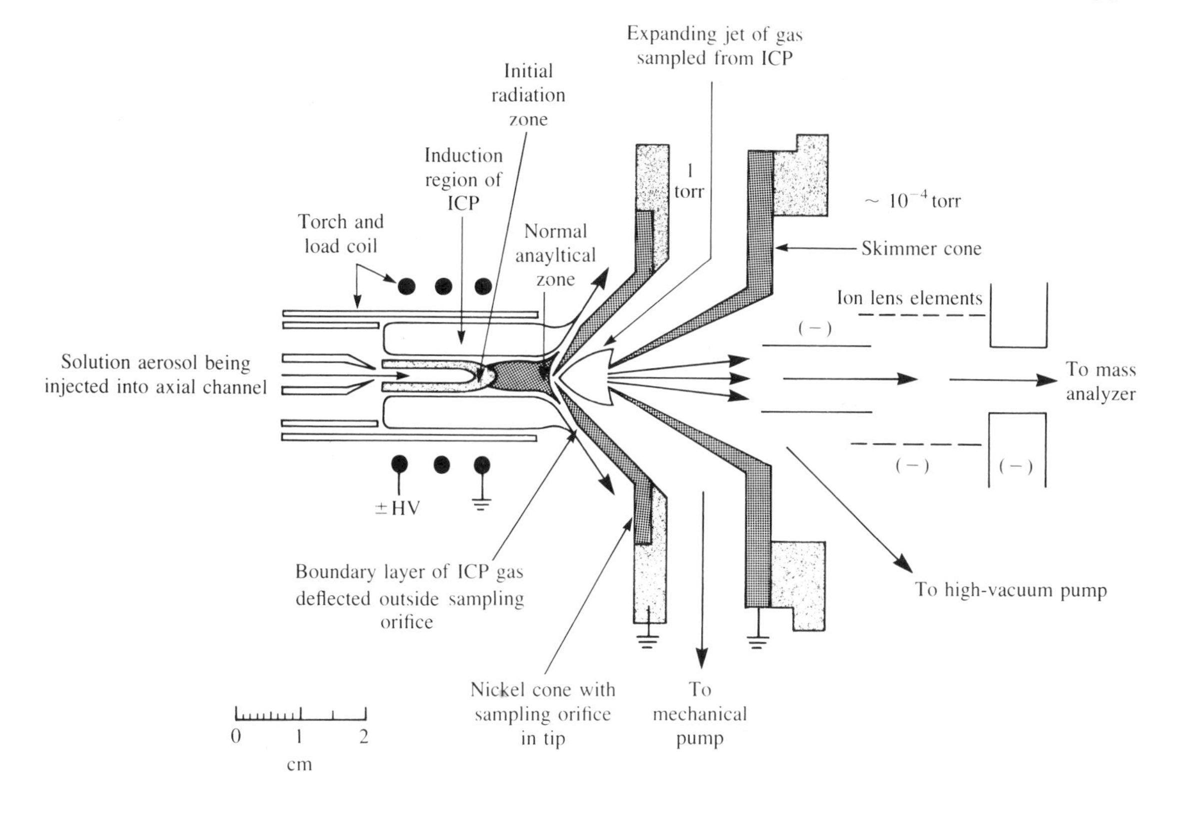

better than in induction-coupled plasma emission spectrometry, especially for elements near the center of the periodic chart that have very rich emission spectra. Under ideal conditions, detection limits of a few tens of parts per trillion are possible for most elements using 10-sec integrations.

16.14 SECONDARY ION MASS SPECTROMETRY (SIMS)

In SIMS an energetic primary ion strikes the sample surface and releases secondary ions. If atomic, these secondary ions are analyzed and detected as such. Molecular ions, however, dissociate to give positive and negative ion mass spectra for the molecules present in the surface. The SIMS spectrum therefore consists of both those secondary ions that are stable to such dissociation and the ionic fragments of those that are not.

SIMS is a rapid, easily used technique that not only affords qualitative identification of all surface elements, including hydrogen, but also permits identification of isotopes and the structural elucidation of molecular compounds present on a surface. The method permits detection sensitivity at the parts-per-million level with a minimum of sample volume (0.2–0.5 nm depth). In addition to depth composition information of about two atomic monolayers, SIMS obtains spatially resolved surface information, such as adsorbed ion images, elemental line scans, and images of surface constituents.

The positive SIMS spectra are extremely sensitive to elements on the left side of the periodic table, whereas the negative SIMS spectra favor the right side. Although SIMS has been applied mainly to metals and semiconductors, it also has been used for the analysis of glasses, rare earth compounds, and minerals. Contributions are being made in the analysis of precipitates, grain boundary segregations, and welded bonds, and to production problems such as temper brittleness, hot workability, hydrogen embrittlement, and hydrogen-induced cracking. SIMS is a popular method for analyzing nonvolatile and thermally labile molecules, including polymers and large biomolecules.

A serious concern in many laboratories is the feasibility of quantitation by SIMS. Although several studies have shown that quantitation may be achieved using stable isotopes as internal standards, the dynamic range appears to be very limited and plagued by competitive and suppressive processes that limit or invalidate quantitation. Generally the secondary ion yield is erratic until a reactive layer forms on the sample and a steady state is reached. The ion yield is then relatively constant, and quantitative analysis is possible.

Instrumentation for SIMS ranges from simple plasma discharge sources coupled with quadrupole mass analyzers to sophisticated ion microprobe mass analyzers. Whatever the level of sophistication, all instruments have a source of primary ions, a mass analyzer, and a sensitive secondary ion detector. All units are enclosed in a high vacuum or an ultrahigh vacuum chamber. Since the ion beam used in SIMS fulfills the excitation as well as the sputtering requirements of the technique, the ion source is a key part of the instrument. An electron-impact ionization source provides a beam of noble gas ions, as described in Section 16.3. The sputtered ions have low energies and are easily analyzed in a quadrupole mass analyzer.

16.15 ION MICROPROBE MASS ANALYZER (IMMA)[34]

The IMMA uses a microfocused primary ion beam from a duoplasmatron source to provide lateral microanalysis with high spatial resolution (2–10 μm) as well as surface analysis and in-depth profiling (Figure 16.24). The impinging ion beam is generated in a duoplasmatron source, which is essentially a low-voltage, low-pressure, hot-cathode arc that is capable of producing either positively or negatively charged ions. This uses a plasma with both electrostatic and magnetic constriction (hence the term *duo*) to produce a high-brightness source of inert or reactive gas ions (usually O^-, F^-, or Ar^+). Ions are extracted from the source through a hole in the anode and accelerated to energies ranging from 5 to 25 keV. Two electrostatic lenses then provide demagnification of the duoplasmatron source image to produce an ion beam diameter of 2–10 μm. The primary beam spot is held stationary for local analysis or moved about the sample surface (rastered) for secondary ion imaging or for producing a flat-bottomed crater for accurate depth profiling.

FIGURE **16.24** Schematic diagram of an ion microprobe mass analyzer. (Courtesy of Applied Research Laboratories.)

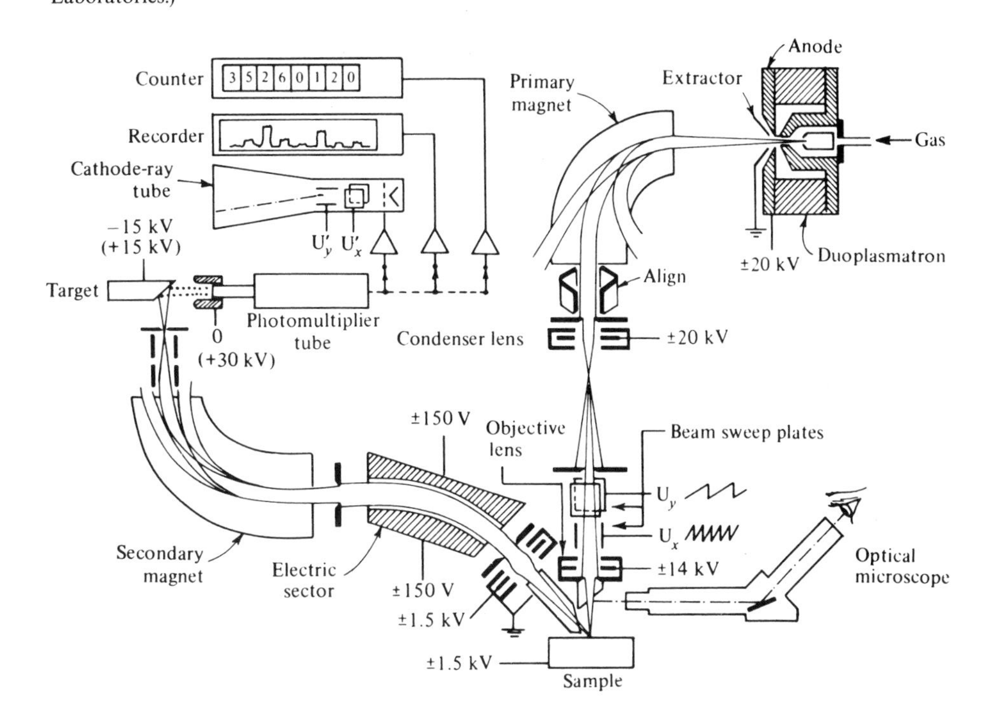

The duoplasmatron produces a variety of ionized species from the sample and, in some cases, ionized molecular fragments. Since it is desirable to have only one type of ion interact with the sample, the primary beam is passed through a mass analyzer placed at the entrance to the electrostatic lens column. The magnetic field is adjusted so that only the desired ion is deflected into the lenses. The composition of the primary beam can be checked by using a sample surface as an electrostatic mirror, thereby deflecting a fraction of the primary beam into the mass spectrometer. In IMMA the magnetic field is the wedge-type with plane but inclined pole pieces.

Sorting out the sputtered secondary sample ions by mass/charge ratio is accomplished in the IMMA system by a two-stage mass spectrometer. The first-stage electrostatic sector sorts the ions on the basis of velocity and brings most of them to a similar speed. Next, the second-stage magnetic sector sorts the ions on the basis of mass/charge ratio. Only those ions that have a discrete selected m/z ratio are passed through the exit slit to the detector. An electrostatic lens increases the angular aperture of the detector.

PROBLEMS

1. For a field strength of 2400 G in 180° magnetic-deflection spectrometer, what electrostatic voltage range suffices for scanning from mass 18 to mass 200? The radius of curvature of the 180° analyzer tube is 12.7 cm.

2. For a drift length of 100 cm in a time-of-flight mass spectrometer, what is the difference in arrival time between ions of $m/z = 44$ and $m/z = 43$ when the accelerating voltage is 2800 V?

3. The spectrum of cetyl palmitate (M.W. 480) was recorded on a linear scale in the presence of $C_{10}F_{19}$ (M.W. 480.970). On the spectrum the molecular ion peak of cetyl palmitate was separated from the $M + 1$ peak by 200.5 mm and from the perfluorocarbon molecular ion peak by 95.3 mm. What is the precise mass of the molecular ion of cetyl palmitate?

4. The parent peak spectrum of tridecylbenzene (260.2504), phenyl undecyl ketone (260.2140), 1,2-dimethyl-4-benzoylnapthalene (260.1201), and 2,2-naphthylbenzothiophene (260.0922) would require what resolution for quantitative analysis based on the parent peak?

5. What resolving power is needed to separate (a) the CH_2N—C_2H_4 doublet at mass 200, (b) the N_2—CO doublet at mass 150, and (c) the CH_2—N doublet at mass 200?

6. A peptide was admitted to a high-resolution mass spectrometer and the parent peak mass was measured relative to the parent peak in the spectrum of dibromobenzene (236.8638). The measured ratio of unknown mass/reference mass was 1.001197 ± 0.000002. Compute the exact weight of the peptide and deduce the molecular formula.

7. From the following exact molecular weights, estimate the empirical formulas assuming only C, H, O, or N is present unless otherwise indicated: (a) 164.0473, (b) 120.0575, (c) 180.0939, (d) 94.0531, (e) 109.0528, (f) 190.9540 (contains Cl), (g) 181.0891, (h) 334.0873 (contains S), and (i) 177.0426.

8. (a) In the high-resolution spectrum of methionine, a quartet of peaks appears at nominal mass 88: 88.0220, 88.0345, 88.0335, and 88.0267. Deduce the fragment ion responsible for each peak. (b) Methionine also gives a doublet at nominal mass 75. One line corresponds to $C_2H_4NO_2$; the other has m/z 75.0267. Outline the process leading to the fragment $C_2H_4NO_2$. Deduce the fragment ion of m/z 75.0267.

9. What is the probable composition of a molecule of mass 142 whose P_{M+1} peak is 1.1% of the parent peak?

10. Deduce the number and type of halogen atoms present in a molecule from the abundance of heavy isotopes and the intensity ratios given in the following table:

	P_M	P_{M+2}	P_{M+4}	P_{M+6}	P_{M+8}
Compound A	30	29	10	1	—
Compound B	13	30	19	6	1
Compound C	5	20	30	19	5
Compound D	23	30	7	—	—
Compound E	18	30	14	2	—

11. From the isotopic abundance information, what can be deduced concerning the empirical formula of each of the following compounds?

m/z	Percent of base peak
90(*P*)	*100.00*
91	5.61
92	4.69

m/z	Percent of base peak
89(*P*)	17.12
90	0.58
91	5.36
92	0.17

m/z	Percent of base peak
206(*P*)	25.90
207	3.24
208	2.48

m/z	Percent of base peak
230(*P*)	1.10
232	2.12
234	1.06

m/z	Percent of base peak
140(*P*)	14.8
141	1.40
142	0.85

m/z	Percent of base peak
151(*P*)	*100.0*
152	10.4
153	32.1
154	2.9

12. The significant portion of the mass spectral data is given for the individual alcohols. Select appropriate analytical masses and write a series of four equations in terms of divisions of base peak due to each of the four alcohols.

	Percent of base peak				*Unknown mixtures*		
m/z	Methyl	Ethyl	*n*-Propyl	Isopropyl	A	B	C
15	35.48	9.44	3.77	10.70			
19	0.29	3.13	0.90	6.51			
27	—	21.62	15.20	15.50			
29	58.80	21.24	14.14	9.49			
31	*100*	*100.00*	*100.00*	5.75			
32	68.03(*P*)	1.14	2.25	—	600	600	2350
33	0.98	—	—	—			
39		—	4.00	5.52	4800	3000	3000
43		7.45	3.18	16.76			
45		37.33	4.39	*100.00*			
46		16.23(*P*)	—	—	1000	1100	698
59			9.61	3.58	4000	2300	5000
60			6.36(*P*)	0.44(*P*)			
Sensitivity: 8.76 (divisions/10^{-3} torr)		17.98	26.51	23.47			

13. The mixture peaks for three unknown mixtures of alcohols are shown in Problem 12. For each mixture compute the mole percent of each alcohol.

14. A material containing only C, H, and O, and in the form of leaflets melting at 40 °C, possesses a rather simple mass spectrum with the parent peak at m/z 184

(10%), the base peak at m/z 91, and small peaks at m/z 77 and 65. Metastable peaks appear at m/e 45.0 and 46.5. Deduce the structure of the compound.

15. The mass spectrum possesses a strong parent peak at m/z 122 (35%) plus peaks at m/z 92 (65%), m/z 91 (100%), and m/z 65 (15%). In addition there are metastable peaks at 46.5 and 69.4 mass units. Deduce the compound's structure.

16. Deduce the structural formula for each of these compounds from the mass spectral data:

$C_4H_8O_2$		$C_4H_6O_2$		*C*		*D*	
m/z	Percent of base peak	m/z	Percent of base peak	m/z	Percent of base peak	m/z	Percent of base peak
27	39.3	15	27.7	29	18	63	22
29	19.8	26	22.4	39	23	64	20
39	14.8	27	68.1	51	29	65	33
41	23.7	29	13.0	65	18	92	82
42	24.7	42	11.8	78	50	93	18
43	22.3	55	*100.0*	91	*100*	120	*100*
45	19.1	58	8.4	105	41	121	34
60	*100.0*	59	5.2	134(*P*)	57.4	152(*P*)	45.0
73	27.1	85	12.3	135	5.80	153	4.1
88(*P*)	1.6	86(*P*)	2.1	136	0.41	154	0.4

17. Deduce the complete structural formula of the compound from the mass spectrum in the figure below.

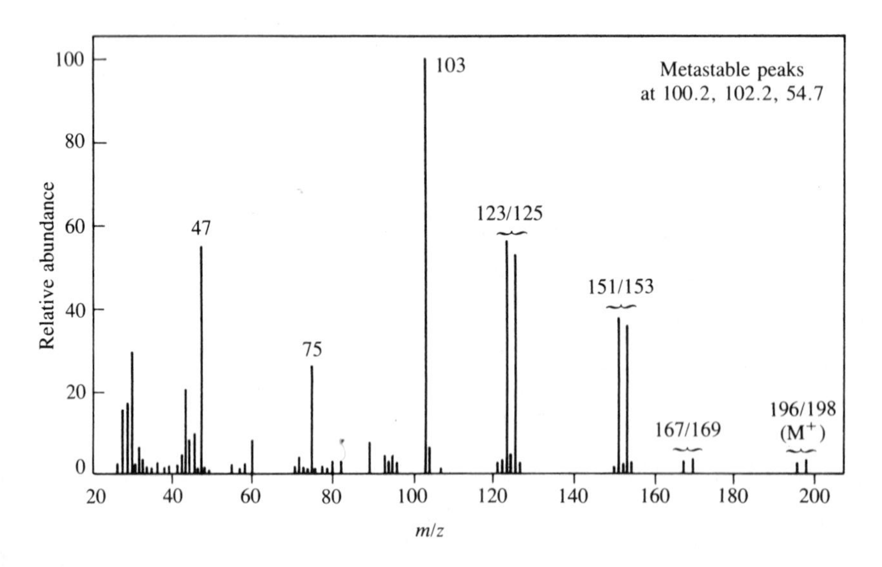

18. Deduce the structural formula of the compound (b.p. 74 °C) whose mass spectrum is shown in the figure on p. 511.

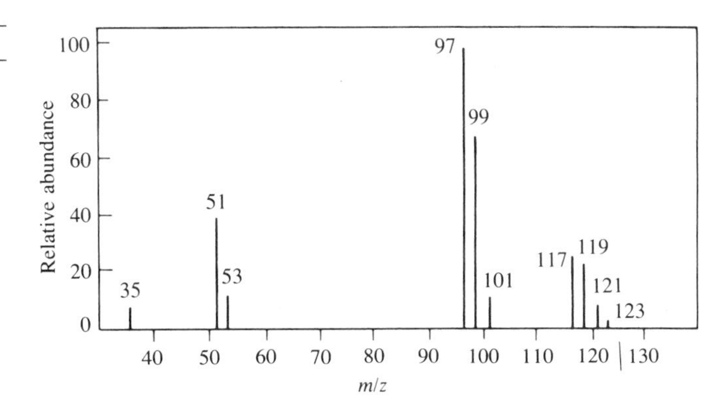

19. What is the characteristic cyclotron frequency of an ion of $m/z = 1000$ when the magnetic field strength is 10,000 gauss?

20. For a 2-m flight tube and an ion energy of 2 keV, what time is required for an ion of 800 daltons to travel from the source to the detector?

BIBLIOGRAPHY

BECKEY, H. D., *Principles of Field Ionization and Field Desorption Mass Spectrometry,* Pergamon, London, 1977.

Eight Peak Index of Mass Spectra, 3rd ed., The Royal Society of Chemistry, Letchworth, Herts., England, 1985.

MCLAFFERTY, F. W., *Interpretation of Mass Spectra,* 3rd ed., University Science Books, Mill Valley, CA, 1980.

MCLAFFERTY, F. W., ed., *Tandem Mass Spectrometry,* John Wiley, New York, 1982.

MILLARD, B. J., *Quantitative Mass Spectrometry,* Heyden, Philadelphia, 1979.

SILVERSTEIN, R. M., G. C. BASSLER, AND T. C. MORRILL, *Spectrometric Identification of Organic Compounds,* 4th ed., John Wiley, New York, 1981.

WATSON, J. T., *Introduction to Mass Spectrometry,* Raven, New York, 1985.

LITERATURE CITED

1. MILBERG, R. M., AND J. C. COOK, JR., "Design Considerations of MS Sources: EI, CI, FI, FD, and API," *J. Chromatogr. Sci.,* **17,** 17 (1979).
2. MUNSON, B., "Chemical Ionization Mass Spectrometry," *Anal. Chem.,* **49,** 772A (1977).
3. DOUGHERTY, R. C., "Negative Chemical Ionization Mass Spectrometry," *Anal. Chem.,* **53,** 625A (1981).
4. ANBAR, M., AND W. H. ABERTH, "Field Ionization Mass Spectrometry: A New Tool for the Analytical Chemist," *Anal. Chem.,* **46,** 59A (1974).
5. REYNOLDS, W. D., "Field Desorption Mass Spectrometry," *Anal. Chem.,* **51,** 283A (1979).
6. SCHULTEN, H. R., AND P. B. MONKHOUSE, "Fast Quantitative Trace Analysis of Metals in Inorganic and Biological Materials," *Am. Lab.,* **15,** 44 (March 1983).
7. MCNEAL, C. J., "Symposium on Fast Atom and Ion Induced Mass Spectrometry of Nonvolatile Organic Solids," *Anal. Chem.,* **54,** 43A (1982).
8. BARBER, M., R. S. BORDOLI, G. J. ELLIOTT, R. D. SEDGWICK, AND A. N. TYLER, "Fast Atom Bombardment Mass Spectrometry," *Anal. Chem.,* **54,** 645A (1982).
9. MAHONEY, J., J. PEREL, AND S. TAYLOR, "Primary Ion Sources for Fast Atom Bombardment," *Am. Lab.,* **16,** 92 (March 1984).

10. MACFARLANE, R. D., "Californium-252 Plasma Desorption Mass Spectrometry," *Anal. Chem.*, **55,** 1247A (1983).
11. DENOYER, E., R. VAN GRIEKEN, F. ADAMS, AND D. F. S. NATUSCH, "Laser Microprobe Mass Spectrometry: Basic Principles and Performance Characteristics," *Anal. Chem.*, **54,** 26A (1982).
12. HERCULES, D. M., R. J. DAY, K. BALASANMUGAM, R. A. DANG, C. C. P. LI, "Laser Microprobe Mass Spectrometry: Applications to Structural Analysis," *Anal. Chem.*, **54,** 280A (1982).
13. COTTER, R. J., AND J. TABET, "Laser Desorption MS for Nonvolatile Organic Molecules," *Am. Lab.*, **16,** 10 (March 1984).
14. COTTER, R. J., "Lasers and Mass Spectrometry," *Anal. Chem.*, **56,** 485A (1984).
15. HERZOG, R. F., "Mattauch-Herzog Mass Spectrograph," *Am. Lab.*, **1,** 15 (May 1969).
16. CAPRIOLI, R. M., W. F. FIES, AND M. S. STORY, "Direct Analysis of Stable Isotopes with a Quadrupole Mass Spectrometer," *Anal. Chem.*, **46,** 453A (1974).
17. BEYNON, J. H., AND A. E. WILLIAMS, *Mass and Abundance Tables for Use in Mass Spectrometry,* Elsevier, Amsterdam, 1963.
18. SILVERSTEIN, R. M., G. C. BASSLER, AND T. C. MORRILL, *Spectrometric Identification of Organic Compounds,* 4th ed., John Wiley, New York, 1981.
19. MCIVER, R. T., JR., "Fourier Transform Mass Spectrometry," *Am. Lab.*, **12,** 18 (November 1980).
20. WILKINS, C. L., AND M. L. GROSS, "Fourier Transform Mass Spectrometry for Analysis," *Anal. Chem.*, **53,** 1661A (1981).
21. BOWERS, W. D., R. L. HUNTER, AND R. T. MCIVER, JR., "FT-MS Uses Image Current Detector to Get High Mass Resolution," *Ind. Res/Dev.*, p. 124 (November 1983).
22. COMISAROW, M. B., AND A. G. MARSHALL, *Can. J. Chem.*, **52,** 1997 (1984).
23. MCLAFFERTY, F. W., "Tandem Mass Spectrometry," *Science,* **214,** 280 (1981).
24. SLAYBACK, J. R. B., AND M. S. STORY, "Chemical Analysis Problems Yield to Quadrupole MS/MS," *Ind. Res/Dev.*, p. 129 (February 1981).
25. COOKS, R. G., AND G. L. GLISH, "Mass Spectrometry/Mass Spectrometry," *Chem. Eng. News,* p. 40 (November 30, 1981).
26. BUSCH, K. L., AND R. G. COOK, "Taking Stock of Mass Spectrometry/Mass Spectrometry," *Anal. Chem.*, **55,** 38A (1983).
27. JOHNSON, J. V., AND R. A. YOST, "Tandem Mass Spectrometry for Trace Analysis," *Anal. Chem.*, **57,** 758A (1985).
28. BORMAN, S. A., "MS/MS Instrumentation," *Anal. Chem.*, **58**(3), 406A (1986).
29. YOST, R. A., AND C. G. ENKE, "Triple Quadrupole Mass Spectrometry for Direct Mixture Analysis and Structure Elucidation," *Anal. Chem.*, **51,** 1251A (1979).
30. YOST, R. A., AND C. G. ENKE, "Structure Elucidation Through Triple Quadrupole MS," *Am. Lab.*, **13,** 88 (June 1981).
31. HOUK, R. S., V. A. FASSEL, G. D. FLESCH, H. J. SVEC, A. L. GRAY, AND C. E. TAYLOR, "Inductively Coupled Argon Plasma as an Ion Source for Mass Spectrometric Determination of Trace Elements," *Anal. Chem.*, **52,** 2283 (1980).
32. DOUGLAS, D. J., AND J. B. FRENCH, "Elemental Analysis with a Microwave-Induced Plasma/Quadrupole Mass Spectrometer System," *Anal. Chem.*, **53,** 37 (1981).
33. HOUK, R. S., "Mass Spectrometry of Inductively Coupled Plasmas," *Anal. Chem.*, **58,** 97A (1986).
34. LIEBL, H., "Ion Microprobe Analyzers: History and Outlook," *Anal. Chem.*, **46,** 22A (1974).

CHROMATOGRAPHY: GENERAL PRINCIPLES

The feature that distinguishes chromatography from most other physical and chemical methods of separation is that two mutually immiscible phases are brought into contact; one phase is stationary and the other mobile. A sample introduced into a mobile phase is carried along through a column (manifold) containing a distributed stationary phase. Species in the sample undergo repeated interactions (partitions) between the mobile phase and the stationary phase. When both phases are properly chosen, the sample components are gradually separated into bands in the mobile phase. At the end of the process, separated components emerge in order of increasing interaction with the stationary phase. The least retarded component emerges first; the most strongly retained component elutes last. Partition between the phases exploits differences in the physical and/or chemical properties of the components in the sample. Adjacent components (peaks) are separated when the later-emerging peak is retarded sufficiently to prevent overlap with the peak that emerges ahead of it.

The separation column is the heart of the chromatograph. It provides versatility in the types of analyses that can be performed. This versatility, due to the wide choice of materials for the stationary and mobile phases, makes it possible to separate molecules that differ only slightly in their physical and chemical properties. Broadly speaking, the distribution of a solute between two phases results from the balance of forces between solute molecules and the molecules of each phase. It reflects the relative attraction or repulsion that molecules or ions of the competing phases show for the solute and for themselves. These forces can be polar in nature, arising from permanent or induced dipole moments, or they can be due to London's dispersion forces. In ion-exchange chromatography, the forces on the solute molecules are substantially ionic in nature but include polar and nonpolar forces as well. The relative polarity of solvents is manifested in their dielectric constant.

17.1 CLASSIFICATION OF CHROMATOGRAPHIC METHODS

The mobile phase can be a gas or a liquid, whereas the stationary phase can be only a liquid or a solid (Table 17.1). When the separation involves predominantly a simple partitioning between two immiscible liquid phases, one stationary and the other mobile, the process is called *liquid-liquid chromatography* (LLC). When physical

TABLE 17.1 CHROMATOGRAPHIC METHODS

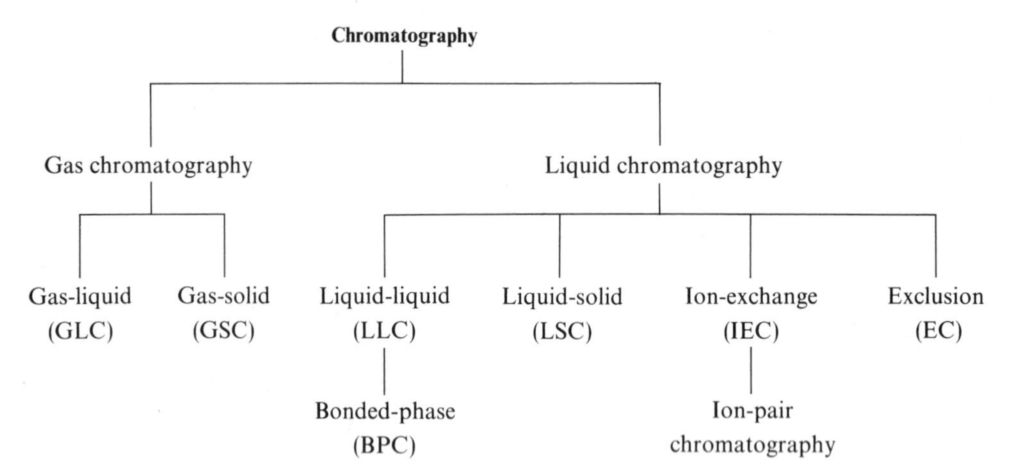

surface forces are mainly involved in the retentive ability of the stationary phase, the process is denoted *liquid-solid* (or adsorption) *chromatography* (LSC).

Two other liquid chromatographic methods differ somewhat in their mode of action. In *ion-exchange chromatography* (IEC), ionic components of the sample are separated by selective exchange with counterions of the stationary phase. The use of exclusion packings as the stationary phase brings about a classification of molecules based largely on molecular geometry and size. *Exclusion chromatography* (EC) is referred to as gel-permeation chromatography by polymer chemists and as gel filtration by biochemists.

When the mobile phase is a gas, the methods are called *gas-liquid chromatography* (GLC) and *gas-solid chromatography* (GSC). Methods involving gas (see Chapter 18) and liquid mobile (see Chapters 19 and 20) phases will be treated individually. Although there are few fundamental reasons for separate treatment, differences in operating technique and equipment warrant individual chapters.

17.2 CHROMATOGRAPHIC BEHAVIOR OF SOLUTES

The chromatographic behavior of a solute can be described in numerous ways. For column chromatography, the retention volume, V_R (or corresponding retention time, t_R), and the partition ratio, k', are the terms most frequently used. By varying the stationary–mobile-phase combinations and various operating parameters, the degree of retention can be varied from nearly total retention to a state of free migration.

Retention Behavior

Retention behavior reflects the distribution of a solute between the mobile and stationary phases. Figure 17.1 shows the separation of two isomeric alkenes. The volume of mobile phase necessary to convey a solute band from the point of injection, through the column, and to the detector (to the apex of the solute peak) is defined as the *retention volume*, V_R. It may be obtained directly from the corresponding *retention time*, t_R, on the chromatogram by multiplying the latter by the *volumetric flowrate*, F_c, expressed as the volume of the mobile phase per unit time:

$$V_R = t_R F_c \tag{17.1}$$

The flowrate, in terms of column parameters, is as follows:

$$F_c = \underbrace{\frac{\pi d_c^2}{4}}_{\substack{\text{cross}\\ \text{section}\\ \text{of empty}\\ \text{column}}} \times \underbrace{\varepsilon_{\text{tot}}}_{\substack{\text{total}\\ \text{porosity}}} \times \underbrace{\frac{L}{t_M}}_{\substack{\text{average}\\ \text{linear}\\ \text{velocity}\\ \text{of eluent}}} = \frac{V_{\text{col}}\,\varepsilon_{\text{tot}}}{t_M} \tag{17.2}$$

where d_c is the column bore, L is the column length, ε_{tot} is the total porosity of the column packing, and V_{col} is the bed volume of the column. The porosity expresses the ratio of the interstitial volume of the packing to the volume of its total mass. For solid packings the total porosity is 0.35–0.45, whereas for porous packings it is 0.70–0.90. In capillary columns the value of ε_{tot} is unity. The *average linear velocity*, u, of the mobile phase,

$$u = \frac{L}{t_M} \tag{17.3}$$

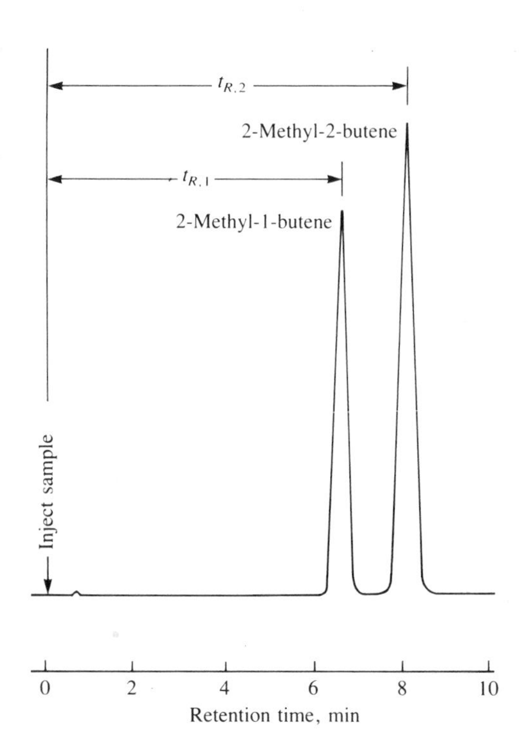

FIGURE 17.1
Separation of 2-methyl-1-butene and 2-methyl-2-butene by gas chromatography on a column packed with 25% SE-30 on Chromosorb W at 41 °C.

is measured by the transit time of a nonretained solute, t_M. In interactive chromatography no material can elute prior to this time. When converted to volume, V_M (or V_o), it represents what is called the dead space, void volume, or holdup volume of a column. It includes the effective volume contributions of the sample injector, any connecting tubing, the column itself, and the detector.

The *adjusted retention volume*, V'_R, or *time*, t'_R, is given by

$$V'_R = V_R - V_M \quad \text{or} \quad t'_R = t_R - t_M \tag{17.4}$$

When the mobile phase is a gas, temperature and pressure must be specified, and retention volumes must be corrected for the compressibility of the gas, since the gas moves more slowly near the inlet than at the exit of the column. The *pressure-gradient correction* (or compressibility) *factor*,* j, is expressed by

$$j = \frac{3[(P_i/P_o)^2 - 1]}{2[(P_i/P_o)^3 - 1]} \tag{17.5}$$

where P_i is the carrier gas pressure at the column inlet and P_o that at the outlet.

Partition Coefficient

When a solute enters a chromatographic system, it immediately distributes between the stationary and mobile phases. If the mobile-phase flow is stopped at any time, the solute assumes an equilibrium distribution between the two phases. The concentration in each phase is given by the *thermodynamic partition coefficient:*

$$K = \frac{C_S}{C_M} \tag{17.6}$$

where C_S and C_M are the concentrations of solute in the stationary and mobile phases, respectively. When $K = 1$, the solute is equally distributed between the two phases. The partition coefficient determines the average velocity of each solute zone—more specifically, the zone center as the mobile phase moves down the column.

For a symmetrical peak, when the peak maximum appears at the column exit, half of the solute has eluted in the retention volume, V_R, and half remains distributed between the volume of the mobile phase, V_M, and the volume of the stationary phase, V_S. Thus,

$$V_R C_M = V_M C_M + V_S C_S \tag{17.7}$$

Rearranging and combining with Equation 17.6, we obtain a fundamental equation in chromatography:

$$V_R = V_M + K V_S \quad \text{or} \quad V_R - V_M = K V_S \tag{17.8}$$

It relates the retention volume of a solute to the column dead volume and the product of the partition coefficient and the volume of the stationary phase. This equation is correct for liquid partition columns, but for adsorption columns, V_S should be replaced by A_S, the surface area of the adsorbent.

* For a derivation of the pressure-gradient correction factor, see W. E. Harris and H. W. Habgood, *Programmed Temperature Gas Chromatography*, John Wiley, New York, 1966, p. 49.

Partition Ratio

The *partition ratio* (or *capacity ratio*), k', is the most important quantity in column chromatography. It relates the equilibrium distribution of the sample within the column to the thermodynamic properties of the column and to the temperature, as will be shown. For a given set of operating parameters, k' is a measure of the time spent in the stationary phase relative to the time spent in the mobile phase. It is defined as the ratio of the moles of a solute in the stationary phase to the moles in the mobile phase:

$$k' = \frac{C_S V_S}{C_M V_M} = K \frac{V_S}{V_M} \tag{17.9}$$

The *volumetric phase ratio*, V_M/V_S, is often denoted by the symbol β. Thus, $k' = K/\beta$. Stated another way, the partition ratio is the additional time a solute band takes to elute, as compared with an unretained solute (for which $k' = 0$), divided by the elution time of an unretained band:

$$k' = \frac{t_R - t_M}{t_M} = \frac{V_R - V_M}{V_M} \tag{17.10}$$

The relation states explicitly how many dead volumes (or t_M) are required to attain V_R (or t_R). Rearranging Equation 17.10 and introducing Equation 17.3, retention times are related to k' by the equation:

$$t_R = t_M(1 + k') = \left(\frac{L}{u}\right)(1 + k') \tag{17.11}$$

As will be shown, values of k' higher than ten waste valuable analytical time. Values less than unity do not provide adequate resolution among early-eluting solutes.

EXAMPLE 17.1

On a 1000-cm wall-coated open tubular column of 0.25-mm bore, the helium carrier gas velocity is 37 cm/sec. The retention time, t_R, for decane is 1.27 min; peak width at half height is 0.88 sec. The retention time for a nonretained compound, t_M, is

$$t_M = \frac{L}{u} = \frac{1000 \text{ cm}}{37 \text{ cm sec}^{-1}} = 27 \text{ sec or } 0.45 \text{ min}$$

The partition ratio, k', is

$$k' = \frac{t'_R}{t_M} = \frac{1.27 - 0.45}{0.45} = 1.82$$

The fraction of time that a solute spends in a particular phase is very close to the fraction of all those particular solute molecules that are instantaneously in the same phase. Thus, the average fraction of time spent by a solute in the mobile phase is

$$\frac{C_M V_M}{C_M V_M + C_S V_S} = \frac{1}{1 + k'} \tag{17.12}$$

Similarly, for the stationary phase,

$$\frac{C_S V_S}{C_M V_M + C_S V_S} = \frac{k'}{1 + k'} \tag{17.13}$$

Relative Retention

The relative retention, α, of two solutes, where solute 1 elutes before solute 2, is given variously by

$$\alpha = \frac{k'_2}{k'_1} = \frac{K_2}{K_1} = \frac{V'_{R,2}}{V'_{R,1}} = \frac{t'_{R,2}}{t'_{R,1}} \tag{17.14}$$

The relative retention is dependent on (1) the nature of the stationary and mobile phases and (2) the column operating temperature. One should always be as selective as possible in choosing a pair of phases for the adjacent solutes most difficult to separate.

17.3 COLUMN EFFICIENCY AND RESOLUTION

Under operating conditions where the partition between the stationary and mobile phases is linear (that is, Henry's law is obeyed), K and k' are independent of the total solute concentration. After 50 or more partitions between the phases, the resultant profile of a solute band closely approaches that given by a Gaussian distribution curve (Figure 17.2). However, as the solute band passes through the chromato-

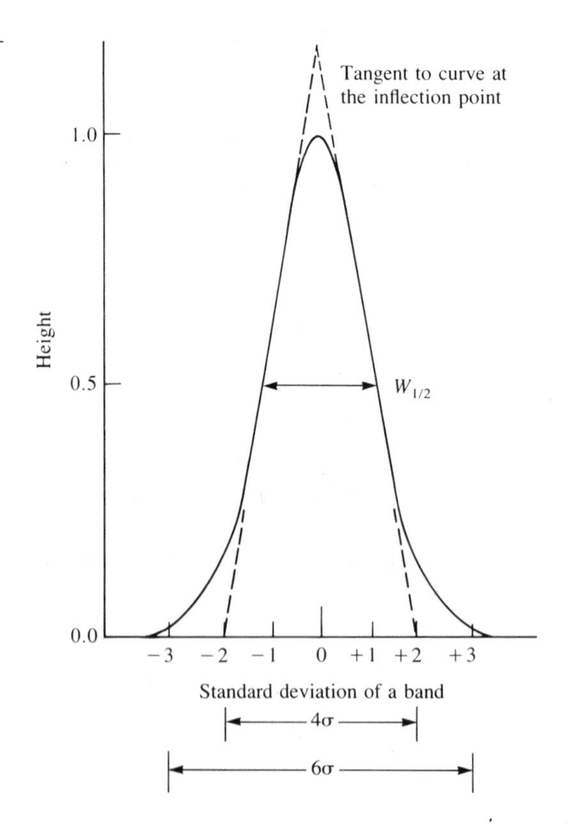

FIGURE 17.2
Profile of a solute band.

graphic column, it broadens and the concentration at the peak maximum decreases. This broadening ultimately affects the resolution of adjacent solute bands, as will be discussed.

Plate Height and Plate Number

An important characteristic of a chromatographic system is its efficiency expressed as a dimensionless quantity called the *effective plate number*, N_{eff}. It reflects the number of times the solute partitions between the two phases during its passage through the column. The effective plate number can be defined from the chromatogram of a single band, as shown in Figure 17.3:

$$N_{\text{eff}} = \frac{L}{H} = \left(\frac{t'_R}{\sigma}\right)^2 \tag{17.15}$$

where L is the column length, H is the plate height, t'_R is the adjusted time for elution of the band center, and σ^2 is the band variance in time units. The width at the base of the peak, W_b (determined from the intersections of tangents to the inflection points with the baseline), is equal to four standard deviations (assuming an ideal Gaussian distribution; see Figure 17.2). Thus, in Equation 17.15, $\sigma = W_b/4$ and

$$N_{\text{eff}} = 16\left(\frac{t'_R}{W_b}\right)^2 \tag{17.16}$$

The upper portion of the peak dictates the tangent line, which minimizes any contribution from a tailing (or fronting) segment of the peak.

Often it is easier to measure the width at half the peak height. Since $\sigma = W_{1/2}/\sqrt{8 \ln 2}$,

$$N_{\text{eff}} = \frac{L}{H} = 5.54\left(\frac{t'_R}{W_{1/2}}\right)^2 \tag{17.17}$$

Measurement of the peak width at half the peak height is less sensitive to peak asymmetry, since tailing often shows up below the measurement location.

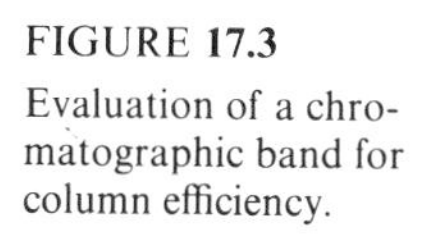

FIGURE 17.3
Evaluation of a chromatographic band for column efficiency.

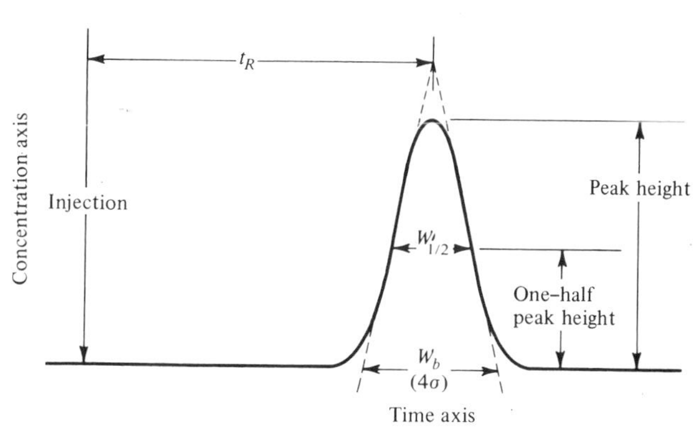

Although this is not recommended, column efficiency is sometimes stated as the number of theoretical plates. In this context no correction is made for the transit time of a nonretained solute.

Plate number is an indication only of how well a column has been packed; it cannot adequately predict column performance under all conditions. It is designed to be primarily a measure of the kinetic contributions to band broadening (see Section 17.4). Other contributions to peak width, such as extracolumn effects and thermodynamic factors (such as peak tailing, discussed next), may play a significant role.

Plate height, H, is the distance a solute moves while undergoing one partition:

$$H = \frac{L}{N_{\text{eff}}} \tag{17.18}$$

Plate height is a good way to express column efficiency in units of length without specifying the length of the column. From a theoretical point of view, plate height can be directly related to the experimental conditions and operating parameters. H is a small number for an efficient column.

EXAMPLE 17.2

Referring back to Example 17.1 and continuing, note that the effective number of plates, N, is

$$N_{\text{eff}} = 5.545\left(\frac{t'_R}{W_{1/2}}\right)^2 = 5.545\left(\frac{49.2 \text{ sec}}{0.88 \text{ sec}}\right)^2 = 17{,}300$$

The plate height, H, is

$$H = \frac{L}{N_{\text{eff}}} = \frac{1{,}000 \text{ cm}}{17{,}300} = 0.058 \text{ cm}$$

Band Asymmetry

Asymmetric bands are a common complaint among chromatographers. Fortunately, the causes are well documented and it is often possible to diagnose the reason for band asymmetry in a particular separation. Symmetrical bands are normally observed only for samples that do not exceed some maximum size, usually 1 mg of sample per gram of stationary phase (0.1 mg/g for pellicular packings). If k' is higher at lower concentrations of solute, then the low-concentration wing of the eluent peak moves more slowly than the high-concentration wing. As an initially symmetric band moves down the column, it becomes skewed and eventually develops a sharp front and a long tail. The reverse type of asymmetry is known as fronting. The solution in this case is to decrease the sample size to the point where retention times and band shape for all bands become symmetrical.

The peak asymmetry factor (AF) is defined as the ratio of the peak half-widths at a given peak height. The lower down the peak the asymmetry is measured, the larger AF is. Because of detector noise, among other factors, an acceptable compromise is to measure the AF at 10% of peak height—that is, the ratio b/a, as shown in Figure 17.4. When the asymmetry factor lies outside the range 0.95–1.15 for a peak of $k' = 2$, the apparent plate number for a column (as calculated by Equation 17.16) is too high. An AF of 1.3 reduces the efficiency by 69% and the resolution by 30%. Foley and Dorsey[1] developed an expression for the column efficiency in terms of graphically measurable parameters when the elution curve is asymmetric:

$$N_{\text{eff}} = \frac{41.7(t'_R/W_{0.1})}{(a/b) + 1.25} \tag{17.19}$$

Many band-tailing problems can be attributed to the wrong combination of sample and column packing. When system mismatch is suspected, a different type of column should be tried.

Heterogeneous retention sites are a problem encountered most commonly with liquid-solid or ion-exchange chromatographic systems. Usually retention sites in liquid-solid or ion-exchange chromatography are not exactly equivalent within a given packing, which results in sites of varying retention affinity. Initially retention occurs on the more active sites. When these sites are not overloaded, there is normal elution of the sample bands. Another approach is to selectively remove the stronger sites—that is, to deactivate the stationary phase as is often done in liquid-solid chromatography.

Asymmetrical peaks may result from actions that occur outside the column, particularly injection problems. They may also arise from a poorly packed column.

Resolution

The degree of separation or *resolution* of two adjacent bands is defined as the distance between band peaks (or centers) divided by the average bandwidth. If retention and bandwidth are measured in units of time, as in Figure 17.5, the resolution, R_s, is given as

$$R_s = \frac{t_{R,2} - t_{R,1}}{0.5(W_2 + W_1)} \tag{17.20}$$

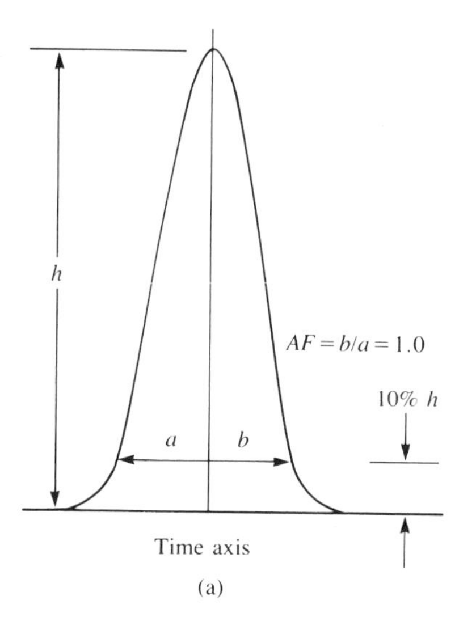

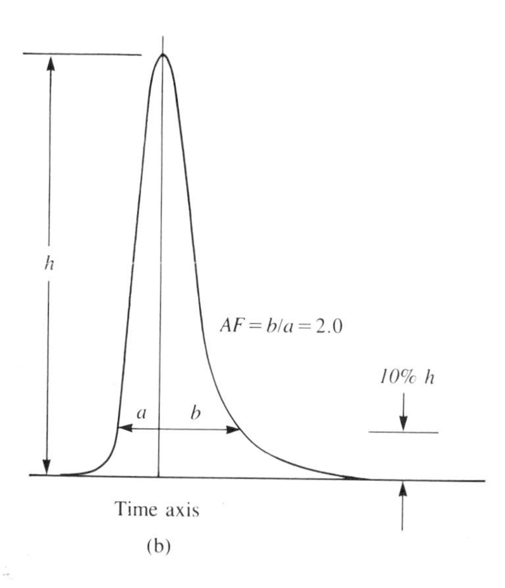

FIGURE 17.4
Peak asymmetry factor: (a) symmetrical band and (b) band tailing present.

FIGURE 17.5
Definition of resolution.

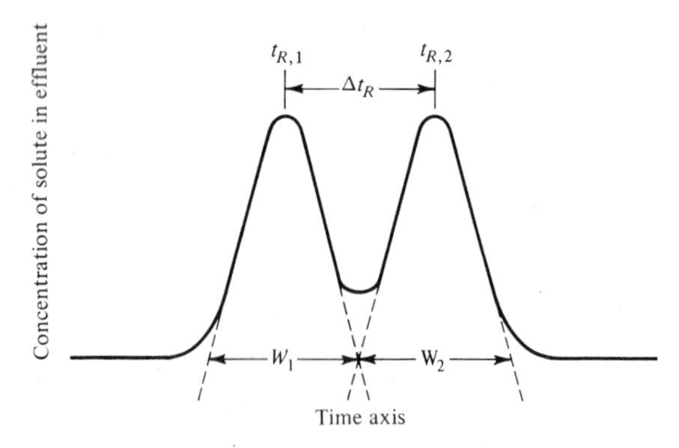

EXAMPLE 17.3

On a column 122 cm long, operated at 160 °C, these retention times (in minutes) were obtained: air peak, 0.90; heptane, 1.22; and octane, 1.43. The base widths of the bands were 0.14 for heptane and 0.20 for octane. What are the relative retention and the resolution for these bands?

$$\alpha = \frac{t'_{R,2}}{t'_{R,1}} = \frac{t_{R,2} - t_M}{t_{R,1} - t_M} = \frac{1.43 - 0.90}{1.22 - 0.90} = 1.66$$

The resolution of the heptane/octane bands (see Equation 17.20) is

$$R_s = \frac{1.43 - 1.22}{0.5(0.20 + 0.14)} = 1.24$$

Solute bands broaden gradually as they migrate through a chromatographic column. Resolution of individual solutes into discrete bands occurs only if the bands broaden to a lesser extent than their peak maxima separate. Values of the baseline bandwidths of adjacent bands are almost constant; that is, $W_1 \simeq W_2$. Since the baseline bandwidth is equal to four standard deviations for a given band, resolution can also be expressed as

$$R_s = \frac{t_{R,2} - t_{R,1}}{4\sigma} \tag{17.21}$$

If inadequate, the resolution of adjacent peaks can be improved either by increasing the separation between peaks or by decreasing individual peak widths. This involves column selectivity when moving peaks farther apart, and column efficiency when attempting to narrow the peak width. Improving the selectivity involves altering the thermodynamics of the chromatographic system. Improving the kinetics of the system increases the efficiency of the separation. As shown in Figure 17.6 (upper chromatogram), a column may have adequate selectivity but exhibit poor efficiency (as compared with the middle chromatogram). The lower chromatogram exhibits excellent efficiency but could have better selectivity; here k' values are probably too low.

Any criterion for resolution will be somewhat arbitrary. For reasonable quantitative accuracy, peak maxima must be at least 4σ (that is, W_b or $2W_{1/2}$) apart. If so, then $R_s = 1.0$, which corresponds approximately to a 3% overlap (cross contamination) of peak areas. A value of $R_s = 1.5$ (for 6σ) represents essentially complete resolution with only 0.2% overlap of peak areas. One caveat; these criteria pertain to roughly equal solute concentrations. Increased resolution may be needed when a band from a major component is adjacent to a band of a minor constituent. In real life there are many instances where baseline resolution for all components may be unachievable. The separation is satisfactory when the least resolved pair of components can be quantitatively determined to an acceptable degree.

Equations 17.20 and 17.21 define resolution in a given situation, but they do not relate resolution to the conditions of separation nor do they suggest how to improve resolution. For these purposes a resolution equation can be derived in a form that explicitly incorporates the terms involving the thermodynamics and kinetics of the chromatographic system. To accomplish this, Equations 17.16 and 17.20 are combined, using W_b as the average baseline width. This gives

$$R_s = \frac{t_{R,2} - t_{R,1}}{4t_{R,2}} N^{1/2} = \frac{N^{1/2}}{4}\left(\frac{1 - t_{R,1}}{t_{R,2}}\right) \tag{17.22}$$

Equation 17.10, expressed in terms of both $t_{R,2}$ and $t_{R,1}$, is substituted into Equation 17.22 to yield

$$R_s = \frac{N^{1/2}}{4}\left(1 - \frac{1 + k'_1}{1 + k'_2}\right) = \frac{N^{1/2}}{4}\left(\frac{k'_2 + k'_1}{1 + k'_2}\right) \tag{17.23}$$

Now from Equation 17.14, the relative retention α is equal to k'_2/k'_1, and the fundamental resolution equation is

$$R_s = \frac{1}{4}\left(\frac{\alpha - 1}{\alpha}\right)\left(\frac{k'}{1 + k'}\right)\left(\frac{L}{H}\right)^{1/2} \tag{17.24}$$

Resolution as expressed by Equation 17.24 is seen to be a function of three separate factors: (1) a column selectivity factor that varies with α, (2) a rate of migration or capacity factor that varies with k' (taken variously as k_2 or the mean value of k_1 and k_2), and (3) an efficiency factor that depends on L/H (or the theoretical plate number). Each factor can be calculated directly from the recorded chromatogram

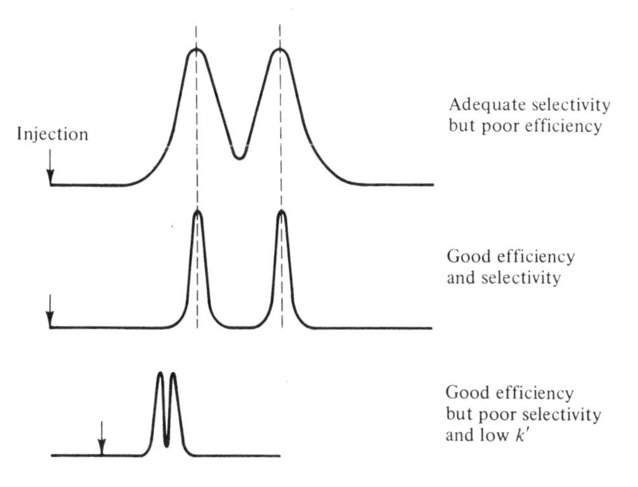

FIGURE **17.6**
Selectivity, efficiency, and partition ratio for columns.

and can be adjusted more or less independently. The first two factors are essentially thermodynamic, whereas the L/H term is mainly associated with the kinetic features of chromatography.

Changes in α and k' are achieved by selecting different stationary and mobile phases, or by varying temperature and (less often) pressure. In addition, k' can be varied by changing the relative amounts of mobile and stationary phases within the column. When optimizing a particular separation, the k' term should be considered first. Maximum resolution in unit time is obtained when $k' = 2$, as shown in Section 17.5. The optimum range of k' values extends from 1 to 10. Furthermore, the first eluted component of a given pair should have a retention time that is twice the passage time of a nonretained solute; that is, $t_R = 2t_M$. Unfortunately, in a complex mixture of many components, it is often possible to optimize separation conditions for only one pair of components. The only effective solution to this problem when working with complex real samples is "k' programming." In gas chromatography an optimum value of k' can be achieved by varying temperature. In liquid chromatography it is usually more profitable to vary systematically the composition of the mobile phase. If resolution is still a problem, an increase in either α or L/H should be explored.

EXAMPLE 17.4

Because the separation of heptane and octane in Example 17.3 is less than the baseline width of the two bands, how much should the column be lengthened from the original 122 cm length?

Since resolution is proportional to the square root of the column length, for a resolution of 1.5,

$$\left(\frac{L}{122\ \text{cm}}\right)^{1/2} = \frac{1.50}{1.24}$$

and $L = 179$ cm.

The first term in Equation 17.24 is very sensitive to changes in the value of α, as shown in Table 17.2. Generally it is desirable to select values of α within the range

TABLE **17.2**

VALUES RELATED TO THE RELATIVE RETENTION

α	$\left(\frac{\alpha}{\alpha - 1}\right)^2$	N_{req} for $R_s = 1.5$ and $k' = 2$	L_{req}, meters for $H = 0.6$ mm
1.01	10,201	826,281	495
1.02	2601	210,681	126
1.03	1177	95,377	52
1.04	676	54,756	33
1.05	441	35,721	21
1.10	121	9801	5.8
1.15	58	4418	2.6
1.20	36	2916	1.7
1.25	25	2025	1.2
1.30	19	1514	1.0

1.05–2.0. For example, an increase in α from 1.05 to 1.10 will improve resolution by a factor of four for the same L/H. When α is quite close to 1, it becomes impractical to operate because the required column length and column inlet pressures become difficult or impossible to achieve.

The L/H term is adjusted to provide maximum efficiency compatible with a reasonably short analysis time. Higher L/H (N) values always provide improved resolution, other factors being equal. Thus resolution may be improved by increasing the column length, but only as the square root of the column length. The plate height may be decreased through improvement in the kinetic features of column operation, perhaps by lowering the flowrate of the mobile phase (but not lower than the minimum in the H/u graph, as will be discussed). Any action that increases the efficiency of the mass transfer of solutes between the stationary and mobile phases will decrease the plate height and thus improve resolution.

17.4 COLUMN PROCESSES AND BAND BROADENING

Various processes take place on a column during a chromatographic separation that contribute to the *peak variance*, σ^2, or band broadening. These will now be discussed.

Theories of band spreading in liquid and gas chromatography are nearly identical. Plate height expresses in simple terms the extent of band broadening and the factors that affect the broadening. It is a function of thermodynamic and kinetic processes within the column. These are (1) flow irregularities that lead to convective mixing, (2) transverse and longitudinal diffusion in the mobile phase, and (3) a finite rate of equilibration of solute between the stationary and mobile phases (mass transfer). Stated as an abbreviated form of the van Deemter equation (introduced originally for gas-liquid chromatography),

$$H = A + \frac{B}{u} + C_{\text{stationary}}u + C_{\text{mobile}}u \tag{17.25}$$

Equation 17.25 and its individual components are shown graphically in Figure 17.7. The average linear velocity of the mobile phase, u, is used because it can be directly related to the speed of analysis, whereas the flowrate depends on the column cross section and the column volume occupied by packing material. Experimentally,

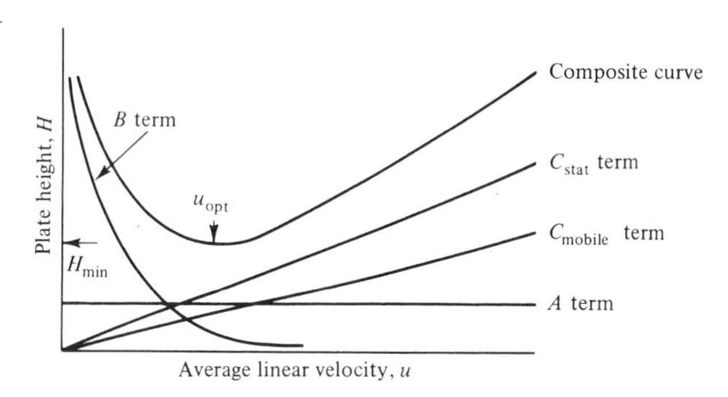

FIGURE 17.7
Typical H/u (van Deemter) curve for a gas chromatographic column.

the average linear velocity is easily determined by injecting a nonretained solute and measuring its passage time through the column. Knowing the column length, we have

$$u = \frac{L}{t_M} \tag{17.26}$$

Eddy Diffusion

The A term, called "eddy diffusion," in Equation 17.25 results from the inhomogeneity of flow velocities and path lengths around packing particles (Figure 17.8a). A is defined as

$$A = \lambda\, d_p \tag{17.27}$$

where d_p is the particle diameter and λ is a function of the packing uniformity and the column geometry. Flow paths of unequal length must exist through any less-than-perfect packing. Some solute molecules of a single species may find themselves swept through the column close to the column wall where the density of packing is comparatively low, especially in small-diameter columns. Other solute molecules pass through the more tightly packed center of the column at a correspondingly lower velocity. Molecules that follow a shorter path elute before those that follow a

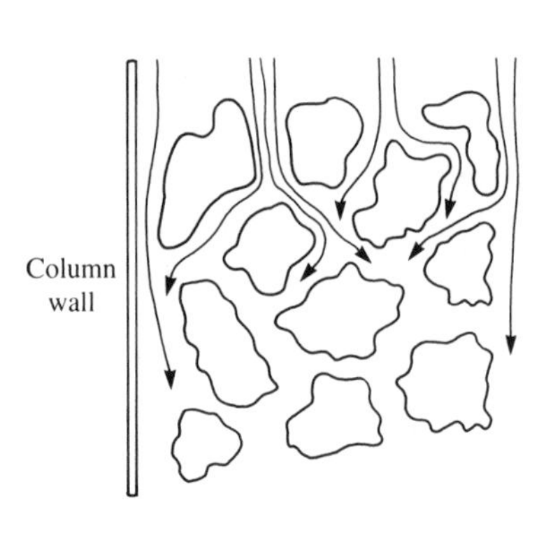

Eddy diffusion
(a)

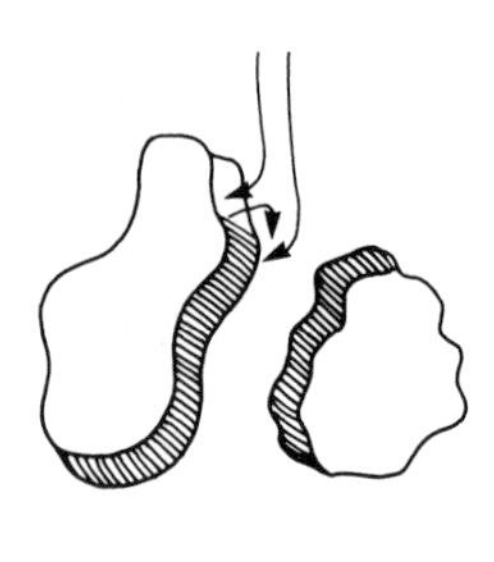

Stationary-phase
mass transfer
(b)

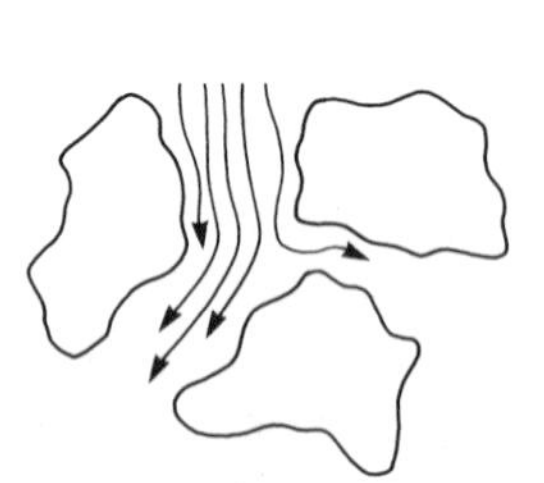

Mobile-phase
mass transfer
(c)

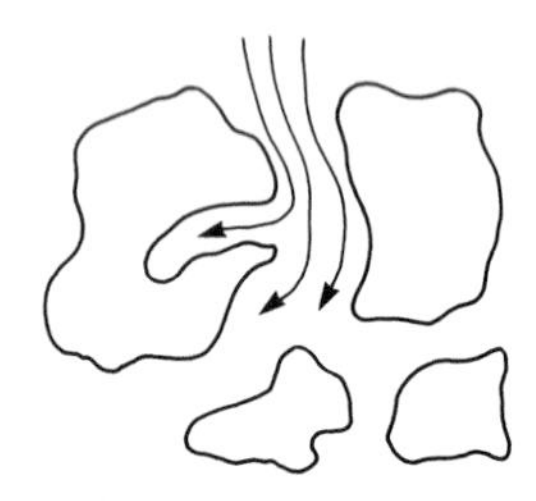

Stagnant mobile-phase
mass transfer
(d)

FIGURE **17.8**
Contributions to band broadening.

series of erratic (and longer) paths. This leads to a broadening of the elution band for each solute.

To minimize the A term, the mean diameter of the particles in a column packing should be as small as possible and packed as uniformly as possible. Of course, the smaller the particles, the higher is the inlet pressure needed to drive the mobile phase through the column and the more difficult it is to pack the column in a uniform manner. However, because of the higher efficiency achieved with smaller-diameter particles, the column length can be shortened somewhat. There has to be a trade-off between particle size, column length, and the pressure required. In gas-liquid chromatography, when films are used on the interior of capillary columns, the A term is zero.

Longitudinal Diffusion

The B term in Equation 17.25 defines the effect of longitudinal, or axial, diffusion—that is, random molecular motion within the mobile phase (not illustrated in Figure 17.8). B is defined as

$$B = 2\gamma D_M \tag{17.28}$$

where γ is an obstruction factor that recognizes that longitudinal diffusion is hindered by the packing or bed structure, and D_M is the solute diffusion coefficient in the mobile phase. In coated capillary columns γ is unity; in packed columns it has a value of about 0.6. The foregoing B term pertains to gas-liquid chromatography. In liquid column chromatography, the ratio D_M/T_M should be used, where T_M, the interparticle tortuosity factor, corrects the diffusion coefficient for the varying size and direction of the interstitial pores in a packing.

The contribution of longitudinal diffusion to plate height becomes significant only at low mobile-phase velocities. Then the high diffusion rates of a solute in the mobile phase can cause the solute molecules to disperse axially while slowly migrating through the column. If this happens, peak broadening occurs.

Mass Transfer

The $C_{\text{stationary}}$ term in Equation 17.25 results from resistance to mass transfer at the solute to stationary phase interface (Figure 17.8b). It is proportional to d_f/D_S, where d_f is the effective thickness of the stationary phase and D_S is the diffusion coefficient of the solute in the stationary phase. Slow molecular movement within the stationary phase means a longer time spent in this phase by a solute molecule, while other molecules are moving forward with the mobile phase. The faster the mobile phase moves through the column and the slower the rate of mass transfer, the broader is the solute band that eventually elutes from the column. Nonviscous liquids should be chosen for the stationary phase so that the diffusion coefficient is not unduly small. Reducing the thickness of the stationary phase is beneficial, although the capacity of the column is lowered.

The C_{mobile} term represents radial mass transfer resistance between adjacent stream lines of mobile phase (Figure 17.8c). It is proportional to the square of the particle diameter of the packing material, d_p^2, and inversely proportional to the diffusion coefficient, D_M, of the solute in the mobile phase. Decreasing the size of the stationary-phase particles is always helpful in decreasing the plate height.

In liquid column, as contrasted with gas-liquid, chromatography, major differences arise from (1) the 10,000-fold decrease in the value of D_M when a liquid constitutes the mobile phase as opposed to a gas, and (2) the presence of stagnant pockets of mobile phase trapped within pores and channels of the stationary phase. Solute molecules move into and out of these pores by diffusion (Figure 17.8d). These molecules are retarded in their forward motion relative to the main band of a given solute and again there is an increase in the molecular spreading. Also the velocity of the mobile phase differs from point to point owing to the perturbation caused by the support particles. Liquid stream lines of the mobile phase near the particle boundaries move slowly, whereas stream lines near the center between particles move more rapidly. Solute molecules transfer constantly by lateral diffusion to a different stream line. Hence, the obstructive path of a solute molecule is due both to diffusion between stream lines and to the necessity for traveling around the stationary-phase particles. Molecular diffusion, coupled with unequal stream lines (multipath effect), gives rise to a convective mixing, or coupled, term:

$$\frac{A}{1 + C_{\text{mobile}}/u^{1/2}} \tag{17.29}$$

The effect of stagnant pools can be minimized in several ways. The internal structure of the packing can be made impervious; an example is surface-coated pellicular packings with a solid core. Reducing the diameter of the particles is very effective. Also, supports can be chosen that have very wide pores so that liquid flows easily in and out, or even through the pore channels.

Plate height in liquid column chromatography can be expressed by the abbreviated equation:

$$H = \underbrace{\frac{B}{u}}_{\text{longitudinal or axial diffusion}} + \underbrace{\frac{A}{1 + C_{\text{mobile}}/u^{1/2}}}_{\text{convective mixing}} + \underbrace{C_{\text{mobile}}u^{1/2}}_{\text{resistance to mass transfer in mobile phase}} + \underbrace{C_{\text{stat}}u}_{\text{resistance to mass transfer in stationary phase}} \tag{17.30}$$

The individual contributions of the four terms in Equation 17.30 are shown graphically in Figure 17.9. In both gas-liquid and liquid column chromatography,

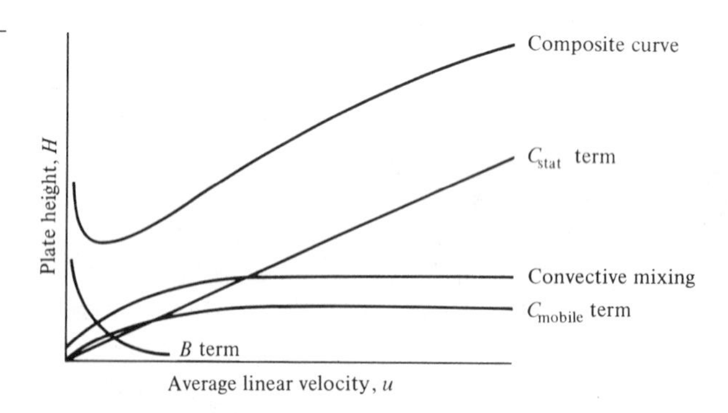

FIGURE **17.9**
Typical H/u curve for a liquid chromatographic column.

longitudinal diffusion (the B term in Equation 17.25 or Equation 17.30) is a significant factor only at mobile-phase velocities less than the minimum in the composite H/u curves. Ideally an operator might wish to utilize a velocity corresponding to the minimum plate height. Actually the plate height in the region to the right of the minimum usually rises only gradually for most gas-liquid or liquid column chromatographic columns. Consequently, a slight loss in column efficiency may be traded for shortened analysis time when velocities somewhat higher than u_{opt} are used.

17.5 TIME OF ANALYSIS AND RESOLUTION

Previous discussions have hinted at an important parameter of concern to the analyst—namely, the retention time required to perform a separation. This is the time needed to get the solute band through one plate, t_p, multiplied by the number of plates required, N_{req}, for the desired resolution. Thus,

$$t_R = t_p N_{req} \tag{17.31}$$

Since t_p is given by the plate height divided by the band velocity, $u/(1 + k')$, where $1/(1 + k')$ is the fraction of time spent by the solute in the mobile phase, Equation 17.31 becomes

$$t_R = N_{req}(1 + k')\left(\frac{H}{u}\right) \tag{17.32}$$

Eliminating N_{req} between Equations 17.24 and 17.32 gives

$$t_R = 16R^2\left(\frac{\alpha}{\alpha - 1}\right)^2\left[\frac{(1 + k')^3}{(k')^2}\right]\left(\frac{H}{u}\right) \tag{17.33}$$

Differentiating Equation 17.33 with respect to k' and placing all other variables into one constant, C, give

$$\frac{dt_R}{dk'} = C\left[\frac{(k')^3 - 3k' - 2}{(k')^3}\right] \tag{17.34}$$

Now t_R is a minimum when $k' = 2$, which is when $t_R = 3t_M$ (see Equation 17.10). There is little increase in analysis time when k' lies between 1 and 10, provided k' has no effect on the other variables.

A twofold increase in the mobile-phase velocity would be expected to halve the analysis time (see Equation 17.33). However, this is not strictly true because H would increase, as shown in Figures 17.9 and 17.10. The ratio H/u can be obtained directly from the experimental plate height/velocity graph. It is the slope of the line drawn from the origin to a point on the graph. This slope decreases as the velocity increases; however, a point of diminishing returns is reached at higher velocities. Nevertheless, it is desirable to use velocities higher than u_{opt}, with a corresponding increase in column length, at least until the column inlet pressure requirement becomes excessive. Although a longer column is needed, the analysis time is shorter.

17.6 QUANTITATIVE DETERMINATIONS

Detectors in chromatography respond either to the concentration of solute or to the mass flowrate. Those that respond to the concentration yield a signal that is proportional to the solute concentration that traverses the detector. An elution peak results when the signal is plotted against time. For such detectors the area under the peak is proportional to the mass of a component and inversely proportional to the flowrate of the mobile phase. It is crucially important that the flow of the mobile phase be kept constant for such detectors if quantitative analysis is to be carried out. In differential detectors that respond to mass flowrate, the peak area is directly proportional to the total mass and there is no dependency on the flowrate of the mobile phase.

In column chromatography the analog signal generated by the detector is graphically recorded in the form of the familiar chromatographic peaks. The area under these peaks can then be integrated in a variety of ways and the resulting data related to the composition of the unknown samples.

Peak Area Integration

Height Times Width at Half-Height. The operations involved are drawing the baseline of the peak, measuring the height from this baseline, positioning the measuring scale parallel to the baseline at half the height, and measuring the width of the peak at this position. The normal (zero signal) baseline is not used because large deviations may be caused by tailing.

Peak Height. Measurement of peak height is inherently simple. The only operations involved are drawing the baseline and measuring the height. The precision is better than measuring the peak area, particularly of narrow peaks. However, peak heights are sensitive to small changes in the technique of sample injection and in operating conditions. Peak height does not always remain directly proportional to sample size. As the latter increases, there is a point at which the peak begins to broaden and no longer increases in height at the same rate.

Ball-and-Disk Integrator. A ball positioned on a rotating flat disk rotates at a speed proportional to its distance from the center of rotation. The ball is positioned on the disk at a distance from the center in the same relationship as the position of the recorder pen to the baseline of the chromatogram. If the disk is rotated at a constant speed (time), the ball rotates at a speed proportional to the position of the recorder pen from zero. This speed is then transmitted to a roller through a second ball, which, by means of a "spiral in" and a "spiral out" cam, actuates the integrator pen at a speed directly proportional to the position of the recorder pen. The drive between the disk and the ball is by traction through an oil film. Although this hydrostatic phenomenon is not clearly understood, the oil film acts similarly to an induction motor where slip is proportional to the driven load.

Reading the integrator trace is done as follows (also refer to Figure 17.10): establish the desired chart time interval from the recorder pen trace of the chromatogram and project directly down to the integrator trace (see the arrows). The value of an interval is obtained by counting the chart gradations crossed by the integrator trace. A full stroke of the "sawtooth" pattern in either direction represents 100 counts. Every horizontal division has a value of 10. Values less than 10 are estimated. In the example, the interval for the main peak is 1083 counts. The pattern can usually be read within two counts. On some models the space between "blips" projecting slightly above the uppermost horizontal line is equivalent to 600 counts, making it possible to record up to 9600 counts per centimeter of chart. The pattern to the right in Figure 17.10 illustrates the method for estimating the baseline correction when the peak baseline does not coincide with the recorder baseline.

Computing Integrator

On-line computer-based data systems provide complete automation. This includes automatic acquisition and reduction of data, storage of calculation methods, and a printout of analytical results. Initially the analog chromatographic signal is digitized by an analog-to-digital converter. The software can then detect the presence of peaks, correct for baseline drift, calculate areas and retention times, determine concentrations of components using stored calibration factors, and generate a complete report of the analysis (Figure 17.11).

Peak area counts are accumulated when the signal leaves the baseline. This departure from the baseline is usually determined by monitoring the slope of the signal. The retention time and signal heights of each peak maximum detected by the program are stored in memory. The termination of a component peak is established when the signal returns to the baseline. During isothermal runs the software can automatically increase the slope sensitivity with time, thus ensuring the program's ability to detect both initial sharp peaks and later low flat peaks with equal precision. In the case of fused peaks, areas can be allocated to each component by dropping perpendiculars from valley points to the corrected baseline (Figure 17.12). In the case of overlapping peaks, special algorithms allot areas to each component.

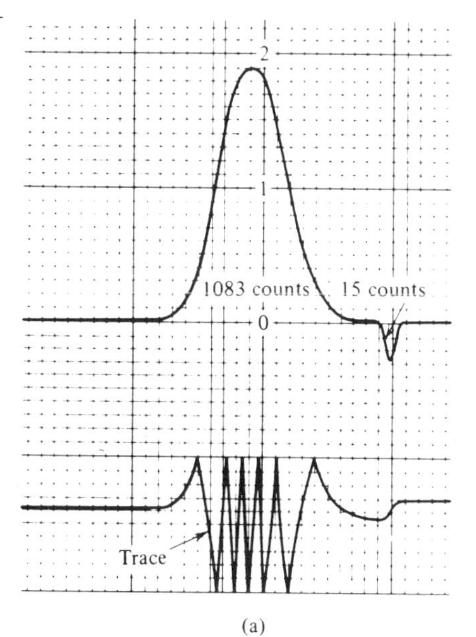

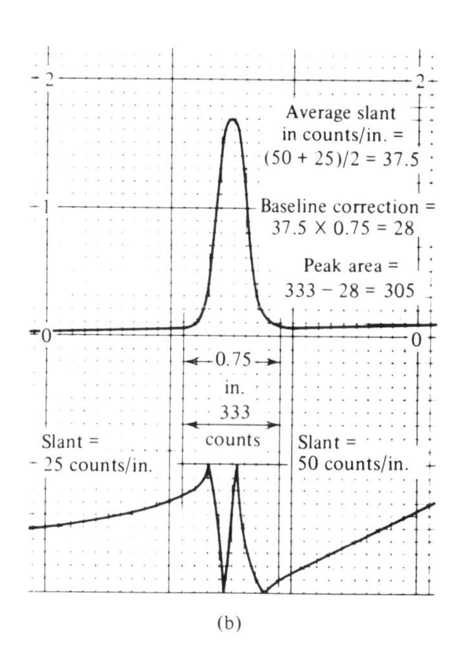

FIGURE **17.10**
(a) Estimation of peak areas with ball-and-disk integrator. (b) Method for handling baseline correction.

Evaluation Methods

The three principal evaluation methods are (1) calibration by standards, (2) area normalization, and (3) internal standard. Each has its place, depending on the nature of the analysis.

Calibration by Standards. When the sample volume is known, calibration by standards is often used. It has the advantage that only the areas of the peaks of interest need to be measured. It does require that the same amount of sample be injected each time. The necessary calibration standard(s) should be run under the same operating conditions as the sample. The percent concentrations are obtained by calculating the ratio of the volume of each component of interest to the sample size. In practice, standard solutions of the component(s) of interest are prepared and injected into the chromatograph. Then for an unknown,

$$X = (\text{area})_X K \tag{17.35}$$

where K is the proportionality constant (slope of the calibration curve).

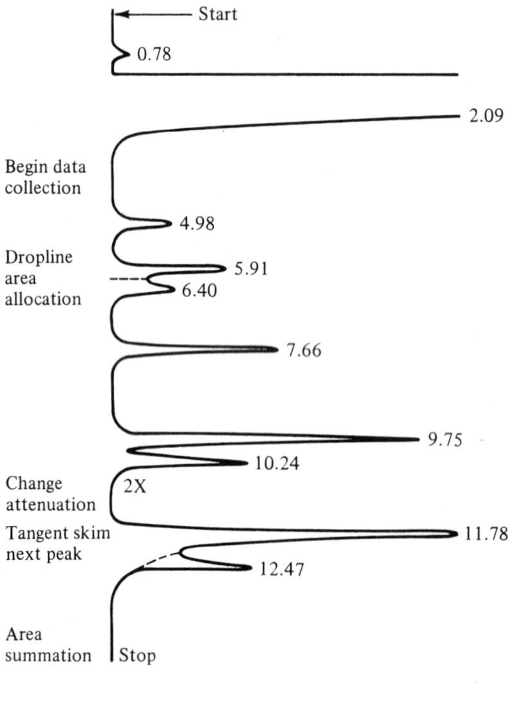

FIGURE **17.11**
Partial printout of a chromatographic run and indicated computations that can be made by computer software programs.

		RT	AREA	CAL	AMT
TEMP 1	150	4.98			
TIME 1	0.0	5.91			
RATE	5.00	6.40			
TEMP 2	250	7.66			
TIME 2	15.0	9.75			
INJ TEMP	250	10.24			
FID TEMP	300	11.78			
CHART SPEED	1.00	12.47			

Relative response factors must be considered when converting area to volume and when the response of a given detector differs for each molecular type or class of compounds. Response factors are best obtained by analyzing standard samples.

Area Normalization. When it is known that the chromatogram represents the entire sample, that all components have been separated, and that each peak has been completely resolved, area normalization may be used for evaluation. To use this method, the area under each individual peak is measured and then divided by its response factor to give the peak's calculated area. Adding together all the peak areas gives the total calculated area. The percent by volume for individual components is obtained by multiplying the individual calculated area by 100 and then dividing by the total calculated area.

Internal Standard. The internal standard method permits the operating conditions to vary from sample to sample and does not require repeatable sample injection. The internal standard has to be a component that can be completely resolved from adjacent peaks, is not present in the unknown mixture, and does not have any interference effects. A known quantity of this standard is chromatographed, and area versus concentration is plotted. A known amount of the standard is then added to the unknown mixture. Any variation in sample size will be immediately apparent

FIGURE 17.12 Computing integrator capabilities. (Courtesy of Spectra-Physics.)

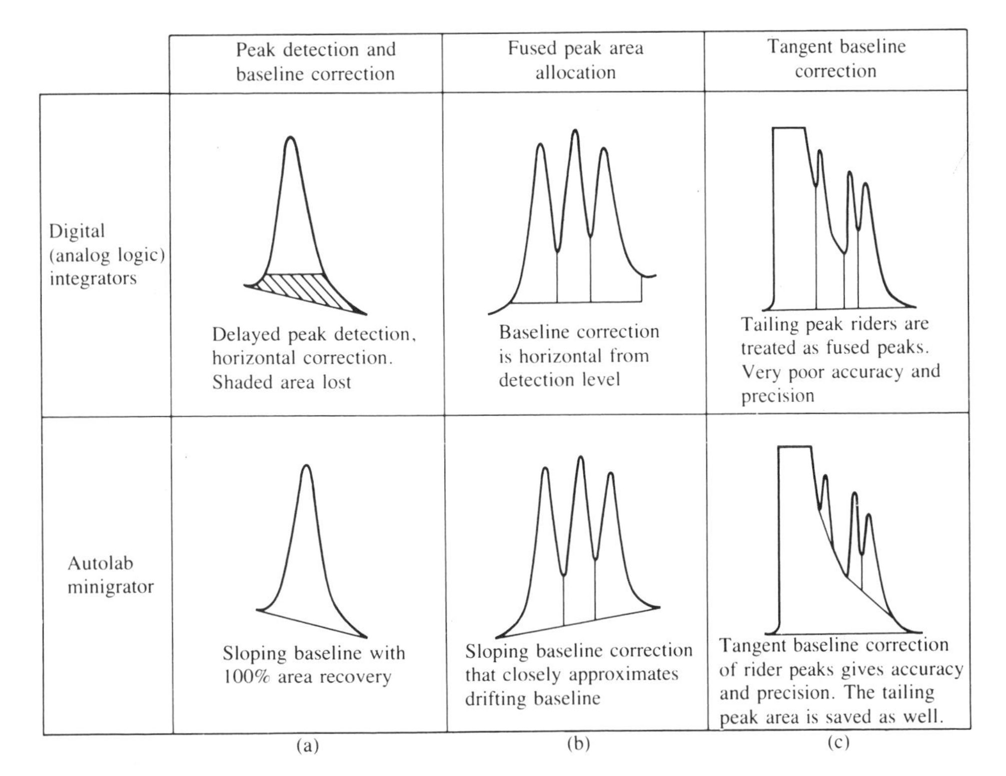

by comparing the peak area of the internal standard in different runs. A correction factor can then be used when determining the exact concentration of the other components.

EXAMPLE 17.5

Assume that 100 mg of internal standard is added to 1.00 g of mixture. Measurement of the resultant chromatogram shows four components (including the internal standard) with areas (in arbitrary units) as follows: $A_1 = 27$, $A_{std} = 80$, $A_2 = 20$, $A_3 = 70$, and area sum $= 197$. The amount of component 3 in the sample is

$$W_3 = W_{std}\left(\frac{A_3}{A_{std}}\right) = 100\text{ mg}\left(\frac{70}{80}\right) = 87.5\text{ mg}$$

$$\text{Percent component 3:}\quad \left(\frac{0.0875\text{ g}}{1.000\text{ g}}\right)100 = 8.75\%$$

Note that component 3 represents less than 9% of the total sample, yet in terms of peak area it appeared to be a major component. This indicates that a large part of the sample does not appear on the chromatogram, as would be the case if the mixture included some inorganic salts.

In Example 17.5 it was assumed that the internal standard and other components responded both to the column and to the detector in the same manner. This might be true if the internal standard was a member of the same homologous series as the components being measured. This is seldom true. However, the ratio of response factors (such as K_{std}/K_3) can be determined experimentally. This value would then be applicable as long as the other requirements were met. When so,

$$\frac{K_{std}}{K_3} = \frac{W_3 A_{std}}{W_{std} A_3}$$

The internal standard method is most often used where a portion of the sample may not elute completely or may be lost in operations preliminary to the chromatographic step. It is important that the internal standard is added before any sample pretreatment.

17.7 RETENTION DATA FOR SAMPLE CHARACTERIZATION

In a fixed chromatographic system the retention time is a constant for a particular solute and, therefore, can be used to identify that solute. Thus, although chromatography is primarily a separation technique, it is possible to identify the separated components of a complex sample by their retention times.

Direct Comparison of Retention Values

Retention times usually vary in a regular and predictable fashion with repeated substitution of some group *i* into the sample molecule as, for example, in a series of homologs, benzologs, or oligomers. Often some function of retention time is linear with the number of repeating groups *i* within the sample molecule—for example, —CH_2— groups for a homologous series. For isocratic elution the retention of the ith member of a homologous series is given by

$$\log t_{r,i} = mN_i + \text{constant} \tag{17.36}$$

where m is a constant and N_i is the number of repeating groups (or the number of carbon atoms) in the homolog. Use of capacity factor k' values is often preferred when retention data are assembled for qualitative comparisons because k' values are not influenced by mobile-phase flowrate or column geometry. Several homologous series are graphed in Figure 17.13.

Retention times can be predicted from the known behavior of other members of a homologous series. The likelihood of a successful match in retention values depends on prior knowledge of the sample and, therefore, the ability to anticipate the presence of specific compounds in the sample. Successful matching of retention values also requires the availability of likely reference compounds.

The method of standard addition can be used to verify the retention value of the compound in question in the actual sample matrix. The retention time of the original sample band should not change after addition of the compound in question if the two compounds are the same.

Chromatographic Cross Check

The reliability of an identification by means of retention times is greatly enhanced by using different solute–stationary phase interactions. By utilizing the selectivity of particular stationary phases in gas chromatography, and particular combinations of mobile and stationary phases in liquid column chromatography, much information can be ascertained about an unknown or a mixture of unknowns. For example, on each of two gas-chromatographic columns, one containing a polar and the other a nonpolar liquid phase, a series of compound classes are run to determine the retentions of each compound (Figure 17.14). By plotting the retention values for the two stationary phases against each other, lines that radiate from the origin are obtained (one for each homologous series, as given by Equation 17.36).

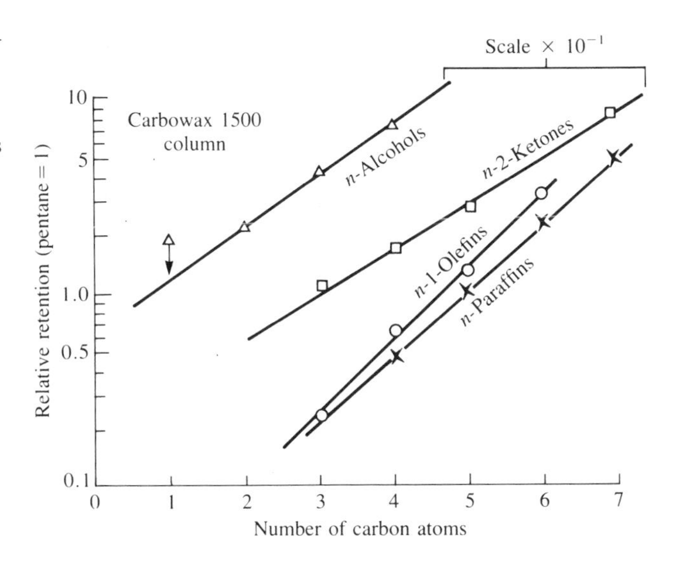

FIGURE **17.13**

Plot of retention time (log scale) versus the number of carbon atoms for several homologous series.

Crowding occurs in the corner near the origin because the points are placed along the lines in a distribution logarithmic to molecular weight. If the logarithms of the retentions are plotted against each other, a corresponding series of approximately parallel lines is obtained with points spaced linearly according to molecular weight.

Retention indexing systems are also valuable for qualitative analysis. As explained in Chapter 18, the Rohrschneider constants are characteristic of the substance being analyzed. Once obtained, these constants are valid for the substance on any column packing.

Identification by Ancillary Techniques

When standard reference compounds are unavailable, recourse can be had to independent structure information from the several spectroscopic techniques

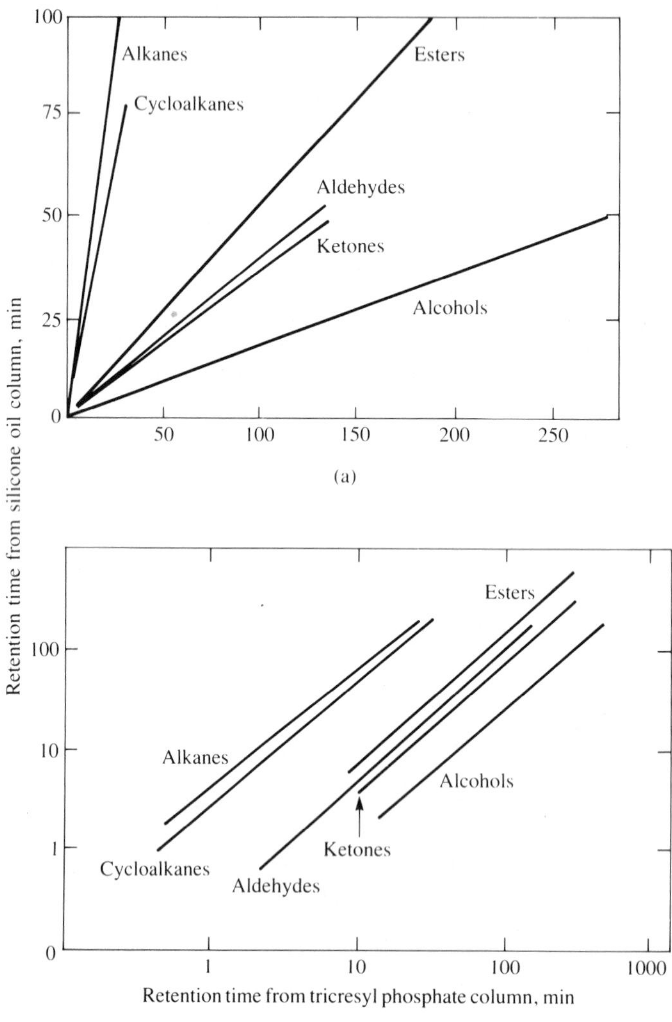

FIGURE 17.14
Two-column plots: (a) linear and (b) logarithmic.

discussed in earlier chapters. In many cases, positive identification of an unknown can be accomplished only by isolating the peak during a chromatographic run and subsequently analyzing it by a supplemental method. Mass spectrometry has been successfully coupled with chromatography, as described in Section 18.8 for gas chromatography and Section 19.9 for liquid column chromatography. For mass spectrometry only about 5 ng of sample is required. The information that can be obtained includes molecular weight and empirical formula, structural information, and confirmation of structure. Infrared spectroscopy has also been coupled with gas chromatography (see Section 18.9) where it aids in identifying functional groups and possible molecular structures. Confirmation is likely if reference spectra are available.

PROBLEMS

1. Consider a 50-cm column with a plate height of 1.5 mm that provides a theoretical plate number of 333 at a flowrate of 3 mL min^{-1}; $V_M = 1.0$ mL. (a) What are the solute retention time and retention volume when k' is 1, 2, 5, and 10? (b) What is the baseline peak width for each of the foregoing values of k'?

2. The data in the table were extracted from the chromatogram of a two-component mixture of x and y. (a) Calculate the capacity ratios of x and y. (b) What parameters affect the capacity ratio, k'? (c) Calculate the separation factor, α. Is this a reasonable value for achieving separation of the components? (d) Calculate the resolution. Is this sufficient to obtain baseline separation?

	Air	x	y
Retention time (sec)	5	25	30
Peak width, baseline (sec)	—	5	4

3. Chromatograms with a standard test mixture were obtained using porous alumina; assume the total porosity is 0.75. The inlet pressure was 22.5 atm for all columns. Operating conditions and retention times are given in the table. (a) Calculate k' for each solute. (b) Calculate the average linear velocity and plate height for each column. (c) Calculate the free column volume and flowrate for each column.

Test substance, t_R (sec)	Column 1	Column 2	Column 3
Nitrobenzene	538	182	91
Anisole	232	76	36
Biphenyl	168	56	26
Toluene	124	40	19
t_M (sec)	104	34	16
L (cm)	50.0	13.5	9.0
N	3200	4450	5000
d_p (μm)	20	10	6.5

4. Keeping all other parameters constant in the resolution equation, (a) graph the effect of k' on resolution and (b) graph the effect of α on resolution.

5. For a typical chromatographic separation giving just-resolved peaks ($R_s = 1.5$), assume that $N = 3600$, $k' = 2$, and $\alpha = 1.15$. Sketch the effects of changing these parameters one at a time to (a) $N = 1600$, (b) $\bar{k}' = 0.8$, and (c) $\alpha = 1.10$.

6. To decrease the plate height and yet increase the resolution, what courses of action are available? What penalties may accrue for each approach?

7. The relative response factors for *p*-dichlorobenzene and *p*-xylene (relative to the value for benzene, assigned unity) were found to be 0.624 ± 0.034 and 0.917 ± 0.018, respectively. Upon integration of chromatographic peaks the results in the table were obtained. Calculate the percent composition of each sample.

	Area		
Sample	**Benzene**	***p*-Xylene**	***p*-Dichlorobenzene**
1	4592	2984	1238
2	512	3527	5495

8. The relative response factors for *p*-xylene and toluene (relative to benzene assigned a value of unity) were found to be 0.570 ± 0.0327 and 0.793 ± 0.0178, respectively. Measurements were made by peak height. Unknown mixtures of these three solutes gave the peak heights (in mm) shown in the table. Calculate the percent composition of each sample.

Sample	**Benzene**	***p*-Xylene**	**Toluene**
1	98	87	86
2	136	82	63
3	148	51	97
4	52	48	81
5	85	35	42

9. A mixture of straight-chain aliphatic alcohols and acetates gave the chromatogram shown in the table. After a separate sample was spiked peaks 1 and 5 were identified as ethanol and 1-butanol. Identify the remaining peaks.

Peak	t'_R **(min)**
1	4.8
2	7.7
3	10.4
4	15.9
5	24.0

BIBLIOGRAPHY

ETTRE, L. S., AND C. HORVATH, "Foundations of Modern Liquid Chromatography," *Anal. Chem.*, **47**, 422A (1975).

GIDDINGS, J. C., "Principles and Theory," *Dynamics of Chromatography*, Part 1, Dekker, New York, 1965.

GILPIN, R. K., "New Approaches for Investigating Chromatographic Mechanisms," *Anal. Chem.*, **57**, 1465A (1985).

HAWKS, S. J., "Modernization of the van Deemter Equation for Chromatographic Zone Dispersion," *J. Chem. Ed.*, **60**, 393 (1983).

LITERATURE CITED

1. FOLEY, J. P., AND J. G. DORSEY, *Anal. Chem.*, **55**, 730 (1983).

GAS CHROMATOGRAPHY

Gas chromatography is the technique of choice for the separation of thermally stable and volatile organic and inorganic compounds. Gas-liquid chromatography (GLC) accomplishes the separation by partitioning the components of a chemical mixture between a moving (mobile) gas phase and a stationary liquid phase held on a solid support. Gas-solid chromatography (GSC) uses a solid adsorbent as the stationary phase. The availability of versatile and specific detectors and the possibility of coupling the gas chromatograph to a mass spectrometer or an infrared spectrophotometer further enhance the usefulness of gas chromatography.

18.1 GAS CHROMATOGRAPHS

A gas chromatograph consists of several basic modules joined together to (1) provide a constant flow of carrier (mobile phase) gas, (2) permit the introduction of sample vapors into the flowing gas stream, (3) contain the appropriate length of stationary phase, (4) maintain the column at the appropriate temperature (or temperature-program sequence), (5) detect the sample components as they elute from the column, and (6) provide a readable signal proportional in magnitude to the amount of each component. The instrument modules are shown schematically in Figure 18.1.

Sample Injection System

The standard mode, suitable for approximately 95% of packed column applications, is direct injection. The sample is injected by a hypodermic syringe through a self-sealing silicone rubber septum onto a glass liner within a metal block, where it is vaporized and swept onto the column (Figure 18.2). The block is heated at a fixed temperature sufficiently high to convert virtually instantly the liquid sample into a "plug" of vapor. Samples, both liquid and gas, can be measured with a calibrated loop, then introduced into the flowing gas stream by means of a valve (Figure 18.3).

Reduction in sample volume is necessary when working with capillary columns. This is accomplished by an injector-splitter, where typically a 1-μL sample is injected but only 0.01 μL enters the capillary; the remainder is vented. This technique prevents column overload but wastes a significant portion of the sample.

When analyses are performed on small amounts of sample with some component concentrations in the parts-per-billion range, too little material is placed on the colum if splitting is used for these samples. Splitless injection is required for these samples. The entire sample, including any solvent, is injected onto the open tubular column through a modified heated flash vaporizer. A large solvent tail is avoided by venting the injection port to the atmosphere at that time (perhaps 30 sec), when most of the solvent and essentially all the sample have entered the column. The proper time until venting is critical; too short a time causes loss of sample components, whereas too long a time causes a solvent peak larger than necessary, which might bury some of the peaks of interest.

Automatic Sampler

An automatic sampler duplicates manual sample measurements and injection. Sample vials are a glass, throw-away type with vapor-tight septum caps. The sampler flushes the syringe with a new sample to remove traces of the previous sample, pumps new sample to wet the syringe completely and eliminate any bubbles, takes in a precisely measured amount of sample, and injects the sample into the gas chromatograph. The automatic samplers are machine reproducible and consistently more precise than a skilled chromatographer; also, unattended operation releases the operator for other duties.

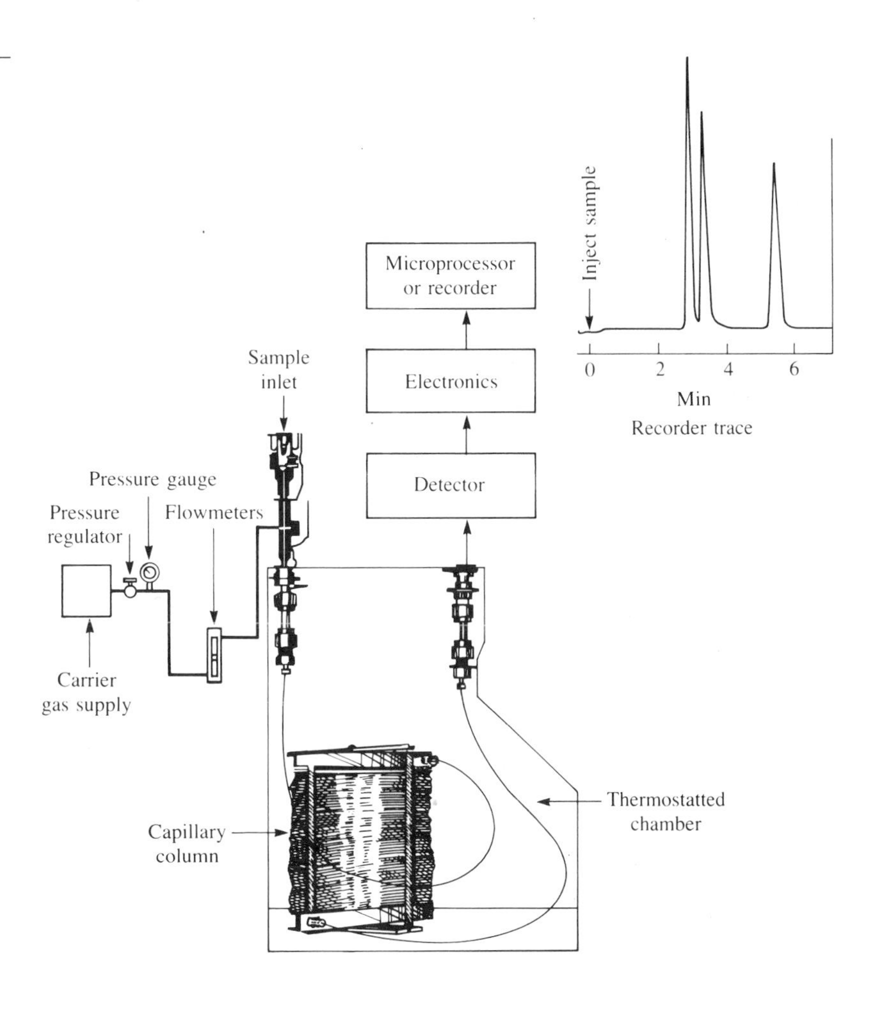

FIGURE **18.1**
Schematic of a gas chromatograph.

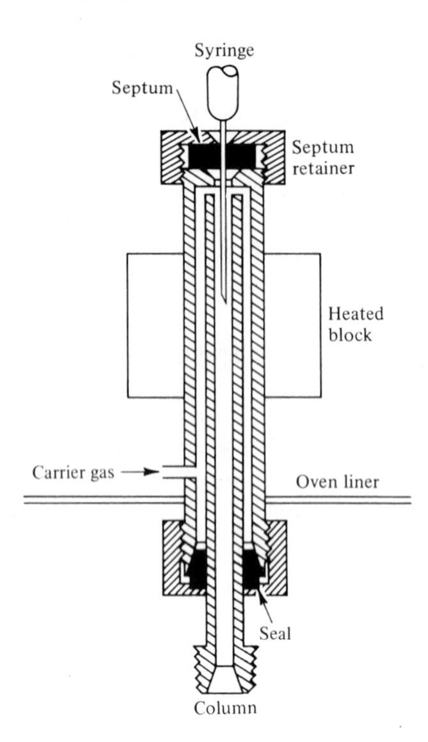

FIGURE **18.2**
Schematic representation of a typical flash vaporizer injection port.

FIGURE **18.3**
(a) Six-port rotary valve and (b) a sliding plate valve.

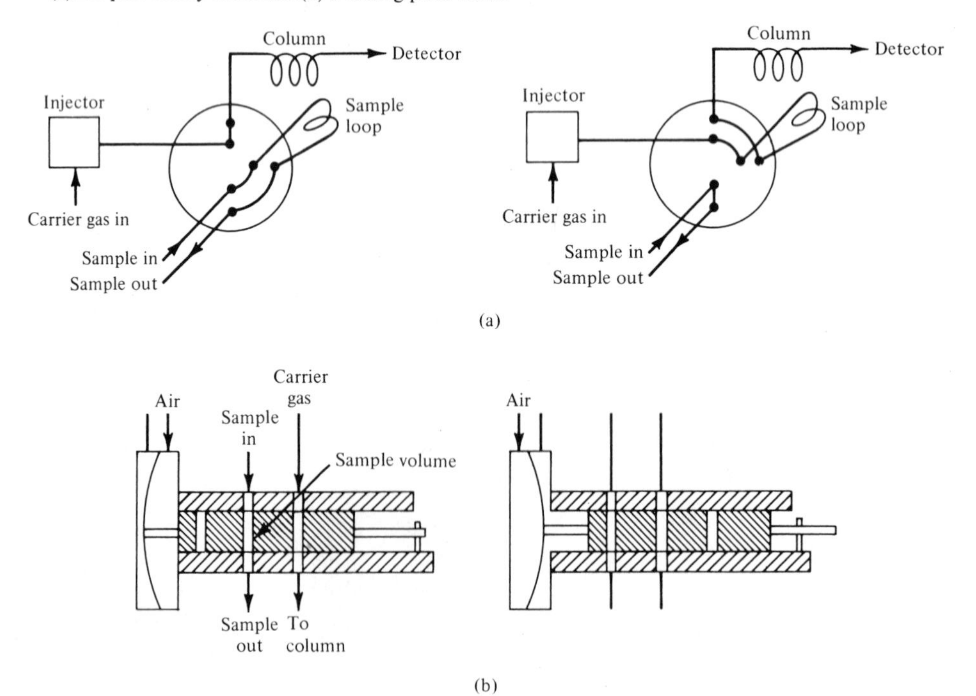

Headspace Sampling[1]

Beyond the conventional GC analysis of gases and low-viscosity liquids, some situations are more effectively handled by headspace sampling. This is true when only the vapor above the sample is of interest, as with perfumes or food products, with volatile organic constituents of samples such as urine, human breath, and environmental samples, and when the sample is a liquid that would normally require some processing before injection, such as blood, waste water, or drinking water. Solvent peaks are much smaller than would be found by injecting the liquid sample itself. Headspace sampling can be done on any sample matrix provided that the partition coefficient allows a sufficient amount of analyte into the gaseous phase.

Purge and Trap

Volatile organic constituents can be purged from the sample and trapped on Tenax-GC contained in an 11-cm tube. Tenax-GC is a porous polymer based on 2,6-diphenyl-*p*-phenylene oxide. Trapped samples can easily be stored or shipped to another site for analysis. Efficient desorption from the Tenax occurs with helium flow at 300 °C. The desorbed volatiles are collected in a precolumn cooled by dry ice. The precolumn is then connected to the GC column, the dry ice is removed, and the analysis is started at room temperature. The precolumn contains the same liquid phase as the regular GC column. Examples of purge and trap are organic constituents of urine, human breath, and environmental air. This technique has been used for the study of volatile metabolites in the urine samples from normal individuals and from individuals with diabetes.

Pyrolysis[2]

Gas-liquid chromatography reaches a practical limit when the amount of energy needed to vaporize the sample is the same as the amount of energy required to break a carbon-carbon bond. The technique of pyrolysis (or controlled thermal fragmentation) extends gas chromatographic analysis to such low-volatility compounds as rubber, polymers, paint films, resins, microorganisms, soils, coals, textiles, and organometallics. Volatile fragments are formed and introduced into the chromatographic column for analysis.

The method lends itself to studies on heat stability and thermal decomposition. It is difficult to predict a pyrolysis pattern before the experiment. Usually trial and error is used to determine which peaks in the chromatogram of the pyrolysis products are related or are unique to the composition in question. In many cases the relationships are obvious and readily predictable on the basis of chemical intuition. For example, pyrolyzed monomeric products from polymers aid in their identification. In a two-step process, the injection port is preheated to perhaps 270 °C. As soon as the sample is inserted, any volatile ingredients are driven off, which serves as a fingerprint of the formulation. Then the pyrolysis step develops the fingerprint of the nonvolatile ingredients. If the top of the sample holder is provided with a septum, known monomers can be injected with a microsyringe for identification of the peaks of an unknown pyrogram.

A pyrolyzer based on the Curie point principle uses a radio-frequency induction heating system that utilizes the Curie points of ferromagnetic metals to achieve good

FIGURE **18.4**

Diagram of a typical Curie-point pyrolyzer. [Reprinted with permission from C. J. Wolf, M. A. Grayson, and D. L. Fanter, *Anal. Chem.*, **52**, 349A (1980). Copyright 1980 American Chemical Society.]

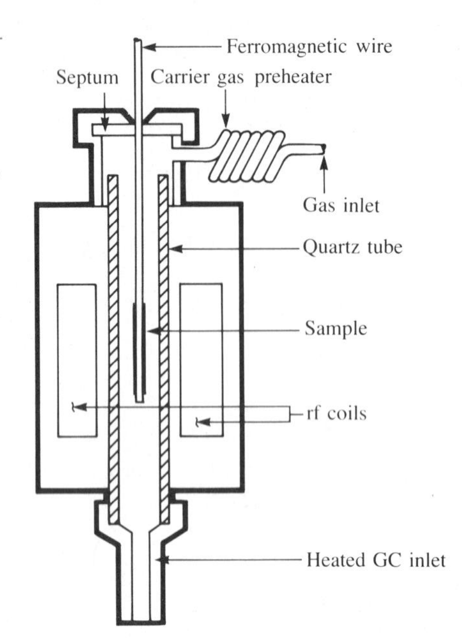

accuracy in temperature control. In these systems a ferromagnetic wire or ribbon, held in a flowing carrier gas stream, is heated inductively in a radio-frequency field to its Curie point (the temperature at which the material loses its magnetic property and ceases to absorb radio-frequency energy). The Curie point of the wire thus provides a predetermined pyrolysis temperature. The temperature rise is very rapid (0.1–0.3 sec). Each magnetic material has its own specific Curie point. Foils of various materials enable an operator to select pyrolysis temperatures from 150 to 1040 °C by simply changing the foil. Material that is soluble in conventional solvents can be applied to the filament or wire in the form of a thin film. Insoluble samples are wrapped in aluminum foil, then rewrapped with the particular magnetic foil and inserted into a sample tube holder (Figure 18.4). The entire assembly is contained in a thermostatically controlled oven to prevent condensation of pyrolyzed products.

Continuous-mode pyrolyzers include tubular furnaces and resistively heated ribbons. The latter type can be operated as either a pulsed or continuous mode system. One system contains both a ribbon and coil filament pyrolyzer. It is based on the concept of a self-sensing platinum element acting as one leg of a Wheatstone bridge configuration, with a digital setting temperature control potentiometer as the other (balancing) leg. The platinum element serves as a temperature sensor, heater, and sample holder simultaneously. Alternatively, a small quartz tube can be inserted into a coil that is resistively heated with the same circuit used to heat the ribbon.

18.2 DERIVATIVE FORMATION[3,4]

Derivatization prior to gas chromatography is often desirable to (1) improve the thermal stability of compounds, particularly compounds that contain polar functional groups, (2) change the separation properties of compounds by the purposeful adjustment of their volatility, and (3) introduce a detector-oriented tag into a molecule. In an ideal derivatization, all the relevant functional groups should be derivatized quantitatively and quickly, preferably in less than 10 min. Then, without further processing, the reaction mixture can be injected directly into the gas chromatograph. Separation of the derivatives should be complete and excess reagent should elute with the solvent peak. A few typical examples will be mentioned.

Silylating agents, such as *N*,*O*-bis(trimethylsilyl)acetamide or the trifluoro analog, *N*,*O*-bis(trimethylsilyl)trifluoroacetamide, convert one active hydrogen in polar groups such as —OH, —COOH, —NH_2, =NH, and —SH to an —O—$Si(CH_3)_3$ group. A typical silylation reaction with an alcohol is shown:

$$H_3C\overset{O}{\overset{\|}{C}}N[Si(CH_3)_3]_2 + -\overset{|}{\underset{|}{C}}-OH \rightarrow -\overset{|}{\underset{|}{C}}-OSi(CH_3)_3 + CH_3\overset{O}{\overset{\|}{C}}NHSi(CH_3)_3 \quad (18.1)$$

The use of perdeuterated derivatives, such as *N*,*O*-bis(trimethylsilyl-d_{18})acetamide, assists materially in interpreting the mass spectra of silylated compounds.

O-Alkylhydroxylamines are used to prepare *O*-alkyloximes of aldehydes and ketones:

$$R_1R_2C{=}O + H_2N{-}O(CH_2)_3CH_3 \xrightarrow{\text{pyridine}} R_1R_2C{=}N{-}O(CH_2)_2CH_3 + H_2O \quad (18.2)$$

The reagents are particularly useful in the analysis of compounds that contain both hydroxyl and carbonyl groups.

The trifluoro group, such as trifluoroacetyl, pentafluoropropionyl, and pentafluorobenzoyl, is commonly used for sensitizing substances to detection by electron capture.

A derivatization method makes feasible the quantitative and qualitative analysis of amino acids. After they are dry, amino acids are converted to butyl ester hydrochlorides by treatment with 1-butanol and hydrogen chloride at 100 °C for 30 min:

$$\underset{\displaystyle NH_2}{\underset{|}{RCHCOOH}} + C_4H_9OH + HCl \rightarrow \underset{\displaystyle NH_3^+\ Cl^-}{\underset{|}{RCHCOOC_4H_9}} + H_2O \quad (18.3)$$

The butyl esters are converted to volatile *N*-trifluoroacetyl butyl esters in sealed tubes:

$$(CF_3CO)_2O + \underset{\displaystyle NH_3^+\ Cl^-}{\underset{|}{RCHCOOC_4H_9}} \rightarrow \underset{\displaystyle NHCOCF_3}{\underset{|}{RCHCOOC_4H_9}} + HCl + CF_3COOH \quad (18.4)$$

Subsequent chromatographic resolution into single peaks by temperature programming is then easily accomplished.

18.3 GAS CHROMATOGRAPHIC COLUMNS

Two basic types of columns are in general use: the packed column and the open tubular or capillary column.

Packed Columns

Packed columns are constructed from tubing of stainless steel, nickel, or glass. Inner diameters may range from 1.6 to 9.5 mm. Length is often 3 m. These columns are packed with an inert support, usually a diatomaceous earth whose internal pore diameters range from 2 μm for material derived from firebrick to 9 μm for materials derived from filter aids. The internal column diameter should be at least eight times the diameter of the support particles. For example, the best particle size is 100/120 mesh (149–125 μm) for 2-mm-bore columns, and 80/100 mesh (177–149 μm) for 4-mm-bore columns.

The support should not take part in the separation. Especially with low liquid loadings (less than 10%) and nonpolar phases, the support must be thoroughly inactivated when polar compounds are to be analyzed. Surface mineral impurities, which can serve as adsorption sites, can be removed by acid washing. If the support is not deactivated at all, it is called nonacid washed. The acid-washed grade performs quite well for the analysis of relatively nonpolar samples. However, the silanol (Si-OH) groups that cover the surface also tend to adsorb solutes. These surface silanol groups can cause peak tailing through hydrogen bonding with polar samples. The problem can be minimized by converting the silanol groups to silyl ethers by treating the column packing with dimethyldichlorosilane (unused silylating agent is removed with methanol). When silanized supports are used, the loading of the stationary phase should be limited to a maximum of 10%.

Special packing material may be needed for particular applications. Very lightly loaded (roughened or texturized) glass beads are used for very rapid analyses well below the boiling point of the sample components. Sieved Teflon supports are used when corrosive substances are handled. In gas-solid chromatography the packing material is an adsorbent, such as silica gel, a bonded-phase support, or a molecular sieve.

Capillary Columns

Capillary columns have an internal diameter of 1 mm or less. These are usually constructed of fused silica (a very high-purity glass), which has a much higher degree of cross linking within the silicon-oxygen matrix than does ordinary glass. The high tensile strength of the silica tubing permits the construction of thin-walled, flexible columns. To protect the thin wall against scratches, a protective coating of polyimide is applied to the outer wall.

There are two main types of capillary columns: (1) packed columns with solid particles over the whole diameter of the column (micropacked) and (2) open tubular columns with an open and unrestricted flow path through the middle of the column. The latter are divided into wall-coated open tubular (WCOT) columns, support-coated open tubular (SCOT) columns, and porous-layer open tubular (PLOT) columns.

The inner surface of the wide-bore (WCOT) capillary column (0.53 mm i.d.) is coated with the stationary liquid phase.[5] Uniform coating of the surface was difficult to obtain until bonded-phase (polymer) columns were introduced. The polymer wets the surface of a fused silica column well, with the result that a very uniform film covers the column walls. A variety of functional groups can be blended into the polysiloxane chain to provide stationary phases of different polarity or selectivity.

The result is a film that is both thermostable and nonextractable. The columns can be flushed with pure solvents to remove contaminants, nonvolatiles, and pyrolysis products.

The major advantage of wide-bore capillaries regardless of the separation is an increase in the speed of analysis (Figure 18.5). Using packed column flowrates, the wide-bore column produces separations with packed column efficiencies but with a total run time roughly three times faster. However, when the flowrate is optimized for the tubing diameter, the wide-bore column produces far superior efficiencies and therefore superior separations, with analysis times approximately equal to those for packed columns. Thus, there are six extremely good reasons for changing from packed columns to wide-bore capillaries: shorter retention times, greater inertness, longer life, lower bleed, higher efficiencies, and greater reproducibility.

Wide-bore capillaries are available in 10- to 30-m lengths. However, the 10-m column closely approximates the capacity and separation of the standard analytical 20 m × 2 mm i.d. packed column with a 3%–5% loading. The ability to manipulate the film thickness reproducibly has led to columns that can be tailored for the analysis of very volatile or very high-boiling mixtures. The customary thickness of the bonded liquid phase is 1–2 μm. Thicker films (3–5 μm) permit the analysis of volatile materials without the need for subambient column temperatures. Films less than 1 μm thick are valuable for analyzing high-molecular-weight compounds with reasonable analysis times.

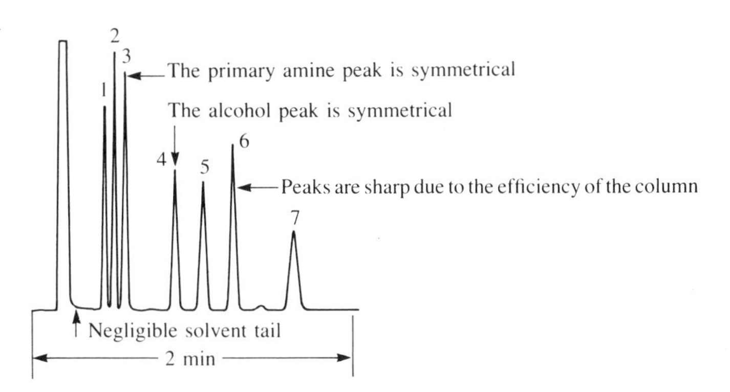

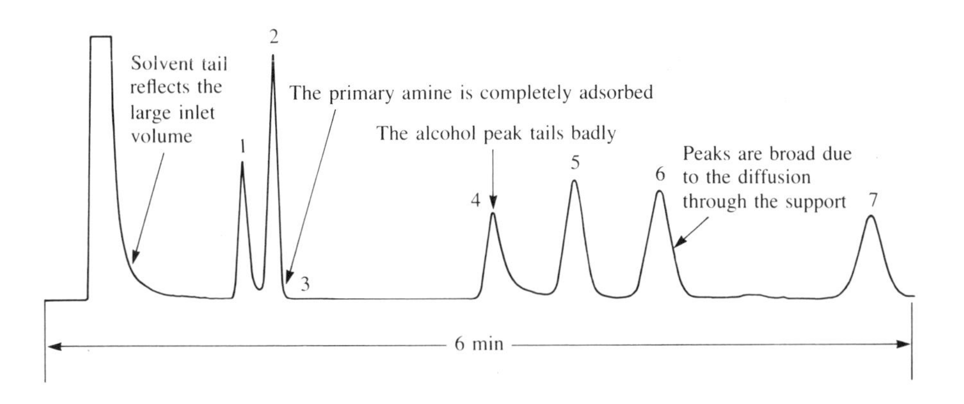

FIGURE **18.5**
Comparison of chromatograms obtained on a wide-bore column (*upper*) and a packed column. Peaks: (1) 4-chlorophenol, (2) dodecane, (3) 1-decylamine, (4) 1-undecanol, (5) tetradecane, (6) acenaphthene, and (7) pentadecane. (Courtesy of J & W Scientific, Inc.)

With helium as the carrier gas, the optimum flowrate for a 0.53-mm i.d. column is approximately 2.5 mL/min. This provides 2100 plates per meter, giving 21,000 theoretical plates with a 10-m wide-bore coated capillary column. This is far superior to the 3000–4000 theoretical plate total column efficiency obtained with a typical 20-m packed column. Nothing is superior to a capillary column for resolving mixtures with many components.

Ovens

Chromatographic columns are coiled and held in a basket that is mounted inside an oven. The column oven must be able to be rapidly heated and cooled. This requires a well-designed and adequate system of air flow. In most designs the air is blown past the heating coils, then through baffles that make up the inner wall of the oven, past the column, and back to the blower to be reheated and recirculated. Ovens are usually constructed of low-mass stainless steel. For temperature programming it is desirable to have a range of temperature program rates from 0.1 to 50 °C /min. Hold times anywhere within the program should be available.

Subambient initial temperatures are useful when working with capillary columns. Temperatures should be maintained within ± 1 °C for isothermal runs and within ± 2 °C of the desired temperature during programming.

18.4 LIQUID PHASES AND COLUMN SELECTION

The function of the stationary liquid phase is to separate the sample components into discrete peaks. In addition, the liquid phase should have reasonable chemical and thermal stability (the upper temperature limit in Table 18.1). Of course, in order to function the stationary phase must remain in the liquid condition and thus the lower temperature limit. Usually operations are kept 10–15 °C below the upper temperature limit. The amount of column bleed (vaporization of the stationary liquid phase) must be minimized to prolong the column life, to prevent any fouling of the detector, and to maintain baseline stability on the chromatogram. Bonding the stationary partitioning liquid to the inner surface of a capillary column or to the column packing permanently anchors the stationary phase to the surface. This provides a liquid phase with no appreciable vapor pressure. Nevertheless, there is an upper temperature limit above which carbon-carbon bonds rupture or chemical reactions (such as hydrolysis with water vapor) apparently occur. Inclusion of a carborane structure, which acts as an energy sink, periodically within the bonded group delays the destruction of the bonded-phase coating and enables some bonded phases to be used up to 500 °C.

Retention Indices

The suitability of a stationary phase for a specific separation depends on the selectivity of the phase. This is a measure of the degree to which polar compounds are retarded relative to their elution on a nonpolar phase. A systematic method for expressing retention data uses the *Kovats retention indices* (RI)[6]. These indices indicate where compounds will appear on a chromatogram with respect to straight-chain alkanes injected with the sample. By definition, the RI for a normal paraffin is

TABLE **18.1** STATIONARY PHASES IN GAS CHROMATOGRAPHY

Liquid phase (chemical type; similar phases)	**Minimum/ maximum temperature (°C)**	*McReynolds constants*					
		x'	y'	z'	u'	s'	Σ
For boiling point separation of broad molecular weight range of compounds							
Squalane (2,6,10,15,19,23-hexamethyltetracosane)	20/150	0	0	0	0	0	0
OV-101 (polydimethylsiloxane; SF 96, SP-2100)	50/350	17	57	45	67	43	229
SE 54 (polydiphenylvinyldimethylsiloxane, 5%/1%/94%)	50/300	33	72	66	99	67	337
OV-3 (polydiphenyldimethylsiloxane, 10%/90%)	0/350	44	86	81	124	88	423
Dexsil 300 (polycarboranemethylsiloxane)	50/500	47	80	103	148	96	474
Dexsil 400 (polycarboranemethylphenylsilicone)	50/500	72	108	118	166	123	587
OV-7 (polydiphenyldimethylsiloxane, 20%/80%; DC 550)	20/350	69	113	111	171	128	592
For unsaturated hydrocarbons and other semipolar compounds							
OV-17 (polydiphenyldimethylsiloxane, 50%/50%)	0/325	119	158	162	243	202	884
Dexsil 410 (polycarboranemethylcyanoethylsilicone)	50/500	72	286	174	249	171	952
OV-215 (polytrifluoropropylmethylsiloxane)	0/275	149	240	363	478	315	1545
OV-225 (polycyanopropylphenylmethylsiloxane; XE 60)	0/265	228	369	338	492	386	1813
For nitrogen compounds							
Poly-A 103 (polyamide)	70/275	115	331	149	263	214	1072
XF-1150 (polycyanoethylmethylsilicone, 50%/50%)	20/200	308	520	470	669	528	2495
Specifically retards compounds with keto groups; for halogen compounds							
OV-210 (polytrifluoropropylsiloxane; QF-1, SP-2401)	0/275	146	238	358	468	310	1520
For alcohols, esters, ketones, and acetates							
SP-2300 (polyethylene glycol; Carbowax 20M, FFAP)	25/275	316	495	446	637	530	2424
Silar-7CP (SP-2310)	0/250	440	638	605	844	673	3200
Silar-10C (SP-2340; tetracyanoethylated pentaerythritol)	25/275	523	757	659	942	801	3682
For fatty acid methyl esters							
Neopentyl glycol succinate (HI-EFF-3BP)	50/230	272	469	366	539	474	2120
Diethylene glycol adipate (SP-2330, HI-EFF-1AP, LAC-1-R-296)	25/275	378	603	460	665	658	2764
Diethylene glycol succinate (HI-EFF-1BP, LAC-3-R-728)	20/200	499	751	593	840	860	3543
OV-275 (polydicyanoallylsilicone)	25/250	781	1006	885	1177	1089	4938
Absolute index values on squalane for reference compounds		653	590	627	652	699	

100 times the number of carbon atoms in the compound regardless of the columns used or the chromatographic conditions. Thus, the RI for pentane is 500, for hexane 600, for heptane 700, and so on. Of course, the type of column and the operating conditions, such as liquid loading and any pretreatment, must be specified. For example, suppose that on a squalane column the retention times for hexane, 1-nitropropane, and octane were 15, 16, and 25 min, respectively. On a graph of $\ln t'_R$ of the alkanes versus their retention indices, a RI of 652 for 1-nitropropane is read off the graph. The number 652 for 1-nitropropane means that it elutes halfway between hexane and heptane on a logarithmic time scale. When the experiment with an OV-275 (dicyanoallylsilicone) column is repeated, the RI for 1-nitropropane is found to be 1177, between undecane and dodecane. This implies that OV-275 will retard 1-nitropropane much more than will squalane; that is, OV-275 is more polar than squalane by 525 units. Consult the entries in Table 18.1 and note the values mentioned in the foregoing discussion.

Within a homologous series, a plot relating the logarithm of the adjusted retention time (or volume) to the number of carbon atoms is linear, provided the lowest member of the series is excluded. Members of homologous series are easily identified from a plot of $\ln(t_R - t_M)$ versus the RI of the paraffins.

EXAMPLE 18.1

The adjusted retention times for these compounds are:

Solute	t'_R(min)	Solute	t'_R(min)
Nonane	0.639	1-Octanol	1.863
1-Heptanol	0.932	Dodecane	5.024
Decane	1.226		

The Kovats retention index for each compound is determined as follows. By definition the retention indices are 900 for nonane, 1000 for decane, and 1200 for dodecane. Substituting the data into the expression and solving the set of simultaneous equations

$$\text{RI} = a + b \ln t'_R$$

give

$$\text{RI} = 966.2 + 143.9 \ln t'_R$$

The retention indices for 1-heptanol and 1-octanol are 956 and 1056, respectively. The predicted values for 1-nonanol and 1-decanol are 1156 and 1256, respectively; and the corresponding adjusted retention times are 3.740 min and 7.492 min, respectively.

Classification of Stationary Phases

Since the chromatographer is interested in the selectivity of a column for a variety of functional groups, it is important to classify each of the stationary phases by its ability to retard specific functional groups. Stationary phases can be classified by the method developed by Rohrschneider[7] and extended by McReynolds.[8] It involves measurement of the RI for selected "index" compounds on a column packed with a particular stationary phase. These RI values are compared with the RI values for the same index compounds obtained on a squalane column. The difference determines the degree to which each index compound is retarded by the stationary phase. It is a measure of solute-solvent interaction due to all intermolecular forces other than London dispersion forces. The latter are the principal solute-solvent effects with squalane.

The overall effects due to hydrogen bonding, dipole moment, acid-base properties, and molecular configuration can be expressed as

$$\Sigma \Delta I = ax' + by' + cz' + du' + es' \tag{18.5}$$

In this expression,

$x' = \Delta I$ for benzene (intermolecular forces typical of aromatic and olefin compounds)

$y' = \Delta I$ for 1-butanol (an electron attractor typical of alcohols, nitriles, acids, nitro compounds, and alkyl mono-, di-, and trichlorides)

$z' = \Delta I$ for 2-pentanone (an electron repeller typical of ketones, ethers, aldehydes, esters, epoxides, and dimethylamino derivatives)

$u' = \Delta I$ for 1-nitropropane (typical of nitro and nitrile compounds)

$s' = \Delta I$ for pyridine

McReynolds constants are listed for selected stationary phases in Table 18.1. Five additional constants and their Kovats indices (in parentheses) are: 2-methyl-2-pentanol (690), 1-iodobutane (818), 2-octyne (841), 1, 4-dioxane (654), and *cis*-hydrindane (1006). All index compounds must be chromatographed under nearly identical conditions of column temperature and stationary-phase loading.

From a practical standpoint, the majority of separations can be made efficiently on one of perhaps six different classes of stationary phases. If the sample components have significantly different boiling points, a nonpolar stationary phase is used (first group in Table 18.1). When some components have very similar boiling points, a stationary phase should be selected that will interact strongly with one (or several) of the components. The interaction desired can be chosen from among the characteristics listed in parentheses for the index compounds in the preceding section. Another application of the table of McReynolds constants is to determine a substitute for a prescribed liquid phase, perhaps a similar phase but usable at a higher temperature.

EXAMPLE 18.2

To elute an alcohol before an ether (both have nearly identical boiling points), a stationary phase with a high z' value (relative to the y' value) is needed. A column with SP-2401 or OV-210 as stationary phase should be tried. Just the reverse is needed if the ether is to be eluted before the alcohol. This suggests an OV-275 stationary phase. In the separation of a benzene/cyclohexane mixture, there is only a 0.5 °C difference in boiling points. Whereas an OV-101 column would fail to separate this mixture, the more polar OV-225 (note x' values in Table 18.1) will retain benzene long enough to provide a clean separation.

Another rule to follow when selecting liquid phases is "like dissolves like." For the separation of alcohols use a polyglycol as a liquid phase, for the separation of hydrocarbons use a hydrocarbon liquid phase, and so on. The polarities of solute and stationary phase should be somewhat alike. Special phases may be needed for nitrogen compounds and halogen compounds.

After a reasonably intelligent choice of liquid phase has been made, the difference between obtaining a good chromatogram and a poor chromatogram usually becomes a simple matter of adjusting the primary operating parameters that the chromatographer can control. These are the type and length of column, detector type, carrier gas type, flowrate, temperature, and sample size. The efficiency of the gas chromatographic process depends on these interrelated factors and they must all be considered collectively (see Section 18.6).

18.5 DETECTORS FOR GAS CHROMATOGRAPHY

A detector, located at the exit of the separation column, senses the presence of the individual components as they leave the column. The detector volume must be small to prevent the remixing of components separated on the column. The electrical analog output of the detector is amplified and then sent directly to a strip chart recorder or converted to a digital signal and sent to a microcomputer system. The computer system can process the data, store them, and display the chromatogram with analytical results on a video screen or recorder. Such systems provide a continuous record of the mass of solute eluted or concentration profiles of eluted solute bands—that is, a chromatogram of concentration versus time.

Thermal Conductivity Detector

The thermal conductivity detector (TCD) uses a heated filament placed in the emerging gas stream. The amount of heat lost from the filament by conduction to the detector walls depends on the thermal conductivity of the gas phase. A diagram of a flow-through TCD is shown in Figure 18.6. Within a cavity in the metal block there extends a tightly coiled filament constructed of tungsten metal, tungsten-rhenium alloy, or tungsten sheathed with gold. The filament is heated to a constant temperature but less than a dull-red condition by a regulated dc current supply. Heat loss from the filament to the metal block is constant when only carrier gas is flowing through the detector. The thermal conductivities of hydrogen and helium are roughly six to ten times greater than those of most organic compounds. Thus the presence of even small amounts of organic materials causes a relatively large decrease in the thermal conductivity of the column effluent. The filament retains more heat, its temperature rises, and its electrical resistance goes up.

The standard detector consists of four identical filaments mounted within one brass block. The filaments make up the arms of a Wheatstone bridge (Figure 18.7a).

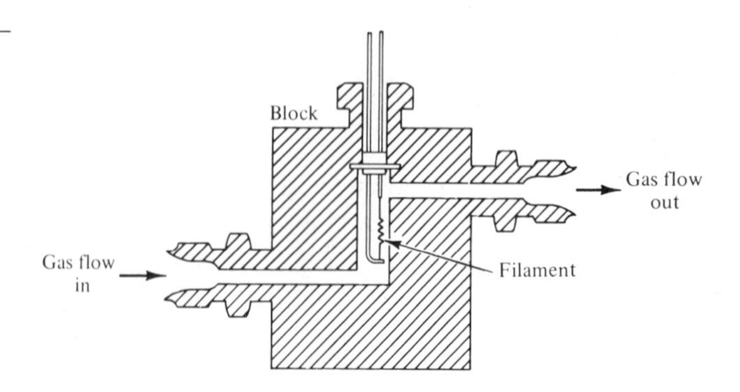

FIGURE **18.6**
Cross-sectional view of a thermal conductivity detector mounted in a heat sink.

In analyses the column effluent is passed through one pair of filaments, and the second pair is placed in the gas stream ahead of the sample injection point. Any imbalance between the pairs of filaments is recorded. In this way the thermal conductivity of the carrier gas is canceled and the effects of variation in flowrate and pressure are minimized. An initial baseline reading is established by passing only carrier gas through both pairs of filaments and then adjusting the setting of the power supply D.

A TCD can be constructed with thermistors. The thermistor is a metal oxide bead with electrical leads attached. The bead exhibits a resistance of 8000-Ω at 25 °C and a negative temperature coefficient of resistance. One bead is mounted in the pure carrier gas stream, the other in the column effluent. The Wheatstone bridge circuit (Figure 18.7b) is completed by a matched pair of 500-Ω, 3-W resistors. Thermistors are not exchangeable either mechanically or electrically with filaments. Thermistors are used for work at ambient or subambient column temperatures.

Flame Ionization Detector

The flame ionization detector (FID) adds hydrogen to the column effluent. Subsequently, the mixture is passed through a jet where it is mixed with external air

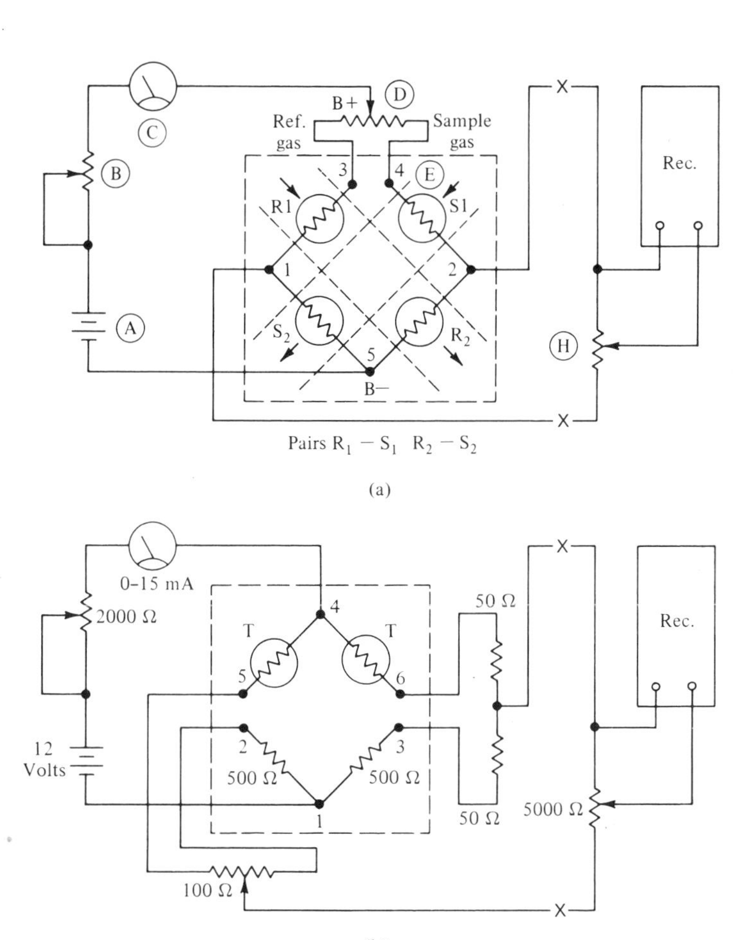

FIGURE **18.7**
Circuitry for (a) thermal conductivity (four-filament) cells and (b) thermistor cells. (Courtesy of Gow-Mac Instrument Co.)

and burned. In the configuration shown in Figure 18.8, the flame jet and a cylinder positioned 0.5–1.0 cm above the tip of the flame constitute the twin electrodes. In another arrangement two parallel plates are mounted above the flame tip. A potential of about 400 V is applied across the two electrodes, which lowers the resistance between the electrodes and causes a current ($\sim 10^{-12}$ A) to flow. The current arises from the ions and free electrons generated in a pure hydrogen/air flame. When ionizable material from the column effluent enters the flame and is burned, the current markedly increases. The current flows through an external resistor, is sensed as a voltage drop, amplified, and finally sent to an output device, a recorder or microcomputer. An opposing voltage equal to the signal from the hydrogen/air flame when only pure carrier gas is passing permits the baseline to be adjusted.

The FID is enclosed within a chimney so that it is unaffected by drafts and can be heated sufficiently to avoid condensation of water droplets from the combustion process. An ignitor coil and flame-out sensor are placed above the jet to reignite the flame if it becomes extinguished.

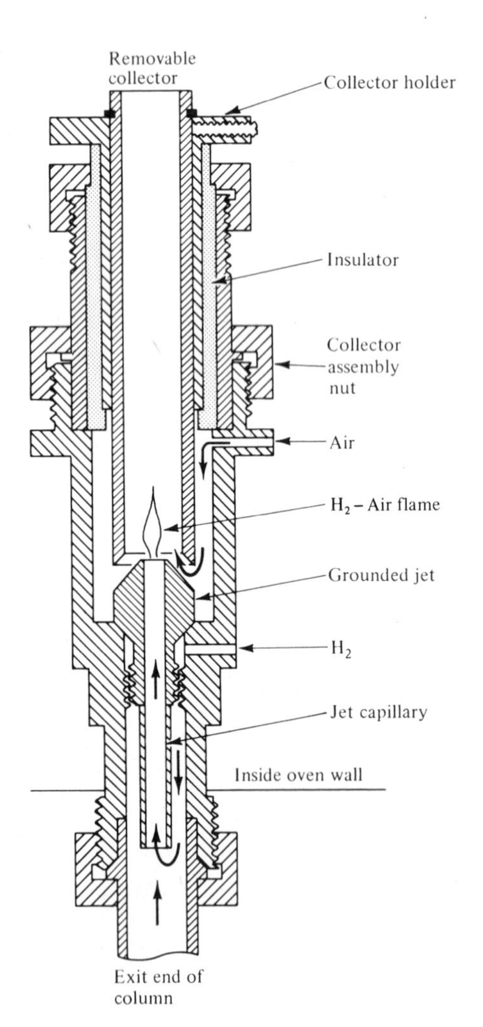

FIGURE **18.8**
Cross section of a flame ionization detector. (Courtesy of Hewlett-Packard Co.)

The FID responds proportionately to the number of $—CH_2—$ groups introduced into the flame. For example, the response to an equimolar amount of butane is twice that to ethane. There is no response from fully oxidized carbons such as carbonyl or carboxyl groups (and thio analogs) and ether groups. The response from carbons attached to hydroxyl groups and amine groups is lower. The insensitivity of the FID to moisture and permanent gases (CO, CO_2, CS_2, SO_2, NH_3, N_2O, NO, NO_2, SiF_4, and $SiCl_4$) is advantageous in the analysis of moist organic samples and in air-pollution studies when small traces of organic materials have to be measured against these permanent gases as background. If desired, CO and CO_2 can easily be converted to CH_4 by reduction with hydrogen over a nickel catalyst and subsequently measured by the detector.

A portable gas chromatograph that can be carried by a shoulder strap has been built around the FID as detector. The dual-mode instrument provides either continuous direct readout of total organic vapor concentration (bypassing the chromatographic column) for screening or survey purposes, or qualitative and quantitative analyses using the GC option. A portion of the sample air is directed through an integral charcoal filter to provide the detector with a supply of pure air. The portable unit carries an 8-hr supply of hydrogen, a small diaphragm air pump, and a readout meter. An isothermal pack offers temperature control of the separation column; the pack encloses a supercooled mixture that, when seeded, provides a constant temperature during its solidification process.

Thermionic Emission Detector

The thermionic emission detector (TED) uses a fuel-poor hydrogen plasma, a low-temperature flame that suppresses the normal flame ionization response of compounds that do not contain nitrogen or phosphorus (Figure 18.9). A nonvolatile bead of rubidium silicate is centered 1.25 cm above the flame tip. In all other respects

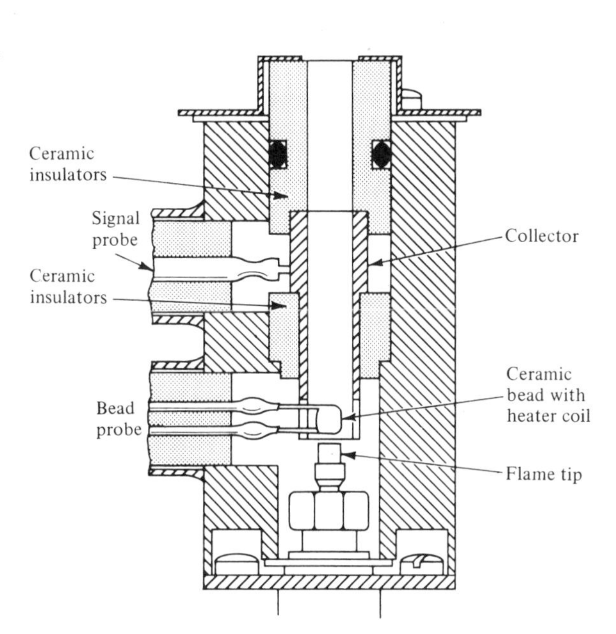

FIGURE **18.9**
Thermionic emission detector. (Courtesy of Varian Associates.)

the physical arrangement resembles a FID. The bead is electrically heated and can be adjusted to between 600 and 800 °C. This option permits adjustment of the bead's temperature independent of the flame as a source of thermal energy.

With a very small hydrogen flow, the detector responds to both nitrogen and phosphorus compounds. Enlarging the plasma size and changing the polarity between the plasma tip and collector cause the TED to respond to only phosphorus compounds.

Electron Capture Detector

In the electron capture detector (ECD) the column effluent passes between two electrodes, as shown in Figure 18.10. One of the electrodes has on its surface a radioisotope that emits high-energy electrons (beta particles) as it decays. These electrons bombard the carrier gas (nitrogen), resulting in the formation of a plasma of positive ions, radicals, and thermal electrons by a series of elastic and inelastic collisions. This process is very rapid (<0.1 μsec). The application of a potential difference to the electron-capture cell allows the collection of the thermal electrons that constitute the detector standing current, or baseline signal, when only carrier gas is passing through the detector. Electron-absorbing compounds in the carrier gas stream react with the thermal electrons to produce negative ions of larger mass. The rate of recombination between negative and positive ions is many times faster than between thermal electrons and positive ions. The decrease in detector current due to removal of thermal electrons by recombination in the presence of electron-capturing compounds forms the quantitative basis of the detector operation.

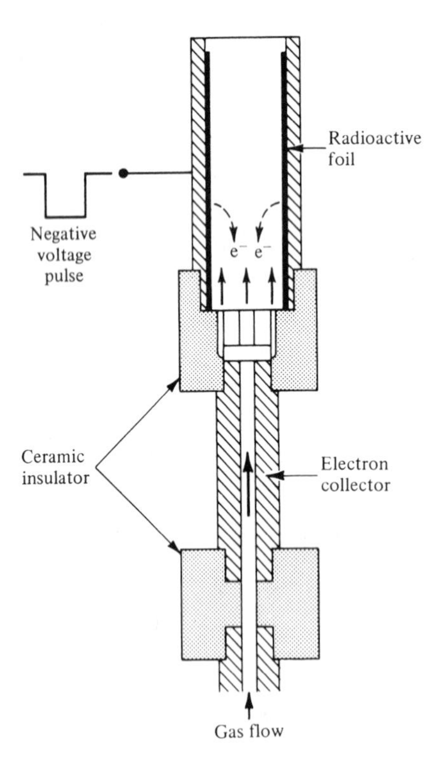

FIGURE **18.10**

Electron capture detector. (Courtesy of Varian Associates.)

Efficient electron collection with nitrogen as the carrier gas is achieved with a displaced coaxial cylindrical cell configuration. A 0.2-mL volume is formed by two closely spaced cylinders in which the carrier gas flow is opposite to the electron flow.

Radioactive sources include tritium adsorbed in titanium or scandium, and nickel-63 as a foil or plated on the interior of the cathode chamber. Nickel-63 is a higher energy source than tritium and can be used up to 400 °C. Although the sensitivity of the nickel detector is about five times less than that of a tritium unit, the standing current remains constant and the detector is much less susceptible to contamination.

To prevent the formation of contact potentials and space charge effects, the ECD voltage is applied as a sequence of narrow pulses with a duration and amplitude (0.6 μsec and 50 V, respectively) sufficient to collect the very mobile electrons but not the heavier, slower negative ions. During the interval between pulses (100–150 μsec) the electron concentration builds up inside the cell. When the pulse is applied, the concentration of electrons is essentially reduced to zero. This mode of operation has several advantages. During most of the time no field is applied; this enables the free electrons to reach thermal equilibrium with the gas molecules. The opportunity for electron capture is maximized, and stable response factors are thereby attained. Negative ion formation occurs in a region where positive ions are also present and recombination can efficiently take place. There is no collection of negative ions. The formation of a space charge is prevented by the brevity of the pulse interval.

The linearity of ECD is a function of the cell design, the composition of the carrier gas, and the method of applying the potential difference to collect the thermal electrons. For the direct current and pulsed ECD, the linear range is about 100, and for pulse-modulated, constant-current ECD about four decades.

Argon mixed with 5%–10% methane is the makeup gas of choice for pulsed ECDs. It is added to the column effluent. At the voltages used, the electron velocity is ten times greater in this mixture than in nitrogen. Through inelastic collisions with methane, the metastable ions formed from argon are eliminated before they can cause undesirable sample ionization.

Other Detectors[9]

The *flame photometric detector* (FPD) is essentially a special type of flame emission filter photometer used primarily for the determination of volatile sulfur or phosphorus compounds.[10] The column effluent passes into a dual hydrogen-enriched, low-temperature flame within a shield. Both air and hydrogen are supplied as makeup gases to the carrier gas. Only the upper flame is viewed. Phosphorus forms an HPO species that emits band emissions at 510 and 526 nm around the sides and base of the flame. Sulfur compounds form a S_2 entity that emits a series of bands centered around 394 nm but also overlapping the phosphorus spectrum. The detector response to phosphorus is linear, whereas the response to sulfur depends on the square of the concentration. It is about 100 times less sensitive to phosphorus than the thermionic emission detector.

The *photoionization detector* (PID) uses ultraviolet radiation from lamps with energies ranging from 9.5 to 11.7 eV to produce ionization of solute molecules.[11,12] The ions are collected at a positively charged electrode and the current is measured.

Compounds whose ionization potentials are lower than the lamp ionizing energy give a response.

The *electrolytic conductivity detector* operates on electrolytical conductivity principles (see Chapter 24). Organic compounds eluting from the GC column are burned in a miniature furnace to form simple molecular species that readily ionize and contribute to the conductivity of deionized water. Changes in electrolytic conductivity are monitored. Analyte ions are removed from the liquid by an ion-exchange column, part of a continuous circulating system that regenerates conductivity-grade water. When the combustion products are mixed with hydrogen gas and hydrogenated over a nickel catalyst at 850 °C in a quartz tube furnace, ammonia is formed from organic nitrogen, HCl from organic chlorides, and H_2S from sulfur compounds.

Comparison of Detectors

In Figure 18.11 gas chromatographic detectors are compared with respect to sensitivity and linearity. The sensitivity values represent those attainable for a reasonable number of compounds, with the minimum detectable quantity being the one that gives a signal-to-noise ratio of 2.

The reason for such widespread use of the flame ionization detector is apparent from Figure 18.11. Other widely used detectors are thermal conductivity, flame ionization, and electron capture. Among these, only the thermal conductivity detector is truly nondiscriminating, although the flame ionization detector responds to almost all organic compounds. Other detectors, which respond selectively or characteristically to a specific property of certain eluted species, may offer a method for discrimination in a chromatogram with a complex maze of peaks or when peaks overlap.

The thermal conductivity detector is simple, rugged, inexpensive, nonselective, and nondestructive and displays a universal response. Being nondestructive, the column effluent can be passed through a TCD and then into a second detector. In this configuration, the TCD serves as a general survey detector, which responds, in some degree at least, to all types of compounds. Since it is a nondestructive detector, it is particularly suitable for fraction collection and preparative gas chromatography. Cavity volume can be varied from 2.5 mL for detectors coupled to packed

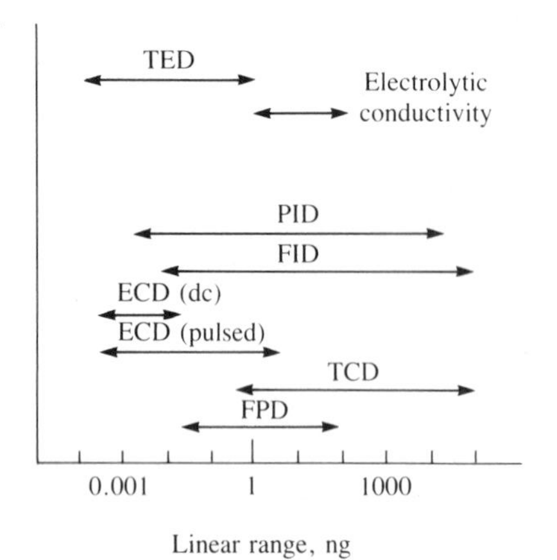

FIGURE **18.11**
Linear dynamic ranges of gas chromatographic detectors: TED, thermionic emission; PID, photoionization; FID, flame ionization; ECD, electron capture; TCD, thermal conductivity; and FPD, flame photometric.

columns, down to less than 30 μL in a detector designed for use with capillary columns. Sensitivity is about 0.3 ng/mL (as butane); linearity is between four and five decades.

The flame ionization detector is the most popular detector because of its high sensitivity, wide range of linearity, great reliability, and nearly universal response. The FID is about 1000 times more sensitive than the TCD. However, whereas the TCD responds to any substance mixed with the carrier gas, the FID responds to only substances that produce charged ions when burned in a hydrogen flame. Fortunately this includes almost all organic compounds. Detection limits for the FID are about 4 pg/sec. Response is linear over seven orders of magnitude. Noise is 2×10^{-14} A. Column operating temperatures can vary from 100 to 420 °C, an obvious advantage in programmed temperature applications.

The thermionic emission detector responds to only a restricted group of compounds—in this instance, to only compounds that contain nitrogen or phosphorus. Compared with the FID, the TED is about 50 times more sensitive for nitrogen and about 500 times more sensitive for phosphorus. The minimum detection limit is 0.06 pg/sec for nitrogen as caffeine. Though depending somewhat on operating conditions and the type of molecule, the selectivity against carbon is always better than 1 in 5000.

The electron capture detector responds to electrophilic species, which gives the detector its specificity (and name). Residual oxygen and water must be removed from the carrier gas and makeup gases. Detection is at the picogram level for polyhalogenated compounds such as pesticides. For the direct current and pulsed ECD, the linear range is about 100, and for pulse-modulated, constant-current ECD about four decades. Steroids, biological amines, amino acids, and various drug metabolites can be converted to perfluoro derivatives prior to electron capture detection. Other applications involve halogenated anesthetics, polynuclear carcinogens, and sulfur hexafluoride. The ECD also responds to anhydrides, peroxides, conjugated carbonyls, nitriles and nitrates, plus ozone and organometallics and sulfur-containing compounds.

For the electrolytic conductivity detector, sensitivity is better than 0.01 ng for sulfur, chlorine, and nitrogen. The linear dynamic range is five orders of magnitude. The detector finds use in the fields of pesticide analysis and herbicide analysis, alkaloids, and pharmaceuticals.

Dual Detectors

Dual-channel detection involves splitting the effluent and passing it through two dissimilar detectors simultaneously. Different detectors, each with its individual selectivity for the same compound, enable an operator to gain information concerning the chemical composition of the sample and also aid in distinguishing components that make up overlapping peaks. Two detectors may be used in series if the first is nondestructive. Often the TCD is used to provide a total signal, since it responds to all compounds. This is followed by a more specific detector. Figure 18.12 illustrates the technique with a sample of aliphatic hydrocarbons and amines. Being a universal detector, the TCD responds to all the components in the sample (upper trace), whereas the more selective TED responds to only the nitrogen-containing components.

FIGURE 18.12
Chromatogram of an aliphatic hydrocarbon and amine sample. (*Upper curve*) response of the TCD; (*lower curve*) response of a TED run in the nitrogen mode. Peaks: (1) heptylamine, (2) dodecane, (3) tridecane, (4) dihexylamine, (5) dicyclohexylamine, and (6) dodecyclamine.

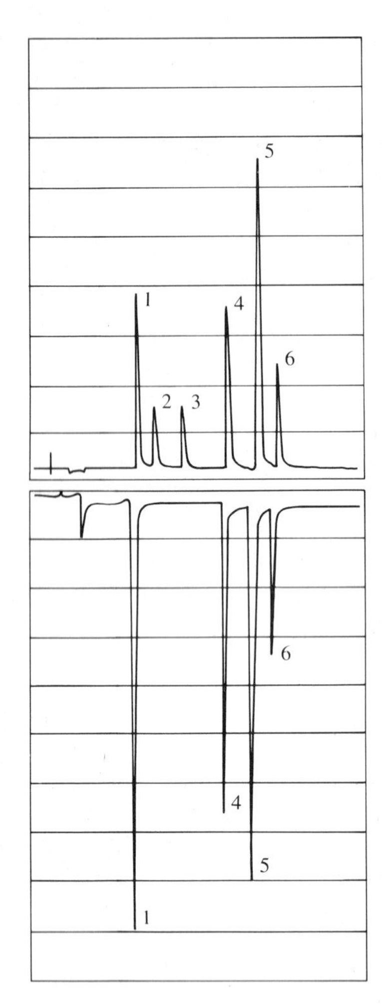

Identical dual detectors can be used in a differential mode, particularly during temperature programming. The effluent from dual columns, one containing the sample and the other just a dummy, eliminates the problem of a drifting baseline or a rising baseline as the upper temperature limit of the liquid substrate is reached in temperature programming.

18.6 OPTIMIZATION OF EXPERIMENTAL CONDITIONS

The three main goals in GLC are resolution, speed, and sample capacity. Each goal can be optimized at the cost of the other two.

For capillary columns the plate height expression can be formulated as

$$H = \frac{2D_M}{u} + \frac{1 + 6k' + 11(k')^2}{24(1 + k')^2}\left(\frac{r^2u}{D_M}\right) + \frac{k'}{6(1 + k')^2}\left(\frac{d_f^2u}{D_S}\right) \qquad (18.6)$$

where r is the radius of the capillary column and the other terms are defined in

Sections 17.3 and 17.4. When film thicknesses are small, the second term in the equation dominates at velocities above the optimum (u_{opt}). This term involves the resistance to mass transfer by the mobile phase. Thus the internal tube diameter should be kept small. Secondarily, plate height is related to k', but optimum values of k' are determined from the resolution equation (Equation 17.24).

For packed columns the plate height is given by the extended form of the van Deemter equation:

$$H = 2\lambda d + \frac{2\gamma D_M}{u} + \frac{2}{3}\frac{k'}{(1 + k')^2}\frac{d_f^2 u}{D_S} + \frac{(k')^2 d_p u}{96 D_M (1 + k')^2} \tag{18.7}$$

where λ is the packing factor ($\simeq 0.5$), γ is the tortuosity factor ($\simeq 0.7$), d_p is the particle diameter, and d_f is the thickness of the droplets of the stationary liquid phase that coat the substrate. At the low reduced velocities used, the coupled form of the A and C_{mobile} (last) terms is not used. When heavily loaded columns are used (for example, 30% liquid loading), the third term on the right-hand side dominates among the C terms. This is because of the slow diffusion in the stationary phase and the film thickness. Columns with 3%–5% loadings are more efficient and have lower C values. Now the fourth term becomes important. Good efficiency and short analysis times require small particle diameters of the support material (diatomaceous earths) and small droplets of the stationary-phase liquid (or thin films of bonded phases). Particles of 100–200 μm represent the best compromise between efficiency and permeability; these require pressure drops of 3–4 atm for columns 2–3 m long.

Once a liquid phase has been selected, k' is inversely proportional to the phase ratio, V_M/V_S. For a packed column the phase ratio is 15–20, whereas for a capillary column it is 60–70. Thus k' is always smaller for a capillary column than for a packed column prepared with the same liquid phase and operated at the same temperature. For example, if $k' = 2$ on a packed column, the value at the same temperature would be 0.5–0.6 on a capillary column. Solutes with short retention times would be difficult to separate on a column with a high phase ratio. The entire advantage of the high plate numbers given by capillary columns is lost if they are operated under conditions in which k' is fractional. For this reason, capillary columns are operated at appreciably lower temperatures than are packed columns. The main advantage of capillary columns is that the permeability is high, and long columns and/or high mobile-phase velocities can be used.

Carrier Gas

The purpose of the carrier gas is to transport the sample through the column to the detector. Selecting the proper carrier gas is very important because it affects both column and detector performance. A van Deemter plot for the three most common carrier gases is shown in Figure 18.13. Virtually the same minimum plate height is achieved with each gas. The difference arises in the optimum linear velocity. Although lower for nitrogen, the curves for helium and hydrogen are flatter. This means little loss in efficiency when linear velocities higher than the optimum are used. The use of higher velocities can shorten the analysis time significantly.

The viscosity dictates the inlet column pressure. The ratio of carrier gas viscosity to the diffusion coefficients of the sample components should be as small as possible. In this respect, hydrogen is the best choice, followed by helium.

The purity of the carrier gas should be at least 99.995%. Impurities such as oxygen or water can cause column and detector deterioration. In temperature programming, impurities can impair the baseline stability of the chromatogram.

Selection of Column Temperature for Isothermal Operation

The solute vapor pressure is described as a function of temperature by the Clausius-Clapeyron equation:

$$\ln p^0 = -\frac{\Delta \bar{H}_v}{RT} + C \tag{18.8}$$

where $\Delta \bar{H}_v$ is the molal heat of vaporization of the bulk solute and C is a constant. When the solute activity coefficient is unity, the *specific retention volume* can be expressed as

$$V_g^0 = \frac{273R}{p^0 M w_L} \tag{18.9}$$

where M is the molecular weight and w_L is the weight of the stationary liquid phase. Combining these two relations, we get

$$\ln V_g^0 = \frac{273R}{M w_L} + \frac{\Delta \bar{H}_v}{RT} + C \tag{18.10}$$

Because the first term on the right-hand side is itself a constant and $\Delta \bar{H}_v + \Delta \bar{H}_s = 0$ when the solute activity coefficient is unity, where $\Delta \bar{H}_s$ is the molal heat of solution,

$$\ln V_g^0 = -\frac{\Delta \bar{H}_s}{RT} + C' \tag{18.11}$$

Plots of $\ln V_g$ versus $1/T$ are linear and the slope yields $-\Delta \bar{H}_s/R$ (or $\Delta \bar{H}_v/R$). These plots have a positive slope for the molal heat of solution, which is invariably negative (-1 to -10 kcal/mol), implying that heat is evolved. Among members of a

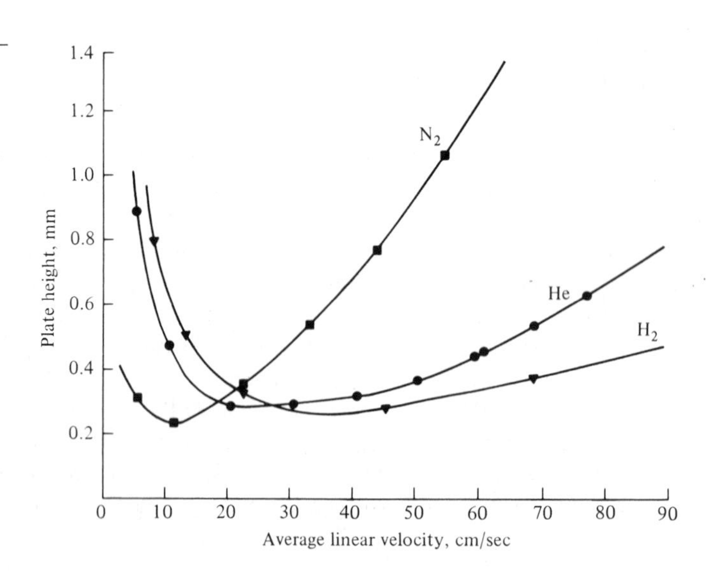

FIGURE **18.13**
Van Deemter graph for various carrier gases flowing through a WCOT glass column coated with OV-101. Column temperature is 175 °C.

homologous series, solutes have the same value of V_g at the column temperature corresponding to their boiling points.

An expression analogous to Equation 18.11 but in terms of the partition coefficient is

$$\ln K = \ln \frac{RT}{V_L} - \frac{\Delta \bar{H}_s}{RT} + C \tag{18.12}$$

where V_L is the molar volume of the stationary phase. Plots of $\ln K$ versus $1/T$ are expected to be curved, since the first term on the right-hand side of the equation is temperature dependent.

The relationship between relative retention and column temperature is

$$\ln \alpha = -\frac{\Delta \bar{H}_2 - \Delta \bar{H}_1}{RT} + \text{const} \tag{18.13}$$

In most cases $\Delta \bar{H}_2 - \Delta \bar{H}_1$ is greater than zero because the final component of an adjacent pair has the higher molal heat of solution in the stationary phase. Thus the relative retention usually decreases with increasing column temperature.

EXAMPLE 18.3

1,2,3-Trimethylbenzene has an adjusted retention time of 21.3 min at 200 °C and 13.3 min at 225 °C. Would it be possible to elute this compound in less than 10 min considering that the liquid phase has a maximum operating temperature of 275 °C?

Set up simultaneous equations based on Equation 18.8 and assume that the vapor pressure and adjusted retention time of a solute are interchangeable. Express temperature in degrees kelvin, and solve for B and C:

$$\ln 21.3 = \frac{B}{473} + C$$

$$\ln 13.3 = \frac{B}{498} + C$$

and $B = 4428$ and $C = -6.305$. Now, there are two ways to answer the question. (1) Insert the maximum permissible operating temperature and solve for the retention time:

$$\ln t'_R = \frac{4428}{548} - 6.305 \quad \text{and} \quad t'_R = 5.9 \text{ min}$$

(2) Insert the desired retention time of 10 min and solve for the column temperature:

$$\ln 10.0 = \frac{4428}{T} - 6.305 \quad \text{and} \quad T = 515 \text{ K (or 241 °C)}$$

To summarize, selecting the column temperature for isothermal operation is a complex problem and usually a compromise is the answer, often dictated by the pair of solutes most difficult to separate. Complex multicomponent samples that cover a wide temperature range of boiling points cannot be satisfactorily chromatographed in a single isothermal run. A run at a moderate column temperature will demonstrate good resolution of the low-boiling compounds, but requires a lengthy period for the elution of the high-boiling material. For the latter the retention time may be unusually long and the peaks so broad that they become indistinguishable from the baseline. Conversely, a chromatogram run at a higher temperature, though providing more rapid elution and better resolution of the higher-boiling materials, has the obvious disadvantage of poor or no resolution of the low-boiling components. The simple solution to this problem is to change the band migration rates during the course of separation by temperature programming.

Temperature Programming

Temperature programming combines the best results of runs at different temperatures. The sample is injected into the chromatographic system with the column temperature below that of the lowest-boiling component of the sample, preferably 90 °C below. Then the column temperature is raised at some preselected heating rate. Earlier peaks, representing low-boiling components, emerge essentially as they would from an isothermal column operated at a relatively low temperature. As the column temperature increases, the high-boiling components are forced through the column at an ever-increasing rate. In an approximate fashion, one can ascertain the proper temperature program.

Earlier it was shown that the fraction of total solute found in the mobile phase is given by Equation 17.12. The partition coefficient is responsible for the rather rapid increase in the amount of solute found in the vapor phase, as shown by Equation 18.12. Since there is nearly always more solute in the liquid phase than in the gas phase—that is, the fraction $1/(1 + k')$ is much less than unity—it is usually a good approximation to ignore $C_M V_M$ in the denominator of Equation 17.12. Remembering that $k' = K/\beta$, then from Equation 18.12, and ignoring the temperature dependency of the first term on the right-hand side, one gets

$$\ln\left(\frac{1}{k'}\right) = -\frac{\Delta\bar{H}_v}{RT} + C' \tag{18.14}$$

Next it is important to determine the average increase in temperature needed to just halve the k' value (which will double the total solute in the vapor phase). If k' is halved by increasing the temperature from T_1 to T_2, then from Equation 18.14, the ratio of k' values is

$$\ln 2 = \frac{-\Delta\bar{H}_v/RT_2}{-\Delta\bar{H}_v/RT_1} = \frac{\Delta\bar{H}_v}{RT}\left(\frac{\Delta T}{T}\right) \tag{18.15}$$

which gives

$$\Delta T = \frac{0.693R(\bar{T})^2}{\Delta\bar{H}_v} \tag{18.16}$$

where $\bar{T}$ is the geometric mean of the two temperatures and $\Delta T = T_2 - T_1$. Trouton's rule is approximately valid, so $\Delta\bar{H}_v/T_b \simeq 23$. Since the chromatographic process is operated near the solute boiling point, $\bar{T} \simeq T_b$. The ratio in Equation 18.16 may also be approximated by 23. Substituting the typical operating temperature (T) into Equation 18.16, we find the temperature increase (ΔT) that typically halves the k' value to be 21 degrees at an operating temperature of 75 °C, 24 degrees at 125 °C, and 30 degrees at 225 °C. The k' value is significant because the more solute that is in the vapor state, the faster the peak migrates. Thus, if $k' = 5$, this means that one-fifth of the solute molecules exist as vapor.

The final temperature should be near the boiling point of the final solute (if the maximum temperature limit of the stationary phase is not exceeded). As the temperature range proceeds, individual compounds automatically select their own ideal temperature in which to migrate and separate within the column. When the column temperature is increased linearly, the members of any homologous series are eluted at approximately equally spaced intervals, as shown in Figure 18.14, rather than proportional to $\ln t'_R$ as in isothermal elution.

EXAMPLE 18.4

A step-function approximation may be used to follow the solute migration. Let us assume that the temperature increases in steps of 23 °C to halve the value of k' each unit time. Furthermore, let us assume that $t_M = 1.0$ time unit and $k' = 60$. In successive steps, where L is the column length, we obtain the following values:

k'	t'_R	Unit distance	Total distance
60	60	$0.02L$	$0.02L$
30	30	$0.03L$	$0.05L$
15	15	$0.07L$	$0.12L$
7.5	7.5	$0.13L$	$0.25L$
3.75	3.75	$0.27L$	$0.52L$
1.88	1.88	$0.53L$	$1.05L$

Only 0.48 of the column length remains for step 6; consequently, the solute will emerge in 0.48/0.53 or 0.90 time unit. The total adjusted retention time is 5.90 time units, which corresponds approximately to a temperature rise of 136 degrees. Note that the solute migrates over half of the column length in the final unit time interval, and one-quarter of the column length in the preceding time interval.

FIGURE **18.14**

Chromatograms of an alcohol mixture: (a) programmed temperature from 100 to 175 °C and (b) isothermal operation at 175 °C.

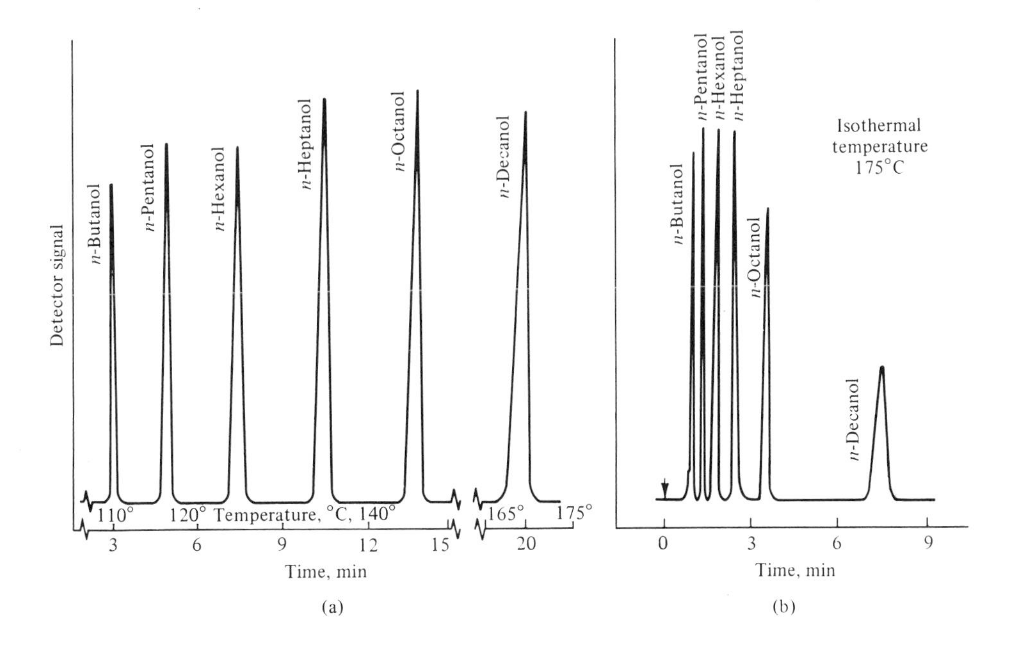

Time of Analysis

The retention time for a solute with a known k' value was given by Equation 17.32. The ratio H/u decreases as the carrier gas velocity increases. A point of diminishing returns is reached when the ascending part of the van Deemter plot starts to become linear. Nevertheless, it is always advantageous to use gas velocities higher than u_{opt}, with a corresponding increase in the column length. Here capillary columns prove valuable. The optimum practical gas velocity can be determined by a plot of N/t_R versus u. Over the range of the first rise and up to the maximum, an increase in column length can compensate for the loss of the overall column efficiency due to the higher gas velocity, but it still keeps the retention time shorter than the value corresponding to u_{opt}. It is always desirable to reduce the time of a gas chromatographic analysis until it is compatible with the time of sample preparation.

EXAMPLE 18.5

The analysis of a benzene–cyclohexane pair at 65 °C on a SCOT column prepared with squalane liquid phase ($\beta = 75$) was conducted (1) at the optimum average velocity (18 cm/sec) and (2) at the optimum practical velocity (60 cm/sec) for helium as the carrier gas. The separation factor was 1.24. For the cyclohexane peak, $k' = 1.41$. The plate height was 0.56 mm for condition 1 and 1.08 mm for condition 2. The desired resolution is 1.5. The plate numbers required for each operating condition along with the column lengths and the analysis times needed will now be calculated. For cyclohexane,

$$N_{req} = 16(1.5)^2\left(\frac{1.24}{1.24 - 1}\right)^2\left(\frac{1.41 + 1}{1.41}\right)^2 = 2800 \text{ plates}$$

$$L = NH = 2800 \times 0.56 \text{ mm} = 1570 \text{ mm} \quad \text{(condition 1)}$$

$$L = NH = 2800 \times 1.08 \text{ mm} = 3030 \text{ mm} \quad \text{(condition 2)}$$

$$t_M = \frac{L}{u} = \frac{157 \text{ cm}}{18 \text{ cm/sec}} = 8.7 \text{ sec} \quad \text{(condition 1)}$$

$$t_M = \frac{L}{u} = \frac{303 \text{ cm}}{60 \text{ cm/sec}} = 5.05 \text{ sec} \quad \text{(condition 2)}$$

$$t_R = t_M(1 + k')$$

$$= 8.7 \text{ sec}(1 + 1.41) = 21.0 \text{ sec} \quad \text{(condition 1)}$$

$$= 5.05 \text{ sec}(1 + 1.41) = 12.2 \text{ sec} \quad \text{(condition 2)}$$

Conclusions: Almost twice the column length is required to maintain the same plate number at the higher carrier gas velocity. However, the analysis time is 72% shorter.

18.7 GAS-SOLID CHROMATOGRAPHY

In gas-solid chromatography (GSC) the columns are packed with interactive solids such as molecular sieves or porous polymers through which the carrier gas flows. Sample injection is by means of a gas sample loop or valve. Normal operation involves the use of columns 1–4 m long and temperatures in the range 40–90 °C.

Adsorbents

Inorganic molecular sieves are naturally occurring or synthetic zeolites. They are frameworks of aluminum and silicon with oxygen that consist of interconnected cavities with precisely uniform openings. Only molecules smaller than these openings are able to enter the cavities. Pore size is very uniform and is dependent on the cation used. Zeolites act as adsorbents but differ from other adsorbents in that they take up molecules that fit into their micropores. A specific affinity based on molecular size and shape is obtained. In addition, normal adsorption based on the physicochemical structure of the components occurs in the micropores as well as in the macropores outside the sieves. Adsorption continues until all the pores are filled. Raising the temperature will displace the molecules from the pores.

Several types of molecular sieves are available. Type 3A, a potassium aluminosilicate, with a pore diameter of 0.3 nm, absorbs molecules such as water and ammonia. Type 4A, the sodium analog of Type 3A, adsorbs all molecules with critical diameters up to 0.4 nm; these include carbon dioxide, sulfur dioxide, hydrogen disulfide, ethane, ethylene, propylene, and ethanol. Type 5A, with calcium replacing the sodium content of Type 4A, has a pore diameter of 0.5 nm. It separates straight-chain hydrocarbons (C_3 through C_{22}) from branched-chain and cyclic hydrocarbons. Type 13X, a sodium aluminosilicate with a different crystalline structure from the preceding types, has a pore diameter of 1 nm. A carbon molecular sieve, Type B, with a pore diameter of 1–3 nm is less polar than squalane. Water is quickly eluted and well separated from methanol, ethanol, and formaldehyde; benzene elutes before hexane. Silica gel columns have a large retention for carbon dioxide.

Column packings may be porous polymers analogous to the materials used in exclusion chromatography. Those made from copolymers of aromatic hydrocarbons provide packings of low to moderate polarity. Polymers made from acrylic esters provide packings of moderate to high polarity. Most gases and low-boiling-point liquids can be separated on one or more of these porous polymers within the temperature range 50–250 °C.

Multicolumn Systems

One problem that makes GSC separations more complicated than GLC separations is that almost all gas mixtures contain some components that cannot be separated on a particular column or cannot pass through a column in a reasonable time. A remedy for the latter problem is the technique known as backflushing. An example will illustrate the technique. In the trace analysis of hydrocarbons in natural gas, interest is usually centered on the C_1 to C_5 hydrocarbons. Consequently, after the pentane peak has eluted from the column, the carrier gas flow is reversed so as to enter the column exit. The C_6 and higher hydrocarbons are backflushed from the upper end of the column through the detector now positioned at the entrance to the column. With proper programming, the column can be backflushed before any component has eluted. This provides a sharper initial combined peak of C_6 and higher hydrocarbons for more precise quantitation (Figure 18.15).

In the "series/bypass" technique (Figure 18.16), a column-switching valve enables column 2 to be bypassed by the carrier gas at selected times. Thus certain components can be temporarily stored in column 2 while separations of the remainder are made on column 1. Consider the analysis of a sample of sulfur-rich hydrogen product gas. Column 1 is a porous polymer and column 2 contains molecular sieve 5A; both columns are operated at 97 °C. In a programmed time interval after sample injection, oxygen, nitrogen, methane, and carbon monoxide will have passed through column 1 and entered column 2, where they are temporarily stored. Now the carrier gas is switched to pass through only column 1. Within 2–10 min, separate peaks for hydrogen, carbon dioxide, ethylene, ethane, acetylene, and hydrogen sulfide will appear (Figure 18.17). Then column 2 is switched back into the gas stream to elute the stored components as distinct peaks in the order enumerated over the next 8-min interval. These column-switching techniques are used extensively in process control GC (see Chapter 26).

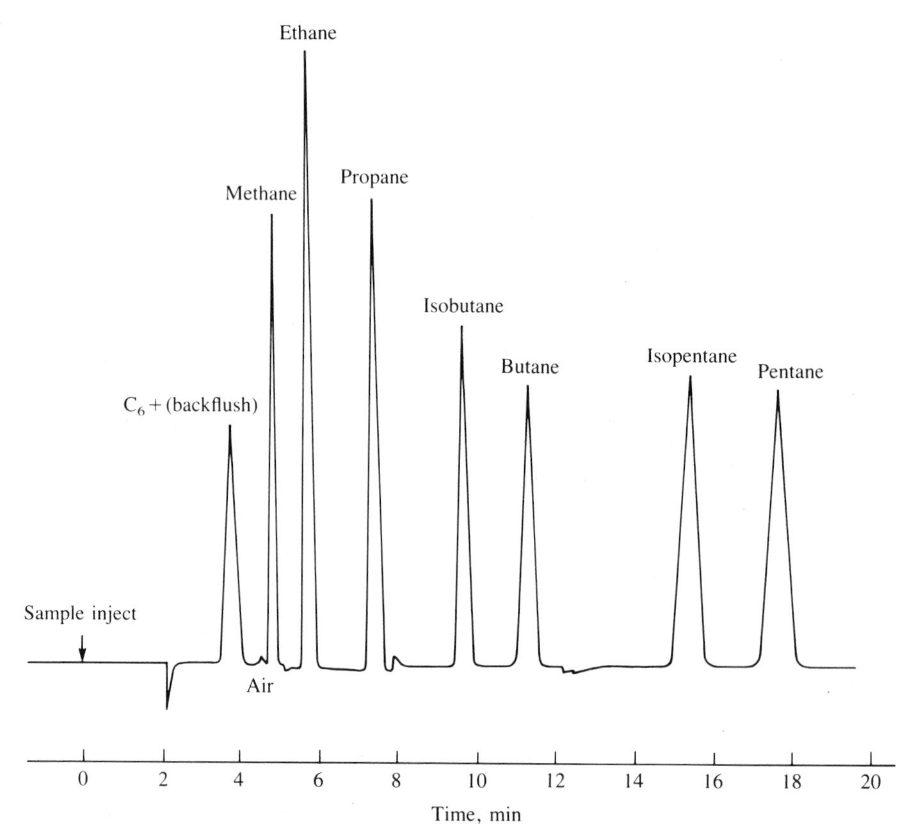

FIGURE **18.15**
Chromatogram of natural gas liquids. Two minutes after sample injection, the C_6 and higher hydrocarbons are backflushed. The forward carrier gas flow is resumed after 2 min of backflushing.

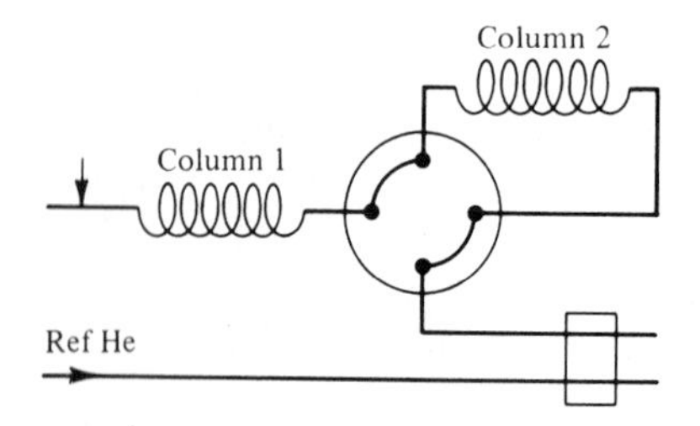

FIGURE **18.16**
Series/bypass arrangement.

FIGURE **18.17** Chromatogram of sulfur-rich hydrogen product gas using the series/bypass arrangement.

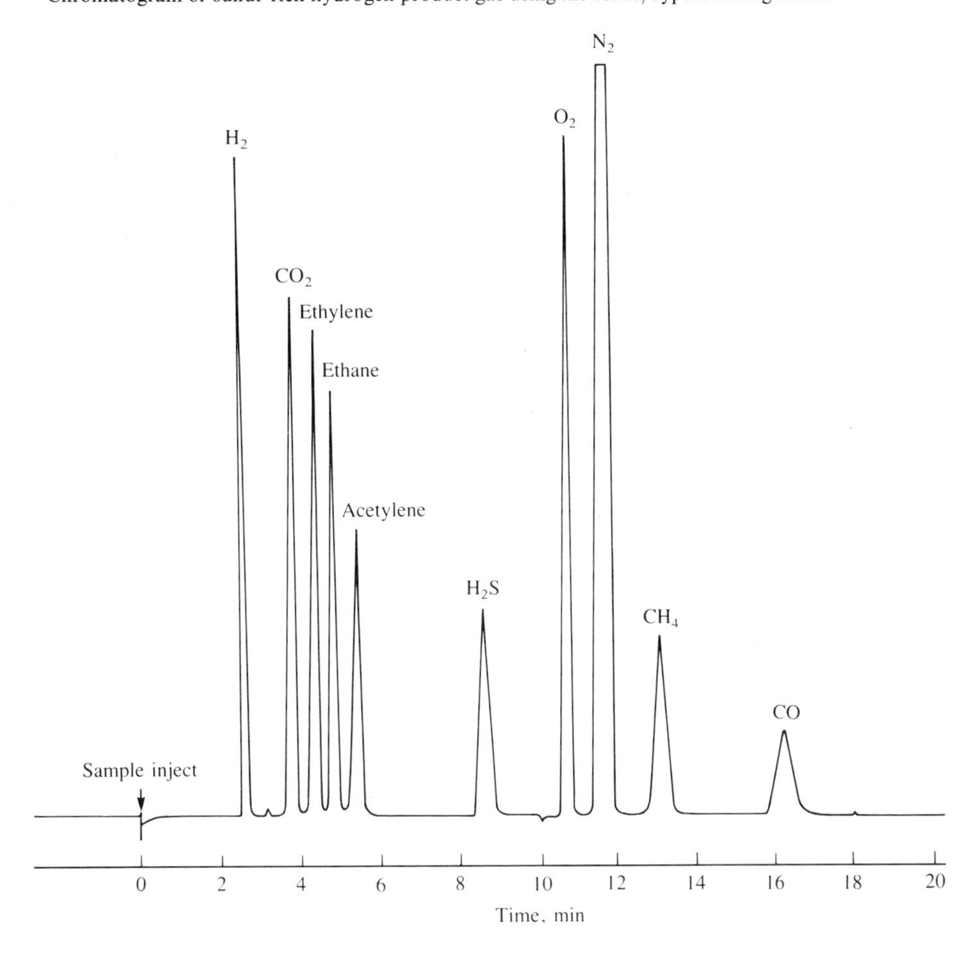

18.8 INTERFACING GAS CHROMATOGRAPHY WITH MASS SPECTROMETRY

The mass spectrometer (see Chapter 16) is a universal detector for gas chromatographs, since any compound that can pass through a gas chromatograph is converted into ions in the mass spectrometer. At the same time, the highly specific nature of a mass spectrum makes the mass spectrometer a very specific gas chromatographic detector. Gas chromatography is an ideal separator, whereas mass spectrometry is excellent for identification. The aim of an interfacing arrangement is to operate both a gas chromatograph and a mass spectrometer without degrading the performance of either instrument. The problem is compatibility. One incompatibility problem is the difference in pressure required for the operation of a gas chromatograph and the mass spectrometer. Whereas the former operates at high pressures, the latter is designed to run under high vacuum. An associated problem is the presence of much carrier gas and little sample in the effluent from the gas chromatograph.

Gas Chromatograph–Mass Spectrometer Interface

The interface must provide the link between the two instruments. Almost all GC–MS interface systems contain an enrichment device (Figure 18.18). However, the high pumping speeds used in mass spectrometers may permit the total effluent from capillary GC columns to be transported to the ion source of the mass spectrometer. When the chemical ionization reagent gas is used as the carrier gas, the effluent can be introduced directly into the mass spectrometer.

Effusion Separator. Since the carrier gas molecules are usually much lighter than those of the sample, they can be removed preferentially by an effusion chamber. Effluent from the gas chromatograph passes through a tube constructed of ultrafine-porosity sintered glass with an average pore size of about 10^{-4} cm. The porous barrier is surrounded by a vacuum chamber. The lighter carrier gas (assumed to be helium) permeates the effusion barrier in preference to the heavier organic molecules. Enrichment is typically five- to sixfold and the yield (sample throughput) is about 27%.

Jet/Orifice Separator. A precisely aligned, supersonic jet/orifice system is effective in removing the carrier gas by effusion. Effluent from the gas chromatograph is throttled through a fine orifice, where it rapidly expands into a vacuum chamber. During this expansion, the faster diffusion rate of helium results in a higher sample concentration in the core of the gas stream, which is directed toward a second jet or orifice aligned with the first jet. Alignment and relative spacing of the expansion and collector orifices are very critical. The distance between jets must be changed for a change in flowrates. Yields are about 25%. An all-glass jet separator is frequently used for packed column operation. The short path through the interface to the ion source reduces dead volume, which gives better peak separation.

Membrane Separator. The membrane separator takes advantage of large differences in permeability between most organic molecules and the carrier gas when both are confronted by a membrane. Effluent from a gas chromatograph enters a cavity that is separated from the mass spectrometer vacuum system by a dimethyl silicone rubber membrane, usually about 0.025–0.040 mm thick. Helium has a low permeability, whereas the organic molecules pass through the membrane and directly into the high vacuum of the mass spectrometer system. Enrichment values are 10- to 20-fold; the yield may be 30%–90%. Major problems with this type of separator are the temperature limits (80–220 °C) and temperature optimization. The upper temperature limit is a serious disadvantage that cuts out a segment of GC–MS work. Each compound has an optimum temperature for membrane enrichment, and thus sample discrimination occurs. There is also a time lag of about 0.1 sec while the sample molecules pass through the membrane. Polar compounds and high-boiling-point compounds tend to be partially adsorbed on the membrane, resulting in the tailing of these materials into the mass spectrometer.

GC–MS Instruments[13]

Two types of mass spectrometers are used for GC–MS work: magnetic sector mass spectrometers and quadrupole mass filters (see Chapter 16). Three characteristics affect the compatibility of GC–MS units: scan speed, sensitivity, and useful dynamic range. Fast peak switching is important for qualitative and quantitative multi-ion selection analyses. In magnetic sector instruments, electromagnets exhibit a kind of inertia, called reluctance, that limits the rate at which a magnet can be forced to change field strength. This phenomenon limits scan rates to 0.1 sec/decade (a decade

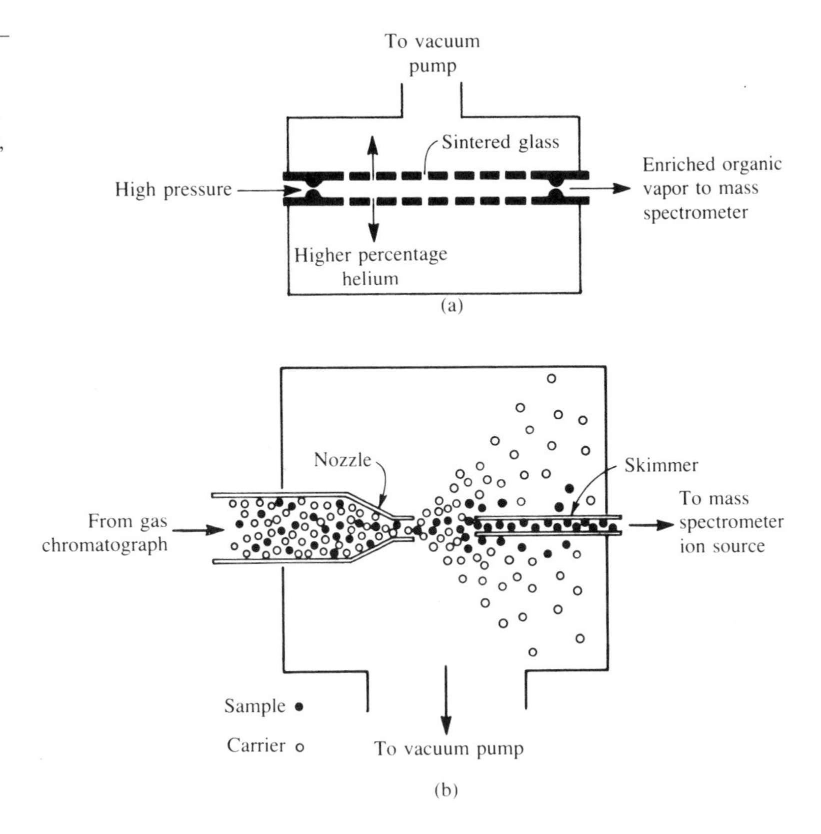

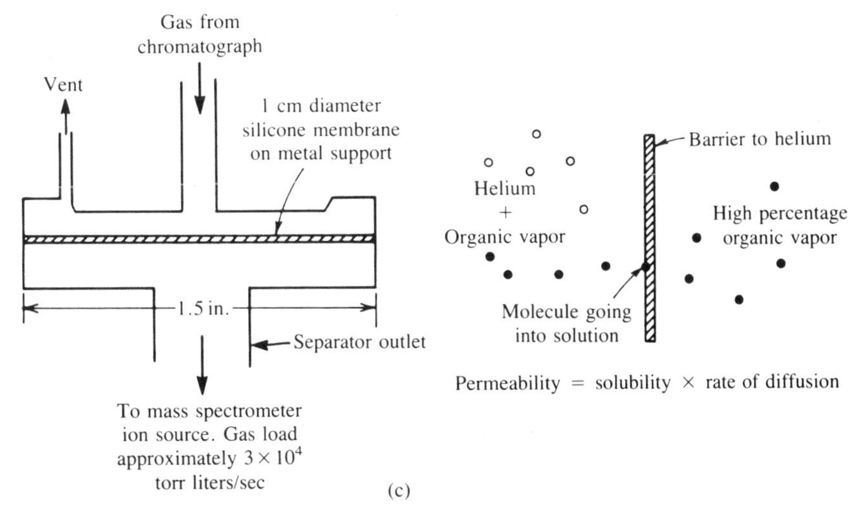

FIGURE **18.18**
Enrichment devices: (a) porous barrier separator or effluent splitter, (b) jet/orifice separator, and (c) molecular separator using a permeable membrane.

is 50–500 mass units, for example). About 0.2 sec is required to reset the magnet between scans. Hence, the scan repetition rate for magnetic sector instruments is 3–4 Hz.

In quadrupole mass filters a maximum scan rate of about 780 mass units per second can be reached. Somewhat faster scan rates can be achieved by increasing the ion energy concurrently with the mass scan ramp in order to maintain a more constant ion velocity and reduce the transit time for heavier ions. The useful upper scan repetition rate is 4–8 Hz. The linear mass scale of the quadrupole mass filter is ideally suited for digital control, a feature that is important for routine analysis with high sample throughput and for unattended operation.

For GC–MS work only nominal masses need to be measured. A low-resolution quadrupole mass spectrometer suffices. The molecular peak and the fragment peaks form a typical pattern that can be interpreted with experience or compared against reference spectra. If precise masses are needed, a high-resolution mass spectrometer is required; this is the domain of magnetic sector instruments.

Mass spectrometers can be used as real-time detectors for gas chromatography. The total ion current is measured and recorded as a function of time. It is a measure of the total number of ions formed from material in the effluent. In selective ion monitoring, the intensities of preselected ions, characteristic of a class or of a particular compound, are monitored throughout the elution cycle. This technique is favored for analyses that require the highest sensitivity, particularly in environmental and biological work. With multiple-ion monitoring, the intensities of two or more preselected ions are recorded as a function of time. To achieve this the analyzer of the mass spectrometer cycles through the group of ions being monitored, switching each into the detector in turn. The intensity of each of the ions is recorded several times per second. Multiple-ion monitoring is useful for deconvoluting overlapping peaks and when assaying a stable isotope incorporated into the sample molecules. Also, the technique can be applied to quantitative studies in which stable isotopes are used as internal standards.

The main advantages of a mass spectrometer as a detector for gas chromatography are its increased sensitivity and its specificity in identifying unknowns or confirming the presence of suspected compounds. The enhanced sensitivity results primarily from the action of the analyzer as a mass filter to reduce background interference and from the sensitive electron multipliers used for detection. The excellent specificity results from characteristic fragmentation patterns that give information about molecular weight and molecular structure.

18.9 INTERFACING GAS CHROMATOGRAPHY WITH INFRARED SPECTROMETRY

Gas chromatography–infrared spectrometry (GC-IR) is a coupled technique in which the gas chromatograph does the separating and the infrared spectrometer does the identifying.[14,15] Usually a capillary GC column is coupled with a Fourier transform infrared spectrometer (FTIR), as shown in Figure 18.19. The modulated IR beam is focused into a heated "light pipe" cell through which the GC effluent is directed. The light pipe is a glass tube with a gold coating on the inside and IR-transmitting windows affixed to each end. The IR beam travels down the light pipe

(2 mm i.d. and 50 cm long) by multiple reflections from the gold coating. The transmittance can be 25%.

Because the detection limits of GC–FTIR systems are usually between 10 and 100 ng, a chromatographic column with 0.3–0.5 mm i.d. and a film thickness of 0.5–1.5 μm gives a suitable compromise between adequate resolution and reasonable sample capacity. Slow column temperature programs must be used. For example, the resolution can be tripled if the temperature program is changed from 5 °C/min to 2 °C/min. The volume of the light pipe is determined by the width of the narrowest peak in the chromatogram. Since it would be awkward to change light pipes continually to match each chromatogram, it is usual to select a typical cell volume and then set the chromatographic conditions so that the half-width volume of each peak is equal to, or slightly greater than, the IR cell volume. For narrow-bore fused-silica columns, cell volumes of 50–100 μL are typical. For wide-bore heavily loaded columns, a volume of about 300 μL is close to the optimum value.

The FTIR instrument is fast enough to obtain several scans over a single chromatographic peak. The accuracy of the frequency scale facilitates subsequent data manipulation, such as signal averaging and spectral subtraction. Since IR spectrometry is nondestructive, the GC effluent can also be directed from the light pipe outlet to another GC detector. The interrelationship of the components of a GC–FTIR system is discussed by Griffiths and colleagues.[14]

In the Cryolect GC–FTIR interface, the GC effluent, helium with 1%–2% argon added, is split into two parts. One part goes to a conventional GC detector. The other part is admitted to the Cryolect interface, where argon atoms and sample molecules condense on a rotating mirrored cylinder that has been cooled to 12 K. Helium atoms do not condense at this temperatures and are pumped away. The frozen matrix, a cryogenically trapped version of the chromatogram, remains on the cylinder as long as the temperature is maintained at 12 K. After the chromatographic run is complete, the operator selects peaks of interest from the chromatogram, and the Cryolect cylinder is positioned by a computer so that each peak of

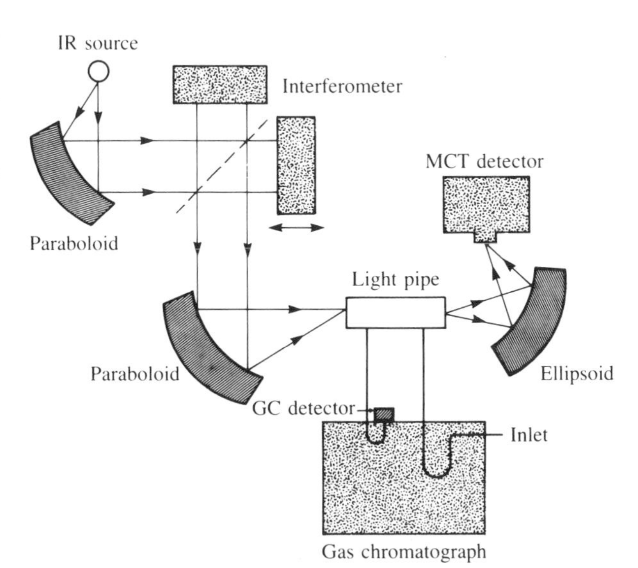

FIGURE **18.19**
Schematic of GC–FTIR instrumentation. [Reprinted with permission from P. R. Griffiths, J. A. de Haseth, and L. V. Azarrage, *Anal. Chem.*, **55**, 3161A (1983). Copyright 1983 American Chemical Society.]

interest can be scanned sequentially by FTIR. Infrared band broadening and band shifting are prevented by the argon matrix, which isolates each analyte molecule, and by the temperature level of the cylinder. Extended observation times make it possible to do signal averaging. Thus, by decoupling the GC and IR, the FTIR generates a high-quality spectrum from a small amount of sample. However, it is not real-time analysis and the IR instrument is not on-line.

The database problem plagues GC-IR. Only about 100,000 compounds are presently in the data collections and more than half of these are not sufficiently volatile to undergo GC separation. Of course, the structural information and functional groups obtained from an infrared spectrum at least provide some information about a compound.

PROBLEMS

1. Often it is tempting to increase the loop volume in order to increase the amount of sample for trace analysis. What problems will arise if the loop volume is increased from 1 to 5 mL when columns with 2-mm bore are used? Assume a flowrate of 30 mL min^{-1} at atmospheric pressure (outlet) and a column inlet pressure of 3 atm.

2. Specific retention volumes (in mL g^{-1}) for some chlorinated hydrocarbons and benzene at several column temperatures on three column substrates are listed in the table. All columns were made from coated Chromosorb packed into columns 1.8 m long with 6.4-mm bore. The paraffin column contained 4.58 g on a 13.93-g support; the tricresyl phosphate column, 4.46 g on 13.71 g; and the Carbowax 4000 column, 4.38 g on 13.1 g. Densities of the liquid phases at 74°/4° in g cm^{-3}: paraffin, 0.768; tricresyl phosphate, 1.128; and Carbowax 4000, 1.081. (a) Compute the heat of vaporization for each solute on each of the stationary liquid phases. (b) Calculate the partition coefficient at each temperature for each solute on each column substrate. (c) Calculate the value of V_g at 150 °C for the solutes on the Carbowax 4000 column. (d) What temperature change halves V_g on each column for each solute?

Temp. (°C)	CH_2Cl_2	$CHCl_3$	CCl_4	$CCl_2{=}CCl_2$	C_6H_5Cl	C_6H_6
Paraffin						
74	29.5	74.3	141	568	677	136
97	17.0	41.5	76.5	273	323	74.3
125	8.08	19.6	34.5	105	124	
Tricresyl phosphate						
74	56.1	137	99.8	359	826	133
97	30.3	68.5	52.7	170	372	69.4
125	13.9	29.6	24.7	69.5	143	
Carbowax 4000						
74	69.6	138	55.8	165	548	89
97	31.9	59.3	27.4	78.7	235	43.3
125	13.5	24.2	13.5	31.1	83.6	

3. The diffusion coefficient, D_M, for butane in helium at 25 °C and 1 atm is

$0.342\ cm^2\ sec^{-1}$. In nitrogen the same constant has a value of $0.0960\ cm^2\ sec^{-1}$. (a) Which terms in Equation 18.7 are affected by this difference? (b) Sketch two graphs of H versus u showing how this difference will change the shape and location of u_{opt} and H_{min}. (c) Which carrier gas will allow faster mobile-phase velocities without a significant loss of efficiency?

4. The data in the table were obtained from a column 360 cm long that was packed with 30/50 mesh Chromosorb P and had various amounts of hexadecane as the liquid substrate (whose density is $0.774\ g\ mL^{-1}$). Column operating temperature was 30 °C. Chart speed was $61\ cm\ hr^{-1}$. Propane was the solute. Carrier gas was hydrogen, except for the final data set when nitrogen was used. (a) Graph the values of effective plate height versus the average linear velocity. (b) Estimate the B and C terms of the van Deemter equation. For the A term use the average particle diameter of the packing (0.045 cm). (c) Estimate the optimum velocity and the minimum plate height for each column. (d) Calculate the retention time for an unretained component at each linear velocity for each column. (e) Calculate k' for each column. (f) The three columns operated with hydrogen as the carrier gas differed only in the amount of liquid stationary phase. How did their mobile-phase velocities differ? Why? (g) Through what range of mobile-phase velocities can 90% of the column efficiency be retained? (h) On the 23% column loading with H_2 carrier gas, a particular separation requires 1500 theoretical plates. What is the fastest mobile-phase velocity at which this can be achieved? How much time will be saved by running at the maximum flowrate instead of at the optimum velocity?

31% Column loading; H_2 carrier gas

u (cm sec^{-1})	V'_R (mm chart)	W_b (mm chart)
1.00	335.4	47.9
1.65	205.0	25.2
3.22	108.1	12.7
6.16	57.3	7.7
8.80	38.5	5.8
13.28	25.0	4.5

23% Column loading; H_2 carrier gas

u (cm sec^{-1})	V'_R (mm chart)	W_b (mm chart)
1.43	157.4	19.2
2.81	78.0	8.1
4.15	54.1	5.2
5.43	40.6	4.0
6.64	32.8	3.3
12.26	18.1	2.1

13% Column loading; H_2 carrier gas

u (cm sec^{-1})	V'_R (mm chart)	W_b (mm chart)
2.49	60.1	6.3
4.78	31.2	2.7
6.86	21.3	1.8
10.51	14.5	1.3

23% Column loading; N_2 carrier gas

u (cm sec^{-1})	V'_R (mm chart)	W_b (mm chart)
0.71	307.2	27.2
1.38	158.9	12.3
2.72	82.4	6.2
6.12	36.2	3.2

5. Values of plate height versus average linear gas velocity for n-heptane ($k' = 2.25$) at 75 °C on a 15-m SCOT column with 0.51-mm bore, and prepared with squalane liquid phase are tabulated below:

u (cm sec^{-1}):	6.5	12.2	19	27	43
H (mm):	1.20	0.91	0.93	1.08	1.48

(a) For each data point, calculate t_R, t'_R, and the theoretical and effective plate numbers. (b) Graph H versus u; calculate u_{opt} and H_{min}. (c) Graph N/t_R versus u and estimate the optimum practical gas velocity.

6. The separation of hydrocarbons was done on a column 1.5 m long with a 2.3-mm bore, packed with *n*-C_8 alkane bonded to porous silica. Column temperature: 25 °C. Detector: flame ionization. Carrier gas: nitrogen at 25 mL min^{-1}. Retention characteristics of selected peaks are shown in the table. (a) Calculate the Kovat's retention indices for all the compounds. (b) Estimate the position of *n*-hexane on the chromatogram. (c) What is the effective plate number for 2-methylpentane? Plate height? (d) What is the resolution between *trans*-2-butene and *cis*-2-butene? (e) Calculate k' for *cis*-2-butene and for 2-methylpentane. (f) To achieve a resolution of 1.5 for the separation of peaks 10 and 11, how many plates would be required? (g) Suggest three ways in which the additional plates could be achieved for part (f). Provide a firm number or basis for each suggested alteration in operating procedure. (h) What is the velocity of the carrier gas through the column?

Compound	Peak	t_R (min)	W_b (min)
Methane	1	0.39	
Propane	4	0.86	
n-Butane	7	1.75	
trans-2-Butene	10	3.12	
cis-2-Butene	11	3.43	0.31
n-Pentane	13	4.31	
2-Methylpentane	17	9.55	0.71

7. A chromatographic column 45.7 m long is operated under two different average linear gas velocities: 28.5 and 49.4 cm sec^{-1}. The respective number of theoretical plates is 81,920 and 47,480. (a) What are the analysis times at each velocity for the methyl oleate peak ($k' = 4.71$)? (b) To what degree can the methyl hexanoate ($k' = 0.12$) and the methyl octanoate ($k' = 0.18$) peaks be resolved on this column? How many theoretical plates would be required for $R = 1.5$ at these retention times? How long a column is required? (c) Repeat part (b) for methyl decanoate ($k' = 0.33$) and methyl dodecanoate ($k' = 0.63$).

8. Determine the time of analysis for each of these columns under the assumed conditions: (a) a packed column with a phase ratio of 20, (b) a WCOT column with a phase ratio of 120, and (c) a SCOT column with a phase ratio of 60. Squalane is the liquid substrate. Assume that $H = 0.60$ mm on each of these columns and that the column temperature is 75 °C. The partition ratio, k', of the second peak on the packed column is to be taken as 1, 2, 3, 6, 12, and 30 in the six comparative cases. Assume a relative retention ratio of 1.10 for the adjacent peaks and that the resolution desired is 1.5. The three columns are operated at the following average linear gas velocities with helium as the carrier gas: packed column, 6 cm sec^{-1}; WCOT column, 12 cm sec^{-1}; and SCOT column, 18 cm sec^{-1}. Note the k' value for the minimum analysis time on each type of column. Graph the analysis time for each comparative case along sets of time scales.

9. The information in the table was obtained from three isothermal chromatograms of n-C_7 to n-C_{11} alkanes run on a column consisting of 10% SE-30 on Chromosorb W (30/60 mesh). Helium was the carrier gas. The column was 122 cm long and the bore was 10 mm. (a) Devise a suitable temperature program for separating the n-alkanes using the step-function approximation method. Estimate the retention temperature for each alkane. What is the maximum heating rate that should be used? (b) Predict the retention temperatures for n-C_{12}, n-C_{13}, and n-C_{14} alkanes. (c) On separate graphs, one for each operating temperature, plot both the theoretical and effective plate numbers of the normal alkanes.

	Column temperature					
	120 °C		**140 °C**		**160 °C**	
Solute	**t_R (min)**	**W_b (min)**	**t_R (min)**	**W_b (min)**	**t_R (min)**	**W_b (min)**
Air peak	0.98		0.95		0.90	
n-C_7	1.72	0.24	1.49	0.18	1.22	0.14
n-C_8	2.28	0.31	1.84	0.25	1.43	0.20
n-C_9	3.26	0.42	2.46	0.34	1.78	0.25
n-C_{10}	5.03	0.65	3.46	0.43	2.30	0.33
n-C_{11}	8.02	0.96	5.07	0.60	3.15	0.43

10. The information in the table was obtained from three isothermal chromatograms of the straight-chain alkyl acetates run on the same column described in Problem 9. Devise a suitable temperature program for separating the n-alkyl acetates using the step-function approximation method. Estimate the retention temperature for each alkane. What is the maximum heating rate that should be used?

	Column temperature		
Solute	**80 °C, t_R (min)**	**100 °C, t_R (min)**	**120 °C, t_R (min)**
Air peak	1.00	0.95	0.90
Methyl	1.50	1.28	1.13
Ethyl	1.95	1.57	1.30
Propyl	2.95	2.08	1.60
Butyl	5.00	3.08	2.13
Pentyl	8.95	4.93	3.08

11. From the van Deemter equation, predict the effect (increase, decrease, no effect, cannot determine) on the plate height in each of these situations, with only one parameter varied at a time: (a) decreasing the particle size of the solid support,

(b) increasing the column temperature, (c) increasing the thickness of the liquid coating, and (d) increasing the gas flowrate.

12. The retention temperature for benzyl acetate is 200 °C under these conditions: flowrate 25.0 mL/min, initial temperature 25 °C, and heating rate 5 degrees/min. Holding all other parameters constant, calculate the heating rate that would be required to obtain a retention temperature of 180 °C.

13. An isopropylbenzene peak has a retention time of 5.36 min at 200 °C and 3.15 min at 225 °C on a Carbopack C/0.1% SP-1000 column that has an efficiency of 2900 theoretical plates. What is the highest column temperature that can be used such that the peak width will not be less than 10 sec?

BIBLIOGRAPHY

BEREZKIN, V. G., *Chemical Methods in Gas Chromatography,* Elsevier, New York, 1983.

COWPER, C. J., AND A. J. DEROSE, *The Analysis of Gases by Gas Chromatography,* Pergamon, New York, 1983.

DROZD, J., *Chemical Derivatization in Gas Chromatography,* Elsevier, New York, 1981.

GROB, R. L., ed., *Modern Practice of Gas Chromatography,* 2nd ed., John Wiley, New York, 1977.

GROB, R. L., AND M. KAISER, *Environmental Problem Solving Using Gas and Liquid Chromatography,* Elsevier, New York, 1982.

JOFFE, B. V., AND A. G. VITENBERG, *Headspace Analysis and Related Methods in Gas Chromatography,* John Wiley, New York, 1984.

LEE, M. L., F. J. YANG, AND K. D. BARTIE, *Open Tubular Column Gas Chromatography: Theory and Practice,* John Wiley, New York, 1984.

MCNAIR, H. M., M. W. OGDEN, AND J. L. HENSLEY, "Recent Advances in Gas Chromatography," *Am. Lab.,* **17,** 15 (August 1985).

MESSAGE, G. M., *Practical Aspects of Gas Chromatography/Mass Spectrometry,* John Wiley, New York, 1984.

Sadtler Research Laboratories, *The Sadtler Capillary GC Standard Retention Index Library and Data Base,* Sadtler, Philadelphia, 1985.

ZLATKIS, A., AND C. F. POOLE, eds., *Electron Capture: Theory and Practice in Chromatography,* Elsevier, New York, 1981.

LITERATURE CITED

1. MCNALLY, M. E., AND R. L. GROB, "Static and Dynamic Headspace Analysis," *Am. Lab.,* **17,** 20 (January 1985); p. 106 (February 1985).
2. WOLF, C. J., M. A. GRAYSON, AND D. L. FANTER, "Pyrolysis Gas Chromatography," *Anal. Chem.,* **52,** 348A (1980).
3. PERRY, J. A., AND C. A. FEIT, "Derivatization Techniques in Gas-Liquid Chromatography," in *GLC and HPLC Determination of Therapeutic Agents,* K. Tsuji and W. Morozowich, eds., Part I, Dekker, New York, 1978.
4. POOLE, C. F., AND A. ZLATKIS, "Derivatization Techniques for the Electron-Capture Detector," *Anal. Chem.,* **52,** 1002A (1980).
5. WIEDEMER, R. T., S. L. MCKINLEY, AND T. W. RENDL, "Advantages of Wide-Bore Capillary Columns," *Am. Lab.,* **18,** 110 (January 1986).
6. ETTRE, L. S., "The Kovats Retention Index System," *Anal. Chem.,* **36,** 31A (1964).
7. ROHRSCHNEIDER, L., *J. Chromatogr.,* **22,** 6 (1966).

8. MCREYNOLDS, W. O., *J. Chromatogr. Sci.*, **8,** 685 (1970).
9. BORMAN, S. A., "New Gas Chromatographic Detectors," *Anal. Chem.*, **55,** 726A (1983).
10. PATTERSON, P. L., R. L. HOWE, AND A. ABU-SHUMAYS, "Dual-flame Photometric Detector for Sulfur and Phosphorus Compounds with Gas Chromatographic Effluents," *Anal. Chem.*, **50,** 339 (1978).
11. DRISCOLL, J. N., ET AL., "Development and Applications of the Photoionization Detector in Gas Chromatography," *Am. Lab.*, **10,** 137 (May 1978).
12. DRISCOLL, J. N., AND J. H. BECKER, "Industrial Hygiene Monitoring with a Variable Selectivity Photoionization Analyzer," *Am. Lab.*, **11,** 69 (November 1979).
13. FENSELAU, C., "The Mass Spectrometer as a Gas Chromatograph Detector," *Anal. Chem.*, **49,** 563A (1977).
14. GRIFFITHS, P. R., J. A. DE HASETH, AND L. V. AZARRAGA, "Capillary GC/FT-IR," *Anal. Chem.*, **55**(13), 1361A (1983).
15. BOURNE, S., G. REEDY, P. COFFEY, AND D. MATTSON, "Matrix Isolation GC/FTIR," *Am. Lab.*, **16,** 90 (June 1984).

HIGH-PERFORMANCE LIQUID CHROMATOGRAPHY: THEORY AND INSTRUMENTATION

Only about 20% of known compounds lend themselves to analysis by gas chromatography either because they are insufficiently volatile and cannot pass through the column or because they are thermally unstable and decompose under the conditions of separation. High-performance liquid chromatography (HPLC) is not limited by sample volatility or thermal stability. HPLC is able to separate macromolecules and ionic species, labile natural products, polymeric materials, and a wide variety of other high-molecular-weight polyfunctional groups. With an interactive liquid mobile phase, a parameter is available for selectivity in addition to an active stationary phase. Chromatographic separation in HPLC is the result of specific interactions between sample molecules with both the stationary and mobile phases. These interactions are essentially absent in the mobile phase of gas chromatography. HPLC offers a greater variety of stationary phases, which allows a greater variety of these selective interactions and more possibilities for separation. Sample recovery is easy in HPLC. Separated fractions are easily collected by placing an open vessel at the end of the column. Recovery is usually quantitative (barring irreversible adsorption on a column), and separated sample components are readily isolated from the mobile-phase solvent. In addition to the usual type of organic compounds, liquid column chromatography can handle separations of ionic compounds, labile naturally occurring products, polymeric materials, and high-molecular-weight polyfunctional compounds.

19.1 OPTIMIZATION OF COLUMN PERFORMANCE[1–4]

Before discussing the instrumental components of a high-performance liquid chromatograph, it is necessary to consider the optimum operating conditions. With two interactive phases, HPLC differs from gas chromatography. The problem of optimization is difficult, not to solve but to define. The many variables in chromatography are linked by equations, but there are more unknowns than equations, so an analytical problem has several degrees of freedom (t_M, ΔP, N). Several sets of experimental conditions can be used to achieve the desired separation of sample components. Some sets are more practical than others—hence, the search for optimization.

The analyst might seek the lowest pressure drop and therefore probably the lowest-cost equipment. Perhaps the major interest is a short analysis time (control laboratories). Or a certain plate count is needed to accomplish a particularly difficult separation. Of course, trade-offs are possible. In the following sections, several approaches will be examined.

First, the proper HPLC system must be selected. Consequently, all the parameters in the equations that depend on the system or on the properties of the mobile and stationary phases are determined and cannot be changed. As discussed in Chapter 17, and explicitly included in Equation 17.24, these are the relative retention, α, the partition coefficient for the solute retained longest, k', and the plate count. The compounds of interest usually need two to ten times longer to transit through the column than the unretained peak, t_M. Of concern also is the mobile-phase viscosity and the diffusion coefficients of the solutes in the mobile phase, all factors that do not seriously concern a gas chromatographic separation. In addition, the type and characteristics of the column packing (porosity, narrow particle size range, good packing procedure, and high-quality packing material) influence the column length and the particle size.

Column Efficiency

What follows is a general treatment of column efficiency as a function of different separation variables. It is based on parameters introduced by Giddings.[5] The *reduced plate height* is

$$h = \frac{H}{d_p} = \frac{L}{Nd_p} \tag{19.1}$$

which simply states the number of particle diameters, d_p, that constitute one plate height. The *reduced velocity* of the eluent is

$$v = \frac{ud_p}{D_M} \tag{19.2}$$

which can be considered as the ratio of the time required to displace solute molecules a distance equal to one particle diameter to the time needed for the same displacement by molecular diffusion. The reduced velocity expresses the balance between mass transport by bulk flow and by diffusion or molecular motion across a single particle. Combined with Equation 17.3, the reduced velocity can also be expressed as

$$v = \frac{Ld_p}{t_M D_M} \tag{19.3}$$

D_M is the diffusion coefficient of the solute in the mobile phase.

The complete equation for the dependence of the reduced plate height on the reduced velocity, as obtained by Knox,[2] is

$$h = \frac{B}{v} + Av^{0.33} + Cv \tag{19.4}$$

The B term is 1.2 for solid-core (pellicular) packings and 2.0 for completely porous column packings. For a well-packed column the A term is about unity. The C term is 0.05 for porous particles, decreasing to 0.003 for pellicular particles. Although no theory accurately describes the dispersion from flow inhomogeneity (convective mixing) in the mobile phase, experiments show it to approximate $Av^{0.33}$.

A logarithmic plot of Equation 19.4 is shown is Figure 19.1. In the intermediate region of velocities, the reduced plate height has a minimum value in the range 2–3 and a reduced velocity in the range 3–5. Here the A term dominates. At low reduced velocities, the B term, arising from axial or longitudinal diffusion, dominates. One usually avoids operating in this region of the h/v curve. At high velocities the C term dominates and is responsible for the increase in the reduced plate height as the velocity increases. The C term contains the contribution from mass transport kinetics and also the contribution from stagnant pockets of mobile phase.

To summarize, three things should be noted in Equation 19.4. First, the reduced plate height is independent of the particle diameter of the column packing. Second, the constants are dependent only on how well the column has been packed. Last, there is no reason or evidence to suggest that changing the column bore has any effect on the plate height–velocity curve. Neither is the optimum particle size nor the column length a function of the column bore. As a result, a plot of h versus v values for any well-packed column falls on the same general curve so long as the same type of packing (porous or pellicular) is used.

The reduced velocity of Equation 19.4 is proportional to $1/D_M$ (Equation 19.3). The solute diffusion coefficient, D_M, is roughly proportional to the (solute molecular weight)$^{-0.4}$, which means that the reduced velocity is proportional to (mol wt)$^{0.4}$. Separations of small molecules usually involve reduced velocities in the range $3 < v < 15$; however, macromolecule ($> 50{,}000$ daltons) separations typically involve reduced velocities greater than 50. This is important with the method of exclusion chromatography (see Chapter 20).

Generation of Required Plate Number in a Reasonable Time

In the first example, the optimization problem is how to generate the required number of plates in a reasonable time while operating the column at the minimum in the h/v plot (Figure 19.1). The column length and particle size of the packing must be ascertained. Then the desired performances (t_M and N) are chosen under the operating conditions of eluent viscosity, η, and column packing, assumed here to be porous. The numerical data used are typical for columns packed with silica particles.

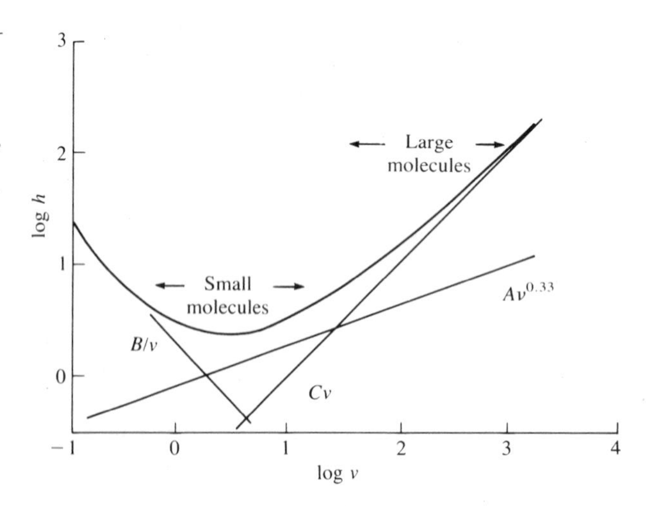

FIGURE 19.1
Logarithmic plot of reduced plate height, h, against reduced velocity, v (Knox equation), with $A = 1$, $B = 2$, and $C = 0.1$.

They include $\eta = 10^{-3}$ N sec m^{-2}, $D = 10^{-9}$ m^2 sec^{-1}, and ϕ, a dimensionless constant called the *specific column resistance*, whose value is in the range 500 (for pellicular packings) to 1000 (for fully porous packings).

The pressure drop, ΔP, across a bed packed with spherical particles of diameter d_p is

$$\Delta P = \frac{\phi \eta L v D_M}{d_p^3} \tag{19.5}$$

Combining this expression with Equations 19.1 and 19.3, one obtains

$$\Delta P = \frac{N^2 h^2 \phi \eta}{t_M} \tag{19.6}$$

Inserting the desired plate count, $N = 2500$ and $t_M = 90$ sec, and using an optimum value of the reduced plate height, $h = 3$, for $\phi = 1000$, Equation 19.6 gives

$$\Delta P = \frac{(2500)^2(3)^2(1000)(10^{-3}\text{ N sec m}^{-2})}{90\text{ sec}}$$

$$\Delta P = 625{,}000\text{ N m}^{-2} \quad (6.2\text{ atm or } 91\text{ psi})$$

The required particle size is given by Equation 19.1:

$$d_p = \frac{L}{Nh} = \frac{L}{(2500)(3)} = \frac{L}{7500}$$

Since normal columns are 100 or 250 mm long, the corresponding particle diameters needed are 13 and 33 μm, respectively.

Repeating the calculations for a 5000 plate count, the pressure drop would be fourfold larger (24.8 atm or 364 psi). The particle sizes required would be half the former sizes: 6.5 and 16.5 μm, respectively, for the columns 100 and 250 mm long. Combinations of column lengths and particle sizes, plus operating pressures for different plate counts and retention times, are given in Table 19.1.

Ascertaining Pressure and Retention Time When Plate Number Is Specified

For the second example, the problem involves ascertaining the operating pressures and retention times when the plate count is specified. No assumptions are made for the values of the reduced plate height and the reduced velocity. The desired separation parameters are $k' = 2$ and $N = 10{,}000$, a situation that corresponds to a resolution of 1.5 and a relative retention of 1.10. For water as eluent, $\eta = 10^{-3}$ N sec m^{-2} and $D_M = 10^{-9}$ m^2 sec^{-1} for many solutes. The well-packed column is prepared from a porous packing for which $\phi = 1000$.

To proceed with the solution, eliminate t_M between Equations 19.3 and 19.6 to give

$$\Delta P = \frac{N h \phi \eta D_M v}{d_p^2} \tag{19.7}$$

Rearrangement gives

$$hv = \frac{\Delta P d_p^2}{\phi \eta N D_M} \tag{19.8}$$

The right-hand side can be evaluated for any ΔP and d_p. Using a pressure drop of 100 atm and a column packed with 3-μm particles gives

$$hv = \frac{(100\text{ atm})(1.01 \times 10^5\text{ N m}^2\text{ atm}^{-1})(3 \times 10^{-6}\text{ m}^2)}{(1000)(10^{-3}\text{ N sec m}^{-2})(10{,}000)(10^{-7}\text{ m sec}^{-1})} = 9$$

Multiplying all terms in Equation 19.4 by the reduced velocity, one obtains

$$hv = 2 + v^{1.33} + 0.05v^2 = 9$$

which can be solved for an approximate value of v and then h. These are $v = 4$ and $h = 2.3$.

Combining Equations 17.3 and 19.5, the result with Equation 19.1 gives

$$t_M = \frac{N^2h^2\phi\eta}{\Delta P} \tag{19.9}$$

$$t_M = \frac{(10000)^2(2.3)^2(1000)(10^{-3}\text{ N sec m}^{-2})}{(100\text{ atm})(1.01 \times 10^5\text{ N m}^{-2}\text{ atm}^{-1})} = 52\text{ sec}$$

TABLE **19.1** TYPICAL PERFORMANCES FOR VARIOUS EXPERIMENTAL CONDITIONS

Performances		***Column parameters***			**Peak bandwidth**
N	**t_M (sec)**	***L* (cm)**	**d_p (μm)**	***P* (atm, psi)**	**4σ (μL)**
2,500	30	2.3	3	18.4 (270)	23
2,500	30	3.7	5	18.4 (270)	37
2,500	30	7.5	10	18.4 (270)	75
5,000	30	4.5	3	74 (1088)	41
5,000	30	7.5	5	74 (1088)	68
5,000	30	15.0	10	74 (1088)	136
10,000	30	9.0	3	300 (4410)	82
10,000	30	15.0	5	300 (4410)	136
10,000	30	30.0	10	300 (4410)	272
10,000	30	9.0	3	300 (4410)	82
10,000	60	9.0	3	150 (2200)	82
10,000	90	9.0	3	100 (1470)	82
15,000	90	2.3	3	223 (3275)	23
15,000	120	2.3	3	167 (2459)	23
11,100	30	10.0	3	369 (5420)	91
11,100	37	10.0	3	300 (4410)	91
11,100	101	10.0	3	100 (1470)	91
27,800	231	25.0	3	300 (4410)	75

Assumed reduced parameters: $h = 3$, $v = 4.5$

From Equation 17.11, the retention time is

$$t_R = (1 + k')t_M$$
$$= (1 + 2)(52 \text{ sec}) = 156 \text{ sec} \quad (\text{or } 2.6 \text{ min})$$

The required column length is obtained from Equation 19.1:

$$L = Nhd_p$$
$$= (10000)(2.3)(3 \times 10^{-6} \text{ m}) = 0.069 \text{ m} \quad (\text{or } 6.9 \text{ cm})$$

Effect of Temperature

In many critical HPLC analyses, variations in temperature can cause significant changes in retention times, making qualitative analysis difficult and affecting the precision of quantitative measurements. Elevated temperatures are advantageous because decreased mobile-phase viscosity, increased mass transfer, and increased sample solubility result in either better resolution or faster analysis.

In Figure 19.2, the same materials are separated at different temperatures while the mobile-phase flowrate and composition are held constant. As the temperature is increased, the k' values decrease and the peaks become sharper. This observation is fairly general for most modes of HPLC. An exception is ion-exchange chromatography in which selectivity changes are frequently observed with changes in temperature.

Limits of Column Performance

Since an analytical separation has several degrees of freedom, various operational variables will be summarized. Also, the influence of the physical properties of the two phases on a separation needs comment.

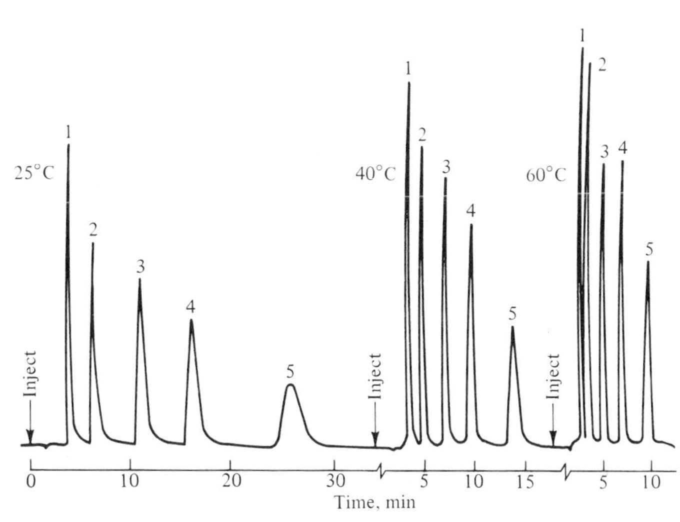

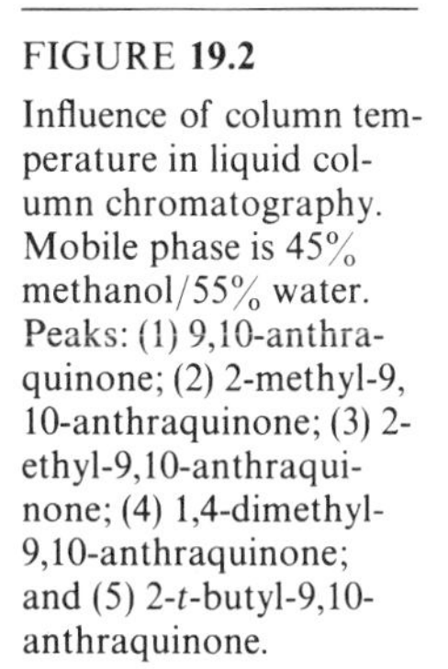
FIGURE 19.2
Influence of column temperature in liquid column chromatography. Mobile phase is 45% methanol/55% water. Peaks: (1) 9,10-anthraquinone; (2) 2-methyl-9,10-anthraquinone; (3) 2-ethyl-9,10-anthraquinone; (4) 1,4-dimethyl-9,10-anthraquinone; and (5) 2-t-butyl-9,10-anthraquinone.

Pressure Drop. As shown by Equation 19.9, the pressure drop varies inversely with the retention time. Although an increase in pressure can increase the maximum attainable plate number, the generation of heat within the column as a result of the work done in forcing the mobile phase through the column at a very high pressure can seriously degrade column performance. Pressures above 5000 psi (340 atm) do not appear worthwhile for most HPLC separations.

Particle Diameter of Stationary Phase. From Table 19.1, for each separation time and operating pressure there exists an optimum particle diameter for which the plate count is maximum. The analytical performances improve dramatically when the particle diameter is reduced, particularly when the column is operated at the optimum velocity. Each time the particle diameter is halved, the pressure drop required is raised by approximately a factor of four. Often, however, the column length can be shortened significantly. There are no practical operating conditions when particles larger than 5 μm would be desirable from the point of view of achieving high plate numbers quickly. For a given column length, the plate count is increased 1.7 times for a 3-μm packing material compared with a 5-μm packing material. Furthermore, within limits, the plate count is also directly proportional to the column length. Of course, one must always keep in mind that the resolution of two peaks is proportional to only the square root of the plate count. Eventually, however, the use of ever finer particles or longer columns of the same-size particles requires pressures that exceed 5000 psi, the practical upper limit. Commercial columns are available with packing materials with particle diameters of 3, 5, and 10 μm.

Column Length. The ratio

$$\lambda = \frac{L}{d_p} \tag{19.10}$$

is the number of particles to the column length and is called the *reduced column length.* Inserted into Equation 19.2 and combined with Equation 19.6, it becomes

$$\Delta P_0 = \frac{\phi \eta \lambda^2}{t_M} \tag{19.11}$$

Equation 19.11 shows that when columns of the same reduced length are eluted with the same eluent to give the same elution time for an unretained solute, the same pressure drop is required. This statement can be verified by the results for $N = 10{,}000$ in Table 19.1. Finer particles are usually paired with shorter columns to achieve a given separation. This necessitates the use of higher inlet pressures to move the mobile phase through the column at the optimum velocity. Both effects eventually result in serious technical difficulties.

Viscosity. A solvent with low viscosity is always preferred in HPLC. While maintaining constant the pressure drop across the column, an increase in the viscosity of the solvent always decreases the flowrate of the mobile phase. The diffusion coefficients of solutes are also affected by the viscosity of the mobile phase. For example, D_M is 0.3×10^{-3} N sec m^{-2} in 1-propanol and 3.0×10^{-3} N sec m^{-2} in hexane.

Extra-Column Band Broadening. A final consideration of importance in optimizing column performance is the effect of extra-column band broadening (the $4\sigma_v$ values of Table 19.1). This spreading includes the dilution factor caused by the injector, any column dead (void) spaces, the volume of connecting tubing before and after the separation column, and the detector volume. Any band spreading from the foregoing factors is added to the random dispersion within the separation column. For symmetrical solute elution bands, these factors are additive as variances; that is,

$$\sigma_{\text{tot}}^2 = (\sigma_{\text{inj}}^2 + \sigma_{\text{trans}}^2 + \sigma_{\text{det}}^2) + \sigma_{\text{column}}^2 \quad (19.12)$$

For the variances that arise from the extra-column system (terms within the parentheses) to be neglected, they must collectively be less than half the column variance at the flow velocities needed for HPLC. With current instrumentation and using standard 8-μL detector cells, it is not generally possible to work with unretained peaks whose peak volumes are less than about 70 μL due to extra-column dispersion produced by the injector/detector system. The base peak width of an unretained solute is obtained by combining Equations 17.2 and 17.1. Assuming a column internal diameter of 4.6 mm and a porosity of 0.75,

$$4\sigma_v = 4(\pi \times 2.3^2 \times 0.75)LN^{-1/2} = 50LN^{-1/2} \quad (19.13)$$

Values of $4\sigma_v$ are included in Table 19.1. When these values are less than 70 μL, special equipment (not always attainable) is needed to keep extra-column effects at an insignificant level. To keep the increase in plate height resulting from remixing of the solute bands in the detector less than 1%, the detector cell volume should be less than or equal to

$$\frac{0.1t'_R F_c}{N^{1/2}}$$

where t'_R is the adjusted retention time (in minutes) of the solute, F_c is the flowrate of the mobile phase (in mL/min), and N is the plate number. For example, the preceding condition is fulfilled with a 10-μL detector cell for a column with 10,000 plates at a flowrate of 0.5 mL/min if the retention time is longer than 20 min. With the use of narrow-bore columns, the detector volume must be 2–3 μL to ensure that the separations achieved within the separation column are not lost through band spreading within the detector.

19.2 GRADIENT ELUTION AND RELATED PROCEDURES

In *isocratic elution* a sample is injected onto a given column and the mobile phase is unchanged throughout the time required for the sample components to elute from the column. No single isocratic elution can separate a complex mixture with adequate resolution in a reasonable time and with good detectability. The isocratic separation of samples with widely varying k' values typically exhibits poor resolution of early-eluting bands, difficult detection of late-eluting bands, and unnecessarily long elution times. To adequately handle samples that have both weakly retained and strongly retained substances, the rates of individual band migrations must be changed during a chromatographic run by solvent programming. Note the analogy with temperature programming in gas chromatography.

Solvent Programming

Solvent programming, also called *gradient elution*, involves changing the mobile-phase composition either stepwise or continuously as elution proceeds.[6] Usually all the sample components are initially retained at the top of the column. After the gradient is begun, the eluent strength of the mobile phase is increased. Eventually the k' value for the earliest eluting component becomes small enough to allow migration of that component along the column at an increasingly faster pace until it exits (analogous to the last stages of a solute's progress down the column in programmed temperature gas chromatography). The same pattern is repeated at a later time for the component eluting next, followed eventually by similar migrations of the remaining components.

Basic equations for gradient elution are expressed in terms of the same parameters as for isocratic elution, discussed in Section 19.1, except that k' is replaced by the average $\bar{k}$ value during gradient elution. The equations are as follows:

$$\text{Retention time:} \quad t_g = t_M \bar{k} \log\left(\frac{2.3k_0}{\bar{k}}\right) \tag{19.14}$$

$$\text{Resolution:} \quad R_s = \left[\left(\frac{1}{4}\right)(\alpha - 1)N^{1/2}\right]\left[\frac{\bar{k}}{1 + \bar{k}}\right] \tag{19.15}$$

$$\text{Bandwidth:} \quad \sigma_g = \frac{V_M(1 + \bar{k})N^{1/2}}{2} \tag{19.16}$$

where k_0 is the value of k' at the beginning of a separation, t_g is the retention time in gradient elution, and σ_g is the gradient bandwidth (one standard deviation). The capacity factor in gradient elution is a function of the gradient time (from start to finish), the flowrate, and the column void volume.

Instantaneous values of $\bar{k}$ during the migration of a particular solute should fall within the range $1 < \bar{k} < 10$, as in isocratic elution. At the time the band leaves the column, $\bar{k}$ for each is fairly small (~ 1). Small $\bar{k}$ values at the time of elution mean narrow bands, and a narrow band means increased sensitivity (in the detector). Compared with isocratic elution, gradient elution provides more nearly equal bandwidths and faster overall separation. A potential twofold or greater increase in sensitivity is achievable for gradient elution versus isocratic elution, even for samples that do not require gradient elution. In the case of bands that would tail in isocratic elution, the decrease in all capacity factor values with time means that the tail of a band moves continually in a region of smaller k' than the band front. Such bands become narrower and increase in sensitivity. Thus, in gradient elution it is possible to obtain both maximum resolution and sensitivity for every solute in the sample.

Gradient Selection

The optimum gradient for a particular separation is selected by trial and error. Of the gradient forms shown in Figure 19.3, the linear form is preferred initially. Of the nonlinear gradient forms, four have fast rates of change at the beginning of the

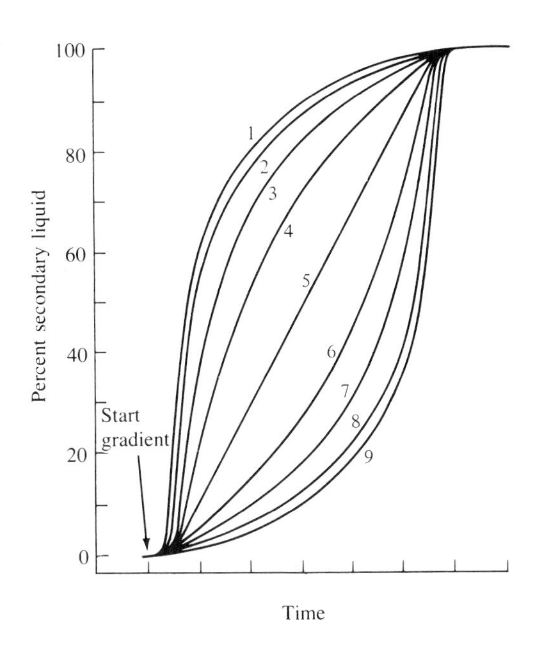

FIGURE **19.3**
Basic gradient forms.

gradient with continually decreasing rates, and four have slow starting rates with continuously increasing rates of change. An interesting feature of gradient 9 is that approximately 50% of the run is isocratic, holding at the initial composition, followed by a built-in flush with the more powerful eluent mixture as the mobile phase.

Stepwise Elution

Stepwise elution is worth considering if the sample contains some fast-moving components followed by some slow-moving solutes but no components with intermediate k' values. Successive samples are injected, with the eluent strength progressively altered by stepwise changes in the composition of the mobile phase, as shown in Figure 19.4. The mobile phase is stepped from one isocratic composition to another. This procedure permits solvent programming information to be obtained in detail. Starting a method development scheme with too powerful an eluent causes almost immediate elution of the sample; peaks 1 and 2 are unresolved in the example when 20% acetonitrile is used. Starting with 12% or 14% acetonitrile and proceeding to 20% acetonitrile would be a solvent program worth trying. The program would follow one of the convex gradient forms (shapes 6–9 in Figure 19.4). The order reversal of peaks 3 and 4 is a problem sometimes encountered in gradient elution.

19.3 DERIVATIZATION

The most common derivatization techniques in HPLC are intended to enhance detectability by ultraviolet absorption, fluorescence, or electrochemistry. Precolumn derivatization is carried out before separation. The derivatized sample is

FIGURE **19.4** Stepwise elution of methyl xanthines on bonded phase C-8 column with decreasing elution strength. The mobile phase contained a phosphate buffer, pH 2.6, and the percent of acetonitrile is marked on the graph. Peaks: (1) theobromine, (2) theophylline, (3) hydroxypropyl theophylline, (4) caffeine, and (5) 8-chlorotheophylline.

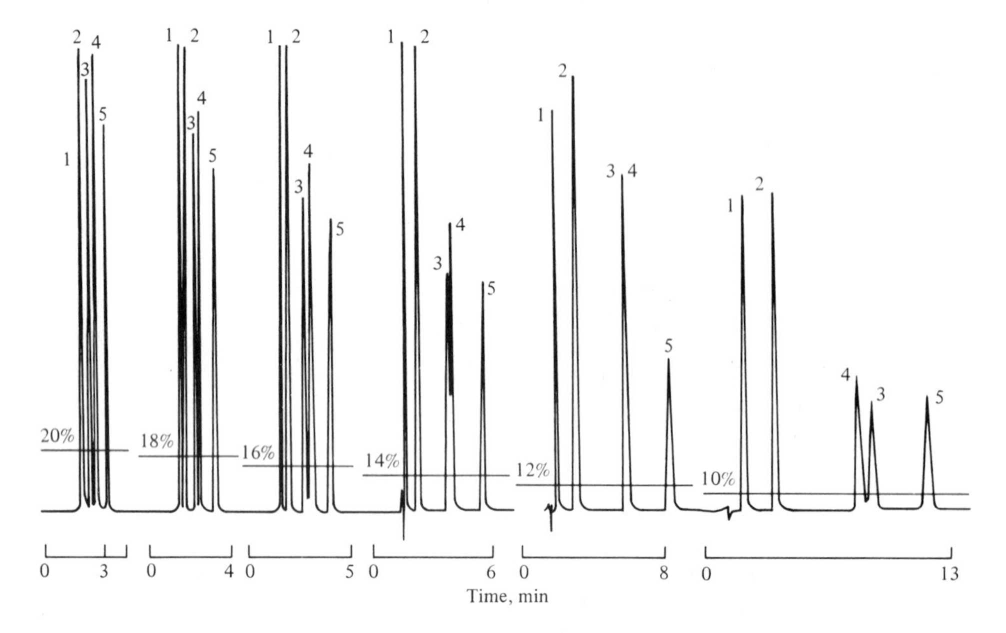

then injected into the column. Postcolumn reactions are carried out after separation. Reactions can be run off-line for long periods, if necessary. The product need not have different detection properties from the reactant or analyte so long as each has different separation properties.

In cases where the detection sensitivity to the 254-nm wavelength in ultraviolet absorption is zero or very low, detection can be enhanced by attaching to the solute a chromophore with high absorption at 254 nm. This might be accomplished, for example, with the following reagents:

Reagents	Reactants
N-Succinimidyl 4-nitrophenylacetate	Amines and amino acids
3,5-Dinitrobenzoyl chloride	Alcohols, amines, and phenols
4-Nitrobenzyloxyamine hydrochloride	Aldehydes and ketones
O-4-Nitrobenzyl-*N,N'*-diisopropylisourea	Carboxylic acids
4-Nitrobenzyl-*N*-propylamine hydrochloride	Isocyanate monomers

All these reagents make reductive electrochemical detection and also fluorescence feasible.

The formation of fluorescent derivatives serves two purposes: it allows the sensitive detection of otherwise nonfluorescent molecules, and it exploits the selectivity of fluorescence by allowing the detection of all compounds with a particular functional group in a sample after derivatization. The following reagents permit fluorescent derivatization:

Reagents	Reactants
4-Bromomethyl-7-methoxycoumarin	Carboxylic acids
7-Chloro-4-nitrobenzyl-2-oxa-1,3-diazole	Amines (primary and secondary) and thiols
1-Dimethylaminonaphthalene-5-sulfonyl chloride	Amines (primary) and phenols
1-Dimethylaminonaphthalene-5-sulfonyl hydrazine	Carbonyls

In postcolumn reactions, effluent from an HPLC column is mixed with a reagent before it enters the detector. The equipment needed to perform postcolumn reactions is relatively simple. A column filled with glass beads mixes the effluent and reagent (added through a tee valve) before the mixture passes into the detector. Reaction times must be short, although a delay tube can be added to the system to allow slow reactions to take place. In postcolumn work there is much less need to convert quantitatively each analyte into a single product; reproducibility is the only requirement.[7]

Postcolumn addition of an alkaline buffer (pH 10.4) increases 20-fold the sensitivity of HPLC analysis for barbiturates (Figure 19.5). As little as 2 ng of drug can be detected by using a postcolumn reaction system because at alkaline pHs the uv absorption maxima for barbiturates shift to longer wavelengths (240 nm). Simply adjusting the mobile phase to an alkaline pH would dissolve gradually a silica-based column packing.

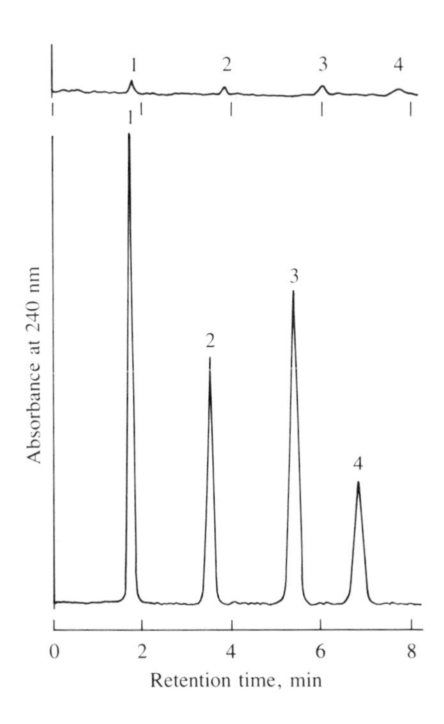

FIGURE **19.5**

Chromatograms of barbiturates without (*upper trace*) and with (*lower*) a postcolumn pH change. Peaks: (1) barbital, (2) butethal, (3) pentobarbital, and (4) secobarbital.

19.4 HPLC INSTRUMENTATION

The general instrumentation for HPLC incorporates the following components, shown in Figure 19.6.

1. There is a solvent reservoir for the mobile phase.
2. The mobile phase must be delivered to the column by some type of pump. To obtain separations either based on short analysis time or under optimum pressure, a wide range of pressure and flows is desirable. The pumping system must be pulse-free or else have a pulse damper to avoid generating baseline instability in the detector.
3. Sampling valves or loops are used to inject the sample into the flowing mobile phase just at the head of the separation column. Samples should be dissolved in a portion of the mobile phase (if possible) to eliminate an unnecessary solvent peak.
4. Ahead of the separation column there may be a guard column or an in-line filter to prevent contamination of the main column by small particulates.
5. To measure column inlet pressure a pressure gauge is inserted in front of the separation column.
6. The separation column contains the packing needed to accomplish the

FIGURE **19.6** General instrumentation for HPLC.

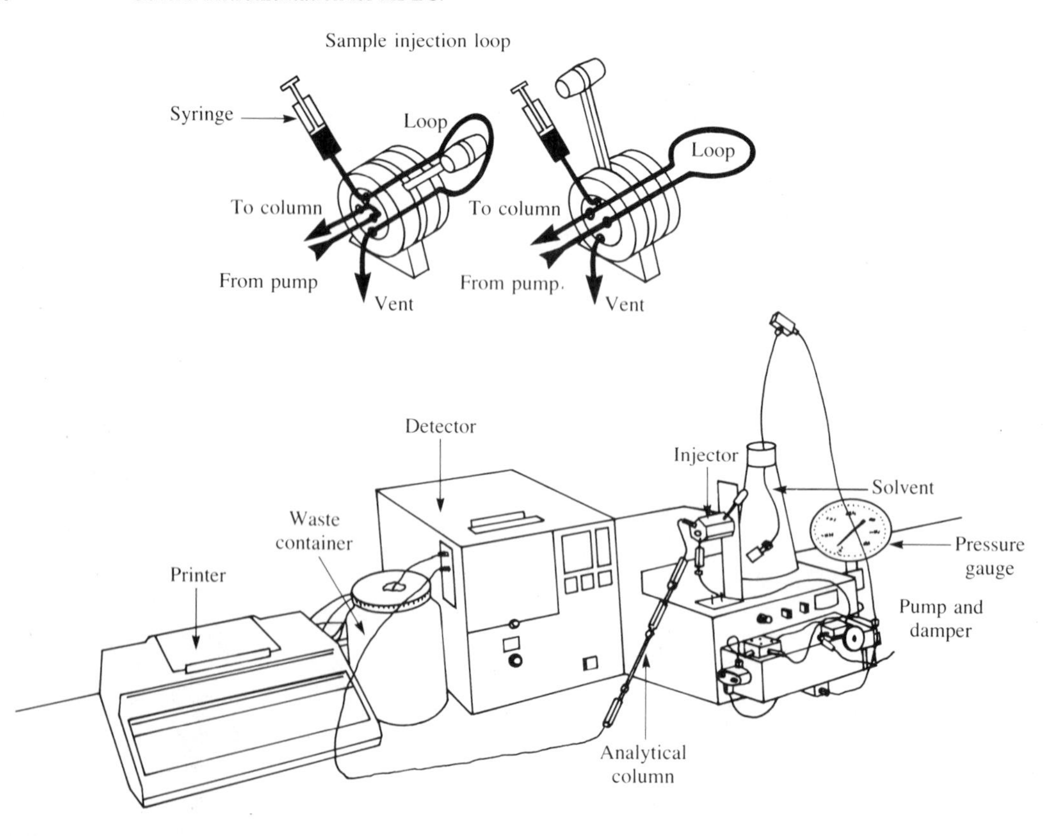

desired HPLC separation. These may be silicas for adsorption chromatography, bonded phases for liquid-liquid chromatography, ion-exchange functional groups bonded to the stationary support for ion-exchange chromatography, gels of specific porosity for exclusion chromatography, or some other unique packing for a particular separation method.

7. A detector with some type of data handling device completes the basic instrumentation.

Additional description and discussion will follow in subsequent sections.

19.5 MOBILE-PHASE DELIVERY SYSTEM

The mobile phase must be delivered to the column over a wide range of flowrates and pressures. To permit the use of a wide variety of organic and inorganic solvents, the pump, its seals, and all connections must be made of materials chemically resistant to the mobile phase. If the system is not pulse-free, some form of pulse damping is needed. A degasser is needed to remove dissolved air and other gases from the solvent. Another desirable feature in the solvent-delivery system is the capability for generating a solvent gradient.

A pump should be able to operate to at least 100 atm (1500 psi), a pressure suited to less expensive chromatographs. However, 400 atm (6000 psi) is a more desirable pressure limit. For many analytical columns only moderate flowrates of 0.5–2 mL/min need to be generated. Microbore columns require flowrates as low as a few microliters per minute. The ease with which solvents may be changed is an important consideration in gradient elution work or when scouting for the optimum solvent. The pump should have a small holdup volume.

Reciprocating Piston Pumps

Of the many types of pump designs currently in use, the reciprocating piston type is the most popular. Pumps of this type are relatively inexpensive and permit a wide range of flowrates. A simple version of this pump is shown in Figure 19.7. A small motor-driven piston moves rapidly back and forth in a hydraulic chamber that may vary from 35 to 400 μL in capacity. By means of check valves, on the backward stroke the piston sucks in solvent from a mobile-phase reservoir. At this time the outlet to the separation column is closed. On the forward stroke the pump pushes solvent out to the column and the inlet from the reservoir is closed. Usually a hydraulic fluid transmits the pumping action to the solvent via a flexible diaphragm. This minimizes solvent contamination and corrosion problems with pump parts. A wide range of flowrates is obtainable by varying either the stroke volume during each cycle of the pump or the stroke frequency. Since flow pulses are produced, some pulse-dampening system must be used.

Dual-head (and triple-head) pumps consist of identical piston-chamber units operated 180° (or 120°) out of phase. Although not free from small pulsations, solvent delivery is smoothed because in the dual-head type one pump is filling while the other is in the delivery cycle. Arranging a piston-driving cycle slightly more than 180° enables gradual takeover periods between alternate delivery by each piston of the pump.

With reciprocating pumps solvent delivery is continuous. There is no restriction on the reservoir size or operating time. These pumps are valuable in equipment used for automatic operation. The pump's small internal volume makes solvent changes rapid and accurate when doing gradient elution and when solvent scouting.

Syringe-Type Pumps

These pumps work through positive solvent displacement by a piston mechanically driven at a constant rate (Figure 19.8). The piston is actuated by a screw-feed drive

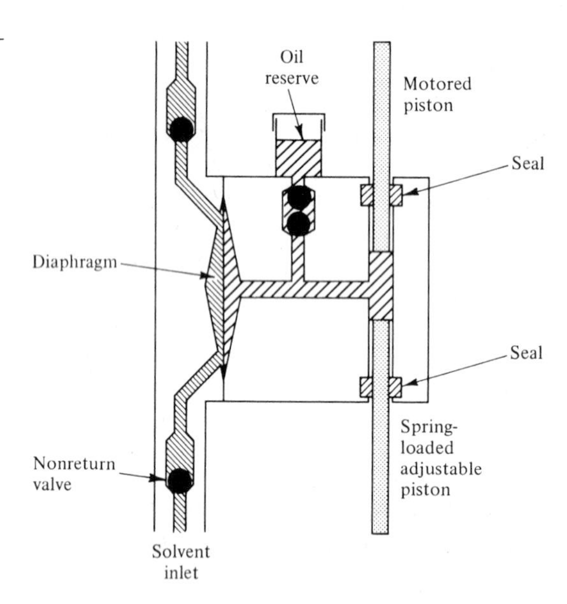

FIGURE **19.7**
Reciprocating pump.

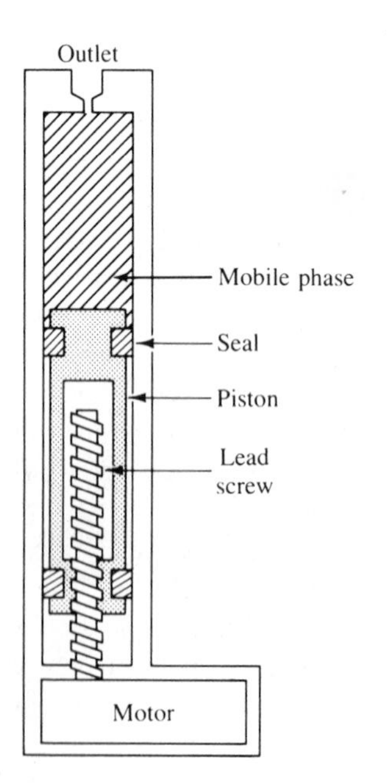

FIGURE **19.8**
Syringe-type pump.

through a gear box, usually run by a digital stepping motor. The rate of solvent delivery is controlled by changing the voltage on the motor. Although the solvent chamber has only a finite capacity (250–500 mL) before it must be refilled, this is adequate for operations with today's small-bore columns. Pulseless flow is achieved along with high-pressure capability (200–475 atm). Tandem operation of two or more pumps permits the use of various types of solvent gradients.

Constant-Pressure Pumps

In these pumps the pressure from a gas cylinder delivered through a large piston drives the mobile phase, as shown in Figure 19.9. Since the pressure on the solvent is proportional to the ratio of the area of the two pistons, usually between 30:1 and 50:1, a low-pressure gas source of 1–10 atm can be used to generate high liquid pressures (1–400 atm). A valving arrangement permits the rapid refill of the solvent chamber whose capacity is about 70 mL. This system provides pulseless and continuous pumping, plus high flowrates for preparative applications. This type is useful for packing columns; however, it is inconvenient for solvent gradient elution.

Pulse Dampers

Several of the detectors used in HPLC are sensitive to variations in flow, most notably refractive index, electrochemical, and conductivity detectors. Even fluorescence and ultraviolet absorbance detectors exhibit increased detector noise and interference with quantitation when flow modulation is excessive in an HPLC system.

A quite simple damping method involves a flexible bellows or a compressible gas in the capped upright portion of a tee tube to take up some of the pulsation energy. When the pump refills, this energy is released to help smooth the pressure pulsations. These types of pulse dampers have large fluid volumes. Pressures in excess of 1000 psi are required for effective operation.

Pulse dampers that use a compressible fluid separated from the mobile phase by a flexible inert diaphragm offer several advantages: easy mobile-phase changeover, effectiveness at low system pressure, wide dynamic range, and minimal dead volume. In this type of pulse damper, the compressible fluid expands when the pump piston retracts, maintaining system pressure and constant solvent flow.

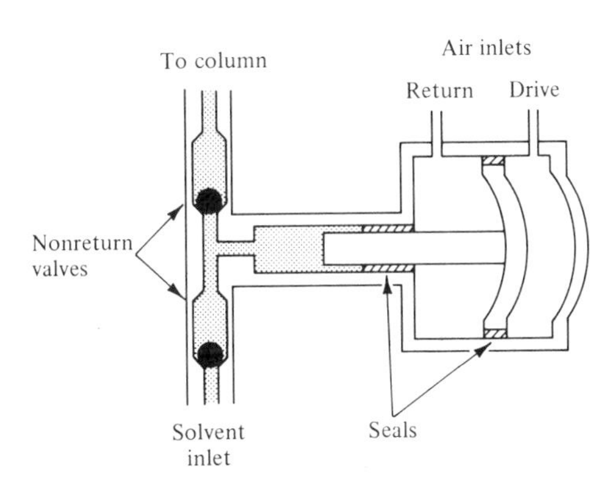

FIGURE **19.9**
Constant-pressure pump.

Electronic pulse dampers provide a small rapid forward stroke of the piston following the pump's rapid refill stroke. The small forward stroke dampens the pulse flow by bringing the solvent up to system pressure before the solvent delivery stroke.

Equipment for Gradient Elution

The methodology and advantages of gradient elution (or flow programming) were discussed in Section 19.2. Changes in the composition of the mobile phase during gradient elution can be stepwise or continuous. Commercial gradient programmers can store and execute up to 50 steps per program, which provides a virtually smooth gradient. A number of programs can be stored. The microprocessor unit permits the creation of any type of gradient (see Figure 19.3), even irregular and step gradients.

In one low-pressure gradient programmer (that is, the gradient is formed ahead of the HPLC pump in use), the solvents are accurately measured by a precise valving system. Up to three solvents can be handled. To promote complete mixing of the solvent with significantly varying viscosities and miscibilities, the micro-mixing vessel is packed with an inert fiber. The fiber creates sufficient turbulence to effectively mix the mobile-phase components while simultaneously causing the release of any generated gas bubbles, which are vented.

Low-pressure systems can handle a series of solvents with good gradient reproducibility to provide a range of elution power. Any solvent-volume change that occurs during mixing is completed before the mixture is pressurized. However, solvents must be thoroughly degassed before mixing. Only one pressurization pump is needed with this type of system, which is a significant cost advantage.

To operate on the high-pressure region of the system, the output from two or more high-pressure pumps is programmed into a low-volume mixing chamber before flowing into the column. The output of each pump is separately controlled by the gradient microprocessor.

19.6 SAMPLE INTRODUCTION

Insertion of the sample onto the pressurized column must be as a narrow plug so that the peak broadening attributable to this step is negligible. The injection system itself should have no dead (void) volume.

Sampling Valves and Loops

Microsampling injector valves are the most widely used sampling devices. Samples are dissolved, if possible, in the mobile phase to avoid an unnecessary solvent peak. The calibrated sample loop is filled by flushing it thoroughly with a sample solution by means of an ordinary syringe. A rotation of the valve rotor places the sample-filled loop into the mobile-phase stream, with subsequent injection of the sample onto the top of the column without significant interruption of flow. Sample loops of different volumes are interchangeable; loop volume is usually 10 or 20 μL.

Low-volume switching valves are also available. The valve uses an internal sample cavity consisting of an annular groove on a sliding rod that is thrust into the flowing stream. The system, shown in Figure 18.3b for gas–liquid chromatography, is applicable to HPLC.

Loop or valve injection is rapid and can operate at pressures up to 470 atm with less than 0.2% error. The system can be located within a temperature-controlled oven for systems that require handling at elevated temperatures (up to about 150 °C).

Stopped-Flow Injection

In stopped-flow injection, the pump is turned off until the inlet column pressure becomes essentially atmospheric. The syringe is then inserted, the sample is injected in the normal fashion, and the pump is turned on. For syringe-type and reciprocating pumps, flow in the column can be brought to zero and rapidly resumed by diverting the mobile phase by means of a three-way valve placed before the injector. This method can be used up to very high pressures.

19.7 SEPARATION COLUMNS

Columns are constructed of heavy-wall, glass-lined metal tubing or stainless steel tubing to withstand high pressures (up to 680 atm) and the chemical action of the mobile phase. The interior of the tubing must be smooth with a very uniform bore diameter. Straight columns are preferred and are operated in the vertical position. Column end fittings and connectors must be designed with zero void volume to avoid unswept corners or stagnant pockets of mobile phase that can contribute significantly to extra-column band broadening. Packing is usually retained by inserting stainless steel frits into the end of the column.

Most column lengths range from 10 to 30 cm; short, fast columns are 3 to 8 cm long. For exclusion chromatography, columns are 50 to 100 cm long.

Standard Columns

Many HPLC separations are done on columns with an internal diameter of 4 to 5 mm. Such columns provide a good compromise between efficiency, sample capacity, and the amount of packing and solvent required. Column packings feature particles that are uniformly sized and mechanically stable. Particle diameters lie in the range 3–5 μm, occasionally up to 10 μm or higher for preparative chromatography.

Radial Compression Columns

Although a decrease in response is associated with increasing the column diameter, there are benefits to using the wider-diameter radial compression columns. A decrease in the overall operating pressure allows the analyst to decrease analysis time by increasing solvent flow. At this point fast liquid chromatography is achieved at the expense of detectability. The radial compression module applies hydraulic pressure (via glycerol in a plastic sleeve) to compress radially a flexible wall cartridge, 10 cm long with an 8-mm bore, held within a plastic holder. Separations are performed while the cartridge is under compression. Compression diminishes the voids and channels, particularly voids between packing particles and the cartridge walls; this increases the column efficiency. When separations are complete, the cartridge can be decompressed, removed from the module, and reused repeatedly without losing efficiency. Low cost permits the dedication of a cartridge to each application; in turn this eliminates the consumption of costly solvents for column purging.

Narrow-Bore Columns[8]

Decreasing the internal diameter of the column by a factor of two increases the signal of a sample component by a factor of four, the square of the change in diameter (Figure 19.10). The linear column velocity of the mobile phase remains essentially the same and the analysis time remains unchanged. When substituting a 2.0-mm bore column for one of 4.0-mm bore, the volumetric flow required to achieve a given eluent velocity is four times less, a substantial saving in solvent costs and fewer problems with solvent waste disposal. The detector response is four times greater. With narrow-bore columns the analyst can opt for mobile phases containing unconventional solvents or high-purity solvents that normally would be precluded because of cost. In narrow-bore columns there is better homogeneity in packing density over the cross-sectional area of the bed and smaller temperature gradients across the column because frictional heat is dissipated better.

When plate counts above 30,000 are required, and therefore long columns are necessary, a number of short columns can be joined together to build up a long column without loss of plate count. This cannot be done with columns of 4.6-mm bore.

If a 2-mm-bore column yields a good response, a 1-mm-bore column will give an even better response if no modifications of the system are necessary. Packed columns with internal diameters of 1.0 mm or less impose rigid limitations on extra-column effects. Detector-injector volume has to be extremely low if band

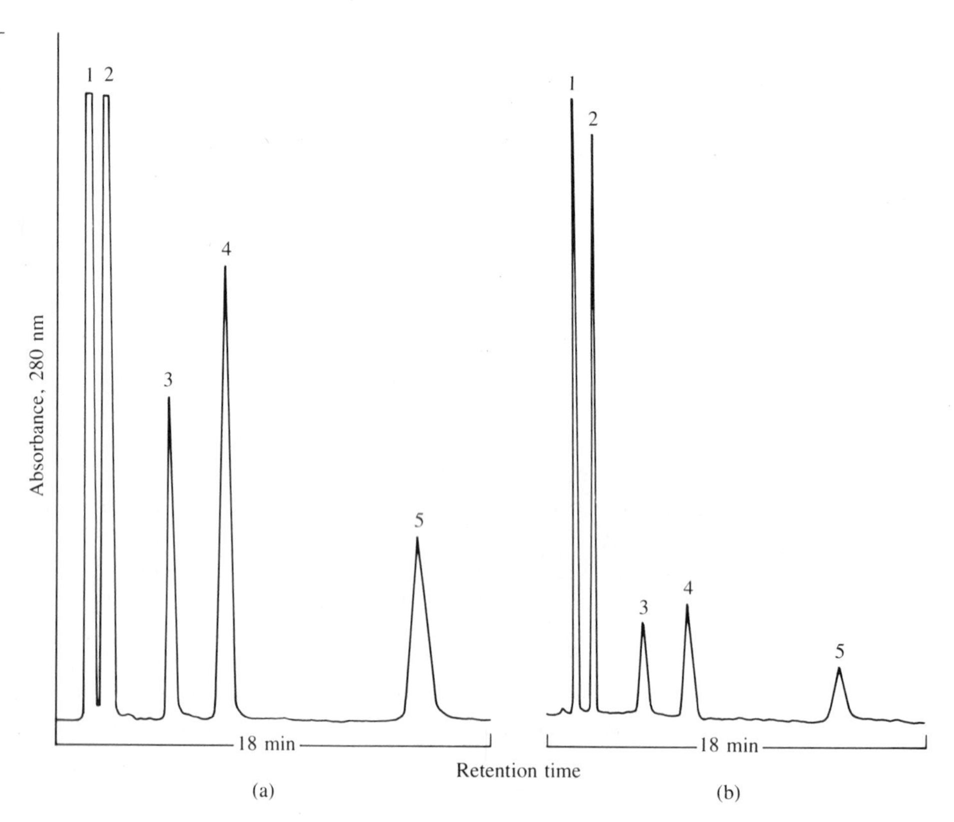

FIGURE **19.10**

Comparison of peak response of (a) a 2 mm × 30 cm column and (b) a 4 mm × 30 cm column, both packed with the same stationary phase. Peaks: (1) ascorbic acid, (2) niacinamide, (3) pyridoxine HCl, (4) riboflavin, and (5) thiamine HCl. Linear velocity is the same for both columns.

broadening is not to become a problem. A standard injection valve cannot be used, or any length of connecting tubing between column and detector. The detector volume has to be greatly reduced (1 μL); as a result the noise level of the detector increases and the concentration sensitivity suffers.

Short, Fast Columns

A short (3–6 cm), conventional-bore column packed with 3-μm particles can save solvent costs, increase sample throughput, and deliver higher sensitivity than conventional-length columns. Typical analysis times are 15–120 sec for isocratic elution and 1–4 min for gradient elution. These columns are useful where analytical speed is essential, as in quality control work.

Guard Columns and In-Line Filters

To prolong the life of analytical columns, guard columns are often inserted ahead of the analytical column where they act as both physical and chemical filters. Guard columns are relatively short (usually 5 cm) and contain a stationary phase similar to that in the analytical column. They protect the analytical column from particulate contamination that arises from a contaminated mobile phase or from degrading sample-injection valves. A guard column extends the lifetimes of the expensive separation column by capturing the strongly retained sample components and preventing them from gradually contaminating the upper layers of the analytical column. Guard columns are by design expendable and are periodically repacked, replaced, or reconditioned.

Guard columns are not an unmixed blessing (Figure 19.11). The approximately 0.08-mL band spreading in the guard column ahead of the 2-mm-bore column

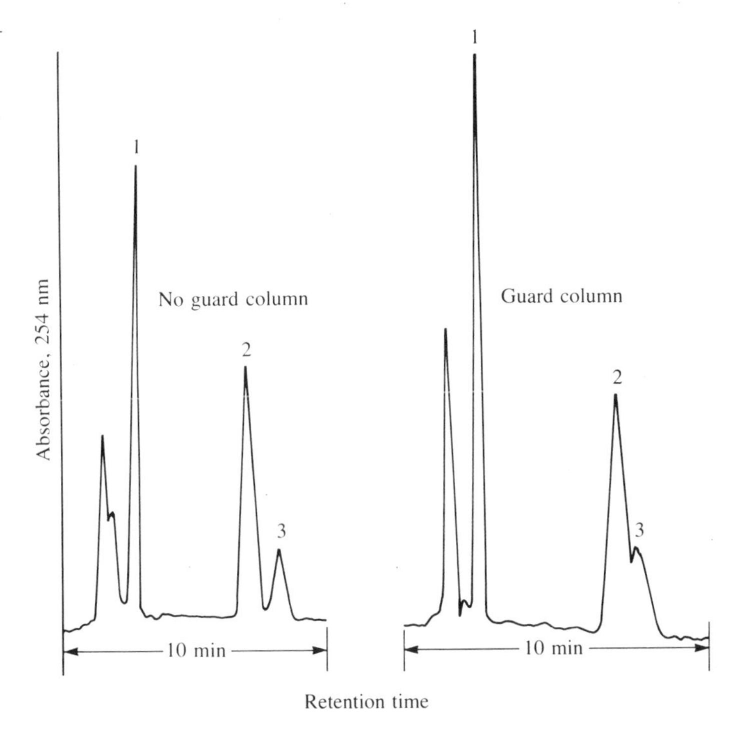

FIGURE **19.11**
Effect on resolution of sample components in a diet beverage when a guard column is used. Peaks: (1) saccharin, (2) caffeine, and (3) benzoic acid.

results in appreciable broadening of solute peaks and degradation of the separation of caffeine and benzoic acid. If the loss in resolution due to the extra-column effect cannot be tolerated, an in-line filter designed to remove particulates should be placed in front of the column.

Temperature Control

Separation columns should be housed within a stable system with temperature variations of less than 0.1 °C when temperature changes must be avoided. Circulating air baths or electrically heated chambers are used to control the column temperature. The solvent is preheated separately before entering the separation column.

19.8 DETECTORS

The sensitive universal detector for HPLC has not been devised yet. Thus it is necessary to select a detector on the basis of the problem at hand and, in doing a variety of separations, it is expected that more than one detector will be needed. Some detectors can be arranged in series.

Bulk property detectors, typified by the refractive index detector, compare an overall change in a physical property of the mobile phase with and without an eluting solute. Although universal, this type of detector tends to be relatively insensitive and requires very good temperature control.

Solute property detectors respond to a physical property of the solute that is not exhibited by the pure mobile phase. These detectors are 1000 or more times sensitive, giving a detection signal for a few nanograms or less of sample. Absorbance, fluorescence, and electrochemical detectors have achieved popularity. Precolumn and postcolumn derivatization expand their applicability.

The *response time*, τ, of the detector is critical and should be at least ten times less than the peak width of a solute in time units. Only then is the peak area not distorted. This condition can be written as

$$\tau \leq \frac{W}{10} = \frac{2t_R}{5\sqrt{N}}$$

Detectors with a time constant of 0.3 sec or less are desirable for high-efficiency small-particle columns.

Detection Limit

To estimate the detection limit using a particular detection system, the band (zone) spreading must be known or estimated (see Equations 19.13 and 19.16). Any connecting tubing must be kept to an absolute minimum. Tubing diameter is most critical because the standard deviation is proportional to the diameter squared but only directly proportional to the length. It should have a bore no larger than 0.25 mm and be no longer than 200 mm; this length would create a volume of 10 μL. Couplings should have zero void volume. Precolumn tubing has a greater effect on the drop in efficiency than postcolumn tubing. This effect results from diffusion of the sample before it reaches the separation column. Nevertheless, the initial effects of the postcolumn tubing are the most detrimental to the separation because any interdiffusion involves the separated components.

Typically the *dilution factor* varies from 5 to 350, meaning that the concentration of the solute peak in the detector is $\frac{1}{5}$ to $\frac{1}{350}$ of the initial sample concentration at injection. This dilution factor naturally affects the detection limits of solutes. For precise quantitation, a tenfold greater concentration than this estimate is needed.

When the system concentration is desired in terms of sample weight, the sample volume must be considered. Here is where the choice of column size enters the picture. Narrow-bore columns dilute small samples less than do large-bore columns. The effect of column parameters and injection volume on detection limits has been examined by Karger, Martin, and Guiochon[9]. For example, consider a porous-layer bead column of 100 cm and 2-mm i.d. If the sample is 5 μL, $N = 900$ plates, and $V_R = 20$ mL, then the dilution is

$$\frac{(5 \times 10^{-6}\ \text{liter})900}{(20 \times 10^{-3}\ \text{liter})2} \quad \text{or } 1{:}335$$

Consider next a porous bead column (particle diameter 10 μm) of 25 cm and 2 mm i.d. With a sample of 5 μL, $V_R = 10$ mL, and $N = 3600$ plates, the dilution is 1:84.

For a detector to be useful in quantitative work, the signal output should be linear with concentration for a concentration-sensitive detector and in mass for a mass-sensitive detector. Also, the detector should have a wide linear dynamic range, perhaps five orders of magnitude, so that major and trace components can be determined in a single analysis.

Ultraviolet-Visible Photometers and Spectrophotometers

Optical detectors based on ultraviolet–visible absorption are the workhorses of HPLC, constituting over 70% of all the detection systems in use. Basically three types of absorbance detectors are available: a fixed-wavelength detector, a variable-wavelength detector, and a scanning (in real time) wavelength detector. These systems, described in detail in Chapter 6, have become so refined that fundamental noise limitations now originate from thermal instabilities in flow cells and in the optical and electronic components. Absorbance detectors may require thermostatting to 0.01 °C to approach the fundamental shot noise limitation of 10^{-6} absorbance unit achievable with high-intensity, fixed-wavelength lamps.

Detector cell volumes on the order of 8 μL per centimeter of optical path length are acceptable for conventional-diameter separation columns. A quartz collimating lens focuses the radiation on the sample and reference cells. Energy that passes through the sample cell is compared with the energy that passes through the reference cell by electronic signal-processing circuits to produce an output that is linear with solute concentration. A 2–3-μL cell is needed when doing work with narrow-bore columns. For fast HPLC, one also needs a rise time of 0.1 sec (0.04 sec time constant) for the detector.

Cell optics must prevent false absorbance, particularly when the uv detector is operated at high sensitivity. False absorbance arises from failure to distinguish between sample peaks and pseudopeaks due to refraction effects and wall reflectance. If the refractive index (RI) within the flow cell changes, the amount of energy that reaches the detector can change because of refractive effects at the cell wall. These RI effects can arise from differences in temperature, flowrate, mobile-phase compositions, or even a RI difference between sample and mobile phases. RI effects cause false absorbances by creating a "dynamic liquid lens" that optically distorts the light beam. This distortion results in absorbance at the cell wall or changes in the amount of the beam striking the detector rather than deflected aside (Figure 19.12). The conical design of the tapered cell continuously eliminates the liquid lens effect as it moves through the cell. The diverging cell walls prevent any distorted light from impinging on the cell wall surfaces. Collimating and masking the light that enters the cell can similarly eliminate these refractive effects, though with some loss in signal strength.

The mobile-phase solvent chosen should absorb only weakly or not at all. For solvents the ultraviolet cutoff is given in Table 7.1. Water, methanol, acetonitrile,

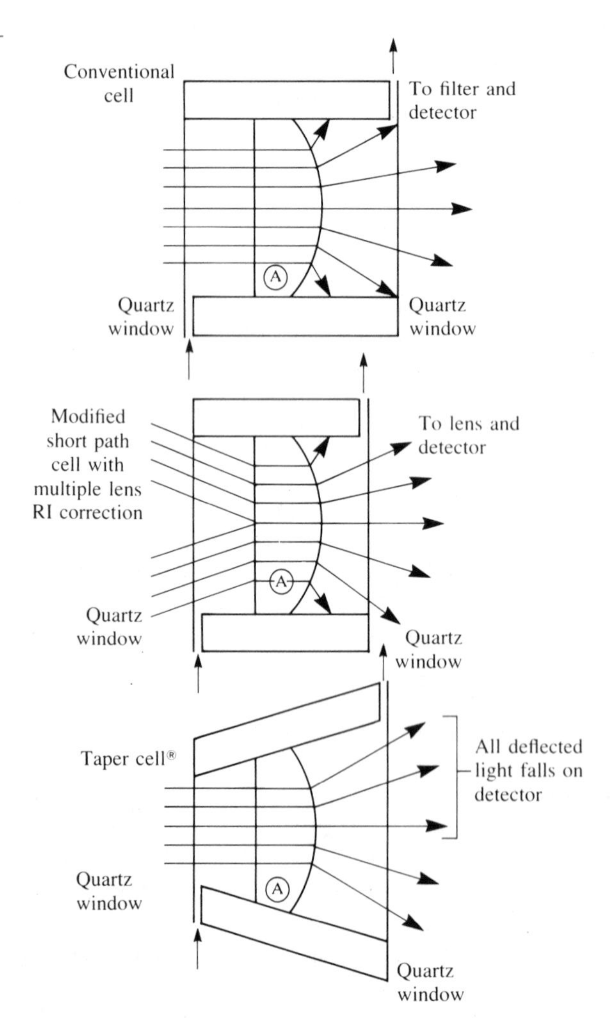

FIGURE **19.12**
The liquid lens effect on absorbance readings with tapered cells and conventional cells. (Courtesy of Waters Associates.)

and hexane all permit operation in the far ultraviolet to at least 210 nm. Functional groups that absorb in the ultraviolet region were discussed in Chapter 7.

Absorbance ratioing is a fast, easy, and accurate method for measuring the purity of a chromatographic peak. These ratios, obtained at any two wavelengths under standard conditions, are specific for each compound and often can be used to identify them and determine their purity.

Photometric detection limits can be estimated if the noise level and the approximate molar absorptivity are known at the operating wavelength. Assuming a path length of 1.00 cm and a noise level of 0.00004 absorbance unit, a concentration detection limit is

$$\frac{2(\text{noise})}{b\varepsilon} = \frac{0.00008}{\varepsilon}$$

in units of mol cm^{-1} liter^{-1}. If ε is 10,000, the minimum detectable concentration is 8 nM/L (or 4 ng/mL for a compound whose molecular weight is 500). If the system concentration is desired in terms of sample weight instead of concentration, the sample volume must be considered and the system dilution factor. For a sample of 5 μL, a dilution factor of 20, and molecular weight of 500, the detection limit is:

$$\frac{(2)(0.00004)(20)(5 \times 10^{-6}\ \text{liter})(500)}{(10{,}000\ \text{liter mol}^{-1}\ \text{cm}^{-1})(1\ \text{cm})} = 0.4\ \text{ng}$$

Fixed-Wavelength Detectors. A fixed-wavelength detector uses a light source that emits maximum light intensity at one or several discrete wavelengths that are isolated by appropriate filters (see Chapter 6). Such a detector offers a minimum of noise but no free choice of wavelength. Using a medium-pressure mercury lamp, wavelengths of 254, 280, 313, 334, and 365 nm can be selected by the use of narrow-bandpass interference filters. Operation in the visible region can be accomplished by using a quartz iodine lamp and appropriate interference filters. Many compounds, including the nucleic acids, have absorption bands, of which a portion encompass the 254-nm wavelength. Placing a special phosphor converter between the mercury lamp and the lens produces an emission band that peaks at 280 nm, a wavelength suitable for detecting proteins. Short-term noise levels are usually less than 0.0001 absorbance unit.

Variable-Wavelength Detector. A variable-wavelength detector is a relatively wide-bandpass ultraviolet-visible spectrophotometer coupled to a chromatographic system. It offers a wide selection of uv and visible wavelengths, but at an increased cost. To obtain a complete spectrum, the eluent flow must be stopped to trap the component of interest in the detector cell while the uv-visible spectral region is scanned.

Scanning-Wavelength Detector. To obtain a real-time spectrum for each solute as it elutes, solid-state diode arrays are required (see Chapter 6). The diode arrays work in parallel, simultaneously monitoring all wavelengths. For example, a complete spectrum can be obtained in as little as 0.01 sec when scanning from 190 to 600 nm.

Simultaneous Monitoring. Rather than scanning the complete spectrum, some instruments can be configured to monitor simultaneously up to nine independent signals (wavelength intervals), permitting multicomponent detection in complex samples. Any one or all of these signals have an associated instrument bandwidth that can be varied between 4 and 400 nm to permit the integration of all signals between two preset wavelengths—a significant increase in detector universality.

Fluorometric Detector

Typical instrument diagrams were shown in Chapter 8. Tapered, square, or cylindrical flow cells have been used. The fluorescence is often collected at a right angle to the excitation beam. However, in this configuration only about one-sixth of the fluorescence is collected. If a concave mirror is placed around the flow cell and the rear of the cell is reflective, about 75% of the emission is collected. With all flow cells, scattered radiation from the excitation source is selectively removed with cutoff or bandpass filters placed before the photomultiplier tube.

Electrochemical (Amperometric) Detectors[10]

The suitability of electrochemical detection depends on the voltammetric characteristics of solute molecules in an aqueous or aqueous-organic mobile phase. Amperometric transducers measure the current (nanoampere level) at a controlled potential as a function of time (see Chapter 23). The flow cell (about 5 μL) is a channel in a thin polyfluorocarbon gasket (50–125 μm) sandwiched between two blocks, one plastic and the other stainless steel (which serves as the auxiliary electrode), as shown in Figure 19.13. Along one side of the channel is positioned a

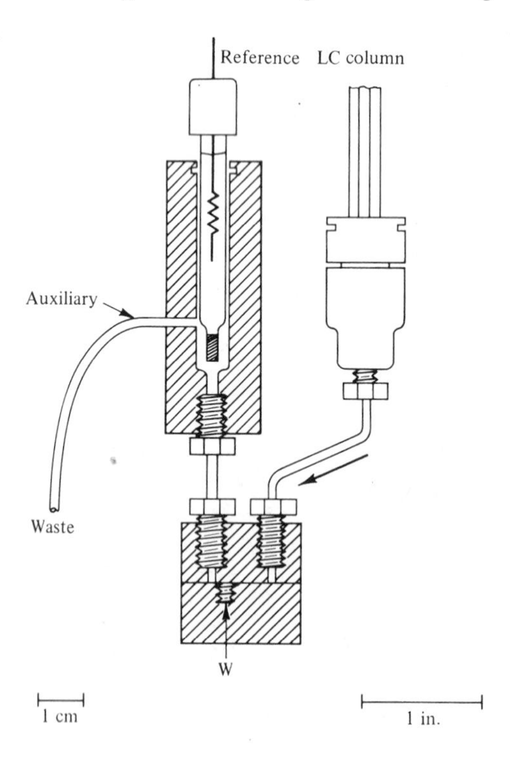

FIGURE 19.13
Thin-layer amperometric detector. (Courtesy of Bioanalytical Systems, Inc.)

working electrode. Farther downstream and connected to the working region by a short length of tubing is a reference electrode (usually Ag/AgCl).

A *dual electrode detector* offers additional specificity. In one configuration, the two working electrodes are placed parallel with the flowing stream. Each electrode is held at a different potential. Two simultaneous chromatograms are generated. The current ratio at the two potential settings is calculated and used for peak confirmation (cf. absorbance ratioing). The second configuration has the two electrodes arranged in series. The upstream working electrode generates an electroactive product from the analyte. That product is subsequently detected at the downstream working electrode. This is a useful strategy if the downstream electrode can be potentiostatted at a value where noise from the electrolysis of the mobile phase (perhaps dissolved oxygen introduced into the eluent by sample injection) is decreased. The series arrangement limits the number of electroactive compounds that are detected; only those compounds that generate an electroactive product that is stable in the time required to reach the second detector electrode are sensed.

Differential Refractometers

A differential refractometer monitors the difference in refractive index between the mobile phase (reference) and the column eluent. It responds to any solute whose refractive index is significantly different from that of the mobile phase. Differential refractometers operate on one of two principles. The *deflection type*, illustrated in Figure 19.14, measures the deflection of a beam of monochromatic light by a double prism. Eluent passes through half of the prism; pure mobile phase passes through, or fills, the other half. An optical mask confines a beam of light from an incandescent tungsten lamp to the face of the sample and reference compartments. The beam, collimated by a lens, passes through the compartments and is reflected back by the mirror through the compartments again. The beam is then focused on a beam-splitter before passing into twin photodetectors. The reference and sample compartments are separated by a diagonal glass divider. If the refractive index of the mobile phase is changed due to the presence of a solute, the beam from the sample compartment is slightly deflected. As the beam changes location on the detector, an out-of-balance signal is generated that is proportional to the concentration of the solute. An optical flat, which deflects the beam from side to side, is used to adjust for a zero output signal when the mobile phase is in both prism compartments. The cell volume is 15–25 μL. Deflection refractive index detectors have the advantage of a wide range of linearity. One cell covers the entire refractive index range.

FIGURE **19.14** Deflection-type refractometer. (Courtesy of Waters Associates.)

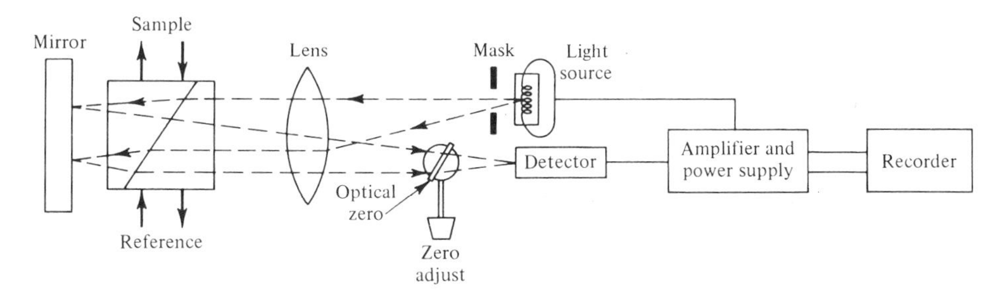

The *reflection-type refractometer* measures the change in percentage of reflected light at a glass-liquid interface as the refractive index of the liquid changes. The instrument design is based on Fresnel's law of reflection, which states that the amount of light reflected at a glass-liquid interface varies with the angle of incidence and the refractive index of the liquid. In the optical path (Figure 19.15) two collimated beams from the projector (light source, masks, and lens) illuminate the reference and sample cells. The cells are formed with a Teflon gasket, which is clamped between the cell prism and a stainless steel reflecting backplate (finely ground to diffuse the light). As the light beam is transmitted through the cell interfaces, it passes through the flowing liquid film and impinges on the surface of the reflecting backplate. This diffuse, reflected light appears as two spots of light that are imaged by lenses onto dual photodetectors. Since the ratio of reflected light to transmitted light is a function of the refractive index of the two liquids, the illumination of the cell backplate is a direct measure of the refractive index of the liquid in each chamber. The cell volume is 3 μL. With mobile phase flowing through both compartments, coarse zero adjustment (equal intensities of light striking the two phototubes) is done by rotating the entire projector assembly. Fine adjustment is done with the optical flat, a glass plate, which can be rotated $\pm 30^\circ$ from normal. This type of refractometer has a relatively limited range. Two different prisms must be used to cover the useful refractive index range (1.33 to 1.63).

Christiansen Effect Detector

The Christiansen effect detector (CED) is a refractive index-dependent detector that passes light through the sample and reference cells in such a way as to observe the difference in the refractive index of the two liquids. Both sample and reference cells are packed with a solid that has the same refractive index as the chromatographic mobile phase. Light is transmitted through the cell as long as the refractive indices of the solid and the liquid remain the same. When a sample is eluted from the column and carried through the cell, the refractive index changes and no longer matches the refractive index of the solid. This change is indicated by a change in the transmission of light and is measured by means of photodetectors as done in dual-beam uv photometers. Solvents with refractive indices greater than 1.44 are matched with various glasses; those with lower refractive indices are matched with various salts or plastics.

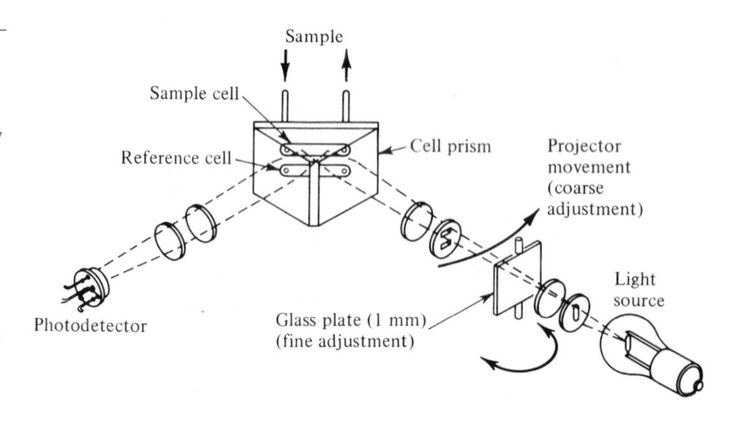

FIGURE **19.15**

Optical diagram of reflection-type (Fresnel) refractometer. (Courtesy of Laboratory Data Control.)

Conductivity Detector

This type of HPLC detector finds use with ion chromatography. The description and general discussion of this detector are found in Section 24.11.

Comparison of HPLC Detectors

Detectors based on the absorption of visible-ultraviolet radiation are virtually insensitive to temperature variations of the test liquid. They may be used with gradients. The advantages of a filter photometer are its relatively low cost and the high sensitivity (nanogram level) achieved for many compounds of chemical and biological interest that absorb ultraviolet or visible light at a fixed wavelength.

A variable-wavelength detector offers a range of wavelengths from 190 to 600 nm, which permits a wavelength to be chosen where the solute absorbance is maximum. It is also possible to select a wavelength that suppresses the absorption of an interfering solute or the mobile phase, yet where the solute has some absorbance. The detector noise levels are greater, usually by a factor of five to ten, than with filter photometers, since the light intensity for each individual wavelength is less.

The simultaneous monitoring of radiation at many wavelengths and the acquisition of the resulting data mean that spectra are available for visual presentation as three-dimensional chromatograms (time vs. wavelength vs. absorbance, Figure 19.16) or storage at any time without disrupting the chromatographic signals. With a single wavelength detector, after one has obtained a chromatogram at the fixed wavelength, one cannot go back and look for other bits of information at other wavelengths. With diode arrays it is possible later to extract data at other wavelengths from the memory. Comparison of absorption spectra with spectra in a user-generated library often gives positive identification of sample components. It is now possible to evaluate a peak for purity by software data manipulation rather than by iteration and refinement of the chromatographic separation. Data

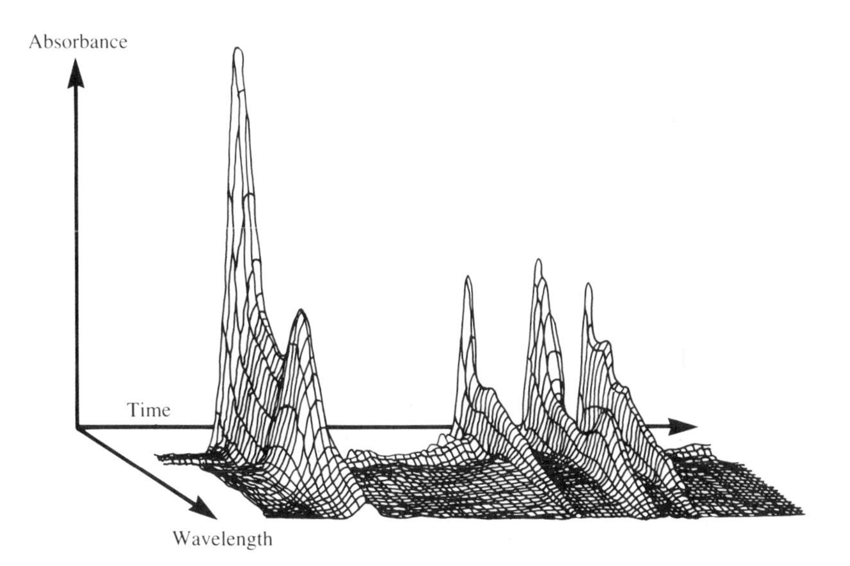

FIGURE **19.16**
Absorbance data as a function of wavelength and time.

manipulation can be a cost-effective substitute for higher chromatographic resolution and can perhaps justify the cost differential between diode arrays ($15,000 to $20,000) and a conventional scanning instrument ($6000 to $9000).

Fluorescence is a means of increasing both the selectivity and the sensitivity of an HPLC analysis (see Chapter 8). Selectivity is enhanced, since not all compounds that absorb radiation fluoresce. Of course, this limits the applicability, although fluorescent derivatives of many nonfluorescing substances can be prepared (see Section 19.3). Typical fluorescing compounds are polynuclear aromatics, steroids, plant pigments, vitamins, alkaloids, catecholamines, aflatoxines, and porphyrins. Sensitivity is improved, since the fluorescence signal is measured against a low background, assuming that the mobile phase does not fluoresce. Fluorescence is frequently 100 to 1000 times more sensitive than absorbance detection (although the two methods are frequently comparable if derivatization is needed) and is approximately 1 ng/mL for strongly fluorescing compounds. The fluorescence detector is not for general use. It is a powerful tool for specific applications with the selective detection of trace components. It may be used with gradients.

Electrochemical detectors have found their greatest applications when polar mobile phases are used. These detectors provide considerable selectivity, since relatively few components in a complex mixture are likely to be electroactive. When applicable, sensitivity is excellent; in many cases a picomole or less can be detected in real time in the eluent. Gradients may be used. Aromatic amines and phenolic compounds are the most important classes of compounds to which electrochemical detection has been applied. These classes include a large number of compounds of biochemical interest.

Differential refractometers can be considered universal, with the exception of a mobile phase that has the same refractive index as a sample component. Limitations of differential refractometers are poor detection sensitivity (typically micrograms), lack of selectivity, and extreme sensitivity to temperature and flow changes. Cell temperature must be controlled within 0.001 °C. Use of a gradient is restricted to a few solvent pairs that have virtually identical refractive indices. The response time is 2 sec.

19.9 INTERFACING HPLC WITH MASS SPECTROMETRY[11–13]

The mass spectrometer is probably the ideal detector for liquid chromatography because it is capable of providing both structural information and quantitative analysis for separated compounds. The problem encountered when interfacing HPLC with mass spectrometry (see Chapter 16) is the mismatch between the mass flows involved in conventional HPLC (about 1 g/min), which are two or three orders of magnitude larger than can be accommodated by conventional mass spectrometer vacuum systems. Another problem is the difficulty of vaporizing involatile and thermally labile molecules without degrading them excessively. Many suggested interfaces involve a considerable compromise in the operating conditions of either the chromatograph or the mass spectrometer. The status of various approaches to overcoming the apparent incompatibility between HPLC and mass spectrometry has been discussed by Arpino and Guiochon[11] and Arpino[12].

Thermospray Method[14,15]

The thermospray technique depends on the thermal generation of a spray and the further separate heat treatment of that spray to yield desolvated ions. The HPLC effluent is fed into a microfurnace, a capillary tube maintained at up to 400 °C that protrudes into a region of reduced pressure (~1 torr) (Figure 19.17). The heat creates a supersonic, expanding aerosol jet that contains a mist of fine droplets of solvent vapor and sample molecules. On their way downstream, the droplets vaporize. The excess vapor is pumped away by an added mechanical pump, which is directly coupled to the ion source. No external ionizing source is required to achieve molecular ions from many nonvolatile solutes at the subnanogram level. Ions of the sample molecules are formed in the spray either by direct desorption or by chemical ionization when used with polar mobile phases that contain appropriate buffers (such as ammonium acetate, 0.0001–0.1M). With weakly ionized mobile phases, a conventional electron beam is used to provide gas-phase reagent ions for the chemical ionization of solute molecules. Chemical ionization spectra are typically accompanied by electron-impact spectra when an electron-beam source is used. The ions formed are led into a quadrupole or magnetic sector mass spectrometer. Flow rates of effluent up to 2 mL/min are permissible.

Monodisperse Aerosol-Generation Interface[16]

The interface is configured in three sections: (1) aerosol generator, (2) desolvation chamber, and (3) two-stage aerosol-beam pressure reducer.

In the aerosol generator the high-pressure effluent from the chromatographic column passes through a small-diameter orifice to form a fine liquid jet. The jet breaks up under natural forces to form uniform drops, which are immediately dispersed with a gas stream introduced at right angles to the liquid flow direction. Coagulation of dispersed drops is minimized by this design.

The solvent evaporates in the desolvation chamber. The chamber is maintained near room temperature by heating gently to replace the latent heat of vaporization necessary for solvent evaporation.

The two-stage aerosol-beam separator consists of two nozzle and skimmer devices. These reduce the pressure from an initial value close to atmospheric

FIGURE **19.17** Schematic of thermospray LC-MS. (Courtesy of VG Instruments, Inc.)

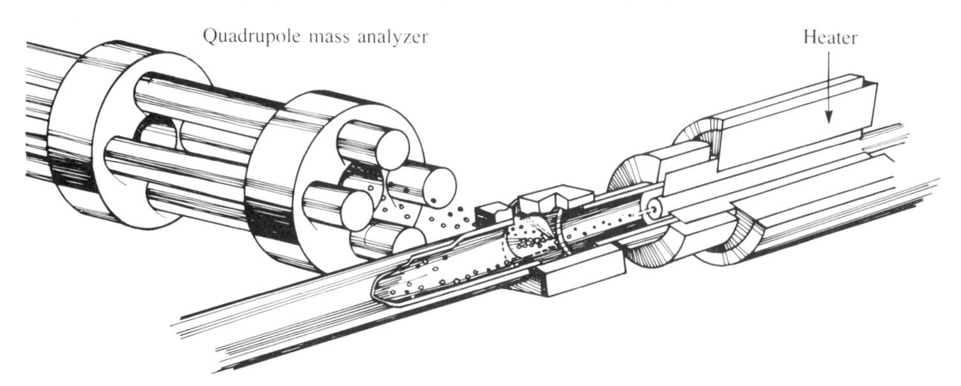

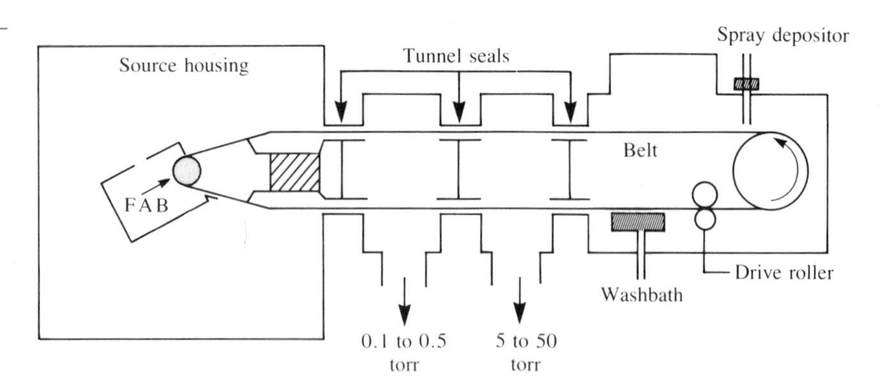

FIGURE **19.18**
Schematic diagram of a moving-belt LC-MS interface. (Courtesy of VG Analytical.)

pressure in the desolvation chamber to a final value close to the pressure in the ion source. The separator also allows solute particles (from the initial aerosol) to be preferentially transferred through the system, while dispersion gas and solvent vapor are pumped away.

Moving-Belt Interface

The schematic diagram of a moving-belt interface is shown in Figure 19.18. The effluent is placed by spray deposition onto a continuous moving belt that is woven from ultrafine quartz fiber. Since the spray deposition is essentially a dry process, no solvent removal is needed. The belt passes through two successive vacuum locks where the pressure is reduced to 0.1 torr before moving into the ion source housing of the fast ion bombardment (FAB) mass spectrometer. After the belt leaves the FAB unit, it exits through the two vacuum chambers (but in the reverse order). Any residual sample or solvent is removed by a wash bath. Earlier versions of the moving belt required a low-temperature infrared heater to evaporate the solvent, a heater to vaporize the sample in the ion chamber of the mass spectrometer, and a powerful clean-up heater to remove any sample residue. This latter type of moving belt interface, though more cumbersome, can be used with magnetic sector or quadrupole instruments and in either the electron-impact or chemical ionization mode.

Comparison of HPLC-MS Interfaces

The development of an interface that allows the use of a wide range of ionization modes is clearly desirable. Electron-impact ionization only recently has been fully exploited for on-line detection largely because of the difficulty in removing solvent from the analyte stream. Only the moving belt and monodisperse aerosol-generation interfaces completely support the electron-impact mode of ionization. Without electron-impact ionization spectra, much structural information is lost along with the ability to identify unknown compounds by comparison with extensive reference libraries of electron-impact spectra.

PROBLEMS

1. What problems may arise from using a poorly packed HPLC column?

2. Prepare a table containing a geometrically similar column set such that for each column the reduced length is 20,000. Do this for particle diameters 1, 2, 3, 4, 5, 7, 10, 12, 15, and 20 μm.

3. Some optimization procedures in LCC originate from the idea of generating the maximum number of plates within the shortest possible time—that is, minimizing t_R/N. (a) Under what conditions can this be achieved? (b) Because of equipment limitations, what does this approach lead to in terms of t_R and ΔP?

4. Express the volume over which the base width of the peak elutes in terms of N, t_M, and k'.

5. For a column exhibiting a plate number of 5000 for a peak whose retention time is 3.0 min, what would be the reduction in plate number for a detector time constant equal to one-tenth of the peak width at baseline (4σ)?

6. Suppose that the volume over which the base width of the peak elutes, V_w, plus the spreading due to the detector volume, V_{det}, is not to be more than 12% greater than V_w alone. Experiment shows that for a typical 8-μL photometer cell into which a 1-μL sample is directly injected by syringe or high-grade injection valve, V_{det} is about 30 μL. (a) Calculate the minimum column volume that will allow these conditions to be met for an unretained solute from a column giving 10,000 plates and packed with 5-μm particles of totally porous silica whose porosity is 0.75. (b) If the column length is 10 cm, what is the minimum column diameter?

7. Estimate the maximum sample volume that should be injected into each of these columns whose porosity is 0.75. Column A: 500 mm by 2 mm, $t_M = 104$ sec, $N = 3200$. Column B: 135 mm by 5 mm, $t_M = 34$ sec, $N = 4450$. Column C: 90 mm by 5 mm, $t_M = 16$ sec, $N = 5000$.

8. Prepare a table giving optimum particle diameters for a column to be operated at a reduced velocity of 3.0. The reduced length is 20,000. Assume t_M values of 10, 30, 100, and 300 sec. Do this for three eluents: 1-propanol ($D_M = 3 \times 10^{-6}$ cm^2 sec^{-1}); water ($D_M = 1.0 \times 10^{-5}$ cm^2 sec^{-1}); and n-hexane ($D_M = 3.0 \times 10^{-5}$ cm^2 sec^{-1}).

9. Calculate the pressure drop required for the individual eluents at each of the values of t_M and d_p tabulated in answer to Problem 8. Do this for solid-core particles ($\phi = 500$) and for porous particles ($\phi = 1000$). Viscosity is, at 25 °C, (in mN sec m^{-2}): 1-propanol, 2.004; water, 0.8903; and n-hexane, 0.313 (at 20 °C).

10. Determine the pressure drop and column length required to maintain $N = 5000$ with $t_M = 100$ sec for particle diameters 2, 3, 4, 5, 7, 10, and 12 μm. Assume that $\phi = 1000$, $\eta = 0.89 \times 10^{-3}$ N m^{-2} sec, and $D_M = 1.0 \times 10^{-5}$ cm^2/sec.

11. Graph the relation between column inlet pressure as a function of column length to obtain a peak with 5000 plates ($k' = 2$) in 300 sec. On the same graph, but with the x-axis extending to the left, graph the column inlet pressure as a function of the packing particle diameter. Use the data previously calculated for Problem 10. Note the trade-off between L and d_p at constant inlet pressure.

12. Calculate the elution time for a nonretained material and the column length required when the pressure drop is constant at 30 or 200 atm and the plate number is 5000, with particle diameters 3, 5, 7, and 10 μm. Assume that $\eta = 10^{-3}$ N sec m^{-2}, $D_M = 1 \times 10^{-5}$ cm^2 sec^{-1}, and $\phi = 500$.

13. Determine the efficiency and plate number of individual columns with particle diameters 3, 4, 5, 7, 10, 12, 15, and 20 μm. All columns are operated at the same pressure drop (20 atm) and elution time ($t_M = 100$ sec). Water is the eluent. Other parameters are $\eta = 10^{-3}$ N sec m^{-2}, $D_M = 1 \times 10^{-5}$ cm^2 sec^{-1}, $\phi = 1000$, $\lambda = 20{,}000$, $A = 1$, $C = 0.05$, and $\gamma = 1$.

14. Prepare a graph of reduced plate height versus reduced velocity from the results of Problem 13. Calculate the physical column length, L, and the average linear velocity, u, for columns packed with particles, 4, 5, and 10 μm in diameter.

15. What is the maximum number of plates that can be obtained from any particle size in a given elution time?

16. Determine the elution time and column length required when the pressure drop is maintained constant at 20 or 100 atm, the plate number is 5000, and the column is packed with particle diameters 3, 4, 5, 7, 10, 15, and 20 μm. All other parameters are the same as in Problem 13.

17. Derive an expression for the maximum number of plates that can be obtained for any particle size given a stated inlet pressure and eluent.

18. To achieve the desired resolution, a change in R_s from 0.8 to 1.25 is required. Assume that the conditions for the starting separation are a 30-cm column of 10-μm particles, reversed-phase LCC (30% acetonitrile: 70% water as mobile phase) at 25 °C, $t_M = 80$ sec, and sample molecular weights are in the range 200–500. Explore each of the possibilites for changing resolution by the required amount: (a) a change in ΔP (or F_c) and (b) a change in L (with ΔP constant). $D_M = 0.56 \times 10^{-5}$ cm^2 sec^{-1}.

19. In gradient elution two sets of conditions may be considered to exist: one to provide maximum resolution, $F_c t/V = 0.70$, and the other to provide minimum analysis time, $F_c t/V = 0.85$. Here t is the time interval for each solvent, and V is the volume of the mixing vessel. The programmer has a 1-min discrimination. Also, $F_c t = 2.5\ V_M$, a condition that ensures that each solvent is used to develop a chromatogram over a k range of 2.5. What should be the volume of the mixing vessel when using a column 25 cm long, a 4.6-mm bore, and with a porosity of 0.72? Do this for flowrates of 1 and 2 mL min^{-1}.

20. The separation of adenosine mono-, di-, and triphosphate nucleotides (AMP, ADP, and ATP) was accomplished in a little over 3 min using 0.4M KH_2PO_4 (plus 3% methanol) and a 15 cm by 2 mm column, packed with 10-μm particles of silica to which was bonded a 3- aminopropylsiloxane phase. The mobile-phase viscosity was 1.4 cPa. The flowrate was 100 mL hr^{-1} at an inlet pressure of 2900 psi. Suggest improvements (with reasons) in the operating procedure.

21. Compare the flowrate through a microbore column of diameter 1.0 mm with the corresponding flow through a 4.6-mm-bore column; both columns are 25.0 cm long. What is the actual flow through the 1.0-mm-bore column over an 8-hr period?

22. For a column packing consisting of 5-μm particles, the reduced plate height translates to about 9000 plates for a column 15 cm long. What would be the plate number for a column 25 cm long packed with the same material?

23. In Problem 22, what is the reduced length of the column 15 cm long? The column 25 cm long?

BIBLIOGRAPHY

HORVATH, C., ed., *High-Performance Liquid Chromatography,* Vol. 1, 1980; Vol. 2, 1980; Vol. 3, 1983; Academic, Orlando, FL.

SCOTT, R. P. W., ed., *Small Bore Liquid Chromatographic Columns: Their Properties and Uses,* John Wiley, New York, 1984.

SIMPSON, C. F., ed., *Techniques in Liquid Chromatography,* Wiley-Heyden, New York, 1982.

SNYDER, L. R., AND J. J. KIRKLAND, *Introduction to Modern Liquid Chromatography,* 2nd ed., Wiley-Interscience, New York, 1979.

YEUNG, E. S., ed., *Detectors for Liquid Chromatography,* Wiley-Interscience, New York, 1986.

LITERATURE CITED

1. GLAJCH, J. L., AND J. J. KIRKLAND, "Optimization of Selectivity in Liquid Chromatography," *Anal. Chem.,* **55,** 319A (1983).
2. KNOX, J. H., "Practical Aspects of LC Theory," *J. Chromatogr. Sci.,* **15,** 352 (1977).
3. MARTIN, M., G. BLU, C. EON, AND G. GUIOCHON, "Optimization of Column Design and Operating Parameters in High Speed Liquid Chromatography," *J. Chromatogr. Sci.,* **12,** 438 (1975).
4. MARTIN, M., C. EON, AND G. GUIOCHON, "Trends in Liquid Chromatography," *Res/Dev.,* p. 24 (April 1975).
5. GIDDINGS, J. C., *Dynamics of Chromatography,* Chap. 2, Dekker, New York, 1965.
6. SNYDER, L. R., M. A. STADALIUS, AND M. A. QUARRY, *Anal. Chem.,* **55,** 1413A (1983).
7. FREI, R. W., H. JANSEN, AND U. A. TH. BRINKMAN, "Postcolumn Reaction Detectors for HPLC," *Anal. Chem.,* **57,** 1529A (1985).
8. KATZ, E., K. OGAN, AND R. P. W. SCOTT, "LC Column Design," *J. Chromatogr.,* **289,** 65–83 (1984).
9. KARGER, B. L., M. MARTIN, AND G. GUIOCHON, "Role of Column Parameters and Injection Volume on Detection Limits in Liquid Chromatography," *Anal. Chem.,* **46,** 1640 (1974).
10. ROSTON, D. A., R. E. SHOUP, AND P. T. KISSINGER, "Liquid Chromatography/Electrochemistry: Thin-layer Multiple Electrode Detection," *Anal. Chem.,* **54,** 1417A (1982).
11. ARPINO, P. J., AND G. GUIOCHON, "LC/MS Coupling," *Anal. Chem.,* **51,** 683A (1979).
12. ARPINO, P. J., *Biomed. Mass Spectrom.,* **9,** 176 (1982).
13. COVEY, T. R., E. D. LEE, A. P. BRUINS, AND J. D. HENION, "Liquid Chromatography/Mass Spectrometry," *Anal. Chem.,* **58,** 1451A (1986).
14. BLAKELEY, C. R., AND M. L. VESTAL, "Thermospray Interface for Liquid Chromatography/Mass Spectrometry," *Anal. Chem.,* **55,** 750 (1983).
15. YANG, L., G. J. FERGUSSON, AND M. L. VESTAL, "A New Transport Detector for High-Performance Liquid Chromatography Based on Thermospray Vaporization," *Anal. Chem.,* **56,** 2632 (1984).
16. WILLOUGHBY, R. C., AND R. C. BROWNER, "Monodisperse Aerosol Generation Interface for Combining Liquid Chromatography with Mass Spectroscopy," *Anal. Chem.,* **56,** 2626 (1984).

20 HIGH-PERFORMANCE LIQUID CHROMATOGRAPHY: METHODS AND APPLICATIONS

20.1 INTRODUCTION

All forms of liquid chromatography are differential migration processes where sample components are selectively retained by a stationary phase. The branches of chromatography were outlined briefly at the beginning of Chapter 17. Liquid chromatographic methods, the topics to be discussed in this chapter, are summarized in Table 20.1 along with the predominant mechanism of each method. Mixed separation mechanisms are frequently used, complicating theoretical interpretations but often enhancing selectivity. The equipment for conducting liquid chromatography is discussed in Chapter 19, also the optimization of column performance. Specific details concerning the particular stationary and mobile phases will be discussed when each HPLC method is treated individually.

Unlike in gas chromatography, the ability to manipulate the character of both the stationary phase and the mobile phase decreases the need for a large number of stationary phases in liquid chromatography. Knowledge of the molecular structure of the sample components can be very helpful in the selection of a liquid chromatographic method. A very general guide for the choice of a method is given in Figure 20.1.

TABLE **20.1** LIQUID CHROMATOGRAPHIC METHODS

Method	Abbreviation	Predominant mechanism
Liquid-solid or adsorption	LSC	Adsorption on surface
Liquid-liquid	LLC	Partition between liquid phases, one mobile and the other stationary
Bonded phase	BPC	Partition and/or adsorption between mobile and bonded phases
Ion pairing	IPC	Separation of ion pairs between mobile and bonded phases
Ion exchange	IEC	Exploitation of charge by adsorption on fixed ionic site via cation or anion exchange
Steric exclusion	EC	Exploitation of size via diffusion into pores by molecules able to enter
Affinity		Use of structure of immobilized ligand to bioselectively bind the desired protein

If the sample is water insoluble, possesses an aliphatic or aromatic character, and has a molecular weight less than 2000 daltons for all of the sample components, adsorption (liquid-solid) chromatography and liquid-liquid chromatography are possibilities. *Adsorption chromatography* works best for class separations or for the separation of isomeric compounds. The technique of *liquid-liquid chromatography* works better for the separation of homologs. Functional groups that are capable of strong hydrogen bonding are retained strongly in adsorption chromatography; however, liquid-liquid chromatography provides an alternate method for the separation of such compounds. These will be samples that possess medium polarity and are soluble in weakly polar to polar organic solvents. In general, the separation of nonelectrolytes in liquid-liquid chromatography is achieved by matching the polarities of the sample and stationary phase and using a mobile phase that has a markedly different polarity.

Ionic groups and ionizable groups suggest the use of *ion-exchange* or *ion-pairing chromatography* when the sample is water soluble and molecular weights are less than 2000 daltons.

When it is known or suspected that the molecular weight exceeds approximately 2000 daltons for some or all of the sample components, then a separation using *exclusion* (gel permeation) *chromatography* is indicated. This method is based on the ability of controlled-porosity substrates to sort and separate sample mixtures according to the size and shape of the sample molecules.

An additional chromatographic procedure depends on the structure of solutes considered as a whole rather than as a function of specific functional groups or of charge or size. *Affinity chromatography* uses immobilized biochemicals as the stationary phase to separate one or a few solutes from hundreds of unretained solutes. The separations exploit the "lock and key" binding that is prevalent in biological systems.

FIGURE **20.1** Guide to liquid chromatography mode selection. (Courtesy of Waters Associates, Inc.)

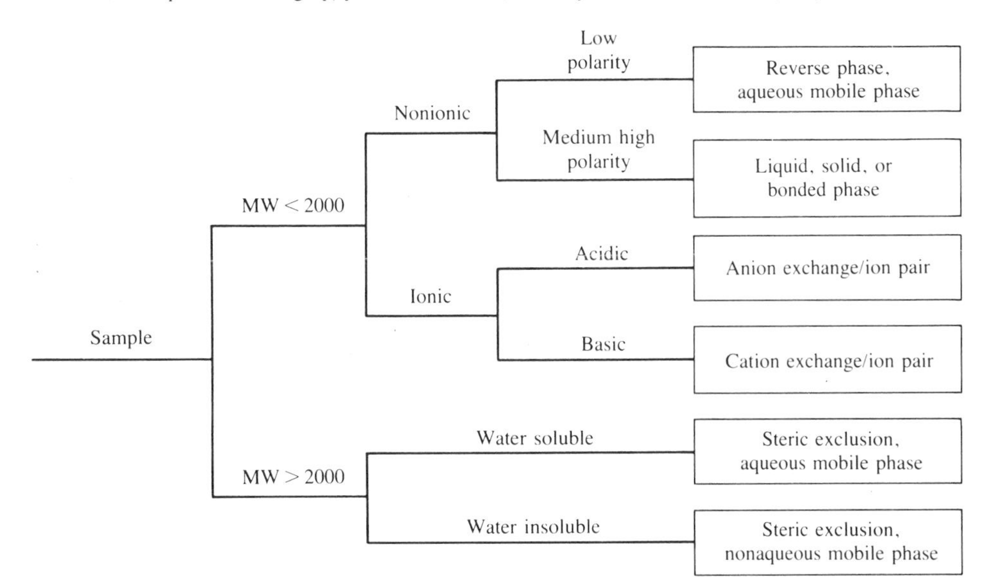

20.2 STRUCTURAL TYPES OF COLUMN PACKINGS

The stationary phase may be either a totally porous particle (or macroporous polymer) or a superficially porous support (porous layer beads or pellicular supports). Either of these types may have a polymer bonded to its surface. These stationary phases are shown in Figure 20.2. Solutes separated by these adsorbent packings are retained almost exclusively on the internal surface of the particle's pores because only a small fraction of the active area is found on the outside of the particles.

Porous Layer Beads

A pellicular or porous layer bead type of packing consists of a solid, spherical glass bead with an average particle diameter of 30–40 μm and a thin, porous outer shell. The outer shell, typically 1–3 μm thick, may be a silica gel layer, a network of small spherical particles bonded to the solid core, or a bonded monomeric or polymeric organic phase. Surface areas of the porous layer beads range from 5 to 15 m^2/g. Columns are easy to pack with these materials because of the dense core, but due to their small surface areas, porous layer beads suffer from limited sample capacity (approximately 0.1 mg/g). Thus the superficially porous column has a sample capacity one-tenth to one-twentieth that of a porous column of equivalent dimensions. If a sample contains seven or more components and/or 0.5 mg or more of the total sample, a porous packing should be used to avoid overloading.

Mass transfer within the stationary phase is greatly improved in a thin porous layer. Consequently, porous layer packings exhibit good efficiency. Longer columns are possible because the pressure drop is lower due to the larger particle size of porous layer supports. Commercial superficially porous packings are offered in several coating thicknesses. Thicker coatings give rise to slower mass transfer but have increased sample capacity.

Porous Particles

The totally porous particle has a large surface area. For silicas, surface areas range from a low of 100 m^2/g to a high of 860 m^2/g, with the average being 400 m^2/g. The mean pore diameter is inversely related to the specific surface area. Silicas with large pores have low specific surface areas, and vice versa. The linear adsorption coefficient of a solute is independent of both these parameters provided that the solute molecule is small enough to enter the pores unimpeded and that the nature of the active surface sites is independent of the pore diameter. Particles can be packed

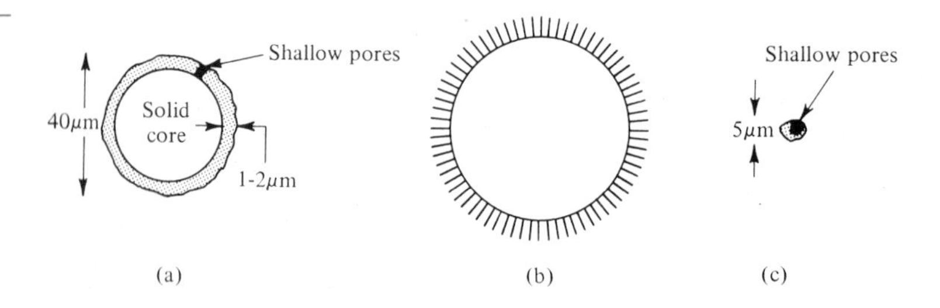

FIGURE **20.2** Stationary phases used in adsorption chromatography: (a) porous layer beads, (b) bonded phase, and (c) porous microparticle.

quite simply into HPLC columns to give efficiencies of up to 800 theoretical plates per centimeter if the 5-μm particle size is used. Columns with larger particles give proportionately smaller numbers of theoretical plates, as was discussed in Section 19.1.

Macroporous Polymers

Macroporous (or macroreticular) styrene-divinylbenzene polymers have large channels, in addition to micropores, which offer the ions easy access to the functional groups of the exchanger. Compared with the classical microreticular cross-linked resins, the macroporous beads do not swell or shrink appreciably with changes in the ionic strength of the mobile phase, or deform at high flow velocities. They are well suited to separations conducted in nonaqueous media.

20.3 ADSORPTION CHROMATOGRAPHY[1]

If the sample is soluble in nonpolar or moderately polar solvents such as hexane, dichloromethane, chloroform, or diethyl ether, then adsorption chromatography is a likely choice. The mechanism for adsorption chromatography involves the interaction between the sample molecule and the stationary phase. This interaction is a competitive situation in which the molecules of the mobile phase and the solute are in competition for discrete adsorption sites on the surface of the column packing. Interaction between a solute molecule and the adsorbent surface is optimum when solute functional groups exactly overlap these adsorption sites. Before considering the adsorption process in more detail, adsorbent and mobile phases must be described.

Adsorbents

Column packings can be described in terms of their chemical composition, whether silica, alumina, or carbon. Silica gel makes up most of the adsorbent column packings, although alumina columns find applications in special situations. Both are general-purpose adsorbents; the silicas are more acidic and therefore good for the separation of basic materials, whereas the aluminas are more basic and therefore good for the separation of acids. Retention and separation on these two adsorbents are generally similar, with the more polar sample components being preferentially held.

Adsorption Processes

Since silica column packings are the great majority of adsorbent columns, the following discussion is limited to silica gel. The main mechanism in adsorption chromatography is the interaction of the hydroxyl groups of silica gel with the polar functional group of a solute molecule (or a solvent molecule). The slightly acidic silanol groups (Si-OH) in silica gel are at the surface and extend out from the surface in the internal channels of the pore structure. The number and topographical arrangement of the several types of hydroxyl groups determine the activity of the adsorbent and thereby the retention of the solutes. These hydroxyl groups can be divided into three types: I, free hydroxyl; II, bound hydroxyl; and III, reactive hydroxyl:

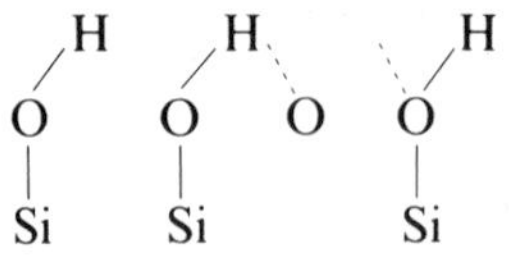

I II III

The activity of the different types of sites increases in this order: bound < free < reactive hydroxyls. These hydroxyl groups interact with polar or unsaturated moieties by hydrogen bonding or dipole interaction.

Current models of the adsorption process assume that the adsorption sites are completely covered by either solute molecules or solvent molecules. Solvent and solute molecules are adsorbed preferentially depending on their relative strength in this competitive interaction. The competition between the solute molecule and the mobile-phase molecule for an active site provides the driving force and selectivity in separations. This is schematically illustrated in Figure 20.3 for possible orientations of phenol on a silica surface in the presence of a weakly adsorbed mobile-phase solvent and a strongly adsorbed solvent. Both the hydroxyl group and the aromatic ring of phenol have adsorption capabilities; the hydroxyl group is more strongly adsorbed due to its greater polarity and its capability to engage in hydrogen bonding. This illustration also points out the role of the mobile phase, which will receive attention in the next section.

Interaction between a solute molecule and the adsorbent surface is best when functional groups overlap adsorption sites. The aliphatic alkane moiety of the solute molecule has little influence on adsorption, and thus there is usually little difference in retention values among molecules that differ only in their aliphatic substituents. In fact, the stronger the mobile-phase interactions with the adsorbent, the less important is the aliphatic part in the adsorption of the solute molecule. Thus, adsorption chromatography is less influenced by molecular weight differences and more by specific functional groups. For compounds of low to moderate polarity, adsorption chromatography often makes possible the separation of complex mixtures into classes of compounds with similar chemical functionality. Examples of group separations are the isolation of polynuclear aromatics from a petroleum sample and the triglycerides from a lipid extract.

FIGURE **20.3**
An adsorbed solute on a silica surface.

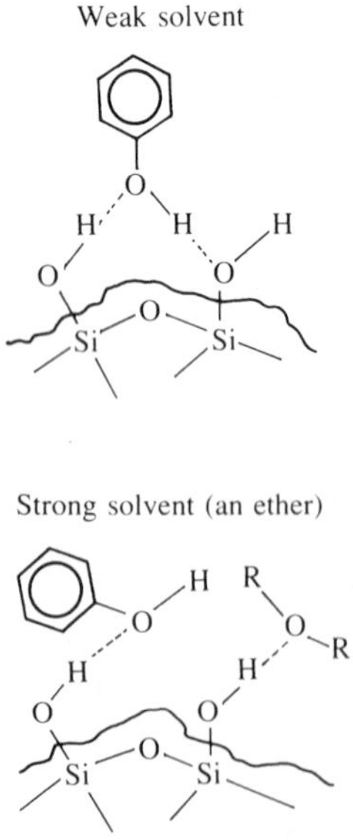

Weak solvent

Strong solvent (an ether)

Mobile Phases

Variations in sample retention for optimum separation in adsorption chromatography are achieved almost exclusively by changes in the composition of the mobile phase. Solvent strength, which controls the k' values of all sample bands, is easily predicted in adsorption chromatography. Snyder defines a solvent strength parameter ε° as the adsorption energy per unit area of adsorbent.[2] Table 20.2 is a list of common solvents used in adsorption chromatography in order of increasing solvent strength. In essence this listing also ranks the adsorption strength of the various functional groups of solute molecules. For a given solute and adsorbent, log k' varies linearly with ε°. An increase in the solvent strength parameter means a stronger eluting solvent and smaller k' values for all sample bands. In terms of the comparable structure of a solute molecule, it means greater retention and the need to select a stronger solvent (or solvent mixture) if elution is to be accomplished in a reasonable period of time.

Initially a solvent can be selected by matching the relative polarity of the solvent to that of the sample components. As a first approximation, a solvent is chosen to match the most polar functional group in the sample—for example, alcohols for the hydroxyl group and ketones or acetates for the carbonyl group. From this first chromatogram, the separation can be refined. If the k' values are too small (sample elutes too rapidly), then a weaker (less polar) solvent is substituted. Conversely, if the sample does not elute in a reasonable time because of high k' values, then a solvent with higher polarity is selected. Two solvents whose solvent strength parameters are, respectively, too small and too large may be blended together in various proportions to allow continuous variation in solvent strength between that of each pure solvent. An increase in solvent ε° by 0.05 unit usually decreases all k' values by a factor of three to four. Thus, the solvents in Table 20.2 allow variation in sample k' values over a range of about 10^{10} for silica and alumina.

TABLE **20.2**

SOLVENT STRENGTH PARAMETER, ε°, AND PHYSICAL PROPERTIES OF SELECTED SOLVENTS

Solvent	$\varepsilon^\circ (SiO_2)$	$\varepsilon^\circ (Al_2O_3)$	**Viscosity, 20°C** ($mN\ sec\ m^{-2}$)	**Refractive index, 20°C**
Pentane	0.00	0.00	0.23	1.358
Hexane		0.00	0.313	1.375
Cyclohexane	−0.05	0.04	0.980	1.426
Carbon disulfide	0.14	0.15	0.363	1.628
Carbon tetrachloride	0.14	0.18	0.965	1.460
1-Chlorobutane		0.26	0.47	1.402
Diisopropyl ether		0.28	0.379	1.368
2-Chloropropane		0.29	0.335	1.378
Benzene	0.25	0.32	0.65	1.501
Diethyl ether	0.38	0.38	0.23	1.353
Chloroform	0.26	0.40	0.57	1.443
Methylene dichloride		0.42	0.44	1.425
Methyl isobutyl ketone		0.43		1.394
Tetrahydrofuran		0.45	0.55	1.407
Acetone	0.47	0.56	0.32	1.359
1,4-Dioxane	0.49	0.56	1.54	1.422
Ethyl acetate	0.38	0.58	0.45	1.370
1-Pentanol		0.61	4.1	1.410
Acetonitrile	0.50	0.65	0.375	1.344
1-Propanol		0.82	2.00 (25°)	1.38
Methanol		0.95	0.60	1.329
Water		Large	1.00	1.333

NOTE: Solvent strength parameter (ε°) values from L. R. Snyder, *Principles of Adsorption Chromatography*, Dekker, New York, 1968.

Binary solvent mixtures offer additional selectivity by fine-tuning the dipole, proton acceptor, and proton donor forces. Relative retention values of sample components can be varied by holding the solvent strength parameter constant and partially exchanging one solvent for another. In practice, this is done with the aid of solvent strength plots such as that in Figure 20.4. In this illustration ε° is plotted across the top. The solvent strengths of various binary solvent combinations with hexane are shown in each of the five horizontal lines below. Each line corresponds to a range from 0% to 100% by volume of that particular solvent in the binary pair. The last line represents the binary mixture of acetonitrile and methanol. For any given solvent strength, Figure 20.4 suggests several binary mixtures. For example, the dashed line in the illustration indicates solvent mixtures with $\varepsilon^\circ = 0.30$. One of these mixtures is 76% by volume of dichloromethane in hexane; another is 49% *tert*-butyl methyl ether in hexane. The presence of ether in the binary mixture would provide donor sites for hydrogen bonding with either the adsorbent or certain solute molecules. Any change in the mobile phase that results in a change in hydrogen bonding between sample and mobile-phase molecules generally results in large changes in relative retention.

Modifiers

Any surface water on silica exists in two forms: (1) chemically combined with the surface as a hydrate, $\geqslant$Si—OH · OH_2, and (2) as molecular water that adsorbs onto strong adsorbent sites. To control adsorbent activity in adsorption chromatography, it is customary to add small amounts of water or other polar modifiers to a nonpolar mobile phase to selectively cover or block the more active sites on the surface, leaving a more uniform population of weaker sites that then retain the

FIGURE **20.4** Binary solvent strengths on silica gel for mixtures of hexane with 2-chloropropane, dichloromethane, *tert*-butyl methyl ether, acetonitrile, or methanol; also for the binary system of acetonitrile with methanol. [Reprinted with permission from D. L. Saunders, *Anal. Chem.*, **46,** 470 (1974). Copyright 1974 American Chemical Society.]

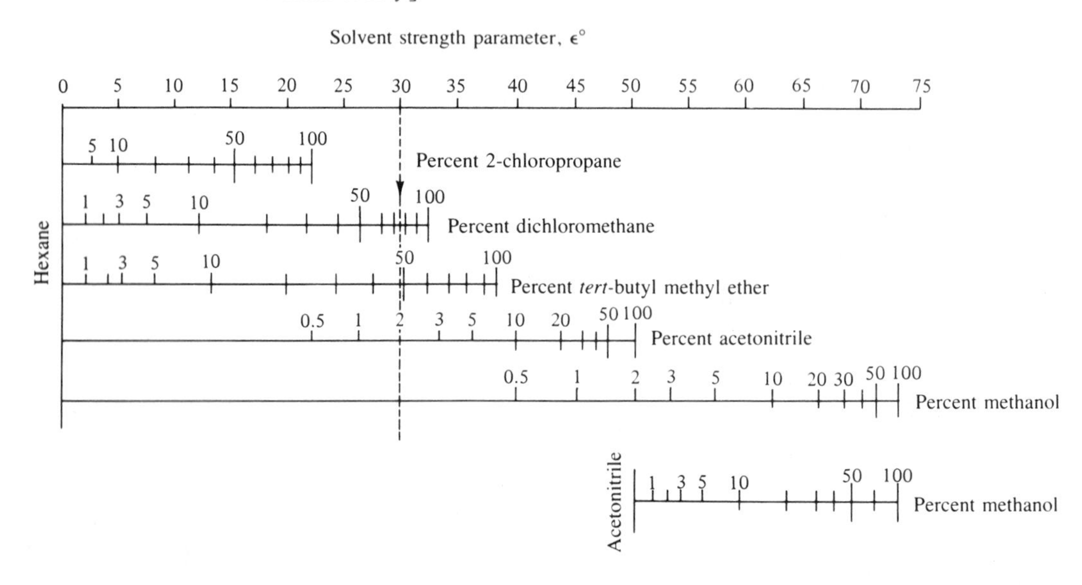

sample. This deactivation of the adsorbent improves subsequent separations. Among the beneficial effects are substantial increases in the sample capacity and an increase in the linear capacity of 5- to 100-fold. The linear capacity of an adsorbent is defined as the maximum weight of sample that can be applied to a gram of adsorbent before the adsorption coefficient falls more than 10% below its linear isotherm value (Henry's law).

The amount of water in the mobile phase should ideally be at about 50% water saturation when packed silica columns are used. In actual practice it is difficult to avoid small changes in the mobile-phase water content. For water-miscible mobile phases or in gradient elution, 0.05%–0.2% acetonitrile or methanol is commonly added as a water substitute.

The addition of modifiers such as 1%–2% of tetrahydrofuran to acetonitrile or methanol sharpens peaks by reducing peak tailing.

Applications

Adsorption chromatography is an analytical method that is particularly suitable for the analysis of nonionizing, water-insoluble, and relatively simple molecules, which are often isomers or very closely related compounds. The range of compound types that can be separated extends from very nonpolar hydrocarbons to strongly polar multifunctional compounds. The order of adsorption follows the general polarity scale for various classes of compounds, increasing from top to bottom of the listings, as outlined in Table 20.2. Not included in the tabulation are sulfides (less polar than ethers), nitrocompounds (more polar than ethers), aldehydes (similar to esters and ketones), amines (similar to alcohols), and sulfones, sulfoxides, amides, and carboxylic acids (between the alcohols and water, with their polarities increasing in the order listed).

Generally the most polar group of a polyfunctional compound governs its adsorption characteristics. Often only one functional group is geometrically positioned with respect to the adsorption site. However, certain polyfunctional solutes are better matched to the adsorbent surface than their isomeric counterparts. Thus the method excels in the separation of positional isomers (Figure 20.5).

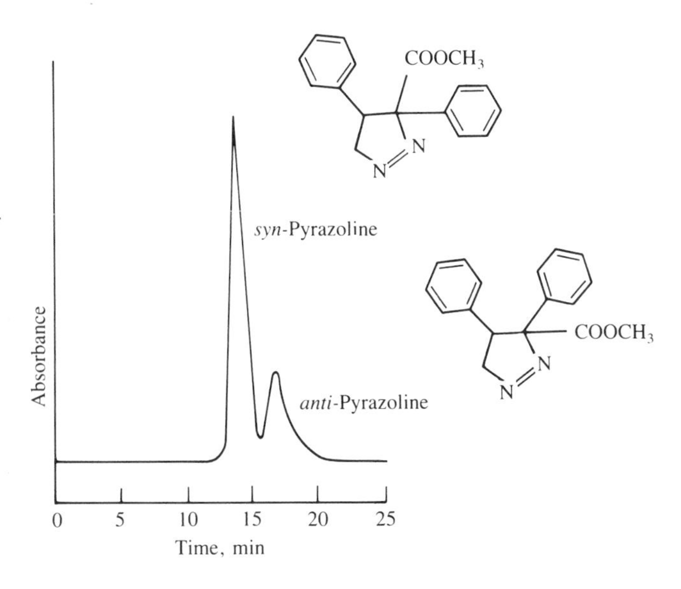

FIGURE **20.5**

Separation of *syn*- and *anti*-pyrazoline isomers on a 2 mm × 10 cm column of pellicular silica. Mobile phase: hexane/dichloromethane (50:50).

FIGURE 20.6

Separation of polynuclear aromatic hydrocarbons on a 4.6 mm × 250 mm column of totally porous, spherical silica. Mobile phase: acetonitrile/water (70:30). Peaks: (1) naphthalene, (2) fluorene, (3) phenanthrene, (4) anthracene, and (5) pyrene.

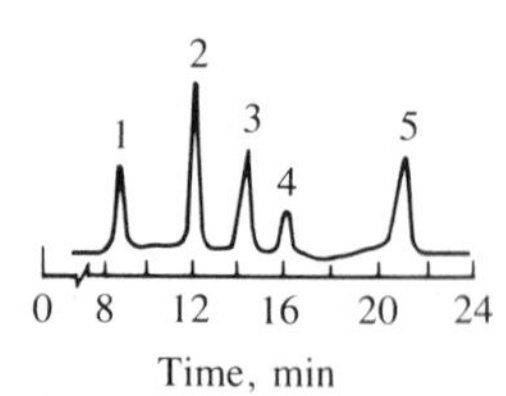

Adsorption chromatography is less influenced by molecular weight differences and more by specific functional groups than, say, liquid-liquid chromatography. Consequently, the separation of compounds that differ only in the degree or type of alkyl substitution, such as members of an homologous series, is usually poor by adsorption. However, even this shortcoming can be used to advantage if the isolation of all members of a particular class of compounds is desired. For example, adsorption chromatography can be used to isolate the polynuclear aromatics from a petroleum sample (Figure 20.6) or the triglycerides from a lipid extract.

The assay of various pharmaceutical products receives heavy emphasis; examples are the separation of vitamin D_3 and its metabolites, vitamins A, D, and E (and compounds closely related to these vitamins), many drugs of abuse (LSD, for example), tricyclic antidepressants, beta-blockers, and the PTH-amino acids. Natural oils and flavor extracts are readily assayed, and the less polar plant pigments, such as carotenoids and porphyrins, have been successfully separated for several decades. Esters are also amenable to separation with glycerides and phthalate esters typical of this class of compounds. Aflatoxins and other mycotoxins have also been separated on silica columns.

20.4 BONDED-PHASE CHROMATOGRAPHY

The most widely used column packings for liquid-liquid partition chromatography are those with chemically bonded, organic stationary phases. They replace the classical packings in which the stationary liquid phase coated a support material (as described for gas-liquid chromatography in Chapter 18). Partition occurs between the bonded phase and a mobile liquid phase.

Preparation of Bonded-Phase Supports[3]

Bonded-phase supports are made from silica by the covalent attachment of an organic hydrocarbon moiety to the surface. Supports include large porous silica gels, porous layer beads, and microparticles. The latter dominate. Bonded-phase packings are quite stable because the stationary phase is chemically bound to the support and cannot be easily removed during use.

The *siloxane* (Si—O—Si—C) type of bond has become the standard for commercial bonded phases. It is formed by reactions of di- (or tri-) alkoxysilane with the surface silanol groups of fully hydroxylated silica gel, as either pellicular or totally porous supports. Equation 20.1 shows in simplified form a typical chlorosilane bonding reaction with a silanol group on the silica gel surface:

$$\geq Si{-}OH + ClSi(CH_3)_2R \rightarrow \geq Si{-}O{-}Si(CH_3)_2R + HCl \quad (20.1)$$

Bonded phases of this type are stable to hydrolysis throughout the pH range 2–8.5

In most cases the actual surface topology of bonded-phase materials on the molecular level is not known. It is unlikely that the octyl or octadecyl hydrocarbonaceous portion would fully extend into the mobile phase. When wetted by the mobile phase, monomeric phases respond rapidly to changes in mobile-phase composition. Lack of wetting causes poor efficiency, with adsorption occurring at the sorbent-solvent interface in addition to the expected liquid-liquid partition equilibrium.

For steric reasons, it is not possible for all silanol groups to react during the chemical bonding step. Consequently, the silica surface consists of a mixture of silanols and hydrophobic chains. Many remaining silanols can be removed by treatment with chlorotrimethylsilane, a small silylating agent that is able to penetrate to the location of the unreacted silanols.

Polymerized Multilayer Bonded Coatings

The polymeric type is prepared from silica using a bifunctional (X_2SiR_2) or trifunctional reactant in the presence of excess water adsorbed on the silica surface. After one or two of the functional groups have reacted with silanol groups on the surface of the gel, the remaining functional groups react with water to form more silanol groups. These new silanol groups in turn react with more reagent to form a cross-linked polymeric structure.

For very small-surface-area silicas, polymerization of the reactants may be required to provide sufficient coverage and adequate sample capacity. For satisfactory chromatography with polymeric layers, the presence of even a small amount of polar solvent appears to be important. The polar solvent apparently penetrates into the polymeric layer, swelling it, and establishing a partitioning phase of solvated polymer.

From a mass transfer point of view, a monomeric layer consisting of "bristles" of alkyl chains is generally preferred to a polymeric layer of disiloxane linkages. The polymer layer must swell before sample molecules are able to penetrate; once within, diffusion in the stationary phase is slow.

Functional Groups in Bonded-Phase Supports

By varying the nature of the organic portion of the bonding silane, the surface polarity of the bonded-phase packing can be altered from an essentially hydrophobic one, consisting of a hydrocarbonaceous layer, to one of various functional groups placed on the outer end of the hydrocarbon moiety. Although several functional groups could be used, a vast number of bonded stationary phases are not necessary because the mobile-phase composition can be altered easily to change the selectivity.

A linear hydrocarbon chain is a popular bonded phase. The alkyl group can be a variety of chain lengths, usually an ethyl (C-2), octyl (C-8), or octadecyl (C-18) group. Bonded octyl packings represent a good compromise for the separation of compounds with low to high polarity and samples with wide-ranging polarities. The octadecyl packing can be used for applications in which maximum retention is required. Figure 20.7 compares chromatograms obtained on columns of octadecyl and octyl packings under the same mobile-phase conditions. A variety of functional groups are represented in the sample. The ethyl group is useful in applications that involve very strongly retained solutes. Bonded alkyl phases permit rapid analyses and rapid reequilibration when the mobile phase is altered, as in solvent programming.

Bonded phases of medium polarity have nitrile functions, such as cyanoethyl ($—CH_2CH_3CN$) groups bonded to the silica surface. Nitrile-bonded phases are useful in separations involving ethers, esters, nitro compounds, double-bond

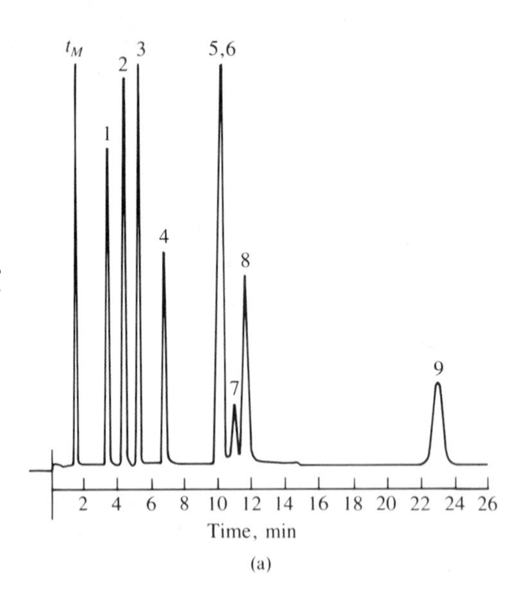

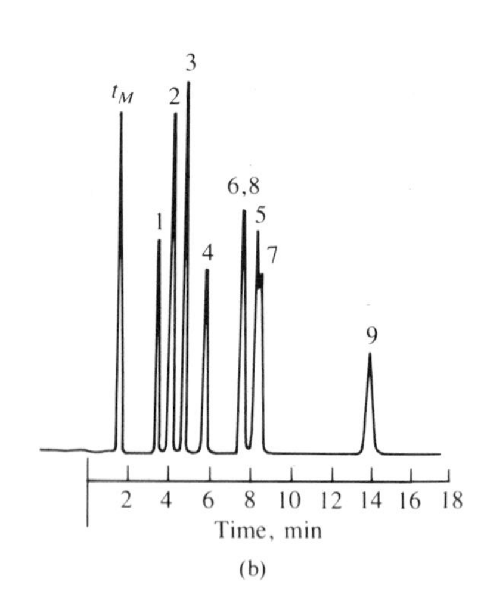

FIGURE **20.7**
Comparison chromatograms obtained on columns of octadecyl (a) and octyl (b) packings under the same mobile-phase conditions (50:50 methanol/water). Peaks: (1) phenol, (2) benzaldehyde, (3) acetophenone, (4) nitrobenzene, (5) methyl benzoate, (6) methoxybenzene, (7) fluorobenzene, (8) benzene, and (9) toluene.

FIGURE **20.8**
Chemical structure of a chiral-urea bonded phase, Supelcosil® LC-(R)-Urea.

isomers, and ring compounds that differ in double-bond content. These are functional groups lying in the middle of Table 20.2.

The bonded phenylsilane functional group provides a material with unique selectivity that allows separations that rely on the interaction of aromatic components with the stationary phase. For certain systems, where $\pi-\pi$ interactions may be important, significantly different separations are obtained with the bonded phenyl phase.

Bonded phases that contain the aminoalkyl functional group are used in applications requiring a highly polar surface (see the separation of sugars in cider, Figure 20.13). This functional group also imparts unusual selectivity. It may function as either a Bronsted acid or base, depending on the solute, or interact with solutes by hydrogen bonding. In a water/acetonitrile mobile phase, polar compounds such as carbohydrates and peptides can be separated. In acidic solution the aminoalkyl group behaves as a weak anion exchanger.

Chiral-phase HPLC columns are a relatively new but invaluable tool for resolving optically active isomers. Figure 20.8 illustrates the chemical structure of a chiral-urea bonded phase. Resolution is determined by the sum of the interactions between functional groups in the optically active sample components and an optically active bonded phase. The polar urea group can form hydrogen bonds or enter into other dipole-dipole interactions. The phenyl group can enter into $\pi-\pi$ interactions with aromatic compounds or other compounds that have π-electron-containing functional groups. The plasma protein α_1-acid glycoprotein has been immobilized on silica microparticles and used as a bonded phase; hydrophilic solutes are retained.

Mobile Phases

The stability of the Si—O—Si bond involved in bonded phases is dependent on the mobile-phase pH. Above about pH 7, the bond is hydrolyzed and serious degradation of the column stationary phase results.

In general, mobile phases with no or little polarity (top of Table 20.2) are used with the polar bonded phases. These are hydrocarbons or hydrocarbons that contain some modifier. Mobile phases for the nonpolar bonded phases are selected from solvents with high polarities (bottom of Table 20.2). Based on initial experiments with a particular column and solvent system considered appropriate for the separation, the mobile-phase composition is adjusted to change the overall retention time and/or to pull specific peak pairs apart. Solvent optimization searches involving three or even four solvents have been described.[4]

Although we speak of a solvent's polarity as if it were a fixed inherent property, it does not accurately reflect the eluting strength for all solutes. This is because solvent strength is the total of all types of molecular interactions acting concurrently—for example, dispersion, orientation, and hydrogen bonding. Solvent strength is particularly high when there is a good interactive match between solvent and solute. Thus the separation of solutes with different functional groups may be improved by the use of ternary mobile phases for the precise control of mobile-phase eluent strength, perhaps in conjunction with solvent programming.

Multiple solutions to a separation problem using ternary mobile phases are shown in Figure 20.9. If the desire of the chromatographer is complete separation of solute peaks in a minimum analysis time, then the mobile-phase mixture indicated for the first chromatogram is the method of choice. If space on the chromatogram is needed for an internal standard, then one of the next two mixtures is optimum. If reaction products or impurities are anticipated in some samples to be analyzed, then one of the last two methods best meets the needs of the chromatographer.

Ion Suppression

When acids or bases are to be separated, the solvent pH has to be controlled by using a buffer. Adjusting the pH on the acid side suppresses the ionization of acids, allowing the acids to be separated by the reverse-phase method. This approach is useful for weak acids and bases in the pH range 2–8. It is advisable to maintain a relatively high buffer concentration to facilitate a rapid establishment of the protonic equilibria. The use of an alkaline eluent is precluded because siliceous bonded phases are not generally stable at pH values higher than 8.

Normal-Phase Liquid-Liquid Chromatography

Normal-phase liquid-liquid chromatography uses a polar stationary phase (often hydrophilic) and a less polar mobile phase. To select an optimum mobile phase, it is best to start with a pure hydrocarbon mobile phase such as heptane. If the sample is strongly retained, the polarity of the mobile phase should be increased, perhaps by adding small amounts of methanol or dioxane. Figure 20.10 shows three aromatic compounds separated on a nitrile-bonded phase; the normal-phase mode was used. Either 2-propanol or tetrahydrofuran is the strong component in the binary mobile phase in this example.

In the normal-phase mode, separations of oil-soluble vitamins, essential oils, nitrophenols, or more polar homologous series have been performed using alcohol/heptane as the mobile phase. Another application is shown in Figure 20.10.

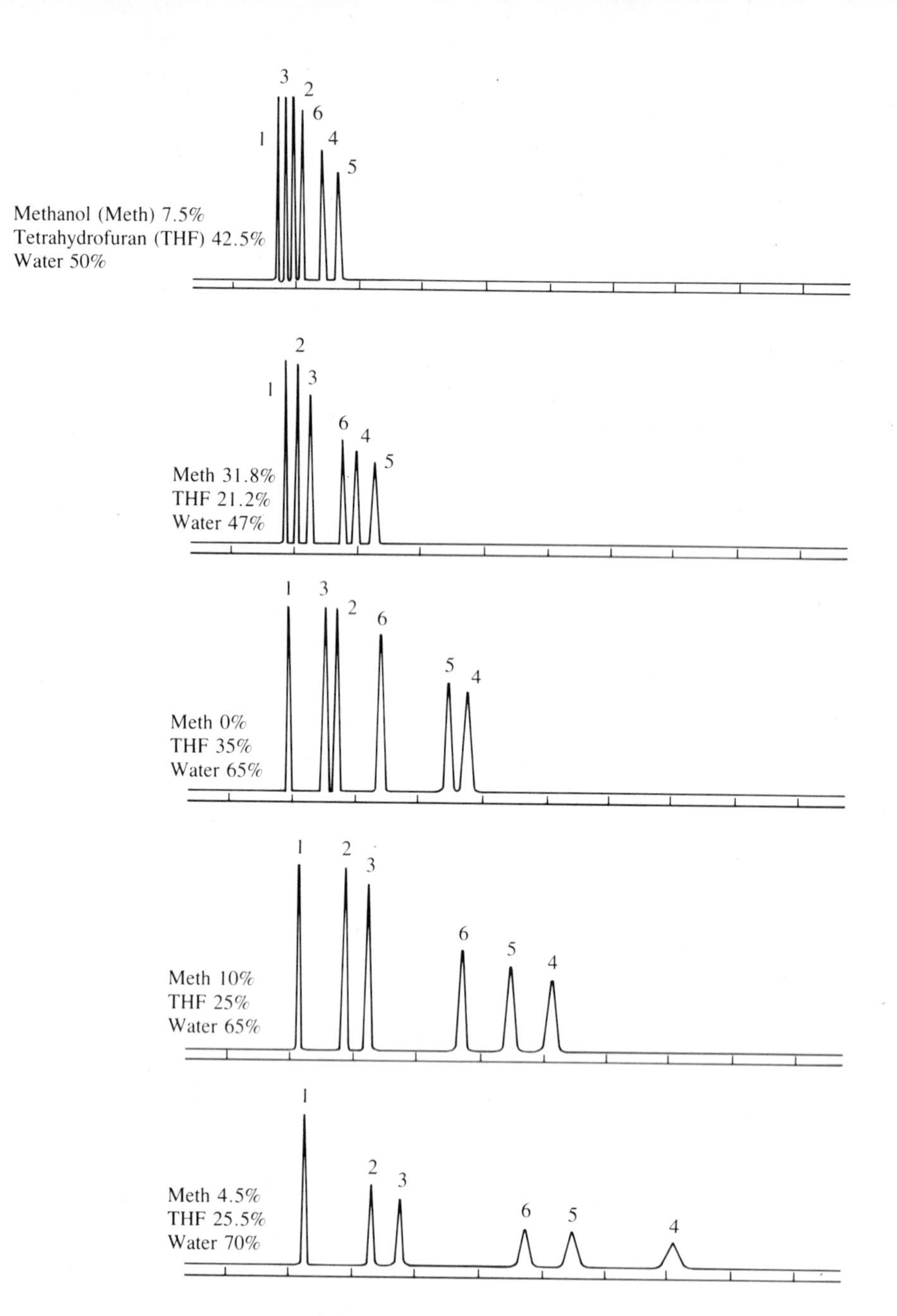

FIGURE **20.9**

Separations representative of five ternary mobile phases. Peaks: (1) benzyl alcohol, (2) phenol, (3) 3-phenylpropanol, (4) 2,4-dimethylphenol, (5) benzene, and (6) diethyl *o*-phthalate. [After R. D. Conlon, "The Perkin-Elmer Solvent Optimization System," *Instrumentation Research,* p. 95 (March 1985). Courtesy of Instrumentation Research.]

20.5 REVERSE-PHASE CHROMATOGRAPHY[5–8]

Reverse-phase chromatography uses a hydrophobic bonded packing, usually with an octadecyl (C-18) or octyl (C-8) functional group and a polar mobile phase, often a partially or fully aqueous mobile phase (Figure 20.11). Polar substances prefer the mobile phase and elute first. As the hydrophobic character of the solutes increases, retention increases. Generally, the lower the polarity of the mobile phase, the higher

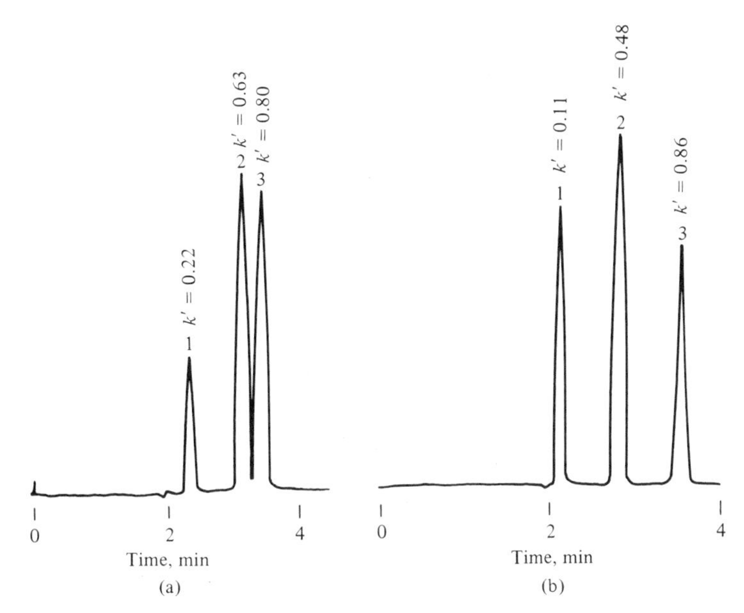

FIGURE **20.10**

Selectivity changes through changes in mobile-phase components in normal-phase liquid chromatography on a nitrile-bonded phase: (a) 25% 2-propanol, 75% cyclohexane; (b) 25% tetrahydrofuran, 75% cyclohexane. Peaks: (1) anisole, (2) nitrobenzene, (3) dimethyl *o*-phthalate.

is its eluent strength. The elution order of the classes of compounds in Table 20.2 is reversed (thus the name reverse-phase chromatography). Hydrocarbons are retained more strongly than alcohols. Also, the eluent strength of the various solvents in reverse-phase chromatography follows approximately the reverse order given in Table 20.2. Thus water is the weakest eluent. Methanol and acetonitrile are popular solvents because they have low viscosity and are readily available with excellent purity. Eluents intermediate in strength between these solvents and water are usually obtained by preparing mixtures. Since the optimum composition of the mobile phase must generally be found by trial and error, it is convenient to start with a 1:1 water/methanol mixture. If the sample components elute at or near the transit time of a nonretained solute, t_M, a lower concentration of methanol is indicated. Changing to acetonitrile, dioxane, or mixtures of 1,4-dioxane/methanol or acetonitrile/2-propanol can often improve selectivity. In reverse-phase chromatography, solvent gradients are generated by a continuous decrease in the polarity of the eluent during the separation—for example, by gradually increasing the organic solvent content in water/methanol or water/acetonitrile mixtures.

FIGURE **20.11**

Silica-based column packing material and mobile-phase solvents for reverse-phase chromatography.

Sample	Column packing	Mobile phase
Low/moderate polarity (soluble in aliphatic hydrocarbons)	Bonded C-18	Methanol/water
Moderate polarity (soluble in methyl ethyl ketone)	Bonded C-8	Acetonitrile/water
High polarity (soluble in lower alcohols)	Bonded C-2	1,4-Dioxane/water

Proposed Mechanism of Reverse-Phase Chromatography

Now let us consider the proposed mechanism of reverse-phase chromatography. When a solute dissolves in water, the strong attractive forces between water molecules become distorted or disrupted. These attractive forces arise from the three-dimensional network of intermolecular hydrogen bonds. Only highly polar or ionic solutes can interact with the water network. Nonpolar solutes are "squeezed out" of the mobile phase but bind with the hydrocarbon moieties of the stationary phase. In reverse-phase chromatography the driving force for retention is *not* the favorable interaction of solute with the stationary phase, but the effect of the mobile-phase solvent in forcing the solute into the hydrocarbonaceous bonded layer. In opposition is the interaction of the solute's polar groups with the mobile phase. As a result, hydrophobic retention involves mainly nonpolar substances or the nonpolar portions of molecules. Hydrophobic retention can be lessened by adding to water any organic solvent that is miscible with water. The less polar the added organic solvent, the greater the effect. Thus one speaks of the added organic modifier as the *stronger* solvent and water as the *weaker* solvent. As an example, better resolution of homologs, for a given stationary phase, is achieved as the water content of the mobile phase is increased.

Trace Enrichment

Trace enrichment is readily adaptable to reverse-phase chromatography. An injection solvent (often water) is used that is significantly weaker than that of the mobile phase. A preconcentration of the solute occurs at the top of the column as the mobile phase at the column entrance is converted to that of the weaker injection solvent. Gradient elution subsequently separates the sample ingredients.

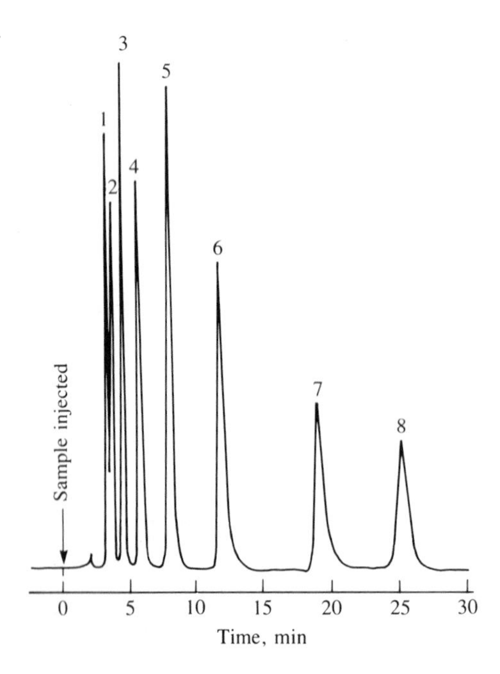

FIGURE **20.12**
Chromatogram of *o*-phthalate esters run on a column with an octadecyl packing and an eluent consisting of methanol/water (90:10). Peaks: (1) dimethyl, (2) diethyl, (3) dipropyl, (4) dibutyl, (5) dipentyl, (6) dihexyl, (7) diheptyl, and (8) dioctyl.

FIGURE 20.13

Chromatograms of sugars in cider run on an aminoalkyl packing with acetonitrile/water (70:30) as eluent. Peaks: (1) solvent, (2) unknown, (3) fructose, (4) mannose, (5) glucose, and (6) sucrose.

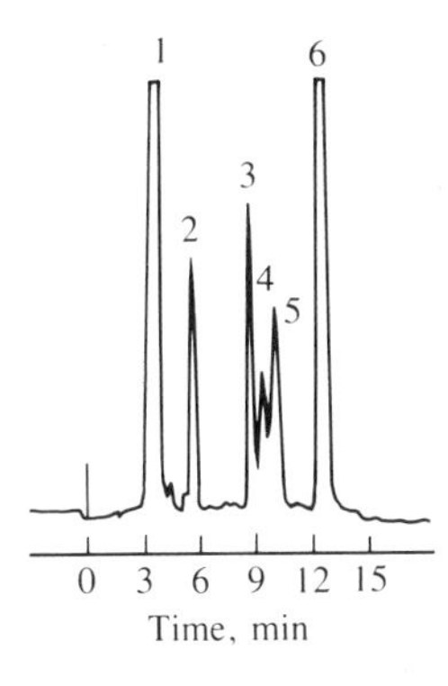

Applications of Reverse-Phase Chromatography

The reverse-phase technique in its various forms is the most widely used mode in HPLC and comprises nearly half of all the liquid chromatographic methods described in the literature. This technique is the most likely to provide optimum retention and selectivity when compounds have no hydrogen-bonding groups or have a predominant aliphatic or aromatic character (see Figure 20.7). The method is well suited for separating solutes based on the size and structure of alkyl groups; this is illustrated in Figure 20.12 for a series of phthalate esters.

In clinical chemistry the quantitative analysis of drugs of abuse is increasingly carried out by this technique. Pharmaceuticals routinely analyzed include barbiturates, antiepileptic drugs, analgesics, and sedatives. Food preservatives, herbicides, and sugars (Figure 20.13) are additional applications for reverse-phase methods. In the pharmaceutical field the use of this technique has been increasing at the expense of adsorption chromatography. Because the water content of the mobile phase can range from 100% to quite low percentages or none at all, a broad spectrum of biomolecules can be chromatographed, lipophilic or ionic, small or large. Lipophilic compounds such as triglycerides, which have very poor solubility in aqueous reverse-phase solvents, often can be separated by a nonaqueous reverse phase using an octadecyl-packed column with polar organic solvents such as acetonitrile or tetrahydrofuran.

20.6 ION-PAIR CHROMATOGRAPHY[9,10]

Ion-pair chromatography, which can be considered a subset of reverse-phase chromatography, can deal with ionized or ionizable species on reverse-phase columns. The method overcomes difficulty in handling certain samples by the other liquid chromatography methods; these are samples that are very polar, multiply ionized, and/or strongly basic. In ordinary reverse-phase HPLC, organic ions show poor peak shapes and inadequate retention. One antidote, ion suppression, was discussed in Section 20.4, but it may not succeed if there is more than one ionizable component in the sample. The ion suppression method is also limited to the pH range 2.0–7.5 by the instability of stationary bonded phases outside this pH range. Ion-exchange packings offer limited choices with little ability to vary selectivity by changing the column packing.

Ion-Pair Reagent

In ion-pair chromatography an ion-pair reagent (a large organic counterion) is added at low concentration (usually 0.005M) to the mobile phase. The ion-pair reagent is itself ionized. One ion of the reagent is retained by the stationary phase, thus providing the otherwise neutral stationary phase with its charge. This charged stationary phase can then retain and separate organic solute ions of the opposite charge by forming a reversible ion-pair complex (a coulombic association species formed between two ions of opposite electrical charge) with the ionized sample as represented by the following equilibrium:

$$RCOO^- + R_4N^+ \rightleftharpoons [R_4N^+, {}^-OOCR]^0 \text{ ion pair} \tag{20.2}$$

Here it is assumed that the solute ion is a carboxylate anion, $RCOO^-$, and that the counter ion is a quaternary ammonium ion, R_4N^+. Thus with a suitable counterion, ionic or ionizable compounds can be converted to electrically neutral compounds that will partition between the mobile and nonpolar stationary phases. At the same time, the stationary phase will not have lost any of its ability to retain and separate nonionized organic substances.

Mechanism of Ion-Pair Chromatography

The exact mechanism for ion-pair chromatography has not been clearly established. There are two fundamental models. The first postulates that the solute molecule forms an ion pair with the counterion in the mobile phase. This uncharged ion pair then partitions into the lipophilic stationary phase (Figure 20.14a). The other mechanism postulates that the counterion partitions into the stationary phase, or is "loaded" onto the bonded reverse-phase packing, with its ionic group oriented at the surface. This produces two possibilities for the material to be chromatographed. It can be attracted to the hydrocarbon portion in the usual reverse-phase manner, or it can interact in an ion-exchange mode (Figure 20.14b). It is quite likely that the true mechanism involves both postulates but is further complicated by adsorption and micelle formation. Whatever the mechanism, ion-pair chromatography

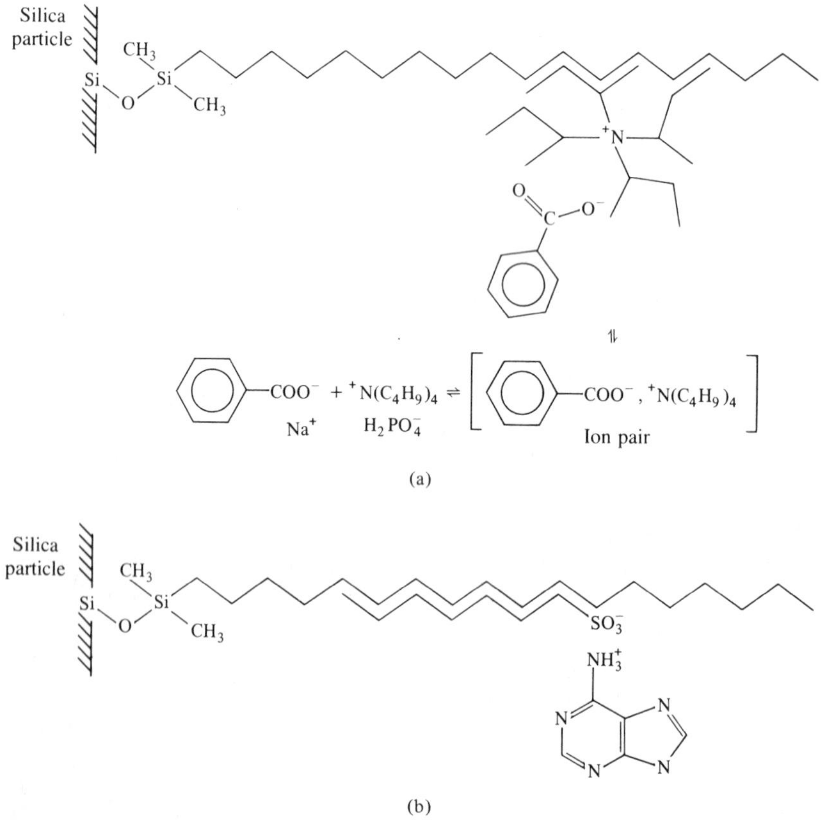

FIGURE **20.14**
Ion-pair chromatography: (a) ion pair partitions into the lipophilic stationary phase (bonded C-18) and (b) counterion (alkyl sulfonate) loaded onto stationary phase.

allows for unique separations not otherwise obtainable by either reverse phase or ion exchange.

Factors Influencing Retention

The extent to which the ionized sample and counterion form an ion-pair complex as well as the binding strength of the complex with the stationary phase affects the retention. Adjusting the pH so that some or all of the sample components are present in their ionic form and choosing the particular counterion and its concentration are under the control of the analyst. The relative size of the lipophobic group on the counterion affects the degree of retention. The longer the alkyl chain, the denser is the ion population in the stationary phase, which, in turn, yields greater retention of given ions. For example, for basic samples, alkyl sulfonates adjusted to a pH of 3.5 are used often as the mobile-phase counterion:

$$RNH_3^+ + RSO_3^- \rightleftharpoons [RNH_3^+, \,^-O_3SR] \tag{20.3}$$

When heptanesulfonic acid is used, as opposed to pentanesulfonic acid, retention for the ionic sample components is increased (Figure 20.15). Increasing the concentration of counterion increases retention up to a limit that is usually set by the solubility of the counterion in the mobile phase, particularly when an organic modifier is used in the system. Changing the reagent concentration is usually reserved to fine-tune a separation.

A range of reagents is needed to match the range of ionic retentions. For the positive organic ions, the sulfonates are available with carbon chain lengths extending upward from C-5. The triethylalkyl quaternary amines provide a corresponding range of reagents for negative organic ions. The molecular shape of the quaternary amine is deliberately unsymmetrical to provide a better association of the long alkyl chain with the paraffinic surface of the bonded C-18 stationary phase.

Alkyl sulfates (for example, dodecyl sulfate) behave similarly to alkyl sulfonic acids but yield different selectivities when used as the counterion. Even perchloric acid forms very strong ion pairs with a wide range of basic solutes.

The control of the pH is a most important parameter. Retention is increased as adjustment of the pH maximizes the concentration of the ionic form of the solutes. Maintaining a pH around 2.0 ensures that both strong and weak bases are in their protonated ionic forms, and any weak acids present are primarily in their nonionic forms. For acidic samples, a quaternary amine is recommended as the counterion. A phosphate buffer keeps the mobile phase at a pH of about 7.5. At this pH both strong and weak acids are in their ionic forms and weak bases are in their nonionic form.

Experimental Conditions

A good first choice of a bonded-phase column is a monolayer C-18 or C-8 column. One starts with a $0.01M$ concentration of a shorter-chain counterion, or if the counterion contains a decyl or longer-chain alkyl groups, one prepares a $0.005M$ solution. The most common solvent combinations are water/methanol and water/acetonitrile. Although acetonitrile offers better column efficiencies due to its lower viscosity, its usefulness is limited by the poor solubility of many ion-pair reagents in it. A reasonable concentration of any buffer required is 0.001–$0.005M$.

Buffer components should be selected with poor ion-pair properties but good solubilities.

Unlike conventional ion exchange, ion-pair partition can separate nonionic and ionic compounds in the same sample. One first optimizes the separation of the nonionic solutes, then selects and adds a counterion to the mobile phase, whereby

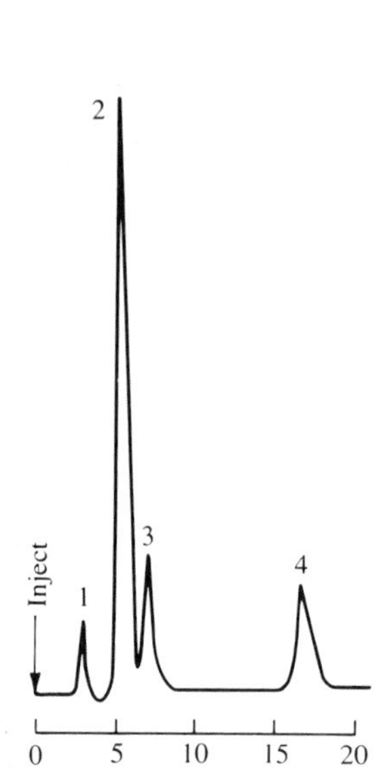

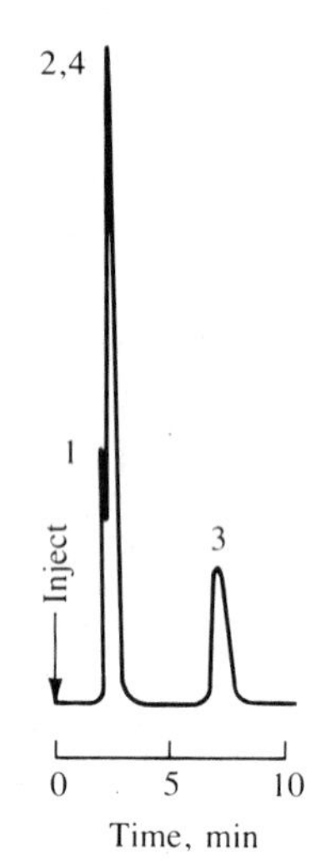

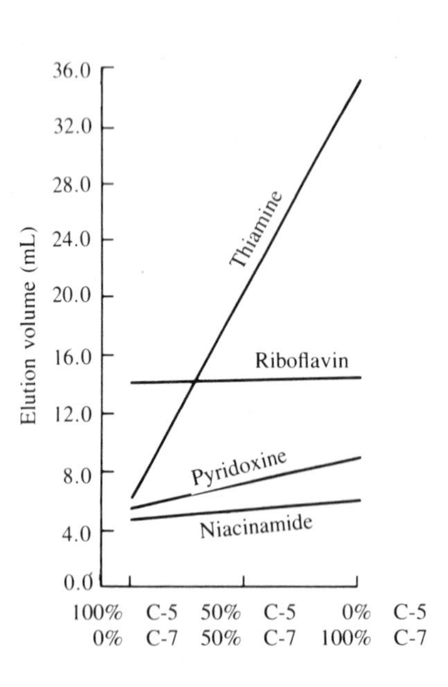

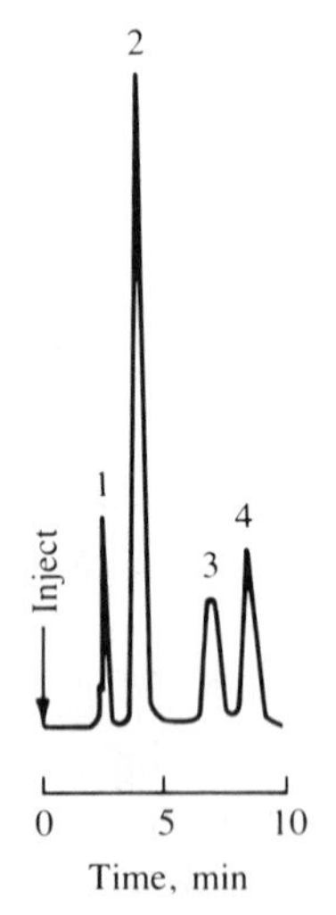

FIGURE **20.15**

Separation of mixtures of ionic and nonionic compounds by reverse-phase ion-pair chromatography. Peaks: (1) niacinamide, (2) pyridoxine, (3) riboflavin, and (4) thiamine. (Courtesy of Waters Associates, Inc.)

the ionic solutes become retained. For example, consider the separation of the water-soluble vitamins. At pH 3.5 thiamine is strongly ionic, pyridoxine and niacinamide are less so, and riboflavin is nonionic. A two-step procedure is required. In the first step, the methanol/water ratio is adjusted to obtain good retention of the nonionic compound, riboflavin. In the second stage, an organic counterion is chosen and added to the eluent to separate the three ionic compounds. The latter exhibit differences in retention with the alkyl chain length of the counterion. The extent of this effect depends on the ease of ionization of the solutes. Thus, thiamine, a quaternary amine, shows the greatest sensitivity to a change in counterion. As shown in Figure 20.15, the optimum separation is achieved with a 50/50 mixture of C-5/C-7 alkyl sulfonic acids. A mixture of counterions added to the mobile phase produces a retention proportional to the concentration of each counterion.

20.7 ION-EXCHANGE CHROMATOGRAPHY

Ion-exchange chromatography is carried out with column packings that have charge-bearing functional groups attached to a polymer matrix. The functional groups are permanently bonded ionic groups associated with counterions of the opposite charge. The most common retention mechanism is the simple exchange of sample ions and mobile-phase ions with the charged group of the stationary phase.

Functional Groups in Column Packings

Some ion-exchange packings bear negatively charged groups and are used for exchanging cationic species. Others, designed for exchanging anionic species, are provided with positively charged groups. The most commonly used functional groups are the sulfonate type for cation exchange and the quaternary amine type for anion exchange (Figure 20.16). Sulfonate exchangers are strongly acidic exchangers, which have the properties of strong acids when in the H-form. Likewise, quaternary ammonium exchangers are strongly basic; when in the OH-form their properties are those of a strong base. Both functional groups are totally dissociated. Therefore, their exchange properties are independent of the pH of the mobile phase; that is, their exchange capacity, which is the number of functional groups available for exchange per mass (or volume) unit of exchanger, is constant and not subject to change with pH.

There are some exchangers whose functional groups have weak acidic or basic properties. A carboxylate group carried by an exchanger permits the exchange of cationic species only when the pH is sufficiently high to permit dissociation of the —COOH site. For the same reason a tertiary amine exchanger has exchanging properties only when in an acidic medium; only then do its functional groups carry a positive charge, since a proton has been bound to the nitrogen atom.

Some functional groups have chelating properties toward certain metallic ions—namely,

$$-N\begin{matrix} \diagup CH_2COOH \\ \diagdown CH_2COOH \end{matrix} \qquad\qquad -PO_3^{2-}$$

Aminodiacetate Phosphonate

FIGURE **20.16**
Functional groups of ion-exchange resins.

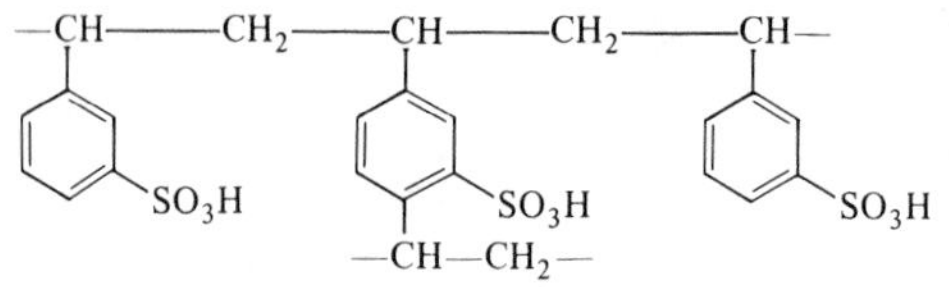

Sulfonic acids cation exchange resin

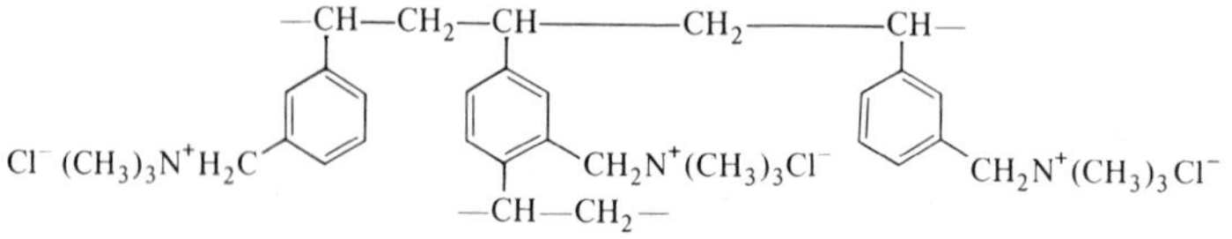

Strong base anion exchange resin

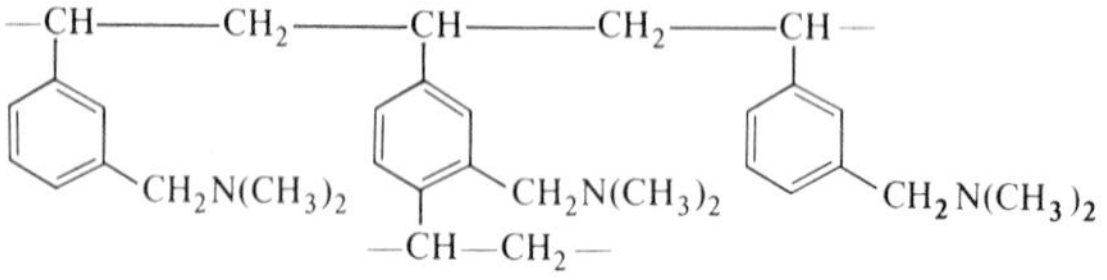

Weak base anion exchange resin

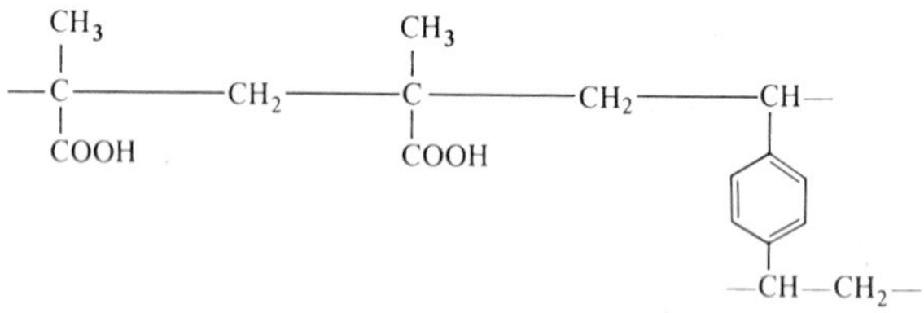

Carboxylic cation exchange resin

These exchangers have considerable affinity for heavy metal cations and, to a lesser extent, for alkaline earth cations. The aminodiacetate exchanger finds use as a column packing for ligand exchange.

Structural Types of Column Packings

Ion-exchange chromatography may be performed on one of several structural types of packings (Figure 20.17). The pellicular type consists of a resin coating, about 1–2 μm thick, on a glass bead whose diameter is 30–40 μm. Superficially porous resins are obtained by coating glass beads with a thin layer of silica microspheres (mean diameter, 0.2 μm) on which is coated or bonded an ion exchanger. This increases the interface between the resin and mobile phase. For either type of packing, the exchange capacity is low: 0.01–0.1 meq/g.

The exchanger may also be bonded to silica microparticles by means of silylation reactions or polymerized into the pores of a superficially porous silica gel. When preparing an ion exchanger by silylation, a vinyl group is chosen for R_3 of

$\geqslant SiOSiR_1R_2R_3$, which leads to a vinylated silica onto which styrene is then polymerized:

$$-CH{=}CH_2 + \underset{\displaystyle C_6H_5}{\underset{|}{CH}}{=}CH_2 \rightarrow -CH-CH_2-\underset{\displaystyle C_6H_5}{\underset{|}{CH}}-CH_2- \qquad (20.4)$$

Afterward the bonded phase is treated with chloromethyl ether and subsequently trimethylamine (or hydroxyethyldimethylamine) to prepare the quaternary amine exchanger:

$$-CH_2-CH_2-\underset{\displaystyle C_6H_5}{\underset{|}{CH}}-CH_2- + ClCH_2OCH_3 \rightarrow -CH_2-CH_2-\underset{\displaystyle C_6H_5CH_2Cl}{\underset{|}{CH}}-CH_2- + CH_3OH$$

$$\xrightarrow{RNH_2} -CH_2-CH_2-\underset{\displaystyle C_6H_5CH_2\overset{+}{N}H_2-R \;\; Cl^-}{\underset{|}{CH}}-CH_2- \quad \text{Weak anion exchanger}$$

$$\xrightarrow{N(CH_3)_2CH_2CH_2OH} -CH_2-CH_2-\underset{\displaystyle C_6H_5CH_2\overset{+}{N}(CH_3)_3 \;\; Cl^-}{\underset{|}{CH}}-CH_2- \quad \text{Strong anion exchanger} \qquad (20.5)$$

Amination of the chloromethyl group with an aliphatic primary or secondary amine yields a weakly basic type of resin that contains the corresponding secondary (as shown) or tertiary amine group at the exchange sites. In the production of strong cation exchangers, the polymer matrix is sulfonated with chlorosulfonic acid to introduce the sulfonic acid group into the benzene rings. Spherical particles are available with diameters of 5 μm; exchange capacities are 0.5–2 meq/g.

The preceding ion-exchange packings have been based on a stationary phase that is hydrophobic in the absence of ionic functional groups. Hydrophilic polymers allow the separation of proteins, nucleic acids, and other large ionic molecules. This hydrophobic character is important because it prevents the denaturation of unstable biological molecules and hydrophobic nonspecific interactions that result in irreversible adsorption. The microporosity of these ion exchangers minimizes

FIGURE **20.17** Structural types of ion-exchange packings: (a) pellicular with ion-exchange film, (b) superficially porous resin coated with exchanger beads, (c) macroreticular resin bead, and (d) surface sulfonated and bonded electrostatically with anion exchanger.

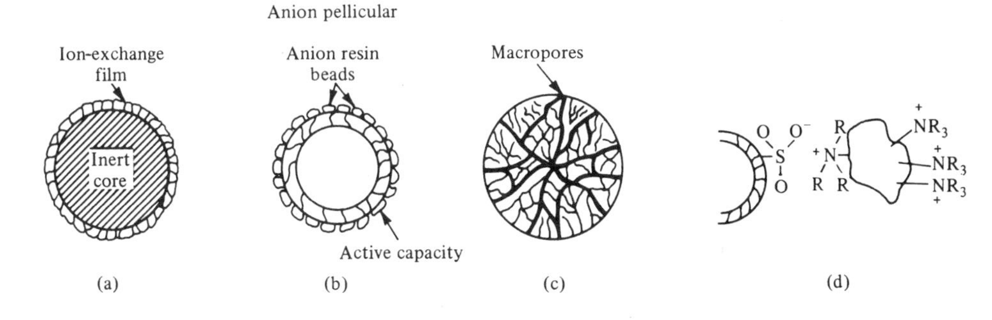

possible exclusion (gel filtration) effects. The chemical structures of different exchangers are shown in Figure 20.18. Each primary monomer has three hydroxymethyl groups that impart hydrophilic character. The exchange capacity ranges from 0.20 meq/mL for the carboxymethyl and sulfonate resin to 0.3 meq/mL for the mixed amine resin.

Exchange Equilibrium

The primary process of ion-exchange chromatography involves adsorption/desorption of charged ionic materials in the mobile phase with a permanently charged (opposite sign) stationary phase. For example, in the case of ionized resin, initially in the H-form, in contact with a solution containing K^+ ions, an equilibrium exists:

$$\text{resin}, H^+ + K^+ \rightleftharpoons \text{resin}, K^+ + H^+ \tag{20.6}$$

which is characterized by the selectivity coefficient, $k_{K/H}$:

$$k_{K/H} = \frac{[K^+]_r[H^+]}{[H^+]_r[K^+]} \tag{20.7}$$

where the subscript r refers to the resin phase. Strictly, the selectivity coefficient is constant only if the activity coefficient ratios in the resin and in the solution phases are constant.

Retention differences are essentially governed by the physical properties of the solvated ions. The resin phase shows a preference for (1) the ion of higher charge, (2) the ion with the smaller solvated radius, and (3) the ion that has greater polarizability. The solvated ionic radius limits the coulombic interaction between

FIGURE 20.18 The chemical structures of polyamide matrix resins: (a) carboxymethyl weak acid exchangers, (b) diethylaminoethyl plus quaternary aminoethyl strong anion exchanger, and (c) sulfonate strong acid exchanger.

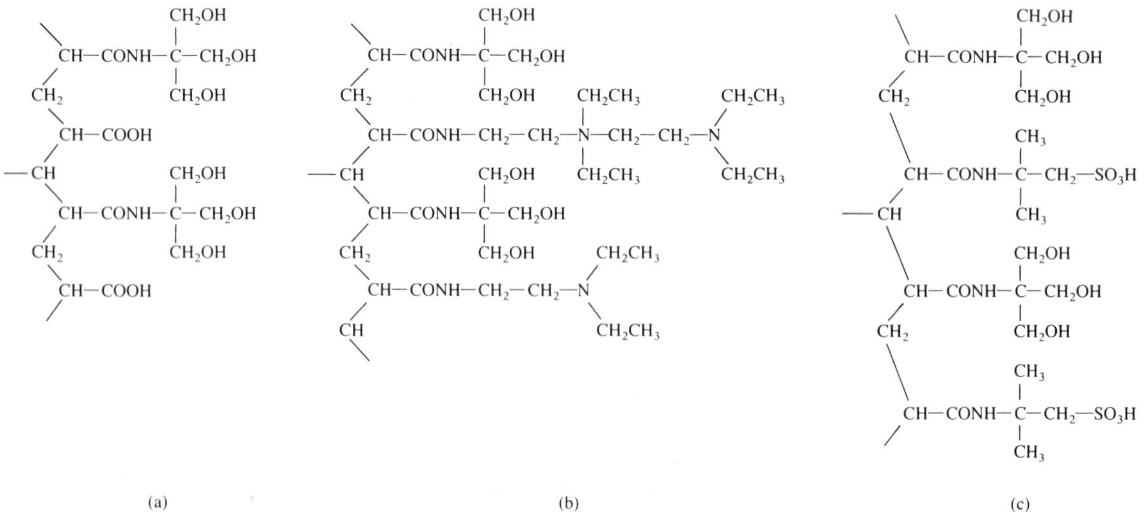

(a) (b) (c)

ions, and the polarizability of the ions determines the van der Waals attraction. Together these factors control the total energy of interaction between oppositely charged species and hence their tendency to exist as ion pairs in the resin matrix. Energy is needed to strip away the solvation shell surrounding ions with large hydrated radii, even though their crystallographic ionic radii may be less than the average pore opening in the resin matrix. This explains the position of the lithium ion among the alkali metal ions and the fluoride ion among the halide ions; each is least retained.

The partitioning of each trace ion, K^+ in the example, between the resin phase and the solution phase can be described by means of a concentration distribution ratio, D_c:

$$(D_c)_K = \frac{[K^+]_r}{[K^+]} \tag{20.8}$$

Combining the equations for the selectivity coefficient and for D_c, one gets

$$(D_c)_K = k_{K/H} \frac{[H^+]_r}{[H^+]} \tag{20.9}$$

Equation 20.9 reveals that the concentration distribution ratio for trace concentrations of an exchanging ion is independent of the respective solution concentration of that ion (neglecting any activity effects). Therefore, the uptake of each trace ion by the resin is directly proportional to its solution concentration. However, the concentration distribution ratios are inversely proportional to the solution concentration of the resin counterion, which is to be expected, since the counterion competes with the trace ion for exchange sites in the resin phase.

To accomplish any separation of two cations (or two anions), it is necessary that one of these cations be taken up by the resin in distinct preference to the other. This is expressed by the separation factor (or relative retention), $\alpha_{K/Na}$, using K^+ and Na^+ as the example. Thus,

$$\alpha_{K/Na} = \frac{(D_c)_K}{(D_c)_{Na}} = \frac{k_{K/H}}{k_{Na/H}} = k_{K/Na} \tag{20.10}$$

If the selectivity coefficient, $k_{K/Na}$, is unfavorable for the separation of K^+ from Na^+, no variation in the concentration of H^+ (the eluent) will improve the separation. The situation is entirely different if the exchange involves ions of different net charges. Now the separation factor does depend on the concentration. For example, the more dilute the counterion concentration in the eluent, the more selective the exchange becomes for polyvalent ions.

Cation Exchange Applications

Acids can be separated according to their strengths on strong anion exchangers, with the weakest acids emerging first, either by elution with a strong acid or, since acid dissociation depends on pH, by gradient elution with buffers of decreasing pH. Amino acids, which add protons to form cations in the pH range below their isoelectric points, can be separated on cation exchangers by gradient elution with buffers of increasing pH; here the more acidic components emerge first and the most basic last (Figure 20.19). Commercial automatic amino acid analyzers have been available for many years.

Inorganic cations can be monitored in foods, such as dietetic foods low in sodium, and urine samples (Figure 20.20). The efficiency of the separation of cations on a cation exchanger can be influenced markedly by the use of complexing agents in the eluent. For example, buffered solutions that contain citrate or ethylenediaminetetraacetate (EDTA) will bring about separations because some cations

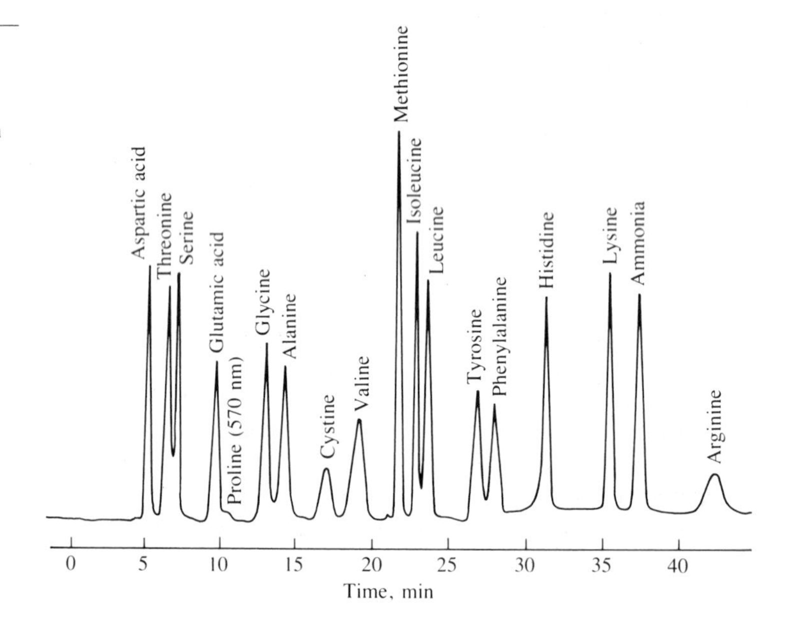

FIGURE **20.19**

Amino acid analysis; photometer senses the amino-acid–ninhydrin complex (postcolumn reaction) at 570 nm.

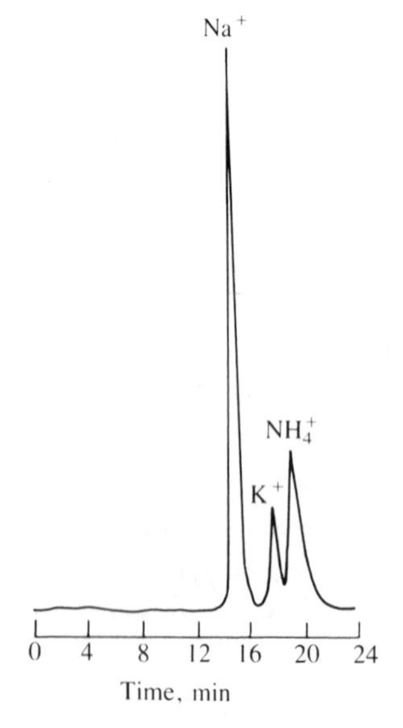

FIGURE **20.20**

Inorganic cations in human urine. Mobile phase: 0.05*M* citric acid (pH 2.4). Sample size: 0.001 mL. Temperature: 30 °C.

are converted into neutral or negatively charged ionic species. The extent of conversion to such species is a function of pH and metal formation constants.

Anion Exchange Applications

Advantage can be taken of weakly basic anion exchange resins to separate acids of different strengths. Acids that have a small dissociation constant are retained to only a slight extent and, if the dissociation constant of the acid is less than the base constant of the exchanger, virtually no retention occurs. In this manner, HCN, carbonic, silicic, and boric acids can be separated from phosphoric, sulfuric, and hydrochloric acids.

The extent of ionization of acids can be controlled by the pH of the mobile phase. An increase in ionization leads to increased retention of the solute. The separation of several nucleotides on a strong anion exchanger is shown in Figure 20.21.

To a substantial extent, practically every metal can be converted to a negatively charged complex ion through suitable masking systems. This fact, coupled with the greater selectivity of anion exchangers, makes anion exchange a logical tool for handling metals. The complexes, negatively charged, are absorbed by the exchanger and eluted by changing the concentration of masking agent in the eluent sufficiently to cause a dissociation of the metal complexes or a decrease in the fraction of the metal present as an ionic complex ion in solution. If the interconversion of complexes is fast, control of the ligand concentration affords a powerful tool to control adsorbability, since the ligand concentration controls the fraction of the metal present as adsorbable complex. The separation of metals that form chloride and fluoride complexes is an example of many such systems. For example, nickel(II), iron(III), and zinc(II) can be separated in a chloride solution on a quaternary amine column packing. Introduced onto the column in a 2.5M HCl solution, the nickel quickly traverses the column because it does not form a chloride complex and is therefore not retained. The iron(III) and zinc ions form $FeCl_4^-$ and $ZnCl_4^{2-}$, respectively. These anionic complexes are retained on the anion exchanger. The tetrachloroferrate(III) complex elutes when the eluent strength is lowered to 0.5M HCl, and the zinc complex elutes with 0.005M HCl. At these lower concentrations of chloride ion, the respective cations no longer form a chloro complex.

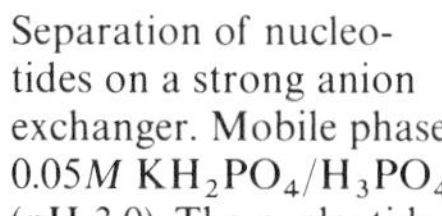

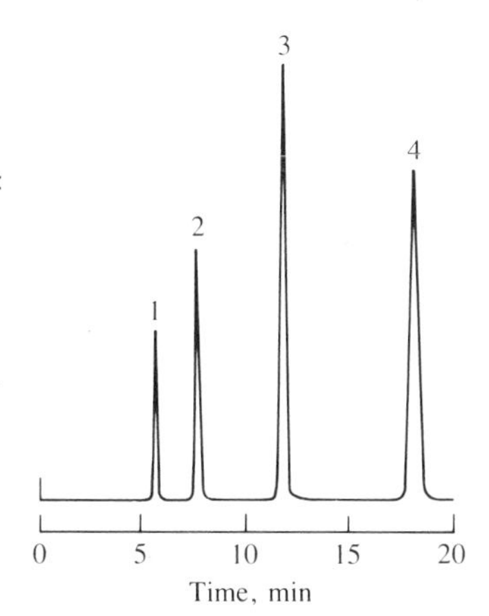

FIGURE 20.21
Separation of nucleotides on a strong anion exchanger. Mobile phase: 0.05M KH_2PO_4/H_3PO_4 (pH 3.0). The nucleotide 5′-monophosphoric acid peaks are (1) cytidine-, (2) adenosine-, (3) uridine-, and (4) guanosine-.

Sugars, and polyhydroxy compounds in general, form with borate ions a series of complexes with varied stabilities that dissociate as weak acids to various degrees:

$$\begin{array}{l} \quad\ \ \mathrm{H} \\ -\mathrm{C}-\mathrm{OH} \\ \quad\ \ | \\ -\mathrm{C}-\mathrm{OH} \\ \quad\ \ | \\ -\mathrm{C}-\mathrm{OH} \\ \quad\ \ | \end{array} + \begin{array}{l} \mathrm{HO} \diagdown \\ \qquad \mathrm{B}-\mathrm{OH} \\ \mathrm{HO} \diagup \end{array} \rightarrow \begin{array}{l} \quad\ \ \mathrm{H} \\ -\mathrm{C}-\mathrm{O} \diagdown \\ \quad\ \ | \qquad\quad \mathrm{B}-\mathrm{O}^{-}\mathrm{H}^{+} \\ -\mathrm{C}-\mathrm{O} \diagup \\ \quad\ \ | \\ -\mathrm{C}-\mathrm{OH} \\ \quad\ \ | \end{array} + 2H_2O \tag{20.11}$$

Disaccharides have less tendency to form complexes and, therefore, can readily be separated from monosaccharides, which form more stable complexes. Moreover, monosaccharides can be separated from each other within the classes of hexoses, pentoses, and tetroses with a borate buffer and pH gradient rising from pH 7 to 10.

The analytical separation of aldehydes and ketones, and the separation of these compounds from alcohols, are based on the fact that the carbonyl group forms addition compounds with the hydrogen sulfite ion:

$$>\mathrm{C}=\mathrm{O} + HSO_3^- \rightarrow >\mathrm{C}\begin{array}{l} \diagup \mathrm{OH} \\ \diagdown SO_3^- \end{array} \tag{20.12}$$

which is strongly adsorbed by anion exchange resins. Alcohols do not form the sulfite-addition product and are therefore not adsorbed. Subsequently, the ketones are desorbed by hot water (since the addition product of ketones is less stable than that of aldehydes at elevated temperature) and the aldehydes are eluted with a NaCl solution.

Ligand Exchange

Cation exchangers that contain counterions, such as Cu^{2+}, Ni^{2+}, or Zn^{2+}, show a unique preference for adsorbing molecules that can act as coordinating ligands. Thus even though they are bound to an exchanger, these metals retain their ability to be the central atom of a coordination compound. Furthermore, the ligand associated with the coordinating metal attached to the resin (R^-) can be replaced by a different ligand, L^-:

$$R^-[Cu(NH_3)_4^{2+}] + 4L^- \rightleftharpoons R^-[CuL_4^{2-}] + 4NH_3 \tag{20.13}$$

No ion exchange takes place; the exchanger acts simply as a support for the coordinating metal ion, itself strongly bound at the primary cation exchange site. This process is called ligand exchange. The chelating resins that have iminodiacetate functional groups attached to a styrene matrix are ideally suited for ligand-exchange work. Divalent metal ions from the transition elements are tightly bound to such exchangers. Consequently, leakage of metal ions from the chelating resin by ordinary ion-exchange reactions with cationic materials in eluting solutions is held to a minimum.

Ligands may be separated from each other by ligand-exchange chromatography. Different ligands have different affinities for the coordinating metal attached to the exchanger. Hence, their migration rates down a column differ and thus separation occurs. For example, purine metabolites have been separated on a chelating resin loaded with copper(II) ion.[11] The eluent was 1*M* NH_3. Caffeine is strongly retained on the same column packing and is well separated from theophylline, theobromine, and most other compounds that appear in coffee, cola beverages, commercial decongestants, and analgesic tablets.

20.8 ION CHROMATOGRAPHY[12–16]

The term *ion chromatography* was coined to describe the analysis of dissolved ions by ion-exchange chromatography using HPLC equipment. What differs from the preceding discussion of ion-exchange chromatography is the nature of the exchange resin.

Column Packings

The column packing consists of a neutral polymer core about 10 μm in diameter. Depending on whether the packing will be used for the separation of anions or cations, the core is lightly sulfonated or aminated, which leads to the formation of a thin surface shell of sulfonic acid or quaternary amine groups. This produces the traditional ion-exchange resin. The new feature is the introduction of a monolayer of polymeric beads, aminated or sulfonated and 100 to 300 nm in diameter, which are bonded electrostatically on the surface of the polymer core. Thus, for an anion exchanger, the packing consists of a rigid inert core, an intermediate layer of sulfonate groups, and a thin layer of anion exchange beads (Figure 20.17d). For a cation exchanger there would be an intermediate layer of aminated groups covered by a thin layer of sulfonated resin beads. Due to the proximity of all the active sites to the eluent-resin interface, this type of exchanger has favorable mass transfer characteristics. The capacity is low, only about 0.020 meq/g of copolymer. Silica-based materials are inappropriate in most applications due to their degradation in the presence of aqueous eluents and their poor selectivity for some ionic species.

Methodology

A typical flow scheme for the two-column method is shown in Figure 20.22. Two columns are used: a separator and a suppressor. Taking as an example the method for anions, the sample is injected at the head of an anion exchange resin bed (separator column). Elution with dilute sodium salicylate accomplishes the chromatographic separation of the species being analyzed. The eluent converts the anions to dissolved sodium salts.

The effluent from the separator column then passes to a bed of cation exchange resin in the hydrogen form (suppressor column) that exchanges hydrogen ions for other cations in the eluent stream. Thus, the suppressor column converts the dissolved sodium salts to their corresponding acids. The sodium salicylate becomes salicylic acid (only slightly ionized). Finally, the effluent from the suppressor column passes through a detector. If the detector is a conductivity cell, the highly conductive anions are detected at high sensitivity in the presence of a low-conductance background of salicylic acid. The suppressor column does postcolumn chemistry to modify the properties of the eluent by removing ions that would cause a large baseline signal in the detector, particularly a conductivity detector. After a period of use, the suppressor column must be regenerated.

The method for cation determination is analogous to that for anions and makes use of a cation exchange column as the separator. HCl is the eluent and the suppressor column is an anion exchange resin in the hydroxyl form.

In a more recent version (single-column method), the eluent is passed down the core of a hollow-fiber cation (or anion) exchange membrane, while a counterflowing stream of acid (or alkali for anion membranes) on the outside continuously supplies hydronium ions to exchange for cations across the membrane. The newest chemical suppressor uses a high-capacity ion-exchange material and a void volume less than 50 μL. Figure 20.23 shows an exploded view of the suppressor, which sandwiches alternating layers of high-capacity ion-exchange screens with ultrathin ion-exchange membranes. The eluent flows lengthwise between the screens and the regenerant flows on both sides of the sandwich. Thus, interruptions for suppressor regeneration have been eliminated. Ion-exchange sites in each screen provide a site-to-site pathway for eluent ions to the membrane. The high capacity of the suppressor unit broadens the choice of eluents and permits the use of much higher eluent concentrations. This expands the range of analyte applications and permits gradient elution to be performed with ion chromatography.

There is also a single-column mode that dispenses with a suppressor column. Short (5 cm) columns are packed with a cross-linked polystyrene material that has been selectively sulfonated to produce a low-capacity ion-exchange resin (approximately 100 μeq/g) specifically designed for ion chromatography. One type of ion-exchange polymer is prepared by selectively sulfonating a large-surface-area macroporous polystyrene matrix. Quaternary amine functional groups, covalently

FIGURE **20.22** Ion chromatography flow scheme for anion analysis using the two-column method. (Courtesy of Dionex Corp.)

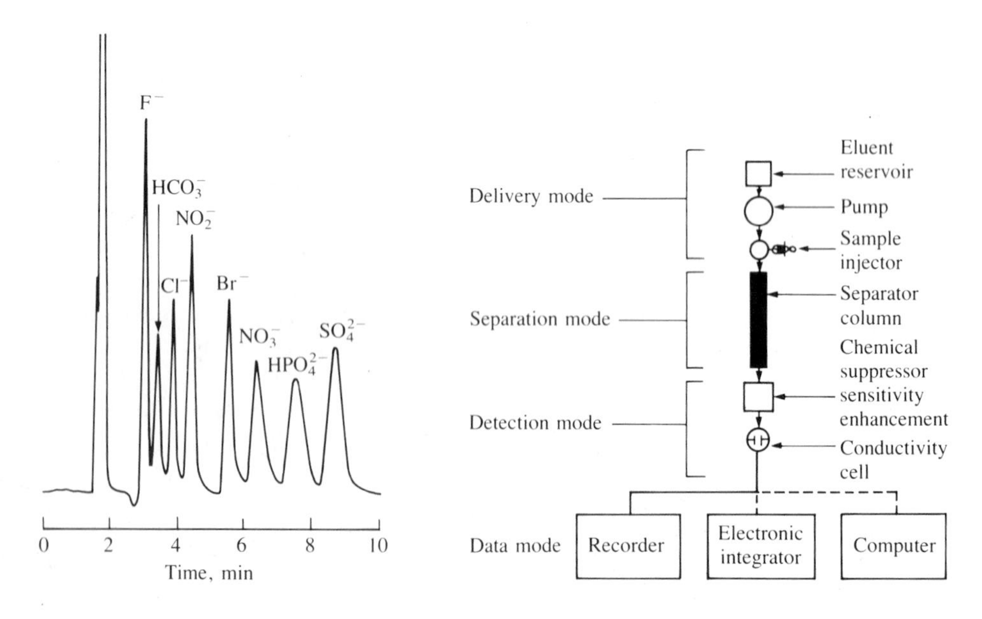

bound to a hydrophilic macroporous matrix, provide the anion exchange resin. This permits the use of low-ionic-strength eluents, which is the key to simplifying the determination, since these columns eliminate the need for an eluent suppression system.

Eluents

Anion eluents may be any species that is anionic above pH 8 and neutral between pH 5 and 8. In addition to salicylate already mentioned, eluents have included these anions: borate, bicarbonate, *p*-cyanophenate, glycine, hydroxide, silicate, and tyrosine. Likewise, cation eluents may be any species that is cationic below pH 5 and neutral between pH 5 and 9. In addition to HCl, these eluents have been used: histidine·HCl, hydroxylamine·HCl, lysine·HCl, triethanolamine·HCl, and HNO_3.

Detectors

In many cases the ionic strength of the eluent can be kept sufficiently low so that conductivity detectors can be used. Modern conductivity detectors are designed with wide signal suppression ranges (i.e., extensive zero-offset capability) and can often be used without suppression with the low concentration of eluents used with low-capacity ion-exchange columns. Conductivity detectors provide the advantage of universal detection. For samples that contain a mixture of ionic and nonionic species, refractometers can be used.

Applications

Matrices that have been analyzed include brine solutions, pulp and paper liquors, soil extracts, pond waters, plating baths, and industrial process, waste, and boiler waters, urine, plasma, and food samples. Among the inorganic ions determined are F^-, Cl^-, PO_4^{3-}, NO_3^-, CrO_4^{2-}, SO_4^{2-}, $C_2O_4^{2-}$, Na^+, K^+, NH_4^+, Mg^{2+}, and Ca^{2+}. Air samples and vapors from Schöniger-type combustions are trapped in an appropriate solvent and diluted as necessary.

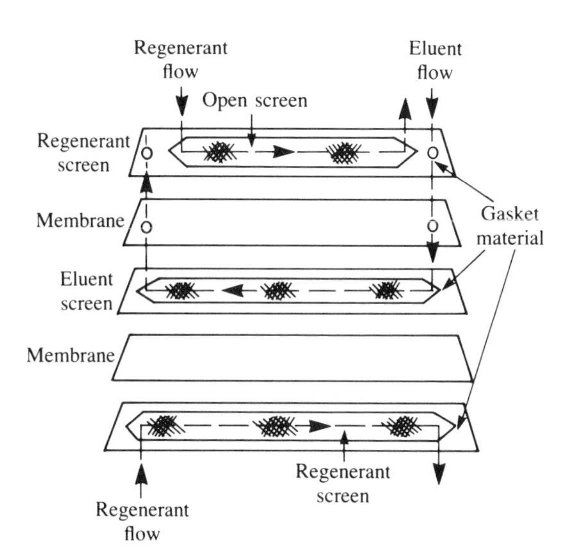

FIGURE **20.23** Exploded view of the micromembrane suppressor. (Courtesy of Dionex Corporation.)

Ion chromatography has also been used to determine transition metals, organic acids, amino acids, and peptides, simply by correctly selecting the proper polymeric column, mobile phase, and detector. Any compound that can be converted to an ionic form is amenable to analysis by this method. The method has the ability to differentiate between species such as iron(II) and iron(III), and to distinguish among polyvalent anions such as pyrophosphate, tripolyphosphate, trimetaphosphate, and tetrapolyphosphate. Carbohydrate analysis is achieved by raising the eluent pH to between 12 and 14. At this pH, carbohydrates are anions and are separated on an anion exchange column. For metals, a combination of separator columns, postcolumn reagent delivery, and visible detection technology permits the detection of multielement transition metals at levels lower than those possible with either inductively coupled plasma or atomic absorption spectroscopy.

20.9 EXCLUSION CHROMATOGRAPHY[17]

Exclusion chromatography, also called *gel permeation chromatography,* is a noninteractive mode of separation. Essentially a maze for molecules, the particles of the column packing have various size pores and pore networks, so that solute molecules are retained or excluded on the basis of their hydrodynamic molecular volume—that is, their size and shape. Strictly speaking, separation in exclusion chromatography is not based on molecular weight.

As the sample passes through the column, the solute molecules are sorted. Very large molecules cannot enter many of the pores, and they also penetrate less into the comparatively open regions of the packing. Thus excluded, they travel mostly around the exterior of the packing and elute at the bed void volume of the mobile phase. Very small molecules diffuse into all or many of the pores accessible to them. With a larger column volume at their disposal, small molecules exit the column last. Between these two extremes, intermediate-size molecules can penetrate some passages but not others and, consequently, suffer retardations in their progress down the column and exit at intermediate times.

Column Packings

Column packings are either semirigid, cross-linked macromolecular polymers or rigid, controlled-pore-size glasses or silicas. The semirigid materials swell slightly, and some care must be taken in their use because these materials are limited to a maximum pressure of 300 psi due to bed compressibility. The styrene–divinylbenzene polymers allow fractionation within the molecular-weight range from 100 to 500 million. Partially sulfonated polystyrene beads are compatible with aqueous systems, the nonsulfonated with nonaqueous systems. Bead diameters are usually 5 μm.

Another class of hydrophilic porous packing is prepared by suspension polymerization of 2-hydroxyethyl methacrylate with ethylene dimethacrylate. The packings can withstand pressures up to 3000 psi and are usable with aqueous systems and with a variety of polar organic solvents.

Porous glasses and silicas cover a wide range of pore-size diameters. For example, one series has the following diameters and the corresponding operating ranges:

Pore-size diameter (nm)	Operating range (daltons)
4	1,000–8,000
10	1,000–30,000
25	2,500–125,000
55	11,000–350,000
150	100,000–1,000,000
250	200,000–1,500,000

These packings are chemically resistant at pH values less than 10 and can be used with aqueous and polar organic solvents. With nonpolar solvents it is desirable to deactivate the surface by silylation to avoid irreversible retention by polar solutes. Porous inorganic packings have distinct advantages over organic exclusion packings. After calibration, columns can be used routinely and indefinitely, with no possibility of sample contamination or biodegradation. The bed volume remains constant at high flowrates and high pressures. Thermal stability permits the use of elevated temperature.

Solvents

Exclusion chromatography requires only a single solvent in which to dissolve and chromatograph the sample. There may be problems caused by the high viscosities exhibited by high-molecular-weight samples. If the viscosity difference between an injected sample and the mobile phase is too great, peak distortion and anomalous changes in elution times may result.

Detectors

The most widely used detectors are the differential refractometer and the spectrophotometric detectors that operate in the ultraviolet and infrared spectral regions. Detectors must be compatible with exclusion columns that are 3–6 m long and have a typical working volume of 1–10 mL. Analysis times are less than 30 min.

A low-angle laser light scattering (LALLS) detector makes it possible to determine absolute molecular weights directly. Although polymer molecules with different molecular weights may elute simultaneously, skewing the molecular weight distributions generated with conventional detectors, the LALLS detector gives correct results because it responds to analyte molecular weight, not just to concentration. When used in conjunction with a viscosimeter and a conventional mass concentration detector (refractive index, ultraviolet, or infrared), it can also provide information on the variation of long-chain branching with molecular weight.

An infrared detector provides information on copolymer composition, branching, and tacticity. (Tacticity refers to the stereoregularity found in certain types of polymers.)

Retention Behavior

The essential behavior of a solute and the characteristics of porous column packings can be discussed in very simple terms. For a packed column of porous particles with a total bed volume, V_t,

$$V_t = V_M + V_S + V_g \tag{20.14}$$

where V_M is the void volume of the mobile phase (that is, the unbound solvent in interstices between the solvent loaded porous particles), as estimated by the elution of a totally excluded solute; V_S is the cumulative internal volume within the porous particles and available to a totally included solute or molecule of solvent (also denoted V_i); and V_g is the volume occupied by the matrix.

If it is assumed that the time taken for a solute molecule to diffuse into a pore is short with respect to the time that the molecule spends in the vicinity of the pore, then the separation process is completely independent of diffusion processes. Under these conditions the retention (elution) volume, V_R, of a solute is the volume of effluent that flows from a column between the sample injection and its emergence in the effluent; that is,

$$V_R = V_M + KV_S \tag{20.15}$$

or, rearranging as the distribution coefficient, one gets

$$K = \frac{V_R - V_M}{V_S} \tag{20.16}$$

which can be stated as a fraction of the internal pore volume that is accessible to the solute. Totally excluded molecules elute in one void volume; that is, $V_R = V_M$, and so $K = 0$. For small molecules that can enter all the pores of the packing, $V_R = V_M + V_S$, and hence $K = 1$. Intermediate-size molecules elute between these two limits, and K ranges from 0 to 1. An elution graph is shown in Figure 20.24. The upper portion is the graph of the logarithm of molecular weight versus retention volume. It is a sigmoid-shaped curve, in which there is a linear range of effective permeation between the limiting values that correspond to exclusion ($K = 0$) and to total permeation ($K = 1$). The maximum elution volume is often only twice the column void volume.

The total number of peaks that can be resolved is limited for any particular packing. However, the various pore sizes available in commercial packings provide selective permeation ranges that permit separating small molecules with molecular weights of less than 100 to large polymers with molecular weights up to 500 million (Figure 20.25). When the probable molecular dimensions or weights of the sample components are known, selecting the particular pore size or exclusion limit of the packing is usually straightforward. For species in the 100–1000 range of molecular weights, there should be differences on the order of 40–50 molecular weight units between components for discrete resolution. Column lengths are determined by the magnitude of the differences. As size differences diminish, longer columns of a given packing are required.

A quick preliminary run can be made using one column packed with 100-nm material and the second with 10-nm pore-size material. Using a flowrate of 3 mL/min, the entire run on each column requires about 4 min per sample. If most of the sample elutes near the exclusion limit with the column packed with 10-nm pore-size material, a column packed with the larger-pore-size material should be

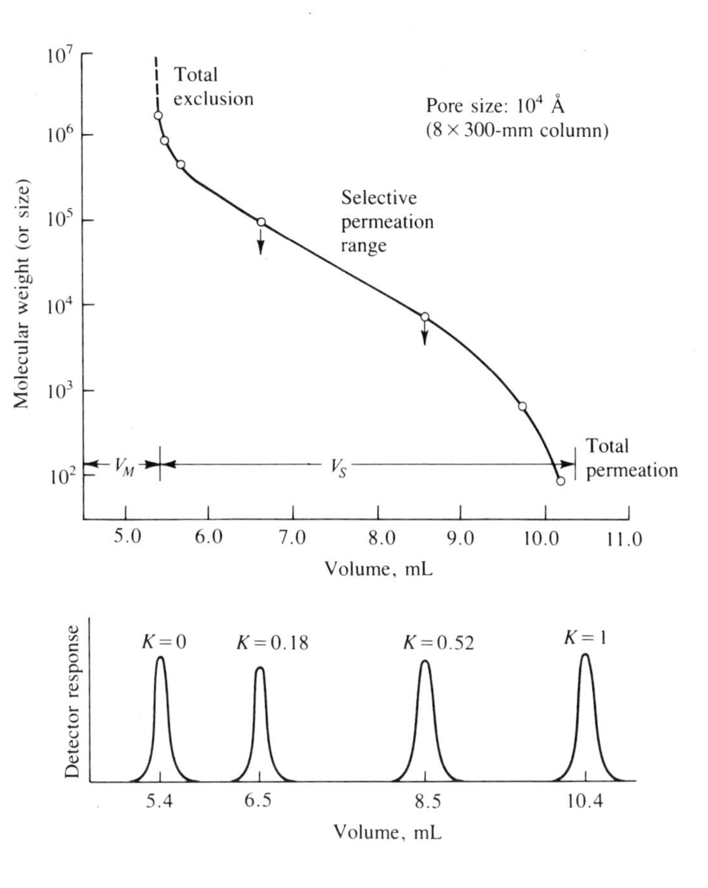

FIGURE **20.24**
Retention behavior in exclusion chromatography.

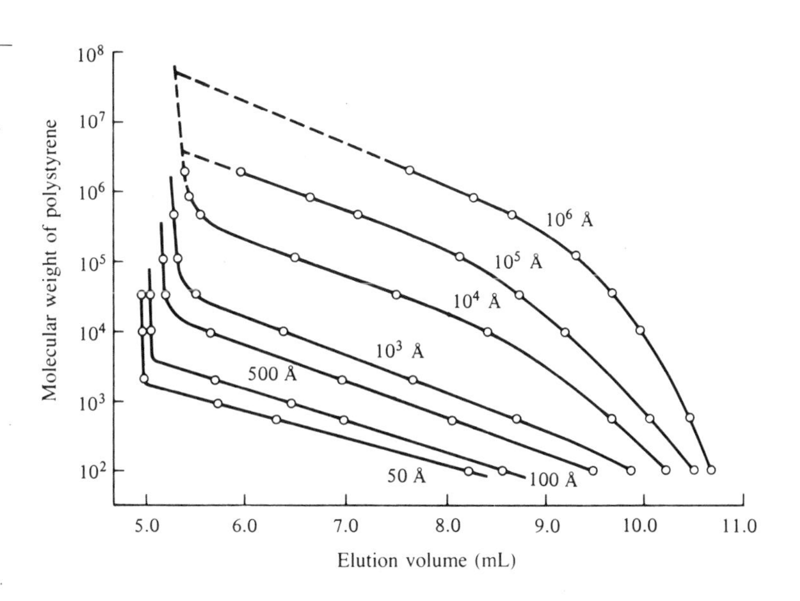

FIGURE **20.25**
Separation range of various pore-size exclusion chromatographic packings.

used. Elution of the majority of the sample halfway between the exclusion and total permeation volumes when using the 100-nm pore-size material suggests a column packing of 50-nm pore size. Near-total permeation indicates a 10-nm packing, and total permeation indicates a packing of 6-nm pore size. If sample components elute over the entire range, a full column set of all four packings should be used.

Sources of Error

Adsorptive sites on the packing can be deactivated by using a polar mobile phase, such as tetrahydrofuran. Or the surface of the exclusion packing may be chemically modified to reduce adsorptivity.

An ion-exclusion effect arises when the pore surface contains charged groups with the same sign as the solute. Charge repulsion may cause limited pore penetration, and therefore smaller elution volumes may be exhibited toward anionic species if the eluent is of insufficient ionic strength to neutralize ion-ion repulsion. This effect can be eliminated by using $0.01 M$ NaCl in the eluent.

Ion inclusion can occur when a polydisperse polyelectrolyte or a mixture of different-sized salts is chromatographed using a porous packing and a low-ionic-strength eluent. The presence of the larger sterically excluded molecules leads to the establishment of a Donnan equilibrium. This leads to a higher included concentration of the smaller charged molecules and a higher elution volume than would be found in the absence of excluded charged molecules. This effect is also suppressed by the addition of sufficient electrolyte to screen the charge on the excluded charged molecule.

Column Calibration

To determine molecular weights for monodisperse species or molecular weight averages and distributions for polydisperse systems, the exclusion columns must be calibrated. This is achieved by eluting the appropriate calibration standards and monitoring the elution volume.[18]

Narrow dispersed standards of polystyrene, polytetrahydrofuran, and polyisopropene are available for use in organic solvents. Samples of dextrans, polyethylene glycols, polystyrene sulfonates, and proteins are available for use in hydrophilic solvents.

Care must be exercised in correlating molecular weight/elution volume data for polymers that have different chemical compositions. Differences in structure and solvent-polymer interactions lead to different hydrodynamic volumes for equivalent molecular weights. The hydrodynamic radius of the molecule is proportional to the logarithm of the product of the molecular weight and intrinsic viscosity—that is, $\ln M[\eta]$. It is a valid parameter for linear polymers. A graph of $\ln M[\eta]$ versus the elution volume produces a unique relationship for determining molecular weights of polymers that are structurally different from those used for calibration of the column.

For polymers, specific parameters of interest are $\bar{M}_w$, the weight-average molecular weight; $\bar{M}_n$, the number-average molecular weight; and $\bar{M}_w/\bar{M}_n$, the dispersivity. A simplified example will illustrate these parameters. Assume that the chromatogram exhibits two peaks of equal concentration that arose from 1 g each of two components, one of molecular weight 150,000 and the other of molecular weight 50,000, and that N is the number of molecules in 1 g of each component. Then

$$\bar{M}_w = \frac{(1\text{ g} \times 150{,}000) + (1\text{ g} \times 50{,}000)}{1\text{ g} + 1\text{ g}} = 100{,}000$$

$$\bar{M}_n = \frac{(0.33 \times 150{,}000) + (1 \times 50{,}000)}{0.33 + 1.0} = 74{,}800$$

$$\frac{\bar{M}_w}{\bar{M}_n} = \frac{100{,}000}{74{,}800} = 1.34$$

$\bar{M}_w$ values are particularly sensitive to the amount of high-molecular-weight material present. An analogous situation exists for $\bar{M}_n$ values at the low-molecular-weight end of the distribution. The dispersivity is essentially a measure of the relative spread in molecular weights in a polymer. An exclusion chromatogram provides a complete, detailed molecular-weight dispersion of a polymer and an analysis of many of the additives that are generally found in finished products. Direct comparison between chromatograms of two or more materials can quickly establish "good" versus "bad" (Figure 20.26), "theirs" versus "ours," "new" versus "old," or product stability such as "virgin" versus "used" in lubricating oils.

Ion-Exclusion Chromatography

Ion-exclusion chromatography is somewhat of a hybrid method. The principal mechanism involved in these separations is ion exclusion, although steric exclusion and reverse-phase partitioning effects have been observed. As a result, separations are dependent on such factors as resin particle size and even isomer distribution in addition to the obvious variables such as cross linking and ionic form. The fact that multiple modes are involved is advantageous, often permitting the separation of species that would otherwise coelute.

Separations are carried out on high-capacity polystyrene-based exchange resins using dilute mineral acid eluents. Acetonitrile may be used as an organic modifier to

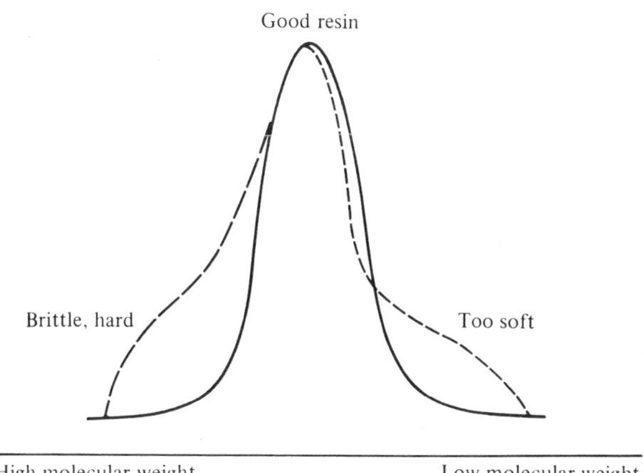

FIGURE **20.26**
Exclusion chromatography used in quality control to evaluate incoming resin. (Courtesy of Waters Associates, Inc.)

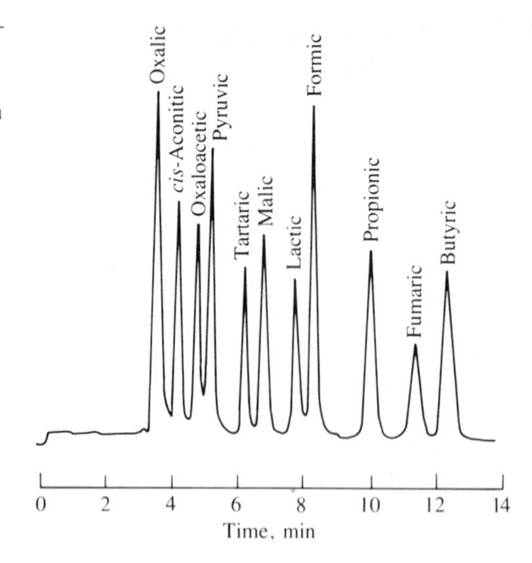

FIGURE **20.27**
Separation of organic acids on anion exclusion column. Eluent: 0.005*M* H_2SO_4. Temperature: 35 °C.

decrease the retention time of relatively nonpolar compounds. For most applications the column must be maintained at elevated temperatures, so a column heating device must be used.

Much work in the food and beverage industry, or physiological samples in which most of the Krebs cycle acids (tricarboxylic acid cycle) are present, is easily done with an anion exclusion column. Wine, beer, fruit juice, and many dairy products are quickly analyzed with minimal sample preparation (usually only filtration or centrifugation). A profile of organic acids is shown in Figure 20.27. Silica columns have been only partially successful in separating organic acids. Another application is the separation of oligosaccharides and sugar alcohols using an anion exclusion column in the calcium ionic form with water as eluent and a temperature of 90 °C. Maltotriose and higher oligosaccharides are separated from mono- and disaccharides by steric exclusion effects.

20.10 AFFINITY CHROMATOGRAPHY[19–21]

The general scheme of affinity chromatography involves the covalent attachment of an immobilized biochemical (called an affinity ligand) to a solid support. When a sample is passed through the column, only the solute(s) that selectively binds to the complementary ligand is retained; the other sample components elute without retention. The separations exploit the "lock and key" binding that is prevalent in biological systems. The retained solute(s) can be eluted from the column by changing the mobile-phase conditions. The major advantage of this technique is its tremendous specificity, which permits rapid isolation (in preparative work) with a good yield in a single step.

Column Matrices

Column matrices encompass a wide variety of materials: organic gels such as beaded agarose (a purified form of the polysaccharide agar), cellulose, dextran, polyacrylamide, combinations of these polymers, and microporous glass beads. The matrix should be sufficiently hydrophilic to avoid the nonspecific binding of solutes and stable to most water-compatible organic solvents. A porous matrix permits a high degree of ligand substitution and is more accessible to larger molecules.

Preactivated agarose supports are available commercially or can be prepared as outlined:

Polysaccharide matrix | Cyanate ester

$$\begin{array}{c} | \\ -\mathrm{C}-\mathrm{OH} \\ | \\ -\mathrm{C}-\mathrm{OH} \\ | \end{array} \xrightarrow{\mathrm{BrCN}} \begin{array}{c} | \\ -\mathrm{C}-\mathrm{O}-\mathrm{CN} \\ | \\ -\mathrm{C}-\mathrm{OH} \\ | \end{array} \xrightarrow{\mathrm{RNH_2}} \begin{array}{c} \qquad\qquad \mathrm{NH} \\ | \qquad\qquad \| \\ -\mathrm{C}-\mathrm{O}-\mathrm{C}-\mathrm{NHR} \\ | \\ -\mathrm{C}-\mathrm{OH} \\ | \end{array}$$

R = alkyl or aryl group

Agarose is a very popular matrix; its porous meshwork can be strengthened by cross-linking it with 1-chloro-2,3-epoxypropane. Other commercial supports emphasize active groups such as $-CO-NHR$, $-CH(OH)CH_2NHR$, $-CH_2NHR$, $-S-S-R$, and $-C_6H_4-N{=}N-C_6H_3(OH)R$.

Affinity Ligands

The affinity ligands can be antibodies, enzyme inhibitors, or other molecules that reversibly and bioselectively bind to the complementary analyte molecules in the sample. A short alkyl chain such as hexamethylenediamine inserted between the ligand and the matrix reduces or eliminates the steric influence of the matrix.

Ligands are either specific or group specific. Specific ligands presumably bind only one particular solute. Advantage can be taken of this high selectivity by using short columns (~ 1 cm) to perform separations by step elution in less than 1 min.

Group-specific ligands bind to certain groups of solutes. These can be separated in one step as a group or separated from each other using isocratic or gradient elution techniques. In the latter case, longer and more efficient columns must be used. Among the more unusual ligands are the triazine dyes. These dyes bind at the nucleotide-binding sites of many enzymes but also bind hydrophobically to many other proteins, including albumin.

Elution Methods

Retained solutes may be eluted either by adding excess ligand to the mobile phase or by adding a compound that interferes with the solute-ligand interactions. The column is then ready for another run after returning to the initial mobile-phase composition.

In biospecific elution, the mobile-phase modifier (called the inhibitor) is a free ligand similar or identical to the immobilized affinity ligand or the solute. The inhibitor competes for sites on the ligand or solutes. For example, if immobilized glucosamine were used to purify a lectin, either glucose or *N*-acetyl-D-glucosamine might be used as the inhibitor.

Nonspecific elution involves the denaturation of either the ligand or analyte by means of pH, chaotropic agents (such as KSCN or urea), or organic solvents, or by changing the ionic strength of the mobile phase. These conditions must be empirically determined.

PROBLEMS

1. What are the major causes of tailing and memory effects in adsorption chromatography?

2. A particular LSC separation specifies that the mobile phase be 10% dichloromethane in hexane. Unfortunately the laboratory supply of dichloromethane is temporarily exhausted. What mobile phase might be substituted that has approximately the same strength?

3. A particular compound is eluted from a silica gel column too rapidly when dioxane is used as the mobile phase. Will methyl isobutyl ketone make the compound move faster or slower?

4. On a C_{18} bonded phase, a particular compound moves too slowly in a methanol/acetonitrile (50/50) solvent. What adjustment should be made in the ratio of solvents?

5. During the separation of carbohydrates using a bonded aminoalkyl functional group, an increase in the water concentration of the acetonitrile/water mobile phase decreases retention. Is the bonded phase acting in the normal or reverse mode?

6. For each of the samples, develop a scheme for ion chromatography by suggesting the separating column, the eluent, and the suppressor column including the chemical stripping action: (a) analysis of fruit juices for Na^+, K^+, and NH_4^+; (b) separation of tetraethylammonium and tetrabutylammonium ions; and (c) determination of Cl^-, SO_4^{2-}, and PO_4^{3-} ions in municipal water supplies and sources.

7. The reverse-phase analysis of a mixture of these diamines, $H_2N(CH_2)_4NH_2$ and $H_2N(CH_2)_3NH_2$, gave two peaks but very poor separation of the peaks when using a methanol-buffered water (60/40, pH 2) mobile phase. To achieve essentially baseline resolution, what modifications in procedure would be needed?

8. Some headache preparations contain aspirin and phenylpropanolamine. Suggest a LCC method for their separation using ion suppression plus ion-pair reverse-phase chromatography.

9. Linear alkyl benzene sulfonates are the major surfactants in household detergents. Therefore the detection of such substances and their separation based on the alkyl chain length in environmental samples are desirable. Reverse-phase techniques using a methanol/water solvent resulted in only two major peaks and no peaks for individual alkyl members. Success is achieved if the ion-pair technique is used. Predict the effect of counterion size on retention considering ammonium, tetramethylammonium, and tetrabutylammonium chlorides.

10. Amphiprotic compounds, such as the monofunctional amino acids, are difficult to chromatograph. (a) What two different approaches could be used if ion exchange is selected? (b) Reverse-phase ion-pair chromatography also offers two approaches. What are they?

11. A cation exchange resin column is saturated with copper(II) ions recovered from rinse waters from plating operations. It is desired to recover the copper and to convert the resin to the H^+ form for reuse. Normally this might be done by washing the column with 3*M* sulfuric acid, but 6*M* hydrochloric acid is found to be superior. Why?

12. An anion exchange column is saturated with chromate ions recovered from an industrial operation. Regeneration of the column with sodium chloride solution is slow and consumes a large volume of reagent. (a) Why is sodium chloride unsatisfactory? (b) Suggest an alternate approach for regeneration of the resin bed.

13. In separate containers are exactly 1-g portions of a cation exchange resin (H^+ form, 4.3 meq/g capacity) and 10 mL of 0.25*M* HCl. To one container is added solute A and to the other solute B, each in concentrations of exactly $10^{-4}M$. After equilibration, 58.84% of A and 32.26% of B remain in the solution phase. (a) Calculate the weight concentration distribution ratio for each solute. (b) Express the separation factor (relative retention) of solute B relative to solute A.

14. Consider the separation of trace quantities of K^+ and Mg^{2+} on a column of pellicular cation exchange resin. The selectivity coefficients, $k_{K/H}$ and $k_{Mg/H}$, are 2.28 and 1.15, respectively. The exchange capacity of the resin bed (1.92 mL) is 1.70 meq/mL when fully swollen. Estimate the individual concentration distribution ratios for these concentrations of hydrogen ion in the aqueous phase; also the relative retention of K/Mg: (a) 0.10*M*, (b) 0.50*M*, (c) 1.00*M*, and (d) 3.00*M*.

15. On a particular size exclusion column, retention volumes for the individual peaks (molecular weight in parentheses) are: #1, 3.55 mL (8,100,000); #2, 4.45 mL (1,800,000); #3, 5.05 mL (500,000); #4, 5.81 mL (25,100); and #5, 5.93 mL (76). The baseline width (4σ) of peak 5 is 0.47 mL. (a) What is the effective operational range, in log molecular weight units, of the column packing? (b) What is the value of V_M? Of V_S? (c) Indicate on the volume axis where $K = 0$ and $K = 1$. (d) Calculate the individual partition coefficients for the components exhibiting peaks #1–#5. (e) Calculate the resolution for peaks #1 and #2. (f) Calculate the plate number from the MW 76 peak; express this also as plates per meter. (g) Given that the total column volume is 7.55 mL, estimate the "void" porosity and the "total" porosity of this particular packing.

16. Suggest a method for calculating V_S in an exclusion chromatographic packing that contains few or no labile protons.

17. After a period of use of a particular column, the solute retention times were observed to be increasing. What operating conditions might be responsible?

18. Suggest a corrective procedure for these observed column operating problems. (a) Severe band tailing occurs particularly at large values of k'. (b) Band tailing increases with an increase in k' for LSC (or ion exchange chromatography). (c) Band tailing decreases for larger values of k'.

19. In a gradient elution separation, what should be changed in regard to solvents A (starting solvent) and B when the following situations develop? (a) The bands are bunched together near t_M. (b) Bands continue to elute for a considerable time after completion of the gradient. (c) To achieve added sensitivity how would the gradient steepness be altered? (d) To achieve increased resolution how would the gradient be altered?

BIBLIOGRAPHY

FRITZ, J. S., "Ion Chromatography," *Anal. Chem.*, **59,** 335A (1987).

FRITZ, J. S., D. T. GJERDE, AND C. POHLANDT, *Ion Chromatography,* A. Huthig, Heidelberg, 1983.

SIMPSON, C. F., ed., *Techniques in Liquid Chromatography,* Wiley-Hayden, New York, 1982.

SNYDER, L. R., AND J. J. KIRKLAND, *Introduction to Modern Liquid Chromatography,* 2nd ed., Wiley-Interscience, New York, 1979.

YAU, W. W., J. J. KIRKLAND, AND D. D. BLY, *Modern Size Exclusion Liquid Chromatography,* Wiley-Interscience, New York, 1979.

LITERATURE CITED

1. SAUNDERS, D. L., "Practical Aspects of Adsorption HPLC," *J. Chromatogr. Sci.*, **15,** 372 (1977).
2. SNYDER, L. R., "Role of the Solvent in Liquid Solid Chromatography—A Review," *Anal. Chem.*, **46,** 1384 (1974).
3. GRUSHKA, E., AND E. J. KIKTA, "Chemically Bonded Stationary Phases in Chromatography," *Anal. Chem.*, **49,** 1004A (1977).
4. HORVATH, C., AND W. MELANDER, "LC with Hydrocarbonaceous Bonded Phases; Theory and Practice of Reversed Phase Chromatography," *J. Chromatogr. Sci.*, **15,** 394 (1977).
5. COOKE, N. H. C., AND K. OLSEN, "Chemically Bonded Alkyl Reversed-Phase Columns: Properties and Use," *Am. Lab.*, p. 45 (August 1979).
6. HALASZ, I., "Columns for Reversed Phase Liquid Chromatography," *Anal. Chem.*, **52,** 1393A (1980).
7. ANTLE, P. E., AND L. R. SNYDER, "Selecting Columns for Reversed-Phase HPLC," *Liq. Chromatogr. HPLC Mag.*, **2,** 840 (1984); **3,** 98 (1985).
8. CONLON, R. D., "The Perkin-Elmer Solvent Optimization System," *Instrumentation Research,* p. 90 (March 1985).
9. TOMLINSON, E., T. M. JEFFERIES, AND C. M. RILEY, "Ion-Pair High-Performance Liquid Chromatography," *J. Chromatogr.*, **159,** 315 (1978).
10. GLOOR, R., AND E. L. JOHNSON, "Practical Aspects of Reverse Phase Ion Pair Chromatography," *J. Chromatogr. Sci.*, **15,** 413 (1977).
11. WOLFORD, J. C., J. A. DEAN, AND G. GOLDSTEIN, "Separation of Oxypurines by Ligand-Exchange Chromatography and Determination of Caffeine in Beverages and Pharmaceuticals," *J. Chromatogr.*, **62,** 148 (1971).
12. HEARN, M. T. W., in *High Performance Liquid Chromatography: Advances and Perspectives,* C. Horvath, ed., Vol. 3, pp. 87–155, Academic, New York, 1983.
13. HEARN, M. T. W., *Advan. Chromatogr.*, **20,** 1 (1980).
14. SMALL, H., "Modern Inorganic Chromatography," *Anal. Chem.*, **55,** 235A (1983).
15. BENSON, J. R., "Modern Ion Chromatography," *Am. Lab.*, **17,** 30 (June 1985).
16. FRANKLIN, G. O., "Development and Applications of Ion Chromatography," *Am. Lab.*, **17,** 65 (June 1985).

17. BORMAN, S. A., "Recent Advances in Size Exclusion Chromatography," *Anal. Chem.*, **55,** 384A (1983).
18. ABBOTT, S. D., "Size Exclusion Chromatography in the Characterization of Polymers," *Am. Lab.*, **9,** 41 (August 1977).
19. WALTERS, R. R., "Affinity Chromatography," *Anal. Chem.*, **57,** 1099A (1985).
20. PARIKH, I., AND P. CUATRECASAS, "Affinity Chromatography," *Chem. Eng. News,* p. 17 (August 26, 1985).
21. DEAN, P. D. G., W. S. JOHNSON, AND F. A. MIDDLE, eds., *Affinity Chromatography,* IRL Press, Oxford, U.K., 1985.

21 INTRODUCTION TO ELECTROANALYTICAL METHODS OF ANALYSIS

Analytical techniques based on electrochemical principles make up one of the three major divisions of instrumental analytical chemistry (see Chapter 1). Each basic electrical measurement of current, resistance, and voltage has been used alone or in combination for analytical purposes. If these electrical properties are measured as a function of time, many additional electroanalytical methods of analysis are possible. A summary of electroanalytical methods is shown in Table 21.1.

The individual techniques are best recognized by their excitation-response characteristics. Less confusion arises when each technique is described by an operational nomenclature that consists of an independent-variable part followed by a dependent-variable part. An example is voltammetry. The name is often preceded by system-specific modifiers such as cyclic voltammetry or square-wave voltammetry.

There are highly refined ways of making reliable electrical measurements in the submicroampere and microvolt range. Reliable analyses at the picogram range are possible. By contrast, selectivity is one of the weakest aspects of electrochemical methods due to poor resolution. However, the combination of electroanalytical methods with chromatography (see Chapters 19 and 20) is a powerful tool for both

TABLE **21.1** SELECTED ELECTROANALYTICAL METHODS

Quantity measured	Variable controlled	Name of method
E	$i = 0$	1. Ion selective potentiometry 2. Null-point potentiometry
E versus volume of titrant	$i = 0$	Potentiometric titrations
Weight of separated phase	E	Controlled potential electrodeposition
i versus E	Concentration	Voltammetry
	t	Linear potential sweep stripping chronoamperometry
	t	Linear potential sweep voltammetry
i versus volume of titrant	E	Amperometric titrations
Coulombs (current × time)	E	Coulometry
$1/R$ (conductance)	Concentration	Conductance measurements
$1/R$ versus volume of titrant		Conductometric titrations

qualitative and quantitative work. Chromatography provides the selectivity and electroanalytical methods provide the sensitivity.

Electroanalytical methods are conveniently divided into two categories: steady state and transient. The steady-state or static methods, such as potentiometry (see Chapter 22), are firmly rooted in the basic concepts of equilibrium and mass action, and the rigor of their presentation in introductory analytical and physical chemistry courses is considered adequate. Steady-state methods entail measurements of the potential difference at zero current. The system defined by the solid-solution interface is not disturbed and equilibrium is maintained. Time is effectively eliminated as a variable, and equilibrium is assured by vigorously stirring the solution with the indicator electrode held stationary, or vice versa. Any concentration gradients at the solution-electrode interface are completely or nearly completely eliminated. In such cases, the indicator electrode potential is related to bulk and surface concentration.

Voltammetry is concerned with the current-potential relationship in an electrochemical cell and, in particular, with the current-time response of an electrode at a controlled potential (see Chapter 23). In a typical voltammetric experiment, the amount of material actually removed or converted to another form is quite small. Polarography is the name applied to dc voltammetry at the dropping mercury electrode. Although the term *polarography* is acceptable in this context, its more general usage is undesirable.

The integrated current (current multiplied by time), or charge (coulombs), is a measure of the total amount of material converted to another form. In controlled-potential coulometry and in controlled-potential electroanalysis (see Chapter 24), this corresponds to the total removal of the reactant solute species.

There is yet another group of methods in which electron-transfer reactions and diffusional transport are unimportant. Charge transport by migration forms the basis for conductometry and conductometric titrations (see Chapter 24).

21.1 ELECTROCHEMICAL CELLS

Electrode potentials are the electromotive force (emf) of electrochemical cells formed by the combination of an individual half-cell with a standard hydrogen electrode, with any liquid-junction potential that arises being set at zero. Thus, when the emf of each half-cell is mentioned, what is actually implied is the emf of the cell:

$$\text{Pt, } H_2(1 \text{ atm})^* \,|\, H^+(m = 1.228) \,\|\, M^{n+}(a = 1), M^\circ \qquad (21.1)$$

standard hydrogen electrode   individual half-cell
liquid junction

In this notation, a vertical line represents a phase boundary and a comma separates two components in the same phase. A double vertical line represents a phase boundary, which may have an associated liquid-junction potential.

The overall chemical reaction that takes place in the cell (Equation 21.1) is made up of two independent *half-reactions,* which describe the real chemical changes at

* One atmosphere equals 1.01325×10^5 pascals (newtons/meter^2).

the two electrodes. Interest often centers on only one of the reactions. The electrode at which the reaction of interest occurs is called the *working* (or *indicator*) electrode. A *reference half-cell* (see Chapter 22) is the other half of the cell. Since the reference electrode has a constant makeup, its potential is fixed. Therefore, any changes in the cell are ascribable to the working electrode (right-hand side of Equation 21.1). In subsequent chapters the methods will involve *observing* or *controlling* the potential of the working electrode with respect to the reference electrode.

By driving the working electrode to more negative potentials, the energy of the electrons is raised, and they eventually reach a level high enough to occupy vacant valence levels on chemical substances in the electrolyte. In this case, electrons flow from electrode to solution. This is a *reduction current*. Similarly, the energy of the electrons can be lowered by imposing a more positive potential. At some point electrons on solutes in the electrolyte find a more favorable energy on the electrode and transfer there. This flow is an *oxidation current*.

The critical potentials at which these processes commence are related to the standard potentials, E°, for the specific chemical substances in the system. In our example (Equation 21.1) these are

$$E^\circ_{\mathrm{H^+,H_2}} \quad \text{and} \quad E^\circ_{M^{n+},M^\circ}$$

for the hydrogen ion/hydrogen gas system and for the metal ion/metal system, respectively. Their difference is the *cell emf*. An applied potential (emf) must exceed the cell emf before any electrochemical reaction is initiated.

Effect of Concentration on Electrode Potentials

Each half-reaction responds to the interfacial potential difference at the corresponding electrode. The potential E of any electrode for an oxidation/reduction system (here abbreviated Ox/Red)

$$\mathrm{Ox} + ne^- = \mathrm{Red} \tag{21.2}$$

is given by the generalized form of the Nernst equation:

$$E = E^\circ - \frac{RT}{nF}\ln\frac{a_{\mathrm{red}}}{a_{\mathrm{ox}}} = E^\circ - \frac{2.3026RT}{nF}\log\frac{a_{\mathrm{red}}}{a_{\mathrm{ox}}} \tag{21.3}$$

where E° is the standard electrode potential, R is the molar gas constant, T is the absolute temperature, n is the number of electrons transferred in the electrode reaction, and a_{ox} and a_{red} are the activities of the oxidized and reduced forms involved in the electrode reaction, respectively. If concentrations are substituted for activities, common logarithms substituted for natural logarithms, and numerical data inserted for the constants, and assuming the temperature is 298 K, the Nernst equation becomes

$$E = E^{\circ\prime} - \frac{0.05916}{n}\log\frac{[\mathrm{red}]}{[\mathrm{ox}]} \tag{21.4}$$

where $E^{\circ\prime}$ is the formal electrode potential defined in terms of concentration rather than activities. A change of one unit in the logarithmic term changes the value of E by $59.16/n$ mV. Those oxidation/reduction systems that respond in this manner are often called *reversible* or *nernstian* because the chemical reactions obey thermodynamic relationships.

Effect of Complex Formation on Electrode Potentials

The effect of reagents that can react with one or both participants of an electrode process is significant. The simplest case involves a single ionic species formed over a range of concentrations of the complexing agent. A typical example is the silver ion/silver metal couple in the presence of aqueous ammonia, where the $Ag(NH_3)_2^+$ complex ion is the major ionic species in the solution phase.

The formation of the silver diammine complex is represented by the equilibrium

$$Ag^+ + 2NH_3 \rightleftharpoons Ag(NH_3)_2^+ \tag{21.5}$$

for which the formation constant is written as

$$K_f = \frac{[Ag(NH_3)_2^+]}{[Ag^+][NH_3]^2} = 6 \times 10^8 \tag{21.6}$$

For the half-reaction involving the silver ion/silver couple, the electrode potential is expressed by

$$E = E^\circ + 0.05916 \log[Ag^+] \tag{21.7}$$

Combining Equations 21.6 and 21.7 yields the potential of a silver electrode in aqueous ammonia systems:

$$E = E^\circ + 0.05916 \log \frac{1}{K_f[NH_3]^2} + 0.05916 \log[Ag(NH_3)_2^+] \tag{21.8}$$

The shift in electrode potential caused by the complexing agent is contained in the second term on the right-hand side of Equation 21.8. Complexation extends the utility of electrodes. The silver electrode can be used for the determination of ammonia, organic amines, and, in fact, any ligand that complexes with the silver ion.

21.2 CURRENT-POTENTIAL RELATIONSHIPS

In general, when the potential of an electrode is moved from its equilibrium (or zero-current) value toward more negative potentials, the substance that is reduced first is the oxidant in the couple with the least negative E°. The current that passes through the solution is carried by the migration of cations to the cathode and of anions to the anode.

Figure 21.1 gives oversimplified, idealized current-potential curves of solutions with varying concentrations of copper(II) ion. Starting with a well-stirred solution of 0.05*M* copper(II) ion and impressing an applied potential across a platinum and a reference electrode, the current-potential curve is traced by curve *OAB*. No current flows until the applied potential is more negative than the most positive electrode potential at which copper plates from a solution that contains 0.05*M* copper(II) ions. At point *A*, copper commences to plate on the platinum electrode and current starts to flow. As the applied potential is gradually made more negative, the current increases linearly (at least for a time) in accordance with Ohm's law. On the other hand, if the applied potential is changed from B to A, the current diminishes gradually to zero (and any deposit of copper dissolves from the platinum electrode). For smaller concentrations of copper, the current-potential traces are given by curves OCD and OEF. In each case, the potential at which copper commences to plate is shifted along the potential axis to a more negative value by 29.5 mV for each tenfold decrease in the concentration of copper(II) ions. The current-potential traces are called *voltammograms.*

Since these oxidation/reduction reactions are governed by the laws of electrolysis enunciated by Faraday, they are called *faradaic* processes. Electrodes at which faradaic processes occur are sometimes called *charge-transfer* electrodes. *Faraday's laws* are (1) the amount of chemical decomposition or the amount of substance deposited on an electrode is proportional to the total charge passed through the electrolyte, and (2) the amounts of different substances liberated at an electrode by the same current flowing for the same time (current × time = coulombs) are proportional to their chemical equivalents.

Under some conditions a given electrode-solution interface shows a range of potentials where no charge-transfer reactions occur because such reactions are thermodynamically or kinetically unfavorable. However, processes such as adsorption and desorption can occur, and the structure of the electrode-solution interface can change with changing potential or solution composition. These processes are called *nonfaradaic.* Although charge does not cross the interface under these conditions, external currents can flow (at least transiently) when the potential, electrode area, or solution composition changes.

An electrode at which there is no charge transfer across the metal-solution interface regardless of the potential impressed by an outside source of voltage is called a *polarizable* electrode. Take, for example, a platinum electrode dipping into a 0.0008M solution of copper(II) ions that is also 0.1M in sulfuric acid. When short-circuited with a saturated calomel reference electrode (a *nonpolarizable* electrode whose potential is constant and fixed by the composition of the solution

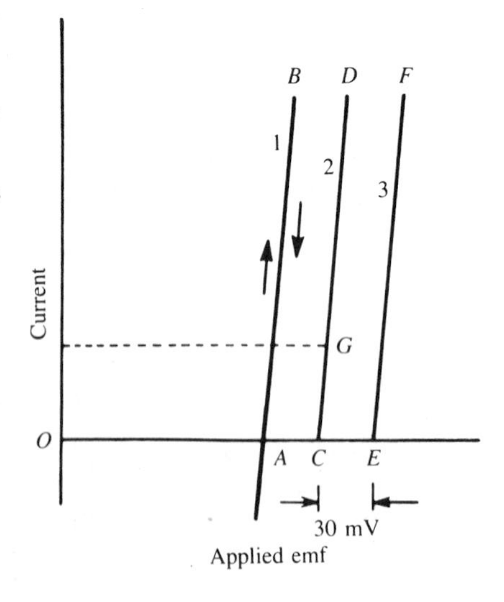

FIGURE **21.1**

Current-potential curves in a well-stirred solution: curve 1: 0.05M copper; curve 2: 0.005M copper; curve 3: 0.0005M copper.

surrounding the mercury), the platinum electrode assumes the potential of the reference electrode (0.2445 V). No current flows. The platinum electrode is polarized, and it remains polarized until a potential is impressed across the two electrodes (with the platinum electrode being the cathode) that exceeds the electrode potential at which copper(II) ions are reduced to metallic copper. At that point metallic copper starts to deposit on the cathode:

$$Cu^{2+} + 2e \rightleftharpoons Cu^{\circ} \qquad (E^{\circ} = 0.337 \text{ V}) \tag{21.9}$$

When the electrode surface is covered with copper, the (working) electrode becomes depolarized. Its potential is now determined by the equation for the reversible oxidation/reduction of the copper(II)/copper system at 25 °C:

$$E = E^{\circ} + \frac{0.05916}{2} \log[Cu^{2+}] \tag{21.10}$$

As long as the working (or reference) electrode is depolarized, passage of current does not cause the electrode potential to deviate from its reversible value.

Stated another way (as shall be seen in voltammetric methods; see Chapter 23), a polarizable electrode is one whose potential is easily changed upon the passage of only a tiny current. By contrast, the potential of a nonpolarizable electrode, such as a reference electrode, remains essentially constant even when an appreciable current flows. Polarized and depolarized are experimental, phenomenological terms, and they do not imply that the electrode reaction is reversible or irreversible in either a chemical or thermodynamic sense.

Returning to Figure 21.1 but using experimental conditions that involve no stirring of the solution and with a microelectrode (an electrode with a very small area of contact with the solution) and impressing an applied potential corresponding to D, the current starts to flow at a value corresponding to D but decreases rapidly to some value G as the concentration of copper(II) ions at the microelectrode surface becomes depleted. The current, represented by G, is characteristic of the rate at which fresh copper(II) ions are supplied to the microelectrode by diffusion. Under these conditions the microelectrode is polarized.

When a current flows across an electrode-solution interface, the electrode potential normally changes from the reversible value it exhibits before the passage of current. The difference between the measured potential (or cell emf) and its reversible value is the *overpotential*. Both cathodic and anodic processes exhibit overpotential, which is affected by many factors. When an anodic process shows an overpotential effect, the applied potential necessary to bring about an electrochemical reaction is always a more positive value than the calculated equilibrium potential, and, for cathodic processes, overpotential requires a more negative applied potential than the calculated value to be used. Although overpotential phenomena complicate the calculation of the applied potential necessary for electrochemical reactions to occur, their effect makes feasible certain separations that would not be expected merely from the consideration of standard electrode potentials.

Complexation may affect the overpotential. The exchange rate of electrons between the electrode and complexed species may be faster or slower than the rate of exchange with the aquated ion. For example, aquonickel ions show an overpotential of about 0.6 V at a mercury surface, whereas complexes of nickel with thiocyanate or

TABLE 21.2 HYDROGEN OVERPOTENTIAL ON VARIOUS CATHODES

Cathode	First visible gas bubbles	Current density	
		$0.01\ A\ cm^{-2}$	$0.1\ A\ cm^{-2}$
Antimony	0.23	0.4	—
Bismuth	0.39	0.4	—
Cadmium	0.39	~0.4*	—
		—	1.2
Copper	0.19	0.4	0.8
Gold	0.017	0.4	1.0
Lead	0.40	0.4	1.2
Mercury	0.80	1.2	1.3
Platinum (bright)	~0	0.09	0.16
Silver	0.097	0.3	0.9
Tin	0.40	0.5	1.2
Zinc	0.48	0.7†	—

SOURCE: J. J. Lingane, *Electroanalytical Chemistry*, 2nd ed., p. 209, Wiley-Interscience, New York, 1958. Reproduced by permission.
NOTE: The electrolyte is $1M\ H_2SO_4$. Overpotentials are given in volts.
* $0.005M\ H_2SO_4$.
† $0.01M\ Zn(C_2H_3O_2)_2$.

pyridine, though actually shifting the equilibrium electrode potential more negative, show a decrease in overpotential that more than compensates for the shift in the equilibrium potential.

The evolution of gases at an electrode is sometimes associated with an overpotential. The anodic overpotential of oxygen evolved on smooth platinum in acidic solutions is approximately 0.4 V. The overpotential involved in the reduction of hydrogen ion (or hydrogen in water) to hydrogen gas, as shown in Table 21.2, is large on metals such as bismuth, cadmium, lead, tin, zinc, and especially mercury. With a mercury cathode, a number of useful separations of metal ion/metal systems from the hydrogen ion/hydrogen gas system become possible.

21.3 MASS TRANSFER BY MIGRATION AND CONVECTION

Electron transfer can be limited by the mass transfer of ions to or from the electrode or by the rate of electron transfer at the electrode-solution interface. In the subsequent discussion it is assumed that the rate of electron transfer at the electrode-solution interface is fast compared with the rate of mass transfer. A *limiting current* results whenever the rate of the electrode reaction is governed by the rate of supply of the electroactive species to the electrode surface by a process whose rate is more or less independent of the electrode potential. There are three general mass-transfer processes by which a reacting species may be brought to an electrode surface: (1) migration of charged ions in an electric field, (2) convection due to motion of the solution or the electrode, and (3) diffusion under the influence of a concentration gradient.

Migration

Electrical migration is restricted to ionic substances or dipolar molecules capable of undergoing directed movement in an electrical field. Contributions to the limiting current by electrical migration depend directly on the transference number (the fraction of the current carried by a given ion) of the ion in the solution and on the electrical potential gradient in the vicinity of the electrode surface. Electrical migration may augment or oppose diffusion depending on whether the electroactive species is a cation or an anion.

In voltammetry the effect of electrical migration can be eliminated by adding some innocuous salt ("supporting electrolyte") in a concentration at least 100-fold greater than the concentration of the substance being determined. A potassium or quaternary ammonium salt is often used. These cations cannot be discharged at the cathode until the impressed potential becomes quite negative. Large numbers of them remain as a cloud around the cathode. This positively charged cloud restricts the potential gradient to a region close to the electrode surface. In the presence of a supporting electrolyte, the transference number of the electroactive species is decreased practically to zero, and the resistance of the solution, and thus the potential (iR) gradient through it, is made desirably small. Under this condition, the limiting current in a quiescent solution is controlled entirely by the diffusional rate of mass transfer.

Convection

Electroactive species may be moved by thermal currents and by density gradients, also whenever the solution is stirred into the path of the electrode (*hydrodynamic transport*). These processes are grouped under the general term *convection*. Although convection can be minimized by using quiescent solutions, in general, diffusion measurements for times longer than about 30 sec are difficult, and even measurements longer than 20 sec may show some convective effects. At longer times the buildup of density gradients and the existence of stray vibrations cause convective disruption of the diffusion layer. Of course, when interest centers around complete removal of some particular electroactive species, efficient stirring and perhaps elevated temperatures should be used.

21.4 MASS TRANSFER BY DIFFUSION

Diffusion, the movement of a chemcial species under the influence of a gradient of chemical potential (that is, a *concentration gradient*), is of paramount concern in voltammetric techniques. Concentration gradients at the electrode surface are dependent on time. In transient or dynamic methods the system is intentionally disturbed from equilibrium by excitation signals consisting of variable potentials or currents. When either a potential or a current is impressed upon the system, the other variable is a function of both the concentration of the electroactive species and time. When the potential reaches a value at which a faradaic process occurs, current flows. The current is the rate at which charge passes through the electrode-solution interface. Current that arises from a faradaic process is a direct measure of the rate of the process and, if the rate is proportional to concentration, it is also a measure of the concentration of electroactive species in the bulk solution.

Fick's First Law

Fick's laws are differential equations describing the flux or movement of a substance and its concentration under diffusion control as functions of time and position. Any concentration distribution can be described from a microscopic model. Consider a particular location x. At time t, assume that $N(x)$ molecules are to the left of this location and that $N(x + dx)$ molecules are to the right. After a time increment, dt, the molecules originally at x will have moved an average distance dx. Because the driving force for the motion of these molecules is a random thermal process, the same number go to the right as go to the left. The net number of molecules that move between x and $(x + dx)$ is given by the difference between the number of molecules on the left and the number on the right. The net movement per unit time, or flux, through an area A is

$$\frac{1}{A}\frac{dN}{dt} = \frac{\text{flux}}{\text{area}} = \frac{\left[\frac{N(x)}{2}\right] - \left[\frac{N(x + dx)}{2}\right]}{A\,dt} \tag{21.11}$$

Multiplying the right-hand side of Equation 21.11 by $(dx)^2/(dx)^2$ converts the number of molecules to concentration, and noting that the concentration of the species is given by $C = N/(A\,dx)$, one obtains

$$\text{flux} = \left[\frac{(dx)^2}{dt}\right]\left[\frac{C(x + dx) - C(x)}{dx}\right] \tag{21.12}$$

The first term on the right-hand side of Equation 21.12 is a function of the average distance the molecules have moved, dx, in a given time, dt. It is a constant, characteristic of the particular system of solute and solvent molecules, and is called the *diffusion coefficient, D*. By allowing dx and dt to approach zero, the differential form, Fick's first law, is obtained:

$$\text{flux per unit area} = -\frac{1}{A}\frac{dN}{dt} = -D\left(\frac{dC}{dx}\right) \tag{21.13}$$

which states that the flux is proportional to the concentration gradient dC/dx, which is decreasing as the electrode surface is approached. The net transfer of solute mass per unit time across a plane intersecting a concentration profile is proportional to the steepness of the profile and in the downhill direction. Molecules are diffusing down their respective concentration gradient, meaning that they diffuse in the direction in which the sign of dC/dx is negative from the molecules' frame of reference.

Fick's Second Law

Fick's second law pertains to the change in the concentration gradient as a function of time. It is derived from the first law by noting that the change in concentration at a location x_2 is given by the differences in the flux into and the flux out of an element of width dx:

$$\left(\frac{dC}{dt}\right)_{x_2} = \frac{D\left(\frac{dC}{dx}\right)_{x_2 - dx} - D\left(\frac{dC}{dx}\right)_{x_2 + dx}}{dx} \tag{21.14}$$

When dx and dt approach zero, and because D is assumed independent of x and t, for planar electrodes Fick's second law becomes

$$\frac{dC}{dt} = D\left[\frac{d^2C}{(dx)^2}\right] \tag{21.15}$$

Concentration Gradients

Now assume that an electroactive species is present and that the potential impressed upon the system causes an electrode reaction to occur. According to Fick's first law, the net rate of diffusion of a species to a unit area of electrode surface A at any time t is proportional to the magnitude of the concentration gradient:

$$\text{flux} = -D\left(\frac{dC}{dx}\right)_{x=0} = \frac{-D(C - C_o)}{\delta} \tag{21.16}$$

where δ is the thickness of the hypothetical diffusion layer about the microelectrode (Figure 21.2). Once the reaction potential has been reached, the concentration of the electroactive species of interest changes from the bulk concentration value, C, observed at positions far removed from the electrode surface, to the concentration, C_o, at the electrode surface. As C_o approaches zero, the rate of diffusion becomes proportional to the concentration in the bulk of the solution.

The concentration in the thin diffusion layer is approximately a linear function of distance from the electrode (Figure 21.2), so that as a first approximation,

$$\frac{dC}{dx} = \frac{C - C_o}{\delta} \tag{21.17}$$

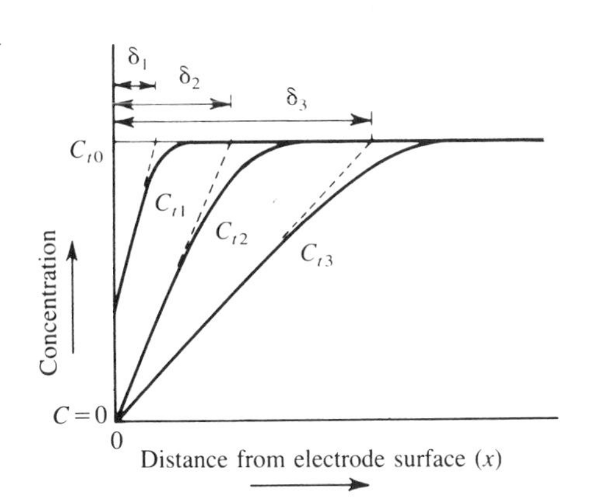

FIGURE **21.2**
Concentration profiles for reacting species at various times after the start of electrolysis under diffusion-controlled conditions. $C_{t0} = C_{\text{bulk}}$; $t_3 > t_2 > t_1$; t_0 = before electrolysis; δ_i refers to thickness of the diffusion layer at time t_i.

Instantaneous Current

Regardless of the current-controlling factor, dN/dt is the number of moles that react at the electrode in unit time. The rate of discharge of ions is equal to i/nF, the quantity of electricity carried by one equivalent of electroactive species. Thus,

$$\frac{i}{nF} = \frac{dN}{dt} \tag{21.18}$$

When the current-controlling factor is the rate of diffusion of the reacting substance through the depleted thin layer of solution in contact with the electrode, the diffusion-limited current at any particular time, $(i_d)_t$, can be expressed in electrochemical parameters by combining Equations 21.13, 21.17, and 21.18:

$$(i_d)_t = \frac{nFAD(C - C_o)}{\delta} \tag{21.19}$$

The applied potential does not have to be increased very much beyond the electrode potential before C_o becomes so small compared with C that $(C - C_o)/\delta$ at any time t becomes essentially equal to C_t/δ, and the expression for the current becomes

$$(i_d)_t = \frac{nFDAC}{\delta} \tag{21.20}$$

The diffusion-limited current is proportional to the concentration and inversely proportional to the thickness of the diffusion layer. A diffusion-limited current decreases with time because of the buildup of the diffusion layer.

Cottrell Equation

When equilibrium is established at a planar microelectrode in an unstirred solution that also contains a large concentration of supporting electrolyte, the rate of discharge of the ions is equal to the rate of diffusion of the electroactive species to the electrode. For times greater than zero and the distance from the electrode approaching infinity, solving these boundary conditions by the use of appropriate mathematical techniques* gives

$$C_{(x,t)} = C \operatorname{erf}\left[\frac{x}{2D^{1/2}t^{1/2}}\right] \tag{21.21}$$

where erf is the error function.

Of primary interest is the variation of the measured current with time. The diffusing material is electrolyzed as soon as it reaches the electrode surface. By taking the partial derivative of Equation 21.21, with $x = 0$, we obtain

$$\left(\frac{\delta C}{\delta x}\right)_{x=0} = \frac{C}{\pi^{1/2}D^{1/2}t^{1/2}} \tag{21.22}$$

* A rigorous derivation of this equation is given by L. B. Anderson and C. N. Reilley in *J. Chem. Educ.*, **44**, 9 (1967), and by A. J. Bard and L. R. Faulkner in *Electrochemical Methods*, John Wiley, New York, 1980, p. 142. Fortunately, the chemist using electrode reactions can accomplish a great deal without more than a cursory appreciation of the mathematics.

Substitution of Equation 21.16 into Equation 21.22 yields the Cottrell equation:[1]

$$i = nFAD\left(\frac{\delta C_{(0,t)}}{\delta x}\right) = \frac{nFACD^{1/2}}{\pi^{1/2}t^{1/2}} \tag{21.23}$$

Its validity was verified in detail by Kolthoff and Laitinen.[2,3]

If the electrode is spherical rather than planar, a spherical diffusion field must be considered. Fick's second law becomes

$$\frac{dC}{dt} = D\left(\frac{\delta^2 C}{(\delta r)^2} + \frac{2}{r}\frac{\delta C}{\delta r}\right) \tag{21.24}$$

where r is the radial distance from the electrode center. The resulting expression for current is

$$i = nFACD\left[\frac{1}{(\pi D t)^{1/2}} + \frac{1}{r}\right] \tag{21.25}$$

The Cottrell equation points out the following facts: (1) The observed faradaic current is directly proportional to the concentration of electroactive material. This relationship is the foundation of quantitative voltammetry. (2) The faradaic current is also directly proportional to the square root of the diffusion coefficient of the electroactive ion. The influence of temperature on the faradaic current is quite marked, particularly because the diffusion coefficient of many ions changes 1%–2% per degree in the vicinity of 25.0 °C, the standard temperature often chosen for voltammetric work. This implies that the temperature of the solution in the electrochemical cell must be controlled to within 0.5 °C. (3) The effect of depleting the electroactive species near the surface is characterized by an inverse $t^{1/2}$ function. This kind of time dependence is encountered in a number of experiments.

Reversible and Irreversible Reactions

For the current-potential relations described, the rate of electron transfer at the electrode surface was assumed to be fast enough to maintain the surface concentrations of reactants and products very close to their equilibrium values; that is, the rate of electron transfer is fast (very rapid electrode kinetics) in comparison with the mass transport process. When this is true, the reaction is called *reversible*. If the energy of activation for the electron-transfer reaction is large and the rate of electron transfer is correspondingly slow, in order to observe the net current, the forward process has to be so strongly activated (by application of an overpotential) that the back-reaction is totally inhibited. The concentrations at the electrode surface are not equilibrium values and the Nernst equation is not applicable. Voltammetric reactions of this type are called *irreversible*.

Electrode processes are not always very facile or very sluggish. In these quasi-reversible cases, the net current involves appreciable activated components from the forward and reverse charge transfers. Most electrode reactions involving the reduction or oxidation of organic compounds are irreversible to some degree.

In essence, there is no such thing as a perfectly reversible electrode reaction. In a practical sense a reaction is called reversible if, within the limits of error of experimental measurement, its behavior follows the Nernst equation under the given experimental conditions. Consequently, a reaction that appears to be reversible under one set of experimental conditions may not behave reversibly under a different set of experimental conditions.

Summary

The diffusion of an electroactive species takes place through a rather sharply defined and relatively thin layer of solution in contact with the electrode. When long reaction times (slow scan rates) are used, the surface concentration initially is the same as the bulk concentration. When the reduction potential of the electroactive species is exceeded, the surface concentration depletes toward zero as the applied potential becomes more negative (assuming the scan is in the negative direction). The final value is governed by the concentration and the diffusion coefficient of the electroactive species at potentials well past the reduction potential. Realistically, the diffusion profile is that shown by the curved portions in Figure 21.2.

When the voltammetric method uses rapid changes in potential, either because the scan rate is fast or because a pulse modulation of some type is used, the slope of the concentration gradient at any particular potential is greater than the slope obtained from slow scans. The bulk concentration is maintained closer to the electrode surface, the number of electroactive particles that arrive at the surface per unit time is greater, and larger current signals result.

21.5 CLASSIFICATION OF INDICATOR ELECTRODES

The simplest situation is a piece of metal in contact with a solution of its ions—for example, silver dipping into a silver nitrate solution. One interfacial equilibrium is involved:

$$Ag^+ + e^- \rightleftharpoons Ag \tag{21.26}$$

The expression for the electrode potential is

$$E = 0.799 - 0.05916 \log \frac{1}{[Ag^+]} \tag{21.27}$$

This is called an electrode of the *first kind*. The metal must be stable with respect to air oxidation, especially at low ion activities. Suitable electrodes are restricted essentially to mercury(I)/mercury and silver(I)/silver.

Electrodes of the *second kind* involve two interfaces, such as a metal coated with a layer of one of its sparingly soluble salts. The underlying electrode must be a reversible electrode of the first kind. Consider a silver wire coated with a thin deposit of silver chloride. At the Ag/AgCl solution interface the electrochemical equilibrium is

$$AgCl(s) + e^- \rightleftharpoons Ag + Cl^- \tag{21.28}$$

In addition, there is a chemical equilibrium,

$$AgCl(s) \rightleftharpoons Ag^+ + Cl^-, \qquad K_{sp} = 1.8 \times 10^{-10} \tag{21.29}$$

Combining these two equations gives the Nernst expression for the chloride ion:

$$E = 0.799 + 0.05916 \log K_{sp} - 0.05916 \log[Cl^-] \tag{21.30}$$

This simplifies to

$$E = 0.2222 - 0.05916 \log[Cl^-] \quad (21.31)$$

Thus when the electrode system is immersed in a filling solution that contains a constant amount of the chloride ion, the electrode potential is constant. Second-kind electrodes can be used as indicators for either the cation or the anion. Limitations on these electrodes are severe. They can be used only over a range of anion activities such that the solution remains saturated (at the interface) with respect to the metal salt coating. Other anions can interfere if they also form an insoluble salt with the cation of the underlying electrode.

Electrodes of the *third kind* involve an electrochemical equilibrium and two chemical equilibria. For example, a small mercury electrode (or gold amalgam wire) in contact with a solution saturated with the mercury chelate of ethylenediaminetetraacetic acid (EDTA, H_4Y) follows the concentration of a second metal ion, M, which also forms a complex with the same chelating agent, Y^{4-}, but a complex that is less stable than the mercury complex. The half-cell established is

$$Hg|HgY^{2-}; MY^{(n-4)+}$$

The electrode potential is given by

$$E = E^\circ + \frac{0.05916}{2} \log \frac{[M^{n+}][HgY^{2-}]}{[MY^{(n-4)+}]} \quad (21.32)$$

Because a fixed amount of HgY^{2-} is present, the electrode potential is dependent on the ratio $[M^{n+}]/[MY^{(n-4)+}]$.

An "inert" electrode, such as gold or platinum, immersed in a solution containing both the oxidized and reduced states of a homogeneous and reversible oxidation/reduction system, is capable of following the ratio of the two oxidation states. An example is a platinum wire in contact with a solution of iron(III) and iron(II) ions. For the half-reaction

$$Fe^{3+} + e^- \rightleftharpoons Fe^{2+}, \qquad E^\circ = 0.771 \text{ V} \quad (21.33)$$

the electrode expression is

$$E = 0.771 - 0.05916 \log \frac{[Fe^{2+}]}{[Fe^{3+}]} \quad (21.34)$$

In many oxidation/reduction systems, the inert electrode is not reversible for one of the half-reactions, as is the case for thiosulfate in iodometric titrations. If the nonreversible system attains a chemical equilibrium quickly with a reversible system (for example, iodine/iodide), however, the latter will serve as the potential-determining half-reaction.

BIBLIOGRAPHY

BARD, A. J., AND L. R. FAULKNER, *Electrochemical Methods, Fundamentals and Applications,* John Wiley, New York, 1980.

GALUS, Z., *Fundamentals of Electrochemical Analysis,* Ellis Harwood, Ltd., Chichester, U.K., 1976.

KISSINGER, P. T., AND W. R. HEINEMAN, eds., *Laboratory Techniques in Electroanalytical Chemistry,* Dekker, New York, 1984.

LINGANE, J. J., *Electroanalytical Chemistry*, 2nd ed., Interscience, New York, 1958.

LITERATURE CITED

1. COTTRELL, F. G., *Z. Physik. Chem.,* **42,** 385 (1902).
2. LAITINEN, H. A., AND I. M. KOLTHOFF, *J. Am. Chem. Soc.,* **61,** 3444 (1939).
3. LAITINEN, H. A., *Trans. Electrochem. Soc.,* **82,** 289 (1942).

22 POTENTIOMETRY

Potentiometric methods embrace two major types of analyses. One involves the direct measurement of an electrode potential from which the activity (or concentration) of an active ion may be derived. The other type involves measuring the changes in the electromotive force (emf) brought about by the addition of a titrant to the sample.

22.1 ELECTROCHEMICAL CELLS

In a potentiometric type of sensor, a membrane or sensing surface acts as a half-cell, generating a potential proportional to the logarithm of the analyte activity (concentration). This potential is measured relative to an inert reference electrode that is also in contact with the sample. Potentiometric measurements are made under conditions of essentially zero current flow so as not to disturb the equilibrium at the sample-membrane interface.

For a complete electrochemical cell from which negligible current is drawn, the emf is given by

$$E_{\text{cell}} = E_{\text{ind}} - E_{\text{ref}} + E_j \tag{22.1}$$

where E_{ind}, E_{ref}, and E_j are the potentials of the indicator electrode, the reference electrode, and the liquid-junction potential, respectively. For conditions in which a current is drawn from the cell, the term for the iR drop (the potential gradient across the solution due to the product of the current flowing and the resistance of the solution) must be included in Equation 22.1. The indicator electrode senses the presence of the analyte in the sample solution, the reference electrode is independent of the sample solution composition, and the liquid (or fluid) junction is an interface between dissimilar solutions. In a properly designed system, E_{ref} is a constant and E_j is either constant or negligible. When these conditions are realized, the indicator electrode can supply information about ion activities.

22.2 REFERENCE ELECTRODES

A reference electrode is an oxidation/reduction half-cell of known and constant potential at a particular temperature. Three main requirements for a satisfactory reference electrode systems are reversibility, reproducibility, and stability, which are interrelated. Its potential must not depart from its equilibrium value when current demands are made upon it, such as the passage of a small current through the cell. This may become important for miniature electrodes and at low concentrations of the potential-determining ion. The problem can be minimized by selecting a reference electrode with large concentrations of the potential-determining ions, with a fairly large active electrode area, and a not-too-small volume of solution. Reproducibility involves two aspects. One is the ability of a particular reference electrode to respond according to the Nernst equation without temperature- (or concentration-) change hysteresis. The other is the feasibility of establishing an easy and standard method of electrode preparation and assembly that will produce electrodes that exhibit a constant potential within an acceptable standard deviation. The third requirement, stability, refers to the useful life of a reference electrode.

A reference electrode consists of three principal parts: (1) an internal element; (2) some filling solution, which constitutes the salt-bridge electrolyte; and (3) an area in the tip of the electrode that permits a slow, controlled flow of filling solution to escape the electrode (called the fluid junction) into the sample under a head of a few inches and where an electrical connection is made with the other components of the electrochemical cell (Figure 22.1).

Internal Elements

The choice of a reference electrode for most applications is between the mercury/mercury(I) chloride (or calomel) half-cell and the silver/silver chloride half-cell as the internal element. Both are electrodes of the *second kind*—that is, anion reversible. The pertinent chemical and electrochemical equilibria were discussed in Section 21.3. When the electrode system is immersed in a filling solution that contains a constant amount of the chloride ion, the electrode potential is constant. The nonpolarizability of reference electrodes of this type is specifically designed into the electrode. Through the use of massive metal electrodes accom-

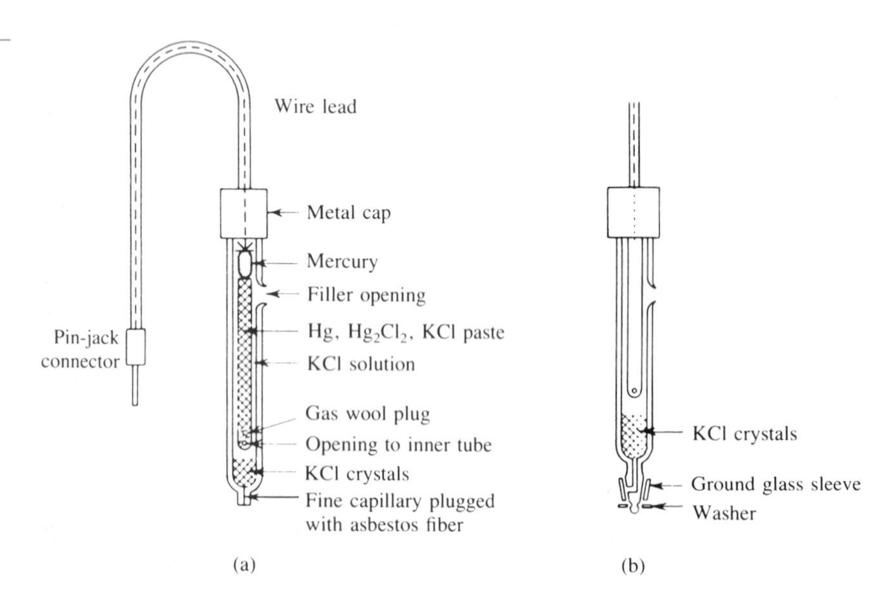

FIGURE 22.1
Calomel reference electrodes: (a) fiber type and (b) sleeve type.

panied by pastes or coatings of the oxidation product of the half-cell reaction, the internal element offers so large an area that polarization is precluded even at currents considerably larger than those typically drawn by high-input impedance operational amplifiers.

Calomel Electrodes

Calomel electrodes comprise an inert or unattackable metal, such as platinum, in contact with mercury and a paste of mercury(I) chloride, mercury, and potassium chloride, moistened with the filling solution. It is enclosed in an inner glass tube and makes contact through a porous plug (e.g., glass wool) with a filling solution of potassium chloride of known concentration (usually 0.1*M*, 1.0*M*, or saturated, 4.2*M*) and saturated with mercury(I) chloride. The saturated calomel electrode (SCE) is often used because it is easy to prepare and maintain. For accurate work other concentrations of KCl are preferred because they reach their equilibrium potentials more quickly and their potential depends less on temperature. The saturated calomel electrode exhibits a perceptible hysteresis following temperature changes, due in part to the time required for a solubility equilibrium to be re-established. At temperatures higher than 80 °C, all calomel electrodes are subject to problems arising from the accelerated disproportionation of mercury(I) to the metal and mercury(II) ions.

Silver/Silver Chloride Electrodes

The silver/silver chloride electrode is made in wire form and cartridge form. In the former, a silver wire is coated with silver chloride by electrolysis or by dipping into the molten salt. The wire is immersed in a filling solution of potassium chloride of known concentration, usually 1.0*M*, and saturated with silver chloride. In the cartridge form, the metal is in contact with a paste of the salt moistened with electrolyte, and all is enclosed in an inner glass tube. Contact with the filling solution in the salt bridge is made through a small aperture. All silver/silver chloride systems demonstrate excellent electrical and chemical stability. The silver/silver chloride electrode should not be used in solutions that contain proteins, sulfide, bromide, iodide, or any other materials that would precipitate (or complex) with the silver found in the filling solution. Strong reducing agents also should be avoided because they can reduce the silver ions to silver metal at the liquid junction.

Liquid Junctions

Electrical contact between the sample and the reference electrode is established by a slow leak of electrolyte (filling) solution through a fluid or liquid junction. The filling solution forms a continuous electrical conduction path from the internal element to the lower tip of the electrode. It also serves to protect the internal element from contamination that would change the reference electrode potential. The leakage rate should be slow, but satisfactory performance depends on a continuous, unimpeded flow of the filling solution out of the electrode through the junction.

At the boundary between two dissimilar solutions (or solids), there is always a fairly high resistance that involves an appreciable ohmic drop when a current is flowing. A junction potential is always involved. It results from the fact that the mobilities of positive and negative ions diffusing across the boundary are unequal.

Because of this difference, one side of the boundary accumulates an excess of ions of one charge type. The junction potential adds to or subtracts from the potential of the reference electrode, depending on which side of the boundary becomes positive. In making emf measurements it is very important that this potential be the same when the reference electrode is in the standardizing and calibration solutions and in the sample solution.

The junction potential is less when the ions of the electrolyte have nearly the same mobilities. In general, a filling solution should be within 5% of being equitransferent; that is, the fractions of the current carried through the solution by the cations and anions should be nearly equal. This accounts for the common use of a potassium chloride solution in salt bridges. A mixture of K^+, Na^+, NO_3^-, and a small amount of Cl^- in the appropriate ratios minimizes the liquid-junction potential. Also the ionic strength of the filling solution should be at least five to ten times greater than the maximum ionic strength expected in the sample and standardizing solutions. This effect counteracts the differences between the transference numbers of the ions in the two half-cell solutions and thus decreases the liquid-junction potential. Approximate liquid-junction potentials are given by Milazzo for several boundary systems.[1] The following data for two rather extreme cases illustrate the magnitude of the junction potential.

	Transference numbers		
Junction	**Cation**	**Anion**	**E_j (mV)**
HCl (0.1*M*) | HCl (0.01*M*)	0.83	0.17	+28
KCl (3.5*M*) | KCl (0.1*M*)	0.49	0.51	−0.2

Transference numbers ordinarily vary slightly with concentration and temperature.

Liquid junctions come in several physical forms. A sleeve or ground junction is designed for samples like slurries, emulsions, suspensions, pastes, and gels, which require a relatively high flow of electrolyte (about 0.1 mL per hour) and a self-cleaning facility. This junction comprises a loose but captive sleeve ground to a tapered section of the electrode stem, which has a hole on the inside tapered portion. Another use is for titrations in nonaqueous solvent systems where the overall resistance at the junction must not be too high. Considerably less filling solution leaks through a porous ceramic, cracked bead, or quartz fiber junction, about 8 μL per hour. One of these latter junctions should be used when contamination of the test solution by the filling solution must be avoided. Double junction, sleeve-type salt bridges overcome problems with leakage of undesirable ions into the sample or compatibility of filling and sample solutions. This type is useful for careful ion-selective analyses.

22.3 THE MEASUREMENT OF pH

The pH scale is a series of numbers that express the degree of acidity of a solution, as contrasted with the total quantity of acid (or base) in some material as found by an alkalimetric (or acidimetric) titration. Sorensen[2] proposed the term *pH*, and defined it as the negative logarithm of the hydrogen ion concentration, expressed in molarity:

$$pH = -\log[H^+] \tag{22.2}$$

The term *pH* is simply a mathematical symbol of convenience, widely accepted and firmly established, but devoid of exact thermodynamic validity. It is the activity of the hydrogen ion that is formally consistent with the thermodynamics of the pH electromotive cell. The activity definition is

$$paH = -\log a_{H^+} \tag{22.3}$$

The activity is the product of concentration and an activity coefficient; that is, $a_{H^+} = f_+[H^+]$. However, single ionic activity coefficients cannot be measured directly; only the mean ionic activity coefficient is available. Now $[H^+]$ and $f_\pm[H^+]$ are often the most useful units for expressing the acidity of aqueous solutions, where $f_\pm$ is the mean ionic activity coefficient. Unfortunately, the established experimental pH method cannot furnish either of these quantities. For those interested in a detailed treatment of the historical development of the concept of pH, the work by Bates, cited in the Bibliography at the end of this chapter, should be consulted.

Operational Definition of pH

The relation by which the emf of a suitable pH-measuring electrode is related to the hydrogen ion concentration was developed by Nernst:

$$E = E^\circ - \left(\frac{RT}{nF}\right)\log[H^+] \tag{22.4}$$

where E° is a potential dependent on the electrode system used, R is the gas constant, T the absolute temperature, F (the faraday) is 96,485 coulombs/mole, and n is the number of electrons involved in the equilibrium. For $n = 1$, the factor $RT/nF = 0.591$ (at 25 °C). Using the Sorenson expression (Equation 22.2) for pH,

$$E = E^\circ + 0.0591 \text{ pH} \tag{22.5}$$

Equation 22.5 establishes the well-known relationship of 59.1 mV per pH unit (at 25 °C) for any electrode system that follows the Nernst equation. Strictly, ion-selective electrodes (which include pH-responsive glass electrodes) do not obey the Nernst equation, although the appropriate equation (see Equation 22.11) has a very similar form. The Nernst equation applies to a redox reaction, whereas the equation for ion-selective electrodes applies to an accumulation of charge at an interface without electron transfer.

Since the glass electrode is the widely accepted detector for pH measurements, it is the only pH-responsive element that will be considered in this text. The National Bureau of Standards (USA) pH scale has been established by assigning pH values to certain standard solutions, with the values chosen to give the maximum possible consistency between precise thermodynamic information, such as dissociation constants of weak acids, and Equation 22.4. In so doing, the pH scale is defined in an operational manner, convenient and easily reproducible but not exact thermodynamically:

$$pH = pH_s + \frac{E - E_s}{2.302RT/F} \tag{22.6}$$

In this definition, E and E_s are, respectively, the emf of an electrochemical cell of the usual design that contains either the unknown solution or a standard reference material of known pH—namely, pH_s:

electrode reversible to hydrogen ions	unknown or standard buffer solution	salt bridge	reference electrode

The pH_s reference materials were assigned values from measurements of the emf of cells that contained hydrogen gas and silver/silver chloride electrodes—that is, cells without a liquid junction:

$$\text{Pt} \,|\, \text{H}_2(1\text{ atm}), \text{H}^+\text{Cl}^-(\text{plus K}^+\text{Cl}^-), \text{AgCl}(s) \,|\, \text{Ag}$$

by the equation

$$E = E^\circ - 0.000198T \log f_{H^+} f_{Cl^-} m_{H^+} m_{Cl^-} \tag{22.7}$$

where E° is the standard potential of the cell. By rearranging Equation 22.7 in terms of the acidity function, $p(a_{H^+} f_{Cl^-})$, one gets

$$p(a_{H^+} f_{Cl^-}) = -\log f_{H^+} f_{Cl^-} m_{H^+} = \frac{E - E^\circ}{0.000198T} + \log m_{Cl^-} \tag{22.8}$$

The pH_s of the chloride-free buffer solution is computed from the equation

$$pH_s = p(a_{H^+} f_{Cl^-})^\circ + \log f^\circ_{Cl^-} \tag{22.9}$$

where $p(a_{H^+} f_{Cl^-})^\circ$ is the value obtained by evaluating $p(a_{H^+} f_{Cl^-})$ at several concentrations of chloride and extrapolation to zero chloride concentration. The activity coefficient of the chloride ion can be estimated for ionic strengths less than 0.1 by the equation

$$-\log f^\circ_{Cl^-} = \frac{A\sqrt{\mu}}{1 + 1.5\sqrt{\mu}} \tag{22.10}$$

where A is a parameter of the Debye-Hückel theory that has a different value at each temperature. The recommended values of pH_s may be found in *Lange's Handbook of Chemistry.*[3] The total uncertainty in pH_s, exclusive of any liquid-junction potentials introduced during the calibration of pH equipment, is estimated as 0.005 pH unit (0 to 60 °C) and 0.008 pH unit (60 to 95 °C). The operational definition of pH is valid for only dilute solutions and for the pH range 2–12 because it is only under these conditions that the liquid-junction potential remains constant between the test solution and the electrolyte in the reference electrode required for the measurement. To detect any serious impairment of the response of a measuring device and electrode assembly outside the pH range 2–12, two secondary standards are included among the pH reference materials. These are potassium tetroxalate and calcium hydroxide solutions.

For accuracy of ± 0.01 pH unit, the temperature should be known to $\pm 2\ ^\circ C$. Not only does the proportionality factor between the cell emf and pH vary with temperature, but dissociation equilibria and junction potentials also have significant temperature coefficients.

The necessity for estimating the individual activity coefficients of the chloride ion in each reference solution deprives the pH_s value of exact fundamental meaning. Nevertheless, the operational definition of pH, chosen in part for its reasonableness but largely for its utility, agrees as closely as possible with the mathematical concepts evolved from the present state of solution theory. Fortunately, a highly precise knowledge of the solution pH is seldom required. Neither is it necessary to know exactly what a particular pH value means. Often in an industrial process it is sufficient to know that at a certain stage a particular pH value is maintained.

pH or pION Measurement System

The measurement of pH (or pION, or simply pI) is one of the most common analytical techniques used in chemistry laboratories. A pH or pI measurement system always consists of four parts: a sensing electrode, a pH (or ion) meter that contains a high-input-impedance operational amplifier (see Chapter 3) to measure and display the cell potential, a reference electrode (see Section 22.2), and the sample being measured. To achieve a reproducibility of ± 0.005 pH unit, an amplifier is needed that is reproducible to at least 0.2 mV. The high electrical resistance (5–500 MΩ) of the glass membrane necessitates measuring circuits with a high-input-impedance voltmeter. Negligible current must be drawn during the measurement if changes in the ion concentration at the electrode surface are to be avoided and no error is to arise from the voltage drop across the inherent resistance of the electrochemical cell. With glass electrodes the current drawn should be 10^{-12} A or less. pH measurements often suffer from the effects of incorrect materials or incorrect maintenance. Certain protocols should be followed for pH measurements and for properly maintaining a pH meter.[4,5]

The pH meter is a voltmeter but with several critical additional functions. Not only does it measure the potential across the pH-sensing and reference electrode system, but it also converts the potential difference measurement at a given temperature into pH terms, and it provides mechanisms to correct for the nonideal behavior of the electrode system. The schematic circuit diagram of a potentiometric type of pH or ion meter that is based on an operational amplifier is shown in Figure 22.2. The operational amplifier not only serves as a high-impedance voltmeter, but also provides stability and automatic operation through the use of the feedback loop. The operational controls on a pH meter are best understood by reference to Figure 22.3. The relationship for the emf of a pH assembly is

$$E = k - KT(\mathrm{pH}) \tag{22.11}$$

Equation 22.11 is the equation of a straight line with slope $-KT$ and a zero-intercept of k. The pH-sensing electrode and its reference electrode must have an isopotential point at 0 V. The isopotential point expresses the zero shift in pH terms and is identified with the pH of a solution in which the emf of the pH assembly does not vary with temperature. This is accomplished by using as the internal solution in the pH-sensing electrode a buffer whose pH change with temperature exactly compensates the temperature changes of the internal and external reference electrodes.

The proper emf/pH slope involves adjusting the KT factor (actually $2.3026RT/F$) to 59.16 mV per pH unit at 25 °C by means of a slope control to rotate the emf/pH slope about the isopotential point (usually pH 7.00). The temperature compensator, which is reserved to correct the slope for the actual temperature of the sample, varies the instrument definition of a pH unit from 54.20 mV at 0 °C to 66.10 mV at 60 °C and 74.04 mV at 100 °C. Many pH meters are equipped to accept an automatic temperature compensator probe, which automatically measures the sample temperature and adjusts the meter sensitivity for the correct temperature.

A single point standardization involves immersing the pH assembly into a standard reference pH buffer whose value lies near the pH expected for the sample. The meter reading is brought into juxtaposition with the pH_s value by adjusting the intercept (standardization, zero, asymmetry) control. This control shifts the

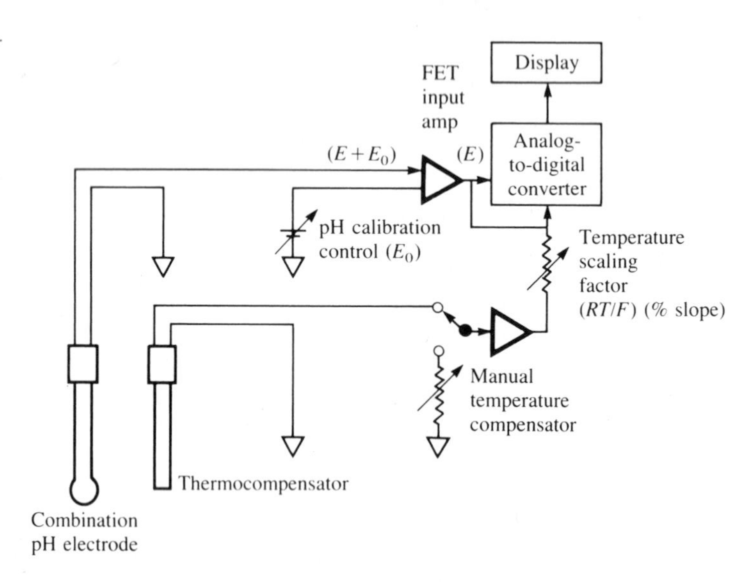

FIGURE **22.2**
Schematic circuit diagram of a pH meter.

FIGURE **22.3**
(a) Typical pH electrode response as a function of temperature (the *slope* control) and (b) operation of the *intercept* control shown schematically.

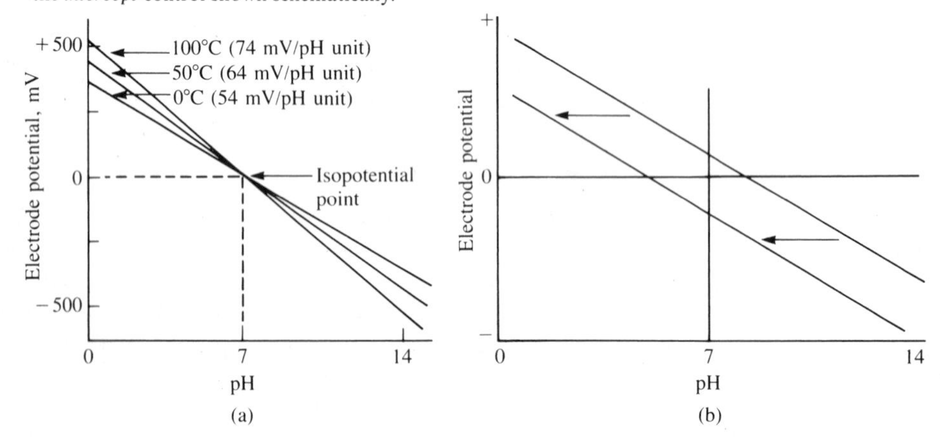

response curve laterally until it passes through the isopotential point (Figure 22.3b). The procedure entails adding or subtracting a dc voltage to correct for the offset voltage (from the isopotential point). Once done, the standardization control must remain undisturbed unless the entire standardization step is repeated. Because both the K and T terms affect the slope of the pH response, standardization and calibration are both carried out at the anticipated temperature of the sample.

Standardization at only one pH value does not assure the validity of readings at other pH values considerably removed from the value of the standard reference buffer. On pH meters that allow two calibration points, a second step involves the calibration of the pH meter with a second standard reference buffer. After the calibration described in the preceding paragraph, the electrode assembly is transferred to either an acidic or basic standard reference buffer so that the two reference buffers bracket the expected pH of the sample. The pH value of the second buffer is set on the meter by adjusting the slope (gain, calibrate) control. What transpires is an adjustment of the amplifier gain to correct for any non-Nernstian behavior of the indicator electrode. This amounts to adjusting the K term in Equation 22.11.

The preceding discussion concerns traditional pH meters in which an amplifier circuit is used to adjust the signal of the electrode so that it can be read directly in pH units.

Microcomputer-based pH meters measure the output of the electrode assembly and subject it to appropriate algorithms based on Equation 22.11. The algorithms perform compensatory calculations equivalent to standardization and calibration. Most pH meters of this type permit standardization and calibration in any order. However, if steps are taken out of order or steps are introduced in the middle of a prescribed operating protocol, the pH meter may not function properly. There is no isopotential point in the meter; the microcomputer typically assumes the isothermal point of the electrode to be pH 7 or 0.0 mV.

Glass-Indicating Electrodes

The classic electrode assembly consists of a glass pH-indicating electrode and a reference electrode. Typical pH-sensitive glass membranes are either sodium/calcium silicate (Corning 015 glass) or lithium silicates with lanthanum and barium ions added. These added ions act as lattice "tighteners" to retard silicate hydrolysis and lessen alkali ion, chiefly sodium ion, mobility. Sodium or lithium ions are the bulk mobile charge carriers under an applied electric field. Upon immersion of the membrane in water, the surface layer becomes involved in an ion-exchange process between the hydrogen ions in the external solution and the sodium or lithium ions of the membrane. The content of H^+ decreases in a complex way with increasing distance into the membrane, whereas Li^+ or Na^+ content increases in such a way that the sum of positive ions, charge carriers, and other cations balances the presumed uniform fixed-site concentration of anions.

The activity of water in the solution appears to play an important role in the development of the pH response of the glass membrane. If the ionic strength of the solution is extremely high, or if a nonaqueous solvent is present, the measured emf deviates from the expected value. All glass electrodes must be conditioned for a time by soaking in water or in a dilute buffer solution, even though they may be used subsequently in media that are only partly aqueous.

The conventional pH electrode has an internal reference electrode (silver/silver chloride or calomel) immersed in a chloride salt buffer (usually a phosphate buffer at pH 7) solution, with the glass membrane separating it from the test solution (Figure 22.4). The body of the glass electrode is a nonconducting glass tube. This is sealed to a bulb made of special conductive glass, which is the pH-sensing membrane. The body is filled with a buffered electrolyte with fixed pH value and ionic concentration. A phosphate buffer is used in most electrodes. An external reference electrode completes the assembly. This design assures that constant potentials are developed on the inner surface of the glass membrane and on the internal reference element. When the electrode assembly is immersed in a solution of pH 7, the sum of these fixed voltages approximately balances the voltage developed on the outer surface of the glass membrane and the external reference electrode.

The combination electrode contains both the pH-sensing and reference electrodes, combined in a single probe body (Figure 22.5).

With glass pH-responsive electrodes, chemical durability and electrical resistance are linked together. So-called universal glass electrodes have an electrical resistance of approximately 100 MΩ at 25 °C, permitting their use at temperatures down to at least 0 °C where resistance increases to about 1000 MΩ. The thick, rugged membrane withstands even rough handling as might be encountered in industrial applications. They exhibit a fast response over the entire 0–14 pH range; the temperature range is from −5 to 110 °C. However, they are not recommended for constant use in solutions with very low or very high pH because electrode life is shortened.

For long-term measurements at extreme pH values, a full-range, high-pH glass electrode should be used. This type of electrode has superior chemical durability, though at the cost of a fourfold increase in electrical resistance and a corresponding elevation of the lower temperature limit of use to 10 °C. This is the glass membrane of choice for measurements in solutions with high alkali metal content and high pH values.

A glass electrode exhibits a reasonably rapid response to rapid and wide changes of pH in buffered solutions. However, valid readings are obtained more

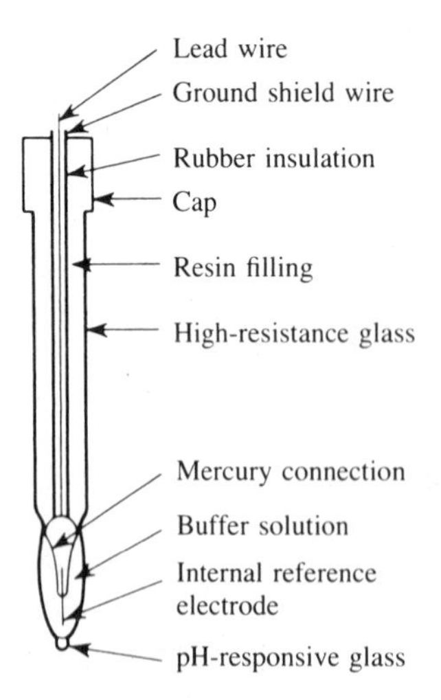

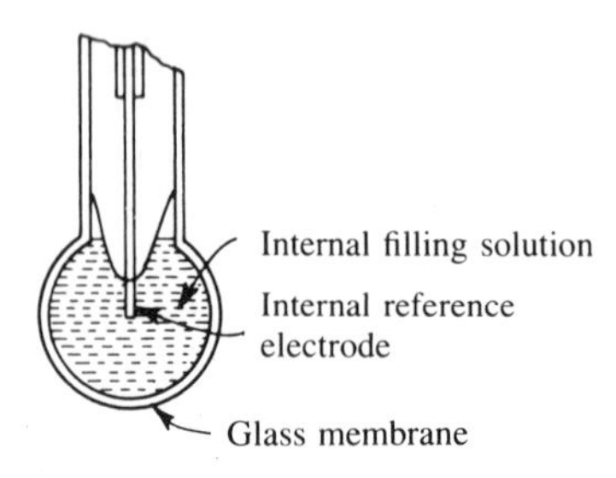

FIGURE **22.4**
Construction of a glass pH-responsive electrode.

slowly in poorly buffered or unbuffered solutions, particularly when changing to these from buffered solutions, as after standardization and calibration. The electrode should be thoroughly washed with distilled water after each measurement and then rinsed with several portions of the next test solution before making the final reading. Poorly buffered solutions should be vigorously stirred during measurement; otherwise, the stagnant layer of solution at the glass–solution interface tends toward the composition of the particular kind of pH-responsive glass. Suspensions and colloidal material should be wiped from the glass surface with a soft tissue.

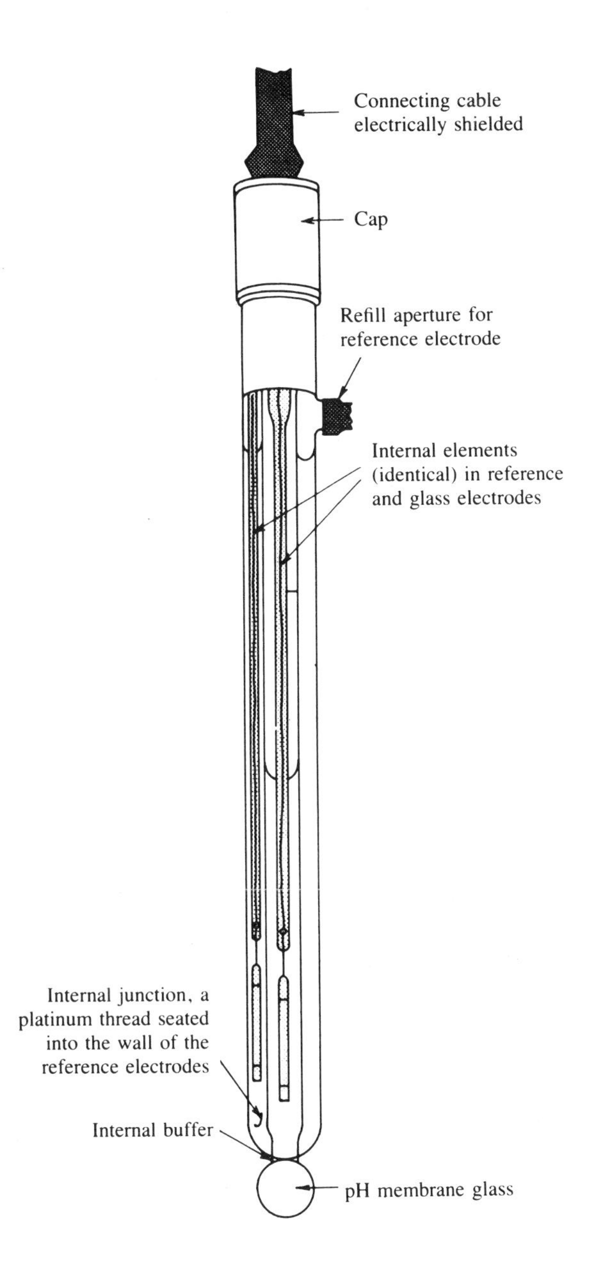

FIGURE **22.5**
Combination pH/reference electrode. (Courtesy of Sargent-Welch.)

Commercial glass electrodes are fabricated in a wide variety of sizes and shapes and for many special applications. Syringe and capillary electrodes require only one or two drops of solution, or even less in ultramicro work, whereas others penetrate soft solids (such as leather) or pastes. The normal-size electrode operates with a volume of solution of 1–5 mL.

22.4 ION-SELECTIVE ELECTRODES

A number of ion-selective electrodes have taken their place beside the historical pH glass electrode discussed in the preceding section. From the analytical viewpoint, ion-selective electrodes are nearly ideal measurement tools because of their ability to monitor selectively the activity of certain ions in solution both continuously and nondestructively. No oxidation/reduction reactions are involved, in contrast to the direct indicating electrodes used in classical potentiometry. All seem to involve an ion-exchange process or a related phenomenon, such as complexation or precipitation, with the active sites on the surface or in the hydrated layer of the electrode membrane. The potential of an ion-selective electrode is actually composed of two or more discrete contributions arising from the various processes at the interfaces and in the bulk of the active membrane material. If a charge separation occurs between ions at an interface, a potential difference is generated across that interface. The problem is to find an interface whose composition will selectively favor one type of ion over all others.

Ion-selective electrodes measure ion activities, the thermodynamically effective free ion concentration. In dilute solutions, ion activity usually approaches the ion concentration. Activity measurements are valuable because the activities of ions determine rates of reactions and chemical equilibria. For example, ion activities are important parameters in predicting corrosion rates, extent of precipitation, formation of complexes, solution conductivities, effectiveness of metal pickling baths and electroplating bath solutions, and physiological effects of ions in biological fluids.

There are three types of membrane sensors: glass, solid state, and solid matrix (liquid ion exchange). Gas-sensing electrodes and biocatalytic electrodes are merely special designs that incorporate one of the three types into the system. Novel approaches in sensor design are discussed by Czaban.[6]

Glass Membrane Electrodes

The various types of ion-selective glass electrodes are all members of a continuum of glass electrodes that belong to the "fixed-site" category of ion-exchange membrane electrodes. This simply means that the active sites on the surface or in the hydrated layer of the glass are not free to move about during the time scale of the measurement. The electrode potential arises from a combination of cation exchange and cation mobility factors that lead to an accumulation of charge at the glass-solution interface. The observed selectivity ratio of these electrodes is the product of the ion-exchange equilibrium constant between the sites and the solution and the mobility ratio of the exchanging ions in the hydrated layer of the glass. Thus, the selectivity properties of a desired electrode can be produced by appropriate adjustment of these parameters by altering the glass composition.

Three subtypes of glass electrodes and their selectivity order to ions can be summarized as follows:

pH type: $H^+ >>> Na^+ > K^+, Rb^+, Cs^+, \ldots >> Ca^{2+}$
Cation-sensitive type: $H^+ > K^+ > Na^+ > NH_4^+, Li^+, \ldots >> Ca^{2+}$
Sodium-sensitive type: $Ag^+ > H^+ > Na^+ >> K^+, Li^+, \ldots >> Ca^{2+}$

The second two subtypes also display considerable response to such univalent cations as thallium(I), copper(I), and alkyl quaternary ammonium ions, but they are primarily unresponsive to other univalent cations and generally quite unresponsive to anions.

As a rule, cation selectivity (over hydrogen ion) can be achieved by adding elements with coordination numbers that are higher than their oxidation numbers—for example, the substitution of aluminum(III) for silicon(IV) in alkali metal-silicate glasses (20% Na_2O:10% CaO:70% SiO_2). Such charge-deficient elements apparently leave the glass with an excess of negatively charged ion sites that attract cations with the proper charge-to-size ratio. Glasses that contain less than about 1% Al_2O_3 yield good pH-responsive electrodes with little metal-ion response. Glasses that have a composition about 27% Na_2O:5% Al_2O_3:68% SiO_2 show a general cation response. Glasses of the composition 11% Na_2O:18% Al_2O_3:71% SiO_2 are highly sodium-selective with respect to other alkali metal ions. More complex glasses that contain other additives frequently yield electrodes with superior mechanical and electrical properties.

The glass electrode construction is the same as shown in Figure 22.4 for pH-responsive electrodes. These electrodes function well in organic solvents and in the presence of lipid-soluble or surface-active molecules. The sodium-responsive glass electrode has extensive use in clinical work.

Solid-State Sensors

The nonglass solid-state sensors replace the glass membrane with an ionically conducting membrane. The electrode body is composed of a chemically resistant epoxy formulation. Bonded to the electrode body is the sensing membrane, which is composed of a single, pure, nonporous material with a homogeneous mirrorlike surface of low microporosity that keeps sample retention to a minimum.

The fluoride electrode is a characteristic solid-state sensor. A cross-sectional view is shown in Figure 22.6. The active membrane is a single crystal of LaF_3, doped with europium(II) to lower its electrical resistance and facilitate ionic charge

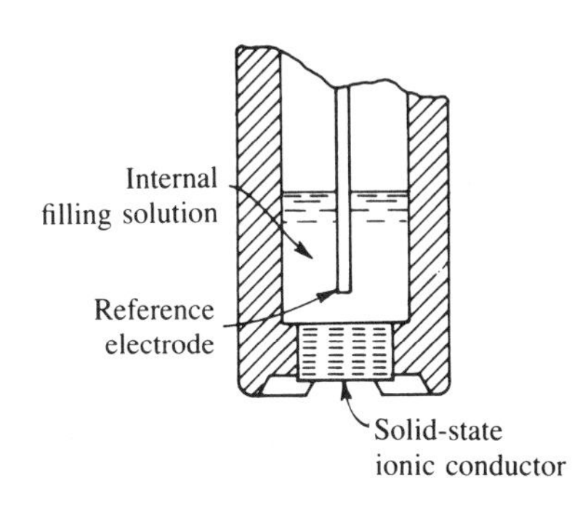

FIGURE **22.6**
Cross-sectional view of solid-state sensor. (Courtesy of Orion Research, Inc.)

transport. Typically, the internal solution is $0.1M$ each in NaF and NaCl. The fluoride ion activity controls the potential of the inner surface of the LaF_3 membrane, and the chloride ion activity fixes the potential of the internal silver/silver chloride wire reference electrode. In contact with the sample, the external surface of the membrane responds to the fluoride ion activity in the sample. The electrochemical cell that incorporates the fluoride electrode is

$$\mathrm{Ag} \mid \mathrm{AgCl}(s), \mathrm{Cl}^-(0.1M), \mathrm{F}^-(0.1M) \mid \mathrm{LaF}_3(s) \mid \text{sample} \parallel \text{reference electrode}$$

It obeys an electrode relationship of the form

$$E = \text{constant} + \frac{RT}{F} \ln \frac{[F^-]_{\text{int}}}{[F^-]_{\text{ext}}} \tag{22.12}$$

If the $[F^-]_{\text{int}}$ is assumed to be constant, Equation 22.12 simplifies to

$$E = \text{constant} + 0.05916\text{pF} \tag{22.13}$$

at 25 °C. Equation 22.13 is only an approximation. Equilibrium selectivity considerations are one of the limiting factors in the use of the lanthanum fluoride crystal as a fluoride-selective electrode. There is an accessibility region with the solution pH as the variable because the usefulness of this electrode is limited by hydroxide ion interference at high pH values (lanthanum hydroxide is less soluble than lanthanum fluoride and so depletes the available lanthanum sites on the crystal surface) and by the formation of hydrogen fluoride species (HF and HF_2^-) at low pH values. The calibration curve of activity versus potential shows that the electrode follows a logarithmic response to fluoride concentrations as low as $10^{-5}M$ and a useful response to at least $10^{-6}M$ fluoride ion. These lower limits are imposed by the solubility of the lanthanum fluoride in the sample solution.

EXAMPLE 22.1

A fluoride-selective electrode has a selectivity ratio of 0.10 relative to the hydroxide ion. At $10^{-5}M$ fluoride ion concentration, what hydroxide ion concentration could be tolerated

Since the fluoride electrode is ten times more sensitive to fluoride ion than to hydroxide ion, a $10^{-4}M$ hydroxide ion concentration will produce a response equal to that from a $10^{-5}M$ fluoride ion. To operate within the stipulated measurement error, the hydroxide ion concentration must be less than $10^{-4} \times 0.02 = 2 \times 10^{-6}M$, which is pH 8.3 or less. In practice, the hydroxide ion concentration is kept constant with buffer solutions. One recommended buffer consists of $0.25M$ acetic acid, $0.75M$ sodium acetate, $1M$ sodium chloride, and $1mM$ sodium citrate [for masking aluminum(III) and iron(III), which interfere by complexing fluoride ion]. The buffer controls overall ionic strength as well as pH.

EXAMPLE 22.2

What should be the lower pH limit when using the fluoride electrode if a 1% error can be tolerated?

The ionization constant for HF is 6.6×10^{-4} (and $pK_a = 3.18$). To never exceed a ratio of $[HF]/[F]^- = 0.01$, the pH must exceed or be 5.18. Comparing the lower pH limit of this example and the upper pH limit of Example 22.1, the need for a buffer that lies within the pH range 5.2–8.3 becomes obvious.

A group of solid-state selective ion electrodes is based on the polycrystalline Ag_2S membrane. By itself, it can be used either to detect silver ions or to measure sulfide ion levels by the mechanism of the solubility product equilibrium already discussed for the fluoride electrode. This electrode transports charge by the movement of silver ions, but the potential is determined by the availability of the sulfide ion or the silver ion in contact with the membrane. As a silver ion detector, it is superior to a silver metal electrode, since it is not attacked by strong oxidizing agents and it is not sensitive to redox couples in the solution. The dynamic range of the electrode extends from saturated solutions down to silver and sulfide levels on the order of $10^{-8} M$. The lower limit of detection reflects the experimental difficulty of preparing extremely dilute solutions of ions without extensive ionic adsorption on and desorption from the surfaces of the containing vessels and the electrodes. For sulfide measurements a special buffer must be mixed with sulfide-containing samples to raise the pH of the sample solution, free sulfide bound to hydrogen, fix the total ionic strength, and retard oxidation of the sulfide.

If this membrane is altered from pure silver sulfide by dispersing within it another metal sulfide, such as CuS, CdS, or PbS, the corresponding metal-selective electrode is obtained. The solubility product of the second metal sulfide must be much larger than that of silver sulfide yet sufficiently small that the level of metal ion in the sample solution produced from the solubility of the metal sulfide is small relative to the levels of the ion that are expected in the sample. In addition the sulfides of the membrane must equilibrate rapidly with the ions of the sample in order for the electrode to have a reasonable time response. Two solid-phase equilibria must now be established. The resulting kinetic complications restrict the number of possible metal selective ion electrodes of this type.

Mixed crystals of $AgX-Ag_2S$ make up the anion-selective electrodes of chloride, bromide, iodide, and thiocyanate, respectively. Again the solubility of the silver halide must be greater than that of silver sulfide. The mixed silver halide electrodes exhibit useful pI ($-\log[\text{ion}]$) ranges up to 5, 6, and 7, respectively, for chloride, bromide, annd iodide ions.

Solid or Liquid Matrix Electrodes

The sensing membrane of the solid matrix electrode uses an ion exchanger permanently embedded in a plastic material that is sealed to the electrode body. The membrane separates the internal filling solution and reference from the external sample solution. Except for the difference in the membrane material, the electrode construction resembles that of a solid-state electrode. In this type of electrode, the "sites" are free to move in the active phase (the membrane). This is an important advantage from the point of view of practical electrodes because it makes possible the design of electrodes that have appreciable selectivity for multivalent ions over univalent ions.

The calcium-selective electrode uses the calcium salt of bis(2-ethylhexyl)-phosphoric acid. The aqueous internal filling solution consists of a fixed concentration of calcium and chloride ions. For the nitrate (and fluoroborate) electrode, a nickel(II)-1,10-phenanthroline ion-pair association site group is used. The corresponding iron(III) ion association complex is used in the perchlorate-selective electrode. These liquid membrane electrodes are responsive exactly to those ions (Ca^{2+}, ClO_4^-, NO_3^-, and BF_4^-) that are extremely difficult to monitor by other techniques.

Membrane electrodes with polyvinyl chloride matrices incorporate different types of neutral carrier or ion-exchange molecules. The binding sites are incorporated into the semirigid, but liquid-state, plasticized polymer membrane. The most successful example is the potassium electrode based on valinomycin. This antibiotic molecule is a doughnut-shaped complex with an electron-rich pocket in the center into which potassium ions are selectively bound. The valinomycin molecule is devoid of charged groups but contains an arrangement of ring oxygens energetically suitable through ion-dipole interaction to replace the hydration shell around potassium. This provides a mechanism for potassium permeation across such normally insulating media. Valinomycin membranes show excellent potassium selectivity—about 3800 times that of sodium and 18,000 times that of hydrogen ions.

Gas-Sensing Electrodes

The construction of a gas-sensing electrode is shown in Figure 22.7. A gas-permeable membrane is used to isolate the analyte from possible interferences in the sample. A thin buffer layer is used to trap the analyte gas and convert it to some ionic species that can be detected potentiometrically. For example, ammonia gas can be monitored via the formation of ammonium ion (with an ammonium ion-selective electrode) or via the shift in pH (with a glass pH-responsive electrode) within the buffer layer. Dissolved ammonia from the sample diffuses through a fluorocarbon membrane until a reversible equilibrium is established between the ammonia level of the sample and the internal solution. Hydroxide ions are formed in the internal filling solution by the reaction of ammonia with water:

$$NH_3 + H_2O \rightleftharpoons NH_4^+ + OH^- \tag{22.14}$$

The hydroxide level of the internal filling solution is measured by the internal sensing element (pH-responsive glass electrode) and is directly proportional to the level of ammonia in the sample. Samples and standards are adjusted to a fixed pH, or to a pH greater than 11. Sensitivity extends from 10^{-6} to 1 M ammonia. Volatile amines may interfere but other common gases do not.

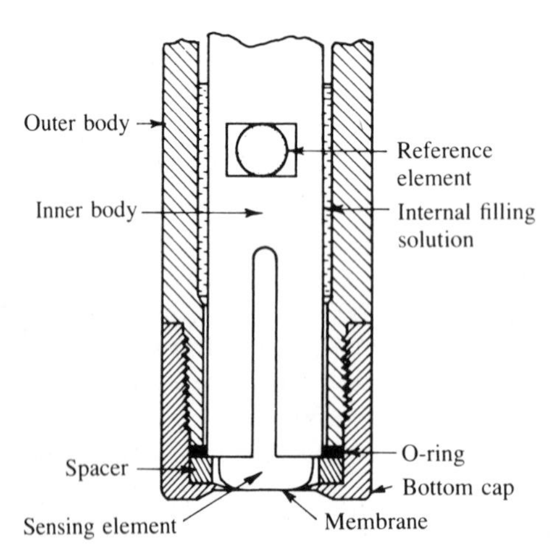

FIGURE 22.7
Construction of a gas-sensing electrode. (Courtesy of Orion Research, Inc.)

Gas-sensing electrodes are available for the measurement of carbon dioxide, nitrite, and sulfur dioxide. For example, the carbon dioxide sensor has either a silicon rubber or a microporous Teflon membrane separating the sample from an internal electrolyte layer of sodium hydrogen carbonate and a combination glass pH electrode. The equilibrium

$$CO_2 + H_2O \rightleftharpoons HCO_3^- + OH^- \tag{22.15}$$

enables a pH-responsive electrode to indirectly follow the carbon dioxide content of a sample. The sulfur dioxide/hydrogen sulfite ratio and the nitrous acid/nitrite equilibria can be followed similarly using the appropriate internal electrolyte layer.

Potentiometric gas electrodes are simple and reliable, but they tend to have a relatively slow response and recovery time (often 30 sec to 5 min).

Biocatalytic Membrane Electrodes

The basic arrangement of a biocatalytic membrane electrode is shown in Figure 22.8. Generally, the membranes are multilayered composites containing one or more biocatalysts, often enzymes, immobilized in a gel layer that coats a conventional ion-selective electrode. For the ammonium-ion-selective electrode, the enzyme urease is fixed in a layer of acrylamide gel held in place around the glass electrode bulb by a nylon mesh or a thin cellophane film. The urease acts specifically upon urea in the sample solution to yield ammonium ions, which diffuse through the gel layer and are sensed by the internal electrode. A pH-responsive glass electrode could replace the ammonium-ion-selective electrode, since the hydrogen ion is affected by the ammonia/ammonium ion ratio.

The enzyme layer thickness and the effective solution substrate diffusion constant are important parameters in determining electrode response times. This latter parameter is directly related to the degree of external solution stirring. The enzyme layer response is best described by enzyme kinetics and diffusion processes.

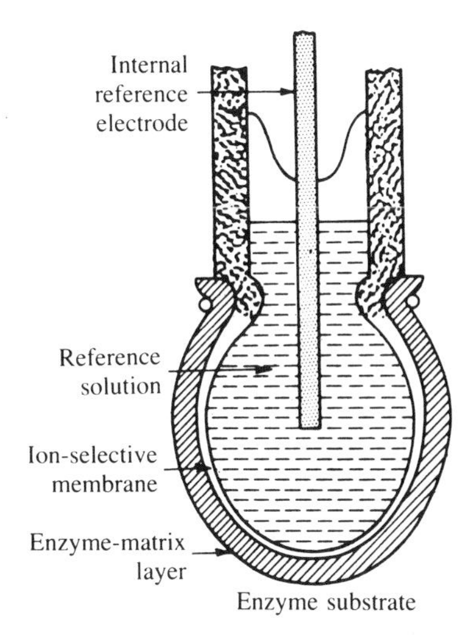

FIGURE **22.8**
Biocatalytic membrane electrode.

There are numerous enzyme–substrate combinations that would yield products measurable with ion-selective electrodes. A guanine-selective electrode has been developed in which a slice of rabbit liver is immobilized at the surface of an ammonia gas sensor. With modifications, biocatalytic membrane electrodes can be reversed to produce an enzyme-sensing electrode.

22.5 ION-ACTIVITY EVALUATION METHODS

The direct measurement technique requires a single emf reading on the sample solution. The sample's millivolt reading from an expanded scale pH/millivolt meter is compared with a previously prepared calibration curve, or with the sample concentration (or activity) read directly from the meter scale of a calibrated specific ion meter. Calibration procedures must use solutions of known activity or concentration, depending on which parameter is required. At less than 10^{-3} to $10^{-4} M$ the two quantities are practically indistinguishable. Direct measurement techniques are useful when samples are essentially pure solutions of the ion to be measured or have a relatively high and constant total ionic strength. To swamp effects caused by variations in total ionic strength, the ionic strength of the sample solution and the calibrating solutions may be adjusted by adding a high level of noninterfering ions.

An approach, similar to the operational definition of pH, can be applied to the problem of measuring the activities of other ions in solution. Standard reference values for pNa and pCl in sodium chloride solutions, pCa in calcium chloride solutions, and pF in sodium fluoride solutions have been determined by Bates and Alfenaar.[7]

In the method of standard addition (or subtraction), the concentration of a specific ion sample is estimated by observing the change in electrode potential when a known incremental (or decremental) change is made in the concentration of the ion in the sample. This approach requires neither preparation of a calibration curve nor calibration of logarithmic scales with standard additions. The ion is added to the test solution in a known amount that changes the total concentration by a known amount, ΔC, but does not change the total ionic strength appreciably or the fraction of the total concentration that is free. Thus, the initial reading is taken on a sample C_1, and the electrode response is

$$E_1 = \varepsilon + S \log(\gamma_1 C_1) \tag{22.16}$$

After the incremental addition,

$$E_2 = \varepsilon + S \log[\gamma_1(C_1 + \Delta C)] \tag{22.17}$$

Combining equations, one gets

$$\Delta E = E_2 - E_1 = S \log \frac{C_1 + \Delta C}{C_1} \tag{22.18}$$

where S is the slope factor or emf/pC slope. Known addition and subtraction methods are particularly suitable for samples with a high unknown total ionic strength. Where the species being measured is especially unstable, known subtraction is preferred over known addition.

22.6 INTERFERENCES

Chemical Interferences

Chemical interference arises when some sample ingredient prevents the probe from sensing the ion of interest. For example, a fluoride electrode can detect only fluoride ions. In an acid solution, however, a fluoride ion forms complexes with the hydrogen ion and is thereby masked from the detector, as shown in Example 22.2. Cations that form stable fluoride complexes interfere similarly. More drastic is the effect of a cyanide ion in contact with a mixed silver halide/silver sulfide membrane. In this case the reaction proceeds virtually to completion with the consumption of silver halide from the membrane.

Electrode Interferences

Electrode interferences arise when the electrode responds also to ions other than the test ion. In a first approximation, the effect of foreign cations on the electrode potential may be fitted by an extended Nikolsky equation:

$$E = \text{constant} + \frac{RT}{z_i F} \ln\left[a_i + \sum K_{ij} a_j^{(z_i/z_j)}\right] \tag{22.19}$$

where a_i is the activity of the primary ion whose charge is z_i, a_j is the activity of an interfering ion with a charge of z_j, and K_{ij} is the selectivity ratio characteristic of a given membrane. The selectivity ratio should ideally be small or zero for minimal interference. The value of the constant in the equation depends on the choice of reference electrodes.

The selectivity ratios can be determined in many ways. The simplest is shown in Figure 22.9. Potential measurements are made in solutions, with the activity of the ion to be measured varying in the presence of a constant background level of the interference. The intercept of the extensions of the response line (slope) with the line defining the plateau in the region of high interference defines a particular intercept activity for the primary ion, a_i. This particular activity is related to the selectivity

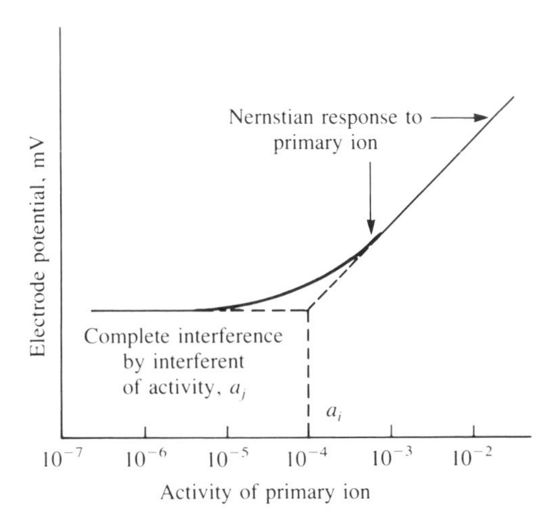

FIGURE 22.9 A method for calculating selectivity ratios.

ratio by the second term in the brackets of Equation 22.19:

$$a_i = K_{ij} a_j^{z_i/z_j} \tag{22.20}$$

where a_j is the constant background level of the interference. Knowing the activity of a_i and a_j then permits solving for the selectivity ratio.

In the separate solution method, the electrode response to various activities of the test ion and, separately, the electrode response of the interference are plotted, as shown in Figure 22.10. When the primary and interference activities are equal (vertical line) and z_i and z_j are equal,

$$\log K_{ij} = z_i \frac{E_2 - E_1}{0.0592} \tag{22.21}$$

Or, for a selected potential (horizontal line),

$$\log a_i = \log(a_j K_{ij}) \tag{22.22}$$

and

$$K_{ij} = \frac{a_i}{a_j} \tag{22.23}$$

In a third method, the interferent is varied with a constant level of primary ion. The interference activity is found from the intercept, as shown in Figure 22.11. This approach is extensively used for ascertaining hydrogen ion interference. The selectivity ratio is

$$K_{M/H} = \frac{a_{(M^{z+})}}{[a_{(H^+)}]^z} \tag{22.24}$$

The interference mechanism that occurs in crystal electrodes is quite different and requires a different method for expressing the selectivity coefficient. Surface reactions can convert one of the components of the solid membrane to a second insoluble compound. As a result, the membrane loses sensitivity to the ion being

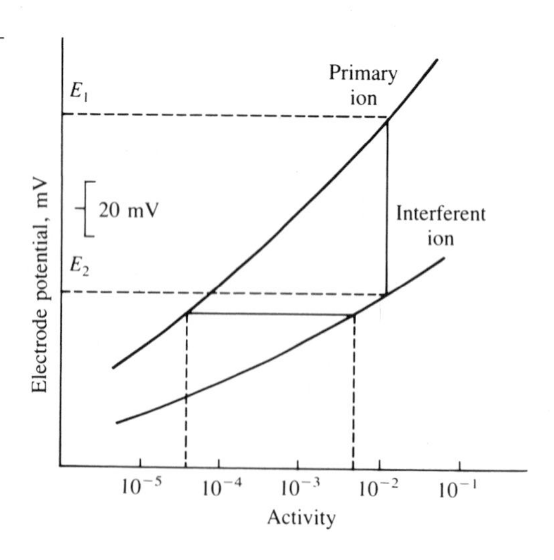

FIGURE **22.10**
Illustration of an electrode response with separate solutions used for the primary ion and the interferent ion.

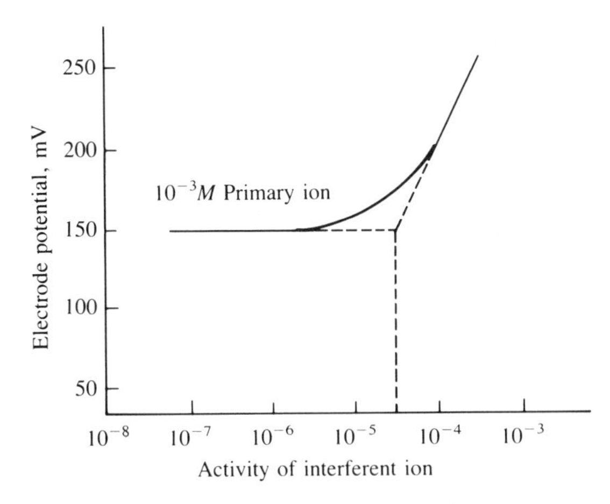

FIGURE **22.11**
Selective-ion electrode response in a solution of the primary ion at varying activities of an interference.

measured. For example, thiocyanate ion can interfere with bromide ion measurements if the reaction

$$SCN^- + AgBr(s) \rightarrow AgSCN(s) + Br^- \tag{22.25}$$

takes place, which will begin if the ratio of thiocyanate ion activity to bromide ion activity exceeds the value given by the ratio of the solubility products of silver thiocyanate to silver bromide, or

$$\frac{1}{k_i} = \frac{a_{SCN}}{a_{Br}} = \frac{1.00 \times 10^{-12}}{5.0 \times 10^{-13}} = 2.0 \tag{22.26}$$

Failure to give an expected response (slope) also occurs when the concentration of a dilute test solution approximates the solubility of the membrane material.

22.7 POTENTIOMETRIC TITRATIONS[8,9]

Less rigorous measurement techniques are involved when following the changes in the cell emf brought about by the addition of a titrant of precisely known concentration to the test solution. The equipment needed to carry out a classical potentiometric titration is illustrated in Figure 22.12. The method can be applied to any titrimetric reaction for which an indicator electrode is available to follow the activity of at least one of the substances involved. A reproducible equilibrium is of little concern. Requirements for reference electrodes are greatly relaxed. In contrast to direct potentiometric measurements, potentiometric titrations generally offer increased accuracy and precision. Accuracy is increased because measured potentials are used to detect rapid changes in activity that occur at the equivalence point of the titration. This rate of emf change is usually considerably greater than the response slope, which limits precision in direct potentiometry. Furthermore, it is the change in emf versus titration volume rather than the absolute value of the emf that is of interest. Thus, the influence of liquid-junction potentials and activity coefficients is minimized.

Location of the Equivalence Point

The critical problem in a titration is the recognition of the point at which the quantities of reacting species are present in equivalent amounts—the *equivalence point*. The titration curve can be followed point by point, plotting as the ordinate successive values of the cell emf versus the corresponding volume of titrant added as the abscissa. Additions of titrant should be the smallest accurately measurable increments that provide an adequate density of points, particularly in the vicinity of the equivalence point. Over most of the titration range the cell emf varies gradually, but near the end point the cell emf changes very abruptly. The resulting titration curve resembles Figure 22.13a. The problem in general is to detect this sharp change in cell emf that occurs in the vicinity of the equivalence point. The equivalence point may be calculated, as outlined in textbooks on analytical chemistry. Usually the analyst must be content with finding a reproducible point, as close as possible to the equivalence point, at which the titration can be considered complete—the *end point*. By inspection the end point can be located from the inflection point of the titration curve. This is the point that corresponds to the maximum rate of change of cell emf per unit volume of titrant added (usually 0.05 or 0.1 mL). The distinctness of the end point increases as the reaction involved becomes more nearly quantitative. Once the cell emf has been established for a given titration, it can be used to indicate subsequent end points for the same chemical reaction.

In the immediate vicinity of the equivalence point the concentration of the original reactant becomes very small, and it usually becomes impossible for the ion or ions to control the indicator electrode potential. The cell emf becomes unstable and indefinite because the indicating electrode is no longer bathed with sufficient quantities of each electroactive species of the desired oxidation/reduction couple. Usually a drop or two of the titrant will suffice to carry the titration through the

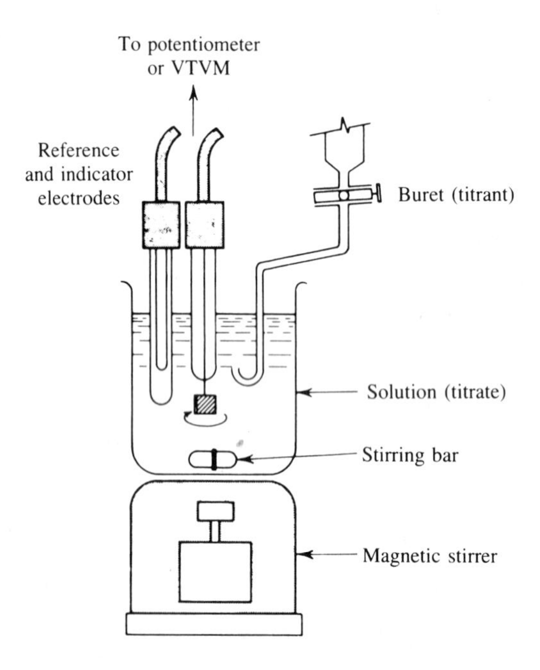

FIGURE **22.12**
Equipment for potentiometric titrations.

equivalence point and into the region stabilized by the electroactive species of the titrant. However, solutions more dilute than $10^{-3}M$ generally do not give satisfactory end points. This is a limitation of potentiometric titrations.

The end point can be more precisely located from the first or second derivative curves. Although either of these methods of selecting the end point is too laborious to do manually for each titration, the determination of derivatives, inflection points, and equivalence points becomes feasible with appropriate software algorithms.

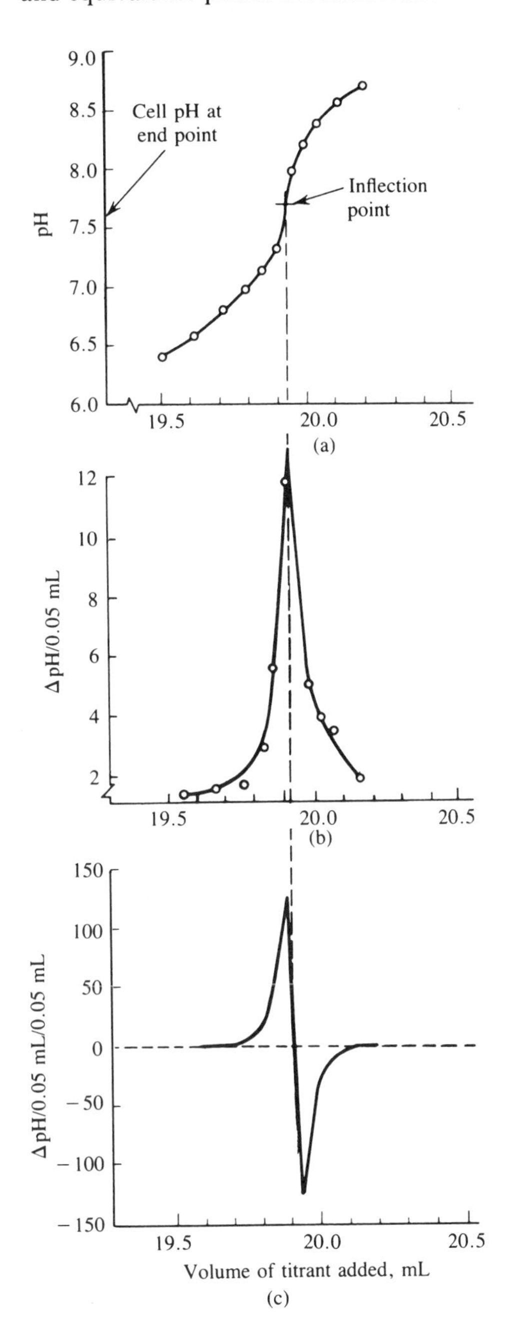

FIGURE **22.13**
Potentiometric titration curves: (a) experimental titration curve, (b) first derivative curve, and (c) second derivative curve.

Automatic Titrators

An entire titration can be performed automatically by titrators equipped with microcomputers and analog-to-digital converters, and using dedicated software. Digitally controlled stepper motors permit precise control of the titrant. A fully automatic unit accepts, serially, samples placed on a turntable. After each titration the turntable rotates, indexes the next sample beneath the electrode assembly, and actuates the titration switch. This type of instrumentation is ideal for performing multiple analyses in which the fundamental analytical procedure remains fixed, as in a quality control situation.

Automatic titrators can also be used to measure the volume of titrant required to maintain the indicator electrode at a constant potential. The volume added is plotted automatically versus time and becomes useful in kinetic studies. Since many enzymatic reactions result in the release or consumption of hydrogen ions, the amount of acid or base required to maintain a constant pH versus time gives a measure of the enzyme activity.

PROBLEMS

1. Determine the stability constant of the silver ammine ion from the cell

$$\text{Ag}\,|\,\text{AgNO}_3\ (0.025M),\ \text{NH}_3\ (1.00M)\,||\,\text{KNO}_3\,||\,\text{AgNO}_3\ (0.010M),$$
$$\text{KNO}_3\ (0.015M)\,|\,\text{Ag}$$

which has an emf of $+0.394$ V at 25 °C.

2. An ion-selective electrode and reference electrode pair were placed in exactly 100 mL of the sample; a reading of 21.6 mV was obtained. After the addition of exactly 10 mL of a standard solution with a concentration of 100 μg/mL, the electrode pair gave a reading of 43.7 mV. The response slope of the indicator electrode was previously determined to be 57.8 mV. What is the sample concentration?

3. Although no nitrite-selective electrode is available, suggest an indirect method to measure nitrite ion activity.

4. A fluoride solid-state electrode has a selectivity coefficient of 0.10 relative to hydroxide ion. At a $10^{-2}M$ fluoride concentration, what hydroxide ion concentration could be tolerated?

5. What should be the lower pH limit when using the fluoride electrode if a 5% error can be tolerated, and samples and standards are not adjusted to the same pH value?

6. Estimate the ratios at which the following ions can be present without impairing the response of the solid-state bromide electrode: chloride, iodide, hydroxide, and cyanide ions.

7. If calcium ion activity is to be measured with a liquid-membrane electrode in samples that contain up to $0.7M$ sodium ion, estimate the minimum level of calcium that can be measured under these conditions. Assume a minimum level of 5% interference. The selectivity ratio for sodium ion is 1.0×10^{-4}.

8. What is the maximum concentration of interfering anions that can be tolerated for a 1% interference level when measuring $10^{-5}M$ BF_4^- with a fluoroborate liquid ion-exchange membrane electrode? The interfering anions and their selectivity ratios are OH^-, 10^{-3}; I^-, 20; NO_3^-, 0.1; HCO_3^-, 4×10^{-3}; and SO_4^{2-}, 1×10^{-3}.

9. What is the total sulfide concentration in 100 mL of sample that gives a potential reading of −845 mV before the addition of 1 mL of 0.1*M* $AgNO_3$ and a reading of −839 mV after the addition?

10. Determine the ionization constant of boric acid. A sample, 0.0309 g in weight, was dissolved in 47.5 mL of distilled water. The temperature was 20 °C; the titrant was 0.1000*M* KOH. The molecular weight of H_3BO_3 is 61.84.

Titrant (mL):	0.5	1.0	1.5	2.0	2.5	3.0	3.5	4.0
pH reading:	8.34	8.68	8.89	9.07	9.26	9.43	9.62	9.84

11. A 0.0611-g sample of benzoic acid was dissolved in 47.5 mL of water and titrated at 20 °C with 0.1000*M* KOH. Determine the ionization constant of benzoic acid.

Titrant (mL):	0.5	1.0	1.5	2.0	2.5	3.0	3.5	4.0
pH reading:	3.38	3.63	3.83	3.99	4.16	4.34	4.53	4.76

12. What is the pH at the end point in the titration of potassium hydrogen phthalate with strong alkali, if the titration is performed under such conditions that the final concentration of phthalate ion is 0.05*M*? The constant *A* in the Debye-Hückel expression is 1.02.

13. Suggest how it might be possible to construct an ion-selective electrode that would be responsive to cyanide ion.

14. For what reasons might cast pellets of the individual silver halides not make satisfactory selective ion electrodes?

BIBLIOGRAPHY

BAILEY, P. L., *Analysis with Ion Selective Electrodes,* 2nd ed., Heyden, London, 1980.

BATES, R. G., *Determination of pH: Theory and Practice,* 2nd ed., John Wiley, New York, 1973.

DURST, R. A., ed., *Ion-Selective Electrodes,* National Bureau of Standards Spec. Publ. 314, Washington, DC, 1969.

FREISER, H., ed., *Ion-Selective Electrodes in Analytical Chemistry,* Vol. 2, Plenum Press, New York, 1980.

KORYTA, J., *Ions, Electrodes and Membranes,* John Wiley, New York, 1982.

KORYTA, J., AND K. STULIK, *Ion-Selective Electrodes,* 2nd ed., Cambridge University Press, Cambridge, U.K., 1983.

MA, T. S., AND S. S. M. HASSAN, *Organic Analysis Using Ion-Selective Electrodes,* Vols. 1 and 2, Academic, London, 1982.

MORF, W. E., "The Principles of Ion-Selective Electrodes and of Membrane Transport," in *Studies in Analytical Chemistry,* Vol. 2, Elsevier, Amsterdam, 1981.

RECHNITZ, G. A., "Bioanalysis with Potentiometric Membrane Electrodes," *Anal. Chem.,* **54,** 1194A (1982).

LITERATURE CITED

1. MILAZZO, G., *Electrochemistry,* Springer-Verlag, Vienna, 1952.
2. SORENSEN, S. P. L., *Compt. rend. trav. lab. Carlsberg,* **8,** 1 (1909).
3. DEAN, J. A., ed., *Lange's Handbook of Chemistry,* 13th ed., pp. 5–92 to 5–108, McGraw-Hill, New York, 1985.
4. FISHER, J. E., "Measurement of pH," *Am. Lab.,* **16,** 54 (June 1984).
5. ROTHSTEIN, F., AND J. E. FISHER, "pH Measurement: The Meter," *Am. Lab.,* **17,** 124 (September 1985).
6. CZABAN, J. D., "Electrochemical Sensors in Clinical Chemistry: Yesterday, Today, Tomorrow," *Anal. Chem.,* **57,** 345A (1985).
7. BATES, R. G., AND M. ALFENAAR, "Activity Standards for Ion-Selective Electrodes," in *Ion-Selective Electrodes,* R. A. Durst, ed., National Bureau of Standards Spec. Publ. 314, Washington, DC, 1969.
8. SERJEANT, E. P., *Potentiometry and Potentiometric Titrations,* John Wiley, New York, 1984.
9. BUCK, R. P., Chap. 2, *Physical Methods of Chemistry,* A. Weissberger and B. W. Rossiter, eds., Part IIA, Vol. 1, Wiley-Interscience, New York, 1971.

23 VOLTAMMETRIC TECHNIQUES

Voltammetry represents a wide range of electrochemical techniques as outlined in Table 23.1. These techniques can be used to study the solution composition through current-potential relationships in an electrochemical cell and with the current-time response of a microelectrode at a controlled potential. The development of polarography (the name applied to dc voltammetry when a dropping mercury microelectrode is used), commencing with the work of Heyrovsky in 1922 (and for which he received the Nobel Prize in 1959), marked a significant advance in electrochemical methodology. The existence of polarized electrodes was recognized and utilized in a practical way.

TABLE **23.1** REPERTOIRE OF VOLTAMMETRIC METHODS

Linear potential sweep (dc) voltammetry
Classical dc polarography at the dropping mercury electrode
Current sampled (Tast) voltammetry
Cyclic voltammetry

Potential step methods
Normal pulse voltammetry
Differential pulse voltammetry
Square-wave voltammetry
Chronocoulometry

Phase-sensitive alternating current voltammetry
Fundamental ac voltammetry

Hydrodynamic methods
Rotating disk, ring, and ring-disk voltammetry
Flow-through electrodes

Linear potential sweep stripping chronoamperometry
Anodic stripping voltammetry
Cathodic stripping voltammetry

Controlled potential in flowing systems
Amperometric titrations
Constant potential amperometric detection

The original dc voltammetric technique suffered from a number of difficulties that made it less than ideal for routine analytical purposes and made the results obtained somewhat difficult to interpret. With the advent of low-cost, fast, stable, operational amplifiers in the early 1960s, some problems were overcome. Investigations of new approaches, such as potential step methods, potential sweep voltammetry, phase-sensitive alternating current voltammetry, hydrodynamic methods, and stripping voltammetry, demonstrated the utility and desirability of the new voltammetric methods.[1]

The advantages of voltammetry quickly demonstrate that it is a potent analytical tool. The foremost advantage is sensitivity. Voltammetry ranks among the most sensitive analytical techniques available; it is routinely used for the determination of electroactive substances in the sub-parts per million range. Analysis times of seconds are possible. The simultaneous determination of several analytes by a single scan is often possible with a voltammetric procedure. Voltammetric techniques have a unique capability to distinguish between oxidation states that may affect a substance's reactivity and toxicology. The theory of voltammetry is well developed, and reasonable estimates of unknown parameters can be made.

In analytical applications the composition of a sample can be investigated using various potential step methods, possibly by ac voltammetry in any of several forms and perhaps anodic and cathodic stripping voltammetry. Chronocoulometry (see Chapter 24) may be a simpler means for obtaining quantitative evaluations. Accordingly, flexibility is required in the format of instrumentation for electrochemical excitation and observation.

23.1 INSTRUMENTATION

The Dropping Mercury Microelectrode

The microelectrode of classical polarography is the dropping mercury electrode. This small polarizable electrode is produced by forcing a stream of mercury through a very fine-bore (0.05–0.08 mm i.d.) glass capillary under the pressure of an elevated reservoir of mercury connected to the capillary by flexible tubing (Figure 23.1). A steady flow of mercury issues from the capillary at the rate of one drop every 2–5 sec. Knockers are often used to detach the drop reproducibly at a preselected time interval. A dropping mercury electrode has several advantages: (1) its surface area is reproducible with any given capillary; (2) the constant renewal of the electrode surface eliminates passivity or poisoning effects; (3) the high overpotential of hydrogen on mercury renders the electrode useful for electroactive species with a reduction potential that is considerably more negative than the reversible potential for the discharge of hydrogen; (4) mercury forms amalgams with many metals and thereby lowers their reduction potential; and (5) the diffusion current assumes a steady value immediately and is reproducible.

The dropping mercury electrode is useful over the range $+0.3$ to -2.8 V versus the saturated calomel electrode (SCE) in aqueous solutions. At potentials more positive than 0.3 V, mercury is oxidized, which results in an anodic wave. The most positive potentials may be attained in the presence of noncomplexing anions that form soluble mercury(I and II) salts—for example, nitrate or perchlorate ions.

Anions that form insoluble mercury salts or stable complexes shift the anodic dissolution potential to more negative values. At potentials more negative than −1.2 V visible hydrogen evolution occurs in 1*M* HCl solutions, and at −2 V the usual supporting electrolytes of alkali salts begin to discharge. The most negative potentials may be attained in solutions in which a quaternary ammonium hydroxide is used as a supporting electrolyte. With tetrabutylammonium hydroxide the limit is −2.7 V.

The Static Mercury Drop Electrode

This electrode uses a hanging mercury drop. The mercury drop is dispensed through a capillary (from a micrometer or solenoid-activated plunger). An entire experiment is performed on one mercury drop held on the capillary tip. That drop is then dislodged and a new drop is dispensed for the next experiment. This type of electrode can also be formed by catching a drop from a capillary and hanging the drop on the tip of the hanging mercury drop electrode's holder. All the characteristics of the dropping mercury electrode that make it a suitable electrode for routine analytical determinations also apply to the hanging mercury drop electrode. It has the added feature that once the drop has expanded to a given area, the electrode area remains constant during subsequent measurements. A static electrode area eliminates the contribution of nonfaradaic current (the charging current from a changing surface area discussed in Section 23.2). This enhances sensitivity. Sensitivity is also increased due to the larger drop size afforded by this type of electrode. Because the mercury drop is renewed reproducibly for each experiment, the condition of the electrode surface is less important as a variable in the analysis. This is not true for solid electrodes.

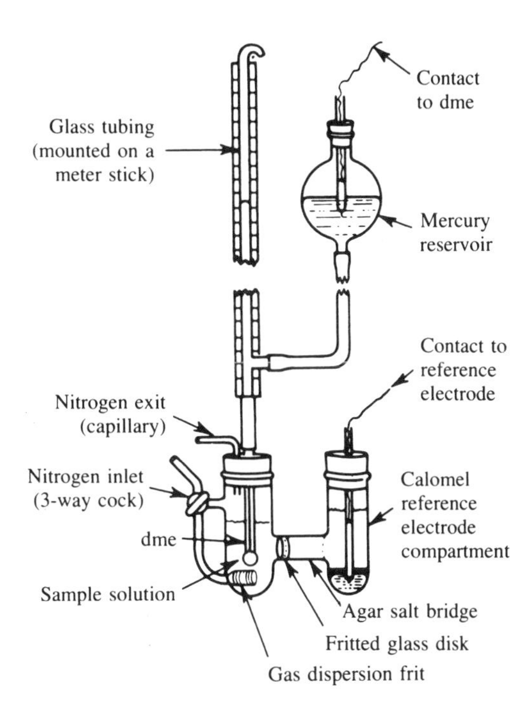

FIGURE **23.1**
Dropping mercury microelectrode and reservoir arrangement for polarographic cell.

Thin-Film Mercury Electrode

A thin-film mercury electrode is prepared by depositing a film of mercury several thousand angstroms thick onto a glassy carbon electrode. This type of electrode is generally used only for anodic stripping voltammetry. Because the layer of deposited mercury is extremely thin, the use of this type of electrode should be limited to analyte concentrations less than $10^{-7}M$ and then only when maximum sensitivity is required. The thin-film electrode offers a large surface-area-to-volume ratio and can be easily rotated or stirred at quite high rates.

Solid Electrodes

The stationary solid electrode renews the diffusion layer on a solid electrode tip by rapidly raising and lowering the electrode at periodic intervals, as selected by the operator. A variety of electrode tips may be used: platinum, gold, glassy carbon, pyrolytic carbon, nickel, titanium, as well as mercury. Glassy carbon electrodes are often used to monitor electro-oxidation reactions, particularly on-line (such as in liquid column chromatography), where nanogram and picogram amounts of suitable materials may be detected at the electrode surface. Electrode fouling is solved by repolishing the exposed electrode surface or using multistep pulsed detection with periodic cleaning pulses.

Rotating Disk and Ring-Disk Electrodes

The rotating disk electrode consists of a disk of electrode material embedded in a rod of an insulating material (glass, Teflon, epoxy resin, or another plastic). A simple form involves a platinum wire sealed in glass tubing with the sealed end ground smooth and perpendicular to the rod axis. The addition of an independent ring surrounding the disk produces a ring-disk electrode. These several electrodes are shown in Figure 23.2.

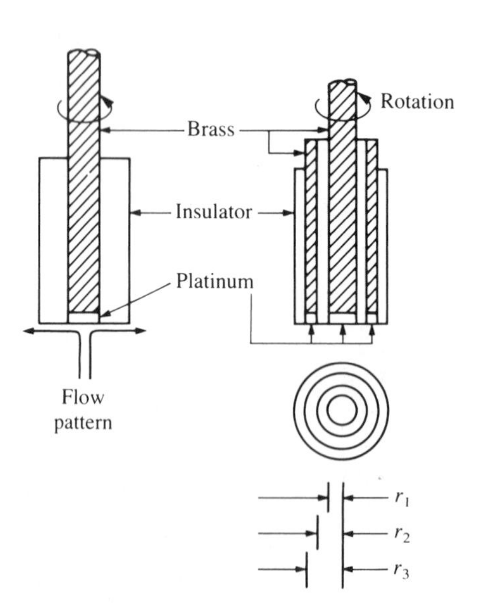

FIGURE **23.2**
Disk, ring, and ring-disk electrodes.

Deaeration of Solutions

Usually when an experiment involves the application of negative potentials, dissolved oxygen must be removed because oxygen is reduced at potentials more negative than about 0.1 V versus SCE. An oxygen voltammogram comprises two waves, each of equal height, that result from the reduction of oxygen to hydrogen peroxide (about -0.1 V vs. SCE in 0.05*M* KCl) and the further reduction of the peroxide to hydroxyl ion (about -0.9 V vs. SCE). Oxygen is removed by bubbling pure nitrogen or argon through the solution via a fine capillary or fritted glass dispersion tube for about 10 min with a capillary or 1 to 2 min with a glass frit. The bubbling of gas through the solution must be stopped prior to experiments that use quiescent solutions. A slight positive pressure of gas is then maintained over the solution. Any traces of oxygen in the nitrogen or argon purge gas should be removed; also the gas should be saturated with solvent vapor to prevent any change in the concentration of the sample.

Drop Detachment

Drop synchronization is essential for reproducible measurements in many voltammetric methods and desirable in other methods. A drop of mercury is detached from the capillary by an electromechanical drop dislodger. One type moves the capillary away from the drop at a fixed time interval; others deliver a sharp knock to the capillary. At the same moment a trigger signal is sent to the time base. Each drop is permitted to grow until its area changes the least, usually after the first 1.5–2.0 sec of drop life. All these operations may be controlled by a computer.

Three-Electrode Potentiostat

A characteristic of modern voltammetric instrumentation is potentiostatic control of the working electrode potential accompanied by the measurement of the current at that electrode. The potentiostat must perform these two functions with electrodes of varying size and surface resistance immersed in solutions of varying conductivity. The potential may be pulsed very rapidly or scanned very slowly, and the resulting cell current may be extremely high or very low. Obviously, a potentiostat must be designed for routine potentiostatic or galvanometric control, for experimental versatility, or for high-current performance.

In a three-electrode potentiostat, a reference electrode is positioned as close as possible to the working (indicator) electrode. The auxiliary (counter) electrode is the third electrode in the electrochemical cell. The function of the potentiostat is to observe the potential of the working electrode (either cathode or anode) against the reference electrode—that is, to sense the potential of the working electrode, since the potential of the reference electrode is constant. The reference–working electrode pair is connected through a circuit that draws essentially no current (Figure 23.3); the input impedance is in excess of 10^{11} Ω. If the voltage of the reference–working electrode pair is less than the dc ramp provided to the scan amplifier, the feedback to the scan amplifier from the operational amplifier control loop will provide a corrective voltage that changes the potential applied across the auxiliary–working electrode pair enough to compensate for the resistance of the cell and electrolyte. Thus, the voltage measured at point *C* should always be the same as that applied to

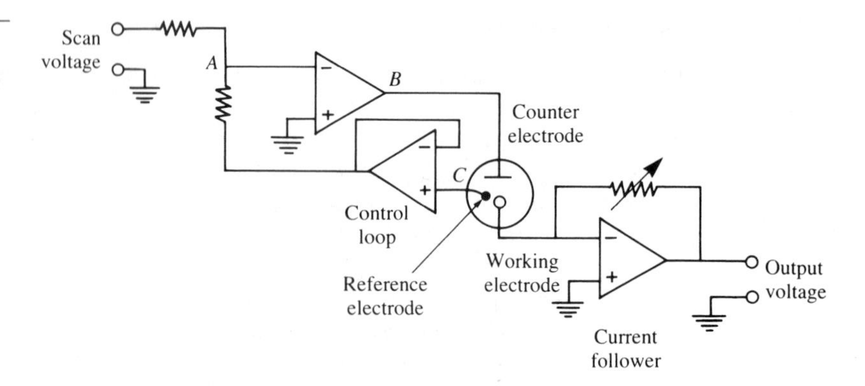

FIGURE 23.3
Schematic diagram of a three-electrode potentiostat.

point A and, if momentarily it is not, the voltage at point B will automatically increase to maintain point C equal to point A.

Current passes between the working electrode and the counter electrode. Although the most direct way to measure the cell current is to place a resistor in series with the counter electrode and measure the voltage drop across it, this voltage is floating because it is not referenced to ground. A floating signal is more susceptible to noise pickup and is more difficult to record. Measurement of the current with a current follower provides a ground-referenced voltage proportional to the cell current.

Although the potentiostat must control the potential at the working electrode surface, the potential is actually measured at the reference electrode. The three-electrode potentiostat automatically compensates for the resistance of the solution between the counter electrode and the reference electrode. This makes it possible to use nonaqueous solvents of high resistance and quite dilute aqueous electrolytes. Distortion of the wave shape and slope of the current-voltage signal is much less pronounced, if not entirely eliminated. Positive feedback circuits, which increase the voltage applied between the working and reference electrodes by an amount proportional to the instantaneous current, may also be used to overcome errors due to very high resistance or improve the frequency response. For extremely small electrodes (ultramicroelectrodes) and extremely small currents, a potentiostat is not essential, and a two-electrode cell suffices. Use of a high-speed signal generator enables much faster experiments than are possible with potentiostats.

23.2 CHARGING OR CAPACITIVE CURRENT

In voltammetry the current that flows across the electrode-solution interface arises from two sources. What one wishes to observe is the *faradaic current,* the current arising from the reduction or oxidation of the species in solution. However, due to the *double layer* that forms at the electrode-solution interface, a significant capacitive current is observed on current-potential plots (Figure 23.4). This is called *nonfaradaic current.* The double layer is caused by the electrostatic attraction or repulsion of cations and anions, respectively, near the electrode surface to balance the charge on the electrode. At the solution-electrode interface, a separation of charge takes place that makes the interface look like a large capacitor to the external circuitry.

With the dropping mercury electrode, the electrode surface area repeatedly grows to a maximum and then suddenly falls to zero as the drop detaches or is dislodged. When a new mercury drop grows, effectively changing the surface area of the interface, or when the potential on the electrode is changed, a current must flow to charge or discharge the capacitor. This current flows in the absence of an accompanying faradaic process. This charging current, i_c (Figure 23.5), appears as a surge at the beginning of each drop when a new capacitor must be charged, then decreases steadily according to the expression

$$i_c = 0.00567C_i(E_z - E)m^{2/3}t^{-1/3} \tag{23.1}$$

where C_i is the integral capacitance of the double layer, m is the mercury flowrate (in mg/sec), and t is the time (in sec). For a given solution there is a potential of zero charge, E_z, at which the double-layer capacitance is minimized. As the potential increases or decreases from this value, capacitance increases and the magnitude of the charging current increases. This charging current is the principal factor that limits the sensitivity of classical voltammetric techniques and its accuracy at low

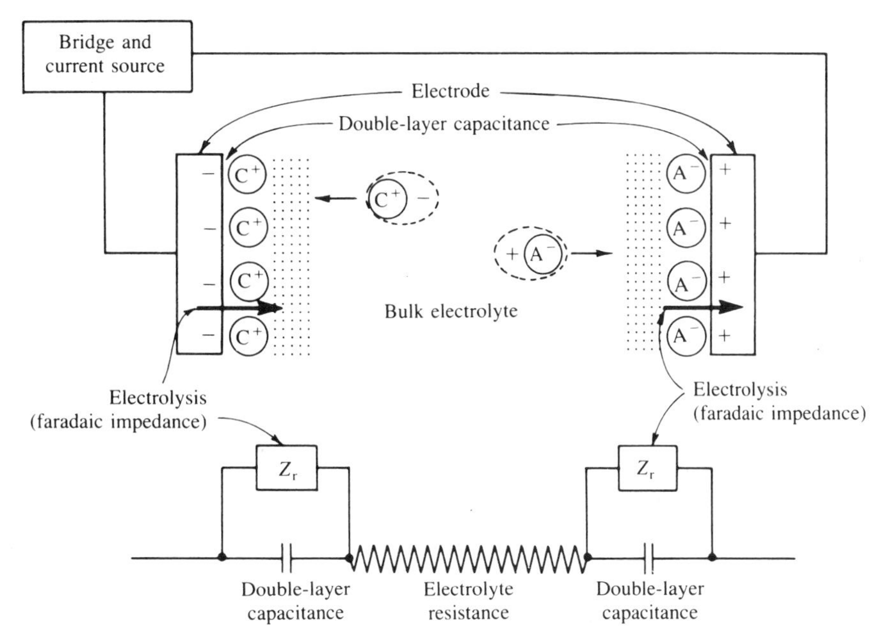

FIGURE **23.4**
Electrolytic cell showing a simplified representation of the double layer at the electrodes, faradaic processes, and migration of ions through the bulk electrolyte. [After J. Braunstein and G. D. Robbins, *J. Chem. Educ.*, **48**, 52 (1971). Reproduced by permission.]

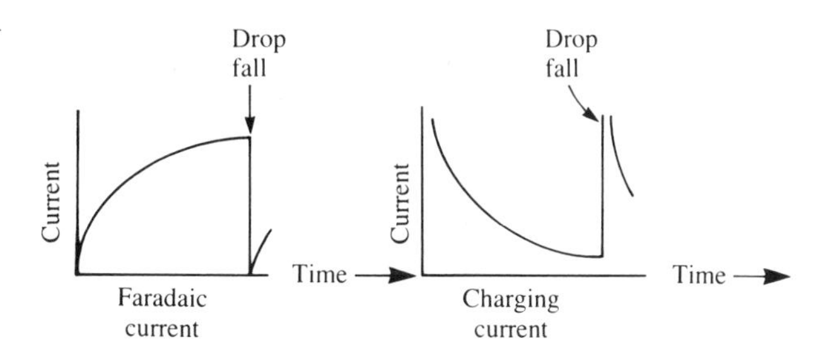

FIGURE **23.5**
Charging (nonfaradaic) and faradaic currents at a dropping mercury electrode.

concentrations. At concentrations around $10^{-5}M$, the charging current is usually larger than the faradaic current for conventional polarographic methods, and the precision of the determination depends principally on how exactly the contribution of the charging current can be estimated, compensated, or minimized at the potential at which the diffusion current is measured.

23.3 LINEAR POTENTIAL SWEEP (DC) VOLTAMMETRY

In its simplest form, dc voltammetry involves applying a linear dc potential ramp between two electrodes, one small and easily polarized (the indicator or working electrode), classically the dropping mercury electrode (polarography), and the other large and relatively immune to polarization (the counter electrode, often a pool of mercury). In polarography the ramp is applied at the rate of a few millivolts per second so as not to change the potential during a drop's life (usually 2–5 sec) more than a few millivolts. The current is measured at the same time in the life of a growing drop. Effectively, a steady state results from the fact that the diffusion-layer thickness is always the same at the time of measurement.

Limiting Current

In direct-current voltammetry using a dropping mercury electrode, it is a formidable problem to develop an expression for the current at an expanding spherical electrode when the surface concentration is held at zero. The original equation was solved by Ilkovič, who assumed semi-infinite linear diffusion to an expanding plane rather than to an expanding sphere. An expression may also be derived by taking the Cottrell equation, substituting the time dependence of the surface area of a growing sphere, and multiplying the result by a correction term to account for the expansion of the drop into the bulk media. The volume of an expanding mercury drop is related to the mercury flowrate, m, and the time, t, since the drop began to form at the capillary tip by

$$\text{volume} = \left(\frac{4}{3}\right)\pi r^3 = \left(\frac{m}{\rho}\right)t \qquad (23.2)$$

where r is the radius of the drop at time t, and ρ is the density of mercury. Solving for the radius of the drop and then calculating the surface area of the drop, one obtains

$$\text{area} = 4\pi\left(\frac{3mt}{4\pi\rho}\right)^{2/3} \qquad (23.3)$$

The substitution of this area into the Cottrell equation (Equation 21.23) gives an approximate expression for the diffusion current:

$$i_d = 4\pi^{1/2}nF\left(\frac{3}{4\pi\rho}\right)^{2/3} D^{1/2}Cm^{2/3}t^{1/6} \qquad (23.4)$$

Expansion of the drop causes the existing diffusion layer to stretch over a still larger sphere. This has the effect of making the layer thinner than it otherwise would be. The concentration gradient at the electrode surface is enhanced, so that larger currents flow. Multiplication of Equation 23.4 by $(\frac{7}{3})^{1/2}$ provides a closer fit to experiment. When the constants are gathered together, the result is the Ilkovič equation:

$$i_d = 708nCD^{1/2}m^{2/3}t^{1/6} \tag{23.5}$$

where i_d is in μA, D in cm^2/sec, C in mM units, m in mg/sec, and t in sec.

The quantities m and t depend on the dimensions of the dropping capillary microelectrode and on the pressure exerted on the capillary orifice due to the height of the mercury column attached to the electrode. A mercury reservoir with a large area is customarily attached to the mercury column to prevent any change in height of the column during a series of analyses (see Figure 23.1). The effects of drop expansion more than counteract depletion of the electroactive substance near the electrode. The current is an increasing function of time, in contrast to the behavior found at a stationary planar electrode. Two important consequences of the increasing current-time function are that the current is greatest and its rate of change is slowest just at the end of the drop's life.

Current-Potential Curves

Consider an overall electrode reaction

$$\text{ox} + ne^- \rightleftharpoons \text{red} \tag{23.6}$$

composed of a series of steps that cause the conversion of the dissolved oxidized species to a reduced form, also in solution. The current is governed by the rates of processes such as mass transfer of the oxidant from the bulk solution to the electrode surface (discussed in Chapter 21) and electron transfer at the electrode surface. Although we shall proceed in a naive way for steady-state voltammetry, the results apply equally well for sampled-current voltammetry. The electrochemical equilibrium may be represented as follows:

$$E = E^\circ - \frac{0.05916}{n}\log\frac{[\text{red}]_i}{[\text{ox}]_i} \tag{23.7}$$

where the subscripts denote concentrations at the electrode-solution interface.

The observed current depends on the rate of diffusion established by the concentration gradient:

$$i = K([\text{ox}] - [\text{ox}]_i)D_{\text{ox}}^{1/2} \tag{23.8}$$

where K includes electrode characteristics. When the current attains the limiting value represented by the diffusion-current plateau, the concentration of oxidant at the electrode-solution interface is essentially zero, and

$$i_d = K[\text{ox}]D_{\text{ox}}^{1/2} \tag{23.9}$$

Solving Equation 23.8 for $[\text{ox}]_i$ and then combining with Equation 23.9, one gets

$$[\text{ox}]_i = \frac{i_d - i}{KD_{\text{ox}}^{1/2}} \tag{23.10}$$

For reaction products that form amalgams with the dropping electrode or are soluble in solution, the concentration of metal amalgam at the surface of the drop is directly proportional to the current on the current-voltage curve, and generally the concentration of reductant formed is proportional to the observed current, so

$$i = K[\text{red}]_i D_{\text{red}}^{1/2} \tag{23.11}$$

Solving for $[\text{red}]_i$ and substituting the result into Equation 23.7 along with Equation 23.10, we can express the potential of an oxidation/reduction system as

$$E = E^\circ - \frac{0.05916}{n} \log \frac{i}{i_d - i} + \frac{0.05916}{n} \log \left(\frac{D_{\text{red}}}{D_{\text{ox}}}\right)^{1/2} \tag{23.12}$$

By definition, the half-wave potential is the point where

$$i = \frac{i_d}{2}$$

At this point the first logarithmic term becomes zero, and

$$E_{1/2} = E^\circ + \frac{0.05916}{n} \log \left(\frac{D_{\text{red}}}{D_{\text{ox}}}\right)^{1/2} \tag{23.13}$$

Equation 23.12 predicts a current-potential sigmoid-shape wave that rises from the baseline to the diffusion-controlled limit over a fairly narrow potential region centered on the half-wave potential (see Figure 23.7, curve *A*). Since the ratio of diffusion coefficients in Equation 23.13 is nearly unity in almost any case, the half-wave potential is usually a very good approximation to E° for a reversible couple. This involves a homogeneous process that is fast enough to be considered always in thermodynamic equilibrium coupled to a nernstian electron-transfer reaction. Reversible systems provide thermodynamic properties, such as standard potentials, free energies of reaction, and various equilibrium constants.

If both oxidant and reductant are initially present and both are in solution, then a voltammetric wave would look like Figure 23.6, curve *B*. The anodic current $(-i_d)_a$ arises when the reduced form is being oxidized at the electrode. Under these conditions the current-potential relation assumes the form

$$E = E_{1/2} + \frac{0.05916}{n} \log \frac{(i_d)_c - i}{i - (i_d)_a} \tag{23.14}$$

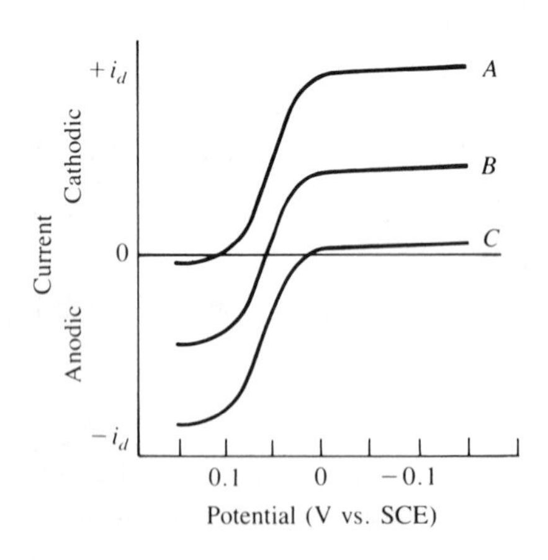

FIGURE 23.6
Voltammograms for a reversible system: *A*, reduction of quinone; *B*, equimolar mixture of quinone-hydroquinone (both reduction and oxidation waves are obtained); and *C*, oxidation of hydroquinone.

Note in Figure 23.6 the sign convention used in voltammetry. A cathodic current is given a positive sign, an anodic current a negative sign.

When a slow chemical reaction (whose rate constant is k) is coupled to a nernstian electron transfer, a reaction layer thickness (u) and a mass-transfer coefficient must be taken into account. Assuming a nernstian charge transfer reaction with an irreversible following reaction involving the reductant, one must modify the last term in Equation 23.12 to yield

$$E_{1/2} = E^\circ - \frac{RT}{nF} \ln\left(\frac{D_{ox}}{D_{red} + uk}\right) - \frac{RT}{nF} \ln \frac{i}{i_d - i} \tag{23.15}$$

When the product $uk \ll D_{red}$, the rate of the following chemical reaction is negligible compared with the rate of mass transfer of reductant away from the electrode surface, and an unperturbed current-potential curve results. In the other limiting case when the product $uk \gg D_{red}$, the effect of the following chemical reaction is to shift the reduction wave in a positive direction without changing the shape. The rigorous treatment of electrode reactions with coupled homogeneous chemical reactions is discussed by Bard and Faulkner.[2]

Experimental Procedure

Voltammograms are usually recorded on strip-chart or x-y recorders as the potential of the dropping mercury electrode is scanned linearly with time. For a reduction, the initial potential is selected so that the reduction of interest will not have begun. The potential is then scanned cathodically through the reduction wave. The result is a sigmoidal curve for each electroactive species in the solution. To obtain the true diffusion current, a correction must be made for the charging or capacitive current, often called the residual current i_r. The most reliable method for making this correction is to evaluate in a separate voltammogram the residual current of the supporting electrolyte alone (Figure 23.7). The value of the residual current at any particular potential of the dropping electrode is then subtracted from the total current observed. Often, an adequate correction can be obtained by extrapolating the residual current portion of the voltammogram immediately preceding the rising part of the current-voltage curve, and taking as the faradaic current the difference between this extrapolated line and the current-voltage plateau. Estimation of residual currents by extrapolation techniques is inaccurate and questionable at low concentrations.

Sometimes the voltammograms show peaks, called *maxima,* that can greatly exceed the limiting currents due to diffusion. Although some maxima are obvious because of the excursion of the current signal above the limiting current plateau before settling back to the plateau, other types of maxima are insidious in that they enhance the plateau current without distorting the sigmoidal wave shape. Maxima are complex phenomena that are reasonably well, but not quantitatively, understood. They arise from convection around the growing mercury drop. Surfactants, such as gelatin or Triton X-100, suppress these maxima and are routinely added in small quantities (0.005% to 0.01%) to all test solutions.

Current Sampled (Tast) Polarographic Voltammetry[3]

In current sampled voltammetry the current is measured (sampled) at a fixed time after the birth of a drop. The sampling period is ended immediately before the drop is dislodged. To control drop life, a time circuit and drop knocker are required to provide reproducible drops and time of measurement. A sample-and-hold circuit (see Chapter 4) samples the current for a selected period each time it receives a trigger from the mechanical drop timer's clock circuit. The sample-and-hold system includes averaging circuitry, which averages the current sampled over the entire aperture period and integrates out noise. It also maintains the current value of the previous sampling interval, which it presents to a recorder as a constant readout until it is replaced at the sampling time during the next drop. This gives the usual sigmoidal voltammogram a staircase appearance (Figure 23.8).

FIGURE **23.7**

Polarographic current-potential curves. Measurement of the faradaic current: (a) subtraction of the signal from the supporting electrolyte and (b) extrapolation of the residual current.

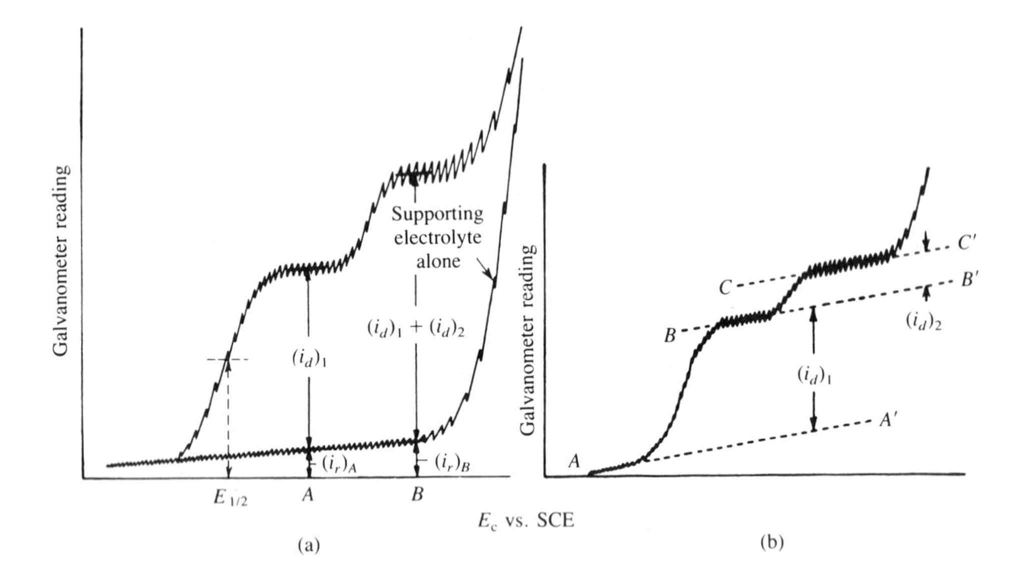

FIGURE **23.8**
Current-sampled voltammogram.

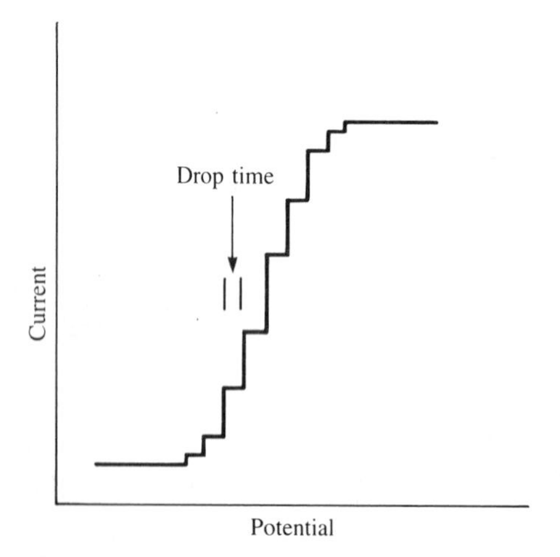

Since the potentiostat is always active and the potential ramp is identical to that used in classical dc voltammetry, the actual current flow at the electrode is the same as that observed in dc voltammetry using controlled drop times. The faradaic component of the limiting current is given by Equation 23.5, whereas the charging component is given by Equation 23.1. The resulting voltammogram is substantially free from the fluctuating pattern of the current response (resembling a noise level that interferes with the quantitation of samples) during the lifetime of each mercury drop. Detection limits are near $10^{-6}\,M$, which are somewhat lower than those of classical dc voltammetry.

23.4 POTENTIAL STEP METHODS[4,5]

The techniques discussed in this section were developed for use with the dropping mercury electrode. All are based on the measurement of current as a function of time (chronoamperometry) after applying a potential pulse. Following a sudden change in applied potential, the capacitative surge decays much more rapidly than does the faradaic current. Clearly one can optimize the ratio of faradaic to charging current, and thus the sensitivity, by sampling the current at the instant just before the drop falls. The techniques accomplish this by different approaches. The availability of relatively low-cost microprocessor-controlled instruments has provided the ability to generate an arbitrary time sequence of potential application and current sampling.

Potential step methods are rarely used for the diagnosis of electrode processes; hence, detailed knowledge of wave shapes is not necessary.

Normal Pulse Voltammetry

In normal pulse voltammetry, also called large-amplitude pulse voltammetry, a series of square-wave voltage pulses of successively increasing magnitude is superimposed upon a constant dc voltage signal (Figure 23.9). This can be done at a dropping mercury electrode or a solid electrode. The electrode is held at a base potential at which negligible electrolysis occurs for most of the life of each mercury drop. After a fixed waiting period, measured from the birth of the drop (timed from the dislodgement of the previous drop), the potential is changed abruptly to value E for about 50 msec. The potential is then returned to the base value. The cycle is repeated with successive drops except that the potential is made a few millivolts higher (stepped) for each drop (or for each pulse when a stationary electrode is used). Near the end of each pulse, and before the drop is dislodged, the current is sampled. This allows time for the charging current to decay to a very low value. During this time interval the faradaic current also decays somewhat. The sampled current is presented as a constant signal to a recorder until the current sample taken in the next drop lifetime replaces it. The output is a stepped sigmoidal-shaped voltammogram of sampled current versus step potential, as shown in Figure 23.9c.

The time delay between pulses must be long enough to restore all concentration gradients to their original state before the next potential pulse is applied. The length of time required depends on the chemical reversibility of the system, being shorter for reversible reactions.

The limiting current is given by the Cottrell equation:

$$i_{\text{lim}} = nFCA\sqrt{\frac{D}{\pi t}} \tag{23.16}$$

where t is the time measured from the pulse rise. For the dropping mercury electrode, A is a function of time, which leads to the Ilkovič equation.

As a diagnostic test for diffusion control, the potential is stepped from the residual current to a potential that is on the current plateau. If the process is diffusion controlled, the current decays as the reciprocal of the square root of time; that is, a plot of log i versus log t yields a straight line with slope -0.5.

Differential Pulse Voltammetry

Although Tast and normal pulse voltammetry give a marked improvement in sensitivity over the classical dc method, they still give the sigmoidal voltammogram. A much more useful variant is differential pulse voltammetry, in which a series of potential pulses (ΔE) of fixed, but small, amplitude (10–100 mV) is superimposed on

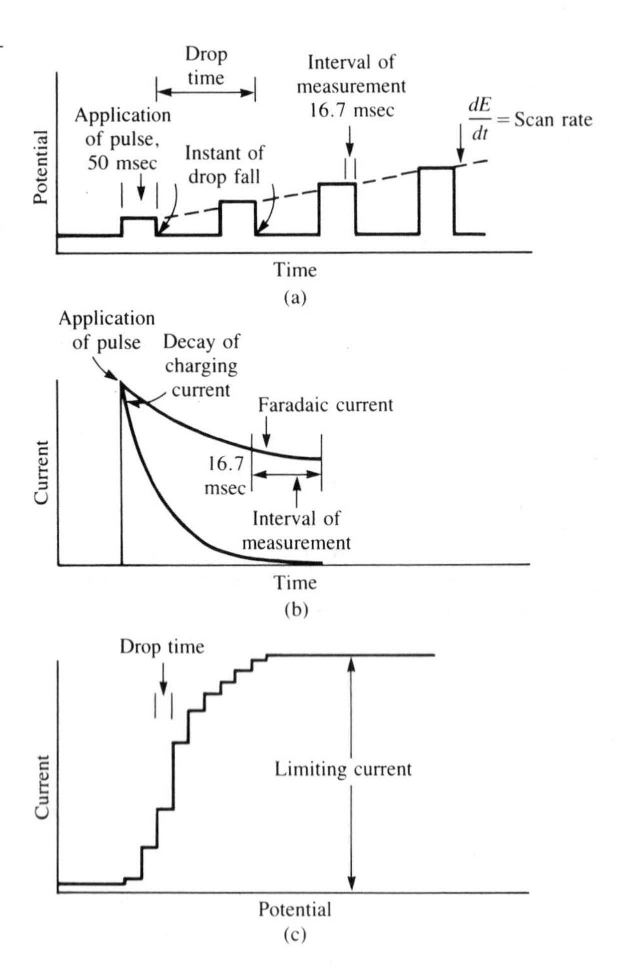

FIGURE 23.9
Normal pulse voltammetry. (a) Schematic showing the pulse application, drop time, and interval of current measurement. (b) Variation of current during the pulse. (c) Stepped voltammogram.

a dc voltage ramp (Figure 23.10). Two current samples are taken during each drop's lifetime. One is taken immediately before applying the potential pulse, and the second is taken late in the pulse (perhaps the last 17 msec of the pulse) and just before the drop is dislodged. For each cycle the first current value is instrumentally subtracted from the second. A plot of this current difference versus applied (dc ramp) potential produces a stepped peak-shaped incremental derivative voltammogram.

The theoretical relationship between peak current, i_p, and the pulse modulation amplitude, ΔE, has been developed by Parry and Osteryoung.[6] The maximum peak current when $\Delta E < RT/nF$ (that is, for pulses of small amplitude) is directly proportional to both the concentration of the electroactive substance and ΔE, and is given by

$$\Delta i_p = \frac{n^2F^2}{4RT} AC(\Delta E)\sqrt{\frac{D}{\pi t}} \tag{23.17}$$

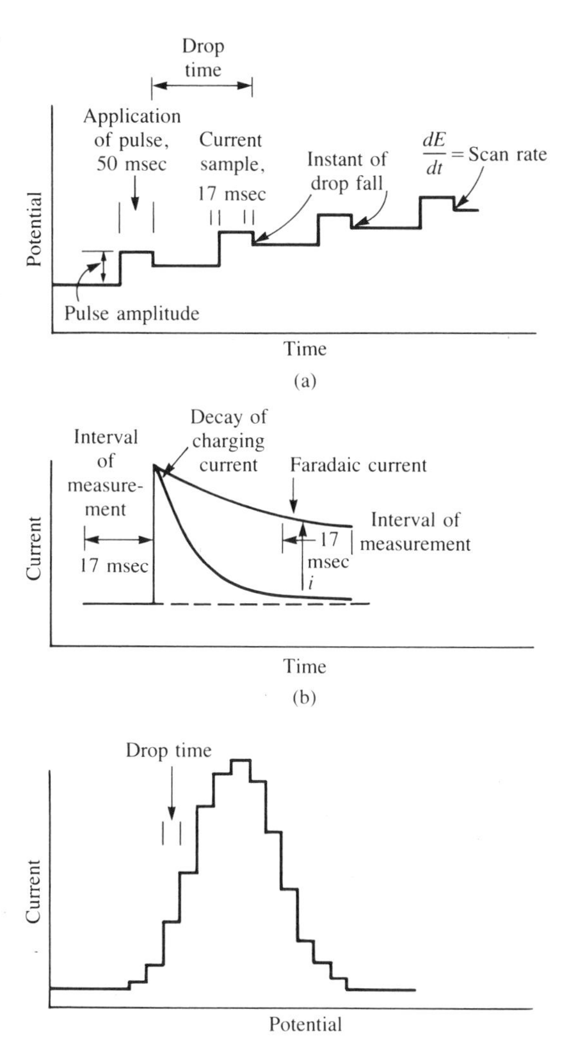

FIGURE **23.10**

Differential pulse voltammetry. (a) A linearly increasing scan voltage upon which a 35-mV pulse is superimposed during the last 50 msec of the drop life. (b) Variation of current during the pulse interval. (c) Net current signal observed between the two intervals of current measurement.

The peak current potential coincides with the half-wave potential.

With pulses of large amplitude, the peak current is given by

$$\Delta i_p = nFAC\sqrt{\frac{D}{\pi t}}\left(\frac{\sigma - 1}{\sigma + 1}\right) \tag{23.18}$$

where $\sigma = \exp[(\Delta E)nF/2RT]$. In the limit, for $\Delta E \gg RT/nF$, $(\sigma - 1)/(\sigma + 1)$ approaches unity and the limiting current is given by the Cottrell equation. The use of a large pulse amplitude allows one to obtain large signals from extremely dilute solutions, albeit with some distortion in the curve shape, because the pulse amplitude is an appreciable portion of the total voltammetric wave. With larger values of modulation amplitude, the peak potential is no longer coincident with the half-wave potential.

Square-Wave Voltammetry[7]

In square-wave voltammetry a symmetrical square-wave pulse is superimposed on a staircase waveform where the forward pulse (point 1 in Figure 23.11) of the square wave (τ is the time for one square-wave cycle or one staircase step in seconds) is coincident with the staircase step. The square-wave frequency in hertz is $1/\tau$. E_{sw} is the height of the square-wave pulse in millivolts, where $2E_{sw}$ is equal to the peak-to-peak amplitude. E_{step} is the staircase step size in millivolts. The current is the difference between the current observed at point 1 (forward current) minus that at

FIGURE **23.11** (a) Excitation signal for square-wave voltammetry. (b) Forward, reverse, and difference currents.

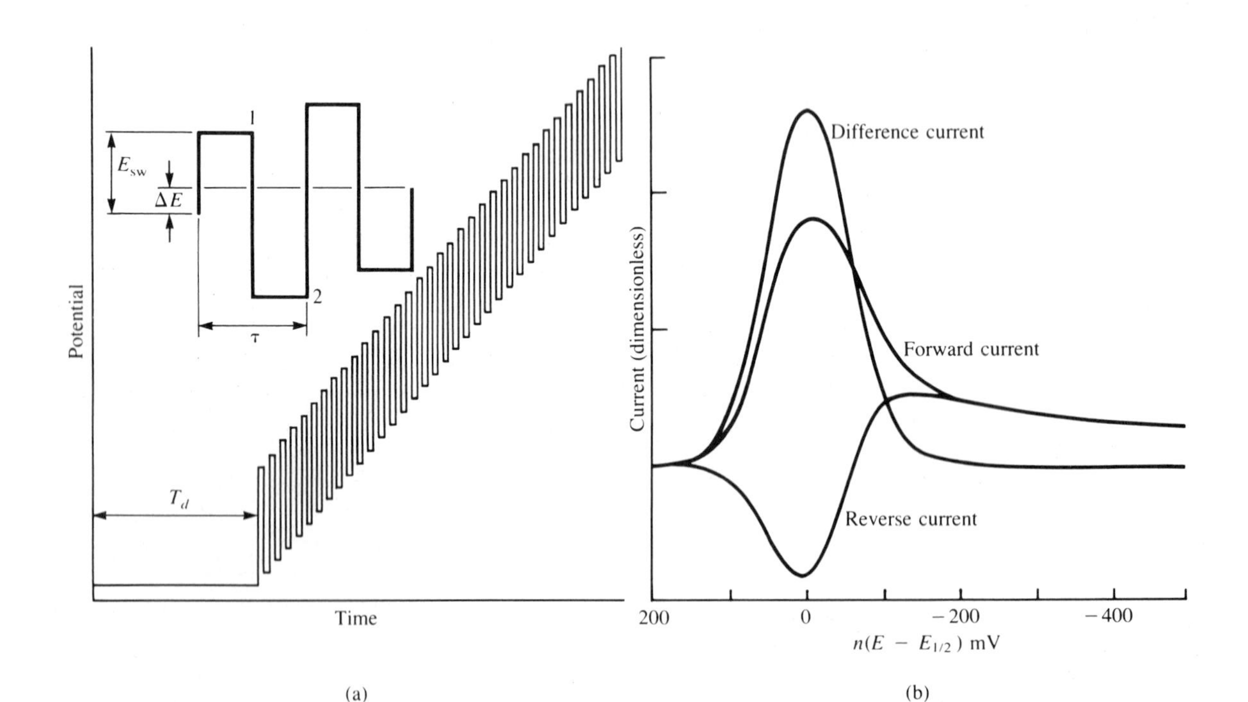

point 2 (reverse current). This complex technique required the power and flexibility of the minicomputer for its development and modern microprocessors for its commercial implementation. These experimental parameters are related by:

$$\text{scan rate (in mV/sec)} = \frac{E_{\text{step}}\ \text{(in mV)}}{\tau\ \text{(in sec)}} \tag{23.19}$$

Frequencies of 1–120 square-wave cycles per second permit scan rates up to 1200 mV/sec. There is a maximum frequency. Although it is possible to scan at a rate faster than the kinetics of the faradaic process will allow, the current will be kinetically controlled rather than diffusion controlled. Under such conditions the peak height may not be directly proportional to the concentration. No peak is observed if the scan rate is significantly faster than the kinetics of the system. Operation at frequencies much greater than 1000 Hz requires careful attention to cell design and electronics, particularly the current capability of the operational amplifiers. About 200 Hz appears to be a good trade-off between sensitivity and stable, trouble-free operation for analytical work.

The current is sampled twice during each square-wave cycle, once at the end of the forward pulse and again at the end of the reverse pulse. Square-wave voltammetry discriminates against the charging current by delaying the current measurement until close to the end of the pulse. The reverse pulse of the square wave occurs halfway through the staircase step. The amplitude of the square-wave modulation is so large ($50/n$ mV is optimum for a reversible process) that the reverse pulse causes reoxidation of the product produced on the forward pulse back to the original state with a resulting anodic current. Thus, the net current at the current-voltage peak is larger than either the forward or reverse current, since it is the difference between them. The peak height is directly proportional to the concentration of the electrochemical species. Concentration levels of parts per billion are easily detectable.

23.5 CYCLIC POTENTIAL SWEEP VOLTAMMETRY[8]

Cyclic voltammetry (also called linear sweep voltammetry) consists of cycling the potential of a stationary electrode immersed in a quiescent solution and measuring the resulting current. The excitation signal is a linear potential scan with a triangular waveform. Symmetrical triangular scan rates range from a few millivolts per second to hundreds of volts per second. This triangular potential excitation signal sweeps the potential of the working electrode back and forth between two designated values called the switching potentials. The triangle returns at the same speed and permits the display of a complete voltammogram with cathodic (reduction) and anodic (oxidation) waveforms one above the other, as shown in Figure 23.12. The current at the stationary working electrode is measured under diffusion-controlled, mass-transfer conditions. Although the potential scan is frequently terminated at the end of the first cycle, it can be continued for any number of cycles. Both the scan rate and the switching potentials are easily varied. Typically, the scan rate is from 20 V/sec to 100 V/sec at macroscopic electrodes, but it can be as high as 10^6 V/sec at ultramicroelectrodes.

In the example (Figure 23.12), the initial potential (0.0 V) applied at point *a* is chosen to avoid any electrolysis of electroactive species in the sample when the experiment is initiated. Then the potential is scanned in the negative direction. When the potential becomes sufficiently negative (−0.6 V) to cause a reduction of an electroactive species at the electrode surface, cathodic current (indicated at *b*) begins to flow. The cathodic current increases rapidly until the surface concentration of oxidant at the electrode surface approaches zero, as signaled by the current, now diffusion controlled, peaking at point *c*. The current then decays with $t^{-1/2}$ according to the Cottrell equation as the solution surrounding the electrode is depleted of oxidant due to its electrochemical conversion to the reduced state. The final rise at point *d* is caused by the discharge of the supporting electrolyte. At the switching potential (−0.9 V) the potential is switched to scan in the positive direction. However, the potential is still sufficiently negative to continue the reduction of the oxidant and so a cathodic current continues for a brief period. Finally the electrode potential becomes sufficiently positive to bring about oxidation of the reductant that had been accumulating adjacent to the electrode surface. At this point an anodic current begins to flow and to counteract the cathodic current. The anodic current increases rapidly until the surface concentration of the accumulated reductant approaches zero, at which point the anodic current peaks (point *f*). The anodic current then decays as the solution surrounding the electrode is depleted of reductant formed during the forward scan. In the course of the cathodic variation in potential, the reduced form of the reactant is produced in the vicinity of the electrode, while the oxidized form is depleted. Given sufficient time, the reduced form would diffuse into the bulk of the solution, but the potential is taken back to the initial value at a rate such that some of the reduced form is still present at the electrode surface and undergoes a process of oxidation back to the form of the couple initially present in the solution.

The important parameters of a cyclic voltammogram are the magnitudes of the anodic peak current, $(i_p)_a$, the cathodic peak current, $(i_p)_c$, the anodic peak potential, $(E_p)_a$, the cathodic peak potential, $(E_p)_c$, and the half-peak potential, $(E_{p/2})_c$. Note that $(E_{p/2})_c$, at which the current is half the peak value, differs from the half-wave

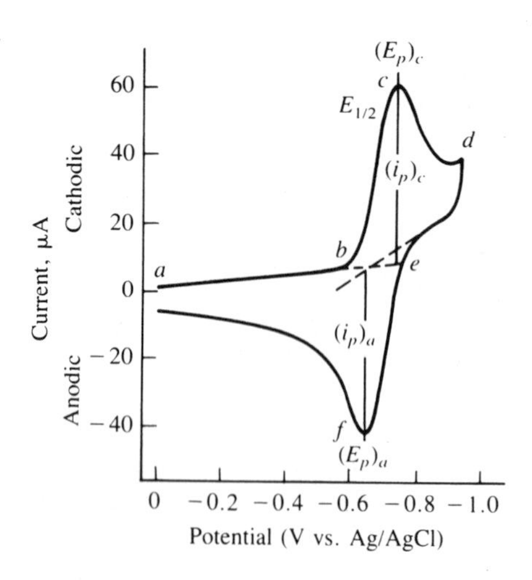

FIGURE **23.12**
Schematic cyclic voltammogram. Scan initiated at 0.0 V versus Ag/AgCl and in negative direction at 20 mV/sec. Scan reversed at −0.9 V.

potential, $E_{1/2}$. The peak current (in amperes) for the oxidant (assuming the initial scan is cathodic) is

$$i_p = n^{3/2}F^{3/2}(\pi v D_{ox}/RT)^{1/2}AC_{ox}\chi(\sigma t)$$

where $\chi(\sigma t)$ is a tabulated function whose value is 0.446 for a simple, diffusion-controlled electron transfer reaction, R is in J $K^{-1}mol^{-1}$, and T is kelvin. At 25 °C this reduces to

$$i_p = (2.69 \times 10^5)n^{3/2}AD_{ox}^{1/2}v^{1/2}C_{ox} \tag{23.20}$$

for A in cm^2, D in cm^2/sec, C in mol/cm^3, and v in V/sec.

To measure accurately peak currents, it is essential to establish the correct baseline. This is not always easy, particularly for more complicated systems. Because the peak may be somewhat broad, so that the peak potential may be difficult to determine, it is sometimes more convenient to report the potential at half the peak height $(E_{p/2})_c$.

For a reversible wave, E_p is independent of the scan rate, and i_p, as well as any other point on the wave, is proportional to $v^{1/2}$. A convenient normalized current function is $i_p/v^{1/2}C$, which depends on $n^{3/2}$ and $D^{1/2}$. For a simple diffusion-controlled reaction, this current function is a constant independent of the scan rate. This constant can be used to estimate n for an electrode reaction if a value of D can be estimated (or the reverse if n is known).

The number of electrons transferred in the electrode reaction for a reversible couple can be determined from the separation between the peak potentials:

$$(E_p)_a - (E_p)_c = \frac{0.057}{n} \tag{23.21}$$

which is valid when the switching potential is at least $100/n$ mV past the cathodic peak potential. The formal potential for a reversible couple is centered between the peak potentials. Electrochemical quasi-reversibility is characterized by a separation of peak potentials greater than indicated by Equation 23.21, and irreversibility by the disappearance of a reverse peak.

On the reverse scan, the position of the peak depends on the switching potential. As this potential becomes more negative, the position of the anodic peak becomes constant at $28.5/n$ mV anodic of the half-way potential. When the switching potential is more negative than $100/n$ mV of the reduction peak, the separation of the two peaks is $57/n$ mV and independent of the scan rate. This is a commonly used criterion of reversibility. Reversibility can also be ascertained by plotting $(i_p)_c$ or $(i_p)_a$ versus the square root of the scan velocity. The plots should be linear with intercepts at the origin. Care must be taken to eliminate any background currents.

23.6 PHASE-SENSITIVE ALTERNATING CURRENT VOLTAMMETRY[9]

Alternating current voltammetry uses a small-amplitude sinusoidal potential modulation of a single frequency that is superimposed on a slowly changing dc potential ramp. The dc potential is effectively constant during the lifetime of each drop if a polarographic method is used. Although not absolutely necessary, fresh drops permit renewal of the diffusion layer. Following the perturbation of the system with the alternating signal, one observes the way in which the electrochemical cell follows the perturbation at steady state.

In a typical experiment, the ramp has a slope of 5 mV/sec or less and the sinusoid is a few millivolts in amplitude at several hundred cycles per second (Figure 23.13). The composite signal is the dc ramp with almost imperceptible sinusoidal wiggles on it. The current is averaged for a number of cycles for each drop just before the drop is dislodged. The direct current component of the total current is blocked out electronically and only the rectified and damped alternating component is displayed as a function of the dc potential. The resulting voltammogram is substantially free of drop growth oscillations.

By looking only at the alternating portion of the current that flows and detecting its amplitude, one is in effect looking at the difference in current that flows between the minimum and maximum applied potentials during the modulation period. The fundamental harmonic ac voltammogram has its maximum amplitude at the half-wave potential. Because the reversible wave attains its maximum slope at the half-wave potential (Figure 23.14), a given ac signal causes the periodic changes in concentration of the electroactive species to be maximal at this potential. At other dc potentials along the wave, the sinusoidal potential variation causes less perturbation of the dc surface concentration and the alternating current is correspondingly less.

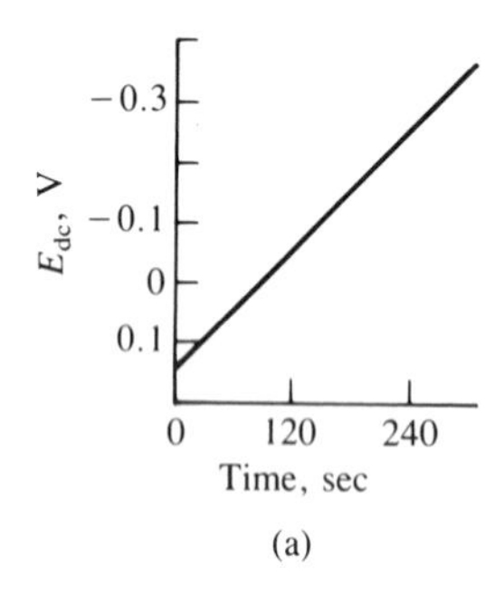

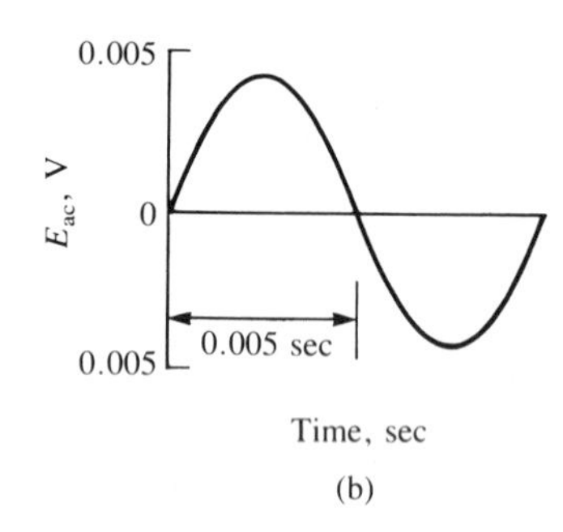

FIGURE **23.13**
Waveforms of the excitation signals applied to the electrochemical cell for phase-sensitive ac voltammetry: (a) the dc voltage ramp and (b) the sinusoidal ac modulation at a single frequency.

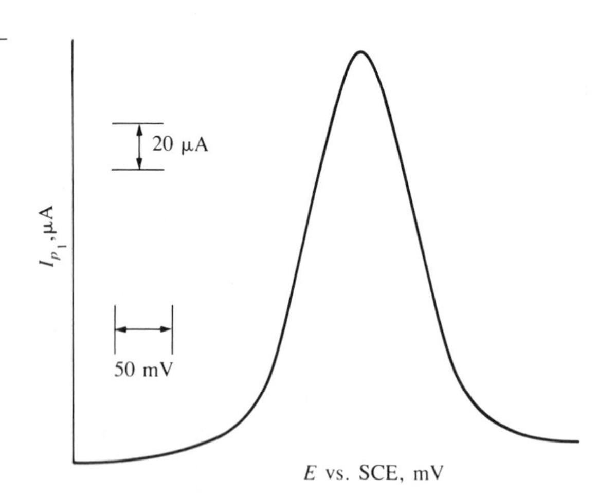

FIGURE **23.14**
Phase-sensitive fundamental harmonic ac voltammogram.

Instrumentation for ac voltammetry incorporates a phase-sensitive detector for measurement of the alternating current. The detector uses the ac potential applied to the cell as a reference and discriminates against those signals that are not in phase with the applied potential and of the same frequency. Detection of only the ac component permits separation of the faradaic and capacitative currents because of the phase difference between them. The total alternating current is the vector sum of a faradaic component and a charging current due to the capacitance of the double layer. Usually the faradaic current has a phase angle of 45° with respect to the applied ac potential, whereas the capacitance current is 90° out of phase with the applied potential. By using a phase-sensitive lock-in amplifier (see Chapter 4), either the faradaic or the capacitative current can be selected while the other is rejected. The capacitative current is important in studies of kinetics and adsorption.

The maximum height of the faradaic alternating current is proportional to the concentration of electroactive material, here assumed to be the oxidant:

$$i_p = \frac{n^2 F^2 A \omega^{1/2} D_{\text{ox}}^{1/2} C_{\text{ox}} \, \Delta E}{4RT} \tag{23.22}$$

where A is the electrode area, ΔE is the amplitude of the voltage signal, and ω is the angular frequency. Because even a moderate cell resistance mixes the phases, successful application of the phase discriminator requires cell resistances to be very low.

23.7 HYDRODYNAMIC METHODS

Methods involving convective mass transport of reactants and products are the subject of this section. These involve systems where the electrode is in motion or where there is forced solution flow past a stationary electrode. These methods take advantage of enhanced sensitivity resulting from the enhanced mass transfer of electroactive substance to the electrode that occurs under hydrodynamic conditions. A steady state is attained rather quickly and measurements can be made with high precision. At steady state, double-layer charging does not enter into the measurement.

Rotating Disk Voltammetry

The construction of a rotating disk electrode was described in Section 23.1. Qualitatively, the spinning disk drags with it the fluid at its surface. The hydrodynamic flow pattern that results from rapid rotation of the disk moves liquid horizontally out of and away from the center of the disk. The fluid at the disk surface is replenished by an (upward) axial flow normal to the surface. Once the velocity profile has been determined, the convective-diffusion relationships can be solved to give the mass transfer-limited current (Levich equation):

$$i_{\text{lim}} = 0.620 nFAD^{2/3} \omega^{1/2} v^{-1/6} C \tag{23.23}$$

where v is the kinematic viscosity (=viscosity/density) in cm^2/sec and ω is the angular frequency of rotation ($2\pi \times$ rotation rate). The rotation rate should be faster than 10 sec^{-1} but slow enough to maintain laminar flow. Turbulent flow must be avoided. The rate at which the electrode potential is scanned must be slow with respect to the rotation rate to allow the steady-state conditions to be achieved.

For a totally reversible reaction, the shape of the current-potential wave is independent of the rotation rate. Deviation of the plot of i versus $\omega^{1/2}$ from a straight line that intersects the origin suggests that some kinetic step is involved in the electron-transfer reaction. When such is the case, the plot will be curved and tend toward the kinetically limited current as the square root of the rotation rate approaches infinity. Plotting $1/i$ versus $1/\omega^{1/2}$ yields $1/i_k$, the kinetic current, upon extrapolation. Determination of the kinetic current at different values of potential then allows determination of the kinetic parameters.

Rotating Ring Voltammetry

The limiting current for a rotating ring electrode is

$$i_{\text{ring, lim}} = 0.620nF\pi(r_3^3 - r_2^3)^{2/3}D^{2/3}\upsilon^{-1/6}\omega^{1/2}C \tag{23.24}$$

where r_2 is the inner radius and r_3 the outer radius of the ring. For a given concentration and rotation rate, a ring electrode will produce a larger current than a disk electrode of the same area.

Rotating Ring-Disk Voltammetry

The addition of an independent ring electrode encircling a disk electrode enables information to be obtained by reversal techniques. The ring functions as a second working electrode. The idea is to generate electrochemically a reactive species at the disk and then electrochemically monitor the species as it is swept past the ring. This method has proved very useful in the study of reaction mechanisms, since electroactive intermediates of coupled homogeneous chemical reactions can be monitored at the ring electrode.

Flow-Through Voltammetry

Flow-through electrodes, both steady-state amperometric and triple pulse types, are widely used as detectors in HPLC (see Section 19.8). In fact, they may be the most widely used hydrodynamic method.

23.8 STRIPPING VOLTAMMETRY[10,11]

Stripping voltammetry, also called stripping chronoamperometry, has the lowest detection limit of any commonly used electroanalytical technique. Basically, electrochemical stripping analysis is a two-step operation. During the first step, analyte is electrolytically deposited onto or into the surface of an electrode typically consisting of a thin film or a drop of mercury or, in some cases, a solid electrode by controlled potential electrolysis (see Chapter 24). This is followed by a reverse electrolysis, or stripping, step, in which the deposited analyte is removed from the electrode. Each electrochemical species strips at a characteristic potential.

The preconcentration or electrodeposition step provides the means for substantially improving the detection limit for the analytical (stripping) step. Since the volume of the mercury electrode or the solid electrode surface is considerably less than the volume of the sample solution in the electrochemical cell, the resulting amalgam or deposit of metal atoms into or onto the electrode may be more concentrated than the original test solution by a factor of up to 1 million.

Anodic Stripping Voltammetry

Anodic stripping voltammetry (ASV) is used primarily to determine the concentration of trace metals that can be preconcentrated at an electrode by reduction. The method is especially effective for metals that dissolve in mercury by forming amalgams. Either a hanging mercury drop electrode or a thin-film mercury electrode can be used. Very electropositive metal ions, such as mercury(II), gold(III), silver, and platinum(IV), are deposited on solid electrodes such as glassy carbon. The geometry of diffusion is critical within or on the electrode surface. A thick planar electrode, such as a mercury pool, is not useful.

In the first step, a deposition potential is chosen that is more negative than the half-wave potential of the metal or metals to be determined. Regions in which particular metal ions are reduced at a mercury electrode are determined from the current-potential curves discussed in previous sections. A suitable potential would lie on the diffusion current plateau of a dc polarogram or a normal pulse voltammogram. For example, copper(II) can be selectively reduced in the potential range 0 to −0.25 V (vs. SCE), whereas copper(II), bismuth, lead, cadmium, and zinc could be reduced simultaneously at a potential of −1.3 V (Figure 23.15). Similarly copper and bismuth could be selectively deposited at −0.4 V. The solution is generally stirred during deposition to maximize analyte-electrode contact.

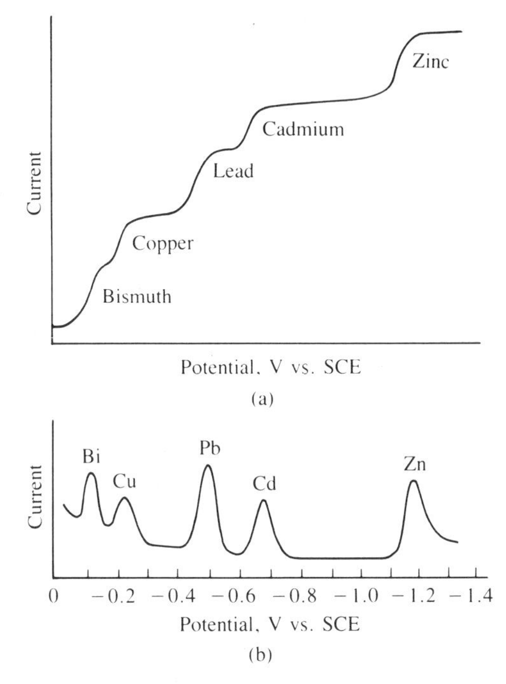

FIGURE **23.15**
(a) Voltammogram of a solution that contains bismuth, copper(II), lead, cadmium, and zinc ions. (b) Anodic stripping voltammogram using differential pulse detection of the same solution.

If more sensitivity is required, the deposition time is simply increased. This increases the degree of preconcentration, making more deposited analyte available at the electrode during the stripping step. However, the deposition step is seldom carried to completion. Usually, only a fraction of the metal ions needs to be deposited, just a sufficient amount to produce a measurable current during the stripping step. However, it is important that the same fraction of metal ion be removed during each experiment. Thus the temperature and stirring of the sample solutions must be kept as constant and reproducible as possible, and the deposition time must be carefully controlled for samples and standards alike.

Following the deposition step, there is a short rest period of 30–60 sec. The deposition potential is still applied to the working electrode, but stirring is halted. This allows convection currents from the stirring to decrease to a negligible level and also gives time for any amalgam to stabilize. If desired, at this time the electrolyte may be changed to one better suited for the stripping process.

In the final stripping step, there is no stirring of the solution. The potential is scanned in the positive direction (typically linear potential sweep voltammetry, differential pulse voltammetry, or square-wave voltammetry). At characteristic potentials the deposited metal atoms are stripped from the electrode back into the solution by oxidation to the ionic form. The potentials of the stripping peaks identify the respective metals, since, ideally, the different metals strip back into solution in reverse sequence to their reduction potentials. The area under the resulting current peaks is proportional to the concentrations of the respective analyte species.

Prior to each anodic stripping experiment, the supporting electrolyte must be conditioned or purified. In the conditioning step, a potential of 0.0 V versus SCE (usually just negative of the oxidation potential of mercury) is applied to the electrode for a controlled time (60–120 sec) to clean the electrode by removing contaminants from the mercury drop or material not removed during the prior stripping step. If a thin-film electrode is being formed in situ, the conditioning potential may be set positive of the oxidation potential of mercury to provide a clean electrode surface for the deposition step. The solution is stirred during the conditioning step. Purging the solution with purified nitrogen gas for 2–10 min eliminates interference from oxygen.

Standard samples and a blank are carried through identical electrodeposition and stripping steps. Often the method of standard addition is used for evaluation. The limit of detection is nearly always governed by the magnitude of the blank value and not by instrumental sensitivity.

Differential pulse anodic stripping has two advantages. First, it discriminates against the capacitive component of the stripping signal. Only a portion of the metal atoms oxidized as a result of the potential step that ends an anodic pulse has a chance to diffuse away from the electrode surface before the potential step that begins the next pulse returns the potential to a value at which the metal is redeposited. Consequently, metal atoms make multiple contributions to the analytical signal and render the detection limit of differential pulse anodic stripping voltammetry lower than that of linear potential sweep anodic stripping voltammetry. The second advantage is that linear sweep methods have continuous interference from charging current as long as the potential is scanned.

Anodic stripping voltammetry can be complicated by intermetallic formation, particularly with a thin-film mercury electrode. This occurs when metals such as zinc and copper are in high concentrations. When such intermetallics are present, the stripping peaks for the constituent metals may be shifted, severely depressed, or absent completely. The formation of intermetallics is less likely to be a problem with a hanging mercury drop electrode because the large electrode volume diminishes the maximum achievable preconcentration. Intermetallic compound formation can render the results of standard addition evaluation incorrect.

Cathodic Stripping Voltammetry

The procedure for cathodic stripping voltammetry follows the same steps as outlined for anodic stripping voltammetry. It involves preconcentration by oxidation with subsequent stripping via a negative potential scan. Cathodic stripping voltammetry is used to determine those materials that form insoluble mercury salts on the electrode surface. At a relatively positive potential, mercury(I) ions are produced at the mercury electrode surface during an anodic preelectrolysis. Materials that precipitate with mercury(I) ions form an insoluble film on the surface of the mercury electrode. After a rest period, a cathodic scan causes the reduction of the salt to mercury and the original anion, giving a cathodic current peak.

Silver can be used as the electrode for the determination of anions that form insoluble silver salts. Materials that can be determined by cathodic stripping voltammetry include arsenate, chloride, bromide, iodide, chromate, tungstate, molybdate, vanadate, sulfate, oxalate, succinate, selenide, sulfide, mercaptans, thiocyanate, and thio compounds. Lead has been determined by cathodically stripping a film of PbO_2 deposited on a SnO_2 electrode.

23.9 VOLTAMMETRIC METHODS COMPARED

The key to classical polarography (slow linear sweep dc voltammetry) is that the current is always measured at the same time in the life of an oscillating drop. Effectively a steady state prevails because the thickness of the diffusion layer is always the same at the time of measurement. The key to the other methods is that the diffusion-layer thickness is a function of time. If repetitive measurements are made at a constant time interval after initiation of the excitation signal in pulse techniques or ac voltammetry, an analogous steady-state result is obtained. In most analyses, aqueous media are used, both for convenience and for compatibility with the chemistry of sample preparation. Other solvents can provide superior working ranges and may merit consideration for new applications.

Slow linear sweep dc voltammetry has the poorest detection limits (about $10^{-5}M$) of any of the methods. The culprit is the charging or capacitive current, due both to drop growth and continually changing the applied potential, whose compensation is difficult at low concentrations.

The potential step methods discriminate against the charging current by not sampling the current until most of the capacitive current has decayed. In addition, the pulse methods gain enhanced sensitivity through the increased faradaic currents obtained by delaying the pulse until near the end of each timer cycle and sampling the current at a short enough time. Also, with digital waveforms, one has a stepped dc ramp (and not a linear dc ramp) so the charging current value attributable to the potential scan is eliminated. All the potential step techniques discussed are directed toward analytical measurements of concentration, so that sensitivity and resolution are the key features of performance. The normal pulse, differential pulse, and square-wave techniques are among the most sensitive for the direct evaluation of concentrations, and they have wide use for trace analysis. Reversibly reducible ions can be detected at concentrations down to $10^{-8}M$, and irreversibly reducible ions can be determined at $10^{-7}M$ concentration. In addition, potential step methods can provide information about the chemical form in which an analyte appears. Oxidation states can be defined, complexation can often be detected, and acid-base chemistry can be characterized. Pulse measurements are sufficiently sensitive that the analyst must pay special attention to impurity levels in solvents and supporting electrolytes.

The system response is not strongly dependent on the electrode kinetics, at least not to the extent expected of ac voltammetry. Consequently, pulse voltammetric techniques may be used for electrochemically irreversible systems.

Differential pulse polarography provides a greatly improved signal-to-noise ratio compared with the Tast or normal pulse method even though the signal is smaller. In fact, the differential mode is less sensitive than the normal mode due to the difference measurement, but the detection limit is better because it discriminates more effectively against charging currents by taking smaller potential steps of constant size. Because the voltammogram is peak-shaped with the signal returning to the baseline, the selectivity is improved.

Square-wave voltammetry materially shortens the observation time in an experiment. A typical experiment that requires 3 min by normal or differential pulse techniques can be performed in a matter of seconds by square-wave voltammetry. The entire voltammogram can be recorded on a hanging mercury drop electrode. If noise reduction through integration of the signal is desired, five scans could be performed in only 30 sec. Since square-wave voltammetry exhibits a rapid response, it can be used for chromatographic detection. Peak heights obtained by square-wave voltammetry are larger than those with the differential pulse technique. This enhanced peak height is a function of the delay time between pulses. In differential pulse voltammetry the delay time is much larger than the pulse time, allowing greater depletion of the analyte concentration near the electrode, whereas in square-wave voltammetry the delay time is equal to pulse time, dependent on frequency, and equal to $\tau/2$. As frequency increases, τ decreases, and the measured faradaic current increases.

The merits of cyclic voltammetry are largely in the realm of qualitative or diagnostic experiments. A cyclic voltammogram is to the electrochemist what the frequency domain spectrum is to the spectroscopist. It gives a picture of a substance's electrochemical behavior while varying the voltage (energy) of an indicator electrode. Cyclic voltammetry is capable of rapidly generating a new oxidation state during the forward scan and then probing its fate on the reverse scan. Peak height is related to both concentration and reversibility of the reaction. Thus, the technique yields information about reaction reversibilities and also offers a very rapid means of analysis for suitable systems. The method is particularly valuable for the investigation of stepwise reactions, and in many cases direct investigation of reactive intermediates is possible. By varying the scan rate, systems that exhibit a wide range of rate constants can be studied, and transient species with half-lives of milliseconds are readily detected. The method can be applied to stationary electrodes as well as to a single mercury drop, and to reactions for which stripping analysis is inapplicable due to highly irreversible electrode processes or the formation of solution-soluble reaction products.

An important characteristic of ac polarography is that it responds to only reversible or quasi-reversible electrode reactions. This limits its applicability but in some cases can be advantageous in avoiding interferences.

Hydrodynamic techniques have very favorable limits of detection when combined with steady-state response. Double-layer charging does not enter the measurement. The rates of mass transfer at the electrode surface in these methods are much faster than the rates of diffusion alone. The effects of transfer kinetics are more pronounced because mass transfer keeps the concentration near the electrode closer to the bulk value. These methods permit the continuous monitoring of flowing liquids (like detectors in HPLC and in amperometric titrations). Dual-electrode techniques can be used to provide the same kind of information that cyclic voltammetry does with a stationary electrode.

The major advantage of stripping voltammetry, as compared with direct voltammetric analysis of the original solution, is the preconcentration (by factors of 100 to more than 10^6) of the material to be analyzed into the small volume of a mercury electrode or onto the surface of an electrode before a voltammetric analysis. As a result, the voltammetric stripping current is less perturbed by charging or residual impurity currents. Combined with differential pulse voltammetry in the stripping step, the technique is especially useful for the analysis of very dilute solutions (down to about $10^{-11}M$).

23.10 VOLTAMMETRIC APPLICATIONS

Determination of inorganic or organic species that are either molecular or ionic can be performed if they undergo oxidation or reduction at a working electrode in the region of potential bounded at the positive limit by the potential at which the solvent, the anion present, or the electrode material is oxidized and at the negative limit by the potential at which the supporting electrolyte, solvent, or in some cases the electrode itself is reduced. Many pertinent organic and inorganic half-wave potentials are listed in the *Handbook of Organic Chemistry*.[12]

Organic Compounds

The types of organic bonds that can be reduced at a dropping mercury electrode are listed in Table 23.2. Often the presence of a single group is insufficient to bring the half-wave potential of the electroactive species to an accessible potential range. For these compounds it is necessary for the material to contain, in addition to the electroactive group, another activating group that affects the electron distribution in the substrate and thereby shifts the half-wave potential of the reduction wave. Both inductive and resonance effects can facilitate the reduction of groups attached to aromatic systems or to double bonds. For example, the reduction potential of the disulfide group linked to a phenyl group is −0.5 V (vs. SCE) and to an alkyl group is

FIGURE 23.16

Square-wave voltammogram of iron(II) and iron(III) in pyrophosphate buffer at pH 9.

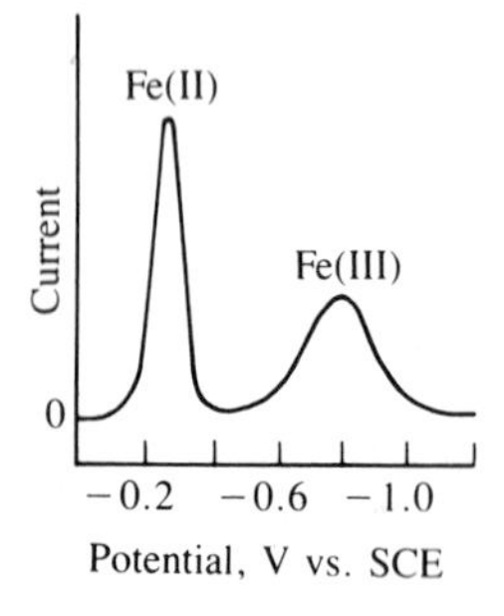

−1.25 V. As the number of condensed aromatic rings increases, the reduction is made easier. Single C—X (where X is halogen) bonds are usually reduced at more negative potentials than C═C—C—X, but at more positive potentials than C—C═C—X. The ease of reduction of the C—X bond increases with the increasing polarizability of the halogen; that is, F < Cl < Br < I. Many organic functional groups can be converted to an electroactive group by appropriate derivatization methods. A number of reagents are listed in Table 23.3.

Inorganic Compounds

Voltammetry can be used to determine the concentration of a particular oxidation state of a metal with multiple oxidation states. For example, iron(II) and iron(III) can be determined simultaneously in a pyrophosphate buffer at pH 9 (Figure 23.16). The oxidation states of iron are of analytical importance in the manufacture of magnetic media or plating nickel-iron alloys. Often a different supporting electrolyte is needed. In chrome-plating baths and in waste-water analysis, chromium(VI) can

TABLE 23.2

REDUCIBLE ORGANIC FUNCTIONAL GROUPS

>C═O	Ketone	—O—O—	Peroxy
—CHO	Aldehyde	—S—S—	Disulfide
>C═C<	Alkene	$—NO_2$	Nitro
Aryl—C≡C—	Aryl alkyne	—NO	Nitroso
>C═N—	Azomethine	—ONO	Nitrite
—C≡N	Nitrile	$—ONO_2$	Nitrate
—N═N—	Azo	—NHOH	Hydroxylamine
—N(O)═N—	Azoxy		

NOTE: Others are dibromides, aryl halide, alpha-halogenated ketone or aryl methane, conjugated alkenes and ketones, polynuclear aromatic ring systems, and heterocyclic double bond.

TABLE 23.3

ORGANIC FUNCTIONAL GROUP ANALYSIS OF ELECTROINACTIVE GROUPS

Functional group	Reagent	Active voltammetric group
Carbonyl	Girard T and D	Azomethine
	Semicarbazide	Carbazide
	Hydroxylamine	Hydroxylamine
Primary amine	Piperonal	Azomethine
	CS_2	Dithiocarbonate (anodic wave)
	$Cu_3(PO_4)_2$ suspension	Copper(II) amine
Secondary amine	HNO_2	Nitrosoamine
Alcohols	Chromic acid	Aldehyde
1,2-Diols	Periodic acid	Aldehyde
Carboxyl	Transform to thiouronium salts	—SH (anodic wave)
Phenyl	Nitration	$—NO_2$

be determined in sodium hydroxide at −0.88 V and chromium(III) in a solution of potassium thiocyanate and acetic acid at pH 3.2 at −0.99 V (both vs. SCE).

The determination of ammonia by differential pulse voltammetry depends on its reaction with excess formaldehyde at pH 4 to produce methyleneimine, which is electrochemically reducible to methylamine. Iodate and bromate are easily distinguishable.

Even some unlikely candidates can be handled by voltammetric techniques. Cyanide, a critical pollutant, forms a strong complex with mercury, which shifts the reduction potential of mercury to −0.1 V. This method was used by the Food and Drug Administration during the Tylenol crises in 1982 and 1986 to confirm the presence or absence of cyanide.

Oxygen Determination

Several compact portable units are available for the determination of dissolved oxygen in various media. The oxygen-sensing probe is an electrolytic cell with a gold (or platinum) cathode separated from a tubular silver anode by an epoxy casting. The anode is electrically connected to the cathode by an electrolytic gel, and the entire chemical system is isolated from the environment by a thin gas-permeable membrane (often Teflon). A potential of 0.8 V is applied between the electrodes. The oxygen in the sample diffuses through the membrane and is reduced at the cathode with the formation of the oxidation product, silver oxide, at the silver anode. The resultant current is proportional to the concentration of oxygen present. The analyzer operates over the range 0.2–50 parts per million of dissolved oxygen. Gases that reduce at −0.8 V will interfere; these include the halogens and sulfur dioxide. Hydrogen sulfide contaminates the electrodes.

Spectroelectrochemistry[13]

Photons are used as nonperturbing probes and as active participants in electrochemical processes. Most optical methods discussed in earlier chapters have been explored. The simplest experiment is to reflect a light beam from or pass the beam through the electrode surface and to measure changes that result from species produced or consumed in the electrode process. The prerequisite is a highly polished reflective electrode, an optically transparent electrode (thin films of a semiconductor, a metal deposited on a transparent substrate, or a fine wire mesh with several hundred wires per centimeter). Transmission measurements may involve the study of absorbance versus time as the electrode potential is stepped or scanned. A rapid spectral scanning system is needed to follow spectral changes over comparatively short time scales. For a completely stable electrochemical reaction product, the integral of its differential absorbance (Beer's law) is equal to Q_d/nFA, where Q_d (the chronocoulometric charge from a diffusing component) is given by the integrated Cottrell equation. Hence,

$$\text{absorbance} = \frac{2\varepsilon_{\text{red}} C_{\text{ox}} D_{\text{ox}}^{1/2} t^{1/2}}{\pi^{1/2}} \tag{23.25}$$

The particular advantage of an optically transparent thin-layer electrode is that bulk electrolysis is achieved in $\lesssim 1$ min, so that for a chemically reversible system the entire solution reaches an equilibrium with the electrode potential.

Spectral data can be gathered on a static solution composition. One can do cyclic voltammetry, bulk electrolysis, and coulometry in the ordinary way while also obtaining spectrochemical information. Spectroelectrochemical methods can be especially useful for unraveling a complex sequence of charge transfers. Charge increments as small as five nanoequivalents have produced significant spectral changes.

Optical signals can also be used to initiate electrochemical reactions. The latter are then followed by any appropriate voltammetric method.

Microelectrodes[14]

So often in every area of analytical chemistry it seems that smaller is better. Electrodes with a radius smaller than 10 μm (one-tenth the diameter of a human hair) exhibit very different voltammetric properties than do electrodes of conventional size because of the effects of convergent diffusion, which enables a steady-state current response to develop on a sub-second time scale. Microelectrodes enable time-independent currents to be easily monitored. This feature simplifies measurements and facilitates their use under conditions where variations in potential are disadvantageous. The currents at these electrodes are extremely small. Because of the small currents, the electrodes can be used in solutions of very high resistance.

The time scale of the measurement and the radius of the electrode affect the type of voltammogram that one obtains in stationary solutions (in vivo measurements). At fast time scales, or at electrodes with a large radius, and when the diffusion layer thickness is much less than the radius, peak-shaped voltammograms are obtained, whereas with a small radius or with long time scales of measurement, the diffusion layer thickness exceeds the radius and convergent diffusion dominates, yielding steady-state voltammograms.

Microelectrodes connected in parallel maintain the features of single electrodes with the advantage of increased current amplitudes. Using a current transducer capable of low-noise subpicoampere measurements and using repetitive potential pulses, concentration changes of $1 \times 10^{-7} M$ for the two-electron oxidation of dopamine in pH 7.4 buffer have been observed.

Because of the small currents involved when microelectrodes are used, nonpolar media such as toluene become usable. Also, work has been reported in frozen and solid media, in solutions without supporting electrolytes and in supercritical fluids. These electrodes enable electrochemical measurements to be made in unusual places and under unusual circumstances.

Amperometric Titrations

When the potential applied across two electrodes is maintained at some constant value, the current may be measured and plotted against the volume of the titrant—hence, the name *amperometric titration*. The current is measured either on the limiting current plateau or somewhere within the diffusion current region of a current-potential curve.

Working Electrode–Reference Electrode. The potential of the indicator electrode is maintained at a constant value with respect to a reference electrode so that a limiting current, which is proportional to the concentration of one or more of the reactants or products of the titration, is measured. A titration curve is obtained by plotting the limiting current as a function of the volume of titrant added. The shape of the titration curve can be predicted from hydrodynamic voltammograms of the solution obtained at various stages of the titration. As an example, consider the titration of a reducible substance (lead ions) with a nonreducible titrant (sulfate ions). A voltammogram of a solution containing lead ions is represented by curve *A* in Figure 23.17. If the applied potential is held at any value on the diffusion-current plateau, the current is represented by i_0. The titrant exhibits no diffusion current at the applied voltage. Successive increments of titrant remove lead ions to form a precipitate of lead sulfate. As the concentration of lead ions decreases, the diffusion current decreases successively to i_1, i_2, i_3, and finally i_r, the residual current characteristic of the supporting electrolyte.

The intersection of the extrapolated branches of the titration curve gives the end point (Figure 23.18). Only three or four experimental points need to be accumulated to establish each branch of the curve. Amperometric titration curves are linear, as opposed to logarithmic, and thus do not rely on data in the vicinity of the end point. This is in marked contrast to the difficulty in establishing the end point in a logarithmic titration curve (as in potentiometry).

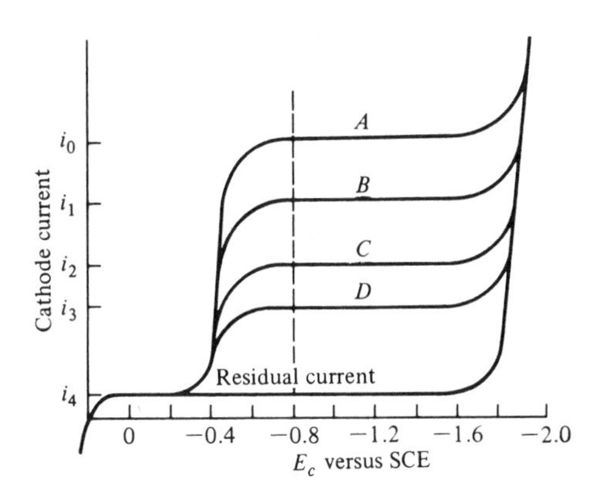

FIGURE **23.17** Successive current-potential curves of lead ion made after increments of sulfate ion were added.

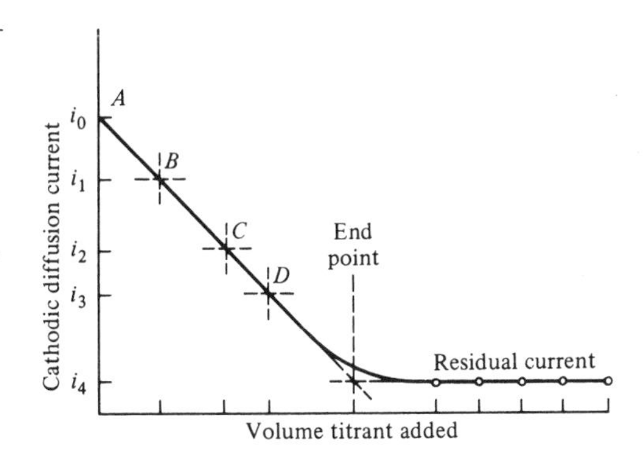

FIGURE **23.18** Amperometric titration curve for the reaction of lead ions with sulfate ions. See Figure 23.17 for corresponding current-potential curves. Performed at −0.8 V vs. SCE.

When both titrant and sample ions give diffusion currents at the chosen voltage, the current will decrease up to the end point and then increase again to give a V-shaped titration curve. If the sample is not electroactive but the titrant is, a horizontal line is obtained that rises after the end point. When possible, the reversed L-shaped curve is preferred.

Automation is easy. The titrator can be programmed to shut off when a specified current level is reached. Titrant is run into a blank until the specified current is reached, sample is added, and the titrant is again added until the specified current level is attained.

Strictly, a correction for dilution is necessary to attain a linear relationship between current and volume of titrant, but by working with a reagent that is tenfold more concentrated than the solution being titrated, the correction becomes negligible. Incompleteness of the reaction in the vicinity of the end point usually does not detract from the results provided the reaction equilibrium is attained quickly. Points can be selected between 0% and 50% and between 150% and 200% of the end-point volume for the construction of the two branches of the titration curve. In these regions the common ion effect represses dissociation of complexes and solubility of precipitates.

FIGURE **23.19**
Two-electrode amperometric titration when titrant electrode reactions are irreversible and with a constant potential difference impressed between two platinum indicator electrodes. The dashed line represents the curve after the end point when the titrant electrode reactions are reversible.

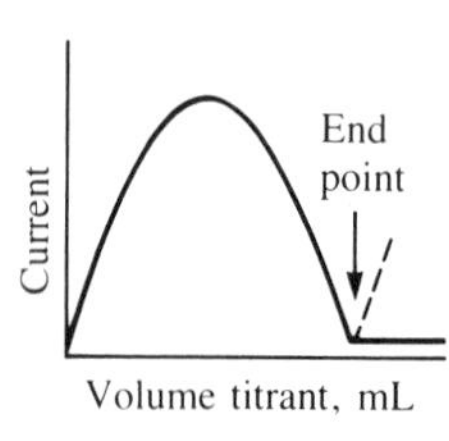

Two Working Electrodes. In a modification of the usual amperometric titration system, the current that results from a small fixed potential impressed between two working electrodes is measured. One electrode functions as an anode and the other as a cathode. The shape of the amperometric titration curve strongly depends on the reversibility of the electrode reactions of the titrant and titrate systems.

When the titrant (sample) involves reversible electrode reactions, a small amount of electrolysis takes place, but no net change in solution composition occurs because the amount of the oxidized form reduced at the cathode is compensated by that formed by oxidation of the reduced form at the anode. If the titrant electrode reactions are irreversible, the titration curve has the shape shown in Figure 23.19. After the equivalence point, the current is zero or close to zero. The method was introduced years ago under the name "dead-stop end point."

When the titrant and its reaction product form a reversible oxidation/reduction system, the current is zero or close to zero only at the end point. Past the equivalence point, the current increases due to the increasing amount of unused electroactive titrant being added to the system.

PROBLEMS

1. When the differential pulse method was used, three individual samples of chloramphenicol in 0.1*M* acetate buffer, pH 4, gave these peak currents measured at −0.27 V versus SCE. What is the concentration of the unknown?

Concentration (μg/mL)	0.48	0.96	*X*
i_{peak} (nA)	2.96	5.92	4.45

2. To exactly 50.00 mL of an unknown zinc solution in a cell were added these increments of 0.100*M* disodium dihydrogen ethylenediaminetetraacetate (EDTA) solution. After each increment of titrant, the resultant faradaic current was measured at −1.20 V versus SCE. What is the concentration of the zinc solution?

EDTA (mL)	0.00	2.00	4.00	6.00	8.00	11.00	14.00
i (mA)	0.300	0.231	0.167	0.107	0.052	0.00	0.00

3. An unknown amount of copper(II) ions produces a faradaic current of 12.3 μA on a normal pulse voltammogram. After 0.100 mL of $1.00 \times 10^{-3} M$ copper(II) ions is added to the original volume of 5.00 mL, the new current is 28.2 μA. Calculate the original amount of copper.

4. Copper(II) ions form a precipitate with α-benzoinoxime in an ammoniacal solution. The supporting electrolyte consists of 0.05M ammonia and 0.1M ammonium chloride. The copper(II) tetraammine(2+) ion is reduced in two steps with half-wave potentials of −0.2 and −0.5 V versus SCE. α-Benzoinoxime gives a reduction wave with a half-wave potential of −1.6 V versus SCE. Deduce the shape of the amperometric titration curve that will be obtained at (a) an applied potential of −0.8 V versus SCE and (b) an applied potential of −1.8 V versus SCE. Which applied potential would be preferred under normal circumstances? Which applied potential would be preferred when nickel and zinc ions are also present in the sample?

5. The measurements shown in the table were made on a reversible dc polarographic wave at 25 °C. The limiting diffusion current was 3.30 μA. (a) What number of electrons were involved in the electrode reaction? (b) What is the half-wave potential for this reaction?

E (V vs. SCE)	i (μA)
−0.519	0.40
−0.520	1.02
−0.531	1.55
−0.538	2.01
−0.548	2.56
−0.560	3.01
−0.568	3.20

6. The data in the table were obtained from cyclic voltammograms of a potassium ferrocyanide solution scanned initially in the anodic direction. Do the data support the concept of a reversible nernstian wave?

v (mV s^{-1})	$(E_p)_a$ (mV)	$(E_{p/2})_a$ (mV)	$(E_p)_c$ (mV)	$(i_i)_a$ (μA)	$(i_i)_c$ (μA)
16.7	311	254	239	9.8	9.6
33.3	316	257	234	14.0	13.7
50.0	314	256	236	17.1	16.9

7. The measurements given in the table were made at 25 °C on the reversible wave for the reduction of a metallic complex ion to the metal amalgam. (a) Calculate the number of ligands associated with the divalent metal ion in the complex. (b) Calculate the formation constant of the complex if the half-wave potential for the reversible reduction of the simple metal ion is +0.081 V versus SCE. Assume that the diffusion coefficients for the complex ion and the metal ion are equal, and that all activity coefficients are unity.

Concentration of ligand (M)	$E_{1/2}$ (V vs. SCE)
0.10	−0.448
0.50	−0.531
1.00	−0.566

8. Consider the amperometric titration of iodine with sodium thiosulfate, using two indicator electrodes. (a) Sketch the current-potential curves for the reversible iodine/iodide system for points at which the titration is 0%, 50%, 100%, and 110% complete. Assume an impressed voltage of 100 mV across two platinum electrodes. (b) Sketch the amperometric titration curve and correlate the several regions on this curve with the corresponding regions on the current-potential curves. (c) How would the amperometric titration curve change if excess iodide ion were initially present?

9. Prepare a sketch of the current-potential curves and the nature of the supporting electrolyte that would be needed during the amperometric titration of an iodide-bromide mixture with a standard silver solution. The indicator electrode is a rotated platinum microelectrode. *Hint:* For success, a complexing agent is needed for one step of the titration.

BIBLIOGRAPHY

BARD, A. J., AND L. R. FAULKNER, *Electrochemical Methods,* Wiley-Interscience, New York, 1980.

BOND, A. M., *Modern Polarographical Methods in Analytical Chemistry,* Dekker, New York, 1980.

LINGANE, J. J., *Electroanalytical Chemistry,* 2nd ed., Wiley-Interscience, New York, 1958.

KISSINGER, P. T., AND W. R. HEINEMAN, eds., *Laboratory Techniques in Electroanalytical Chemistry,* Dekker, New York, 1984.

KOLTHOFF, I. M., AND J. J. LINGANE, *Polarography,* 2nd ed., Wiley-Interscience, New York, 1952 (2 volumes).

Southampton Electrochemistry Group, *Instrumental Methods in Electrochemistry,* John Wiley, New York, 1985.

LITERATURE CITED

1. FLATO, J. B., "The Renaissance in Polarographic and Voltammetric Analysis," *Anal. Chem.,* **44,** 75A (1972).
2. BARD, A. J., AND L. R. FAULKNER, Chap. 11, *Electrochemical Methods,* John Wiley, New York, 1980.
3. BOND, A., *Modern Polarographic Methods in Analytical Chemistry,* Dekker, New York, 1980.

4. BURGE, D. E., "Pulse Polarography," *J. Chem. Educ.,* **47,** A81 (1970).
5. BORMAN, S. A., "New Electroanalytical Pulse Techniques," *Anal. Chem.,* **54,** 698A (1982).
6. PARRY, E. P., AND R. A. OSTERYOUNG, *Anal. Chem.,* **37,** 1634 (1964).
7. OSTERYOUNG, J. G., AND R. A. OSTERYOUNG, "Square Wave Voltammetry," *Anal. Chem.,* **57,** 101A (1985).
8. HEINEMAN, W. R., AND P. T. KISSINGER, "Cyclic Voltammetry: Electrochemical Equivalent of Spectroscopy," *Am. Lab.,* **15,** 29 (November 1982).
9. CURRAN, D. J., "Sinusoidal ac Voltammetric Methods," *Am. Lab.,* **14,** 27 (June 1981).
10. PETERSON, W. M., AND R. V. WONG, "Fundamentals of Stripping Voltammetry," *Am. Lab.,* **13,** 116 (November 1981).
11. COPELAND, T. R., AND R. K. SKOGERBOE, "Anodic Stripping Voltammetry," *Anal. Chem.,* **46,** 1257A (1974).
12. DEAN, J. A., ed., *Handbook of Organic Chemistry,* McGraw-Hill, New York, 1987.
13. HEINEMAN, W. R., F. M. HAWKRIDGE, AND H. N. BLOUNT, Chap. 1, *Electroanalytical Chemistry,* A. J. Bard, ed., Vol. 13, Dekker, New York, 1984.
14. WIGHTMAN, R. M., "Microvoltammetric Electrodes," *Anal. Chem.,* **53,** 1125A (1981).

24 ELECTROSEPARATIONS, COULOMETRY, AND CONDUCTANCE METHODS

24.1 ELECTROSEPARATIONS WITH CONTROLLED-ELECTRODE POTENTIAL

Electroseparation is electrolysis in which a quantitative reaction or, at the very least, an appreciable amount of electro-oxidation or electroreduction takes place at an electrode. Bulk (or exhaustive) electrolytic methods are characterized by a large ratio of electrode area to solution volume and by mass-transfer conditions as effective as possible. Although bulk electroseparations are generally characterized by large currents and time scales of experiments of minutes or longer, the basic principles governing electrode reactions described in the previous chapters still apply.

Completeness of an Electrode Process

Since the potential of the working electrode is the basic variable that controls the degree of completion of an electrolytic process in most cases, controlled-potential techniques are usually the most desirable for bulk electrolysis. The extent of completion of a bulk electrolytic process can often be predicted for reversible reactions from the applied electrode potential and the Nernst equation. For the deposition of a solid, when more than a monolayer of solid is deposited on an inert electrode (such as platinum), the activity of the solid is constant and equal to unity at the completion of the electrolysis. The Nernst equation yields

$$E = E^\circ + \frac{RT}{nF} \ln[C_i(1 - x)] \tag{24.1}$$

where C_i is the initial concentration of the oxidized form and x is the fraction of the oxidized form reduced at the electrode potential, E.

For example, for 99.9% completeness of reduction of an oxidant to the metallic state, the potential of the working electrode at 25 °C should be

$$E = E^\circ + \frac{0.0592}{n} \log[C_i(1 - 0.999)] \tag{24.2}$$

or $178/n$ mV more negative than E°. The current through the system steadily decreases as the deposition proceeds. However, the maximum permissible current is used at all times so that the electrolysis proceeds at the maximum rate.

Any overpotential term is added to Equation 24.2 and takes the sign of the electrode. The potential range for a successful separation can best be found by determining the current-potential curve on a microelectrode under the same conditions (concentration, supporting electrolyte, temperature) considered for the separation (see Chapter 23).

Equipment for Electrolytic Separations

Bulk electrolysis using controlled-potential techniques requires a potentiostat with large output current and voltage capabilities. Stable reference electrodes must be carefully positioned to minimize uncompensated resistance effects. The auxiliary electrode is positioned to provide a fairly uniform current distribution across the surface of the working electrode. A schematic arrangement is shown in Figure 24.1. Sometimes the auxiliary electrode is placed in a separate compartment isolated from the working-electrode compartment by some type of separator.

In flow electrolytic methods the solution to be electrolyzed flows continuously through a porous working electrode with a large surface area. The counter electrode is interleaved with the working electrode and insulated from it with separators, unless a divided cell is necessary.

EXAMPLE 24.1

Would it be possible to separate lead quantitatively from a solution 0.01 M in lead nitrate with a pH of 1?

For a quantitative removal, the lead concentration should be lowered to $10^{-6}M$. The corresponding cathode potential would be

$$E = -0.126 + \frac{0.05916}{2} \log 10^{-6} = -0.304 \text{ V}$$

A possible interference would be the evolution of hydrogen, which would commence at

$$E = 0.0 + 0.05916 \log 0.1 = 0.059 \text{ V}$$

if it were not for the hydrogen overpotential of about -0.40 V on a lead-coated cathode (see Table 21.2). Thus, hydrogen would not interfere until the cathode attained a potential of -0.459 V, which is sufficient to lower the lead concentration well below the specified amount.

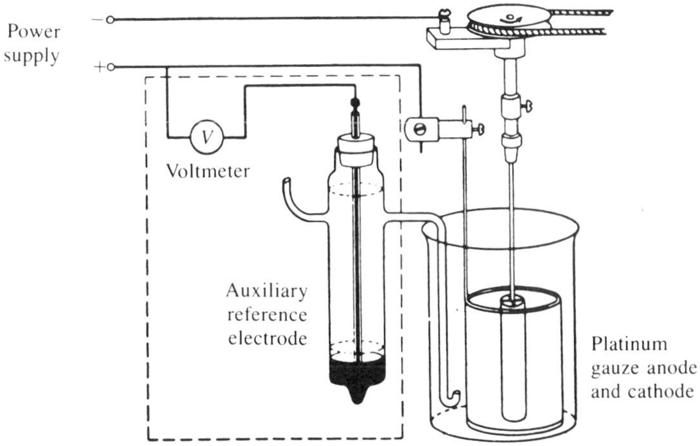

FIGURE **24.1**
Electrolytic cell for bulk electrolysis.

Composition of the Electrolyte

When the separation between the formal potentials of two oxidation/reduction systems is less than $0.178/n$ mV, changing the supporting electrolyte to one that complexes one or both of the metals often improves the separation. The addition of a masking agent to form a complex with a metal ion usually shifts the deposition potential more negative. Take, for example, copper(II), lead(II), and tin(IV) ions in a $0.1M$ sodium tartrate solution adjusted to pH 5.0 and containing hydrazine. Copper is deposited at -0.05 V and lead at -0.35 V. After the solution is acidified to destroy the tartrate complex of tin(IV), the tin is deposited at -0.40 V. Tin(II) and lead(II) have almost identical deposition potentials in a chloride medium, but oxidation of tin to the tetravalent state and then complexation of tin(IV) with tartrate ions enables lead to be deposited in the presence of tin.

Suitable oxidation/reduction systems that are preferentially reduced at the cathode, or oxidized at the anode, may be used to limit and maintain a constant potential at the particular electrode. The name *potential buffer* is applied to this type of supporting electrolyte because of its functional resemblance to pH buffers. The successful deposition of copper from chlorocuprate(I) ions and the prevention of the competing oxidation of these ions to copper(II) ions at the anode are due to the buffering action of hydrazine. The oxidation of hydrazine, for which the electrode potential is 0.17 V, takes place in preference to that of chlorocuprate(I) ions, for which the formal electrode potential is 0.51 V. As long as hydrazine is present in excess, the anodic oxidation of copper(I) (and also chloride) ions is prevented.

EXAMPLE 24.2

Under what pH conditions would it be possible to initiate the deposition of zinc onto a copper-clad electrode from a solution that is $0.01M$ in zinc ions? To what pH should the solution be adjusted to permit the quantitative removal of zinc ions?

The deposition of zinc will commence at

$$E = -0.76 + 0.0296 \log(0.01) = -0.82 \text{ V}$$

and the potential will need to be controlled at -0.94 V ($-0.76 - 0.178$) to lower the zinc concentration to 0.01% of its initial concentration.

Turning to the expression for the evolution of hydrogen, it is possible to calculate the minimum electrolyte pH to permit the deposition of zinc to commence, assuming the overpotential of hydrogen on a copper-clad electrode to be 0.40 V:

$$E = 0.0 + 0.05916 \log[H^+] + (-0.40) = -0.82 \text{ V}$$

from which the pH is calculated to be 7.

Although it might be expected that the pH would have to be raised to about 9.1 to remove the zinc completely,

$$E = -0.76 + 0.0296 \log 10^{-6} = -0.94 \text{ V}$$

$$-0.94 = 0.0 - 0.0592\text{pH} - 0.40 \quad \text{and} \quad \text{pH} = 9.1$$

that would be true only if the amount of zinc is insufficient to coat completely the exposed electrode surface. As soon as the electrode becomes coated with zinc, the overpotential of hydrogen rises to the value on a zinc surface (0.7 V):

$$-0.94 = 0.0 - 0.0592\text{pH} - 0.70 \quad \text{and} \quad \text{pH} = 4.1$$

In practice, an ammoniacal buffer is used (pH = 9.2), partly to take advantage of the superior nature of the deposit from zinc ammine ions ($E^\circ = -1.04$ V). Now, under these conditions and assuming that the ammoniacal buffer is 1 M in ammonia, the cathode potential would need to reach -1.22 V for the quantitative removal of zinc:

$$E = -1.04 + 0.0296 \log 10^{-6} = -1.22 \text{ V}$$

Under these conditions, the pH should not be less than

$$-1.22 = 0.0 - 0.0592\text{pH} - 0.70 \quad \text{and} \quad \text{pH} = 8.8$$

a condition that is met by the ammoniacal buffer used.

The Mercury Cathode

Probably the most widely used type of electroreduction involves a mercury cathode to which a constant potential is applied. Current is forced through the cell without any attempt to control the potential of the working electrode (although it is possible to control the latter with a three-electrode potentiostat). Consequently selectivity is poor. This method has extensive use for the removal of elements that interfere in various instrumental methods. The method is not used to determine any of the metals deposited.

Two factors set mercury apart from other electrode materials. Many metals are soluble in mercury, coupled in some cases with the formation of compounds (amalgams) of the metal and mercury. Because the activity of the metal in the amalgam is less than that of the solid metal, the free energy required for reduction is less and the deposition potential is correspondingly less negative than when a platinum electrode is used. Deposition of metals is also aided by the fact that the hydrogen overpotential on mercury is particularly large. As a result, the deposition, even from a fairly acid solution, is possible for metals such as iron, nickel, chromium, and zinc without coevolution of hydrogen.

The cell designed by Melaven, shown in Figure 24.2, is in common use. The cathode consists of 35–50 mL of pure mercury in a modified separatory funnel. The apparatus has a conical base fitted with a three-way stopcock. One arm of the stopcock is connected to a leveling bulb that controls the level of the mercury in the cell; the other permits removal of the electrolyte. The anode is a

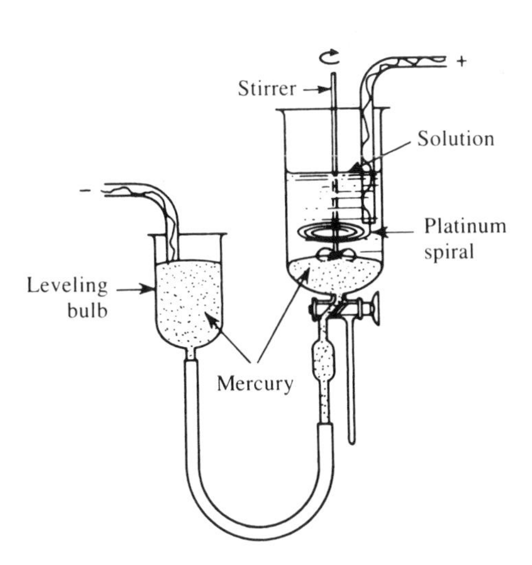

FIGURE **24.2**
Mercury cathode cell. [Reprinted with permission from A. D. Melaven. *Ind. Eng. Chem., Anal. Ed.*, **2**, 180 (1930). Copyright 1930 American Chemical Society.]

platinum wire formed into a flat spiral. Agitation is accomplished with a magnetic stirring bar that floats on the mercury surface or by letting the impeller blades of a mechanical stirrer be only partially immersed in the mercury. The supporting electrolyte is usually a 0.1–0.5M solution of sulfuric acid or perchloric acid.

A commercial unit has been devised in which a magnetic circuit provides the stirring, with the electrolyte and the mercury becoming the two independent rotors of a dc motor. In addition, the magnetic field immediately removes deposited ferromagnetic materials from the mercury-solution interface and retains them beneath the surface of the mercury. Circulating tap water removes the heat developed by the resistance of the electrolyte.

Electrography

The electrographic method is a useful microanalytical tool for accurately identifying and determining substances.[1] This method consists of anodically dissolving a minute amount of the test substance onto a piece of highly absorbent paper or, for a more accurate rendition, gelatin-coated paper that has been soaked in a suitable electrolyte. The test sheet is held under pressure between the sample surface, the anode, and a suitable cathode surface. The latter may be a flat square electrode for flat surfaces, a long narrow electrode for use on metal ribbons, or sponge rubber covered with aluminum foil for uneven sample surfaces. The unit is connected to several dry cells and current is allowed to flow for several seconds (Figure 24.3). While the current flows, ions leave the surface of the specimen and migrate into the permeable sheet. Their presence can be made manifest, if they are colorless, by treating the test sheet with selective reagents. Distinctive identifying colors result and appear in an exact chemical and physical image of the surface. In general, 50 μg of most metals will produce brilliantly colored products when the reaction is confined to an area of 1 cm^2. These conditions require a current of 15 mA and an exposure time of 10 sec.

The test sheet may be moistened with only a neutral electrolyte, such as sodium nitrate or chloride, or it may be impregnated with a reagent for the metal or metals to be detected. With a neutral electrolyte the print must be further developed by immersion in a developing reagent(s) that forms a reaction product with a distinctive color. Individual patterns can be secured by developing successive prints with different selective reagents. The prints are sharp and permit many fine features to be detected.

The electrographic method is applicable only to materials that are conductors of the electric current. It can be applied for the inspection of lacquer coating and plated metals for pinholes and cracks in their surface. It can be used for many alloy identifications and the distribution of metal constituents within an alloy. In the

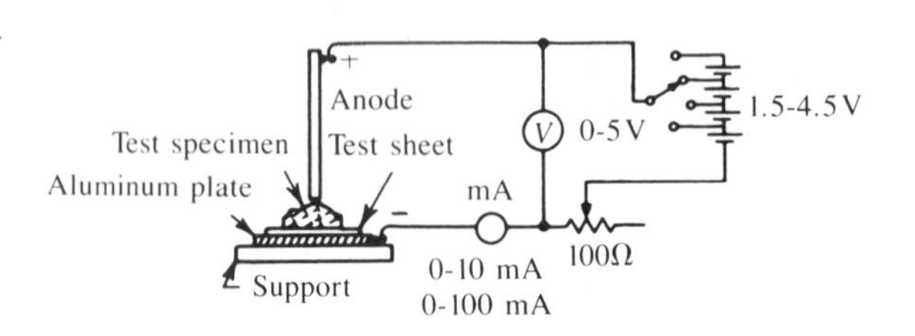

FIGURE **24.3**
Schematic arrangement of equipment and electrical circuit for electrographic analysis.

biological field the method is applicable to the localization of those constituents that are normally present within the tissue in an ionic state. Portable field kits have found extensive use in inspection and sorting work and in mineralogical field work.

24.2 CONTROLLED-POTENTIAL COULOMETRY

The fundamental requirements of coulometric analysis are that only one overall reaction of known stoichiometry may take place and that it proceed with 100% current efficiency. There may be no side reactions of different stoichiometry. The method is particularly useful and accurate in the range from milligram quantities down to microgram quantities and, therefore, in trace analysis. In practice, sensitivity is limited only by problems of sample handling and end point detection.

Coulometric methods eliminate the need for burets and balances, and the preparation, storage, and standardization of standard solutions. Procedures can be automated readily and are especially adaptable to remote operation and control. In a sense the electron becomes the primary standard. Coulometric methods produce reagents in solution that would otherwise be difficult to use, volatile reactants such as chlorine, bromine, or iodine, or unstable reactants such as titanium(III), chromium(II), copper(I), or silver(II).

General Principles

In controlled-potential coulometry the total number of coulombs consumed in an electrolysis is used to determine the amount of substance electrolyzed. A three-electrode potentiostat maintains a constant electrode potential by continuously monitoring the potential of the working electrode as compared with a reference electrode. The current is adjusted continuously to maintain the desired potential.

Consider the case when one or more of the ions present can undergo reduction or oxidation within the potential frame extending from the reduction of hydrogen ions to the region where water is oxidized to oxygen. Controlling the potential of the working electrode allows for selectivity of the redox reaction. The mini-potential frame for a particular redox reaction is represented by the Nernst equation (Equation 21.4). To lower the oxidant to 0.001% of its initial concentration, or $10^{-6}C_i$, requires a potential change of $E^\circ - 0.355/n$. In the reverse situation, a potential change of $E^\circ + 0.355/n$ is needed to convert the reduced form to the oxidized state. A second redox system whose mini-potential frame overlaps this system would constitute an interference.

To conduct controlled-potential coulometry, current-potential diagrams must be available for the oxidation/reduction systems to be determined and also for any other system that can react at the working electrode. Current-potential diagrams are obtained by plotting current against the cathode-reference electrode potential (rather than the cathode-anode potential, which would include the large and variable iR drop in the cell). The necessary data can be obtained by setting the potentiostat to one cathode-reference potential after another in sequence, allowing only enough time at each setting for the current indicator to balance. Alternatively, the reduction (or oxidation) is performed in the usual manner except that periodically throughout the electrolysis the potential is adjusted to a value that stops the current flow. The net charge transferred up to this point and the electrode potential are noted, and the electrolysis is then continued. Curves plotted from a series of points, called coulograms, establish the optimum electrode potentials because they relate the extent of reaction with electrode potentials under actual titration conditions and with actual electrode material (Figure 24.4).

The coulogram in Figure 24.4 involves a mixture of antimony(V) and antimony(III) in a supporting electrolyte that contains 6*M* HCl plus 0.4*M* tartaric acid. Plateaus are centered at −0.21 and −0.35 V versus SCE. To proceed with the experiment, the supporting electrolyte is prereduced at −0.35 V. Then the sample is introduced and the system deoxygenated. Finally the reduction starts at −0.21 V for antimony(V) to antimony(III), followed by the reduction at −0.35 V for antimony(III) to the metal. Initially, the electrolysis may proceed at a constant rate. Most coulometers have an upper limit to the current that they can furnish and, therefore, in the early stages may not be able to carry out the electrolysis at a rate sufficiently high to attain the limiting current. Eventually the potential of the working electrode reaches the limited value, in this case −0.21 V. Then the potentiostat takes over, and the current through the cell gradually decreases until all antimony(V) has been reduced to antimony(III). This procedure is then repeated at −0.35 V.

Factors Governing Current

If a reaction is 100% current efficient—that is, has only one electrochemical reaction route—the passage of one faraday of electricity (96,487 C) involves the reaction of one equivalent weight of substance. The relationship between the weight in grams, *W*, the number of coulombs, *Q*, the molecular weight, *M*, the number of faradays involved in the reaction of 1 mole, *n*, and the value of one faraday in coulombs, *F*, is given by the equation:

$$W = \frac{QM}{nF} \tag{24.3}$$

In controlled-potential coulometry, the current changes continuously and *Q* is given by the integration of time versus current:

$$Q = \int_0^t i\,dt \tag{24.4}$$

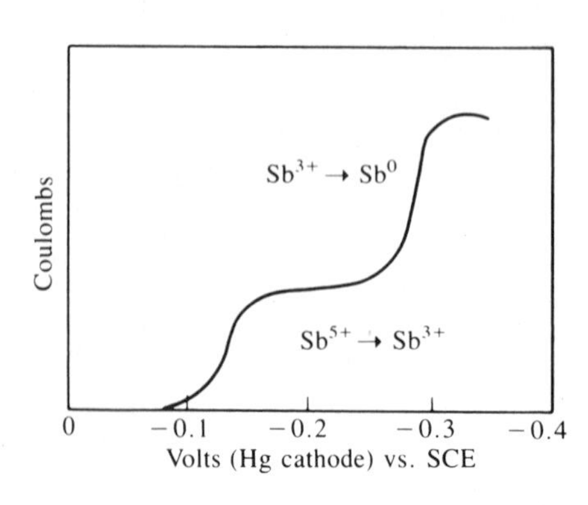

FIGURE **24.4**
Electrolytic reduction of antimony(V) by a two-step process in 6*M* HCl plus 0.4*M* tartaric acid. [After L. B. Dunlap and W. D. Shults, *Anal. Chem.*, **34,** 499 (1962). Courtesy of American Chemical Society.]

Most controlled-potential coulometric experiments are conducted under conditions in which the current is controlled by diffusion. When so, the relationship between current, i_t, at any time, t, and concentration, C_t, is given by

$$i_t = \frac{nFADC_t}{\delta} \tag{24.5}$$

where A is the area of the working electrode (in cm^2), D is the diffusion coefficient of the electroactive species (in $cm^2\ min^{-1}$), and δ is the thickness of the Nernst diffusion layer (in cm). From Faraday's law, the rate of change of concentration with time is given by the relationship

$$\frac{dC_t}{dt} = -\frac{i_t}{nFV} \tag{24.6}$$

where V is the volume of the solution (in cm^3). Substitution of Equation 24.5 into Equation 24.6 and integration yield these results for the concentration at any time t as a function of the initial concentration, C_0:

$$C_t = C_0 e^{-kt} \tag{24.7}$$

and current as a function of the initial current, i_0,

$$i_t = i_0 e^{-kt} \quad \text{or} \quad 2.3 \log \frac{i_0}{i_t} = kt \tag{24.8}$$

The constant k is given by $DA/V\delta$. The practical value of these relations is that they provide a logical basis for the optimum design of cells and the choice of experimental conditions to obtain rapid electrolysis. Note that the rate of electrolysis (i.e., the decadic change in concentration) is independent of the initial concentration of reactant.

EXAMPLE 24.3

If the initial silver concentration in a sample is $1 \times 10^{-3} M$, what time should be required for a controlled-potential coulometric procedure, assuming a diffusion-layer thickness of 2×10^{-3} cm, a diffusion coefficient of $4.2 \times 10^{-3}\ cm^2\ min^{-1}$, a solution volume of 20 mL, and an electrode area of 15 cm^2?

From the constant in Equation 24.8 (right side),

$$k = \frac{0.434(4.2 \times 10^{-3}\ cm^2\ min^{-1})(15\ cm^2)}{(2 \times 10^{-3}\ cm)(20\ cm^3)} = 0.684\ min^{-1}$$

$$t_{1/2} = \frac{0.693}{k} = \frac{0.693}{0.684\ min^{-1}} = 1.014\ \text{min (or 60.8 sec)}$$

Thus a controlled-potential coulometric electrolysis is like a first-order reaction, with the concentration and the current decaying exponentially with time during the electrolysis (Figure 24.5) and eventually attaining the background level (residual current of the supporting electrolyte). The reaction would be complete (within 0.1%) after ten half-lives (see Chapter 14), or 10.14 min (608 sec).

The number of coulombs, Q, that pass up to time t is obtained by integration:

$$Q_t = \int_0^t i\,dt = \int_0^t i_0 e^{-kt}\,dt = \left(\frac{i_0}{k}\right) - \left(\frac{i_t}{k}\right) \tag{24.9}$$

Equation 24.9 is useful for estimating the total number of coulombs required for the complete reaction before the reaction is actually completed. Q_t is read at several values of t, preferably in the range of 90%–99% completion, and then these values of Q_t are plotted versus i_t. Q_∞ is determined by extrapolation of the straight line so obtained to the coulomb axis. The limiting value of Q is i_0/k.

Instrumentation

Four instrumental units are involved: a dc current supply, a potentiostat, an electrolytic cell, and a coulometer. A mercury pool is most often used as a working electrode for electrolyses that involve reduction processes. Oxidations can be performed at a platinum working electrode, often cylindrical. Because the current is continuously changing, decreasing from a relatively large value at the beginning to essentially zero at the completion of the reaction, the charge transfer during this process must be integrated by a coulometer. A potentiostat is necessary to control the potential of the working electrode within 1–5 mV of the desired value (see Section 23.1). The electrolysis is terminated when the current has diminished to 0.1% or less of its initial value or when the current becomes equal to the residual current as measured on a sample of supporting electrolyte alone.

Current integration can be performed digitally by a 50-kHz voltage-to-frequency converter and counter system. The integrator is preceded by a chopper-stabilized current-to-voltage converter with current ranges of 1 μA to 1 A available in decade steps. The frequency output of this converter is then digitally counted to yield the current integral.

A technique called *digital normalization* can be used to evaluate the background current. While an electronic logarithmic display of the digitally recorded data is viewed, a constant number of counts is subtracted from each data channel until all the data points in the decay curve fall on a straight line. The summation of the counts subtracted is equal to the background current contribution. Software programs carry out the normalization automatically.

Applications of Controlled-Potential Coulometry

By controlling the potential of the electrode at a suitable value, it is possible to reduce a metal completely to a lower valency state. Then at a more positive potential, the metal can be oxidized quantitatively to a higher valency state. For example, at -0.15 V versus SCE with a mercury electrode, the reductions of

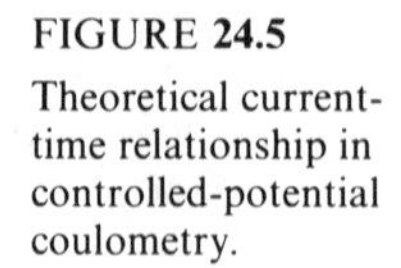

FIGURE **24.5**
Theoretical current-time relationship in controlled-potential coulometry.

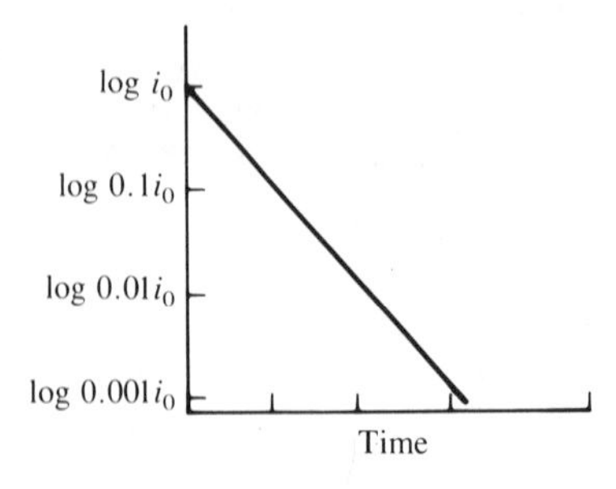

uranium(IV to III) and chromium(III to II) occur simultaneously. If pre-electrolysis is carried out at −0.55 V, only uranium(III) is oxidized to uranium(IV). Although the reaction does not occur with 100% current efficiency, it is complete. When all the uranium(III) has been oxidized, chromium(II) is determined by oxidation to chromium(III) at −0.15 V.

Indirect methods are possible. In the determination of plutonium in the presence of iron, the first step is the reduction of plutonium(VI) to plutonium(IV) and then to plutonium(III) ($E^\circ = 0.74$ V) and the partial reduction of iron(III) to iron(II) ($E^\circ = 0.56$ V) at a platinum cathode at 0.50 V in a 0.5M sulfuric acid electrolyte. When this is followed by electro-oxidation (at 0.92 V) of the mixture to plutonium(IV) and iron(III), the net reaction is the reduction of plutonium(VI) to plutonium(IV). Interference caused by the presence of iron is thereby avoided.

Controlled-potential coulometry suffers from the disadvantage of requiring relatively long electrolysis times, although it proceeds virtually unattended with automatic instruments. However, optimum conditions for successive reactions are easily obtained. No indicator electrode system is necessary, since the magnitude of the final current is sufficient indication of the degree of completion of the reaction. Although the concentration limits vary for each individual case, the upper limit is about 2 meq and the lower limit is about 0.05 meq. The latter limit is largely set by the magnitude of the residual current and the many factors that affect it.

24.3 CONSTANT-CURRENT COULOMETRY

In situ generation of known amounts of chemical reagents from a titrant precursor forms the basis of constant-current coulometry. A constant current is maintained throughout the reaction period. As long as the impressed current is less than the limiting current at a given bulk concentration, the electrode reaction proceeds with 100% current efficiency. The quantity of unknown present is given by the number of coulombs (the product of current and time) of electricity used.[2,3]

Primary Coulometric Titrations

The primary titration technique is attempted only with electrodes of silver metal, silver/silver halide, or mercury amalgams. All these electrodes are the source of the electrogenerated species. The substance to be determined reacts directly at the electrode, or with a reactant electrogenerated from the working electrode. Since the potential of the working electrode is not controlled, this class of titrations is limited generally to reactants that are nondiffusible. These include mercury amalgams, silver ions generated by anodization of silver metal, and the halides liberated by reduction of the appropriate silver/silver halide electrode.

One major area of application involves the electrode material itself participating in an anodic process—for example, in the reaction of mercaptans, sulfhydryl groups, and ionic halide ions with silver ions generated at a silver anode. For chloride samples, the initial reaction may be

$$Cl^- + Ag \rightarrow AgCl(s) + e \tag{24.10}$$

followed by

$$Ag \rightarrow Ag^+ + e \tag{24.11}$$

as soon as the limiting current (supply of chloride ions to the anode) has become smaller than the current impressed upon the electrolytic cell. At this point the silver ion generated anodically diffuses into the solution, and precipitation occurs with the chloride ions left in solution. Of course, the result of the two reactions is identical. The end point of the titration is determined amperometrically. Combustion by the oxygen flask method precedes the titration step for nonionic halides in organic compounds. Mercaptan samples are dissolved in a mixture of aqueous methanol and benzene to which aqueous ammonia and ammonium nitrate are added to buffer the solution and to supply sufficient electrolyte to lower the solution resistance.

EXAMPLE 24.4

If a constant current of 10.00 mA passes through a chloride solution for 200 sec, what weight of chloride will have reacted with the silver anode?

The net charge involved is

$$Q = it = (10.00 \times 10^{-3}\ \text{A})(200\ \text{sec}) = 2.00\ \text{C}$$

or

$$\frac{2.00}{96{,}487} = 2.075 \times 10^{-5}\ \text{equivalents of chloride}$$

Since $n = 1$ for the reaction illustrated by Equation 24.11, in weight the amount of chloride ion, present as AgCl, is $(2.075 \times 10^{-5})(35.45) = 0.735 \times 10^{-3}$ g (or 0.735 mg).

Secondary Coulometric Titrations

In the majority of applications, the secondary coulometric titration technique is used, in which an oxidation/reduction buffer serves as the titrant precursor. An active intermediate from the titrant precursor must first be generated with 100% efficiency by the electrode process. The intermediate must then react rapidly and completely with the substance being determined. Some end point detection technique must be used to indicate when the coulometric generation should be stopped.

Almost any titrant can be generated from an appropriate titrant precursor. An excess of a titrant precursor is always added to the supporting electrolyte. To avoid unwanted reactions, the standard potential of the titrant precursor must lie between the potential window of the unknown analyte and the potential at which the supporting electrolyte or another sample constituent undergoes an electrode reaction.

A knowledge of current-potential curves aids the analyst in choosing the titrant precursor (see Chapter 23). For example, consider the coulometric determination of iron(II) in the presence of cerium(III) at constant current. Pertinent current-potential curves are shown in Figure 24.6. To complete the titration within a reasonable time, usually 10–200 sec, a finite current must be selected, say i_0. The initial potential (anodic) is V_0 for the initial concentration of iron(II) present in the solution. At the beginning, iron(II) is oxidized directly at the anode to iron(III). As the concentration of iron(II) decreases with the progress of the oxidation, the current transported by the iron(II) ions decreases. Since a constant current is being imposed on the system, the anodic potential would drift ultimately to the decomposition potential of the solvent system (in this case, water). If i_0 is selected sufficiently small to delay the onset of the undesired anodic oxidation, the time required for a determination becomes too long for practical consideration.

The addition of a titrant precursor with an E° that lies between the potentials at which iron(II) and water are oxidized permits the titration of iron(II) to reach 100% completion. No interfering electrode reaction can occur so long as the potential of the anode is prevented from reaching the value that would initiate the decomposition of water. Such potential drift is limited by having a precursor of the secondary titrating agent present in relatively high (greater than 10:1) concentration. With a large excess of cerium(III) present, as soon as the limiting current of iron(II) falls below the value of the current forced through the cell, the cerium(III) commences to undergo oxidation to cerium(IV) at the anode in increasing amounts until it may be the preponderant anode reactant. Since the cerium(IV) formed reacts instantly and stoichiometrically with the iron(II) to form iron(III) and re-form cerium(III), the total current ultimately used in attaining the oxidation of iron(II) is the same as would have been required for the direct oxidation. Because there is a relatively inexhaustible supply of cerium(III), the anode potential is stabilized at a value (about 1.3 V in 0.5*M* sulfuric acid) less than the decomposition potential of water (about 1.7 V). The end point is signaled by the first persistence of excess cerium(IV) in the solution and may be detected potentiometrically with a platinum-reference electrode pair or spectrophotometrically at a wavelength at which cerium(IV) absorbs strongly.

FIGURE **24.6** Current-potential curves pertinent to the coulometric titration of iron(II) with cerium(III) as the precursor system.

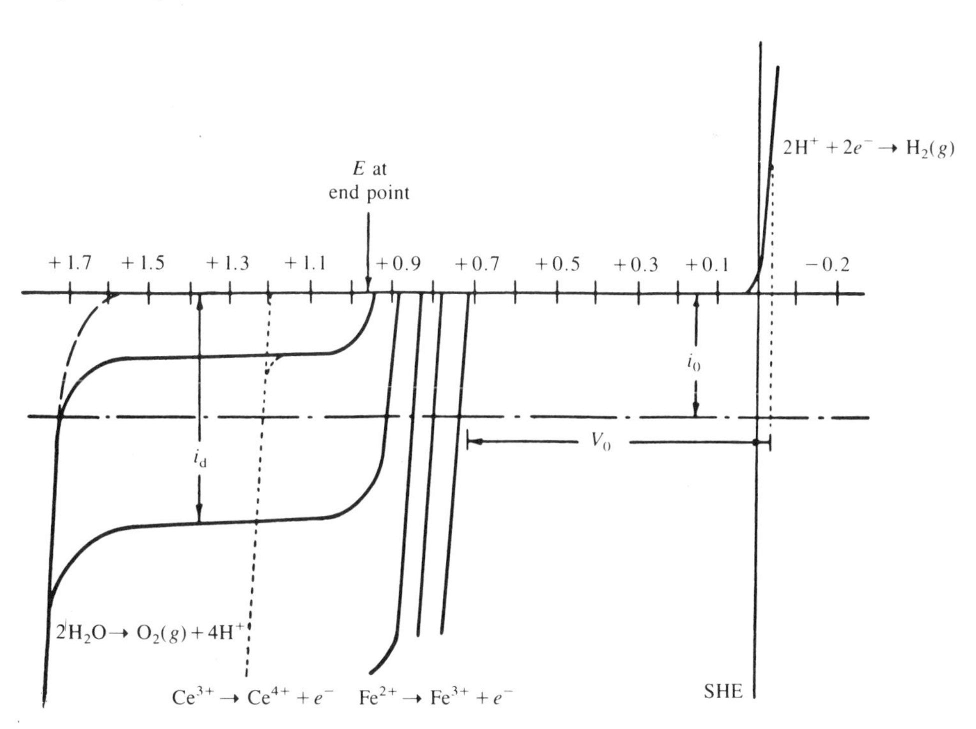

Several sample aliquots can be titrated in the same precursor solution before it becomes too dilute to function properly. Errors due to impurities in the supporting electrolyte or in the titrant precursor can be avoided by performing a pretitration in the same supporting electrolyte.

Instrumentation for Constant-Current Coulometry

Only a knowledge of the current and elapsed time is needed to determine the number of coulombs involved in the desired reaction. Both current and time can be measured with good accuracy and with relatively simple equipment. The instrumentation required consists of a galvanostat (see Chapter 3) to force a constant current through the generator cell and some device to measure the electrogeneration time. A constant-current source can be constructed from a series combination of a battery and a resistance network, as shown in Figure 24.7, along with the schematic

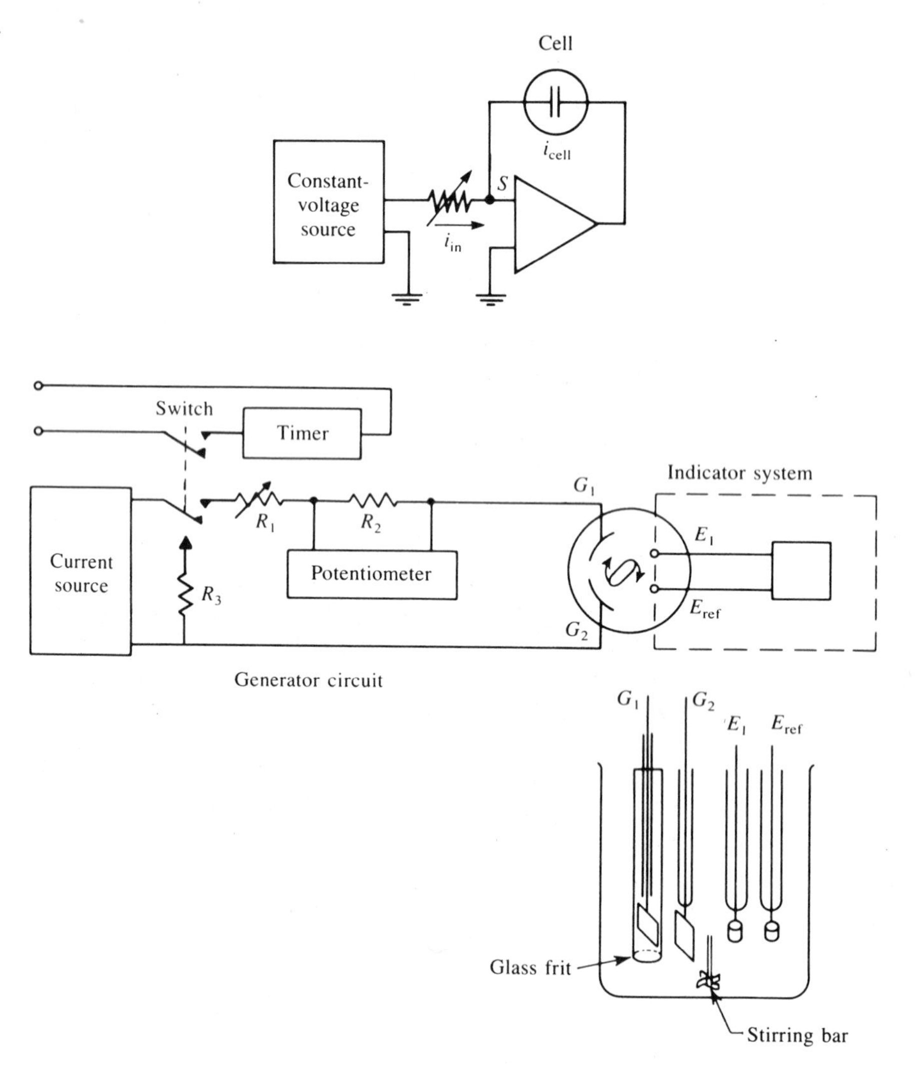

FIGURE 24.7
Schematic of equipment for constant-current coulometry and a constant-current source using either a battery and series resistor or an operational amplifier (galvanostat). R_1 is the series (ballast) resistor; R_2 is the precision resistor; G_1 and G_2 are the generator electrodes (one isolated behind a porous frit barrier); and E_1 and E_{ref} are the electrodes for the end point detector system.

arrangement of the generation cell and detection system. The galvanostat may simply be a high-gain, high-current operational amplifier with the two-electrode coulometry cell placed in its feedback loop. Constant current flowing through the amplifier's input resistor from a precision constant-current source causes a current of equal magnitude but opposite polarity to flow through the cell to the summing point (S) of the amplifier. Resistor R_3 serves as a dummy cell so that the current generator will see a finite load when the cell is out of the circuit. The current can be indicated approximately by a calibrated milliammeter and measured precisely by means of the voltage drop across a precision resistor (one of several selected by a current-selector switch) incorporated directly in series with the electrolytic cell. The voltage drop across the resistor can be measured with a recording potentiometer with a precision of about 0.002%. Time measurements accurate to 0.01 sec are normally made with a precision electric stop clock. For greater precision the manual switch should be replaced by a solid-state electronic gate, and the stop clock by a high-frequency thermostatted crystal oscillator with appropriate counting electronics, perhaps an electronic accumulating scaler capable of measuring a 1-sec interval within better than one part in a million. Power supplies of commercial units provide a dc voltage with a maximum of about 300 V, sufficient for work with high-resistance electrolytes.

Electrode materials used for titrant generation at positive potentials are generally gold or platinum, fabricated in the form of wires, foils, or gauze. The high hydrogen overpotential characteristics of mercury make it unique for titrant generation at negative potentials. The electrode area, generally 1–10 cm^2, must be sufficient to keep the current density sufficiently low and thereby keep electrode polarization within the limits necessary for 100% current efficiency. In most instances the auxiliary electrode must be isolated by placing it in a physically separate compartment, electrically connected through a low-leakage salt bridge.

The end point detection system may be colored indicators, provided the indicator itself is not electroactive, or it may be a potentiometric, amperometric, or spectrophotometric system. Potentiometric and spectrophotometric indication has use in acid-base and oxidation/reduction titrations, whereas amperometric procedures are applicable to oxidation/reduction and ion-combination reactions and, in particular, to these systems as the sample solutions become more dilute.

Applications of Secondary Coulometric Titrations

Uncommon titrants such as chromium(II), silver(II), copper(I), chlorine, titanium(III), uranium(V), and bromine, which would be difficult or impossible to prepare and store as standard solutions because of their high redox strength or instability, are very easily generated *in situ*. This is one of the virtues of secondary coulometric methods.

Electrolytic generation of hydroxyl ion has some advantages over conventional methods. Very small amounts of titrant can be prepared, and in a carbonate-free condition. To analyze dilute acid solutions, such as would result from adsorption of acidic gases, the cathode reaction generates the hydroxyl ion. In the initial stages it is possible for the hydrogen ion to react directly at the cathode, but in the vicinity of the end point, the secondary generation predominates. The anode reaction must also be considered. If a platinum anode is used, it must be isolated in a separate compartment, since hydrogen ions are liberated at its surface. Alternatively, a silver anode may be used within the electrolytic cell so long as excess bromide ions are present, for then the anode reaction is

$$Ag^{\circ} + Br^{-} \rightarrow AgBr(s) + e^{-} \tag{24.12}$$

and the electrogenerated silver ions are fixed as a coating of silver bromide on the electrode surface. Fresh surface must be exposed for each analysis.

Halogens generated internally, and particularly bromine, have found widespread application, especially in organic analysis, because substitution reactions, addition reactions, and others are possible. In contrast with certain difficulties encountered in the use of bromate-bromide mixtures by conventional volumetric procedures, coulometry is much simpler. Bromates are not soluble in many organic solvents, and many organic samples are not soluble in water. However, sodium and lithium bromides are quite soluble in various organic solvents in which brominations can be conducted.

Another important titrant is Karl Fischer reagent, which is used for the direct titration of water and indirectly for the determination of many organic materials that contain certain functional groups that will produce or consume water in a quantitative manner. Iodine is electrochemically generated from an iodide salt in anhydrous methanol that contains sulfur dioxide and Hydranal (the trade name for a mixture of amine solvents developed to replace pyridine with its unpleasant smell). The coulometric Karl Fischer titration allows the determination of microgram amounts of water and is particularly advantageous for organic liquids that contain only traces of moisture and for the determination of moisture in gases.

The complexing ability of ethylenediaminetetraacetic acid (EDTA, H_4Y) has been exploited in the coulometric titration of metal ions. The method depends on the reduction of the mercury(II) or cadmium chelate of EDTA:

$$CdY^{2-} + 2e^{-} \rightleftharpoons Cd + Y^{4-} \tag{24.13}$$

and the titration of the metal ion (for example, magnesium) to be determined by the anion of EDTA that is released:

$$Mg^{2+} + Y^{4-} \rightleftharpoons MgY^{2-} \tag{24.14}$$

The end point can be ascertained amperometrically by the appearance of free cadmium ions. If the direct reaction of a metal ion with EDTA is too slow, excess EDTA anion can be generated and then the excess back-titrated by cadmium generated at a cadmium-amalgam electrode.

External generation of a titrant should be considered when conditions conducive to the optimum generation of reactant and to rapid reaction with the analyte are not compatible. Such is the case in the titration of azo dyes with titanium(III) generated by the reduction of titanium(IV). At room temperature the reaction rate of titanium(III) with the dye is slow. Yet on raising the temperature, hydrolysis of titanium(IV) and bubble formation at the generating electrode surface lead to low current efficiencies. If the titanium(III) is generated at room temperature and then delivered to the hot dye solution via a capillary delivery tube, however, optimum conditions prevail for each step. A double-arm electrolytic cell with separate anode and cathode delivery tubes is used for the external generation of titrant.[4]

FIGURE **24.8**
Cathodic reduction of tarnish films on copper: curve 1, copper(I) oxide; curve 2, copper(I) sulfide.

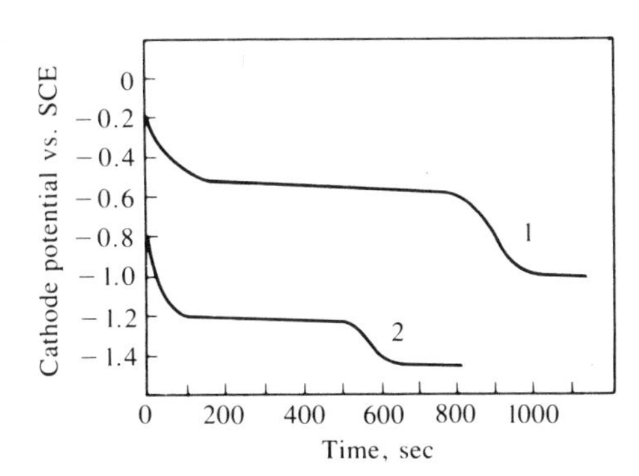

The removal (stripping) of deposits is used to measure the thickness of plated metals and of corrosion or tarnish films. In the case of oxide tarnish on the surface of metallic copper, the specimen is made the cathode and the copper oxide is reduced slowly with a small, but constant, known current to metallic copper. When the oxide film has been completely reduced, the potential of the cathode changes rapidly to the discharge potential of hydrogen. The equivalence point is taken as the point of inflection of the voltage-time curve, as illustrated in Figure 24.8, curve 1. From the known current, i, expressed in milliamperes, and the elapsed time, t, in seconds, the film thickness, d, in nanometers, can be calculated from the known film area, A, in cm^2, and the film density, ρ, according to the equation

$$d = \frac{10^4 Mit}{AnF\rho} \tag{24.15}$$

where M is the molecular weight of the oxide comprising the film and F is the Faraday. From mixed films of oxide and sulfide on a metal, such as copper, two-step potentials would be obtained.

Analogous anodic dissolution is used to determine the successive coatings on a metal surface. Iron is sometimes clad with a tin undercoating for adhesion and a copper-tin surface layer for protection from corrosion. The two coatings exhibit individual step potentials. In a similar manner, duplex nickel coatings on a copper-plated steel substrate can be measured; the step potential difference between bright nickel and semibright nickel is between 110 and 150 mV. Accuracy in these corrosion studies is limited by the residual current and by the fact that the last traces of deposit may not dissolve uniformly from the surface.

24.4 CONDUCTANCE METHODS

One of the oldest and in many ways simplest of the electrochemical methodologies is the measurement of electrolytic conductance. The ions of the electrolyte transport charge (conduct current) through the bulk of a solution. Due to thermal energy, molecules of a solvent and solvated ions are continually colliding, and thus move in a random manner with frequent changes of speed and direction. When an electrical field is applied to the solution, ions experience a force that attracts them toward the oppositely charged electrode.

Practical applications are of three types: direct analysis, stream monitoring, and conductometric titrations.

Electrolytic Conductivity

Electrolytic conductivity is a measure of the ability of a solution to carry an electric current. Solutions of electrolytes conduct an electric current by the migration of ions under the influence of a potential gradient. The ions move at a rate dependent on their charge and size, the microscopic viscosity of the medium, and the magnitude of the potential gradient. Like a metallic conductor, they obey Ohm's law. Thus, for an applied potential, E, maintained constant but at a value that exceeds the deposition potential of the electrolyte, the current, i, that flows between the electrodes immersed in the electrolyte varies inversely with the resistance of the electrolytic solution, R. The reciprocal of the resistance, $1/R$, is called the *conductance*, S, and is expressed in reciprocal ohms, or mhos (in SI nomenclature the reciprocal ohm takes the name siemens and the symbol S).

The standard unit of conductance is *specific conductance*, κ, which is defined as the reciprocal of the resistance in ohms of a 1-cm cube of liquid at a specified temperature. The units of specific conductance are $\Omega^{-1}\ \text{cm}^{-1}$, or S/cm. The observed conductance of a solution depends inversely on the distance, d, between the electrodes and directly on their area, A:

$$\frac{1}{R} = S = \kappa \frac{A}{d} \tag{24.16}$$

Analytical applications depend on the relation between the conductance and the concentration of various ions and their specific ionic conductance. The electrical conductance of a solution is a summation of contributions from all the ions present. It depends on the number of ions per unit volume of the solution and on the velocities (mobilities) with which these ions move under the influence of the applied electromotive force. As a solution of an electrolyte is diluted, the specific conductance decreases. There are fewer ions to carry the electric current in each cubic centimeter of solution. However, in order to express the ability of individual ions to conduct, a function called the *equivalent conductance* is used. It may be derived from Equation 24.16, where A is equal to the area of two large parallel electrodes set 1 cm apart and holding between them a solution that contains one equivalent of solute. If C_s is the concentration of the solution in gram equivalents per liter, then the volume of solution in cubic centimeters per equivalent is equal to $1000/C_s$, so that Equation 24.16 becomes

$$\Lambda = 1000 \frac{\kappa}{C_s} \tag{24.17}$$

At infinite dilution the ions theoretically are independent of one another and each ion contributes its part to the total conductance; thus,

$$\Lambda_\infty = \Sigma(\lambda_+) + \Sigma(\lambda_-) \tag{24.18}$$

where λ_+ and λ_- are the ionic conductances of cations and anions, respectively, at infinite dilution. Values for the limiting ionic conductances of many ions in water at 25 °C are available in reference handbooks.[5]

There are practical limits of measured electrolytic resistance for any desired accuracy and sensitivity. The optimum appears to lie in the vicinity of 500–10,000 Ω when errors are not to exceed ±0.1%. In solutions of low conductance, the electrode area should be large and the plates close together; for highly conducting solutions, the area should be small and the electrodes far apart. For a given cell with fixed electrodes, the ratio d/A is a constant, called the *cell constant*, θ. It follows that

$$\kappa = \frac{1}{R}\frac{d}{A} = \frac{\theta}{R} = S\theta \tag{24.19}$$

For conductance measurements a cell is calibrated by measuring R when the cell contains a standard solution of known specific resistance, and the cell constant is then computed.

EXAMPLE 24.5

When a certain conductance cell was filled with a 0.0100M solution of KCl, whose specific conductance is 0.001409 mho/cm (or Ω^{-1} cm^{-1} or S/cm) at 25 °C, it had a resistance of 161.8 Ω, and when filled with 0.0050M NaOH it had a resistance of 190 Ω. The cell constant is

$$\theta = (0.001409 \text{ mho/cm})(161.8\ \Omega) = 0.2280 \text{ cm}^{-1}$$

The specific conductance of the sodium hydroxide solution is

$$\kappa = \frac{\theta}{R} = \frac{0.2280 \text{ cm}^{-1}}{190\ \Omega} = 0.00120\ \Omega^{-1}\text{ cm}^{-1}$$

and the equivalent conductance is

$$\Lambda = \frac{(1000)(0.00120\ \Omega^{-1}\text{ cm}^{-1})}{0.005 \text{ equiv/L}} = 240 \text{ cm}^2 \text{ equiv}^{-1}\ \Omega^{-1}$$

The conductivity of solutions is quite temperature dependent. An increase in temperature invariably results in an increase in the ionic conductance, and for most ions this amounts to 2% to 3% per degree. For precise work, conductance cells must be immersed in a constant-temperature bath. A practical means of providing temperature compensation is to introduce into the measuring circuit a resistive element that will change with temperature at the same rate as the solution being tested.

Conductance Cells

Electrolytic conductance measurements usually involve determining the resistance of a segment of solution between two parallel electrodes by means of Ohm's law. In conductometric measurements, the electrodes are placed in intimate contact with the solution and an alternating potential is applied. Polarization at the surface of these electrodes is avoided by increasing the macroscopic surface area through platinization and/or operation at sufficiently high frequency. Under these conditions the contribution of nonfaradaic currents is so large that the electrode process itself is not governing and the potential drop across the electrode-solution interfaces is constant, and assumed the same for the bulk solution.

Various types of conductance cells are commercially available. The dip cell is the simplest to use whenever the liquid to be tested is in an open container (Figure 24.9). It is merely immersed in the solution to a depth sufficient to cover the electrodes and the vent holes. A pair of individual platinum electrodes on glass wands is useful in conductance titrations. Pipet cells permit measurements with small volumes of solution, as little as 0.01 mL in some designs. A conductance cell is calibrated by using a solution of known conductivity, usually a potassium chloride solution.[5]

Any smooth metal surface can serve as an electrode at an operating frequency of 3000 Hz. For industrial on-line applications, stainless steel electrodes have desirable rigidity and low cost. This facilitates the fabrication of cells with tubular geometry. The smooth, continuous internal diameter minimizes mixing due to eddy currents in the flow passageway, which may be as small as 0.8 mm in diameter. The active volume of the cell is about 2 μL; the nominal cell constant is 80 cm^{-1}. In small cells, where the electrodes are close together, the cell constant is in fact not a constant but varies with concentration. Thus, for wide-range measurements using the micro cells, a calibration is needed over the entire conductance range.

Some industrial flow cells have three electrodes. In flow cells of constant 1 cm^{-1} and higher, the electrodes must be spaced farther apart than the bore of the cell. This means that they do not face each other perpendicular to the flow, but are aligned instead along the major axis. If small ac gradients exist from the electrical leakage of pumps or stirrers, as they do in many sampling lines, these are picked up by the two electrodes in their alignment, amplified, and appear as an error in the readout device. By providing a third in-line electrode and then connecting the outer two in common, all pickup is eliminated. A further problem with on-line instrumentation is also avoided—namely, the contribution to any measured resistance from the shunt path (whether electrolytic, metallic, or a combination) that arises as a result of any junction of the incoming and outgoing sample lines.

Conductivity Meters[6,7]

Some of the more important phenomena associated with the application of a potential between electrodes immersed in a liquid electrolyte are indicated in Figure 23.4 for an idealized system. To eliminate the effects of faradaic processes, measurements are made with an alternating current. If an alternating potential is applied, alternating current flows through the double-layer capacitances at the two electrodes and the resistance of the solution between them, including the interelectrode capacitance in parallel with the cell. Each double-layer capacitance provides such an easy path for alternating current that a potential cannot build up across the corresponding faradaic impedance to the point where faradaic current can flow; that is, the deposition potential is not exceeded. Platinizing the electrodes increases the double-layer capacitance manyfold.

Wheatstone Bridges. In the classical mode of conductance measurements, some variation of the Wheatstone bridge is used. The bridge circuit must contain not only

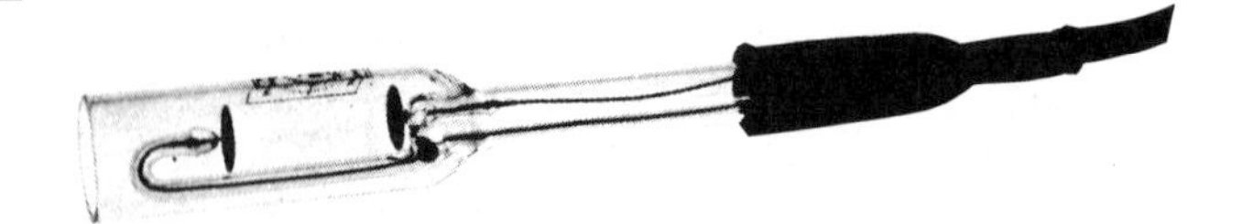

FIGURE **24.9**
Dip-type conductance cell.

resistance, but also capacitance (or inductance) to balance the capacitive effects in the conductance cell. The introduction of a variable capacitance (or inductance) into the bridge circuit permits compensation of the phase shift between current and voltage caused by the capacitance in the electrolytic cell. Parallel resistance-capacitance balancing arms are used more frequently than a series arrangement because smaller capacitance values are needed, and small capacitances can be obtained with higher accuracy and less frequency dependence than large ones.

A typical commercial conductivity bridge is designed to measure electrolytic conductance in microsiemens (or micromhos) and resistance in ohms. Commercial instruments are manufactured in many levels of sophistication, but the principle of operation is the same, as illustrated in Figure 24.10. A built-in generator provides bridge current at frequencies of 100, 1000, and 3000 Hz. Generally the lower frequency is preferred when the measured resistance is high, and the higher frequency when the measured resistance is low. The generator supplies a sinusoidal drive voltage to the bridge arm as well as a reference voltage for the phase detector. In the resistance mode, the bridge is an equal-arm bridge with cell and decade resistance in adjacent arms. At balance the decade resistance equals the cell resistance. The bridge is balanced with the aid of a phase-sensitive detector and a null meter by adjusting the readout dial resistor, coupled to a mechanical counter. The range of the instrument is changed by switching in different multiplier resistors for each range. The cell constant of the conductivity cell should be selected to maintain the measured resistance between 100 Ω and 1.1 MΩ. In the conductance

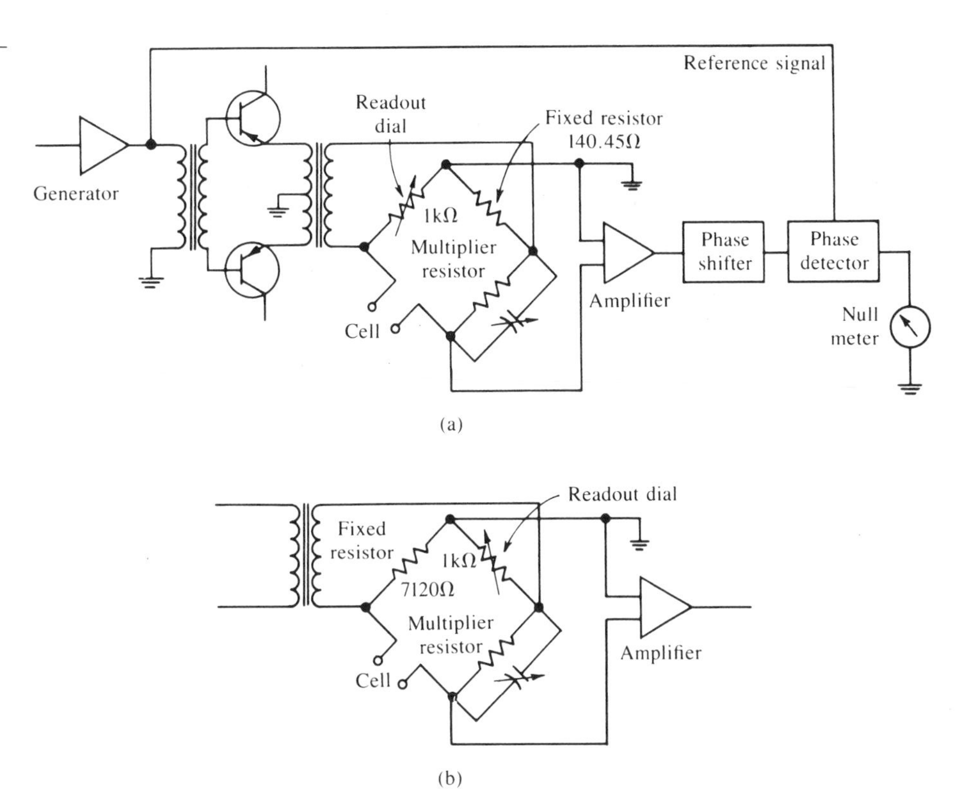

FIGURE **24.10**
Simplified schematic of a conductivity bridge: (a) resistance mode and (b) conductance mode. (Courtesy of Beckman Instruments, Inc.)

mode (Figure 24.10b), the decade resistance is placed in the arm opposite the cell. All other circuits remain the same and perform the same functions.

It is essential to introduce some reactive component into the bridge to obtain a true balance in view of the capacitance present in the conductance cell. This is usually done by placing an adjustable capacitance (8–200 pf) in parallel with the balancing resistor.

Operational Amplifier Units. The Wheatstone bridge can be replaced by operational amplifier circuitry (see Chapter 3), as shown in Figure 24.11. If the cell is connected in place of R_1, the result is a current $i = E/R_1 = ES_1$, which is proportional to the conductance of the cell.

For resistance measurements, a fixed resistance R_1 is used in the input current-generating circuit and the cell is connected in place of R_2 in Figure 24.11. Since this same current passes through the cell and the operational amplifier maintains point S at the common potential, the output voltage is equal to $-iR_{\text{cell}}$. This results in an output voltage directly proportional to the resistance of the cell.

If the feedback resistor R_1 is replaced with a semiconductor thermistor, the circuit has the ability to correct for changes in the solution temperature of the cell. A pulsed voltage source eliminates undesirable joule heating in the thermistor and its surroundings.

Differential Meters. Differential conductivity measurements offer advantages for several applications, such as monitoring a flowing stream to which a reagent or contaminant is added or measuring a small but discrete conductance change in the presence of a steadily changing conductance (gradient elution liquid chromatography). In the first example, the conductance changes are measured by two cells. One cell is upstream and one downstream from the point where the additive enters the stream. In gradient elution chromatography a second cell that monitors the gradient can provide a flat baseline, except when a discrete change occurs due to elution of the sample.

Electrodeless Conductivity Meters. Electrodeless conductivity systems have considerable industrial importance. No electrodes are in contact with the solution and therefore there is no chance for fouling. Coupled coils, as shown in Figure 24.12, are toroidally wound and are operated at a frequency of about 20 kHz. An oscillator energizes the primary toroid (toroidal transformer), which along with the secondary toroid is immersed in the solution to be examined. The current induced in this loop

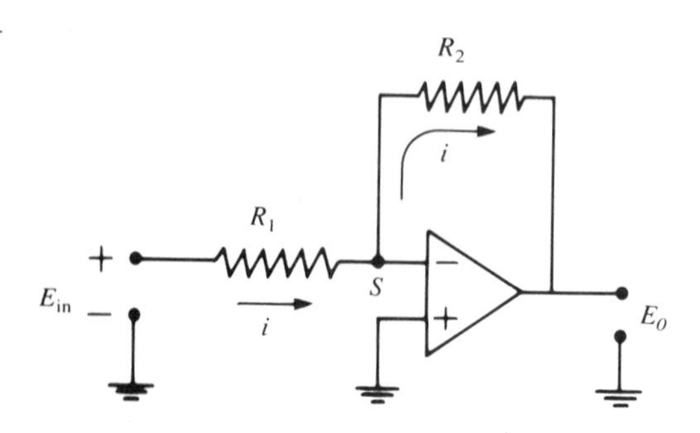

FIGURE **24.11**
Operational amplifier used for resistance and conductance measurements.

of solution energizes the second toroidal transformer. The greater the conductivity of the solution, the greater is the current induced in the secondary toroidal transformer. This current is handled by a bandpass amplifier, which drives a meter in a direct-reading instrument. Circuit parameters are chosen so that the meter deflection is proportional to the conductance, and the instrument thus reads out linearly with the solution conductivity. In practice, the toroidal coils are arranged parallel to one another and mounted coaxially close together in a watertight corrosion-resistant housing.

The lower portion of Figure 24.12c shows a self-balancing transformer bridge circuit. This circuit has an additional loop, called the standard loop, that is formed of wire and contains a variable resistor. This standard loop is so wound through the two toroidal transformers that it bucks any signal induced by the solution loop in the second toroidal transformer. When the resistances in the solution loop and the standard loop are equal, the currents through the two loops are equal. The fluxes that they induce in the second toroidal transformer cancel, and no signal reaches the amplifier. If the resistance of the solution loop now changes, a signal is impressed on the phase-sensitive amplifer, which will drive the motor in the proper direction to readjust the resistance of the standard loop and restore balance. A pointer attached to the variable resistor is realigned against its scale to reflect the change in the

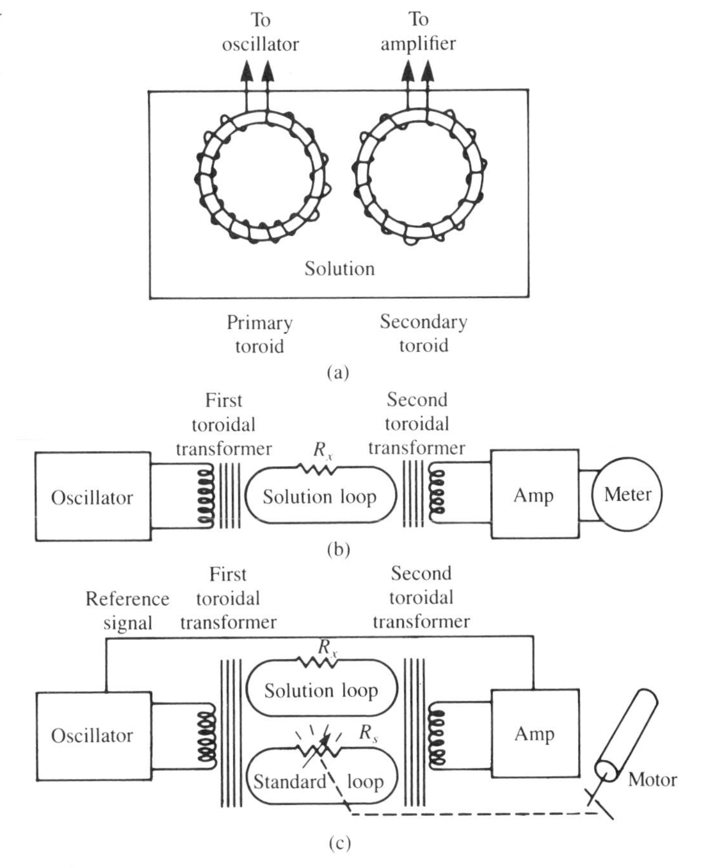

FIGURE **24.12**
Electrodeless conductivity systems: (a) schematic of toroids (toroidal transformers), (b) direct-reading conductance meter, and (c) self-balancing bridge circuit.

solution loop resistance. Up to three settable alarm switches may also be actuated by the motor.

The ruggedness of the electrodeless conductivity equipment enables measurements to be made on corrosive, abrasive-containing, and other hostile fluids. One example is the on-line analysis of oleum that contains 2%–7% free SO_3.

Direct Concentration Determinations

Despite the lack of specificity in conductance measurements, the technique is useful in giving a measure of the overall ionic content of a solution, such as determining the purity of potable and product-effluent waters, and monitoring solutions that contain only a given electrolyte type. In general, the success of a measurement depends on relating the property of the sample that is to be estimated to the conductance of some highly conducting ion. For example, free caustic (NaOH) that remains in scrubbing-tower solutions can be estimated by observing the decrease in conductance of the solution. This is possible even in the presence of the salts that are formed upon neutralization of the alkali because the conductance of the hydroxyl ion is approximately fivefold greater than that of any other anion. Similarly, the unusually high conductance of the hydrogen ion permits an estimation of the free acid content in acid pickling baths. The changing conductivity resulting from the adsorption of gaseous combustion products in suitable solutions is frequently used for the determination of carbon, hydrogen, oxygen, and sulfur individually and in organic and inorganic compounds. The change in conductivity is always measured in relation to an identical solution that is not in contact with the combustion products. On the other hand, when checking the purity of distilled or deionized water, steam distillates, rinse waters, or boiler waters, or in the regeneration of ion-exchanger columns, it is the total salt content that is sought.

For all these purposes very compact and inexpensive conductance bridges are available with scales calibrated directly in the desired units of measurement. For example, instruments can be supplied for direct indication, such as 1–12 lb of sodium carbonate per 100 gallons, 0–40 parts per thousand salinity, and 0.4%–12% sodium hydroxide.

EXAMPLE 24.6

The scale of a conductivity bridge is inscribed from 0.005% to 2.0% sulfuric acid in approximately a logarithmic manner. For these solutions the specific conductance ranges from about 0.00044 to 0.176 mho/cm. What range of resistances are involved and what cell constant is compatible?

The resistance values range from

$$R = \frac{\theta}{0.00044 \text{ mho/cm}} = 2270\theta$$

to

$$R = \frac{\theta}{0.176 \text{ mho/cm}} = 5.68\theta$$

A suitable cell constant would be 10.0 cm^{-1}; the resistance readings would range from 57 to 22,700. A cell constant of 20 cm^{-1} would also be suitable. A smaller cell constant would provide too low a resistance for the stronger acid solutions. A cell with a constant of 10 cm^{-1} would have electrodes of moderate area and some distance apart, perhaps 0.5 cm^2 in area and spaced 5 cm apart.

Conductometric Titrations

In conductometric titrations the variation of the electrical conductivity of a solution during the course of a titration is followed. It is not necessary to know the actual specific conductance of the solution. Specificity is achieved by replacing a given ion with a second one. A conductometric titration is devised so that the ionic species to be determined can be replaced by another ionic species of significantly different conductance. The end point is obtained by the intersection of two straight lines that are drawn through a suitable number of points obtained by measurement of the conductivity after each addition of titrant.

As an illustration, suppose that an approximately 0.001*M* solution of HCl is progressively titrated with 0.1*M* NaOH, so that there is essentially no change in volume. By the formation of water, the highly conducting hydronium ion ($\lambda_+ = 350$) is replaced by a less highly conducting sodium ion ($\lambda_+ = 50$), and the conductivity falls linearly, reaching a minimum when the solution consists of only NaCl. Continued titration then results in increased conductivity, largely because the hydroxyl ion, which is a very good conductor ($\lambda_- = 198$), is no longer being consumed. Figure 24.13 illustrates the titration curve and also the relative contribution of each ion in the process. The falling branch represents the conductance of the hydrochloric acid still present in the solution, together with that of the sodium chloride already formed. Unused NaOH and previously formed NaCl constitute the conductance of the rising branch of the titration curve. Usually three or four readings are taken to establish each branch of the titration curve.

Hydrolysis, dissociation of the reaction product, or appreciable solubility in the case of precipitation reactions gives rise to curvature in the vicinity of the end point. At a sufficient distance away from the end point (from 0% to 50% and between 150% and 200% of the equivalent volume of titrant), sufficient common ion is present to repress these effects. By extrapolating these portions of the two branches, the position of the end point can be determined.

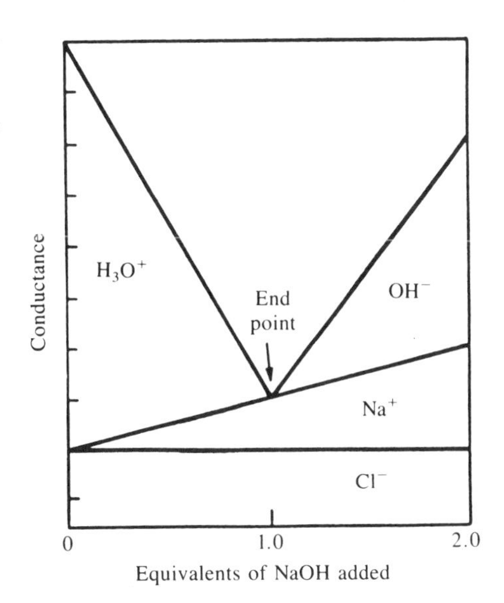

FIGURE **24.13**
Titration of hydrochloric acid with sodium hydroxide.

The acuteness of the angle at the point of intersection of the two branches is a function of the individual ionic conductances of the reactants. In Figure 24.13, the falling branch is steep because it involves the replacement of hydrogen ion. Similarly, the rising branch is relatively steep also, but not as steep as the falling branch because the conductance of the hydroxyl ion is considerably less than the corresponding value for the hydrogen ion.

The titrant should be at least ten times as concentrated as the solution being titrated in order to keep the volume change small. If necessary a correction may be applied; all conductance readings are multiplied by the term $(V + v)/V$, where V is the initial volume and v is the volume of titrant added up to the particular conductance reading.

The major applications are to acid-base titrimetry and to certain precipitation and other ion-combination titrations. When applicable, conductometric titration is particularly valuable at low titrant concentrations, about $0.0001M$. On the other hand, because every ion present contributes to the electrolytic conductivity, there should be no large amounts of extraneous electrolytes. This usually precludes conductometric redox titrations principally because such titrations are usually made in acid or other strongly conducting media. Under optimum conditions the end point can be located with a relative error of approximately 0.5%.

PROBLEMS

1. At what value should the cathode potential be controlled if one desires to separate silver from a $0.005M$ solution of Cu^{2+} ions? If the initial silver concentration is $0.05M$, how long should the deposition take, assuming $\delta = 2 \times 10^{-3}$ cm, $D = 7 \times 10^{-5}$ cm^2 sec^{-1}, $V = 200$ mL, and $A = 150$ cm^2?

2. A solution is initially $0.01M$ in silver ion and $0.5M$ in copper(II) ions. (a) What cathode potential is needed theoretically for the complete deposition of silver? (b) What cathode potential may be required considering concentration polarization? (c) How much silver remains in the solution when the cathode potential has been brought to 0.45 V versus SHE?

3. In an electrolytic determination of bromide ion from 100 mL of solution, the silver anode, after electrolysis was completed, was found to have gained 0.8735 g. (a) Calculate the molarity of bromide in the original solution. (b) Calculate the potential of the silver electrode at the beginning of the electrolysis, assuming the solubility product of AgBr is 4×10^{-13}.

4. A solution that is $0.01M$ in zinc sulfate, and buffered at pH 4 with an acetate buffer, is to be electrolyzed using a copper-clad cathode. The overpotential of hydrogen on copper is 0.75 V at the current density to be used, and that of oxygen on the platinum anode is 0.50 V. (a) Calculate the cell potential of the solution when electrolysis commences, assuming that the iR drop is 0.5 V. (b) Will the cell potential change as the electrolysis proceeds? (c) How much zinc will remain in solution at the point when hydrogen gas begins to be liberated?

5. At a current density of 0.01 A cm^{-2} the overpotential of hydrogen gas on cadmium is 0.4 V. Would it be possible to deposit cadmium quantitatively in a solution buffered at pH = 2?

6. The cathode potential is controlled at 0.05 V less negative than the value at which tin would be deposited from a 0.005*M* solution. (a) Calculate the molarity of copper ions that remain in a sulfate solution. (b) Estimate the quantity of undeposited copper in a solution 1.0*M* in HCl and containing hydrazine hydrochloride.

7. If lead is used as the soluble anode in the internal electrolysis of a lead solution containing a small amount of copper, what would be the final concentration of copper in solution if the lead concentration is 0.2*M*?

8. By means of suitable calculations, show why zinc can be successfully plated onto a copper-clad electrode from a solution buffered at pH = 10.5, whereas the deposition would not occur without the concurrent evolution of hydrogen gas if smooth platinum electrodes were substituted. Assume a current density equivalent to the appearance of the first visible gas bubbles.

9. Suggest an electrographic method for detecting the presence of copper filings on an ax blade suspected of being used to cut telephone cables. The method should be adaptable to courtroom demonstration before a jury of lay people.

10. For a typical laboratory deposition of 0.200 g copper onto a platinum electrode (area 160 cm^2) from 200 mL of 0.5*M* tartrate solution adjusted to pH 4.5 and containing hydrazine, calculate the time required to reduce the copper concentration (a) to 1% of its original value and (b) to 0.1%. Starting at an initial value of 2.6 A, the current decreased to 1.3 A after 2 min, to 0.65 A after 4 min, and to 0.33 A after 6 min.

11. The initial current is 90.0 mA and decreases exponentially with $k =$ 0.0058 sec^{-1}; the titration time is 714 sec. How many milligrams of uranium(VI) are reduced to uranium(IV)?

12. These results were obtained during the titration of three successive 1.00-mL aliquots of As_2O_3 solution with electrically generated iodine at pH 8 and using amperometric indication of the end point. Graph the results and determine the normality of the As_2O_3 solution. The microequivalents of iodine generated are followed by the amperometric signal in microamperes: Pretitration—0.00 microequivalents = 0.4 μA; 5.10 = 0.7; 9.90 = 1.3; and 15.0 = 1.7. First aliquot—15.0 = 0.4; 50.0 = 0.4; 100.0 = 0.4; 149.5 = 1.3; 154.5 = 2.0; 160.0 = 2.6; 164.9 = 3.0. Second aliquot—164.9 = 0.4; 200.0 = 0.4; 250.0 = 0.4; 273.8 = 1.0; 277.4 = 2.2; 280.8 = 2.7; 286.1 = 3.2. Third aliquot—286.1 = 0.4; 350.0 = 0.4; 400.0 = 0.4; 402.0 = 0.8; 406.3 = 2.6; 411.0 = 3.1; 416.1 = 3.8.

13. Calculate the concentration of acid in a 10.0-mL aliquot that required a generation time of 165 sec for the appearance of the pink color of phenolphthalein. The voltage drop across a 100-Ω resistor was 0.849 V.

14. Sketch the current-potential curves that would pertain to each of these coulometric systems: (a) the titration of acids with electrically generated hydroxyl ion in a potassium bromide electrolyte and using a silver anode; (b) the generation of excess bromine in a potassium bromide electrolyte, followed by the generation of

copper(I) to react with the unused bromine; and (c) the titration of zinc with generated ferrocyanide ions.

15. In the coulometric determination of permanganate ion by generating iron(II) from iron(III), the permanganate was all reduced to manganese(II) by a constant current of 2.50 mA acting for 10.37 min. Calculate the molarity of the permanganate if the initial volume was 25.00 mL.

16. Assuming that the coulograms for iron and vanadium, shown in the figure, are reversible, outline a constant-current procedure for the determination of iron(II) in the presence of vanadium(IV).

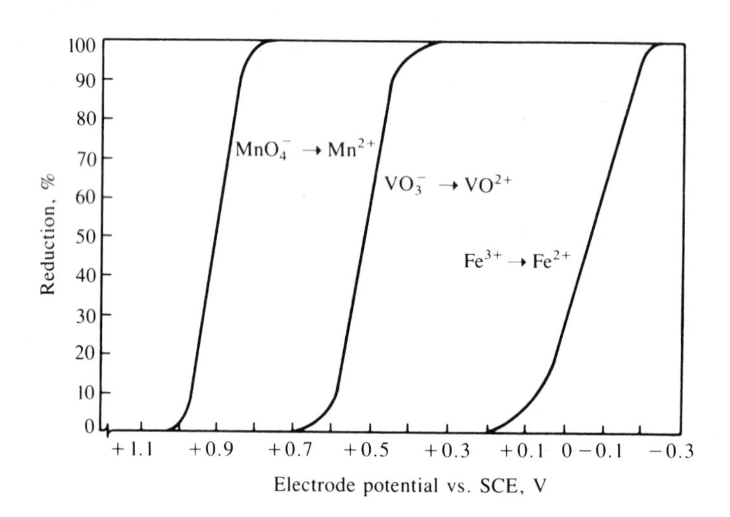

17. From the information in Problem 16, outline a procedure for the determination of the amounts of vanadium(V) and vanadium(IV) in a mixture that contains the two oxidation states.

18. How long should a constant current of 100.0 mA be passed through a solution to prepare 100 mL of a solution of 0.0100*M* Ni^{2+} using an anode of pure nickel?

19. A cell constant of 20.0 cm^{-1} is recommended for a commercial conductivity bridge designed to span the range from 1% to 18% HCl. The corresponding conductance ranges from 0.0630 to about 0.750 Ω^{-1}. What range of resistance values is involved?

20. A meter scale is to be inscribed from 2 to 1000 ppm Na_2SO_4, and the midpoint of the logarithmic scale will correspond to 40 ppm. Suggest a compatible set of instrument parameters—that is, resistance range and cell constant.

21. Individual instruments are to be designed for each of these systems. Compute the resistance range and a compatible cell constant. Use handbooks to locate necessary conductance values, and assume average distilled water has a specific conductance of $2 \times 10^{-6}\ \Omega^{-1}\ cm^{-1}$. (a) 0%–5% HCl; (b) 0.5%–5% NH_3;

(c) 0–60 ppm sodium formate; (d) 0–40 ppm salinity (as NaCl); (e) 96%–99.5% H_2SO_4; (f) 0.1%–10% CrO_3.

22. The equivalent conductance of a 0.002414*N* acetic acid solution is found to be 32.22 at 25 °C. Calculate the degree of dissociation of acetic acid at this concentration, and calculate the ionization constant.

23. The specific conductance at 25 °C of a saturated solution of barium sulfate was $4.58 \times 10^{-6}\ \Omega^{-1}\ cm^{-1}$, and that of the water used was 1.52×10^{-6}. What is the solubility of $BaSO_4$ at 25 °C in moles per liter and in grams per liter? Calculate the solubility product.

24. In the titration of 100 mL of H_2SO_4 in glacial acetic acid with 0.500*M* sodium acetate in the same solvent, the following specific conductance ($\times 10^6$) data were obtained at the indicated buret readings:

0.50 mL = 2.95	3.50 mL = 4.78	7.00 mL = 3.20
1.00 mL = 3.30	4.00 mL = 4.73	7.50 mL = 3.20
1.50 mL = 3.65	4.50 mL = 4.40	8.00 mL = 3.47
2.00 mL = 4.00	5.00 mL = 4.04	8.50 mL = 3.82
2.50 mL = 4.35	5.50 mL = 3.76	9.00 mL = 4.18
3.00 mL = 4.65	6.00 mL = 3.43	9.50 mL = 4.50

What is the molarity of the sulfuric acid solution?

25. The following relative conductance readings were obtained during the titration of a mixture containing an aliphatic acid and an aromatic sulfonic acid. The titrant was 0.200*N* NH_3. Readings have been corrected for titrant volume.

0.00 mL = 2.01	3.20 mL = 1.19	5.00 mL = 1.51
1.00 mL = 1.75	3.50 mL = 1.26	6.00 mL = 1.52
2.00 mL = 1.47	4.00 mL = 1.41	8.00 mL = 1.53
2.50 mL = 1.33	4.20 mL = 1.47	
3.00 mL = 1.19	4.50 mL = 1.51	

Calculate the number of equivalents of each acid present in the mixture.

26. Using equivalent conductance values, sketch the general form of the titration curve in each of the following cases: (a) titration of $Ba(OH)_2$ with HCl, (b) titration of NH_4Cl with NaOH, (c) titration of silver nitrate with potassium chloride, (d) titration of silver acetate with lithium chloride, (e) titration of sodium acetate with HCl, (f) titration of a mixture of a sulfonic acid and a carboxylic acid with NaOH, and (g) titration of $KH_3(C_2O_4)$ with NH_3.

BIBLIOGRAPHY

BARD, A. J., ed., *Electroanalytical Chemistry, a Series of Advances,* Dekker, New York; a series of monographs published usually yearly commencing in 1966.

KISSINGER, P. T., AND W. W. HEINEMAN, eds., *Laboratory Techniques in Electroanalytical Chemistry,* Dekker, New York, 1984.

LINGANE, J. J., *Electroanalytical Chemistry,* 2nd ed., Wiley-Interscience, New York, 1958.

LITERATURE CITED

1. HERMANCE, H. W., AND H. V. WADLOW, "Electrography and Electrospot Testing," pp. 500–520, *Standard Methods of Chemical Analysis*, 6th ed., F. J. Welcher, ed., Vol. 3, Part A, Van Nostrand, New York, 1966.
2. EWING, G. W., "Titrate with Electrons," *Am. Lab.*, **13,** 16 (June 1981).
3. CLEM, R. G., "Coulometry Present," *Ind. Research,* p. 50 (September 1973).
4. DEFORD, D. D., J. N. PITTS, AND C. J. JOHNS, *Anal. Chem.*, **23,** 938 (1951).
5. DEAN, J. A., ed., *Lange's Handbook of Chemistry*, 13th ed., pp. 6–36, McGraw-Hill, New York, 1985.
6. EWING, G. W., "The Measurement of Electrolytic Conductance," *J. Chem. Educ.*, **51,** A469 (1974).
7. STORK, J. T., "Two Centuries of Quantitative Electrolytic Conductivity," *Anal. Chem.*, **56,** 561A (1984).

THERMAL ANALYSIS

Thermal analysis includes a group of techniques in which specific physical properties of a material are measured as a function of temperature (Table 25.1). The production of new high-technology materials and the resulting requirement for a more precise characterization of these substances have increased the demand for thermal analysis techniques. Current areas of application include environmental measurements, composition analysis, product reliability, stability, chemical reactions, and dynamic properties. Thermal analysis has been used to determine the physical and chemical properties of polymers, electronic circuit boards, geological materials, and coals. An integrated, modern thermal analysis instrument (Figure 25.1) can measure transition temperatures, weight losses, energies of transitions, dimensional changes, modulus changes, and viscoelastic properties.

Thermal analysis is useful in both quantitative and qualitative analyses. Samples may be identified and characterized by qualitative investigations of their thermal behavior. Information concerning the detailed structure and composition of different phases of a given sample is obtained from the analysis of thermal data. Quantitative results are obtained from changes in weight and enthalpy as the sample is heated. The temperatures of phase changes and reactions as well as heats of reaction are used to determine the purity of materials.

TABLE **25.1** SUMMARY OF THERMAL ANALYSIS TECHNIQUES

Technique	Quantity measured	Typical application
Differential scanning calorimetry (DSC)	Heats and temperatures of transitions and reactions	Reaction kinetics, purity analysis, polymer cures
Differential thermal analysis (DTA)	Temperatures of transitions and reactions	Phase diagrams, thermal stability
Thermogravimetric analysis (TGA)	Weight change	Thermal stability, compositional analysis
Thermomechanical analysis (TMA)	Dimension and viscosity changes	Softening temperatures, expansion coefficients
Dynamic mechanical analysis (DMA)	Modulus, damping, and viscoelastic behavior	Impact resistance, mechanical stability
Evolved gas analysis (EGA)	Amount of gaseous products of thermally induced reaction	Analysis of volatile organic components of shale

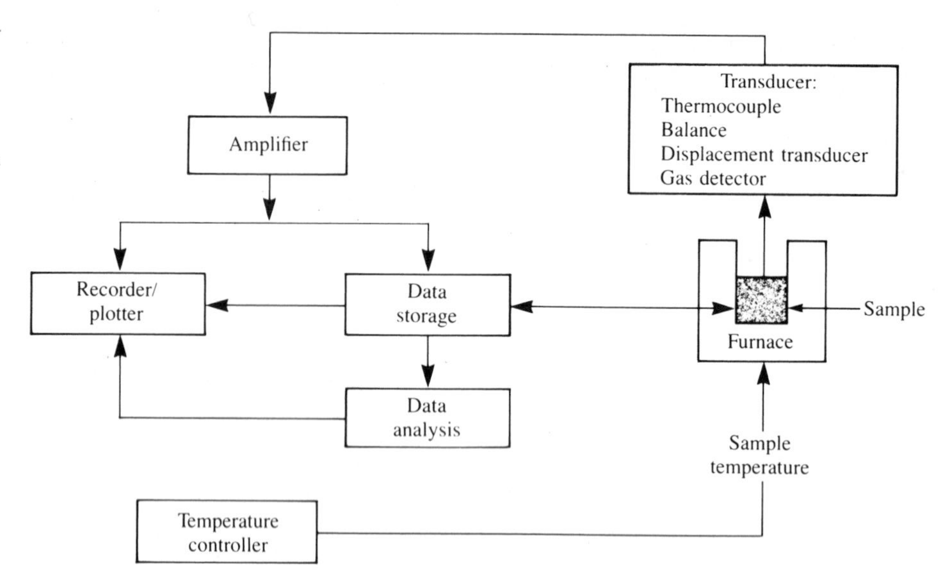

FIGURE 25.1
A complete thermal analysis system. (Courtesy of DuPont Clinical and Instruments Systems Division.)

The use of microprocessors has both enhanced and simplified the techniques of thermal analysis.[1] The sample is heated at a programmed rate in the controlled environment of the furnace. Changes in selected properties of a sample are monitored by specific transducers, which generate voltage signals. The signal is then amplified, digitized, and stored on a magnetic disk along with the corresponding direct temperature responses from the sample. The data may also be displayed or plotted in real time. The microcomputer is used to process the data with a library of applications software designed for thermal analysis techniques. The multitasking capabilities of some computer systems allow a single microcomputer to operate several thermal analyzers simultaneously and independently.

A major advantage of microcomputer systems in thermal analysis is that the operator seldom, if ever, needs to repeat an analysis because of an improper choice of ordinate scale sensitivity. The software does this rescaling after all the data have been collected. In some systems both axes are automatically rescaled after the last data point has been received. When the amount of time necessary to obtain thermal data is considered, the advantage is obvious. For example, a differential scanning calorimetric run at 10 °C/min from room temperature to 1100 °C takes 100 min, and thus the rerun of the sample would take longer than 2 hr including the time required for cooling and sample reloading.

25.1 DIFFERENTIAL SCANNING CALORIMETRY AND DIFFERENTIAL ANALYSIS

Differential scanning calorimetry (DSC) has become the most widely used thermal analysis technique. In this technique, the sample and reference materials are subjected to a precisely programmed temperature change. When a thermal transition (a chemical or physical change that results in the emission or absorption of heat) occurs in the sample, thermal energy is added to either the sample or the reference containers in order to maintain both the sample and reference at the same temperature (Figure 25.2a). Because the energy transferred is exactly equivalent in magnitude to the energy absorbed or evolved in the transition, the balancing energy yields a direct calorimetric measurement of the transition energy. Since DSC can measure directly both the temperature and the enthalpy of a transition or the heat of a reaction, it is often substituted for differential thermal analysis as a means of determining these quantities except in certain high-temperature applications.

In differential thermal analysis (DTA), the difference in temperature between the sample and a thermally inert reference material is measured as a function of temperature (usually the sample temperature) (Figure 25.2b). Any transition that the sample undergoes results in liberation or absorption of energy by the sample with a corresponding deviation of its temperature from that of the reference. A plot of the differential temperature, ΔT, versus the programmed temperature, T, indicates the transition temperature(s) and whether the transition is exothermic or endothermic. DTA and thermogravimetric analyses (measurement of the change in weight as a function of temperature; see Section 25.2) are often run simultaneously on a single sample.

Instrumentation

A typical DSC cell uses a constantan (Cu-Ni) disk as the primary means of transferring heat to the sample and reference positions and also as one element of the temperature-sensing thermoelectric junction[2] (Figure 25.3). The sample and a reference are placed in separate pans that sit on raised platforms on the disk. Heat is transferred to the sample and reference through the disk. The differential heat flow to the sample and reference is monitored by the chromel/constantan thermocouples

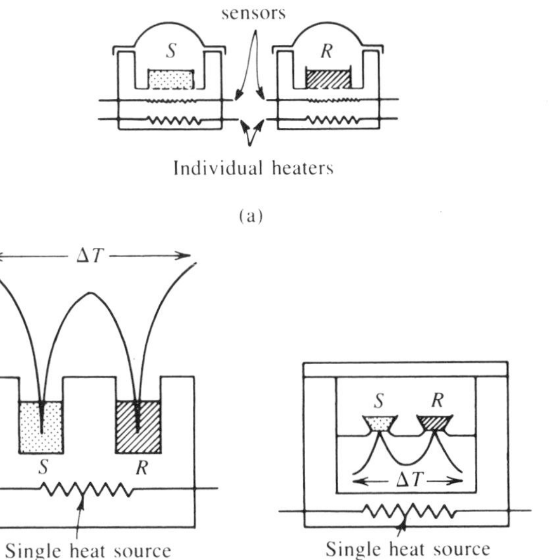

FIGURE **25.2**
Arrangement of temperature sensors in (a) DTA and (b) DSC.

formed by the junction of the constantan disk and the chromel wafer covering the underside of each platform. Chromel and alumel wires connected to the underside of the wafers form a chromel/alumel thermocouple, which is used to directly monitor the sample temperature. Constant calorimetric sensitivity is maintained by computer software, which linearizes the cell-calibration coefficient. DSC provides maximum calorimetric accuracy from -170 to 750 °C. Sample sizes range from 0.1 to 100 mg.

The change in enthalpy, ΔH, of the sample is equal to the difference between the heat flow to or from the sample, Q_s, and the heat flow to or from the reference material, Q_r (ΔH is used to indicate that the heat flow is the change in enthalpy):

$$\Delta H = Q_s - Q_r \tag{25.1}$$

According to the thermal analog of Ohm's law,

$$Q = \frac{T_2 - T_1}{R_{\text{th}}} \tag{25.2}$$

the heat flow is proportional to the driving force (the temperature difference between temperatures T_1 and T_2) and inversely proportional to the thermal resistance, R_{th}. Combining the preceding relationships yields

$$\Delta H = Q_s - Q_r = \frac{T_c - T_s}{R_{\text{th}}} - \frac{T_c - T_r}{R_{\text{th}}} \tag{25.3}$$

where T_c is a constant temperature external to the sample and reference, T_s is the sample temperature, and T_r is the reference temperature. The system is so designed that the two T_c and two R_{th} values are identical. Thus Equation 25.3 reduces to

$$\Delta H = -\frac{T_s - T_r}{R_{\text{th}}} \tag{25.4}$$

The measured signal is the voltage from the thermocouple or thermopile, which is proportional to the temperature difference, $(T_s - T_r)$.

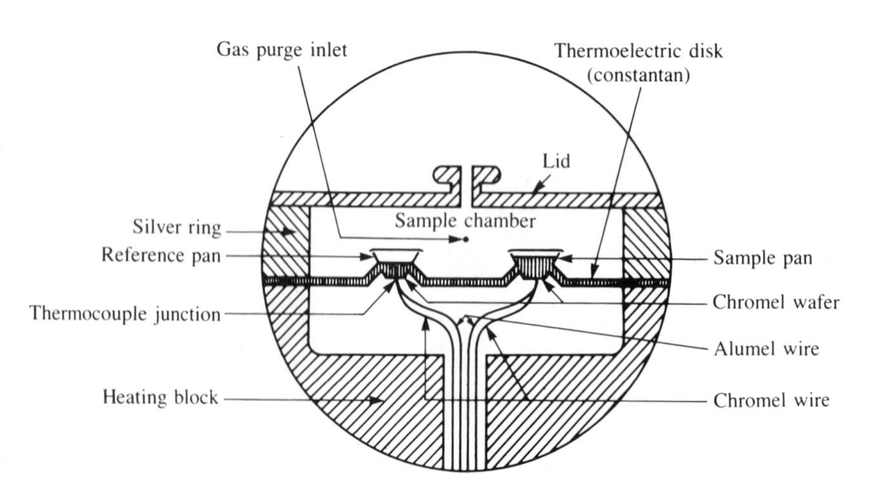

FIGURE 25.3
DSC cell cross section. (Courtesy of DuPont Clinical and Instruments Systems Division.)

Integration of the area under a DSC curve provides a direct measurement of ΔH for thermally induced transitions according to the equation

$$A = -k'm\,\Delta H \tag{25.5}$$

where A is the area; k' the instrument constant, which is independent of the temperature; m the mass; and H the enthalpy of the reaction or transition. Precise determinations require enlargement of the thermogram for accurate area measurements and a time base display rather than the temperature-based thermogram. Also, precise quantitative heat capacity measurements are typically made at high-sensitivity settings. The data stored on disks are recorded at maximum sensitivity. Thermograms can be replotted over a temperature range and on a scale selected by the operator. Temperature expansion to 0.2 °C/cm is possible. When this feature is combined with a high calorimetric sensitivity of 0.01 mW/cm, thermal occurrences that produce very small amounts of heat can be recorded. For maximum accuracy, a baseline is obtained and subtracted from the sample thermogram to determine the heat capacity. Heats of transitions are calculated from stored data by the applications software of the microcomputer.

In DTA, the furnace contains a block with identical and symmetrically located chambers (Figure 25.4). The sample is placed in one chamber and a reference material, such as α-Al_2O_3, is placed in the other chamber. A thermocouple is inserted into the center of the material in each chamber. The furnace and sample

FIGURE **25.4** Schematic diagram of the DuPont differential thermal analyzer. (Courtesy of E. I. DuPont de Nemours, Inc.)

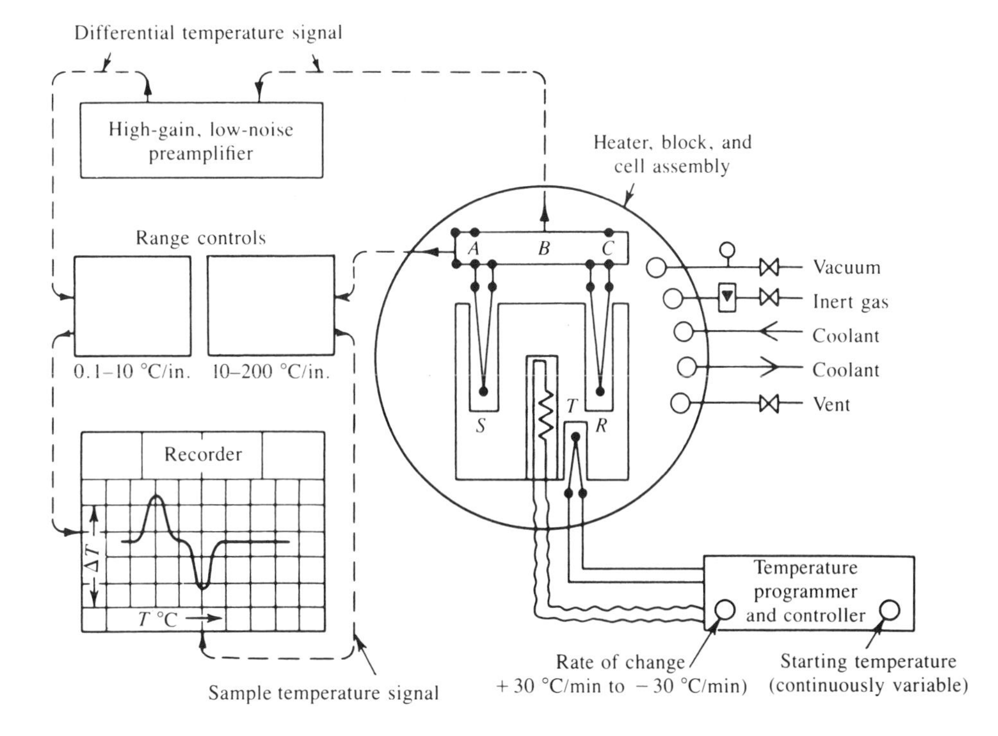

blocks are then heated by a microprocessor-controlled heating element. The difference in temperature between sample and reference (S, R) thermocouples, connected in series opposition, is continuously measured. After amplification (about 1000 times) by a high-gain, low-noise dc amplifier for microvolt-level signals, the difference signal is recorded as the y-axis. The temperature of the furnace is measured by an independent thermocouple and recorded as the x-axis. Because the thermocouple is placed in direct contact with the sample, DTA provides the highest thermometric accuracy of all thermal methods. DTA can be used in the temperature range from −190 to 1600 °C. Sample sizes are similar to those used in DSC.

Although the area of a DTA peak is proportional to the heat of reaction and the mass of the sample, it is inversely proportional to the sample's thermal diffusivity, which is a function of grain size and compactness. This inverse relationship prevents DTA peak areas from being used to provide direct calorimetric measurements. It is necessary to calibrate a DTA instrument for each type of sample and to carefully control experimental parameters to obtain useful thermodynamic data.

Pressure DSC

The ability to control the pressure of the atmosphere above the sample is often useful in DSC analysis. Applications of this technique include studies of pressure-sensitive reactions, evaluation of catalysts, and resolution of overlapping transitions. Figure 25.5 illustrates the latter application in studying the curing of phenolic resins.[1] When these resins are cured, they can liberate water and ammonia simultaneously during the exothermic curing reaction. Since both the vaporization of the above components and the curing reaction are exothermic, the two processes are in thermal competition and result in the top curve of Figure 25.5. There are no easily determined transitions in this scan. The dependence of vaporization on

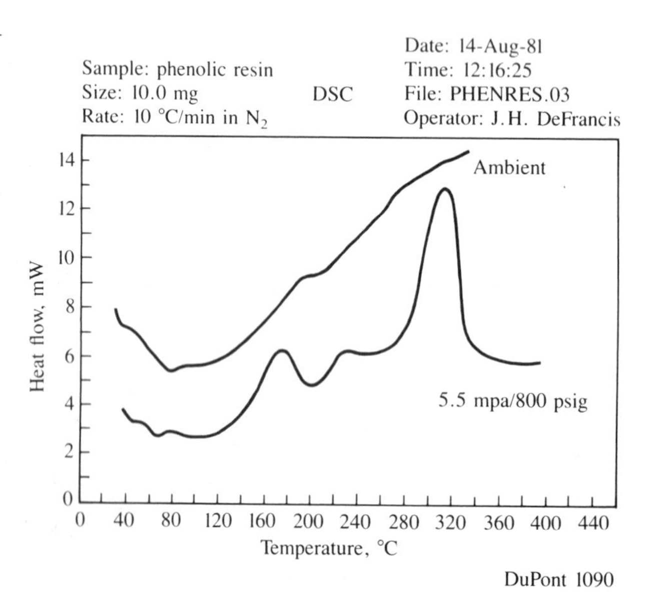

FIGURE **25.5**

Cure of a phenolic thermoset by pressure DSC. [After P. Gill, *Am. Lab.*, **16**(1), 39 (1984). Reproduced with permission.]

pressure can be used to resolve the two transitions. Elevation of the pressure (Figure 25.5, lower curve) causes the processes associated with the vaporization of water and ammonia to occur at higher temperatures and permits observation of the transitions associated with the two-stage curing process.

25.2 THERMOGRAVIMETRY[3]

Thermogravimetry (TG) or thermogravimetric analysis (TGA) provides a quantitative measurement of any weight changes associated with thermally induced transitions. For example, TG can record directly the loss in weight as a function of temperature or time (when operating under isothermal conditions) for transitions that involve dehydration or decomposition. Thermogravimetric curves are characteristic of a given compound or material due to the unique sequence of physical transitions and chemical reactions that occur over definite temperature ranges. The rates of these thermally induced processes are often a function of the molecular structure. Changes in weight result from physical and chemical bonds forming and breaking at elevated temperatures. These processes may evolve volatile products or form reaction products that result in a change in weight of the sample. TG data are useful in characterizing materials as well as in investigating the thermodynamics and kinetics of the reactions and transitions that result from the application of heat to these materials. The usual temperature range for TG is from ambient to 1200 °C in either inert or reactive atmospheres.

In TG the weight of the sample is continuously recorded as the temperature is increased. Samples are placed in a crucible or shallow dish that is positioned in a furnace on a quartz beam attached to an automatic recording balance. Figure 25.6 shows a TG instrument that contains a taut-band suspension electromechanical transducer.[2] The horizontal quartz beam is maintained in the null position by the current flowing through the transducer coil of an electromagnetic balance. A pair of photosensitive diodes acts as a position sensor to determine the movement of the beam. Any change in the weight of the sample causes a deflection of the beam, which is sensed by one of the photodiodes. The beam is then restored to the original null position by a feedback current sent from the photodiodes to the coil of the balance. The current is proportional to the change in weight of the sample.

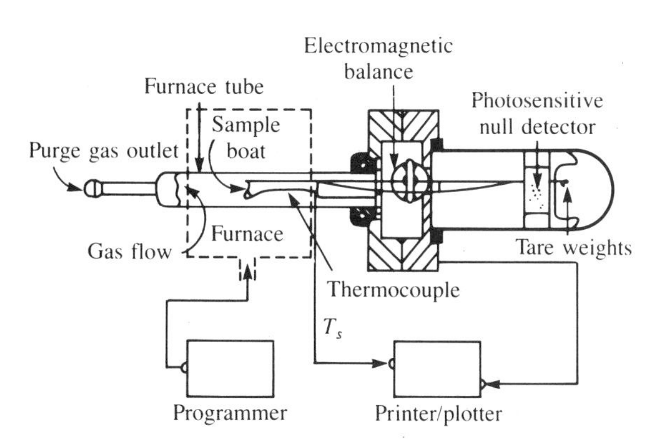

FIGURE **25.6**
Schematic of thermogravimetric analyzer. (Courtesy of DuPont Clinical and Instruments Systems Division.)

Linear heating rates from 5 to 10 °C/min are typical. Sample sizes range from 1 to 300 mg. Computer software allows the computation of $\Delta w/\Delta t$, which is important in kinetic interpretations of reactions and processes. TG has been used in the kinetic analysis of polymer stability, compositional analyses of multicomponent materials, atmospheric analyses and corrosion studies, moisture and volatiles determinations, and accelerated tests of aging.

25.3 EVOLVED GAS DETECTION AND ANALYSIS[3]

The analysis of the purge gas exit stream from differential thermal analysis, differential scanning calorimetry, and thermogravimetric analyzers is useful in establishing mechanisms and stoichiometric relationships of thermal decompositions. In evolved gas analysis (EGA) the absolute identities of the gaseous components are determined, whereas in evolved gas detection (EGD) the presence of only a single, preselected component of the evolved gas is sensed. An appropriate analyzer may be coupled to a thermogravimetric system for performing either EGA or EGD. The resulting hyphenated methods are powerful analytical tools. Two analyzers that have been successfully coupled to TG systems are mass spectrometers (MS) and flame ionization detectors (FID).[4] The TG-MS or TG-MS-MS combination is used for evolved gas analysis, whereas the TG-FID combination provides evolved gas detection. These hyphenated methods are used in studies of the volatile organic pyrolysis products of oil shales.

EXAMPLE 25.1

The simultaneous TG-DTA-MS analysis of copper sulfate-pentahydrate is shown in Figure 25.7. The quadrupole mass spectrometer detects the temperature-dependent intensity changes of water during dehydration to 300 °C. The evolution of sulfur dioxide during the two-step sulfate decomposition to copper(II) oxide between 600 and 900 °C is also clearly seen. Electron-impact ionization in the MS is responsible for the production of sulfur monoxide (SO) radical ions.

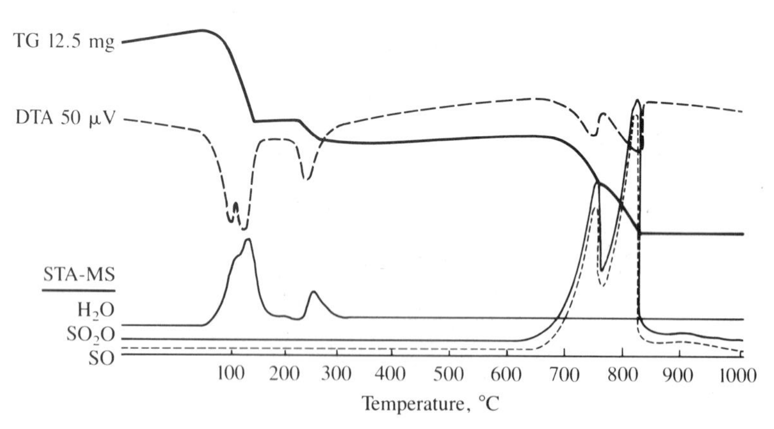

FIGURE **25.7** Simultaneous TG–DTA–MS analysis of $CuSO_4 \cdot 5H_2O$. (Courtesy of Netzsch Geratebau, GmbH.)

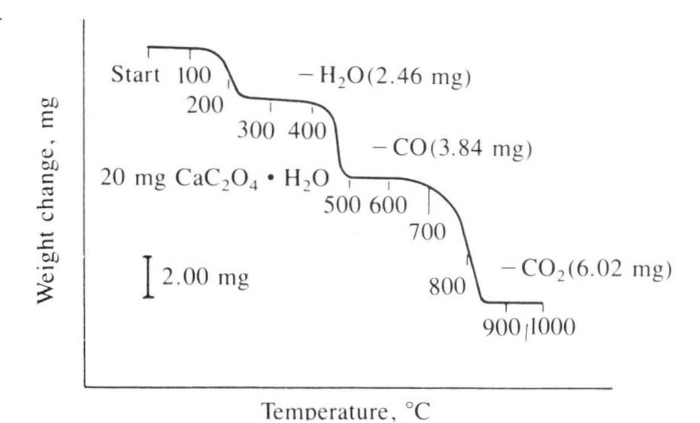

FIGURE **25.8** Thermogravimetric evaluation of calcium oxalate monohydrate. The heating rate is 6 °C/min.

25.4 METHODOLOGY OF THERMOGRAVIMETRY, DIFFERENTIAL SCANNING CALORIMETRY, AND DIFFERENTIAL THERMAL ANALYSIS

Thermogravimetry (TG)

The weight-change TG curve for calcium oxalate monohydrate is shown in Figure 25.8. Water is evolved beginning slightly above 100 °C. At about 250 °C the curve breaks at the stoichiometry corresponding to that of the anhydrous salt. Further heating gives definite weight plateaus for the carbonate (from 500 to 600 °C) and finally the oxide (hotter than about 870 °C). Exact locations of the weight plateaus are dependent on the heating rate (a slower heating rate shifts values to lower temperatures) and the ambient atmosphere around the sample particles. The curve is quantitative in that calculations can be made to determine the stoichiometry of the compound at any given temperature.

Thermal analysis is affected by the experimental conditions. Deviations caused by instrumental factors include furnace atmosphere, size and shape of the furnace and sample holder, sample holder material and its resistance to corrosive attack, wire and bead size of the thermocouple junction, heating rate, speed and response of the recording equipment, and location of the thermocouples in the sample and reference chambers. Another set of factors that influence the results depends on the sample characteristics; these include layer thickness, particle size, packing density, amount of sample, thermal conductivity of the sample material, heat capacity, the ease with which gaseous effluents can escape, and the atmosphere surrounding the sample.

Thermogravimetry, a valuable tool in its own right, is perhaps most useful when it complements differential thermal analysis studies. Virtually all weight-change processes absorb or release energy and are thus measurable by DTA or DSC, but not all energy-change processes are accompanied by changes in weight. This difference in the two techniques enables a clear distinction to be made between physical and chemical changes when the samples are subjected to both DSC (or DTA) and TG tests.

EXAMPLE 25.2

Thermogravimetry can also be used to determine the composition of complex materials such as carbon black-filled rubber. Figure 25.9 shows the results of rapidly heating a rubber sample in an inert atmosphere of nitrogen from room temperature to 950 °C and then quickly changing the atmosphere to air. Heating the sample in an inert environment results in the

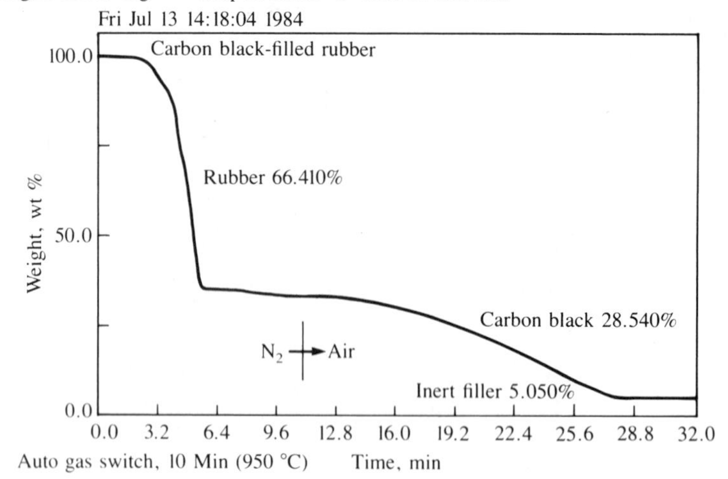

FIGURE **25.9**
TG scan of carbon black-filled rubber material. (Courtesy of Perkin-Elmer Corp.)

pyrolytic decomposition of the rubber and gives the first major weight loss observed in the scan. The addition of air at 950 °C causes the carbon black to undergo combustion to produce the second loss in weight, leaving in the balance pan only inert filler. All the major components of the rubber sample were determined quantitatively from a single TG scan.

Differential Scanning Calorimetry (DSC) and Differential Thermal Analysis (DTA)

In general, each substance gives a DSC or DTA curve in which the number, shape, and position of the various endothermic and exothermic features serve as a means of qualitative identification of the substance. When an endothermic change occurs, the sample temperature lags behind the reference temperature because of the heat in the sample. The initiation point for a phase change or chemical reaction is the point at which the curve first deviates from the baseline. When the transition is complete, thermal diffusion brings the sample back to equilibrium quickly. The peak (or minimum) temperature indicates the temperature at which the reaction is completed. When the break is not sharp, a reproducible point is obtained by drawing one line tangent to the baseline and another tangent to the initial slope of the curve.

Various behaviors deduced from a DTA curve are shown in Figure 25.10. The heat capacity at any point is proportional to its displacement from the blank

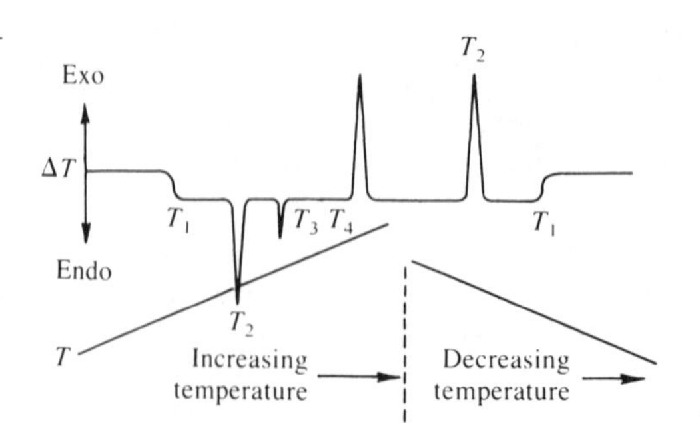

FIGURE **25.10**
DTA curve of a hypothetical substance.

baseline. A broad endotherm indicates a slow change in heat capacity. A "second-order or glass" transition, observed as a baseline shift (T_1), denotes a decrease in order within the system. This is the temperature at which a polymer changes from a brittle, glasslike material to a tough, resilient material. The lower the glass transition temperature, the lower the temperature at which the polymer is useful in applications, such as adhesives or impact-resistant structures. In a thermoset, a high glass transition temperature indicates incomplete cure of the resin; in a thermoplastic, a high glass transition temperature indicates the use of the wrong plasticizer or incomplete reaction in the formation of the polymer itself. Endotherms generally represent physical rather than chemical changes. Sharp endotherms (T_3) are indicative of crystalline rearrangements, fusions, or solid-state transitions for relatively pure materials. Broader endotherms (T_2) cover behavior ranging from dehydration and temperature-dependent phase behaviors to the melting of polymers. Exothermic behavior (without decomposition) is associated with the decrease in enthalpy of a phase or chemical system. Narrow exotherms usually indicate crystallization (ordering) of a metastable system, whether it be supercooled organic, inorganic, amorphous polymer, or liquid, or annealing of stored energy resulting from mechanical stress. Broad exotherms denote chemical reactions, polymerization, or curing of thermosetting resins. Exotherms with decomposition can be either narrow or broad depending on the kinetics of the behavior. Explosives and propellants are sharpest, and the "unzipping" of polyvinyl chloride is rapid, whereas oxidative combustion and decomposition are generally broad.

On cooling one would expect the reverse of features observed on the heating cycle (Figure 25.10). Since T_4 does not recur on cooling, the reaction is obviously nonreversible (perhaps a pyrolytic decomposition). Instead of taking the system up to T_4, the cooling cycle should be started before that temperature. As it cools, the substance is seen to lose its transition peak at T_3. Judging from the area under the T_2 peak, the transition energy of T_3 has been added to T_2. This indicates a metastable condition at T_3, with the retained energy being released in one large step at a lower temperature. Further along the cooling curve, the glass transition at T_1 falls properly into place to complete the cycle.

Determining the significance of thermoanalysis curves is not always a straightforward task. A reference library of curves of specific interest to a particular laboratory is vital. Computerized systems offer the capability of storage and quick retrieval of data. Thermoanalysis data on commercial products or thermal transition points for pure substances reported in the literature are of little value for the comparison with a dynamically scanned thermal profile. Complementary techniques are valuable. The correlation of thermally evolved gaseous products with differential scanning calorimetry or differential thermal analysis transitions using evolved gas analysis or evolved gas detection often assists in elucidating the decomposition mechanism (see Section 25.3). Running thermal decompositions in inert, oxidative, or special atmospheres can often provide valuable clues from changes in the curves.

EXAMPLE 25.3

The TG and DTA curves of manganese phosphinate monohydrate are shown in Figure 25.11. The weight-loss data (TG curve) from a 200-mg sample run under vacuum and with the analysis of effluent gases showed the loss of 1 mole of water at 150 °C, 1 mole of phosphine at

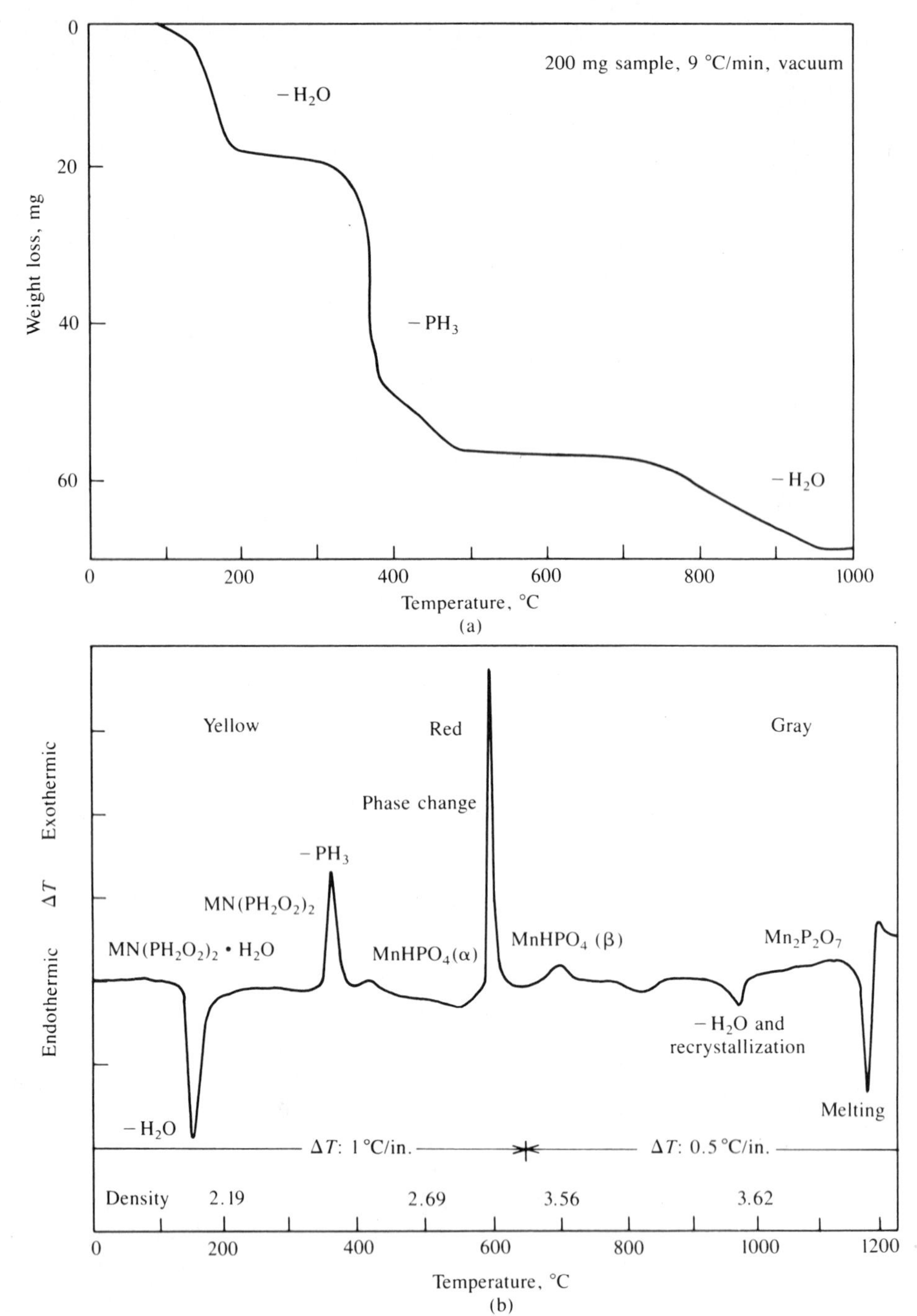

FIGURE 25.11
(a) TG and (b) DTA curves for $Mn(PH_2O_2)_2 \cdot H_2O$. (Courtesy of American Instrument Co.)

360 °C, and the slow loss of another mole of water starting around 800 °C. After comparison with the DTA curve, two major peaks remain unidentified: the large exotherm at 590 °C and the endotherm at 118 °C, plus several smaller thermal features. Thermogravimetric data obtained from runs performed under vacuum and in a nitrogen atmosphere failed to show any loss associated with these peaks. Each sample was measured for its real density. The resulting

data are shown on the DTA curve. Undoubtedly the sharp DTA exotherm at 590 °C represents a phase change. The relatively small endotherm starting above 900 °C must represent a recrystallization exotherm following the elimination of water, which is superimposed on the latter endotherm. The peak at 1180 °C is due to melting. With this information the thermal decomposition reactions and phase changes are:

$$Mn(PH_2O_2)_2 \cdot H_2O(s) \rightarrow Mn(PH_2O_2)_2(s) + H_2O(g)$$

$$Mn(PH_2O_2)_2(s) \rightarrow MnHPO_4(s) + PH_3(g)$$

$$\alpha\text{-}MnHPO_4(s) \rightarrow \beta\text{-}MnHPO_4(s)$$

$$2MnHPO_4(s) \rightarrow Mn_2P_2O_7(s) + H_2O(g) \text{ (and recrystallization)}$$

$$Mn_2P_2O_7(s) \rightarrow Mn_2P_2O_7(l)$$

Thermal studies with polymers can predict a product's performance in use—that is, its stiffness, toughness, or stability.[5] Melting-point, phase-transition, pyrolysis, and curing temperatures are accurately measured. Once a polymer has been broadly classified by other methods, curves are often used to establish, by comparison with known reference materials, the degree of polymerization, the thermal history of the sample, crystal perfection and orientation, the effect of different coreactants and catalysts, the percentage of crystalline polymer, and the extent of chain branching. For example, curves for a low-molecular-weight, nonlinear, branched-chain polymer show a continuous series of rather broad and low-melting-point endotherms, whereas a high-molecular-weight, stereoregular, linear polymer reveals a single narrow and higher-melting-point endotherm. If a polymer has been incompletely cured, the heating cycle may reveal an exotherm at a temperature close to the one used for the polymerization reaction. An exotherm just below the melting temperature indicates "cold crystallization," which results if a sample is quenched quickly after being melted. On reheating, crystallites form rapidly and exothermically just prior to remelting of the polymer. Annealing temperatures are similarly revealed as exotherms.

If the molecular weight or density of a polymer has been established by appropriate (often lengthy) methods, subsequent determination of its melt temperature (a 15-min process) can be related to molecular weight or density. Product quality is maintained subsequently by simply examining curves of polymer materials to obtain molecular weights or densities from an appropriate calibrated graph.

Instead of using the traditional method of preparing a derivative from the organic sample and a reagent, the sample is heated with a specific reagent at a programmed heating rate in a selected atmosphere. The DTA or DSC curve shows the derivative-forming reaction, the physical transitions of the sample or reagent (whichever is in excess), and the physical transitions of the intermediates and final products. When one reactant is volatile and in excess, a rerun usually shows only the derivative characteristics.[6]

The area of exotherms or endotherms is used to calculate the heat of the reaction or the heat of a phase transition. Suitable calibration is necessary with DTA equipment, but the values are given directly with DSC instruments.

EXAMPLE 25.4

Polyethylene is a semicrystalline thermoplastic that when heated undergoes a process of melting. This melting destroys the crystal structure of the polymer and is an endothermic process. Although plastics usually melt over a temperature range, the melting point is defined

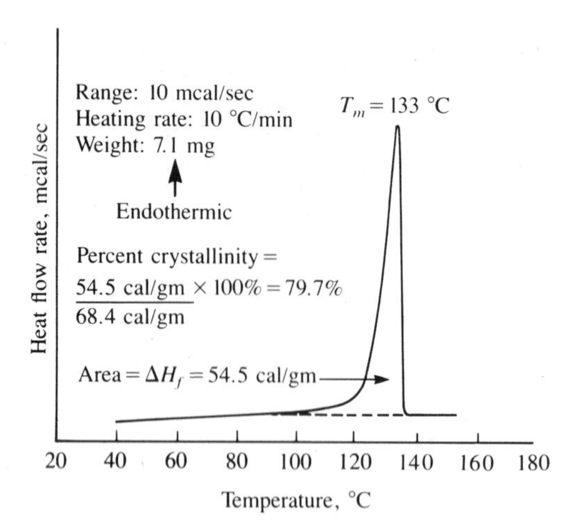

FIGURE 25.12
DSC analysis of polyethylene: melting point and percent crystallinity. (Courtesy of Perkin-Elmer Corp.)

as the temperature at which the melting is complete. The melting point of a plastic is an important property because it is the minimum temperature for processing the plastic and the maximum temperature for applications where structural integrity is required. A second important property for the characterization of semicrystalline thermoplastics is the percent crystallinity. Many physical properties that give the plastic its useful attributes are dependent on the percent crystallinity.

Both the melting point and the percent crystallinity are obtained from a single differential scanning calorimetry scan (Figure 25.12). The melting point is taken as 133 °C, where the melting is approximately complete. Since differential scanning calorimetry is directly quantitative, the peak area is equal to the heat of fusion, H_f, in units of calories per gram. The percent crystallinity is determined by assuming that the heat of fusion is proportional to the percent of crystallinity of the sample. Thus if the heat of fusion of 100% polyethylene has been determined to be 68.4 cal/g and the measured heat for the sample is 54.5 cal/g, the percent crystallinity is 79.7%, as shown in Figure 25.12.

25.5 THERMOMECHANICAL ANALYSIS

Thermomechanical analysis (TMA) provides measurements of penetration, expansion, contraction, and extension of materials as a function of temperature. The typical apparatus, diagrammed in Figure 25.13, is a probe connected mechanically to the core of a linear variable differential transformer (LVDT). The core is coupled to the sample by means of a quartz probe that contains a thermocouple for measurement of the sample temperature. Any movement of the sample is translated into a movement of the transformer core and results in an output that is proportional to the displacement of the probe, and whose sign is indicative of the direction of movement. The temperature range is from that of liquid nitrogen to 850 °C.

In the penetration and expansion modes, the sample rests on a quartz stage surrounded by the furnace. Under no load, expansion with temperature is observed. The thermal coefficient of linear expansion is calculated directly from the slope of the resulting curve. A weight tray attached to the upper end of the probe allows a predetermined force to be applied to the sample to study variations under load.

Probes with a small tip diameter and a loaded weight tray are used when the sensitive detection of softening temperatures, heat-distortion temperatures, and glass transitions are of interest. Larger tip diameters and zero loading are used in the expansion mode when coefficients of expansion and dimensional changes due to stress relief are the objects of investigation. Sample sizes may range from a 2.54-μm coating to a 1.3-cm-thick solid. Sensitivities down to a few micrometers are observable.

For the measurement of samples in tension, the sample stage and probe are replaced by a sample holder system consisting of stationary and movable hooks

FIGURE **25.13** (a) Thermomechanical analyzer. (Courtesy of Perkin-Elmer Corp.) (b) Probe configurations. (Courtesy of E. I. DuPont de Nemours, Inc.)

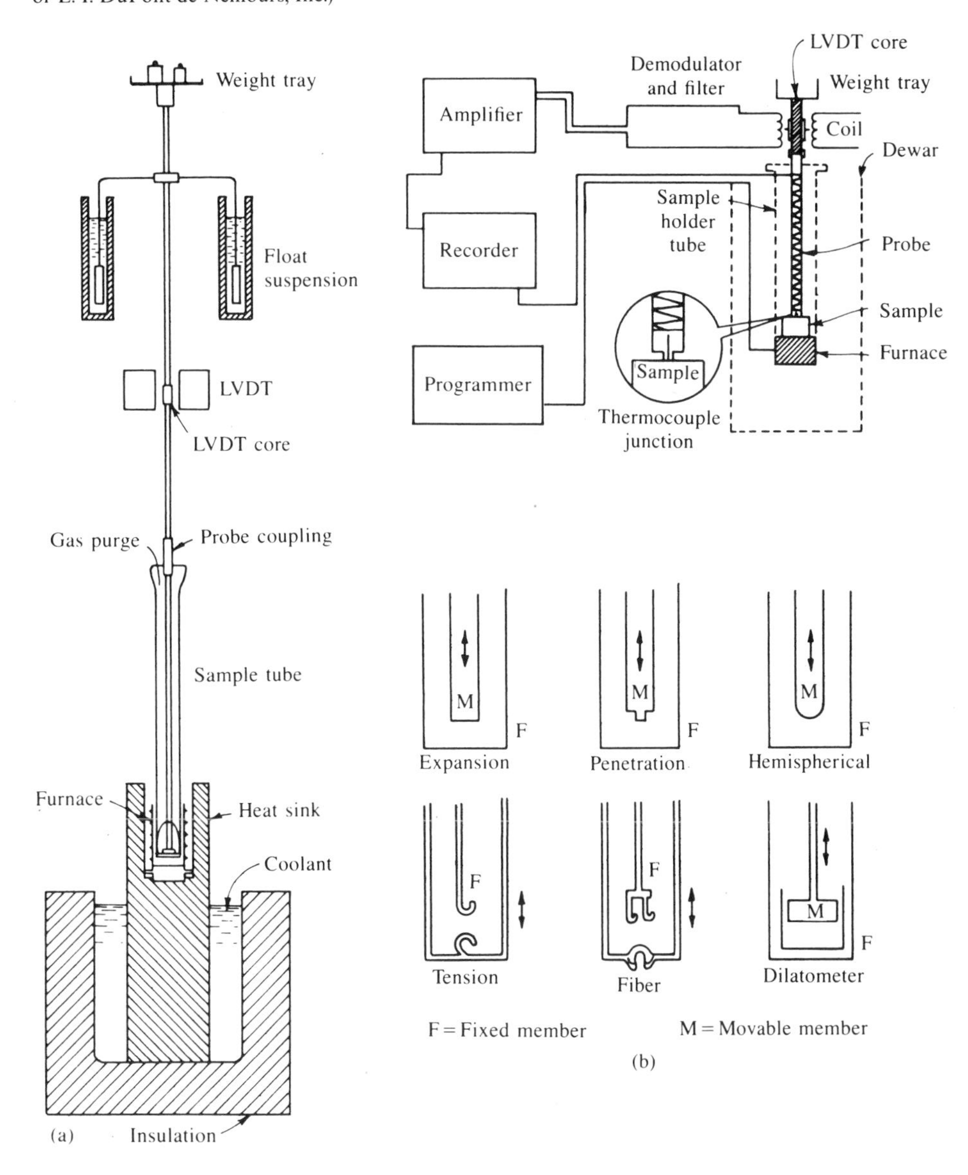

FIGURE 25.14
Thermomechanical (expansion) behavior and differential scanning calorimetry of herring oil.

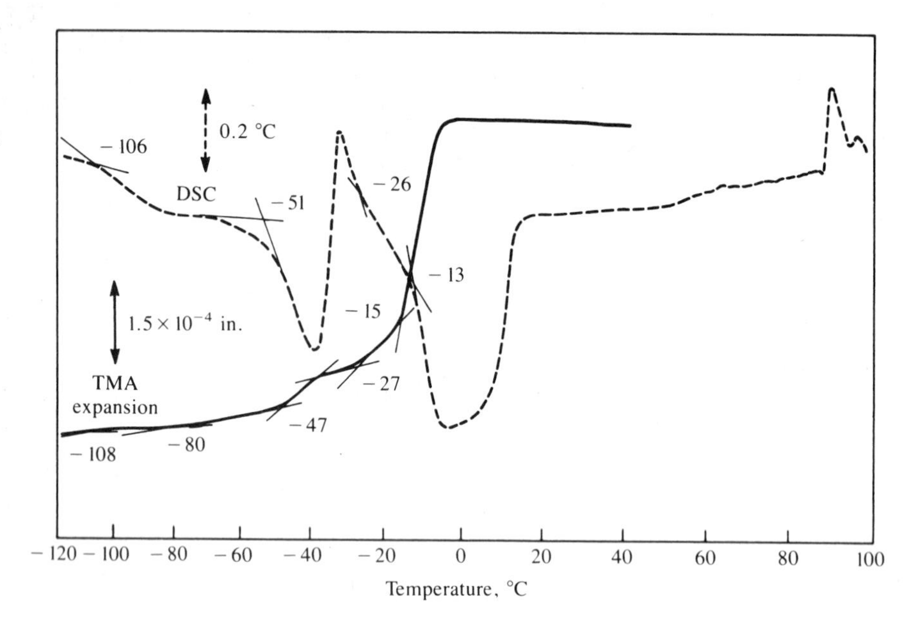

constructed of fused silica. This permits extension measurements on films and fibers. Holes about 0.6 cm apart are punched into injection-molded pieces or solution-cast or extruded films; also a fiber fused into a loop can be used for this test. The double-hook probe is designed to grasp a pair of aluminum spheres that are crimped onto either end of a fiber sample. Measurements made with these probes are related to the tensile modulus of a sample.

Volume-expansion characteristics of samples are measured by placing the sample in a quartz cylinder fitted with a flat-tipped quartz probe in a cylinder-piston arrangement. Sample volume changes are translated into linear motion of the piston.

EXAMPLE 25.5

The expansion and heat capacity behavior of herring oil are shown in Figure 25.14. The expansion characteristics show changes at -108, -80, -47, -27, and -15 °C, when the material is apparently fluid. These volume changes confirm the changes in heat capacity measured by DSC at -106, -51, -26, and -13 °C, and emphasize the need for using more than one mode of thermal analysis to illustrate the thermal response of a system.

25.6 DYNAMIC MECHANICAL ANALYSIS

Dynamic mechanical analysis is the most sensitive thermal analytical technique for detecting transitions associated with the movement of polymer chains. The technique involves measuring the resonant frequency and mechanical damping of a material forced to flex at a selected amplitude. Mechanical damping is the amount of energy dissipated by the sample as it oscillates, whereas the resonant frequency defines Young's (elastic) modulus or stiffness. The loss modulus and the ratio of loss modulus to elastic modulus are calculated from the raw frequency and/or damping data. In general, modulus and frequency, as well as damping, change more dramatically than heat capacity or thermal expansion during secondary transitions. For example, dynamic mechanical analysis is helpful in determining the effectiveness of reinforcing agents and fillers used in thermoset resins. Both DMA and TMA are used to determine physical properties important in the production of multilayer printed circuit boards.

25.7 THERMOMETRIC TITRIMETRY AND DIRECT-INJECTION ENTHALPIMETRY[7]

Thermometric enthalpy titrations (TET) and direct injection enthalpimetry (DIE) are the two main types of enthalpimetric analysis. The application of TET and DIE to the determination of an analyte is contingent upon the knowledge of two fundamental quantities: the stoichiometry of the reaction with thermometric enthalpy titrations and the heat of the reaction with DIE.

Instrumentation

The equipment consists of a reagent addition system (motor-driven, automated buret), an adiabatic reaction cell, stirring device, thermistor, Wheatstone bridge circuit, and a recorder or other output device (Figure 25.15). To minimize heat transfer between the solution and its surroundings, the titrations are performed under as near adiabatic conditions as possible in an insulated beaker or Dewar flask of 100- to 250-mL capacity that is closed with a stopper provided with holes for the buret tip, a glass stirrer, and the thermistor. The titrant is delivered at flowrates of 0.1–1.0 mL/min. To obviate volume corrections and to minimize temperature variations between the titrant and sample, the titrant concentration is usually 100 times greater than that of the reactant. Amounts of the sample are selected so that a volume of titrant not exceeding 1–3 mL is required.

Because the temperature changes during a titration range between 0.1 and 0.2 °C, the accuracy of the temperature measurement must be about 10^{-4} °C. For a thermistor that has a resistance of 2 kΩ and a sensitivity of -0.04 ohm ohm^{-1} deg^{-1} Celsius in the 25 °C temperature range, a change of 0.01 °C corresponds to an imbalance potential of 0.157 mV. Temperatures of the titrant and sample should be within 0.2 °C before a titration is begun. A small heating element, located inside the

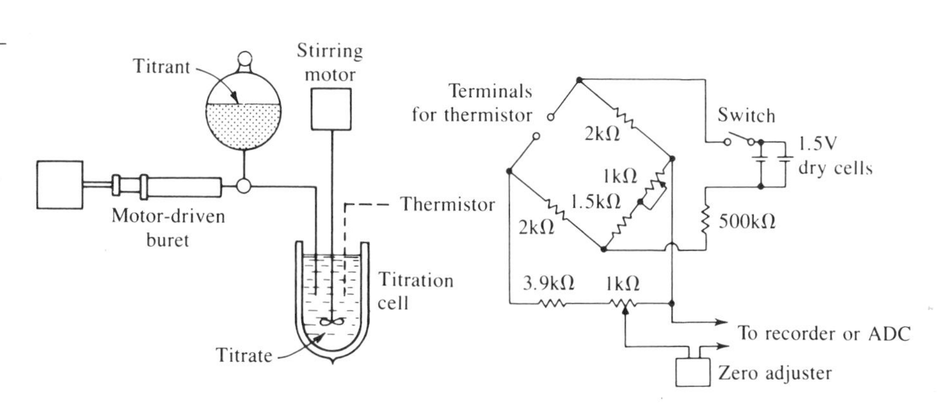

FIGURE **25.15** Schematic titration assembly and bridge circuit for conducting thermometric titrations. [After H. W. Linde, L. B. Rogers, and D. N. Hume, *Anal. Chem.*, **25**, 494 (1953). Courtesy of American Chemical Society.]

titration vessel, can be used to warm the sample to the temperature of the titrant or as a calibrating device when estimating the heats of reaction or mixing.

In a differential thermometric apparatus, temperature-sensing elements are placed in both the sample and blank (pure solvent plus titrant) solutions. The sensitivity is improved and extraneous heat effects, such as stirring and heats of dilution, are minimized.[8]

For conventional liquid-phase DIE, it is important to introduce the reagent instantaneously as a single plug. Since the injected reagent is usually in excess and the heat capacity of the system is easily determined after each injection, precise volume measurements are not required. Thus a manually operated syringe is adequate. The same adiabatic conditions required for thermometric enthalpy titrations are necessary for direct-injection enthalpimetry.

Methodology

Contrary to various types of potentiometric titrations that depend solely on the equilibrium constant, K, and therefore on the free energy of the reaction, ΔG, or

$$-\Delta G = RT \ln K \tag{25.6}$$

where R is the gas constant and T is the kelvin temperature of the reaction, thermometric titrations depend only on the enthalpy of the reaction, ΔH, or

$$\Delta H = \Delta G - T\,\Delta S \tag{25.7}$$

where ΔS is the entropy of the reaction. Thus, a thermometric titration may be feasible when all "free energy" methods fail. This point is clearly illustrated in Figure 20.16, where the thermometric titration curves for HCl and H_3BO_3 are shown. In contrast to the potentiometric curve, the thermometric titration curve has a well-defined end point for the weak acid. The change in temperature of the titration curve is dependent on the heat of reaction of the system, according to the equation

$$\Delta T = \frac{N\,\Delta H}{Q} \tag{25.8}$$

where N represents the number of moles of water formed in the neutralization, ΔH is the molar enthalpy of neutralization, and Q is the heat capacity of the system. In practice,[9] H and Q are constant throughout the reaction so that ΔT is proportional to N.

On the thermometric titration curve shown in Figure 25.16, point A occurs at the beginning of the temperature readings, and line AB is a trace of the temperature of the solution before the addition of titrant. If the line AB shows a marked slope, it is an indication of excessive heat transfer between the solution and its surroundings. At point B the addition of titrant begins; line BC shows the gradual evolution of heat of the reaction. Point C is the end point. Line CD may slope either up or down. The linear portions of the curves are extrapolated to give the initial and equivalence points, and the distance between them is measured along the volume (or time) axis of the graph to determine the volume of titrant consumed in the reaction. The vertical line BB' is the temperature difference (ΔT) used to evaluate the enthalpies (Equation 25.8).

FIGURE 25.16 (a) Potentiometric and (b) thermometric titration curves for hydrochloric and boric acids with 0.2610M sodium hydroxide.

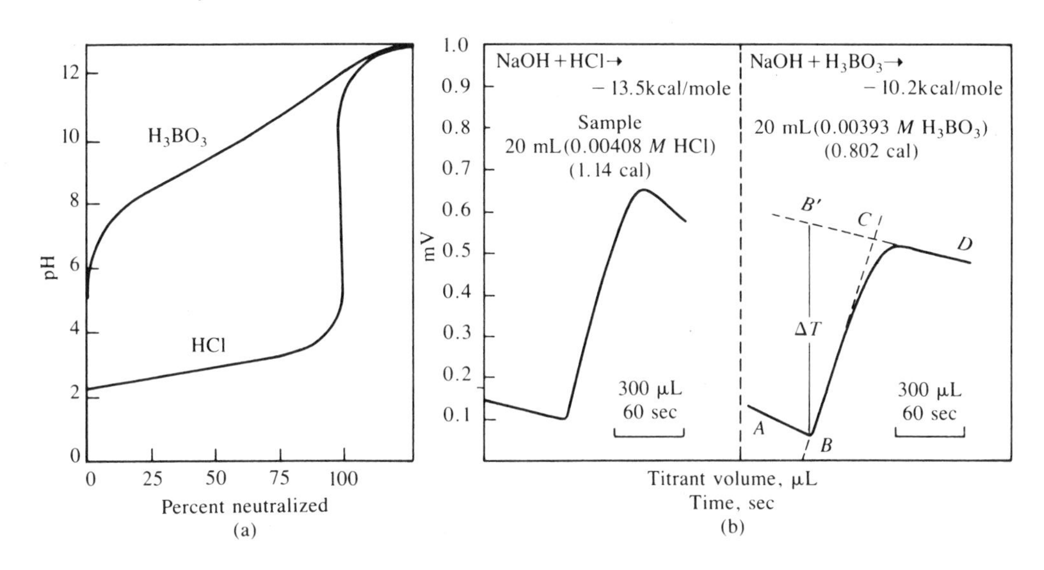

FIGURE 25.17 Direct-injection enthalpimetric (DIE) thermogram.

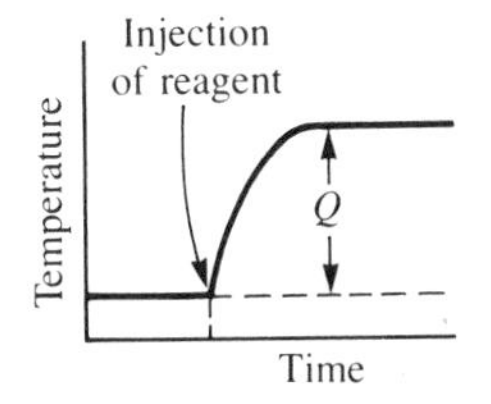

A typical direct-injection enthalpimetry thermogram is shown in Figure 25.17. By measuring the difference in the temperature of the solution before and after the injection of an appropriate reagent, this technique can determine the heat evolved or absorbed by the reaction and hence the amount of analyte present in the solution.

Applications

Applications of thermometric titrimetry include the determination of the concentration of an unknown substance, the reaction stoichiometry, and the thermodynamic quantities ΔG, ΔH, and ΔS. The first application is perhaps the most useful to the analytical chemist. Precision and accuracy of measurements depend largely on the enthalpy of the reaction involved and range from 0.2% to 2%. About 0.0001M is the lowest limit of concentration that can be successfully titrated in the more favorable cases.

All acids with $Ka \geq 10^{-10}$ can be titrated thermometrically in 0.01M solution with a precision of 1% if the heat of neutralization is 13 ± 3 kcal/mole. The extension to acids too weak to titrate potentiometrically is clearly demonstrated by the curves in Figure 25.16. Good end points are obtained for other weak acids and bases, even in emulsions and thick slurries.

Nonaqueous systems are well suited for thermometric titrations, although attention must be paid to the heat of mixing of solvents and dilution. The lower specific heat of many organic solvents introduces a favorable sensitivity factor. Under strictly anhydrous conditions, even diphenylamine, urea, acetamide, and acetanilide are readily titratable with perchloric acid in glacial acetic acid.[10] Lewis bases, such as dioxane, morpholine, pyridine, and tetrahydrofuran, have been titrated with the Lewis acid $SnCl_4$ in the solvents CCl_4, benzene, and nitrobenzene.[11]

Thermometric titrations are very useful in titrating acetic anhydride in acetic acid–sulfuric acid acetylating baths, water in concentrated acids by titration with fuming acids, and free anhydrides in fuming acids. In fact, methods based on the heats of reaction offer one of the few approaches to the analysis of concentrated solutions of these materials.[12]

Good results are obtained in precipitation and ion-combination reactions such as the halides with silver or mercury(II), and cations such as Mn(II) with EDTA and oxalate. Silver titration of halides has been done at elevated temperatures in molten salts. Enthalpimetric analysis has been used in enzyme assay and immunological determinations as well as in the analysis of alkaloid drugs such as codeine phosphate and morphine sulfate.

When the titration reaction is appreciably incomplete in the vicinity of the equivalence point, actual titration curves exhibit a curvature from which equilibrium constants and corresponding free energies can be calculated. The temperature rise that occurs during an exothermic reaction can be used to determine constituents. For example, benzene has been determined rapidly and with good precision in the presence of cyclohexane by measuring the heat of nitration when a standard nitrating acid mixture is added to the sample; the temperature rise is a direct function of the benzene present. In a similar manner, heats of reaction are used to estimate the heats of successive steps in the formation of metal-ammine complexes,[13] the heats of chelation,[14,15] and the heats of reaction in fused salts under virtually isothermal conditions.[12]

PROBLEMS

1. Formulate the solid-state reaction of sodium bicarbonate when heated. It decomposes between 100 and 225 °C with the evolution of water and carbon dioxide. The combined loss of water and carbon dioxide totaled 36.6% by weight, whereas the weight loss due to carbon dioxide alone was found to be 25.4%.

2. A definite relationship exists between the decomposition temperature of $CaCO_3$ and the equilibrium partial pressure of CO_2. A series of thermograms was obtained with a dynamic flow of CO_2 in the pressure range from 40 to 600 torr. Since pure CO_2 was used, the partial pressure is equivalent to the system pressure. Estimate the heat of dissociation from the following data:

Initial decomposition temperature (°C)	926	895	840	802	759	749
Pressure of CO_2 (torr)	600	400	200	100	50	40

3. Ascertain the glass transition of a polycarbonate resin from the following heat capacity measurements:

Temperature range (K)	Specific heat	Temperature range (K)	Specific heat
400.0–402.5	0.345	412.5–415.0	0.373
402.5–405.0	0.349	415.0–417.5	0.385
405.0–407.5	0.355	417.5–420.0	0.417
407.5–410.0	0.361	420.0–422.5	0.449
410.0–412.5	0.367		

4. The heat of fusion of a mixture of AgCl-AgBr can be used for the analysis of Cl-Br mixtures because these ions form ideal solid solutions in all proportions. ΔH_{fusion} (in cal/g) was found to be 12.1 for pure AgBr and 22.0 for pure AgCl. What weight percent of AgCl is present in a mixture that has the following values of heat of fusion: (a) 14.4, (b) 20.0, (c) 16.9, (d) 16.05, and (e) 19.6?

5. On a DSC curve obtained with 10.2 mg of dotriacontane and an external 12.1-mg standard of indium whose $\Delta H_{\text{fusion}} = 6.8$ cal/g, the following areas were obtained: chain rotation (65 °C), 158 units; fusion of dotriacontane (72 °C), 439 units; and fusion of indium, 93 units. Calculate the transition energies of dotriacontane.

6. A mixture of 95% Ar and 5% O_2 was passed through a DTA oven that was heated at 10 °C/min. A sample of 1.000 g of UO_2 registered an exotherm at 360°C with a peak area of 25.6 cm^2. When a current of 2.1 A at 3.6 V was passed for 30 sec, a calibration peak of 15.6 cm^2 was obtained. Determine the energy liberated in the reaction $3UO_2 + O_2 \rightarrow U_3O_8$.

7. In the accompanying figure curve *A* is the weight-loss thermogram from pure $CaCO_3$, curve *C* shows a similar trace from $MgCO_3$, and curve *B* is the thermogram of a limestone sample. (a) Derive an expression for the direct quantitative analysis of CaO and MgO. (b) Write equations for the solid-state decomposition of $MgCO_3$. (c) Calculate the percent CaO and MgO in the limestone sample.

PROBLEM 7

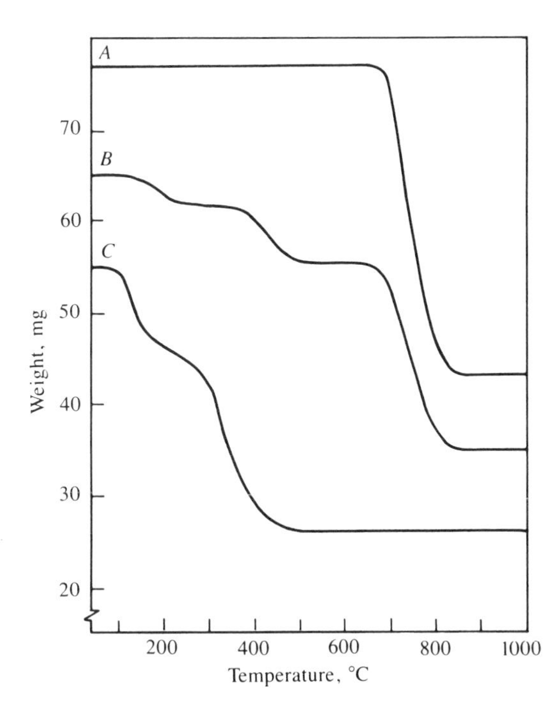

8. The decomposition reactions of a 100-mg sample of nickel oxalate dihydrate vary in different atmospheres. In both flowing and stationary air, successive weight losses of 19 and 39 mg were observed. However, in flowing CO_2 and in flowing N_2

the successive weight losses were 19 and 49 mg. The same temperature program was used in the four runs. Write the decomposition reactions.

9. In fused lithium nitrate-potassium nitrate at 158 °C, the shape of the titration curve obtained in a 8.6×10^{-4} molal solution of potassium chloride with 1.40-molal silver nitrate showed that precipitation at the equivalence point was about 20% incomplete. In contrast, precipitation of a 1.17×10^{-2} molal solution of KCl was 98.5% complete. Estimate the molal solubility product of AgCl in the eutectic salt melt.

10. State the measured quantity and describe the technique used to obtain this quantity for each of the following: (a) DTA, (b) DSC, (c) TG, (d) TMA, and (e) EGA.

11. Estimate the values of ΔH and sketch the hypothetical titration curve for a mixture of calcium and magnesium ions titrated with EDTA. Thermodynamic characteristics at 25 °C of chelation equilibria with EDTA are given here.

Cation	$pK_{stability}$	ΔS° (entropy unit)
Ca^{2+}	−11.0	+31
Mg^{2+}	− 9.1	+60

12. A simultaneous DTA-TGA curve for manganese hydrogen carbonate in a porous crucible is shown in the figure (solid lines). (a) What are the transitions involved at each peak on the DTA trace, and what are the products at each TGA plateau? (b) Another laboratory, using a controlled atmosphere with 13 atm CO_2, obtained the curves shown (dashed line). Why is the initial oxide different?

PROBLEM 12

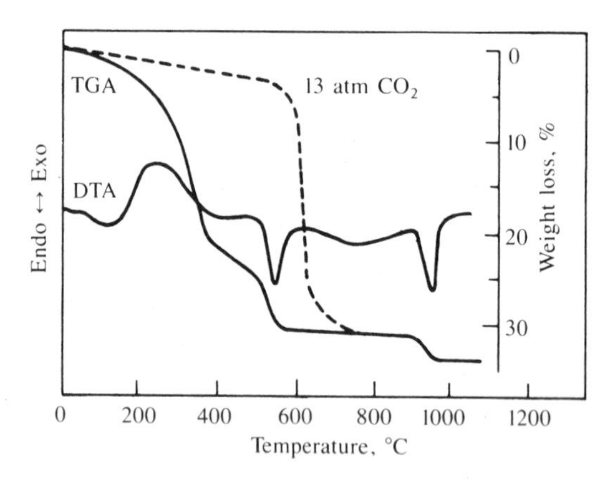

13. From the thermomechanical penetration curve shown on p. 783, deduce the nature of the two transitions.

14. Three successive runs on the same sample of a fiberglass mat impregnated with an uncured epoxy resin are shown in the figure on p. 783. The scan in run A was stopped at 90 °C, and the sample was cooled and rerun (run B). Run C is the sample from run B after cooling. Discuss the features in the DSC scans.

PROBLEM 13

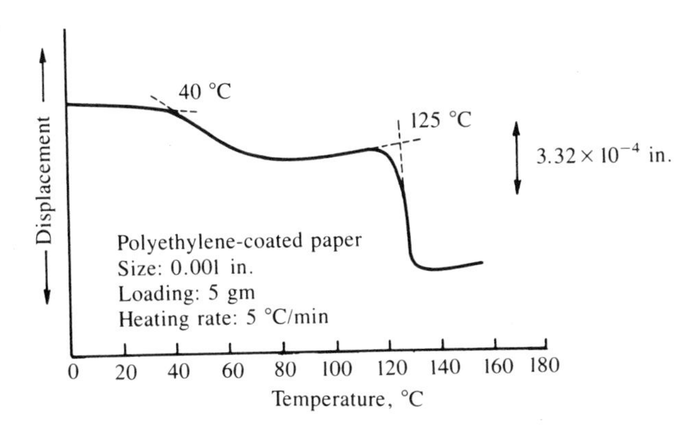

PROBLEM 14

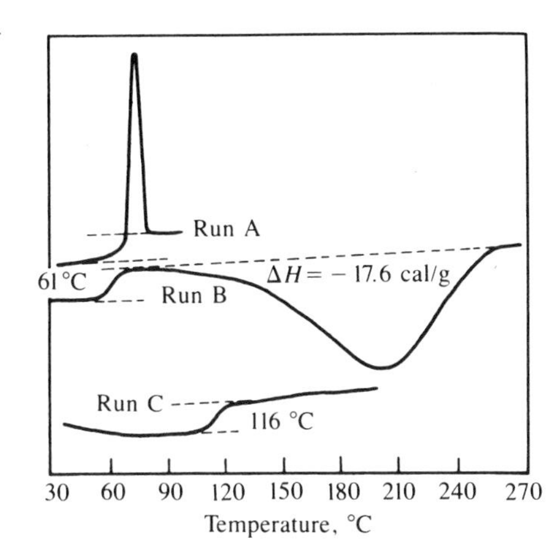

15. The figure shows the micro DTA curves of the two-phase system involving picryl chloride and hexamethylbenzene; construct the phase diagram.

PROBLEM 15

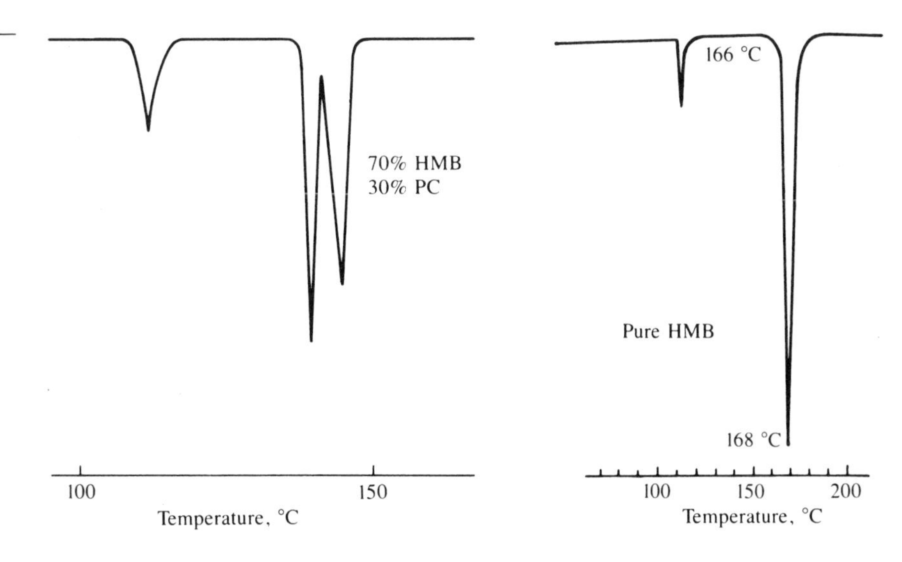

PROBLEM **15**
(continued)

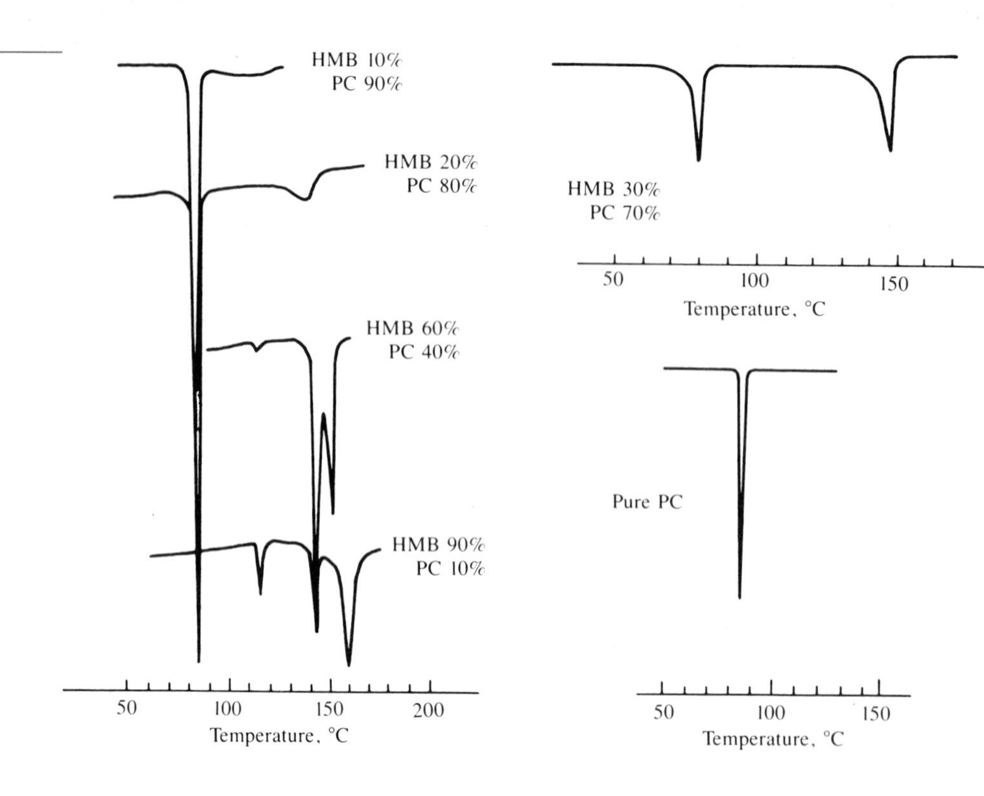

BIBLIOGRAPHY

BRENNAN, W., et al., *Am. Lab.*, **18**(1), 82 (1986).

GILL, P., *Am. Lab.*, **17**(1), 34 (1985).

KOLTHOFF, I. M., P. J. ELVING, AND C. MURPHY, eds., *Treatise on Analytical Chemistry*, 2nd ed., Part I, Vol. 12, "Thermal Methods," John Wiley, New York, 1983.

MEISEL, T., AND K. SEYBOLD, "Modern Methods of Thermal Analysis," *Crit. Rev. Anal. Chem.*, **12**, 267 (1981).

SVEHLA, G., ed., *Wilson and Wilson's Comprehensive Analytical Chemistry*, Vol. XII, "Thermal Analysis," Elsevier, New York, 1984.

VAUGHN, G., *Thermometric and Enthalpimetric Methods*, Van Nostrand, New York, 1973.

WENDLANDT, W., *Thermal Methods of Analysis*, 3rd ed., John Wiley, New York, 1986.

LITERATURE CITED

1. GILL, P., *Am. Lab.*, **17**(1), 34 (1985).
2. GILL, P., *Am. Lab.*, **16**(1), 39 (1984).
3. EARNEST, C., *Anal. Chem.*, **56**(13), 1471A (1984).
4. GOHLKE, R., AND H. LANGER, *Anal. Chem.*, **37**(10), 25A (1965); **37**, 433 (1965).
5. TURI, E., *Thermal Characterization of Polymeric Materials*, Academic, New York, 1982.
6. CHIU, J., *Anal. Chem.*, **34**, 1841 (1962).
7. JORDAN, J., *Anal. Chem.*, **48**, 427A (1976).
8. TYSON, B., W. MCCURDY, AND C. BRICKER, *Anal. Chem.*, **32**, 1640 (1961).
9. JORDAN, J., *Record of Chem. Prog.*, **19**, 193 (1958).

10. KEILY, H., AND D. HUME, *Anal. Chem.*, **36,** 543 (1964).
11. CIOFFI, F., AND S. ZENCHELSKY, *J. Phys. Chem.*, **67,** 357 (1963).
12. JORDAN, J., et al., *Anal. Chem.*, **32,** 651 (1960).
13. PAULSEN, I., AND J. BJERRUM, *Acta Chem. Scand.*, **9,** 1407 (1955).
14. JORDAN, J., AND T. ALLEMAN, *Anal. Chem.*, **29,** 9 (1957).
15. JORDAN, J., et al., *Anal. Chem.*, **31,** 1439 (1959).

26 PROCESS INSTRUMENTS AND AUTOMATED ANALYSIS

26.1 INTRODUCTION

This chapter will discuss automatic analyzers and automated instrument systems used primarily for routine analysis. The chapter is divided into two parts: on-line process analyzers and laboratory instruments. The number of methods available in each area is growing at a rapid rate, stimulated by advances in both sensor (detector) and microcomputer technologies. No attempt has been made to provide exhaustive coverage of these areas. Rather, examples have been selected to illustrate some of the important techniques used for analyses involving continuous analysis and large sample throughputs. The features of total laboratory automation using networks of instruments and computers were discussed in Chapter 4.

At the outset it is necessary to distinguish between the characteristics of automatic and automated devices. According to the current definitions of the International Union of Pure and Applied Chemistry (IUPAC),[1] both devices are designed to replace, refine, extend, or supplement human effort and facilities in the performance of a given process. The unique feature of automated devices is the feedback mechanism, which allows at least one operation associated with the device to be controlled without human intervention. For example, an automatic photometer might continuously monitor the absorbance of a given component in a process stream, generating some type of alarm if the absorbance exceeds a preset value. By contrast, an automated photometer system could transmit absorbance values to a control unit that adjusts process parameters (temperature, amount of additional reagent, and so on) to maintain the concentration of the measured component within preset limits. In spite of this fundamental difference, the terms *automatic* and *automated* are often interchanged.

Development of Process Control and Laboratory Analyzers

For the past 50 years both automatic and automated instruments have been used to monitor and control process stream environments. Nonselective properties of the process stream, such as density, viscosity, and conductivity, have long been used to monitor and regulate process conditions, such as temperature, pressure, and flowrate. Selective determination of chemical components in process streams was originally performed on grab samples, which were subsequently analyzed in control laboratories by off-line techniques with obvious shortcomings in time, economy, and human error. Increasing surveillance and control of the individual steps in industrial processes required rapid, selective analyses. In the 1950s and 1960s, a variety of continuous analyzers with the capability for selective determination of chemical composition began to emerge. The more successful of these analyzers used refractometry, gas chromatography, potentiometry, and infrared spectroscopy. These analyzers provided a dynamic rather than historic determination of the composition of starting materials, intermediates, products, and contaminants. Data from these analyzers, fed back to controllers, permitted better regulation of stream composition than did earlier nonselective methods. Most recently, process control analyzers have become the sensors for large computer-based process control systems.

The developments in process stream analyzers are paralleled by the appearance of automatic laboratory instruments such as titrators, elemental analyzers, and continuous flow analyzers. Increased demand for medical testing provided the impetus for the development of sophisticated automatic clinical analyzers.

During the 1970s progress in solid-state electronics and in computer technology greatly influenced the design of both process stream and laboratory analyzers. Instrument size was reduced, while reliability increased. These analyzers have one or more of the following capabilities: routine analysis and monitoring, acquisition, processing, and storage of data; on-line process control; and generation of data necessary for records and reports. The application of this technology to industrial process control was slower than application to laboratory analyzers for two reasons: the expense involved in replacing existing equipment and apprehension concerning the reliability of this new equipment.

Automation Strategy

Automated (or automatic) analyses are usually implemented by one of two methods, continuous or discrete (batch). In continuous flow analysis, samples are analyzed from a flowing stream with any necessary operations, such as filtering and reagent addition, performed prior to the measurement of chemical and physical properties. Actual determinations are made using flow-through sample cells. In discrete analysis, samples are placed in individual containers for the duration of the analysis. Preliminary operations of dilution, reagent addition, and mixing are performed on each sample at different locations within the analyzer. Each treated sample is then presented in sequence to the sensing device. A batch of samples is usually preloaded for processing by this technique. Although both techniques are used in process control and laboratory analyses, continuous analysis is more often utilized in automated process control applications because of its faster response time.

Not every process or analytical method lends itself to automation. Analyses involving gaseous or liquid samples have most often been successfully automated, whereas applications involving solid samples have generally proved most difficult. In fact, even though a wide variety of methods exist for automation, there will most likely always remain a hard core of complex chemical analyses that defy automation or are too costly to automate.

Figure 26.1 illustrates the importance of automated analysis in the production of high-quality drinking water and the treatment of waste water. Fast detection and correction of abnormalities are important factors in reducing treatment costs and keeping a treatment facility in compliance with regulatory agencies.

FIGURE **26.1** Typical treatment plants and monitoring points for water analysis. PC is potentiometric concentration. (Courtesy of Hach Co.)

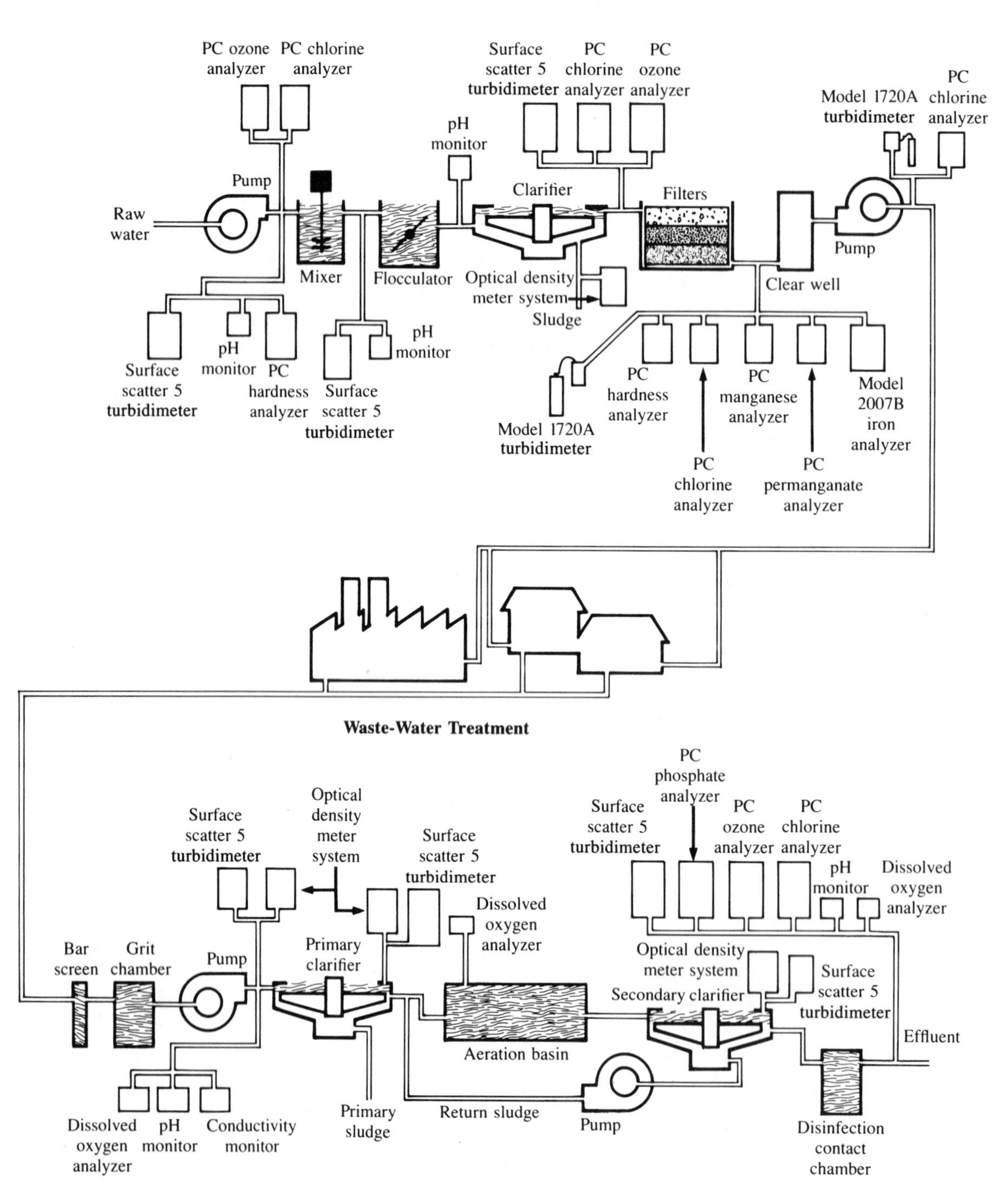

26.2 INDUSTRIAL PROCESS ANALYZERS[2]

Process control analyzers have properties that in many instances differ markedly from those of analogous laboratory instruments. Rapid analysis is essential if the results are to be useful in high-speed processes. Analyzers must be simple in design and easy to maintain. This simplification is often achieved at the expense of analytical versatility. Gas chromatography is widely used for process stream analysis partly because these analyzers are composed of simple, easily maintainable components. The ability to withstand harsh plant environments (mechanical shocks and vibrations, dust, and sometimes weather) is another important consideration.

Process control analyzers often compromise selectivity to increase sensitivity to changes in the chemical composition of the process stream. Nondispersive infrared analyzers are more sensitive toward a specific component in many situations, but interfering components cause greater problems than they do in dispersive laboratory spectrometers. Decreasing the size of the analyzers facilitates their installation and reduces cost. Electrical components are contained in explosion-proof housing for safe operation in areas where combustion hazards may occur. Reliability is of primary importance in an environment where downtime is expensive. An analyzer that costs several thousand dollars cannot be permitted to delay a process involving millions of dollars in materials, equipment, and labor.

In deciding between continuous and discrete analysis methods, rapid and more sensitive analyses favor the continuous technique, whereas increased selectivity is usually achieved by processing individual samples in batches. In addition, continuous analyzers are usually simpler to design, operate, and maintain. The features of continuous analysis are illustrated schematically in Figure 26.2. A sample is obtained at a controlled rate from the main process stream or other source of material. When necessary, provision is made for preparing the initial sample for analysis by further operations, such as the addition of reagents or filtration. The actual measurement is made by an appropriate sensor, and the results of the

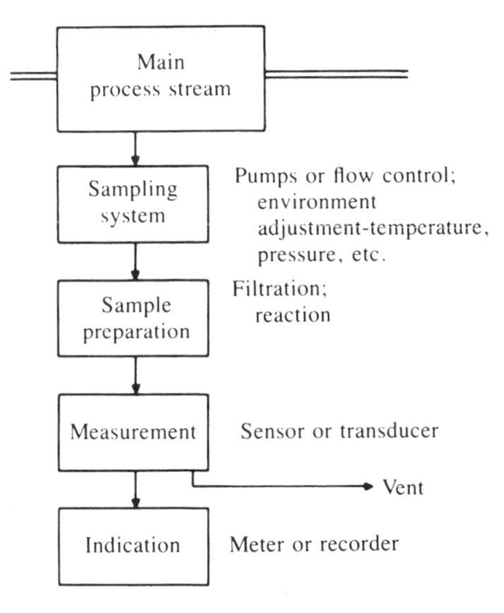

FIGURE **26.2**
Features of continuous analysis.

determination are indicated in terms of the concentration of the desired sample component. This information is directly displayed by a readout device such as a meter or a strip chart recorder, or transmitted to a computer for further processing and storage.

One class of primary sensors consists of devices that measure physical variables, such as temperature, pressure, fluid flowrate, and liquid level. The other broad category of primary sensors, often referred to as on-stream or on-line analyzers, determines chemical composition. In many cases they achieve this by measuring some physical property quantitatively related to chemical composition, such as the absorption of infrared, ultraviolet, or visible radiation, thermal or electrical conductivity, dielectric constant, paramagnetic susceptibility, density, or refractive index.

Most significant among the factors in the development of on-stream analyzers for chemical process control is the difficulty in obtaining samples and preparing them properly for analysis. Sampling often represents as much as 90% of the total analysis problem.[3] Chemical process streams may be hot and under pressure, supersaturated with certain dissolved salts, highly corrosive, laden with fibrous or particulate matter, or even full of radioactive materials. Even when acceptable samples are obtained, they may need to be cooled, filtered, or otherwise processed before analysis. All these operations must be performed on either a continuous or an automatic semicontinuous basis. Transport lag must be eliminated and dead volume reduced if events within the analyzer are to be representative of the process stream being monitored. Special bypass pumping devices are often needed to keep fresh sample rapidly supplied to the input of the analyzer. Extreme care is required in the continuous analysis of trace quantities. Leakage of a sample stream at a fitting or connection causes serious back-diffusion of the atmosphere into the stream; examples are dissolved oxygen in power plant condensate streams and water in dry gas streams.

Corrosive samples or materials under extremes of pressure or temperature restrict the latitude in the design of analyzers, especially in the area of sample cells, for various optical methods of analysis. Gas-handling components should always include a small protective filter, preceded by a major filter if the gas contains suspended matter that requires removal. If a gas sample has a water vapor concentration high enough to cause condensation within the analyzer, or if moisture is an interferent, stream drying equipment must be installed.

If several separate process streams are reasonably identical in major constituents and the concentrations of the desired components do not vary over too large a range, all streams are connected through solenoid-operated, three-way valves to a common manifold that leads to the measurement cell. Each stream is directed to the cell in a sequence determined either by hard-wired electronic control logic or by a software program resident in a computer interfaced to the control valves. This method of sampling provides rapid analysis for several streams; the exact number depends on the required sampling rate.

In summary, process control instrumentation must provide quick, reliable results when operated in a harsh environment by semiskilled technicians. Thus, in many cases, the design of these instruments is quite different from those found in the controlled environment of a general-purpose analytical laboratory.

26.3 METHODS BASED ON BULK PROPERTIES

A number of instrumental methods are used to continuously measure bulk properties of process streams. These properties include pH, density, viscosity, conductance, capacitance, and combustibility. Selected examples illustrate these types of analyses.

In a binary system, when the difference between the values of a given physical property of the components is large, the composition of the system is determined easily by monitoring this property. A calibration curve of the property's signal versus composition yields the correlation. For example, water in many organic materials is readily determined because water has a dielectric constant of 80, whereas most organic materials have values between 1 and 10. Thus, the moisture content of the paper web can be monitored during manufacture at speeds of up to 3000 ft/min, as shown in Figure 26.3. The measuring head is an electrical capacitor, which uses the paper web as a part of the measuring circuit. The dielectric constant of dry paper is about 3.

In many process streams a pseudobinary situation exists in which the components of the stream fall into two responsive groups. If the signal difference between groups is large compared with the differences among individual components of each group, a successful "group-type" analysis is possible. Group analysis for aromatics, diolefins, ketones, and aldehydes in process streams is accomplished with ultraviolet instruments that use a suitable source and filters. The utilization of conductance measurements to monitor the ionic content of water has important applications in boiler operations, cooling-tower losses, and the manufacture of pulp and paper.

Simple, rugged, continuous flow photometric analyzers are used to monitor either liquid or gaseous process streams. The dual path–dual frequency analyzer provides a high level of stability with minimum noise and maintenance (Figure 26.4). This analyzer measures either visible or ultraviolet radiation transmitted through the sample at two frequencies. The analyte intensity, M_s, is selected to be the wavelength that is most strongly absorbed by the component of interest in the

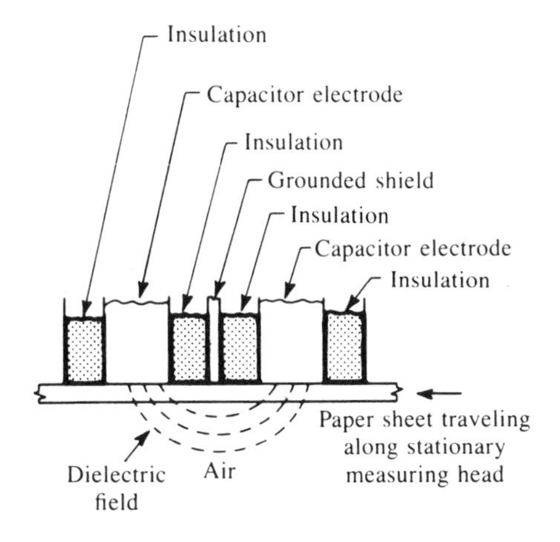

FIGURE 26.3
Schematic diagram of measuring head for control of moisture in paper web. (Courtesy of Foxboro Co.)

sample. The reference wavelength is chosen for maximum intensity, R_s, by the sample. It is used to compensate for nonspectral variations in the optical system. In addition to the sample beam, a compensation beam is used to monitor variations in the intensity of the source at both the analyte, M_c, and reference, R_c, frequencies (Figure 26.4). Thus four separate detectors produce electrical outputs M_s, R_s, M_c, and R_c, each of which is proportional to the energy focused on the detector. The concentration of the component of interest in the sample, C, is given by Beer's law:

$$C = \log_{10}\left(\frac{R_s \times M_c}{R_c \times M_s}\right) \tag{26.1}$$

Examination of Equation 26.1 confirms that neither sample cell contamination nor variation in the source intensity affects the value of C. Typical determinations include Cl_2, SO_2, or H_2S in stack gas; aromatics in hydrocarbons, dienes, polyenes, and polyacetylenes; and aromatic hydrocarbons, phenols, or inorganic (copper, cobalt, and nickel) salts in aqueous solutions.

A refractometer is the instrument of choice whenever the sample to be analyzed is a simple binary mixture. If the range of compositions is broad, analyzers that use density measurements are applicable. When the concentration range is narrow and an analysis of the liquid phase of a suspension or slurry is required, refractometry is again the method of choice. The critical-angle refractometer measures the index of refraction due to the interface between the prism and the process stream and therefore requires no penetration of the stream by the light from the source. This analyzer performs equally well on process streams that are dark or turbid and on those liquid streams that are clear and transparent. In addition, the measurement is not affected by large amounts of solid matter or gas bubbles contained in the streams. Streams that are viscous or must be held at high temperatures or pressures in order to prevent unwanted reactions are analyzed directly by the critical-angle refractometer. The probe sensing head (Figure 26.5) is easily inserted into any stream by threading it into a standard fitting in the process vessel, tank, or pipeline.

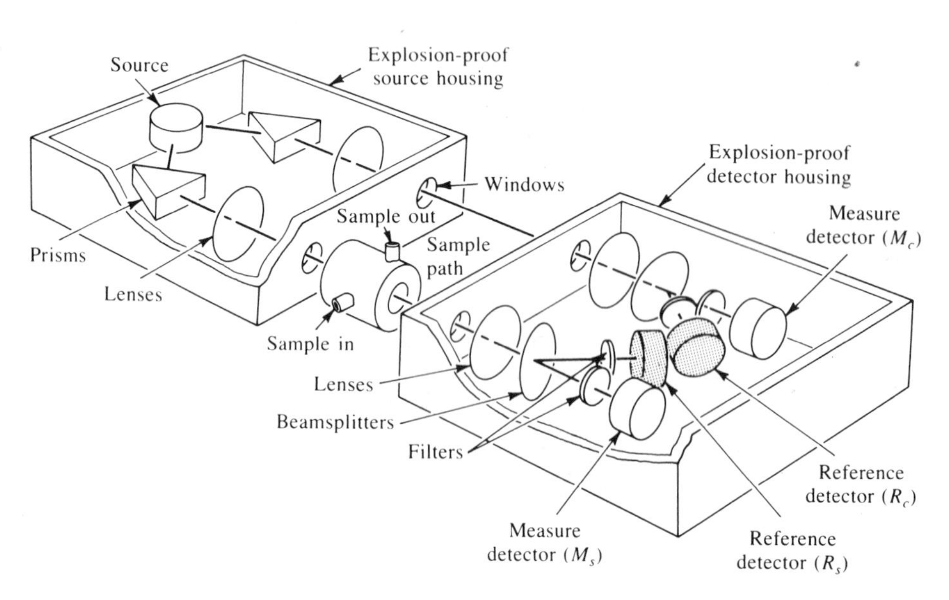

FIGURE **26.4**
Dual path–dual frequency photometric process analyzer. (Courtesy of Anacon, Inc.)

Fiber optics are used to guide the light from the source to the prism. An automatic washing jet is provided to prevent film buildup on the prism surface. A thermistor probe inserted into the stream monitors the temperature and provides the data necessary to correct for temperature changes. It may also automatically correct the refractive index for variations in temperature if linked to a computer with appropriate software.

Although gas-density detectors are used to measure concentration changes in the effluent from gas chromatography columns, they also have applications in continuous, direct monitoring of process gas streams. These detectors continuously measure the density or average molecular weight of binary or multicomponent gas streams.

At constant temperature and pressure, 1 mole of any pure, ideal gas occupies the same volume as 1 mole of any other ideal gas. Consequently, the density of an ideal gas is a direct linear function of the molecular weight of that gas. Although this is strictly true only for ideal gases, nearly all real gases behave like ideal gases at temperatures near room temperature and pressures near atmospheric. Two designs of the gas-density detector are in commercial use: the original Martin design and the Nerheim design. Both function on the same basic operational principles but differ in sensing elements, configuration, and simplicity.

The Nerheim configuration (Figure 26.6) illustrates the principle of operation: With the conduit network mounted vertically, the reference (carrier) gas enters at A, splits into two streams, and exits at D. Two flow meters, B_1, B_2, are installed, one in

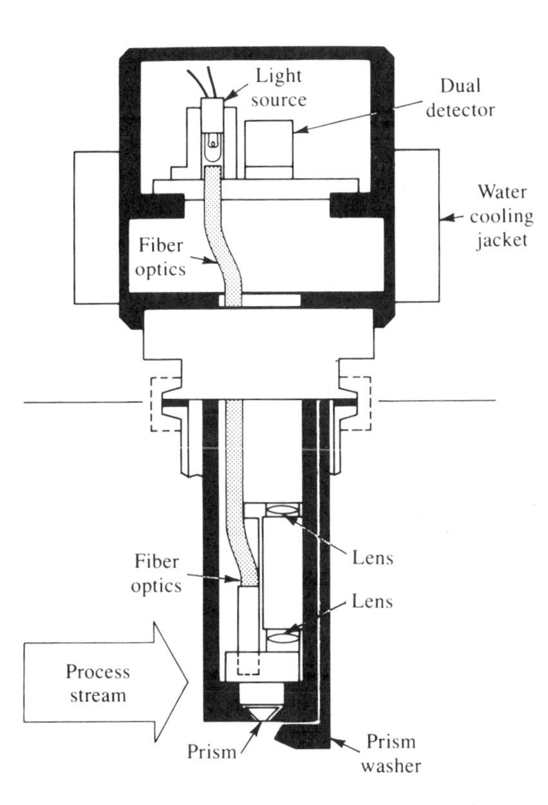

FIGURE **26.5**
Probe head for critical-angle process control refractometer. (Courtesy of Anacon, Inc.)

each stream, and are wired into a Wheatstone bridge. When the flow is balanced, the detector elements, which form a matched pair, are equally cooled and the bridge is balanced, thus giving a baseline (zero) trace. The detector elements are either hot wires or thermistors, depending on the desired operating temperature. These are connected via an electrical bridge to a recording potentiometer. The sample gas (or effluent from a chromatographic column) enters at C, splits into two streams, mixes with the reference gas in the horizontal conduits, and exits at D. The sample gas never comes into contact with the detector element, thus avoiding problems caused by corrosion or carbonization.

If the sample gas has the same density as the reference, there will be no unbalance of reference streams or of detector elements. When the sample gas carries transient trace impurities that are heavier than the reference gas, the density of the heavier molecules causes a net downward flow, partially obstructing the flow A-B_2-D, with a temperature rise of element B_2, and permitting a corresponding increase in the flow A-B_2-D, with a temperature decrease of element B_1. In a similar manner, lighter molecules cause a net upward flow and the reverse is true—namely, a temperature rise of element B_1 and a decrease of element B_2, with a signal of opposite polarity from the first case. Bridge unbalance is linear over a broad range and directly related to the gas-density difference. Calibration for individual components is eliminated because the response depends on a predictable relationship, the difference in molecular weights of component and reference gas:

$$\Delta p = k \frac{n_s(M_s - M_r)P}{RT} \tag{26.2}$$

where k is a constant whose value depends on cell geometry, viscosity, the flow-measuring system, and the thermal conductivity of the gases; M_s and M_r are the molecular weights of the sample and reference gases; and n_s is the mole fraction of the component in the sample.

The sensitivity of the gas-density detector with the thermistor sensor is comparable to that of a thermistor-type, thermal-conductivity cell and requires no amplification at room temperature. The hot-wire sensor, which has one-sixth the sensitivity of thermistors at 25 °C, may require low-level amplification at 100 °C. The low noise level of the detectors permits effective use to 300 °C, although the

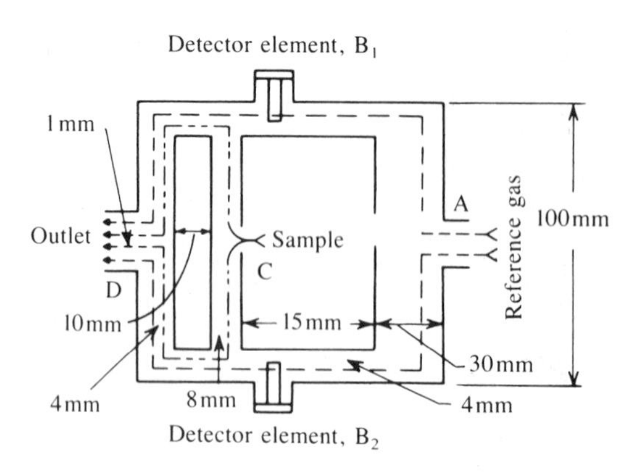

FIGURE **26.6**
Nerheim gas-density balance. Schematic view as mounted in a vertical plane. (Courtesy of Gow-Mac Instrument Co.)

sensitivity decreases rapidly with increasing temperature for thermistors, being at 250 °C one-fifth that at 50 °C. The Nerheim design has an effective sample volume of approximately 5 mL.

26.4 INFRARED PROCESS ANALYZERS[4]

Infrared instruments are widely used in process stream analyses. Infrared analyzers fall into three categories: dispersive, nondispersive, and bandpass optical filter.

A dispersive infrared instrument is patterned after a double-beam spectrometer (see Chapter 11). Radiation at two *fixed* wavelengths passes through a cell containing the process stream to provide a continuous measurement of the absorption ratio. The two wavelengths are chosen so that the component of interest in the process sample stream absorbs at one wavelength. At the other wavelength, the sample either does not absorb or exhibits a small, constant absorption. The ratio of absorbance readings is converted electronically to read the concentration of the component of interest and recorded. This type of instrument handles liquid systems as well as gas streams, and has the ability to analyze quite complex mixtures. Tunable laser diodes, fabricated from lead-salt semiconductors, are sometimes used as monochromatic infrared sources for trace gas monitoring. The detector may be the radiation thermocouple used in laboratory instruments. Dispersion infrared analyzers are used in process applications when the necessary selectivity cannot be achieved by nondispersive methods, in cases involving liquid process streams, and in analyses at wavelengths longer than 10 μm (less than 1000 cm^{-1}).

The popular method for the selective analysis of gaseous process streams is nondispersive infrared absorption. These rugged, reliable nondispersive analyzers are commonly used as sensing components of automated process-control loops. Since these instruments do not require monochromatic radiation, their design and operation are simpler than those of dispersive instruments. Excellent sensitivity is achieved due to the strong signal power resulting from all of the radiation from the infrared source passing through the sample.

In practice, the nondispersive analyzer is really a filter photometer with the filter usually composed of the vapors of the component of interest. Interference filters constructed of nongaseous materials are also used successfully in the design of these compact analyzers. An analyzer that contains an interference filter has the selectivity approaching that of a dispersive instrument while maintaining the simplicity of a colorimeter.

In the positive-filter type of analyzer, the radiation from two matched energy sources is passed through a chopper that blocks both sources simultaneously with equal on-off periods (Figure 26.7). This is done to minimize variations in the signal due to fluctuations in the ambient temperature and source intensity. One beam of radiation then passes through the sample cell, which is a highly polished tube that conducts radiant energy by multiple reflection. The gaseous sample from the process stream flows continuously through this cell. The other beam of radiation passes through a similar cell that contains a nonabsorbing reference gas such as nitrogen. Both beams of radiation then pass into the detector, a sealed compartment filled with the pure form of the gas being analyzed. A nonabsorbing pressurizing gas may be added to the detector to optimize its sensitivity. The detector is divided into halves separated by a flexible diaphragm. When some of the component of interest appears in the sample stream, the sample side of the detector receives less radiant energy and the diaphragm expands outward from the reference side into the sample side. The diaphragm is positioned so that it forms a capacitor plate in an electrical circuit. Thus the difference in radiant energy received by the two compartments of the detector is transformed into an electrical signal, which is proportional to the concentration of the component of interest in the sample stream.

If, for example, the detector of this nondispersive analyzer is filled with carbon dioxide, the carbon dioxide absorbs infrared radiation at its characteristic wavelengths, principally 4.2 μm (Figure 26.7). If carbon monoxide is present in the sample stream, it does not interfere with the analysis of carbon dioxide because its absorption bands do not overlap those of carbon dioxide. Thus, an analyzer with a detector filled with carbon dioxide detects primarily carbon dioxide and is not affected by the presence of other components, such as carbon monoxide, unless these components are present in high concentrations or the component absorption bands overlap those of carbon dioxide. Where interference is a problem, standard optical filters, gas cell filters, and/or window material can be placed in both radiation beams to eliminate or reduce the interference to an insignificant level. While these filtering techniques increase the selectivity of the analysis, they also reduce the sensitivity of the analyzer toward the component of interest.

Although this type of nondispersive instrument is both sensitive and selective, it suffers from a major limitation: a separate analyzer is required to monitor each component of interest. In many applications it is desirable to monitor simultaneously several components of a process stream or of ambient air. To perform these analyses at increased sensitivities and with a minimum of instrumentation, the design of the nondispersive photometer is modified. The pneumatic detector is

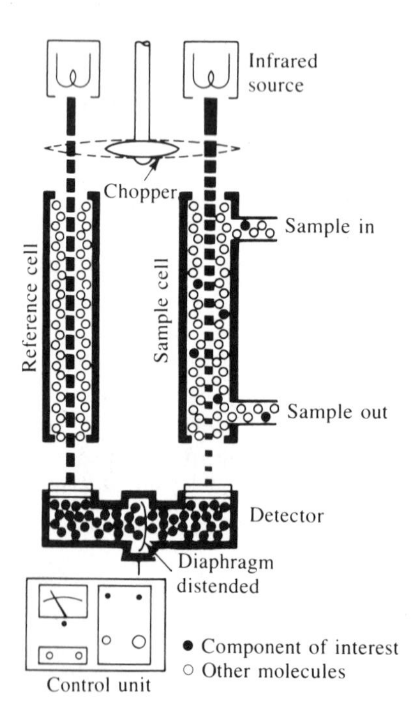

FIGURE **26.7**
Positive-filter nondispersive process-stream infrared analyzer. (Courtesy of Beckman Instrument Co.)

replaced by a more sensitive broadband, solid-state detector. The functions of the chopper, gas-filled reference cell, and the component in the detector compartments are combined in a filter wheel. This wheel consists of pairs of gas-filled filters, alternately positioned in the single optical beam of the infrared source (Figure 26.8). One cell is filled with a pure sample of the component of interest, the next with nitrogen. When a filter that contains a given component of interest is in the optical path, the amount of this component present in the sample cell has little effect on the detector, since the component gas in the filter has already absorbed most of the radiation. This filter is called the reference cell. When the filter cell that contains the nitrogen is rotated into the optical beam, little energy is absorbed by the filter cell (known as the sensitive cell), and thus the amount of absorption depends on the concentration of the component in the flow-through sample cell. If e_r is the detector output signal when the reference filter is placed in the optical path and e_s is the output when the filter that contains the component of interest is in the optical path, then the ratio e_r/e_s is proportional to the concentration of the component of interest.

The first electronic circuitry developed for this type of nondispersive analyzer was entirely analog. Current instruments use digital circuitry with microcomputer control. Replacement of analog hardware with microcomputer software leads to easier maintenance, increased reliability, and increased flexibility in both instrument design and operation (see Chapter 4).

The versatile, portable Wilks miniature infrared analyzer (MIRAN) produced by Foxboro Analytical replaces the pairs of gas-filled cells with a variable filter wheel. A single component is monitored by setting the wheel at the appropriate wavelength, or the analyzer can scan through discrete portions of the spectrum to measure the concentrations of several different sample components. A number of different sampling accessories are available, including variable-path gas cells (Figure 26.9), liquid flow-through multiple internal reflection (MIR) cells, and an automatic film sample handler. More sophisticated models perform multicomponent analyses using a microprocessor to control analytical parameters, perform data reduction, and generate reports. The MIRAN analyzer is also the heart of a

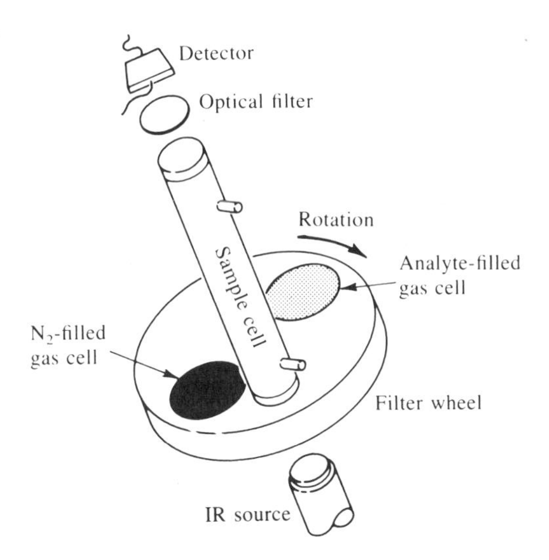

FIGURE **26.8**
Filter wheel nondispersive infrared (NDIR). (After R. J. Bibbero, *Microprocessor in Instruments and Control*, p. 180, John Wiley, New York, 1977.)

multipoint, multicomponent environmental air-monitoring system that can analyze samples in up to 24 locations for as many as ten components. The system produces data in a form acceptable to the Occupational Safety and Health Agency (OSHA) and the Environmental Protection Agency (EPA).

Components frequently determined by infrared analyzers in industrial processes and in ambient air-monitoring stations are listed in Table 26.1; these include vapors of substances that are liquids at room temperatures. For example, using the measurement of carbon dioxide in industrial processes, there are many applications, such as the control of combustion, the manufacture of cement, and the production of ethylene oxide, phthalic anhydride, and ammonia. By contrast, these infrared instruments cannot selectively measure similar substances, such as butane in propane.

Fourier transform infrared (FTIR) instruments are used for continuous real-time monitoring of solids, liquids, and gaseous manufacturing processes. The process parameters monitored include chemical composition, degree of cure (polymers), impurity levels, and film thickness. Process FTIR instrumentation combines the FTIR optical system (see Chapter 8), high-speed array data processing data systems (see Chapter 2), and chemometric process modeling software (see Chapter 2). During on-line operation, the FTIR analyzer automatically performs measuring procedures established by the modeling process. These include spectral acquisition, data point selection and scaling, composition of output, and triggering of any necessary alarms. The speed of the FTIR technique permits complete measurements to be made as rapidly as once a second.

Another variation of infrared spectroscopy, near-infrared reflectance analysis (NIRA), has found wide application in the routine analysis of highly absorbing materials such as coal, grains, and some pharmaceuticals (Figure 26.10). The reflected radiation from the sample surface at a series of wavelengths is followed by measurements of reflectance of a standard reference surface at the same wavelengths. Thus the reflectance from the sample is reported relative to the standard reflector. In this method, little radiation is absorbed and the angle of reflectance is equal to the angle of incidence. The incident radiation penetrates the surface of a sample a small distance and can transfer vibrational energy to the bonds of the molecules in the sample. The energy is transferred when the frequency of the incident radiation is the same as the frequency (fundamental or overtone) of the chemical bond. A series of samples that contain known amounts of a given analyte are scanned to find a correlation between the analyte concentration and the absorption of NIR radiation. With a mathematical correlation transform, an equation can be developed to determine the concentrations of analyte. This method requires little, if any, sample preparation and can handle both liquid and solid samples.

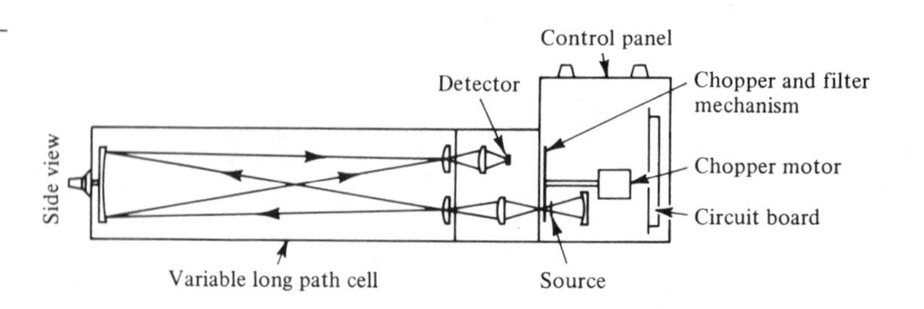

FIGURE **26.9**
Single-beam infrared analyzer with variable-path gas cell. (Courtesy of Foxboro Analytical Co.)

TABLE **26.1**

APPLICATIONS OF INFRARED SPECTROSCOPY TO PROCESS STREAMS

Gas	Analytical wavelength, μm	Minimum detectable concentration (ppm) 20-m path	Maximum concentration 1.0 absorbance (ppm or %) 0.75-m path
Cyclohexane	3.4	0.04	4000
CO_2	4.25	0.08	7500
N_2O	4.5	0.03	2100
CO	4.65	1.2	8.3%
COS	4.85	0.02	1100
NO	5.3	1.5	8.2%
CH_3COCH_3	5.75	0.1	4000
SO_2	7.4	0.1	3000
Vinyl acetate	8.2	0.06	1000
1–4 Dioxane	8.8	0.2	2100
CH_3CH_2OH	9.4	0.4	5000
NH_3	10.75	2.2	2.1%
Freon 11	11.8	0.06	300
CCl_4	12.6	0.05	250
CH_2Cl_2	13.3	0.4	1900

SOURCE: Courtesy of Wilks Scientific Corp.
NOTE: Data for MIRAN (miniature infrared analyzers) dispersive gas analyzer with variable path gas cell (0.75 to 20 m).

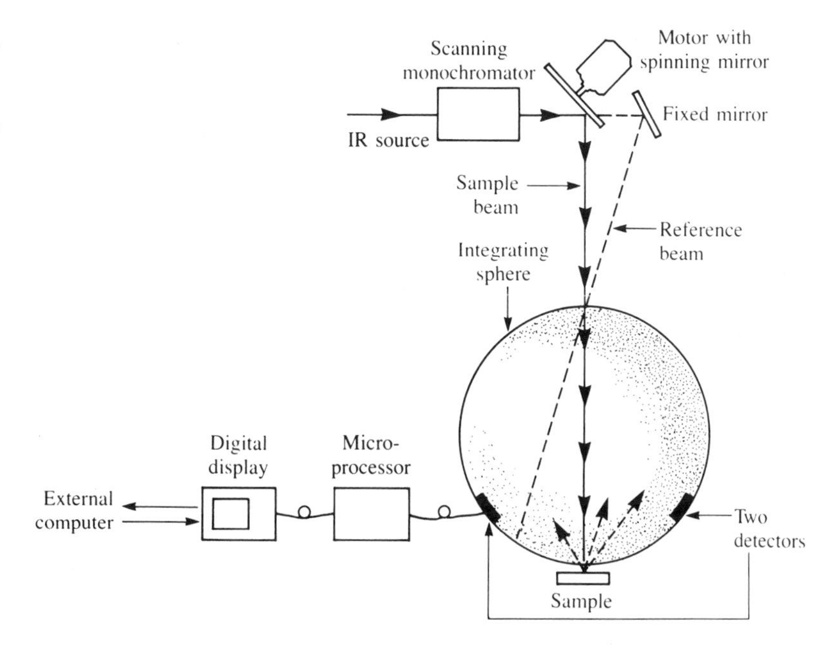

FIGURE **26.10**
Instrumentation for near-infrared reflectance analysis. (Courtesy of Technicon Corp.)

Specific applications include the moisture content of coal, cosmetics and detergent powders, the protein content of cereals and grain, and hydrogenation of unsaturated fats and oils.

26.5 OXYGEN ANALYZERS

The determination of oxygen is important in monitoring processes as varied as respiration (blood gas) and combustion (industrial flue gas). Energy conservation and environmental emission standards have forced fuel-consuming industries to improve combustion performance. The oxygen content in flue gas is a good indicator of combustion efficiency.

Methods used to measure oxygen are classified as either physical or chemical. Physical methods use the paramagnetic property of oxygen or thermal conductivity as the basis for quantitative determinations. Chemical methods include potentiometry and catalytic combustion. The choice of a particular method is determined by whether the measurement is for oxygen in gas samples or dissolved oxygen in liquids. The presence of interfering substances and the required limits of detection also are important factors in selecting a method. Several of these methods are briefly described in the following sections.

Magnetic Susceptibility

Oxygen, nitric oxide, and nitrogen dioxide are unique among the ordinary gases in that they are paramagnetic; that is, they are attracted into a magnetic field. Most gases are slightly diamagnetic and repelled out of a magnetic field. Oxygen is several times more paramagnetic than nitric oxide or nitrogen dioxide. The values of the volume susceptibilities are 146.6, 65.2, and 4.3×10^{-9}, respectively. Advantage is taken of this property of oxygen in gaseous oxygen analyzers.

The instrument shown in Figure 26.11 uses the magnetic properties of oxygen along with thermal conductivity for the measurement of oxygen in a gas. The gas sample is passed across the bottom of a gas cell that contains an electrically heated

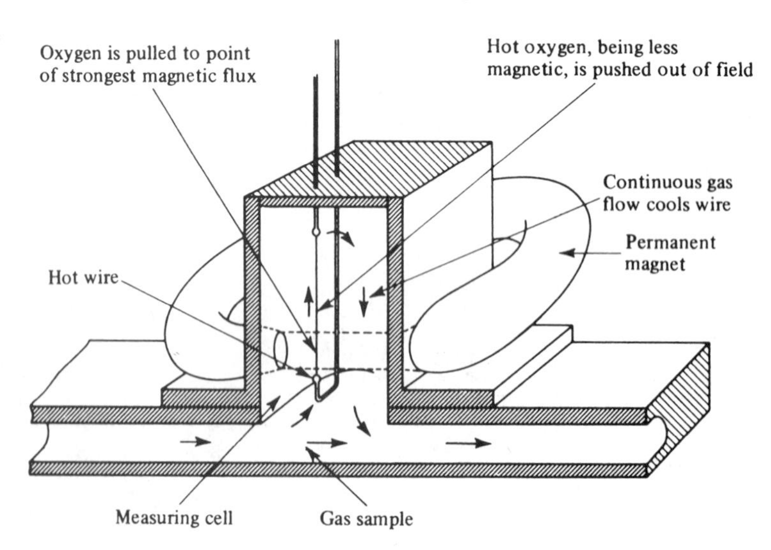

FIGURE **26.11**
Schematic of Hays Magno-Therm oxygen recorder: schematic operation of analyzing cell. (Courtesy of Hays Corp.) .

wire. A strong magnetic flux from a permanent magnet is directed across the wire. The oxygen is pulled into the region around the hot wire by the magnetic flux and is heated by the wire. Oxygen loses its magnetic susceptibility in inverse proportion to the square of the absolute temperature, and therefore the heated, relatively nonparamagnetic gas is continually displaced by the cooler, more paramagnetic oxygen moving in from below. A flow of gas proportional to the amount of oxygen present is set up around the hot wire. The hot wire is cooled, and its resistance is thereby decreased. The resistance of the wire in the analysis cell is compared with the resistance of a similar wire in a comparison cell by means of a Wheatstone bridge circuit. The comparison cell contains the sample of gas as does the measuring cell, but it does not have a magnetic flux around the wire. Thus all variables are canceled except the cooling due to the oxygen present. The zero setting of the instrument is checked by swinging the magnet away from the measuring cell without interrupting the gas flow. The overall accuracy of the instrument is claimed to be 0.25% oxygen (up to 20%) and 2.5% of the range up to 100%.

Electroanalytical Methods

The Hersch galvanic cell provides an electrolytic method for the determination of low concentrations of oxygen in gas streams. The gas stream passes through an electrolytic cell, which consists of a silver cathode and an anode of active lead or cadmium. The electrodes are separated by a porous tube saturated with an electrolyte of potassium hydroxide. Oxygen in the gas sample is reduced to hydroxyl ions at the silver cathode. The metallic lead or cadmium in turn is oxidized to lead(II) ions or cadmium hydroxide. The magnitude of the cell current is a measure of the oxygen in the sample; a sensitivity of 1 ppm is attainable. Acidic substances are removed in advance by scrubbing the gas stream with an alkali hydroxide. Calibration is achieved by periodically generating known amounts of oxygen in a separate electrolysis cell.

Oxygen in flue gases is monitored by probe-type analyzers that are inserted directly into high-temperature flue streams. Zirconium(IV) oxide (ZrO_2), stabilized by traces of materials such as CaO or Yb_2O_3, has a crystal structure that contains cation vacancies or holes. At temperatures of around 1500 °C, oxygen anions tend to migrate into the holes, causing the vacancy (hole) to be displaced. If platinum electrodes are placed on opposite sides of the ZrO_2 lattice and the lattice is maintained at a high temperature, the voltage difference between the electrodes indicates the difference in partial pressure of oxygen between the two sides (Figure 26.12). If the oxygen partial pressure is controlled on the reference side, a millivolt signal proportional to the difference between the oxygen pressures is generated. At 816 °C the magnitude of the signal is given by the Nernst equation:

$$E = 0.054 \log_{10}\left(\frac{P_r}{P_s}\right) \tag{26.3}$$

where P_r and P_s are the partial pressures of oxygen on the reference and sample sides, respectively. An output voltage of 54 mV/decade of oxygen pressure difference is obtained at 816 °C.

This oxygen sensor is operated at or above flue gas temperatures and provides a strong output signal specific to the oxygen content of the gas. No sample preparation is required. The sensor is rugged enough for service in the severe environment of a hot flue and can be repaired easily in the field. Sensor calibration is performed without removing the probe from the flue stream. The temperature of the sensor must be controlled to minimize changes and gradients.

A unique method of sampling liquid streams for the measurement of dissolved oxygen is shown in Figure 26.13. A sample is induced to flow past the oxygen

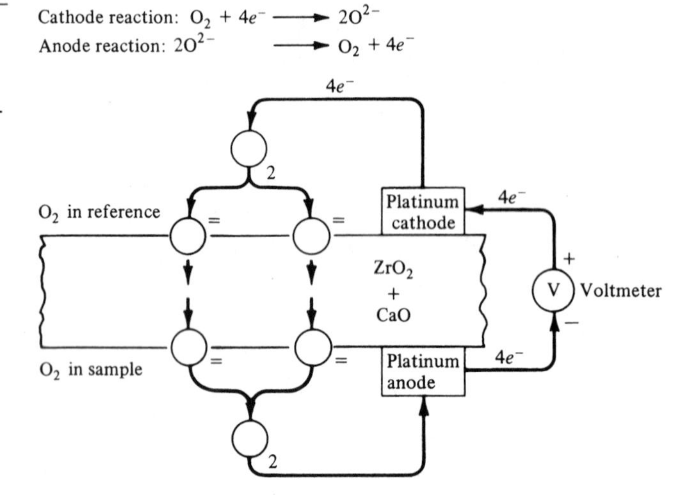

FIGURE **26.12**
Zirconia oxygen analyzer. (Courtesy of Milton Roy Co., Hayes-Republic Division.)

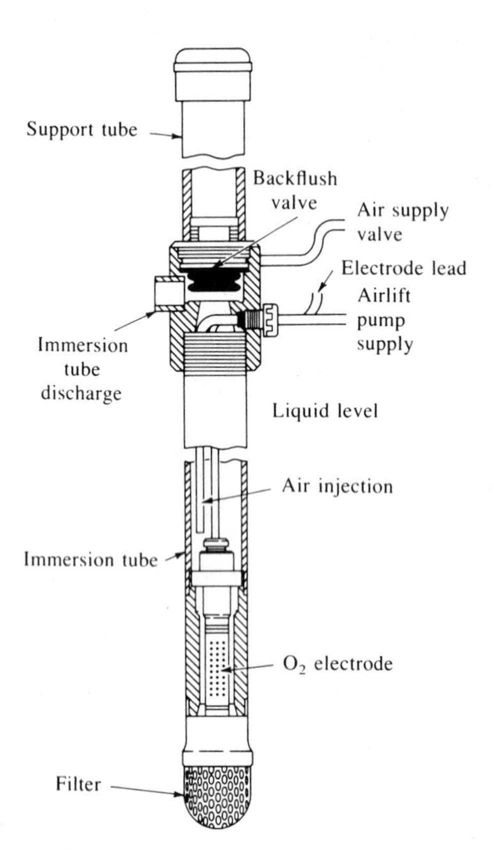

FIGURE **26.13**
Dissolved oxygen electrode system. (Courtesy of Ionics, Inc.)

electrode by air injected into the immersion tube above the electrode. The resultant air/liquid mixture in the immersion tube is less dense than that of the sample stream. Thus the incoming liquid from the stream forces the liquid already in the immersion tube up the tube to a point of hydrostatic equilibrium. Since the immersion tube outlet is below the point of hydrostatic equilibrium, the incoming sample flow is maintained as long as air injection is continued. The sample stream is drawn through a filter and past the oxygen electrode to continuously present it with fresh sample. The injected air does not affect the measurement because it is injected above the sensing point.

If the filter becomes clogged, a backflushing method is available. The immersion tube discharge is closed. If the air continues to flow into the immersion tube, it displaces the liquid, which must flow back through the filter, dislodging trapped debris. After all the liquid has been flushed out of the immersion tube, the electrode is surrounded by damp air, which provides a 100% saturated sample for calibration purposes.

26.6 ON-LINE POTENTIOMETRIC ANALYZERS

Potentiometric, single-component analyzers that give continuous, real-time analyses are available for a variety of inorganic ions and dissolved gases. These systems use rugged specific ion electrodes to obtain laboratory accuracy in the environment of an industrial plant. The design of these analyzers is simple and straightforward (Figure 26.14). In this system the sample is first filtered to prevent clogging of the

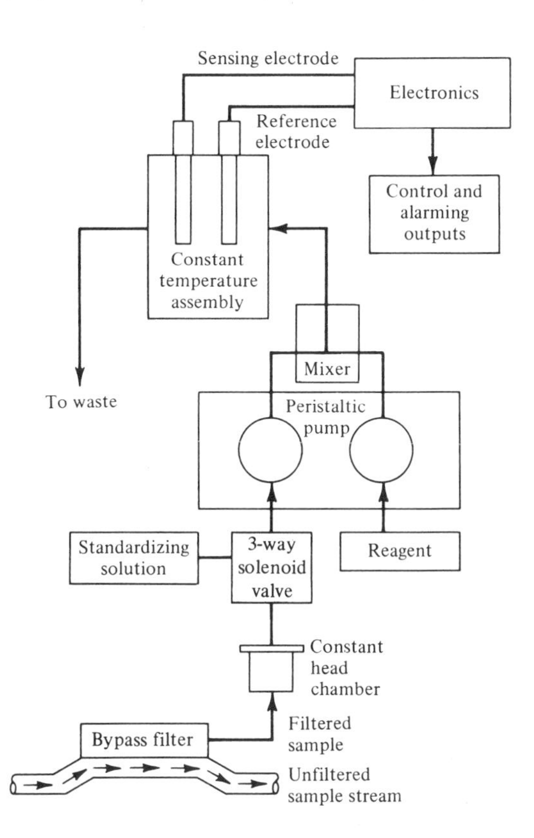

FIGURE **26.14**
On-line potentiometric analyzer. (Courtesy of Orion Industrial.)

electrode surfaces and then mixed with a reagent to minimize interferences. A unique bypass filter uses the flow of the sample stream to keep the filter surface clean. After mixing, the sample/reagent solution is pumped through a thermostatted electrode assembly for measurement. The sample's color and turbidity do not affect accuracy.

The electrode signals are transmitted to the electronic unit where their voltage difference is amplified. Both voltage and current outputs are provided for remote reading and control. A panel meter gives continuous readings, while a strip chart recorder gives a permanent record of the sample stream concentration. The results are utilized for a local alarm process or sent to a host computer for further analysis and process control. Applications include quality control measurements of chloride ion in the manufacture of ascorbic acid, monitoring cyanide in effluent from electroplating operations, and control of water fluoridation.

26.7 CHEMICAL SENSORS

Advances in fiber optics and microelectronics have led to the development of two types of small, relatively inexpensive chemical sensors that are well suited for automated analysis and process control applications: optodes and microsensors. An optode is a device that consists of a chemical reagent phase positioned at the end of a fiber optic. The analyte interacts with the reagent phase to produce changes in the optical properties of the reagent that are probed and detected through the fiber optic (Figure 26.15). Optical properties such as absorbance, reflectance, and luminescence can be measured. For example, a pH sensor consisting of an immobilized dye whose color or fluorescence varies with pH has been developed.[5]

Some advantages of optodes over electrodes are no electrical interference, no "reference electrode" required, easy replacement or substitution of reagent phase, and the use of a single spectrometer with several optodes. Optode limitations include interference due to ambient light, lack of long-term stability of reagent phases, slow response times, dependence of optical intensities on the amount of reagent phase, and a limited dynamic concentration range compared with electrodes.

The second type of chemical sensors, known as microsensors, are microfabricated devices that produce reproducible electronic signals as a result of a chemical stimulus.[6,7] These devices can be divided into two categories. One includes devices that experience the internal transport of electrical charge as a result of interaction with an analyte. The second group includes all other microsensors that produce electrical signals as a result of a variety of phenomena such as pyroelectric enthalpimetry, acoustic wave propagation, and potentiometric gas sensing.

Two components are essential for the operation of a chemical microsensor: a microfabricated probe and a chemically selective coating. Because the coating must be in physical contact with the medium that contains the analyte, the performance of the sensor is determined by the coating chemistry.

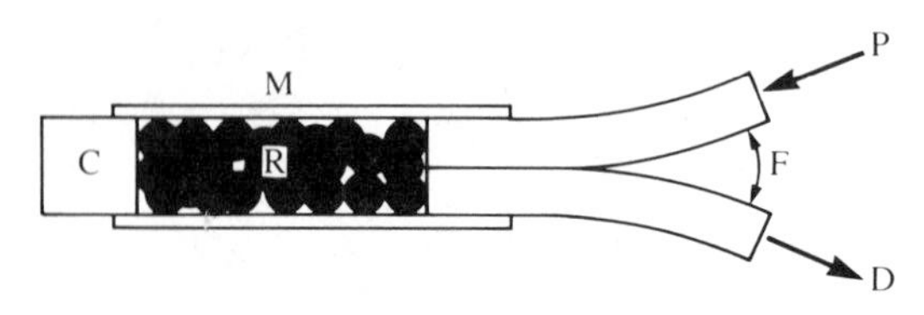

FIGURE **26.15**
A pH sensor based on absorbance. P = probe radiation, F = optical fibers, R = reagent phase immobilized on polyacrylamide spheres, M = cellulose membrane, and D = detected radiation. The reagent phase includes polystyrene spheres to redirect incident probe radiation. A filter wheel is used to sequentially determine transmitted light at two wavelengths, one where the base form of reagent absorbs and the other where no absorption occurs. The cap, C, serves to confine the reagent and prevent incident radiation from entering the sample. The same arrangement is also used for an oxygen sensor based on fluorescence quenching. [After R. Seitz, *Anal. Chem.*, **56**(1), 20A (1984). With permission.]

The CHEMFET is an example of the first class of microsensors (Figure 26.16). In this device, the metallic gate of a MOSFET (see Figure 3.5) is replaced with a chemical coating and a reference electrode. If, for example, the coating is palladium, the CHEMFET becomes a sensitive detector for hydrogen gas in concentrations less than 1 ppm. Other gases such as CO, NH_3, and H_2S can be detected with CHEMFETs. A major problem is the difficulty of encapsulating CHEMFETs so that moisture and contaminants do not introduce instabilities in their performance.

A successful application of chemical sensors is the miniature gas chromatography system shown in Figure 26.17.[8] Many of the components are fabricated

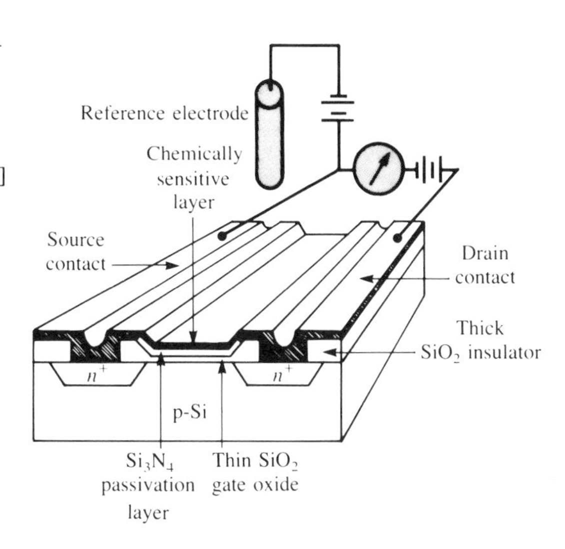

FIGURE **26.16**
A CHEMFET sensor. [After H. Wholtjen, *Anal. Chem.*, **56,** 90A (1984). With permission.]

FIGURE **26.17**
A miniature gas chromatograph. (a) 3 silicon GC wafer. (b) Schematic representation of the micro GC. [After S. Saadat and S. Terry, *Am Lab.*, **16**(5), 90 (1984). With permission.]

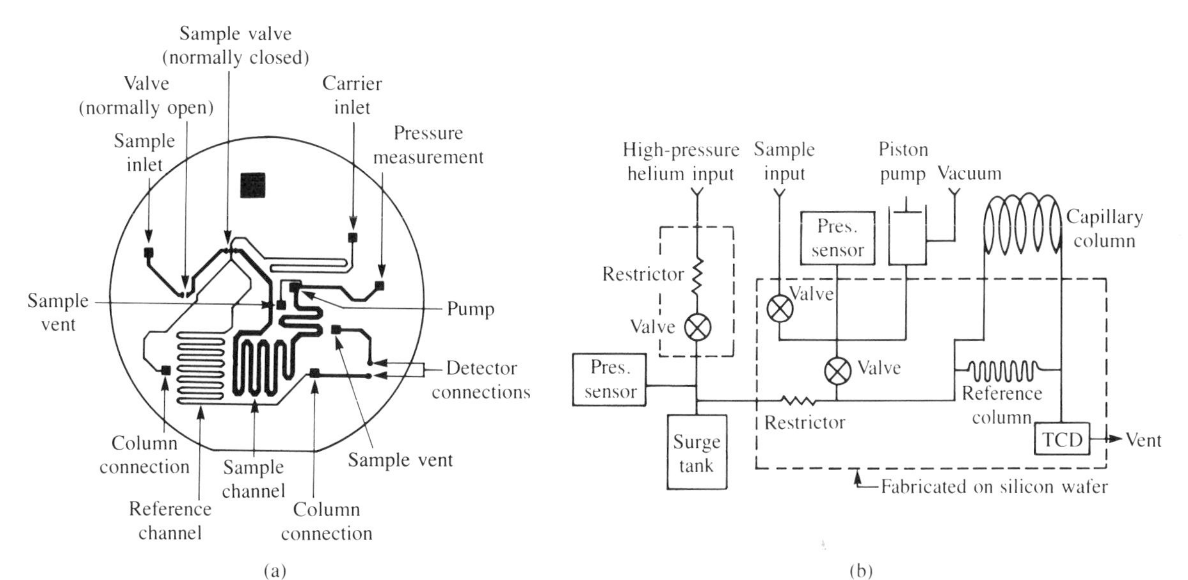

on a silicon wafer: the injection port, the thermal conductivity detector, and all interconnecting plumbing. The extremely small volumes of these components make them ideal for use with short, micropacked columns or small-diameter capillary columns. These columns can be optimized for rapid, efficient separations (less than 45 sec) of small (approximately 80 nL) gas samples at ambient temperature.

26.8 PROCESS GAS CHROMATOGRAPHY[9]

The first process gas chromatograph was put into service in 1954. Since that time these chromatographs have become the most widely used process analyzers in the petrochemical and refining industries. Although the similarities between laboratory gas chromatographs (see Chapter 18) and their process counterparts are apparent, differences important to process applications are less obvious. These differences include the physical appearance of the chromatograph, techniques used to acquire data, and most importantly, the way the analytical results are used.

On-line analyzers are installed for providing analytical results with a response time comparable to the changes in the process being monitored. This information is then used to take appropriate control measures. The process gas chromatograph is designed to accomplish this by operating continuously on-line and automatically analyzing one or more components of process streams in a cyclic, repetitive manner. A single analyzer typically monitors several streams (Figure 26.18).

FIGURE 26.18 Process control gas chromatography system.

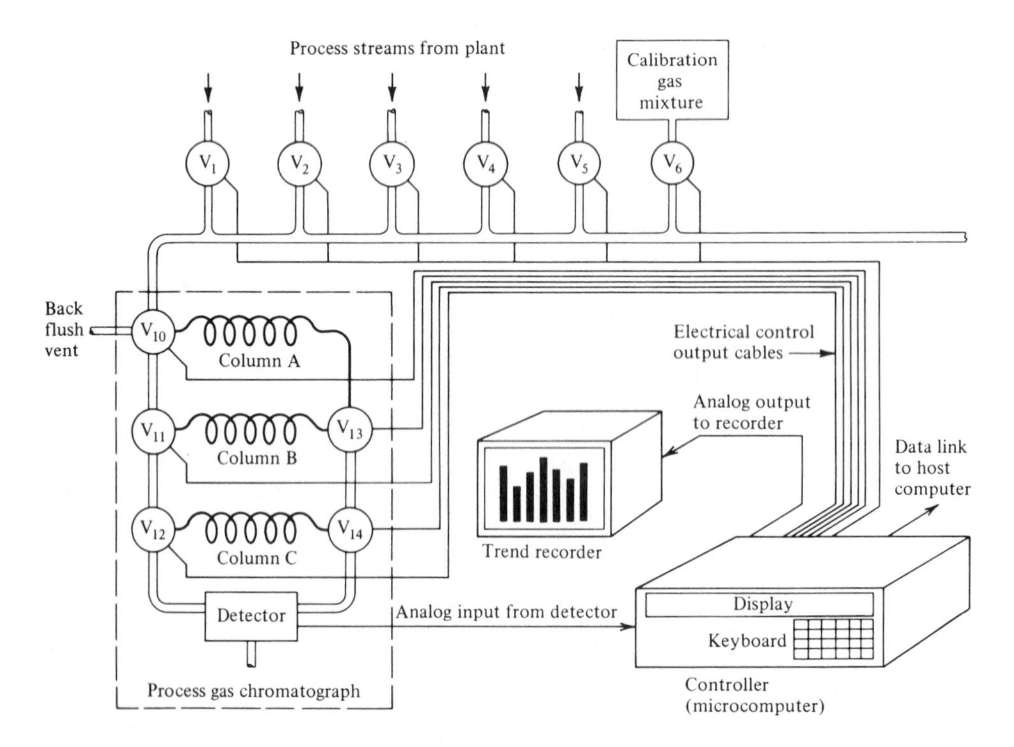

Differences between Process and Laboratory Chromatographs

Among the characteristics unique to process gas chromatography is the method of sample handling. The sample is transferred directly from the sampled stream to the chromatographic column(s) with no human intervention. Sample valves and connecting lines are maintained at elevated temperatures, thus allowing samples that contain large amounts of water vapor or other condensables to be transferred to the analyzer without altering the sample composition. Many sampling situations that are difficult or impossible using laboratory chromatographs are routine for process control analyzers. The inlet septum, common to laboratory chromatographs, is replaced by a sample valve. This valve is a critical component in the process chromatograph, for it must repetitively and reproducibly transfer a precisely measured volume of the sample into the carrier gas, which flows through the analyzer columns. If the sample is in liquid form, the liquid sample valve must deliver a measured volume of sample to a vaporizer, where it is first volatilized and then passed to the carrier gas. In addition to handling samples from one or more process streams, the valve system must have a means for periodically providing a sample of known composition from a reference tank for the purpose of calibration.

As in laboratory chromatography, column parameters are optimized to separate all components of interest in the shortest possible time. Analysis time is critical and must be minimized for process control applications. All components of the sample must be quantitatively accounted for and removed from the column(s) during each cycle. Residual components either accumulate and change the properties of the column, or they elute during a later cycle and interfere with the analysis. The system should be as simple as possible to operate, adjust, and maintain. Personnel assigned to process analyzers usually do not have training comparable to that of the personnel assigned to laboratory chromatographs. Process gas chromatographs are used in environments where downtime must be prevented.

Because of the requirements of process analysis, column technology has developed in different directions from that used in laboratory practice. Isothermal, multicolumn analysis methods are emphasized, whereas little or no use is made of techniques such as temperature programming and derivatization, which are used in laboratory separations. A large number of designs using valves to switch the sample among several columns have been developed. One widely used configuration is the stripper (Figure 26.19), which consists of two columns in series: a stripper (or precut) and an analysis column. Provision is made for backflushing the stripper column through a separate vent in order to reject unwanted components while providing carrier gas to the analysis column. The desired components are further separated on the analysis column while the stripper is being backflushed. This configuration is used to remove moisture, an impurity often found in hydrocarbon samples. Use of a stripper column ensures that unknown components of higher molecular weight are quantitatively removed from the system during each cycle and will not appear unexpectedly during a subsequent analysis (see Section 18.7).

The thermal conductivity detector is the primary detector for process gas chromatography. It is reliable, simple, easy to maintain, inexpensive, and universal in its response to components. Increased demands for monitoring and process control require greater sensitivity and selectivity inherent in ionization detectors.

The flame ionization detector has proved to be the most practical alternative to thermal conductivity detectors.

Process Chromatography Systems

On-line digital computers have revolutionized the methods used in both laboratory and process gas chromatography. In addition to providing real-time analysis of chromatograms and postrun calculation of results, the computer can control the operation of the process analyzer system by controlling both stream sampling and column switching valves. This computer control improves the efficiency and reliability of the system as well as reducing maintenance costs. Initially large minicomputer systems were used to control as many as 30 or 40 on-line analyzers. This approach was useful in applications, such as ethylene production, that use large numbers of chromatographs. Smaller plants that require only a few analyzers had difficulty justifying the large initial capital investment of the minicomputer system. This problem is being solved as microcomputers replace minicomputers. Individual chromatographic analyzers that contain microcomputers have the full capabilities of the minicomputer at drastically reduced costs. A complete chromatographic analyzer system with microcomputer controller is shown in Figure 26.18. Tasks performed by the computer include program control (timing of operations in analysis), peak detection, baseline correction, resolution of overlapping peaks, postrun calculations, calibration, scaling of analog outputs, and serial communications with display devices and the host computer. In addition to these basic tasks, the computer can perform maintenance tests, such as monitoring oven temperature and detector currents, and can control the sampling system.

High-performance liquid chromatographs (HPLC) have been adapted for use as process control analyzers. The problems associated with sampling, maintenance of column conditions, and general reliability are much more difficult to overcome in process HPLC than in process gas chromatography. Nevertheless, HPLC process

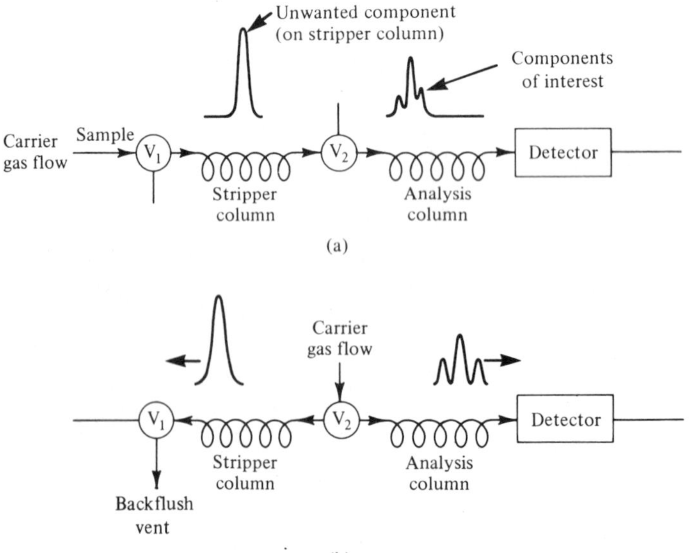

FIGURE **26.19**
Use of a stripper column in process control gas chromatography. (a) Initial sample configuration. (b) Backflushing of stripper column while components of interest move to detector.

analyzers are used in many industries such as food and pharmaceuticals. In combination with ultrafiltration membranes to eliminate high-molecular-weight components, HPLC provides an analyzer for some processes important to the biotechnology industry.

Applications of process chromatographs frequently encountered are for open- or closed-loop process control and environmental monitoring. In open-loop control the operator adjusts the process conditions based on the chromatographic results, In closed-loop control analyzer data are sent either to hardware electronic controllers or to a process control computer, which controls the process automatically. Monitoring of harmful components, such as vinyl chloride, in the atmosphere of factories that either produce or use them is required by OSHA regulations. A single analyzer can obtain samples from a number of locations throughout the plant, pump them to the chromatograph equipped with a flame ionization detector, and record concentrations as small as 0.1 ppm.

26.9 CONTINUOUS ON-LINE PROCESS CONTROL[10]

Automated control has become a major factor in the efficient operation of large-scale chemical processes. Production plants are facing difficulties of rapidly rising costs of energy and raw materials, environmental regulations, and varying personnel attitudes. Sensitive control systems are required to reduce the margin of operational errors. Present control specifications require these systems to be highly flexible, yet easy to operate and maintain.

In process control applications, a continuous analyzer is attached to a sampling line and thereafter automatically and continuously obtains a signal proportional to the instantaneous concentration of a selected component in the flowing stream. The information provided is then automatically used to set the process environment controllers and to take any corrective action necessary to control the process. Thus, continuous stream analyzers take over the function of the control laboratory with an increase in speed and efficiency.

A number of steps are involved in setting up on-stream control facilities. The analyst, in close collaboration with the project engineer, must determine the analytical task or tasks to be performed in order to follow a process as effectively as possible. The number of constituents monitored and the number and location of checkpoints must be decided. The operator should never be delayed with a mass of information, much of which may be useless. The guiding principle is to provide the operator with the least amount of data needed to produce the desired product. Economic considerations and worker requirements for installation, calibration, and maintenance must not be overlooked. Different analytical methods may be desirable at each checkpoint. As process control improves, production costs decrease; however, analysis costs rise. The selection of an optimum analytical method depends on minimizing the total of these costs.[11]

Design Features

When an analyzer is selected for a given problem, the first step is to make certain that the particular analysis can be made by the instrument and that it has sufficient sensitivity to determine the component of interest in the range of concentrations expected. Although similar in operating principles to their laboratory cousins, process stream analyzers differ in some important respects. Moving the automatic instrument sensors from the laboratory to the plant confronts instrument designers with major problems. Design criteria must incorporate these features: (1) reliability, (2) operational simplicity, (3) readout as foolproof as possible, (4) ease of maintenance, and (5) flexibility for future growth.

First and foremost is reliability. Instrument downtime means plant process downtime, or operation without control, which is very costly. Hence, long-term stability and reliability are essential characteristics. The availability of modular plug-in-type construction shortens necessary repairs and goes a long way toward making the instrumentation as reliable as any of the links in the process. Operational simplicity implies a minimum of controls and infrequent attention by the operator. Preferably once every shift, a cursory check is run. Thorough overhaul and inspection are done only during normal process downtime.

Readout and control functions must be made as foolproof as possible. Digital readout devices are utilized extensively. The environment in which the analyzer is used differs from the relative calm of the laboratory. Analyzers must withstand wide ambient temperature fluctuations and heavy vibrations, and they must not create explosion hazards. Often the units are completely sealed to operate independently of outside conditions and to withstand the onslaught of monkey-wrench mechanics.

Closed-Loop Control

An automated control system typically consists of the sensing element with an amplifier, the controller, and a final control element. These elements interact with one another and form the closed loop as illustrated in Figure 26.20. All elements are of equal importance.

Automated process control systems depend on deviations from a predetermined set-point signal for their operation. The controller acts on the difference between the set-point signal and the signal from the sensing element by sending an output signal to the final control element, which produces a correction in the

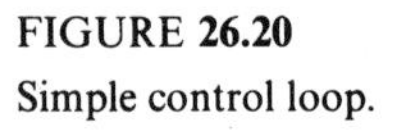
FIGURE 26.20
Simple control loop.

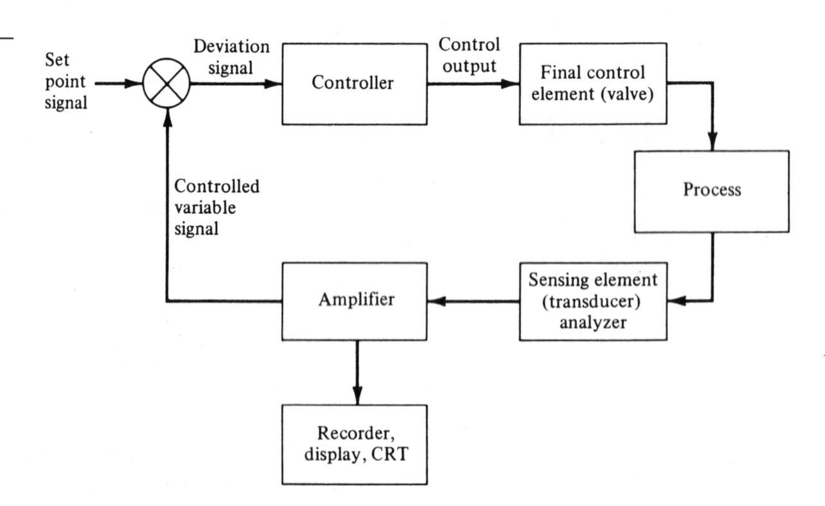

process. The corrective cycle terminates when the modified signal from the sensing element equals the set-point signal.

A time lag is involved in each step of this feedback control loop. This time lag is composed of the time lapses before corrective action can be initiated and the dead time (time delay between corrective action and the corresponding signal change detected by the sensor).[12]

The combination of current computer technology with on-stream sensors provides the ultimate in flexible, precise control of processes. To properly use this technology, the process under consideration must be studied to understand the reactions that take place, to determine how the flowing stream responds dynamically to changes in valve position, and to work out the mathematics required to control the process. Equations are written to express the desired amount of influence from a given variable. In predictive control, the desired correction is made not on the basis of set points previously determined by the operator or process engineer, but on the basis of continuing calculations made by the computer. When a computer completes the closed control loop (Figure 26.20) and should the operating conditions suddenly vary, the computer can rapidly make the necessary calculations so that several of the set points are properly adjusted to optimize the yield of the desired product. Computers are also used to evaluate measurement results and to search for useful correlations not usually discernible by operators or engineers.

26.10 AUTOMATIC CHEMICAL ANALYZERS

The increased demand for medical services beginning in the late 1950s resulted in growing workloads for clinical laboratories. Furthermore, new, specialized tests, such as analyses of serum enzymes and protein fractions, have been added to the list of routine determinations. This increase in the number of both samples and available tests has hastened the development of automatic chemical analyzers. Since the introduction of automatic instruments into clinical laboratories, these techniques have also been applied successfully to other areas such as environmental, pharmaceutical, food, and agricultural analyses, where large numbers of samples must be quickly analyzed.

Almost any repetitive analytical determination can be automated. Steps common to most analyses are: (1) sampling, (2) any required separation of sample components, (3) the addition of one or more required reagents and subsequent mixing, (4) detection of the analyte of interest by an appropriate sensor, and (5) collection, storage, and presentation of analytical data. Current automatic analyzers combine these steps to assay simultaneously up to 20 components in a 0.5-mL serum sample, and they can produce up to 3000 separate assays per hour.

An important difference between automatic and manual systems is that reactions do not have to be carried to completion in the automatic analyzer. Since conditions in a given automatic analyzer are unvarying and known, standard samples are subjected to the same treatment as samples of unknown concentration. Answers are supplied for most analyses from 10 to 100 times faster than with manually operated methods—that is, in minutes rather than hours, or even days. Furthermore, the results are not affected by operator fatigue or momentary inattentiveness.

Automatic clinical analyzers are classified as one of three types based on their method of operation: discrete, continuous flow, or centrifugal.

Discrete Analyzers

A discrete analyzer handles each sample as a separate entity and usually only one assay is made per sample. In some systems both the sample and the required reagents are metered into discrete reaction vessels, test tubes, or specially designed cells (Figure 26.21a). Other systems use cuvettes that contain prepacked quantities

FIGURE **26.21** Single test analyzer cell (STAC) and mixing procedure. (Courtesy of Technicon Corp.)

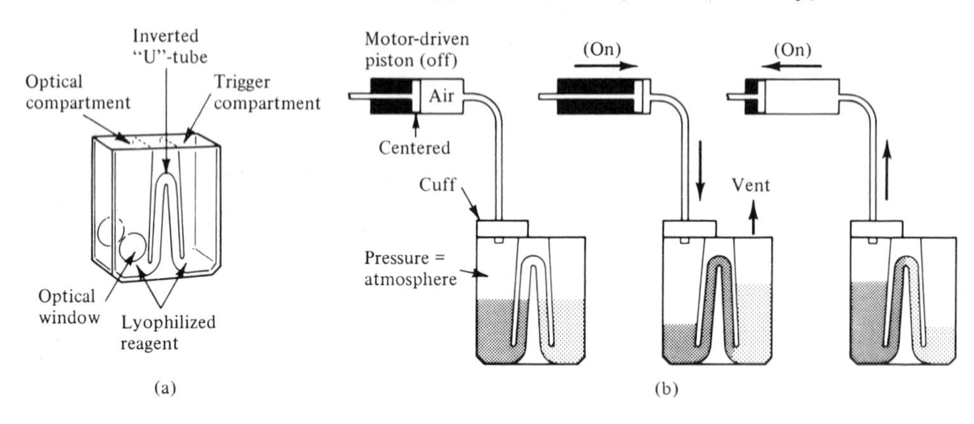

FIGURE **26.22** Automatic clinical analyzer (ACA). (Courtesy of E. I. DuPont de Nemours.)

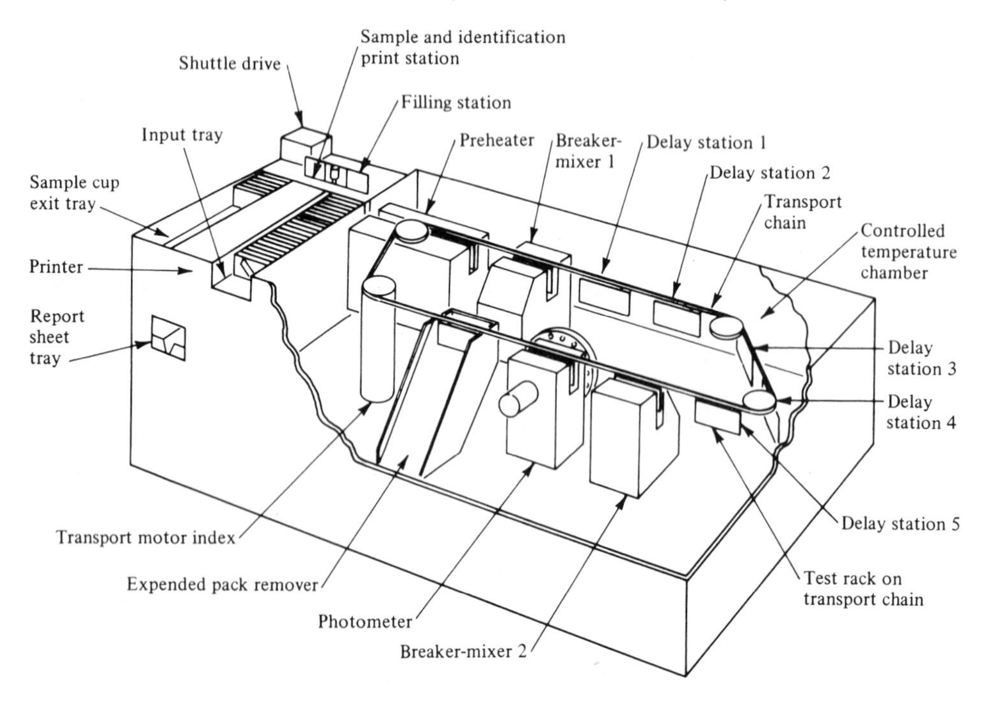

of the reagent required for a given sample. It is necessary to add only the sample. In both cases mixing and analysis take place in the test tube, sample cell, or cuvette. These containers are placed on a conveyor belt or turntable and the mixture is agitated (Figure 26.21b). The mixture in the container is passed through a water bath for temperature control (Figure 26.22). Additional reagents, if necessary, are added prior to measurement at the sensor station. Some systems include a component designed to wash the reaction vessels and recycle them back into the analysis train. Commercial analyzers of this type are generally capable of running a variety of different tests.

The Technicon Single Test Analyzer (STAC) (Figure 26.23) uses a computer to monitor and control the operation of the analyzer module (Figure 26.24) and to

FIGURE **26.23** STAC analyzer unit. (Courtesy of Technicon Corp.)

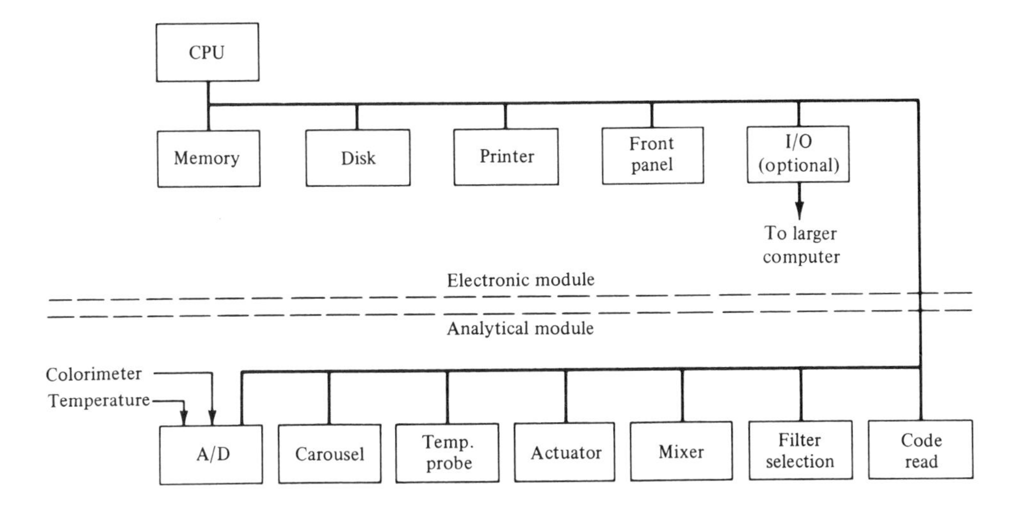

FIGURE **26.24** STAC computer-analyzer system. (Courtesy of Technicon Corp.)

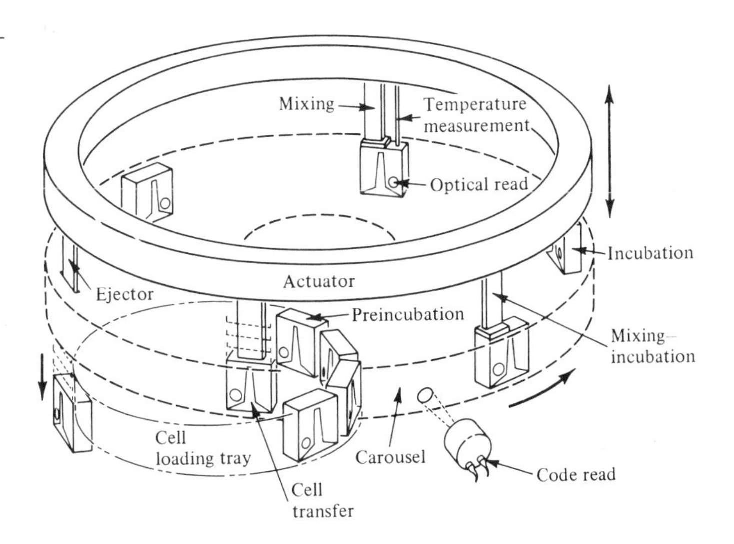

acquire and process the desired analytical data. Built-in detection assists the operator in diagnosing system malfunctions. Flexibility in programming mixing times and photometer reading patterns optimizes individual enzyme determinations to provide the best method at the lowest cost.

Discrete analyzers allow established manual techniques to be directly emulated by automatic operations. Only the test requested is performed on a given sample. This selectivity is not possible with continuous flow analyzers. The discrete analyzer uses smaller amounts of expensive reagents. A disadvantage associated with the use of discrete analyzers is their mechanical complexity compared with that of the other two types of analyzers. The use of prepackaged reagents in the sample containers adds to the expense of this method.

Discrete Analysis with Dry Reagents[13]

The key component in this technique is a slide composed of dry-film layers. All the reagents necessary to perform a specific bioassay and to screen out substances that might interfere with the analysis are contained on the slide. The only liquid required is 10 μL of sample. The effect of the analyte on the reagent is monitored by diffuse reflectance spectroscopy or, less frequently, fluorescence spectroscopy.

When a specific test is ordered, a slide is automatically dispensed from the cartridge that holds the slides for the specified test. A small drop of the patient's sample is applied to the slide by the analyzer. The drop is quickly and uniformly distributed by an isotropically porous spreading layer of the slide. This layer can perform several additional functions. It can retard proteins and other large molecules and allow the protein-free filtrate to pass into the reagent layer, where in most slides a reaction occurs. In some slides, such as those used for kinetic enzymes, total protein, and albumin, reactions take place in the spreading layer. In either case, a colored reactant is produced in an amount proportional to the quantity of the analyte in the patient's sample.

Complex sequential reactions, which are impossible to run in solutions or with other dry methods, can be performed on the slides. Multilayer coatings (with physical and chemical properties such as filtration, selective adsorption, and reactant stabilization or kinetic optimization) offer domains for multiple reactions within a single slide. In some layers interfering substances can be retained or altered (e.g., in the bilirubin assay, hemoglobin is retained in the spreading layer), whereas in other layers reactions or detection signals may be enhanced.

The slide used for blood urea nitrogen (BUN) determinations contains a transparent support layer, two reagent layers separated by a semipermeable membrane, and a spreading and reflectance layer (Figure 26.25). A drop of undiluted patient's serum is applied to the porous spreading and reflectance layer, which spreads the sample uniformly and rapidly through the capillary structure of the layer. The sample then enters the first reagent layer where the urea is converted to ammonia and carbon dioxide by the enzyme urease. The semipermeable membrane blocks the hydroxyl ions and allows the ammonia into the second reagent layer. In this layer the ammonia causes a change in the color of the pH indicator, which is detected by a reflectance meter mounted below the slide. The results are expressed as the concentration of blood urea nitrogen within 7 min.

Kinetic analysis can be performed on slides similar to those used for colorimetric end point assays. The enzyme analysis slides contain all the substrates and cofactors necessary and are constructed to provide an optimum pH for the specific analysis. The enzyme reactions produce a change in the optical density of the reflected light at a wavelength specific for the reaction being run. The change in light density becomes linear with time and proportional to the activity of the analyte. The change in light density is measured 54 times during a 5-min thermostatted incubation period to determine the linear portion of the curve. Data from the linear portion of the curve are used to determine the rate of the reaction. Finally, the rate is converted to enzyme activity using calibration values from reference materials.

Slides used for electrolyte analysis are designed for a single analyte (Figure 26.26). The slides contain two identical electrodes, sample and reference, and an ionophore specific for the electrolyte being measured. In the slide for potassium determinations, each electrode contains a silver/silver chloride electrode, a reference layer containing potassium chloride, and an ion-selective membrane containing

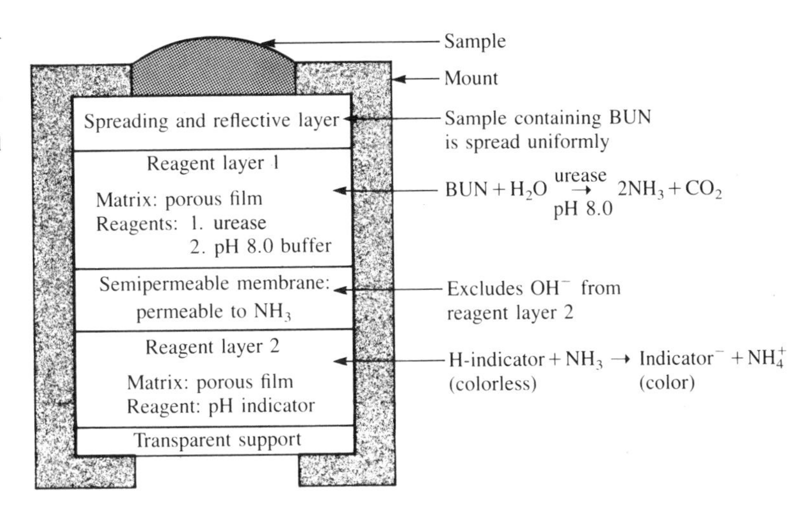

FIGURE **26.25**
Cross section and chemistry of Eastman Kodak Ektachem slide for blood urea analysis. [After B. Walter, *Anal. Chem.*, **55,** 504A (1983). With permission.]

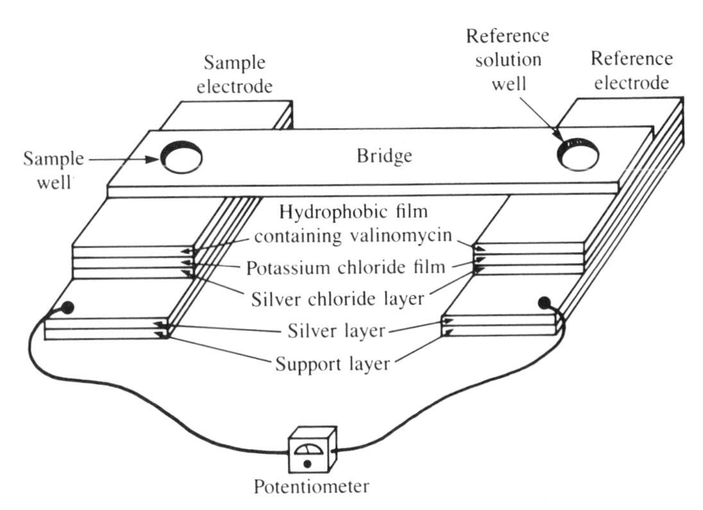

FIGURE **26.26**
Diagram of Eastman Kodak Ektachem slide for potassium analysis by electrochemical means. [After B. Walter, *Anal. Chem.*, **55,** 506A (1983). With permission.]

valinomycin. Potassium ions are sequestered by this ionophore at the expense of other cations. Fluid from the patient and reference fluid of known potassium ion concentration are applied automatically to the two electrodes. The fluids flow toward each other through a paper bridge and a stable liquid junction is formed. An electrometer measures the potential difference between the two half-cells. This difference potential is sent to the microprocessor, which calculates and reports the potassium concentration. One test requires about 3 min. Unlike conventional electrode measurements, the sample is referenced directly to a standard solution. Single-use, ion-selective electrodes are inexpensive, require no preconditioning or membrane change, and lose no sensitivity due to protein buildup.

Continuous Flow Methods

There are two methods in which the samples are allowed to react with reagents in a continuously flowing system and the amount of product measured after a specific interval of time. In the continuous flow analysis (CFA) method, the sample is injected into the flowing reagent stream.[14,15] Intermixing of individual samples in the stream is prevented by air bubbles placed between samples. The flow-injection analysis (FIA) method uses the laminar flow present in narrow-bore tubing to mix the sample with the reagent, thus eliminating the necessity of air bubble partitioning.[16,17] The two methods complement each other. Flow-injection analysis is best suited for analyses that require less than 30 sec and the sequential addition of only one or two reagents. Continuous flow analysis is preferred for analysis times longer than 2 min and the sequential addition of three or more reagents.

The continuous flow analysis method is illustrated by the AutoAnalyzer, which consists of a series of modules, each performing one specific function in the programmed sequence of events that constitute the analysis (Figure 26.27a). Modules can be interchanged and rearranged for different analytical methods. All aspects of sampling, separation, reagent addition, mixing, temperature control, sensing, readout, and recording are totally automatic. A sampling capillary dips into each sample on the sampling tray, aspirates its contents for a timed interval, and then feeds it into the analyzer (Figure 26.27b). Between samples the capillary is raised to aspirate air for another timed interval. This step is followed by aspiration of an intersample liquid wash solution and another air aspiration before the next sample is introduced into the system.

The heart of the analyzer is the proportionating pump. It can deliver 12 or more separate fluids (reagents, diluents, air, and so on) simultaneously while varying their flowrates in any ratio from 1:1 to 79:1. The pump consists of two parallel stainless-steel roller chains with spaced roller thwarts that press continuously against a spring-loaded platen (Figure 26.27c). Variable flowrates are obtained by using flexible tubing with different inside diameters.

At each point in the system where two liquids come together, there is a mixing coil of sufficient length to provide the required mixing time. In clinical analyzers, dialyzable analytes diffuse through a dialyzer membrane into a second segmented stream (Figure 26.27a). This mixture reacts as it passes through a thermostatted reaction coil. After the sample has reacted, the analyte is monitored by a suitable sensing device, such as a narrow-bandpass ultraviolet/visible spectrometer, a fluorometer (for special cases that require increased selectivity or sensitivity), a flame

FIGURE 26.27 (a) Typical single-channel flow schematic, (b) details of sampler, and (c) proportioning pump of the AutoAnalyzer. (Courtesy of Technicon Corp.)

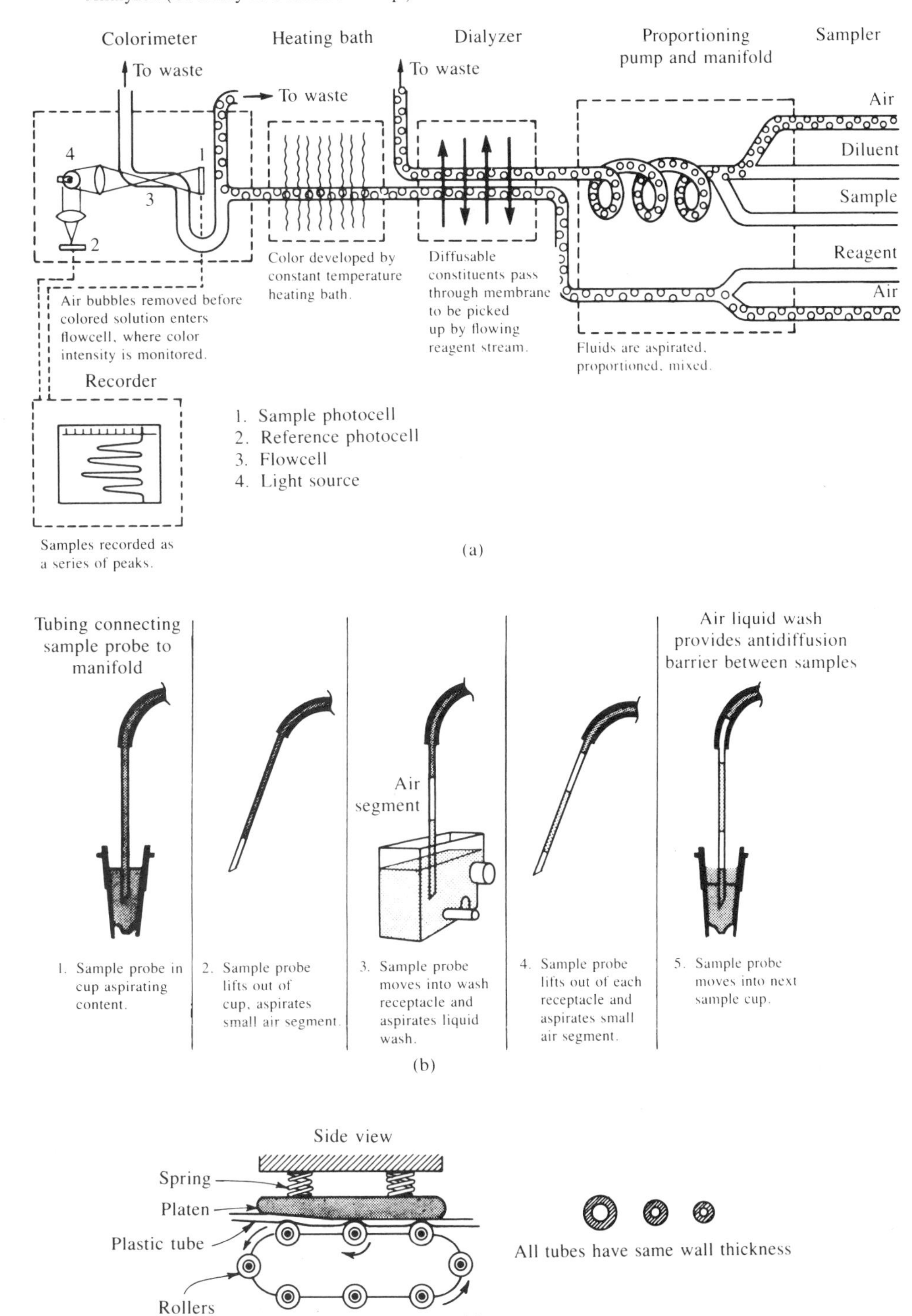

emission photometer (for Na^+ and K^+), or ion-selective electrodes. The results are printed on a continuous strip chart or sent to a computer interfaced to the analyzer. In the latter case, these results are stored in a computer memory and may be accessed by one or more devices: a video terminal for surveying the data quickly, a printer that generates a report in the required format, or a larger computer for further processing and storage in a larger data file. Calibration of the system is checked periodically by interspersing standards of known concentration.

The performance of a continuous flow analyzer is determined from the shape of the signal produced by the sensor. Three principal parameters influence the quality of this signal: sample dispersion, mixing, and flow stability. Sample dispersion is minimized by the segmenting air bubbles and wash solutions that serve as physical barriers to contamination. The bubbles also act as wiping agents in the tubular systems. Excessive sample dispersion leads to the overlap of adjacent samples. This effect is countered by increasing the intersample wash time at the expense of either sampling time (volume) or total sample throughput (sampling time not reduced).

The sample and all added reagents must be completely mixed in each sample segment. Incomplete mixing results in noisy, unstable detector output. Important mixing parameters are liquid viscosity, density, flowrate, internal diameter of tubing, helix coil diameter, and length of sample segment in the tube. Flow stability results in constant proportioning of the sample to reagents for all segments from one sampling cycle to the next. An incorrectly proportioned sample stream also appears as a noisy detector output. Variation in liquid or air rates, compression of intersample air bubbles, a short sample segment, and blockages in the sampling probe or manifold tubes result in flow instability in a continuous flow analyzer.

The addition of computers to continuous flow analyzer systems has increased the accuracy and precision of analyses as well as producing higher sampling rates. Monitoring and analysis of detector output signals by the computer flag results caused by the following errors: malfunction in sample or reagent dispenser, insufficient sample or reagent, incomplete mixing, and carryover from a high-concentration sample to an adjacent low-concentration sample. When the detector output curves do not meet preselected "normal" criteria, the computer flags the assays for rerunning. At the same time the computer analyzers record abnormalities and alert the operator as to what corrective action is necessary. The Technicon SMAC/SDM high-speed computer-controlled biochemical analyzer and disk-based data processing system allow assay results to be stored on magnetic disk (64 analyses for each of 1500 patients) and also transferred electronically to remote locations. Twenty different analyses may be selected from 26 available methods. Only a 0.45-mL sample is required for the selected assays.

Multichannel flow analyzers offer advantages to laboratories that have high volumes of multiassay samples. A typical physician's request for the analysis of a single serum sample might include sodium, potassium, chloride, carbon dioxide, glucose, urea nitrogen, albumin, and total protein. One computer-controlled analyzer can run approximately 240 complete analyses (1920 individual tests) per hour. By contrast, for laboratories with lighter workloads or irregular periods of activity, the discrete analyzer offers the advantages of quick startup and the ability to specify only certain types of tests.

The basic system for automated flow-injection analysis is shown in Figure 26.28. The sample is aspirated from a container in the tray into the loop of a sample insertion valve. Upon actuation of the valve, the sample is injected into an unsegmented continuous stream of solvent that is quickly mixed with a reagent stream by conduction and diffusion. The product of the quantitative reaction then passes through the detector. Peak height and peak area measurements are made and concentrations determined by comparison with measurements made on standard solutions. Flow-injection analysis systems perform routine replicate analyses at rates of 100 or more samples per hour. In many cases, results are obtained within 15 sec after sample injection, since the method is based on kinetic rather than equilibrium production concentrations. Results from typical flow-injection analysis assays are shown in Figure 26.29. Several parameters are critical to the success of the method: small-bore tubing (usually 0.5 mm i.d.) must be used, flowrates (1–14 mL/min) must be precisely controlled, and the volume available for mixing must be minimized. Almost any detector used with HPLC systems can be used with flow-injection analysis systems. Microcomputers have been added as important components of flow-injection analysis to provide precise, reliable control and data acquisition capabilities.

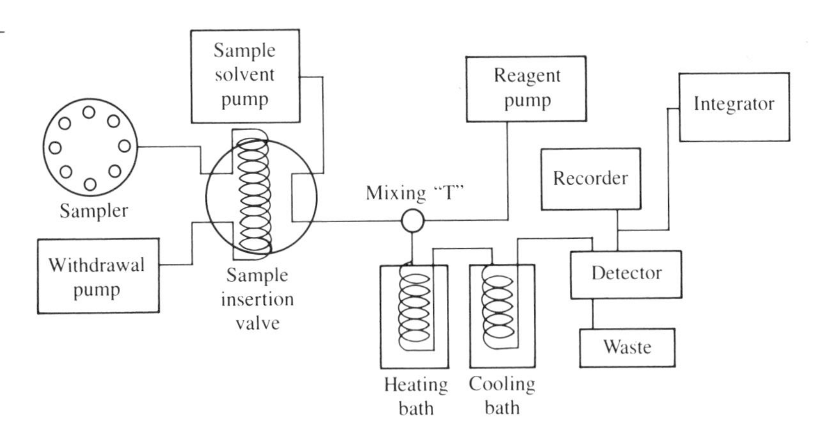

FIGURE **26.28**
Schematic of an automated FIA system. [After K. Stewart, *Anal. Chem.*, **55,** 932A (1983). With permission.]

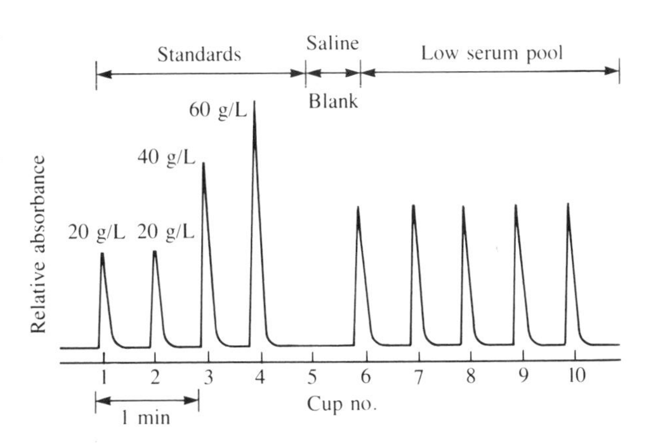

FIGURE **26.29**
Recorder tracing of an FIA determination of serum albumin with bromcresol green. [After K. Stewart, *Anal. Chem.*, **55,** 932A (1983). With permission.]

Centrifugal Analyzers

In both discrete and continuous flow analyzers, the samples are presented serially to the detector, with each sample spending a fixed amount of time in the detector. In situations such as kinetic assays of serum enzymes, where it is necessary to monitor the detector output for extended time periods, the sample throughput (samples per hour) becomes a function of the time that each sample spends in the detector. The use of discrete or continuous flow analyzers requires a trade-off between the time that the sample is exposed to the detector and the rate of sample throughput. Either exposure times must be shortened or throughput rates must be reduced. Centrifugal analyzers offer an alternative approach by effectively multiplexing the detector by rapidly rotating a disk that contains samples past a single detector. Sample exposure time is increased without decreasing sample throughput, since several samples are being monitored during a single extended exposure time.

In this type of analyzer, centrifugal force provides the means for mixing the sample and reagent and for transferring this reaction mixture to the cell. Figure 26.30 shows the cross section of the sample disk and detector compartment for one instrument. In use, samples and reagents are placed in cups arranged in concentric circles in a transfer disk. The cups in the various circles are aligned radially and arranged so that samples and reagents move through a transfer cavity when the rotor accelerates. After mixing, each sample solution is pulled into a cuvette, where its transmittance or luminescence is measured. As the sample disk rotates, the cuvettes pass sequentially past the detector. The reading for each sample is referenced to a reagent blank contained in one of the cuvettes. Thus the analyzer functions as a multichannel double-beam spectrophotometer that produces a complete set of sample readings with each revolution of the disk (100 msec per set at 600 rpm). Typically measurements from eight consecutive rotations are averaged for each sample. For extended kinetic determinations, sets of eight measurements are averaged over longer time periods (seconds or minutes), and these values are displayed, printed, or stored for use in obtaining the final assay.

Computers built into the instrument control the operation of the analyzer and functions associated with the acquisition, averaging, manipulation, display, and storage of data (Figure 26.31). The sample disk is configured to run the same test on many different samples or different assays on the same sample. After the samples

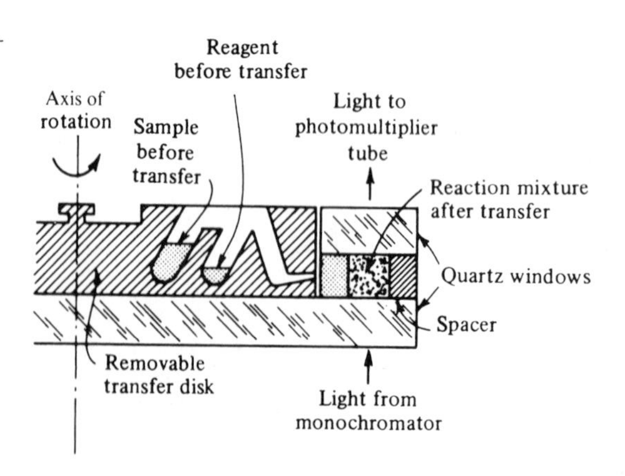

FIGURE **26.30**
Cross section of sample disk and centrifugal analyzer section.

FIGURE 26.31 Centrifugal analyzer system. (Courtesy of American Instrument Co.)

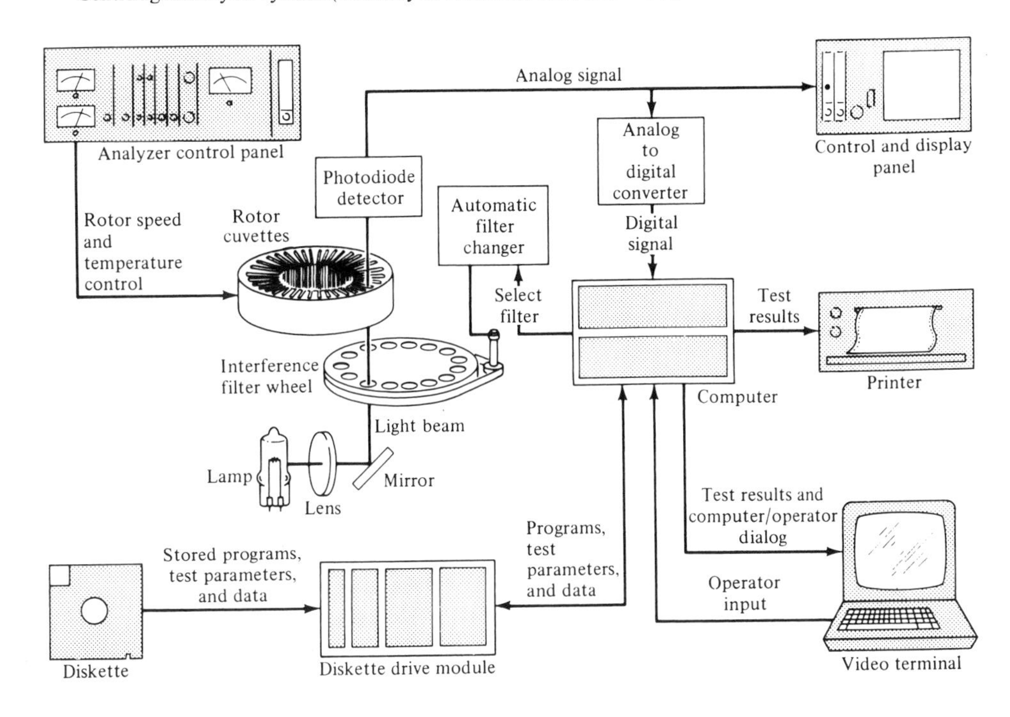

have been analyzed, the disk compartments are cleaned automatically. The disks are usually loaded with samples and reagents outside the centrifugal analyzer. This is done manually or by an automatic dispensing module. Sample volumes run from 5 to 100 μL for each assay.

Important advantages of this approach are virtually simultaneous measurement of samples and standards under the same conditions, optimization of analysis time for solutions in which sample-reagent reactions are slow, simple mechanical design, and the ability to change the procedure easily by changing reagents in the disks and the monochromator setting of the photometric detector. The major disadvantage of centrifugal analyzers is their batch mode of operation, which necessitates reloading of disks with both reagents and samples for each group of samples analyzed.

26.11 AUTOMATIC ELEMENTAL ANALYZERS

Current instrumentation for carbon, hydrogen, and nitrogen (CHN) analyzers is based on one of two general procedures. One involves the separation of carbon dioxide, nitrogen, and water by a gas-liquid chromatographic column. The other involves separation by means of specific absorbants for water and for carbon dioxide with the resulting change in composition of the gas mixture being measured. Thermal conductivity is the detection method in both techniques. Results are calculated by analyzing standard samples and an occasional blank.

The sample (0.1–3 mg range) is either burned under static conditions in a pure oxygen atmosphere at 900 °C, whereby the sample boat can subsequently be removed for weighing any residue, or mixed with cobalt(III) oxide [or a mixture of the oxides of manganese(IV) and tungsten(VI)] to provide the oxygen and heated to 900 °C. Combustion converts the carbon in the sample to carbon dioxide and carbon monoxide, and hydrogen to water. Nitrogen is released as the free gas, along with some oxides. A stream of helium carries the combustion gases into a reaction furnace operated at 750 °C. Here a chemical change completes the simultaneous oxidation and reduction of the sample gases. Hot copper reduces the nitrogen oxides to nitrogen and removes the oxygen. Copper oxide converts the carbon monoxide to carbon dioxide. If needed, a magnesium oxide layer in the middle of the furnace removes fluorine, and a silver-wool plug at the exit removes chlorine, iodine, and bromine and also any sulfur or phosphorus compounds that result from the combustion of the sample. In CH analyzers, oxides of nitrogen are removed a little later in the train with manganese dioxide.

Separation of the combustion gases by gas chromatography involves the following sequence: the gases pass through a charge of calcium carbide where water vapor is converted to acetylene. A nitrogen cold-trap freezes the sample gases and isolates them in a loop of tubing. A valve seals off the combustion train, which is then ready for another sample. The chromatographic stage is begun by lowering the cold-trap and heating the injection loop. Another stream of dry helium gas carries the gases (as a plug) into the chromatographic column where the three gases—N_2, CO_2, and C_2H_2—are completely separated. Most organic samples can be completely burned in 10–12 min, and the chromatographic separation requires another 10 min. Figure 26.32 gives the schematic diagram of a CHN analyzer of this type.

The other general procedure involves separation by means of specific adsorbants for water and carbon dioxide. Three pairs of thermal conductivity cells (glass-coated platinum filaments) are used in series for detection: one pair each for water, carbon dioxide, and nitrogen (Figure 26.33). A magnesium perchlorate trap between the first pair of cells absorbs any water from the gas mixture before it enters the second pair of cells. The differential signal, measured before and after the trap, is proportional to the amount of hydrogen (measured as water removed) in the sample. Similarly a soda-asbestos trap between the second and third pairs of cells results in a signal proportional to the carbon (carbon dioxide) in the sample. The last pair of cells detects nitrogen by comparing the helium-nitrogen mixture with pure helium.

For oxygen analysis, the combustion furnace is replaced by a quartz pyrolysis tube that contains platinized carbon. The reduction furnace is replaced by a tube that contains copper oxide. The operating temperatures are the same as for CHN analysis, but the oxygen supply is shut off. Samples are handled exactly as they are in the CHN analysis mode, with the sample initially heated in a helium atmosphere so that any oxygen in the sample forms carbon monoxide. This carbon monoxide is converted in the copper oxide tube to carbon dioxide, which is then detected and measured in precisely the same manner used for the carbon analysis.

Sulfur analysis can be performed if the combustion furnace is replaced with a tube that contains tungsten(VI) oxide packing plus a dehydrating reagent. The

FIGURE **26.32** Schematic diagram of the Fisher carbon-hydrogen-nitrogen analyzer. (Courtesy of Fisher Scientific Co.)

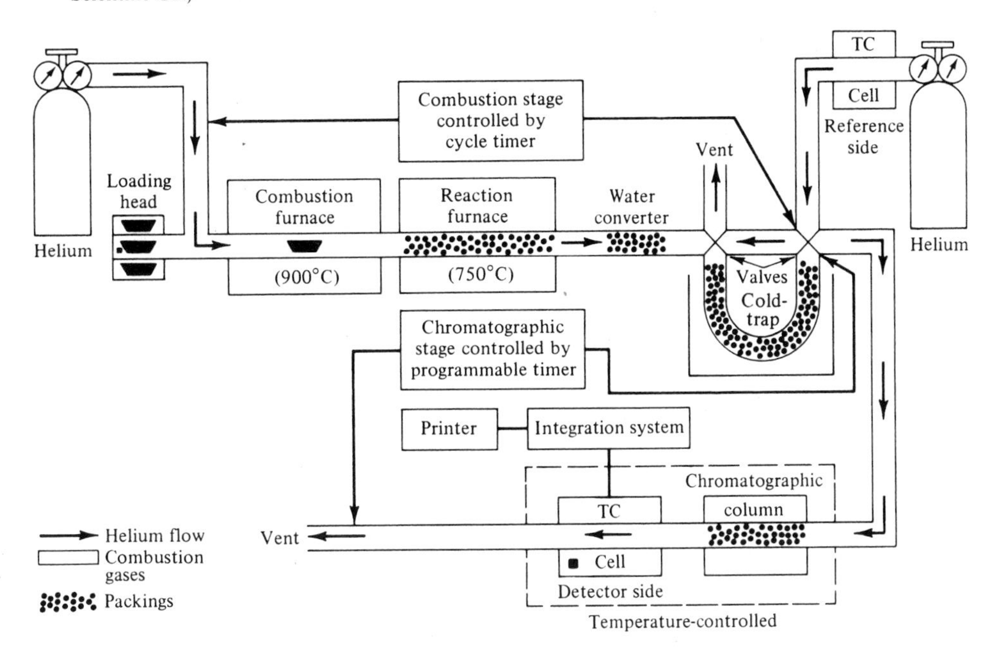

FIGURE **26.33** Flow schematic of the Perkin-Elmer Model 240B elemental analyzer. (Courtesy of Perkin-Elmer Corp.)

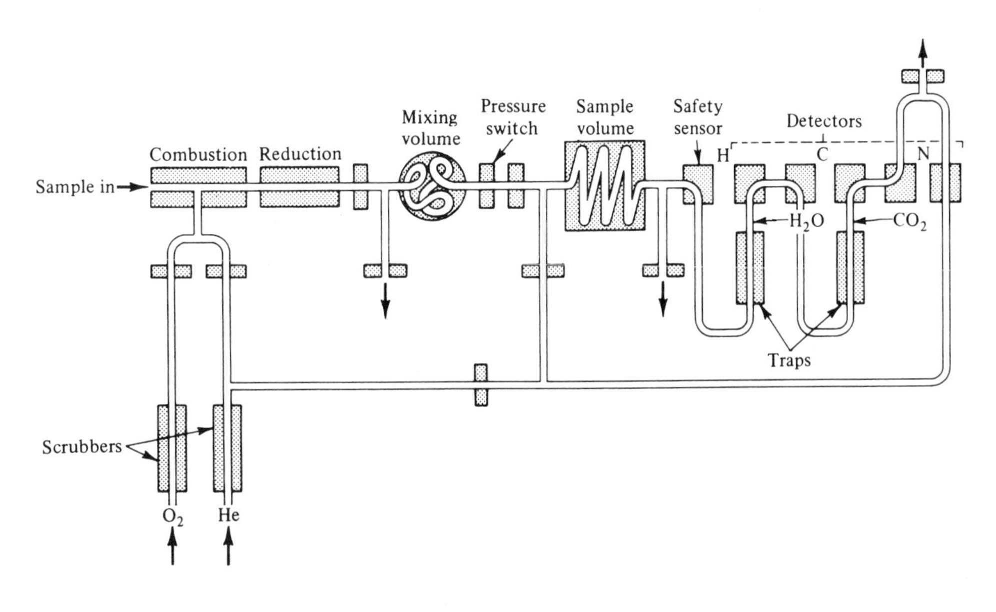

FIGURE 26.34 Model 2400 CHN Analyzer. (Courtesy of Perkin-Elmer Corp.)

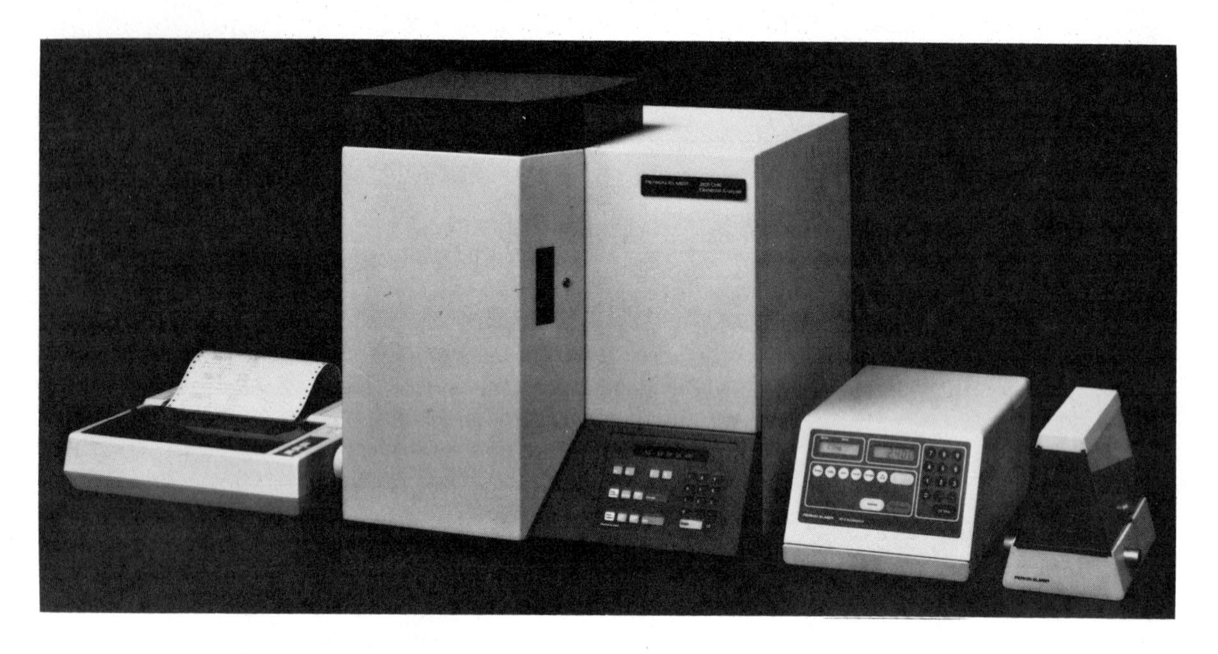

water trap is removed from the bridge area and replaced with a trap containing silver oxide to adsorb the SO_2 produced in the combustion tube.

In the first generation of CHN analyzers, the sample throughput was limited by the semiautomatic design, which required each individual sample to be placed into the analyzer by the operator. Second-generation elemental analyzers resolved this limitation by adding automatic sampling and data handling modules to produce a fully automatic analyzer system. The Perkin-Elmer Model 2400 analyzer automatically feeds and computes up to 60 samples without operator attention (Figure 26.34). Sample containers are loaded into a disk-shaped magazine, automatically fed through the analyzer, and removed at the end of each analysis. The system includes a laboratory data station (microcomputer system) for the control of instrument parameters, acquisition and processing of data, graphical display of results, and storage of information. A robot arm can be interfaced to the system to automatically manipulate the samples.

The typical analytical report for each sample includes the sample identification number, data, sample weight, sample gas responses, weight percentages, and simplest formula. This automation results in increased sample throughput, reduction of human error, and flexibility of analyzer operation and data processing to meet a variety of applications.

26.12 LABORATORY ROBOTS[18–20]

In most analytical laboratories concerned with automated analysis there is a large deficiency, the automation of physical manipulations. These manipulations may include several of the following operations: sample identification, extraction of aliquots, dilution, concentration, extraction, crushing, grinding, weighing, and addition of reagents. Important considerations in the use of laboratory robots are coordinate systems, drive systems, hands, sensors (sight and touch), and programming. In addition to reducing the cost of performing such operations by a factor of four compared with human labor, the use of robots frees the chemist from many tedious tasks.

The typical laboratory robot consists of a body, an arm, and a hand or gripper. Movement of the body or arm allows the hand to sweep out a space that can be viewed as Cartesian, cylindrical, or spherical. Joints may be added to the arm to allow more flexibility. At a minimum, the hand must contain a pair of fingers to grasp an object. The parts of the robot are moved by electrical servo or stepper motors. Servo motors are more reliable under mechanical torque, whereas stepper motors provide low cost and an easy interface to digital microcomputers. Regardless of the type of motor used, the mechanical properties of the drive train between the motor and the robot component are important to precise, reliable robot operation.

The robot is only one component of a robotic laboratory work station (Figure 26.35); other components are a microcomputer and the instrumentation required for the analysis. The three components are linked by a hardware-software interface and by the devices that are required to give the robot precise and reliable mechanical coupling to the elements at the work station. Because of the limited sensory input available with current robots, precise mechanical interfaces between the robot and

FIGURE **26.35** Diagram of a robotic laboratory work station. (Courtesy of Zymark Corp.)

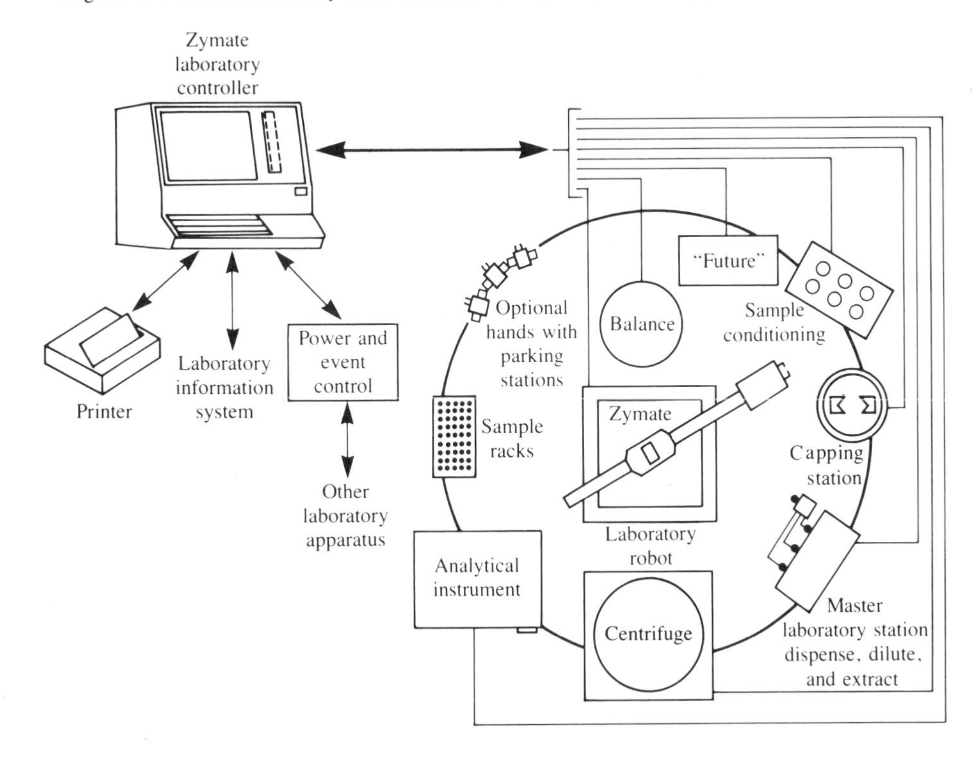

FIGURE **26.36** Laboratory robotic work station. (Courtesy of Zymark Corp.)

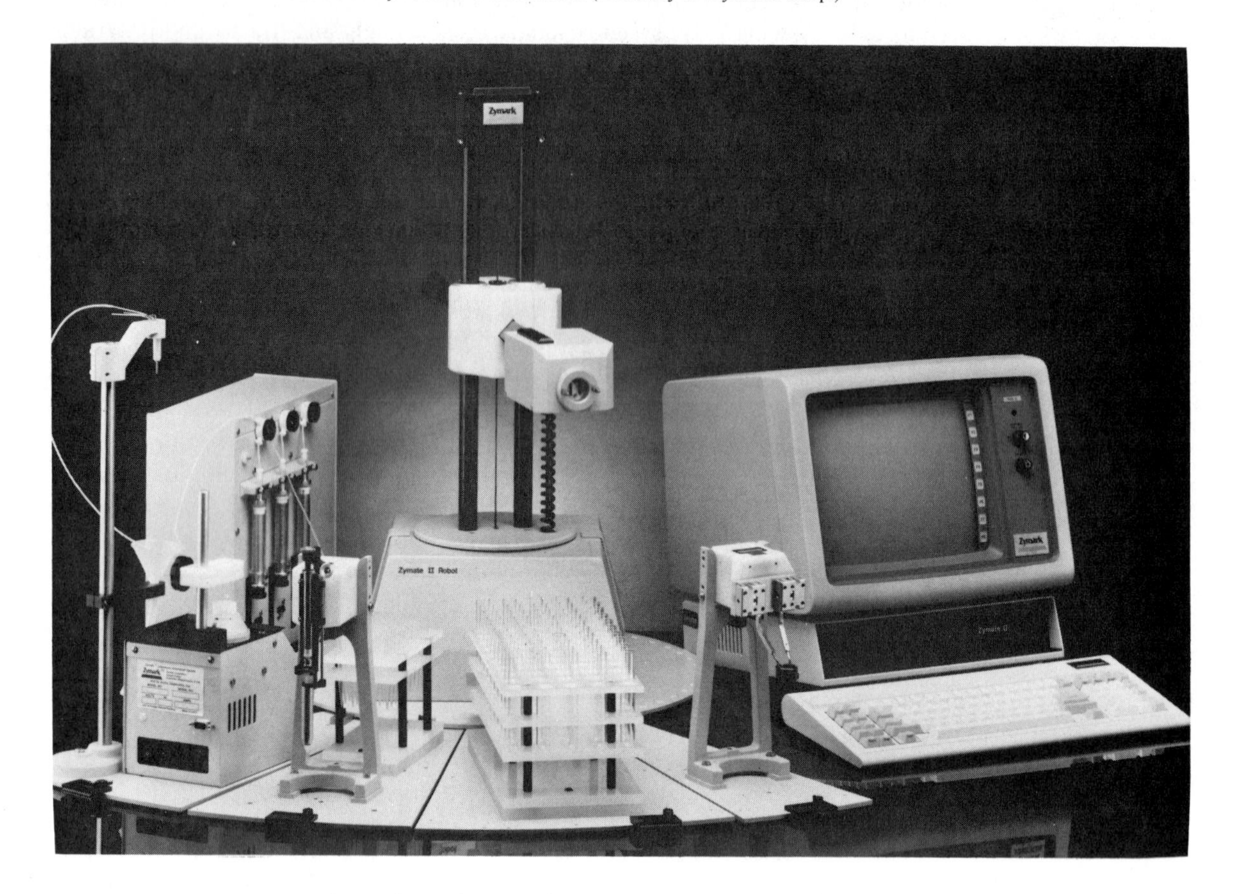

each work station component are necessary. The environment around the robot must be designed so that the robot can find and grasp the desired component. Mechanical devices used to accomplish the required tasks include a mobile base for the robot, gripper fingers or hands for handling test tubes, beakers, or flasks, sample racks and empty container magazines, and holders for electrodes and pipet tips.

Although most commercially available robots can be programmed using a hand-held training pendant, the capabilities of many robots can be extended by interfacing them to microcomputers (Figure 26.36). Comprehensive robotic work station software can be developed using high-level languages such as BASIC or FORTH.

PROBLEMS

1. Describe a simple *automatic* instrument for monitoring the pH in the production of disodium phosphate from soda ash and phosphoric acid. Explain how this instrument could be used as part of an *automated* system for the continuous production of disodium phosphate.

2. In the separation of aromatics from saturates, using a refractive index to follow a process, what precision can be expected? The aromatics have a refractive index of about 1.50, and the saturates about 1.40.

3. What advantages do nondispersive infrared analyzers offer over dispersive instruments? What disadvantages are associated with nondispersive infrared analyzers?

4. Explain the major differences between process chromatographs and laboratory chromatographs.

5. Suggest a method for handling the following on-stream process situations: (a) fractionating-tower control in the separation of cyclohexane and *n*-hexane; (b) fractionating column for the determination of butane in isobutane; (c) control of the end point in making shortenings and margarine from materials such as soybean and cottonseed oils; and (d) changes in the concentration of individual solutions of inorganic salts, HCl, or H_2SO_4.

6. Design a method for monitoring boiler stack gas in pulp mills for soda, both to aid in minimizing soda losses and to help control air pollution. Keep in mind that light transmission or scattering is handicapped by rigid sample-handling requirements and high emission rates. In addition, highly conductive dissolved gases that are also present offset the advantages of measuring the electrical conductivity in scrubbed gas samples.

7. A Hersch galvanic cell is used to monitor oxygen in flue gases at 816 °C. If the partial pressure of oxygen on the reference side is 200 mm, what is the partial pressure of oxygen in the flue gas?

8. The readings in the table are obtained at the four detectors of the dual path–dual frequency photometric analyzer (Figure 26.4) in the on-line analysis of SO_2 in stack gas using a sample cell with a 50-cm path length. Calculate the concentration of SO_2 in the unknown stack gas sample.

		Radiation intensity	
Detector (Figure 26.4)	**Wavelength (nm)**	**Calibration mixture (38 ppm SO_2)**	**Unknown stack gas**
M_s	280	45	32
R_s	280	98	95
M_c	365	97	92
R_c	365	99	93

9. Compare the three types of automatic chemical analyzers with respect to the mode of operation, advantages, and disadvantages.

10. Discuss several limitations inherent in the gas chromatographic system of analysis used in some CHN analyzers.

11. A CHN analyzer was calibrated by burning 1.657 mg of dimethylglyoxime. The total signal was C = 17,500 units, H = 1062 units, and N = 4128 units. The unknown compound, a 2.021-mg sample, gave these signals: C = 40,760 units, H = 1078 units, and N = 3195 units. Calculate the percent carbon, hydrogen, and nitrogen.

BIBLIOGRAPHY

BORER, J., *Instrumentation and Control for the Process Industries,* Elsevier, New York, 1985.

CALLIS, J., D. ILLMAN, AND B. KOWALSKI, *Anal. Chem.,* **59,** 624A (1987).

FRANT, M., AND R. OLIVER, *Anal. Chem.,* **52,** 1252A (1980).

HUSKINS, J., *General Handbook of On-Line Process Analyzers,* John Wiley, New York, 1982.

JOHNSON, C., *Process Control Instrument Technology,* John Wiley, New York, 1977.

MIX, P., *The Design and Application of Processor Analyzer Systems,* John Wiley, New York, 1984.

YOUNG, D., *Automation in Fundamentals of Clinical Chemistry,* Chap. 4, N. Tietz, ed., Saunders, Philadelphia, 1976.

LITERATURE CITED

1. IRVING, H., H. FREISER, AND T. S. WEST, *Compendium of Analytical Nomenclature,* Pergamon, New York, 1978.
2. FRANT, G., AND R. OLIVER, *Anal. Chem.,* **52,** 1251A (1980).
3. NICHOLS, G., *Anal. Chem.,* **53,** 489A (1981).
4. FRANT, G., AND M. LABUTTI, *Anal. Chem.,* **52,** 1331A (1980).
5. SEITZ, R., *Anal. Chem.,* **56,** 16A (1984).
6. WHOLTJEN, H., *Anal. Chem.,* **56,** 87A (1984).
7. DESSY, R., *Anal. Chem.,* **57,** 1189A (1985).
8. SAADAT, S., AND S. TERRY, *Am. Lab.,* **16**(5), 90(1984).
9. VILLABOBAS, R., *Anal. Chem.,* **47,** 983A (1975).
10. HUSKINS, J., *Chem. Eng. News,* **7** (June 7, 1984).
11. LEEMANS, F., *Anal. Chem.,* **43,** 36A (1971).
12. HALL, G., et al., "Principles of Automatic Control," Chap. 18, *Process Instruments and Control Handbook,* 2nd ed., D. Considine, ed., McGraw-Hill, New York, 1974.
13. WALTER, B., *Anal. Chem.,* **55,** 499A (1983).
14. SNYDER, L., et al., *Anal. Chem.,* **48,** 942A (1976).
15. MOTTOTA, H., *Anal. Chem.,* **53,** 1313A (1981).
16. STEWART, K., *Anal. Chem.,* **55,** 931A (1983).
17. RUZICKA, J., *Anal. Chem.,* **55,** 1041A (1983).
18. DESSY, R., *Anal. Chem.,* **55,** 1101A (1983).
19. MCGRATTAN, B., AND D. MACERO, *Am. Lab.,* **16**(9), 19 (1984).
20. OWENS, G., AND R. DEPALMA, *Trends in Analytical Chemistry,* **4,** 32 (1985).

ANSWERS TO PROBLEMS

CHAPTER 2

1. Sensitivity is the ratio of the change of instrument response to the corresponding change in analyte concentration. Detection limit is the analyte concentration that produces a signal that is 3 standard deviations greater than the blank signal. Sensitivity can vary over the range of measured concentrations, whereas the detection limit is measured in the region where the signal disappears into the noise.

2.

Measurement 1		Measurement 2	
1.52	± 0.05	0.94	± 0.03
−1.38	± 0.07	−0.81	± 0.02
0.14	3(0.05)	0.13	3(0.03)

Since the signal due to the analyte (0.14) is less than three times the value of the standard deviation of the measurement (0.15), the detection limit has been reached in measurement 1. **3.** **(a)** Thermal, shot; **(b)** thermal; **(c)** reducing the frequency does not necessarily reduce noise; however, in the case of environmental noise, moving the frequency of the signal away from the frequency of the noise reduces noise. Moving the signal to lower frequencies increases the contribution of flicker noise. **4.** The S/N ratio is proportional to the square root of the integration time; thus, the improvement in the S/N ratio would be

$$\sqrt{\frac{5}{1}} \cong \frac{2.24}{1.00} = 2.24$$

5. **(a)** Thermal, shot, environmental; **(b)** random noise due to thermal, shot, or flicker; **(c)** environmental. **6.** Active filters can move the signal frequency away from environmental noise frequencies and can also discriminate among the signal and nonenvironmental types of noise (thermal, shot, and flicker). **7.** Software boxcar averaging is best suited for applications in which the signal amplitude is changing slowly, whereas ensemble averaging is better suited to signals that are changing rapidly. **8.** **(a)** The S/N ratio is proportional to the number of scans. The improvement in the S/N ratio in going from a single scan to 200 repetitive scans is $\sqrt{200} = 14.1$. **(b)** The time required to make repetitive scans limits the usefulness of ensemble averaging. **9.** The use of fast Fourier transformations allows appropriate digital filter software functions to be applied to the signal in order to reduce noise components. Data contained in a frequency-amplitude format are transformed to time-amplitude format. The resulting waveform is then multiplied by an appropriate mathematical function to obtain the desired frequency response. Finally an inverse Fourier transformation is carried out to convert the filtered time-amplitude signal back to the original frequency-amplitude format. **10.** FFT requires less time to produce a spectrum with an S/N ratio equivalent to the S/N ratio in the same spectrum using ensemble averaging. **11.** Indeterminate (random) errors due to nonenvironmental noise sources (thermal, shot, flicker). **12.** An analyst obtained the following results on two NBS standard reference copper ore samples by repeated analysis.

Sample x: 27.52% Cu by weight
Sample y: 3.41% Cu by weight

The values for percent copper determined by NBS were sample x: 27.62% and sample y: 3.30%.

$$\%E_x = \frac{.10}{27.52} \times 100 = 0.363\%$$

$$\%E_y = \frac{.11}{3.41} \times 100 = 3.226\%$$

Since sample y contains less copper the absolute error is larger. **13.** Each additional significant figure in a given measurement adds to the expense of the measurement. Higher quality reagents, more expensive instruments, and increased analysis time contribute to the increased expense.

14. $$\frac{x}{29} = \frac{x+25}{53} = \frac{x+50}{78}, \qquad x(\text{avg.}) = 29.7\ \mu\text{g/mL}$$

15. $$C_f = \frac{0.42/18{,}500}{1.18/45{,}000} = 0.866$$

$$\text{Conc. alcohol} = 0.866\,\frac{24{,}200 \times 1.50}{60{,}000} = 0.52\ \text{g}$$

$$\%\ \text{wt. alcohol} = 100 \times \frac{0.52}{6.54} = 7.95\%$$

16. Specific activity of the tracer = 150 cpm/mg. Specific activity of isolated sample = 4 cpm/0.2 mg = 20 cpm/mg. Weight of aureomycin = (1 mg)[(150/20) − 1] = 6.5 mg.

CHAPTER 3

1. (a) Bipolar—must have two or more junctions between n-type and p-type semiconductive materials in order to control current flow (Figures 3.1 and 3.3). MOS—include metals and metal oxides in addition to semiconductive material in their construction. The arrangement and operation of these devices are shown in Figures 3.4 and 3.5. **(b)** and **(c)** The packing density of transistors on MOS chips is about four to ten times greater than that for bipolar chips. Bipolar devices have faster response times than MOS devices. **2.** **(a)** The *integral* of a chromatographic signal is proportional to the area under a given peak. **(b)** Location of the *inflection* point of the first derivative of potentiometric titration data determines the equivalence point of the titration. **(c)** A *current-to-voltage converter* is required to convert the output of an amperometric titration (current) to voltage. **(d)** *Interrogration* of a constant voltage produces a voltage ramp signal. **(e)** *Current-to-voltage converters* generate the voltages to be added by the *summation circuit.*

3. $\frac{12}{1000} = 12 \times 10^{-3}\ \text{V} = 12\ \text{mV}$

4.

A	*B*	*C*	*T*
0	0	0	1
1	0	0	1
0	1	0	1
0	0	1	1
1	0	1	1
0	1	1	1
1	1	0	1
1	1	1	0

5. **(a)** Both contain flip-flops, whereas only counters contain additional logic gates. **(b)** An asynchronous counter counts by rippling a count through a sequence of flip-flops. A synchronous counter changes all the flip-flops simultaneously in a given count. **(c)** A *d*-latch will move the logic level on its input to the Q output when the latch receives a clock pulse. This latch has no other external control input. The J–K flip-flop moves logic level from input to output when it receives a clock pulse but also has additional control inputs that make this device more versatile than the *d*-latch. **(d)** An incremental counter counts pulses from zero up, whereas a decremental counter counts down to zero from a preset value. **(e)** A decade counter counts from 0 to 9 before repeating, whereas a modulus 16 counter counts from 0 to 15. **6.** **(a)** Logic 1, logic 0, and high impedance (effectively disconnected); **(b)** The data bus must be capable of communicating with many devices but only with one device at any instant. Devices other than the one being addressed may be effectively disconnected from the bus.

CHAPTER 4

1. The central processing unit (CPU) controls the overall operation of the computer and has three components: ALU (arithmetic logic unit) performs all arithmetic and logic operations; CU (control unit) manages the interpretation and execution of instructions and controls the flow of data on the buses; and registers are temporary storage locations that can be rapidly accessed by the CU and ALU. Memory consists of main memory (RAM and ROM) and mass memory (floppy disks, hard disks, etc.), which provide both permanent and temporary storage of programs and data. I/O devices are keyboards, video displays, printers, plotters, and instrument I/O (data acquisition and control lines). **2.** Data bus carries digital data to and from devices using parallel transmission. Address bus carries the address of the external device to be connected to the data bus for transmission. Control bus carries the control and status signals required for transmission of data.

3. **(a)** A microprocessor usually contains only the CPU with limited amounts of memory (registers), whereas a microcomputer includes memory (RAM, ROM), disks, keyboard, and facilities for connecting other I/O devices to the CPU. **(b)** In serial mode the data are transmitted sequentially, one bit after another, and require only two wires for transmission, whereas in parallel mode the data are transmitted simultaneously on parallel wires, usually in groups of bits (one byte). **(c)** Multiplexing involves selecting one output from several possible inputs, whereas demultiplexing involves sending one input to one of several possible outputs. **(d)** An ADC converts analog signals to digital format, whereas a DAC reverses the process. **(e)** Both devices sample an analog signal and hold it until an ADC can make the conversion. If the sampling time is long compared with the holding time, the device is known as a track-and-hold device. If the sampling time is shorter than the holding time, the device is called a sample-and-hold amplifier.

4. $136_8 = 94_{10}$, $\quad 5.00\text{ V}\,\dfrac{94}{255} = 1.84\text{ V}$

5. $\dfrac{6.23}{10.0} \times 1024 = 638_{10} = 1176_8 = 1001111110_2$

6. **(a)** $1/2^4 = 1/16$ or 1 part in 16 resolution. **(b)** $1/2^{12} = 1/4096$ or 1 part in 4096 resolution. **7.** **(a)** Fast conversion times that are independent of the magnitude of the analog signal. **(b)** Better for smaller analog signals (millivolts), slow but can most accurately convert signals with small amplitudes. **(c)** Excellent accuracy but slow conversion rate. **8.** **(a)** Universal asynchronous transmitter/receiver—facilitates serial data transmission between computers and instruments. **(b)** Programmable peripheral interface chip—a chip containing parallel connections for linking the computer to several instruments; the configuration of the chip can be programmed in software. **(c)** Buffers—parallel registers that usually hold at least one byte (8 bits) of data in temporary storage. **9.** **(a)** Direct communications with CPU—no intermediate software required; tedious for human programmers but allows fast execution of programs; all software programs are ultimately converted to machine code for execution. **(b)** Next most direct means of programming requires a software program known as an assembler to convert instructions in assembly language to machine language. **(c)** Easier for human programmers, one instruction may generate several machine language instructions, requires software known as compilers or interpreters to convert instructions (source code) to machine language (object code). **(d)** Program dedicated to performing specific tasks such as word processing, spreadsheet calculations, or data management; contains powerful, easy-to-understand functions; usually written in a high-level language that is transparent to the user. **10.** **(a)** Both generate object code from source code; an interpreter translates one line at a time, while a compiler operates on the entire program. Interpreters are well suited for interactive programming, whereas compilers generate more efficient source code. **(b)** All computer systems require software known as an operating system to run high-level languages. An applications package is dedicated to performing specific tasks and requires an operating system to function. **(c)** When a number of devices have data to be sent to a computer, one of two strategies is usually used. Each device may be "polled" sequentially in a continuous loop, or devices may "interrupt" the current operation of the computer to report in. **11. (a)** $157_8 = 6F_{16} = 1101111_2$. **(b)** $011110010_2 = F2_{16} = 242_{10}$. **(c)** $010011001001_2 = 2311_8 = 1225_{10}$. **(d)** $1245_{10} = 010011011101_2 = 2335_8 = 4dd_{16}$. **12.** A local area network (LAN) can be used to connect instruments to computers. These intermediate computers may have the capability for laboratory information management (LIM). The intermediate computers may then pass data and reports on to larger computers at remote locations (see Figures 4.28–4.30).

CHAPTER 6

1. From 20.0° (violet) to 39.05° (red). **2. (a)** Assuming normal incidence, 741 grooves/mm; **(b)** 108.4 nm; **(c)** 22.95°; **(d)** 312.8 grooves/mm; **(e)** 523.7 nm; **(f)** 123.2 grooves/mm. **3. (a)** 0.148 μm; **(b)** 0.167 μm; **(c)** 0.210 μm; **(d)** 0.272 μm. **4.** First-order passband is at 1.250 μm; third order at 416.7 nm. **5. (a)** From the second order at 750 nm and the fourth at 375 nm. **(b)** A cutoff filter lying midway between 500 and 750 nm would eliminate 750 nm; a cut-on filter lying midway between 500 and 375 nm would eliminate 375 nm. Also a first-order interference filter peaked at 500 nm would be excellent. **(c)** 543 nm. **6.** From 4.3% to 1.3%. Reflectance is 33%, and since there is no absorption, transmittance is 67%. **7.** In both cases A and B the slit distribution is trapezoidal with a bandwidth equal to the larger slit and a bandpass equal to 1.5 times the larger slit. In the last cases, the slit distribution is rectangular and the bandwidth and bandpass are equal. **8. (a)** 999 grooves in the first order; 250 grooves in the fourth order; **(b)** just under 0.59 nm bandpass or 0.29 nm FWHM; **(c)** 0.18 nm. **9.** No. If bandwidth = 8.0 nm, bandpass = 16.0 nm, and at the baseline an overlap of 1.9 nm exists. **10. (a)** 5.05 nm/mm. **(b)** Yes; $\Delta\lambda = 0.050$ nm bandwidth or 0.10 nm bandpass. The lines are separated by 0.18 nm. **11.** Effects of dark current are additive, affecting both sample and reference beam powers: $A = \log (P_0 + P_D)/(P + P_D)$. A factor would have to affect P_0 and P by the same fraction of each—that is, be multiplicative rather than additive.

CHAPTER 7

1. Stray radiation is probably from second-order reflection of 758-cm^{-1} radiation because CaF_2 is opaque at 738 cm^{-1} as well as 379 cm^{-1}. If the stray radiation were due to the third-order reflection of 1137-cm^{-1} radiation, then both plates, which transmit 1137-cm^{-1} radiation, would have shown the 379-cm^{-1} stray radiation.

2.

True absorbance	*Observed absorbance for stray radiation at* 0.1%	1.0%	5.0%
0.500	0.499	0.491	0.457
1.000	0.996	0.963	0.845
1.500	1.487	1.375	1.104
2.000	1.959	1.703	1.243

3. $\varepsilon = 3120$ liter $mol^{-1}\,cm^{-1}$. **4.** For Ti: $x = 5.40A_{400} - 3.39A_{460}$; for V: $y = 15.98A_{460} - 7.95A_{400}$. Sample 1, 0.053% Ti, 0.044% V; sample 2, 0.059% Ti, 0.396% V; sample 3, 0.099% Ti, 0.182% V; sample 4, 0.198% Ti, 0.188% V; sample 5, 0.294% Ti, 0.194% V. **5.** Tyrosine at 294 nm and tryptophan at 280 nm. **6.** First end point at 6.80 mL; second end point at 9.80 mL. Sodium acetate, 0.0749*M* and chloraniline, 0.0331*M*. **7.** First rise extrapolated back to zero absorbance gives system blank, 0.23 mL. Net titrant volume: 2.70 – 0.23 = 2.47 mL; 88.2 μg of magnesium present. **8.** For *p*-nitrophenol at 407 nm where anion absorbs, $pK_a = 7.01$; at 317 nm where undissociated acid absorbs, $pK_a = 6.96$. For papaverine (protonated cation), $pK_a = 6.40$ (average). **9.** Bromophenol blue $pK_a = 4.06$. Methyl red: $pK_a = 4.93$. Bromocresol purple: $pK_a = 6.29$.

10.

Concentration (*M*)	dC/C (thermal noise)	dC/C (shot noise)*
0.0050	0.0332	0.0310
0.0100	0.0192	0.0167
0.0150	0.0146	0.0119
0.0200	0.0126	0.0096
0.0250	0.0116	0.0082
0.0300	0.0111	0.0073
0.0400	0.0109	0.0063
0.0500	0.0115	0.0058
0.0600	0.0126	0.0055
0.0800	0.0163	0.0055
0.1000	0.0226	0.0057
0.1100	0.0270	0.0060

* Assuming $k = 0.004$, a reasonable estimate for high-quality spectrometers.

11.

Concentration (*M*)	dC/C
0.0600	0.00255
0.0700	0.00237
0.0800	0.00231
0.0900	0.00230
0.100	0.00231
0.110	0.00234
0.120	0.00240
0.130	0.00246
0.140	0.00250

12.

Concentration (*M*)	dC/C
0.0100	0.0171
0.0200	0.0113
0.0300	0.0103
0.0400	0.0104
0.0500	0.0117
0.0600	0.0141
0.0700	0.0189
0.0800	0.0307
0.0900	0.0705

13.

Concentration (*M*)	dC/C
0.0600	0.0030
0.0700	0.0021
0.0800	0.0022
0.0900	0.0029

14. (a) $dC/C = 5.12dT$ by ordinary method (thermal noise) or $1.41k$ (shot noise). Using 5.0-mg reference for 100% T, $dC/C = 0.389dT$ (thermal noise) or

0.369k (shot noise), a 13.2-fold increase (thermal noise) or 3.09-fold increase (shot noise). **(b)** $dC/C = 0.149dT$ by maximum precision method (thermal noise) or 0.127k (shot noise); a 35-fold and 11-fold increase, respectively. **15. (a)** FeL_3^{2+}; **(b)** 3.27×10^{17} (in a medium of approximately pH 4). **16. (a)** FeL_3^{2+}; **(b)** $\varepsilon = 12{,}600$. **17.** Ratio of slopes: 0.353/0.117 = 3/1; FeL_3^{2+}. **18.** Degree of dissociation = 0.333; $K_f = 8.5 \times 10^9$.

CHAPTER 8

1. 1.87 μg. **2.** Equation 8.6 is valid to 8 μg/50 mL. **3. (a)** With mercury lamp, lines at 303 or 313.5 nm fall near excitation maximum; emission filter should reject wavelengths shorter than about 330 nm but pass longer wavelengths. **(b)** Use peak maximum near 295 nm for excitation and fluorescence emission at 350 nm. **4.** Excitation at 340 nm produces only phenanthrene emissions. Excitation at 275 nm maximizes the naphthalene fluorescence at 315 nm, which is almost free from phenanthrene emission. Phenanthrene fluorescence at 370 nm is almost free from naphthalene fluorescence. Phosphorescence offers no advantage. **5. (a)** Fluorescence of window material; scatter from scratches, bubbles, or etched surfaces. **(b)** Concentration quenching and scatter interference. **6. (a)** When excitation wavelength is varied, Rayleigh scatter will vary exactly as the excitation, whereas Raman shift is a constant energy shift. Any fluorescence peaks vary only in intensity, with the wavelength maxima remaining unchanged. **(b)** Limits sensitivity by producing a high blank reading, which drowns out weak fluorescence signals. **(c)** Change excitation wavelength; increase resolution to separate Raman scatter from the sample; use a polarizer; change solvents. **(d)** Raman shifts of the four solvents at excitation wavelengths are:

	Excitation wavelengths			
Solvent	**254 nm**	**313 nm**	**365 nm**	**436 nm**
Cyclohexane	274	344	408	499
Water	278	350	416	511
Ethanol	274	345	409	500
Chloroform	275	346	410	502

The wavelengths listed for each solvent are the Raman band maxima (in nm). **7.** 0.106 μg/liter. **8.** For aluminum, fluorescent peaks are at 417.1, 440.0, and 467.6 nm, giving $^1S^* \rightarrow {}^1S_0$ transitions to $v = 0, 2, 4$ ground-state vibrational levels (23,975, 22,727, and 21,386 cm^{-1}, respectively). Phosphorescence peaks occur for $v = 0, 1, 2, 3, 4,$ and 5 and involve transitions from the triplet state of 20,790, 20,202, 19,531, 18,857, 18,182, and 17,482 cm^{-1}, respectively. [See also *J. Chem. Phys.*, **17**, 1182 (1949).] **9.** Lifetime for a, $\tau = 1.6 \times 10^{-8}$ sec and for b, $\tau = 2.0 \times 10^{-7}$ sec. Demodulation factors are: for a, 0.71; for b, 0.08.

CHAPTER 9

1. Cement A: 0.14% Na_2O, 0.62% K_2O. Cement B: 0.37% Na_2O, 0.43% K_2O. Cement C: 0.22% Na_2O, 0.55% K_2O. **2.** Emission from molecular band systems of CaO and CaOH. Blank would be larger because reading would increase approximately with the square of the bandpass of the filter or monochromator. **3. (a)** 55 μg/mL; **(b)** 170 μg/mL; **(c)** 130 μg/mL. **4. (b)** Ionization of strontium atoms in flame. **(c)** Ionization in (b) is repressed by large excess of calcium atoms. Addition of calcium also contributes a significant background emission due to CaOH. **5.** 0.715 μg/200 mL. **6.** Water: 27.0 μm; 50% methanol–water: 18.7 μm; 40% glycerol–water: 29.4 μm.

7.

	Droplet diameter (μm)		
Flowrate (mL/min)	**Water**	**50% EtOH–H_2O**	**MIBK**
1.0	16.1	12.0	10.9
2.0	18.3	16.3	13.0
3.0	21.0	21.8	15.7
5.0	28.0	35.9	22.5

[See also *Applied Optics*, **7,** 1353 (1968).] **8. (a)** 0.0094; **(b)** 0.094; **(c)** 0.29. **9. (a)** 0.35; **(b)** 0.75; **(c)** 0.97. **10.** To suppress the ionization of 0.23 μg/mL of sodium to 0.01 (1%), add:

Element	At 2500 K	At 2800 K
Cs	0.44 μg/mL	0.74 μg/mL
Rb	1.05 μg/mL	1.6 μg/mL
K	1.04 μg/mL	1.4 μg/mL
Li	22.50 μg/mL	19.5 μg/mL

11. (a) $S/N = 13.3$ (or 6.7 peak-to-peak) for 2 μg/mL; **(b)** Sensitivity = 0.22 μg/mL; **(c)** Detection limit = 0.30 μg/mL.

12.

	Advantages	Limitation
Flame AAS	Relatively inexpensive, fast	Analyte must be in solution
Electrothermal AAS	Lower detection limits, handles solid samples directly	Less precision, slower
FES	Complements AAS, fast, inexpensive	Limited to alkali and alkaline earth metals
AFS	Good for trace analysis, faster than electrothermal AAS	No instrumentation is made commercially

13. (a) Correction of measurements from sample solutions with measurements made on blank solutions, simple to implement, difficult to prepare representative blank solutions. **(b)** Broad-band continuum source can be used with single-beam instruments, fast, necessary to change continuum sources for visible and uv measurements. **(c)** Zeeman correction requires a single source and is fast; magnetic field may destabilize the emission signal from the lamp source. **14. (a)** The process used to convert liquid samples to a stream of small droplets. **(b)** The type of burner in which the lines of flow of the input gases change smoothly, if at all, in both time and space; provides a constant, reproducible flame. **(c)** A compound that is thermally stable (does not decompose at high temperatures)—for example, CaO. **(d)** An easily ionized element that is added to the sample solution to suppress ionization of the analyte. **(e)** A thin graphite plate surrounded by a graphite cylinder; the sample is placed on the plate for electrothermal atomization. **(f)** A device for producing volatile hydrides of certain metals (e.g., As, Ge, Sn, and Te) to improve the AAS detection limit for these metals. **(g)** The central wavelength of the narrow band used for the quantitative determination of an element in AAS. **(h)** Results from the motions of the atoms radiating from the hot cathode of the source, causes a broadening of the emission lines of the spectrum, which in turn reduces the detection limit and sensitivity of elements in AAS. **15. (a)** Spectral line interferences due to inadequate resolution of the resonance line of the analyte from the lines of other elements or from a molecular band; can be corrected by using a better monochromator or removing the interfering substance prior to the measurement. **(b)** Vaporization interferences—a situation in which the vaporization of the matrix material controls the rate of release of analyte to the flame. Thus, the composition of the matrix determines the amount of analyte measured; can minimize this interference in the determination of many elements by increasing the flame temperature. **(c)** Ionization interferences—ionization of elements with low ionization potentials reduces the observed concentration of the element in the flame. Addition of an easily ionized element as a buffer to suppress the ionization of the analyte helps to minimize this type of interference. **16. (a)** Drying—evaporization of the solvent and removal of any volatile matrix components. **(b)** Ash—volatilization of higher-boiling-point matrix components and pyrolysis of matrix materials. **(c)** Atomization—converts analyte into free atoms.

CHAPTER 10

1. Cadmium, magnesium, and zinc. **2.** On double logarithmic paper, graph the reciprocal of the ratio (because these are density readings): Pb reading/Mg reading, against the lead concentration: **(a)** 0.26 mg/mL; **(b)** 0.34 mg/mL; **(c)** 0.41 mg/mL. **3.** Answer from graphical solution:

$$0.0035\ \mu g/mL \times 10\ (\text{dilution factor}) = 0.035\ \mu g/mL$$

4. $124.5 - 8.2 = 116.3$ units

$$1.4 - 0.02 = 1.38\ \text{ppm}$$

$$116.3/1.38 = 84.3\ \text{units/ppm}$$

$$94.5 - 8.2 = 86.3\ \text{units}$$

$$(86.3\ \text{units}) \times (1\ \text{ppm}/84.3\ \text{units}) = 1.02\ \text{ppm} = 1.0\ \text{ppm}$$

5. dc arc—useful in qualitative analysis of trace elements; results are nonreproducible, since the arc tends to wander over the sample. ac arc—better reproducibility but less sensitivity than dc arc. ac spark—good precision and stability; not as sensitive as dc arc source. **6.** ICP—better precision, ease of handling liquid samples and control source. Electrical sources have fewer parameters to control and in general are less expensive than ICP systems. Electrical sources are well suited for qualitative and semiquantitative analyses of solid samples. **7.** Less sample reaches atomizer in ICP; time required to wash out analyte spray chamber is longer for ICP due to the slower flowrates involved. Parameters affecting drop size must be more carefully controlled in ICP. **8.** Paschen-Runge—covers a large range of wavelengths, nonlinear dispersion, slow optical speed. Wadsworth mounting—fast optical speed, lower linear dispersion than Paschen-Runge, sharp intense lines, mounting takes up more space than P-R. Echelle—excellent dispersion and resolution over a wide range of wavelengths, facilitates the detection of low-intensity lines. **9.** Photo—integrates intensity of impinging radiation over the entire time of exposure; records a large number of wavelengths simultaneously; nonlinear response, film processing required; longer exposure time required for weak lines. Photomultiplier tubes—linear response, good dynamic intensity range, quick data acquisition, excellent precision and sensitivity; one detector required for each wavelength. Diode arrays—simultaneous measurement of intensities at a large number of wavelengths, easy to interface an array to a computer system.

CHAPTER 11

1. 910 cm^{-1}. **2.** First overtone at 1108 cm^{-1}; second overtone at 1662 cm^{-1}. NaCl for overtone bands; KBr if fundamental band is also to be observed. **3.** **(a)** 3060 cm^{-1}; **(b)** 3293 cm^{-1}; **(c)** 1128 cm^{-1}; **(d)** 1465 cm^{-1}; **(e)** 2143 cm^{-1}; **(f)** 1744 cm^{-1}. **4.** **(a)** 1.91 μg/mL; **(b)** 1.30 μg/mL; **(c)** 0.351 μg/mL; **(d)** 0.283 μg/mL; **(e)** 0.391 μg/mL; **(f)** 1.71 μg/mL. **5.** 1.72%. **6.** **(a)** $1.72 \times 10^{-4} M$; **(b)** $4.30 \times 10^{-4} M$; **(c)** $1.46 \times 10^{-3} M$; **(d)** $1.06 \times 10^{-4} M$; **(e)** $5.06 \times 10^{-5} M$. **7.** **(a)** Cell 1, 0.085 cm; cell 2, 0.13 cm; cell 3, 0.044 cm; cell 4, 0.022 cm; **(b)** 0.031 mm. **8.** **(a)** *meta*, **(b)** *para*, **(c)** *ortho*. **9.** *p*-Bromotoluene. **10.** $(CH_3)_3—C—COCH_3$. **11.** About 3.3%. **12.** 14.5% of compound 1, 10.1% of compound 2, and 75.4% of compound 3. **13.** $X = 0.53\%$, $Y = 1.40\%$.

CHAPTER 12

1. There is failure to appreciate the magnitude of the decrease in the efficiency of the photomultiplier tube coupled with the decrease in grating efficiency over the Raman range of 2960 cm^{-1} when the spectrum is excited by a He-Ne laser. **2.** **(a)** 16 cm^{-1} per sec or 960 cm^{-1} per minute; **(b)** 632.8 nm $\pm$ 0.55/2 nm is a range from 15,809.5 cm^{-1} to 15,795.8 cm^{-1}, thus 13.7 cm^{-1} slit width. Therefore, the slit width must be 0.6 mm. **3.** It is reduced by a factor of approximately 22.7. **4.** 0.35 to 1.

5.

Benzene (Δ cm^{-1})	Wavelength of Raman lines (nm) for exciting lines 632.8	488.0	514.5	568.2	647.1
606	658.1	502.9	531.1	588.5	673.5
850	668.8	509.1	538.0	597.0	684.8
991	675.1	512.8	542.1	602.1	691.4
1176	683.7	517.7	547.6	608.9	700.4
1584	703.3	528.9	560.2	624.4	721.0
1605	704.3	529.5	560.8	625.2	722.1
3047	783.9	573.2	610.2	687.2	806.0
3063	784.9	573.8	610.7	687.9	807.1

6. The C—H stretching frequencies at 3047 and 3063 cm^{-1} from the shorter wavelength exciting line would overlap the spectrum from the longer wavelength exciting line in each case. **7.** N_2O has no center of symmetry while CS_2 does. The structures must be N—N—O and S—C—S (type of bonds not intended to be indicated). **8.** The molecule has a center of symmetry. If planar, the molecule would have three unpolarized Raman lines; if tetrahedral, it would have only one. The spatial configuration is tetrahedral.

9.

Unknown	Volume percent 1,2,3-	1,2,4-	1,3,5-
A	33.3	33.2	33.3
B	40.0	26.0	34.1
C	25.0	37.0	38.0

10. Benzoyl chloride. **11.** Dimethylacetylene.

CHAPTER 13

1. 0.2066Å; $Z = 70$. **2. (a)** L_{II} to K_I; **(b)** M_{III} to K_I; **(c)** M_V to L_{III}.

3 and 4.

Element	K edge (eV)	L_{III} edge (eV)	$K\alpha_1$ (Å)
Mg	1.299	0.050	9.92
Mn	6.540	0.639	2.10
Cu	8.980	0.933	1.54
Mo	20.0	2.523	0.709
W	69.6	10.20	0.209
Pt	78.4	12.06	0.187

5. For L_{III} spectra, $Z \leq 92$; for K spectra, $Z = 39$. **6.** Operating voltage, 1300–1700 V. **7.** Operating voltage, 1470–1545 V. **8.** K edges of iodine and argon, respectively.

9.

Crystal	Ti	Co	Ag	Pt
NaCl	58.3°	37.0°	11.4°	3.8°
EDDT	36.4°	23.4°	7.3°	2.4°
Gypsum	20.8°	13.5°	4.2°	1.4°

10. Use of pulse height analyzer, regulation of voltage applied to source, use of appropriate filter, incorporation of monochromator into system.

11.

U (wt. %)	Slope (counts sec^{-1} $\%^{-1}$ U)	Time (1.96 σ) (min)
2	165	2.57
5	110	1.77
10	67	1.80
15	45	2.08
20	31	2.87

12. NaCl crystal; $2\theta = 143°41'$; tube voltage, 3.5 kV; counting time: background, 28 sec; sample, 84 sec for 0.4% S. **13. (a)** Set baseline at 6.5 V and window for 6 V. **(b)** Set baseline at 15.5 V and window for 9.5 V. **(c)** Set baseline at 17.5 V and window for 4.5 V. **(d)** Set baseline at 27 V and leave window open (or set window for 37 V). **(e)** Set baseline at 8.2 V and window for 11.8 V. **14. (a)** Use analyzing crystal of higher dispersion and/or finer collimation. Topaz will resolve these lines with a $\Delta\ 2\theta$ of 3.0°, while LiF would have a $\Delta\ 2\theta$ of only 1.5°. The use of Fe $K\alpha_1$ radiation would excite Cr spectra (K edge at 2.070 Å) but not the Mn $K\alpha$ spectra (K edge at 1.895 Å). **(b)** Maintain the X-ray tube voltage less than 10.4 kV, where $\lambda_{max} = 1.192$ Å, L absorption edge for Re = 1.177 Å and Zn K edge = 1.283 Å. One could also use the K absorption edge of Ga (1.195 Å) as a selective filter. **(c)** Use a thin sheet of Ni as a selective filter.

15. (a)

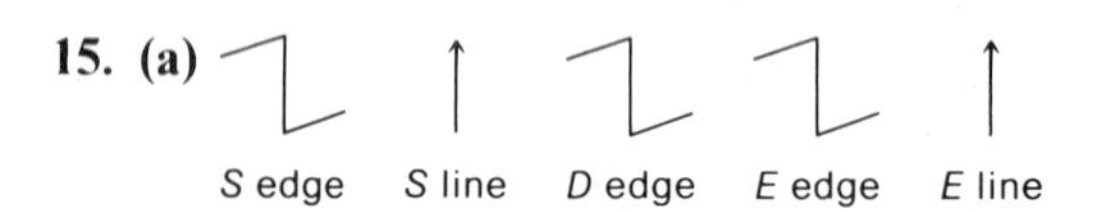

(b) Exchange position of S edge and line with E edge and line.

(c)

E edge   E line   D edge   D line   S edge   S line

(d) Exchange position of D edge and line with S edge and line.

16. (a) Sr: 16.09 kV; V: 17.05 kV. **(b)** Sr: $K\alpha_1 = 25.2°$ and $K\beta = 22.4°$; Y: $K\alpha_1 =$ 23.8° and $K\beta = 21.2°$. **(c)** A: 0.251%; B: 0.180%; C: 0.059%. **17.** $\Delta\lambda = 0.002$ Å, $\Delta\theta = 0.028°$. Use a filter to reduce Zr $K\alpha_1$ intensity or, better, use a proportional or semiconductor detector with pulse height discrimination.

18.

Top layer			Bottom layer		
2θ	**λ (Å)**	**Line**	**2θ**	**λ (Å)**	**Line**
27.6°	2.53	Ti $K\beta$	34.0°	3.10	Ca $K\beta$
30.0°	2.75	Ti $K\alpha$	36.0°	3.28	Ca $K\alpha$
58.0°	5.15	Ti $K\beta$ (2nd order)	71.0°	6.17	Ca $K\alpha$ (2nd order)
63.0°	5.55	Ti $K\alpha$ (2nd order)			

19. **(a)** 16 cm^2/g; **(b)** 52 cm^2/g; **(c)** 47 cm^2/g. **20.** **(a)** Decrease in intensity is 64% for 1-cm cell length versus air; **(b)** a 44% decrease versus an octane blank. **21.** Decrease in intensity is 97.6% for 1-cm cell length versus air or 58.3% versus an octane blank. **22.** Sample 1, Ni $K\alpha_1$ on Mo $K\alpha_1$. Sample 2, Cu $K\alpha_1$ on Mo $K\alpha_1$. Sample 3, Au $L\beta_1$ on Ni $K\alpha_1$. Sample 4, Pd $K\alpha_1$ on Ni $K\alpha_1$.

23.

2θ:	111.0°	100.2°	57.8°	48.8°	45.1°	44.0°	40.4°
λ(Å):	3.320	3.090	1.947	1.664	1.545	1.509	1.391
Element:	Ni $K\alpha_1$(2nd)	Cu $K\alpha_1$(2nd)	Fe $K\alpha_1$	Ni $K\alpha_1$	Cu $K\alpha_1$	Ni $K\beta$	Cu $K\beta$

24. See G. L. Clark, ed., *Encyclopedia of Spectroscopy*, pp. 704–711, Van Nostrand Reinhold, New York, 1960. **25.** Elements with $Z < 25$, especially good for Cl, P, and S in hydrocarbon matrices. **26.** 1.66 Å, Ni $K\alpha_1$; 2.10 Å, Mn $K\alpha_1$; 3.32 Å, Ni $K\alpha_1$(2nd); 4.20 Å, Mn $K\alpha_1$(2nd); 6.64 Å, Ni $K\alpha_1$(4th). **27.** A fully extended planar carbon chain, since the C—C distance is essentially this value. **28.** No; the recorded peaks would overlap by about 136 eV.

CHAPTER 14

1. 5.62×10^9 disintegrations/min.

2.

	A/A_0 fraction		
Nuclide	**14 days**	**30 days**	**60 days**
^{32}P	0.506	0.233	0.054
^{131}I	0.299	0.0754	0.0057
^{198}Au	0.0272	4.45×10^{-4}	1.98×10^{-7}

3.

Number of counts	P.E. (%)	1σ (%)	2σ (%)
3,200	1.19	1.77	3.54
6,400	0.84	1.25	2.50
8,000	0.75	1.12	2.24
25,600	0.42	0.63	1.26
102,400	0.21	0.31	0.62

4. Dead time is 197 μsec; loss is 704 counts/min. **5.** **(a)** 50%; **(b)** 80%; **(c)** 96%; **(d)** 99.9% efficiency. **6.** For a 99% counting efficiency, the counting rate should not exceed 1% of the reciprocal value of the dead time. **(a)** 40,000 counts/sec; **(b)** 10,000 counts/sec; **(c)** 2000 counts/sec; **(d)** 37 counts/sec. **7.** $A_0 =$ 2250 counts/min. The half-life is 18 min. The isotope is bromine-80. $B_0 = 440$ counts/min. The half-life is 264 min. The isotope is metastable bromine-80. **8.** $E_{max} = 0.34$ MeV. The aluminum half-thickness is 11 mg/cm^2. **9.** 200 mg. **10.** In air: 162 cm for strontium-90 and 910 cm for yttrium-90. In iron: 0.25 mm for strontium-90 and 1.39 mm for yttrium-90. **11.** 71 min. **12.** 0.26 μg. **13.** 2.8 mg. **14.** 62 sec for copper; 96 sec for manganese; 9.0 days for cobalt; impossible for nickel unless the cooling period is shortened or the neutron flux is increased tenfold. **15.** $A_0 = 1.23 \times 10^6$ disintegrations/min on the reference date; 1.68 picomoles of thymidine per microgram of DNA. **16.** For sodium, whose beta is 1.39 MeV, the range is 634 mg/cm^2. For potassium, whose beta is 3.52 MeV, the range is 1740 mg/cm^2. Use an aluminum absorber: 634 mg/cm^2 to remove the sodium activity. **17.** 10.2% error. **18.** Nuclides observed: sodium-22, calcium-47, iron-59, cobalt-60, zinc-65, strontium-85, rubidium-86, zirconium-95, niobium-95, ruthenium-103, antimony-124, cesium-134, barium-140, and lanthanum-140.

CHAPTER 15

1. Bearing in mind the 1:2 relationship of the enolic hydrogen to the keto methylene group: % enol = [37.0]/[(37.0 + 1/2 × 19.5)] × 100 = 79.1%. **2.** Since benzophenone contains 5.53% hydrogen, the sample contains (0.8023/0.3055) (184/228) (5.53%) = 11.72% **3.** Since excess phenol is used, the chains must be terminated at both ends by phenol groups, and the average number of methylene bridges per chain is one less than the number of phenolic nuclei. The average number of aromatic and methylene protons per molecule (chain) is $(3n + 2)$ and $2(n - 1)$, respectively, where n is the number of monomer units. One extra proton is present on each terminal phenol group. Thus, $(3n + 2)/2(n - 1) = 30/18$ and $n = 16$. Average molecular weight is $1684 = (16 \times 92) + (15 \times 14) + 2$. **4. (a)** The three signals are attributable to the methyl group in *o*-cyanotoluene, the methylene group on *o*-cyanobenzyl chloride, and the methine group in *o*-cyanobenzal chloride. **(b)** The relative molar proportions are 13/3, 20/2, and 10/1, and the proportions by weight are 1.0, 3.0, and 3.7.

5. 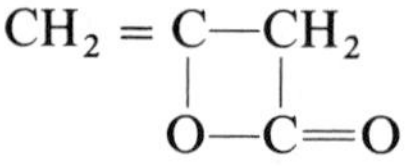
```
CH2 = C—CH2
      |   |
      O—C=O
```
6. In compound I the CH_3 resonance would be a doublet with $J = 6$ Hz, showing that there was a proton attached to the adjacent carbon atom. In compound II the CH_3 resonance, in addition to being farther downfield, would be only weakly coupled with protons on carbons once removed and would have $J \cong 1$ Hz. **7.** $HPO(OH)_2$; $H_2PO(OH)$. **8.** First structure. **9.** Second structure.

10. 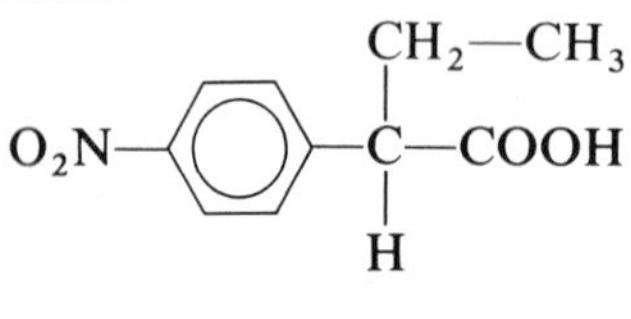
```
                   CH2—CH3
                   |
O2N—(benzene ring)—C—COOH
                   |
                   H
```
11. Ditolyl disulfide.

12. $C_6H_5-CH_2-C(CH_3)_2-Cl$ (benzene ring$-CH_2-\overset{CH_3}{\underset{CH_3}{C}}-Cl$) **13.** $CH_3-CH_2-\overset{H}{\underset{Br}{C}}-\overset{O}{\overset{\|}{C}}-O-CH_2-CH_3$

14. $CH_3-CH_2-O-\underset{O}{\underset{\|}{C}}-CH_2-CH_2-\underset{O}{\underset{\|}{C}}-O-CH_2-CH_3$

15. On the delta scale, benzene is located at $(418.4/60) = 6.97$. The olefinic protons are located at $6.97 - (56.9/60) = 6.03$ and at $6.97 - (87.5/60) = 5.52$. Each is weakly coupled with the other ($J = 1$ Hz) and with the methyl protons. The methyl protons at $6.97 - (304.6/60) = 1.90$ are coupled to two nonequivalent olefinic protons. Singlet is CH_3-O group. **16.** Use 4x multiplier yielding an output of 20 MHz plus a 0.7-MHz incremental oscillator; mix the two outputs in a single sideband modulator and select the lower sideband: $20 - 0.7 = 19.3$ MHz. **17.** Varying H_0 while keeping the observing rf field, ν_1, and the perturbing rf field, ν_2, constant would mean that ν_2 would have the correct frequency to resonate with the nuclei to be decoupled at only one value of H_0. If, as H_0 is varied with ν_1 remaining constant, ν_2 is also varied by a feedback mechanism so that the ratio of ν_2 to H_0 remains constant, a magnetic sweep can be used for decoupling experiments. See the EM 360 diagram, Figure 15.6. **18. (a)** 1.988; **(b)** 0.126. **19. (a)** Compound 1 = 51.50 mole percent, compound 2 = 39.79 mole percent, and compound 3 = 8.71 mole percent; **(b)** weight percentages: compound 1 = 71.4%, compound 2 = 20.2%, compound 3 = 8.4%.

CHAPTER 16

1. 2490 V for mass 18 and 224 V for mass 200. **2.** 9.03 μsec for mass 44 and 8.9 μsec for mass 43. **3.** 480.493. **4.** For the pair—tridecylbenzene and phenyl undecyl ketone, the resolution required is 7140. For the pair—1,2-dimethyl-4-benzoylnaphthalene and 2,2-naphthylbenzothiophene, the resolution required is 9320. **5. (a)** 10,000; **(b)** 6500; **(c)** 10,000. **6.** The mass is 237.1473(5) with the uncertainty for the fourth place in parentheses. The empirical formula is $C_{12}H_{19}N_3O_2$. **7. (a)** $C_9H_8O_3$; **(b)** C_8H_8O; **(c)** $C_{14}H_{12}$; **(d)** $C_5H_6N_2$; **(e)** C_6H_7NO; **(f)** $C_6H_3Cl_2NO_2$; **(g)** $C_{13}H_{11}N$; **(h)** $C_{17}H_{18}O_5S$; **(i)** $C_9H_7NO_3$. **8. (a)** The fragment ions are C_3H_6NS, C_4H_8S, $C_3H_6NO_2$, and $^{13}CC_3H_7S$, respectively. **(b)** The hydrogen atom located three carbon atoms away from the carbonyl group is transferred to the carbonyl oxygen via a McLafferty rearrangement with the simultaneous cleavage of the C=2, C=3 bond. **(c)** The fragment ion of mass 75.0267 is $CH_3SCH_2CH_2^+$. **9.** Only one carbon atom is indicated plus a heavy monoisotopic element; thus, CH_3I is the structure. **10.** Compound A has three chlorine atoms. Compound B has five bromine atoms. Compound C has four bromine atoms. Compound D has one chlorine atom and one bromine atom. Compound E has two chlorine atoms and one bromine atom. **11.** $m/z = 90(P)$: one sulfur atom is present from the isotopic abundance. Now $5.61 - 0.78 = 4.58$, and $4.58/1.08 = 4$ carbon atoms (or mass 58). The difference between the parent peak of 90 and the mass of one sulfur atom, divided by 14, equals four methylene groups plus two hydrogen atoms. The empirical formula is $C_4H_{10}S$. The compound could be a dialkyl sulfide or an alkyl mercaptan. $m/z = 89(P)$: one chlorine atom and one nitrogen atom (note the odd mass of the parent peak) are present; the residual mass is 40. Since the mass at $P + 1$ indicates not more than two carbon atoms, the probable empirical formula is C_2ClNO. $m/z = 206(P)$: two sulfur atoms are present plus a residual mass of 142. $P + 1$ is 12.5%, which indicates that not more than ten carbon atoms are present [12.5 − 2(0.78) = 10.9]. The empirical formula is $C_{10}H_{22}S_2$. $m/z = 230(P)$: two bromine atoms are present plus a residual mass of 72. Now 72/14 = 5 methylene groups plus two hydrogen atoms. The empirical formula is $C_5H_{12}Br_2$. $m/z = 140(P)$: the peak at $P + 1$ is 9.54% and that at $P + 2$ is 5.77% of the parent peak. The latter indicates that one sulfur atom is present. Not more than seven carbon atoms are present (9.54 − 0.76 = 8.78 and 7 × 1.1 = 7.7). An unusually large residuum remains. Table 16.1a uncovers the possibility that two nitrogen atoms might be present. $m/z = 151(P)$: one chlorine atom and one nitrogen atom are present. This leaves a residual mass of 102, and 102/12 = 8 carbon atoms plus six hydrogen atoms. Since the base peak is the parent peak, we may be dealing with an aromatic compound. **12.** The appropriate equations are

$$
\begin{aligned}
68.03x_1 + 1.14x_2 + 2.25x_3 &= M_{32} \\
4.00x_3 + 5.52x_4 &= M_{39} \\
16.23x_2 &= M_{46} \\
9.61x_3 + 3.58x_4 &= M_{59}
\end{aligned}
$$

13.

Unknown	Methanol (%)	Ethanol (%)	1-Propanol (%)	2-Propanol (%)
A	0.9	8.2	11.7	79.0
B	2.5	13.5	6.9	77.0
C	7.1	7.8	52.7	31.4

14. The metastable peak at 45.0 indicates that the parent ion decomposes to mass 91 with the loss of neutral fragments that total mass 93. The metastable peak at 46.5 indicates that the fragment at mass 91 decomposes to mass 65 plus a neutral fragment of mass 26. These data, coupled with the peak intensities, indicate that one is dealing with a ring compound possessing the structure $C_6H_5{-}CH_2{-}$ plus mass 93, which is probably C_6H_5O. The compound is tolyl phenyl ether. **15.** The strength of the parent indicates that a ring structure may be involved. The metastable peak at 69.4 indicates that the decomposition route proceeds from mass 122 to mass 92 plus a neutral fragment of mass 30. The metastable peak at 46.5 indicates that the decomposition of the fragment ion of mass 91 proceeds to mass 65 plus a neutral fragment of mass 26. The ion peak at mass 92, an even mass arising from an even mass molecular ion, suggests a rearrangement reaction. Coupled with the possible transition from mass 122 to mass 91, which involves a loss of mass 31, the presence of a CH_2OH is suggested. The latter group could participate in a

McLafferty rearrangement that would account for the loss of CH_2O (mass 31). The compound is 2-phenylethanol. **16.** $C_4H_8O_2$: The empirical formula indicates either one ring or one double bond. The base peak at mass 60 is typical of an acid and arises from a McLafferty rearrangement with a loss of a neutral fragment of mass 28 that would be $CH_2{=}CH_2$ (CO is seldom lost from a molecular ion). Butyric acid is the compound. $C_4H_6O_2$: The empirical formula indicates two double bonds and two rings, or one double bond and one ring. From the parent peak one hydrogen is lost from mass 86 to mass 85, and either CH_3O or CH_2OH is lost from mass 86 to mass 55 (the base peak). The base peak at mass 55 suggests an unsaturated resonance structure such as $C{=}C{-}C{=}O$. Coupled with the loss of a methoxy group, the suggested structure is methyl acrylate. Compound C: The isotopic abundance indicates not more than nine carbon atoms. The strength of the parent peak suggests an aromatic ring, which is reinforced by the peaks at mass 91 and mass 92. The latter is the result of a McLafferty rearrangement, which must involve a carbonyl group (actually the CHO group rather than an ethyl group). The neutral fragment lost to attain the fragment ion of mass 105 could be either an ethyl group or a formyl group. The compound is 3-phenylpropionaldehyde. Compound D: The isotopic abundance indicates not more than eight carbon atoms. The strength of the parent peak suggests an aromatic system, which is reinforced by the rearrangement peak at mass 92. The rearrangement transition from the parent peak to the base peak involves the loss of a neutral fragment of mass 32, whereas the fission transition results in the loss of mass 31. These neutral fragments could be CH_3O ($m/z = 31$) and CH_2OH ($m/z = 32$). For the latter, the donor hydrogen must be six bonds away from the electron pair on an oxygen. Coupled with the structure, $HO{-}C_6H_4{-}C{=}O$, typical for a peak at mass 121, there must be a hydroxy group *ortho* to the side chain. The compound is methyl salicylate (or methyl 2-hydroxybenzoate. **17.** The isotopic clusters indicate the presence of one bromine atom in the parent molecule and at fragment ions of masses 167, 151, and 123. From the metastable peaks:

m/z 102.2: $153^+ \rightarrow 125^+ + 28$

m/z 100.2: $151^+ \rightarrow 123^+ + 28$

m/z 54.7: $103^+ \rightarrow 75^+ + 28$

In each instance the neutral fragment could be CO or $CH_2{=}CH_2$. Other possible fragmentation steps are

$196^+ \rightarrow 167^+ + 29$ loss of CHO or C_2H_5

$196^+ \rightarrow 151^+ + 45$ loss of $HOCH_2CH_2$ or C_2H_5O

$196^+ \rightarrow 103^+ + 93$ loss of CH_2Br (isotope cluster absent)

The pattern of neutral fragment loss strongly suggests the ethoxy group, and probably two groups. These are being lost as an ethoxy group, the ethyl portion, or an ethylene molecule; these losses are typical of an ethoxy group. If there are a $-CH_2Br$ group and two C_2H_2O groups, only 13 mass units are unaccounted for. The latter must be a methine group, which is the juncture point for the other groups.

The structure is $BrCH_2CH(OC_2H_5)_2$. **18.** The isotopic cluster at mass 117 indicates the presence of three chlorine atoms, that at mass 97 indicates two chlorine atoms, and that at mass 51 indicates one chlorine atom. Mass 117, less three chlorine atoms, leaves a residual mass of 12, which implies a fragment ion, Cl_3C—, and not the parent peak. Mass 97, less two chlorine atoms, leaves a residual mass of 27, which suggests CH_3CCl_2—. Mass 51 appears to be CH_4Cl, whose origin must involve complex fragmentation. Putting together the pieces, one deduces that the parent is CH_3CCl_3. **19.** 72.2 kHz. **20.** 89 μsec.

CHAPTER 17

1.

k'	1	2	5	10
t_R (sec)	40	60	120	220
V_R (mL)	2	3	6	11
W_b (mL)	0.44	0.66	1.32	2.41

2. (a) For x, $k' = 4$, and for y, $k' = 6.5$. **(b)** The nature of the stationary and mobile phases and their relative amounts, column temperature, and the ratio L/u. **(c)** $\alpha = 1.2$; yes. **(d)** $Rs = 1.1$, which is not sufficient to obtain a baseline separation. **3. (a)** k' values: toluene, 0.19; biphenyl, 0.61; anisole, 1.23; and nitrobenzene, 4.08. **(b), (c), and (d)** See table below.

	Column 1	Column 2	Column 3
Average linear velocity (cm/sec)	0.48	0.40	0.56
Plate height (mm)	0.16	0.030	0.018
Reduced velocity	32	13.3	12.2
Reduced plate height	7.8	3.0	2.8
Reduced column length	25,000	13,500	13,800
Free column volume, V_M (mL)	1.18	1.99	1.33
Flowrate, F_c (mL/min)	0.68	3.51	4.97

4. (a) $Rs = \text{const}[k'/(1 + k')]$. **(b)** $Rs = \text{const}[(\alpha - 1)/\alpha]$. Graph these equations.

5.

N	k'	α	Rs
3600	2	1.15	1.31
1600	2	1.15	0.87
3600	2	1.10	0.87
3600	0.8	1.15	0.91

6. (a) Decrease the mobile-phase velocity; this means a longer analysis time. One can increase the mobile-phase velocity provided that the column length is increased proportionately. **(b)** Decrease the stationary film thickness to a minimum but not to the point where adsorption on the naked support begins to adversely affect the plate height. When possible, choose as films liquids with a low viscosity so that the diffusion coefficient of the solute(s) will not be unduly small. **(c)** Decrease the size of the support particles (if a packed column is being used). Permeability decreases and higher inlet pressures are needed to force the mobile phase through the column packing. However, because of the higher efficiency, the column length can be decreased, thus decreasing the needed pressure drop somewhat. **(d)** In HPLC use a mobile phase that has a low viscosity; this also means that a lower inlet pressure is required. **(e)** Use a support wherein the likelihood of large stagnant pockets of mobile phase are minimized; these might be superficially porous materials or microspheres with small pores. **7.** Sample 1: total adjusted response is 9830; 46.7% benzene, 33.1% *p*-xylene, and 20.2% *p*-dichlorobenzene. Sample 2: total adjusted response is 13,164; 3.9% benzene, 29.2% *p*-xylene, and 66.9% *p*-dichlorobenzene.

8. Sample	**Benzene**	***p*-Xylene**	**Toluene**
1	27.4%	42.5%	30.2%
2	37.8%	40.0%	22.2%
3	41.2%	24.8%	34.0%
4	21.7%	35.0%	43.3%
5	42.7%	30.7%	26.6%

CHAPTER 18

1. The volumetric flow through the column and sample loop is 10 mL/min. If the loop volume is increased to 5 mL, it will take 30 sec to sweep only the sample into the column. Thus no peak can be narrower than 30 sec. **2. (a)** ΔH, kcal/mole:

	CH_2Cl_2	$CHCl_3$	CCl_4	$CCl_2{=}CCl_2$	C_6H_5Cl	C_6H_6
Tricresyl phosphate	7.50	8.23	7.50	8.82	9.42	7.18
Paraffin	6.96	7.16	7.57	9.07	9.12	6.67
Carbowax 4000	8.81	9.35	7.62	8.97	10.1	7.95
(c) V_g (150 °C)	6.98	12.0	7.54	16.2	39.9	11.1

(b) As a typical example and using 0.001/°C as the coefficient of variation for the density, at 74 °C and for $CHCl_3$, $K = 74.3 \times 0.768 = 57.1$. At 97 °C, $K = 41.5[0.768 - 0.001\ (97 - 74)] = 30.9$. **(c)** See table above. **(d)** The change in temperature required to halve V_g varies from 19 °C at the lower temperature to 28 °C at the higher temperature on the Carbowax 4000 column. **3. (a)** The B term is proportional to D_M and the C_{mobile} term is a function of $1/D_M$. **(b)** See Figure 18.15. The optimum mobile-phase velocity is greater for helium than for nitrogen. **(c)** Helium.

4. (a) and (c) See table below.

Column loading*	u (cm/sec)	H (cm)	t_M (min)	Column loading	u (cm/sec)	H (cm)	t_M (min)
31% (H_2)	1.00	0.46	6.00	23% (H_2)	1.43	0.335	4.20
	1.65	0.34	3.63		2.81	0.243	2.14
	3.22	0.31	1.86		4.15	0.208	1.45
	6.16	0.41	0.97		5.43	0.218	1.10
	8.80	0.52	0.68		6.64	0.228	0.90
	13.28	0.73	0.45		12.26	0.303	0.49
13% (H_2)	2.49	0.247	2.41	23% (N_2)	0.71	0.176	8.45
	4.78	0.169	1.26		1.38	0.135	4.35
	6.86	0.161	0.88		2.72	0.127	2.21
	10.51	0.170	0.57		6.12	0.176	0.98

* The carrier gas is shown in parentheses.

(b), (d), (e), and (g) See table below.

Column loading*	B term	C term	u_{opt} (cm/sec)	H_{min} (cm)	k'	u range (cm/sec) for 90% efficiency
31% (H_2)	0.357	0.0479	2.73	0.307	5.61	1.6 to 4.6
23% (H_2)	0.376	0.0190	4.45	0.214	3.64	3.1 to 7.5
13% (H_2)	0.458	0.0072	7.99	0.160	2.44	4.2 to >13
23% (N_2)	0.084	0.019	2.11	0.125	3.61	1.3 to 4.0

* The carrier gas is shown in parentheses.

(f) As the column loading decreased, the optimum mobile-phase velocity increased because the effective liquid film thickness decreased. **(h)** 7.5 cm/sec; 0.59 normal time (or 1.68 times faster). **5. (a) and (c)** See table below.

u (cm/sec)	t_M (min)	t_R (min)	t'_R (min)	N_{eff}	N_{theor}	N/t
6.5	3.85	31.7	27.9	12,500	14,100	445
12.2	2.05	16.9	14.9	16,500	18,600	1100
19	1.32	10.9	9.54	16,100	18,200	1680
27	0.93	7.64	6.71	13,900	15,700	2050
43	0.58	4.80	4.22	10,100	11,400	2380

(b) $u_{opt} = 14.5$ cm/sec; $H_{min} = 0.90$ mm. **6. (a)** RI $= 371 + 94.31nt'_R$. RI values: propane, 300; butane, 400; *trans*-2-butene, 466; *cis*-2-butene, 476; pentane, 500; and 2-methylpentane, 580. **(b)** $t'_R = 11.33$ for hexane. **(c)** $N_{eff} = 2660$; $H = 0.56$ mm. **(d)** $Rs = 1.0$. **(e)** k' is 7.8 for *cis*-butene-2 and 23.5 for 2-methylpentane. **(f)** $N_{req} = 3460$. **(g)** Increase the column length 2.25 times. Adjust the carrier gas velocity to the optimum value. Change to helium as the carrier gas. **(h)** 6.4 cm/sec. **7. (a)** 15.2 min and 8.81 min. **(b)** $Rs = 3.72$ and 2.75, which are more than adequate.

At the lower velocity a 7.45-m column (13,300 plates) is required. At the higher velocity, a 13.6-m column (14,130 plates) suffices. **(c)** $Rs = 14.5$ and 11.2. Column lengths of 0.49 m and 0.82 m, respectively, are sufficient.

8.

Packed column

k'	N_{req}	L (cm)	t_R (sec)
1	17,400	1045	348
2	9800	588	294
3	7740	465	310
6	5930	356	415
12	5110	307	664
30	4650	279	1440

WCOT column

k'	N_{req}	L (cm)	t_R (sec)
0.17	212,700	12,760	1241
0.33	69,800	4190	465
0.5	39,200	2350	294
1	17,400	1045	174
2	9800	588	147
5	6270	376	188

SCOT column

k'	N_{req}	L (cm)	t'_R (sec)
0.33	69,800	4188	310
0.67	27,200	1630	151
1	17,400	1045	116
2	9800	588	98
4	6800	408	113
10	5270	316	193

For the minimum analysis time on each column, $k' = 2$. **9. (a)** Start the program at room temperature and use a heating rate of 18–23 °C/min. **(b)** Predicted retention temperatures are in the range: 178 to 180 °C for dodecane, 196 to 200 °C for tridecane, and 204 to 220 °C for tetradecane. **(c) and part of (a)** See table below.

		Heptane	Octane	Nonane	Decane	Undecane
(120 °C)	N_{theor}	822	866	964	958	1120
	N_{eff}	152	281	472	621	860
(140 °C)	N_{theor}	1100	866	837	1040	1140
	N_{eff}	144	200	316	545	750
(160 °C)	N_{theor}	1220	820	810	780	860
	N_{eff}	84	112	198	290	440
Retention temp. (°C)		84	102	124	143	163

10. (a) Start the program at room temperature, if possible. The methyl ester is best eluted isothermally. Use a heating rate of 16 °C/min. The retention temperatures are 72 °C for ethyl acetate, 96 °C for propyl acetate, 107 °C for butyl acetate, and 127 °C for pentyl acetate. **11. (a)** Decrease; **(b)** decrease; **(c)** increase; **(d)** cannot determine unless the optimum mobile-phase velocity is known. Plate height will increase on decreasing the gas flowrate below the optimum value and also on increasing the gas flowrate above the optimum value. **12.** 241 °C.

CHAPTER 19

1. Resettling of the packing with use that could create a void at the top of the column that would lead to broad peaks with poor symmetry. **2.** $L = 20{,}000d_p$.

d_p (μm)	1	2	3	4	5	7	10	12	15	20
L (cm)	2	4	6	8	10	14	20	24	30	40

3. (a) The higher the pressure, the more plates there are per unit time whatever the column. **(b)** This approach leads to too high an inlet pressure and too short an analysis time. **4.** $V_w = W_t V_{col}\varepsilon_{total}/t_M$; $V_w = 4V_{col}\varepsilon_{total}(1 + k')N^{-1/2}$. **5.** 16% decrease. **6. (a)** 2.0 mL; **(b)** 5 mm. **7.** Column A: 10.4 μL; column B: 14.9 μL; column C: 9.3 μL.

8.

	Optimum particle diameters (μm)				
Eluent	**t_M (sec): 10**	**30**	**60**	**120**	**300**
1-Propanol	0.7	1.2	1.6	2.3	3.7
Water	1.2	2.1	3.0	4.2	6.7
Hexane	2.1	3.7	5.2	7.4	11.7

9.

	Pressure drop (atm)				
Eluent	**t_M (sec): 10**	**30**	**60**	**120**	**300**
1-Propanol	804	268	134	67	26.8
Water	352	117	59	29.3	11.7
Hexane	123	41.1	20.6	10.3	4.1

These values are for porous particles; for solid-core particles, divide all the tabulated values by 2.

10.

	Particle diameter (μm)								
	2	**3**	**4**	**5**	**7**	**9**	**10**	**11**	**12**
L (cm)	3.67	4.13	4.76	5.65	8.54	14.0	18.0	24.1	31.7
ΔP (atm)	33.1	18.6	14.0	12.6	14.7	23.9	31.9	47.3	68.7

11.

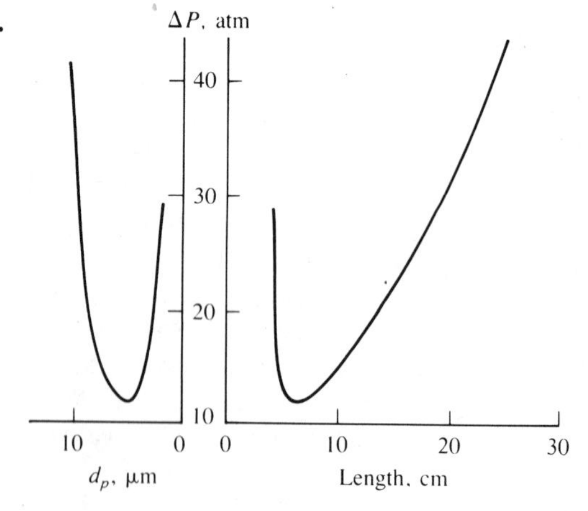

12.

ΔP	30 atm				200 atm			
d_p (μm)	3	5	7	10	3	5	7	10
t_M (sec)	22.5	34.5	51	81	8.7	17.4	28.7	50
L (cm)	3.5	7.3	12.4	22.3	5.6	13.3	23.9	45

13 and 14.

	Particle diameter (μm)							
	3	4	5	7	10	12	15	20
N	8280	8880	8510	7100	5280	4410	3440	2420
v	1.8	3.2	5.0	9.8	20.0	28.8	45	80
h	2.42	2.25	2.35	2.82	3.79	4.54	5.81	8.27
L (cm)	6	8	10	14	20	24	30	40
u (cm/sec)	0.060	0.080	0.10	0.14	0.20	0.24	0.30	0.40
H (mm)	0.0073	0.009	0.012	0.020	0.038	0.054	0.087	0.165

15. At high mobile-phase velocities the equation for reduced plate height becomes approximately $h/v = C$. Under conditions required to obtain the maximum number of plates, the pressure drop tends to infinity.

16.

	Particle diameter (μm)						
	3	4	5	7	10	15	20
20 atm							
L (cm)	3.9	4.5	5.8	9.1	16	30	49
t_M (sec)	85	63	66	84	122	201	298
100 atm							
L (cm)	3.8	5.9	8.4	15	27	54	90
t_M (sec)	16	21	28	43	70	128	200

17. At low mobile-phase velocities, $hv = B$. Under conditions required to obtain the maximum number of plates, t_M becomes infinite. **18. (a)** The reduced velocity is 67 and the reduced plate height is 7.44 for the conditions when $Rs = 0.8$. When $Rs = 1.25$, the reduced plate height is 3.05 and the reduced velocity is 12.0 (also 1.0 on the low-velocity side of the h/v minimum). The pressure drop is 5.6-fold less. Of course, the elution time for a nonretained solute is 5.6-fold longer. **(b)** The column length must be increased 2.44-fold; correspondingly, the analysis time is 13.6 times longer. **19.** For maximum resolution the volume is 11.4 mL, whereas the volume is 9.4 mL for a minimum time of analysis. These values remain the same for both flowrates. However, the time interval is 8 min at the lower flowrate and 4 min at the higher flowrate. **20.** At higher operating temperatures, the viscosity and therefore the pressure drop are reduced, whereas the efficiency is increased. **21.** 21 times less; 23 mL over an 8-hr period. **22.** 15,000 plates. **23.** 30,000; 50,000.

CHAPTER 20

1. It is due primarily to insufficient deactivation of the silica or alumina surface of the column. **2.** Hexane and pentane are identical in eluent strength. Either a volume ratio of 33:67 for an isopropyl chloride/hexane mixture or 9:91 for a diethyl ether/hexane mixture would provide the same solvent strength parameter. Pure carbon disulfide also has the same value. **3.** Slower. **4.** Increase the "stronger" solvent content, in this case acetonitrile, in steps of 10% to a 40:60 volume ratio. **5.** The bonded phase functioned in the normal mode because more polar solvent (water) decreased the competitive interaction between sugar and the polar bonded phase. **6. (a)** Use surface sulfonated styrene/divinylbenzene resin in the separating column, 0.01*M* HCl as eluent, and a strong anion-exchange resin (in the OH form) in the stripping column. The latter column retains the chloride ion and releases the hydroxyl counterion, which combines with the hydrogen ion (from HCl) to form water. **(b)** Use a surface sulfonated styrene/divinylbenzene resin in the separating column, 0.002*M* silver nitrate plus 0.0004*M* nitric acid as eluent, and a dual column stripper with a strong anion-exchange resin in the Cl form in the first column and a strong anion-exchange resin in the OH form in the second column. The first stripper column removes the nitrate and the released chloride ion forms a precipitate of silver chloride. The second stripper column removes the nitrate ion from the nitric acid and neutralizes the hydrogen ion with the released hydroxyl ion. **(c)** Use an anion-exchange resin in the carbonate form in the separating column, 0.003*M* hydrogen carbonate plus 0.0024*M* carbonate as the eluent, and a cation-exchange resin in the hydrogen form in the suppressor column. The latter column reacts with the hydrogen carbonate (or carbonate) to form carbon dioxide and water. **7.** A convenient way to improve this separation of amines is either to change the solvent composition in order to neutralize the solute (pH modification or addition of a pairing agent) or to enhance the hydrophobic effect in order to increase the retention time (more water in the mobile phase). **8.** Use a solvent consisting of methanol and water (in the volume ratio 45:55) and containing heptyl sulfonic acid (pH 3.5) to ensure that the base is completely ionized and the weak acid is not ionized. In this manner aspirin can be separated by ionic suppression while the ionized base is separated by reverse-phase ion-pair partition. **9.** A small counterion would be most effective. Actually the ammonium ion and the tetramethylammonium ion yield very similar resolution, although the latter displays a significant loss of resolution because its larger size dominates the retention of the ion pair. The latter yields larger k' values than does the ammonium ion. **10. (a)** Use a cation-exchange column and a solution at low pH and base the separation on the alkylammonium groups. Also one could use an anion-exchange column and a solution at high pH and base the separation on the carboxyl groups. The first method is preferred because the site of ionization is closer to the side chain of the amino acid, particularly if steric effects exert any influence on the separation. **(b)** By reverse-phase ion-pair partition the amino acids can be chromatographed using either a quaternary amine (to pair with the carboxyl group) in a solution adjusted to pH 7.5 or an alkyl sulfonic acid (to pair with the alkylammonium group) in a solution with a low pH. **11.** Hydrochloric acid forms tetrachlorocuprate(II)(2−) and trichlorocuprate(II)(1−) species with copper(II), which will not bind to the cation-exchange sites. **12.** Use a reducing agent, perhaps hydrogen peroxide, in a dilute regenerant consisting of a HCl—NaCl mixture. Chromium(III) will be formed and, being positively charged, will not be retained on the anion-exchange resin. **13. (a)** The distribution ratio is 7.0 for component A and 21.0 for component B. **(b)** $\alpha = 3.0$.

14. $[H^+]$, aqueous phase	$(D_c)_K$	$(D_d)_{Mg}$	Relative retention
0.10	0.074	1.4×10^{-3}	0.1
0.50	0.015	5.6×10^{-5}	0.5
1.00	0.0074	1.4×10^{-5}	1.0
3.00	0.0025	1.6×10^{-6}	3.0

15. **(a)** In log mol wt units, the range is 5–7. **(b)** $V_M = 3.55$ mL; $V_S = 2.38$ mL. **(c)** $V = 3.55$ mL ($K = 0$) and $V = 5.93$ mL ($K = 1$). **(d)** Number 1, 0.10; number 2, 0.38; number 3, 0.63; and number 4, 0.95. **(e)** $Rs = 1.39$ (peaks 1 and 2). **(f)** 10,200 plates/meter. **(g)** Void porosity is 0.47; total porosity is 0.79. **16.** Inject an isotopically labeled solvent. V_S corresponds to the subtraction of the void volume from the elution volume of the solvent. **17.** Low mobile-phase flow-rate, low column temperature, improper gradient, column activity increasing (system not equilibrated with the proper content of water or organic modifier), mobile phase may be removing water from an adsorbent column (in LSC), or incorrect mobile phase is being used. **18.** **(a)** Change to another type of column packing; avoid the use of cationic or basic samples on silica. Try ion-pair chromatography. **(b)** Tailing should decrease if a smaller sample is injected. The addition of water or another polar modifier to the mobile phase is useful. The use of gradient elution is always beneficial. **(c)** Replumb the system to reduce extra-column band broadening. Use a detector with smaller volume or faster response. Use a column with a larger volume. **19.** **(a)** Decrease the eluting strength of the initial mobile phase. **(b)** Increase the strength of the final solvent. **(c)** Increase the gradient steepness if added sensitivity is desired. **(d)** Decrease the gradient steepness if increased resolution is desired.

CHAPTER 22

1. 1.12×10^7. **2.** 6.07 μg/mL. **3.** Add bromine to the sample to oxidize the nitrite ion to nitrate ion. Measure the amount of bromide produced. **4.** The fluoride electrode is ten times more sensitive to fluoride ion than it is to hydroxide ion. Therefore, a $0.001M$ hydroxide concentration will double the response from a $0.0001M$ fluoride concentration. To operate within a stipulated measurement error, the hydroxide concentration must be less than the product of 0.001 and the stipulated error. **5.** pH = 4.48. **6.** Ratio of interferant to bromide ion: chloride, 330; iodide, 1.7×10^{-4}; hydroxide, 4000; cyanide, 4×10^{-4}. **7.** $9.8 \times 10^{-4}M$. **8.** Hydroxide, $0.0001M$; iodide, $5 \times 10^{-9}M$; nitrate, $1 \times 10^{-6}M$; hydrogen carbonate, $2.5 \times 10^{-5}M$; and sulfate, $1 \times 10^{-8}M$. **9.** $1.33 \times 10^{-3}M$. **10.** $pK_a = 9.26$. **11.** $pK_a = 4.16$. Be sure to apply a correction for the concentration of hydrogen ions.

CHAPTER 23

1. 0.72 μg/mL. **2.** $0.0200M$. Be sure to correct each measured diffusion current for the dilution caused by the titrant. **3.** $1.49 \times 10^{-5}M$. **4.** At −0.8 V (vs. SCE) the titration curve is L-shaped, whereas at −1.7 V the curve is V-shaped. Under normal circumstances the titration is conducted at −0.8 V; also when nickel and zinc are present to avoid the waves of these metals ammine complexes. **5.** $n = 2$; $E_{1/2} = -0.532$. **6.** Yes. E_p is independent of the scan rate; i_p is proportional to $v^{1/2}$; and the difference between the anodic and cathodic peak potentials is 0.058 V, which is within 1.7% of the theoretical difference. **7.** Four ligands are associated with the central metal atom. The formation constant is 7.2×10^{21}.

8.

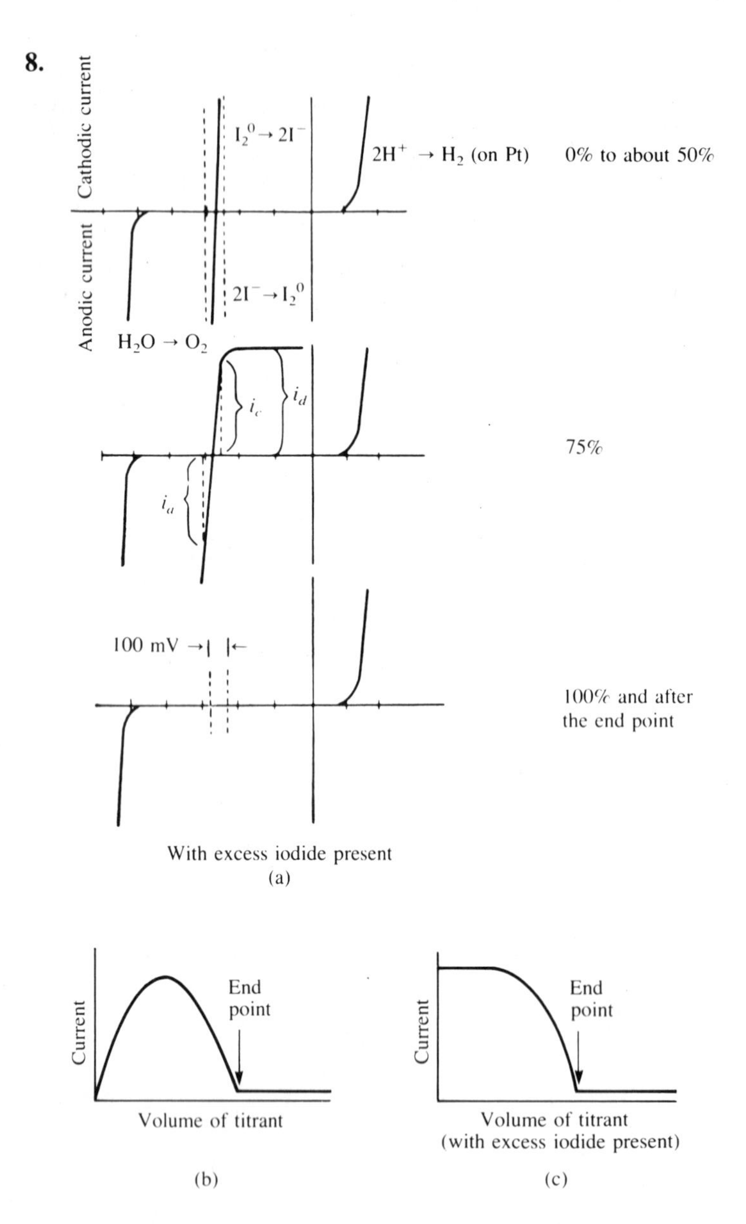

9. For the iodide titration use a solution that is 0.3M in ammonia and adjusted to pH 9. The potential is −0.2 V versus SCE. For the bromide titration use a solution that is 0.8M in nitric acid. The potential range is from +0.2 to −0.2 V (vs. SCE).

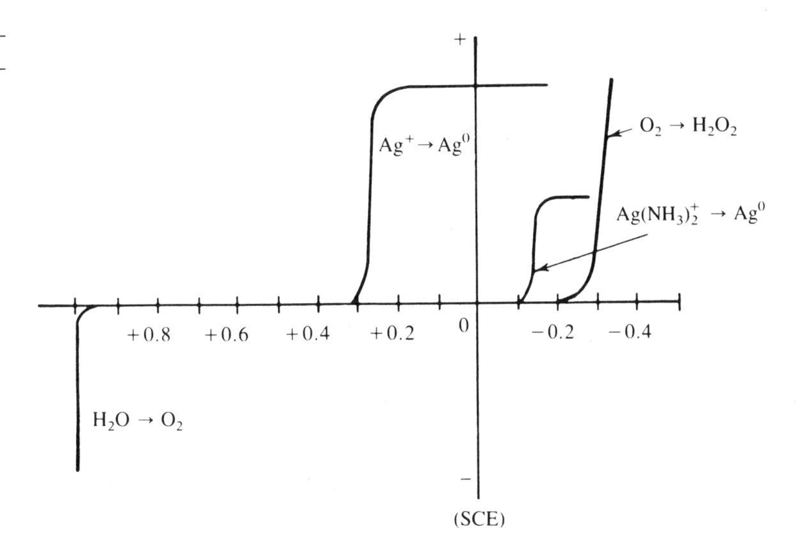

CHAPTER 24

1. (a) The cathode potential should exceed 0.26 V versus SHE. **(b)** To remove 99.9% of the silver, 10.2 min are required. **2. (a)** 0.50 V versus SHE; **(b)** 0.38 V versus SHE; **(c)** $1.2 \times 10^{-6}M$. **3. (a)** $0.109M$; **(b)** 0.128 V versus SHE. **4. (a)** The cathode potential (controlled by the zinc ion/zinc metal system) is −0.82 V versus SHE. Hydrogen gas will not be liberated until the electrode potential is −0.99 V versus SHE. The anode potential is 1.97 V. The cell potential required to initiate the anode–cathode reaction is 3.29 V. **(b)** Yes. **(c)** Hydrogen gas is liberated from the zinc surface at −0.94 V versus SHE, and the zinc remaining in solution at this cathode potential is $8.3 \times 10^{-7}M$. **5.** Not completely. Hydrogen begins to evolve at −0.52 V versus SHE. At this cathode potential 9.9 mg of cadmium remain in solution. **6. (a)** Tin would begin to deposit when the cathode potential is −0.21 V versus SHE. The copper concentration at −0.16 V is $2.8 \times 10^{-17}M$. **(b)** The copper reduction proceeds stepwise and the reduction from the trichlorocuprate(I)(2−) ion to copper is the controlling electrode reaction. The residual copper concentration is $2.1 \times 10^{-6}M$. **7.** Since the anode and cathode are connected directly to each other—that is, shorted—the cathode potential is limited to −0.147 V (the anodic potential of the lead system). The residual copper concentration is $7.9 \times 10^{-17}M$. **8.** On smooth platinum the evolution of hydrogen begins at −0.62 V, which is less negative than the electrode potential of the zinc system. On a copper-coated electrode the evolution of hydrogen begins at −0.81 V and plating of zinc could begin. Once the cathode is coated with zinc, the larger hydrogen overpotential on this surface delays the evolution of hydrogen until −1.10 V. The residual zinc concentration is $3.3 \times 10^{-12}M$. **9.** Use the electrographic method with a sodium carbonate/sodium nitrate electrolyte. Wash the test sheet with dilute acetic acid and develop the green copper complex with a 1% alcoholic solution of α-benzoinoxime. **10. (a)** 13.3 min; **(b)** 19.9 min. **11.** 18.8 mg.

12. Mean value: $0.133N$. **13.** $0.00145N$. **14. (a)** Cathode: $2H_2O + 2e^- = H_2 + 2OH^-$; anode: $Ag + Br^- = AgBr + e^-$. **(b)** Cathode: $2H_2O + 2e^- = H_2 + 2OH^-$; anode: $2Br^- = Br_2 + 2e^-$. Excess unused bromine reacted with $CuCl_3^{2-}$ generated at the cathode: $Cu^{2+} + 3Cl^- + e^- = CuCl_3^{2-}$; anode compartment must be isolated. **(c)** Cathode: $Fe(CN)_6^{3-} + e^- = Fe(CN)_6^{4-}$; anode: $2H_2O = 4H^+ + O_2 + 4e^-$. **15.** $1.29 \times 10^{-4}M$. **16.** Use the reversible vanadium(V/IV) system as a potential buffer. Iron(III) will be formed either by direct oxidation at the anode or indirectly by oxidation with the vanadium(V) formed at the anode. The end point can be ascertained by the reaction between the first persistence of vanadium(V) that reacts with diphenylamine sulfonic acid to give a blue color. **17.** Control the anode potential at 0.75 V versus SCE and measure the coulombs that are involved in the oxidation of vanadium(IV). Reverse the electrode potential and control the cathode value at 0.30 V versus SCE and measure the coulombs required for the reduction of the original and newly generated vanadium(V). The difference in coulombs is the amount of the original vanadium(V). **18.** 1930 sec. **19.** 26.7 to 318 Ω. **20.** The resistance should range from $546 \times \Theta$ to $273{,}000 \times \Theta$. If $\Theta = 0.2$, resistance readings are 100–54,600 Ω with 2740 Ω at midscale. **21. (a)** $R = 2\Theta$. Θ should be 50 in order to provide a resistance of 100 Ω (for 5% HCl) and 85,000 Ω (for $0.001M$, the assumed low concentration). **(b)** Resistance varies from 512 to 1620 times Θ; Θ could be 0.2. **(c)** The concentration ranges from $8.82 \times 10^{-4}M$ to perhaps 1/600 of this value (for 0.1 μg/mL). Resistance varies from 108 to 64,800 Ω for $\Theta = 0.01$. Θ could also be 0.1. **(d)** Case (c) repeated essentially. **(e)** $\Theta = 0.01R$; $\Theta = 50$ is used. **(f)** The resistance varies from 2.5 to 250 times Θ. $\Theta = 20$ is used. **22.** $\alpha = 0.0824$; $K_a = 1.79 \times 10^{-5}$. **23.** The solubility is $1.07 \times 10^{-5}M$; $K_{sp} = 1.15 \times 10^{-10}$. **24.** $0.0176M$. There are individual end points for each replaceable hydrogen. **25.** 0.630 mEq of the sulfonic acid and 0.230 mEq of the carboxylic acid.

26.

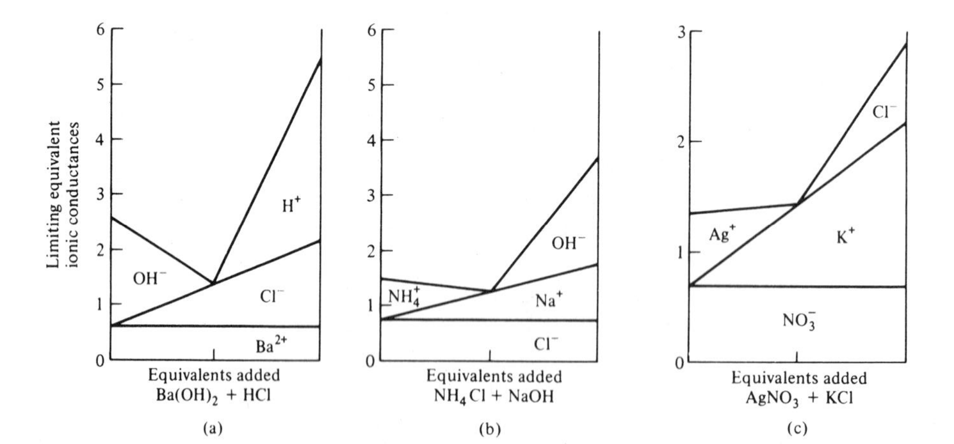

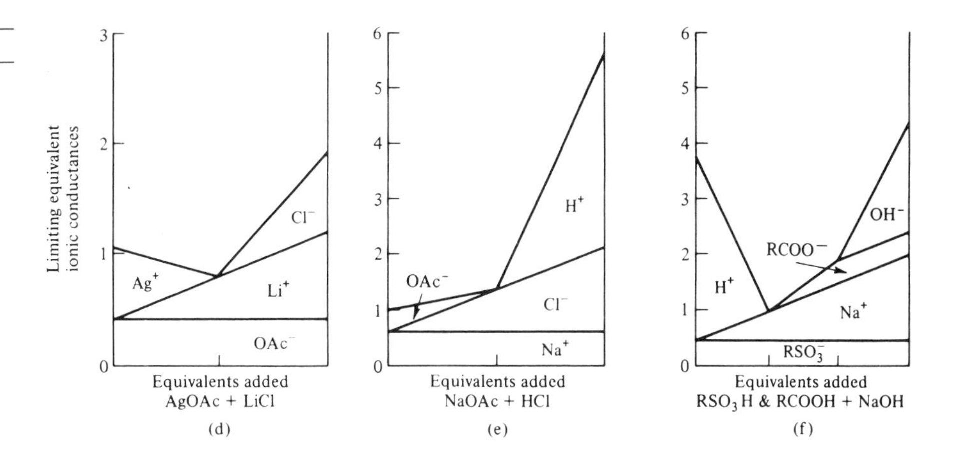

(g) The graph resembles the one for part (f) except that there is no sharp rise after the second end point, since excess ammonia contributes no conductance. The first end point occurs after one equivalent of ammonia has been added; the second end point after three equivalents have been added.

CHAPTER 25

1. $2NaHCO_3 \rightarrow Na_2CO_3 + CO_2 + H_2O$. **2.** $\Delta H = -(\text{slope})\ (2.303)\ (1.987) =$ 39.0 kcal/mole, where slope is found from plotting log p_{CO_2} versus $1/T$ (in K). **3.** Plot specific heat versus temperature; T_g, given by intersection of two linear segments, is 415 K. **4. (a)** 24.0%; **(b)** 80.0%; **(c)** 49.0%; **(d)** 40.0%; **(e)** 76.0%. **5.** Chain rotation, 13.7 cal/g; fusion, 38.0 cal/g. **6.** 89.1 cal/g of UO_2. **7. (a)** $w_{CaO} = (w_{600°} - w_{900°})\ (56/44)$; $w_{MgO} = 1.5(w_{300°} - w_{600°})\ (40.6/44)$. **(b)** $3MgCO_3 \rightarrow MgO \cdot 2MgCO_3 + CO_2$. $MgO \cdot 2MgCO_3 \rightarrow 3MgO + 2CO_2$. **(c)** 40.0% CaO and 13.5% MgO. **8.** In all cases, the first loss is the 2 moles of hydrated water. In an oxidizing atmosphere the final product is NiO; in CO and N_2, it is nickel metal. **9.** $K_{sp}^{158°} = 3.0 \times 10^{-8}$. **10.** DTA—difference in temperature of sample and temperature of thermally stable reference material versus temperature of reference material. DSC—amount of heat added to sample or reference material to keep both at the same temperature versus the temperature. TG—mass of the sample versus the temperature of the sample. TMA—movement of the sample versus temperature. EGA—analysis of gas evolved at a specific temperature or temperature range. **11.** Titration curve exhibits a rising portion (Ca^{2+} reacting) followed by a descending portion (Mg^{2+} reacting). For Ca: $\Delta H = -5.7$ kcal/mole; for Mg: $\Delta H = 5.6$ kcal/mole. **12. (a)** $MnCO_3 \xrightarrow{400°} MnO_2 \xrightarrow{550°} Mn_2O_3 \xrightarrow{900°} Mn_3O_4$. **(b)** In CO_2 atmosphere, decomposition is delayed; Mn_2O_3 is formed at 600°. Formation of MnO_2 requires presence of oxygen. **13.** Glass transition at 40 °C; melting of coating at 125 °C. **14.** The sharp endothermic peak at 70 °C is superimposed upon a glass transition at 61 °C, followed by the exothermic curing reaction, which appears to be complete near 260 °C. Cured sample manifests a glass transition at 116 °C. **15.** At a mole ratio of 1:1 there occurs an intermolecular substance; 110 °C is the transition point of hexamethylbenzene.

CHAPTER 26

1. A rugged continuous-flow-type pH electrode assembly is mounted in the line connecting the mixing tank and the initial filter. Phosphoric acid and soda ash are mixed at approximately 85 to 100 °C and the hot solution is filtered to remove iron and aluminum phosphates as well as silica. The signal from the electrode assembly is amplified and sent to a recorder. In an automated system, the amplified electrode signal would also be used to actuate a diaphragm motor valve controlling the addition of phosphoric acid. (See D. M. Considine, ed., *Process Instruments and Controls Handbook*, 2nd ed., pp. 6–96, McGraw-Hill, New York, 1974.) **2.** Measurements to the fourth decimal place provide a precision of about 0.1%. **3.** The design of nondispersive analyzers is less complicated; thus, these instruments are easier to operate and maintain. The sensitivity of nondispersive analyzers is usually better. Nondispersive analyzers are not generally as selective or flexible as dispersive units and are thus dedicated to sampling light intensity at a limited number of predetermined wavelengths. **4.** In process gas chromatography, the sample is always transferred from the process stream to the analyzer without human intervention. Process instruments are operated in a harsh plant environment by personnel with minimal analytical training. Process chromatography utilizes multicolumn techniques, such as stripping and backflushing, which are not typically used in laboratory chromatography. The data from process instruments are presented in formats, such as bar graphs, not generally used for laboratory data. The data are used immediately to control the monitored process. In process chromatography, the same components are usually in each sample (one knows what to expect). **5.** All can be handled by refractometric methods. [See *J. Chem. Educ.*, **45**, A470 (1968).] **6.** Gas sample is drawn through the probe by means of a steam-operated aspirator. Stream-gas mixture is condensed in a cooling chamber and the resulting condensate, containing all the sodium ion entrained in the original gas sample, is separated from the gas and passed through a rotameter ahead of the sodium analyzer (selective-ion electrode). Gas flow measured in second rotameter provides a multiplier, which may easily be corrected for absolute humidity, whereby the recorded sodium data can be reported on a dry gas basis. **7.** Restate the problem as follows: A Hersch galvanic cell is used to monitor oxygen in flue gas at 816 °C. If the partial pressure of oxygen on the reference side is 600 mm, what is the partial pressure of oxygen in the flue gas if the voltage reading is 75 mV?

$$75 \text{ mV} = 54(\log 600 - \log P_s) \quad \text{(See Equation 26.3)}$$

$$1.38 = 2.78 - \log P_s$$

$$\log P_s = 1.39$$

$$P_s = 25 \text{ mm}$$

8. $$38\text{ ppm} = A \log\left(\frac{98 \times 97}{99 \times 45}\right) \quad \text{(see Equation 26.1)}$$

$$K = 111.1\text{ ppm} \qquad \text{calibration}$$

$$\text{conc. } SO_2 = 111.1 \log\left(\frac{95 \times 92}{93 \times 32}\right)$$

$$\text{conc. } SO_2 = 53\text{ ppm}$$

9.

Analyzer	**Mode of operation**	**Advantages**	**Disadvantages**
Discrete	Batch, serial processing of sample	Runs selected test on limited number of samples, quick start-up	Expensive prepackaged reagents and cells, mechanically complex
Continuous flow	Continuous processing of large number of samples for same series of tests	Fast, simple equipment, reagent cost minimized	Difficult to vary test doing a series of samples, tubing is critical
Centrifugal	Sample mixed with reagent by centrifugal force	Fast, all samples and controls measured under the same conditions	Loading samples is time-consuming, can run only a limited number of tests.

10. All the parameters of the gas chromatograph must be optimized and controlled. The method depends on the use of small samples of about 0.6 mg, which allows for rapid combustion and the introduction of the products into the chromatographic column without excessive dilution by the carrier gas. **11.** 78.99% C; 5.78% H; 15.29% N.

APPENDIXES

APPENDIX A COMMON TECHNIQUES OF ANALYTICAL CHEMISTRY

MOLECULAR IDENTIFICATION AND ANALYSIS*

Method	Principal applications	Molecular phenomenon	Advantages in qualitative analysis
Infrared spectroscopy	Structure determination and identity of organic and inorganic compounds; generally quantitative analysis	Excitation of molecular vibrations by light absorption	Identification of functional groups; largest file of reference spectra available; virtually no sample limitation; impurity detection
Raman spectroscopy	Structure determination and identity of organic compounds, symmetry of molecular groups in solid state	Excitation of molecular vibrations by light scattering	Identification of functional groups (usually different from those identified by ir); use of water solutions
Mass spectrometry	Structure determination and identity of organic compounds; analysis of trace volatiles in nonvolatiles	Ionization of molecule and cracking of molecule into fragment ions	Precision molecular wt. (molecular ion), masses of integral parts of molecule (fragment ions), very high sensitivity, impurity detection
Nuclear magnetic resonance	Structure determination and identity of organic compounds, molecular conformation	Reorientation of magnetic nuclei in a magnetic field	Determines chemical type and number of atoms (^{1}H, ^{19}F, ^{13}C, ^{31}P, etc.), molecular configuration and conformation, detects impurities, uses neat sample, including water solutions
Ultraviolet and visible spectrophotometry	Quantitative analysis, esp. as end methods in chemical analysis schemes	Excitation of valence electrons	Special applications
Gas chromatography	General multicomponent quantitative analysis of volatile organics, highly efficient separation technique	Partitioning between vapor phase and substrate	Separates materials for examination by other techniques

SOURCE: Modern Methods of Chemical Analysis, © 1985 The Dow Chemical Company, Midland, MI with permission.
* This table compares only the more widely used techniques. Methods of identification and analysis of compounds discussed elsewhere in this book but not included in this table can be considered to be specialized techniques.

Advantages in quantitative analysis	**Average sample desired for qualitative analysis**[†]	**Method limitations**	**Sample limitations**
Widely applicable	10 μg to 10 mg	Medium sensitivity,[‡] no direct information about size of molecule	Avoid aqueous solutions if possible
Special applications	0.01 mg	Low sensitivity;[‡] no direct information about size of molecule	Sample must not fluoresce, avoid turbid materials, some restrictions on colored material
High sensitivity	<1 mg	Cannot always identify isomeric structures	Difficult to analyze non-volatile samples
Standards not required; no chemical or physical alteration of a neat sample	10 mg	Medium sensitivity; most useful information from a limited number of elements (H, C, F, P, Si, Sn, N)	Liquid or soluble solid preferred (wide variety of solvent choices)
High precision, high sensitivity, simplicity	0.01 mg	Low specificity; little information on molecular structure	Soluble in uv-transparent solvent (wide variety of choices)
Widely applicable to volatile materials, multi-component analyses; high sensitivity in special cases	1 mg	Identifies materials only in special cases	$> \sim 1$ torr vapor pressure at sample inlet temperature

(*continued*)

[†] The amount of sample listed in this column is an estimate of the average minimum sample required. Usually a larger sample is preferred in order to do the best possible analysis. On the other hand, successful identifications can often be done with much smaller amounts.

[‡] *Sensitivity* as used here indicates ability to determine a small amount of one material in the presence of large amounts of other material(s).

MOLECULAR IDENTIFICATION AND ANALYSIS

Method	Principal applications	Molecular phenomenon	Advantages in qualitative analysis
Combined gas/liquid chromatography–mass spectrometry	Identification and analysis of mixtures	Combines separation efficiency of GC/LC with sensitivity and specificity of mass spectrometry	Applicable to identity of sub-ppm components in mixtures
Liquid chromatography (including ion exchange and thin layer)	Separation technique for less volatile and ionic materials, multi-component quantitative analysis	Partitioning between liquid solution and substrate	Separates materials for examination by other techniques
Size exclusion Chromatography	Separation according to hydrodynamic volume, determination of polymer molecular weight distribution	Solute size-dependent partitioning between packing pore volume and interstitial volume	Separates materials for examination by other techniques
X-ray diffraction	Identification of solid compounds, crystallite size, phase changes, and crystallinity	Diffraction of X rays from crystal planes	High specificity for crystalline solids; can distinguish isomers, polymorphs, conformers, and different hydrated structures; compound specific
Chemical reaction methods	Variety of specialized quantitative analysis applications	Stoichiometry of chemical reactions	Special applications

Advantages in quantitative analysis	Average sample desired for qualitative analysis[†]	Method limitations	Sample limitations
Specific identification of GC-LC peaks being determined	<1 mg	Cannot always identify isomeric structures	Difficult to analyze non-volatile samples
Multicomponent analyses of less volatile materials	10 mg	Method development is time-consuming	None
Determines molecular weight distribution	500 mg	Can require calibration of the molecular weight–hydrodynamic volume relationship	Material must be soluble in a suitable solvent
Useful for quantitating isomers, polymorphs, and conformer mixtures	0.1 mg	Detection and sensitivity dependent on crystallinity and crystallite size	Applicable to crystalline solids and polymers
High precision for assay analyses, absolute calibration	1000 mg	Time-consuming, interferences often a problem	None

ATOMIC IDENTIFICATION AND ANALYSIS*

Method	Principal applications	Atomic phenomenon	Advantages in qualitative analysis
Atomic emission spectroscopy	Qualitative analyses for 25 elements using simultaneous detection; quantitative analyses for all elements emitting in uv-visible spectrum	Light emission from excited electronic states of atoms	General for all metallic elements; simultaneous analysis of metallic elements
Atomic absorption spectroscopy	Precision quantitative analysis for a given metal, trace analysis for a given metal	Absorption of atomic resonance line	(Not applicable)
X-ray fluorescence	General qualitative and semiquantitative survey of all elements atomic no. ≥ 5(B); precision quantitative analysis of elements, esp. heavier nonmetals (P, Cl, Br, I); trace analysis	Re-emission of X rays from excited atoms	General for all elements atomic no. ≥ 5(B); minimum sample preparation
Neutron activation	Precision quantitative analysis of most elements; trace and ultratrace element analysis; general qualitative analysis of most elements	Counting of radioactive species produced by neutron reactions	Minimum sample preparation
Chemical reaction methods	Variety of specialized quantitative analysis applications	Stoichiometry of chemical reactions	Special applications

SOURCE: Modern Methods of Chemical Analysis, The Dow Chemical Company, Midland, MI (1985) with permission.

* This table compares only the more general techniques, omitting those that are applicable to only one or a few elements. For each method, the sensitivity and precision with which a given element can be determined can vary considerably. Therefore, this table can be considered only as a very rough guide to a proper choice for a given analysis.

† The amount of sample listed in this column is only a rough index. Sample required will vary enormously depending on whether the problem is identity of a major elemental component or precise determination of a trace element.

‡ None of these methods handles gases easily and conveniently (although it could be done in special cases). For elemental analysis in gases, mass spectrometry is the best general choice.

Advantages in qualitative analysis	Average sample required[†]	Method limitations	Sample limitations[‡]
General for all metallic elements; high sensitivity in many cases	0.25–2 g	Limited sensitivity for halogens and other nonmetals; not well suited for quantitative microanalysis	Most organic liquid and solid samples require wet digestion prior to analysis
Fast, reliable analysis for a given element; high sensitivity in some cases; simplicity	100 mg	Metals analyzed individually, not simultaneously; usually not applicable to nonmetallic elements	Element being analyzed must be in a solution (many solvent choices)
General for all elements atomic no. ≥ 13(Al), high sensitivity in some cases; simplicity; minimum sample preparation	500 mg (nondestructive)	Nonsensitive to elements of atomic no. <5(B), precision limited by nonuniformity of sample	Applicable principally to solids and nonvolatile liquids
Highest sensitivity for many elements, high confidence level, only general instrumental method capable of N, O, and F analysis	100 mg (nondestructive)	Sensitivity varies considerably among elements (but sensitive to amounts <1 μg for most elements); multicomponent analyses present some problems	Applicable to solids and liquids
High precision for assay analysis, absolute calibration	1000 mg	Time-consuming	None

APPENDIX B SOURCES OF LABORATORY EXPERIMENTS

BUNCE, S., ed., *Annotated List of Laboratory Experiments from the Journal of Chemical Education, 1957–1984*, ACS Educational Division, Office of College Chemistry, Washington, DC, 1985. MicroCHEMLAB, computer searchable, is also available from Project Seraphim, Department of Chemistry, Eastern Michigan University, Ypsilanti, MI. Organized by principal areas of chemistry including analytical. Good source of current experiments.

GILLESPIE, A. M., JR., *A Manual of Fluorometric and Spectrophotometric Experiments*, Gordon and Breach Science Publishers, New York, 1985. Good experiments with supplementary materials.

KEALEY, D., *Experiments in Modern Analytical Chemistry*, Chapman and Hall, New York, 1986. Collection of standard experiments for both wet chemical and instrumental analysis.

SAWYER, D. T., W. R. HEINEMAN, AND J. M. BEEBE, *Chemistry Experiments for Instrumental Methods*, John Wiley, New York, 1984. Excellent source of experiments for commonly used instrumental methods. Clearly written background materials as well as laboratory procedures.

Instrument manufacturers' technical bulletins often address educational applications as well as providing ideas for laboratory experiments.

APPENDIX C POTENTIALS OF SELECTED HALF-REACTIONS AT 25 °C

A summary of oxidation/reduction half-reactions arranged in order of decreasing oxidation strength and useful for selecting reagent systems.

Half-reaction			$E°$ (V)
$F_2(g) + 2H^+ + 2e^-$	=	$2HF$	3.06
$O_3 + 2H^+ + 2e^-$	=	$O_2 + H_2O$	2.07
$S_2O_8^{2-} + 2e^-$	=	$2SO_4^{2-}$	2.01
$Ag^{2+} + e^-$	=	Ag^+	2.00
$H_2O_2 + 2H^+ + 2e^-$	=	$2H_2O$	1.77
$MnO_4^- + 4H^+ + 3e^-$	=	$MnO_2(s) + 2H_2O$	1.70
$Ce(IV) + e^-$	=	$Ce(III)$ (in $1M$ $HClO_4$)	1.61
$H_5IO_6 + H^+ + 2e^-$	=	$IO_3^- + 3H_2O$	1.6
Bi_2O_4 (bismuthate) $+ 4H^+ + 2e^-$	=	$2BiO^+ + 2H_2O$	1.59
$BrO_3^- + 6H^+ + 5e^-$	=	$\frac{1}{2}Br_2 + 3H_2O$	1.52
$MnO_4^- + 8H^+ + 5e^-$	=	$Mn^{2+} + 4H_2O$	1.51
$PbO_2 + 4H^+ + 2e^-$	=	$Pb^{2+} + 2H_2O$	1.455
$Cl_2 + 2e^-$	=	$2Cl^-$	1.36
$Cr_2O_7^{2-} + 14H^+ + 6e^-$	=	$2Cr^{3+} + 7H_2O$	1.33
$MnO_2(s) + 4H^+ + 2e^-$	=	$Mn^{2+} + 2H_2O$	1.23
$O_2(g) + 4H^+ + 4e^-$	=	$2H_2O$	1.229
$IO_3^- + 6H^+ + 5e^-$	=	$\frac{1}{2}I_2 + 3H_2O$	1.20
$Br_2(l) + 2e^-$	=	$2Br^-$	1.065
$ICl_2^- + e^-$	=	$\frac{1}{2}I_2 + 2Cl^-$	1.06
$VO_2^+ + 2H^+ + e^-$	=	$VO^{2+} + H_2O$	1.00
$HNO_2 + H^+ + e^-$	=	$NO(g) + H_2O$	1.00
$NO_3^- + 3H^+ + 2e^-$	=	$HNO_2 + H_2O$	0.94
$2Hg^{2+} + 2e^-$	=	Hg_2^{2+}	0.92
$Cu^{2+} + I^- + e^-$	=	$CuI(s)$	0.86
$Ag^+ + e^-$	=	Ag	0.799
$Hg_2^{2+} + 2e^-$	=	$2Hg$	0.79
$Fe^{3+} + e^-$	=	Fe^{2+}	0.771
$O_2(g) + 2H^+ + 2e^-$	=	H_2O_2	0.682
$2HgCl_2 + 2e^-$	=	$Hg_2Cl_2(s) + 2Cl^-$	0.63
$Hg_2SO_4(s) + 2e^-$	=	$2Hg + SO_4^{2-}$	0.615
$Sb_2O_5 + 6H^+ + 4e^-$	=	$2SbO^+ + 3H_2O$	0.581
$H_3AsO_4 + 2H^+ + 2e^-$	=	$HAsO_2 + 2H_2O$	0.559
$I_3^- + 2e^-$	=	$3I^-$	0.545
$Cu^+ + e^-$	=	Cu	0.52
$VO^{2+} + 2H^+ + e^-$	=	$V^{3+} + H_2O$	0.337
$Fe(CN)_6^{3-} + e^-$	=	$Fe(CN)_6^{4-}$	0.36
$Cu^{2+} + 2e^-$	=	Cu	0.337
$UO_2^{2+} + 4H^+ + 2e^-$	=	$U^{4+} + 2H_2O$	0.334

(continued)

APPENDIX C *(continued)*

Half-reaction			$E°$ (V)
$Hg_2Cl_2(s) + 2e^-$	$= 2Hg + 2Cl^-$		0.2676
$BiO^+ + 2H^+ + 3e^-$	$= Bi + H_2O$		0.32
$AgCl(s) + e^-$	$= Ag + Cl^-$		0.2222
$SbO^+ + 2H^+ + 3e^-$	$= Sb + H_2O$		0.212
$CuCl_3^{2-} + e^-$	$= Cu + 3Cl^-$		0.178
$SO_4^{2-} + 4H^+ + 2e^-$	$= SO_2(aq) + 2H_2O$		0.17
$Sn^{4+} + 2e^-$	$= Sn^{2+}$		0.15
$S + 2H^+ + 2e^-$	$= H_2S(g)$		0.14
$TiO^{2+} + 2H^+ + e^-$	$= Ti^{3+} + H_2O$		0.10
$S_4O_6^{2-} + 2e^-$	$= 2S_2O_3^{2-}$		0.08
$AgBr(s) + e^-$	$= Ag + Br^-$		0.071
$2H^+ + 2e^-$	$= H_2$		0.0000
$Pb^{2+} + 2e^-$	$= Pb$		−0.126
$Sn^{2+} + 2e^-$	$= Sn$		−0.136
$AgI(s) + e^-$	$= Ag + I^-$		−0.152
$Mo^{3+} + 3e^-$	$= Mo$	*approx.*	−0.2
$N_2 + 5H^+ + 4e^-$	$= H_2NNH_3^+$		−0.23
$Ni^{2+} + 2e^-$	$= Ni$		−0.246
$V^{3+} + e^-$	$= V^{2+}$		−0.255
$Co^{2+} + 2e^-$	$= Co$		−0.277
$Ag(CN)_2^- + e^-$	$= Ag + 2CN^-$		−0.31
$Cd^{2+} + 2e^-$	$= Cd$		−0.403
$Cr^{3+} + e^-$	$= Cr^{2+}$		−0.41
$Fe^{2+} + 2e^-$	$= Fe$		−0.440
$2CO_2 + 2H^+ + 2e^-$	$= H_2C_2O_4$		−0.49
$H_3PO_3 + 2H^+ + 2e^-$	$= HPH_2O_2 + H_2O$		−0.50
$U^{4+} + e^-$	$= U^{3+}$		−0.61
$Zn^{2+} + 2e^-$	$= Zn$		−0.763
$Cr^{2+} + 2e^-$	$= Cr$		−0.91
$Mn^{2+} + 2e^-$	$= Mn$		−1.18
$Zr^{4+} + 4e^-$	$= Zr$		−1.53
$Ti^{3+} + 3e^-$	$= Ti$		−1.63
$Al^{3+} + 3e^-$	$= Al$		−1.66
$Th^{4+} + 4e^-$	$= Th$		−1.90
$Mg^{2+} + 2e^-$	$= Mg$		−2.37
$La^{3+} + 3e^-$	$= La$		−2.52
$Na^+ + e^-$	$= Na$		−2.714
$Ca^{2+} + 2e^-$	$= Ca$		−2.87
$Sr^{2+} + 2e^-$	$= Sr$		−2.89
$K^+ + e^-$	$= K$		−2.925
$Li^+ + e^-$	$= Li$		−3.045

APPENDIX D SELECTED HALF-WAVE POTENTIALS FOR INORGANIC SUBSTANCES

(Courtesy of EG and G, Princeton Applied Research.)
Key:
All potentials are referenced to the saturated calomel electrode.
* Potentials that have an asterisk are differential pulse peak potentials. All other potentials are half-wave potentials.
[ ] Where a working electrode other than the dropping mercury electrode is used, the electrode material is indicated by brackets.

Ag(I): Ions in bold type are those that can also be determined using differential pulse anodic or cathodic stripping voltammetry.
( ) Potential values in parentheses indicate that the electrode reaction is an oxidation.
Acetate(4.5): Numbers in parentheses next to a supporting electrolyte are the pH values for that electrolyte.
Abbreviated electrolytes are explained following the table.

Ion	Supporting electrolytes	Supporting electrolytes are listed in approximate order of preference. These supporting electrolytes were selected from commonly available literature sources and are not meant to be all-inclusive.		
Ag(I)	0.1*M* KNO_3: 0.10[Pt]	NH_3—NH_4Cl: −0.24		
Al(III)	SVRS-Ac(4.5): −0.46*	0.1*M* TMAC: −1.75		
As(III)	1*M* HCl: −0.42,* −0.84*	1*M* H_2SO_4: −0.43, −0.81	1*M* NaOH: −0.3	H_2SO_4—NaCl: −0.20
As(V)	$HClO_4$-pyrogallol: −0.11			
Au(I)	0.1*M* KOH: −1.16			
Au(III)	1*M* HCl: 0.37[GCE]			
Ba(II)	0.1*M* LiCl: −1.92			
Bi(III)	1*M* HCl: −0.09	Tartrate(4.4): −0.14	NH_4Cit(3): −0.19	
Br^-	0.1*M* KNO_3: (0.12)	KNO_3—MeOH: (0.15) [Ag]		
BrO_3^-	H_2SO_4—KNO_3: −0.41	0.1*M* KCl: −1.78		
Cd(II)	NH_4Cit(3): −0.63*	Acetate(4.5): −0.65	1*M* HCl: −0.64	NH_4Tart(9): −0.69*
Ce(III)	2*M* K_2CO_3: (−0.16)			
Ce(IV)	2*M* K_2CO_3: −0.16			
Cl^-	0.1*M* KNO_3: (0.25)	KNO_3—MeOH: (0.28) [Ag]		
ClO^-	0.5*M* K_2SO_4(7): 0.08			
ClO_2^-	1*M* NaOH: −1.0			
CN^-	Borate(9.75): (−0.27)*	0.1*M* NaOH: (−0.36)		
Co(II)	NH_3—NH_4Cl: −1.30	Py-PyHCl: −1.06	5*M* $CaCl_2$: −0.82	NH_4Cit(9): −1.39
Cr(III)	KSCN(3.2): −0.75*			
Cr(VI)	NH_3—NH_4Cl: −0.30*	1*M* NaOH: −0.84*	NH_4Tart(9): −0.24	
Cu(I)	NH_3—NH_4Cl: (−0.22), −0.50			
Cu(II)	NH_4Cit(3): −0.06*	Acetate(4.5): −0.07	1*M* HCl: −0.22	NH_4Tart(9): −0.36*
Eu(III)	0.1*M* NH_4Cl: −0.67			
Fe(II)	Oxalate(4): (−0.23)*	$Na_4P_2O_7$(9): (−0.37)*		
Fe(III)	Oxalate(4): −0.23*	NaOH-TEA: −1.01*	NH_4Tart(9): −1.45*	$Na_4P_2O_7$(9): −0.99
Ga(III)	1*M* NaSCN(2): −0.83			
Ge(II)	6*M* HCl: −0.45			
Hg(II)	1*M* HCl: 0.44[Au]			
H_2O_2	PO_4—Cit(7): (0.18), −1.0			

(continued)

Ion	Supporting electrolytes	Supporting electrolytes are listed in approximate order of preference. These supporting electrolytes were selected from commonly available literature sources and are not meant to be all-inclusive.		
I^-	0.1*M* KNO_3: (−0.03)	KNO_3—MeOH: (0.0) [Ag]		
In(III)	Acetate(4.5): −0.71	1*M* HCl: −0.56		
IO_3^-	Phosphate(6.4): −0.79	1*M* KCl: −1.16		
IO_4^-	K_2SO_4—H_2SO_4: −0.12			
Ir(IV)	1*M* HCl: 0.65[GCE]			
K(I)	0.1*M* TBAOH: −2.14			
Li(I)	0.1*M* TBAOH: −2.33			
Mn(II)	NH_3—NH_4Cl: −1.66	NH_4Tart(9): −1.55*		
Mo(VII)	0.3*M* HCl: −0.26, −0.63	H_3Cit: 0.04, −0.44		
Na(I)	0.1*M* TBAOH: −2.12			
Nb(V)	8*M* HCl: −0.46, −0.70			
Ni(II)	NH_3—NH_4Cl: −1.10	NH_4Tart(9): −0.98*	Py-PyHCl: −0.75	
NH_2OH	1*M* NaOH: (−0.43)			
N_3^-	0.1*M* KNO_3: (0.25)			
NO_2^-	DPA—SCN(1): −0.54*	2*M* Cit(2.5): −1.06*	U(VI)-Ac-KCl(2): −0.98*	
NO_3^-	U(VI)—Ac-KCl(2): −0.98*			
O_2	0.1*M* KNO_3: −0.05, −0.90			
Pb(II)	NH_4Cit(3): −0.48*	Acetate(4.5): −0.50	1*M* HCl: −0.44	NH_4Tart(9): −0.52*
Pd(II)	NH_3—NH_4Cl: −0.75			
Rb(I)	0.1*M* TBAOH: −2.03			
Rh(III)	NH_3—NH_4Cl: −0.93			
Ru(IV)	1*M* $HClO_4$: 0, 0.02, −0.34			
S^{2-}	0.1*M* NaOH: (−0.78)*			
Sb(III)	6*M* HCl: −0.23*	1*M* HCl: −0.15		
Sb(V)	6*M* HCl: −0.23*			
Se(IV)	1*M* HCl: −0.10, −0.40			
Sn(II)	1*M* HCl: (−0.1), −0.47	NH_4Cit(3): (−0.21), −0.54	NH_3Tart(9): (−0.53), −0.77	1*M* NaOH: (−0.73), −1.22
Sn(IV)	HCl—NH_4Cl: −0.25, −0.52	1*M* HCl: −0.1, −0.47		
SO_3^{2-}	Acetate(5): (−0.62)*			
$S_2O_3^{2-}$	Acetate(5): (−0.21)*			
Te(IV)	NH_4Tart(9): −0.71	NH_3—NH_4Cl: −0.67		
Ti(III)	H_2Tart: (−0.44)			
Ti(IV)	0.1*M* HCl: (−0.81)			
Tl(I)	Acetate(4.5): −0.47	1*M* HCl: −0.48	Ac-EDTA(4.5): −0.50	
U(VI)	0.1*M* HCl: −0.18, −0.94			

Ion	Supporting electrolytes	Supporting electrolytes are listed in approximate order of preference. These supporting electrolytes were selected from commonly available literature sources and are not meant to be all-inclusive.		
V(V)	H_2SO_4-KSCN: −0.52*			
W(VI)	10*M* HCl: 0, −0.60			
Zn(II)	NH_4Cit(3): −1.05*	Acetate(4.5): −1.1	NH_3—NH_4Cl: −1.35	NH_4Tart(9): −1.24*

Supporting electrolytes: The list below includes only those supporting electrolytes that were abbreviated for the table.

1. Ac-EDTA(4.5): 0.1*M* sodium acetate −0.1*M* acetic acid −0.1*M* disodium ethylenediamine tetraacetic acid, pH 4.5
2. Acetate(4.5): 0.1*M* sodium acetate −0.1*M* acetic acid, pH 4.5
3. Acetate(5): 0.1*M* sodium acetate + acetic acid to pH 5
4. Borate(9.75): 0.1*M* boric acid + NaOH to pH 9.75
5. 2*M* Cit(2.5): 2*M* citric acid + NaOH to pH 2.5
6. DPA—SCN(1): 1.3 × 107^{-4}*M* diphenylamine (DPA) −0.01*M* NaSCN −0.04*M* $HClO_4$
7. H_3Cit: saturated citric acid
8. HCl—NH_4Cl: 1.0*M* HCl − 4*M* NH_4Cl
9. $HClO_4$-pyrogallol: 2*M* $HClO_4$ −0.5*M* pyrogallol
10. H_2SO_4—KNO_3: 0.1*M* H_2SO_4 −0.2*M* KNO_3
11. H_2SO_4—KSCN: 0.1*M* H_2SO_4 −0.1*M* KSCN
12. H_2SO_4—NaCl: 2*M* H_2SO_4 −2*M* NaCl
13. H_2Tart: saturated tartaric acid
14. KNO_3—MeOH: 0.1*M* KNO_3 in 50% methanol
15. KSCN(3.2): 0.2*M* NaSCN −0.2*M* acetic acid, pH 3.2
16. K_2SO_4—H_2SO_4: 0.16*M* K_2SO_4 −1*M* H_2SO_4
17. NaOH—TEA: 0.3*M* triethanolamine −0.2*M* NaOH
18. $Na_4P_2O_7$(9): 0.2*M* sodium pyrophosphate + H_3PO_4 to pH 9
19. NH_3—NH_4Cl: 1*M* NH_3 −0.1*M* NH_4Cl
20. NH_4Cit(3): 0.1*M* citric acid + NH_4OH to pH 3
21. NH_4Cit(9): 0.1*M* citric acid + NH_4OH to pH 9
22. NH_4Tart(9): 0.1*M* tartaric acid + NH_4OH to pH 9
23. Oxalate(4): 0.1*M* oxalic acid + NaOH to pH 4
24. Phosphate(6.4): 0.2*M* sodium dihydrogen phosphate + NaOH to pH 6.4
25. PO_4-Cit(7): 0.1*M* sodium dihydrogen phosphate −0.1*M* sodium citrate adjusted to pH 7.0
26. Py-PyHCl: 0.1*M* pyridine −0.1*M* pyridine-HCl
27. SVRS-Ac(4.5): 0.1*M* acetate buffer, pH 4.7 − 1.4 × 10^{-4}, solochrome violet RS − 12% ethanol
28. Tartrate(4.4): 0.1*M* Na_2 tartrate, pH 4.4
29. 0.1*M* TBAOH: 0.1*M* tetrabutylammonium hydroxide
30. 0.1*M* TMAC: 0.1*M* tetramethylammonium chloride
31. U(VI)-Ac-KCl(2): 20 ppm U(VI) − 0.2*M* KCl − 0.1*M* acetic acid

References:

1. EG&G, Princeton Applied Research, Application Briefs and Application Notes.
2. Donald T. Sawyer and Julian L. Roberts, Jr., *Experimental Electrochemistry for Chemists*, John Wiley, New York, 1974.
3. Louis Meites, *Polarographic Techniques*, Interscience, New York, 1955.
4. Louis Meites, *Handbook of Analytical Chemistry*, McGraw-Hill, New York, 1963.

APPENDIX E PROTON-TRANSFER REACTIONS OF MATERIALS IN WATER AT 25 °C

Substance	pK_1	pK_2	pK_3	pK_4
Acetic acid	4.76			
Ammonium ion	9.24			
Anilinium ion	4.60			
Arsenic acid	2.20	6.98	11.5	
Arsenous acid	9.22			
Ascorbic acid	4.30	11.82		
Benzoic acid	4.21			
Boric acid: meta-	9.24			
tetra-	4	9		
Bromocresol green	4.68			
Bromocresol purple	6.3			
p-Bromophenol	9.24			
Bromophenol blue	3.86			
Bromothymol blue	7.1			
Carbonic acid ($CO_2 + H_2O$)	6.38	10.25		
Chloroacetic acid	2.86			
Chlorophenol red	6.0			
Chromic acid		6.50		
Citric acid	3.13	4.76	6.40	
Cresol purple (acid range)	1.51			
(base range)	8.32			
Cresol red	8.2			
Dichloroacetic acid	1.30			
Ethanolammonium ion	9.50			
Ethylammonium ion	10.63			
Ethylenediaminetetraacetic acid (EDTA)	2.0	2.67	6.27	10.95
Ethylenediammonium ion	6.85	9.93		
Ferrocyanic acid			2.22	4.17
Formic acid	3.75			
Glycine (protonated cation)	2.35	9.78		
Hydrazinium ion	−0.88	7.99		
Hydrocyanic acid	9.21			
Hydrofluoric acid	3.18			
Hydrogen peroxide	11.65			
Hydrogen sulfide	6.88	14.15		
Hydroquinone	10.0	12.0		
Hydroxylammonium ion	5.96			

Substance	pK_1	pK_2	pK_3	pK_4
N,*N*-bis(2-hydroxyethyl) glycine (bicine)(protonated cation)	8.35			
tris(hydroxymethyl)aminomethane (TRIS)(protonated cation)	8.08			
N-2-Hydroxyethylpiperazine-*N*′-2-ethanesulfonic acid (HEPES)	7.55			
N-tris(hydroxymethyl)methylglycine (TRIS)(protonated cation)	8.08			
Hypochlorous acid	7.50			
Methyl orange	3.40			
Methyl red	4.95			
2-(*N*-Morpholino)ethanesulfonic acid (MES)	6.15			
Nitrous acid	3.35			
Oxalic acid	1.27	4.27		
1,10-Phenanthrolinium ion	4.86			
Phenol	9.99			
Phenol red	7.9			
Phenolphthalein	9.4			
Phenylacetic acid	4.31			
Phosphoric acid: ortho	2.15	7.20	12.36	
pyro	1.52	2.36	6.60	9.25
o-Phthalic acid	2.95	5.41		
Pyridinium ion	5.21			
Salicylic acid	3.00	12.38		
Succinic acid	4.21	5.64		
Sulfamic acid	0.988			
Sulfuric acid		1.92		
Sulfurous acid ($SO_2 + H_2O$)	1.90	7.20		
Tartaric acid: meso-	3.22	4.81		
Thymol blue	8.9			
Thymolphthalein	10.0			
Triethanolammonium ion	7.76			
Vanillin	7.40			
Veronal	7.43			

APPENDIX F CUMULATIVE FORMATION CONSTANTS FOR METAL COMPLEXES AT 25 °C

	log K_1	**log K_2**	**log K_3**	**log K_4**	**log K_5**	**log K_6**
Ammonia						
Cadmium	2.65	4.75	6.19	7.12	6.80	5.14
Cobalt(II)	2.11	3.74	4.79	5.55	5.73	5.11
Cobalt(III)	6.7	14.0	20.1	25.7	30.8	35.2
Copper(I)	5.93	10.86				
Copper(II)	4.31	7.98	11.02	13.32	12.86	
Nickel	2.80	5.04	6.77	7.96	8.71	8.74
Silver(I)	3.24	7.05				
Zinc	2.37	4.81	7.31	9.46		
Chloride						
Copper(I)		5.5	5.7			
Copper(II)	0.1	−0.6				
Tin(II)	1.51	2.24	2.03	1.48		
Tin(IV)						4
***Citrate* (L^{3-} *anion*)**						
Cadmium	11.3					
Cobalt(II)	12.5					
Copper(II)	14.2					
Iron(II)	15.5					
Iron(III)	25.0					
Nickel	14.3					
Zinc	11.4					
Cyanide						
Cadmium	5.48	10.60	15.23	18.78		
Copper(I)		24.0	28.59	30.30		
Nickel				31.3		
Silver(I)		21.1	21.7	20.6		
Zinc				16.7		

	log K_1	log K_2	log K_3	log K_4	log K_5	log K_6
Ethylenediamine-N,N,N′,N′-tetraacetic acid						
Calcium	11.0					
Copper(II)	18.7					
Iron(II)	14.33					
Iron(III)	24.23					
Magnesium	8.64					
Mercury(II)	21.80					
Zinc	16.4					
1,10-phenanthroline						
Cadmium	5.93	10.53	14.31			
Cobalt(II)	7.25	13.95	19.90			
Copper(II)	9.08	15.76	20.94			
Iron(II)	5.85	11.45	21.3			
Iron(III)	6.5	11.4	23.5			
Nickel	8.80	17.10	24.80			
Zinc	6.55	12.35	17.55			

APPENDIX G A COMPARISON OF TRACE ELEMENT DETECTION LIMITS FOR ATOMIC SPECTROSCOPIC METHODS*

Element name	Symbol	Atomic number	*Method*†						
			FES	FAAS	EAAS	ICPS	DCPS	AFS	LAFIS
Aluminum	Al	13	3	20	0.01	0.2	2	0.6	
Antimony	Sb	51	200	0.1	0.08	10	3	0.1	
Arsenic	As	33	2000	0.02	0.08	2	45	0.1	
Barium	Ba	56	1	8	0.04	0.01	2	2	0.2
Beryllium	Be	4	100	1	0.003	0.003	0.5	1	
Bismuth	Bi	83	1000	0.02	0.1	10	75	2	2
Boron	B	5	50	700	15	0.1	5	2000	
Bromine	Br	35				‡			
Cadmium	Cd	48	300	0.5	0.0002	0.07	0.5	0.001	
Calcium	Ca	20	0.1	0.5	0.01	0.0001	0.2	0.08	0.1
Carbon	C	6				44			
Cerium	Ce	58	150			0.4		500	
Cesium	Cs	55	0.02	8	0.04				
Chlorine	Cl	17				‡			
Chromium	Cr	24	1	2	0.004	0.08	1	1	2
Cobalt	Co	27	5	2	0.008	0.1	1	2	
Copper	Cu	29	3	1	0.005	0.04	2	0.03	100
Dysprosium	Dy	66	20	50		4	5	300	
Erbium	Er	68	20	40	0.3	1	5	500	
Europium	Eu	63	0.2	20	0.5	0.06		20	
Fluorine	F	9				‡			
Gadolinium	Gd	64	120	1000	8	0.4		800	
Gallium	Ga	31	5	50	0.01	0.6	38	0.9	0.07
Germanium	Ge	32	400	50	0.1	0.5		100	
Gold	Au	79	500	6	0.01	0.9	3	5	
Hafnium	Hf	72		2000		10			
Holmium	Ho	67	10	40	0.7	3		100	
Indium	In	49	1	20	0.02	0.4	38	0.2	0.006
Iodine	I	53				‡			
Iridium	Ir	77	400	500	0.5	30	60		
Iron	Fe	26	10	3	0.01	0.09	3	3	2
Lanthanum	La	57	5	2000	0.5	0.1	2		
Lead	Pb	82	0.2	10	0.007	1	23	10	0.6
Lithium	Li	3	0.001	0.3	0.01	0.02	1	0.4	0.001
Lutetium	Lu	71	200	700		0.1			
Magnesium	Mg	12	1	0.1	0.0002	0.003	0.2	0.1	0.1
Manganese	Mn	25	1	0.8	0.0005	0.01	0.5	0.4	0.3
Mercury	Hg	80	150	0.001	0.2	1	75	0.003	
Molybdenum	Mo	42	10	10	0.02	0.2	0.5	12	
Neodymium	Nd	60	200	600		0.3	38	2000	
Nickel	Ni	28	10	2	0.05	0.2	2	2	8

Element name	Symbol	Atomic number	Method†						
			FES	**FAAS**	**EAAS**	**ICPS**	**DCPS**	**AFS**	**LAFIS**
Niobium	Nb	41	60	1000		0.2	38	1000	
Nitrogen	N	7				‡			
Osmium	Os	76	2000	80	2	0.4			
Oxygen	O	8				‡			
Palladium	Pd	46	40	10	0.05	2	9	40	
Phosphorus	P	15	100		0.3	15	75		
Platinum	Pt	78	2000	40	0.2	0.9	26	300	
Potassium	K	19	0.01	1	0.004	30	0.3	0.8	1
Praseodymium	Pr	59	500	2000		10	53	1000	
Rhenium	Re	75	200	200	10	6	23		
Rhodium	Rh	45	10	2	0.1	30	45	100	
Rubidium	Rb	37	0.02	0.3			15		
Ruthenium	Ru	44	300	70		30	30	500	
Samarium	Sm	62	50	500		1	83	100	
Scandium	Sc	21	10	20	6	0.4		10	
Selenium	Se	34		0.02	0.05	1	45	0.06	
Silicon	Si	14		20	0.005	2	15	300	
Silver	Ag	47	2	0.9	0.001	0.2	2	0.1	1
Sodium	Na	11	0.01	0.2	0.004	0.1	0.05	0.1	0.04
Strontium	Sr	38	0.1	2	0.01	0.002	2	0.3	
Sulfur	S	16	1600	20	10	30			
Tantulum	Ta	73		9		5	75		
Tellurium	Te	52	600	0.002	0.03	15	75	0.08	
Terbium	Tb	65	200	600		0.1		500	
Thallium	Tl	81	2	9	0.01	40		4	0.09
Thorium	Th	90				3			
Thulium	Tm	69	4	10		0.2		100	
Tin	Sn	50	100	10	0.03	3	23	10	2
Titanium	Ti	22	30	10	0.3	0.03	1	2	
Tungsten	W	74	200	500		0.8	30	2000	
Uranium	U	92	100		30	1.5	150		
Vanadium	V	23	7	20	0.1	0.06	8	30	
Ytterbium	Yb	70	0.2	5	0.1	0.02		10	
Yttrium	Y	39	40	50	10	0.04	2	500	
Zinc	Zn	30	1000	0.8	0.0006	0.1	2	0.0003	
Zirconium	Zr	40	1000	350		0.06	8		

SOURCE: Applied Spectroscopy, **37** (5), 411 (1983), with permission.

* All concentrations are in nanograms per milliliter.

† Abbreviations: FES, flame emission spectroscopy; FAAS, flame atomic absorption spectroscopy; EAAS, electrothermal atomic absorption spectroscopy; ICPS, inductively coupled plasma spectroscopy; DCPS, direct-current plasma spectroscopy; AFS, atomic fluorescence spectroscopy; LAFIS, laser-assisted flame ionization spectroscopy.

‡ Can be determined, but exact detection limits are not established.

INDEX